ANNUAL REVIEW OF BIOCHEMISTRY

ANNUAL REVIEW OF BIOCHEMISTRY

VOLUME 51, 1982

ESMOND E. SNELL, *Editor*
University of Texas at Austin

PAUL D. BOYER, *Associate Editor*
University of California, Los Angeles

ALTON MEISTER, *Associate Editor*
Cornell University Medical College

CHARLES C. RICHARDSON, *Associate Editor*
Harvard Medical School

ANNUAL REVIEWS INC.   4139 EL CAMINO WAY   PALO ALTO, CALIFORNIA 94306 USA

ANNUAL REVIEWS INC.
Palo Alto, California, USA

International Standard Serial Number: 0066-4154
International Standard Book Number: 0-8243-0851-4
Library of Congress Catalog Card Number: 32-25093

Annual Reviews Inc. and the Editors of its publications assume no responsibility for the statements expressed by the contributors to this *Review*.

PRINTED AND BOUND IN THE UNITED STATES OF AMERICA

Annual Review of Biochemistry
Volume 51, 1982

CONTENTS

vi   CONTENTS

SOME RELATED ARTICLES IN OTHER *ANNUAL REVIEWS*

From the *Annual Review of Genetics,* Volume 15 (1981)

Directed Mutagenesis, D. Shortle, D. DiMaio, and D. Nathans

Simian Virus 40 as a Eukaryotic Cloning Vehicle, J. T. Elder, R. A. Spritz, and S. M. Weissman

Transposable Elements in Prokaryotes, N. Kleckner

From the *Annual Review of Medicine, Volume 33 (1982)*

The Thalassemia Syndromes: Models for the Molecular Analysis of Human Disease, E. J. Benz, Jr., and B. G. Forget

From the *Annual Review of Microbiology,* Volume 36 (1982)

Mechanism of Incorporation of Cell Envelop Proteins in Escherichia coli, S. Michaelis and J. Beckwith

Phycobilisomes: Structure and Dynamics, A. N. Glazer

Colicins and Other Bacteriocins with Established Modes of Action, J. Konisky

Metabolic Acquisitions Through Laboratory Selection, R. P. Mortlock

Microbial Envelope Proteins Related to Iron, J. B. Neilands

Low-Molecular-Weight Enzyme Inhitibors of Microbial Origin, H. Umezawa

From the *Annual Review of Neuroscience,* Volume 5 (1982)

Inherited Metabolic Storage Disorders, R. O. Brady

The Molecular Forms of Cholinesterase and Acetylcholinesterase in Vertebrates, J. Massoulié and S. Bon

From the *Annual Review of Physiology,* Volume 44 (1982)

Processing of Endogenous Polypeptides by the Lungs, J. W. Ryan

Structure and Function of the Calcium Pump Protein of Sarcoplasmic Reticulum, N. Ikemoto

Turnover of Acetylcholine Receptors in Skeletal Muscle, D. W. Pumplin and D. M. Fambrough

ERRATA

ANNUAL REVIEW OF BIOCHEMISTRY, Volume 50 (1981)

In *Proteolipids,* by Milton J. Schlesinger

On page 196, in the first paragraph, the statement that a lysine hexapeptide from brain myelin proteolipoprotein contained fatty acid is in error. Only the 13-residue Arg-peptide was found to have bound fatty acid.

In *The Biochemistry of the Complications of Diabetes Mellitus,* by M. Brownlee and A. Cerami

On page 415, myoinositol is described as an isomer of sorbitol when in fact it is a cyclic, nonreducing isomer of glucose.

W. Engelhardt.

Ann. Rev. Biochem. 1982. 51:1–19
Copyright © 1982 by Annual Reviews Inc. All rights reserved

LIFE AND SCIENCE

Wladimir A. Engelhardt

Institute of Molecular Biology, Academy of Sciences of the USSR,
Moscow, USSR

CONTENTS

Childhood. University. First Steps in Science

Many of my predecessors in this section of the Reviews of Biochemistry, the prefatory chapters, must have experienced the same difficulty that I feel confronting me now. One has to steer between Scylla and Charybdis—find the correct way between giving a scientific review article, summarizing the scientific contributions, and giving a strictly autobiographical sketch of the events that constituted one's life course, without overloading it with scientific details. I have tried my best to avoid the two extremes. I will be happy if I have succeeded in achieving the necessary blend.

I was born in 1894 in Moscow, not in my parent's home, but in the obstetric department of Moscow's Central Children's Hospital, where my father temporarily worked. Two months later I was taken to Jaroslavl, a district town on the Volga, where our permanent home was. Here I spent my childhood and college years, till the age of 25, so that I regard Jaroslavl as being practically my home town. As though I had some indirectly introduced genes relating me to medicine, I kept in touch with this sphere for a considerable time, in different ways, besides being born in medical surroundings. My grandfather on my mother's side was the chief surgeon and head of the local hospital of Jaroslavl. My father was the head of the department of obstetrics and gynecology of the same hospital. After college

I entered the medical faculty of Moscow University, graduated as an MD and served two years as a military doctor. My first scientific appointment was in the Biochemical Institute of the Commissariat of Public Health, my first professorship was the chair of Biochemistry of the medical faculty of Kazan University (later Medical Institute), and my first academic nomination was as a member of the newly formed Academy of Medical Sciences of the USSR. But looking back at my life's course, I cannot remember a single case where I healed a person of any disease. The effect of medical genes seems to continue into the next generation: my younger daughter works as a scientist in the laboratory of the Oncological Center, Moscow, and one of my granddaughters is a postgraduate at the chair of normal and pathological neuropsychology of Moscow State University.

Perhaps the first signs of my inclination toward a scientific profession became perceptible at a fairly early stage. It so happened that during primary school my classmates, fond of giving nicknames, as probably all boys in the world do, honored me with the title "Wolodya-Outcheny," which means scientist, Wolodya being the diminutive of Wladimir. This probably was due to my liking for playing around with all kinds of simple apparatus—electric bells, primitive toys—and gadgets such as those that show the attraction of light objects to a rubbed haircomb or glass-rod, etc. I even recollect my first "invention:" a small test tube with a light bead of elder-marrow in it and two wires attached to the ends of the tube. I declared that this apparatus could serve as a "cunny-scope," showing the degree of cunning of a person, according to the displacement of the bead when the wires were applied to definite parts of the body. Needless to say that the effect was produced by dexterously rubbing one end of the tube with a piece of silk cloth by which the "apparatus" was held. Thus I started my scientific career as a quack. I hope that my later behaviour has amended this regrettable qualification.

The next landmark of my career was in the field of chemistry, in which I became interested in college. As every beginner I was attracted by the handling of explosive materials. Nitrogen iodide was the object of choice, for when dry it explodes when even slightly touched. I carried a piece, damp and therefore safe, the size of a wheat-corn, to enlighten my classmates. But I dropped the damned stuff near the teacher's pulpit during the lesson on orthodox religion. At the very end of the lesson, the boy in charge of pronouncing the closing prayer stepped on the ill-fated piece, producing no harm but great noise.

It was a period when attempts on the life of tzars and members of the royal family were fashionable, so the priest who gave us the lesson rushed out of the classroom to the headmaster's study shouting: "an attempt has been made on my life . . ." Efforts by my parents were necessary to prevent my being expelled from the school. But not the highest mark for conduct stood in my final diploma which I had to present when applying for entrance to the Petersburg Polytechnical School where I intended to study electrical engineering. I did not pass the competition. This was a blow to my expectations. As a matter of fact it seems that the ominous chemical experience diverted my interests from chemistry to the physical field, particularly electricity. Still at college, with no help or advice from teachers, I constructed entirely from scrap collected on the market a complete radio set, transmitter, and receiver, with a crystal detector or coherer of iron filings; it exchanged signals all through our flat, passing closed doors and walls, to the bewilderment and great admiration of my parents. A high-frequency Tesla transformer followed, with mirror galvanometer, thermoelectric battery etc. I even published in a semipopular journal, a small article describing the replacement of the usual vibration interrupter of a small Rumcorf coil by a miniature mercury switch, which doubled the efficiency. That was my first scientific publication!

Having missed the Polytechnical School, I entered the mathematical faculty of Moscow University, where the requirements were less rigid, expecting that a good knowledge of mathematics would certainly be useful if I continued as an engineer. At school I was regarded as fairly good in mathematics, but a few months sufficed to destroy this illusion. I felt completely incapable of grasping the very first principles of higher mathematics, exposed by the eminent professor Lusin. After one completely fruitless term, I started to attend courses in chemistry and even passed one of the examinations, but changed my mind once more and finally settled down as a student of the faculty of medicine.

I hardly attended the lectures or the practical courses, but spent all my days working in the biochemical laboratories of a few different chairs. How I managed to obtain my medical diploma remains a mystery to me. In the evenings I attended lectures of a more general biological character, delivered at the Popular Shanyavsky University on immunology, and especially on physico-chemical biology, by professor Nikolai Koltzov.

My true inclination developed rapidly: the study of biological phenomena by means of exact sciences such as chemistry and physics. Perhaps this was the result of two fascinating books recommended by Koltzov, and of his brilliant lectures. The books were by Fisher: *Oedema* and *Nephritis*. I not only studied in the laboratory the uptake of water by tissue slices, but during my summer vacations, when I worked as volunteer in a small hospital in Jaroslavl, I tried to apply my modest knowledge to the treatment of renal diseases. Unfortunately the results were not exactly favorable for the patients and I was severely forbidden to continue my experiments on humans.

At home, electrical apparatus gradually became replaced by a self-made thermostat, heated by a kerosene lamp, and dangerous explosives gave place to harmless test tubes with protozoa or yeast suspensions. I succeeded in demonstrating that a mistake was present in Koltzov's conclusion on the nature of the influence of pH on phagocytosis by a unicellular organism. The effect was measured by the rate of uptake of India-ink suspension, and I established that changes of the electrical charge of the lamp-black particles were responsible for the changes observed and not the behavior of the organisms. When I reproduced an experiment at Koltzov's laboratory and told him about my explanation, he was quite excited and said that this was the correct way to tackle an experimental problem, not to be afraid of disagreeing even with authorities.

Civil war ravaged the country. The first socialist state stood alone, deprived of any external supply, against the adherents of the broken-down tzarist regime, the so–called "white forces" supported by armies of many western countries. It was a struggle of ideas against military force, and the victory was at last won by the ideas.

Immediately after graduating from the University I was called for military service as a doctor, and spent two years on the Southern front as head of a field hospital of a cavalry division. I went from the Don to the Crimea, and ended in the Caucasus, after the explusion of the English occupation forces.

In those years the war was as much against typhoid fever as against military forces—"war against lice"—and medical personnel were the first to fall victims. By pure wonder I escaped the fatal end. I withstood a severe infection under most unfavorable conditions: on a stretcher in an unheated wagon, during retreat before an advancing White-Cossak army. A good friend of mine kept me alive with huge doses of camphor, the only drug available, through a week of unconsciousness. After a month's recovery I was again on active service, now, fortunately, having become immune.

In 1921 the civil war came practically to an end, I was dismissed from military service, returned to Moscow, and from that time on began my scientific career.

By happy chance I was admitted to the newly organized Biochemical Institute of Public Health. I wonder whether there was not some kind of favorable influence from Koltzov, who was at that time director of another, closely related neighboring institute. Our Institute was under professor Alexei Bach, an eminent chemist and political figure. He was well-known for his research on the enzymatic mechanisms of biological oxidation, where a central role was ascribed to the formation of peroxides as the initial, all-important step. Bach's attitude even when dealing with young beginners, was not to "lead by the hand" by everyday instructions, but to give first of all general advice, indicating the main directions in which to proceed. At the beginning, Bach suggested that I study the properties of immune antienzymes, particularly the antiphenolases. One part of Bach's advice I have always remembered. After having done reasonable work on immune antibodies against enzynes (antienzymes) I once told Bach that I had a theory concerning the nature of immunity. What did he think about my developing it? "Dear boy," he told me, "if I were paid for producing theories, I could sit all my life inventing new and better ones. Good theories come from good facts. You had better turn to your bench, to your bench, my boy."

I have never had regular training in biochemistry and related subjects. In this respect, in scientific education and studies, I have to regard myself as a self-made investigator, not having spent any time in a traditional school. Of course I have been influenced by scientists of the older generation, but by more or less fragmentary contacts rather than in a systematic fashion. More exactly, there was just one very brief period when, after several years of work at the Biochemical Institute, I spent in 1927 about two months in Peter Rona's laboratory at the Charité hospital in Berlin. Here again the atmosphere was extremely liberal. Everyone could choose the subject on which he preferred to work; Rona himself was always ready to render assistance, but no strict program was ever prescribed. His lectures were excellent, meticulously prepared and accompanied by impressive experiments. Here I became familiar with the peculiar German way of expressing admiration during some especially intricate lecture demonstration: it was by stamping the feet. I regarded it as a special kind of applause by the lower extremities, the hands being busy taking notes or drawing sketches of the apparatus.

Rona's laboratory had a worldwide reputation. I even remember a jocular interpretation of the abbreviation, FRS, as meaning "Frühere Rona's Schüler." It was a kind of breeding pool from which many future excellent researchers emerged. For many of them it was the way to a scientific paradise—some of the Kaiser-Wilhelm Institutes, the KWI as they were known. Future Nobel prize winners were among the students there, as Fritz Lipmann, Hans Krebs, Max Perutz, E. B. Chain. There one could make close aquaintance with a broad diversity of interesting and able people. I remember with great pleasure working at the same bench, side by side, with David Nachmansohn; on the upper floor Hans Weber was working on muscle proteins. Rona arranged visits to KWI for me to meet Carl Neuberg, Otto Warburg, Karl Lohmann, and Otto Meyerhof. Some of these contacts, even if initially of short duration, grew later into life-long friendships.

The work on antienzymes did not bring results of any importance from the point of view of giving new insight into the nature of enzymes or their action. But it led to a discovery, of minor importance by itself, but which nevertheless appeared later to contain elements of broader significance. I have in mind the use of what I later called "the principle of fixed partner" (1). The starting point were observations that showed that immune antibodies can produce their reaction with the antigen even when transferred from solution into a heterogeneous state, in adsorbed form on some suitable carrier such as kaolin or aluminum hydroxide. The use of this principle permitted establishment of the antigenic properties of haemoglobin. Under usual conditions, immunization with this protein did not produce antibodies; if examined in ordinary solution no precipitation or changes in its specific, oxygen-binding properties were observed. But if the serum of an animal immunized with haemoglobin was adsorbed on a colloidal carrier, this suspension had the remarkable property of binding hemoglobin and removing it from solution. Similar conditions were observed in experiments with yeast invertase. Whereas other enzymes, such as oxidases, produced antibodies that inhibited their enzymatic activity, invertase seemed to be devoid of antigenic properties. But here again, when the serum from an immunized animal was used in the form of an adsorbate, in the fixed state, complete removal of the enzyme from solution resulted after centrifugation. It was easy to show that in this bound state, after reaction with the "fixed partner," the enzyme still exhibited its original catalytic action. These experiments might be regarded as precursors of the now so popular use of "immobilized" forms of enzymes or other biologically active substances.

Discovery of Oxidative Phosphorylation

Immunological studies, attractive as they were, seemed at the time to have led to a kind of blind alley. My interest shifted to the study of metabolic processes involving phosphoric acid, which in the meantime had become a central point of significance. Its participation in anaerobic carbohydrate metabolism attracted the greatest interest, due to the fundamental work of Embden, Lohmann, Meyerhof, and Parnas after the breakthrough discoveries of Harden and Young.

In 1929 I accepted the invitation from the Kazan University to occupy the chair of biochemistry. The laboratory had to be organized completely anew, since it lacked even the simplest modern equipment. All I possessed for my personal work after a year of effort was a poor imitation of a Warburg respirometer, made in the university's modest workshop, and the simplest kind of colorimeter, an "Authenrieth" wedge apparatus, only later replaced by a small Duboscq model. But the teaching took very little of my time; I had plenty of it for working on my modest bench and for meditation.

The results were highly satisfying—the discovery, that respiration in cells can produce a synthesis of ATP. At that time it was well known that ATP is synthesized in the course of anoxidative breakdown of glucose, as a result of fermentation and glycolysis. In retrospect it appears almost impossible that nothing whatever was known about a possible participation of ATP (or of phosphate in general) in the processes of the other great predominant source of energy, namely respiration. The explanation is simple: it was the lack of appropriate experimental objects. A very limited number of biological objects served at that time as materials of choice for the study of the two fundamental sources of energy in living objects: fermentation and respiration. They were yeast, on the one hand, and muscle and liver tissue on the other. Experimenters neglected the valuable advice given by the remarkable biologist, Dr. Krogh, in his address to one of the Physiological Congresses. He said that Nature has been generous toward naturalists, by creating some special object particularly suited for the study of each of the more important problems. The condition of success for a scientist attacking a new problem is to find and use the appropriate object.

As a paradox, the first indication of participation of ATP in the processes of cellular respiration came not from experiments where respiration proceeded, but from those where it was absent.

By lucky chance it so happened that I had chosen the kind of cells for my study that might be regarded as specially suited for investigating the possible participation of ATP in respiratory processes. These were the nucleated avian blood cells. Their structure is of the simplest kind: they have only one strictly defined function to fulfill, i.e. to keep in a closed space a highly concentrated solution of hemoglobin, in which swims a large nucleus. Evidently, thanks to the latter feature, they possess a very intense respiration, in contrast to the non-nucleated mammalian erythrocytes. And most important, they have a very high content of ATP, of almost exactly the same order as muscle tissues.

This ATP content remains practically constant as long as respiration goes on. But as soon as the latter is interrupted, by cyanide poisoning or removal of oxygen by repeated evacuation and flushing with pure nitrogen, rapid dephosphorylation of ATP is observed, accompanied by a corresponding increase of inorganic phosphate. Within an hour at 37° the acid-labile phosphate has disappeared, hydrolyzed to inorganic phosphate. How was the stability of the ATP content under aerobic conditions to be interpreted? Obviously, the choice had to be made between two explanations. Either the splitting of ATP was stopped by oxidative conditions, or it did proceed, but was compensated for by a steadily proceeding reverse process: reesterification of the inorganic phosphate that had been set free.

The preference was decidedly the second alternative, which assumes a continuing "phosphate cycle," for no example was known of an inhibition of hydrolytic processes by oxygen. This would have remained an indirect conclusion, were it not unambiguously supported by the direct demonstration of esterification of inorganic phosphate, formed during a period of anaerobiosis, after reestablishment of aerobic conditions. Thus aerobic esterification of inorganic phosphate, with formation of ATP was proved beyond any doubt (2, 2a). I called the process "respiratory resynthesis" of ATP; it corresponds to the nowadays generally accepted term oxidative phosphorylation.

The experiments permitted at least a rough quantitative evaluation of the efficiency of the process, represented in the form of the P/O quotient. This value was found to be about 1.0, of the order of magnitude established later in numerous cases.

With the discovery of oxidative phosphorylation, ATP was immediately raised to a pivotal role in bioenergetics, especially when very soon the work of Wladimir Belitzer, and of Herman Kalckar established the important fact that not only the primary attack on the sugar molecule was accompanied by the binding of inorganic phosphate, but also the oxidation of several of the intermediates in the long chain of steps leading to the oxidative breakdown of hexose. It became commonplace to regard ATP as the storage form, the energetic currency by which the energy produced by fermentation and respiration is made available for fulfillment of all physiological functions. The stored chemical energy is set free by the action of the corresponding enzyme, the ATPase.

A peculiar property of ATPase should be mentioned here, which became apparent during our further studies of the ATP metabolism of nucleated blood cells. In the absence of respiratory resynthesis, their ATP content is hydrolyzed within about one hour. This means that the ATPase activity is small. But if the cells are hemolyzed, the ATP is split almost instantly, in an explosive way. When my collaborator Tatiana Venkstern and I studied this peculiar behaviour (3) it appeared that there were two kinds of ATPase activity. The smaller part is present in the internal contents of the cell, and is responsible for the slow breakdown or turnover of ATP in normal, intact blood cells. The other, many times larger part, is firmly localized on the exterior surface of the cells and its activity is directed outwards. It is this latter ATPase—we called it "ecto-ATPase"—that splits the ATP of the cells when they are lyzed. A simple experiment shows its presence. If we add some ATP to a suspension of avian bloodcells in normal saline, all the added triphosphate is rapidly hydrolyzed. But the inner content of ATP remains completely unattacked at the initial, normal level. Circumstances did not permit us to continue this study. The biological role of the ecto-ATPase represents an intriguing mystery. We have to suppose that the enzyme never meets its substrate, for there is no measurable amount of ATP in the blood plasma. Perhaps it is a peculiar ontogenetic relic from the process of erythropoiesis?

Nature of the Pasteur Effect

Cellular respiration may be regarded as possessing a dual role (7). Oxidative phosphorylation is obviously its most important manifestation, a powerful generator of high-energy phosphate bonds that serve as the immediate source of chemical energy for all physiological functions. The other role is of, so to say, vicarial significance, of less direct character. This function is represented by the so-called Pasteur effect. It governs the interrelation between respiratory and anaerobic (glycolytic or fermentative) metabolism. Under aerobic conditions the wasteful fermentative breakdown of the carbohydrate is suppressed. Thus through the action of the Pasteur effect the fate of a hexose molecule becomes decided—whether it will follow the respiratory or the fermentative pathway.

It was natural that my interest was attracted by this, at the time, hardly explored problem—the nature of the Pasteur effect. The most plausible explanation of its mechanism was to assume that the suppression of fermentation under aerobic conditions was due to the oxidative inactivation of some member of the enzymatic machinery of fermentation and glycolysis. The leading idea of our approach, which I undertook in collaboration with my postgraduate student Nikolai Sakov, was to investigate the sensitivity toward oxidation of the different enzymes that take part at the first stages of the anaerobic breakdown of glucose.

Accordingly, the effect of redox dyes of different redox potential was tested on the enzymes responsible for the initial stages of the glycolytic pathway.

Of the several enzymes tested (hexokinase, isomerase, aldolase etc) all, with one exception, showed complete insensitivity toward the redox dyes over the whole range of concentration or E'_o values. The only exception, significantly, showed a striking difference. This was the enzyme, phosphofructokinase, which leads to the formation of the starting point of fermentation, hexose-1,6-diphosphate, by transferring a phosphate residue from ATP onto the preceding stage, fructose-6-phosphate.

Phosphofructokinase appeared to be highly sensitive to the action of redox dyes, with positive potential within the range of 0.05 and 0.250 V. Complete inhibition was obtained also with a variety of oxidizing agents: iodine, quinine, hydrogen peroxide, dehydroascorbic acid etc. All the reduced forms were without effect.

Evidently, the effect of these agents, completely alien to the normal catalytic system of the cell, even if highly suggestive, was only of an indirect kind. But an impressive proof of the validity of the findings was obtained when an exactly similar effect was found using the major physiological oxidizing system, cytochrome and its oxidase. In the presence of a suitable intermediate carrier, oxidized cytochrome by itself taken in stoichiometric amount, inhibited the phosphofructokinase. But, most important, the inhibition could be obtained with minute, catalytic amounts of cytochrome in the presence of cytochrome oxidase. In air, almost complete inhibition is observed, whereas in nitrogen no inhibition occurs. This experiment can well be regarded as the closest modeling of the Pasteur effect under the most simplified conditions.

We have to regard hexose-6-monophosphate as the turning point at which the further fate of the hexose molecule is decided. If the first C atom is phosphorylated, with formation of the Harden-Young ester, the molecule is predestined to be split into two equal parts, the trioses, and then to follow a series of fermentation steps; we called this the dichotomic pathway. If on the contrary, the first carbon atom, instead of being phosphorylated is oxidized, hexose enters the oxidative pathway, which may proceed by consecutive splitting off of single-carbon products; we called this the apotomic pathway.

Sakov was perhaps the most brilliant of my pupils; he had a keen mind and was an excellent experimenter. His lot was tragic. The work on the Pasteur effect had a deeply regretful issue. The experiments were finished in the spring of 1941. The war broke out; Sakov was soon called to military service. My family and I were evacuated from Moscow to Frunze, the capital of the Kazakh republic in Central Asia. I carried the laboratory protocols and notebooks with me, waiting for news from Sakov. They came only after a long silence, and were sad. A note from the armed forces informed me that Sakov had perished on the battlefields of Stalingrad.

The paper was published posthumously. There was no possibility of sending it to a foreign journal, and it appeared only in the Russian journal *Biokhimiya* (4) and thus remained almost completely unknown, except that some of the initial stages of this research, already showing the important points, were mentioned in a few lines in an article by Dean Burk in the Cold Spring Harbor Annual Report (5). The interpretation we gave to the mechanism of the Pasteur effect was "rediscovered" after exactly twenty years, by Janet Passoneau and O. Lowry, and published under the title "*Phosphofructokinase and the Pasteur effect*" (6). I reported the work at a conference convened in Paris in 1946, in commemoration of the fiftieth anniversary of Pasteur's death. Apparently the Proceedings remained unpublished; at least I received neither reprints nor information about publication. With a lapse of over thirty years I gave an account of the main results at a conference on bioenergetics, held at the American Academy of Arts and Sciences, in Boston, in 1973 (7).

Myosin and Adenosinetriphosphatase

Obviously, muscle is the most appropriate object for the study of problems of bioenergetics. With the discovery by Einar Lundsgaard of the "alactacid contraction" it became clear that the immediate source of energy for the work of muscle is the splitting of ATP and that consequently, great importance should be attributed to the enzyme that catalyzes this splitting, adenosinetriphosphatase.

There was every reason to regard this enzyme as fulfilling a key role in the function of muscle and to study its properties. But, strangely enough, this did not happen for a considerable time. The reason was peculiar—no ATP-splitting activity was found in aqueous extracts of muscle tissue. And it was these extracts that were the favorite material of muscle biochemists. Intense work was proceeding, directed toward obtaining the different enzymes that constitute the glycolytic complex, in soluble form. "Water-soluble enzymes" formed a distinguished group; the discovery of a new enzyme belonging to it was the predominant aim of experimental efforts, a kind of hypnotic spell. But adenosinetriphosphatase was not found in these extracts and remained neglected, an enzymological Cinderella, although it was known that if ATP is added to minced muscle it is rapidly split. But the extracted residue from which all the water-soluble enzymes have been removed was generally regarded as presenting no further interest and was discarded.

In collaboration with my wife, Militza Nikolaevna Lyubimova, my former postgraduate student, we undertook the study of the apparently evasive ATP-splitting activity. We turned our attention to the residue that remains after the extraction of the water-soluble enzymes. The very first experiment, prosaic in its simplicity, brought the unequivocal answer; an extremely high enzymatic activity exists in the "insoluble" part of muscle tissue, after the rather mild conditions of aqueous or saline extraction that are usually applied.

At this stage our merit was modest. Figuratively speaking, we simply took out from the waste-bucket the stuff that others rejected as not deserving attention. Much more significant was the next step, when we tried to isolate the ATP-splitting activity, that is, to separate the enzyme from the different water-insoluble proteins to which the function of contractility belongs.

Having once violated the canons by studying the residue instead of the extract, we continued to move along a heretical way, by using, instead of water or saline, concentrated salt solutions, which were known to extract the main contractile protein, myosin. It was natural to start with the removal of this protein, which was known to be the preponderant component of the insoluble protein fraction. Great was our astonishment, when after treating the residue with solutions of higher ionic strength, as used for the isolation of myosin, we found the full enzymatic activity present in the myosin-containing extract. This result was exactly opposite to our expectations. We intended to remove the bulk of "structural" protein and find the enzyme in some minute, individual fraction. On the contrary, the whole activity was found in the myosin fraction itself; all methods known at that time for the isolation of myosin invariably yielded products carrying the whole enzymatic activity. Moreover, the well-known great heat lability of myosin was found to be characteristic also for the enzymatic properties. Unexpected as it was, and apparently unlikely, the conclusion had to be drawn that the enzymatic, adenosinetriphosphatase activity belonged to myosin itself.

In our first publication in *Nature* (8) we ventured to introduce an abbreviated name for the enzymatic property of myosin, calling it ATPase instead of the cumbersome adenosinetriphosphatase, but the editors rejected it. By now it has won general acceptance and is used here.

Naturally, the fact that enzymatic properties were ascribed to a highly specialized protein that carries out structural and mechanical functions and constitutes a large percentage of the dry weight of our body, seemed highly improbable. No wonder that, regardless of strong experimental evidence, certain sceptical authors, some very eminent, raised objections and expressed strong doubts. But all these objections could easily be shown to be untenable and the initial statement remained unaltered. We had no reason to complain on the verdict of the most exacting judge—time. At present even the precise localization of the catalytic activity within the molecule of myosin is known.

The importance of the fact that myosin possesses ATPase activity needs hardly to be stressed. It means that the structural contractile substance that performs the mechanical work of muscle, itself provides the moving force for this work, by splitting ATP and liberating its chemical energy.

One can now wonder how it happened, that for such a long time the study of muscle proceeded by two completely independent, separate lines. On the one hand was the study of the building material of muscle, the proteins, which constitute the physical basis of the living machine. On the other hand was the study of chemical processes and the corresponding enzymes, which provide the moving force for the accomplishment of mechanical work. The finding that chemical and mechanical properties are combined in myosin, the establishing of its ATPase properties, served to fill out the gap that separated the two approaches, introduced a certain kind of unity, and replaced the former dualism.

But this was not the whole story. A new aspect, of not minor importance, was introduced when it appeared that interaction of ATP and myosin is of a bivalent, reciprocal character, going beyond the simple relation between an enzyme and its substrate. Namely, it was found that ATP changes the physical properties of myosin, for which there was every reason to expect some fundamental role in the mechanical effects involved in muscular contraction. Work on this line started almost simultaneously in two places, by the group of Joseph Needham in Cambridge and in our laboratory.

Evidently, the motivation was the same in both groups: if myosin acts on ATP and liberates its energy, does not ATP in its turn act on myosin, and produce changes in its physical properties that could play a role in the performance of work?

Needham and co-workers studied the physical properties (i.e. the viscosity and flow birefringence) of myosin solutions (9). In both cases addition of ATP produced strong effects, which disappeared as soon as the added ATP became split by the ATPase action of myosin. This clearly demonstrated the bilateral character of the ATP-myosin interaction as observed at or near the molecular level. Needham aptly named myosin a contractile enzyme (10).

I started with experiments with monomolecular layers of myosin, using a Langmuir trough with a rotating disc in the plane of the surface. The measurements of the viscosity of the monolayer were very impressive, and I remember the excitement of Needham, when on a visit to the USSR, he came to my laboratory and I demonstrated the apparatus (again, as in younger years, self-made partly from scrap) to him. But the effects of ATP were poorly reproducible, and I soon abandoned this approach, switching over to an object of higher structural order, myosin threads. These can easily be prepared by squirting a concentrated myosin solution into distilled water. They possess a small, but measurable degree of strength that can be registered by a torsion balance.

ATP has a well-reproducible, strong effect that greatly changes the extensibility of the threads. This appeared to depend on the ATPase activity: after treatment of the threads with very dilute silver nitrate solution, which completely inhibits ATPase activity, no effect of ATP on the mechanical properties of the thread could be observed (11).

These observations of mine were even reflected in an auto-epigram, which I attached to an amiable cartoon drawn by a friend of mine on the occasion of the award of the State Prize, which my wife and I received for our work on myosin. The drawing represented my wife holding onto my shoulders, hanging from a thin thread, which grows, downwards, into a solid rope.

The epigram, originally in Russian, and later translated by me into English, ran thus:

Those who envy will say: "on a thinnest thread
Holds the fame of this new laureate."
And I pray to the Almighty: "My only hope
Is that the myosin thread be as strong as a rope."

The subsequent discovery, by Bruno Straub in Hungary, of the second muscle protein, actin, intimately associated with myosin, brought new details to this field. Later investigations in our laboratory and others of spermatozoa, ciliae of unicellular organisms, protoplasmic flow, etc, have led to the conclusion that similar fundamental principles apparently hold true for biological motility in general. These experiments led us to formulate the principle of mechanochemistry of muscle (and other motile objects). Its essence is the assumption of a reciprocal interrelation between the source of chemical energy, ATP, on the one hand, and the basic component of contractile structures, myosin (or actomyosin, or myosin-like proteins), on the other. ATP provides the energy, and at the same time alters the mechanical properties of the contractile protein, myosin. The latter by its catalytic, enzymatic property, liberates the energy of the high-energy bonds of ATP, and at the same time undergoes changes of its physical state, which are responsible for the production of mechanical work.

Many years earlier Meyerhof gave a very concise formulation, based on clear thermodynamic considerations, of a principle well-known and almost obvious, which must be expected to be operative in a machine where transformation of energy takes place. These are his words: "In any machine, if it is not a heat engine, the energy-producing chemical process must interfere with the structural base of the machine, in order to produce the changes which are the source of the work performed." Is it not astonishing, that this succinct statement was not applied, for more than a decade, to the phenomenon of muscle movement? The discovery of the ATPase properties of myosin provided an excellent place to test the validity of the above-mentioned principle.

Administrative Work

Years went on, and duties unrelated to research work, accumulated, of an organizational, administrative, and social character. Managing human affairs, even of a scientific nature, required increasing amounts of time that couldn't be compensated for by more work at the bench, as had been my custom. In all the investigations mentioned above I took an active part by experimenting with my own hands.

It is to Archibald Hill that I owe my introduction into the attractive, but time-consuming engagement in scientific public affairs. It was during his presidency of the International Council of Scientific Unions (ICSU) that I was invited to become a member of its Bureau, and in later years Vice-President. This gave me fascinating opportunities to meet and often come in close contact with a great variety of scientists, both in my own field of biochemistry, and in the worldwide scientific community represented in this international organization. I remember with great affection my years with the ICSU. After Hill, Rudolf Peters took over the presidency and his unforgettable charm remains with me over the years. He masterfully managed to keep an excellent atmosphere during the meetings of the ICSU Bureau, despite the sharp controversies that sometimes arose. I recollect vividly the clash of opinions between myself and Lloyd Berkner, of the USA, when I strongly opposed what I considered to be his intentions to make ICSU dependent on "Big Business," as represented by various American trusts and foundations. I was happy when my point of view became adopted and the really international and independent character of ICSU was supported by the other members. The wise leadership of Peters helped to keep up the fruitful, friendly collaboration that always permeated our work.

With modesty I feel that I can attribute to my efforts an important result achieved during these years: the acceptance of the German Democratic Republic as an ICSU member. The opposition was tacit but strong, and my struggle lasted long; it seems that I exhibited unexpected diplomatic abilities. Anyhow the result was achieved, to my great satisfaction. It was the first time that the GDR, of the socialist fraternity, became a member of an international, powerful organization.

My experimental work came to a complete standstill when the Academy of Sciences of the USSR assigned me a difficult charge relating to its diverse biological research institutes, by appointing me to the post of Academic Secretary in Biology. These were the years when Lyssenko was still in full power, and the situation of biological science within the Academy and in the country as a whole was, in many respects, in a highly unfavorable state, to say the least. For several years I carried out my duties as best I could, but evidently did not fulfill expectations. I was finally dismissed.

An excellent recompense for my dismissal was given me in charging me to organize a new institute for the study of the physical and chemical basis of life phenomena. The duties of Director of this Institute of Molecular Biology I continue to carry out. The Director's study and writing desk have replaced my laboratory bench; a skillful secretary helps me instead of a talented collaborator. I do what I can to keep abreast of the work going on in several teams of younger and middle-aged personnel of the Institute, and to provide the necessary material and human conditions for carrying out their research and to ensure a favorable atmosphere of friendship, collaboration, and mutual respect.

The organization of our Institute marked a turning point in the development of physico-chemical biology in our country. It was the first and only center of its kind at that time. We were successful in attracting a number of able chemists; biologists were easier to find, and physicists were fewer in number, but very eager to bring all the help necessary. My leading idea was to have something like a 30:30:30 ratio of chemists, physicists and biologists. We had to start from practically zero level, but the enthusiasm was great and the support from the Academy generous. The Institute organized the publication of a large series of monographs on most important aspects of molecular biology. It was a great relief to me and a happy chance for the growth and successive development of the Institute that a considerable number of excellent scientists of the intermediate and younger generations joined the staff at the early stages of its existence. Particularly fruitful was the renewed partnership with my former pupils, some of whom, as for instance Alexander Bayev or Alexander Braunstein, had been among my postgraduate students in a rather remote past, and now formed the backbone of the newly created Institute. We owed to Bayev and his group the first outstanding success in the field of nucleic acid study. It resulted in establishing the full primary structure of the second nucleic acid to become known after the first achievement in this line, the work of Holley on alanine tRNA. Two years later Bayev's group gave the structure of the valine tRNA. Continuing his work at the Institute in the field of genetic engineering, Bayev is at present Past President of the International Union of Biochemistry, under the ICSU.

The formation of the Institute of Molecular Biology occurred first under another name: "Institute of Physico-Chemical and Radiation Biology." This was a kind of camouflage, because at that time the very name molecular biology sounded somehow suspicious. Only after a few years, when the reputation of the Institute became well-established and the general situation changed for the better, did I ask our President of the Academy, M. Keldysh, on the occasion of an anniversary of some kind, whether he would like to make me a present. There was some anxiety in the expression on his face when he asked what I would particularly like to have. I told him: "Just to change the name of our Institute to what it should be." With evident relief he immediately agreed, and at the next meeting of the Academy's Presidium the matter was settled and the Institute obtained its present name.

The strained situation with even the concept of molecular biology appeared clearly during the preparation of the XVth International Congress of Biochemistry, which was held in Moscow in 1961. There was a section on problems of molecular biology, where Max Perutz and myself were the joint organizers. We intended to name it simply "molecular biology section." But when I proposed this at a meeting of the Moscow group of our organizing committee, strong objections were raised by the more orthodox members: "What is molecular biology? We do not know such a science or branch." To satisfy my opponents (I was in a minority of one!) I invented a euphemistic name, "biological functions at the molecular level." This was accepted.

The Institute is now twenty years of age. Affiliated with the institute is a "Scientific Council on problems of molecular biology." It consists of about 20 members, who represent the main centers of research in the field of molecular biology, which have grown in the Soviet Union during these last decades. At the annual meeting of the Council or of its Bureau the results obtained during the period are summed up, and plans for future activity, steps needed to avoid unnecessary duplication or to fill undesirable gaps in the treatment of studies on urgent problems that present special interest and importance are discussed. Special attention is paid to the effective planning of the publication of the corresponding scientific literature, the organization of reference editions etc. Of great significance for the progress of the young science was the organization of the specialized journal "Molecularnaja Biologia," where the scientific production of the country is published. The journal appears also in English translation in the USA.

In order not to break my bonds to concrete science, I engaged in a project named "Revertase," designed to develop on a broader base studies of the enzyme of the so-called reverse transcription. The project serves to coordinate and support research going on in research centers in Moscow, Kiev, Novosibirsk, Riga, and others in the USSR, and also in the German Democratic Republic, Czechoslovakia, Poland. Winter schools on various relevant topics are organized yearly by the Council in Moscow or other cities. Directing these schools gives me the refreshing opportunity to keep in closer contact with the youngest generation of scientists, on whom the future growth and development of our science depends.

Usually in biographical articles, even in shorter notes of a questionnaire type, some place is given to the topic of "Hobbies." I could mention only a single one, in my younger years, mountaineering. Mostly accompanied by my wife, who bravely shared the difficulties, we visited first the mountain ridges of the central Caucasus, later followed high passes and glaciers of the Pamirs, and then the mountain crests in Tien-Tchang, on the border of China. Fresh in my memory is the day, when, in company with the Swiss yachtswoman-writer-mountaineer, Ella Maillart, we stood on top of a steep ridge, over 4 thousand meters in altitude, which formed the frontier between the USSR and China. We stood with one foot on Soviet soil, the other on Chinese territory, with the hazy sky extending over the great desert of Takla Makan before us. Mountain friendships are among the strongest. I was happy when I was able to pay a visit to Ella at her chalet in a valley near the Matterhorn. I attended a conference in Arolla, the same Valais canton in Switzerland, and a friend, Guido Pontecorvo, arranged a trip by car to Chandolin, to meet Ella after a lapse of some forty years. Great was our joy!

One is well justified to regard this century as crucial for both my country and for the world as a whole. My fatherland has passed through several wars: the Russian-Japanese, when a vast but feeble empire was defeated by a small, almost negligible neighbor; the First World War, when the country was again defeated, but also liberated of the rotten tzarist regime; the civil war of the young new state against the remnants of the old regime, which were supported by armed forces of about a dozen interventionist and occupational countries: the English in the Crimea and the Caucasus, whose aim was the rich oilfields, The French in Archangel, the Japanese in the Far East, the Austrians Rumanians on the west front, and so on. And finally we suffered the terrible Second World War. We had two revolutions in 1905; that drowned in blood by Nicolas II, and the victorious Great October Revolution. This changed the world and created a new society in which, I believe, human values—brotherhood, social equality and freedom from exploitation—replaced the "anti-values" of previous generations—wealth among pauperism and profit as a dominant directive force of the whole social structure. In the new society pecuniary wealth has disappeared and has given place to social, spiritual, and intellectual values.

There have been years of great hardships and struggles, also of great triumphs. I think that both kinds of experiences helped me to develop my strong optimism. Recently my friends, cineasts, were showing a film "Faustian Tale," in which I took part. They asked me in their professional language whether, if presented with such a possibility, I would change the film of my life to something different. I replied that fate has been benevolent to me, and that I have no complaints with the scenario that it chose for me. My wish would only be for a prolongation of the lives of my mother, who died in middle age because of a street accident; and of my wife, my true and devoted life-companion, whom I lost a few years ago. For the rest I have no complaints; the reel could well be turned again.

A beautiful verse by our great lyricist, Tyutchev, is always in my mind. It is with his lines that I close my article. I could not find in Moscow an English edition of Tyutchev's poetry, and I am not a Marshak, the excellent translator of verse from English into Russian and vice versa. So I have to rely on the indulgence of my readers when I give a word for word unrhymed translation, knowing well that much of the charm of Tyutchev is lost:

> Blessed be those who visit our world
> At its crucial, fateful moments.
> They have been invited by the celestial powers
> To partake at their convivial feast.

Literature Cited

1. Engelhardt, W. A., Kisselev, L. L., Nezlin, R. S. 1970. *Monatsh. Chem.* 101:1510–17
2. Engelhardt, W. A. *Biochem. Z.* 1930. 227:16–38
2a. Engelhardt, W. A. 1932. *Biochem. Z* 251:343–68
3. Wenkstern, T. W., Engelhardt, W. A. 1959. *Folia Haematol.* 76:362–71
4. Engelhardt, W. A., Sakov, N. 1943. *Biokhimiya* 8:9–36
5. Burk, D. *Cold Spring Harbor Symp. Quant. Biol.* 1939. 7:420–59
6. Passoneau, J. V., Lowry, O. H. 1962. *Biochem. Biophys. Res. Commune* 7:10–15
7. Engelhardt, W. A. *Mol. Cell. Biochem.* 1974. 5:25–33
8. Engelhardt, W. A., Ljubimova, M. N. 1939. *Nature* 144:668–69
9. Needham, J., Kleinzeller, A., Miall, M., Dainty, M., Needham, D. M. 1942. *Nature* 150:46–49
10. Needham, D. M. 1972. *Machina carnis.* Cambridge: Cambridge Univ. Press
11. Engelhardt, W. A. 1946. *Adv. Enzymol.* 6:147–91

Ann. Rev. Biochem. 1982. 51:21–59

ENZYMOLOGY OF OXYGEN

Bo G. Malmström

Department of Biochemistry and Biophysics, University of Göteborg
and Chalmers University of Technology, S-412 96 Göteborg, Sweden

CONTENTS

INTRODUCTION

The enzymology of oxygen as a field of scientific inquiry is more than one hundred years old. In the period 1858–1886 Traube in Germany carried out investigations, partly summarized in his monograph *Oxydationsfermente* (1874), which led him to suggest that biological oxidation involves the activation of molecular oxygen by intracellular enzymes [for a detailed documentation of the historical notes, see (1, 2)]. This concept was soon provided with a firm experimental basis through the studies of Yoshida (1883) in Japan and Bertrand (1894) in France on the enzymes involved in the darkening and hardening of latex from Oriental lacquer trees. Bertrand gave the name laccase to the enzyme responsible for the oxidation of the phenolic substances (urushiol, laccol) in the latex. He later found laccase in several mushrooms, which also contained a distinctly different phenol oxidase, named tyrosinase. Bertrand introduced the term oxidase for these enzymes and suggested that they are catalytic metalloproteins.

21

0066-4154/82/0701-0021$02.00

In the beginning of this century Bach and Chodat proposed that biological oxygen activation involves a reaction with an acceptor molecule, catalyzed by an "oxygenase," followed by oxidation of the organic substrate, catalyzed by a "peroxidase." Their hypothesis did not receive any experimental support, however, and it was soon abandoned, as a completely different picture emerged from investigations largely by Thunberg, Wieland, Warburg, and Keilin. Their work demonstrated that biological oxidation of organic substrates generally involves dehydrogenation, the hydrogen acceptor being an organic coenzyme, e.g. NAD. Dioxygen reoxidizes the reduced co-enzyme. This occurs in the respiratory chain, whose terminal component, cytochrome oxidase, catalyzes the reduction of dioxygen to two molecules of water.

It was soon found that many other oxidases reduce dioxygen to the level of peroxide only, but in no case was the insertion of atoms from dioxygen into the organic substrate demonstrated. In those cases in which oxygen was incorporated into organic compounds, it was assumed to be derived from water through the addition to a double bond. Not until the 1950s was it unequivocally demonstrated that there exist true oxygenases, which catalyze the insertion of oxygen atoms from dioxygen in organic substrates. This occurred in pioneering experiments of Hayaishi and Mason, using oxygen isotopes.

There are now more than 200 enzymes known that have molecular oxygen (dioxygen) as one of their substrates (3). These are divided into two main categories: (*a*) **oxygenases** and (*b*) **oxidases.** Oxygenases catalyze reactions in which atoms of oxygen are incorporated into organic substrates. They are in turn grouped into two classes: (*a*) **dioxygenases,** which catalyze the insertion of both atoms of dioxygen into the organic substrate, and (*b*) **monoxygenases** (mixed-function oxidases), which catalyze the insertion of one atom of dioxygen; the other one is reduced to water. In oxidase-catalyzed reactions the dioxygen molecule functions as an electron acceptor only, and is reduced to O_2^-, H_2O_2, or $2H_2O$.

Just a list of the relevant enzymes, with a few key references to each one, would easily fill the allocated space for this review. It is therefore understandable that the whole area has only been treated once in this series, by Mason in 1965 (4). Limited parts of it have, however, been reviewed later: porphyrin enzymes in 1967 (5), flavoproteins in 1967 and 1970 (6, 7), enzymic hydroxylation in 1969 (8), oxygenases in 1975 (9), chemical models for redox enzymes in 1978 (10), and cytochrome P-450 in 1980 (11).

This review attempts a survey of the whole field. It does not, however, provide a detailed documentation of the recent work on all the individual enzymes. Instead the treatment is entirely focused on work that has yielded significant information on the molecular mechanism of dioxygen activation and reduction. To this end a few key enzymes that illustrate the major types of reactions with dioxygen are discussed in some detail, whereas I may not even mention those enzymes for which knowledge is limited to the nature of the overall reaction, the identity of the prosthetic group, or the reaction with the reducing substrate. Oxidases are stressed partly because oxygenases have been covered occasionally (9–11), but also because oxidases represent the author's main research interest. One oxygenase, cytochrome P-450, was extensively reviewed two years ago (11), but a short discussion of dioxygen binding and activation in this enzyme is still included here, because a major purpose of this chapter is to see if some general principles of dioxygen activation by enzymes can be extracted, and cytochrome P-450 is a particularly well-characterized enzyme with respect to the dioxygen reaction. Enzymes that have dioxygen as a product rather than as a substrate (e.g. catalase and superoxide dismutase) are not discussed. Oxygen-binding proteins are also outside the scope of the chapter. As O_2 activation involves a binding step, some information on proteins and on model systems interacting reversibly with O_2 is, however, introduced at appropriate points.

As the topic has not been reviewed in its entirety since 1965 (4), the discussion is not limited to papers published within the last few years, although the emphasis is on recent literature.[1] In many cases, particularly in the sections on dioxygen chemistry and model compounds, reference is, however, made to important review articles rather than to original research papers. Several comprehensive reviews of oxygen enzymology have appeared in recent years (3, 12–26).

The problem of mechanism is introduced by a brief consideration of the chemistry of dioxygen, followed by a discussion of chemical models for oxygen-binding proteins and oxygen-activating enzymes. This is followed by a survey of oxygen activation by oxygenases and oxidases. The main review is preceded by a short summary of some principles of dioxygen activation.

PERSPECTIVES AND SUMMARY

In this section some of the points discussed in detail in the main parts of the review are briefly summarized without documentation.

All enzymes activating and reducing dioxygen are conjugated proteins. The prosthetic groups involved are flavins, copper, and iron (3). Many of the enzymes are heme proteins, but nonheme prosthetic groups are common in oxygenases and are also found in some oxidases. Some flavoenzymes

---

[1]This review was finished in July 1981, thus contributions in nonEuropean journals after May 1981 could not be considered.

contain molybdenum as well as iron-sulfur centers, but these groups do not appear to be directly involved in dioxygen reduction.

The dioxygen reaction usually occurs at a prosthetic group. No single reaction principle can be discerned for all prosthetic groups, but in a broad sense three types of dioxygen activation only are involved:

1. combination of O_2 with a fully reduced flavin to yield a flavin peroxide, whose further reactions are controlled by specific effects in the active site;
2. binding of O_2 to a reduced metal center with varying degrees of electron transfer from the metal to O_2;
3. activation by an oxidized metal center of the organic substrate, which can then react directly with dioxygen bound in the active site.

There are two main factors making dioxygen kinetically inert: (a) its triplet ground state, which imposes spin restrictions on certain of its reactions, and (b) the negative standard reduction potential for the one-electron reduction of O_2 to the superoxide radical, O_2^-, which makes this a nonspontaneous process with electron donors that have higher potentials. Reduced flavins can easily overcome both restrictions. First, a one-electron oxidation of a fully reduced flavin produces a flavin radical, so that spin is conserved when O_2 is the acceptor. Second, flavins usually have low reduction potentials, so that their reduction of O_2 to O_2^- will be associated with a negative free-energy change. With nonenzymic systems, generation of O_2^- in the reduced flavins with O_2 is readily demonstrated. One might thus expect that flavin-containing oxidases producing H_2O_2 actually reduce O_2 to O_2^- only in a one-electron transfer, H_2O_2 being formed in a subsequent dismutation reaction. Even if xanthine oxidase, for example, can produce O_2^-, there is evidence that in catalytic turnover with excess reducing substrate, O_2 is directly reduced to H_2O_2 in a two-electron process. The flavin-containing oxygenases do not seem to form O_2^- either; the active dioxygen species is instead a flavin hydroperoxide.

The second category of O_2 activation is the most common, and it is also best understood from extensive model studies as well as from direct characterizations of oxygen intermediates in the active sites of several enzymes. The first step in the activation involves binding of O_2 to a low-valence form of the metal prosthetic group (Fe^{2+}, Cu^{1+}). This entails internal electron transfer from the metal to oxygen, the extent of transfer in a given protein being one factor determining whether this is going to function as a reversible oxygen binder, as an oxygenase, or as an oxidase. The extent of electron transfer is controlled in a subtle manner by variations in protein ligands and by specific effects of the protein conformation. This adjustment of the reactivity through the protein structure is understood in the most general terms only, and there has been limited success in reproducing it in model systems. Small heme compounds mostly react irreversibly with O_2, even if it has been possible to make O_2-binding model compounds by the introduction of certain steric restraints. In proteins the heme group can, however, catalyze oxygenation reactions, for example in cytochrome P-450, as well as direct electron-transfer to O_2, for example in cytochrome c oxidase.

Dioxygen activation by metal prosthetic groups may involve mono- as well as bimetallic centers. In the few oxidases reducing O_2 to $2H_2O$, the O_2-reducing sites are always bimetallic, for example Cu_B-cytochrome a_3 in cytochrome oxidase. One reason for this is undoubtedly that, for thermodynamic reasons discussed in a later section, bimetallic centers can bind O_2 at higher reduction potentials than monometallic sites, and these enzymes have high potentials. Other reasons are the possibility of a two-electron reduction of O_2, with H_2O_2 as the first reduced intermediate, thus bypassing the thermodynamically unfavorable formation of O_2^-, and the increased capability to stabilize the variety of intermediates that must be formed in the four-electron reduction of O_2.

The third category of O_2 activation, which does not involve the metal prosthetic group directly, seems to be limited to reactions in which there is a formal two-electron reduction of O_2. Galactose oxidase appears to be an example, but this type of activation is also observed in some oxygenases, for example protocatechuate dioxygenases.

In addition to the types of O_2 activation described so far, a number of other effects may increase the reactivity of O_2 in enzymes. Increased hydrophobicity in the active site may be an important factor; in some model systems there is, for example, a need for the use of aprotic solvents. In addition, the binding of O_2 creates favorable entropic conditions for the reaction. Finally, metal-O_2 complexes are generally diamagnetic, so that the spin restriction has been removed; however, formation of a diamagnetic complex instead puts energetic requirements on the O_2-binding step.

For a few enzymes—for example, cytochrome P-450, laccase, and cytochrome c oxidase—there is particularly detailed information available on the mechanism of the dioxygen reaction. Even in these cases the mechanisms are rather sketchy, and some steps are incompletely understood. This applies particularly to the breaking of the O–O bond and to the proton transfers involved in the formation of H_2O. The mechanistic knowledge for these enzymes is, on the other hand, considerably more advanced than for the few "simpler" catalysts devised by coordination chemists, where information on reaction intermediates is at best indirect. Despite their complexity, enzymes, thanks to their specificity and to the binding of reactants in active sites with a unique structure, often allow a more rigorous mechanistic characterization than small-molecule catalysts. Knowledge of enzyme mechanisms may indeed provide major clues to a rational design of new catalysts for O_2 reactions.

THE CHEMISTRY OF DIOXYGEN

The chemical properties of O_2, with particular reference to its role in biological electron-transfer reactions, have been reviewed repeatedly (27–35). Consequently, only a brief summary and discussion of some recent work is given here.

One reason for the prominent role of O_2 in biological redox reactions is its high reduction potential, i.e. it is thermodynamically a good oxidizing agent. In that respect it is not unique; fluorine, for example, is an even more powerful oxidant. But whereas fluorine and other halogens are very reactive, O_2 is kinetically quite inert. At physiological temperatures it will therefore react only in the presence of a catalyst, and this is an important aspect of its biological fitness (28). Through the specificity of enzymes the chemical selectivity required in life processes can thereby be achieved. At the same time it is this kinetic inertness that poses the mechanistic problems forming the focus of discussion in this chapter.

The low reactivity of O_2 is intimately related to its peculiar electronic structure (36). Dioxygen has a triplet ground state, $^3\Sigma$, whose molecular orbital description is $(\sigma_s)^2(\sigma_s^*)^2(\sigma_p)^2(\pi_p)^4 (\pi_p^*)^1(\pi_p^*)^1$. As shown, the two unpaired electrons, which give the molecule its paramagnetism, are found in two degenerate antibonding π_p^* orbitals. The bond order is 2, with one σ and one π bond; the bond length is 1.21 Å. When electrons are added to the π_p^* level to give O_2^- and O_2^{2-}, the bond order changes to 1.5 and 1, respectively, and the bond length to 1.28 and 1.49 Å, respectively. It is thus seen that addition of electrons weakens the O–O bond. As is discussed in some detail below, binding of O_2 to a transition metal ion is associated with electron transfer from the metal ion to O_2. The consequent weakening of the O–O bond in the complex is probably an important feature in O_2 activation by metalloenzymes. In model systems there is, however, no simple correlation between bond length and reactivity (33).

There are no unpaired electrons in H_2O_2 and H_2O, so that the reduction of O_2 involves a forbidden spin change, which is a major reason for the kinetic inertness. There are two ways to overcome this restriction. In a radical chain mechanism, in which the first products are O_2^- and an organic free radical, spin is conserved ($S = 1$ for reactants as well as products). Such radical mechanisms are common in model systems, but evidence suggests that they do not occur in enzymes. Spin can also be conserved if the electron donor is a transition metal ion, which itself has unpaired electrons, as for example in the reaction Fe^{2+} ($S = 2$) + O_2 ($S = 1$) → Fe^{3+} ($S = 5/2$) + $O_2^-(S = \frac{1}{2})$. This is often stated to be the reason that oxidases contain transition metals. It is, however, forgotten that enzymic reduction of O_2

does not occur in bimolecular collision reactions. Instead the initial step is the binding of O_2 to a metal ion in the active site, and the complex so formed is expected to be diamagnetic. Thus the problem of spin restriction is shifted from the electron transfer to the binding step in the catalytic sequence. This imposes specific demands on the electronic structure of the O_2-binding center, as is discussed in the next section.

The reduction potentials for the successive steps in a stepwise reduction of O_2 are other important properties governing O_2 reactivity. The potentials (in volts) at 25°C and pH 7 are summarized in Scheme 1 (28, 30, 37, 38):

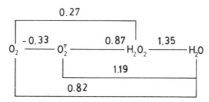

Scheme 1

The value of the reduction potential for the O_2/O_2^- couple is difficult to determine experimentally, and results ranging from +0.005 V (39) to –0.59 V (28) have been reported. Two recent estimates (37, 38) have, however, given the value –0.33 V. The value is important, as it imposes a kinetic restriction on the one-electron reduction of O_2 to O_2^-, depending on the reduction potential of the electron donor. Consider the following reaction (Scheme 2), in which M is a metal and the numerical values are reduction potentials:

$$- 0.33 \text{ V}$$
$$M^{2+} + O_2 \longrightarrow M^{3+} + O_2^-$$
$$+ 0.27 \text{ V}$$

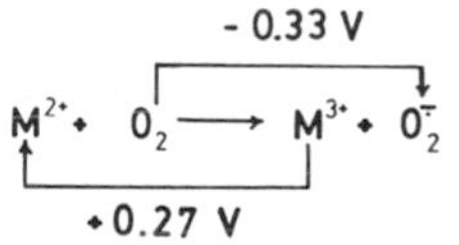

Scheme 2

The reaction from left to right is associated with a change in standard potential of –0.60 V, which corresponds to a positive standard free energy change (ΔG°) of 58 kJ·mol^{-1}. This would also be the minimum activation energy in any mechanism having this reaction as an elementary step. The value 0.27 V was chosen, because it represents the mean reduction potential for the hemes in cytochrome oxidase. The reduction of O_2 to O_2^-, by reduced cytochrome oxidase would thus be associated with a considerable energy barrier. The corresponding barrier would be as high as 107 kJ.mol^{-1} for fungal laccase, which has a reduction potential of 0.78 V. A one-electron reduction of O_2 by reduced laccase would consequently be expected to be extremely slow.

The argument given so far may not be relevant for the enzymic mechanisms, because in these O_2 is not reduced in bimolecular collision reactions. A sufficiently strong binding of the reduction product, O_2^-, could thus eliminate the barrier. It has, in fact, been claimed (38) that the restriction no longer applies to cytochrome oxidase if the enzyme has an affinity for O_2^- of 10^4–10^5 M^{-1}, which is quite a reasonable binding constant. The same potentials as given here were used, therefore this value must involve a miscalculation; the required affinity really is 10^{10} M^{-1}. Even if such a strong interaction is not impossible, it would make the next reductive step correspondingly less favorable, i.e. 0.27 V instead of 0.87 V. This is a valid objection, as cytochrome oxidase must reduce O_2 all the way to the level of H_2O. With laccase the barrier is so large, that no reasonable binding constant can eliminate it. In fact, with this enzyme, even a two-electron reduction to H_2O_2 is still associated with an appreciable increase in free energy.

The kinetic consequences of the thermodynamic restrictions just discussed are the reason that transition metal ions often reduce O_2 directly to H_2O_2 in a two-electron process (29). The same arguments could also explain why the O_2-reducing sites in enzymes, such as cytochrome oxidase and laccase, are bimetallic complexes that can function as two-electron redox centers.

In enzymes having redox centers with a reduction potential lower than -0.33 V, the reduction of O_2 to O_2^- would, of course, be a spontaneous process. In flavoproteins it may thus be expected that the primary reduction product of O_2 will be O_2^-, which then forms H_2O_2 in a dismutation reaction. Xanthine oxidase can, for example, be shown to generate O_2^-, but apparently this occurs only under conditions when a considerable amount of partially reduced enzyme is formed (40). Under turnover conditions with excess-reducing substrate the fully reduced flavin is produced, and this yields H_2O_2 without the intermediate formation of O_2^-.

CHEMICAL MODELS FOR PROTEINS BINDING AND ACTIVATING DIOXYGEN

There has been a great interest among coordination chemists in metal complexes that can bind or react with O_2. Most simpler complexes that bind O_2 reversibly, do not, however, closely resemble biological O_2 carriers. Although considerable success has been achieved in the synthesis of O_2-binding iron porphyrins (35), there are no simple bimetallic complexes of copper or nonheme iron binding O_2, despite the importance of such centers in biology (e.g. hemocyanin and hemerythrin). The most common metal in synthetic O_2-binding complexes is cobalt, which is not found in any oxygenase or oxidase. Nevertheless, the simpler systems may reveal certain principles governing the interaction between metals and O_2, and hence a brief discussion is presented here. Several reviews dealing with synthetic O_2 carriers are available (33, 35, 41–46).

Binding is associated with electron transfer[2] from the metal ion (M^n) to dioxygen: $M^n + O_2 \rightarrow M^{n+1} \cdot O_2^-$. Therefore, the metal is always a transition element, which can exist in more than one valence state (43), and in the O_2–binding complex the metal is in its low-valence form (Co^{2+}, Fe^{2+}). With simple complexes a second metal ion often combines with $M^{n+1} \cdot O_2^-$ to form M^{n+1}-O-O--M^{n+1} (35).

In mononuclear complexes the tendency to bind O_2 is related to the reducing power of M^n, i.e. it is inversely related to the standard reduction potential of the complex, as has been established by changing the metal with the same ligand (e.g. tetraphenylporphyrin) or by changing the ligands with the same metal (e.g. Co^{2+}) (35, 47). Thus, if the potential is too high, i.e. the complex is an oxidant rather than a reductant, there will be no binding; whereas, if it is too low, the binding will be irreversible. One important factor behind this tendency is related to the spin restriction associated with the formation of the complex, which is generally diamagnetic. The reason for this is that there is no way of accommodating the two unpaired electrons of triplet O_2 in the ground state of the molecular orbital scheme for the complex, because of its low symmetry. Thus, only singlet O_2 will bind. In the absence of photochemical excitation, binding will therefore only occur if the driving force for electron transfer from the metal ion to O_2 is large, i.e., the reduction potential is sufficiently low. Of course, the relevant potential is really that of the O_2 complex itself, so the relation cannot be expected to be strict (47).

There is the same type of relation between potential and O_2 affinity in binuclear complexes, but binding occurs at higher absolute values of the reduction potential (35, 47). This can be explained by the fact that once the spin restriction has been overcome in forming the first M–O bond, then the second O–M bond can be very strong. Thus, the reaction forming the second bond drives the energy-requiring formation of the first one. This effect may be one reason why high-potential oxidases, such as laccase, have binuclear O_2-reducing centers.

Since the time of the classical paper on oxyhemoglobin by Pauling & Coryell (48), there has been considerable controversy regarding the manner in which O_2 is bound in metal complexes; for example, whether the bound species can formally be regarded as O_2, O_2^-, or O_2^{2-} (35, 49). Measurements of the O–O stretching frequencies in metal-O_2 complexes have, however, shown that these are similar to those found for compounds containing ionic

---

[2]The discussion in this section is intentionally oversimplified. Strictly speaking, of course, bound O_2 is not present as O_2^- or O_2^{2-} in the complex.

superoxide (O_2^-), or peroxide (O_2^{2-}), which suggests a substantial transfer of electron density from the metal to oxygen (35). This conclusion is also supported by Mössbauer and photoelectron spectroscopy (35), and the optical spectra of metal-O_2 complexes have been interpreted on the same basis (50). Thus, the structures derived from spectroscopic measurements are consistent with those suggested by the thermodynamic arguments given earlier.

In oxygenases and oxidases the reactivity of the bound O_2 is probably to a large degree determined by the extent of electron transfer from the metal. This is regulated by the nature of the protein ligands and by subtle effects from the protein conformation. For example, in cytochrome P-450 the axial protein ligand to the heme iron is believed to be a thiol group (51). This will increase the tendency for electron donation from the heme iron, an effect that could partly explain why this protein is a hydroxylating agent rather than just an O_2 binder. In oxidases, carefully balanced protein effects must adjust the redox properties of the metal centers, so that not only O_2 but also oxygen intermediates are bound without the formation of irreversible complexes.

In addition to the effects discussed so far, entropy changes are important in determining the O_2 affinity of a metal complex. This is generally explained by the formation of the strongly polar $M^{n+1} \cdot O_2^-$ bond causing solvent reorganization, which thermodynamically is manifested as a negative contribution to the entropy of binding (35). In a protein the hydrophobic metal site may, however, already be organized to accommodate a polar metal-oxygen group. Thus, the entropy of binding may be more favorable. In addition, the hydrophobic environment in the active site may be a factor in preventing irreversible oxidation, as aprotic solvents have been seen to have this effect in model complexes (41).

Attempts to make simple complexes that not only bind O_2 but also mimic oxygenases and oxidases have been largely disappointing. There is a vast body of work on metal-catalyzed oxygen insertion (35, 52, 53), but most of the successful catalysts have been found to operate by mechanisms that must differ from those of the biological systems. In many cases, for example, the model systems involve metal-catalyzed free-radical processes. There are, however, a few systems that may have biological relevance. For example, Co^{2+} complexes of *bis*(salicylidene)ethylenediamine and of tetraphenylporphyrin have been shown to catalyze the insertion of dioxygen in the oxidative cleavage of indoles (35, 52). The reaction may mimic that catalyzed by indoleamine 2,3-dioxygenase, as with tryptophan as a substrate the products are the same as those formed in the natural system. Recent studies of ferryl complexes in oxygen atom transfer (52–54) may be of interest in relation to the mechanism of cytochrome P-450 (51).

There are no simple metal complexes that catalyze the reduction of O_2 to $2H_2O$ as efficiently as oxidases, e.g. laccase. In most cases, if a third electron is added to a complex of the type $M^{n+1}-O^--O^--M^{n+1}$, a metal ion, and not the peroxide, becomes reduced, and the complex decomposes. Some moderately effective catalysts have, however, been synthesized. An interesting recent example is a cobalt porphyrin complex that catalyzes the electrochemical reduction of O_2 to $2H_2O$ (55). It was found that H_2O, rather than H_2O_2, was produced with bimetallic complexes only, and the distance between the two cobalt porphyrins was critical. This was taken as evidence for a μ-peroxo intermediate, similar to those postulated in the laccase and cytochrome oxidase reactions.

OXYGENASES

Flavin-Containing Oxygenases

There are very few flavin-containing dioxygenases (3), and very little information about their dioxygen activation. For a number of monoxygenases, on the other hand, the reaction mechanism has been characterized in considerable detail. This is particularly true for p-hydroxybenzoate hydroxylase, which is the focus of this section.

Good reviews of the roles played by flavins in biological redox reactions have been written by Hemmerich & Massey, together (56, 57) as well as individually (58, 59). Flavin-enzymes catalyze dehydrogenation (including transhydrogenation), electron-transfer, and O_2-activating reactions. These activities are in many cases combined; for example, in some oxygenases the flavin is reduced in a dehydrogenase reaction prior to reacting with O_2. The activation of O_2 often involves the rapid formation of a flavin C (4a) peroxide:

1.

The reactions of such flavin peroxides with O_2 in nonenzymic systems have been discussed by Bruice (60, 61). There are, however, notable differences between the models and the enzymes (58). Natural O_2 activation has $2e^-$ or $4e^-$ stoichiometry and yields H_2O_2 in the oxidases, or oxene (31) and H_2O

in the oxygenases, without the formation of radical intermediates. Artificial flavin-activation, on the other hand, shows $1e^-$ stoichiometry and gives O_2^-, together with blue flavin radicals. These blue radicals are neutral, whereas the red radicals observed in some oxidases (though not formed in natural turnover) are anionic.

p-Hydroxybenzoate hydroxylase catalyzes the hydroxylation of the aromatic substrate with the concomitant dehydrogenation of NADPH (56). Extensive steady-state kinetic data (62) are consistent with a ping-pong mechanism. First a ternary complex of the enzyme with p-hydroxybenzoate and NADPH is formed in a random reaction. In this complex the flavin is reduced, with the release of $NADP^+$. The complex between the reduced enzyme and the substrate then reacts with O_2, which yields 3,4-dihydroxy-benzene and the oxidized enzyme.

The reaction of O_2 with the complex between the reduced enzyme and p-hydroxybenzoate has been studied in detail by the stopped-flow technique (62, 63). Three intermediates have been detected. Intermediate I is formed in a second-order reaction with O_2 without any oxygen transfer, and is presumably a C(4a) adduct (Formula 1). This structure is supported by the spectral similarity with a compound made by adding a synthetic C(4a) adduct to the apoprotein (64), and with C(4a) peroxides having N(5) blocked in aprotic solvents (65). A similar intermediate has also been proposed in bacterial luciferase (65).

The spectrum of intermediate III is similar to that of I (63). It is probably a C(4a)-hydroxyflavin, which regenerates the oxidized flavin by the elimination of H_2O. The real clue to the hydroxylation reaction would seem to center around intermediate II, but unfortunately its structure is still a matter of considerable debate. Massey and his colleagues (63) have suggested a structure with one flavin ring open between C(4a) and N(5), but this proposal has been criticized on the ground that the high extinction characterizing II would not be expected (58). As an alternative it was suggested that the actual hydroxylating agent is a transiently formed C(10a) peroxide and that consequently the ring opening in II occurs between C(10a) and N(10). Such a mechanism is, however, made very unlikely by the recent crystal structure determination (66) of the enzyme in a complex with its substrate. This shows the N(5) edge of the isoalloxazine ring oriented towards the p-hydroxybenzoate.

The aromatic ring is not generally hydroxylated by H_2O_2, whereas this reacts very readily with the thiol group of p-mercaptobenzoate (67). With the thiol compound used as a substrate the rate of formation of intermediate II is significantly enhanced (67), consistent with the idea that the reduction I→II involves the oxygenation of the substrate. This reaction does, on the other hand, not occur at all when the prosthetic group in the enzyme is replaced by 1-deaza-FAD (68). Presumably 1-deaza-FAD-4a-peroxide is a poorer donor of electrophilic oxygen (oxene) than the native flavin peroxide.

Mechanisms involving peroxidation of the substrate, without the formation of structures of the type discussed so far for intermediate II, have been put forth (60, 61, 69). They seem, however, to be ruled out by recent work with fluorinated derivatives of p-hydroxybenzoate as substrates (70). The alternative mechanism would require that the fluorine atom be eliminated as F^+ rather than as F^-, as observed (70).

In summary, it would appear that the mechanism originally suggested by the Massey group (63) is the most tenable one for flavin-containing monoxygenases generally. A key step is the formation of a flavin-4a-peroxide, whose physical and chemical properties are, however, modified by protein effects not well understood so far. Further work with modified flavins (68) may be expected to give important clues to this problem.

Metal-Containing Oxygenases

DIOXYGENASES   The metal-containing dioxygenases generally contain iron as a prosthetic metal (3, 71). This may be part of a heme group, as in tryptophan 2,3-dioxygenase and indolamine 2,3-dioxygenase, or it may be in a nonheme form, as in protocatechuate 3,4-dioxygenase and lipoxygenase (71). The nonheme iron found in these enzymes is unusual, as in some enzyme forms it gives an EPR signal at $g = 6$ (72, 73) generally associated only with heme-bound high-spin Fe^{3+}.

A natural hypothesis for the mechanism of iron-containing dioxygenases is that the metal has to be reduced to the Fe^{2+} state before O_2 becomes bound, and that O_2 activation involves electron transfer from the metal to the bound O_2. Such a mechanism has been established for some dioxygenases (71), and indoleamine 2,3-dioxygenase (74) is here discussed as an example. In other dioxygenases, however, the iron does not seem to undergo a valency change, and the metal may, in fact, not be directly involved in O_2 activation. Protocatechuate 3,4-dioxygenase (75) is used to illustrate this category, as it has been extensively investigated in recent years.

Indolamine 2,3-dioxygenase catalyzes the oxidative ring cleavage of various indoleamine derivatives, including tryptophan, but is distinct from tryptophan 2,3-dioxygenase (71). It requires methylene blue and ascorbic acid as cofactors for maximum activity in vitro, although the ascorbic acid can be replaced by different systems generating O_2^- (71, 74). These observations can be explained on the assumption that the oxygenated intermediate contains a $Fe^{3+} \cdot O_2^-$ complex. As the resting enzyme has the iron in the Fe^{3+} state, this complex can be formed in a reaction with O_2^- but not with O_2. When KO_2 is used in the assay mixture under anaerobic conditions, the amount of product formed is, however, much less than under aerobic conditions (76), which suggests that the product is released from the Fe^{2+} enzyme, which can then form the oxygenated intermediate in a reaction with O_2 rather than with O_2^-. Such a mechanism, summarized in Figure 1, is supported by extensive kinetic studies on the binding of O_2^- and O_2 to the enzyme (77).

Further mechanistic details have been obtained by measurements of the interaction of tryptophan (a substrate) and CO (an inhibitor) with the oxidized as well as with the reduced enzyme (77, 78). It was found that the Fe^{2+} enzyme binds tryptophan first, followed by O_2. The Fe^{3+} enzyme, on the other hand, binds O_2^- first and then tryptophan (see Figure 1). The binding of tryptophan to the Fe^{2+} enzyme causes only a small decrease in the rate of CO binding (77), which shows that the active site of indoleamine 2,3-dioxygenase has distinctly different properties compared to those of two other heme-containing oxygenases, tryptophan 2,3-dioxygenase and cytochrome P-450$_{cam}$. These enzymes do not, in fact, have a stable oxygenated form in the absence of substrate. Indoleamine 2,3-dioxygenase is, on the other hand, similar to horseradish peroxidase with respect to the effect of the substrate on CO binding (77). The so-called compound III of peroxidase can be formed in a reaction of O_2^- with the Fe^{3+} enzyme or of O_2 with the Fe^{2+} enzyme (see 77). It would thus appear that the heme environments in horseradish peroxidase and indoleamine 2,3-dioxygenase are related. This analogy may provide clues to the oxygenase mechanism.

In 1972 Hayaishi's group (79) demonstrated a spectral shift when the enzyme-substrate complex (E·S) of protocatechuate 3,4-dioxygenase reacts with O_2, and ascribed this to the formation of a ternary complex E·S·O_2. This was first considered to be similar in nature to the other oxygenated intermediates that the group had discovered (71). Later, extensive EPR and Mössbauer studies (73, 80) of the resting enzyme, of the E·S and the

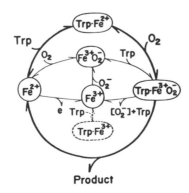

Product

Figure 1   Catalytic cycle for indolamine 2,3-dioxygenase (78).

presumed ternary complex, and of enzyme-product and enzyme-inhibitor complexes revealed, however, that the iron remains high-spin Fe^{3+} in all the forms of the enzyme. There are, of course, changes in the ligand field environment in the various complexes, as reflected in the zero-field splitting parameters. The so-called ternary complex is unique among biological iron compounds in having high-spin Fe^{3+} in an "axial" environment characterized by a negative zero-field-splitting parameter D (73).

Dioxygen has so far not been found to form complexes with high-spin Fe^{3+}, and consequently the suggestion was made that the function of the metal center in protocatechuate 3,4-dioxygenase may be to activate the organic substrate rather than O_2 (80). This idea gained support from the spectroscopic studies of the various enzyme complexes (73, 80). A mechanism was proposed (75, 80) in which there is first a keto-enol tautomerization of the substrate, induced by an interaction of one of its OH groups with Fe^{3+}. The activated keto form of the substrate is assumed to react with O_2 to form the semiquinone and O_2^-. The O_2^- then binds to Fe^{3+}, and the two radicals couple to form a peroxy intermediate, which is assumed to be identical with the optical oxygenated intermediate. This suggestion was supported by a comparison (80) with oxyhemerythrin, which also is characterized by a strong charge-transfer transition around 500 nm.

Even if the mechanism just described seems very attractive, much subsequent work (81–84) has shown some of its features to be incorrect. Experiments in which the reaction was quenched before completion and the organic compounds bound to enzyme were extracted, indicated that the so-called oxygenated intermediate is really a binary complex between enzyme and product (81). This interpretation is consistent with the effect of pH on the kinetics of the breakdown of the intermediate (82). In addition, resonance-Raman measurements (83) have invalidated the analogy with oxyhemerythrin, as no peroxide $\rightarrow Fe^{3+}$ charge-transfer band was observed. No evidence could, in fact, be found for any O_2 vibrations in the "ternary" complex, and it was concluded that O_2 is attached only to the substrate but not to Fe^{3+}. Recent resonance-Raman data (84) for enzyme-inhibitor complexes support the conclusion that the postulated $E \cdot S \cdot O_2$ complex is really a rather stable enzyme-product complex. Thus the organic substrate is probably sufficiently activated in its interaction with Fe^{3+} in the active site to allow a rapid reaction with O_2 to give the enzyme-bound product. Clearly, the role of the metal in oxygenases is not always to activate the kinetically inert O_2 molecule.

MONOXYGENASES   Among the metal-containing monoxygenases, cytochrome P-450 is undoubtedly best characterized with respect to the

O_2-activating reactions. The section on cytochrome P-450 is short, however, as this enzyme was the subject of an entire chapter in the 1980 volume (11).

Cytochrome P-450   The proceedings (85) of a symposium on cytochrome P-450, held in June 1980, supplement the reviews published in the same year (11, 51). The enzyme catalyzes the hydroxylation of a wide variety of organic substrates with the concomitant dehydrogenation of NADPH. Serious studies of the mechanism of O_2 activation were initiated in 1972 (86) by the optical identification of an oxygenated form of this hemoprotein. For this to be formed, a substrate must be bound and the enzyme must be reduced to the Fe^{2+} state. The oxygenated species has a spectrum quite similar to those of oxyhemoglobin and oxymyoglobin.

The ability of the oxygenated enzyme to act as a hydroxylating species is, of course, intimately connected with the specific interactions of the heme group with protein residues in the active site. A key feature is undoubtedly the nature of the fifth iron ligand. On the basis of the unusual EPR spectrum this was early proposed (87) to be a thiolate anion of a cysteine residue. This suggestion has received support from extensive model studies (51,88) but it still lacks a direct protein chemical confirmation. As an alternative, imidazole has been suggested (89) as the fifth ligand; the unique spectral properties are ascribed to interactions with nearby aromatic residues. Magnetic circular dichroism (MCD) measurements on the oxygenated species (90) seem to disprove an axial histidine ligand, however, as there are substantial spectral differences compared to oxymyoglobin. The sixth ligand in the resting enzyme is probably H_2O and should be easily exchangeable; this would account for the low-to-high spin conversion on the binding of substrate (11, 51).

The reduction of the Fe^{3+} enzyme is a one-electron process (91) characterized by a half-reduction potential of -326 mV (92). This very low potential can probably be ascribed to the thiol ligation, and it favors the binding of O_2 to the Fe^{2+} form with concomitant electron transfer to give $Fe^{3+} \cdot O_2^-$, as discussed in a previous section. Such a structure is supported by the observation (93) that O_2^- is released from the oxyenzyme in the absence of a source of additional reducing equivalents. The interpretation of this experiment is, however, complicated by the claim (94) that the NADPH-cytochrome P-450 reductase used in the assay can itself generate O_2^-.

A simplified reaction cycle for cytochrome P-450 is given in Figure 2. The resting form of some iso-enzymes actually have an equilibrium between low- and high-spin Fe^{3+} (95). An extensive EPR investigation (96) suggests that the protein can exist in two conformations. One has high-spin Fe^{3+}, and this state has a higher affinity for the substrate, in agreement with the spin change on substrate binding.

The steps included in Figure 2 are those for which there is good evidence (11, 51) but not those that are most interesting in relation to the mechanism of O_2 activation. As indicated earlier, $(RH)Fe^{2+}$ (O_2) is probably converted to $(RH)Fe^{3+}$ (O_2^-) before receiving the second electron, to give $(RH)Fe^{3+}$ (O_2^{2-}). This intermediate can only formally be represented as an Fe^{3+} peroxide, as the electron distribution and protonation state are not known (11, 51). The least understood step is the conversion of the peroxide intermediate to yield the hydroxylating species, after the breaking of the O–O bond with the release of one molecule of H_2O. Essentially two types of mechanism have been considered (11, 51): (a) a heterolytic cleavage of the peroxide bond, also called the oxenoid pathway, or (b) a homolytic cleavage, the so-called quasi-Fenton pathway. Much recent evidence (11, 51, 97) tends to favor the second alternative. However, the oxenoid mechanism may operate when iodosobenzene is used as an exogenous oxygen donor (98). The reaction then occurs by a different mechanism compared to normal hydroxylation, as solvent O atoms become incorporated into the product. This is explained by the postulated formation of an $Fe^{5+}=O$ intermediate, similar to compound I in peroxidase, which can exchange O atoms with H_2O.

Important mechanistic information has been derived from studies (11, 51, 97, 99) in which organic hydroperoxides, peracids, or peroxyesters are substituted for NADPH and O_2. With peroxyphenylacetic acid the reaction produces benzyl alcohol (97), which cannot be explained in terms of a peroxidase-like mechanism. Instead of a perferryl ion $(Fe^{5+}=O)$ intermediate of the compound I type, an oxyradical and a thiyl-Fe^{3+} hydroxide complex ($S\cdot Fe^{3+}OH^-$) are proposed as the active oxygen species. The specific function of the thiolate ligand to the heme would thus be to serve as a one-electron donor to the peroxide via the iron. The results are undoubtedly applicable to the NADPH/O_2-dependent reactions as well, as these

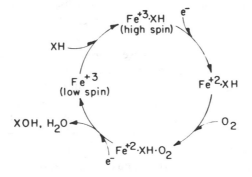

Figure 2 Simplified catalytic cycle for cytochrome P-450 (89). XH is identical with RH in the text.

certainly proceed via a peroxide intermediate. In fact, recent data (100) suggest the formation of a protein-bound peroxy acid in the normal mechanism of O_2 activation by cytochrome P-450.

Other metal-containing monoxygenases   The information so far available on the mechanism of O_2 activation in metal-containing monoxygenases other than cytochrome P-450 is scanty. This group includes, however, the few monoxygenases having copper rather than iron as the prosthetic group, notably tyrosinase and dopamine β-hydroxylase, and consequently a brief discussion is in place.

Tyrosinase is an interesting enzyme in that it is both a monophenol hydroxylase and a diphenol oxidase (101). Although it was shown to be a copper-containing enzyme more than four decades ago (see 101), it has only recently been established, through work on the *Neurospora* enzyme, that it contains two copper ions per active subunit (102). Most of the copper in the enzyme as isolated is not detectable by EPR (101, 103), and it was therefore long assumed that the metal in the native enzyme is Cu^{1+}. Later it was shown, however, that the active site contains a bimetallic copper center (104), and the absence of an EPR signal could then be explained on the basis of an antiferromagnetic coupling between two Cu^{2+} ions.

The addition of stoichiometric amounts of H_2O_2 to tyrosinase results in the formation of a compound with an optical spectrum very similar to that of oxyhemocyanin (101, 105). Recent chemical and spectroscopic studies (106, 107) have established the close relationship between tyrosinase and hemocyanin; the oxygenated sites in both cases are proposed to have the following structure:

$$
\begin{array}{c}
N \diagdown \quad \diagup O\text{-}O^{2-} \diagdown \quad \diagup N \\
Cu^{2+}Cu^{2+} \\
N \diagup \quad \diagdown R \diagup \quad \diagdown N
\end{array}
\qquad 2.
$$

Tyrosinase in which the Cu^{2+} pair has been substituted by two Co^{2+} ions, has been prepared (108). In view of the many bimetallic Co^{2+} complexes with an ability to bind O_2 that have been synthesized (35), it is interesting that an oxy form of cobalt tyrosinase cannot be made (108). Spectroscopic data indicate a tetrahedral coordination, however, whereas O_2-binding Co^{2+} complexes generally have an octahedral corrdination.

In the catalytic cycle, the Cu^{2+} pair in the resting tyrosinase is first reduced, and the reduced enzyme then combines with O_2 to form oxytyrosinase (101, 107). The activation of O_2 is assumed to involve electron transfer from the Cu^{1+} ions to O_2, so that this is really bound in the form of peroxide (see Formula 2). Possibly oxytyrosinase is the hydroxylating species, since peroxide complexes are known to be able to oxygenate aromatic compounds (31).

Dopamine β-hydroxylase is a tetrameric molecule containing four Cu^{2+} ions (109). There is some evidence (110) that it contains two bimetallic copper sites, but these must be very different from the tyrosinase sites, as all Cu^{2+} ions are detectable by EPR (111). The Cu^{2+} ions can be reduced by the cosubstrate ascorbate (111, 112); the reduction occurs in a series of one-electron transfers (113, 114).

A mechanism has been proposed (115, 116) in which two Cu^{2+} ions per active site accept one electron each before the binding of O_2 and substrate. Later results (111, 117, 118) have, however, provided strong evidence against such a ping-pong mechanism. Instead they support a sequential mechanism in which the reduced enzyme contains one electron per active site, so that the catalytic cycle cannot be completed without excess reductant.

In terms of the sequential mechanism just described it is easy to understand why an oxy form similar to oxytyrosinase has not been observed with dopmaine β-hydroxylase. An oxygenated enzyme cannot be made by the addition of O_2^- (119), as with indoleamine 2,3-dioxygenase (77, 78), even if O_2^- can stimulate the reaction by acting as the reducing cosubstrate (119). The mechanism of O_2 activation in this enzyme is thus largely unknown, and the work reviewed emphasizes the danger in postulating mechanisms on the basis of analogies with related enzymes.

OXIDASES

Oxidases that Reduce Dioxygen to Hydrogen Peroxide

FLAVIN-CONTAINING OXIDASES   Flavins (FAD or FMN) are the most common prosthetic groups in oxidases that donate two electrons only to O_2, reducing it to H_2O_2 (3). Some of these oxidases are simple flavoproteins, i.e. they contain no other prosthetic group than the flavin, whereas a few are complex metalloflavoproteins with molybdenum and iron sulfur centers in addition to the flavin. To the first category belong, for example, D- and L-amino acid oxidase, glucose oxidase, some monoamine oxidases, and the Old Yellow Enzyme (120). Lactate oxidase can also be put in this group despite the fact that the products of its reaction are acetate, CO_2, and H_2O, as there is evidence (121) that $FMNH_2$ in the reduced enzyme reacts with O_2 to give H_2O_2, which remains bound in the active site; in fact, with glyoxylate as substrate the products are oxalate and H_2O_2 (122). The most studied metalloflavoprotein oxidases are xanthine oxidase and aldehyde oxidase (123).

The two groups of flavin-containing oxidases have quite different reaction patterns, and this is reflected in a number of chemical properties. It has already been mentioned that the simple flavoprotein oxidases under some conditions form red, anionic flavin radicals (57, 58, 120), but the xanthine oxidase and sulfite oxidase radicals are blue and neutral (123). Another characteristic of simple flavoprotein oxidases is their ability to form a flavin-sulfite adduct (124). The properties of their 8-mercaptoflavin derivatives (125) can also be used to distinguish them from other flavoproteins.

Reduced flavoprotein oxidases react with O_2 to yield the oxidized enzyme and H_2O_2 with rates very similar to those observed for the oxidation of reduced free flavins (120). This might suggest that the enzyme and the model reactions have similar mechanisms, which, in fact, they do not (57, 58, 120). Probably the first step in both cases is the formation of an adduct between reduced flavin and O_2, as already discussed in the section on flavin-containing oxygenases. The further reactions of this adduct are, however, distinctly different. In the models (60, 61) it undergoes a homolytic cleavage to give a flavin radical and O_2^-, thus initiating an autocatalytic reaction, whereas the red radicals of the oxidases are not formed during normal catalytic turnover (57, 58, 120).

Even if flavin radicals are not catalytic intermediates in flavoprotein oxidases, the red, anoinic radicals do generally react rapidly with O_2 (56, 61). In methanol oxidase the anion radical is, however, remarkably inert (126). The reactive radical of lactate oxidase can be stabilized by the binding of pyruvate (127), whereas this greatly increases the reaction of the reduced enzyme with O_2 (121). The latter effect has been suggested (127) to be due to a flattening of the bent dihydroflavin molecule on complex formation, as the planar molecule is more reactive with O_2.

There is little knowledge of the way in which the protein modifies the reactivity of the flavin-O_2 adduct, so that it undergoes a heterolytic cleavage in the oxidases, giving H_2O_2 and oxidized flavin, compared to the homolytic reaction in model systems or in electron transferases, such as flavodoxin. One suggestion (58, 128) is that the $N(1)C(2)O$ region (see Formula 1) is stabilized by hydrogen bonding with the protein, and that this would favor the formation of a $C(10a)$ peroxide, which can undergo heterolytic cleavage, as illustrated in Figure 3. In electron transferases, on the other hand, the hydrogen bonding is thought to occur at the $N(5)H$, and a $C(4a)$ peroxide is formed.

Xanthine oxidase, in its reaction with xanthine and O_2, generates significant amounts of O_2^- in addition to H_2O_2 (129), which shows that its mechanism of O_2 reduction differs from that of the simple flavoprotein oxidases. As O_2^- radicals readily dismutate, with the formation of H_2O_2, one may think that the enzyme in its reaction with O_2 operates according to a homolytic mechanism similar to that of the model flavin systems. Detailed investigations (40, 130) of the reaction of reduced xanthine oxidase with O_2 show, however, that this is only partly true.

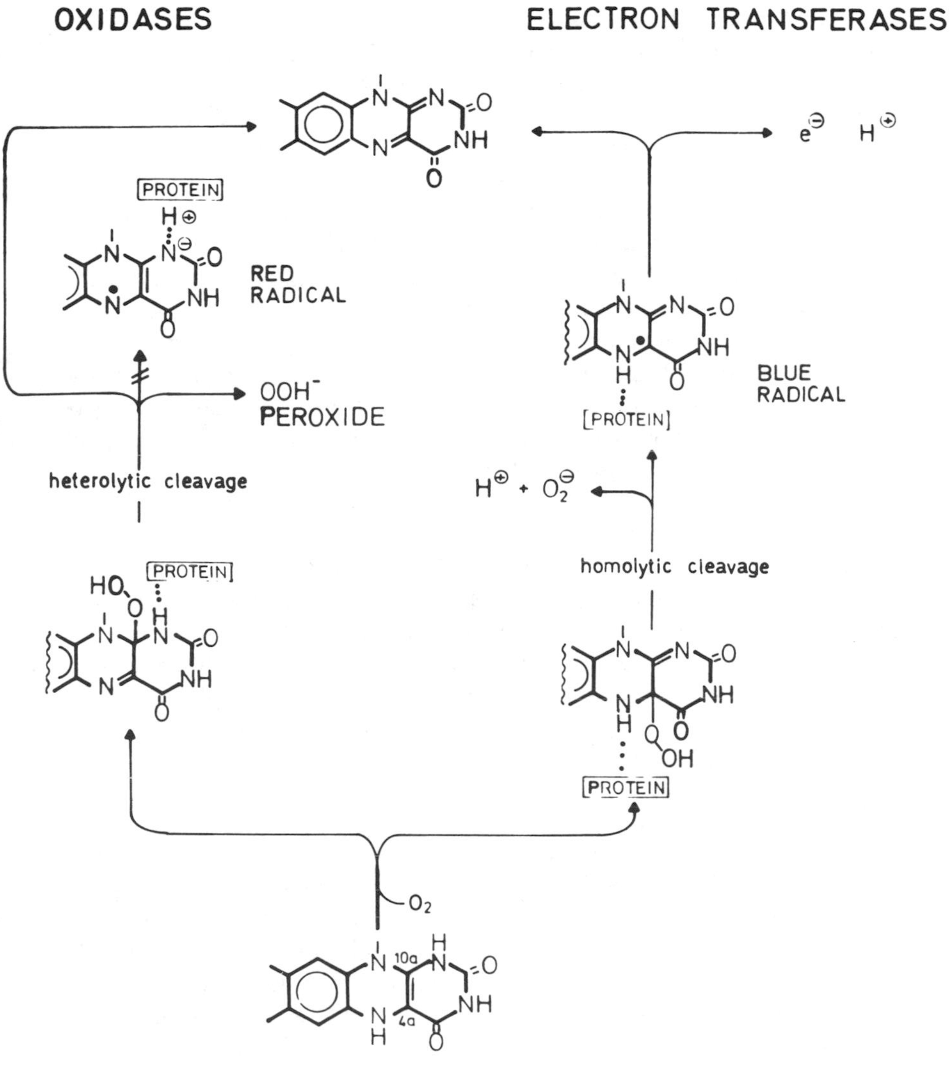

OXIDASES   ELECTRON TRANSFERASES

Figure 3 Suggested reactions of flavins with O_2 in oxidases and electron transferases (57).

Each independent functional unit in the dimeric xanthine oxidase molecule contains 1 Mo, 1 FAD, and 2 Fe_2S_2 centers (123). The metals are not directly involved in the O_2 reaction, which occurs at the FAD site (131). There is a rapid redox equilibrium between the various sites (40), and reduced metal sites become reoxidized by transferring their electrons to the flavin site because of its high standard reduction potential compared to the other sites (132).

Oxidized xanthine oxidase can accept six electrons, two each at the molybdenum and flavin sites, and one each at the two Fe_2S_2 centers (123). The reaction of the fully reduced enzyme with O_2 is biphasic, the slower phase representing reoxidation of the one-electron reduced enzyme (40). Extensive kinetic measurements (40, 130) indicate that the overall reoxidation process proceeds in a sequence of steps, which generate intermediates containing a successively decreasing number of reducing equivalents: $6\rightarrow4\rightarrow2\rightarrow1\rightarrow0$. The first two steps represent two-electron oxidations giving H_2O_2, probably via the heterolytic cleavage of a flavin peroxide intermediate, as in the simple flavoprotein oxidases. At the two-electron reduced stage the enzyme switches,however, from an oxidase to an electron transferase, according to the scheme in Figure 3. As the last two steps then are one-electron processes, they must, of course, generate O_2^-. The final step may be slow because it involves the reaction of O_2 with the blue flavin radical, which is known to be rather inert toward O_2 (133), but a more important factor is probably that in the one electron–reduced enzyme there is very little flavin reduction (132).

In catalytic turnover with excess reducing substrate the O_2^- formation would be expected to be minimized, as some of the enzyme molecules would be re-reduced before reaching the two-electron reduced state. It would thus appear that O_2 activation in all flavin-containing oxidases and oxygenases involves mainly the formation of a flavin peroxide that is split heterolytically. In this way H_2O_2 is produced directly without the intermediate formation of O_2^-.

METAL-CONTAINING OXIDASES   Copper is more prevalent than iron in those metal-containing oxidases that reduce O_2 to H_2O_2 (3). Some amine oxidases contain an organic prosthetic group, probably pyridoxal phosphate, in addition to copper (134, 135). Several reviews are available (17, 20, 22, 26, 134). One enzyme in the group, galactose oxidase, is also the subject of a few specific surveys (136, 137). Not much is known of the mechanism of O_2 activation, a statement that will be supported by a brief discussion of galactose oxidase and plasma amine oxidase.

The most controversial aspect of galactose oxidase research has been the claim (138) that the two-electron redox process catalyzed by this enzyme is mediated by the copper, which is supposed to be reduced to Cu^{1+} by the substrate and then oxidized to Cu^{3+} by O_2, with the formation of H_2O_2. Even if the existence of Cu^{3+} complexes is well documented (139), such a mechanism appears excluded on thermodynamic grounds alone (134, 137). The main basis for the hypothesis was the observation (138) that the addition of $Fe(CN)_6^{3-}$ decreases the Cu^{2+} EPR signal of the resting enzyme with a concomitant increase in activity. EPR experiments at 9 and 35 GHz around 10°K have, however, shown that the titration with $Fe(CN)_6^{3-}$ leads to the appearance of an EPR signal from an $S = 1$ species, which was interpreted to be a spin-coupled Fe^{3+}-Cu^{2+} pair [unpublished experiments of M. J. Ettinger, B. Reinhammar, and R. Aasa, quoted in (137)].

Not only is the purported role of Cu^{3+} in the mechanism excluded, but it is, in fact, unlikely that copper is directly involved in the O_2 reaction. Galactose cannot reduce Cu^{2+} (140), and indeed the Cu^{2+} is redox inactive in the absence of mediators (137). Galactose binds to Cu^{2+} in the enzyme, however (141). NMR relaxation measurements with ^{19}F and ^1H (142) have allowed a comparison of the stability constants with kinetically determined parameters. No competition between O_2 and bound F^- was observed. The results suggest that the role of the metal is to interact with the substrate, and that this reacts directly with O_2 without any role of the metal in the electron transfer.

A similar situation probably exists for plasma benzylamine oxidase, as there is no evidence for reduction or oxidation of Cu^{2+} in any intermediate (143). It has been observed (144) that Cu_{aq}^{2+} and some low-molecular-weight Cu^{2+} complexes with the ability to catalyze O_2^- dismutation, decrease the rate of benzylamine oxidation. This was taken as evidence for a role of copper in the one-electron reduction of O_2 to O_2^- despite the failure to observe a Cu^{1+} intermediate. Cu^{2+} will indeed inhibit most enzymes, not only those having O_2^- as a catalytic intermediate. Other results (145) give strong support to a mechanism in which the amine substrate undergoes a transamination with the pyridoxal prosthetic group; the pyridoxamine formed is then oxidized by O_2 with the formation of H_2O_2 and NH_3. Thus, again the catalytic role of the metal is to activate not O_2 but the organic substrate.

Oxidases that Reduce Dioxygen to Water

BLUE OXIDASES   Ascorbate oxidase, ceruloplasmin, and laccase are a group of copper-containing enzymes, known collectively as the blue oxidases (134) because of the cerulean color imparted to them from an unusually strong charge-transfer absorption band around 600 nm (146). Despite their peripheral biological roles they have received a good deal of attention from chemists for at least two reasons. First, like cytochrome oxidase, they catalyze the four-electron reduction of O_2 to $2H_2O$, perhaps by analogous mechanisms, but unlike the oxidase they are water-soluble enzymes easily prepared in a homogeneous form amenable to rigorous physico-chemical investigations. Second, their metals are in "unique" coordination environments (147), as reflected in a number of anomalous properties compared to small Cu^{2+} complexes, and this has roused the interest of inorganic chemists (30, 148). As a result of the great attraction of the blue oxidases for many investigators there has been a surge of review articles in the last decade (134, 148–156).

The blue oxidases are more complex with regard to their prosthetic group composition than the enzymes considered so far, with the exception of the metalloflavoprotein enzymes. Oxygenases and most oxidases reducing O_2 to H_2O_2 have single prosthetic groups, even when they contain more than one metal ion, as in tyrosinase, or, in the case of multisubunit proteins, several identical prosthetic groups in independent active sites. The blue oxidases, on the other hand, are structurally and functionally asymmetric (157), with four (in laccase) or more (in ascorbate oxidase and ceruloplasmin) copper ions in at least three distinct coordination environments (149, 150). These are generally referred to as type 1, 2, and 3 copper ions, respectively, and can be characterized by the following operational criteria:

1. Type 1 Cu^{2+} has a strong blue color (extinction co-efficient $>10^3$ M^{-1} cm^{-1}) and an EPR spectrum with a narrow hyperfine splitting in the $g_{||}$ region ($A_Z < 10$ mT).
2. Type 2 Cu^{2+} has normal EPR parameters but has an anomalously high affinity for certain anions, e.g. F^-, which act as inhibitors.
3. Type 3 copper ions are EPR nondetectable but have a strong absorption band in the near-ultraviolet spectral region; this band disappears on reduction of the enzymes.

Type 1 Cu^{2+} ions are also found in some small proteins with a single metal ion, such as azurin and plastocyanin, and for these the structural basis of the anomalous properties is known (156, 158). Key features are a thiolate ion from a cysteine residue as one ligand, functioning as the electron donor in the charge-transfer transition (146, 158), and a flattened tetrahedral ligand geometry. Sequence homologies (154, 159), as well as the similarities in physical properties, suggest an essentially identical coordination in the blue oxidases.

The type 3 copper ions, which constitute the O_2-reducing sites in these enzymes, consist of an antiferromagnetically coupled Cu^{2+}-Cu^{2+} pair, as in tyrosinase and hemocyanin (104). In laccase the coupling is so strong that the pair does not display any paramagnetism in the temperature range 40–300° K (160, 161), but its presence can be diagnosed by the reactions of the enzyme with NO (162). Under some circumstances it is possible to produce a type 3 Cu^{2+} EPR signal (163, 164) by differential reduction of one of the ions, which definitely demonstrates that the copper ions are in the Cu^{2+} state despite being EPR undetectable in the oxidized enzymes.

The catalytic mechanism, especially the O_2 reduction, is best character-ized for laccase from the tree *Rhus vernicifera* or from the fungus *Polyporus versicolor,* and consequently laccase is the focus of attention in this section. Steady-state kinetics (165) have established that laccase follows a ping-pong mechanism in which the enzyme must be reduced by the organic substrate before O_2 reacts. Electrons are taken up from reductants one at a time (166); the primary electron acceptor is the type 1 Cu^{2+}. The single electron in the reduced type 1 site cannot be transferred to the type 3 pair before the type 2 Cu^{2+} has also been reduced, because the type 3 site functions as a cooperative two-electron acceptor (167, 168). Thus, contrary to a recent claim (169), the one-electron reduced enzyme cannot react with O_2, as was, in fact, established by kinetic experiments already in 1969 (170).

The type 2 Cu^{2+} does not react with the substrate until the type 1 copper has been reduced (166), in agreement with much evidence (171–173) that this site is not available in the oxidized enzyme. Some results (171–174) suggest that the type 2 and 3 ions are in the same cavity, which is unavailable to bulk solvent when the type 1 copper is oxidized.

Reduction of the type 3 copper pair occurs by an intramolecular transfer of two electrons from the reduced type 1 and 2 sites (166, 175). In anaerobic reduction experiments (166) this step appears too slow to be part of turnover, however. This dilemma is resolved by the finding (166, 175, 176) that the resting enzyme is in an inactive form, which is converted to the active form in the first redox cycle during the presteady state. The inactive form has an OH^- coordinated to the type 2 Cu^{2+} (175); it should be noted that binding of F^- to type 2 Cu^{2+} also prevents the intramolecular electron transfer (166, 170).

The first demonstration of an optical intermediate in the reaction of a reduced blue oxidase with O_2 was made with ceruloplasmin (177). Shortly afterward a similar intermediate was observed with fungal laccase (178). As its optical spectrum closely resembled that of a complex made by adding H_2O_2 to the oxidized enzyme (179), it was suggested to be a peroxy species. In this, H_2O_2 was thought to be bound to type 2 Cu^{2+}, on the basis of a slight shift in a low-field hyperfine line in the EPR spectrum (178). Later extensive investigations (169, 176, 180–182) of the tree laccase-H_2O_2 derivative strongly implicated the type 3 copper ions as the H_2O_2-binding site, however. Among other observations was an increase in the magnetic susceptibility (182), ascribed to a decrease in the antiferromagnetic coupling of the type 3 copper pair. The effects on the type 2 Cu^{2+} was suggested to be due to an interaction between the type 2 and 3 sites, in agreement with many other observations (168, 172–174).

The circular dichroism spectrum of the laccase-H_2O_2 complex (169) indicates a structure similar to that in oxytyrosinase and oxyhemocyanin (107). The complex was proposed (180) to be identical with the optical intermediate, which has a similar spectrum (183), particularly since it could also be made by reacting the two-electron reduced enzyme with O_2 (176). Undoubtedly, a peroxy laccase is an intermediate in the reaction, but other data (175, 183) suggest that it accepts an additional electron from the reduced type 1 copper ion so rapidly that the first transient observable is a three-electron reduced oxy species.

The strongest evidence that the optical intermediate is not an H_2O_2 complex is its kinetic identification (183) with the paramagnetic intermediate demonstrated by EPR (184, 185). With the aid of ^{17}O-enriched O_2 (185) this has unambiguously been shown to be an oxygen radical, and this represents the first really direct demonstration of a true oxygen intermediate in an oxidase. On the basis of its unusual relaxation properties the radical has been suggested to be O^- (185), in agreement with the finding that it should be a species in which three electrons have been transferred to O_2 (175, 183). This highly reactive radical must undoubtedly be stabilized by electron delocalization onto the type 3 copper pair (155, 176, 185).

The paramagnetic intermediate appears to be reduced too slowly (184) to be a species in the catalytic turnover, when starting from the fully reduced enzyme. This is, however, a consequence of the fact that the last electron needed to reduce O^- to H_2O must in this case be taken from reduced type 2 copper, which is a slow process. In a turnover situation type 1 Cu^{2+} is immediately re-reduced and can then rapidly donate an electron to the type 3 site (175, 186). The EPR signal of the intermediate can, in fact, be observed during the steady state (175, 185). The use of the fully reduced enzyme to investigate the O_2 activation by an oxidase, a common practice, may thus represent an artificial situation that gives results not directly applicable to the true catalytic cycle.

It was suggested that type 2 copper plays a role in the reduction of O_2 (169, 179), particularly in the stabilization of oxygen intermediates. An investigation (186) with type 2–depleted laccase (187), however, eliminated this possibility. In the modified enzyme the reduction of type 1 Cu^{2+} proceeds normally, but the intramolecular electron transfer to the type 3 site is prevented, just as it is on binding of F^- to type 2 Cu^{2+} (166, 170). The type 3 copper pair in type 2–depleted laccase can be reduced, however, if it is left anaerobically with a reducing agent for about 20 h. When this three electron–reduced laccase is mixed with O_2, the paramagnetic intermediate is formed and is stable for hours. It is, on the other hand, rapidly reduced via the type 1 copper on the addition of an external reducing agent. It would thus appear that the sole function of the type 2 copper is to donate the second electron required for the initial reduction of the type 3 copper pair, but that it plays no direct role in the O_2 reduction.

On the basis of the extensive results summarized here it is possible to formulate the catalytic cycle shown in Scheme 3, in which the copper ions represent type 1, 2, and 3 in the order written:

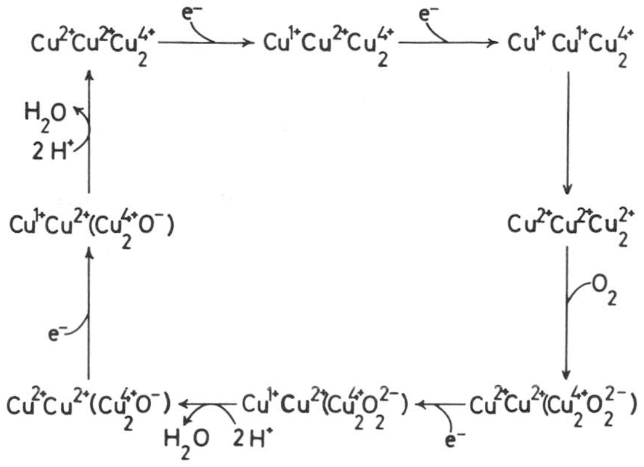

$$Cu^{2+}Cu^{2+}Cu_2^{4+} \xrightarrow{e^-} Cu^{1+}Cu^{2+}Cu_2^{4+} \xrightarrow{e^-} Cu^{1+}Cu^{1+}Cu_2^{4+}$$

$$Cu^{1+}Cu^{2+}(Cu_2^{4+}O^-) \qquad Cu^{2+}Cu^{2+}Cu_2^{2+}$$

$$Cu^{2+}Cu^{2+}(Cu_2^{4+}O^-) \xleftarrow[H_2O \ \ 2H^+]{} Cu^{1+}Cu^{2+}(Cu_2^{4+}O_2^{2-}) \xleftarrow[e^-]{} Cu^{2+}Cu^{2+}(Cu_2^{4+}O_2^{2-})$$

Scheme 3

An important feature in this mechanism is the fact that the initial reaction of O_2 is with the two electron–reduced type 3 site, so that O_2 can be reduced directly to O_2^{2-}, thus bypassing the thermodynamically unfavorable formation of O_2^-. Further reduction then proceeds in discrete one-electron steps mediated by the type 1 copper. This leads to the formation of an oxygen radical intermediate, which remains bound in the type 3 copper site.

CYTOCHROME OXIDASE The most complex of all the enzymes discussed in this chapter is cytochrome c oxidase, which not only catalyzes the four-electron reduction of O_2 to $2H_2O$, like the blue oxidases, but also couples the electron-transfer reactions to the translocation of protons across an energy-transducing membrane (188). It can also be regarded as the biologically most important of the enzymes having O_2 as a substrate, at least from a quantitative point of view, as the O_2 consumption by all the oxygenases and the other oxidases put together is almost negligible compared to that in the respiratory chain. Consequently, many investigators have found it worthwhile to try to overcome the experimental obstacles posed by its complexity, and the literature on cytochrome oxidase is voluminous. Fortunately there are a number of reviews available (19, 32, 189–194), and hence very few references to original papers published before 1979 are given here. A specific survey of the O_2 reaction with the reduced enzyme has been written (195), and a brief discussion of the mechanism of O_2 reduction is included in a review (188) mainly devoted to the problem of proton translocation.

The functional unit of cytochrome oxidase, like that of laccase, contains four metal-ion prosthetic groups (32, 192). Two of these are heme a bound in different ways to the protein, cytochrome a and a_3, whereas two are copper ions in separate chemical environments, Cu_A and Cu_B. The mechanism of electron transfer from reduced cytochrome c to O_2, the topic of this section, has generally been studied by following changes in the spectroscopic properties of the metal ions on reduction or binding of ligands. Thus, the mechanistic discussion must be preceded by a summary of the spectroscopic characteristics of various forms of the enzyme and their structural implications. EPR spectra are stressed, as these generally allow assignments to specific sites, which is seldom possible by optical measurements alone (32, 192). In recent years magnetic circular dichroism (MCD) has also been a valuable tool in ascribing chemical changes to a particular prosthetic group (192).

In oxidized cytochrome oxidase two components only, cytochrome a and Cu_A, are detectable by EPR (192). The EPR signal from cytochrome a ($g = 3.03$, 2.21, 1.45) (196) is typical for a low-spin heme Fe^{3+}. The absorption at g 2, generally ascribed to Cu_A^{2+}, is quite anomalous, having properties more like a radical signal (32, 196). This led to the suggestion (197) that it represents $Cu^{1+} \cdot S^{\cdot}$, an idea that has received support from X-ray absorption edge spectroscopy showing that one of the coppers is in the Cu^{1+} state (198). On the other hand, other investigators (199) suggested that this was an artifact due to reduction induced by the synchrotron radiation; but a later careful reinvestigation (200) has, in fact, demonstrated that the oxidase sample remains intact under the conditions used. Amino-acid sequence studies (201) have shown that cytochrome oxidase contains a site homologous to the type 1 Cu^{2+} site in the blue proteins, so one would expect one of the copper ions to have type 1 properties. Perhaps the energetic balance in this highly unusual Cu^{2+} coordination is so fine that what is the excited state in the charge-transfer transition of the blue proteins, i.e. $Cu^{1+} \cdot S^{\cdot}$, becomes the ground state for Cu_A. Sulfur ligation to copper in cytochrome oxidase has been demonstrated by extended X-ray absorption fine structure (EXAFS) measurements (202), and there is some evidence that two sulfur atoms are bound to Cu_A.

The absence of any EPR signals from oxidized cytochrome a_3 and Cu_B is generally ascribed to an antiferromagnetic coupling between these components (32, 192), although there has been a suggestion (203) that it is due to the presence of Fe^{4+} and Cu_B^{1+} in the oxidized enzyme. Much recent work (204–210) gives, however, additional support to the original model. A g 6 signal from high-spin Fe^{3+} appears on partial reduction (196), and this has been unambiguously assigned to cytochrome a_3 (204). Under other conditions, an EPR signal from Cu_B^{2+} can be observed (163, 205, 206). Extensive EPR studies (206) of the reactions of NO with cytochrome oxidase show that the oxidized enzyme contains cytochrome a_3^{3+} and Cu_B^{2+} at an estimated distance of 3.4 Å. Even if resonance-Raman measurements (207) indicate that the heme of cytochrome a_3 is in a unique environment, they give no evidence (208) for an unusual valence state of the iron in this cytochrome. MCD studies (209, 210) of cytochrome a_3-CN^- derivatives are also consistent with a coupling between cytochrome a_3 and Cu_B, but do not by themselves exclude the alternative model (203).

The ligand mediating the antiferromagnetic coupling between cytochrome a_3 and Cu_B is possibly a thiolate ion, on the basis of EXAFS studies (211). A bridging imidazole is ruled out (206, 212). On the other hand, histidine has unambiguously been identified as the axial fifth ligand of cytochrome a_3 by elegant experiments (213) with NO derivatives of yeast cytochrome oxidase that had ^{15}N-histidine incorporated.

The nature of the Cu_B site remains controversial. On the basis of optical (214) and X-ray edge absorption data (199) it has been proposed that this, rather than Cu_A, is a blue type 1 Cu^{2+}. This assignment depended, however, on the attribution of a considerable absorption of Cu_B^{2+} at 830 nm, whereas this near-infrared band has generally been associated with Cu_A. Several later investigations (210, 215, 216) have shown that Cu_B^{2+} does not contribute significantly to the 830-nm band. In addition, experiments (217) with Ag^{1+} and Hg^{2+} provided evidence against Cu_B being a type 1 copper. The EPR parameters of Cu_B^{2+} (163, 205) also show that this is not a type 1 ion; it is instead very similar to the type 3 copper in laccase.

The binding of CO to reduced cytochrome a_3 has been a classical approach in investigations of the site of the O_2 reaction in cytochrome oxidase (189). The present concept is that this site is actually a cytochrome a_3–Cu_B pair, and the CO compound of cytochrome a_3 appears to behave as a two-electron acceptor (216, 218). A claim (219) that a species with an absorption at 607 nm (214) is a CO compound with Cu_B oxidized was later retracted (220); instead it probably is the product of the reaction of O_2 with the so-called mixed-valence oxidase (i.e. enzyme with only cytochrome a_3 and Cu_B reduced) (214, 221).

In addition to the EPR signals mentioned so far, a number of less well understood signals can be observed under special conditions. On reoxidation of reduced enzyme with O_2 a signal at $g = 5$, 1.78, 1.69 develops (222, 223), and this can sometimes be seen in the oxidized protein as prepared (224). Experiments with $^{17}O_2$ (222, 223) indicated that oxygen is not involved in the structure giving this signal. The g 5 species changes on standing to one giving an unusual signal at g 12 (225). Both signals are

ascribed to cytochrome a_3 in different forms of the oxidized enzyme (225).
A normal low-spin Fe^{3+} signal at g 2.6, observed on partial reduction (196),
particularly at high pH, can be assigned to a cytochrome $a_3^{3+} \cdot OH^-$
complex (226).

Kinetic experiments have given further evidence that there exist more
than one oxidized form of cytochrome oxidase. Thus, the resting enzyme
is converted to a more active form, called "pulsed" oxidase, on reduction
and reoxidation with O_2 (227, 228). The intramolecular electron transfer
from the primary acceptors, cytochrome a and Cu_A, to the cytochrome
a_3–Cu_B site, is increased in the pulsed oxidase (227–230). The structural
difference has been suggested (206, 227) to be that the pulsed species retains
one of the oxygen atoms as a μ-oxo bridge between cytochrome a_3 and
Cu_B, but this idea seems ruled out by recent experiments with $^{18}O_2$ (231).
It seems likely (223) that pulsed oxidase is identical with the g 5 form of
the enzyme. The latter can, however, only be prepared with O_2 as the
oxidant (222, 223), whereas it is claimed (232) that pulsed oxidase can be
made by oxidation with ferricyanide. On the basis of MCD and resonance-
Raman spectra of three different "oxygenated" forms of the oxidase it has
been concluded (233) that in pulsed oxidase, cytochrome a_3 in an intermediate spin state ($S = \frac{3}{2}$) is coupled to Cu_B, whereas it is high-spin cytochrome a_3 in the resting form.

At room temperature the reaction of fully reduced (234) or "mixed-
valence" cytochrome oxidase (235) is too rapid for any reaction intermediates involving the $[Cu_B a_3 O_2]$ unit to be detected. It was therefore a major
advance when Chance and his colleagues (236) introduced the low-temperature trapping technique to stabilize possible intermediates. With this
method they (237) as well as others (221, 238) found a number of intermediates when the solubilized enzyme reacts with O_2, starting either from the
fully reduced or the mixed valence state:

$$\text{Fully reduced} + O_2 \longrightarrow \underset{(A_1)}{I} \longrightarrow II \longrightarrow \underset{(B)}{III} - - \rightarrow$$

$$\text{Mixed valence} + O_2 \longrightarrow \underset{(A_2)}{I_M} \longrightarrow II_M \longrightarrow \underset{(C)}{III_M} - - \rightarrow$$

Scheme 4

The designations used here are those of Clore et al (221, 238), but the
nomenclature of the other group (214, 236, 237) is shown in parenthesis.
In discussing the evidence for the structure of the intermediates and their
place in the catalytic cycle, the scheme in Figure 4 is used. This also includes

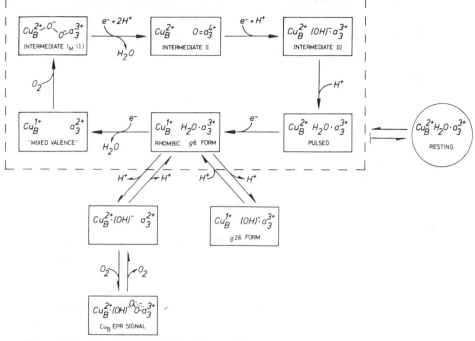

Figure 4   Reactions of the $[a_3Cu_B]$ unit in cytochrome oxidase (240). Those species that form the catalytic cycle are enclosed by a dashed line.

many of the enzyme forms discussed earlier, and arguments are given for placing some of them outside the cycle.

The Chance group has suggested (211, 236, 237) that the successive intermediates involve electron redistributions solely within the $[Cu_Ba_3O_2]^{3+}$ unit, but no additional electron transfers to this unit from reduced cytochrome a and Cu_A. Quantitative EPR measurements (221, 238) have, however, established that this is true only in the case of the mixed-valence oxidase, in which no electron transfer from cytochrome a and Cu_A is possible, because these components are already oxidized. In the reaction of the fully reduced enzyme (238) with O_2, on the other hand, intermediates II and III are formed from I in two sequential one-electron transfers.

Intermediate I has recently been stated (211) to be an O_2 adduct, in which O_2 is bound to reduced cytochrome a_3 only. The structure suggested in Figure 4 would, however, appear more likely. First, the formal valence of III is that of fully oxidized $[Cu_Ba_3]$ and fully reduced O_2. As it is formed by the addition of two electrons to I, this should already have O_2 at the peroxide level. Second, experience with simpler bimetallic complexes binding O_2 suggests that O_2 in intermediate I is bound as O_2^{2-}, as discussed in the section on models.

The structure for intermediate II given in Figure 4 has been proposed earlier (206). It has recently been established by the discovery (239) of a Cu_B^{2+} EPR signal with unusual properties associated with this intermediate. The hyperfine coupling shows that the unpaired electron is mainly associated with a copper nucleus. The signal is, however, characterized by unusual relaxation properties and shape, which can only be explained by an interaction with a nearby paramagnetic ion. As the formal valence state of II is $[Cu_B a_3 O_2]^{2+}$ (238) and the signal stems from Cu_B^{2+}, the rest of the unit should be $[a_3 O_2]^0$. Undoubtedly the addition of a third electron in the formation of intermediate II from I results in a breaking of the O–O with the release of one molecule of H_2O, so that the cytochrome part of the O_2-reducing unit should be written $[FeO]^{2+}$. This can have several possible electron distributions ($Fe^{2+} \cdot O$, $Fe^{3+} \cdot O^-$, or $Fe^{4+}=O$), but only one of them, namely the ferryl ion ($Fe^{4+}=O$), would be expected to be paramagnetic.

The suggestion (211, 237) that intermediate III is a peroxy oxidase with the same structure as here proposed for intermediate I must be wrong, as III is formed by the addition of one electron to II (238) with the structure $[Cu_B^{2+} O-a_3^{4+}]$. There seems to be no doubt that the two groups suggesting different structures are studying the same compound, as III (B) is a stable end product at the temperature used (173 K). It is apparent that EPR is more useful than visible spectra for assigning valence states to the metal components in oxygen intermediates.

The state of the $[Cu_B a_3]$ unit in intermediate III is not identical with that in the fully oxidized enzyme, in either the pulsed or the resting conformation, even if the formal valence is the same. This is shown by the absence of the 655-nm band (240) in III. The pulsed or g 5 form, which has been shown (223, 233) to be formed later than III in the reoxidation of the reduced enzyme, does, on the other hand, have this absorption band. The H_2O molecule proposed to be bound to cytochrome a_3 in the g 5 form would be expected to exchange rapidly, in agreement with the $^{18}O_2$ experiments (231) quoted earlier.

To investigate which of the many species detectable by EPR are really part of the catalytic cycle, experiments have been performed in which the enzyme reaction was stopped by rapid-freeze quenching during turnover with ascorbate, trimethyl-p-phenylenediamine (TMPD), cytochrome c, and O_2 in the reaction mixture (241). The steady-state level of cytochrome a was varied from nearly fully oxidized (in slow turnover) to nearly fully reduced (in rapid turnover) by varying the TMPD concentration. Apart from weak g 5 signals, no other of the many EPR signals described than those from cytochrome a^{3+} and Cu_A^{2+}, at varying levels depending on the turnover rate, were detected. This means that either the EPR-detectable species are outside the catalytic cycle, or they are poorly populated during the steady state. Of course, an EPR-detectable state must be formed in the re-reduction of the $[Cu_B a_3]$ unit, unless this is a concerted two-electron process, as in laccase (167, 168). The appearance of g 6 signals on partial reduction under anaerobic conditions (196) indicates, however, that this is not the case. It is, on the other hand, possible that the process is kinetically, if not thermodynamically, concerted, i.e. as soon as one electron has been transferred to the unit, the second one is added very rapidly. This suggestion receives support from some old kinetic observations (242).

Earlier, the form labeled "Cu_B EPR signal" has been suggested to be in the cycle (205), but the g 6 species would appear more likely. The main reason for this is the observation (229) that a g 6 signal is rapidly formed on partial anaerobic reduction of the pulsed enzyme, which demonstrates that Cu_B is reduced before cytochrome a_3. Thus, O_2 will not react until the unit has received two electrons, and intermediate I (or rather I_M) is formed directly. In the absence of the second reducing equivalent, under aerobic conditions, the Cu_B EPR species is formed, in agreement with experimental observations (205), by a displacement of the equilibria in Figure 4, because of the high affinity of cytochrome a_3^{2+} for O_2.

Under very slow turnover in the absence of TMPD a considerable amount of g 5 signal is seen (241). This is consistent with the scheme, as intermediate III would then build up because of the low rate of electron flow into the $[Cu_B a_3]$ unit.

A study of the CO complex of the reduced $[Cu_B a_3]$ unit by Fourier-transform infrared spectroscopy (212) has led to the proposal of a nonpolar surrounding of cytochrome a_3 and a flexible structure for Cu_B. This is consistent with the observations (205, 239) that the two species giving Cu_B^{2+} EPR signals have slightly different hyperfine splittings. The nonpolar site may be an important factor in the O_2 activation, as discussed earlier for the model systems.

The mechanism of O_2 reduction given in Figure 4 is very similar to that earlier described for laccase. It would appear that a common feature of the few enzymes reducing O_2 to $2H_2O$ is a bimetallic O_2-reducing site. This is initially reduced by two electrons transferred intramolecularly from the two primary electron acceptors. In this way O_2 is rapidly reduced to O_2^{2-} without the intermediate formation of O_2^-. The paramagnetic intermediates formed by the addition of a third electron have different structures in the two enzymes, however, because of the different chemical nature of their bimetallic sites. Cytochrome oxidase utilizes the ability of a heme group to form a ferryl ion ($Fe^{4+}=O$), whereas laccase probably has an O^- radical bound to the two type 3 copper ions. A notable feature of the proposed cytochrome oxidase mechanism is that neither the fully reduced nor the fully oxidized enzyme is part of the catalytic cycle.

Thus, we know a good deal about the manner in which O_2 is activated and reduced not only in cytochrome oxidase but also in a few other oxidases and oxygenases. Many aspects of the O_2 reaction mechanisms are, however, incompletely understood. This applies, for example, to the protonation steps involved in the formation of H_2O in enzymes such as cytochrome P-450, laccase, and cytochrome oxidase. It is also impossible as yet to define those specific features of the unique bimetallic centers in laccase and cytochrome oxidase that allow the further reduction of the peroxide intermediates to yield the paramagnetic oxygen intermediates. The construction of bimetallic model complexes mimicking these steps should offer a challenge to the bioinorganic chemist.

ACKNOWLEDGMENTS

Part of the review was written while I was a Sherman Fairchild Distinguished Scholar, 1980–81, at the California Institute of Technology. I want to acknowledge many helpful discussions with my colleagues there, particularly Drs. F. C. Anson, S. I. Chan, R. R. Gagné, and H. B. Gray. I am indebted to Drs. H. Beinert, S. I. Chan, G. M. Clore, H. B. Gray, K. Lerch, V. Massey, B. Reinhammar, and M. T. Wilson for providing me with manuscripts prior to publication. Drs. V. Massey and T. Vänngård have given valuable comments on parts of the manuscript. I would also like to thank Dr. D. L. Roth at Caltech and Dr. H. Csopak at Chalmers Library for help with a computer search of the literature. My own investigations have been supported by grants from the Swedish Natural Science Research Council.

Literature Cited

1. Keilin, D. 1966. *The History of Cell Respiration and Cytochrome.* London: Cambridge Univ. Press. 416 pp.
2. Fruton, J. S. 1972. *Molecules and Life,* pp. 262–396. New York: Wiley-Intersci. 579 pp.
3. Keevil, T., Mason, H. S. 1978. *Methods Enzymol.* 52:3–40
4. Mason, H. S. 1965. *Ann. Rev. Biochem.* 34:595–634
5. Caughey, W. S. 1967. *Ann. Rev. Biochem.* 36:611–44
6. Wellner, D. 1967. *Ann. Rev. Biochem.* 36:669–90
7. Neims, A. H., Hellerman, L. 1970. *Ann. Rev. Biochem.* 39:867–88
8. Hayaishi, O. 1969. *Ann. Rev. Biochem.* 38:21–44
9. Gunsalus, I. C., Pederson, T. C., Sligar, S. G. 1975. *Ann. Rev. Biochem.* 44:377–407
10. Walsh, C. 1978. *Ann. Rev. Biochem.* 47:881–931
11. White, R. E., Coon, M. J. 1980. *Ann. Rev. Biochem.* 49:315–56
12. Hess, B., Staudinger, Hj., eds. 1968. *Biochemie des Sauerstoffs.* Heidelberg: Springer. 360 pp.
13. Ochiai, E.-I. 1975. *J. Inorg. Nucl. Chem.* 37:1503–9
14. Boyer, P. D., ed. 1975. *Enzymes* 12:1–647(3rd ed.)
15. Hayaishi, O., ed. 1974. *Molecular Mechanisms of Oxygen Activation.* New York: Academic. 678 pp.

16. Hayaishi, O., ed. 1962. *Oxygenases.* New York: Academic. 588 pp.
17. Yasunobu, K. T., Mower, H. F., Hayaishi, O., eds. 1976. *Iron and Copper Proteins.* New York: Plenum. 596 pp.
18. Singer, T. P., Ondarza, R. N., eds. 1977. *Mechanisms of Oxidizing Enzymes.* Amsterdam: Elsevier/North-Holland. 313 pp.
19. King, T. E., Orii, Y., Chance, B., Okunuki, K., eds. 1979. *Cytochrome Oxidase.* Amsterdam: Elsevier/North-Holland. 426 pp.
20. Beinert, H. 1980. *Coordin. Chem. Rev.* 33:55–85
21. Spiro, T. G., ed. 1980. *Metal Ion Activation of Dioxygen.* New York: Wiley. 247 pp.
22. Spiro, T. G., ed. 1981. *Copper Proteins.* New York: Wiley. 363 pp.
23. Caughey, W. S., ed. 1979. *Biochemical and Clinical Aspects of Oxygen.* New York: Academic. 866 pp.
24. King, T. E., Mason, H. S., Morrison, M., eds. 1981. *Oxidases and Related Redox Systems.* Oxford: Pergamon. 1225 pp.
25. Ho, C., ed. 1982. *Interaction Between Iron and Proteins in Oxygen and Electron Transport.* Amsterdam: Elsevier/North-Holland. In press
26. Smith, B. E., Knowles, P. F. 1980. In *Inorganic Biochemistry,* ed. H.A. O. Hill, 1:271–316. London: Chem. Soc. 442 pp.
27. George, P., Griffith, J. S. 1959. In *The Enzymes,* eds. P. D. Boyer, H. Lardy, K. Myrbäck, Vol. 1, pp. 347–89. New York: Academic. 785 pp.
28. George, P. 1965. In *Oxidases and Related Redox Systems,* eds. T. E. King, H. S. Mason, M. Morrison, 1:3–33. New York: Wiley. 1144 pp.
29. Taube, H. 1965. *J. Gen. Physiol.* 49:29–50
30. Bennett, L. E. 1973. *Prog. Inorg. Chem.* 18:1–176
31. Hamilton, G. A. 1974. See Ref. 15, pp. 405–51
32. Malmström, B. G. 1973. *Q. Rev. Biophys.* 6:389–431
33. Valentine, J. S. 1973. *Chem. Rev.* 73:235–45
34. Hill, H. A. O. 1979. In *Oxygen Free Radicals and Tissue Damage,* pp. 5–11. Amsterdam: Excerpta Medica. 381 pp.
35. Jones, R. D., Summerville, D. A., Basolo, F. 1979. *Chem. Rev.* 79:139–79
36. DeKock, R. L., Gray, H. B. 1980. *Chemical Structure and Bonding,* pp. 232–44. Menlo Park, Calif: Benjamin. 491 pp.
37. Wood, P. M. 1974. *FEBS Lett.* 44:22–24
38. Ilan, Y. A., Czapski, G., Meisel, D. 1976. *Biochim. Biophys. Acta* 430:209–24
39. Jacq, J., Bloch, O. 1970. *Electrochim. Acta* 15:1945–66
40. Olson, J. S., Ballou, D. P., Palmer, G., Massey, V. 1974. *J. Biol. Chem.* 249:4363–82
41. Collman, J. P., Halbert, T. R., Suslick, K. S. 1980. See Ref. 21, pp. 1–72
42. Vogt, L. H. Jr., Faigenbaum, H. M., Wiberley, S. E. 1963. *Chem. Rev.* 63:269–77
43. Basolo, F., Hoffman, B. M., Ibers, J. A. 1975. *Acc. Chem. Res.* 8:384–92
44. Vaska, L. 1976. *Acc. Chem. Res.* 9:175–83
45. Collman, J. P. 1977. *Acc. Chem. Res.* 10:265–72
46. Mimoun, H. 1980. *J. Mol. Catal.* 7:1–29
47. Wong, C.-L., Switzer, J. A., Balakrishnan, K. P., Endicott, J. F. 1980. *J. Am. Chem. Soc.* 102:5511–18
48. Pauling, L., Coryell, C. D. 1936. *Proc. Natl. Acad. Sci. USA* 22:210–16
49. Goddard, W. A. III, Olafson, B. D. 1979. See Ref. 23, pp. 87–122
50. Lever, A. B. P., Gray, H. B. 1978. *Acc. Chem. Res.* 11:348–55
51. Coon, M. J., White, R. E. 1980. See Ref. 21, pp. 73–123
52. Groves, J. T. 1980. See Ref. 21, pp. 125–62
53. Groves, J. T. 1979. *Adv. Inorg. Biochem.* 1:119–42
54. Chin, D.-H., La Mar, G. N., Balch, A. L. 1980. *J. Am. Chem. Soc.* 102:5945–47
55. Collman, J. P., Denisevich, P., Konai, Y., Marrocco, M., Koval, C., Anson, F. C. 1980. *J. Am. Chem. Soc.* 102:6027–36
56. Massey, V., Hemmerich, P. 1975. See Ref. 14, pp. 191–252
57. Massey, V., Hemmerich, P. 1981. See Ref. 24, pp. 370–91
58. Hemmerich, P., Wessiak, A. 1979. See Ref. 23, pp. 491–510
59. Massey, V. 1979. See Ref. 23, pp. 477–88
60. Bruice, T. C. 1980. *Acc. Chem. Res.* 13:256–62
61. Bruice, T. C. 1981. See Ref. 24, pp. 412–30
62. Husain, M., Massey, V. 1979. *J. Biol. Chem.* 254:6657–66
63. Entsch, B., Ballou, D. P., Massey, V. 1976. *J. Biol. Chem.* 251:2550–63
64. Ghisla, S., Entsch, B., Massey, V., Hu-

sein, M. 1977. *Eur. J. Biochem.* 76:139–48

65. Kemal, C., Bruice, T. C. 1976. *Proc. Natl. Acad. Sci. USA* 73:995–99
66. Wierenga, R. K., De Jong, R. J., Kalk, K. H., Hol, W. G. J., Drenth, J. 1979. *J. Mol. Biol.* 131:55–74
67. Entsch, B., Ballou, D. P., Husain, M., Massey, V. 1976. *J. Biol. Chem.* 251: 7367–79
68. Entsch, B., Husain, M., Ballou, D. P., Massey, V., Walsh, C. 1980. *J. Biol. Chem.* 255:1420–29
69. Kemal, C., Bruice, T. C. 1979. *J. Am. Chem. Soc.* 101:1635–38
70. Husain, M., Entsch, B., Ballou, D. P., Massey, V., Chapman, P. J. 1980. *J. Biol. Chem.* 255:4189–97
71. Hayaishi, O., Nozaki, M., Abbott, M. T. 1975. See Ref. 14, pp. 119–89
72. Slappendel, S., Veldink, G. A., Vliegenthart, J. F. G., Aasa, R., Malmström, B. G. 1981. *Biochim. Biophys. Acta* 667: 77–86
73. Que, L. Jr., Lipscomb, J. D., Zimmermann, R., Münck, E., Orme-Johnson, N. R., Orme-Johnson, W. H. 1976. *Biochim. Biophys. Acta* 452:320–34
74. Hayaishi, O. 1981. See Ref. 24, pp. 769–85
75. Wood, J. M. 1980. See Ref. 21, pp. 163–80
76. Hayaishi, O., Hirata, F., Ohnishi, T., Henry, J.-P., Rosenthal, I., Katok, A. 1977. *J. Biol. Chem.* 252:3548–50
77. Taniguchi, T., Sono, M., Hirata, F., Hayaishi, O., Tamura, M., Hayashi, K., Iizuka, T., Ishimura, Y. 1979. *J. Biol. Chem.* 254:3288–94
78. Sono, M., Taniguchi, T., Watanabe, Y., Hayaishi, O. 1980. *J. Biol. Chem.* 255: 1339–45
79. Fujisawa, H., Hiromi, K., Uyeda, M., Okuno, S., Nozaki, M., Hayaishi, O. 1972. *J. Biol. Chem.* 247:4422–28
80. Que, L. Jr., Lipscomb, J. D., Münck, E., Wood, J. M. 1977. *Biochim. Biophys. Acta* 485:60–74
81. Nakata, H., Yamauchi, T., Fujisawa, H. 1978. *Biochim. Biophys. Acta* 527: 171–81
82. May, S. W., Phillips, R. S. 1979. *Biochemistry* 18:5933–39
83. Keyes, W. E., Loehr, T. M., Taylor, M. L., Loehr, J. S. 1979. *Biochem. Biophys. Res. Commun.* 89:420–27
84. Que, L. Jr., Epstein, R. M. 1981. *Biochemistry* 20:2545–49
85. Gustafsson, J.-Å., Carlstedt-Duke, J., Mode, A., Rafter, J., eds. 1980. *Biochemistry, Biophysics and Regulation of Cytochrome P-450.* Amsterdam: Elsevier/North-Holland. 626 pp.

86. Peterson, J. A., Ishimura, Y., Griffin, B. W. 1972. *Arch. Biochem. Biophys.* 149:197–208
87. Mason, H. S., North, J. C., Vanneste, M. 1965. *Fed. Proc.* 24:1172–80
88. Tang, S. C., Koch, S., Papaefthymiou, G. C., Foner, S., Frankel, R. B., Ibers, J. A., Holm, R. H. 1976. *J. Am. Chem. Soc.* 98:2414–34
89. Peterson, J. A. 1979. See Ref. 23, pp. 227–62
90. Dawson, J. H., Cramer, S. P. 1978. *FEBS Lett.* 88:127–30
91. Peterson, J. A., White, R. E., Yasukochi, Y., Coomes, M. L., O'Keefe, D. H., Ebel, R. E., Masters, B. S. S., Ballou, D. P., Coon, M. J. 1977. *J. Biol. Chem.* 252:4431–34
92. Guengerich, F. P., Ballou, D. P., Coon, M. J. 1975. *J. Biol. Chem.* 250:7405–14
93. Kuthan, H., Tsuji, H., Graf, H., Ullrich, V., Werringloer, J., Estabrook, R. W. 1978. *FEBS Lett.* 91:343–45
94. Bösterling, B., Trudell, J. R. 1981. *Biochem. Biophys. Res. Commun.* 98: 569–75
95. Chiang, Y.-L., Coon, M. J. 1979. *Arch. Biochem. Biophys.* 195:178–87
96. Lipscomb, J. D. 1980. *Biochemistry* 19:3590–99
97. White, R. E., Sligar, S. G., Coon, M. J. 1980. *J. Biol. Chem.* 255:11108–11
98. Heimbrook, D. C., Sligar, S. G. 1981. *Biochem. Biophys. Res. Commun.* 99: 530–35
99. Koop, D. R., Hollenberg, P. F. 1980. *J. Biol. Chem.* 255:9685–92
100. Sligar, S. G., Kennedy, K. A., Pearson, D. C. 1980. *Proc. Natl. Acad. Sci. USA* 77:1240–44
101. Vanneste, W. H., Zuberbühler, A. 1974. See Ref. 15, pp. 371–404
102. Lerch, K. 1976. *FEBS Lett.* 69:157–60
103. Bouchilloux, S., McMahill, P., Mason, H. S. 1963. *J. Biol. Chem.* 238:1699–1777
104. Mason, H. S. 1976. See Ref. 17, pp. 464–69
105. Jolley, R. L., Evans, L. H., Makino, N., Mason, H. S. 1974. *J. Biol. Chem.* 249:335–45
106. Kuiper, H. A., Lerch, K., Brunori, M., Finazzi Agrò, A. 1980. *FEBS Lett.* 111:232–34
107. Himmelwright, R. S., Eickman, N. C., LuBien, C. D., Lerch, K., Solomon, E. I. 1980. *J. Am. chem. Soc.* 102:7339–44
108. Rüegg, C., Lerch, K. 1981. *Biochemistry* 20:1256–62
109. Skotland, T., Ljones, T. 1979. *Eur. J. Biochem.* 94:145–51

110. Blackburn, N. J., Mason, H. S., Knowles, P. F. 1980. *Biochem. Biophys. Res. Commun.* 95:1275–81
111. Ljones, T., Flatmark, T., Skotland, T., Petersson, L., Bäckström, D., Ehrenberg, A. 1978. *FEBS Lett.* 92:81–84
112. Walker, G. A., Kon, H., Lovenberg, W. 1977. *Biochim. Biophys. Acta* 482: 309–22
113. Diliberto, E. J. Jr., Allen, P. L. 1981. *J. Biol. Chem.* 256:3385–93
114. Skotland, T., Petersson, L., Bäckström, D., Ljones, T., Flatmark, T., Ehrenberg, A. 1980. *Eur. J. Biochem.* 103: 5–11
115. Friedman, S., Kaufman, S. 1966. *J. Biol. Chem.* 241:2256–59
116. Goldstein, M., Joh, T. H., Garvey, T. Q. III, 1968. *Biochemistry* 7:2724–30
117. Skotland, T., Ljones, T., Flatmark, T. 1978. *Biochem. Biophys. Res. Commun.* 84:83–88
118. Klinman, J. P., Humphries, H., Voet, J. G. 1980. *J. Biol. Chem.* 255:11648–51
119. Henry, J.-P., Hirata, F., Hayaishi, O. 1978. *Biochem. Biophys. Res. Commun.* 81:1091–99
120. Bright, H. J., Porter, D. J. T. 1975. See Ref. 14, pp. 421–505
121. Lockridge, O., Massey, V., Sulivan, P. A. 1972. *J. Biol. Chem.* 247:8097–8106
122. Massey, V., Ghisla, S., Kieschke, K. 1980. *J. Biol. Chem.* 255:2796–2806
123. Bray, R. C. 1975. See Ref. 14, pp. 299–419
124. Massey, V., Müller, F., Feldberg, R., Schuman, M., Sullivan, P. A., Howell, L. G., Mayhew, S. G., Matthews, R. G., Foust, G. P. 1969. *J. Biol. Chem.* 244:3999–4006
125. Massey, V., Ghisla, S., Moore, E. G. 1979. *J. Biol. Chem.* 254:9640–50
126. Geissler, J., Hemmerich, P. 1981. *FEBS Lett.* 126:152–56
127. Choong, Y. S., Massey, V. 1980. *J. Biol. Chem.* 255:8672–77
128. Massey, V., Hemmerich, P. 1980. *Biochem. Soc. Trans.* 8:246–56
129. Knowles, P. F., Gibson, J. F., Pick, F. M., Bray, R. C. 1969. *Biochem. J.* 111:53–58
130. Hille, R., Massey, V. 1981. *J. Biol. Chem.* 256:9090–95
131. Komai, H., Massey, V., Palmer, G. 1969. *J. Biol. Chem.* 244:1692–1700
132. Cammack, R., Barber, M. J., Bray, R. C. 1976. *Biochem. J.* 157:469–78
133. Massey, V., Palmer, G., Ballou, D. 1973. In *Oxidases and Related Redox Systems*, eds. T. E. King, H. S. Mason, M. Morrison, pp. 25–43. Baltimore: Univ. Park Press, 883 pp.
134. Malmström, B. G., Andréasson, L.-E., Reinhammar, B. 1975. See Ref. 14, pp. 507–79
135. Finazzi Agrò, A., Guerrieri, P., Costa, M. T., Mondovì, B. 1977. *Eur. J. Biochem.* 4:435–40
136. Bereman, R. D., Ettinger, M. J., Kosman, D. J., Kurland, R. J. 1977. In *Bioinorganic Chemistry-II*, ed. K. N. Raymond, pp. 263–80. Washington DC: Am. Chem. Soc. 448 pp.
137. Bereman, R. D., Kurland, R. J., Ettinger, M. J., Kosman, D. J. 1981. See Ref. 24, pp. 219–31
138. Hamilton, G. A., Dyrkacz, G. R., Libby, R. D. 1978. *J. Am. Chem. Soc.*
139. Margerum, D. W., Wong, L. F., Bossu, F. P., Chellappa, K. L., Czarnecki, J. J., Kirksey, S. T. Jr., Neubecker, T. A. 1977. See Ref. 136, pp. 281–303
140. Kosman, D. J., Bereman, R. D., Ettinger, M. J., Giordano, R. S. 1973. *Biochem. Biophys. Res. Commun.* 54: 856–61
141. Kwiatowski, L. O., Siconoffi, L., Weiner, R. E., Giordano, R. S., Bereman, R. D., Ettinger, M. J., Kosman, D. J. 1977. *Arch. Biochem. Biophys.* 182:712–22
142. Marwedel, B. J., Kurland, R. J. 1981. *Biochim. Biophys. Acta* 657:495–506
143. Grant, J., Kelly, I., Knowles, P., Olsson, J., Pettersson, G. 1978. *Biochem. Biophys. Res. Commun.* 83:1216–24
144. Dooley, D. M., Coolbaugh, T. S. 1980. *Biochem. Biophys. Res. Commun.* 96: 823–30
145. Lindström, A., Pettersson, G. 1978. *Eur. J. Biochem.* 84:479–85
146. Dooley, D. M., Rawlings, J., Dawson, J. H., Stephens, P. J., Andréasson, L.-E., Malmström, B. G., Gray, H. B. 1979. *J. Am. Chem. Soc.* 101: 5038–46
147. Malmström, B. G. 1970. *Pure Appl. Chem.* 24:393–406
148. Holwerda, R. A., Wherland, S., Gray, H. B. 1976. *Ann. Rev. Biophys. Bioeng.* 5:363–96
149. Malkin, R., Malmström, B. G. 1970. *Adv. Enzymol.* 33:177–244
150. Fee, J. A. 1975. *Struct. Bonding (Berlin)* 23:1–60
151. Vänngård, T. 1972. In *Biological Applications of Electron Spin Resonance*, eds. H. M. Swartz, J. R. Bolton, D. C. Borg, pp. 411–47. New York: Wiley. 569 pp.
152. Reinhammar, B. 1979. *Adv. Inorg. Biochem.* 1:91–118
153. Reinhammar, B., Malmström, B. G. 1981. See Ref. 22, pp. 110–49

154. Reinhammar, B. 1982. In *Copper Proteins,* ed. R. Lontie, Vol. II. Boca Raton, Fa: CRC In press
155. Rotilio, G. 1979. In *Metalloproteins,* ed. U. Weser, pp. 1–28. Stuttgart: George Thieme. 286 pp.
156. Malmström, B. G. 1978. In *New Trends in Bio-Inorganic Chemistry,* eds. R. J. P. Williams, J. R. R. F. Da Silva, pp. 59–77. London: Academic. 489 pp.
157. Malmström, B. G. 1969. In *Symmetry and Function of Biological Systems at the Macromolecular Level,* eds. A. Engström, B. Standberg, pp. 153–63. New York: Wiley. 436 pp.
158. Gray, H. B., Solomon, E. I. 1981. See Ref. 22, pp. 1–39
159. Malmström, B. G. 1980. In *The Evolution of Protein Structure and Function,* eds. D. S. Sigman, M. A. B. Brazier, pp. 87–96. New York: Academic. 350 pp.
160. Petersson, L., Ångström, J., Ehrenberg, A. 1978. *Biochim. Biophys. Acta* 526: 311–17
161. Dooley, D. M., Scott, R. A., Ellinghaus, J., Solomon, E. I., Gray, H. B. 1978. *Proc. Natl. Acad. Sci. USA* 75:3019–22
162. Martin, C. T., Morse, R. H., Kanne, R. M., Gray, H. B., Malmström, B. G., Chan, S. I. 1981. *Biochemistry.* 20: 5147–55
163. Reinhammar, B., Malkin, R., Jensen, P., Karlsson, B., Andréasson, L.-E., Aasa, R., Vänngard, T., Malmström, B. G. 1980. *J. Biol. Chem.* 255:5000–3
164. Reinhammar, B. 1981. *J. Inorg. Biochem.* 15:27–39
165. Petersen, L. C., Degn, H. 1978. *Biochim. Biophys. Acta* 526:85–92
166. Andreásson, L.-E., Reinhammar, B. 1976. *Biochim. Biophys. Acta* 445: 579–97
167. Fee, J. A., Malkin, R., Malmström, B. G., Vänngard, T. 1969. *J. Biol. Chem.* 244:4200–7
168. Reinhammar, B. 1972. *Biochim. Biophys. Acta* 275:245–59
169. Farver, O., Goldberg, M., Pecht, I. 1980. *Eur. J. Biochem.* 104:71–77
170. Malmström, B. G., Finazzi Agrò, A., Antonini, E. 1969. *Eur. J. Biochem.* 9:383–91
171. Brändén, R., Deinum, J. 1977. *FEBS Lett.* 73:144–46
172. Brändén, R., Deinum, J., Coleman, M. 1978. *FEBS Lett.* 89:180–82
173. Goldberg, M., Vuk-Pavlović, S., Pecht, I. 1980. *Biochemistry* 19:5181–89
174. Morpurgo, L., Graziani, M. T., Finazzi Agrò, A., Rotilio, G., Mondovì, B. 1980. *Biochem. J.* 187:361–66
175. Andréasson, L.-E., Reinhammar, B. 1979. *Biochim. Biophys. Acta* 568: 145–56
176. Goldberg, M., Farver, O., Pecht, I. 1980. *J. Biol. Chem.* 255:7353–61
177. Manabe, T., Manabe, N., Hiromi, K., Hatano, H. 1972. *FEBS Lett.* 23: 268–70
178. Andréasson, L.-E., Brändén, R., Malmström, B. G., Vänngård, T. 1973. *FEBS Lett.* 32:187–89
179. Brändén, R., Malmström, B. G., Vänngård, T. 1971. *Eur. J. Biochem.* 18: 238–41
180. Farver, O., Goldberg, M., Lancet, D., Pecht, I. 1976. *Biochem. Biophys. Res. Commun.* 73:494–500
181. Farver, O., Goldberg, M., Pecht, I. 1978. *FEBS Lett.* 94:383–86
182. Farver, O., Pecht, I. 1979. *FEBS Lett.* 108:436–38
183. Andréasson, L.-E., Brändén, R., Reinhammar, B. 1976. *Biochim. Biophys. Acta* 483:370–79
184. Aasa, R., Brändén, R., Deinum, J., Malmström, B. G., Reinhammar, B., Vänngård, T. 1976. *FEBS Lett.* 61: 115–19
185. Aasa, R., Brändén, R., Deinum, J., Malmström, B. G., Reinhammar, B., Vänngård, T. 1976. *Biochem. Biophys. Res. Commun.* 70:1204–9
186. Reinhammar, B., Oda, Y. 1979. *J. Inorg. Biochem.* 11:115–27
187. Graziani, M. T., Morpurgo, L., Rotilio, G., Mondovì, B. 1976. *FEBS Lett.* 70: 87–90
188. Wikström, M., Krab, K., Saraste, M. 1981. *Ann. Rev. Biochem.* 50:623–55
189. Lemberg, M. R. 1969. *Physiol. Rev.* 49:48–121
190. Nicholls, P., Chance, B. 1974. See Ref. 15, pp. 479–534
191. Caughey, W. S., Wallace, W. J., Volpe, J. A., Yoshikawa, S. 1976. *Enzymes* 13:299–344 (3rd ed.)
192. Malmström, B. G. 1979. *Biochim. Biophys. Acta* 549:281–303
193. Malmström, B. G. 1980. See Ref. 21, pp. 181–207
194. Brunori, M., Antonini, E., Wilson, M. T. 1981. In *Metal Ions in Biological Systems,* ed. H. Siegel, Vol. 13, pp. 187–228. New York: Marcel Dekker 394 pp.
195. Clore, G. M. 1981. *Rev. Inorg. Chem.* 2:343–60
196. Aasa, R., Albracht, S. P. J., Falk, K.-E., Lanne, B., Vänngård, T. 1976. *Biochim. Biophys. Acta* 422:260–72
197. Peisach, J., Blumberg, W. E. 1974. *Arch. Biochem. Biophys.* 165:691–708

198. Hu, V. W., Chan, S. I., Brown, G. S. 1977. *Proc. Natl. Acad. Sci. USA* 74: 3821–25
199. Powers, L., Blumberg, W. E., Chance, B., Barlow, C. H., Leigh, J. S. Jr., Smith, J., Yonetani, T., Vik, S., Peisach, J. 1979. *Biochim. Biophys. Acta* 546: 520–38
200. Brudvig, G. W., Bocian, D. F., Gamble, R. C., Chan, S. I. 1980. *Biochim. Biophys. Acta* 624:78–89
201. Steffens, G. J., Buse, G. 1979. See Ref. 19, pp. 79–90
202. Scott, R. A., Cramer, S. P., Shaw, R. W., Beinert, H., Gray, H. B. 1981. *Proc. Natl. Acad. Sci. USA* 78:664–67
203. Seiter, C. H. A., Angelos, S. G. 1980. *Proc. Natl. Acad. Sci. USA* 77:1806–8
204. Beinert, H., Shaw, R. W. 1977. *Biochim. Biophys. Acta* 462:121–30
205. Karlsson, B., Andréasson, L.-E. 1981. *Biochim. Biophys. Acta* 635:73–80
206. Brudvig, G. W., Stevens, T. H., Chan, S. I. 1980. *Biochemistry* 19:5275–85
207. Woodruff, W.-H., Dallinger, R. F., Antatis, T. M., Palmer, G. 1981. *Biochemistry* 20:1332–38
208. Babcock, G. T., Callahan, P. M., Ondrias, M. R., Salmeen, I. 1981. *Biochemistry* 20:959–66
209. Thomson, A. J., Johnson, M. K., Greenwood, C., Gooding, P. E. 1981. *Biochem. J.* 193:687–97
210. Johnson, M. K., Eglinton, D. G., Gooding, P. E., Greenwood, C., Thomson, A. J. 1981. *Biochem. J.* 193:699–708
211. Powers, L., Chance, B., Ching, Y., Angiolillo, P. 1981. *Biophys. J.* 34: 465–98
212. Alben, J. O., Moh, P. P., Flamingo, F. G., Altschuld, R. A. 1981. *Proc. Natl. Acad. Sci. USA* 78:234–37
213. Stevens, T. H., Chan, S. I. 1981. *J. Biol. Chem.* 256:1068–71
214. Chance, B., Saronio, C., Leigh, J. S. Jr. 1979. *Biochem. J.* 177:931–41
215. Beinert, H., Shaw, R. W., Hansen, R. E., Hartzell, C. R. 1980. *Biochim. Biophys. Acta* 591:458–70
216. Boelens, R., Wever, R. 1980. *FEBS Lett.* 116:223–26
217. Brudvig, G. W., Chan, S. I. 1979. *FEBS Lett.* 106:139–41
218. Lindsay, J. G., Wilson, D. F. 1974. *FEBS Lett.* 48:45–49
219. Nicholls, P. 1978. *Biochem. J.* 175: 1147–50
220. Nicholls, P., Chanady, G. A. 1981. *Biochim. Biophys. Acta* 634:256–65
221. Clore, G. M., Andréasson, L.-E.,

Karlsson, B., Aasa, R., Malmström, B. G. 1980. *Biochem. J.* 185:155–67
222. Shaw, R. W., Hansen, R. E., Beinert, H. 1978. *J. Biol. Chem.* 253:6637–40
223. Shaw, R. W., Hansen, R. E., Beinert, H. 1979. *Biochim. Biophys. Acta* 548: 386–96
224. Jensen, P., Aasa, R., Malmström, B. G. 1981. *FEBS Lett.* 125:161–64
225. Brudvig, G. W., Stevens, T. H., Morse, R. H., Chan, S. I. 1981. *Biochemistry* 20:3912–21
226. Lanne, B., Malmström, B. G., Vänngård, T. 1979. *Biochim. Biophys. Acta* 545:205–14
227. Antonini, E., Brunori, M., Colosimo, A., Greenwood, C., Wilson, M. T. 1977. *Proc. Natl. Acad. Sci. USA* 74:3128–32
228. Brunori, M., Colosimo, A., Rainoni, G., Wilson, M. T., Antonini, E. 1979. *J. Biol. Chem.* 254:10769–75
229. Rosén, S., Brändén, R., Vänngård, T., Malmström, B. G. 1977. *FEBS Lett.* 74:25–31
230. Petersen, L. C., Cox, R. P. 1980. *Biochim. Biophys. Acta* 590:128–37
231. Shaw, R. W., Rife, J. E., O'Leary, M. H., Beinert, H. 1981. *J. Biol. Chem.* 256:1105–7
232. Brunori, M., Colosimo, A., Sarti, P., Antonini, E., Wilson, M. T. 1981. *FEBS Lett.* 126:195–98
233. Carter, K. R., Antalis, T. M., Palmer, G., Ferris, N. S., Woodruff, W. H. 1981. *Proc. Natl. Acad. Sci. USA* 78:1652–55
234. Greenwood, C., Gibson, Q. H. 1967. *J. Biol. Chem.* 242:1782–87
235. Greenwood, C., Wilson, M. T., Brunori, M. 1974. *Biochem. J.* 137: 205–15
236. Chance, B., Saronio, C., Leigh, J. S. Jr. 1975. *J. Biol. Chem.* 250:9226–37
237. Chance, B., Saronio, C., Leigh, J. S., Jr., Ingledew, W. J., King, T. E. 1978. *Biochem. J.* 171:787–98
238. Clore, G. M., Andréasson, L.-E., Karlsson, B., Aasa, R., Malmström, B. G. 1980. *Biochem. J.* 185:139–54
239. Karlsson, B., Aasa, R., Vänngård, T., Malmström, B. G. 1981. *FEBS Lett.* 131:186–88
240. Beinert, H., Hansen, R. E., Hartzell, C. R. 1976. *Biochim. Biophys. Acta* 423: 339–55
241. Wilson, M. T., Jensen, P., Aasa, R., Vänngård, T., Malmström, B. G. 1982. *Biochim. J.* In press
242. Antonini, E., Brunori, M., Greenwood, C., Malmström, B. G. 1970. *Nature* 228:936–37

Ann. Rev. Biochem. 1982. 51:61–87

DNA REPAIR ENZYMES

Tomas Lindahl

Imperial Cancer Research Fund, Mill Hill Laboratories, London NW7, England

CONTENTS

Perspectives and Summary

Preventing an unacceptably high mutation rate is a challenge to the cell. Enzymatic mechanisms of great intricacy have evolved to ensure that DNA is replicated with high fidelity, and mismatch repair activities exist to remove the rare misincorporated residues that have escaped proofreading during replication. Potentially mutagenic alterations also arise in nonreplicating DNA by spontaneous hydrolysis under physiological conditions and by exposure to radiation and chemical mutagens. The spontaneous decay of DNA is greater than generally recognized; the loss of bases due to depurination and depyrimidination of DNA amounts to several thousand

61

0066-4154/82/0701-0061$02.00

residues per genome per day for a mammalian cell. Moreover, premutagenic deamination of about one hundred cytosine residues and a few adenine residues also occurs under the same conditions. The recent observation that S-adenosylmethionine acts as a weak DNA alkylating agent indicates that nonenzymatic DNA methylation takes place in vivo. The sites methylated in this fashion are different from those modified by DNA methylases, and the amount of DNA damage could be similar to that caused by the total spontaneous hydrolytic decay (1).

Repair activities that recognize these various alterations in DNA, and also other important forms of damage, such as the most common radiation-induced lesions, presumably appeared very early during evolution as they seem to be universally distributed in living cells. The major pathways of DNA repair are, in fact, surprisingly similar in *Escherichia coli* and in mammalian cells. It would appear that the same kinds of repair enzymes, which presumably evolved to avoid mutations to lethality or auxotrophy in primitive organisms, may also be employed to prevent certain deleterious events unique to more complex structures, such as the persistence of lesions that might cause transformation to malignancy; no novel DNA repair pathways have been detected in higher cells for these purposes. (Cells from long-lived organisms such as man have more effective mechanisms of DNA repair than those from mice and other short-lived animals.) The universally occurring repair activities seem to serve efficiently to counteract transformation, as exemplified by the relatively harmless effects of sunlight on normal humans when compared to its strongly carcinogenic effects in xeroderma pigmentosum patients. Further, repair enzymes that function to correct radiation damage and unavoidable DNA decay in the intracellular milieu also provide protection against many potential, chemical mutagens recently introduced into the human diet and environment. For instance, the occurrence of sulfur dioxide as an air pollutant, and the use of its neutral aqueous form, bisulfite, as a component of many common beverages, is a recent phenomenon. The latter agent causes one of the same mutagenic DNA alterations produced by spontaneous hydrolysis, that is, deamination of cytosine to uracil (2), and is only a very weak mutagen, since the DNA damage it causes is efficiently repaired.

The most important DNA repair pathways, in *E. coli* as well as in human cells, depend on the excision of an altered residue or group, and many different enzymes initiating such reactions have been discovered recently. The two main types of activities reside within DNA glycosylases, which cleave the base-sugar bond of a nonconventional nucleotide residue, and nucleases, which incise DNA by the specific cleavage of a phosphodiester bond adjacent to a damaged residue. Repair activities of these kinds possibly comprise the largest group of enzymes acting on DNA in the cell. So far, about 15 different enzymes have been discovered that apparently serve exclusively in the early stages of DNA repair, and this no doubt represents only a fraction, perhaps one half, of those needed to fulfill the functions required for initiating the removal of commonly occurring lesions from DNA. In this review, the properties of the enzymes hitherto known to be involved in these processes are summarized.

In addition to the various types of excision mechanisms, post-replication repair occurs that, at least in bacteria, depends on recombination between daughter strands, and error-prone inducible repair functions have been demonstrated. As these latter processes have not yet been reproduced in cell-free systems, a discussion of the enzymes possibly involved in them is deferred. Further, factors postulated to make the DNA in mammalian cells accessible (or nonaccessible) to repair enzymes are not discussed. It has often been proposed that the DNA in nucleosomes could be refractory to repair, but this seems unlikely from a physiological point of view, and the experimental evidence is not convincing.

The present review covers enzymes that specifically recognize different kinds of altered DNA, and special emphasis is laid on progress made over the last three years (see 3). The more general activities that catalyze exonucleolytic excision, gap-filling, and ligation are not described. A summary of the latter has been included in a recent review by Grossman (4), and a general and comprehensive review of different repair pathways has already appeared in this series (5). A number of other reviews covering various aspects of DNA repair have also appeared (6–9), and the informative article by Friedberg et al (8) on inherited human diseases associated with defective DNA repair is especially noteworthy.

DNA Glycosylases

Several enzymes termed DNA glycosylases, which catalyze the cleavage of base-sugar bonds in DNA, have been found recently, and most of them are widely if not universally distributed. They have the common property of acting only on altered or damaged nucleotide residues. So far, attempts to find DNA glycosylases that might correct mismatched bases or remove 5-methylcytosine from DNA have failed. The physical and general biochemical properties of the different DNA glycosylases are similar; the enzymes are of relatively low molecular weight, with reported values between 18,000 and 31,000. Further, they do not require cofactors such as divalent metal cations, and apparently act by simple hydrolytic cleavage of the glycosyl bond. Double-stranded DNA is the preferred substrate; with the notable exception of uracil-DNA glycosylase, there is little or no activity against damaged single-stranded DNA, and no DNA glycosylase cleaves mononucleotides. Since these activities resemble each other also from the point of view that they have similar purification properties and appear to be present only in small amounts in cells (again with the exception of uracil-DNA glycosylase), it becomes an increasingly arduous task to show that each newly discovered member of this class is not identical to a previously known DNA glycosylase. Nevertheless, there appears to be little overlap between the various activities, although all of them have not yet been characterized in detail. Each enzyme has a narrow substrate specificity, and may have evolved as a defence against a single type of DNA lesion. Thus, two different DNA glycosylases are required to remove deaminated cytosine and deaminated adenine from DNA, and these enzymes cannot replace each other. Further, two distinct DNA glycosylases serve to remove 7-methylguanine and its imidazole ring–opened derivative from alkylated DNA, and neither of these two activities can release guanine that contains a bulky substituent group, such as an aflatoxin B_1 residue, at the 7 position. The properties and the substrate specificities of the DNA glycosylases strongly indicate that they are all involved in repair processes. However, direct evidence for this notion is only available in the cases in which enzyme-deficient mutant cells have been isolated and shown to be anomalously sensitive to DNA-damaging agents. It is clear, though, that the apurinic or apyrimidinic site generated by the release of a nonconventional base can be rapidly corrected by an excision-repair process that involves the replacement of a single nucleotide residue or at most a small number of such residues.

URACIL-DNA GLYCOSYLASE   The uracil-DNA glycosylase from human cells has recently been extensively purified and characterized (10, 11). It has a molecular weight of about 30,000, similar to the 25,000 found for the *E. coli* enzyme (12). The human enzyme, however, has a 20 times higher K_m (about 1 μM) for dUMP residues in DNA than the bacterial enzyme. Human uracil-DNA glycosylase is found mainly in the cell nuclei (13). Mitochondria also contain uracil-DNA glycosylase, but whether this minor activity is distinct from the nuclear enzyme is not known (14). The nuclear enzyme appears to be induced during cell proliferation in contrast to the mitochondrial activity (14a). Both the *E. coli* and human uracil-DNA glycosylase are product-inhibited by free uracil, with a K_i of about 2.10^{-4} M, and somewhat surprisingly, initial velocity measurements at different uracil concentrations indicate a noncompetitive form of inhibition (10, 12). One possible explanation of these data would be that the enzyme acts in a processive fashion, and that the free base interferes with the progression of the enzyme along the DNA. A similar model (15) to explain this kind of unusual product inhibition of a DNA glycosylase postulates separate enzyme domains for binding, with release of the base residue preceding dissociation of the enzyme from DNA. Many different uracil analogues, including derivatives having modifications at positions 1–4, have been tested and show no detectable inhibition of the enzyme. Most derivatives modified at positions 5 and 6 are also ineffective, but 6-aminouracil and 5-azauracil inhibit almost as well as unsubstituted uracil, and 5-fluorouracil is a weak inhibitor (11). These data suggest that the enzyme is able to remove the latter derivaties if incorporated into DNA, and this has already been shown in the case of 5-fluorouracil (16–18), which is excised by either the *E. coli* or human uracil-DNA glycosylase, albeit 20 times more slowly than unsubstituted uracil. In addition, 5-fluoro-dUTP is a substrate for the cellular dUTPase, so the mechanisms for avoiding incorporation of 5-fluorouracil into DNA are almost the same as those employed to prevent uracil incorporation (16–18). These findings may be related to the cytotoxic and chemotherapeutic effect of 5-fluorouracil on mammalian cells. Another cytotoxic compound, methotrexate, causes a great increase in the intracellular dUTP level and a corresponding decrease in dTTP by inhibition of dihydrofolate reductase. This leads to an increased frequency of misincorporation of uracil into DNA, accompanied by increased fragmentation of the newly synthesized DNA because of the concerted action of uracil-DNA glycosylase and an endonuclease for apurinic and apyrimidinic sites (AP endonuclease) (19). It is possible that this process is responsible for the observed cytotoxicity.

An entirely different kind of inhibitor of uracil-DNA glycosylase is a small protein (mol wt 18,000) induced in *Bacillus subtilis* by the uracil-containing DNA phage PBS2. This protein apparently binds stoichiometrically to the host enzyme. It has been extensively purified (20) and shown to inhibit uracil-DNA glycosylases from several other sources, including *E. coli*, yeast, and human cells. On the other hand, DNA glycosylases acting on hypoxanthine or 3-methyladenine in DNA are resistant to inhibition by the PBS2-induced protein (21). A similar uracil-DNA glycosylase inhibitor is induced as an immediate early function by phage T5 in *E. coli* (22). Since T5 DNA does not normally contain uracil, the reason for the occurrence of this inhibitor is obscure.

The physiological role of uracil-DNA glycosylase is to correct for deaminated cytosine residues in DNA. This has been clarified by the isolation and characterization of enzyme-deficient *E. coli* mutants, *ung* (23), and also by the discovery that 5-methylcytosine residues are hot spots of spontaneous base substitutions (24). In the latter case, 5-methylcytosine would be deaminated to thymine, which cannot be removed by the uracil-DNA glycosylase. Direct evidence that the enzyme serves to remove spontaneously deaminated cytosine residues from DNA in vivo has been obtained recently by the demonstration that in an *E. coli ung* strain, the rate of spontaneous transitions at cytosine residues was raised to that found at 5-methylcytosine residues in wild-type cells (25). Thus, in an *ung* mutant, unmodified as well as modified cytosine residues are hot spots of mutation. On the other hand, incorporation of uracil instead of thymine in DNA does not appear to be markedly mutagenic, since *E. coli ung* mutants also deficient in dUTPase (*dut*) have as much as 15–20% of their thymine residues replaced by uracil, but retain their viability (26).

HYPOXANTHINE-DNA GLYCOSYLASE   This enzyme presumably acts in an analogous fashion to uracil-DNA glycosylase; it removes spontaneously deaminated adenine residues (27). However, no bacterial mutant deficient in hypoxanthine-DNA glycosylase has been isolated, so there is only circumstantial evidence to suggest a role for the enzyme in vivo. The latter includes the observation that just as cytosine is deaminated to uracil, adenine in DNA is slowly converted to hypoxanthine by hydrolysis at neutral pH (28). Further, the enzyme does not release derivatives similar to deaminated adenine such as xanthine or alkylated purine bases from DNA. This narrow substrate specificity indicates that it specifically recognizes dIMP residues. A mammalian hypoxanthine-DNA glycosylase has recently been purified from calf thymus (28). It is similar to the enzyme from *E. coli,* but, in contrast, shows markedly higher activity in the presence of 0.1 M KCl. The calf thymus enzyme has a molecular weight of 31,000 and is not product inhibited by free hypoxanthine. As with uracil-DNA glycosylase, the hypoxanthine-DNA glycosylase acts equally well on matched or mismatched purine-pyrimidine base pairs; that is, hypoxanthine is removed from a double-stranded polydeoxynucleotide having either C or T residues in the complementary chain.

3-METHYLADENINE-DNA GLYCOSYLASE I   One of the major alkylation products in DNA that has been treated with simple methylating agents is 3-methyladenine. The alkyl group of this derivative is located in the minor groove of the DNA helix, and in this regard 3-methyladenine differs from more innocuous methylated bases such as 7-methylguanine, 5-methylcytosine, and N^6-methyladenine. 3-Methyladenine in DNA is not well tolerated by cells, and it is released both in bacteria and in mammalian cells very rapidly after formation by a DNA glycosylase. Since this alkylation lesion is formed in similar amounts by weakly mutagenic and carcinogenic agents such as methyl methanesulfonate and dimethyl sulfate, and more strongly mutagenic compounds such as N-methyl-N'-nitro-N-nitrosoguanidine (MNNG) and N-methyl-N-nitrosourea, it seems likely that 3-methyladenine is a potentially lethal or inactivating lesion rather than a strongly mutagenic one. Traces of 3-methyladenine are continuously formed by the nonenzymatic alkylation of DNA by S-adenosylmethionine, under physiological conditions (1).

A distinct DNA glycosylase, which removes 3-methyladenine from DNA, has been found in *E. coli* (29, 30). It is now termed 3-methyladenine-DNA glycosylase I, because a glycosylase of overlapping specificity (3-methyladenine-DNA glycosylase II) has been recently discovered. A 3-methyladenine-DNA glycosylase has also been found in *Micrococcus luteus* (31, 32). The *E. coli* enzyme (molecular weight 20,000) shows a narrowly defined substrate specificity in that it does not catalyze the release of other alkylated purines such as 7-methylguanine, 7-methyladenine, 3-methylguanine, N^6-methyladenine, or 1-methyladenine from DNA. On the other hand, the ethylated base derivative analogous to 3-methyladenine, 3-ethyladenine, is released efficiently by the enzyme. 3-Methyladenine-DNA glycosylase I is similar to the uracil-DNA glycosylase from the point of view that no endonucleolytic activity has been detected in the highly purified enzyme. Thus, the end products of the reaction of the enzyme with alkylated DNA are free 3-alkyladenine and DNA that contains apurinic sites; no chain cleavage occurs at the latter sites even in the presence of high concentrations of enzyme. Another similarity is that the enzyme is product inhibited in a noncompetitive fashion, in this case by free 3-methyladenine.

Mutants of *E. coli* deficient in 3-methyladenine-DNA glycosylase I, *tag*, have been isolated (33). One mutant shows a temperature-sensitive phenotype and contains an anomalously heat-labile enzyme, which indicates that the mutation is in the structural gene. In general, the *tag* mutants show two characteristic properties. First, after exposure to a high dose of a methylating agent, they are unable to eliminate 3-methyladenine rapidly from their DNA; they differ in this regard from wild-type cells. Second, the mutants are more sensitive to killing by alkylating agents such as methyl methanesulfonate in short-term experiments, and they also show impaired host cell reactivation of alkylated phage lambda and T7 (33, 33a). On the other hand, the *tag* mutants show normal resistance to agents such as ultraviolet (UV) light, X-rays, and nitrous acid. Initial attempts to map the *tag* mutants were complicated by the inducibility of a second 3-methyladenine-DNA glycosylase in *E. coli* (see below) and by the inadvertent isolation of a double mutant also deficient in the adaptive response to alkylating agents. It has recently been determined that the *tag* gene is located close to the gene for streptomycin resistance, at 70–74 min on the standard *E. coli* K-12 map (E. Seeberg, personal communication).

3-METHYLADENINE-DNA GLYCOSYLASE II The *E. coli tag* mutants are not totally deficient in 3-methyladenine-DNA glycosylase activity; the cell extracts retain 5–10% of the wild-type enzyme level. This residual activity is due to another enzyme, 3-methyladenine-DNA glycosylase II, which is present at similar levels in *tag* mutants and in wild-type cells. It differs from the *tag*$^+$ gene product in that it is a considerably more heat-stable enzyme and shows no product inhibition by free 3-methyladenine (33). A more detailed study of the general properties and substrate specificity of 3-methyladenine-DNA glycosylase II (which has a molecular weight of 27,000, significantly larger than the *tag*$^+$ gene product) has been performed recently (15). The highly purified but nonhomogeneous enzyme catalyzed the liberation of 3-methyladenine and 3-methylguanine equally efficiently from alkylated DNA, and in addition released 7-methyladenine and 7-methylguanine, albeit at about a twentyfold slower rate. The analogous ethylated derivatives were similarly liberated from DNA that had been treated with an ethylating agent. However, O^6-methylguanine residues or the imidazole ring–opened derivative of 7-methylguanine could not be released. The activities found to release 3-methyladenine and 7-methylguanine from DNA chromatographed together on several different columns and showed identical rates of heat inactivation, so they were most likely due to the same enzyme. A DNA glycosylase, present in small amounts in *E. coli* cell extracts, which can release 7-methylguanine from alkylated DNA, has been reported by Laval et al (34). This latter activity is presumably due to 3-methyladenine-DNA glycosylase II.

When *E. coli* are exposed to small amounts of alkylating agents such as MNNG, they gain increased resistance to the mutagenic and killing effects of a subsequent challenge with the reagent. This inducible repair pathway, termed the adaptive response, is not dependent on the *recA*$^+$ gene product (35–37). While the adaptation that allows mutation resistance is largely due to the induction of a repair function for O^6-methylguanine (38), the adapted resistance to killing of the cell shows somewhat different features. In particular, the latter is dependent on a functional *polA*$^+$ gene product (37), which implies that excision-repair of some alkylated residue that is a potentially lethal lesion occurs more efficiently in adapted than in normal cells. This lesion may well be identical with 3-methyladenine (or 3-methylguanine). It has been found recently that 3-methyladenine-DNA glycosylase II can be induced and occurs at a 20-fold increased level in adapted *E. coli*, whereas 3-methyladenine-DNA glycosylase I appears to be constitutively expressed and is present at an unchanged level of activity after adaptation (1). The expression of 3-methyladenine-DNA glycosylase II is dependent on a functional *alk*$^+$ gene, which maps 44 min, but the mechanism of the induction process remains unclear (G. B. Evensen and E. Seeberg, personal communication).

The apparently broad substrate specificity of 3-methyladenine-DNA glycosylase II is at first sight puzzling, since most other DNA glycosylases show such a restricted specificity. It seems possible that the unique feature recognized by this enzyme is a positively charged purine residue, rather than methylation or ethylation at some specific site. If this notion is correct, the enzyme should also be able to release unsubstituted adenine and guanine from DNA at low pH values. Unfortunately, the low activity and rapid inactivation of the enzyme under such conditions, as well as the acid-catalyzed depurination of the DNA substrate, make attempts at a practical demonstration difficult. The recognition of a common denominator such as an alteration in charge would make it easy to understand how a single enzyme can remove several different types of alkylated bases without attacking unsubstituted DNA.

A 3-methyladenine-DNA glycosylase has been partly purified from several mammalian tissues (39–41a). The enzyme is present in the cell nuclei and appears to be very labile. Consequently, it has not been highly purified, but partly purified preparations also release 7-methylguanine at a slow rate. In this regard, it therefore seems to resemble the E. coli 3-methyladenine-DNA glycosylase II rather than I. The mammalian activities against 3-methyladenine and 7-methylguanine co-chromatographed and showed the same heat lability (41). It is not known at present if mammalian cells resemble E. coli in having two different DNA glycosylases that release 3-methyladenine, or if only a single enzyme exists that preferentially releases 3-methyladenine over 7-methylguanine. Another possibility, not yet ruled out, is that in addition to a 3-methyladenine-DNA glycosylase, a distinct 7-methylguanine-DNA glycosylase occurs; two groups have reported on the presence of an activity of this kind in crude cell extracts (42, 43). After treatment of mammalian cells with methylating agents in vivo, 3-methyladenine is liberated much more rapidly from DNA than 7-methylguanine (44).

FORMAMIDOPYRIMIDINE-DNA GLYCOSYLASE   7-Methylguanine residues, the most abundant lesions in alkylated DNA, are susceptible to alkali-catalyzed cleavage of the imidazole ring. Thus, in methylated DNA incubated at high pH, the 7-methylguanine residues are converted to 2,6-diamino-4-oxy-5-N-methylformamido-pyrimidine. This ring-opening reaction occurs only at a very slow rate at neutral pH, so substituted formamidopyrimidine is a minor secondary alkylation lesion derived from 7-methylguanine. Nonenzymatic cleavage of the glycosyl bonds of 7-methyldeoxyguanosine residues in DNA to yield apurinic sites occurs about 300 times more rapidly at pH 7.4 than the purine ring-opening reaction (L. Breimer, unpublished information). Nevertheless, the latter reaction may be of significance because of the large amounts of 7-methylguanine generated by the action of methylating agents on DNA.

E. coli cell extracts contain a unique DNA glycosylase that effectively catalyzes the release of the substituted formamidopyrimidine from alkylated, alkali-treated DNA. The partly purified enzyme has a molecular weight of about 30,000, and does not release intact 7-methylguanine from DNA (45). Further, the enzyme does not catalyze the release of a similarly ring-opened guanine residue carrying an aflatoxin B_1 adduct rather than a methyl group, nor can it be induced in cells by MNNG treatment (T. Lindahl, unpublished data). It is not presently known if this enzyme has as its sole function the removal of the secondary alkylation product, which presumably is a dangerous lesion with miscoding or noncoding properties, or if it can also serve to remove unsubstituted guanine and adenine residues with opened imidazole rings. The latter derivatives are the major purine lesions in DNA exposed to ionizing radiation (46), and in addition, ring-opened adenine residues may be generated as spontaneous hydrolytic lesions (47). A formamidopyrimidine-DNA glycosylase with properties similar to the bacterial enzyme has been detected in rodent liver cell extracts (43).

UREA-DNA GLYCOSYLASE   Ionizing radiation causes ring opening and fragmentation of pyrimidine residues in DNA (48). Thus, urea and N-substituted urea derivatives are generated as remnants of pyrimidines. These residues, which may remain attached to deoxyribose in DNA, are unlikely to retain any coding information and presumably have to be repaired. In a search for DNA glycosylases acting on such species, a DNA glycosylase that cleaves a urea-deoxyribose bond was identified (49). The substrate used was a $KMnO_4$-treated polydeoxynucleotide containing a few fragmented thymine residues. This enzyme seems unable to release N-substituted urea derivatives such as formylpyruvylurea and formylurea, and it is presently not known if other DNA glycosylases exist for this purpose. The urea-DNA glycosylase appears to be a distinct glycosylase; it is present at similar levels in *E. coli* wild-type cells and in *ung* and *tag* mutants, and the partly purified enzyme has chromatographic properties different from several other DNA glycosylases. A similar urea-DNA glycosylase has also been observed in crude extracts of several types of mammalian cells (L. Breimer, unpublished).

THYMINE GLYCOL-DNA GLYCOSYLASE   Ring-saturated pyrimidine residues are common base lesions in DNA exposed to ionizing radiation or UV light. Such residues may also be generated, without introducing other forms of damage, by treatment of single-stranded DNA with osmium tetroxide. A DNA glycosylase activity is present in *E. coli* that catalyzes the release of thymine glycol (5',6'-dihydroxydihydrothymine) and a compound tentatively identified as 5,6-dihydrothymine from OsO_4 treated DNA (50). It is not known if the enzyme also would liberate analogous cytosine residues. This glycosylase activity appears to be associated with *E. coli* endonuclease III, an AP endonuclease discussed below, as judged from the fractionation properties, reaction optima, and heat sensitivities of the two activities. Thus, a single enzyme may in this case account both for the release of the damaged base and the subsequent cleavage of the DNA molecule at the apyrimidinic site.

PYRIMIDINE DIMER-DNA GLYCOSYLASE   In organisms particularly resistant to UV light, such as *M. luteus* and phage T4–infected *E. coli*, a small "UV endonuclease," which specifically cleaves DNA at pyrimidine dimers, has been detected. The enzyme seems to have no counterpart in uninfected *E. coli*, or in mammalian cells. Recent studies on the mechanism of action of this enzyme have yielded the surprising result that it cleaves initially one of the two glycosyl bonds within a pyrimidine dimer in DNA (50–58). The initial partial unhooking of the pyrimidine dimer at its 5' side was first recognized from the anomalous properties and alkali lability of the resulting oligonucleotide observed by sequencing UV-irradiated, enzyme-treated DNA (51). Several laboratories have verified this mode of action by demonstrating the release of free thymine from the DNA product by photo-reversal of thymine dimers at enzyme-cleaved sites, either by irradiation or by treatment with a photoreactivating enzyme. In bacteriophage T4, the pyrimidine dimer-DNA glycosylase (molecular weight 18,000) is the prod-uct of the *denV* gene. The enzyme also has a low intrinsic AP endonuclease activity, and is consequently able to cleave the DNA at the 3' side of the apyrimidinic site generated at a pyrimidine dimer by the glycosylase activ-ity. It is now evident that the two activities reside within the same protein. In addition to cofractionation of the activities until a single homogeneous protein was obtained, an amber mutant in the *denV* gene was isolated, and both enzyme activities were partly recovered on infection of an amber suppressor host strain (54). Further, infection of *E. coli xth* mutants, which are deficient in the major host AP endonuclease, with T4 *denV*+ phage yielded significantly higher AP endonuclease activity in cell extracts than that found in extracts from cells infected with T4 *denV* mutants (57). The DNA glycosylase function of the enzyme appears much more active than the AP endonuclease. In vitro, the two activities do not appear to act in a concerted fashion, so that many uncleaved apyrimidinic sites are generated in UV-irradiated DNA treated with a small amount of the *denV*+ gene product (53, 54). The AP endonuclease activity of the enzyme may there-fore only be an accessory activity whose role can be easily replaced by one of the AP endonucleases of the host cell. This notion is supported by the properties of the enzyme synthesized by a UV-resistant *denV* mutant. In this case, the DNA glycosylase activity was retained, but the AP endonuclease activity had been inactivated (57). The T4 *denV*$^+$ gene has recently been cloned in the plasmid pBR322, and the expression of such cloned DNA in *E. coli uvrA* and *uvrB* mutants renders them less sensitive to UV light (58). The T4 enzyme has also been introduced into human cells from xeroderma pigmentosum patients by concomitant treatment of cells with the purified enzyme and UV-inactivated Sendai virus. In this remarkable experiment the enzyme was shown to be able to correct the incision defect of the cells and permit repair to take place at pyrimidine dimers (59).

AP Endonucleases

Apurinic and apyrimidinic (AP) sites in DNA may be generated by several routes, and two of the most important ones involve spontaneous hydrolysis and the action of DNA glycosylases (3). Such defects are repaired very efficiently, but in view of their frequent occurrence in DNA they may nevertheless contribute to occasional inactivation or mutagenesis of cells (60). It seems unlikely, however, that the generation of AP sites in DNA plays a major role in spontaneous mutagenesis, since *E. coli xth* mutants do not show increased spontaneous mutation frequency, although they are deficient in the major endonuclease for AP sites and presumably have an impaired ability to remove these sites from DNA (61, 62).

The correction of AP sites in DNA is initiated by endonucleases that cleave DNA specifically at these sites. Several enzymes of this type have been detected in *E. coli,* and recent extensive studies by Linn and co-workers (63–66) show that two classes of enzymes are present, those that cleave DNA at the 3' side of the AP site (class I enzymes) and those that cleave at the 5' side (class II enzymes). In either case, a 5'-phosphate and a 3'-OH group are generated at the cleavage site. Class II enzymes yield a terminal phosphate bound to a deoxyribose residue that does not carry a base and account for most of the AP endonuclease activity present in cell extracts from either *E. coli* or mammalian cells.

A method devised by Clements et al (67) has been employed in several laboratories (64, 68, 69) to define the site of cleavage by different AP endonucleases. Alternating poly(dA-dT) was synthesized in the presence of small amounts of [α-^{32}P]dUTP, which led to the replacement of a few dTMP residues with radioactive dUMP residues. Treatment with uracil-DNA glycosylase was then used to convert the latter to apyrimidinic sites. Cleavage of the resulting polymer with class II AP endonucleases yielded the radioactive phosphate residues in a phosphatase-sensitive form, while this was not the case for class I enzymes. In a different approach to define the terminal structures at the site of cleavage, it was observed that the 3'-OH termini generated by class II enzymes acted as good primers for *E. coli* DNA polymerase I, while this was not the case for the termini generated by class I enzymes (63). Thus the 3'→5' exonuclease function of DNA polymerase I removes a 3'-terminal deoxyribose residue without an attached base only very slowly. It is presently unclear if the two different kinds of AP endonucleases may act in concert, that is, if cleavage by a class I enzyme is followed by cleavage with a class II enzyme, or vice versa, to release the deoxyribose-5'-phosphate residue from an AP site (69a). This seems feasible since cells contain both kinds of activities. Alternatively, a sugar-phosphate residue may be excised by one of the cellular exonucleases after initial cleavage of the damaged DNA by an AP endonuclease. There may be no stringent borderline between class I and II AP endonucleases in all cases, because a homogeneous AP endonuclease from human placenta (68) has recently been observed to be able to incise DNA at either side of an AP site, although one of the modes of cleavage is preferred to the other one. A note of caution is expressed concerning the classification as an AP endonuclease of any protein that increases the rate of chain cleavage at apurinic sites, since basic proteins such as cytochrome *c* and pancreatic ribonuclease show some activity in this regard, as do polyamines. Further, it has been shown recently that at high concentrations the tripeptides Lys-Trp-Lys and Lys-Tyr-Lys catalyze chain cleavage at AP sites relatively effectively (69b, 69c). The general properties of AP endonucleases from different sources have been reviewed previously (3), and recent progress is summarized below.

E. COLI EXONUCLEASE III   This is the major AP endonuclease of *E. coli* (62). It has also been called endonuclease VI (70). Although the enzyme was first discovered and characterized as an exonuclease, its AP endonuclease function may be of greater physiological relevance (61). The AP endonuclease activity of exonuclease III is of the class II type, and in addition the enzyme has associated RNase H, 3'→5' exonuclease, and phosphatase activities. This useful reagent enzyme has recently been cloned, and overproducing *E. coli* strains have been obtained (71). *E. coli* mutants deficient in exonuclease III, *xth,* show two known phenotypic differences from wild-type cells; they are slightly sensitive to methyl methanesulfonate (71a) and unusually sensitive to hydrogen peroxide (B. Demple, personal communication).

E. COLI ENDONUCLEASE III   The apparent broad specificity of this enzyme for a variety of different types of damaged DNAs (72, 73) first appeared puzzling, but has now been largely resolved by the demonstration that it has an associated DNA glycosylase activity that releases ring-saturated thymine (50). Endonuclease III, a class I AP endonuclease, is a small enzyme (mol wt $\sim$ 23,000) that does not require Mg^{2+} for activity (63). The AP endonuclease activities associated with certain DNA glycosylases have been consistently of the class I type, while the major AP endonucleases without any demonstrable glycosylase activity have been of the class II type (66).

E. COLI ENDONUCLEASE IV   This is an enzyme activity first discovered in extracts of *E. coli xth* mutants that catalyzes the cleavage of DNA at AP sites in the same fashion as exonuclease III; that is, it appears to be a class II AP endonuclease (63, 74). *E. coli* endonuclease IV differs from the AP endonuclease function of exonuclease III in that it has no associated exonuclease activity, and it is present in similar amounts in wild-type cells and in *xth* mutants (74). The presence of this enzyme in *E. coli* may explain why *xth* mutants are resistant to most DNA-damaging agents, if it is assumed that endonuclease IV may substitute for the AP endonuclease activity of exonuclease III. *E. coli* mutants deficient in endonuclease IV have not been isolated, so a direct test of this hypothesis is not presently possible.

E. COLI ENDONUCLEASE V   This endonuclease, of unclear biological function, has been characterized in some detail (75, 76). It is much more active on certain forms of damaged DNA than on native DNA, which suggests that it may be a DNA repair enzyme; but in contrast to the enzymes described above it will also attack intact DNA in an endonucleolytic fashion. In vitro, endonuclease V efficiently degrades DNA in which a large proportion of thymines have been replaced by uracils. It also attacks several other kinds of damaged DNA effectively in vitro and catalyzes the formation of both single-strand and double-strand breaks in undamaged DNA at a slow rate (76a). It seems unable to catalyze some of these reactions in vivo to any significant extent, however, since DNA containing large amounts of uracil is not degraded in *E. coli ung* mutants (26). Further, endonuclease V does not cleave circular PM2 DNA molecules that contain small amounts of uracil residues any faster than circles lacking such residues, in marked contrast to the action of uracil-DNA glycosylase. Endonuclease V cleaves DNA preferentially at the 3' side of AP sites; no associated DNA glycosylase activity has been detected. Possibly the enzyme recognizes some distorted form of secondary structure in DNA, perhaps associated with a reduction in base stacking interactions. A better understanding of the physiological role and specificity of this intriguing enzyme will almost certainly have to depend on the isolation of endonuclease V–deficient mutants. For comparison, the different exonucleolytic, endonucleolytic, and DNA-unwinding activities of the $recBC^+$ gene product would appear bewildering in the absence of any genetic information on its function.

Another strange enzyme activity recently discovered in *E. coli* extacts resides in an endonuclease (mol wt 55,000) that cleaves single-stranded DNA at apyrimidinic sites (E. Friedberg, personal communication). The enzyme does not act on double-stranded DNA containing either apyrimidinic or apurinic sites. Its physiological role has not been defined.

MAMMALIAN AP ENDONUCLEASES   The dominant AP endonuclease in several different types of mammalian cells appears to be a class II enzyme (65, 68, 69), and there is no convincing evidence for the presence of more than one such enzyme (3, 68). This mammalian enzyme resembles *E. coli* endonuclease IV in its specificity. Thus, no mammalian AP endonuclease with an associated exonuclease activity has been found. The human AP endonuclease has been purified to homogeneity from HeLa cells (65) and from placenta (68). It is a monomeric protein (mol wt 32,000–41,000, depending on the method of determination) and requires Mg^{2+} for activity. In addition to this class II enzyme, an AP endonuclease of the class I type is present in human fibroblast extracts. This activity has so far only been partly purified and characterized; attempts to inhibit it with antibodies against the class II enzyme have yielded equivocal results (64). Confirmation is needed that this class I type AP endonuclease activity is absent in fibroblasts from xeroderma pigmentosum group D (64), which is an unexpected observation, since the gene product missing in this complementation group of xeroderma pigmentosum is required for incision at pyrimidine dimers (59, 77) and presumably is analogous to one of the *E. coli uvr*$^+$ gene products. Two AP endonucleases from mouse cells have been described (77a); one acts only at AP sites while the other has a broader substrate specificity and may be similar to the *E. coli* endonuclease III or V.

DO PURINE INSERTASES EXIST?   There are several reports of an enzyme that can directly reinsert purines at apurinic sites in DNA, both in *E. coli* (78) and in human fibroblasts (79, 80). Such a proposed activity would be reminiscent of certain tRNA modifying enzymes such as the one that removes an unsubstituted guanine residue within the anticodon region by cleavage of a glycosyl bond and inserts a hypermodified guanine at the same site (81). However, the report that a certain *E. coli* DNA insertase activity requires purine deoxynucleoside triphosphates as donors has not been reproducible. M. Sekiguchi and H. Kataoka (submitted for publication) have established that while an *E. coli* activity reminiscent of an "insertase" could be observed using partly depurinated DNA as a substrate, the residue added at an apurinic site was a single purine mononucleotide and not a free purine. Further, the apparent insertase activity was absent in *E. coli polA* mutants, although homogeneous DNA polymerase I lacked ability to insert free purines into depurinated DNA. In conclusion, the results indicated that short-patch excision repair with replacement of a single nucleotide residue could account for the observations initially reported as a DNA insertase activity in *E. coli.*

The activity from human cells has been studied in greater detail and therefore requires more serious consideration. It appears to be due to a DNA-binding protein (molecular weight 120,000) that does not require Mg^{2+} for activity and employs free purines rather than deoxynucleoside triphosphates as donors. There are two puzzling aspects of this activity that need to be resolved before the notion of a DNA insertase can be accepted. First, there is no apparent energy source for the insertion reaction. We can postulate that the increase in base stacking interactions allowed in the repaired DNA can provide this energy, or alternatively that the enzyme itself occurs in an activated form such as an enzyme-adenylate complex. Second, the activity seems to prefer guanine over adenine. Because of its virtual insolubility in neutral aqueous solution, free guanine can bind very tightly to DNA in a noncovalent fashion, and this may have occurred during measurements of possible insertase activity.

UVR⁺ Endonuclease

DNA lesions such as pyrimidine dimers and polycyclic hydrocarbon adducts, which cause major helix distortions, are removed by an excision-repair process initiated by a complex endonuclease activity. In *E. coli,* this multisubunit enzyme is the product of the *uvrA⁺*, *uvrB⁺*, and *uvrC⁺* genes. A similar endonuclease is presumably present in eukaryotes, and the gene products missing in human cells of various complementation groups of xeroderma pigmentosum and in certain UV-sensitive yeast mutants are believed to be analogous to the *E. coli uvr⁺* gene products, although there is as yet no direct evidence for this. The *E. coli uvr⁺* endonuclease function can be obtained in active form from gently lysed cells, and cell-free extracts from different *uvr* mutants complement each other. This endonuclease requires Mg^{2+} and ATP for activity (82). It does not seem to have any associated DNA glycosylase activity, and in the case of pyrimidine dimers, it acts by cleavage of a phosphodiester bond adjacent to a dimer (50). However, it is not known whether the enzyme incises DNA at the 5' or 3' side of the dimer, or if a 5'- or 3'-terminal phosphate is generated. The substrate for the enzyme in UV-irradiated DNA can be removed by treatment of the DNA with photoreactivating enzyme prior to exposure to the *uvr⁺* endonuclease, which indicates that the enzyme is recognizing pyrimidine dimers and not some other type of UV-induced damage (83). It is known from in vivo data that the uvr^+ endonuclease attacks DNA at a variety of bulky lesions, and it has recently been shown that the partly purified enzyme incises DNA at photochemically bound psoralen residues and at N-acetoxy-2-acetamidofluorene adducts in vitro (83, 84). Thus, there is satisfactory agreement between the in vivo and in vitro properties of this complex enzyme. While the endonuclease attacked DNA at psoralen monoadducts as well as at psoralen-induced cross-links, no psoralen-base adducts were released in free form from DNA that contained radioactive psoralen residues, which again proves indirectly the absence of an intrinsic DNA glycosylase activity in the enzyme (84).

The different subunits of the uvr^+ endonuclease do not remain attached to each other in cell extracts, but they can be separately purified as three proteins of high molecular weight and identified by a complementation assay (82). Because of the low amounts and instability of the active proteins in cell extracts and the relatively complicated assay procedure, progress with purification has been slow. Recently, the different E. coli uvr^+ genes have been separately cloned on small multicopy plasmids, and techniques have been devised for their selective expression, which promises more rapid progress in this area. The development of a procedure to identify the proteins encoded by recombinant plasmids, the maxi-cell method (85, 86), has been of considerable importance in this regard. In this technique, E. coli recA uvrA double mutants carrying a multicopy plasmid such as pBR322 are irradiated with UV light. While many of the plasmids remain intact, this leads to degradation of chromosomal DNA and cell death. Following addition of cycloserine to prevent the outgrowth of any bacterial survivors, the cells are incubated and then labeled with [^{35}S] methionine. Only plasmid-coded proteins are radioactively labeled under these conditions, and such proteins can then be easily isolated in a radiochemically pure and biologically active form. A plasmid carrying the $uvrA^+$ gene coded for a protein of mol wt 114,000 as estimated by SDS gel electrophoresis. This was identified as the $uvrA^+$ gene product by its absence in cells carrying plasmids in which the cloned uvrA gene had been inactivated by the integration of an insertion sequence (86). The molecular weight of the biologically active $uvrA^+$ gene product has been found to be 100,000–130,000 (84, 86), so this protein is a monomer. By similar techniques, the $uvrB^+$ (87) and $uvrC^+$ (88) genes have been cloned and the proteins identified. The molecular weight of the $uvrB^+$ protein is 84,000, while that of the $uvrC^+$ protein is 68,000.

The different subunits of the uvr^+ endonuclease show no detectable endonucleolytic activity by themselves. The $uvrA^+$ protein, however, is a DNA-binding protein that also shows ATPase activity (84, 88a). It may be that the $uvrA^+$ protein is the subunit that initially recognizes the lesions in

DNA and binds at damaged sites, and that the $uvrB^+$ and $uvrC^+$ proteins then interact with the $uvrA^+$ subunit and catalyze the chain cleavage. The helical region destabilized by the $uvrA^+$ protein binding may determine the patch size of the region subsequently excised. The role of the cofactor, ATP, in the process is obscure. The $uvrB^+$ protein may have an additional repair function in the cell, since it seems to be present in larger amounts than either the $uvrA^+$ or $uvrC^+$ protein (86–88). Moreover, $polA$ $uvrB$ mutants are nonviable, whereas $polA$ $uvrA$ mutants have been constructed (89). The products of the uvr^+ genes have previously been regarded as being constitutively expressed, but the synthesis of $uvrA^+$ protein can in fact be induced severalfold by exposure of E. coli to UV light (90). This induction process is dependent on functional $recA^+$ and $lexA^+$ genes (91).

Photolyase

The photoreactivating enzyme, which is the product of the *phr* gene in *E. coli* (92, 93), catalyzes the direct monomerization of pyrimidine dimers in DNA without any associated excision. The *E. coli* enzyme (molecular weight 37,000) seems to contain small amounts of carbohydrate and RNA (94). Visible light (340–400 nm) is required for the monomerization process. The nature of the light-absorbing cofactor has not been identified, in spite of many attempts.

Photoreactivation may also occur in the absence of enzymes, and can be promoted by tryptophan-containing peptides (95). Thus, as discussed above for AP endonucleases, it is important to distinguish between distinct enzymes, which catalyze the process efficiently, and proteins that promote the reaction in a relatively unspecific manner (96). Photolyase is found in bacteria and lower eukaryotes. Despite extensive debate, it is not yet clear whether enzymatic photoreactivation occurs in mammalian cells.

Transmethylase for O^6-Methylguanine

The major mutagenic DNA lesion in cells exposed to simple alkylating agents such as MNNG or N-methyl-N-nitrosourea is O^6-methylguanine (38, 97, 98). On replication, frequent incorporation of thymine instead of cytosine residues occurs, leading to the accumulation of transition mutations. The persistence and repair of this form of DNA damage has been intensively studied because carcinogenesis induced by methylating and ethylating agents apparently is correlated with defective or insufficient repair of O^6-alkylguanine in the target cells and organs (99, 100). With regard to bacteria, E. coli cells usually have limited capacity for removing O^6-methylguanine from DNA, but an inducible repair function is expressed after the exposure of cells to low concentrations of alkylating agents. This repair pathway, termed the adaptive response, allows rapid and error-free repair of O^6-methylguanine residues (38, 101, 102). The adaptive response can be induced in *recA* mutants and is consequently different in nature from inducible error-prone repair processes. The signal for adaptation is not known, but a comparison of the relative efficiencies of various alkylating agents as inducers indicates that it is due to an O-alkylated rather than an N-alkylated residue, and possibly resides within O^6-methylguanine itself (P. Karran, personal communication). An important and unexpected finding has been that the repair activity is expended in its reaction with alkylated DNA and thus acts only once (103). *E. coli* mutants defective in the adaptive response have been isolated, and have been essential for the detailed analysis of the pathway. These mutants are either unable to express an adaptive response, *ada,* or express the response in a constitutive fashion, Adc (104, 105). Both types of mutations have been mapped at 47 min on the chromosomal map by P1 transduction (B. Sedgwick, *J. Bacteriol.* In press).

The development of an in vitro assay for the adaptive response (106) has allowed a biochemical analysis of the process. Cell extracts from adapted *E. coli* cause the specific disappearance of O^6-methylguanine from alkylated DNA in a reaction independent of divalent metal ions. A striking feature of this reaction is that the methyl groups are not released as low-molecular-weight material, but remain bound to an acid-precipitable macromolecule. The receptor has been identified as the methyltransferase[1] itself, which removes an alkyl group from the O^6 position of guanine onto one of its own cysteine residues. Thus, the reaction products are an S-methylcysteine in the protein and unsubstituted guanine in the DNA (107, 108). It would consequently appear that no excision of base or nucleotide residues from DNA is associated with the repair of O^6-methylguanine. The transferase is specific for methyl residues only at the O^6-position of guanine in DNA, and no mobilization of methyl groups from 7-methylguanine or 3-methyladenine occurs. However, the same protein repairs O^6-ethylguanine in ethylated DNA with the concomitant formation of an S-ethylcysteine residue, although this analogous reaction proceeds more slowly than with a methylated substrate (109). The reaction between the methyltransferase and O^6-methylguanine in DNA exhibits features that are in good agreement with the properties of the adaptive response in vivo. Thus, the transferase is consumed in the reaction (111). This suicide inactivation is not associated

---

[1]While an enzyme, strictly defined, should not be consumed in the reaction with its substrate, the present reaction has many features of the interaction between a suicide enzyme inactivator and its target enzyme (110). Further, it has not been ruled out that the transferase activity might not be slowly regenerated under some conditions. For these reasons, and to distinguish this reaction from the direct methylation of proteins by treatment with alkylating agents, we designate the protein induced during the adaptive response a methyltransferase.

with any detectable alteration in the size of the protein, since both the inactive methylated protein (in 6 M guanidium hydrochloride) and the active unmethylated species show a molecular weight of about 17,000. It would thus appear that the inactivation is due to blocking of a reactive cysteine residue in the protein, without any accompanying dissociation or degradation (107). While S-methylcysteine has been found previously in proteins directly exposed to alkylating agents (112, 113), it has not been detected in enzymatically methylated proteins; in the latter case, methyl groups are usually bound to either lysine, arginine, histidine, or glutamic acid residues (114). Certain chemical mutagens, however, in a detoxifying reaction catalyzed by a glutathione S-transferase, have been found to be bound to the cysteine residue of glutathione. Glutathione S-transferase can convert the *cis* isomer of dimethyl 1-carbomethoxy-1-propen-2-yl phosphate (Phosdrin) to its O-demethylated derivative, with the simultaneous formation of S-methylglutathione (115). It is not known at present if the active site of the methyltransferase has any structural similarity to glutathione. The removal of the methyl group from O^6-methylguanine in DNA is unusual in that it represents an example of enzymatic methyl group transfer to a protein in a situation where S-adenosylmethionine was not employed as the methyl donor. In this context, it should also be mentioned that it is not without precedent to find a protein-modifying enzyme that uses itself as the main target of modification. For example, a number of protein kinases that catalyze autophosphorylation are known, and the major acceptor for poly(ADP-ribose) in mammalian cell nuclei is the poly(ADP-ribose) synthetase itself (116, 117).

Uninduced *E. coli* B or *E. coli* K-12 cells growing in conventional media contain low but detectable amounts of the methyltransferase that acts on O^6-methylguanine in DNA, corresponding to about 20 protein molecules per cell (1). In contrast, adapted cells, or Adc mutants, contain 3000–10,000 molecules per cell. The Adc mutants show a high frequency of spontaneous reversion, which indicates that persistent, high levels of the protein are unfavorable to the cell (105). An adaptive response to alkylating agents, associated with the induction of a methyltransferase for O^6-methylguanine residues in DNA, has also been demonstrated in *M. luteus* (S. Riazuddin, personal communication). Mammalian cells have recently been found to contain a very similar methyltransferase, and a partly purified activity from mouse liver that employs a cysteine residue as its own methyl group receptor has been shown to be consumed in the reaction with alkylated DNA (118). O^6-Ethylguanine in ethylated DNA is apparently also repaired in this fashion in mammalian cells (119, 120). No obvious or striking induction of the mammalian activity has been achieved by treatment of tissue culture cells with alkylating agents, however, and it is presently unclear if this methyltransferase can be induced in mammalian cells (121–123; P. Karran, unpublished information). The most intriguing property of the mammalian O^6-methylguanine methyltransferase is that it no longer seems active in some human tumors and tumor cell lines, in particular in lines from cells transformed with DNA tumor viruses, whereas normal diploid fibroblasts and other control cells express this repair function (124–126). The tumor cells lacking the repair activity, termed Mer⁻ or mex⁻ cells, are markedly sensitive to simple alkylating agents, and it is an exciting possibility that a further clarification of the mechanisms involved may lead to a better rationale for the treatment of certain tumors with alkylating agents.

Mismatch Repair

A number of mismatch repair systems occur universally in cells, and such functions are important in view of the vastly increased mutation frequency observed in certain bacterial strains defective in this form of DNA correction. Nevertheless, little is presently known about the biochemistry of mismatch repair, and this process may be regarded as a major remaining mystery of DNA metabolism. While it is easy to see that bacterial strains lacking uracil-DNA glycosylase, or having a defective DNA polymerase (127), might exhibit a moderately increased frequency of spontaneous mutation, the very high mutation rate (up to 10^4 times that of wild-type cells) found in *E. coli* strains such as *mutH, mutL, mutS, mutU* (*uvrDE*), *mutD* (*dnaQ*), and *mutT* (128) is not understood. The multitude of gene products involved implies that a correction system of considerable complexity exists.

Mismatch repair activities should be able to discriminate between a newly synthesized DNA strand, which might contain replication errors, and the parental template strand. For this reason, the suggestion (129) that methylation of the parental strand might serve to instruct the mismatch correction system has kindled considerable interest. Several types of evidence have been obtained for some kind of involvement of DNA methylation in mismatch repair. First, Marinus's isolation of *E. coli* mutants deficient in DNA methylation and his demonstration that strains defective in the adenine DNA methylase, *dam,* had a moderately increased spontaneous mutation frequency (130) implied that a connection exists between mismatch repair and methylation. The major weakness in this argument is that it is not known whether the *dam⁺* gene product is the methylase itself, or if it is a control function that regulates the expression of several different gene products, including the methylase. The DNA adenine methylase employs S-adenosylmethionine as methyl donor and has been found to methylate adenine at the N^6 position within the DNA tetranucleotide sequence, GATC (131). Apparent revertants of *E. coli dam* mutants, isolated as 2-aminopurine resistant clones, are second-site mutations in the *mutH, mutL,* and *mutS* genes, which indicates a close connection between the *dam*⁺ product and several mutator genes (132, 133).

In a different experimental approach to investigate the role of methylation in mismatch repair, transfection experiments with phage λ heteroduplex DNA molecules containing one methylated and one unmethylated strand have been performed. The data showed that for the correction of some mismatches, the unmethylated strand was preferentially repaired in wild-type cells, while in various mutator strains no strand bias was observed (132, 134). However, similar transfection experiments with phage T7 argue against a general role of methylation in mismatch correction (135). In the latter system, repair of phage heteroduplex DNA containing mismatched bases, though markedly reduced, was not totally eliminated in *E. coli mutH, mutL, mutS,* and *mutU* mutants. The observed repair did not seem to involve a T7 phage-coded function but rather depended upon the host cell machinery. In wild-type host cells, the H strand of the viral DNA was preferentially corrected, although T7 DNA contains very few methyl groups in either strand, and the GATC sequence is not methylated at all. Thus, the *dam*⁺-controlled methylase does not seem to act on T7 DNA, and the mismatch repair system presumably recognized some structural feature of the viral H strand other than methylation. It is still possible that these differences are more apparent than real and may be accounted for by the different modes of replication of λ and T7 DNA. Mismatch repair activities have been observed in several other microorganisms. Thus, pneumococcus mutants, *hex,* have been isolated that are defective in mismatch repair of heteroduplex DNA (136). It seems likely the *hex*⁺ gene product is analogous to the product of one of the *E. coli* strong mutator genes.

Two *E. coli* mutator genes with properties different from those described above have been found. The *mutD* strains show a strong mutator phenotype only when the growth medium has been supplemented with thymidine, and transversions as well as transitions and frameshifts have been observed (137). Only the latter two types of mutations are seen in *mutH, mutL, mutS,* and *mutU* strains. It has been proposed that the *mutD*⁺ gene product (25,000 mol wt) may be a subunit of DNA polymerase III, since this enzyme purified from a *mutD* strain had altered chromatographic properties, anomalously low exonuclease activity, and reduced ability to discriminate between dATP and its 2-aminopurine analogue (138, 138a). The other unusual mutator gene, *mutT,* differs from other strains in that mutants exhibit a greatly increased frequency only of the specific transversion $A \cdot T \rightarrow C \cdot G$ (139). Transversion mutations usually result from purine·purine, rather than from pyrimidine·pyrimidine mismatches in DNA, with one of the purine residues being accommodated within the double helix in its *syn* conformation (140, 141). It would consequently appear that *mutT* mutants may be defective in a repair enzyme that specifically recognizes A·G mismatches and removes the G residue in DNA. Alternatively, the *mutT*$^+$ gene may code for a factor that serves to prevent the introduction of a purine deoxynucleoside triphosphate opposite to a DNA purine residue during replication (128).

Conclusions

The present wide-ranging studies on DNA repair activities will most probably reveal many previously unknown enzymes over the next few years, some of which may act on DNA in novel and unexpected ways. Even with regard to a relatively well-characterized group of chemical mutagens such as the simple methylating agents, the mechanisms of repair of several alkylated pyrimidines have not yet been biochemically characterized, although it is known that lesions of this kind are actively removed from DNA in vivo (142). Moreover, since both bacteria and growing mammalian cells contain long stretches of single-stranded DNA (143, 144), and since precursor deoxynucleoside triphosphates have been found to react readily with alkylating agents (145), significant alkylation probably occurs at sites normally protected by hydrogen bonding in double-stranded DNA. Modification at such sites would be expected to result in the formation of miscoding lesions. For example, 1-alkyladenine is readily produced by the action of alkylating agents on adenine residues in monomeric form or in single-stranded DNA (146). It seems likely that repair functions exist to deal with damage of this kind, although none has yet been reported. Many other types of potentially mutagenic DNA lesions, such as purine N oxides, have not been investigated so far with regard to possible repair. Further, many DNA repair enzymes no doubt exist that act on the major base lesions introduced by exposure of cells to ionizing radiation. A start has been made to investigate such enzymes (49, 50), but nothing is known about the removal of important X-ray lesions (48) such as 5-hydroxy-5-methylhydantoin and formylpyruvylurea. A clarification of the latter, presently hypothetical, repair mechanisms may well be a prerequisite for the definition of the specific repair activities apparently lacking in anomalously radiation-sensitive cells such as those obtained from ataxia-telangiectasia patients (147, 147a). (With regard to ionizing radiation, the induction of poly(ADP-ribose) synthesis in mammalian cell nuclei in response to chain breaks in DNA (148) suggests that this polymer may play a role in DNA repair, but the mechanism is not understood.)

In the past, it has often been technically difficult to observe the removal of a specific altered base from DNA in the presence of several other forms of damage. Liquid chromatography techniques (149) have in many cases now removed this experimental hurdle. Another practical problem in the elucidation of repair activities has been that DNA repair enzymes, with the exception of the ones acting at important spontaneous lesions such as apurinic sites and deaminated cytosine residues, seem to be present in low amounts in cells. Twenty enzyme molecules per cell represents a typical value in *E. coli*. This is probably a reflection of the fact that, under most conditions, cells are exposed to relatively low levels of DNA-damaging agents. Consequently, they are usually well equipped to deal with a small number of a certain kind of DNA lesion, whereas sudden exposure to a high dose of a mutagen might cause many mutations due to transient saturation of the repair capacity. In some cases, cells have the ability to adapt to new environmental conditions by the induction of repair enzymes. In *E. coli,* the *uvrA*$^+$ gene product (90), the methyltransferase acting on O^6-methylguanine (102), the 3-methyladenine-DNA glycosylase II (1), a repair activity apparently recognizing an unknown lesion introduced by long-wave UV light (150), and an elusive factor that seems to allow error-prone replication over potentially lethal lesions such as pyrimidine dimers (151) are all inducible. It seems likely that the expression of several other repair functions may also be dependent on exposure to the damaging agent, and if the conditions of induction can be experimentally established it will clearly be easier to define biochemically the activities involved. In addition, cloning of the genes for various DNA repair enzymes into multicopy plasmids is presently being carried out in many laboratories, and this should permit easier access to these enzymes in purified form.

Literature Cited

1. Lindahl, T., Rydberg, B., Hjelmgren, T., Olsson, M., Jacobsson, A. 1981. In *Molecular and Cellular Mechanisms of Mutagenesis,* ed. J. F. Lemontt, W. M. Generoso. New York: Plenum. In press
2. Shapiro, R., Braverman, B., Louis, J. B., Servis, R. E. 1973. *J. Biol. Chem.* 248:4060–64
3. Lindahl, T. 1979. *Prog. Nucleic Acid Res. Mol. Biol.* 22:135–92
4. Grossman, L. 1981. *Arch. Biochem. Biophys.* 211:511–22
5. Hanawalt, P. C., Cooper, P. K., Ganesan, A. K., Smith, C. A. 1979. *Ann. Rev. Biochem.* 48:783–836
6. Lehmann, A., Karran, P. 1981. *Int. J. Cytol.* 72:101–46
7. Hall, J. D., Mount, D. W. 1981. *Prog. Nucleic Acid Res. Mol. Biol.* 25:53–126
8. Friedberg, E. C., Ehmann, U. K., Williams, J. I. 1979. *Adv. Radiat. Biol.* 8:85–174
9. Roberts, J. J. 1978. *Adv. Radiat. Biol.* 7:211–436
10. Caradonna, S. J., Cheng, Y. C. 1980. *J. Biol. Chem.* 255:2293–300
11. Krokan, H., Wittwer, C. U. 1981. *Nucleic Acids Res.* 9:2599–2613
12. Lindahl, T., Ljungquist, S., Siegert, W., Nyberg, B., Sperens, B. 1977. *J. Biol. Chem.* 252:3285–94
13. Sekiguchi, M., Hayakawa, H., Makino, F., Tanaka, K., Okada, Y. 1976. *Biochem. Biophys. Res. Commun.* 73:293–99
14. Anderson, C. T. M., Friedberg, E. C. 1980. *Nucleic Acids Res.* 8:875–88
14a. Gupta, P. K., Sirover, M. A. 1981. *Cancer Res.* 41:3133–36
15. Thomas, L., Yang, C. H., Goldthwait, D. A. 1982. *Biochemistry.* In press
16. Warner, H. R., Rockstroh, P. A. 1980. *J. Bacteriol.* 141:680–86
17. Ingraham, H. A., Tseng, B. Y.,

Goulian, M. 1980. *Cancer Res.* 40:998–1001

18. Caradonna, S. J., Cheng, Y. C. 1980. *Mol. Pharmacol.* 18:513–20

19. Goulian, M., Bleile, B., Tseng, B. Y. 1980. *Proc. Natl. Acad. Sci. USA* 77:1956–60

20. Cone, R., Bonura, T., Friedberg, E. C. 1980. *J. Biol. Chem.* 255:10354–358

21. Karran, P., Cone, R., Friedberg, E. C. 1981. *Biochemistry* 20:6092–96

22. Warner, H. R., Johnson, L. K., Snustad, D. P. 1980. *J. Virol.* 33:535–38

23. Duncan, B. K., Rockstroh, P. A., Warner, H. R. 1978. *J. Bacteriol.* 134:1039–45

24. Coulondre, C., Miller, J. H., Farabaugh, P. J., Gilbert, W. 1978. *Nature* 274:775–80

25. Duncan, B. K., Miller, J. H. 1980. *Nature* 287:560–61

26. Warner, H. R., Duncan, B. K., Garrett, C., Neuhard, J. 1981. *J. Bacteriol.* 145:687–95

27. Karran, P., Lindahl, T. 1978. *J. Biol. Chem.* 253:5877–79

28. Karran, P., Lindahl, T. 1980. *Biochemistry* 19:6005–10

29. Lindahl, T. 1976. *Nature* 259:64–66

30. Riazuddin, S., Lindahl, T. 1978. *Biochemistry* 17:2110–18

31. Laval, J. 1977. *Nature* 269:828–32

32. Shackleton, J., Warren, W., Roberts, J. J. 1979. *Eur. J. Biochem.* 97:425–33

33. Karran, P., Lindahl, T., Ofsteng, I., Evensen, G. B., Seeberg, E. 1980. *J. Mol. Biol.* 140:101–27

33a. Dodson, L. A., Masker, W. E. 1981. *J. Bacteriol.* 147:720–27

34. Laval, J., Pierre, J., Laval, F. 1981. *Proc. Natl. Acad. Sci. USA* 78:852–55

35. Samson, L., Cairns, J. 1977. *Nature* 267:281–83

36. Jeggo, P., Defais, M., Samson, L., Schendel, P. 1977. *Mol. Gen. Genet.* 157:1–9

37. Jeggo, P., Defais, M., Samson, L., Schendel, P. 1978. *Mol. Gen. Genet.* 162:299–305

38. Schendel, P. F., Robins, P. 1978. *Proc. Natl. Acad. Sci. USA* 75:6017–20

39. Brent, T. P. 1979. *Biochemistry* 18:911–16

40. Ishiwata, K., Oikawa, A. 1979. *Biochim. Biophys. Acta* 563:375–84

41. Cathcart, R., Goldthwait, D. A. 1981. *Biochemistry* 20:273–80

41a. Gallagher, P. E., Brent, T. P. 1981. *Biochem. Biophys. Res. Commun.* 101:956–62

42. Singer, B., Brent, T. P. 1981. *Proc. Natl. Acad. Sci. USA* 78:856–60

43. Margison, G. P., Pegg, A. E. 1981. *Proc. Natl. Acad. Sci. USA* 78:861–65

44. Medcalf, A. S. C., Lawley, P. D. 1981. *Nature* 289:796–98

45. Chetsanga, C. J., Lindahl, T. 1979. *Nucleic Acids Res.* 6:3673–84

46. Hems, G. 1960. *Radiat. Res.* 13:777–87

47. Garrett, E. R., Mehta, P. J. 1972. *J. Am. Chem. Soc.* 94:8542–47

48. Téoule, R., Bert, C., Bonicel, A. 1977. *Radiat. Res.* 72:190–200

49. Breimer, L., Lindahl, T. 1980. *Nucleic Acids Res.* 8:6199–6211

50. Demple, B., Linn, S. 1980. *Nature* 287:203–8

51. Haseltine, W. A., Gordon, L. K., Lindan, C. P., Grafstrom, R. H., Shaper, N. L., Grossman, L. 1980. *Nature* 285:634–41

52. Radany, E. H., Friedberg, E. C. 1980. *Nature* 286:182–85

53. Seawell, P. C., Smith, C. A., Ganesan, A. K. 1980. *J. Virol.* 35:790–97

54. Nakabeppu, Y., Sekiguchi, M. 1981. *Proc. Natl. Acad. Sci. USA* 78:2742–46

55. Gordon, L. K., Haseltine, W. A. 1980. *J. Biol. Chem.* 255:12047–50

56. Warner, H. R., Christensen, L. M., Persson, M. L. 1981. *J. Virol.* 40:204–10

57. McMillan, S., Edenberg, H. J., Radany, E. H., Friedberg, R. C., Friedberg, E. C. 1981. *J. Virol.* 40:211–23

58. Lloyd, R. S., Hanawalt, P. C., 1981. *Proc. Natl. Acad. Sci. USA* 78:2796–2800

59. Tanaka, K., Sekiguchi, M., Okada, Y. 1975. *Proc. Natl. Acad. Sci. USA* 72:4071–75

60. Kunkel, T. A., Shearman, C. W., Loeb, L. A. 1981. *Nature* 291:349–51

61. Weiss, B., Rogers, S. G., Taylor, A. F. 1978. In *DNA Repair Mechanisms,* ed. P. C. Hanawalt, E. C. Friedberg, C. F. Fox, New York: Academic. pp. 191–94

62. Rogers, S. G., Weiss, B. 1980. *Methods Enzymol.* 65:201–11

63. Warner, H. R., Demple, B. F., Deutsch, W. A., Kane, C. M., Linn, S. 1980. *Proc. Natl. Acad. Sci. USA* 77:4602–6

64. Mosbaugh, D. W., Linn, S. 1980. *J. Biol. Chem.* 255:11743–52

65. Kane, C. M., Linn, S. 1981. *J. Biol. Chem.* 256:3405–14

66. Linn, S., Demple, B., Mosbaugh, D. W., Warner, H. R., Deutsch, W. A. 1981. In *Chromosome Damage and Repair,* ed. E. Seeberg, K. Kleppe, New York: Plenum. pp. 97–112

67. Clements, J. E., Rogers, S. G., Weiss, B. 1978. *J. Biol. Chem.* 253:2990–99

68. Shaper, N. L., Grossman, L. 1981. *J. Biol. Chem.* In press

69. Verly, W. G., Colson, P., Zochi, G., Goffin, C., Liuzzi, M., Buchsenschmidt, G., Muller, M. 1981. *Eur. J. Biochem.* 118:195–201

69a. Gordon, L. K., Haseltine, W. A. 1981. *J. Biol. Chem.* 256:6608–16

69b. Pierre, J., Laval, J. 1981. *J. Biol. Chem.* 256:10217–20

69c. Behmoaras, T., Toulmé, J. J., Hélène, C. 1981. *Nature* 292:858–59

70. Verly, W. G. 1981. In *DNA repair. A Laboratory Manual of Research Procedures,* ed. E. C. Friedberg, P. C. Hanawalt, Vol. 1, Part A., New York: Dekker pp. 237–51

71. Rogers, S. G., Weiss, B. 1980. *Gene* 11: 187–95

71a. Ljungquist, S., Lindahl, T., Howard-Flanders, P. 1976. *J. Bacteriol.* 126: 646–53

72. Radman, M. 1976. *J. Biol. Chem.* 251: 1438–45

73. Gates, F. T., Linn, S. 1977. *J. Biol. Chem.* 252:2802–7

74. Ljungquist, S. 1977. *J. Biol. Chem.* 252:2808–14

75. Gates, F. T., Linn, S. 1977. *J. Biol. Chem.* 252:1647–53

76. Demple, B., Gates, F. T., Linn, S. 1980. *Methods Enzymol.* 65:224–31

76a. Demple, B., Linn, S. 1982. *J. Biol. Chem.* In press

77. Zelle, B., Lohman, P. H. M. 1979. *Mutat. Res.* 62:363–68

77a. Nes, I. F. 1981. *FEBS Lett.* 133: 217–20

78. Livneh, Z., Elad, D., Sperling, J. 1979. *Proc. Natl. Acad. Sci. USA* 76:1089–93

79. Deutsch, W. A., Linn, S. 1979. *Proc. Natl. Acad. Sci. USA* 76:141–44

80. Deutsch, W. A., Linn, S. 1979. *J. Biol. Chem.* 254:2099–2103

81. Katze, J. R., Farkas, W. R. 1979. *Proc. Natl. Acad. Sci. USA* 76:3271–75

82. Seeberg, E. 1978. *Proc. Natl. Acad. Sci. USA* 75:2569–73

83. Seeberg, E. 1981. *Mutat. Res.* 82:11–22

84. Seeberg, E. 1981. *Proc. Nucl. Acid Res. Mol. Biol.* 26:217–26

85. Sancar, A., Hack, A. M., Rupp, W. D. 1979. *J. Bacteriol.* 137:692–93

86. Sancar, A., Wharton, R. P., Seltzer, S., Kacinski, B. M., Clarke, N. D., Rupp, W. D. 1981. *J. Mol. Biol.* 148:45–62

87. Sancar, A., Clarke, N. D., Griswold, J., Kennedy, W. J., Rupp, W. D. 1981. *J. Mol. Biol.* 148:63–76

88. Yoakum, G. H., Grossman, L. 1981. *Nature* 292:171–73

88a. Kacinski, B. M., Sancar, A., Rupp, W. D. 1981. *Nucleic Acids Res.* 9:4495–4508

89. Morimyo, M., Shimazu, Y. 1976. *Molec. Gen. Genet.* 147:243–50

90. Kenyon, C. J., Walker, G. C. 1980. *Proc. Natl. Acad. Sci. USA* 77:2819–23

91. Kenyon, C. J., Walker, G. C. 1981. *Nature* 289:808–10

92. Youngs, D. A., Smith, K. C. 1978. *Mutat. Res.* 51:133–37

93. Sancar, A., Rupert, C. S. 1978. *Mutat. Res.* 51:139–43

94. Snapka, R. M., Sutherland, B. M. 1980. *Biochemistry* 19:4201–8

95. Hélène, C., Charlier, M., Toulmé, J., Toulmé, F. 1978. See Ref. 61, pp. 141–46

96. Sutherland, B. M. 1978. See Ref. 61, pp. 113–22

97. Sklar, R., Strauss, B. 1980. *J. Mol. Biol.* 143:343–62

98. Newbold, R. F., Warren, W., Medcalf, A. S. C., Amos, J. 1980. *Nature* 283: 596–99

99. Goth, R., Rajewsky, M. 1974. *Proc. Natl. Acad. Sci. USA* 71:639–43

100. Lewis, J. G., Swenberg, J. A. 1980. *Nature* 288:185–87

101. Cairns, J. 1980. *Nature* 286:176–78

102. Cairns, J., Robins, P., Sedgwick, B., Talmud, P. 1981. *Prog. Nucleic Acid Res. Mol. Biol.* 26:237–44

103. Robins, P., Cairns, J. 1979. *Nature* 280: 74–76

104. Jeggo, P. 1979. *J. Bacteriol.* 139:783–91

105. Sedgwick, B., Robins, P. 1980. *Molec. Gen. Genet.* 180:85–90

106. Karran, P., Lindahl, T., Griffin, B. 1979. *Nature* 280:76–77

107. Olsson, M., Lindahl, T. 1980. *J. Biol. Chem.* 255:10569–71

108. Foote, R. S., Mitra, S., Pal, B. C. 1980. *Biochem. Biophys. Res. Commun.* 97: 654–59

109. Sedgwick, B., Lindahl, T. 1982. *J. Mol. Biol.* In press

110. Abeles, R. H., Maycock, A. L. 1976. *Acc. Chem. Res.* 9:313–19

111. Lindahl, T. 1981. See Ref. 66, pp. 207–18

112. Craddock, V. M. 1965. *Biochem. J.* 94: 323–30

113. Bailey, E., Connors, T. A., Farmer, P. B., Gorf, S. M., Rickard, J. 1981. *Cancer Res.* 41:2514–17

114. Paik, W. K., Kim, S. 1980. *Protein Methylation.* New York: Wiley-Intersci. 282 pp.

115. Morello, A., Vardanis, A., Spencer, E. Y. 1967. *Biochem. Biophys. Res. Commun.* 29:241–45

116. Jump, D. B., Smulson, M. 1980. *Biochemistry* 19:1024–30

117. Ogata, N., Ueda, K., Kawaichi, M., Hayaishi, O. 1981. *J. Biol. Chem.* 256: 4135–37
118. Bogden, J. M., Eastman, A., Bresnick, E. 1981. *Nucleic Acids Res.* 9:3089–3103
119. Renard, A., Verly, W. G. 1980. *FEBS Lett.* 122:271–74
120. Renard, A., Verly, W. G., Mehta, J. R., Ludlum, D. B. 1981. *Fed. Proc.* 40:1763
121. Montesano, R., Bresil, H., Planche-Martel, G., Margison, G. P., Pegg, A. E. 1980. *Cancer Res.* 40:452–58
122. Samson, L., Schwartz, J. L. 1980. *Nature* 287:861–63
123. Margison, G. P. 1981. *Carcinogenesis* 2:431–34
124. Day, R. S., Ziolkowski, C. H. J., Scudiero, D. A., Meyer, S. A., Lubiniecki, A. S., Girardi, A. J., Galloway, S. M., Bynum, G. D. 1980. *Nature* 288:724–27
125. Erickson, L. C., Laurent, G., Sharkey, N. A., Kohn, K. W. 1980. *Nature* 288:727–29
126. Sklar, R., Strauss, B. 1981. *Nature* 289:417–20
127. Engler, M. J., Bessman, M. J. 1978. *Cold Spring Harbor Symp. Quant. Biol.* 43:929–35
128. Cox, E. C. 1976. *Ann. Rev. Genet.* 10:135–56
129. Wagner, R., Meselson, M. 1976. *Proc. Natl. Acad. Sci. USA* 73:4135–39
130. Marinus, M. G., Morris, N. R. 1975. *Mutat. Res.* 28:15–26
131. Lacks, S., Greenberg, B. 1977. *J. Mol. Biol.* 114:153–68
132. Glickman, B. W., Radman, M. 1980. *Proc. Natl. Acad. Sci. USA* 77:1063–67
133. McGraw, B. R., Marinus, M. 1980. *Mol. Gen. Genet.* 178:309–15
134. Meselson, M., Pukkila, P., Rykowski, M., Peterson, J., Radman, M., Wagner, R., Herman, G., Modrich, P. 1980. *J. Supramol. Struct.* Suppl. 4, p. 311

135. Bauer, J., Krämmer, G., Knippers, R. 1981. *Mol. Gen. Genet.* 181:541–47
136. Claverys, J. P., Roger, M., Sicard, A. M. 1980. *Mol. Gen. Genet.* 178:191–201
137. Ehrlich, H. A., Cox, E. C. 1980. *Mol. Gen. Genet.* 178:703–8
138. DiFrancesco, R. A., Hardy, M. R., Bessman, M. J. 1981. *Fed. Proc.* 40: 1763
138a. Horiuchi, T., Maki, H., Maruyama, M., Sekiguchi, M. 1981. *Proc. Natl. Acad. Sci. USA* 78:3770–74
139. Yanofsky, C., Cox, E. C., Horn, V. 1966. *Proc. Natl. Acad. Sci. USA* 55: 274–81
140. Topal, M. D., Fresco, J. R. 1976. *Nature* 263:285–89
141. Fersht, A. R., Knill-Jones, J. W. 1981. *Proc. Natl. Acad. Sci. USA* 78:4251–55
142. Bodell, W. J., Singer, B., Thomas, G. H., Cleaver, J. E. 1979. *Nucleic Acids Res.* 6:2819–29
143. Piwnicka, M., Maciejko, D., Piechowska, M. 1981. *J. Bacteriol.* 147:206–16
144. Bjursell, G., Gussander, E., Lindahl, T. 1979. *Nature* 280:420–23
145. Topal, M. D., Hutchinson, C. A., Baker, M. S., Harris, C. 1980. In *Mechanistic Studies of DNA Replication and Genetic Recombination*, ed. B. Alberts, pp. 725–34. New York: Academic
146. Bodell, W. J., Singer, B. 1979. *Biochemistry* 18:2860–63
147. Smith, P. J., Paterson, M. C. 1980. *Biochem. Biophys. Res. Commun.* 97:897–905
147a. Inoue, T., Yokoiyama, A., Kada, T. 1981. *Biochim. Biophys. Acta* 655:49–53
148. Durkacz, B. W., Omidiji, O., Gray, D. A., Shall, S. 1980. *Nature* 283:593–96
149. Lawley, P. D., Warren, W. 1981. See Ref. 70, pp. 129–42
150. Peters, J., Jagger, J. 1981. *Nature* 289: 194–95
151. Villani, G., Boiteux, S., Radman, M. 1978. *Proc. Natl. Acad. Sci. USA* 75: 3037–41

Ann. Rev. Biochem. 1982. 51:89–121
Copyright © 1982 by Annual Reviews Inc. All rights reserved

CHROMATIN

Tibor Igo-Kemenes, Wolfram Hörz, and Hans G. Zachau[1]

Institute für Physiologische Chemie, Physikalische Biochemie und Zellbiologie der Universität München, D-8000 München 2, Goethestrasse 33, Federal Republic of Germany

CONTENTS

PERSPECTIVES AND SUMMARY

Many facets of the sequence organization of the eukaryotic genome have become known in recent years thanks to rapid advances in recombinant DNA and sequencing techniques. Also many basic features and details of chromatin structure are clear today. However, an understanding of the principles that govern gene expression and its regulation awaits further progress in chromatin research.

In the nucleus, the DNA is compacted with the help of histones such that it is accessible for polymerases and numerous effector molecules. At the first level of organization, uniform in all eukaryotic organisms, the DNA is

[1]Work of the authors was supported by Deutsche Forschungsgemeinschaft, Forschergruppe "Genomorganisation."

0066-4154/82/0701-0089$02.00

wrapped around histone cores, which gives rise to beaded chains. The structure of the beads or nucleosomes is known in principle, although many details are still missing; practically nothing is known on the dynamics of the structure. The process of assembling chromatin in vitro from its constituents has been widely studied. It is presently possible to reconstruct nucleosome cores, but not even the next step, the reconstruction from pure components of the beaded chain with proper spacing of the cores, has been achieved yet. The question of a specific positioning of nucleosomes relative to the DNA sequence, and the possible biological importance of a "phased" arrangement are presently stimulating much research and discussion.

At the second level of chromatin organization, the beaded chain is folded into 250 Å fibers. It is not certain whether the beaded chains in these fibers are organized in a continuous helical fashion or in discontinuous globular structures. At a third level, the 250 Å fibers are arranged into loops or domains in both metaphase chromosomes and interphase nuclei. Such domains might well turn out to be not only structural units but also related to units of function. Elements that distinguish the active from the inactive state of chromatin are in the center of interest. Although many details are known, no consistent picture of the structure of transcribing chromatin has emerged, and we are far from understanding the process of activation. A well-studied example of inactive chromatin is satellite DNA–containing chromatin, which can now be isolated in fairly homogeneous form. This may allow analysis of the mechanism(s) underlying heterochromatin formation.

Chromatin composition and structure depend critically on the mode of isolation from cells or nuclei. It therefore has to be kept in mind that chromatin is an operationally defined term.

Chromatin research has been periodically reviewed in this series (1–3) and in other books (4–10). Reviews of special aspects of chromatin are cited in the respective sections of this article. In order to avoid overlaps with existing reviews we concentrate on the recent literature. The sections on nucleosomes and the 250 Å fibers are a continuation of McGhee & Felsenfeld's article (3), and that on active chromatin takes up the thread where it has been left by Mathis et al (9). The sections on nucleosome phasing, chromatin domains, and satellite DNA–containing chromatin have no direct predecessors, but even there we do not attempt to cover the complete literature. [For reviews of the chemistry of chromatin proteins see (11–14), and for chromatin replication (15–17).] We mention only briefly the extensive literature on cytological and ultrastructural work because of a forthcoming summary article by Laemmli in Volume 52 of this series. Quite naturally, the extent to which the various topics are dealt with in the present article reflects the interests and personal experience of the authors.

It might be fair to say that the period covered in our review was not characterized by a particular breakthrough in chromatin research but several important aspects of chromatin structure and function have been clarified and many have progressed toward their elucidation.

STRUCTURE OF THE NUCLEOSOME

Nucleosome has become the generally accepted term for the chromatin subunit. From nuclease digestion experiments it has become clear that the nucleosome is composed of a well protected nucleosome core and a nuclease sensitive linker region of variable length. The core consists of 146 base pairs (bp) of DNA wrapped around a globular histone bead containing two each of the histones H2A, H2B, H3, and H4. The beaded structures seen in electron micrographs under low salt conditions are core particles containing more or less of the linker DNA.

The basic structure of the nucleosome core is now firmly established. X-ray diffraction analysis of crystals from core particles has progressed to a resolution of 5 Å (19) and is complemented by neutron diffraction work on crystals using contrast variation (20). In addition, a low resolution three-dimensional density map of DNA-free histone octamers has been produced from electron microscopy by using an image reconstruction method (21). Based on these results and those of chemical cross-linking of histones to core DNA carried out in a number of laboratories, notably that of Mirzabekov (7, 22), a proposal for the spatial arrangement of histones in the nucleosome core has been formulated (21, 23). It is clear that answers to many questions concerning the structure of chromatin will depend on further progress in the crystallographic analyses.

With respect to the interaction of other proteins with the nucleosome core, there is general agreement that histone H1 is located at the region where DNA enters and exits the core particle (24–27). The central globular domain of H1 is able to close two full turns of DNA around the histone octamer (28). The corresponding particle containing 166 bp of DNA has been termed chromatosome (25). The globular domain of H1 can be cross-linked with a zero-length cross-linking agent to H2A (29). In other experiments with the same agent, H1 was found to be cross-linked to each of the core histones, H3 being somewhat favored (30). In any case, H1 must be in close contact with the histone octamer. Interaction of H1 with the nucleosome core is also indicated by the finding that H1 depletion has a profound effect on the distribution of cuts within the core DNA generated late in the digestion of chromatin with micrococcal nuclease (31, 32). Pauses early in digestion, which result in discrete DNA fragments between 166 and 200 bp, may be related to the presence of H1 and H5 (33–35). The thermodynamic parameters of the salt-dependent dissociation of H1 and H5 from nondigested chromatin were determined (36).

Polyacrylamide gels make it possible to identify H1-containing mononucleosomes and demonstrate that such particles can be reconstituted from H1-depleted nucleosomes by the addition of this histone (37). H1 blocks the abrupt unfolding of nucleosome cores that is otherwise observed when the NaCl concentration is below 1–2 mM (38).

The occurrence of an H1 subfraction denoted H1° correlates with the absence of DNA synthesis (39). H1° could also be shown to accumulate in growth-inhibited cultured cells (40). There is an analogy in this respect to histone H5; this histone replaces H1 in transcriptionally inactive avian erythrocytes. Sequence analyses have indeed revealed significant homology between H1° and H5 (41, 42). Like H5, H1° replaces H1 in the nucleosomal linker region (43). Its relative abundance is unchanged, however, in a chromatin fraction enriched for putatively active nucleosomes, which might be interpreted as evidence against a suppressive effect of H1° on transcription (43). The heterogeneity of the H1 class is also indicated by the finding of an uneven distribution of H1 subfractions in different chromatin subunits released by micrococcal nuclease (44).

The high mobility group (HMG) proteins are at present the most thoroughly characterized nonhistone proteins in chromatin (14, 45). Micrococcal nuclease–prepared mononucleosomes and subnucleosomal particles with bound HMG 14 and HMG 17 [or H6, the equivalent in fish (46)] can be demonstrated as distinct particles in polyacrylamide gels (47–54) and also by immunochemical methods (55). Two molecules of HMG 14 and/or 17 can be bound to a mononucleosome (47, 51, 52). Major sites of interaction are near the ends of the DNA associated with the histone octamer (51) or at the immediately adjoining linker DNA (49). At the latter location, HMG 14 and/or 17 may take the place of H1 (49), even though mononucleosomes with two HMG 17 molecules and one H1 can be isolated (47). Interaction of HMG 14 with DNA at low ionic strength occurs through the N-terminal half of the molecule (56). HMG 14 and HMG 17 have a preferential affinity for single-stranded DNA (57). The binding of HMG 14/17 to mononucleosomes is reversible; particles containing these proteins can be reconstituted from HMG-depleted nucleosomes (51). The role of the HMG proteins in transcriptionally active chromatin is discussed below.

Among the numerous histone modifications, the conjugation of ubiquitin with H2A to give protein A24 (58) and, as recently discovered, also with H2B (59), is the most extensive one known. Protein A24 is considered a likely candidate for regulatory processes especially in view of its rapid turnover (60). No clear picture has emerged however. Increased transcription in nucleolar chromatin is accompanied by a decrease in A24 content

(61), but it is similarly depleted in the totally inactive chicken erythrocyte nucleus (62) where A24 lyase activity is also lost (63). One nucleosome core can accommodate two molecules of A24 in place of H2A, as determined in reconstitution experiments (64). The resulting particles were found to differ only minimally from standard core particles as far as nuclease sensitivity and HMG 14/17 binding are concerned (64). Protein A24 is not found in metaphase chromosomes, which led to the suggestion that its removal may trigger mitosis (65). Future investigations may benefit from the finding that, by gel electrophoresis, mononucleosomes can be separated that either contain or lack A24 (54, 66). Free ubiquitin is also present in the nucleus and was known as HMG 20 for a while until its identification (67). Ubiquitin is not only an ubiquitously occurring protein but also appears to serve different functions; it has recently been identified as the ATP-dependent proteolysis factor of rabbit reticulocytes (68).

There is a growing appreciation for the existence of tissue specific nonallelic histone variants and their potential importance in the function of chromatin (69). Variants of histones H2A, H2B, and H3, differing in a few amino acids from the standard histones, have been demonstrated in a number of species (70,71). In addition there are histone-like minor components, for example, protein D2 from *Drosophila* (72), that are similar to both H2A and H2B and occur at a frequency of roughly 1 molecule per 5 nucleosomes. Even though it has long been known that there are cases where gross changes in the histone population are observed in the course of development, for example in sea urchin (73), no correlation has yet been determined between the occurrence of minor histone variants and functional aspects of the chromatin structure. The presence in sea urchins of an H2B variant in active tissues as well as in the completely inactive sperm cell (74) seems to disprove such a correlation. Differences might well be restricted, however, to a few strategic sites, which would presently escape detection.

NUCLEOSOME ASSEMBLY

The reconstruction of distinct chromosomal regions from purified constituents would help greatly the study of many structural aspects and also the function of individual components. As a first step in this direction the reconstruction of nucleosome cores is being investigated in many laboratories. The traditional approach has been dialysis of DNA and histones against gradients of salt in a reconstitution process (reviewed in 3, 6, 75, 76). It is possible to combine DNA and histones at physiological ionic strength in what has been called a nucleosome assembly process (for review see 77). However, the terms reconstitution and assembly are sometimes used interchangeably. The assembly process is facilitated by so-called assembly factors (77), but they do not seem to be absolutely required (78). The most thoroughly characterized assembly factor is nucleoplasmin, an acidic, thermostable protein first isolated from *Xenopus* eggs (79), which is present in great abundance in the nucleoplasm of many cell types (80–82). This protein appears to form complexes with the histones, thereby preventing nonspecific aggregate formation between DNA and histones and facilitating an ordered assembly process (82, 83). The acidity of nucleoplasmin may be crucial for its function. This is in keeping with the observation that polyglutamic acid and polyaspartic acid facilitate nucleosome assembly in in vitro experiments (84). Another chromatin assembly activity from *Drosophila* extracts ascribed originally to topoisomeraseI (85) is now ascribed to a high molecular weight RNA present in the original enzyme preparations (86).

The nucleosome spacing generated in the in vitro assembly experiments, in the absence or presence of histone H1, is 140–165 bp (78, 83, 86, 87), i.e. it differs from the spacing of approximately 200 bp in native chromatin. Only by the use of unfractionated cell homogenates from either *Xenopus* eggs (88) or *Drosophila* embryos (89) has it been possible to obtain nucleosomes spaced at about 195 bp. Those systems utilize a stored histone pool that also contains H1.

Reconstitution experiments with DNA and histones H3 and H4 have demonstrated that these histones can, by themselves, form nucleosome-like particles (3). It has been concluded from cross-linking experiments and subsequent electron microscopy that tetrameric and octameric histone complexes with 145 bp DNA have a diameter of 80–90 Å and thus have similar compaction as standard core particles (90). Particles resembling nucleosomes have also been observed in the electron microscope upon incubating closed-circular DNA with calf thymus HMG1 and HMG 2, but these proteins do not confer upon the DNA protection against nucleases as histones do (91). Nucleosome-like complexes have also been formed on single-stranded DNA fragments, 140–160 nucleotides long (92, 93). The reconstituted particles contain an octamer of the four histones, and it was suggested that during replication and transcription of chromatin, histone octamers may become transiently bound to single strands of DNA.

Core particles reconstituted with synthetic polydeoxyribonucleotides of defined sequence may prove especially suitable for physicochemical investigations and crystallization. The results of such reconstitution experiments were reported almost simultaneously by several authors. Poly (dA-dT) can be folded by histones to yield material that can be digested by pancreatic DNase to multiples of about 10.5 bp, and by micrococcal nuclease to 146 bp core particles (94–96). Because of the absence of sequence heterogeneity, the digestion patterns yield fine structural details not usually seen with core particles. The copolymer, poly(dG-dC), could also be reconstituted into nucleosomes (94, 96), but with slightly different properties (94), while the homopolymers, poly(dA)·poly(dT) and poly (dG)·poly(dC), did not form chromatin-like structures (94, 96).

The question of sequence specificity in the binding of histones to DNA in reconstitution experiments continues to attract attention, especially in view of the interest in nucleosome phasing (see below). In reconstitution experiments with SV40 DNA, histone octamers were deposited, with a certain preference at four regions on the DNA (97); see also 98). Chao et al (99) have demonstrated two preferred positions for binding of histones to *lac* operator DNA. Repressor binding and the response to an inductor are hardly diminished in the DNA histone complexes as compared to free DNA. This result indicates that in the reconstituted nucleosome cores the repressor binding surface of the DNA must face outward to be available to the repressor (100, 101). In a different study (102) it was found that after reconstitution with histones the *lac* operon still served as efficient template for transcription and still responded to the catabolite activator protein.

Nonrandom deposition of histones on pBR322 DNA and on *Xenopus* 5S genes has been inferred from DNase I and *E.coli* exonuclease III digestion experiments with the reconstituted material (103). Identical results were obtained with both the traditional dialysis against gradients of salt and the factor-mediated assembly. It is difficult at present to assess the significance of such sequence specificity. We need to know more about the mechanism(s) underlying nonrandom deposition of histones on DNA and whether the positions of the nucleosomes on eukaryotic DNAs after reconstitution correspond to those in native chromatin.

NUCLEOSOME PHASING

How nucleosomes are arranged relative to the DNA sequence has received much attention lately (reviewed in 104, 105). One possibility is that they are arranged randomly. Alternatively they may be located in unique positions that are identical in a population of cells of the same kind; or they may occupy a small number of distinct positions. Such arrangements have been termed phasing. This term is sometimes reserved for situations in which the nucleosomes are regularly positioned with respect to a repeating DNA sequence; the word positioning is then used for unique locations of nucleosomes on nonrepetitive DNA (10). In this review we use only the term phasing, but further progress may make distinctions in nomenclature necessary. Phasing has to be clearly distinguished, however, from spacing.

The term spacing refers to the distance between adjacent nucleosome cores without regard to specific DNA sequences. Data for the average spacing or nucleosome repeat length were reported for many species and cell types (reviewed in 2, 3). Recently, additional evidence for changes in the average nucleosome spacing during early sea urchin embryogenesis has been reported (106). With the advent of blot hybridization techniques, different nucleosome repeat lengths within the same cell could be characterized. The ribosomal 5S RNA genes of *Xenopus* red blood cells were shown to have a repeat length of 175 bp as compared to 189 bp in bulk chromatin (107, 108). The nucleosomal repeat length of the DNA of rat satellite chromatin was found to differ significantly from the 195 bp average of bulk rat liver chromatin. The repeat length of satellite chromatin was determined to be 185 bp (109), which is exactly half of the length of the satellite DNA repeat (110). It is still unclear what determines nucleosome spacing. Histones H1 and H5 (2, 3), and more recently histone H2B (111) have been inferred to influence the spacer length. Results of digestion experiments of Pospelov et al with a nuclease from *Serratia marcescens* (112) are in keeping with the observation (2, 3) that the lengths of spacer DNA are always integral multiples of roughly 10 bp. It is not yet clear, however, whether this is true for all spacers or just for particular fractions of chromatin.

Interest in phasing stems from the idea that preferential or specific nucleosome locations may be important in chromosome mechanics and could be one means of regulating gene expression and/or replication. While the issue was controversial a few years ago, many recent reports agree that, in the systems investigated, there exists a preference of the nucleosomes for certain positions.

Early experiments with viral chromatin pointed to a random location of nucleosomes on the DNA (e.g. 113); but the possibility of a number of defined nucleosome locations was already suggested (114). In the first experiments with cellular chromatin, Prunell & Kornberg (115) studied the reassociation of exonuclease III–treated DNA from nucleosomal core particles of rat liver and concluded that the relation of nucleosomes to nucleotide sequences is random for most single copy sequences. Experiments with DNA fragments from DNase I digestions were similarly interpreted (116). On the basis of experiments with rat 5S RNA, Baer & Kornberg (117) favored a random distribution of nucleosomes for the 5S RNA genes, although the data may not be incompatible with complex phasing models. Digestion experiments with nuclei containing double labeled DNA were interpreted to mean that nucleosomes are arranged randomly with regard to DNA base composition (118).

In only one instance has a precise register been reported between nucleosomes and a repetitive DNA; that is in the α-satellite DNA–containing chromatin of African green monkey cells (119, 120). The conclusion was based on micrococcal nuclease digestion of chromatin. Also in later work a unique phasing was proposed although in a completely different register (120a). Experiments in other laboratories led to the opposite conclusion (121, 122). The finding of preferential cleavage sites for micrococcal nuclease in control experiments with protein free α-satellite DNA (122) offered an alternative explanation for the earlier results (119, 120). By experiments using a number of nucleases, a simple phase relationship in α-satellite DNA–containing chromatin was excluded, but the possibility of more complicated phase relationships in this chromatin was noted (122).

Current views on preferential nucleosome location are based on the results of three different approaches. Chromatin containing satellite DNA, 5S RNA genes, and tRNA genes have been studied by preparing nucleosomal core particles with micrococcal nuclease, or by trimming dinucleosomes with exonuclease and by subsequent mapping of the ends of the DNA relative to known restriction sites. Chromatin containing protein coding genes was mainly investigated with the indirect end labeling technique (see below) or, in a third approach, by probing directly for the accessibility of restriction sites to the respective nuclease. Many conclusions rest on the appearance of certain sharp bands in gel electrophoretic separations. Hence the importance of the control experiments with the respective protein-free DNAs, which in several cases were not run in a convincing manner. Particularly the preference of micrococcal nuclease for certain DNA sequences rather than for AT-rich regions, which has recently been demonstrated (123, 124), may have been underestimated by some authors. It was also pointed out that preferential micrococcal nuclease cleavage sites occur approximately every 200 bp in the nontranscribed portion of a *Drosophila* heat shock locus, while in a prokaryotic DNA such sites are not regularly spaced (125).

Satellite DNA–containing chromatin has the technical advantage of multiplying manyfold any preferred nucleosome site. 370-bp repeat units of rat satellite I DNA containing two nucleosomes were isolated and trimmed. The resulting DNA yielded, on HindIII digestion, three distinct fragments, which suggested that the nucleosomes are located in either of two alternative defined sites (126). The positions of the nucleosomes could be mapped with an uncertainty of only a few base pairs. Part of the satellite DNA–containing chromatin, however, is organized in another, still unknown way. Chromatin comprising the 5S RNA genes of *Xenopus* (oocyte type) (127) and *Drosophila* (128) was investigated by characterizing core particle DNA with the appropriate probes. 200 bp upstream of the *Xenopus* 5S RNA genes, a site highly sensitive to nuclease was found and, on the genes themselves, the histone octamers seem to be arranged in at least four different defined modes or frames [but see (108)]; presumptive promoter regions were found in or near nucleosome linkers. The latter is true also for the *Drosophila* 5S RNA genes, where the histone octamers apparently are positioned in two distinct frames. In chicken chromatin, nucleosomes were reported to be phased on three tRNA genes; this was based on restriction patterns of cloned and uncloned DNA of tetranucleosomes from partial micrococcal nuclease digests (129). A detailed study using micrococcal nuclease indicated that nucleosomes are phased on a tRNA gene cluster in *Xenopus* erythrocyte chromatin (130); in liver and cultured kidney cell chromatin, however, nucleosomes were found either randomly arranged or in many different phases.

Results with protein coding genes are similar in principle. The genes for the heatshock proteins hsp 70 and hsp 83 of *Drosophila* were studied by Wu (131) who described the elegant approach of indirect end labeling. The same approach had been used before by Nedospasov & Georgiev (132) in the analysis of SV40 minichromosomes. In the case of *Drosophila* DNA, the appearance of regularly spaced bands in the micrococcal nuclease pattern indicated an ordered nucleosomal arrangement upstream of the hsp 83 gene (131). On the basis of restriction patterns of core particle DNA, Levy & Noll (133) favored a nucleosome alignment in at least three frames for the repressed state of the hsp 70 and 83 genes. The indirect end labeling technique was successfully applied also to the chicken β-globin genes by Weintraub & Groudine (134) and to the *Drosophila* histone genes by Samal et al (135). Nucleosomes were found to be in preferential locations on and around the embryonic β-globin gene of red blood cells, where the gene was in the nonexpressed state. In brain cells, however, where the β-globin genes are also in the nonexpressed state, the nucleosomal distribution on these genes appears to be random. On the histone gene cluster, nucleosomes were reported to be precisely positioned, at least on the nontranscribed spacer.

With a third approach, nonrandom nucleosome arrangements were found on the constant gene region of the immunoglobulin kappa light chain of the mouse (136), which in liver cells is not rearranged and not active. The conclusion is based on BspRI digestion experiments with nuclei, in which certain cleavage sites were found to be much more accessible than others. The preferential nucleosome location is lost in the rearranged and active form of the gene region, as it is present in a myeloma.

SV40 and polyoma minichromosomes are interesting models of cellular chromatin (137) but because of inherent difficulties of the systems a number of results have remained equivocal. A short but functionally important region of the SV40 genome, including the origin of replication, control regions for late transcription, and a major binding site for the viral T antigen show pronounced accessibility to endogenous nucleases (138, 139), micrococcal nuclease (140), DNase I (138), and restriction nucleases (141, 142). Subsequent electron microscopic studies together with nuclease digestion experiments have shown that the sensitive region is free of histone octamers (143, 144). However, according to electron microscopy, the region at the origin of replication is exposed only in 20–25% of the molecules in a minichromosome preparation (143, 144). This may also explain the observation that in some experiments this region is found only partially accessible to restriction nucleases (145, 146). In virion-derived minichromosomes the region is not found preferentially accessible to DNase I (147) and partially disrupted virions may have been present in some minichromosome preparations. Moreover, the histone H1 and nonhistone protein contents of the minichromosomes depend on the mode of isolation (148, 149). With respect to the nucleosome phasing on SV40 DNA outside of the gap region at the origin the situation is unclear. Preferential location of nucleosomes has been reported (132) but, on the basis of different experiments, other authors concluded that the nucleosomes are randomly distributed along the SV40 DNA (150). Whether this difference and also some of the other discrepancies with respect to nuclease accessibilities (140–142, 145, 146, 151, 152) are due to the differing composition of the minichromosome preparations needs to be clarified.

What is the structural basis of nonrandom nucleosome arrangements? One of the ways to answer the question may be to continue the above described nucleosome reconstitution experiments with defined DNAs (97–99, 103, 153–155) in order to find out whether or to what extent determinants of DNA recognition reside in the histone octamer itself or in histone H1. Such determinants may be accentuated by histone modification and the occurrence of histone variants. Regulatory proteins and other nonhistone proteins may bind with high preference to certain DNA sequences and then direct the placement of nucleosomes. From such "nucleating" nucleosomes the assembly of nucleosomes may proceed by a passive spreading mechanism if one assumes defined, but not necessarily equal linker lengths in the respective region. Such spreading may operate only over a short range, but it may be renewed by other preferentially bound nucleosomes. Higher orders of chromatin structure could modulate the positioning of nucleosomes, and ordered nucleosomes, in turn, could determine some features of this structure. Recently, Trifonov (156) speculated that histone octamers are directed into preferential positions by a bending of the DNA, which may be caused by a 10.5-bp periodicity in the occurrence of possibly "wedge shaped" dinucleotide pairs in eukaryotic DNA (157, 158). Weintraub considered the restrictions that DNA sequence-specific recognition proteins place on the mode of packaging of DNA into chromatin (159). Kornberg (105) quoted unpublished computation work of Stryer & Kornberg and pointed out that, subject to some constraints concerning e.g. specifically bound proteins, random deposition of histone octamers may result in a degree of order considered by ourselves and others to be equivalent to at least one type of phasing.

A functional significance of nucleosome phasing, which is not proven yet, may be as multifaceted as its structural basis. Regular nucleosome arrangements on satellite DNA may, e.g. by modulating chromatin superstructure, affect chromosome recognition and pairing. On protein coding sequences nucleosome phasing could be related to the interconversion between the active and inactive states of a gene. Particularly interesting, of course, are regulatory sequences which, together with other aspects, are discussed below in the context of DNase I hypersensitivity. At present, research on nucleosome phasing is on the level of nuclease digestion and blot hybridization. Improvements in the clarity of the experimental results and new approaches may be necessary for a full answer to the open questions.

HIGHER ORDER STRUCTURE

The 250 Å Fiber

The nucleosome provides only a first level of condensation of the DNA. Inside the cell the nucleosomal chain must itself be further compacted, probably in a hierarchy of superstructures culminating in the compactness of metaphase chromosomes (3, 6, 160). The so-called thick fiber, with a diameter of 250–300 Å represents a second level of chromatin condensation. Although there is general agreement that histone H1 is an absolute requirement for its formation from extended chains of nucleosomes, and that the ionic environment plays a crucial role, there is some debate as to the exact nature of the 250-Å fiber. One line of evidence, based mostly on electron microscopy and neutron diffraction, points toward continuous coiling of the nucleofilament to give what has been termed a solenoid with 6–7 nucleosomes per turn (161–164). The results of hydrodynamic studies are in full agreement with this model (165–167) as are studies by small angle X-ray scattering (168).

An alternative model, also supported mostly by electron microscopy, envisions the 250-Å fiber to be a chain of repeating globular supranucleosomal units, originally named superbeads (169–177). Biochemical studies supporting the superbead model suffer from the fact that nuclease digestion experiments do not yield periodic DNA patterns that would indicate a supranucleosomal repeat unit. It is only at the nucleoprotein level that particles composed of 6–12 nucleosomes can be detected (170, 178). Sucrose gradient centrifugations at appropriate salt concentrations are used in such experiments. The stability of the particles is dependent on histone H1 (179, 180). The observation of multimeric particles in such experiments (170), which would have provided evidence for a repeating superbead structure, could not be substantiated in later studies (181) and might have been due to contaminating ribonucleoprotein particles (182).

It is impossible to decide at present to what extent supranucleosomal units are disassembly products of a continuous structure (181, 183, 184) generated reproducibly in the preparation of the specimens for electron microscopy (185). On the other hand, definitive proof that the truly native structure is all continuous is equally hard to come by.

A puzzling aspect of the higher order structure of chromatin is the so-called linking number paradox: even though DNA is coiled twice around each histone octamer in chromatin, there appears to be close to one topological turn of DNA per nucleosome. Different solutions to this problem have been proposed. Stein (186) has argued from reconstitution experiments with SV40 DNA that H1 and H5 also contribute to the supercoiling and might thus resolve the paradox. Worcel et al (187) have pointed out that a dinucleosome repeat unit with a spacer crossover could produce a linking number of -1. In DNase I digestion experiments of nuclei a periodicity equivalent to a dinucleosomal repeat length has actually been observed (188, 189). The possibility of intersecting spacers has also been considered by Grigoryev & Joffe (190). Lutter (191) and Klug & Lutter (192) have approached the problem from a different angle and have reexamined the number of base pairs per turn of DNA on the nucleosome. They interpret the nuclease cutting data (191, 193) as being consistent with a value of 10.0 bp rather than 10.4 bp, found for DNA in solution. A change from 10.4 to 10.0 in this value, upon association of the DNA with the histone core, would resolve the linking number problem without the necessity of introducing further stipulations (192).

The transition from the 100 Å to the 250 Å chromatin fiber leads to a change in the susceptibility of the DNA to DNase II. While cleavage in the extended conformation occurs in the classical 200-bp mode, a 100-bp periodicity is obtained from the H1-mediated condensed state (194). This is not due, however, to additional cleavage in the center of the nucleosome but instead around positions 20 and 125 of the nucleosomal core DNA; this is of interest considering the internal structure of the core particle (195). DNase II has also been successfully used to study structural transitions of metaphase chromosomes (196). This nuclease complements other methods as a probe for condensed and extended states of chromatin.

A detailed model for the 250-Å fiber has been proposed by McGhee et al (197) based on electric dichroism data. It specifies that the faces of the nucleosome cores must be oriented fairly close to parallel to the fiber axis. Similar measurements with differently prepared chromatin led Lee et al (198) to propose a model in which the faces of the nucleosome cores are more steeply inclined to the fiber axis.

A decondensed chromatin fraction that can be extracted by buffer containing ammonium sulfate has been analyzed in detail by physicochemical methods (199, 200). From a reduction of the molecular weight to about one half upon removal of H1 from this fraction, Lindigkeit et al (201) have proposed an alternative model for the 250-Å fiber consisting of two closely packed nucleosome chains arranged side by side and stabilized by histone H1.

Answers concerning the folding of the nucleosome chain may come with an understanding of the role of histone H1. Cross-linking studies have been one approach to this problem. H1 homopolymers can be obtained from H1-containing chromatin but not from free H1, which suggests a lattice of H1 present in chromatin (3). Itkes et al (202) have found that, upon cross-linking, about 5% of the H1 polymers consist of integral multiples of an elementary structure of 12 H1 molecules, which they called clisone, reminiscent of a superbead unit. Regarding the question of what parts of the neighboring H1 molecules are close enough in intact nuclei to be cross-linked, they found that the amino- and carboxyl-terminal halves of H1 were cross-linked in all combinations (203). Thomas & Khabaza (204) found that in cross-linking experiments hexamers of H1 are the largest polymer. They noticed that the H1 cross-linking pattern depended very little on ionic strength, which may severely limit this assay for probing the higher order structure of chromatin. The differences between the results of the two H1 cross-linking studies (202, 204) may be related to differences in the reaction and extraction procedures used.

The concept that the 250-Å fiber constitutes a dynamic structure is supported by two independent studies that show that at physiological salt concentrations there is rapid and complete equilibration between H1 molecules in segments of chromatin (205, 206). This is not the case for H5, which is known to bind more tightly to chromatin (206).

Chromatin Domains

Both in interphase nuclei and in metaphase chromosomes the 250-Å chromatin fibers appear to be folded into loops (207, 208) or domains (209). Stretches of chromatin comprising 35–85 kilobase pairs (kb) of DNA (207–209; but 210) are believed to be anchored in a supporting structure of the nucleus that has been termed matrix (211–215), lamina (216, 217), envelope (218, 220), nuclear membrane (221), ghost (222), cage (210) or, in the case of metaphase chromosomes, scaffold (223–225). The domain model has been formulated on the basis of three experimental approaches: centrifugation, electron microscopy, and nuclease digestion studies.

Centrifugation studies with lysed mouse (226, 227), hamster (228), HeLa (229, 230), *Drosophila* (207), and yeast (231) cells revealed supercoiled circular DNA structures with a characteristic biphasic response of sedimentation rates to increasing concentrations of intercalating agents. The sedimentation properties were affected by ionic detergents [(207, 229) but see (226, 227)], proteases (207, 226–231), or RNases (207), which suggests that proteins and/or RNAs are involved in the formation of domains.

Electron microscopic investigations on histone-depleted interphase nuclei from mouse cells suggest that a structural skeleton of nonhistone proteins organizes the DNA into loops of approximately 54 kb (232). Transmission electron microscopy of interphase nuclei revealed DNase sensitive fibers cross-linked to nonsensitive structures (222). Three-dimensional reconstruction from electron micrographs of serial sections (218) and freeze-fracture electron microscopy (233) suggest that most chromatin is located in large domains in contact with the nuclear envelope. In spread preparations of histone-depleted metaphase chromosomes from HeLa cells, Paulson & Lämmli (208) observed loops extending from a central protein scaffold. These and other electron microscopic studies have led to the radial loop model of chromosome structure (234) in which the loops are tightly packed around the central scaffold on a helical path. In the unit fiber model the condensation of the 250 Å fibers to chromosome structures, which requires compaction by a factor of about 200, is accomplished by introduction of another level of spiralization of the fibers (235–237) and folding of the thus created supersolenoids. In *Bombyx mori* meiotic prophase chromosomes, the loops appear to be separated by short regions with 5–10 nucleosomes comprising about 0.08% of the DNA of the nucleus (238). The loops of lampbrush chromosomes, which may be related to the domains, have been thoroughly reviewed (239–241) and are not discussed here.

A third line of evidence pointing to the domain structure of interphase chromatin comes from restriction nuclease and micrococcal nuclease digestion studies on rat liver nuclei (209, 242). The quantitative analysis of soluble and insoluble chromatin fractions upon nuclease digestion and extraction of nuclei is best interpreted in terms of the domain model, according to which the insoluble fraction remains attached to the nuclear matrix, while the soluble fraction is excised from the loops by the action of the DNases. Micrococcal nuclease induces an increase in the effective spherical volume of chicken erythrocyte nuclei; the observed nuclear expansion was correlated with the introduction of double-strand chain scission and subsequent relaxation of the domains (243).

The distribution of DNA sequences and the location of the various proteins within chromatin domains are beginning to be explored. There are first experiments pertaining to whether there are specific DNA sequences at the base of the loops, and what involvement they may have in the formation of domains. Repetitive DNA has been found enriched in the insoluble fraction after nuclease digestion (244, 245–247), which would point in this direction. Cook & Brazell (210) found that DNA sequences coding for human α- and β-globin genes are located nonrandomly with respect to the domain topology. A similar result was obtained for viral sequences in SV40-transformed mouse cells (248). The nature of the proteins involved in the formation of domains (e.g. 249–252) is far from clear. It may be that the domain-stabilizing proteins are different in interphase and metaphase, since the lamina, which is believed to be associated with the interphase chromatin, is depolymerized during metaphase and reformed at the end of the cell cycle (253).

Numerous experimental results point to the domains or loops as an essential element of higher order chromatin structure. The interest in the domains stems from the notion that they may be related to units of replication (reviews 254, 255) and/or, as the loops in lampbrush chromosomes, to units of transcription (e.g. 232, 238, 256). If domains are functional units, their specific structure, possible rearrangements, and the time course of their activation may play a role also in developmental regulation.

ACTIVE CHROMATIN

Electron Microscopy of Transcribed Chromatin

Many intriguing questions on the ultrastructure of active chromatin, discussed in previous reviews (3, 9, 241, 257–259), are still open. There is general agreement that beaded (nucleosomal) structures are absent from highly active gene regions. But the notion that, in spite of this, histones are present has gained further support (260), at least in the case of nonribosomal genes. Progress with respect to the dimensions of the chromatin strands and the degree of DNA compaction was made in Daneholt's lab (261) and by Olins et al (262) by studying serial sections of a *Chironomus* Balbiani ring. It appears that, on the active genes, the reversible transition from beaded structures to uniform fibers is caused by an unfolding of the polynucleosome filament rather than by histone depletion.

The absence of nucleosomes in Miller spreads of active chromatin regions was partly explained by an enhanced susceptibility of active nucleosomes, as compared to inactive ones, to commonly used detergents (263) and/or to forces generated during the spreading process (262). This interpretation is based on experiments with chromatin from mouse embryos and from *Chironomus* salivary glands, respectively. Other authors, however, feel that the absence of nucleosomes in their experiments with *Physarum* chromatin is not due to the use of a detergent (264). They found that the nonbeaded appearance extends beyond the transcribed region itself. The ultrastructural results were related to changes in micrococcal nuclease and DNase I digestion patterns [(265, 266); but see (267)].

The transcribed genes for ribosomal RNA in amphibian oocytes are not compacted at all (268) or only by a factor of 1.4 (269). The nontranscribed spacers have a beaded structure that, according to one study, reflects a low degree of compaction (268); another study (269) considers it to be a supranucleosomal structure with a compaction factor of at least 20. A detailed study on *Xenopus laevis* nucleolar chromatin supports the view that active ribosomal genes are not compacted into nucleosomes (269a). In *Tetrahymena pyriformis* the rRNA genes themselves lack compact nucleosomes (270–272), while the distal spacers appear to be compacted into nucleosomes (268).

Nuclease Sensitivity of Active Chromatin

A number of interesting features have been discovered, both with respect to the nuclease sensitivity of active gene regions and with respect to nuclease hypersensitive sites in the neighborhood of the genes (see next section). The advances were made possible by progress in the blot hybridization techniques, which are replacing solution hybridization methods in the analysis of nuclease sensitivity of chromatin. While in most blotting experiments nitrocellulose paper (273) was used (e.g. 131, 274), there are a few reports (275, 276) in which diazobenzyloxymethyl-cellulose paper (277) was applied. The latter method has the advantage that the paper with the covalently attached DNA fragments can be used repeatedly and with different probes.

DNase II, the first nuclease to be used in the study of active chromatin (278; reviewed in 9, 279) is now largely replaced by other nucleases: DNase I, which recognizes features of the internal organization of nucleosomes, appears to degrade active chromatin faster because of its altered nucleoprotein structure; micrococcal nuclease and restriction nucleases differentiate between nucleosome cores and linker regions, and, in the case of active chromatin, may recognize differences related to the higher orders of structure.

DNase I was introduced by Weintraub & Groudine in a study of transcribed and nontranscribed chicken globin genes (280). It has been known for a while that DNase I does not simply distinguish active from inactive genes but that, for instance, in some cells genes that have been transcribed retain their increased DNase I sensitivity (9). An inactive gene that in a later state of development will be transcribed in a cell type is already DNase I sensitive. Changes in the chromosomal structure of embryonic and adult β-globin genes during erythropoiesis were investigated by monitoring the disappearance of specific restriction bands upon DNase I digestion of chicken embryo nuclei (281). In embryonic red cells the genes for both the embryonic and the not yet expressed adult β-globin genes were shown to be very nuclease sensitive. In older embryonic cells, the embryonic β-globin gene, which at that stage is inactive, retained an intermediate sensitivity.

DNase I sensitivity is not restricted to the coding regions of the genes but extends far upstream and downstream (256, 275, 281–283). In the work of Bellard et al (275) the nontranscribed regions in the vicinity of the ovalbumin gene in hen oviduct cells exhibit enhanced DNase I and micrococcal nuclease sensitivity. A total of 25 kb upstream and downstream from the ovalbumin gene has been probed and found to be uniformly sensitive. In the work of Lawson et al (282) the probed region included also the ovalbumin-related X and Y genes and long stretches between these genes. A total of about 30 kb has been found to be DNase I sensitive, but the authors estimate that probably the whole 54 kb of the ovalbumin domain that has been characterized is in a DNase I–sensitive chromatin conformation. At least 7.8 and 5.8 kb of DNA flanking the 5' and 3' ends of the ovomucoid gene, respectively, exhibit a DNase I sensitivity that seems to be indistinguishable from that observed for the transcribed regions (282). Chromatin structures of intermediate DNase I sensitivity have been defined by determining the rates of nuclease degradation (256, 275, 281, 284, 285). The sensitivity of β-globin and ovalbumin genes to micrococcal nuclease and DNase I was measured in chicken erythrocyte and oviduct nuclei (275). The actively transcribed ovalbumin gene was found to be more sensitive in oviduct nuclei than the globin gene in erythrocyte nuclei (275) where it had been expressed and still contained bound RNA polymerase molecules (286). Both genes were digested more rapidly than the inactive reference genes, i.e. the ovalbumin gene in the erythrocyte and the globin gene in the oviduct. The problems of intermediate DNase I sensitivity and of the sensitivity of adjacent noncoding regions were also investigated by Stalder et al (256). In chicken red cell nuclei the coding regions of α- and β-globin genes were found to be organized in a highly sensitive structure, while adjacent noncoding regions were organized in a moderately sensitive structure. All these regions were resistant, under the digestion conditions used, in cells that never express globin genes. The preferential DNase I sensitivity extends to at least 8 kb on the 3' side of the β-globin gene and to 6 or 7 kb on the 5' side.

Zasloff & Camerini-Otero (287) have extended the use of DNase I as a tool for chromatin structure by studying early single strand nicks in active and inactive gene regions. They showed that in chicken erythrocyte nuclei the β-globin gene is about 25 times more sensitive to nicking than is the nontranscribed albumin gene or an average DNA sequence. The sites of initial DNase I nicking were found to be clustered within the transcribed sequence of the β-globin gene. The chromatin structure of *Drosophila melanogaster* hsp 70 genes in the repressed and actively transcribing states has been investigated by Levy & Noll (276). The coding regions in the repressed state were found to be more resistant to micrococcal nuclease digestion than the bulk of the chromatin. In the active state the coding regions became more susceptible, were degraded 30 times more rapidly than in the repressed state, and yielded products without any apparent nucleosomal periodicity. The structural alterations were confined to the coding region itself. In both states the 3' flanking region apparently maintained a structure similar to bulk chromatin. Changes in the higher order structure, and modification or removal of nucleosomes from transcribed DNA were discussed as possible reasons for the structural alterations. In contrast, transcribed chromatin from yeast (288) and the expressed herpes thymidine kinase gene (289) introduced into mouse cells were found to be in compact nucleosomes.

The protamine genes of trout testes (290), the ribosomal RNA genes of *Tetrahymena* (291), a retrovirus gene in chicken cells (292), and the genes for the kappa light chains of mouse immunoglobulins (283) were also found to be DNase I–sensitive when in the active state. The constant region of the latter genes were also studied with the restriction nuclease, Bsp RI (136). Higher sensitivity was found in the active than in the inactive state. This sensitivity may make it possible to isolate active chromatin fractions with proteins still attached to long fragments of DNA.

Preferential DNase I sensitivity of active chromatin in nuclei has been exploited to label this fraction using the standard nick-translation reaction with DNA polymerase (293). The labeled chromatin fraction was preferentially released from nuclei on limited micrococcal nuclease treatment, but the enrichment of the fraction for active genes has yet to be determined.

DNase I sensitivity as a property of active gene regions has been correlated with undermethylation of DNA [reviews on DNA methylation (294–296)]. The state of methylation of the sequence CCGG can be easily monitored by comparing its susceptibility to restriction nucleases (297); Hpa II does not cleave it when the middle C is methylated, while Msp I also acts on the methylated sequence (298). Undermethylation by this criterion seems to be related to DNase I sensitivity in the chicken globin (299, 300), ovalbumin (301), and a retroviral system (292). In addition, undermethylation is correlated with gene activity in many systems (e.g. 298–307). It is unclear whether undermethylation is a reason or a consequence of an altered chromatin structure. Moreover it is not known whether methylation or undermethylation affects many sites or only those few that can be monitored with the restriction nucleases.

DNase I–Hypersensitive Sites in Chromatin

A new feature of chromatin structure was discovered by Wu et al (274, 308), when they mildly digested with DNase I nuclei from *Drosophila melanogaster* tissue culture cells; in these cells the heatshock genes are inactive at 25° and active at 35°. Electrophoretic separation of the digested DNA from cells grown at the low temperature, and blot hybridization with probes from the hsp 70 and 83 heatshock gene regions (and other probes) revealed discrete fragments of a few kb to more than 20 kb. These fragments arose from cleavage at sites that are hypersensitive to DNase I and located in specific positions of the genome. The sites are more sensitive by an order of magnitude than active or potentially active gene regions (131) and about two orders of magnitude more accessible than inactive chromatin regions. Some of the sites were mapped by the indirect end-labeling technique and localized in the vicinity of the heatshock genes (131). Preferred cuts were found at two sites adjacent to the 5' ends of the coding regions of three hsp 70 genes at the cytogenetic locus 87c. Preferred cuts were also localized at the 5' ends of two hsp 70 genes at locus 87A. Three sites were mapped near or at the 5' terminus of the presumed precursor to the hsp 83 mRNA and two sites downstream of the hsp 83 gene. The DNase I hypersensitivity at the 5' side of the hsp 70 genes is present before induction and is maintained after heatshock.

Near the genes for the four small heatshock proteins hsp 22, hsp 23, hsp 26, and hsp 28, DNase I–hypersensitive sites were found in duplicate (309), major ones at or very close to the 5' ends of the genes and secondary ones approximately 300 bp upstream. The size of the hypersensitive regions was estimated to be 50–100 bp.

The α- and β-globin gene clusters of chicken were studied also with respect to DNase I hypersensitivity (256, 299, 300). In precursor cells, not yet capable of globin synthesis, no DNase I–hypersensitive sites were found (299), but such sites became readily detectable in nuclei of red blood cells of 5-day and 14-day embryos, which produce embryonic and adult hemoglobin, respectively. Near the 5' end of the embryonic β-globin gene a site was found that at the later stage was replaced by two sites approximately 2 and 6 kb upstream from the adult β-globin gene (256). In the α-globin domain the hypersensitive site at the 5' side of the U gene, the first one to be expressed during the very early stages of development, is present when the gene is expressed, but disappears after it has been turned off (300).

Other hypersensitive sites upstream and downstream of the α_A and α_D genes seem to persist from the fifth day, when α_A is expressed, to the fourteenth day of development, when α_A is turned off and α_D is expressed (300).

A tissue specificity of DNase I hypersensitivity was found in the rat preproinsulin II gene system (310). A 250–350 bp region at the 5' terminus of the gene is hypersensitive in a rat insulinoma, while in nonexpressing tissues such as kidney, spleen, and brain no such sites were found.

Hypersensitive sites were also detected in the long terminal repeats of Rous sarcoma virus coded sequences when productively integrated into chicken DNA, but not when in inactive locations (292). In the histone gene repeat of *Drosophila melanogaster* cells, the 5' termini of all five genes are highly susceptible to DNase I (135). Extrachromosomal chromatin of *Tetrahymena pyriformis* containing ribosomal DNA genes has DNase I–hypersensitive regions within the nontranscribed central spacer region and near the 3' end of the genes (271). In the macronucleus of the ciliate *Stylonichia mytilus* the inverted repeat sequences at the ends of the gene-size DNA fragments exhibit a preferential sensitivity to DNase I digestion (311).

The structural basis and the functional significance of DNase I hypersensitivity are far from clear. The 50–350 bp long hypersensitive sites may be nucleosome free, as was found for the 250–400 bp region at the origin of replication of the SV40 minichromosome (143, 144). This notion is supported by recent experiments in which a 115-bp DNA fragment is excised by MspI from the hypersensitive region upstream of the adult β-globin gene from chicken erythrocyte nuclei (311a). It is, of course, tempting to relate the hypersensitive sites to the regulatory sequences that are being defined in the vicinity or within eukaryotic genes (reviewed in 312, 313). At the borders of nucleosome-free regions there may be precisely positioned nucleosomes that directly or through a number of (phased) nucleosomes could influence the chromatin structure at the region of transcription initiation. In a more general way one may say that the hypersensitive sites are probably related to a modulation of chromatin structure by binding of sequence and tissue-specific proteins. But one does not know, for instance, whether the generation of hypersensitive sites is a primary or secondary effect in gene activation. In at least one case, the chicken ovalbumin gene region, no hypersensitive site was found (301), which is explained by the fact that here an area of several hundred bp is highly nuclease sensitive (M. Bellard and P. Chambon, personal communication) Also, some but not all sites persist after expression has been turned off (256, 299), which indicates the rather complex nature of DNase I hypersensitivity.

Nonhistone Proteins Associated with Expressed Chromatin Regions

The vast literature on nonhistone proteins and histone modification has been well reviewed (9, 45, 314–316; see also 317–320). A recent book on the high mobility group (HMG) proteins (14) contains also chapters on their role in active chromatin. In the following we concentrate on four aspects: proteins preferentially released on nuclease digestion of chromatin; proteins of isolated nucleosomal particles from active chromatin regions; reconstitution experiments with HMG proteins; and protein localization on chromosomes by immunological methods.

The many studies include: partial digestions of nuclei or chromatin with DNase I (290, 321, 322), DNase II (323–326), micrococcal nuclease (48, 49, 54, 327–331), or endogenous nucleases (332), and subsequent chromatin fractionation in the presence of monovalent (53, 326, 333, 334) or divalent cations (323, 326, 335). Preferentially released chromatin fractions were in several cases found to be enriched in DNA sequences of genes being transcribed in the tissue examined (48, 50, 290, 321, 333); generally, the preferentially released fractions contained increased amounts of HMG proteins (49, 290, 321, 325, 326, 328, 330, 333), other nonhistone proteins (324, 331, 336–338), phosphorylated HMG proteins (339), highly acetylated histones (334, 340–342), ADP-ribosylated histones and nonhistone proteins (343), RNA polymerase (322, 327, 344), and nascent RNA (327).

In a chromatin fraction obtained by homogenization of mouse cell nuclei and subsequent counter current distribution, increased amounts of HMG 14 and HMG 17, the H2A histone variant M1, RNA polymerase B, and nascent RNA have been detected; histone H1 was found in reduced amounts and HMG 1 and HMG 2 seemed not to be present in this fraction (345, 346). In related studies, mononucleosomes and subnucleosomal particles that were preferentially released by micrococcal nuclease digestion were fractionated by differential salt extractions (49, 53, 326 reviewed in 347) and/or polyacrylamide gel electrophoresis (48, 53, 54, 348). With the latter technique, particles containing HMG proteins separate well from other particles and were shown to be enriched in expressed DNA sequences (50, 51). The HMG proteins 1 and 2 as well as the related trout HMG T are believed to be located in linker regions (328, 330, 335, 349; see also 350). The association of HMG 14 and 17 with the nucleosome core is discussed above.

The results described in the two foregoing paragraphs suggest that HMG proteins 14 and 17 and some other nonhistone proteins are preferentially located on actively transcribed chromatin. But it should be kept in mind that this conclusion is largely based on indirect evidence. It may also be mentioned that Mathew et al (348) found in nucleosome fractions from chicken erythrocyte nuclei no direct correlation between HMG protein content and globin gene enrichment. Moreover it was shown that the composition of nucleosomal subfractions depends on an intact nuclear structure and is altered upon disruption of nuclei prior to nuclease treatments (351).

The notion that HMG proteins are associated with active chromatin regions (also potential or former ones) is supported by reconstitution experiments. Weisbrod et al (352, 353) showed that upon removal of HMG from chicken erythrocyte chromatin, the region containing globin gene lost its enhanced DNase I sensitivity, and regained it on reconstitution with HMG proteins 14 and 17. This finding was confirmed by others (51, 354). In such experiments DNase I sensitivity is conferred, for instance, upon the globin genes and not the ovalbumin gene of red blood cells (352). But genes that are transcribed at different rates have about the same affinity for HMG 14 and HMG 17 (51, 353). The association of HMG 14 and HMG 17 with HMG-depleted active nucleosomes was found to be specific in the presence of a ten to twenty fold excess of inactive nucleosomes (353). It was therefore concluded that active nucleosomes have at least one more unique but still unknown feature that distinguishes them from inactive nucleosomes and directs the HMG binding in reconstitution experiments.

Taking advantage of the specificity of HMG 14 and HMG 17 interaction with actively transcribed chromatin, HMG proteins were covalently bound to agarose and used to purify active nucleosomes (355). This new method extends the limited range of chromatin fractionation procedures.

Immunological methods of analyzing the distribution of chromosomal proteins and RNA polymerase have been thoroughly summarized (356–359). The identification of chromocenter (centromer) associated proteins or ribonucleoproteins (360–362), the localization of ribonucleoproteins on *Drosophila* chromosomes (363, 364), and the localization of HMG 1 (365) and ecdysterone (366) on *Chironomus* chromosomes have demonstrated the potential of these methods. The use of monoclonal antibodies further improved the studies of distribution of nonhistone proteins in active chromosome regions (367, 368) and in band and interband structures (367).

SATELLITE DNA–CONTAINING CHROMATIN

The terms constitutive heterochromatin and satellite DNA–containing chromatin have been used interchangeably since there is in most organisms a good correlation between heterochromatin and highly repetitive DNA (reviewed in 369–373). It must be kept in mind, however, that heterochromatin is operationally defined by cytological criteria as that part of the chromatin that does not decondense during interphase, whereas satellite DNA–containing chromatin is defined by its DNA constituent. A property of satellite DNA–containing chromatin that strengthens the correlation with heterochromatin is the higher degree of condensation, which permits its isolation by differential centrifugation after sonication of cell nuclei (374, 375).

The molecular basis for this higher degree of condensation of satellite DNA–containing chromatin is unknown. As far as the organization at the nucleosomal level is concerned, satellite DNA–containing chromatin from mouse and rat was shown to be very much like the rest of the chromatin (184, 376–379). The possibility that a specific phase relationship between DNA repeat and nucleosome spacing could contribute to the condensed state is discussed above. One might expect that the high compaction of satellite DNA–containing chromatin would reduce its susceptibility to nucleases. This was found not to be the case, however, for heterochromatin from *D. ordii* (380) and *M. musculus* (381, 382). It is still possible, though, to enrich satellite DNA–containing chromatin after digestion with e.g. micrococcal nuclease, because the compact state, which is the basis for the separation, does not depend on the continuity of the DNA but on noncovalent interactions that are highly sensitive to the ionic environment (382).

A different approach to the preparation of satellite DNA–containing chromatin not exploiting its high degree of condensation became possible through the use of restriction nucleases. Most satellite DNAs are either completely resistant to certain restriction nucleases or give regular patterns consisting of small fragments, quite different from the continuous fragment distribution found with nonrepetitive eukaryotic DNA. After the digestion of nuclei with restriction nucleases (383), satellite DNA–containing chromatin can be separated accordingly either as high or as a low molecular weight fraction from the rest of the chromatin, as demonstrated for calf (384), rat (242), African green monkey (385), and mouse chromatin (386–388).

The availability of purified satellite DNA–containing chromatin fractions has allowed a search for proteins unique to this chromatin. It was reported that in satellite DNA–containing chromatin from African green monkey cells, histone H1 is replaced by a set of nonhistone proteins (385); a loss of H1 from the fraction by rearrangement or degradation was not excluded, however. A chromatin fraction containing rat satellite DNA I had a typical histone composition (109) and reduced amounts of HMG 14 and 17 proteins (389). Nonhistone proteins, possibly specific for the satellite DNA, were present in low amounts (389).

Blumenfeld demonstrated that in *Drosophila* the amount of heterochromatin in different tissues correlated with a phosphorylated variant of histone H1 (390). In *Drosophila* egg extracts a protein has been discovered that binds tightly and with high specificity to the *Drosophila* 1.688 satellite DNA (391). The presence of a satellite DNA–binding protein was also implied by in situ DNA binding experiments with *Chironomus thummi thummi* chromosomes (392).

It appeared to be a general rule that satellite DNAs are not transcribed in vivo; at least, RNA transcripts had not been detected in the cell. There is now, however, evidence from in situ hybridization experiments for the transcription of highly repeated DNA sequences during the lampbrush stage in *Triturus cristatus carnifex* (393) and of both strands of satellite I DNA in the newt *Notophtalmus* (394). In both cases it appears that transcription begins at a gene region, fails to terminate at the end of the gene, and continues without interruption into the adjacent satellite DNA region. Transcripts were also detected of rat satellite I DNA in hepatoma tissue culture cells (395), but the biological significance of this finding is unclear.

In *D. melanogaster,* where there is a variety of different satellite DNAs, it was shown by in situ hybridization that each chromosome has its own distinctive pattern of heterochromatic sequences (369). A similar situation has been demonstrated in chromosomes from rye (396). It is difficult to say if such patterns are the results of specific constraints or if they are incidental. By genetic means it has been demonstrated that in *Drosophila* the topology of gene regions relative to heterochromatic areas can play a role for their expression (reviewed in 397), but the question as to the function of repetitive DNA and heterochromatin remains an enigmatic one. In the light of the continuing debate on "selfish DNA" (398, 399) a better understanding may come with a knowledge not just of the structure of repetitive DNA itself but of its organization within the cell nucleus.

Literature Cited

1. Elgin, S. C. R., Weintraub, H. 1975. *Ann. Rev. Biochem.* 44:725–74
2. Kornberg, R. D. 1977. *Ann. Rev. Biochem.* 46:931–54
3. McGhee, J. D., Felsenfeld, G. 1980. *Ann. Rev. Biochem.* 49:1115–56
4. Busch, H., ed. 1978–1981. *The Cell Nucleus* Vols. 4–9. New York: Academic
5. Garrett, R. A. 1979. In *International Review of Biochemistry. Chemistry of Macromolecules II B,* ed. R. E. Offord, 25:179–203. Baltimore: Univ. Park
6. Lilley, D. M. J., Pardon, J. F. 1979. *Ann. Rev. Genet.* 13:197–233
7. Mirzabekov, A. D. 1980. *Quart. Rev. Biophys.* 13:255–95
8. Lewin, B. 1980. *Gene Expression,* Vol. 2. New York: Wiley. 2nd ed. 1159 pp.
9. Mathis, D., Oudet, P., Chambon, P. 1980. *Prog. Nucleic Acid Res. Mol. Biol.* 24:2–49
10. VanHolde, K. 1982. *Chromatin.* Heidelberg:Springer In Press
11. Isenberg, I. 1979. *Ann. Rev. Biochem.* 48:159–91
12. von Holt, C., Strickland, W. N., Brandt, W. F., Strickland, M. S. 1979. *FEBS Lett.* 100:201–18
13. Sperling, R., Wachtel, E. J. 1981. *Adv. Protein Chem.* 34:1–60
14. Johns, E. W. ed. 1982. *The HMG Chromosomal Proteins.* NY:Academic
15. DePamphilis, M. L., Wassarman, P. M. 1980. *Ann. Rev. Biochem.* 49:627–66
16. Seale, R. L. 1981. *Prog. Nucleic Acid Res. Mol. Biol.* 26:123–34

17. Kelly, T. J., Challberg, M. D. 1982. *Ann. Rev. Biochem.* 51:901–34
18. Deleted in proof
19. Finch, J. T., Brown, R. S., Rhodes, D., Richmond, T., Rushton, B., Lutter, L. C., Klug, A. 1981. *J. Mol. Biol.* 145:757–69
20. Bentley, G. A., Finch, J. T., Lewit-Bentley, A. 1981. *J. Mol. Biol.* 145:771–84
21. Klug, A., Rhodes, D., Smith, J., Finch, J. T., Thomas, J. O. 1980. *Nature* 287:509–16
22. Mirzabekov, A. D., Belyavsky, A. V., Bavykin, S. G., Shick, V. V. 1980. *BioSystems* 12:265–71
23. Richmond, T. J., Klug, A., Finch, J. T., Lutter, L. C. 1981. In *Proc. SUNYA Conversation in the Discipline Mol. Stereodyn.* 2nd. New York: Adenine
24. Noll, M., Kornberg, R. D. 1977. *J. Mol. Biol.* 109:393–404
25. Simpson, R. T. 1978. *Biochemistry* 17:5524–31
26. Thoma, F., Koller, T., Klug, A. 1979. *J. Cell Biol.* 83:403–27
27. Moyne, G., Freeman, R., Saragosti, S., Yaniv, M. 1981. *J. Mol. Biol.* 149:735–44
28. Allan, J., Hartman, P. G., Crane-Robinson, C., Aviles, F. X. 1980. *Nature* 288:675–79
29. Boulikas, T., Wiseman, J. M., Garrard, W. T. 1980. *Proc. Natl. Acad. Sci. USA* 77:127–31
30. Ring, D., Cole, R. D. 1979. *J. Biol. Chem.* 254:11688–95
31. Smerdon, M. J., Lieberman, M. W. 1981. *J. Biol. Chem.* 256:2480–83
32. Ishimi, Y., Ohba, Y., Yasuda, H., Yamada, M. 1981. *J. Biochem. Tokyo* 89:1881–88
33. Muyldermans, S., Lasters, I., Wyns, L., Hamers, R. 1980. *FEBS Lett.* 119:93–96
34. Muyldermans, S., Lasters, I., Wyns, L., Hamers, R. 1981. *Nucleic Acids Res.* 9:3671–80
35. Allan, J., Cowling, G. J., Harborne, N., Cattini, P., Craigie, R., Gould, H. 1981. *J. Cell Biol.* 90:279–88
36. Kumar, N. M., Walker, I. O. 1980. *Nucleic Acids Res.* 8:3535–51
37. Nelson, P. P., Albright, S. C., Wiseman, J. M., Garrard, W. T. 1979. *J. Biol. Chem.* 254:11751–60
38. Burch, J. B. E., Martinson, H. G. 1980. *Nucleic Acids Res.* 8:4969–87
39. Panyim, S., Chalkley, R. 1969. *Biochem. Biophys. Res. Commun.* 37:1042–49
40. Pehrson, J., Cole, R. D. 1980. *Nature* 285:43–44
41. Smith, B. J., Walker, J. M., Johns, E. W. 1980. *FEBS Lett.* 112:42–44
42. Pehrson, J. R., Cole, R. D. 1981. *Biochemistry* 8:2298–2301
43. Smith, B. J., Johns, E. W. 1980. *Nucleic Acids Res.* 8:6069–79
44. Gorka, C., Lawrence, J. J. 1979. *Nucleic Acids Res.* 7:347–59
45. Goodwin, G. H., Walker, J. M., Johns, E. W. 1978. In *The Cell Nucleus,* ed. H. Busch, 6:181–219, New York: Academic
46. Dixon, G. H. 1982. See Ref. 14, pp. 149–92
47. Albright, S. C., Wiseman, J. M., Lange, R. A., Garrard, W. T. 1980. *J. Biol. Chem.* 255:3673–84
48. Bakayev, V. V., Smatchenko, V. V., Georgiev, G. P. 1979. *Nucleic Acids Res.* 7:1525–40
49. Goodwin, G. H., Mathew, C. G. P., Wright, C. A., Venkov, C. D., Johns, E. W. 1979. *Nucleic Acids Res.* 7:1815–35
50. Albanese, I., Weintraub, H. 1980. *Nucleic Acids Res.* 8:2787–805
51. Sandeen, G., Wood, W. I., Felsenfeld, G. 1980. *Nucleic Acids Res.* 8:3757–78
52. Mardian, J. K. W., Paton, A. E., Bunick, G. J., Olins, D. E. 1980. *Science* 209:1534–36
53. Hutcheon, T., Dixon, G. H., Levy-Wilson, B. 1980. *J. Biol. Chem.* 255:681–85
54. Levinger, L., Barsoum, J., Varshavsky, A. 1981. *J. Mol. Biol.* 146:287–304
55. Tahourdin, C. S. M., Neihart, N. K., Isenberg, I., Bustin, M. 1981. *Biochemistry* 20:910–15
56. Cary, P. D., King, D. S., Crane-Robinson, C., Bradbury, E. M., Rabbani, A., Goodwin, G. H., Johns, E. W. 1980. *Eur. J. Biochem.* 112:577–80
57. Isackson, P. J., Reeck, G. R. 1981. *Nucleic Acids Res.* 9:3779–91
58. Goldknopf, I. L., Busch, H. 1978. See Ref. 45, pp. 149–80
59. West, M. H. P., Bonner, W. M. 1980. *Nucleic Acids Res.* 8:4671–80
60. Seale, R. L. 1981. *Nucleic Acids Res.* 9:3151–58
61. Ballal, N. R., Kang, Y.-J., Olson, M. O. J., Busch, H. 1975. *J. Biol. Chem.* 250:5921–25
62. Goldknopf, I. L., Wilson, G., Ballal, N. R., Busch, H. 1980. *J. Biol. Chem.* 255:10555–58
63. Goldknopf, I. L., Cheng, S., Andersen, M. W., Busch, H. 1981. *Biochem. Biophys. Res. Commun.* 100:1464–70
64. Kleinschmidt, A. M., Martinson, H. G. 1981. *Nucleic Acids Res.* 9:2423–31

65. Matsui, S.-I., Seon, B. K., Sandberg, A. A. 1979. *Proc. Natl. Acad. Sci. USA* 76:6386–90
66. Levinger, L., Varshavsky, A. 1980. *Proc. Natl. Acad. Sci. USA* 77:3244–48
67. Walker, J. M., Goodwin, G. H., Johns, E. W. 1978. *FEBS Lett.* 90:327–30
68. Wilkinson, K. D., Urban, M. K., Haas, A. L. 1980. *J. Biol. Chem.* 255:7529–32
69. Zweidler, A. 1980. In *Gene Families of Collagen and Other Structural Proteins,* eds. D. J. Prockop, P. C. Champe, Amsterdam: Elsevier
70. Urban, M. K., Franklin, S. G., Zweidler, A. 1979. *Biochemistry* 18:3952–60
71. Benezra, R., Blankstein, L. A., Stollar, B. D., Levy, S. B. 1981. *J. Biol. Chem.* 256:6837–41
72. Palmer, D., Snyder, L. A., Blumenfeld, M. 1980. *Proc. Natl. Acad. Sci. USA* 77:2671–75
73. Davidson, E. H. 1976. *Gene Activity in Early Development,* pp. 87–129. New York: Academic. 2nd ed.
74. Zalenskaya, I. A., Zalensky, A. O., Zalenskaya, E. O., Vorob'ev, V. I. 1981. *FEBS Lett.* 128:40–42
75. Bekhor, I. 1978. In *The Cell Nucleus,* ed. H. Busch, 5:137–66. New York: Academic
76. Steinmetz, M., Streeck, R. E., Zachau, H. G. 1978. See Ref. 75, pp. 167–83
77. Laskey, R. A., Earnshaw, W. C. 1980. *Nature* 286:763–67
78. Ruiz-Carrillo, A., Jorcano, J. L., Eder, G., Lurz, R. 1979. *Proc. Natl. Acad. Sci. USA* 76:3284–88
79. Laskey, R. A., Honda, B. M., Mills, A. D., Finch, J. T. 1978. *Nature* 275:416–20
80. Krohne, G., Franke, W. W. 1980. *Proc. Natl. Acad. Sci. USA* 77:1034–38
81. Krohne, G., Franke, W. W. 1980. *Exp. Cell Res.* 129:167–89
82. Mills, A. D., Laskey, R. A., Black, P., DeRobertis, E. M. 1980. *J. Mol. Biol.* 139:561–68
83. Earnshaw, W. C., Honda, B. M., Laskey, R. A., Thomas, J. O. 1980. *Cell* 21:373–83
84. Stein, A., Whitlock, J. P. Jr., Bina, M. 1979. *Proc. Natl. Acad Sci. USA* 76:5000–4
85. Germond, J.-E., Rouviere-Yaniv, J., Yaniv, M., Brutlag, D. 1979. *Proc. Natl. Acad. Sci. USA* 76:3779–83
86. Nelson, T., Wiegand, R., Brutlag, D. 1981. *Biochemistry* 20:2594–601
87. Fulmer, A. W., Fasman, G. D. 1979. *Biochemistry* 18:659–68
88. Laskey, R. A., Mills, A. D., Morris, N. R. 1977. *Cell* 10:237–43
89. Nelson, T., Hsieh, T.-S., Brutlag, D. 1979. *Proc. Natl. Acad. Sci. USA* 76:5510–14
90. Thomas, J. O., Oudet, P. 1979. *Nucleic Acids Res.* 7:611–23
91. Mathis, D. J., Kindelis, A., Spadafora, C. 1980. *Nucleic Acids Res.* 8:2577–90
92. Palter, K. B., Foe, V. E., Alberts, B. M. 1979. *Cell* 18:451–67
93. Palter, K. B., Alberts, B. M. 1979. *J. Biol. Chem.* 254:11160–69
94. Simpson, R. T., Kuenzler, P. 1979. *Nucleic Acids Res.* 6:1387–1415
95. Bryan, P. N., Wright, E. B., Olins, D. E. 1979. *Nucleic Acids Res.* 6:1449–65
96. Rhodes, D. 1979. *Nucleic Acids Res.* 6:1805–16
97. Wasylyk, B., Oudet, P., Chambon, P. 1979. *Nucleic Acids Res.* 7:705–13
98. Hiwasa, T., Segawa, M., Yamaguchi, N., Oda, K. 1981. *J. Biochem. Tokyo* 89:1375–89
99. Chao, M. V., Gralla, J. D., Martinson, H. G. 1979. *Biochemistry* 18:1068–74
100. Chao, M. V., Gralla, J. D., Martinson, H. G. 1980. *Biochemistry* 19:3254–60
101. Chao, M. V., Martinson, H. G., Gralla, J. D. 1980. *Biochemistry* 19:3260–69
102. Holland , L. J., McCarthy, B. J. 1980. *Biochemistry* 19:2965–76
103. Gottesfeld, J. 1982, personal communication
104. Zachau, H. G., Igo-Kemenes, T. 1981. *Cell* 24:597–98
105. Kornberg, R. 1981. *Nature* 292:579–80
106. Savic, A., Richman, P., Williamson, P., Poccia, D. 1981. *Proc. Natl. Acad. Sci. USA* 78:3706–10
107. Humphries, S. E., Young, D., Carroll, D. 1979. *Biochemistry* 18:3223–31
108. Gottesfeld, J. M. 1980. *Nucleic Acids Res.* 8:905–21
109. Omori, A., Igo-Kemenes, T., Zachau, H. G. 1980. *Nucleic Acids Res.* 8:5363–75
110. Pech, M., Igo-Kemenes, T., Zachau, H. G. 1979. *Nucleic Acids Res.* 7:417–32
111. Zalenskaya, I. A., Pospelov, V. A., Zalensky, A. O., Vorob'ev, V. I. 1981. *Nucleic Acids Res.* 9:473–87
112. Pospelov, V. A., Svetlikova, S. B., Vorob'ev, V. I. 1979. *Nucleic Acids Res.* 6:399–418
113. Cremisi, C., Pignatti, P. F., Croissant, O., Yaniv, M. 1976. *J. Virol.* 17:204–11
114. Ponder, B. A. J., Crawford, L. V. 1977. *Cell* 11:35–49
115. Prunell, A., Kornberg, R. D. 1978. *Cold Spring Harbor Symp. Quant. Biol.* 42:103–8
116. Prunell, A. 1979. *FEBS Lett.* 107:285–87

117. Baer, B. W., Kornberg, R. D. 1979. *J. Biol. Chem.* 254:9678–81
118. Nelson, P. P., Albright, S. C., Garrard, W. T. 1979. *J. Biol. Chem.* 254:9194–99
119. Musich, P. R., Maio, J. J., Brown, F. L. 1977. *J. Mol. Biol.* 117:657–77
120. Brown, F. L., Musich, P. R., Maio, J. J. 1979. *J. Mol. Biol.* 131:777–99
120a. Musich, P. R., Brown, F. L., Maio, J. J. 1982. *Proc. Natl. Acad. Sci. USA.* 79:118–22
121. Singer, D. S. 1979. *J. Biol. Chem.* 254:5506–14
122. Fittler, F., Zachau, H. G. 1979. *Nucleic Acids Res.* 7:1–13
123. Hoerz, W., Altenburger, W. 1981. *Nucleic Acids Res.* 9:2643–58
124. Dingwall, C., Lomonossoff, G. P., Laskey, R. A. 1981. *Nucleic Acids Res.* 9:2659–73
125. Keene, M. A., Elgin, S. C. R. 1981. *Cell* 27:57–64
126. Igo-Kemenes, T., Omori, A., Zachau, H. G. 1980. *Nucleic Acids Res.* 8:5377–90
127. Gottesfeld, J. M., Bloomer, L. S. 1980. *Cell* 21:751–60
128. Louis, C., Schedl, P., Samal, B., Worcel, A. 1980. *Cell* 22:387–92
129. Wittig, B., Wittig, S. 1979. *Cell* 18:1173–83
130. Brian, P. N., Hofstetter, H., Birnstiel, M. 1981. *Cell.* 27:459–66
131. Wu, C. 1980. *Nature* 286:854–60
132. Nedospasov, S. A., Georgiev, G. P. 1980 *Biochem. Biophys. Res. Commun.* 92:532–39
133. Levy, A., Noll, M. 1980. *Nucleic Acids Res.* 8:6059–68
134. Weintraub, H., Groudine, M. 1982. Personal communication
135. Samal, B., Worcel, A., Louis, C., Schedl, P. 1981. *Cell* 23:401–9
136. Pfeiffer, W., Zachau, H. G. 1980. *Nucleic Acids Res.* 8:4621–38
137. Moyne, G., Katinka, M., Saragosti, S., Chestier, A., Yaniv, M. 1981. *Prog. Nucleic Acid Res. Mol. Biol.* 26:151–67
138. Scott, W. A., Wigmore, D. J. 1978. *Cell* 15:1511–18
139. Waldeck, W., Foehring, B., Chowdhury, K., Gruss, P., Sauer, G. 1978. *Proc. Natl. Acad. Sci. USA* 75:5964–68
140. Sundin, O., Varshavsky, A. 1979. *J. Mol. Biol.* 132:535–46
141. Varshavsky, A. J., Sundin, O. H., Bohn, M. J. 1978. *Nucleic Acids Res.* 5:3469–77
142. Varshavsky, A. J., Sundin, O., Bohn, M. 1979. *Cell* 16:453–66
143. Saragosti, S., Moyne, G., Yaniv, M. 1980. *Cell* 20:65–73
144. Jakobovits, E. B., Bratosin, S., Aloni, Y. 1980. *Nature* 285:263–65
145. Shelton, E. R., Wassarman, P. M., DePamphilis, M. L. 1980. *J. Biol. Chem.* 255:771–82
146. Das, G. C., Allison, D. P., Niyogi, S. K. 1979. *Biochem. Biophys. Res. Commun.* 89:17–25
147. Hartmann, J. P., Scott, W. A. 1981. *J. Virol.* 37:908–15
148. Fey, G., Hirt, B. 1974. *Cold Spring Harbor Symp. Quant. Biol.* 39:235–41
149. Varshavsky, A., Bakayev, V. V., Chumakov, P. M., Georgiev, G. P. 1976. *Nucleic Acids Res.* 3:2101–13
150. Shelton, E. R., Wassarman, P. M., DePamphilis, M. L. 1978. *J. Mol. Biol.* 125:491–514
151. Nedospasov, S. A., Shakhov, A., Georgiev, G. P. 1981. *FEBS Lett.* 125:35–38
152. Bakayev, V. V., Nedospasov, S. A., Georgiev, G. P. 1981. *Mol. Biol. USSR* 15:939–49
153. Steinmetz, M., Streeck, R. E., Zachau, H. G. 1975. *Nature* 258:447–50
154. Tatchell, K., Van Holde, K. E. 1979. *Biochemistry* 18:2871–80
155. Simpson, R. T., Stein, A. 1980. *FEBS Lett.* 111:337–39
156. Trifonov, E. N. 1981. In *International Cell Biology 1980–1981,* ed. H. Schweiger, pp. 128–38. Heidelberg: Springer
157. Trifonov, E. N. 1980. *Nucleic Acids Res.* 8:4041–53
158. Trifonov, E. N., Sussman, J. L. 1980. *Proc. Natl. Acad. Sci. USA* 77:3816–20
159. Weintraub, H. 1980. *Nucleic Acids Res.* 8:4745–53
160. Georgiev, G. P., Nedospasov, S. A., Bakayev, V. V. 1978. See Ref. 45, pp. 3–34
161. Finch, J. T., Klug, A. 1976. *Proc. Natl. Acad. Sci. USA* 73:1897–1901
162. Thoma, F., Koller, T., Klug, A. 1979. *J. Cell Biol.* 83:403–27
163. Suau, P., Bradbury, E. M., Baldwin, J. P. 1979. *Eur. J. Biochem.* 97:593–602
164. Thoma, F., Koller, T. 1981. *J. Mol. Biol.* 149:709–33
165. Butler, P. J. G., Thomas, J. O. 1980. *J. Mol. Biol.* 140:505–29
166. Thomas, J. O., Butler, P. J. G. 1980. *J. Mol. Biol.* 144:89–93
167. Osipova, T. N., Pospelov, V. A., Svetlikova, S. B., Vorob'ev, V. I. 1980. *Eur. J. Biochem.* 113:183–88
168. Brust, R., Harbers, E. 1981. *Eur. J. Biochem.* 117:609–15
169. Renz, M., Nehls, P., Hozier, J. 1977. *Proc. Natl. Acad. Sci. USA* 74:1879–83
170. Straetling, W. H., Mueller, U., Zent-

graf, H. 1978. *Exp. Cell Res.* 117: 301–11
171. Franke, W. W., Scheer, U., Trendelenburg, M. F., Zentgraf, H., Spring, H. 1978. *Cold Spring Harbor Symp. Quant. Biol.* 42:755–72
172. Meyer, G. F., Renz, M. 1979. *Chromosoma* 75:177–84
173. Pruitt, S. C., Grainger, R. M. 1980. *Chromosoma* 78:257–74
174. Scheer, U., Sommerville, J., Mueller, U. 1980. *Exp. Cell Res.* 129:115–26
175. Klingholz, R., Straetling, W. H., Schaefer, H. 1981. *Exp. Cell Res.* 132:399–409
176. Zentgraf, H., Mueller, U., Scheer, U., Franke, W. W. 1981. See Ref. 156, pp. 139–151
177. Subirana, J. A., Munoz-Guerra, S., Martinez, A. B., Perez-Grau, L., Marcet, X., Fita, I. 1981. *Chromosoma* 83:455–71
178. Renz, M. 1979. *Nucleic Acids Res.* 6:2761–67
179. Jorcano, J. L., Meyer, G., Day, L. A., Renz, M. 1980. *Proc. Natl. Acad. Sci. USA* 77:6443–47
180. Straetling, W. H. 1979. *Biochemistry* 18:596–603
181. Straetling, W. H., Klingholz, R. 1981. *Biochemistry* 20:1386–92
182. Muyldermans, S., Lasters, I., Wyns, L., Hamers, R. 1980. *Nucleic Acids Res.* 8:2165–72
183. Rattner, J. B., Hamkalo, B. A. 1978. *Chromosoma* 69:373–79
184. Miller, F., Igo-Kemenes, T., Zachau, H. G. 1978. *Chromosoma* 68:327–36
185. Labhart, P., Koller, T. 1982. *Eur. J. Cell Biol.* In press
186. Stein, A. 1980. *Nucleic Acids Res.* 8:4803–20
187. Worcel, A., Strogatz, S., Riley, D. 1981. *Proc. Natl. Acad. Sci. USA* 78:1461–65
188. Burgoyne, L. A., Skinner, J. D. 1981. *Biochem. Biophys. Res. Commun.* 99:893–99
189. Khachatrian, A. T., Pospelov, V. A., Svetlikova, S. B., Vorob'ev, V. I. 1981. *FEBS Lett.* 128:90–92
190. Grigoryev, S. A., Joffe, L. B. 1981. *FEBS Lett.* 130:43–46
191. Lutter, L. C. 1981. *Nucleic Acids Res.* 9:4251–66
192. Klug, A. C., Lutter, L. C. 1981. *Nucleic Acids Res.* 9:4267–83
193. Lutter, L. C. 1979. *Nucleic Acids Res.* 6:41–56
194. Hoerz, W., Miller, F., Klobeck, G., Zachau, H. G. 1980. *J. Mol. Biol.* 144:329–51
195. Hoerz, W., Zachau, H. G. 1980. *J. Mol. Biol.* 144:305–27
196. Fittler, F., Ibel, K., Hoerz, W. 1981. *FEBS Lett.* 132:341–43
197. McGhee, J. D., Rau, D. C., Charney, E., Felsenfeld, G. 1980. *Cell* 22:87–96
198. Lee, K. S., Mandelkern, M., Crothers, D. M. 1981. *Biochemistry* 20:1438–45
199. Fenske, H., Schmidt, G., Karawajew, L., Heymann, S., Boettger, M., Billwitz, H., Eichhorn, I., Lindigkeit, R. 1980. *Acta Biol. Med. Ger.* 39:343–54
200. Boettger, M., Karawajew, L., Fenske, H., Grade, K., Lindigkeit, R. 1981. *Mol. Biol. Rep.* 7:231–34
201. Lindigkeit, R., Boettger, M., Mickwitz, C.-U., Fenske, H., Karawajew, L., Karawajew, K. 1977. *Acta Biol. Med. Ger.* 36:275–79
202. Itkes, A. V., Glotov, B. O., Nikolaev, L. G., Preem, S. R., Severin, E. S. 1980. *Nucleic Acids Res.* 8:507–27
203. Nikolaev, L. G., Glotov, B. O., Itkes, A. V., Severin, E. S. 1981. *FEBS Lett.* 125:20–24
204. Thomas, J. O., Khabaza, A. J. A. 1980. *Eur. J. Biochem.* 112:501–11
205. Caron, F., Thomas, J. O. 1981. *J. Mol. Biol.* 146:513–37
206. Lasters, I., Muyldermans, S., Wyns, L., Hamers, R. 1981. *Biochemistry* 20:1104–10
207. Benyajati, C., Worcel, A. 1976. *Cell* 9:393–407
208. Paulson, J. R., Laemmli, U. K. 1977. *Cell* 12:817–28
209. Igo-Kemenes, T., Zachau, H. G. 1977. *Cold Spring Harbor Symp. Quant. Biol.* 42:109–18
210. Cook, P. R., Brazell, I. A. 1980. *Nucleic Acids Res.* 8:2895–2906
211. Berezney, R., Coffey, D. S. 1974. *Biochem. Biophys. Res. Commun.* 60: 1410–17
212. Comings, D. E., Wallach, A. S. 1978. *J. Cell Sci.* 34:233–46
213. Long, B. H., Huang, C. Y., Pogo, A. O. 1979. *Cell* 18:1079–90
214. Herlan, G., Eckert, W. A., Kaffenberger, W., Wunderlich, F. 1979. *Biochemistry* 18:1782–88
215. Dijkwel, P. A., Mullenders, L. H. F., Wanka, F. 1979. *Nucleic Acids Res.* 6:219–30
216. Aaronson, R. P., Blobel, G. 1975. *Proc. Natl. Acad. Sci. USA* 72:1007–11
217. Stick, R., Hausen, P. 1980. *Chromosoma* 80:219–36
218. Murray, A. B., Davies, H. G. 1979. *J. Cell Sci.* 35:59–66
219. Krohne, G., Franke, W. W., Ely, S.,

D'Arcy, A., Jost, E. 1978. *Cytobiol.* 18:22–38
220. Richardson, J. C. W., Maddy, A. H. 1980. *J. Cell Sci.* 43:253–67
221. Comings, D. E. 1980. *Human genetic* 53:131–43
222. Riley, D. E., Keller, J. M., Byers, B. 1975. *Biochemistry* 14:3005–13
223. Adolph, K. W., Cheng, S. M., Paulson, J. R., Laemmli, U. K. 1977. *Proc. Natl. Acad. Sci. USA* 74:4937–41
224. Adolph, K. W., Cheng, S. M., Laemmli, U. K. 1977. *Cell* 12:805–16
225. Campbell, A. M., Briggs, R. C., Bird, R. E., Hnilica, L. S. 1979. *Nucleic Acids Res.* 6:205–18
226. Ide, T., Nakane, M., Anzai, K., Andoh, T. 1975. *Nature* 258:445–47
227. Nakane, M., Ide, T., Anzai, K., Ohara, S., Andoh, T. 1978. *J. Biochem. Tokyo* 84:145–57
228. Hartwig, M. 1978. *Acta Biol. Med. Ger.* 37:421–32
229. Cook, P. R., Brazell, I. A. 1975. *J. Cell Sci.* 19:261–79
230. Levin, J. M., Jost, E., Cook, P. R. 1978. *J. Cell Sci.* 29:103–16
231. Pinon, R., Salts, Y. 1977. *Proc. Natl. Acad. Sci. USA* 74:2850–54
232. Hancock, R., Hughes, M. E. 1982, personal communication
233. Lepault, J., Bram, S., Escaig, J., Wray, W. 1980. *Nucleic Acids Res.* 8:265–78
234. Marsden, M. P. F., Laemmli, U. K. 1980. *Cell* 17:849–58
235. Bak, A. L., Zeuthen, J., Crick, F. H. C. 1977. *Proc. Natl. Acad. Sci. USA* 74:1595–99
236. Bak, P., Bak, A. L., Zeuthen, J. 1979. *Chromosoma* 73:301–15
237. Sedat, J., Manuelidis, L. 1978. *Cold Spring Harbor Symp. Quant. Biol.* 42:331–50
238. Rattner, J. B., Goldsmith, M., Hamkalo, B. A. 1981. *Chromosoma* 82:341–51
239. Macgregor, H. C. 1977. In *Chromatin and Chromosome Structure*, eds.H. J. Li, R. Eckhardt pp. 339–57. New York: Academic
240. Davidson, E. H. 1976. See Ref. 73, pp. 322–78
241. Scheer, U., Spring, H., Trendelenburg, M. F. 1979. In *The Cell Nucleus*, ed. H. Busch, 7:3–47. New York: Academic
242. Igo-Kemenes, T., Greil, W., Zachau, H. G. 1977. *Nucleic Acids Res.* 4:3387–3400
243. Hyde, J. E. 1982. *Exp. Cell. Res.* In press
244. Jeppesen, P. G. N., Bankier, A. T. 1979. *Nucleic Acids Res.* 7:49–67
245. Razin, S. V., Mantieva, V. I., Georgiev, G. P. 1979. *Nucleic Acids Res.* 7:1713–35
246. Mantieva, V. L., Razin, S. V., Georgiev, G. P. 1979. *Mol. Biol. USSR* 13:1360–68
247. Razin, S. V., Mantieva, V. L., Georgiev, G. P. 1978. *Nucleic Acids Res.* 5:4737–51
248. Nelkin, B. D., Pardoll, D. M., Vogelstein, B. 1980. *Nucleic Acids Res.* 8:5623–33
249. Jeppesen, P. G. N., Bankier, A. T., Sanders, L. 1978. *Exp. Cell Res.* 115:293–302
250. Hyde, J., Igo-Kemenes, T., Zachau, H. G. 1979. *Nucleic Acids Res.* 7:31–48
251. Avramova, Z., Dessev, G., Tsanev, R. 1980. *FEBS Lett.* 118:58–62
252. Werner, D., Zimmermann, H. P., Rauterberg, E., Spalinger, J. 1981. *Exp. Cell Res.* 133:149–57
253. Gerace, L., Blobel, G. 1980. *Cell* 19:277–87
254. Berezney, R. 1981. See Ref. 156, pp. 214–24
255. Hand, R. 1978. *Cell* 15:317–25
256. Stalder, J., Larsen, A., Engel, J. D., Dolan, M., Groudine, M., Weintraub, H. 1980. *Cell* 20:451–60
257. Franke, W. W., Scheer, U., Spring, H., Trendelenburg, M. F., Zentgraf, H. 1979. See Ref. 241, pp.49–95
258. McKnight, S. L., Martin, K. A., Beyer, A. L., Miller, O. L. Jr. 1979. See Ref. 241, pp. 97–122
259. Daskal, Y., Busch, H. 1978. In *The Cell Nucleus*, ed. H. Busch, 4:3–45. New York: Academic
260. Scheer, U., Sommerville, J., Bustin, M. 1979. *J. Cell Sci.* 40:1–20
261. Andersson, K., Bjoerkroth, B., Daneholt, B. 1980. *Exptl. Cell Res.* 130:313–26
262. Olins, A. L., Olins, D. E., Franke, W. W. 1980. *Eur. J. Cell Biol.* 22:714–23
263. Petrov, P., Raitcheva, E., Tsanev, R. 1980. *Eur. J. Cell Biol.* 22:708–13
264. Scheer, U., Zentgraf, H., Sauer, H. W. 1981. *Chromosoma* 84:279–90
265. Johnson, E. M., Campbell, G. R., Allfrey, V. G. 1978. *Science* 206:1192–94
266. Jalouzot, R., Briane, D., Ohlenbusch, H. H., Wilhelm, M. L., Wilhelm, F. X. 1980. *Eur. J. Biochem.* 104:423–31
267. Annesley, M., Davies, K. E., Kumar, N. M., Walker, I. O. 1981. *Nucleic Acids Res.* 9:831–39
268. Scheer, U. 1980. *Eur. J. Cell. Biol.* 23:189–96
269. Pruitt, S. C., Grainger, R. M. 1981. *Cell* 23:711–20

269a. Labhart, P., Koller, T. 1982. *Cell.* In press
270. Borkhardt, B., Nielsen, O. F. 1981. *Chromosoma* 84:131–43
271. Borchsenius, S., Bonven, B., Leer, J. C., Westergaard, O. 1981. *Eur. J. Biochem.* 117:245–50
272. Cech, T. R., Karrer, K. M. 1980. *J. Mol. Biol.* 136:395–416
273. Southern, E. M. 1975. *J. Mol. Biol.* 98:503–17
274. Wu, C., Bingham, P. M., Livak, K. J., Holmgren, R., Elgin, S. C. R., 1979. *Cell* 16:797–806
275. Bellard, M., Kuo, M. T., Dretzen, G., Chambon, P. 1980. *Nucleic Acids Res.* 8:2737–50
276. Levy, A., Noll, M. 1981. *Nature* 289:198–203
277. Alwine, J. C., Kemp, D. J., Stark, G. R. 1977. *Proc. Natl. Acad. Sci. USA* 74:5350–54
278. Gottesfeld, J. M., Garrard, W. T., Bagi, G., Wilson, R. F., Bonner, J. 1974. *Proc. Natl. Acad. Sci. USA* 71:2193–97
279. Gottesfeld, J. M. 1979. In *Chromatin Structure and Function,* Part B., ed. C. A. Nicolini, pp. 541–60. New York: Plenum
280. Weintraub, H., Groudine, M. 1976. *Science* 193:848–56
281. Stalder, J., Groudine, M., Dodgson, J. B., Engel, J. D., Weintraub, H. 1980. *Cell* 19:973–80
282. Lawson, G. M., Tsai, M.-J., O'Malley, B. W. 1980. *Biochemistry* 19:4403–11
283. Storb, U., Wilson, R., Selsing, E., Walfield, A. 1981. *Biochemistry* 20:990–96
284. Felber, B. K., Gerber-Huber, S., Meier, C., May, F. E. B., Westley, B., Ryffel, G. U. 1981. *Nucleic Acids Res.* 9: 2455–74
285. Gerber-Huber, S., Felber, B. K., Weber, R., Ryffel, G. U. 1981. *Nucleic Acids Res.* 9:2475–94
286. Gariglio, P., Bellard, M., Chambon, P. 1981. *Nucleic Acids Res.* 9:2589–98
287. Zasloff, M., Camerini-Otero, R. D. 1980. *Proc. Natl. Acad. Sci. USA* 77:1907–11
288. Lohr, D. E. 1981. *Biochemistry.* 20:5966–72
289. Camerini-Otero, R. D., Zasloff, M. A. 1980. *Proc. Natl. Acad. Sci. USA* 77:5079–83
290. Levy-Wilson, B., Kuehl, L., Dixon, G. H. 1980. *Nucleic Acids Res.* 8:2859–69
291. Giri, C. P., Gorovsky, M. A. 1980. *Nucleic Acids Res.* 8:197–214
292. Groudine, M., Eisenman, R., Weintraub, H. 1981. *Nature* 292:311–17

293. Gazit, B., Cedar, H. 1980. *Nucleic Acids Res.* 8:5143–55
294. Razin, A., Riggs, A. D. 1980. *Science* 210:604–10
295. Ehrlich, M., Wang, R. Y.-H. 1981 *Science* 212:1350–57
296. Doerfler, W. 1981. *J. Gen. Virol.* 57:1–20
297. Bird, A., Southern, E. M. 1978. *J. Mol. Biol.* 118:27–48
298. Waalwijk, C., Flavell, R. A. 1978. *Nucleic Acids Res.* 5:3231–36
299. Groudine, M., Weintraub, H. 1981. *Cell* 24:393–401
300. Weintraub, H., Larsen, A., Groudine, M. 1981. *Cell* 24:333–44
301. Kuo, M. T., Mandel, J. L., Chambon, P. 1979. *Nucleic Acids Res.* 7:2105–13
302. van der Ploeg, L. H. T., Flavell, R. A. 1980. *Cell* 19:947–58
303. McGhee, J. D., Ginder, G. D. 1979. *Nature* 280:419–20
304. Mandel, J. L., Chambon, P. 1979. *Nucleic Acids Res.* 7:2081–2103
305. Shen, S. T., Maniatis, T. 1980. *Proc. Natl. Acad. Sci. USA* 77:6634–38
306. Sutter, D., Doerfler, W. 1980. *Proc. Natl. Acad. Sci. USA* 77:253–56
307. Desrosiers, R. C., Mulder, C., Fleckenstein, B. 1979. *Proc. Natl. Acad. Sci. USA* 76:3839–43
308. Wu, C., Wong, Y.-C., Elgin, S. C. R. 1979. *Cell* 16:807–14
309. Keene, M. A., Corces, V., Lowenhaupt, K., Elgin, S. C. R. 1981. *Proc. Natl. Acad. Sci. USA* 78:143–46
310. Wu, C., Gilbert, W. 1981. *Proc. Natl. Acad. Sci. USA* 78:1577–80
311. Lipps, H. J., Ehrhardt, P. 1981. *FEBS Lett.* 126:219–22
311a. McGhee, J. D., Wood, W. I., Dolan, M., Engel, J. D., Felsenfeld, G. 1981. *Cell.* 27:45–55
312. Breathnach, R., Chambon, P. 1981. *Ann. Rev. Biochem.* 50:349–83
313. Hentschel, C. C., Birnstiel, M. L. 1981. *Cell* 25:301–13
314. Johnson, E. M., Allfrey, V. G. 1978. In *Biochemical Actions of Hormones,* ed. G. Litwack, 5:1–53. New York: Academic
315. Glotov, B. O., Trakht, I. N., Grozdova, I. D., Nicolaev, L. G., Itkes, A. V., Severin, E. S. 1980. *Adv. Enzyme Regul.* 18:261–73
316. Smulson, M. 1979. *Trends Biochem. Sci.* 4:225–28
317. Ajiro, K., Borun, T. W., Cohen, L. H. 1981. *Biochemistry* 20:1445–54
318. Hofmann, K. W., Arfmann, H.-A., Bode, J., Arellano, A. 1980. *Int. J. Biol. Macromol.* 2:25–32

319. Cousens, L. S., Gallwitz, D., Alberts, B. M. 1979. *J. Biol. Chem.* 254:1716–23
320. Davie, J. R., Saunders, C. A., Walsh, J. M., Weber, S. C. 1981. *Nucleic Acids Res.* 9:3205–16
321. Hamana, K., Zama, M. 1980. *Nucleic Acids Res.* 8:5275–88
322. Kastern, W. H., Eldridge, J. D., Mullinix, K. P. 1979. *J. Biol. Chem.* 254:7368–76
323. Pashev, I. G., Mencheva, M. M., Markov, G. G. 1980. *Biochim. Biophys. Acta* 607:269–76
324. Georgieva, E. I., Pashev, I. G., Tsanev, R. G. 1981. *Biochim. Biophys. Acta* 652:240–42
325. Billett, M. A., 1981, *Cell Bio. Int. Rep.* 5:55-58
326. Levy-Wilson, B., Connor, W., Dixon, G. H. 1979. *J. Biol. Chem.* 254:609–20
327. Dimitriadis, G. J., Tata, J. R. 1980. *Biochem. J.* 187:467–77
328. Teng, C. S., Andrews, G. K., Teng, C. T. 1979. *Biochem. J.* 181:585–91
329. Goodwin, G. H., Wright, C. A., Johns, E. W. 1981. *Nucleic Acids Res.* 9:2761–75
330. Kuehl, L., Lyness, T., Dixon, G. H., Levy-Wilson, B. 1980. *J. Biol. Chem.* 255:1090–95
331. Defer, N., Crepin, M., Terrioux, C., Kruh, J., Gros, F. 1979. *Nucleic Acids Res.* 6:953–66
332. Sargan, D. R., Butterworth, P. H. W. 1981. *Cell Biol. Int. Rep.* 5:59–66
333. Levy-Wilson, B., Dixon, G. H. 1979. *Proc. Natl. Acad. Sci. USA* 76:1682–86
334. Nelson, D., Covault, J., Chalkley, R. 1980. *Nucleic Acids Res.* 8:1745–63
335. Jackson, J. B., Pollock, J. M., Rill, R. L. 1979. *Biochemistry* 18:3739–48
336. Djondjurov, L., Ivanova, E., Tsanev, R. 1979. *Eur. J. Biochem.* 97:133–39
337. Gates, D. M., Bekhor, I. 1980. *Science* 207:661–62
338. Norman, G. L., Bekhor, I. 1981. *Biochemistry* 20:3568–78
339. Levy-Wilson, B. 1981. *Proc. Natl. Acad. Sci. USA* 78:2189–93
340. Levy-Wilson, B., Watson, D. C., Dixon, G. H. 1979. *Nucleic Acids Res.* 6:259–74
341. Egan, P. A., Levy-Wilson, B. 1981. *Biochemistry* 20:3695–3702
342. Perry, M., Chalkley, R. 1981. *J. Biol. Chem.* 256:3313–18
343. Levy-Wilson, B. 1981. *Arch. Biochem. Biophys.* 208:528–34
344. Hatayama, T., Omori, K., Inoue, A., Yukioka, M. 1981. *Biochim. Biophys. Acta* 652:245–55
345. Faber, A. J., Cook, H., Hancock, R. 1981. *Eur. J. Biochem.* 120:357–61
346. Gabrielli, F., Hancock, R., Faber, A. J. 1981. *Eur. J. Biochem.* 120:363–69.
347. Comings, D. E., Peters, K. E. 1981. In *The Cell Nucleus* ed. H. Busch, 9:89–118. New York: Academic
348. Mathew, C. G. P., Goodwin, G. H., Wright, C. A., Johns, E. W. 1981. *Cell Biol. Int. Rep.* 5:37–43
349. Peters, E. H., Levy-Wilson, B., Dixon, G. H. 1979. *J. Biol. Chem.* 254:3358–61
350. Romani, M., Vidali, G., Tahourdin, C. S. M., Bustin, M. 1980. *J. Biol. Chem.* 255:468–74
351. Levy-Wilson, B. 1979. *Nucleic Acids Res.* 7:2239–54
352. Weisbrod, S., Weintraub, H. 1979. *Proc. Natl. Acad. Sci. USA* 76:630–34
353. Weisbrod, S., Groudine, M., Weintraub, H. 1980. *Cell* 19:289–301
354. Gazit, B., Panet, A., Cedar, H. 1980. *Proc. Natl. Acad. Sci. USA* 77:1787–90
355. Weisbrod, S., Weintraub, H. 1981. *Cell* 23:391–400
356. Silver, L. M., Elgin, S. C. R. 1978. See Ref. 75, pp. 215–62
357. Stollar, B. D. 1978. In *Methods in Cell Biology,* eds. G. Stein, J. Stein, L. T. Kleinsmith, 18:105–22, New York: Academic
358. Bustin, M. 1979. *Curr. Topics Microbiol. Immunol.* 88:105–42
359. Hnilica, L. S., Chiu, J.-F., Hardy, K., Fujitani, H., Briggs, R. 1978. See Ref. 75, pp. 307–31
360. Will, H., Bautz, E. K. F. 1980. *Exp. Cell Res.* 125:401–10
361. Moroi, Y., Peebles, C., Fritzler, M. J., Steigerwald, J., Tan, E. M. 1980. *Proc. Natl. Acad. Sci. USA* 77:1627–31
362. Moroi, Y., Hartman, A. L., Nakane, P. K., Tan, E. M. 1981. *J. Cell Biol.* 90:254-59
363. Jamrich, M., Christensen, M. E., Le Stourgeon, W. M. 1979. *J. Supramol. Struct.* 3:74
364. Christensen, M. E., Le Stourgeon, W. M., Jamrich, M., Howard, G. C., Serunian, L. A., Silver, L. M., Elgin, S. C. R. 1981. *J. Cell Biol.* 90:18–24
365. Kurth, P. D., Bustin, M. 1981. *J. Cell Biol.* 89:70–77
366. Gronemeyer, H., Hameister, H., Pongs, O. 1981. *Chromosoma* 82:543–59
367. Saumweber, H., Symmons, P., Kabisch, R., Will, H., Bonhoeffer, F. 1980. *Chromosoma* 80:253–88
368. Howard, G. C., Abmayr, S. M., Shinefeld, L. A., Sato, V. L., Elgin, S. C. R. 1981. *J. Cell Biol.* 88:219–25

369. Appels, R., Peacock, W. J. 1978. *Int. Rev. Cytol.* 8:69–126. (*Suppl.*)
370. John, B., Miklos, G. L. G. 1979. *Int. Rev. Cytol.* 58:1–114
371. Brutlag, D. L. 1980. *Ann. Rev. Genet.* 14:121–44
372. Singer, M. *Int. Rev. Cytol.* 1982. In press
373. Jelinek, W. R., Schmid, C. W. 1982. *Ann. Rev. Biochem.* 51:813–44
374. Brown, S. W. 1966. *Science* 151:418–21
375. Yunis, J. J., Yasmineh, W. G. 1971. *Science* 174:1200–9
376. Bokhon'ko, A., Reeder, R. H. 1976. *Biochem. Biophys. Res. Commun.* 70:146–52
377. Hoerz, W., Igo-Kemenes, T., Pfeiffer, W., Zachau, H. G. 1976. *Nucleic Acids Res.* 3:3213–26
378. Gottesfeld, J. M., Melton, D. A. 1978. *Nature* 273:317–19
379. Igo-Kemenes, T., Miller, F., Zachau, H. G. 1979. *FEBS Proc. Meet* 51:219–29
380. Bostock, C. J., Christie, S., Hatch, F. T. 1976. *Nature* 262:516–19
381. Kuo, M. T., 1979. *Chromosoma* 70:183–94
382. Horvath, P., Hoerz, W. 1982. *FEBS Lett.* 134:25–28
383. Pfeiffer, W., Hoerz, W., Igo-Kemenes, T., Zachau, H. G. 1975. *Nature* 258:450–52
384. Lipchitz, L., Axel, R. 1976. *Cell* 9:355–64
385. Musich, P. R., Brown, F. L., Maio, J. J. 1977. *Proc. Natl. Acad. Sci. USA* 74:3297–3301

386. Bernstine, E. G. 1978. *Exp. Cell Res.* 113:205–8
387. Mazrimas, J. A., Balhorn, L., Hatch, F. T. 1979. *Nucleic Acids Res.* 7:935–46
388. von Oefele, K. 1979. Darstellung von Heterochromatinfraktionen aus Mäuseleberzellkernen mit Hilfe von Restriktionsnucleasen. Doctoral thesis. Universitaet Muenchen, FRG. 42 pp.
389. Mathew, C. G. P., Goodwin, G. H., Igo-Kemenes, T., Johns, E. W. 1981. *FEBS Lett.* 125:25–29
390. Billings, P. C., Orf, J. W., Palmer, D. K., Talmage, D. A., Pan, C. G., Blumenfeld, M. 1979. *Nucleic Acids Res.* 6:2151–64
391. Hsieh, T.-S., Brutlag, D. L. 1979. *Proc. Natl. Acad. Sci. USA* 76:726–30
392. Schmidt, E. R., Keyl, H.-G. 1981. *Chromosoma* 82:197–204
393. Varley, J. M., MacGregor, H. C., Nardi, I., Andrews, C., Erba, H. 1980. *Chromosoma* 80:289–307
394. Diaz, M. O., Barsacchi-Pilone, G., Mahon, K. A., Gall, J. G. 1981. *Cell* 24:649–59
395. Sealy, L., Hartley, J., Donelson, J., Chalkley, R., Hutchison, N. 1981. *J. Mol. Biol.* 145:291–318
396. Bedbrook, J. R., Jones, J., O'Dell, M., Thompson, R. D., Flavell, R. B. 1980. *Cell* 19:545–60
397. Hilliker, A. J., Appels, R., Schalet, A. 1980. *Cell* 21:607–19
398. Doolittle, W. F., Sapienza, C. 1980. *Nature* 284:601–3
399. Orgel, L. E., Crick, F. H. C. 1980. *Nature* 284:604–7

Ann. Rev. Biochem. 1982. 51:123–54

THE CONFORMATION, FLEXIBILITY, AND DYNAMICS OF POLYPEPTIDE HORMONES

Tom Blundell and Stephen Wood

Laboratory of Molecular Biology, Department of Crystallography, Birkbeck College, University of London

CONTENTS

PERSPECTIVES AND SUMMARY

Polypeptide hormones are manufactured and often stored in specialized endocrine cells and are circulated to specific target tissues. In many cases the hormone message is known to be conveyed to the target cell via a cell surface receptor for the hormone, and rapid degradation, often mediated by

123

0066-4154/82/0701-0123$02.00

receptor binding and internalization, seems to be a common feature. We clearly need to define the conformation of polypeptide hormones during this complex life cycle if we are to understand the various processes at the molecular level.

Analysis of conformation, flexibility, and dynamics presents a range of challenging problems both experimentally and conceptually, as Schwyzer was the first to realise (1). Small polypeptide hormones of less than 30 amino acids will generally be flexible, and their conformations will depend on their concentration, on the solvent, and on other molecules in solution (1, 2). For instance, glucagon has little defined structure in dilute aqueous solutions (3), but secondary structure is stabilized by self-association at high concentrations (2, 4) and by the presence of nonaqueous solvents and lipid micelles (5). The conformations of some small hormones such as oxytocin may be limited by disulfide bridges and consequently they may have better defined mainchain conformations, but even these are flexible and comprise several conformers in equilibrium in aqueous solutions (6). Many larger polypeptide hormones such as insulin, and probably glycoprotein hormones and growth hormones, have "globular" structures with hydrophobic cores (7). Their preferred conformations are probably retained in aqueous solutions, in oligomeric forms, and in the crystalline state, although even insulin shows large changes under extreme conditions such as high salt concentrations, and the conformation and flexibility in dilute solutions has been the subject of some debate (8, 9). The smallest polypeptide hormone with a hydrophobic core appears to be pancreatic polypeptide, with 36 amino acids (10), but many larger polypeptides, such as β lipotropin, have more flexible structures. Clearly, stability of the tertiary structure of the globular polypeptide hormones makes experimental data in vitro more easily relevant to events in vivo.

The conformational analysis of a polypeptide hormone cannot be considered complete until a detailed structure has been defined in crystals and in solutions of different properties corresponding to the environment in storage granules, in circulation, and in the receptor complex. It would be most advantageous to study at high resolution the granules themselves and the isolated receptor complex, but these both present great experimental difficulties. Perhaps the best that has been done so far is with glucagon: a low resolution electron microscopal and diffraction study has been made of granules (11); X-ray analysis has defined the conformation in the crystalline state (2, 12); NMR has been used to detail the conformation and dynamics of monomers in dilute aqueous solutions (3) and in hormone micelle complexes (5), and of trimers in more concentrated aqueous solutions (4); and a range of other spectroscopic probes such as circular dichroism (13) and tryptophan fluorescence (14) have given supplementary information on the conformation in aqueous and nonaqueous solutions and in the presence of lipid micelles.

X-ray crystallography can provide the most detailed time-averaged conformational description of a polypeptide hormone in the crystalline state, given that crystals are obtained. For the larger hormones, like insulin, where the crystals contain considerable amounts of solvent, the method of multiple isomorphous replacement has been quite successful. Some of the smaller hormones, like oxytocin, produce crystals containing little solvent and provide problems at the limit of the more traditional methods of structure analysis of small molecules. Recently, the refinement of protein structures at high resolution and the definition of local disorder and molecular flexibility have been receiving increased attention from protein crystallographers. Pancreatic polypeptide (PP) has now been refined at 0.98 Å resolution, which allows a detailed analysis of the thermal vibrations. These indicate a flexible C terminus and a more ordered globular region (10). Glucagon on the other hand is relatively disordered even in the crystals with a limiting resolution of ~ 3.0 Å (12).

Spectroscopic methods provide important complementary information on the solution conformation of polypeptides, particularly with respect to the dynamics. Development of high resolution NMR techniques have allowed an extensive assignment of resonances even in molecules as complex as glucagon (3, 4). Chemical shifts and nuclear Overhauser effects have been analyzed to define proximity of side groups, and distance geometry algorithms have been used to generate conformations consistent with these data (3). The recent analysis of glucagon bound to deuterated lipid micelles is very encouraging (5) and may represent a method that can be extended to hormone receptor complexes.

For the larger hormones, circular dichroism provides an overall impression of secondary structure composition and flexibility. Calculation of optical activity from crystal atomic coordinates may permit the extraction of more detailed information from circular dichroism measurements (8). Spectroscopic methods provide particularly important data to supplement prediction of structures from amino acid sequence. Predictions based on sequence information only give indications of secondary structure, but more detailed predictions have been achieved by model building members of protein families where detailed structural information is available for at least one member. This type of approach has proved particularly useful for insulin-related growth factors where scarcity of materials has limited other experimental investigations (16).

Much of our knowledge of the tertiary structures of polypeptide hormones, especially of the larger molecules such as the growth hormones and glycoproteins, derives from studies of accessibility to chemical probes and cross-linking agents. Unfortunately, space limitation permits these approaches to be discussed only in relation to the results of spectroscopic and diffraction techniques, which provide the central theme of this review.

Although hormones are defined in terms of their endocrine origins and their action via the circulation, it is now evident that there are no distinct frontiers between them and locally acting regulatory factors, the so-called paracrine agents, neurotransmitters, and growth factors (17). Indeed, insulin may act as a growth factor by binding to somatomedin C (insulin-like growth factor) receptors; vasoactive intestinal polypeptide (VIP) is primarily a neurotransmitter in the central nervous system rather than a gut hormone; and somatostatin, while acting as a local inhibitory agent both in the pancreas and in the pituitary, also has action on distant tissues through the circulation. For these reasons we consider some polypeptide paracrine agents, growth factors, and neurotransmitters in our discussion. We avoid too much emphasis on classification due to origin—pancreatic hormones, pituitary hormones, etc—but try to discuss the molecules in terms of families related by sequence homologies and conformational preferences (17).

INSULIN-LIKE POLYPEPTIDES

Insulin Structures in Crystals

The conformation of insulin has been studied using a wide selection of techniques, the most penetrating of which have been X-ray analyses of the crystal structures of different oligomeric forms. The first successful analyses were carried out on rhombohedral porcine 2Zn insulin crystals in which insulin hexamers are assembled from three equivalent dimers (18, 19). Each dimer is coordinated to two zinc ions lying 16 Å apart on a threefold axis, and each zinc is coordinated by three equivalent B10 histidines. There are two crystallographically independent molecules (I and II) within each dimer. In the medium resolution analysis several parts of the mainchain and many sidechains were poorly defined, but these difficulties have been largely overcome by high resolution refinement at 1.5 Å resolution by Dodson et al (20, 21), at 1.1 Å resolution by Sakabe et al (23–26), and at 1.8 Å resolution by Liang et al (19). Errors in atomic positions in these structures range between 0.03 Å for many well-defined, usually buried mainchain atoms and 1.0 Å for some surface sidechains and solvent molecules. The structures of bovine (S. P. Wood and S. Bedarkar, unpublished results) and human (semisynthesized) 2Zn insulin have also been studied by medium resolution X-ray analysis, and crystals of recombinant DNA synthesized human insulin are under study (A. Cleasby, J. E. Pitts, S. P. Wood, and T. L. Blundell, unpublished results). In the presence of high chloride ion concentrations insulin crystallizes in a rhombohedral form containing four zinc ions bound to the insulin hexamer in which the differences between molecules I and II are greater than in 2Zn insulin (21, 27, 28). The structure of this form was determined initially at 2.8 Å resolution using isomorphous replacement (27) and refined at 1.5 Å resolution (21, 28). A cubic crystal form containing symmetrical insulin dimers and no zinc ions can be prepared from slightly alkaline solutions, and this structure has been determined using the methods of molecular replacement (29). Insulin from the hagfish does not bind zinc ions and produces tetragonal crystals of dimers containing a twofold axis. This structure was determined initially by isomorphous replacement to 3.1 Å resolution, and the refinement to 1.9 Å is in progress (21, 30).

A comparison of the conformations of the protomers (see Figure 1) in various crystallographic forms reveals much about the flexibility of the insulin molecule that may be relevant to its structure in solution and its biological properties (21). Each protomeric structure is a fairly compact globular structure with two essentially nonpolar surfaces that are involved in intermolecular contacts leading to the formation of dimers and hexamers. With the exception of molecule I of the 4Zn hexamers, each protomer involves right-handed helical segments A2–A8, A13–A19, and B9–B19, extended regions B1–B7 and B24–B30, and turns hinging on glycines at B8, B20, and B23. The structures of molecule II of both the 2Zn and the 4Zn hexamers and the hagfish insulin resemble each other closely (21), and this structure may therefore represent a stable conformer.

The conformational differences between molecules I and II of the 2Zn form appear to be concerted, and presumably derive from intermolecular interactions in the crystals, for the hagfish dimeric structure shows that they could be identical with perfect twofold symmetry. The different crystallographic environments of A8, A10, and B5 lead to: differences in the helices A2–A7; a rotation around the A5–A6 bond of 40°; a change of the A7–B7 conformation; and movement of the B-chain C-terminal residues, of which B25 Phe adopts a different conformation in molecule I (21, 28).

In 4Zn insulin the asymmetry between the two protomers of the dimer is greater, particularly toward the B-chain N terminus of molecule I, where the B-chain α helix starts at residue 2 rather than at residue 9 as in molecule II (21, 28). This allows B5 and B10 of adjacent dimers each to bind one zinc ion, thus generating three sites around the threefold axis in addition to the one zinc site on the threefold axis coordinated by three histidines at B10 of molecule II. In fact there is disorder in this part of the molecule, with a percentage of the histidines at B10 coordinating a partially occupied zinc ion on the threefold axis. This is consistent with the observation of transitions between the 2Zn and 4Zn insulin forms in the crystalline state (31).

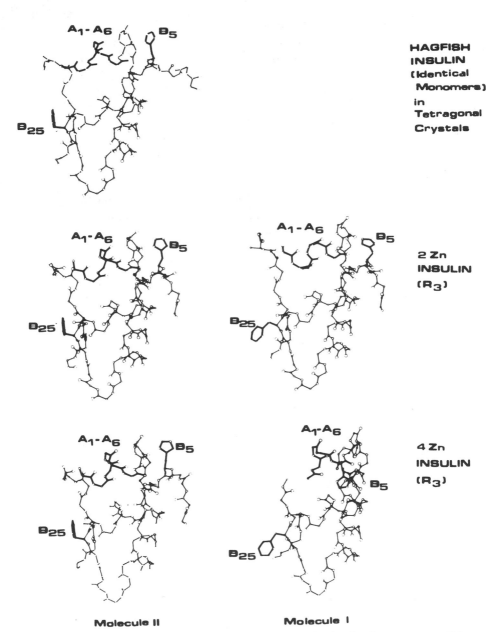

Figure 1 Comparison of five independent insulin molecules showing the arrangement of the N terminus of the A chain (A1–A6), B5 histidine, and B25 phenylalanine. Reproduced by permission of Dr. G. Dodson from (21).

The hydrophobic core of the insulin molecule in all five conformers (Figure 1) receives a rather similar contribution from B-chain sidechains; but A-chain contributions are more varied. With the exception of B25 phenylalanine, the dimer-forming residues in all the insulins are rather similar in spite of the conformational changes occurring elsewhere, and this highlights the strength and specificity of this interaction. Certain regions of the insulin structure are disordered even in the crystals. For instance, B22 Arg, a residue apparently important to structure and activity, is now seen to occupy two positions, forming a salt bridge with either A21 carboxylate as proposed earlier, or with A17 glutamate sidechain carboxylate.

The comparison of the five conformers (2Zn I and II, 4Zn I and II, and hagfish) refined structures has highlighted areas of flexibility in the insulin molecule. Although it is possible that molecule II of 2Zn insulin has a unique stability, the capacity of the insulin molecule to change its conformation may be relevant to the generation of biological activity at the receptor.

In the very high resolution (1.1 Å) refinement of the 2Zn insulin structure at 4°C by Sakabe et al (23–26), some hydrogen atoms bound to atoms with high thermal parameters could not be seen in the electron density difference maps, but many were defined by positive regions, as shown in Figure 2 for B10 His. Detailed analysis of the hydrogen bonding network of insulin has shown that in the B-chain helix the bond lengths vary between 2.8 and 3.2

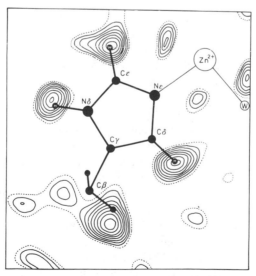

Figure 2   A difference electron density map of the imidazole ring plane on B10 histidine in molecule 1 of 2Zn insulin hexamers. The phases were calculated without contributions of the hydrogen atoms on the basis of the structure refined at 1.1 Å resolution. Reproduced by permission of Drs. N. and K. Sakabe from (25).

Å, and the rms difference between molecules of the dimer is 0.04 Å in bond length and less than 2° in the defining angles. Consistently shorter bond lengths for buried as opposed to solvent exposed hydrogen bonds indicate that the B-chain helix tends to curve against the interior of the molecule. In the β sheet of the dimer, the two interior of the four hydrogen bonds are shorter and straighter, which indicates some cooperativity in the hydrogen bonding. About 150 water molecules lie within 3.5 Å of protein atoms and make up the first hydration shell; 73% of these are hydrogen bonded to polar insulin atoms.

Insulins with reduced receptor affinities due to addition of thiazolidine (32, 33), t-butoxycarbonyl (32, 33), and glutamyl [(34); S. Bedarkar, unpublished] groups at A1 Gly have been crystallized in the 2Zn crystal form and studied by difference Fourier and refinement techniques. The addition of these groups appears to cause rotation and some disordering of the residues A2–A5, and very small movements in the adjacent B-chain residues. Larger changes appear in insulin crosslinked between A1 and B29 with diaminosuberimidate (36). Several insulins with shortened B-chain C termini have been crystallized (37–39), and crystal forms of despentapeptide insulin have been solved by molecular replacement (39), which indicates a similar general tertiary fold for this molecule, together with some movement of the B-chain N-terminal residues. This is the first monomeric insulin to be studied by X-ray techniques.

Dimeric and 2Zn hexameric forms of proinsulin have been crystallized (40). Preliminary molecular replacement studies on large crystals of the dimeric form suitable for high resolution X-ray analysis indicate a twofold axis (T. L. Blundell, G. Godley, J. E. Pitts, S. P. Wood, I. J. Tickle and G. L. Taylor, unpublished results).

Insulin Structure in Solution

Circular dichroism spectroscopy has been used widely in the study of insulin conformation in solution. However, many early interpretations have been confused by the effects of self-association in both the near UV (due to restriction of rotation of aromatic sidechains) and the far UV (where changes in β sheet and perhaps helix affect the spectra). Clearly insulins at equivalent states of association must be compared (42). Helix estimations (43) are probably made with most confidence, and for insulin these agree fairly well with the crystal structure. Strickland & Mercola (44) and more recently, Wollmer and co-workers (8, 45) have used the crystal atomic coordinates to calculate the tyrosyl circular dichroism of insulin resulting from electric dipole-electric dipole coupling employing either a distributed monopole or a dipole approximation to obtain the interaction energies. The calculations generated the correct sign for the 275 nm band observed experimentally and provided a fairly good estimate of the magnitude in different states of association. Experimentally this band is enhanced about fourfold in the hexamer compared to the monomer.

However, there are still considerable technical barriers in the study of the circular dichroism of insulin at the high dilutions characteristic of monomer formation. Recently Pocker & Biswas (9) have reported circular dichroism spectra for insulin at concentrations of 1 μM in the near UV and down to 60 nM in the far UV using an instrument in which a high frequency elasto-optic modulator is used in place of the pockel cells used in older instruments and in which the dynode voltage is artificially suppressed during recordings at extreme dilutions. On dilution to an insulin concentration of 60 nM the far UV spectrum corresponds to 28% helix content compared with 45% estimated from the crystal structure. Assuming that the concentrations are not confused by absorption of insulin on to surfaces and that the measurement techniques are valid, this may represent further evidence for flexibility in the insulin structure.

A similar change in the circular dichroism is observed in guinea pig and porcupine insulins, which form only monomers (42, 46, 47). However, this is not unexpected, as B22 Arg is replaced by a glutamic acid group in these insulins (see below). In casiragua insulin (48), which is also monomeric, the far UV circular dichroism is consistent with a structure containing a percentage of helix closer to that of the crystal structure.

NMR spectroscopy has not made major contributions to our understanding of insulin conformation in solution, probably as a result of the complexity of the dimeric and hexameric species and their low solubility at physiological pH. However, Bradbury & Brown have described NMR studies of the amino groups (49), and Williamson & Williams (50) have reported a 270 MHz proton NMR study of insulin oligomers and zinc binding. Having assigned key reference resonances of methyl groups of isoleucines, valines, and leucines and aromatic protons they were able to follow conformational changes on the assembly of 2Zn insulin hexamers and 4Zn insulin hexamers and even predict a 6Zn hexamer that has yet to be identified crystallographically. ^{13}C NMR measurement of insulin carbamylated with ^{13}C enriched potassium cyanate (51) showed two closely spaced carbamyl glycine lines in conditions where dimers were the dominant species in solution. This finding has been held to reflect the imperfect twofold axis seen in crystals, but may be due to different conformations of monomeric and dimeric forms coexisting under the conditions of the experiment.

Hystricomorph Insulins

Many insulin and insulin-like polypeptides have not been available in sufficient quantities for detailed crystal or solution studies of their conformations. However, the amino acid sequences show a conservation of features necessary to achieve an insulin-like fold in spite of often very low total homology (17), and recent developments in interactive computer graphics allow detailed prediction of their conformations.

The sequences of hystricomorph insulins show unusual substitutions that are of interest, as they lead to reduction of potency (42, 46–48). Models of guinea pig and porcupine (M. Bajaj, R. Horuk, S. P. Wood and T. L. Blundell, unpublished) and casiragua (53) and coypu (53) insulins show that porcine insulin-like dimers and zinc hexamers cannot form. Furthermore, substitution of B22 Arg by glutamate in porcupine and guinea pig insulins would introduce a negative charge in close proximity to the A21 carboxylate, the repulsion between which may lead to the difference of conformation indicated by circular dichroism. Consistent with this idea, the circular dichroism spectra of guinea pig and porcupine insulins become more porcine-insulin-like at low pH when the charge is neutralized (46).

Insulin-Like Growth Factors and Relaxins

Insulin-like growth factors (I and II) have also been modeled (see Figure 3) on the basis of their homology with insulin, which includes identical disulfide dispositions, hydrophobic cores, and glycines equivalent to B8, B20, and B23 (17, 54; S. Bedarkar, T. L. Blundell and R. E. Humbel, unpublished results). The shorter C peptides (8 and 12 residues) have high β-turn potential and are easily accommodated, connecting residues equivalent to A1 and B30 of insulin without affecting the general features of the fold. The A-chain extensions of 8 (I) and 6 (II) residues must also lie on the surface, and here again, there is probably a β turn. Surface residues give rise to extensive charge networks. The lack of B10 His, the existence of B17 Phe, and the charge at B4 Glu make it unlikely that insulin-like growth factors form zinc insulin hexamers. However, the unusual hydrophobic surface patch, especially that involving B14 Ala, B17 Phe, B18 Val, and A13 Leu of insulin-like growth factor I, would be available for interaction with a binding or carrier protein. The models show that the antigenic regions of insulin, which usually include B1–B4 and A8–A10, are quite different in the insulin-like growth factors, which explains their nonsuppressibility with anti-insulin antibodies.

Although the primary structure of pig ovary relaxin indicates a two-chain, disulfide bonded protein with little homology to insulin, the main-chain A1–20 and B6–21 can adopt a conformation identical to that of insulin, so that intrachain and interchain disulfides occupy equivalent spatial positions to those of insulin (56, 57). Residues B6 Leu and B14 Ala, which are close together in insulin, have their respective positions reversed to B6 Ala and B14 Leu in porcine relaxin. A similar effect occurs with A2

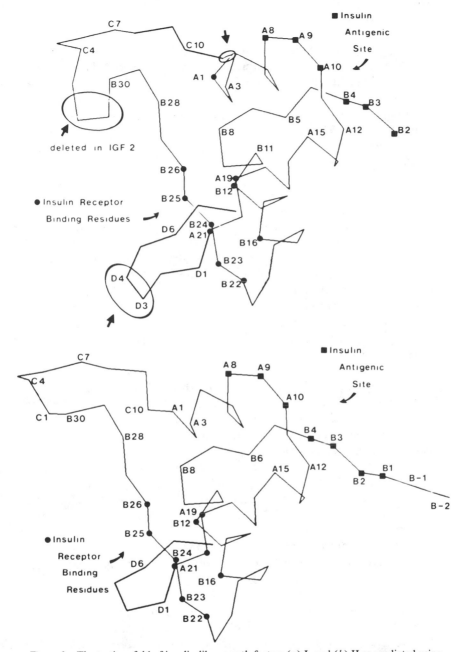

Figure 3 The tertiary fold of insulin-like growth factors (*a*) I, and (*b*) II as predicted using interactive computer graphics model building and the atomic coordinates of porcine insulin (17, 54). The α carbons only are shown. The numbering is as porcine insulin for the A- and B-chain regions. The connecting peptide residues are numbered with a prefix C and the residues at the extension of the A chain with D.

Leu and A16 Ile of relaxin. These pairs of sidechains point toward the core from opposite sides, thus maintaining the core volume. B15 Trp is almost completely buried in the core, and the tryptophan at B24 probably compensates for the lack of other aromatic residues equivalent to B25 Phe and B26 of insulin. Circular dichroism has confirmed the possibility of an insulin-like conformation for relaxin (58). The sequences of rat and sand tiger shark relaxins show numerous differences from porcine relaxin but features crucial to the insulin fold are maintained (59, 60).

Insulin Structure in Storage and at the Receptor

In the B-cell storage granule, the high concentration of insulin and the presence of zinc and calcium ions favor the presence of zinc insulin hexamers. Electron micrographs of storage granules from several species indicate crystals of many morphologies with sharp edges and regular repeats, which reflects the diversity of crystal forms grown in vitro. For instance, the rat and mouse produce two insulins of different sequence; in vitro, one crystallizes in a rhombohedral form (Rat I), while the other (Rat II), produces octahedral crystals with space group $P4_232$ with a cubic cell dimension of $\sim 67 \text{ Å}$ (61). In the electron microscope, sections of granules consistent with both forms are observed, and in some cases a 50 Å repeat (hexamer diameter) is observed. Granules of high quality from the grass snake, after fixing and sectioning, are clearly rhombic dodecahedra, and probably contain zinc insulin hexamers packed with cubic symmetry (62, 63). Such a packing could give rise to the 90 Å repeat seen in micrographs.

The exact metal ion concentrations and pH within the storage granule are not certain, and variation of these factors could have important consequences for the state of the stored protein. For instance, if the pH were around 7, then zinc insulin hexamers would bind extra zinc ions at a number of sites on the hexamer surface and possibly also at the B13 glutamate residues in the central channel, sites that have been defined by X-ray analysis of crystals. It has also been demonstrated that calcium ions can crosslink adjacent hexamers (through A4 and B30) and stabilize the crystal lattice, and this may represent a further means of rapidly removing freshly prepared insulin into the solid storage form (63).

The formation of amorphous or crystalline aggregates would be an effective form of concentration of the hormone and would ensure that hydrophobic surface regions were unavailable for interaction with the membranes of the vesicles (2, 7, 17). Aggregated forms of insulin are also less susceptible to proteolytic attack during prolonged storage. Some species, however, do not bind zinc ions or form hexamers, and in some of these animals (for instance, guinea pigs and porcupines) B22 arginine, which is a susceptible site for proteolytic degradation in the insulin monomer, has been substituted. Although most of the amino acid substitutions in guinea pig, coypu, and casiragua insulin would stabilize the monomeric form by making it more hydrophilic, they lead to a great loss of affinity for insulin receptors. When released into the circulation, the insulin granules experience a rapid dilution to $\sim 10^{-10}$ M, where the insulin monomer is the prevalent form. It therefore seems that the conformation of the insulin monomer is important for receptor binding, perhaps because some region of its surface has a capacity to recognize complementary regions on the cell surface receptor (7, 17, 18). Analysis of the hormone receptor affinity constant, the kinetics of the association and dissociation processes, and comparative studies of various hormone analogues have been employed to probe the nature of the interaction. The first proposal for the nature of the receptor binding region involved an insulin conformer similar to that defined by X-ray analysis, with receptor interactions mediated by a largely invariant surface region, part of which is involved in dimer formation (18). The model has been more clearly defined by extending the analogy of receptor binding and insulin dimerization; the latter process involves a complementarity in shape, charge, hydrophobicity, and hydrogen bonding capacity, as one might expect of the receptor interaction (32). The insulin surface region is larger than the region involved in dimer formation, which is consistent with thermodynamic analysis of the interaction (64). It includes a central hydrophobic region (B24 Phe, B25 Phe, B26 Tyr, B12 Val, B16 Tyr, and A19 Tyr) and polar side groups (A1 Gly, A4 Glu, A5 Gln, A21 Asn, B9 Ser, B10 His, B13 Glu, B21 Glu, B22 Arg, and B27 Thr) on the periphery. Even if the insulin molecule is somewhat less-ordered in solution than in the crystal, the ease of attainment of the crystal structure that occurs on dimerization could well find a parallel in the receptor docking process, and this would explain the observed importance of the tertiary structure and dimer-forming residues for full receptor affinity. The model that assumes a receptor binding region larger, but inclusive of the region involved in dimer formation, requires a decrease in receptor binding if the dimerization is reduced, but the converse is not necessarily true.

For insulin-like growth factors monomeric species are likely to bind the receptor. Part of the receptor binding region of insulin, especially that involved in dimer formation, is retained in insulin-like growth factors, but much is inaccessible due to the A-chain extension and the connecting peptide (17, 33). As this effect is less marked in insulin-like growth factor II, its stronger binding to the insulin receptor is predicted by the model. For relaxin, extensive sequence changes throughout the proposed binding region explain the absence of insulin-like biological activity. The receptor binding region of insulin-like growth factors may involve some residues in common with insulin, as insulins show a small but measurable ability to bind growth factor receptors. Certain hystricomorph rodent insulins are more potent growth factors than expected from their poor ability to bind insulin receptors (65). Work on definition of residues and conformers important for growth promoting effects will clearly be an area of intense activity in the near future.

GROWTH HORMONE FAMILY

Comparison of the amino acid sequences of growth hormone (somatotropin), prolactin, and placental lactogen (chorionic somatomammotropin) shows extensive homologies, which indicates that they can be usefully considered as members of a protein family (for review see 66). They are single chain proteins with molecular weights in the range 20,000–22,000. Growth hormone and placental lactogen have two equivalent intrachain disulfide bridges, while prolactin has a third. Study of the structure of these hormones has been limited so far to circular dichroism, fluorescence (67), sequence based predictive methods, and chemical accessibility measurements. Several crystal forms of placental lactogen have been reported (33), and recently the quality of one of these forms has been significantly improved, which indicates that a crystal structure should be available soon (68). Circular dichroism measurements suggest a high proportion of α helix (45–60%) (67) that persists in a range of conditions, which indicates stable globular structures. Prediction methods indicate rather less helix ($\sim$37%) and likely regions of β sheet and β turns (69, 70).

Enzymatic digestion experiments have been performed in efforts to identify active fragments of growth hormone. Plasmin digestion removes a hexapeptide (comprising residues 135–140) without affecting activity. Subsequent breakage of the disulfide bonds yields two peptides (comprising residues 1–134 and 141–191) that have little or no activity. However, full activity has been restored by noncovalent recombination of these fragments (71). Recently this approach has been extended to the preparation of recombinants of growth hormone and placental lactogen fragments (72) containing at least one reformed disulfide bond. The amino-terminal fragment (residues 1–134) was found to dominate the circular dichroism and biological characteristics of either recombinant that were close to the native values. However, cleavage of placental lactogen into two large fragments at the single tryptophan residue 86, while maintaining the disulfide bond of the above recombinants, led to a complete loss of activity and major structural changes (73).

Extracts of human pituitary glands contain a lower molecular weight variant of growth hormone where residues 32–46 are absent. Comparative

NMR spectroscopy, circular dichroism, spectrophotometric titration, and tyrosine nitration experiments (74) indicate that in whole growth hormone, residues 32–46 are mainly helical. If the secondary structures of the residues in common to the two molecules are identical then the difference spectrum is consistent with the contribution expected for residues 32–46 in human growth hormone from a Chou & Fasman analysis (70).

GLYCOPROTEIN HORMONES

This family of hormones comprises the gonadotrophins: lutropin, follitropin, and thyrotropin from the anterior pituitary and chorionic gonadotrophin (75). They are relatively large hormones (molecular weight 28,000–38,000) and comprise two subunits (α and β). The α chain is slightly smaller (96 residues) and is common to all the members of the group from one animal. Hormone specificity resides in the β subunit, but even here, considerable homology is observed. There are several carbohydrate attachment sites distributed over both subunits, and the sugar accounts for 16–45% of the weight of the hormones. The α subunits contain five disulfide bonds and the β subunit six, but the exact pairing of the cysteine residues is not yet clear. The noncovalent interaction between subunits is very strong, and therefore the associated form is probably present in circulation and involved in receptor binding.

Much information has been gathered on the properties of the isolated subunits that contributes to the picture we have of the conformation of these hormones (76, 77). The structural and functional integrity of the α subunit of lutropin can be regained following complete reduction and reoxidation of the disulfide bonds, but this is not the case for the β subunit. Sequential enzymatic digestion of the α subunit of lutropin yields a "core" containing all the disulfide bonds, somewhat reminiscent of snake venom α neurotoxins. Circular dichroism measurements have been reported by many workers, and there is agreement that β sheet ($\sim$30%) is the predominant feature of secondary structure. Sequence based predictions indicate a much higher β structure content ($\sim$60%) and a considerable β-turn contribution. This abundance of β sheet and turns, coupled with hydrodynamic indications of a compact structure and the high proline and disulfide content, all contribute to a general view of a globular protein. Circular dichroism studies of tyrosine exposure on subunit dissociation, IR spectroscopy on shielding of exchangeable peptide hydrogens on association, and NMR estimates of the pKa's of imidazole protons have been reported (reviewed in 76, 77), but a detailed crystal structure analysis is needed to add cohesion to these data.

PANCREATIC POLYPEPTIDE

Crystal Structure

The structure of the 36 amino acid avian (turkey) pancreatic polypeptide was determined by X-ray analysis, initially at 2.1 Å resolution using isomorphous replacement and anomalous scattering measurements. Subsequently the phase information was extended to 1.4 Å resolution, and the structure has been refined at 0.98 Å resolution, providing a detailed description of the molecule and its association mechanism (10, 78–80). The molecule comprises two well-defined secondary structures lying approximately antiparallel, their connection including two β turns (Figure 4). Residues 1–8, including prolines at 2, 5, and 8, form a polyproline-like helix, the proline rings of which all contribute toward hydrophobic interactions, interdigitating with nonpolar sidechains of an α helix formed by residues 14–32. The polyproline-like helix bends gently around the α helix in a way that optimizes contacts of Pro 2 with Val 30 and Tyr 27; of Pro 5 with Tyr 27, Leu 24, and Phe 20; and of Pro 8 with Phe 20 and Leu 17. The aromatic

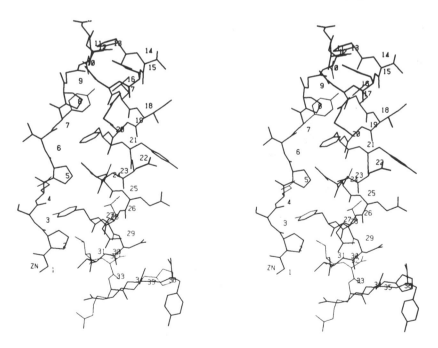

Figure 4 A stereo view of the structure of avian pancreatic polypeptide (80). Prolines 2, 5, and 8 are part of a polyproline-helix type conformation, and they interdigitate with the hydrophobic sidechains of the α helix in residues 18–32. The C terminus is flexible.

sidechains of Tyr 7 and Phe 20 produce a further hydrophobic contact. Residues 33 and 34 form a turn, orienting the less-well-defined and highly flexible C-terminal residues away from the helix axis.

An extensive hydrophobic surface region of the molecule, which involves both the helices, is in contact with a twofold symmetry-related molecule in the crystal, which defines a dimer. The axes of the two α helices are separated by about 11 Å and make an angle of about 150°, which allows sidechains of one helix to fit the grooves between sidechains of the second. The remainder of the surface of the molecule is polar.

Dimers of avian pancreatic polypeptide coordinate zinc ions in the crystal. Each zinc ion has five ligands including the N-terminal nitrogen and carbonyl oxygen of Gly 1 from one dimer, the amide of Asn 23 from a second, and Nε of the imidazole ring of His 34 from a third dimer. A water molecule is the fifth ligand in a distorted trigonal bipyramidal geometry of coordination.

Only sixteen residues of avian pancreatic polypeptide are conserved in the bovine molecule, but computer graphics model building and predictive methods show that a rather similar structure can be achieved (79). The major difficulty is experienced in the replacement of His 34 by proline. This change may be expected to alter the flexibility and conformation of the C-terminal residues. The bovine molecule would be expected to form dimers, but the substitutions at His 34 and Asn 23 of avian pancreatic polypeptide indicate that zinc-linked oligomers are unlikely. Similar conclusions apply to the human pancreatic polypeptide molecule.

Structure in Solution

Centrifuge and gel filtration, and UV spectra experiments have been reported for the avian and bovine polypeptides in solution (81–83) that indicate that pancreatic polypeptides form dimers. At neutral pH, avian pancreatic polypeptide shows a $K_D \simeq 5 \times 10^{-8}$ M; the strength of this association and its temperature dependence is consistent with the extensive nonpolar interactions indicated in the crystal dimer. The association is weaker for the bovine polypeptide; this permits a study of the near UV circular dichroism, which shows a concentration-dependent diminution between 10^{-4} and 10^{-6} M, indicative of the release of aromatic residues from restricted environments in the dimer. Tyr 7, Tyr 20, and Tyr 27 would all be expected to experience such effects on disruption of the dimer. Difference UV absorption measurements over this concentration range also show perturbation of one or more tyrosine residues. In the far UV the spectra for both polypeptides indicate a considerable amount of helix. Of particular importance is the finding that for bovine polypeptide the helix signal is maintained at high dilution where there is an appreciable monomer population.

In conclusion, it appears that in spite of the small size of this molecule, it can adopt an essentially globular structure, major elements of which are retained in the monomeric state in solution. This presumably is the species involved in binding to target tissue receptors, but aggregated states, in particular zinc linked oligomers, may be more important at sites of storage in the endocrine cell.

GLUCAGON FAMILY

Pancreatic glucagon (29 amino acids) is a member of a large family of pancreatic polypeptides found, amongst other places, in the alimentary tract and the central nervous system. This family also includes secretin (the first hormone to be discovered), vasoactive intestinal polypeptide, and glucose-dependent insulinotropic peptide (formerly known as gastric inhibitory peptide). Evidence for the existence of further homologous intestinal peptides indicates that the family may be very extensive (17).

Crystal Structure of Glucagon

Although porcine glucagon crystals are most easily obtained at high pH, where glucagon is most soluble, Sasaki et al (12) showed that these crystals undergo a phase change on lowering the pH to between 6 and 7 that involves a retention of cubic symmetry, but a shortening of the cubic cell parameter from 47.9 Å to 47.1 Å. Similar crystals are obtained by careful crystallization at pH 6.5; less-ordered crystals with a cell parameter of 48.7 Å form at lower pH (~3) (84, 85). These observations reflect the flexibility of glucagon (85).

In the cubic crystals, residues 6–28 are approximately α helical, although Gly 4 endows the N terminus with flexibility (12, 85). The helical conformer brings together the hydrophobic groups in two patches comprising Phe 6, Tyr 10, Tyr 13, and Leu 14 toward the N terminus, and Ala 19, Phe 22, Val 23, Trp 25, Leu 26, and Met 27 toward the C terminus. These hydrophobic patches are responsible for the intermolecular interactions that stabilize the cubic crystals involving a series of trimers that are not mutually exclusive, each of which contains a perfect threefold axis. Trimer 1 involves heterologous hydrophobic interactions between N- and C-terminal hydrophobic patches, while trimer 2 (Figure 5) has contacts between the C-terminal hydrophobic patches alone. Although trimer 1 involves a greater decrease of accessibility of the helical conformer (18.4%) compared with trimer 2 (14%), trimer 1 involves a greater necessary formation of helix (between residues 6 and 26) compared with trimer 2, which requires helix formation only between residues 16 and 27. Trimer 1, which involves charged interactions between Asp 9 and Asp 21 and positively charged guanidinium groups, as well as a buried tyrosyl hydroxyl group (Tyr 10), would not be expected to be stable at very low or very high pH, an observation that is consistent with the greater solubility of glucagon in an alkaline solution (85).

Solution Structure of Glucagon

In dilute aqueous solution, glucagon exists as a monomer (3). A largely unstructured and flexible chain is indicated by recent proton NMR studies on dilute aqueous solutions, which show narrow line widths and a single set of chemical shifts that, with the exception of the aromatic tyrosines, are independent of temperature and phosphate concentration (3, 4). A mainly

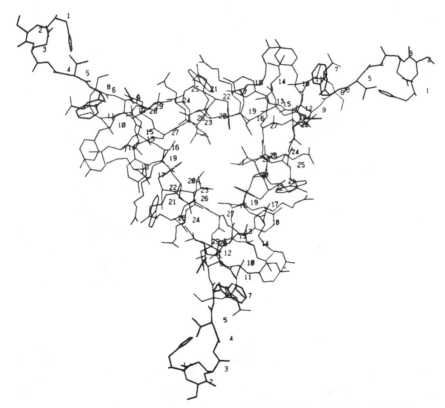

Figure 5 A glucagon trimer defined by X-ray analysis has hydrophobic contacts between C-terminal residues (85). In solution, NMR indicates that the helical region exists only in the C-terminal residues involved in the contacts.

flexible and nonspherical structure is also indicated by viscosity measurements (86), although not by the earlier results (87); the intrinsic viscosity is found to be insensitive to denaturing agents such as urea (86). Also, most NMR chemical shifts are not affected by 8 M urea and are similar to those expected for random coil, except those of Val 23 methyl resonances, Val 23 Hβ, and Leu 26 Hγ. These have been attributed (3) to a proximity between the sidechain for Val 23, with the aromatic ring of Trp 25 in about 20% of the population of conformers; a similar interaction occurs in a peptide fragment corresponding to residues 22–26; these residues cannot have a helical conformation. There may also be some structure in a percentage of the conformers involving Tyr 10 and Tyr 13, which unfold with increasing temperature (86).

The concentration dependence of most spectroscopic probes indicates that glucagon self-associates, with an accompanying change in secondary structure. Circular dichroism and optical rotary dispersion (13) indicate that conformers of 35% α-helical content are induced in concentrated solutions. Fluorescence studies of rhodamine 6G dye bound to glucagon at pH 10.6 (88), and optical detection of magnetic resonance (14) favor preferential formation of secondary structure in the C-terminal region close to the tryptophan (Trp 25). Proton NMR studies in concentrated solutions at high pH also suggest that aggregation involves C-terminal residues 22–29, and the concentration dependence of the chemical shifts can be fully accounted for by a two-state model involving a monomer-trimer equilibrium (4).

The constants for trimerization (Kα) at pH 10.6 for glucagon in D$_2$O with 0.2 M sodium phosphate, are 1.08 ($\pm$.42) $\times$ 10^6 M^{-2} at 30°C, and 5.86 ($\pm$1.07) $\times$ 10^4 M^{-2} at 50°C (4). Similar values are obtained from the concentration dependence of the circular dichroism at different temperatures (89). Analysis of the temperature dependence of the association constants (4, 89) and calorimetry (90) show that glucagon association is characterized by large negative values of both $\Delta H°$ and $\Delta S°$, whose temperature dependence decreases with increasing temperature. $\Delta Cp°$ has a large negative value, which indicates hydrophobic interactions (89). These data are consistent with a dominant negative entropy of a coil-helix transition on association that is greater than the positive entropy change resulting from hydrophobic interactions. The negative enthalpy change makes the overall free energy change favorable. The data indicate a trimeric structure similar to trimer 2 that involves hydrophobic interactions, with a well-defined structure only in the C-terminal region. This is supported by evidence from nuclear Overhauser enhancement indicating no interactions between residues widely separated in the glucagon sequence of the kind found in trimer 1 (86).

Although the evidence is clear that trimer 1 does not exist in acidic or alkaline solutions, a small percentage of such conformers may exist in neutral conditions. The increase in oligomer formation on decreasing the pH corresponds to the pK of the tyrosine, and the same tyrosine involvement in self-association is observed at lower pH. However, Wagman et al (4) find no evidence for an abnormal pK for the tyrosines, which would be expected of trimer 1.

On standing or warming acid solutions, long fibrils form; IR spectra indicate that these are antiparallel β sheet. The ability of glucagon to form both α-helical and β-sheet conformers is reflected in the prediction of secondary structure (91) that the sequence favors both α-helix and β-sheet formation, and suggests that the conformation is delicately balanced between these two conformations.

Micelle-Bound Glucagon

As glucagon is amphipathic in a helical conformation, hydrophobic surfaces favor helix formation. Indeed fluorescence and circular dichroism studies (92–94) show that chloroethanol, detergents, and surfactant micelles bind glucagon, with a concomitant formation of ordered secondary structure that is possibly α helix and probably mainly in the C-terminal region.

More recently, Wüthrich and co-workers (5) have undertaken a complete conformational analysis of glucagon bound to micelles, using high resolution NMR techniques. They show that 1 glucagon molecule binds to 40 detergent molecules (perdeuterated dodecylphosphocholine) with a well-defined and extended conformation. Electron paramagnetic resonance and NMR studies indicate that the molecule is parallel to the micelle surface (95). The combined use of proton-proton Overhauser enhancements and distance geometry algorithms using upper limits for selected proton-proton distances define possible conformations of the glucagon molecules bound to the micelle (5). Several structures meet the distance criteria for the segment Ala 19 to Met 27. Although certain distances are inconsistent with the existence of the helical conformer found in the crystals, it is possible to build a helical conformer with these interatomic distances by moving the sidechain of Met 27, which is restricted by intermolecular interactions in trimer 2, and by rotating some sidechains through small angles that are not precluded at 3.0 Å resolution by the X-ray analysis. The details of the conformer bound to the micelles will be further defined as the NMR analysis proceeds.

Secretin

The difficulties in obtaining quantities of natural secretin suitable for structural studies have meant that synthetic secretin and analogues have been used (96–101). The circular dichroism of secretin between pH 5 and pH 8 varies little and indicates perhaps a 30% α-helical conformation. In more acid solutions the ellipticity at 225 nm decreases, which indicates some loss of structure, but the helix is stabilized by lowering the temperature. Circular dichroism of synthetic analogues indicates that four residues at the N terminus do not contribute toward stabilization of the conformation. Ion pair interactions between Glu 9 and Asp 15 and the arginyl groups appear to be critical as the helical character decreases on modifying the carboxylates to carboxamides (98, 99). However, it is apparent that the helical stretch is most likely in the C-terminal half, as it is in glucagon (98). This would bring the hydrophobic residues Leu 19, Leu 22, Leu 23, Leu 26, and Val 27 into a hydrophobic patch that may interact with hydrophobic groups such as Phe 6, Leu 10, and Leu 13 in the same molecule. This structure may be stabilized by ion pair interactions. However, intermolecular interactions as in glucagon may also occur.

Glucagon in Storage Granules and at the Receptor

Glucagon is stored in granules that are usually amorphous in character. However, the glucagon granules of teleosts are crystalline rhombic dodecahedra (11, 102, 103) with cell dimensions that vary between 41 and 48 Å. Values smaller than that observed for porcine glucagon crystals bathed in their solvent of crystallization at pH 6.5 are expected as a result of the dehydration necessary for electron microscopy. These crystalline granules almost certainly have a structure similar to that of the crystals studied by X-ray analysis, and contain helical glucagon trimers. Amorphous granules may also contain trimers (2). At neutral pH, glucagon is very insoluble, and the high rate of precipitation makes crystals very difficult to obtain. Thus it is not surprising that the granules are often amorphous in mammalian A cells, even though conditions (and possibly the chemical sequence) may be conducive to crystallization in the teleost.

Amorphous or crystalline granules containing trimers are an effective way of concentrating the glucagon molecules for storage. The existence of trimers increases the thermodynamic stability and makes the hormone less available to degradation by proteolytic enzymes. Thus there is a strong parallel in the roles of zinc insulin hexamers of insulin and trimers and higher oligomers of glucagon in the storage of these hormones (17).

Glucagon storage granules, like those of insulin, become unstable at high dilutions in circulation, and circulating glucagon must exist largely as monomers with little defined secondary structure. Study of the data on receptor binding and activity of glucagon suggests that almost the entire molecule is required for full biological potency. Des-His[1]-glucagon and other glucagons modified at the N terminus are partial agonists with reduced affinity for the glucagon receptor, which suggests that this region might be involved in hormone action, whereas the major part of the molecule, including the C-terminal hydrophobic region, might be responsible for enhancing receptor affinity (2). Evidence that interaction with the receptor is entropy-driven suggests a model in which a helical conformer, stabilized by hydrophobic interactions, is induced or selected at the receptor from a population of conformers.

NEUROHYPOPHYSEAL HORMONES

Many of the recent conformational studies of neurohypophyseal hormones concern developments of the model for oxytocin, originally derived by Urry & Walter (104) on the basis of proton NMR studies in dimethyl sulfoxide. A β turn leads to a hydrogen bond between the mainchain carbonyl of Tyr 2 and the mainchain NH of Asn 5, and the 20-membered ring is closed by a right-handed disulfide bridge. The C-terminal residues form a further β turn with a hydrogen bond between the carbonyl of Cys 2 and NH of Gly 9. It was proposed that the structure is made compact by folding the C-terminal residues so that the terminal amide is close to the protonated amino terminus and the sidechain of Asn 5 hydrogen bonds to NH of Leu 8. This model has been held to be consistent with hydrogen deuterium exchange studies of oxytocin in dimethylsulfoxide, with the temperature dependence of the peptide proton chemical shifts, with the coupling constants for the NH–CH dihedral angles, and with solvent perturbation studies (105). Calculations show that this conformation is one of several low energy conformers, most of which show no preference for hydrogen bonds (106). In fact, NMR studies carried out in aqueous solutions (107) and at different pHs show that the molecule is flexible and sensitive to the environment. Alternative structures that may coexist in solution have been predicted using interactive computer graphics model building techniques (108).

NMR, circular dichroism, and laser Raman spectroscopic studies (109, 110) show that the gross conformations of the mainchains of the tocin and pressin 20-membered rings, including the disulfide bridges, appear to be similar, although in vasopressin there is the possibility of interaction between Tyr 2 and Phe 3 aromatic rings, which become more closely associated in water than in dimethylsulfoxide. The vasopressin molecule, however, appears to be more flexible than oxytocin, and the C-terminal residues are less tightly associated with the 20-membered ring structure. This is reflected in the easier solvent perturbation in 8-lysine-vasopressin than in oxytocin. Part of the flexibility appears to derive from the presence of two positive charges, the α-amino group of Cys 1 and the ϵ-amino group of Lys 8; if one of these is deleted, the conformation tightens up. Conformational changes also occur in neurohypophyseal hormones when the pH is lowered to deprotonate the α-amino group.

Earlier discussion of the relation of conformation to the biological action of neurohypophyseal hormones considered a relatively rigid structure binding to the receptor (111). Modifications were considered to be of three types. Group A included those modifications that disturbed the conformation, such as replacement of Asn 5, the sidechain of which is postulated to be involved in an intramolecular interaction with Leu 8 NH, replacement of Ile 3 with proline (112), replacement of the sulfur by selenium, or even replacement of the two half cystines by alanines. These modifications lead to a change in all bioactivities. Group B includes modifications such as at positions 3 or 8, which maintain the mainchain conformation but tend to have differential effects on either "tocin" or "pressin" receptor binding. Group C includes modifications that maintain conformation and affinity but decrease intrinsic activity leading to antagonist action, for example, in 2–O–ethyltyrosine oxytocin. More recently, Hruby and collaborators (6) have emphasized the importance of flexibility and the dynamics of receptor interactions. A ^{15}N NMR study of oxytocin in water has shown that although the tocin ring is rigid on the time scale of 10^{-10} sec, there is dynamic interconversion between conformers, with only Gln 4 and Asn 5 peptide nitrogens in relatively restricted backbone conformations (113). ^{13}C NMR (114), circular dichroism (115), and laser Raman spectroscopy (115) in aqueous solutions show that the antagonist [1-penicillamine]-oxytocin, which has two methyl groups substituted on the Cys 1 sulfur, is relatively more rigid than oxytocin itself. Thus the antagonist action may arise from the inability of the modified oxytocin molecule, once bound to the receptor, to explore those conformations that are important in eliciting a biological response.

Although flexibility of the neurohypophyseal hormone structures makes solution studies and therefore NMR studies of critical importance, some ambiguities in conformational parameters would certainly be clarified by a crystal structure analysis. Although oxytocin in various complexes, deamino oxytocin, deamino-1-seleno-oxytocin, and deamino-6-seleno-oxytocin, have all been crystallized, the X-ray analyses have proved complicated. Recently, new three-dimensional X-ray data sets have been measured in our laboratory with a view to using the sulfurs as heavy atoms, and anomalous scatterers to solve the phase problem. X-ray analysis of the crystalline C-terminal tetrapeptide Cys -Pro-Leu-Gly-NH$_2$ revealed a Gly-9-NH to Cys-6-CO hydrogen bond, as predicted in the model of Walter & Urry (104), although solutions contain a percentage of *cis* Cys-Pro peptide not found in oxytocin.

SOMATOSTATIN

The conformation of the hypothalamic tetradecapeptide, somatostatin, has been investigated using various spectroscopic techniques, but no crystallographic results have been published. The first tentative model presented by Holladay & Puett (116) was based on circular dichroism spectra and their dependence on guanidinium hydrochloride concentrations, which indicate the presence of ordered secondary structure in the monomer, and secondary structure prediction methods, which show residues 6–12 to have a high β -structure potential. The model consists of an elongated hairpin loop with several residues in antiparallel β structure. One end of the molecule is hydrophobic, and the aromatic rings partially shield the peptide-peptide hydrogen bonds. Subsequently, similar techniques were used to investigate perturbations of the proposed β turn by changes in amino acid sequence (117). Long-lived (D-Trp[8])-somatostatin showed a distinct conformational difference from somatostatin. Laser Raman spectroscopy allows estimation of disulfide torsion angles and indicates β-sheet structure, but does not distinguish between inter-and intramolecular hydrogen bonding, as good spectra were only recorded at very high peptide concentrations ($\sim$9%) (118).

NMR studies on fragments (119, 120), analogues (121), and whole somatostatin (122) have been hindered by severe line broadening, aggregation, and signal assignment problems and it is clear that a complete assignment of observed resonances to a 14 amino acid molecule is still not a trivial process. Hallenga et al (119, 120) have reported NMR spectra for peptide fragments in the region 10–13 and 9–14 and noted conformational effects that are also present in the whole hormone. The inequivalence of Thr 10 and Thr 12 resonances was attributed to a shielding of Thr 10 sidechain by the Phe 11 aromatic ring, a conformer favored in dimethylsulfoxide. Semiempirical energy calculations indicate an averaged conformation in which semiextended and folded structures are important, but only a portion of the folded forms are capable of explaining some chemical shift differences. Subsequently, two almost complete NMR assignments of somatostatin have been presented that employ a wide selection of assignment techniques (122, 123). However, there are some significant discrepancies between the assignments. Both studies suggest that somatostatin could exist in aqueous solution in a conformational equilibrium between several low energy conformations. The amides of residues 8, 9, and 10 must be hydrogen bonded or solvent shielded, and while types I and II β turns are excluded, participation of β-II turns in the conformational equilibrium is possible. In particular, a β-II turn from Phe-7-Thr-10 would permit a close approach of Lys 9 and Trp 8 sidechains, thus explaining the upfield shift of the Lys 9 γ-CH$_2$ resonances. Substitution of D-tryptophan at residue 8 should stabilize this conformation, and for this analogue a much larger upfield shift of the Lys 9 resonance is observed (121). On the basis of studies with this analogue it has been suggested that the biologically active conformation of somatostatin at its receptor is not the predominant conformer in solution, but rather a conformer favored by D-tryptophan substitution. It is known that this substitution greatly prolongs the life of this analogue in vivo, but its high potency persists in in vitro assays where degradation is minimized.

Veber and co-workers (124) have extended the study of analogues of this type by introducing covalent conformational constraints and eliminating apparently unimportant residues. The analogue cyclo(Aha-Cys-Phe-DTrp-Lys-Thr-Cys) where Aha is 7-aminoheptanoic acid, has equivalent or higher activity than somatostatin in inhibiting growth hormone, insulin, and glucagon release. The diminished molecular flexibility of this analogue has been defined by NMR measurements. Using a computer modeling system, the feasibility of replacing the Cys-Aha-Cys bridge by various dipeptides in the conformations defined for the various types of β turn was tested. Suitable cyclic hexapeptides were synthesized and bioassayed, and NMR spectra were recorded. The characteristic upfield γ-CH$_2$ protons of Lys 9 were apparent in most of the analogues, but the activity was augmented substantially by closing the ring with Phe-Pro, which provided hydrophobicity and a residue with restricted conformational preferences. The NMR spectrum of this hexapeptide indicated a high degree of molecular rigidity, suiting conformational analysis. The analogue was almost twice as potent as somatostatin in inhibiting growth hormone release and was degraded slowly. Thus nine of the amino acids of native somatostatin were replaced with a single proline residue, which favors a distinct conformer with an enhanced receptor affinity.

ENDORPHIN AND ADRENOCORTICOTROPIN FAMILIES

The sequences of ACTH, melanocyte stimulating hormone, β lipotropin and β endorphin (the C-terminal 31 residues of β lipotropin) are contained within their pituitary precursor, proopiomelanocortin. [Met^5]-enkephalin, a natural opiate of the brain, is contained in residues 61–65 of β lipotropin, although this peptide is synthesized in its own precursor. The polypeptides of this group all show great conformational flexibility in aqueous solutions; but secondary structure can be induced in the hormones and in fragments by nonaqueous solvents, lipid micelles, and salts.

The studies of Schwyzer and his colleagues (1) have some time ago established the flexibility of melanocyte stimulating hormone (MSH) and ACTH in aqueous solutions. Although α helix is not easily induced in the whole molecule, theoretical predictions and circular dichroism indicate that fragments such as [Pro^1, Ala^2, Ala^3, Val^4]-ACTH (10) can be highly helical, especially in trifluoroethanol (125).

The peptides [Leu^5]-enkephalin (Tyr Gly Gly Phe Leu) and [Met^5]-enkephalin (Tyr Gly Gly Phe Met) bind to the same receptor site as the rigid opiate compounds of the morphine and oripavine family. Their small size has allowed synthesis of many analogues and the production of large quantities for NMR, circular dichroism, Raman spectroscopy, and X-ray crystallographic analysis. The conformation is very flexible, and the existence of a preferred conformer depends critically on whether the molecule is uncharged, cationic, or zwitterionic; on the concentration; and on the solvent. ^{1}H and ^{13}C NMR and Raman spectroscopy (126–129) show that the zwitterionic form favors a type I β turn at residues Gly-3-Phe-4 of [Leu^5]-enkephalin, one of several low-energy forms indicated by conformational analysis (130–133). This interpretation has, however, been disputed. The cationic form is in an unfolded state in dimethyl sulfoxide (134). In aqueous solutions the molecule exists as an ensemble of conformers with Tyr 1 relatively rigidly fixed, although there is little evidence for a folded structure (135–138).

Attempts at crystal structure analysis have confirmed the existence of several different conformers. When crystallized from alcohol [Leu^5]-enkephalin (S. Bedarkar, T. L. Blundell, L. Hearn, B. Morgan and I. J. Tickle, unpublished results) is arranged in hydrogen bonded sheets probably comprised of head-to-tail β-sheet dimers with an approximate twofold axis (space group P2; a = 11.46, b = 15.59, c = 16.73 Å, β = 92.2°; Z = 4). The exact nature of the hydrogen bonding has yet to be defined. The [Met^5]-enkephalin forms similar sheets in a related crystal cell (space group P2; a = 11.61, b = 17.99, c = 16.52Å, β = 91.2°; Z = 4). These may correspond to the associated forms observed in solution at high concentrations (134–140). However, in aqueous ethanol, quite different crystals form, with four molecules in the asymmetric unit, each containing a distorted β turn at residues Gly-2-Gly-3 (141, 142). The four molecules are arranged in two pairs, with similar tyrosine sidechain positions. Very small conformational differences distinguish the molecules with similar tyrosine sidechain orientations (142).

The variety of conformers observed confirms the existence of molecular flexibility. Identification of the receptor bound conformer awaits study of the enkephalin-receptor complex, but some understanding has been obtained by comparing conformations with morphine and oripavine structures (143) and by conformational study of analogues with restricted flexibility, such as a cyclic enkephalin analogue in which the COOH-terminal carboxyl group of Leu 5 is cyclized to the γ-amino moeity of α, γ -diaminobutyric acid substituted in position 2 (144). This analogue prevents the realization of many of the conformational features ascribed to native enkephalin in solution or in the crystalline state, and its high in vitro activity suggests a model for the enkephalin bound at the receptor.

Circular dichroism spectra of β endorphin and β lipotropin (145–147) in water show little evidence of secondary structure, and intrinsic viscosities and sedimentation coefficients of the two polypeptides indicate that neither is compact or globular. However, methanol, trifluorethanol, dioxane, or sodium dedecylsulfate promote the formation of up to 50% α-helical structure. Similar α-helical structures are induced by aqueous solutions of various lipids including cerebroside sulfate, ganglioside GM1, phosphatidyl serine, and phosphatidic acid, but not cerebroside or phosphatidyl choline (148). The helices are broken up by Ca^{2+} ions. It is suggested that similar interactions with lipids may be of importance in allowing orientation of the NH_2-terminal pentapeptide (Met-enkephalin) for binding and activation of the receptor.

CONCLUSIONS

The interaction of a polypeptide hormone with its receptor must involve a complementarity in terms of hydrophobicity, charge distribution, and hydrogen bonding capacity of the two molecular surfaces. In this respect, compared to a flexible molecule, a more rigid hormone molecule might appear to be advantageous in that there is a smaller loss of entropy on formation of the receptor hormone complex. This must be the main advantage of the relatively stable conformers of the globular polypeptide hormones such as insulin, glycoprotein hormones, and growth hormones, and of cyclic peptides such as oxytocin. Furthermore, the high biological potency of the cyclic analogues of somatostatin and enkephalin may be due to their being held in a rigid conformation complementary to the receptor. However, studies of polypeptide hormones indicate that many smaller molecules exist as an ensemble of conformers, especially in aqueous solutions, and even the larger molecules such as insulin can undergo conformational changes on modification of the environment. Flexibility may have several advantages. First it may accelerate the hormone receptor recognition process by allowing more interactions to be explored; second, conformational changes in the receptor and the hormone may be required once the hormone is bound in order to activate the biological response. The greater flexibility of the N terminus of glucagon, which is important for full potency, and the discovery that oxytocin antagonists are often more rigid than the native hormone may be reflections of this process. Third, flexibility of the hormone may be important for its clearance from the circulation or for its deactivation by proteolysis.

However, the molecular biology of polypeptide hormones is complex; it involves biosynthesis, transport within the endoplasmic reticulum, storage, circulation, receptor binding, and clearance. The roles of conformation in these processes may be varied, reflecting different ways of optimizng through evolution their overall usefulness.

Literature Cited

1. Schwyzer, R. J. 1973. *J. Mondial Pharm.* 3:254–60
2. Blundell, T. L. 1979. *Trends Biochem. Sci.* 4:80–83
3. Boesch, C., Bundi, A., Oppliger, M., Wüthrich, K. 1978. *Eur. J. Biochem.* 91:209–14
4. Wagman, M. E., Dobson, C. M., Karplus, M. 1980. *FEBS Lett.* 119:265–70
5. Braun, W., Boesch, C., Brown, L. R., Gō, N., Wüthrich, K. 1981. *Biophys. Biochim. Acta.* In press
6. Meraldi, J. P., Hruby, V. J., Brewster, A. I. 1977. *Proc. Natl. Acad. Sci. USA* 74:1373–77
7. Blundell, T. L. 1979. *Trends Biochem. Sci.* 4:51–54
8. Wollmer, A., Strassburg, W., Hoenjet, E., Glatter, U., Fleischhauer, J., Mercola, D. A., De Graaf, R. A. G., Dodson, E. J., Dodson, G. G., Smith, D. G., Brandenburg, D., Danho, W. 1980. *Insulin: Chemistry, Structure and Function of Insulin and Related Hormones,* pp. 27–38. Berlin: Walter De Gruyter. 752 pp.
9. Pocker, Y., Biswas, S. B. 1980. *Biochemistry* 19:5043–49
10. Tickle, I. J., Glover, I. D., Pitts, J. E., Wood, S. P., Blundell, T. L. 1981. *Acta Crystallogr.* Sect 1, Subsect. 4–5 (Suppl.)
11. Lange, R. H., Kobayashi, K. 1980. *J. Ultrastruct. Res.* 72:20–26
12. Sasaki, K., Dockerill, S., Adamiak, D. A., Tickle, I. J., Blundell, T. L. 1975. *Nature* 257:751–57
13. Srere, P. A., Brooks, G. C. 1969. *Arch. Biochem. Biophys.* 129:708–10
14. Ross, J. B. A., Rousslang, K. W., Deranleau, D. A., Kwiram, A. L. 1977. *Biochemistry* 16:5398–402
15. Deleted in proof
16. Blundell, T. L., Bedarkar, S., Rinderknecht, E., Humbel, R. E. 1978. *Proc. Natl. Acad. Sci. USA* 75:180–84
17. Blundell, T. L., Humbel, R. E. 1980. *Nature* 287:781–87
18. Blundell, T. L., Dodson, G. G., Hodgkin, D. C., Mercola, D. A. 1972. *Adv. Protein Chem.* 26:279–402
19. Peking Insulin Structure Research Group. 1981. *Structural Studies on Molecules of Biological Interest,* pp. 501–8. Oxford: Clarendon Press. 610 pp.
20. Dodson, E. J., Dodson, G. G., Hodgkin, D. C., Reynolds, C. D. 1979. *Can. J. Biochem.* 57:469–79
21. Cutfield, J. F., Cutfield, S. M., Dodson, E. J., Dodson, G. G., Reynolds, C. D., Vallely, D. 1981. See Ref. 19, pp. 501–8
22. Deleted in proof
23. Sakabe, K., Sakabe, N., Sasaki, K. 1980. *Water and Metal Cations in Biological Systems,* pp. 117–27. Tokyo: Japan Scientific Societies Press
24. Sakabe, N., Sakabe, K., Sasaki, K. 1978. *Proinsulin, Insulin and C-Peptide,* pp. 73–83. Amsterdam: Excerpta Medica
25. Sakabe, N., Sakabe, K., Sasaki, K. 1981. See Ref. 19, pp. 509–26
26. Sakabe, N., Sasaki, K., Sakabe, K. 1981. *Acta Crystallogr.* 02.7–01 (Suppl.)
27. Bentley, G. A., Dodson, E. J., Dodson, G. G., Hodgkin, D. C., Mercola, D. A. 1976. *Nature* 261:166–69
28. Dodson, E. J., Dodson, G. G., Reynolds, C. D., Vallely, D. 1980. See Ref. 8, pp. 9–16
29. Dodson, E. J., Dodson, G. G., Lewitova, A., Sabesan, M. 1978. *J. Mol. Biol.* 125:387–96
30. Cutfield, J. F., Cutfield, S. M., Dodson, E. J., Dodson, G. G., Emdin, S. F., Reynolds, C. D. 1979. *J. Mol. Biol.* 132:85–100
31. Bentley, G., Dodson, G. G., Lewitova, A. 1978. *J. Mol. Biol.* 126:871–75
32. Pullen, R. A., Lindsay, D. G., Wood, S. P., Tickle, I. J., Blundell, T. L., Wollmer, A., Krail, G., Brandenburg,

D., Zahn, H., Gliemann, J., Gammeltoft, S. 1976. *Nature* 259:369–73
33. Pullen, R. A., Jenkins, J. A., Tickle, I. J., Wood, S. P., Blundell, T. L. 1975. *Mol. Cell. Biochem.* 8:5–20
34. Friesen, H-J., Brandenburg, D., Diaconescu, C., Gattner, H-G., Naithani, V. K., Nowak, J., Zahn, H., Dockerill, S., Wood, S. P., Blundell, T. L. 1977. *Proc. Am. Peptide Symp., 5th,* pp. 136–46
35. Deleted in proof
36. Dodson, G. G., Cutfield, S., Hoenjet, E., Wollmer, A., Brandenburg, D. 1980. See Ref. 8, pp. 17–26
37. Peking Insulin Structure Research Group. 1976. *Sci. Sin.* 19:358–63
38. Lu, Z., Yu, R. 1980. *Sci. Sin.* 13:1592–99
39. Bi, R. C., Cutfield, S., Dodson, E. J., Dodson, G. G. 1981. *Acta Crystallogr.* 01.3–01 (Suppl.)
40. Rosen, L. S., Fullerton, W. W., Low, B. W. 1972. *Arch. Biochem. Biophys.* 152:569–733
41. Deleted in proof
42. Wood, S. P., Blundell, T. L., Wollmer, A., Lazarus, N. R., Neville, R. W. J. 1975. *Eur. J. Biochem.* 55:531–42
43. Hennessey, J. P., Johnson, W. C. 1981. *Biochemistry* 20:1085–94
44. Strickland, E. H., Mercola, D. 1976. *Biochemistry* 15:3875–83
45. Wollmer, A., Fleischhauer, J., Strassburger, W., Thiele, H., Brandenburg, D., Dodson, G., Mercola, D. 1977. *Biophys. J.* 20:233–43
46. Horuk, R., Blundell, T. L., Lazarus, N. R., Neville, R. W. J., Stone, D., Wollmer, A. 1980. *Nature* 286:822–24
47. Horuk, R., Wood, S. P., Blundell, T. L., Lazarus, N. R., Neville, R. W. J., Raper, J. H., Wollmer, A. 1980. *Hormones and Cell Regulation,* 4:123–39. Amsterdam: Elsevier/North Holland Biomed. Press
48. Horuk, R., Wood, S. P., Blundell, T. L., Lazarus, N., Neville, R. 1980. *Actual. Chim. Therap.* 7:15–25
49. Bradbury, J. H., Brown, L. R. 1977. *Eur. J. Biochem.* 76:573–80
50. Williamson, K. L., Williams, R. J. P. 1979. *Biochemistry* 18:5966–72
51. Led, J. J., Grant, D. M., Horton, W. J., Sundby, F., Vilhelmsen, K. 1975. *J. Am. Chem. Soc.* 97:5997–6002
52. Deleted in proof
53. Blundell, T. L., Horuk, R. 1981. *Hoppe-Seylers Z. Physiol. Chem.* 362:727–37
54. Blundell, T. L., Bedarkar, S., Rinderknecht, E., Humbel, R. E. 1978. *Proc. Natl. Acad. Sci. USA* 75:180–84
55. Deleted in proof
56. Bedarkar, S., Turnell, W. G., Blundell, T. L., Schwabe, C. 1977. *Nature* 270:449–51
57. Isaacs, N., James, R., Niall, H. 1978. *Nature* 271:278–81
58. Schwabe, C., Harmon, S. J. 1978. *Biochem. Biophys. Res. Commun.* 84: 374–80
59. Gowan, L. K., Reinig, J. W., Schwabe, C., Bedarkar, S., Blundell, T. L. 1981. *FEBS Lett.* 129:80–82
60. Bedarkar, S., Blundell, T. L., Gowan, L. K., Schwabe, C. 1981. *New York Acad. Sci.* 637: In press
61. Wood, S. P., Tickle, I. J., Blundell, T. L., Wollmer, A., Steiner, D. F. 1978. *Arch. Biochem. Biophys.* 186:175–83
62. Cao, Q., Li, T., Peng, X., Zhang, Y. 1980. *Sci. Sin.* 23:1309–13
63. Pitts, J. E., Wood, S. P., Horuk, R., Bedarkar, S., Blundell, T. L. 1980. See Ref. 8, pp. 673–82
64. Waelbroeck, M., van Obberghen, E., De Meyts, P. 1979. *J. Biol. Chem.* 254: 7736–45
65. King, G. L., Kahn, R. 1981. *Nature* 292:644–46
66. Wallis, M. 1975. *Biol. Rev.* 50:35–98
67. Aloj, S., Edelhoch, H. 1971. *J. Biol. Chem.* 246:5047–52
68. Moffat, K. 1980. *Int. J. Peptide Protein Res.* 15:149–53
69. Chen, C. H., Sonenberg, M. 1977. *Biochemistry* 16:2110–18
70. Hartman, P. G., Chapman, G. E., Moss, T., Bradbury, E. M. 1977. *Eur. J. Biochem.* 88:363–71
71. Li, C. H., Bewley, T. A. 1976. *Proc. Natl. Acad. Sci. USA* 73:1476–79
72. Russell, J., Sherwood, L. M., Kowalski, K., Schneider, A. B. 1981. *J. Biol. Chem.* 256:296–300
73. Russell, J., Katzhendler, J., Kowalski, K., Schneider, A. B., Sherwood, L. M. 1981. *J. Biol. Chem.* 256:304–7
74. Chapman, G. E., Rogers, K. M., Brittain, T., Bradshaw, R. A., Bates, O. J., Turner, C., Cary, P. D., Crane-Robinson, C. 1981. *J. Biol. Chem.* 256:2395–2401
75. Pierce, J. G., Parsons, T. F. 1981. *Ann. Rev. Biochem.* 50:465–95
76. Guidice, L. C., Pierce, J. G. 1978. *Structure and Function of Gonadotrophins,* Ch. 4. New York: Plenum
77. Garnier, J. 1978. See Ref. 76, Ch. 17
78. Wood, S. P., Pitts, J. E., Blundell, T. L., Tickle, I. J., Jenkins, J. A. 1977. *Eur. J. Biochem.* 78:119–26
79. Pitts, J. E., Blundell, T. L., Tickle, I. J., Wood, S. P. 1979. *Proc. Am. Peptide Symp., 6th,* pp. 1011–16

80. Blundell, T. L., Pitts, J. E., Tickle, I. J., Wood, S. P., Wu, C-W. 1981. *Proc. Natl. Acad. Sci. USA* 78:4175–79
81. Noelken, M. E., Chang, P. J., Kimmel, J. R. 1980. *Biochemistry* 19:1838–43
82. Chang, P. J., Noelken, M. E., Kimmel, J. R. 1980. *Biochemistry* 19:1844–49
83. Pitts, J. E. 1980. *Structures and function of pancreatic polypeptide hormones.* PhD thesis. Univ. Sussex, UK. 210 pp.
84. Blundell, T. L., Dockerill, S., Pitts, J. E., Wood, S. P., Tickle, I. J. 1978. *Membrane Proteins,* pp. 249–57. Oxford: Pergamon
85. Blundell, T. L. 1981. *Glucagon.* Berlin: Springer. In press
86. Wagman, M. 1981. *Proton NMR studies of glucagon association in solution.* PhD thesis. Harvard Univ. Cambridge, Mass. 403 pp.
87. Epand, R. M. 1971. *Can. J. Biochem.* 49:166–69
88. Formisano, S., Johnson, M. L., Edelhoch, H. 1978. *Biochemistry* 17:1468–73
89. Formisano, S., Johnson, M. L., Edelhoch, H. 1978. *Proc. Natl. Acad. Sci. USA* 74:3340–44
90. Johnson, R. E., Hruby, V. J., Rupley, J. A. 1979. *Biochemistry* 18:1176–79
91. Chou, P. Y., Fasman, G. D. 1975. *Biochemistry* 14:2536–41
92. Gratzer, W. B., Beaven, G. H., Rattle, H. W. E., Bradbury, E. M. 1968. *Eur. J. Biochem.* 3:276–83
93. Epand, R. M., Jones, A. J. S., Sayer, B. 1977. *Biochemistry* 16:4360–68
94. Wu, C. S. C., Yang, J. T. 1980. *Biochemistry* 19:2117–22
95. Brown, L. R., Boesch, C., Wüthrich, K. 1981. *Biochim. Biophys. Acta.* In press
96. Van Zon, A., Beyerman, H. C. 1976. *Helv. Chim. Acta* 59:1112–19
97. Jäger, E., Filippi, B., Knof, S., Lehnert, P., Moroder, L., Wünsh, E. 1979. *Hormone Receptors in Digestion and Nutrition,* pp. 25–35 Amsterdam: North-Holland
98. Bodanszky, M., Fink, M. L. 1976. *Bioorg. Chem.* 5(3):275–82
99. Bodanszky, M., Fink, M. L., Funk, K. W., Said, S. I. 1976. *Clin. Endocrinol.* 195S–205S (Suppl.)
100. Bodanszky, M. 1976. *J. Am. Chem. Soc.* 98:974–77
101. Yanahara, N., Yanahara, C., Kubota, M., Sakagami, M., Otsuki, M., Baba, S., Shiga, M. 1979. *Peptides,* pp. 539–42. Rockford, Ill: Pierce Chem. Co.
102. Lange, R. H. 1976. *Endocrine Gut and Pancreas,* pp. 167–78. Amsterdam: Elsevier/North-Holland Biomed. Press
103. Lange, R. H. 1979. *Eur. J. Cell. Biol.* 20:71–75
104. Urry, D. W., Walter, R. 1971. *Proc. Natl. Acad. Sci. USA* 68:956–58
105. Walter, R., Glickson, J. D. 1973. *Proc. Natl. Acad. Sci. USA* 70:1199–1203
106. Kotelchuck, D., Scheraga, H. A., Walter, R. 1972. *Proc. Natl. Acad. Sci. USA* 69:3629–33
107. Brewster, A. I. R., Hruby, V. J. 1973. *Proc. Natl. Acad. Sci. USA* 70:3806–9
108. Honig, B., Kabat, E. A., Katz, L., Levinthal, C., Wu, T. T. 1973. *J. Mol. Biol.* 80:277–95
109. Wyssbrod, H. R., Fischman, A. J., Live, D. H., Hruby, V. J., Argawal, N. W., Upson, D. A. 1979. *J. Am. Chem. Soc.* 101:4037–43
110. Tu, A. T., Lee, J., Deb, K. K., Hruby, V. J. 1979. *J. Biol. Chem.* 254:3272–78
111. Walter, R., Schwartz, I. L., Darnell, J. H., Urry, D. W. 1971. *Proc. Natl. Acad. Sci. USA* 68:1355–59
112. Deslauriers, R., Smith, I. C. P., Levy, G. C., Orlowski, R., Walter, R. 1978. *J. Am. Chem. Soc.* 109:3912–17
113. Live, D. H., Wyssbrod, H. R., Fischman, A. J., Agosta, W. C., Bradley, C. H., Cowburn, D. 1979. *J. Am. Chem. Soc.* 101:474–79
114. Hruby, V. J., Deb, K. K., Spatola, A. F., Upson, D. A., Yamamoto, D. 1979. *J. Am. Chem. Soc.* 101:202–12
115. Hruby, V. J., Deb, K. K., Fox, J., Bjornason, J., Tu, A. T. 1978. *J. Biol. Chem.* 253:6060–67
116. Holladay, L. A., Puett, D. 1976. *Proc. Natl. Acad. Sci. USA* 73:1199–1202
117. Holladay, L. A., Rivier, J., Puett, D. 1977. *Biochemistry* 16:4895–900
118. Han, S. L., Rivier, J. E., Scheraga, H. A. 1980. *Int. J. Peptide Protein Res.* 15:355–64
119. Knappenberg, M., Brison, J., Dirkx, J., Hallenga, K., Deschrijver, P., van Binst, G. 1979. *Biochim. Biophys. Acta* 580:266–76
120. Hallenga, K., van Binst, G., Knappenberg, M., Brison, J., Michel, A., Dirkx, J. 1979. *Biochim. Biophys. Acta* 577:82–101
121. Arison, B. H., Hirschmann, R., Veber, D. F. 1978. *Bioorg. Chem.* 7(5):447–51
122. Buffington, L., Garsky, V., Massiot, G., Rivier, D., Gibbons, W. A. 1980. *Biophys. Res. Commun.* 93(2):376–84
123. Hallenga, K., van Binst, G., Scarso, A., Michel, A., Knappenberg, M., Dremier, C., Brison, J., Dirkx, J. 1980. *FEBS Lett.* 119:47–52
124. Veber, D. F., Freidinger, R. M., Perlow, D. S., Paleveda, W. J. Jr., Holly, F.

W., Strachan, R. G., Nutt, R. F., Arison, B. H., Homnick, c., Randall, W. C., Glitzer, M. S., Saperstein, R., Hirschmann, R. 1981. *Nature* 292: 55–56

125. Mutter, H., Mutter, M., Bayer, E. 1979. *Z. Naturforsch. Teil. B* 34:874–85
126. Stimpson, E. R., Meinwald, Y. C., Scheraga, H. A. 1979. *Biochemistry* 18:1661–71
127. Khaled, M. A., Urry, D. W., Bradley, R. J. 1979. *J. Chem. Soc. London Perkin Trans. II* 12:1693–99
128. Niccolai, N., Garsky, V., Gibbons, W. A. 1980. *J. Am. Chem. Soc.* 102: 1517–20
129. Jones, C. R., Garsky, V., Gibbons, W. A. 1977. *Biochem. Biophys. Res. Commun.* 76:619–25
130. Isogai, Y., Nemethy, G., Scheraga, H. A. 1977. *Proc. Natl. Acad. Sci. USA* 74:414–18
131. Momany, F. A. 1977. *Biochem. Biophys. Res. Commun.* 75:1098–1103
132. DeCoen, J. L., Humblet, C., Koch, M. H. J. 1977. *FEBS Lett.* 73:38–42
133. Balodis, Y. Y., Nikifarovich, G. V., Grinsteine, I. V., Vegner, R. E., Chipens, G. I. 1978. *FEBS Lett.* 86:239–42
134. Higashijima, T., Kobayashi, J., Nagai, U., Miyazawa, T. 1979. *Eur. J. Biochem.* 97:43–57
135. Bleich, H. E., Day, A. R., Freer, R. J., Glasel, J. A. 1979. *Biochem. Biophys. Res. Commun.* 87:1146–53

136. Fischman, A. J., Riemen, M. W., Cowburn, D. 1978. *FEBS Lett.* 94:236–40
137. Kobayashi, J., Higashijima, T., Nagai, U., Miyazawa, T. 1980. *Biochim. Biophys. Acta* 621:190–203
138. Kobayashi, J., Nagai, U., Higashijima, T., Miyazawa, T. 1979. *Biochim. Biophys. Acta* 577:195–206
139. Deleted in proof
140. Khaled, M. A., Long, M. M., Thompson, W. D., Bradley, R. J., Brown, G. B., Urry, D. W. 1977. *Biochem. Biophys. Res. Commun.* 76:224–29
141. Smith, G. D., Griffin, J. F. 1978. *Science* 199:1214–16
142. Blundell, T. L., Hearn, L., Tickle, I. J., Palmer, R. A., Morgan, B. A., Smith, G. D., Griffin, J. F. 1979. *Science* 205:220
143. Gorin, F. A., Balasubramanian, T. M., Barry, C. D., Marshall, G. R. 1978. *J. Supramol. Struct.* 9:27–39
144. Di Maio, J., Schiller, P. W. 1980. *Proc. Natl. Acad. Sci. USA* 77:7162–66
145. St.-Pierre, S., Gilardeau, C., Chretien, M. 1976. *Can J. Biochem.* 54:992–98
146. Yang, J. T., Brewley, T. A., Chen, G. C., Li, C. H. 1977. *Proc. Natl. Acad. Sci. USA* 74:3235–38
147. Hollosi, M., Kajtar, M., Grat, L. 1977. *FEBS Lett.* 74:185–88
148. Wu, C. S. C., Lee, N. M., Loh, H. H., Yang, J. T., Li, C. H. 1979. *Proc. Natl. Acad. Sci. USA* 76:3656–59

Ann. Rev. Biochem. 1982. 51:155–183

COMPONENTS OF BACTERIAL RIBOSOMES,

H. G. Wittmann[1]

Max-Planck-Institut für Molekulare Genetik, D-1000 Berlin-Dahlem, Germany

CONTENTS

PERSPECTIVES AND SUMMARY

Research on the structure and function of ribosomes is being undertaken because: (*a*) Ribosomes play an essential role in the biosynthesis of proteins, and an understanding of this complicated process at the molecular level is not possible without a detailed knowledge of the ribosomal structure; (*b*) Ribosomes are the only cell organelles that are present in all organisms, therefore their protein and RNA components are excellent subjects for evolutionary studies; (*c*) The hundreds of mutants with altered or even

[1]I thank Drs. R. Brimacombe, J. Dijk, V. A. Erdmann, J. Littlechild, K. H. Nierhaus, P. Wills, B. Wittmann-Liebold, and A. Yonath for reading the manuscript and helpful comments.

155

0066-4154/82/0701-0155$02.00

missing ribosomal proteins that have been isolated and characterized are of great help in genetic, structural, and functional studies, (*d*) The size and complexity of the ribosomes (50–60 components in prokaryotes and 70–80 components in eukaryotes) constitute a challenge, not only to isolate and study the chemical and physical properties of the individual proteins and RNA components, but also to understand at the molecular level the complicated processes that take place on the ribosome. These processes include specific RNA-protein recognition, assembly, conformational changes, and interaction with mRNA, tRNAs, and factors. On the other hand ribosomes, in contrast to many other multicomponent systems or organelles, are not so complex that a successful attack on their structure and function at the molecular level is out of the range of currently available methodology. In summary, ribosomes provide an excellent system for investigating phenomena of general biochemical interest.

Studies on ribosomes have mainly concentrated on: (*a*) isolation and characterization of the numerous ribosomal components; (*b*) elucidation of the ribosomal architecture; and (*c*) investigation of ribosomal functions, i.e. protein biosynthesis, at the molecular level. Only *a* has so far been realized to any extent: The ribosomal proteins of two bacteria (*Escherichia coli* and *Bacillus stearothermophilus*) and of rat liver have been isolated and characterized. The primary structures of all 53 ribosomal proteins and the three RNAs from *E. coli* are now known, and reliable models for the secondary structure of ribosomal RNAs are available. Rapid progress has also been made in investigating the secondary and tertiary structure of the ribosomal proteins from the two bacteria above, culminating in the recent crystallization and X-ray structure analysis of several individual proteins.

This review describes the primary, secondary, and tertiary structure of isolated ribosomal components. Further, it covers the specific binding of ribosomal proteins to homologous and heterologous rRNAs and deals with evolutionary aspects by comparing the structure of ribosomal proteins and RNAs from different organisms. Unfortunately, a summary of the enormous progress made in elucidating ribosomal architecture and function at a molecular level cannot be included here for reasons of space.

RIBOSOMAL PROTEINS

Primary Structure

The complete amino acid sequences of all 53 different proteins from the *E. coli* ribosome (Table 1) have been elucidated mainly by Wittmann-Liebold and her co-workers during the last decade. Since the previous compilations of the amino acid sequence data for the *E. coli* ribosomal proteins (1–3) the primary structures of the following proteins have been determined: protein S1 (4), S2 (5), S10 (6), S11 (7), S14 (M. Yaguchi and H. G. Wittmann, unpublished data), L2 (8), L4 (9), L9 (R. Kamp and B. Wittmann-Liebold, unpublished data), L22 (10), L29 (11), and EF-Tu (12, 13).

According to the sequence data, the number of amino acids contained in the 21 proteins of the 30S subunit from *E. coli* strain K12 is 3108 with a total molecular weight of 350,000. The 50S subunit correspondingly contains 32 proteins with 4228 amino acids and has a molecular weight of 460,000 (taking into account that proteins L7/L12 are present in four copies per 50S subunit). Adding the molecular weights of the 30S proteins to that of the 16S RNA (14, 15) a chemical molecular weight of 850,000 for the 30S subunit is obtained . A similar calculation for the 50S subunit, summing

Table 1 Number of amino acid residues and molecular weights of ribosomal (and related) proteins from *E. coli*

Protein	Residues	Mol wt	Protein	Residues	Mol wt
S1	557	61,159	L10	165	17,737
S2	240	26,613	L11	141	14,874
S3	232	25,852	L12	120	12,178
S4	203	23,137	L13	142	16,019
S5	166	17,515	L14	123	13,541
S6	135	15,704	L15	144	14,981
S7K	177	19,732	L16	136	15,296
S7B	153	17,131	L17	127	14,364
S8	129	13,996	L18	117	12,770
S9	128	14,569	L19	114	13,002
S10	103	11,736	L20	117	13,366
S11	128	13,728	L21	103	11,565
S12	123	13,606	L22	110	12,227
S13	117	12,968	L23	99	11,013
S14	97	11,063	L24	103	11,185
S15	87	10,001	L25	94	10,694
S16	82	9,191	L26 = S20	86	9,553
S17	83	9,573	L27	84	8,993
S18	74	8,896	L28	77	8,875
S19	91	10,299	L29	63	7,274
S20	86	9,553	L30	58	6,411
S21	70	8,369	L31	62	6,971
L1	233	24,599	L32	56	6,315
L2	269	29,416	L33	54	6,255
L3	209	22,258	L34	46	5,381
L4	201	22,087	IF-1	71	8,119
L5	178	20,171	IF-3	181	20,695
L6	176	18,832	EF-Tu	393	43,225
L7	120	12,220	NS1	90	9,226
L9	147	15,531	NS2	90	9,535

the molecular weights of the proteins, the 23S RNA (16), and the 5S RNA (17) gives a chemical molecular weight of 1.45×10^6 for the 50S particle. The sum of the molecular weights for the two subunits yields a chemical molecular weight of 2.3×10^6 for the 70S *E. coli* ribosome. The values of the molecular weights for the 30S, 50S, and 70S particles determined by sequence analyses are considerably lower than the corresponding values determined by physical studies, which are $0.9-1.0 \times 10^6$ for the 30S, $1.6-1.8 \times 10^6$ for the 50S, and $2.6-2.9 \times 10^6$ for the 70S particles (reviewed in 18). However, the calculation of the chemical molecular weights does not include the presence of salt ions, e.g. Mg^{2+} and K^+, and of spermidine in the ribosome. If these are included, the chemical and the physical molecular weights of the ribosomal particles agree within an experimental error of 10%.

Most of the *E. coli* ribosomal proteins are rich in basic amino acids; e.g. protein L33 contains 22 mole percent lysine and L34 has 24 percent arginine. Proteins L7/L12, S20, and L10 are rich in alanine (23%, 22%, and 20%, respectively), protein L29 in leucine (19%), proteins S6 and L7/L12 in glutamic acid (15% and 13%, respectively), and proteins S17 and L23 in valine (15% and 14%). Many ribosomal proteins contain few aromatic residues; e.g. the sum of phenylalanine, tyrosine, and tryptophan is less than 2 mole percent in proteins S10, S13, L7/L12, L22, L29, and L30.

Modified amino acids have been identified in nine *E. coli* ribosomal proteins; e.g. acetylation of the N terminus in proteins S5, S18, and L7 and methylation in proteins S11, L3, L7/L12, L11, L16, and L33. The most heavily methylated protein is L11, which contains nine methyl groups at three positions of the protein chain [see (2) for details]. The modifications occur after translation, and the addition of an acetyl group to the same amino acid (alanine) at the same position (the N terminus) of proteins S5 and S18 is catalyzed by different enzymes (19). On the other hand, the methylation of different amino acids at positions 1,3, and 39 of protein L11 is probably by the same enzyme (20, 21).

In addition to the primary structure of all the *E. coli* ribosomal proteins (Table 1), the complete amino acid sequences of 15 ribosomal and 2 related proteins from 9 other organisms have also been determined (Table 2). Furthermore, partial sequences are available for many other ribosomal proteins (3, 27–32). In particular, the N-terminal regions are known, since these can be determined by automated sequence analysis of the intact proteins in a sequenator. The results from the sequence analyses of ribosomal proteins from different organisms can be summarized as follows:

1. Complete or partial primary structures are now available for many basic and acidic ribosomal proteins from *B. stearothermophilus* and *B. subtilis,* and can be compared to those from *E. coli.* In most cases there is

Table 2 Completely sequenced ribosomal (and related) proteins from organisms other than *E. coli*

Organisms	Protein	aa-number	Ref	Correspondence to *E. coli*
Bacillus stearothermophilus	BL10	177	22	EL6
B. stearothermophilus	BL17	147	23	EL9
B. stearothermophilus	BL29	104	a	EL24
B. stearothermophilus	BL34	62	b	EL30
B. stearothermophilus	IF-3	183	b	IF-3
B. stearothermophilus	BSb	90	b	NS2
B. subtilis	BL9	122	24	L7/L12
Micrococcus lysodeikticus	MA1	118	25	L7/L12
M. lysodeikticus	MA3	128	c	
Streptomyces griseus	A	126	c	L7/L12
MRCC 11227	A	122	d	L7/L12
Halobacterium cutirubrum	HL20	112	e	L7/L12
Saccharomyces cerevisiae	YP-A1	110	c	L7/L12
S. cerevisiae	YP44	103	26	
S. cerevisiae	YP55	88	f	
Artemia salina	eL12	111	27	L7/L12
Rat (liver)	P3	104	g	L7/L12

[a] K. Ashman and M. Kimura, unpublished.
[b] M. Kimura and K. Appelt, unpublished.
[c] T. Itoh, unpublished.
[d] P. Falkenberg, M. Yaguchi, C. Roy, and A. T. Matheson, unpublished.
[e] G. Oda, M. Yaguchi, C. Roy, L. P. Visentin, A. T. Matheson, and P. Falkenberg, unpublished.
[f] T. Itoh and B. Wittmann-Liebold, unpublished.
[g] A. Lin, B. Wittmann-Liebold, and I. G. Wool, unpublished.

a high degree of structural homology. This is not only true for the comparison between the ribosomal proteins from *B. subtilis* and *B. stearothermophilus,* but also for comparisons between proteins from the Gram-negative *E. coli* on the one hand and the two Gram-positive *Bacillus* species on the other. Based on the partial and especially on the complete primary structures, there is little difficulty in correlating those proteins from *E. coli* and *Bacillus* ribosomes that are homologous (22, 24, 31; see also Table 2).

2. The "A-proteins," which are acidic ribosomal proteins corresponding to proteins L7/L12 from the *E. coli* ribosome, can be isolated relatively easily. In consequence, A proteins have already partially or completely been sequenced from bacteria, plants, yeast, and animals [Table 2; summarized in (28 and 33)]. There is a strong homology not only among the A-proteins from eubacteria (24, 28) but also between proteins L7/L12 from *E. coli* and an acidic ribosomal protein from spinach chloroplasts (M. Bartsch, M. Kimura, A. R. Subramanian, unpublished information). On the other hand there is only very weak, if any, homology between the amino acid sequences

of A-proteins from eubacteria and eukaryotes, whereas strong homology has been found between cytoplasmic ribosomal proteins isolated from eukaryotic species, such as yeast, brine shrimp, wheat, and rat (3, 33).

3. Based mainly on comparative studies on ribosomal RNAs from various organisms Woese & Fox (34) postulated three kingdoms: eubacteria, archaebacteria, and eukaryotes. To the archaebacteria belong the extreme halophiles, e.g. *Halobacterium cutirubrum,* the methanogens, e.g. *Methanobacterium thermoautotrophicum,* as well as other bacteria, such as *Sulfolobus* and *Thermoplasma.* When the A-proteins from *Halobacterium* and *Methanobacterium* were isolated and partially sequenced it was found that they show a strong homology (56%) to each other (35). Their primary structures are distinct from the A-proteins of eubacteria, particularly from Gram-positive bacteria, and they are more closely related to eukaryotic A-proteins. For instance the degree of homology (about 40%) between the A-proteins from *Methanobacterium* and yeast is about the same as between the two eukaryotes *Artemia* and yeast. These and other data suggest that at least some ribosomal proteins of eukaryotes may have evolved from archaebacteria (28). It is proposed that a primitive progenote separated initially into eubacteria and a common ancestor of archaebacteria and eukaryotes, which later separated into the two kingdoms (28, 33, 36).

Secondary Structure

Two different approaches have been used to gain information concerning the secondary structure of ribosomal proteins, namely prediction of the secondary structures based on the known amino acid sequences of the proteins, and measurement of the circular dichroism (CD) of individual proteins in solution. The occurrence of conformational domains (α-helix, extended structure, turn, random coil) along the protein chain has been predicted by four methods for all the *E. coli* ribosomal proteins whose primary structure is known (2, 37–39). Secondary structure predictions have also been made for a few ribosomal proteins, such as S4 (40), S15 (41), and L7/L12 (42, 43). It remains however an open question as to how far the prediction rules that are derived from studies with other proteins, e.g. enzymes, can be applied to the ribosomal proteins which interact not only with themselves, but also with RNA chains within the ribosomal particles.

A preliminary answer can be given by comparing the predicted secondary structure of proteins L7/L12 with the structure that has been derived for the C-terminal fragment of L7/L12 by X-ray analysis (44). In this case there is good agreement for β-turns, while the prediction for the β-sheets is inconsistent with the structure obtained from the X-ray study. It remains to be seen whether this disagreement is an isolated case for specific reasons or whether it points to a more general phenomenon. For many ribosomal proteins the predicted β-structure values are in general lower than those obtained from CD-studies (45; M. Dzionara and B. Wittmann-Liebold, unpublished) whereas there is a much better agreement for the α-helical regions.

Circular dichroism and optical rotatory dispersion measurements have been performed both with mixtures of proteins extracted from the ribosome or its subunits, and later with several individual ribosomal proteins in solution. Since these studies were reviewed (2) an extensive CD investigation has been made on 17 proteins from the small subunit and 18 proteins from the large subunit of the *E. coli* ribosome (45). These proteins have been isolated by mild procedures using salt extraction and column chromatography without urea (46), and the CD-spectra were analyzed by a new procedure that is expected to give more reliable results, especially for β-sheets. According to this study most of the 30S proteins are rich in α-helix and contain a small amount of β-sheet, whereas the proteins from the large subunit are more diverse, especially in their α-helix content; some of them have a low content of α-helix, and almost all the 50S proteins are relatively rich in β-sheet. It was concluded that the proteins isolated by the mild procedure mentioned above have an unique and well-defined secondary structure in solution. By comparison of the CD-spectra from this study (45) with those obtained for proteins S4 (40, 47), S7 (48), and S15 (41), which were isolated in the presence of urea at pH 7, it can be concluded that under appropriate conditions (e.g. renaturation) a secondary structure can be obtained with the urea-treated proteins that is similar to that of salt-extracted proteins isolated in the absence of urea.

Tertiary Structure

In order to obtain information about the tertiary structure of individual ribosomal proteins, several physical and chemical methods have been used, namely limited proteolysis, proton magnetic resonance, crystallization of proteins followed by X-ray analysis, fluorescence spectroscopy (49–51), and microcalorimetry (52–54). Little progress has been reported using the latter two techniques since the time of a recent review (2), and the following discussion is therefore confined to the first three methods.

LIMITED PROTEOLYSIS   Proteins isolated by salt extraction and chromatograpy in the absence of urea (46) were treated with proteolytic enzymes at 0°C. They can be divided into four groups, according to their resistance towards proteolytic attack (55): (*a*) Proteins S15, S16, S17, and L30 are completely resistant to various proteases under the conditions applied. All four of these proteins show proton magnetic resonance (PMR) spectra with pronounced structure. (*b*) Proteins S1, S3, S4, S14, S20, S21,

L1, L11, L16, L17, L19, L25, and L28 are degraded to large fragments by the proteolytic treatment. The size of the fragments corresponds to 65–90% of the intact protein and is in general independent of the protease used, which shows that it is the protein structure and not the enzyme that governs the digestion pattern. (c) Proteins S5, S7, S8, L3, L6, L23, L24, and L32 produce relatively small fragments (25–60% of the protein length) by the protease treatment. (d) Proteins S2, S6, S9, L2, L27, L29, and L33 are completely degraded by the proteases and therefore seem to have an open and unfolded structure, which makes them highly susceptible to proteolytic attack. However, one of these proteins, namely L29, has been crystallized (56), and most proteins in this group also have a high content of α-helix or β-sheet (45), which indicates that they possess a well-defined secondary structure. The reasons for this apparent discrepancy are not yet established.

Fragments arising from proteolytic treatment of the proteins belonging to the second and third groups can be isolated and used for further structural and functional studies. Additional fragments can also be obtained by chemical cleavage of a protein chain with reagents such as CNBr, which causes a specific cleavage after methionine residues. In this way several fragments from protein S1 have been isolated (57), and the biological functions of these and other enzymatically derived fragments have been elucidated (57–59). It has been shown that the N-terminal region (positions 1–193) of protein S1, which consists of 557 amino acid residues (4), corresponds to the ribosome binding domain, and the middle region (224–309) to the nucleic acid binding domain of S1. The C-terminal region, containing about 150 amino acids, seems to have no important function, since *E. coli* mutants with an S1 molecule lacking this region (60) are able to grow almost normally.

In addition to protein S1, enzymatically or chemically derived fragments have also been isolated from other proteins, namely S4 (61–63), S8 (64; K. Paterakis and J. Littlechild, unpublished data), S20 (K. Paterakis, P. Woolley, and J. Littlechild, unpublished data), L7/L12 (54, 65, 66), L11 (2, 67), and L18 (68). The fragments were studied for their capacity to bind to ribosomal RNAs (61, 64, 68) and to ribosomal cores (65, 66), or used for PMR studies (63, 67). These studies will be discussed below in the appropriate chapters.

PROTON MAGNETIC RESONANCE (PMR)   PMR spectroscopy, especially with instruments of high resolution, is a powerful technique for investigating the tertiary structure of protein molecules. Because of the restricted immobilization of amino acid side chains, the resonances in the specta of proteins with a well-defined tertiary structure are generally broadened as opposed to denatured proteins. The interaction between apolar methyl groups of hydrophobic amino acid side chains and aromatic amino acids, especially tryptophan, can be revealed by the appearance of ring current–shifted resonances in the high field region of the PMR spectrum (0–1 ppm) and by perturbed aromatic resonances in the 6–9 ppm region. Furthermore, information about the environment of histidine residues in the tertiary structure of the protein molecule can be obtained.

From an extensive PMR study (55) on 15 proteins from the small and 18 proteins from the large subunit of *E. coli,* isolated without urea and lyophilization (46), the following results have been obtained: (*a*) A number of proteins (S4, S7, S8, S15, S16, S17, L6, L11, L16, L19, L25, L28, and L30) show ring current–shifted methyl resonances in the 0–1 ppm region and perturbed aromatic spectra in the 6–9 ppm region, indicating a well-folded tertiary structure for these proteins. Among them are also the four proteins (S15, S16, S17, and L30) that are completely resistant to proteolytic attack under mild conditions (see above). (*b*) On the other hand, the PMR spectra of seven proteins (S5, S6, S20, L2, L27, L29, and L33) gave no evidence for a compact folding of their protein chains. Five of these (S6, L2, L27, L29, and L33) are completely degraded by mild treatment with proteases. This would indicate an unfolded structure, although L29 has been crystallized (56). Several proteins that show very pronounced shifted resonances, e.g. S16, L11, and L30, are good candidates for a further detailed structural analysis.

PMR studies on ribosomal proteins L11 (67, 69) and L25 (70), isolated in the presence of urea, show that these proteins can be renatured under appropriate conditions, since their PMR spectra show evidence for a well-defined tertiary structure and are very similar, but not identical, to those of salt-extracted proteins. However, as exemplified with protein L11 (69), the success of the renaturation, as judged from the PMR spectrum, depends on minor details of the method used to return the protein to nondenaturing conditions. When salt-extracted proteins are treated with urea or subjected to high temperature or low pH they become denatured, as revealed by their PMR spectra. In many cases, after removal of the urea, the original spectra are obtained, which demonstrates a renaturated state of the protein (2, 55, 71). This renaturation is not always quantitative, i.e. not all of the denatured protein molecules return to their native configuration.

By isolation and PMR analysis of fragments produced by enzymatic or chemical cleavage of the protein it is possible to identify which part of the protein chain has a well-formed tertiary structure. It has for instance been shown that the C-terminal fragment of protein S4 (2) and the N-terminal region of L11 (2, 67) contain most of the structure observed in the intact molecule. The study of fragments also facilitates the assignment of PMR signals to amino acid residues in certain positions (53).

CRYSTALLIZATION   The most direct, although the most difficult, way to get information about the tertiary structure of proteins is crystallization followed by X-ray structural analysis. It was not until a few years ago that any progress in this field had been made. The N- and C-terminal fragments of protein L7/L12 have now been crystallized, and the crystals diffract to 4 Å and 2.6Å, respectively (72). The recent X-ray analysis of the C-terminal fragment (44) shows that it has a compact, plum-shaped tertiary structure with three α-helices and three β-sheets.

Besides the L7/L12-fragments, only protein L29 has so far been crystallized from the *E. coli* ribosome (56). On the other hand, attempts to crystallize ribosomal proteins from the thermophile *B. stearothermophilus* have been more successful. Protein BL17 was the first intact ribosomal **protein to give useful crystals (73). The X-ray diffraction analysis of this** protein is already well under way, with a resolution of better than 3 Å. The complete native data have been collected, and data from two isomorphous heavy atom derivatives of BL17 are being measured (R. Reinhardt and O. Epp, unpublished information). Furthermore, two other *B. stearothermophilus* ribosomal proteins (BL10 and BL34) have recently been crystallized, and the crystals reflect to 4 Å and 3 Å, respectively (56). From the amino acid sequence analysis it is clear that BL10 corresponds to EL6 from *E. coli* and BL34 to EL30 (see Table 2).

Attempts to crystallize intact ribosomal subunits have also been more successful with *B. stearothermophilus* than with *E. coli* ribosomes. While only small helical arrays of subunits have so far been obtained with *E. coli* (74, 75), it was possible to get relatively large three-dimensional crystals containing undegraded and biologically active 50S subunits of *B. stearothermophilus* (76). An electron micrograph of a section through such a crystal is shown in Figure 1. These crystals are presently being used for three-dimensional structure analysis (77).

Shape

The shape of many *E. coli* ribosomal proteins has been studied by a variety of techniques, such as X-ray and neutron scattering, as well as by hydrodynamic methods, i.e. by the combination of sedimentation, diffusion, and viscosity measurements. Information on the shape of 15 proteins (S1, S2, S3, S4, S5, S6, S7, S8, S13, S15, S16, S17, S18, S20, and S21) from the small subunit and 10 proteins (L1, L3, L6, L7/L12, L9, L11, L18, L24, L25, and L30) from the large subunit is now available and has been recently reviewed (1, 2). Since these compilations, studies on the shape of the following proteins have been published: S2 (78), S3 (79), S4 (47, 80), S6 (81), S15 (41), S17 (79), S20 (81), L1 (82), L3 (83), L7/L12 (43), L9 (82), L24 (83), L25 (82), and L30 (82).

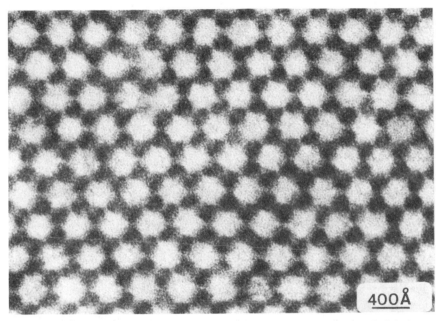

400Å

Figure 1:   Electron micrograph of a section through a three-dimensional crystal of the large subunits from *stearothermophilus* ribosomes (A. E. Yonath, J. Müssig, B. Tesche, S. Lorenz, V. A. Erdmann, and H. G. Wittmann, unpublished information).

The reported parameters for the shape of the individual proteins differ considerably. The main discrepancy exists between the finding of Serdyuk et al (41, 47, 48, 80, 84) that proteins S4, S7, S8, S15, and S16 have a compact globular structure, and the finding of several other groups [(summarized in (1 and 2)] that at least some of these proteins, e.g. S4, have an elongated shape in solution. The reasons for this discrepancy can probably be explained by differences in the procedures used for the isolation of the proteins and/or in the measurement and interpretation of the experimental data (see below).

The first ribosomal proteins whose shapes were determined had been taken from the same protein preparations that were isolated and used for the determination of the primary structures. They were extracted from the ribosomal subunits by acetic acid and isolated by column chromatography in the presence of 6 M urea at pH 3.5–4.5 followed by lyophilization (85, 86). By comparative CD and PMR studies it was shown (87) that these proteins have less secondary and tertiary structure than those that are salt-extracted and isolated under conditions avoiding acetic acid, urea, and lyophilization; i.e. the urea-treated proteins were partially denatured. However, proteins isolated in the presence of urea can be renatured under appropriate conditions, as demonstrated by the finding that the CD- and PMR-spectra of the renatured proteins are very similar to those of the salt-extracted proteins (2, 41, 42, 45, 47, 48, 55, 67, 69, 70, 71, 87). The extent of the renaturation is sensitive to minor variations in the renaturation procedure and also depends on the particular protein; i.e. with some proteins the renaturation is greater than with others under identical conditions. Therefore, an important reason for the discrepancy between the various results on the shape of the proteins is the quality of the protein sample under study, i.e. whether, and to what extent, a particular protein has been denatured during its isolation and to what extent the denaturation has been reversed before the measurement.

If differences in the qualities of a protein sample were the only reason for the discrepancy, one would expect that proteins with the same (or at least very similar) CD- and PMR-spectra would have the same shape. However, different results have also been obtained in such cases. For instance the CD- and PMR-spectra of protein S4 isolated in the complete absence of urea (45, 55) and of protein S4 isolated in the presence of urea followed by dialysis against urea-free buffers (47) are almost identical; i.e. the secondary and tertiary structures of the two S4-protein samples are very similar if not identical, according to these studies. However, the radii of gyration as measured in both cases by small angle X-ray scattering are 42 Å (88) and 19.5 Å (80). Since this large discrepancy cannot be explained by differences in the protein structure (see above) it must be due to differences in the measurements and/or their interpretation. Possible sources of these differences are: determination of protein concentrations, aggregation of the protein, measurement inside or outside of the Guinier range, choice of measured points for slope determination, and desmearing of the data, etc.

In view of this controversy about the radii of gyration and the shape of ribosomal proteins it would be helpful to have independent data from other sources besides hydrodynamic studies. To this end gel filtration experiments have been carried out with most of the *E. coli* 50S ribosomal proteins at very low concentrations, and their Stokes radii, which can be measured accurately under these conditions, have been determined (89). A wide range of gross conformations have been found: several proteins are compact and symmetric, e.g. L17, L25, L28, L29, and L30, whereas others, e.g. L2, L3, L9, L11, L15, L23, L27, L32, and L33, are extended or expanded. The shapes of the remaining proteins, e.g. L1, L4, L5, L6, L13, L16, L19, and L24, are intermediate.

Some additional and independent information with respect to the radius of gyration of ribosomal proteins in situ (i.e. proteins incorporated in the ribosomal particle) has recently been obtained from neutron scattering studies (90). From the mapping algorithm of 12 *E. coli* 30S proteins (S1,

S3, S4, S5, S6, S7, S8, S9, S10, S11, S12, and S15) their radii of gyration were estimated, and it was concluded that most of these proteins are compact, whereas proteins S1 and S4 are more elongated. Unfortunately, the experimental error inherent in these estimates is still large, especially for the small proteins. However, this approach may in the future give more accurate and useful information concerning the radii of gyration of ribosomal proteins in situ, as more estimates of the distances between proteins become available.

A new and more direct method for determining the shape of proteins in situ using neutron scattering has recently been developed and applied to proteins within the 50S subunit. The principle of the "contrast variation technique" is to grow two *E. coli* cultures in 76% and 84% D_2O, respectively, and to reconstitute a 50S particle using ribosomal RNA from the cells grown in 76% D_2O and ribosomal proteins from the cells grown in 84% D_2O. In this way a 50S particle is obtained that (in contrast to the native 50S subunit) is homogenous for the neutron beam. If in such a 50S particle, one of the proteins is incorporated in a fully protonated form by appropriate reconstitution procedures, the radius of gyration of this particular protein can be determined from its neutron scattering curve. In this way it has, for instance, been determined that protein L4 has a radius of gyration of 20 ± 2 Å in situ corresponding to an axial ratio of 3:1 (K. H. Nierhaus, R. Lietzke, R. May, V. Nowotny, H. Schulze, K. Simpson, P. Wurmbach, and H. Stuhrmann, manuscript submitted). It can be expected that the application of this method will lead to the elucidation of the shape of many proteins within the ribosomal particle and circumvent the difficulties that are encountered in studies on the shape of ribosomal proteins in solution. Furthermore, a comparison of reliable data about the shape of a protein in situ and in the isolated state will allow an answer to the question as to whether proteins have the same shape within and outside the ribosomal particle.

RIBOSOMAL RNAs

Although roughly two thirds of the mass of bacterial ribosomes consist of ribosomal RNAs (rRNAs), until recently relatively little was known of their structure and function. In spite of considerable efforts to elucidate the primary structure of the 16S and 23S RNAs (91), it was not until a few years ago that a breakthrough in their structure determination was made possible due to the application of modern techniques for rapid RNA and DNA sequence analysis. As is also true for ribosomal proteins, the knowledge of the primary structure of rRNAs is an essential prerequisite for the elucidation not only of their structural features, such as their secondary structure and their spatial arrangement within the ribosomal particle, but also of their role in the function of the ribosome. It is now evident that rRNAs participate in a number of processes taking place on the ribosome during protein biosynthesis, e.g. binding of mRNA and tRNA molecules, association of the ribosomal subunits, and conferring of resistance to the action of antibiotics.

In the following three sections, the recent progress in determining the structure of bacterial 5S, 16S, and 23S RNA molecules as well as in elucidating the interaction between these and other ribosomal components is summarized.

5S RNA

5S RNA FROM E. COLI   Since the discovery (92) and the primary structure determination (17) of the 5S RNA, which is part of the large subunit of the *E. coli* ribosome, numerous chemical and physical studies have been carried out to obtain further information about the structural features and the functional role of this relatively small RNA molecule. Because the early investigations of this molecule have been extensively reviewed (93, 94), only subsequent results are considered here.

In comparison to the 16S and 23S RNAs, the 5S RNA with 120 nucleotides is a small molecule that interacts with only three ribosomal proteins. Therefore, its secondary and tertiary structure as well as its interaction with these three binding proteins can be more readily investigated than is feasible for the much larger 16S and 23S RNAs. A great variety of chemical and physical techniques have been used to obtain an insight into the structural features of the 5S RNA molecule: chemical modification with nucleotide specific reagents, e.g. kethoxal and monoperphthalic acid (95–99); binding of oligonucleotides to 5S RNA (100, 101); cross-links within the 5S RNA molecule (102, 103), between 5S RNA and proteins (103), and between the proteins bound to the 5S RNA (104); enzymatic cuts with specific ribonucleases (105–109); melting (110–112) and refolding of the RNA (113); circular dichroism (111, 112, 114); UV-absorption (112, 114); infrared spectroscopy (115, 116); Raman spectroscopy (117–119); fluorescence (99); proton nuclear magnetic resonance (120); small angle X-ray diffraction (121); neutron scattering (122); hydrodynamic studies (112); electron microscopy (123); and slow tritium exchange (124).

Based on these chemical and physical studies and on evolutionary comparisons, a number of models for the secondary structure of the 5S RNA have been proposed (107, 109, 120, 125–130) in addition to those previously reviewed (see 94). The majority of these models modifies that of Fox & Woese (131), which was derived from comparative sequence studies on 5S RNAs from various organisms and which can be regarded as a model with the minimal secondary structure common to the 5S RNAs of many species. Its essential feature is the occurrence of four helices: the "molecular stalk" (positions 1–10 and 110–120), the "weak tuned helix" (18–23 and 60–65), the "common arm base" (31–34 and 48–51) and the "prokaryotic loop" (82–86 and 90–94).

Three groups have proposed models for the tertiary structure of the 5S RNA molecule based mainly on small angle X-ray diffraction (121), on hydrodynamic studies (112), and on enzymatic cleavage with specific nucleases (109, 132). The most recent of these models (132) is depicted in Figure 2. It shows an interaction between the regions 41–44 and 74–77 that is characteristic for this particular tertiary structure model.

A comparison of the 5S RNA models proposed by various groups reveals a general agreement on important structural features but also a number of differences, both in the secondary structure of the 5S RNA and in the protein binding sites. Reasons for this discrepancy are probably: (a) the quality of the RNA and protein samples used for the experiments, i.e. the extent to which the 5S RNA and/or the binding proteins were denatured (it is known that the E. coli 5S RNA molecule can occur in a native (A

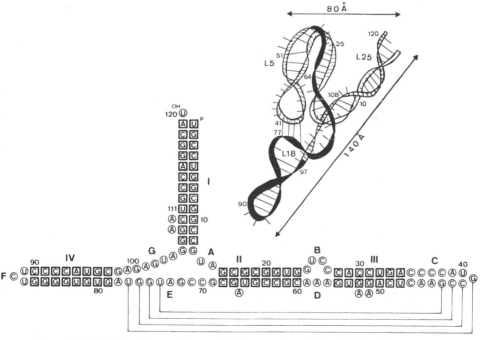

Figure 2 Models for the secondary and tertiary structure of the *E. coli* 5S RNA molecule (132).

form) or in a somewhat denatured (B form) conformation); (b) the buffer conditions used in the experiments, such as concentration of Mg^{2+} and other ions, pH, and temperature; (c) the emphasis the different groups lay on the results from their own studies or those of other groups.

In spite of numerous approaches little is known about the function of the 5S RNA. The results of several experimental but circumstantial approaches seem to support the hypothesis that 5S RNA is involved in the tRNA binding to the ribosomal A site (reviewed in 94). More recent results from 50S reconstitution experiments in the absence of 5S RNA are consistent with this hypothesis (133). However, convincing evidence, e.g. by UV-cross-links between the TψCG region of tRNA and the CGAA region of 5S RNA within the ribosome, is still missing.

As shown previously (for review see 94) protein L5, L18, and L25 bind to the 5S RNA. Recently, a fourth protein, L31a, was found to bind (104). The binding of L5, L18, and L25 was studied in some detail (68, 97, 105, 106, 108, 111, 114, 134–136). It was shown that the proteins bind cooperatively and, under appropriate conditions, nearly stoichiometrically, and that the 5S RNA undergoes a conformational change as a result of the protein binding, as revealed by circular dichroism measurements and other methods.

Conformational switches between different secondary structures of the 5S RNA molecules, possibly triggered by ribosomal proteins, have been assumed to contribute to the movement of the ribosome relative to mRNA (113, 125, 128). Although this hypothesis is interesting and supported by some data (113) more experimental results are needed to strengthen it. The same is true for the suggestion, based on a comparison of RNA structures, that base-pairing between 5S RNA in the large subunit and 16S RNA in the small subunit may be involved in the reversible association of the two ribosomal subunits (137).

5S RNA FROM OTHER BACTERIA   The large number of 5S RNAs from different organisms with known primary structures [(see 138) for a recent compilation] enables the construction of secondary structure models and of a phylogenetic tree for this molecule. The most extensive study in this direction was carried out by Hori & Osawa (36, 139–141) who divided the 5S RNA sequences into four groups: (a) the eubacterial 116-N-type with 116 (or sometimes 117) nucleotides, which occurs in Gram-positive bacteria; (b) the eubacterial 120-N-type with 120 nucleotides, which occurs in Gram-negative bacteria. There are characteristic differences in the secondary structure proposed for these two eubacterial types; (c) the metabacterial type to which the 5S RNA from *Halobacterium cutirubrum* belongs. (Unfortunately there is no uniform nomenclature: in the taxonomy of Woese

& Fox (34) the bacterial group to which *Halobacterium* belongs are called archaebacteria and in that of Hori & Osawa (141) metabacteria.) This type with 121 nucleotides resembles more the eukaryotic than the eubacterial type; (*d*) the eukaryotic type, which has 116–118 nucleotides in plants and 120 nucleotides in other eukaryotes. It differs from the two eubacterial types in several characteristic features of the 5S RNA secondary structure (36, 141).

As shown by reconstitution tests, it is possible to incorporate 5S RNAs from bacteria belonging to different families into the 50S subunit of *B. stearothermophilus,* but this cannot be achieved with 5S RNAs from eukaryotes (94, 142). These results reveal important differences between pro- and eukaryotic 5S RNA molecules. Further reconstitution experiments were performed with artificial 5S RNAs, constructed by combining parts of *Bacillus* 5S RNA molecules that differ in the molecular stalk, i.e. the base-paired region near the 5' and 3' ends. It was found that conservation of the base pairing within this region is not essential for the biological activity of the 5S RNA, at least in the poly(U) system. However, it is important for the rate of reconstitution, i.e. ribosomal assembly (143).

Based on a sequence comparison of 17 prokaryotic 5S RNAs and on helical energy filtering calculations, a general model for the secondary structure of prokaryotic 5S RNA molecules has recently been proposed. This "wishbone" model contains eight double helical regions and corresponds to the lowest energy homologous secondary structure in the 17 prokaryotic 5S RNAs used for the comparison (130).

Studies on the 5S RNA, the binding proteins, and the protein-RNA complex from the extreme halophilic archaebacterium *H. cutirubrum* (144, 145) showed these components to have several unique features, and to be more closely related to the equivalent structures in the eukaryotic rather than those in the eubacterial 5S RNA-protein complexes. The ribosomes from *H. cutirubrum* are only stable in high concentrations of salt, e.g. 3–4 M KCl and 100 mM Mg^{2+}, and their proteins are much more acidic than those from eubacteria. Two proteins specifically bind to the 5S RNA, namely HL13, which corresponds to EL18 (and possibly EL25) from *E. coli,* and HL19, which corresponds to EL5 (146). Similar to the ribosomes themselves, high salt concentrations are also required for the stability of the 5S RNA-protein complex (147).

The 5S RNA from *H. cutirubrum* has 121 nucleotides, i.e. one residue more than that of eubacteria, and no 5'-terminal mononucleoside diphosphate (148). It seems to contain two stable hairpin loops in place of the single "prokaryotic loop" in most other bacteria. Binding of protein HL13 (but not of HL19) results in a change in the secondary structure of both the 5S RNA and the protein (149). CD spectra, ethidium bromide binding, and limited nuclease digestion experiments (149, 150) revealed remarkable structural similarities between the 5S RNA-protein complexes from *H. cutirubrum* and *E. coli.* Considering the large differences in the primary structures of the RNA and the proteins from these two bacteria, as well as in the stability of the 5S RNA protein complexes under various salt conditions, it can be concluded that certain key structural features important for the biological function of the 5S RNA-protein complex have been conserved.

16S RNA

PRIMARY AND SECONDARY STRUCTURES   The elucidation of the primary structure of the *E. coli* 16S RNA by rapid DNA (14) or RNA (15) gel sequencing techniques provided an unambiguous interpretation of previous data on the structure of this molecule and on those RNA sites to which ribosomal protein bind specifically. Furthermore, it stimulated new chemical and physical investigations to determine the secondary (and possibly) tertiary structure of the 16S RNA.

Since studies on the structure of the 16S RNA have been reviewed in some detail (151) the emphasis in this chapter is on results described in more recent papers. The development of models for the secondary structure of the 16S RNA is based on a variety of techniques. They include: (*a*) chemical modification of the RNA (152) with reagents such as kethoxal, glyoxal, bisulfite, and *m*-chloroperbenzoate, which are specific for certain bases; (*b*) treatment with nucleases (153–155) that are single- or double-strand specific, or that cleave only after certain nucleotides; (*c*) cross-linking between different regions of the RNA strand (156–158); (*d*) isolation and sequence analysis of double-stranded regions after treatment with nuclease S1 (159); and (*e*) comparison of rRNA sequences from different organisms (160–164).

Several groups have proposed secondary structure models for the *E. coli* 16S RNA using one or more of these methods. Noller, Woese, and their co-workers (151, 160, 165) based their model mainly on chemical modification studies and on the phylogenetic comparison of the 16S RNA oligonucleotide catalogues derived from bacteria; especially informative was the comparison between the 16S RNA structure of *E. coli* and *B. brevis.* Brimacombe and his co-workers (162, 164, 166) mainly used results obtained from intramolecular RNA cross-links and the analysis of base-paired regions isolated after treatment of 16S RNA with the single-stranded RNA-specific nucleases S1 or T1. Furthermore, they compared the structure of the RNA from the small ribosomal subunits from *E. coli* (14, 15), maize chloroplasts (161), and mouse and human mitochondria (167, 168), and from yeast (169), and *Xenopus laevis* cytoplasm (170). The model of Stiegler et al (163, 171, 172) was mainly derived from studies on the cleavage of 16S RNA with single- and double-stranded RNA-specific nucleases as well as from sequence comparison, especially between the 16S RNAs from *Proteus vulgaris* and *E. coli*. Each of the three groups mainly relied on their own results for construction of their secondary structure model, but also discussed, and partly used, the findings of the other two groups for the refinement of their model.

There is substantial agreement among the models proposed by the three groups, although they differ in detail (164, 165, 172). This is not unexpected, since the groups mainly use their own results and emphasize other results differently. However, there is a large discrepancy between these three models and that of Cantor and his co-workers (156, 157). The latter is based on intramolecular RNA cross-linking by photo-activated psoralen derivatives, followed by electron microscopic identification of the cross-linked 16S RNA regions. The discrepancy remains unexplained.

The secondary structure of the *E. coli* 16S RNA (164, 165, 172) has a number of interesting features: (*a*) There are several long-range interactions, i.e. regions that are far (sometimes several hundred nucleotides) apart in the primary structure, but connected to each other by base-pairing. (*b*) The secondary structure model can be divided into four distinct and well-defined structural domains that are separated by three sets of long-range interactions. (*c*) There are no extended completely double-stranded regions; the largest consists of about ten base pairs. However, several such short stems are often arranged one after the other in a row and are separated by small unpaired loops or bulges. This probably confers an increased flexibility on the 16S RNA molecule, necessary for the compact folding of the RNA strand during assembly. (*d*) The extent of base-pairing deduced from the models (approximately 50%) is lower than that determined by physical techniques (60–70%). It is possible that tertiary structure interactions contribute to the higher value. (*e*) There are no knots in the RNA strand, which is in disagreement with conclusions drawn from cross-linking with psoralen derivatives (see above). (*f*) Each of several RNA regions can base-pair with more than one other region of the 16S RNA strand (151, 162). These multiple interactions, so-called "switches", may be involved in conformational changes in the 16S RNA occurring at different functional stages of the ribosome during the protein synthesis process. Although no direct experimental evidence for this hypothesis is so far available, considerable support for such switches comes from the isolation and analysis of base-paired stems (159, 166) and from the comparison of the secondary structure of rRNAs from different species, e.g. *E. coli* and yeast (162, 164). These results are consistent with the existence of switches.

As discussed above, phylogenetic sequence comparison is an important and valuable tool for constructing and confirming secondary structure models. Comparison of the models of RNA in the small ribosomal subunit of divergent species gives information about the evolutionary changes that have taken place. This is illustrated in Figure 3, in which the secondary structures of *E. coli* 16S RNA, yeast 18S RNA, and human mitochondrial 12S RNA are compared (164). It is obvious that the secondary structure of the RNA molecules has been strongly conserved throughout evolution, and that the secondary structure is in general more conserved than the primary structure of these RNAs. The different lengths of the RNA strands (954, 1542, and 1789 nucleotides in 12S, 16S, and 18S RNA, respectively) are caused by insertions or deletions ("amputations") of certain structural elements, such as stems and loops, or by shrinkage of whole domains.

Strong support for the conclusion that the secondary structure of the RNA is often more conserved during evolution than its primary structure comes from the finding that a base-paired stem in the *E. coli* 16S RNA also remains a stem in the rRNA of other organisms, e.g. yeast, although the nucleotide sequences within the corresponding regions of the two species differ drastically. This conservation is achieved by "compensating base changes;" for instance an A-U base pair in a given stem of the 16S RNA from *E. coli* is replaced by a G-C pair in the corresponding site of the stem in the yeast 18S RNA. In this way the secondary structural feature, i.e. the base-paired stem in the corresponding regions of the RNA molecules, remains the same, although the sequence is different. Examples are known where the majority of the base pairs within a given stem are such compensating pairs, apparently to conserve the structurally and/or functionally important stem character of the particular RNA region.

PROTEIN BINDING SITES ON THE 16S RNA   During the last decade, a wealth of information concerning the nature and the binding sites of those 30S ribosomal proteins (S4, S7, S8, S15, S17, and S20) that bind independently and specifically to the 16S RNA has been accumulated. Many chemical and physical methods have been used to investigate the binding of these ribosomal proteins to the 16S RNA: (*a*) After binding of a given protein the unprotected RNA regions are digested by treatment with nucleases, and the protected region is analyzed. (*b*) The RNA strand is cleaved into several large fragments that are then tested for their capacity to bind proteins. (*c*) Similarly, the protein chain is cleaved into two or more fragments that are tested for binding to the RNA strand. (*d*) Partially unfolded subunits are treated with nucleases, and the resulting RNA-protein complexes are isolated and analyzed for the RNA region and the proteins that they contain. (*e*) Analysis of cross-links, caused by UV-irradiation or chemical **reagents, which allows the identification of those regions of the binding**

HUMAN MITOCHONDRION 12S rRNA

E.COLI 16S rRNA

YEAST 18S rRNA

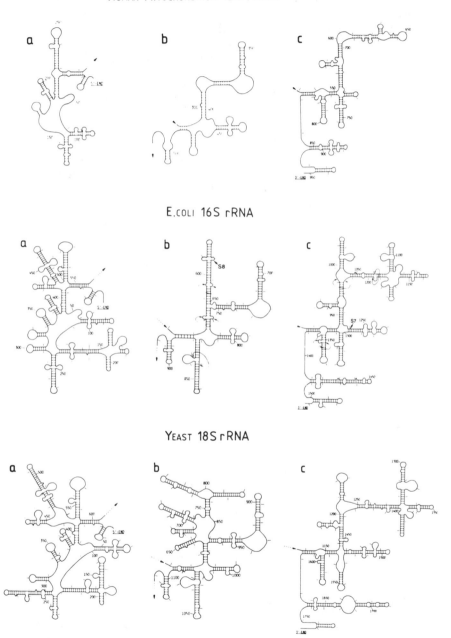

Figure 3   Comparison of the secondary structure of RNAs from the small ribosomal subunits of human mitochondrion, *E. coli* and yeast (164). Section *a* represents the 5'-region, *b* the middle region, and *c* the 3'-region of the RNA molecules.

protein and the RNA that are in close contact to each other. (*f*) The nucleotides that are modified by single strand–specific reagents, such as kethoxal, are monitored before and after binding of a protein to the RNA. (*g*) Individual amino acids (e.g. cysteine, tryptophan, tyrosine, and arginine) are chemically modified to see whether this has an effect on the binding of the particular protein. (*h*) The binding capacity of a mutationally altered protein is compared with that of the wild-type protein. (*i*) The binding site of a protein on the RNA strand is visualized by electron microscopy (*j*) Binding constants are determined for the binding proteins by membrane filter and other assays. (*k*) Physical techniques, such as small angle X-ray and neutron scattering, hydrodynamic methods, and CD-measurements, allow conclusions about possible conformational changes induced by the binding of the protein to the RNA. (*l*) Spectroscopic studies on the natural fluorescence of ribosomal proteins caused by their aromatic amino acids give information about changes in the environment of the amino acid after binding of the protein.

These studies (reviewed in 173, 174) have revealed that proteins S4, S17, and S20 bind to the 5'-proximal one third of the 16S RNA strand, proteins S8 and S15 (together with S6 and S18) to its central region, and protein S7 (together with S9, S13, and S19) to the 3' region. The binding sites are structurally diverse and differ drastically in size, ranging from 40–60 nucleotides for proteins S8 and S15 to approximately 500 nucleotides for protein S4.

The availability of a reliable model for the secondary structure of the *E. coli* 16S RNA (see above) enables the experimental data on the binding of proteins to RNA to be correlated with this model, and allows the identification of those structural features, e.g. double-stranded stems or single-stranded loops, to which the proteins bind (this is [summarized in (165)]). Isolation and analysis of the fragments produced by mild nuclease treatment of the complex between 16S RNA and a binding protein, e.g. S4, show that in general the same structural domains are obtained that are postulated by the secondary structure model. These results not only give additional support to the model, but also indicate that the proteins bind to already existing structural features of the RNA that are then stabilized by the bound protein. This is consistent with results from recent low angle scattering and hydrodynamic studies, which were unable to reveal conformational changes of the RNA induced by the bound protein (114, 175–177).

23S RNA

PRIMARY AND SECONDARY STRUCTURES   The primary structure of the *E. coli* 23S RNA was determined completely by sequencing the DNA in the rrnB operon (16) or to an extent of about 70% by RNA sequencing techniques (178). As with 16S RNA (see above) knowledge of the primary structure of the 23S RNA, which consists of 2904 nucleotides, is a prerequisite for a better interpretation and understanding of data on its structure and on the binding of ribosomal proteins to this RNA molecule. It also allows a comparison to be made between the 23S RNA of *E. coli* and the corresponding ribosomal RNA of other species.

Sequence comparisons revealed that the 5' region of *E. coli* 23S RNA (with approximately 160 nucleotides) corresponds to the 5.8S RNA of various eukaryotes (179–181), and that its 3' region corresponds to the 4.5S RNA contained in the large subunit of chloroplast ribosomes (182–184). There is a 90% homology between the sequences of fragments comprising 110 nucleotides isolated from the 3' ends of the 23S RNAs from *E. coli* and *P. vulgaris,* while the homology between the corresponding fragments from *E. coli* and *Aeromonas punctata* is only 60% (183). By comparison of the complete primary structure of the 23S RNAs from maize chloroplasts and *E. coli* 70% homology was found (185). Mature 23S RNA from prokaryotes appears to be homologous with the precursor of 25–28S RNAs from eukaryotes (186).

Secondary structure models have recently been proposed not only for the 23S RNA of *E. coli* but also for the RNAs from the large ribosomal subunits of maize chloroplast and of mouse and human mitochondria (187). They were based on a combination of results from different studies: (*a*) isolation and analysis of base-paired fragments obtained after treatment of the *E. coli* 23S RNA with ribonuclease T1; (*b*) analysis of intra-RNA cross-links; (*c*) sequence comparison between the four RNAs mentioned. Similar to the *E. coli* 16S RNA, the secondary structures of the large subunit RNAs from these four species have well-defined structural domains. There are more than 450 compensating base changes (explained above) between the 23S RNAs from *E. coli* and maize chloroplasts, as well as more than 100 between the RNAs from mouse and human mitochondria. Furthermore, some switches, i.e. multiple interactions between different regions within the same RNA molecule, were also found.

PROTEIN BINDING SITES ON THE 23S RNA   Most of the studies on the binding of ribosomal proteins to 23S RNA have been carried out with the components from *E. coli*. The results have recently been extensively reviewed (173, 174), and therefore only a brief summary is given here. In early studies (188, 189) ten *E. coli* 50S proteins (L1, L2, L3, L4, L6, L13, L16, L20, L23, and L24) were found to bind to the homologous 23S RNA. When proteins isolated in the absence of urea were tested, it was revealed that proteins L11, L15 (190) and a complex of L7/L12+L10 (191) could also bind. The binding of this latter complex to 23S RNA is probably mediated by the L10 component (192, 193). In addition to the proteins mentioned above, the L9, L17, L18, L22, and L29 (194) proteins were found to bind to 23S RNA under the conditions used for the reconstitution of the 50S subunit. Thus approximately 20 proteins, i.e. more than 50% of the proteins in the large subunit, are able to bind to the 23S RNA to differing extents. The number of binding proteins and their binding strength apparently depend on the quality of the protein and RNA samples as well as on the ionic conditions for the binding tests. Differences in these parameters would explain the different results obtained in the binding studies described.

By means of nuclease protection studies, the binding sites on the 23S RNA strand have been localized for the L1 (195–199), L11 (200), L20 (195, 201), L23 (195, 201), and L24 (202–205) binding proteins. Similarly to the binding sites on the 16S RNA, the sizes of the protected regions vary drastically and range from 40–50 nucleotides for protein L20 to 480 nucleotides for protein L24. Further information about the localization of binding sites has been obtained from studies where the 23S RNA is enzymatically cleaved into three fragments (13S, 8S, and 12S), and where these fragments are then tested for their capacity to bind proteins (206–208). It was found that proteins L4, L7/L12+L10, and L11 bind to the 5'-proximal 13S fragment, proteins L2 and L13 to the 8S fragment in the middle of the 23S RNA strand, and proteins L3 and L16 to the 3'-terminal 12S region. Proteins L14 and L16 bind to a large 18S fragment that corresponds to the RNA region spanned by the fragments 8S+12S.

The determination of precise contact points between proteins and the 23S RNA within the 50S subunits is possible after cross-linking by UV irradiation or treatment with specific bifunctional reagents. The cross-linking procedure is then followed by the identification of the cross-linked amino acid and nucleotide, respectively. In this way uracil at position 615 of the 23S RNA was found to be cross-linked to tyrosine at position 35 of protein L4 (209). Additional cross-links between five 50S proteins (L6, L21, L23, L27, and L29) and the 23S RNA have recently been localized (210). Protein L6 is cross-linked within an oligonucleotide corresponding to positions 2473–2481, protein L21 to positions 540–548, L23 to 137–141, L27 to 2332–2337, and L29 to 99–107. Cross-links reflect the topographical neighborhood between a given protein and an RNA region within the subunit, but do not necessarily reflect a specific protein-RNA interaction, although this is likely, especially in those cases where the cross-link is extremely short (e.g. with UV cross-links). *E. coli* ribosomal proteins not only bind to the rRNA of *E. coli* but can also interact with rRNAs from a wide range of other bacteria, such as enterobacteria and *Bacillus* (211, 212) as well as *Clostridium, Anacystis,* and *Chromatium* (213). Some of the *E. coli* proteins, especially S8, S15, and L1, can bind to the 23S RNA of archaebacteria (214), and protein L1 even binds to the large subunit RNA of the eukaryotic slime mold *Dictyostelium discoideum* (215).

The binding sites for *E. coli* protein L1 on the large subunit rRNA from *P. vulgaris* (216), *B. stearothermophilus* (217), and *D. discoideum* (215) have been sequenced and compared to the corresponding region in *E. coli* 23S RNA. Similarities in both the primary and especially the secondary structure were revealed. This finding points to a conservation throughout evolution of the structural features of the RNA region that are essential for the specific recognition and binding of protein L1. It is possible that similar analyses and comparisons of binding sites for other ribosomal proteins will lead to a better understanding of the molecular processes that govern protein-RNA recognition in ribosomes.

Literature Cited

1. Brimacombe, R., Stöffler, G., Wittmann, H. G. 1978. *Ann. Rev. Biochem.* 47:217–49
2. Wittmann, H. G., Littlechild, J. A., Wittmann-Liebold, B. 1979. In *Ribosomes: Structure, Function and Genetics,* ed. G. Chambliss, G. R. Craven, J. Davies, K. Davis, L. Kahan, M. Nomura, pp. 51–88. Baltimore: Univ. Park Press
3. Wittmann-Liebold, B. 1980. In *Genetics and Evolution of RNA Polymerase, tRNA and Ribosomes,* ed. S. Osawa, H. Ozeki, H. Uchida, T. Yura, pp. 639–54. Amsterdam: Elsevier/North Holland Biomed. Press
4. Schnier, J., Kimura, M., Foulaki, K., Subramanian, A. R., Isono, K., Wittmann-Liebold, B. 1981. *Proc. Natl. Acad. Sci. USA.* In press
5. Wittmann-Liebold, B., Bosserhoff, A. 1981. *FEBS Lett.* 129:10–16
6. Yaguchi, M., Roy, C., Wittmann, H. G. 1980. *FEBS Lett.* 121:113–16
7. Kamp, R., Wittmann-Liebold, B. 1980. *FEBS Lett.* 121:117–22
8. Kimura, M., Mende, L., Wittmann-Liebold, B. 1982. *FEBS Lett.* In press
9. Kimura, M., Wittmann-Liebold, B. 1980. *FEBS Lett.* 121:317–22
10. Wittmann-Liebold, B., Greuer, B. 1980. *FEBS Lett.* 121:105–12
11. Wittmann-Liebold, B., Kamp, R. 1980. *Biochem. Int.* 1:436–45
12. Arai, K., Clark, B. F. C., Duffy, L., Jones, M. D., Kaziro, Y., Laursen, R. A., L'Italien, J., Miller, D. L., Nagarkatti, S., Nakumura, S., Nielsen, K. M., Petersen, T. E., Takahashi, K., Wade, M. 1980. *Proc. Natl. Acad. Sci. USA* 77:1326–30
13. Jones, M. D., Petersen, T. E., Nielsen, K. M., Magnusson, S., Sottrup-Jensen, L., Gausing, K., Clark, B. F. C. 1980. *Eur. J. Biochem.* 108:507–26
14. Brosius, J., Palmer, M. L., Kennedy, P. L., Noller, H. F. 1978. *Proc. Natl. Acad. Sci. USA* 75:4801–5
15. Carbon, P., Ehresmann, C., Ehresmann, B., Ebel, J. P. 1978. *FEBS Lett.* 94:152–56
16. Brosius, J., Dull, T. J., Noller, H. F. 1980. *Proc. Natl. Acad. Sci. USA* 77:201–4
17. Brownlee, G. G., Sanger, F., Barrell, B. G. 1967. *Nature* 215:735–36
18. van Holde, K. E., Hill, W. E. 1974. In *Ribosomes,* ed. M. Nomura, A. Tissières, P. Lengyel, pp. 53–91. Long Island, NY: Cold Spring Harbor Lab.
19. Isono, K., Cumberlidge, A. G., Isono, S., Kitakawa, M., Schnier, J., Hirota, Y. 1980. See Ref. 3, pp. 329–40
20. Colson, C., Lhoest, J., Urlings, C. 1979. *Mol. Gen. Genet.* 169:245–50
21. Isono, K. 1979. See Ref. 2, pp. 641–69
22. Kimura, M., Rawlings, N., Appelt, K. 1982. *FEBS Lett.* 136:58–64
23. Kimura, M., Dijk, J., Heiland, I. 1980. *FEBS Lett.* 121:323–26
24. Itoh, T., Wittmann-Liebold, B. 1978. *FEBS Lett.* 96:392–94
25. Itoh, T. 1981. *FEBS Lett.* 127:67–70
26. Itoh, T., Wittmann-Liebold, B. 1978. *FEBS Lett.* 96:399–402
27. Amons, R., Möller, W. 1980. See Ref. 3, pp. 601–8
28. Yaguchi, M., Matheson, A. T., Visentin, L. P., Zuker, M. 1980. See Ref. 3, pp. 585–99
29. Itoh, T., Higo, K. I., Otaka, E., Osawa, S. 1980. See Ref. 3, pp. 609–24
30. Matheson, A. T., Nazar, R. N., Willick, G. E., Yaguchi, M. 1980. See Ref. 3, pp. 625–37
31. Higo, K. I., Itoh, T., Kamazaki, T., Osawa, S. 1980. See Ref. 3, pp. 655–66
32. Wittmann-Liebold, B., Geissler, A. W.,

180 WITTMANN

Lin, A., Wool, I. G. 1979. *J. Supramol. Struct.* 12:425–33

33. Matheson, A. T., Möller, W., Amons, R., Yaguchi, M. 1979. See Ref. 2, pp. 297–332
34. Woese, C. R., Fox, G. E. 1977. *Proc. Natl. Acad. Sci. USA* 74:5088–90
35. Matheson, A. T., Yaguchi, M., Balch, W. E., Wolfe, R. S. 1980. *Biochim. Biophys. Acta* 626:162–69
36. Hori, H., Osawa, S. 1980. See Ref. 3, pp. 539–51
37. Dzionara, M., Robinson, S. M. L., Wittmann-Liebold, B. 1977. *Hoppe-Seylers Z. Physiol. Chem.* 358:1003–19
38. Wittmann-Liebold, B., Robinson, S. M. L., Dzionara, M. 1977. *FEBS LETT.* 81: 204-13
39. Dzionara, M., Robinson, S.M.L., Wittmann-Liebold, B. 1977. *J. Supramol. Struct.* 7:191–204
40. Allen, S. H., Wong, K. P. 1978. *Biophys. J.* 23:359–73
41. Gogia, Z. V., Venyaminov, S. Y., Bushuev, V. N., Serdyuk, I. N., Lim, V. I., Spirin, A. S. 1979. *FEBS Lett.* 105:63–69
42. Gudkov, A. T., Behlke, J., Viturin, N. N., Lim, V. I. 1977. *FEBS Lett.* 82:125–29
43. Luer, C. A., Wong, K. P. 1979. *Biochemistry* 18:2019–27
44. Leijonmark, M., Eriksson, S., Liljas, A. 1980. *Nature* 286:824–26
45. Dijk, J., Littlechild, J., Freund, A. M., Pouyet, J., Daune, M., Provencher, S. W. 1982. *J. Biol. Chem.* In press
46. Dijk, J., Littlechild, J. A. 1979. *Methods Enzymol.* 59:481–502
47. Serdyuk, I. N., Gogia, Z. V., Venyaminov, S. Y., Khechinashvili, N. N., Bushuev, V. N., Spirin, A. S. 1980. *J. Mol. Biol.* 137:93–107
48. Gogia, Z. V., Venyaminov, S. Y., Bushuev, V. N., Spirin, A. S. 1981. *FEBS Lett.* 130:279–82
49. Gérard, D., Lemieux, G., Laustriat, G. 1975. *Photochem. Photobiol.* 22:89–95
50. Brochon, J. C., Wahl, P., Vachette, P., Daune, M. P. 1976. *Eur. J. Biochem.* 65:35–39
51. Chu Y. G., Cantor, C. R. 1979. *Nucleic Acids Res.* 6:2363–79
52. Khechinashvili, N. N., Koteliansky, V. E., Gogia, Z. V., Littlechild, J., Dijk, J. 1978. *FEBS Lett.* 95:270–72
53. Gudkov, A. T., Khechinashvili, N. N., Bushuev, V. N. 1978. *Eur. J. Biochem.* 90:313–18
54. Gudkov, A. T., Behlke, J. 1978. *Eur. J. Biochem.* 90:309–12

55. Littlechild, J., Malcolm, A. L., Paterakis, K., Ackermann, T., Dijk, J. 1982. *J. Biol. Chem.* In press
56. Appelt, K., Dijk, J., Reinhardt, R., Sanhuesa, S., White, S. W., Wilson, K. S., Yonath, A. 1981. *J. Biol. Chem.* 256:11787–90
57. Subramanian, A. R., Rienhardt, P., Kimura, M., Suryanarayana, T. 1981. *Eur. J. Biochem.* 119:245–49
58. Suryanarayana, T., Subramanian, A. R. 1979. *J. Mol. Biol.* 127:41–54
59. Giorginis, S., Subramanian, A. R. 1980. *J. Mol. Biol.* 141:393–408
60. Subramanian, A. R., Mizushima, S. 1979. *J. Biol. Chem.* 254:4309–12
61. Changchien, L. M., Craven, G. R. 1976. *J. Mol. Biol.* 108:381–401
62. Newberry, V., Yaguchi, M., Garrett, R. A. 1977. *Eur. J. Biochem.* 76:51–61
63. Littlechild, J., Morrison, C. A., Bradbury, E. M. 1979. *FEBS Lett.* 104: 90–94
64. Bruce, J., Firpo, E. J., Schaup, H. W. 1977. *Nucleic Acids Res.* 4:3327–40
65. Koteliansky, V. E., Domogatsky, S. P., Gudkov, A. T. 1978. *Eur. J. Biochem.* 90:319–23
66. Gudkov, A. T., Tumanova, L. G., Gongadze, G. M., Bushuev, V. N. 1980. *FEBS Lett.* 109:34–38
67. Tumanova, L. G., Gudkov, A. T., Bushuev, V. N., Okon, M. S. 1981. *FEBS Lett.* 127:241–44
68. Newberry, V., Brosius, J., Garrett, R. A. 1978. *Nucleic Acids Res.* 5:1753–66
69. Kime, M. J., Ratcliffe, R. G., Moore, P. B., Williams, R. J. P. 1980. *Eur. J. Biochem.* 110:493–98
70. Kime, M. J., Ratcliffe, R. G., Moore, P. B., Williams, R. J. P. 1981. *Eur. J. Biochem.* 116:269–76
71. Littlechild, J. A. 1980. *FEBS Lett.* 111:51–55
72. Liljas, A., Eriksson, S., Donner, D., Kurland, C. G. 1978. *FEBS Lett.* 88:300–04
73. Appelt, K., Dijk, J., Epp, O. 1979. *FEBS Lett.* 103:66–70
74. Clark, M. W., Hammon, M., Langer, J. A., Lake, J. A. 1979. *J. Mol. Biol.* 135:507–12
75. Lake, J. A. 1979. See Ref. 2, pp. 207–36
76. Yonath, A. E., Müssig, J., Tesche, B., Lorenz, S., Erdmann, V. A., Wittmann, H. G. 1980. *Biochem. Int.* 1:428–35
77. Leonard, K. R., Arad, T., Tesche, B., Erdmann, V. A., Wittmann, H. G., Yonath, A. E. 1981. *EMBO Symp. Ribosomes, Heidelberg* (Abstr.)
78. Georgalis, Y., Giri, L., Littlechild, J. 1981. *Biochemistry* 20:1061–64

79. Franz, A., Georgalis, Y., Giri, L. 1979. *Biochim. Biophys. Acta* 578:365–71
80. Serdyuk, I. N., Sarkisyan, M. A., Gogia, Z. V. 1981. *FEBS Lett.* 129:55–58
81. Georgalis, Y., Giri, L. 1980. *Biochem. Int.* 1:105–12
82. Giri, L., Franz, A., Dijk, J. 1979. *Biochemistry* 18:2520–25
83. Giri, L., Dijk, J. 1979. *Arch. Biochem. Biophys.* 193:122–29
84. Serdyuk, I. N., Zaccai, G., Spirin, A. S. 1978. *FEBS Lett.* 94:349–52
85. Hindennach, I., Stöffler, G., Wittmann, H. G. 1971. *Eur. J. Biochem.* 23:7–11
86. Hindennach, I., Kaltschmidt, E., Wittmann, H. G. 1971. *Eur. J. Biochem.* 23:12–16
87. Morrison, C. A., Bradbury, E. M., Littlechild, J., Dijk, J. 1979. *FEBS Lett.* 83:348–52
88. Österberg, R., Sjöberg, B., Garrett, R. A., Littlechild, J. 1978. *Nucleic Acids Res.* 5:3579–87
89. Wills, P., Georgalis, Y., Dijk, J. 1981. *EMBO Symp. Ribosomes, Heidelberg* (Abstr.)
90. Ramakrishnan, V. R., Yabuki, S., Sillers, I. Y., Schindler, D. G., Engelman, D. M., Moore, P. B. 1981. *J. Mol. Biol.* 153:739–60
91. Fellner, P. 1974. See Ref. 18, pp. 169–91
92. Rosset, R., Monier, R. 1963. *Biochim. Biophys. Acta* 68:653–56
93. Monier, R. 1974. See Ref. 18, pp. 141–68
94. Erdmann, V. A. 1976. *Progr. Nucleic Acids Res. Mol. Biol.* 18:45–90
95. Nishikawa, K., Takemura, S. 1977. *J. Biochem.* 81:995–1003
96. Noller, H. F., Garrett, R. A. 1979. *J. Mol. Biol.* 132:621–36
97. Garrett, R. A., Noller, H. F. 1979. *J. Mol. Biol.* 132:637–48
98. Larrinua, I., Delihas, N. 1979. *Proc. Natl. Acad. Sci. USA* 76:4400–4
99. Digweed, M., Erdmann, V. A., Odom, O. W., Hardesty, B. 1981. *Nucleic Acids Res.* 9:3187–98
100. Lewis, J. B., Doty, P. 1977. *Biochemistry* 16:5016–25
101. Wrede, P., Pongs, O., Erdmann, V. A. 1978. *J. Mol. Biol.* 120:83–96
102. Wagner, R., Garrett, R. A. 1978. *Nucleic Acids Res.* 5:4065–75
103. Wagner, R., Hancock, J., Chiam, C. L., Kühne, C. 1980. In *Biological Implications of Protein-Nucleic Acid Interactions,* ed. J. Augustyniak, pp. 30–43. Amsterdam: Elsevier/North-Holland Biomed. Press
104. Fanning, T. G., Traut, R. R. 1981. *Nucleic Acids Res.* 9:993–1004
105. Zimmermann, J., Erdmann, V. A. 1978. *Mol. Gen. Genet.* 160:247–57
106. Zimmermann, J., Erdmann, V. A. 1978. *Nucleic Acids Res.* 5:2267–68
107. Vigne, R., Jordan, B. R. 1979. *J. Mol. Evol.* 10:77–86
108. Douthwaite, S., Garrett, R. A., Wagner, R., Feunteun, J. 1979. *Nucleic Acids Res.* 6:2453–70
109. Pieler, T., Appel, B., Kluwe, D., Schuster, L., Bölker, M., Hering, L. B., Symanowski, R., Erdmann, V. A. 1980. *Eur. J. Cell Biol.* 22:132
110. Weidner, H., Crothers, D. M. 1977. *Nucleic Acids Res.* 4:3401–14
111. Fox, J. W., Wong, K. P. 1978. *J. Biol. Chem.* 253:18–20
112. Fox, J. W., Wong, K. P. 1979. *J. Biol. Chem.* 254:1775–77
113. Kao, T. H., Crothers, D. M. 1980. *Proc. Natl. Acad. Sci. USA* 77:3360–64
114. Bear, D. C., Schleich, T., Noller, H. F., Garrett, R. A. 1977. *Nucleic Acids Res.* 4:2511–26
115. Appel, B., Erdmann, V. A., Stulz, J., Ackermann, T. 1979. *Nucleic Acids Res.* 7:1043–58
116. Stulz, J., Ackermann, T., Appel, B., Erdmann, V. A. 1981. *Nucleic Acids Res.* 9:3851–61
117. Chen, M. C., Giege, R., Lord, R. C., Rich, A. 1978. *Biochemistry* 17:3134–38
118. Luoma, G. A., Marshall, A. G. 1978. *J. Mol. Biol.* 125:95–105
119. Luoma, G. A., Marshall, A. G. 1978. *Proc. Natl. Acad. Sci. USA* 75:4901–5
120. Burns, P. D., Luoma, G. A., Marshall, A. G. 1980. *Biochem. Biophys. Res. Commun.* 96:805–11
121. Österberg, R., Sjöberg, B., Garrett, R. A. 1976. *Eur. J. Biochem.* 68:481–87
122. Lorenz, S., Erdmann, V. A., May, R., Stöckel, P., Strell, I., Hoppe, W. 1980. *Eur. J. Cell Biol.* 22:134
123. Tesche, B., Schmiady, H., Lorenz, S., Erdmann, V. A. 1980. *Electron Microsc.* 2:534–35
124. Farber, N. M., Cantor, C. R. 1981. *J. Mol. Biol.* 146:223–39
125. Weidner, H., Yuan, R., Crothers, D. M. 1977. *Nature* 266:193–94
126. Erdmann, V. A., Appel, B., Digweed, M., Kluwe, D., Lorenz, S., Lück, A., Schreiber, A., Schuster, L. 1980. See Ref. 3, pp. 553–68
127. Hori, H., Osawa, S. 1980. See Ref. 3, pp. 539–51
128. Jagadeeswaran, P., Cherayil, J. D. 1980. *J. Theor. Biol.* 83:369–75

129. Garrett, R. A., Douthwaite, S., Noller, H. F. 1981. *Trends Biochem. Sci.* 6:137–39
130. Studnicka, G. M., Eiserling, F. A., Lake, J. A. 1981. *Nucleic Acids Res.* 9:1885–904
131. Fox, G. E., Woese, C. R. 1975. *Nature* 256:505–7
132. Pieler, T., Erdmann, V. A. 1982. *Proc. Natl. Acad. Sci. USA.* In press
133. Dohme, F., Nierhaus, K. H. 1976. *Proc. Natl. Acad. Sci. USA* 73:2221–25
134. Spierer, P., Zimmermann, R. A. 1978. *Biochemistry* 17:2474–79
135. Spierer, P., Bogdanov, A. A., Zimmermann, R. A. 1978. *Biochemistry* 17: 5394–98
136. Spierer, P., Wang, C. C., Marsh, T. L., Zimmermann, R. A. 1979. *Nucleic Acids Res.* 6:1669–82
137. Azad, A. A. 1979. *Nucleic Acids Res.* 7:1913–29
138. Erdmann, V. A. 1979. *Nucleic Acids Res.* 6:R29–R44
139. Hori, H. 1976. *Mol. Gen. Genet.* 145:119–23
140. Hori, H., Osawa, S. 1979. *Proc. Natl. Acad. Sci. USA* 76:381–85
141. Hori, H., Osawa, S. 1979. See Ref. 5, pp. 333–55
142. Wrede, P., Erdmann, V. A. 1977. *Proc. Natl. Acad. Sci. USA* 74:2706–9
143. Raué, H. A., Lorenz, S., Erdmann, V. A., Planta, R. J. 1981. *Nucleic Acids Res.* 9:1263–69
144. Matheson, A. T., Yaguchi, M., Nazar, R. N., Visentin, L. P., Willick, G. E. 1978. In *Energetics and Structure of Halophilic Microorganisms,* ed. S. R. Caplan, M. Ginzburg, pp. 481–501. Amsterdam: Elsevier/North Holland Biomed. Press
145. Matheson, A. T., Nazar, R. N., Willick, G. E., Yaguchi, M. 1980. See Ref. 3, pp. 625–37
146. Smith, N., Matheson, A. T., Yaguchi, M., Willick, G. E., Nazar, R. N. 1978. *Eur. J. Biochem.* 89:501–9
147. Willick, G. E., Williams, R. W., Matheson, A. T., Sendecki, W. 1978. *FEBS Lett.* 92:187–89
148. Nazar, R. N., Matheson, A. T., Bellemare, G. 1978. *J. Biol. Chem.* 253:5464–69
149. Willick, G. E., Nazar, R. N., Ngugen, T. V. 1980. *Biochemistry* 19:2738–42
150. Nazar, R. N., Willick, G. E., Matheson, A. T. 1979. *J. Biol. Chem.* 254:1506–12
151. Noller, H. F. 1979. See Ref. 2, pp. 3–22
152. Herr, W., Chapman, N. M., Noller, H. F. 1979. *J. Mol. Biol.* 130:433–49
153. Müller, R., Garrett, R. A., Noller, H. F. 1979. *J. Biol. Chem.* 356:3873–78
154. Carbon, P., Ehresmann, C., Ehresmann, B., Ebel, J. P. 1979. *Eur. J. Biochem.* 100:399–410
155. Ehresmann, C., Stiegler, P., Carbon, P., Ungewickell, E., Garrett, R. A. 1980. *Eur. J. Biochem.* 103:439–46
156. Cantor, C. R. 1979. See Ref. 2, pp. 23–49
157. Cantor, C. R., Wollenzien, P. L., Hearst, J. E. 1980. *Nucleic Acids Res.* 8:1855–72
158. Zwieb, C., Brimacombe, R. 1980. *Nucleic Acids Res.* 8:2397–411
159. Ross, A., Brimacombe, R. 1979. *Nature* 281:271–76
160. Woese, C. R., Magrum, L. J., Gupta, R., Siegel, R. B., Stahl, D. A., Kop, J., Crawford, N., Brosius, J., Guttel, R., Hogan, J. J., Noller, H. F. 1980. *Nucleic Acids Res.* 8:2275–93
161. Schwarz, Z., Kössel, H. 1980. *Nature* 283:739–42
162. Brimacombe, R. 1980. *Biochem. Int.* 1:162–71
163. Carbon, P., Ebel, J. P., Ehresmann, C. 1981. *Nucleic Acids Res.* 9:2325–33
164. Zwieb, C., Glotz, C., Brimacombe, R. 1981. *Nucleic Acids Res.* 9:3621–40
165. Noller, H. F., Woese, C. R. 1981. *Science* 212:403–11
166. Glotz, C., Brimacombe, R. 1980. *Nucleic Acid Res.* 8:2377–95
167. Eperon, I. C., Anderson, S., Nierlich, D. P. 1980. *Nature* 286:460–67
168. Van Etten, R. A., Walberg, M. W., Clayton, D. A. 1980. *Cell* 22:157–70
169. Rubtsov, P. M., Musakhanov, M. M., Zakharyev, V. M., Krayev, A. S., Skryabin, K. G., Bayev, A. A. 1980. *Nucleic Acids Res.* 8:5779–94
170. Salim, M., Maden, B. E. H. 1981. *Nature* 291:205–8
171. Stiegler, P., Carbon, P., Zuker, M., Ebel, J. P., Ehresmann, C. 1980. *CR Acad. Sci. Paris Ser. D* 291:937–40
172. Stiegler, P., Carbon, P., Zuker, M., Ebel, J. P., Ehresmann, C. 1981. *Nucleic Acids Res.* 9:2153–72
173. Zimmermann, R. A. 1979. See Ref. 2, pp. 135–69
174. Garrett, R. A. 1979. *Int. Rev. Biochem.* 25:121–77
175. Österberg, R., Sjöberg, B., Garrett, R. A., Littlechild, J. 1978. *Nucleic Acids Res.* 5:3579–87
176. Allen, S. H., Wong, K. P. 1979. *J. Biol. Chem.* 254:1775–77
177. Serdyuk, I. N., Shpungin, J. L., Zaccai, G. 1980. *J. Mol. Biol.* 137:109–21

178. Branlant, C., Krol, A., Machatt, M., Ebel, J. P. 1979. *FEBS Lett.* 107:177–81
179. Nazar, R. N. 1980. *FEBS Lett.* 119:212–14
180. Walker, W. 1980. *FEBS Lett.* 126:150–51
181. Jacq, B. 1981. *Nucleic Acids Res.* 9:2913–32
182. MacKay, R. M. 1981. *FEBS Lett.* 123:17–18
183. Machatt, M. A., Ebel, J. P., Branlant, C. 1981. *Nucleic Acids Res.* 9:1533–49
184. Edwards, K., Bedbrook, J., Dyer, T., Kössel, H. 1981. *Biochem. Int.* 2: 533–38
185. Edwards, K., Kössel, H. 1981. *Nucleic Acids Res.* 9:2853–69
186. Cox, R. A., Kelly, J. M. 1981. *FEBS Lett.* 130:1–6
187. Glotz, C., Zwieb, C., Brimacombe, R., Edwards, K., Kössel, H. 1981. *Nucleic Acids Res.* 9:3287–306
188. Stöffler, G., Daya, L., Rak, K. H., Garrett, R. A. 1971. *Mol. Gen. Genet.* 114:125–33
189. Garrett, R. A., Müller, S., Spierer, P., Zimmermann, R. A. 1974. *J. Mol. Biol.* 88:553–57
190. Littlechild, J., Dijk, J., Garrett, R. A. 1977. *FEBS Lett.* 74:292–94
191. Dijk, J., Littlechild, J., Garrett, R. A. 1977. *FEBS Lett.* 77:295–300
192. Dijk, J., Garrett, R. A., Müller, R. 1979. *Nucleic Acids Res.* 6:2717–29
193. Pettersson, I. 1979. *Nucleic Acids Res.* 6:2637–46
194. Marquardt, O., Roth, H. E., Wystup, G., Nierhaus, K. H. 1979. *Nucleic Acids Res.* 6:3641–50
195. Branlant, C., Krol, A., Sriwidada, J., Ebel, J. P., Sloof, P., Garrett, R. A. 1975. *FEBS Lett.* 52:195–201
196. Sloof, P., Garrett, R. A., Krol, A., Branlant, C. 1976. *Eur. J. Biochem.* 70: 447–56
197. Branlant, C., Krol, A., Sriwidada, J., Ebel, J. P., Sloof, P., Garrett, R. A. 1976. *Eur. J. Biochem.* 70:457–69
198. Branlant, C., Korobko, V., Ebel, J. P. 1976. *Eur. J. Biochem.* 70:471–82
199. Branlant, C., Krol, A., Ebel, J. P. 1980. *Nucleic Acids Res.* 8:5567–78

200. Schmidt, F. J., Thompson, J., Lee, K., Dijk, J., Cundliffe, E. 1981. *J. Biol. Chem.* 256:12301–5
201. Branlant, C., Sriwidada, J., Krol, A., Ebel, J. P. 1977. *Nucleic Acids Res.* 4:4323–45
202. Branlant, C., Krol, A., Sriwidada, J., Fellner, P., Crichton, R. R. 1973. *FEBS Lett.* 35:265–73
203. Branlant, C., Sriwidada, J., Krol, A., Ebel, J. P. 1977. *Eur. J. Biochem.* 74:155–70
204. Sloof, P., Hunter, J. B., Garrett, R. A., Branlant, C. 1978. *Nucleic Acids Res.* 5:3503–13
205. Krol, A., Machatt, M. A., Branlant, C., Ebel, J. P. 1978. *Nucleic Acids Res.* 5:4933–47
206. Spierer, P., Zimmermann, R. A., Mackie, G. A. 1975. *Eur. J. Biochem.* 52:459–68
207. Spierer, P., Zimmermann, R. A., Branlant, C. 1976. *FEBS Lett.* 68:71–75
208. Spierer, P., Wang, C. C., Marsh, T. L., Zimmermann, R. A. 1979. *Nucleic Acids Res.* 6:1669–82
209. Maly, P., Rinke, J., Ulmer, E., Zwieb, C., Brimacombe, R. 1980. *Biochemistry* 19:4179–88
210. Wower, I., Wower, J., Meinke, M., Brimacombe, R. 1981. *Nucleic Acids Res.* 9:4285–4302
211. Garrett, R. A., Rak, K. H., Daya, L., Stöffler, G. 1971. *Mol. Gen. Genet.* 114:112–24
212. Daya-Grosjean, L., Geisser, M., Stöffler, G., Garrett, R. A. 1973. *FEBS Lett.* 37:17–20
213. Thurlow, D. L., Zimmermann, R. A. 1978. *Proc. Natl. Acad. Sci. USA* 75: 2859–63
214. Zimmermann, R. A., Thurlow, D. L., Finn, R. S., Marsh, T. L., Ferrett, L. K. 1980. See Ref. 3, pp. 569–84
215. Gourse, R. L., Thurlow, D. L., Gerbi, S. A., Zimmermann, R. A. 1981. *Proc. Natl. Acad. Sci. USA* 78:2722–26
216. Branlant, C., Krol, A., Machatt, M. A., Ebel, J. P. 1981. *Nucleic Acids Res.* 9:293–307
217. Stanley, J., Sloof, P., Ebel, J. P. 1978. *Eur. J. Biochem.* 85:309–16

Ann. Rev. Biochem. 1982. 51:185–217

PROTON ELECTROCHEMICAL GRADIENTS AND ENERGY-TRANSDUCTION PROCESSES[1]

Stuart J. Ferguson

Department of Biochemistry, University of Birmingham, Birmingham
B15 2TT, United Kingdom

M. Catia Sorgato

CNR Unit for Study of Mitochondrial Physiology, Istituto di Chimica
Biologica, Università di Padova, 35100 Padova, Italy

CONTENTS

[1]The following abbreviations are used: $\Delta\psi$, membrane potential; ΔpH, difference in
pH across membrane; $\Delta\mu_H^+$, electrochemical proton gradient; Δp, protonmotive force =
$\Delta\mu_H^+/F$; ΔE, redox potential span; ΔG_p, phosphorylation potential or free energy of ATP
synthesis.

0066-4154/82/0701-0185$02.00

I. PERSPECTIVES AND SUMMARY

An electrochemical proton gradient can act as an intermediate between two reactions; this is simply illustrated with an example. If the oxidation of NADH by the respiratory chain of a bacterium is linked to the translocation of protons from the cytoplasm to the external aqueous phase, then an electrochemical gradient of protons will develop across the cell membrane. Return of protons into the cell, and therefore down their electrochemical gradient, via a catalyst can in turn drive a second reaction (e.g. ATP synthesis) thermodynamically uphill; thus coupling between two distinct reactions is achieved. This description corresponds in a very elementary way, without refinement of specifying stoichiometry or mechanism, to the chemiosmotic theory of energy coupling first enunciated by Peter Mitchell in 1961 (1). An electrochemical proton gradient, $\Delta \mu_{H^+}$, has a precise definition, containing both an electrical and a concentration gradient term. Therefore it is important to avoid using the term proton gradient because it can be taken to mean either the total electrochemical gradient or just the concentration gradient of protons.

Investigation of electrochemical proton gradients requires their measurement, usually by relatively indirect means, and determination of the stoichiometry and mechanism of proton translocation associated with any given reaction. Neither of these tasks is easy, but there are grounds for optimism. The question of stoichiometry can be approached by determining how many protons may be driven against a measured gradient by a reaction for which the Gibbs free energy change is known, or by asking the reverse question: how many protons must move down the gradient in order to account for observed displacement of a reaction from its usual uncoupled equilibrium position? Other contemporary reviews (2–8) present complementary material and discuss in more detail possible mechanisms of proton translocation; on this aspect we restrict ourselves to a brief account of how some simple bacterial systems generate a gradient.

An electrochemical proton gradient as an intermediate in energy coupling is a thermodynamic description, and therefore does not explicitly suggest how rates of reactions may depend on the magnitude of the gradient. But there probably are relationships between kinetics and the size of the intermediate, and accordingly we discuss this with respect to both the rates of electron transfer and of ATP synthesis. In the latter case we outline how understanding the relationship ought to provide important clues about the mechanism utilized by ATP synthase enzymes to couple the movement of protons to the condensation of P_i with ADP.

Within individual sections we frequently focus on data from a single system, particularly mitochondria, but with minor adaptation the same arguments often apply to analogous processes in bacteria and chloroplast thylakoids. Indeed, the principles surrounding the use of proton electrochemical gradients are now recognized to apply to a variety of processes in several parts of cells.

Proton electrochemical gradients are frequently regarded as the sole factor involved in coupling electron transfer to ATP synthesis or other energy-linked reactions, but we also review observations that do not fit into this picture. Therefore we assess the feasibility of alternative mechanisms in which protons flow within or along the membrane between coupled reactions rather than, as in the chemiosmotic theory, via the bulk aqueous phases.

Some nomenclature needs clarification. A proton electrochemical gradient, $\Delta \mu_{H^+}$, has units of free energy, and thus the term protonmotive force, Δp (or protonic potential difference) was coined by Mitchell for when, as is usual, the gradient is expressed in the electrical units of mV; we follow this convention. Δp usually comprises two components, a membrane potential $\Delta \psi$, and a pH gradient ΔpH, but there are experimental conditions where only one component is significant. For example with very high and equivalent buffering capacity on two sides of a membrane this will be $\Delta \psi$, whereas in illuminated thylakoids the movement of anions in the same direction as translocated protons can collapse $\Delta \psi$ leaving only ΔpH. The only necessary relation between $\Delta \psi$ and ΔpH is that their sum total cannot exceed the maximum attainable value of Δp, which in turn is dictated by the thermodynamics of the reaction that is responsible for generating Δp.

II. THE PROBLEM OF MEASURING Δp

Measurements of each of the two components of Δp, $\Delta \psi$, and ΔpH, are a prerequisite for a complete understanding of the coupling of two reactions via Δp. The most common method for $\Delta \psi$ measures the distribution of a permeant ion between the external and internal aqueous phases, which are separated by a membrane. Assuming that the ion reaches its electrochemical equilibrium across the membrane, $\Delta\psi$ is calculated from its internal and external concentrations, using the Nernst equation. A cation is used when the internal phase is negative relative to the external, and vice versa. ΔpH is frequently estimated from the distribution of a weak acid or base across a membrane, assuming that only the unprotonated species is permeant. The theoretical basis of these methods, descriptions of the full range of experimental procedures and suitable solutes, and details of alternative methods using spectroscopic indicators are described elsewhere (9–11). Here we are concerned with the accuracy of the solute uptake methods, including comparisons with three independent methods, using carotenoid band shifts, ^{31}P NMR, and microelectrodes.

Does Solute Binding or Failure to Equilibrate Cause Errors?

In the uptake of a solute, it is not immediately apparent whether all the solute accumulated by, e.g., a mitochondrion is dissolved in the internal aqueous space, or whether a significant fraction is bound to phospholipid and protein. Only the concentration of the dissolved solute can be related to $\Delta\psi$ or ΔpH. To test for binding, the ratio of internal to external concentration of a solute is usually measured over a range of solute concentrations and with different amounts of biological material (e.g. 12). A constant ratio in this procedure is taken as evidence that binding is negligible, but it is rarely possible to exclude unequivocally that an almost linear region of a binding curve is not being covered. For mitochondria, which are osmotically active, an alternative procedure is possible because the internal volume can be altered by changing the external osmolarity. By this procedure it has been shown that the accumulation ratios of Rb^+ (in the presence of valinomycin) and lactate are independent of a fivefold change in matrix volume, strong evidence that the extent of binding, which would be independent of matrix volume, was minimal (13). In contrast, there was an apparent uptake of methylamine by mitochondria that was independent of the external osmolarity and could thus be attributed to binding rather than accumulation (13). Real uptake of methylamine is difficult to reconcile with accumulation of lactate, which indicates an alkaline matrix that should exclude a weak base such as methylamine. The use of lactate rather than acetate as the weak acid for the mitochondrial ΔpH determinations has been recommended, as there is evidence that acetate can be incorporated into acetyl-CoA; thus measurements of the uptake of [^{14}C]-acetate can give overestimates for ΔpH (13, 14).

A widely quoted value of Δp in respiring mitochondria is 228 mV as reported by Nicholls (15), which is significantly larger than the 165–175 mV reported by some other workers (9, 10), but comparable to the first reported value (16). It has been suggested that 228 mV is an overestimate owing to a 2.5-fold underestimate of the mitochondrial matrix volume (9), but this is not supported by determinations (13) of the matrix volume under conditions similar to those used by Nicholls. The 228 mV may be an overestimate because of the unusually large contribution from ΔpH of $\sim$ 80 mV (1.4 pH units) under conditions that were not explicitly designed to increase ΔpH at the expense of $\Delta\psi$. Two possible methodological causes are the use of acetate distribution as the indicator of ΔpH, and of methylamine as marker for the extramitochondrial space on the filters that were used to separate mitochondria from reaction medium. Any binding of methylamine to mitochondria would have led to an overestimate of the extra- mitochondrial space and thus an underestimate of the matrix volume, leading in turn to an overestimate of both ΔpH and $\Delta\psi$, whilst metabolic incorporation of acetate could have contributed to an overestimate for ΔpH (cf Halestrap 14). Despite the original evidence (15) that none of these factors was significant, further checks seem advisable before a value of 228 mV for respiring mitochondria is accepted.

The lipophilic phosphonium salts, tetraphenylphosphonium and triphenylmethylphosphonium, are now very widely used indicators of $\Delta\psi$ in both mitochondria and cells, but the lipophilic character which renders them permeant to membranes might be expected to cause extensive binding to lipids and hydrophobic proteins. Undoubtedly there is binding but it is not difficult to determine its extent using lysed mitochondria or heat-treated bacteria when $\Delta\psi$ will be zero. The crucial question is whether this will provide the appropriate correction to apply when $\Delta\psi$ is large, in which case the internal concentration of the ion is high, and thus more internal but not external binding sites may be occupied. There are however indications that in some cases increased binding at high $\Delta\psi$ is not significant, since for both Ehrlich cells (17) and *Streptococcus lactis* (18) the extent of phosphonium ion uptake, after correction for binding to cells with zero $\Delta\psi$, indicated the same potential as was expected from imposed K^+ – diffusion potentials. On the other hand for mitochondria the binding correction appears to increase with increasing $\Delta\psi$ (19), whilst with erythrocytes the $\Delta\psi$ indicated from the phosphonium ion distribution correlated well with inwardly directed Cl^- – diffusion potential, although there were discrepancies with outwardly directed K^+-diffusion potentials (20). The results described suggest on the whole that phosphonium salts are able to equilibrate with $\Delta\psi$ in accordance with the Nernst equation, and that simple corrections for non-specific binding can be made. Parenthetically, the same may not be true of the tetraphenylboron anion for which a complex binding correction is necessary (21).

Triphenylmethylphosphonium sometimes equilibrates with $\Delta\psi$ only in the presence of catalytic amounts of tetraphenylboron. Recent work with phospholipid vesicles has explained this effect. Within the membrane, the positive and negative lipophilic ions form an ion-pair that moves more rapidly than the phosphonium ion alone, whilst tetraphenylboron can rapidly recycle across the membrane (22). However, work with model membranes can sometimes be misleading. SCN^- is frequently used as a permeant ion for measuring $\Delta\psi$, but appears to be almost impermeant toward black lipid membranes (23). The higher dielectric constant of biological membranes relative to membranes made of pure phospholipid may be the basis for the greater permeability of biological membranes to SCN^- (23.) Although several minutes are needed for SCN^- to reach its equilibrium distribution across a membrane (24), this does not preclude its use as an indicator for $\Delta\psi$; equilibrium distributions can be reached whatever the rate. The extent of binding of an ion to a phospholipid vesicle may not be comparable to its binding to a biological membrane. Singer (25) has recently proposed that amphipathic ions bind to vesicles more strongly than to "real" membranes.

Carotenoid Band Shifts Indicate Larger $\Delta\psi$ Than Ion Uptake

The electrochromic response of a population of carotenoid molecules is a frequently used and characterized indicator for $\Delta\psi$ in membranes of certain photosynthetic bacteria (7, 8). The carotenoids indicate very much larger $\Delta\psi$ values than permeant ion uptake, in this case SCN^-, for illuminated chromatophores;[2] 300 mV compared with 140 mV under conditions where ΔpH is small (26). The proposal (27) that this discrepancy arises because an inadequate light intensity was used in the experiments with SCN^-, which require a higher concentration of chromatophores than is needed for measurement of carotenoid absorbance shifts, has been disproved (12, 28). A comparison of carotenoid shifts with butyltriphenylphosphonium uptake in cells of *Rhodopseudomonas capsulata* (29) has also shown that upon illumination the former method gives a value for $\Delta\psi$ of 290 mV, whereas the extent of permeant ion uptake is consistent with a $\Delta\psi$ of 160 mV.

There are extensive discussions elsewhere (7, 8, 29) on why the carotenoids indicate higher values for $\Delta\psi$, but no satisfactory explanation has emerged. The possibility that the carotenoids might respond to a local change separation in the photosynthetic reaction centers can now probably

---

[2]Chromatophores are inside-out vesicles that catalyze photophosphorylation. They are prepared by disrupting the invaginated plasma membranes of certain photosynthetic bacteria.

be eliminated as the $\Delta\psi$-indicating carotenoids are associated with a light-harvesting protein rather than with the reaction center (7, 8). Uncertainty about carotenoid shifts arises because their extent is related to a value of $\Delta\psi$ by calibrating with K^+ – diffusion potentials. In this way it is possible to impose $\Delta\psi$ values only in the range –150 mV to +150 mV; thus it is necessary to assume linearity in the extrapolation of the calibration procedure to convert larger shifts into $\Delta\psi$ values exceeding 150 mV (7). A suggestion that this calibration procedure may require correction (30) to allow for the effects of changes in surface charge has not been confirmed by more recent work (29). The carotenoid shift probably overestimates $\Delta\psi$, and may indicate $\Delta\psi$ as well as charge separation within the membrane dielectric. If the carotenoids were to report the correct $\Delta\psi$ in the photosynthetic systems, then the ion-uptake methods would provide substantial underestimates not only in these cases, but presumably also throughout the other systems where they have been applied. Future work must elucidate the basis for the differences between the carotenoid and permeant ion methods.

^{31}P NMR Can Measure ΔpH

The phosphorus nucleus of inorganic phosphate, or of the γ-phosphate group of ATP, will resonate in a high magnetic field at a frequency that depends on the extent of protonation of the phosphate. Hence, provided a pK_a value can be determined (31) for a dissociating proton in a given environment (e.g. cell cytoplasm or mitochondrial matrix), then the resonance frequency can give a readout of pH, although the low sensitivity of ^{31}P NMR means that dense suspensions of mitochondria or cells are required. To slow down metabolism and attain reasonable levels of oxygenation it is necessary to work between 0°C and 4°C. Under these conditions in isolated rat liver cells the contribution of ΔpH to the mitochondrial Δp was 20 mV or less (32), except when the cells were perturbed by treatments designed to increase ΔpH at the expense of $\Delta\psi$. Earlier work had shown that uptake of a weak acid and the ^{31}P NMR method indicated similar values for ΔpH ($\sim$ 30 mV) in isolated respiring mitochondria (33), and more recently it was found using ^{31}P NMR that also ATP hydrolysis generated a ΔpH of approximately 30 mV (34). These findings with the NMR method, despite its difficulties, reinforce our earlier argument that ΔpH is unlikely to be as large as 80 mV (15) unless movement of ions is permitted to enhance ΔpH. NMR and the uptake of benzoic acid have been compared as indicators of ΔpH in cells of *Rhodopseudomonas sphaeroides* (35). At the concentration of cells necessary to detect the internal P_i resonance it was not possible to provide a saturating intensity of illumination, and thus it was necessary to extrapolate the NMR measurements to estimate ΔpH under saturating conditions (35). Despite this problem the two independent methods gave essentially identical values for ΔpH, which strongly supports, together with results from chromaffin granules (see Section VIII), the contention that distribution of weak acids or bases is a reliable procedure for estimating ΔpH.

Use of Microelectrodes to Determine $\Delta \psi$

$\Delta \psi$ would best be measured with the microelectrode techniques of electrophysiology, but the small size of "normal" bacteria and mitochondria precludes their use. However, there are ways for obtaining unusually large cells of *Escherichia coli* (3–5 μm in diameter) (36). Microelectrodes were successfully introduced into such cells and indicated a $\Delta \psi$ of ~ 150 mV. A similar value was obtained in parallel studies with the tetraphenylphosphonium ion (36), although it seems that ascorbate plus phenazine methosulfate was present as an electron donor to the respiratory chain only in the measurements of phosphonium ion uptake, and that no test was made of whether the $\Delta \psi$ reported by the microelectrode was dissipated by an uncoupler of oxidative phosphorylation. Therefore a slight uncertainty surrounds the otherwise very important conclusion that electrode and permeant ion methods give identical estimates for $\Delta \psi$. It would be wrong to assume that in contrast to other methods, measurements with microelectrodes are automatically error free. The introduction of the electrode punctures the membrane of the cell or organelle, and if the resealing processes are inadequate the permeability to ions can increase, and the measured $\Delta \psi$ may be less than the real value.

In contrast to the work with *E. coli*, microelectrodes have never detected significant respiration-dependent values for $\Delta \psi$ in enlarged mitochondria (37). We are not competent to comment on the electrophysiological aspects of this work, but note that Tedeschi and colleagues have performed a large number of experiments to show that the electrode does enter the mitochondrial matrix and that under such conditions the mitochondrion is able to synthesize ATP [extensively reviewed in (37)]. The latter observation can only be reconciled with the absence of an appreciable $\Delta \psi$ if the presence of the electrode renders ΔpH the main component of Δp (37), or if only a low $\Delta \psi$ or Δp is required to account for the amount of ATP synthesized. In concluding from electrode measurements that $\Delta \psi$ in mitochondria is negligible, Tedeschi is forced to adopt several unacceptable arguments including the untenable proposition that the cytoplasmic membrane of a respiratory bacterium does not function in the same way as the mitochondrial inner membrane (37). *Paracoccus denitrificans,* for example, which can maintain a $\Delta \psi$ of approximately 150 mV (38), has a complement of respiratory chain components that is almost identical to that of a rat liver mitochondrion. A second argument of Tedeschi is that permeant ions cannot be used to detect or to measure $\Delta\psi$; he suggests (37) that as the entry of a permeant cation into mitochondria is matched almost stoichiometrically by an efflux of protons, the cation uptake cannot be an electrophoretic response to $\Delta\psi$, but rather is an exchange for protons. However, owing to the relatively low electrical capacitance of the inner mitochondrial membrane only a few protons need move out of the mitochondrion before $\Delta\psi$ is established. Thus when a permeant cation, even at very low (μM) concentration, is added externally to mitochondria, the number of positive charges moved into the mitochondrion, as the ion attempts to attain its electrochemical equilibrium distribution, would be sufficient to reduce $\Delta\psi$. But, as the proton translocating respiratory chain is operating continuously, sufficient extra protons, which otherwise would have returned down their electrochemical gradient to the matrix via leaks, will remain in the external phase and thus maintain $\Delta\psi$ at approximately the same value as before the permeant ion was added. Thus the net efflux of an approximately stoichiometric number of protons is the expected observation if a permeant cation enters mitochondria electrophoretically. When higher concentrations of the permeant ion are used, there is a greater accumulation of protons in the external phase (and depletion from the internal phase) and consequently ΔpH increases at the expense of $\Delta\psi$.

Concluding Remarks on the Magnitude of Δp

The agreement between values of $\Delta\psi$ and ΔpH measured using uptake of solutes, ^{31}P NMR, and, in *E. coli,* microelectrodes, together with the evidence that binding of solutes is either negligible or can be corrected for, suggests that all these independent methods are giving reasonably accurate estimates for the sizes of the components of Δp.

A reasonable value for Δp in mitochondria that are respiring but not synthesizing net ATP is 200 mV (comprising $\Delta\psi = 170$ mV and ΔpH $= 30$ mV), although further work is needed to resolve the discrepancies between the current range of values. For a respiratory bacterium with an external pH of 7.5, Δp appears to lie between 150 and 180 mV, with $\Delta\psi$ approximately 150 mV and ΔpH between 0 and 30 mV. These lower values for bacteria probably arise because Δp in bacteria is driving energy-linked reactions, including ATP synthesis (38).

III. MECHANISMS FOR GENERATING Δp

Δp can be generated by two types of mechanism: through direct pumping of protons across a membrane, as, for example, catalyzed by the mitochondrial ATPase; or by electron-transfer chains, in which the electron may be the charged species that moves across a membrane. In the latter case a

Δp would develop because a hydrogen-carrying molecule first takes up protons and is reduced by an electron donor at one side of the membrane; after electroneutral movement it releases the protons and is oxidised at the opposite side. The hydrogen movement generates ΔpH, whilst the subsequent return of the electron across the membrane is the step that generates $\Delta\psi$. This corresponds to Mitchell's original proposal of a looped mitochondrial respiratory chain (39). It is clear from recent reviews that we are some way from understanding the details of the first mechanism, and it is not yet certain whether the second mechanism operates in mitochondria (2, 3).

Simple Bacterial Systems

The organization of bacterial electron transfer chains has been much less studied than the mitochondrial counterpart, but already there are examples of oxidoreduction reactions that automatically establish a Δp by a variant of the second type of mechanism. In these a proton-releasing reaction in one aqueous phase is coupled to a proton-consuming reaction in a second aqueous phase via a flow of electrons across the membrane that separates the two phases. Thus Δp is generated by a combination of an electrogenic reaction to give $\Delta\psi$, and protolytic reactions whose contribution to ΔpH will depend on the buffering capacity of the two aqueous phases. Some examples of these reactions follow.

ELECTRON TRANSFER FROM FORMATE OR HYDROGEN TO FUMARATE   In the bacterium *Vibrio succinogenes* the oxidation of formate by fumarate is an important reaction for generating ATP. Indications are that NAD-independent formate oxidation occurs on the periplasmic side of the cell membrane, and that fumarate is reduced at the cytoplasmic surface (40). Thus for each molecule of formate oxidised, two electrons cross the membrane, leaving two protons that are released into the external aqueous phase, whilst two protons for fumarate reduction are taken up from the cytoplasm (40). Although the scheme for *V. succinogenes* is appealingly simple a different organization has been proposed for the same reaction in *E. coli* (41). In the latter both formate oxidation and fumarate reduction are suggested to occur at the cytoplasmic surface (42, 43) and the generation of Δp (44) is attributed to the operation of a single redox loop in which hydrogen atoms from formate oxidation are carried to the periplasmic side, where protons are released and from where electrons return across the membrane to the site of fumarate reduction. Energetically there is little to be gained by using the redox loop mechanism rather than the simple electron translocating mechanism suggested for *V. succinogenes,* but a cytoplasmic site may be advantageous for oxidizing formate that is generated by intracellular metabolism. All bacteria may not have adopted the same mechanisms for generating Δp, but if formate dehydrogenase in *V. succinogenes* spans the membrane, as it does in *E. coli* (45), then some of the evidence that formate is oxidized periplasmically may be rendered ambiguous.

Two protons have been shown to be released at the periplasmic side of *E. coli* when H_2 is oxidised by hydrogenase, and when the electron acceptor is fumarate a single electron translocating step is proposed to generate $\Delta\psi$ (46). When instead, nitrate is the final electron acceptor, it is proposed that this electron translocating step is followed by a single complete redox loop involving ubiquinone, and that no proton pumps are linked to these electron transfer reactions in *E. coli* (47).

ELECTRON TRANSFER FROM METHANOL TO OXYGEN   Bacteria that grow on methanol as carbon source possess a methanol dehydrogenase that has a pyrrolo-quinoline quinone as cofactor and feeds electrons to c-type cytochromes that in turn are oxidised by aa_3-type cytochromes. For *Paracoccus denitrificans* there is evidence that the dehydrogenase, which can be isolated as a water soluble protein, is located at the periplasmic surface (48). Thus protons released upon oxidation of methanol will be released into the external phase [as observed with ferricyanide as electron acceptor from cytochrome *c* (49)], whilst it is probable that the electrons will be carried across the membrane by cytochrome aa_3 to combine with protons and O_2 on the cytoplasmic side (48). This simple electron translocating scheme for generating $\Delta\psi$ predicts the appearance of two protons in the external phase for each methanol molecule oxidised by oxygen, as observed in some bacteria (48, 50). However, in *P. denitrificans* a stoichiometry of four protons has been observed (49), and thus it is possible that the electron translocation mechanism may be supplemented by a proton pump associated with cytochrome aa_3, although there is no evidence yet for this from experiments with the reconstituted purified protein (51).

ELECTRON TRANSFER FROM Fe^{2+} TO OXYGEN   *Thiobacillus ferro-oxidans* secures its energy supply for growth from the oxidation of Fe^{2+} by O_2. This reaction has been shown to involve the inward movement of electrons from the periplasmic site of Fe^{2+} oxidation to the cytoplasmic site of O_2 reduction (52). The role of this reaction in generating $\Delta\psi$ is evident when the change in Δp upon initiating electron transfer is examined. This bacterium grows when the external pH is 2, and in the absence of Fe^{2+} oxidation there is a substantial ΔpH (cytoplasm alkaline) equivalent to 242 mV, while $\Delta\psi$, possibly due to a proton diffusion potential, has an opposing value of 184 mV (cytoplasm positive) leaving a net Δp of 58 mV (53). Electron transfer reduces $\Delta\psi$ to 10 mV (cytoplasm positive) which, together with a contribution of 266 mV from ΔpH under these conditions, provides a net Δp of 256 mV (53). This establishes that a chemiosmotic mechanism is compatible with cell growth when the external pH is several units lower than intracellular pH.

GENERAL REMARKS   The existence of these relatively simple mechanisms for generating Δp shows how charge separation can take place across the whole width of a membrane exactly as expected from chemiosmotic theory, and it is difficult to reconcile these observations with suggestions that energy coupling occurs through localized charge separation or without the involvement of bulk aqueous phases (Section IX). It would, however, be an oversimplification to infer that proton-consuming reactions are always at the cytoplasmic side of bacterial membranes, as, for example, there is strong evidence that nitrite reduction occurs at the periplasmic surface in two species of bacteria (54–56).

IV. THERMODYNAMIC RELATIONSHIPS BETWEEN SPANS OF REDOX POTENTIAL AND Δp

A current uncertainty about the mitochondrial electron transfer chain is the number of protons that are translocated per pair of electrons passing from either succinate or NADH to oxygen. The technical problems in making these measurements and the suggested stoichiometries are reviewed elsewhere (3, 5), but these stoichiometries cannot be considered in isolation from the magnitude of Δp and the driving force for proton translocation which, for a two electron reaction, is twice the difference in redox potential ($2 \times \Delta E$) between the reductant and the oxidant couples. The general relationship at equilibrium between ΔE and Δp is given in Table 1. Thus, as discussed by Wikström et al (2), under conditions where the redox span from succinate to O_2 is 2×760 mV, and if Δp is 200 mV (section II), it is difficult to accommodate the translocation of eight protons per electron pair, especially if allowance is made for the partial irreversibility of O_2 reduction (2) which would tend to prevent complete equilibration of ΔE with Δp. However, these considerations do not necessarily rule out that, under the conditions of direct measurement (e.g. for O_2 pulse measurements), when Δp is small, the translocation stochiometry is indeed eight, because as discussed in Section VI the effective stoichiometry may be closer to the mechanistic one at smaller Δp.

Not all redox spans in either mitochondria or bacteria can be treated as simply as the succinate-O_2 reaction. In the latter there is no overall movement of electrons across the membrane because succinate oxidation and O_2 reduction are thought to occur at the same (matrix) side. Thus the

Table 1 Summary of important thermodynamic relationships

Protonmotive force:
$$\Delta p^a = \Delta\psi - 2.3RT\ \Delta pH/F^b$$

Phosphorylation potential:
$$\Delta Gp = \Delta G^{0'} + RTln[ATP]/[ADP]\ [P_i]$$

Equilibrium relationship between ΔGp and Δp excepting extramitochondrial (i.e. cytosolic in eukaryotes) ΔGp:
$$\Delta Gp = H^+/ATP \cdot F^b \cdot \Delta p$$

Equilibrium relationship between extramitochondrial ΔGp and Δp:
$$\Delta Gp = (H^+/ATP + x^c) \cdot F^b \cdot \Delta p$$

Usual equilibrium relationship between redox potential span (ΔE) over a segment of electron transfer chain that translocates n H^+ per $2e \cdot^d$:
$$2\Delta E = n \cdot \Delta p$$

---

[a] The sign of Δp depends on which side of the membrane is taken as the reference; our preference is to assign a positive value when Δp is capable of driving an endergonic reaction. The signs of $\Delta\psi$ and ΔpH are not always stated consistently; recall that ΔpH adds to $\Delta\psi$ when the relatively positive phase is also relatively acid.

[b] Faraday constant.

[c] $x = 1$ and represents the combined effect of the exit of OH^- from mitochondria in electroneutral exchange for P_i together with exchange of ATP^{4-} for ADP^{3-}, which is equivalent to entry of one proton into the mitochondrial matrix.

[d] Does not necessarily apply when the number of translocated protons is not equal to the number of translocated charges (see Section IV).

number of translocated charges also equals the number of translocated protons. In the cytochrome c to O_2 reaction there is the movement of one charge across the membrane as an electron flows from cytochrome c to O_2, and also (see 3, 5), the movement of a second charge as one proton is pumped in the opposite direction. Therefore the redox reaction moves two charges but only one proton, and thus, provided the individual redox reactions are defined in the phase where they occur, and if equilibrium were ever reached, ΔE (for one electron) should be equated with $2\Delta\psi + \Delta pH$. Wikström (2) prefers a different but thermodynamically equivalent analysis for this reaction in which he formally regards both the redox reactions as occurring in the same external phase (i.e. the cytoplasmic side of the membrane). The proton required for O_2 reduction is considered to be formally translocated in addition to the pumped proton. Thus at equilibrium the redox reaction, as defined by Wikström, would equate with $2\Delta\psi + 2\Delta pH$, where the redox potential of the O_2/H_2O couple is calculated using the external pH, whereas in our treatment the internal pH is used. For equal concentrations of O_2 and of both reduced and oxidised cytochmome c the two treatments predict the same equilibrium values for $\Delta\psi$ and ΔpH.

There is widespread agreement that the transfer of two electrons from succinate to cytochrome c is associated with the translocation of four protons, but only two positive charges, out of mitochondria (2). The discrepancy between the number of protons and charges arises because cytochrome c is on the opposite site of the membrane to succinate dehydrogenase. Following our preference for the basis of evaluating ΔE, this span of the chain will thus reach equilibrium when ΔE (for one electron) $= \Delta\psi + 2\Delta pH$, which in turn means that ΔE can be greater than Δp ($= \Delta\psi + \Delta pH$) if ΔpH contributes to Δp. As with the cytochrome c-O_2 reaction, ΔE can alternatively be defined with respect to the cytoplasmic (external) side of the mitochondrial membrane, in which case ΔE (with a different value) would be equated with $\Delta\psi + \Delta pH$ (2). However, for submitochondrial particles and chromatophores, in which the external phase is topologically equivalent to the mitochondrial matrix, both methods of analysis would equate ΔE with $\Delta\psi + 2\Delta pH$ for electron flow through the cytochrome bc_1 complex. These points need to be understood clearly in order to describe situations where substrate oxidation and the physiological terminal electron accepting reaction occur on opposite sides of a membrane, as is the case for nitrite reduction in some bacteria (54–56).

Unequal P:2e Ratios for Different Segments of the Mitochondrial Respiratory Chain?

The distinction between the numbers of translocated charges and protons affects the yield of ATP. According to the above descriptions the passage of two electrons from succinate to cytochrome c is associated with the movement of two charges, whereas the flow of two electrons from cytochrome c to oxygen results in the movement of four charges. As ATP synthesis in mitochondria is associated with the movement in the reverse direction of a given number (see Section V) of charges (protons) through the ATP synthase, a higher stoichiometry (by a factor of two) of ATP production should be associated with the cytochrome c to O_2 reaction, even though the number of translocated protons is larger (four per two electrons compared with two) for the succinate to cytochrome c reaction. The discrepancy between the yield of ATP from the two reactions becomes even greater if the proposals of two translocated protons per electron flowing from cytochrome c to O_2 are accepted (57, 58), although according to this scheme an equal number of protons (four per two electrons) would be translocated by the two reactions. The number of charges moved would now differ by a factor of three.

That ATP synthesis stoichiometries are unequal for these two segments of the mitochondrial respiratory chains is supported by experimental evidence (5, 59) and is consistent with a much smaller redox span for succinate-cytochrome c than for cytochrome c-O_2. Furthermore, as Δp is a delocalized energy pool between electron transfer and ATP synthesis, there is no requirement for those stoichiometries to be whole numbers. It is also of interest that steady-state swelling of mitochondria, as a result of respiration-driven uptake of potassium phosphate or calcium acetate, depended on the respiratory span used. Electron flow from succinate to ferricyanide, even at high rates, was able to maintain only a fraction of the swelling that was seen during electron flow from succinate or ascorbate to oxygen (60).

V. THERMODYNAMIC RELATIONSHIPS BETWEEN FREE ENERGY OF ATP SYNTHESIS AND Δp

The relationship between Δp and ΔG_p (Table 1) is important because it places limits on the stoichiometry of proton translocation per molecule of ATP synthesized or hydrolyzed. Alternatively, if the latter is known from independent experiments, the comparison of Δp with ΔG_p provides a test of the thermodynamic competence of Δp. With the exception of mitochondria, a stoichiometry obtained in this way will relate to the protons that are translocated by the ATP synthase enzyme, and is expressed as H^+/ATP (Table 1). In the case of mitochondria, the external ΔG_p is related to Δp not only by H^+/ATP but also by an extra term (Table 1), which expresses the net movement of protons into the matrix associated with the combined process of phosphate transport and ATP-ADP exchange. The internal (matrix) ΔG_p is related to Δp by H^+/ATP alone.

Measurement of Δp and ΔG_p can only give a value for H^+/ATP (or $x + (H^+/ATP)$) if equilibrium between Δp and ΔG_p is attained. When Δp is driving ATP synthesis this condition may in principle be achieved, although in practice it is possible that kinetic factors prevent ΔG_p from reaching its maximum value, in which case comparison with Δp gives a minimum value for H^+/ATP. When ATP hydrolysis is generating Δp, a system probably never reaches equilibrium owing to dissipation of Δp, and an overestimate for H^+/ATP will ensue. Moreover in both cases the ATP synthase enzyme might not operate as a fully coupled proton translocator, which has the same consequence for the estimate of H^+/ATP as failure to reach equilibrium (61).

Values for H^+/ATP

MITOCHONDRIA Under some reaction conditions the values of Δp and extramitochondrial ΔG_p are consistent with a value of three for $(x + (H^+/ATP))$ (15, 62–64), which, taken together with the generally accepted value of $x = 1$, suggests that mitochondrial ATP synthase translocates two protons for each molecule of ATP synthesized. However, this may be an underestimate: First, the values of ΔG_p that have been compared with Δp are often less than the maximum values of ΔG_p previously measured for mitochondria (65, 66); second, there is some evidence that the adenine nucleotide translocator imposes a kinetic limitation, thus preventing full equilibration of ΔG_p with Δp (67); and finally, slight overestimation of the value of Δp cannot be excluded.

Direct determinations of H^+/ATP by simultaneously measuring ATP hydrolysis and proton translocation can have technical difficulties. The most recent study (57) suggests a value of three, in contradiction to the estimate of two that was made in earlier work. But the value of H^+/ATP must be related not only to Δp but also to the P/O ratio, which for succinate oxidation has been suggested to be only ~ 1.4 (68). Only if this is the case can a translocation of 6 protons (which is widely accepted as a plausible value) per 2 electrons passing from succinate to O_2 be consistent with H^+/ATP equals 3 and x equals 1. In summary, a final adjudication on the value of H^+/ATP cannot yet be made from results of experiments in mitochondria.

SUBMITOCHONDRIAL PARTICLES   The maximum value of ΔG_p sustained by respiring particles is very similar to the intramitochondrial ΔG_p and much less than the extramitochondrial ΔG_p (69–71). As the particles are largely of the opposite orientation to intact mitochondria, these results are in full agreement with the concept that the elevated value of ΔG_p measured outside mitochondria is a consequence of the proton movement associated with adenine nucleotide and phosphate transport (Table 1).

Measurements of Δp in particles are valuable because they provide results that are complementary to those from mitochondria, and comparison of ΔG_p with Δp relates to H^+/ATP for the ATP synthase. The values of Δp, and the relative contributions from $\Delta \psi$ and ΔpH, vary widely; the discussion of some of the underlying reasons is given by Branca et al (61) who found that the H^+/ATP value depended on the reaction conditions chosen. The overall conclusion was that for the ATP synthase in submitochondrial particles the minimum H^+/ATP value is three. An alternative approach to the estimation of H^+/ATP is to measure ATP-dependent Δp at a range of values of ΔG_p in the absence of electron transport. When this is done using a new procedure to clamp ΔG_p, a value of $H^+/ATP = 3$ emerges (72). Direct measurements of ATP-driven uptake of protons into particles suggested an H^+/ATP equals two (73), and therefore further work is needed to resolve the discrepancy between the thermodynamic and kinetic methods for determining H^+/ATP values.

BACTERIA   It is straightforward to measure the total contents of ATP, ADP, and phosphate in a bacterium, although the relative amounts may be perturbed slightly during the necessary acid-quenching of the cells. However, the ΔG_p value obtained in this way might be erroneous because ΔG_p, which should be calculated using only the unbound concentrations, would be, for example, overestimated if ATP is preferentially bound to cellular components. Nevertheless, in a respiring bacterium (excluding for the moment alkalophiles), a ΔG_p value of the order of 11 kcal mol^{-1} is a reasonable estimate, which, together with a value of between 150 mV and 170 mV for Δp, suggests a value for H$^+$/ATP of more than two. This is a minimum value because Δp is not likely to reach equilibrium with ΔG_p, since ATP is constantly utilized in the cell.

It is possible to prepare from bacteria inverted membrane vesicles that catalyze oxidative- or photophosphorylation. These vesicles are able to generate high values of ΔG_p (12, 74–77), comparable in some cases to the extramitochondrial values. Measurement of Δp made with lipophilic anions and weak bases gives values that are inconsistent with an H$^+$/ATP of two and that suggest that this stoichiometry must be at least three if not higher (12, 76, 77). Only when Δp was measured using either the carotenoid band shift or an oxonol dye (78) for $\Delta \psi$ determination, and 9-amino acridine fluorescence quenching for ΔpH, was the data consistent with an H$^+$/ATP of two (75). However, the use of the oxonol dye with *Rhodospirillum rubrum* chromatophores (78) required a calibration of the dye against the carotenoid band shift in another bacterium, which means that any overestimation of $\Delta \psi$ by the carotenoid shift would inevitably be duplicated by the oxonol dye (77, 79).

Evidence that H$^+$/ATP equals two for a bacterial ATPase has come from experiments in which the movement of charges, which were assumed to be protons, across chromatophore membranes was measured and correlated with ATP synthesis following a single turnover of the electron transfer chain (27). Subsequently it was recognized that there may have been a thermodynamic discrepancy, because even when estimated on the basis of the carotenoid band shift, $\Delta \psi$ (ΔpH was negligible) was inadequate to account for net ATP synthesis under the prevailing value of ΔG_p (80). Therefore the interpretation that H$^+$/ATP equals 2 has an element of uncertainty.

The consequence of an equal H$^+$/ATP for mitochondrial and bacterial ATP synthases is that for a given value of Δp, bacterial membranes would generate a lower ΔG_p at equilibrium than is found outside mitochondria, because only in the latter is there a contribution to the development of a high ΔG_p value from adenine nucleotide translocation and phosphate transport (Table 1). However, considering that there is little difference in the two experimental values of ΔG_p, then either Δp or H$^+$/ATP is higher for bacterial systems. From the available experimental evidence the latter is more probable, although further data are required to resolve this matter, and indeed to determine whether all bacterial ATP synthases have identical stoichiometries (see Section V: alkalophilic bacteria).

CHLOROPLAST THYLAKOIDS   The values of Δp [almost exclusively ΔpH in the steady state (81)], and ΔG_p are commensurate with an H^+/ATP of three (81, 82), and a similar value has been estimated both from the extent of proton uptake into thylakoids driven by ATP hydrolysis (82), and from measurements of the proton efflux associated with ATP synthesis (83). Thus although some measurements made under conditions of limited turnover of the ATP synthase have given a value of 2.5 (84) the weight of evidence from independent approaches points strongly to an H^+/ATP of 3.

GENERAL REMARKS   The similarities of the proton translocating ATPases from mitochondria, bacteria, and thylakoids, might suggest that they have identical H^+/ATP ratios, although there is no stringent reason for believing that enzymes from different phylogenetic positions should adhere to a single pattern of behavior. Therefore the experimental evidence that indicates a stoichiometry of two for the mitochondrial but three for the thylakoid enzyme is not necessarily contradictory. If the proton translocation stoichiometries do indeed differ, this will have interesting consequences for structure-function relationships when the ATP synthases from different sources are better characterized.

Variations in Δp Are Not Always Reflected in ΔG_p

Provided that Δp equilibrates with ΔG_p, any alteration of the value of Δp should be paralleled by a proportional change in ΔG_p. Over a limited range of Δp values, this was the behaviour observed with chromatophores from *R. capsulata* (75) and mitochondria from brown fat (62), whilst vesicles from *P. denitrificans* exhibited only a small deviation from the expected pattern (28). However, in several studies with liver mitochondria, addition of an uncoupler, or altering the reaction conditions in a variety of other ways, reduced Δp considerably more than ΔG_p (19, 63, 85). On the basis of these results it was considered that Δp might not be the intermediate in oxidative phosphorylation, and a localized flow of protons was suggested to couple electron transfer to ATP synthesis (63, 85). More explicitly it was suggested that there was a resistance to proton flow into the bulk aqueous phase, and therefore dissipation of Δp by stimulating the movement of protons from the bulk phase by addition of an uncoupler would not be accompanied by a directly proportional effect on the "real" driving force for ATP synthesis (85). However, this view now appears untenable, as addition of a respiratory chain inhibitor, which should not stimulate the dissipation of Δp, also causes a greater reduction in Δp than in ΔG_p (64).

The discovery of bacteriorhodopsin as a simple light-driven proton pump that occurs in distinct patches of the membrane of *Halobacterium halobium,* and therefore could have no direct interactions with the ATP synthases, gave a final impetus to the acceptance of the role of Δp in energy coupling. Yet recent work with cells of *H. halobium* (86) has apparently shown that under some conditions, including the presence of an uncoupler, there is poor correspondence between changes in Δp and ATP synthesis. Indeed, if correction is made for the binding of the phosphonium cation that was used for $\Delta \psi$ determinations, substantial ATP synthesis is observed upon illumination of cells under conditions in which the measured Δp was zero (86). Thus Michel & Oesterhelt concluded that protons pumped by bacteriorhodopsin do not equilibrate with the bulk phase, but rather flow along the surface of the membrane until they reach an ATP synthase enzyme (86). This is equivalent to the mechanism that was discussed above to account for the independent variation of Δp and ΔG_p in mitochondria, and we return in Section IX to this problem, which is also extensively discussed by Westerhoff et al (64).

An alternative explanation for the different behavior of Δp and ΔG_p would be if the ATP synthase enzymes were able to increase their mechanistic stoichiometry H^+/ATP as Δp decreases. Although this might be difficult to envisage, Mitchell has already entertained this concept (87), and Cockrell et al (65) inferred a variable H^+/ATP for mitochondrial ATPase from measurements of ATP-driven K^+ transport. More recently there has been evidence that the plasma membrane ATP-driven proton pump of *Neurospora crassa* alters H^+/ATP from two to one depending on the prevailing metabolic conditions (88). Thus, however unattractive, the notion of variable stoichiometry cannot be dismissed.

Alkalophilic Bacteria

It has long been suspected that these bacteria, which grow at high external pH, might pose an interesting challenge to the role of Δp. The argument has run that because cells will tend to maintain an internal pH of around seven, then when the external pH is as high as eleven the contribution from ΔpH would subtract 4 pH units or 240 mV from the total Δp. Thus only if $\Delta \psi$ were to be well in excess of 240 mV would Δp be sufficient to drive ATP synthesis and active transport processes. Studies of three species of alkalophilic bacteria have now shown that the internal pH can be as high as 9.5 (89–91), and thus just as bacteria have evolved strategies for survival at high temperatures so cellular metabolism can adapt to relatively high pH. In *Bacillus alkalophilus,* for example, the internal pH was nine when the external pH was eleven, but as $\Delta\psi$ was 135 mV Δp was only 15 mV (89). This raises obvious problems for the synthesis of ATP, unless the cellular ΔG_p is exceptionally low or the H^+/ATP is very high (Table 1). Work with *B. alkalophilus* has now been extended to measurements of ATP synthesis and Δp in right-side-out vesicles loaded with ADP and P_i during their preparation (92), and in which respiration generated a $\Delta\psi$ of 122 mV (92). As the external pH was, at 10.5, 1.5 units higher than the internal pH, Δp was only 32 mV and not thermodynamically competent to account for the observed ΔG_p unless H^+/ATP was at least 6. This work also showed that ATP synthesis was sensitive to the usual uncouplers, and was not dependent on the Na^+ gradient that has been shown to participate in some active transport processes in this bacterium (92). Thus the expected challenge to the role of Δp has emerged, but in preliminary work with *B. pasterii* larger values of $\Delta\psi$, and thus Δp were reported, although an external pH of ten was the highest studied (90). A possible source of underestimation of $\Delta\psi$ in *B. alkalophilus* may, in principle, be an overestimation of the internal volume of the cells, but the adoption of a lower volume would also increase the estimate of the internal concentration of the methylamine that was used to indicate ΔpH. The consequence is that an increase in $\Delta\psi$ would be matched by an equal increase in the opposing ΔpH, and Δp would remain unaltered.

Further work on alkalophiles may reveal whether these bacteria, which appear to drive outwardly directed respiratory chain-linked proton translocation with no special features that would distinguish them from bacteria in general (93), adhere to fundamental chemiosmotic principles, but with higher than usual stoichiometries of proton translocation, or whether they will prove the exception to a general pattern of energy coupling and even perhaps prompt a wider reappraisal of the role of Δp.

VI. RELATIONSHIPS BETWEEN RATES OF ELECTRON TRANSFER AND Δp

Effect of Processes that Dissipate Δp

During steady-state respiration Δp remains constant, because the rate of proton translocation driven by the mitochondrial electron transfer chain exactly matches the rate of proton reentry into the mitochondrial matrix. Provided the same number of protons is always translocated for each pair of electrons passing through a section of the chain, the stimulation of respiration upon initiation of ATP synthesis, or upon addition of a protonophore uncoupler, can be qualitatively understood, because in both cases the proton reentry rate is increased. However, a quantitative description requires an understanding of changes in Δp, so that it can be determined how the magnitude of Δp is related to the respiration rate. In mitochondria from both brown adipose tissue and liver, increases of up to sevenfold in respiration rate upon adding an uncoupler are accompanied by linear, but relatively small, decreases in Δp (94, 95). With adipose mitochondria other means of reducing Δp had exactly equivalent effects on the respiration rate (94). The situation is not so simple in liver mitochondria, because the stimulation of respiration by the onset of ATP synthesis apparently is accompanied by a smaller diminution in Δp than is seen when respiration is increased to the same extent with an uncoupler (16, 67, 95, 96). This result seems to pose problems for the understanding of respiratory control in terms of Δp alone and even for the role of Δp as the intermediate of oxidative phosphorylation (95, 96), and therefore stringent tests to exclude experimental problems are required.

As Mitchell and Moyle (16) pointed out, and Hansford (97) has reemphasised, respiratory control may depend not on Δp alone, but rather on the difference between $n\Delta p$ and twice the span of redox potential difference (2 ΔE) over a segment of the respiratory chain that translocates n protons upon the passage of two electrons. Therefore under conditions of maximum respiratory control the values of $2\Delta E$ and $n\Delta p$ would be approximately equal (16), and increases in the respiratory rate would be related to the extent of disequilibrium between $n\Delta p$ and $2\Delta E$, although, a priori, there is no reason why the relationship should be linear. It is at present difficult to assess this proposal since, with the exception of a study of blowfly mitochondria (98), in which, atypically, the respiratory chain becomes more oxidised during uncontrolled respiration, parallel measurements of Δp and ΔE have not been made.

An alternative view of respiratory control envisages that the rate of respiration depends on the concentration of reduced cytochrome a_3, because the reaction of this species with oxygen is the only irreversible step in the respiratory chain and therefore it is the prime candidate for the site of control (99). A low concentration of reduced cytochrome a_3, and therefore a low respiration rate occurs when the respiratory chain on the reducing side of cytochrome a_3 is at near equilibrium with the extramitochondrial ΔG_p and is relatively oxidized [(99); but contrast (98)]. However, as Δp is the probable intermediate between the respiratory chain and ATP synthesis, it is likely to be the factor that is in near-equilibrium, and should thus be more closely related to the respiratory rate than ΔG_p, as indeed some (62, 67, 94), but not all (19), of the experimental evidence suggests. Sufficient data are not available to decide whether the relationship between Δp and the concentration of reduced cytochrome

a_3, or alternatively the concentration of an oxygenated intermediate that participates in an irreversible partial step (2,100) in the cytochrome aa_3 reaction, correlates with respiratory control, but this warrants consideration.[3]

Effect of Inhibitors of Electron Transfer

Whereas we have a qualitative understanding of the relationship between decreases in Δp and increases in electron transport rate, that between decreases in electron transport after addition of an inhibitor and changes in Δp is less clear. If the mitochondrion or other energy transducing unit were to have a constant resistance to the reentry of protons via passive leaks (we consider only conditions where there is no utilization of Δp), then the rate of reentry might be predicted to depend linearly (by analogy with Ohm's law) on the driving force Δp. Therefore any reduction in the electron transfer rate would be accompanied by a parallel decrease in the rate of proton entry, the size of Δp, and the extent of respiratory control, provided the stoichiometry of proton translocation is invariant. Indeed a study (101) with submitochondrial particles of the decrease in extent of respiratory control as the respiratory rate was reduced by titration with inhibitors, appeared to fit well to this model, although Δp was not measured. However, this work predated the recognition that in mitochondria (15), submitochondrial particles (102), bacteria (75, 103), and thylakoids (104) substantial extents of inhibition of electron transfer have little or no effect on Δp. Nevertheless, despite a constant Δp, the decrease in respiratory control with increasing inhibitor concentration in *P. denitrificans* vesicles (103) closely resembled the pattern in submitochondrial particles.

The insensitivity of Δp to the electron transfer rate means that the requirement for equal rates of proton translocation and reentry via leaks in a steady state is met either because there is a mechanism whereby the reentry rate can vary at constant Δp, or because the effective or phenomenological stoichiometry of proton translocation increases as the electron transfer rate decreases. Although the latter explanation has been recognised as a possibility (15, 102), the idea of a variable proton leakage rate or conductance has been most favored, largely because of the difficulty, especially from the standpoint of proton translocating redox loops (39), of imagining

[3]Note that an irreversible step in the respiratory chain is not compatible with the recent contention (100a) that the overall reaction from NADH to O_2 closely approaches equilibrium with the ATP synthesis reaction. However, the evidence for an irreversible step within the overall reaction catalyzed by cytochrome aa_3 has been stengthened by the demonstration that under appropriate conditions H_2O can be oxidized as far as a peroxide or equivalent derivative of cytochrome aa_3 but no further (100b).

how the extent to which electron flow is coupled to proton translocation can be varied. Nevertheless, Pietrobon et al (105) have suggested that the very small decrease in Δp upon titrating succinate oxidation in mitochondria with malonate, or antimycin, is best explained if electrons can sometimes pass through proton translocating segments of the respiratory chain without any proton translocation occurring; a process they term redox slip. Therefore as the respiratory rate is slowed, the proportion of redox slips would correspondingly decrease so as to maintain a constant rate of proton translocation and thus a relatively constant Δp without any requirement for variable rates of proton reentry (leaks).

A possible implication of redox slips is that, during controlled steady-state respiration, when Δp is high, the effective or phenomenological stoichiometry of proton translocation may not be as high as the stoichiometry observed in O_2 pulse experiments where Δp is low. However, it would be the lower effective stoichiometry that ought to be used in thermodynamic comparisons of Δp with redox spans (see Section IV). It is also of interest here that an increase in proton translocation stoichiometry has been reported for thylakoids when the light intensity, and therefore the electron transfer rate, is decreased (83).

At present a final choice between variable proton leaks and redox slips, or a combination of the two, as the basis for an approximately constant Δp cannot be made, although reversible increases in conductivity of some membranes have been reported (106). On the other hand, there is evidence from work with *Streptococcus lactis* that the magnitude of proton leaks is independent of Δp (107). The response of Δp to decreases in rates of electron transfer is related to the problem of respiratory control, because the relation between Δp and the rate of proton leakage will in turn determine the rate of proton translocation in a steady state. Further studies with inhibitors, which in general cause increased oxidation in a respiratory chain after their site of action, should also assist in deciding whether $2\Delta E - n\Delta p$ is an important factor in controlling the respiration rate.

There is no thermodynamic basis for a necessary dependence of Δp on the rate of electron transfer, and indeed there are plausible physiological reasons why an organelle or organism should maintain a constant Δp. For example, the rate of electron transport in a bacterium may vary quite widely according to fluctuations in the supply of substrate and/or electron acceptor, but if Δp also fluctuated the consequence could be drastic changes in cellular ion and metabolite concentrations. The related implications of a constant proton leakage rate in photosynthetic organisms growing at low irradiances have been considered by Raven & Beardall (108).

VII. RELATIONSHIPS BETWEEN RATE OF ATP SYNTHESIS AND Δp

The concept that ATP synthesis is driven by Δp, and is independent of reactions that generate Δp, is not accompanied by any a priori expectation as to the exact relationship of Δp to the rate of ATP synthesis. It is axiomatic that equivalent changes in Δp, whatever their cause, be reflected by identical alterations in the rate of ATP synthesis. Accordingly, equivalent attenuations of Δp in thylakoids, either by reducing the light intensity or by partial uncoupling, had equal effects on the rate of ATP synthesis (109). When similar experiments were done with bacterial chromatophores a quite different picture was obtained, since, whereas progressive addition of an uncoupler increasingly inhibited the rate of ATP synthesis and diminished Δp (75), reduction of light intensity or addition of antimycin, caused very much smaller decreases in Δp for equivalent extents of inhibition of ATP synthesis (75). These findings have led to an extensive study in chromatophores of the factors controlling the rate of ATP synthesis in both pre-steady-state and steady-state conditions, and the general conclusion is that the provision of Δp, although essential for ATP synthesis, is not sufficient (6, 7, 110). The synthesis of ATP is seen to be tightly coordinated to the flow of electrons through the proton translocating segment of the electron transfer chain. One possibility is that this coordination is observed because protons flow from the electron transfer chain to the ATP synthase without fully delocalizing into the aqueous phase (6, 7, 80). It should also be recognized that these observations cannot be separated from the considerations of Section VI; it is the failure of Δp to decrease when electron flow is inhibited that creates the unexpected relationships between Δp and the rate of ATP synthesis.

It is not only in chromatophores that substantial extents of inhibition of electron transfer are paralleled by a proportional decrease in the rate of ATP synthesis, whilst the magnitude of Δp is either unaltered or decreases only very slightly, but also in mitochondria (111), submitochondrial particles (112), and bacterial vesicles (103). These findings of an independence of the rate of ATP synthesis from the magnitude of Δp have again led to suggestions (28, 111, 112) that protons flow relatively directly from the redox reactions to ATP synthases without fully equilibrating with Δp, or that the rate of ATP synthesis is kinetically controlled by electron transfer events (28, 111, 112). Some of the data could be explained without invoking such factors if the rate of ATP synthesis were to be highly sensitive to very small changes in Δp. This argument flounders because, for an equal decrease in the rate of ATP synthesis, uncoupling causes larger changes in Δp than inhibition of electron flow, although in contrast to chromatophores this discrepancy is small in bacterial vesicles (28), whilst in mitochondria the corresponding comparison (111) is difficult to interpret because the rate of ATP synthesis can also be altered by changes in the rate of adenine nucleotide translocation upon partial uncoupling.

The exact relationship between Δp and the rate of ATP synthesis is not only important for understanding the overall coordination between electron transfer and ATP synthesis (discussed above), but also for providing clues to the mechanism of the ATP synthase. Partial uncoupling of chromatophores revealed a linear relationship between Δp and the rate of ATP synthesis (75), with a 100 mV decrease in Δp causing a 50% inhibition of ATP synthesis; and logarithmic-linear relationships have been observed in thylakoids (109). However, linearity clearly does not account either for the results of the experiments with inhibitors of electron flow in chromatophores, or for other sets of data with thylakoids (113).

Of interest are two recent, related suggestions for the behavior of ATP synthase. Jencks (114) has proposed that there are two states of the ATP synthase, only one of which can catalyze ATP synthesis. An equilibrium between these states is displaced toward the catalytically competent form of the enzyme to an extent that will depend on the magnitude of Δp. But the dependence is not linear and is closer to the profile for a pH titration of a weak acid or base. Rumberg & Heinze (113) have formulated for the thylakoid enzyme a more explicit model in which translocated protons, three per ATP synthesized, bind sequentially but equivalently to the enzyme, and thus displace an equilibrium in favor of its active form. This model, which fits well with their experiments in which ΔpH was the sole component of Δp (113), predicts that the rate of ATP synthesis can be a steep function of Δp, because steps of the binding of protons from the internal phase and subsequent proton release to the external phase, when the enzyme returns to the inactive state, will be related to the cube of proton concentrations that are experienced by the inward and outward facing parts of the enzyme. For example, under some conditions a decrease in Δp from 180 mV to 175 mV would be sufficient to cause a 30% decrease in the rate of ATP synthesis. A model of this kind also predicts threshold values of Δp below which the ATP synthesis rate is virtually zero, even if ΔG_p is sufficiently low that there is no thermodynamic barrier to ATP synthesis.

As ΔpH is the sole component of Δp in thylakoids, changes in ΔpH must be accompanied by changes in the pH of the internal and/or external phases. Therefore it has to be assumed that changes in rates of ATP synthesis are a consequence of the binding and release of the translocated protons by the enzyme, and not of the influence of pH on the catalytic functional groups (113). The possibility of interference from the latter effect would be eliminated in a system in which $\Delta \psi$ is the sole component of Δp, but then the question arises whether $\Delta\psi$ can influence proton binding to and release from the ATP synthase in equivalent fashion to ΔpH. One possibility is that the proton channel segment (Fo) of the ATP synthase is a proton well so that at the interface between Fo and the catalytic segment (F_1) $\Delta\psi$ is proportionally converted into a high local concentration of protons (39), and thus the ATP synthase responds identically to $\Delta\psi$ or ΔpH of equivalent size. In *P. denitrificans* vesicles, where $\Delta\psi$ is the only component of Δp, small changes in Δp upon partial uncoupling cause large changes in the rate of ATP synthesis (28), and the data can be approximately fitted to a model related to that for the thylakoid enzyme (115). This result agrees with the contention that $\Delta\psi$ and ΔpH act equivalently on the ATP synthase, as was also concluded from measurements of rates of ATP synthesis in *S. lactis* (116).

A further feature of the thylakoid enzyme scheme (113) is that the highest values of Δp usually encountered (approximately 200 mV) are only just sufficient for the ATP synthase to reach its maximum rate (V_{max}). This aspect might provide the basis for understanding why an inhibitor of electron transfer and an inhibitor of the ATP synthase act additively on the overall reaction of oxidative phosphorylation or ATP-driven reversed electron flow [116a], even though Δp is a delocalised pool intermediate. If the rate of ATP synthesis is a linear function of the concentration of ATP synthase enzyme and a complex function of Δp (113), then whatever the rate of generation of Δp by electron transfer it can be predicted that the ATP synthase inhibitor sensitivity of the rate of ATP synthesis will be constant. In contrast there are examples, for instance in electron transport (117, 118), where reduced sensitivity to an inhibitor that acts after an intermediate pool is observed when the input to the pool is decreased. The invariant but unexplained sensitivity of adenine nucleotide translocation to atractyloside, irrespective of the flux rate (118), may have an analogous explanation to that advocated for the ATP synthase, because the mitochondrial $\Delta\psi$ is probably only just sufficient to ensure the operation of the translocator at its maximum rate (119).

Although models that relate Δp to the rate of ATP synthesis can suggest how small changes in Δp cause large changes in the rate of ATP synthesis, they cannot account for the same rate of ATP synthesis being associated with very different values of Δp in chromatophores (75), or why there is no change in Δp and yet a large change in the ATP synthesis rate (112). These problems need to be understood, otherwise the related and important steps of elucidating the mechanism of the ATP synthase from studies of the relation between Δp and the rate of ATP synthesis may not have such a secure basis as expected from chemiosmotic principles.

VIII. THE WIDESPREAD ROLE OF Δp

Our concentration on the behavior of Δp as the intermediate between redox reactions and ATP synthesis must not obscure the equally important appearances of Δp as the immediate driving force for other processes, including transport in mitochondria and bacteria. Amongst these processes is bacterial motility, and it is of particular interest that the relationship between Δp and the rate of bacterial motility has been studied (120, 121) (an analogy with our discussion on Δp and the rate of ATP synthesis). A striking development has been the realization that Δp is generated in eukaryote organelles other than mitochondria and chloroplasts, and we illustrate this with particular reference to one system.

Chromaffin granules

These are the storage sites for catecholamines in chromaffin cells of adrenal medulla. Work reviewed by Njus & Radda (122) has established that these granules have an ATPase that has features closely in common with the mitochondrial enzyme (123), is accessible to cytosolic ATP, and pumps protons from the cytosol into the interior of the granules. It is of general interest that the acidification of the interior can be directly followed from the ^{31}P NMR signal of internal ATP and that pH measured in this way agrees well with the value obtained from the uptake of methylamine (124, 125).

ATP hydrolysis drives a carrier-mediated and uncoupler-sensitive uptake of catecholamines into the granules, and evidence now indicates that protonated catecholamines carrying one positive charge are transported in exchange for two protons that leave the granule (126, 127). The thermodynamic implication of this mechanism is important because the driving force for transport will not be Δp but rather $\Delta \psi + 2\Delta pH$ (126, 127) as one net charge but two protons move out of the granule (compare discussion of redox spans in Section IV). A consequence of this mechanism is that even in the absence of ATP hydrolysis there is a small driving force for catecholamine uptake, because the internal pH of the granules is usually lower than the external pH owing to the presence of an internal acidic protein, chromagranin A. The membrane is sufficiently permeable to protons (on a time scale of minutes) for the development of an outwardly directed proton diffusion potential (i.e. one that is negative inside in contrast to positive when the ATPase is pumping protons), so that $-\Delta \psi = \Delta pH$, which leaves ΔpH as a net driving force for limited catecholamine uptake (126). The latter observation confirms the proposed relative stoichiometry of proton and catecholamine movement.

It is probable that chromaffin granules will prove to be a prototype for other storage systems in eukaryotes. In the yeast *Saccaromyces cerevisiae* basic amino acids are stored in vacuoles, and recently vacuolar membrane vesicles have been prepared (128). These have an inwardly directed ATP-driven proton pump that is estimated to generate Δp of 170 mV, which in turn can drive arginine transport (128). The vacuoles of some types of plant cells may also possess an electrogenic proton pumping ATPase (129).

There is increasing evidence that lysosomes have an inwardly directed ATP-driven proton pump (130, 131) but there is no direct indication yet that this can generate $\Delta\psi$, although attempts to detect $\Delta\psi$ (131) were made in a medium containing 100 mM Cl^-, which would probably cause dissipation of $\Delta\psi$ [cf chromaffin granules (122)]. There is speculation that an ATPase in lysosomes serves both to maintain an acidic interior and to generate $\Delta\psi$, which is involved in driving the uptake of molecules that are destined for degradation.

Even this brief survey, and we have not space to discuss new developments in ATP-dependent movement of protons across plasma membranes of *N. crassa* (132) and yeast (133), shows that proton electrochemical gradients are at work to drive many function in cells.

IX. COMMENTS ON ALTERNATIVES TO Δp AS THE INTERMEDIATE IN ENERGY COUPLING

This review gives examples, including the thermodynamics of ATP synthesis by alkalophiles and *H. halobium* and the relationship between the rate of ATP synthesis and Δp, that suggest that factors in addition to, or instead of, Δp play a role in energy transduction. Such examples have led Kell to formulate the electrodic view in which protons are envisaged to flow along the surface of a membrane within the Stern-Grahame layer, from which they are unable to equilibrate with a bulk phase (134). However, taking refuge in this electrodic or any other model in which charge separation or protons are confined to the membrane (85, 135, 136) is uncomfortable, because an imposed Δp in submitochondrial particles (137) or thylakoids (138) drives a rapid initial rate of ATP synthesis without any significant lag. Thus unless another interpretation can be placed upon those experiments, the crucial kinetic competence of Δp has been established, which means in turn that any protons within, or on the surface of, the membrane can equilibrate rapidly with the bulk aqueous phases, and therefore problems of energetic shortfall in Δp are not resolved. Also, in conditions where $\Delta\psi$ is zero, an enhancement of the buffering capacity of the internal aqueous phase of illuminated thylakoids leads to increased lags before the onset of

ATP syntheis (139, 140), exactly as would be expected if Δp, in the form of ΔpH, between aqueous phases is the intermediate in ATP synthesis.[4] In the case of thylakoids, consideration of the surface properties of the membrane has caused Barber (141) to conclude that an electrodic-type of mechanism probably does not operate.

On the other hand, the observation of $\Delta\psi$ between the aqueous phases on either side of a membrane is not necessarily incompatible with a role for charge separation restricted within membranes, because, if any such local charge separation has a component normal to the plane of the membrane, it will generate $\Delta\psi$ between the aqueous phases. This is illustrated by an experiment with chromatophores in which conditions were arranged so that illumination drove only the migration of electrons from reaction centers to primary acceptors, and thus the locations of the separated charges were known (142). The localized potential between the two charges was considered to be relayed to the bulk aqueous phase by a rapid capacitative coupling, by which is meant a reorientation of dipoles both within the membrane dielectric between the point charges and the interfaces as well as in the aqueous phases (142). Thus again events within a membrane are shown to be rapidly communicated to the bulk phases. However, the bulk phase $\Delta\psi$ established by charge separation or proton movement within the membrane would be less than had the charges been separated across the total width of the membrane. It follows that measurements between bulk phases (e.g. with permeant ion) would underestimate the true driving force for ATP synthesis if the charge separations involved were confined to the membranes.

Despite the reservation expressed in this section about the feasibility of alternatives to Δp, there is no escape from the unremitting requirement that intermediates in reactions must be thermodynamically competent, and thus, as long as there is good experimental data indicating that Δp does not meet this requirement (see Section V), the possibility, however difficult to imagine, that Δp is not the primary form in which energy is conserved will continue to warrant attention.

[4]Lemasters & Hackenbrock (140a) have argued that the quantity of uncoupler-sensitive ATP synthesis observed after mitochondrial electron transfer has been rapidly blocked with an inhibitor is consistent with a relatively large capacity for the intermediate of oxidative phosphorylation. Furthermore, when ADP was added to mitochondria immediately after inhibiting electron transfer the initial rate of ATP synthesis was comparable to that observed during steady state respiration. These data were considered to be fully compatible with Δp being an intermediate, and incompatible with models in which protons flow between electron transfer and ATP synthase reactions without equilibrating with the bulk aqueous phases (140a).

ACKNOWLEDGMENTS

We thank Baz Jackson, John McCarthy, and Mårten Wikström for valuable discussions, Maurizia Cuccia for splendid secretarial assistance, and many colleagues for sending us reprints and preprints. Work in the authors' laboratories has been supported by the Science and Engineering Research Council (SERC) (UK), Consiglio Nazionale delle Ricerche (CNR) (Italy), and NATO (Grant 1771).

Literature Cited

1. Mitchell, P. 1961. *Nature* 191:144–48
2. Wikström, M., Krab, K., Saraste, M. 1981. *Ann. Rev. Biochem.* 50:623–55
3. Wikström, M., Krab, K. 1980. *Curr. Top. Bioenerg.* 10:51–101
4. Conover, T. E., Azzone, G. F. 1981. *Mitochondria and Microsomes* ed. C. P. Lee, G. Schatz, G. Dallner, pp. 481–518 Reading, Mass: Addison-Wesley.
5. Wikström, M., Krab, K. 1979. *Biochim. Biophys. Acta* 549:177–222
6. Ort, D. R., Melandri, B. A. 1982. *An Integrated Approach to Plant and Bacterial Photosynthesis* ed. Govindjee, New York: Academic. In press
7. Baccarini-Melandri, A., Casadio, R., Melandri, B. A. 1982. *Curr. Top. Bioenerg.* In press
8. Junge, W., Jackson, J. B. 1982. See Ref. 6. In press
9. Rottenberg, H. 1979. *Methods Enzymol.* 55:547–69
10. Van Dam, K., Wiechmann, A. H. C. A. 1979. *Methods Enzymol.* 55:225–29
11. Price, N. C., Dwek, R. A. 1979. *Principles and Problems in Physical Chemistry for Biochemists,* pp. 124–29. Oxford: Oxford Univ. Press. 296 pp. 2nd ed.
12. Kell, D. B., Ferguson, S. J., John, P. 1978. *Biochim. Biophys. Acta* 502:111–26
13. Beavis, A. D. 1977. *Ornithine and Citrulline Transport in Mitochondria.* PhD thesis. Bristol Univ., UK 302 pp.
14. Halestrap, A. P. 1978. *Biochem. J.* 172:389–98
15. Nicholls, D. G. 1974. *Eur. J. Biochem.* 50:305–15
16. Mitchell, P., Moyle, J. 1969. *Eur. J. Biochem.* 7:471–84
17. Heinz, E., Geck, P., Pietrzyk, C. 1975. *Ann. NY Acad. Sci.* 264:428–41
18. Kashket, E. R., Blanchard, A. G., Metzger, W. C. 1980. *J. Bacteriol.* 143:128–34
19. Holian, A., Wilson, D. F. 1980. *Biochemistry* 19:4213–21
20. Cheng, K., Haspel, H. C., Vallano, M. L., Osotimehin, B., Sonenberg, M. 1980. *J. Membr. Biol.* 56:191–201
21. Casadio, R., Venturoli, G., Melandri, B. A. 1981. *Photobiochem. Photobiophys.* 2:245–53
22. Stark, G. 1980. *Biochim. Biophys. Acta* 600:233–37
23. Dilger, J. P., McLaughlin, S. G. A., McIntosh, T. J., Simon, S. A. 1979. *Science* 206:1196–98
24. Kell, D. B., John, P., Sorgato, M. C., Ferguson, S. J. 1978. *FEBS Lett.* 86:294–98
25. Singer, S. J. 1981. *Biochem. Soc. Trans.* 9:203–6
26. Ferguson, S. J., Jones, O. T. G., Kell, D. B., Sorgato, M. C. 1979. *Biochem. J.* 180:75–85
27. Petty, K. M., Jackson, J. B. 1979. *FEBS Lett.* 97:367–72
28. McCarthy, J. E. G., Ferguson, S. J. 1981. *Vectorial Reactions in Electron and Ion Transport in Mitochondria and Bacteria,* ed. F. Palmieri et al, pp. 349–57, Amsterdam: Elsevier. 429 pp.
29. Clark, A. J., Jackson, J. B. 1981. *Biochem. J.* 200:389–97
30. Symons, M., Nuyten, A., Sybesma, C. 1979. *FEBS Lett.* 107:11–14
31. Jacobson, L., Cohen, J. S. 1981. *Biosci. Rep.* 1:141–50
32. Cohen, S. M., Ogawa, S., Rottenberg, H., Glynn, P., Yamane, T., Brown, T. R., Shulman, R. G., Williamson, J. R. 1978. *Nature* 273:554–56
33. Ogawa, S., Rottenberg, H., Brown, T. R., Shulman, R. G., Castillo, C. L., Glynn, P. 1978. *Proc. Natl. Acad. Sci. USA* 75:1796–1800
34. Ogawa, S., Shen, C., Castillo, C. L. 1980. *Biochim. Biophys. Acta* 590:159–69
35. Nicolay, K., Lolkema, J., Hellingwerf, K. J., Kaptein, R., Konings, W. N. 1981. *FEBS Lett.* 123:319–23

36. Felle, H., Porter, J. S., Slayman, C. L., Kaback, H. R. 1980. *Biochemistry* 19:3585–90
37. Tedeschi, H. 1980. *Biol. Rev. Cambridge Philos. Soc.* 55:171–206
38. McCarthy, J. E. G., Ferguson, S. J., Kell, D. B. 1981. *Biochem. J.* 196:311–21
39. Mitchell, P. 1976. *Biochem. Soc. Trans.* 4:399–430
40. Kröger, A., Dorrer, E., Winkler, E. 1980. *Biochim. Biophys. Acta* 589: 118–36
41. Jones, R. W. 1980. *FEMS Microbiol. Lett.* 8:167–71
42. Garland, P. B., Downie, J. A., Haddock, B. A. 1975. *Biochem. J.* 152: 547–59
43. Boonstra, J., Konings, W. N. 1977. *Eur. J. Biochem.* 78:361–68
44. Hellingwerf, K. J., Bolscher, J. G. M., Konings, W. N. 1981. *Eur. J. Biochem.* 113:369–74
45. Graham, A., Boxer, D. H. 1981. *Biochem. J.* 195:627–37
46. Jones, R. W. 1980. *Biochem. J.* 188:345–50
47. Jones, R. W., Lamont, A., Garland, P. B. 1980. *Biochem. J.* 190:79–94
48. Alefounder, P. R., Ferguson, S. J. 1981. *Biochem. Biophys. Res. Commun.* 98:778–84
49. Van Verseveld, H. W., Krab, K., Stouthamer, A. H. 1981. *Biochim. Biophys. Acta* 635:525–34
50. Dawson, M. J., Jones, C. W. 1981. *Biochem. J.* 194:915–24
51. Ludwig, B. 1980. *Biochim. Biophys. Acta* 594:177–89
52. Ingledew, W. J., Cox, J. C., Halling, P. J. 1977. *FEMS Microbiol. Lett.* 2:193–97
53. Cox, J. C., Nicholls, D. G., Ingledew, W. J. 1979. *Biochem. J.* 178:195–200
54. Wood, P. M. 1978. *FEBS Lett.* 92:214–18
55. Meijer, E. M., van der Zwaan, J. W., Stouthamer, A. H. 1979. *FEMS Microbiol. Lett.* 5:369–72
56. Alefounder, P. R., Ferguson, S. J. 1980. *Biochem. J.* 192:231–40
57. Alexandre, A., Reynafarje, B., Lehninger, A. L. 1978. *Proc. Natl. Acad. Sci. USA* 75:5296–300
58. Azzone, G. F., Pozzan, T., Di Virgilio, F. 1979. *J. Biol. Chem.* 254:10206–12
59. Brand, M. D., Harper, W. G., Nicholls, D. G., Ingledew, W. J. 1979. *FEBS Lett.* 95:125–29
60. Baum, H. 1978. *The Molecular Biology of Membranes*, ed. S. Fleischer, Y.

Hatefi, D. H. MacLennan, A. Tzagoloff. pp. 243–62. New York: Plenum
61. Branca, D., Ferguson, S. J., Sorgato, M. C. 1981. *Eur. J. Biochem.* 116:341–46
62. Nicholls, D. G., Bernson, V. S. M. 1977. *Eur. J. Biochem.* 75:601–12
63. Azzone, G. F., Pozzan, T., Masari, S., Pregnolato, L. 1978. *Biochim. Biophys. Acta* 501:307–16
64. Westerhoff, H. V., Simonetti, A. L. M., van Dam, K. 1981. *Biochem. J.* 200: 193–202
65. Cockrell, R. S., Harris, E. J., Pressman, B. C. 1966. *Biochemistry* 5:2326–35
66. Slater, E. C., Rosing, J., Mol, A. 1973. *Biochim. Biophys. Acta* 292:534–53
67. Küster, U., Letko, G., Kunz, W., Duszynsky, J., Bogucka, K., Wojtczak, L. 1981. *Biochim. Biophys. Acta* 636:32–38
68. Hinkle, P. C., Yu, M. L. 1979. *J. Biol. Chem.* 254:2450–55
69. Ferguson, S. J., Sorgato, M. C. 1977. *Biochem. J.* 168:299–303
70. Thayer, W. S., Tu, Y. L., Hinkle, P. C. 1977. *J. Biol. Chem.* 252:8455–58
71. Bashford, C. L., Thayer, W. S. 1977. *J. Biol. Chem.* 252:8459–63
72. Sorgato, M. C., Branca, D., Simon, S., Stefani, L., Ferguson, S. J. 1981. See Ref. 28, pp. 407–10
73. Thayer, W. S., Hinkle, P. C. 1973. *J. Biol. Chem.* 248:5395–402
74. Eilermann, L. J. M., Slater, E. C. 1970. *Biochim. Biophys. Acta* 216:226–28
75. Baccarini-Melandri, A., Casadio, R., Melandri, B. A. 1977. *Eur. J. Biochem.* 78:389–402
76. Kell, D. B., John, P., Ferguson, S. J. 1978. *Biochem. J.* 174:257–66
77. Cirillo, V. P., Gromet-Elhanan, Z. 1981. *Biochim. Biophys. Acta* 636: 244–53
78. Bashford, C. L., Baltscheffsky, M., Prince, R. C. 1979. *FEBS Lett.* 97:55–60
79. Ferguson, S. J., Sorgato, M. C., Kell, D. B., John, P. 1979. *Biochem. Soc. Trans.* 7:870–74
80. Petty, K. M., Jackson, J. B. 1979. *Biochim. Biophys. Acta* 547:474–83
81. Avron, M. 1978. *FEBS Lett.* 96:225–32
82. McCarty, R. E. 1978. *The Proton and Calcium Pumps*, ed. G. F. Azzone, M. Avron, J. C. Metcalfe, E. Quagliariello, N. Siliprandi, pp. 65–70. Amsterdam: North Holland
83. Rathenow, M., Rumberg, B. 1980. *Ber. Bunsenges. Phys. Chem.* 84:1059–62
84. Witt, H. T. 1979. *Biochim. Biophys. Acta* 505:355–427
85. Van Dam, K., Wiechmann, A. H. C. A., Hellingwerf, K. J., Arents, J. C.,

216   FERGUSON & SORGATO

Westerhoff, H. V. 1978. *FEBS Proc. Meet.* 45:121–32
86. Michel, H., Oesterhelt, D. 1980. *Biochemistry* 19:4607–14
87. Mitchell, P. 1969. *Molecular Basis of Membrane Function*, ed. D. C. Tosterson, pp. 483–518. Englewood Cliffs NJ: Prentice-Hall
88. Warncke, J., Slayman, C. L. 1980. *Biochim. Biophys. Acta* 591:224–33
89. Guffanti, A. A., Susman, P., Blanco, R., Krulwich, T. A. 1978. *J. Biol. Chem.* 253:708–15
90. Hoddinot, M. H., Reid, G. A., Ingledew, W. J. 1978. *Biochem. Soc. Trans.* 6:1295–98
91. Guffanti, A. A., Blanco, R., Benenson, R. A., Krulwich, T. A. 1980. *J. Gen. Microbiol.* 119:79–86
92. Guffanti, A. A., Bornstein, R. F., Krulwich, T. A. 1981. *Biochim. Biophys. Acta* 635:619–30
93. Haddock, B. A., Cobley, J. G. 1976. *Biochem. Soc. Trans.* 4:709–11
94. Nicholls, D. G. 1979. *Biochim. Biophys. Acta* 549:1–29
95. Azzone, G. F., Pozzan, T., Massari, S., Bragadin, M. 1978. *Biochim. Biophys. Acta* 501:296–306
96. Padan, E., Rottenberg, H. 1973. *Eur. J. Biochem.* 40:431–37
97. Hansford, R. G. 1980. *Curr. Top. Bioenerg.* 10:217–78
98. Johnson, R. N., Hansford, R. G. 1977. *Biochem. J.* 164:305–22
99. Owen, C. S., Wilson, D. F. 1974. *Arch. Biochem. Biophys.* 161:581–91
100. Wilson, D. F., Owen, C. S., Holian, A. 1977. *Arch. Biochem. Biophys.* 182:749–62
100a. Lemasters, J. J. 1980. *FEBS Lett.* 110:96–100
100b. Wikström, M. K. F. 1981. *Proc. Natl. Acad. Sci. USA* 78:4051–54
101. Hinkle, P. C., Tu, Y.-S. L., Kim, J. J. 1975. *Molecular Aspects of Membrane Phenomena*, ed. H. R. Kaback, H. Neurath, G. K. Radda, R. Schwyzer, W. R. Wiley, pp. 222–32. Berlin: Springer
102. Sorgato, M. C., Ferguson, S. J. 1979. *Biochemistry* 18:5737–42
103. Kell, D. B., John, P., Ferguson, S. J. 1978. *Biochem. Soc. Trans.* 5:1292–95
104. Schonfeld, M., Neumann, J. 1977. *FEBS Lett.* 73:51–54
105. Pietrobon, D., Azzone, G. F., Walz, D. 1981. *Eur. J. Biochem.* 117:389–94
106. Benz, R., Conti, F. 1981. *Biochim. Biophys. Acta* 645:115–23
107. Maloney, P. C. 1979. *J. Bacteriol.* 140:197–205

108. Raven, J. A., Beardall, J. A. 1981. *FEMS Microbiol. Lett.* 10:1–5
109. Portis, A. R., McCarty, R. E. 1976. *J. Biol. Chem.* 251:1610–17
110. Melandri, B. A., Venturoli, G., De Santis, A., Baccarini-Melandri, A. 1980. *Biochim. Biophys. Acta* 592:38–52
111. Yaguzhinskii, L. S., Krasinskaya, I. P., Dragunova, S. F., Zinchenko, V. P., Yevtodiyenko, Yu. V. 1979. *Biophys. USSR* 24:1100–3
112. Sorgato, M. C., Branca, D., Ferguson, S. J. 1980. *Biochem. J.* 188:945–48
113. Rumberg, B., Heinze, Th. 1980. *Ber. Bunsenges. Phys. Chem.* 84:1055–1059
114. Jencks, W. P. 1980. *Adv. Enzymol.* 51:73–106
115. McCarthy, J. E. G. 1981. *Bioenergetics of Paracoccus denitrificans* PhD thesis. Birmingham Univ., UK. 198 pp.
116. Maloney, P. C., Schattschneider, S. 1980. *FEBS Lett.* 110:337–40
116a. Baum, H., Hall, G. S., Nalder, J., Beechey, R. B. 1971. *Energy Transduction in Respiration and Photosynthesis*, ed. E. Quagliariello, S. Papa, C. S. Rossi, pp. 747–755. Bari, Italy: Adriatica Editrice
117. Kröger, A., Klingenberg, M. 1973. *Eur. J. Biochem.* 39:313–23
118. Stubbs, M., Vignais, P. V., Krebs, H. A. 1978. *Biochem. J.* 172:333–42
119. Krämer, R., Klingenberg, M. 1871. See Ref. 28, pp. 291–98
120. Khan, S., MacNab, R. M. 1980. *J. Mol. Biol.* 138:599–614
121. Shioi, J., Matsuura, S., Imae, Y. 1980. *J. Bacteriol.* 144:891–97
122. Njus, D., Radda, G. K. 1978. *Biochim. Biophys. Acta* 463:219–44
123. Apps, D. K., Schatz, G. 1979. *Eur. J. Biochem.* 100:411–19
124. Njus, D., Sehr, P. A., Radda, G. K., Ritchie, G. A., Seeley, P. J. 1978. *Biochemistry* 17:4337–43
125. Pollard, H. B., Shindo, H., Creutz, C.-E., Pazoles, C. J., Cohen, J. S. 1979. *J. Biol. Chem.* 254:1170–77
126. Knoth, J., Handloser, K., Njus, D. 1980. *Biochemistry* 19:2938–42
127. Phillips, J. H., Apps, D. K. 1980. *Biochem. J.* 192:273–78
128. Ohsumi, Y., Anraku, Y. 1981. *J. Biol. Chem.* 256:2079–81
129. Lüttge, U., Smith, J. A. C., Marigo, G., Osmond, C. B. 1981. *FEBS Lett.* 126:81–84
130. Mego, J. L. 1979. *FEBS Lett.* 107:113–16
131. Schneider, D. L. 1981. *J. Biol. Chem.* 256:3858–64

132. Scarborough, G. A. 1980. *Biochemistry* 19:2925–31
133. Malpartida, F., Serrano, R. 1981. *Eur. J. Biochem.* 116:413–17
134. Kell, D. B. 1979. *Biochim. Biophys. Acta* 549:55–99
135. Fillingame, R. H. 1980. *Ann. Rev. Biochem.* 49:1079–1113
136. Williams, R. J. P. 1978. *FEBS Lett.* 85:9–19
137. Thayer, W. S., Hinkle, P. C. 1975. *J. Biol. Chem.* 250:5336–42
138. Smith, D. J., Stokes, B. O., Boyer, P. D.

1976. *J. Biol. Chem.* 251:4165–71
139. Vinkler, C., Avron, M., Boyer, P. D. 1980. *J. Biol. Chem.* 255:2263–67
140. Davenport, J. W., McCarty, R. E. 1980. *Biochim. Biophys. Acta* 589:353–57
140a. Lemasters, J. J., Hackenbrock, C. R. 1980. *J. Biol. Chem.* 255:5674–80
141. Barber, J. 1980. *Biochim. Biophys. Acta* 594:253–308
142. Packham, N. K., Greenrod, J. A., Jackson, J. B. 1980. *Biochim. Biophys. Acta* 592:130–42

Ann. Rev. Biochem. 1982. 51:219–50

INTERMEDIATE FILAMENTS:
A Chemically Heterogeneous, Developmentally Regulated Class of Proteins

Elias Lazarides[1]

Division of Biology, California Institute of Technology,
Pasadena, California 91125

CONTENTS

[1]I thank John Ngai for his critical reading of the manuscript and his many helpful comments.

219

Perspectives and Summary

Electron microscopy has identified a major filamentous system in the cytoplasm of higher eukaryotic cells that is distinct from actin filaments, microtubules, and myosin filaments. Filaments of this system from muscle cells grown in tissue culture exhibit no crossreaction with the actin binding proteolytic fragment of myosin, heavy meromyosin, and a characteristic aggregation into cytoplasmic filament bundles when the myogenic cells are exposed to the antimitotic drug Colcemid (1). As their characteristic diameter of 100 Å is between that of actin filaments (60 Å) and microtubules (250 Å) in nonmuscle cells, and myosin filaments (150 Å) in skeletal muscle cells, this newly discovered filamentous system is said to be composed of intermediate filaments (1). Intermediate filaments were at first widely regarded as disaggregation products of either microtubules or myosin filaments and as a result they were neglected despite numerous observations of their existence in the cytoplasm of a variety of cell types. The development of new biochemical and immunofluorescent techniques for the study of cytoplasmic filaments has recently established intermediate filaments as a distinct fibrous system composed of chemically heterogeneous subunits (2). Their subunit structure defines five major subclasses of intermediate filaments that can be distinguished both biochemically and immunologically: (a) keratin (tono) filaments, found in epithelial cells and cells of epithelial origin; (b) desmin filaments, found predominantly in smooth, skeletal, and cardiac muscle cells as well as in certain nonmuscle cells; (c) vimentin filaments, found in most differentiating cells, in cells grown in tissue culture, and in certain fully differentiated cells; (d) glial filaments, thus far detected only in glial cells and cells of glial origin; and (e) neurofilaments, thus far detected only in neurons. In the course of studies that established the chemical heterogeneity of intermediate filaments, two major observations emerged. The first is that the subunits of two chemically distinguishable intermediate filaments can coexist in the same cell. Depending on the subunit type, the coexisting subunits can be components of distinct filaments in the cytoplasm, as is the case with the keratins and vimentin. Alternatively, they may be components of the same filament, i.e. form copolymers, as is the case with desmin and vimentin. The second observation is that the subunit composition of a given filament subclass can vary both with regard to the ratio of the major subunits that comprise this class of filaments and to the accessory proteins that associate with these subunits. Such variations become evident when the composition of a given class of filaments is examined in different cell types or in different stages of differentiation of a given cell type. This review highlights the most recent studies that have led to these conclusions and focuses on certain observations that shed light on the cytoplasmic function of these filaments. From these studies it is becoming evident that intermediate filaments comprise a class of proteins that are chemically heterogeneous and developmentally regulated, whose expression and function is tailored to the differentiated state of the cell.

Keratin Filaments: A Family of Proteins

The characteristic component of intermediate filaments in epithelial cells and cells of epithelial origin are the keratins. The stratified squamous epithelium is the most common covering or lining surface of the animal body. It may be of ectodermal origin (the epidermis) or of endodermal origin (the esophageal epithelium). The dominant cell type of these epithelia is the keratinocyte, which contains abundant 80-Å filaments in the form of filament bundles, composed of keratin proteins. Cells of this type cultivated from different stratified squamous epithelia are rather similar (3); the keratins account for about 30% of the cellular protein. During the differentiation of the mammalian epidermis, keratins are actively synthesized by the inner, living cell layers of the epidermis, and represent its major differentiation product. Eventually they form the bulk of the outer, fully differentiated, dead, stratum corneum layer, during which time they become cross-linked with disulfide bonds (see below).

Keratins are insoluble in neutral aqueous buffers, but can be extracted as their constituent polypeptides by denaturing solvents (e.g. 8 M urea). A reducing agent is also required for the efficient solubilization of keratins from the terminally differentiated cells. Electrophoresis in denaturing gels separates the keratins into a family of polypeptides with molecular weights ranging from 40,000 to 65,000 (4–7 and references therein). Keratins extracted in the presence of urea from the living cells of epidermis (4–6) or from cultured human keratinocytes (7) are composed of six to seven major polypeptides that have molecular weights of 47,000 to 58,000. Dialysis of the urea extracts against a dilute salt solution results in the spontaneous formation of filaments without any obvious requirement for a specific cofactor(s) such as divalent cations and nucleoside triphosphates (6, 7). The reconstituted filaments show the same polypeptide composition, morphology, and X-ray diffraction patterns as native filaments (4–7). Extraction and reconstitution experiments suggest that most keratins of living cells exist in a reduced form that favors polymerization and become increasingly cross-linked by intermolecular disulfide bonds as differentiation proceeds (6–8).

The keratins comprise a highly complex family of polypeptides. Electrophoretic analysis has shown that the keratins of stratum corneum and those of the living layers of the epidermis exhibit differences in the number of polypeptide subunits and in their apparent molecular weights (7, 9, 10). There are also size differences between the keratins of stratum corneum and those of cultured epidermal cells (9–11). The different keratins are most likely distinct polypeptides rather than proteolytic fragments of a larger precursor, because they exhibit immunological differences and can be shown to be translated in vitro by distinct mRNA species (7, 9–11). Different subsets of the entire keratin family are also expressed by the various stratified squamous epithelial tissue types in vivo (12, 13). Stratified squamous epithelia of internal organs (e.g. esophageal, tongue, and buccal epithelia), which do not form a typical stratum corneum, do not make the higher-molecular-weight keratins (63–67 K) characteristic of the outer layers of epidermis. Their keratins are also different from those of cultured keratinocytes even though they have similar molecular weights (11). These differences show that the behavior of the keratins in a given type of epidermis is only one of the possible programs for keratin gene expression in epithelia. Such differences appear to be general among different mammalian species, as they have been observed with human, rabbit, mouse, and bovine epithelial cells (9–13). In addition, several lines of evidence indicate that the program for keratin synthesis can be altered when the environment of the cell is changed. Keratinocytes of epidermal, corneal, nasopharyngeal, or tracheal origin, which would otherwise synthesize different keratins, have been found to synthesize similar keratins when cultured under common conditions (11, 14). Even epithelia cells not derived from stratified squamous epithelia, such as HeLa cells and PtK_2 cells (15, 16), synthesize some keratins resembling those of cultured keratinocytes. Altered expression of keratin genes is also related to malignant growth properties. Certain human squamous cell carcinomas (malignant variants of the human keratinocyte) grown in tissue culture, express a smaller keratin in addition to the keratins found in normal epidermal and buccal keratinocytes, which by various biochemical and immunological criteria is a distinct keratin polypeptide rather than a proteolytic fragment of a higher-molecular-weight keratin (17). All these results show that the keratins comprise a complex multigene family, subsets of which are differentially expressed in epithelia according to type- and stage-specific developmental programs.

Despite this extreme heterogeneity among different keratin polypeptides, the keratins share a number of striking similarities that imply substantial amino acid sequence homology and classify these proteins as a family of related polypeptides. Perhaps the strongest evidence thus far stems from the observation that mammalian keratins from a variety of species are immunologically related. Although the keratins within a species or among species can be distinguished immunologically, antibodies raised against total human epidermal keratins (18, 19) and bovine epidermal keratins (16, 20–24) show wide crossreactivities with keratins in a variety of epithelial cells of the respective species and in other species, including human, rabbit, mouse, rat, cow, and amphibia. These crossreactive antibodies have been very useful for demonstrating that a given epithelium contains keratin-related polypeptides and that keratins are specific to epithelial cells and cells of epithelial origin, since they are not present in cells of other embryonic origin (see Table 1). Furthermore, these antibodies have allowed a direct demonstration that the identifiable intermediate filaments in the cytoplasm of epithelial cells such as liver hepatocytes, pancreatic acinar cells, and myoepithelial cells are indeed keratins and none of the other classes of intermediate filaments (18, 19, 23, 25). An exception to this has been the cells of the mammalian iris and lens epithelium. These cells' filaments do not crossreact with antibodies to mammalian keratins (19), and detailed examination of the intermediate filaments in these cells has shown them to be composed of vimentin (26; see below). Such widely crossreacting keratin antibodies have also been useful in demonstrating that cells of epithelial origin (keratinocytes) grown in tissue culture maintain their ability to express keratins that can be visualized by immunofluorescence as a cytoplasmic network of filaments (18, 20, 23, 24). In contrast to the effect that Colcemid has on the distribution of other classes of intermediate filaments, the distribution of epithelial keratin filaments is unaffected by this drug (18,20) (presumably because these filaments terminate and are anchored to desmosomes). Similarly the distribution of keratin filaments is unaffected by cytochalasin B, which induces a disorganization of actin filaments (18, 20). Maintenance of the epithelial character of tumorigenic cells grown in tissue culture can be assayed by their continued ability to express keratins, as shown by immunofluorescence (15, 27). Such antibodies therefore can be used as an assay to determine whether a given tumor growth in an animal is of epithelial origin (28).

In addition to the immunological similarities among keratins, the amino acid compositions of individual keratin polypeptides from a given type of epithelium (e.g. stratum corneum) are remarkably similar, even though these molecules differ substantially in molecular weight (9). Polypeptide fragments produced by partial enzymatic hydrolysis show strong homologies among all the keratins of human stratum corneum, as well as those of human cultured epidermal cells and rodent cells (10). Furthermore, keratins isolated from different epithelial cells exhibit remarkably similar in vitro polymerization properties even though the number and molecular weights of the reconstituted polypeptides are different (5, 6). All these results suggest that the keratin polypeptides share one or more regions of amino acid homology as well as regions of extensive amino acid sequence diversity. The regions of homology among the keratins may be responsible for the conserved chemical properties of the proteins, including antigenic relatedness, conserved polymerization, and similar X-ray diffraction pat-

Table 1 Presence of intermediate filament subunits in different cell types

Cell type	Vimentin	Desmin	Glial filaments (GFAP)	Neuro-filaments	Keratins
Adult muscle					
Skeletal, cardiac, smooth	+	+	−	−	−
Purkinje fibers	−	+	−	−	−
Vascular smooth muscle	+	+	−	−	−
Aortic smooth muscle	+	−	−	−	−
Myotubes	+	+	−	−	−
Myoblasts	+	+	−	−	−
Avian red blood cells	+	−	−	−	−
Leukocytes	+	−	−	−	−
Macrophages	+	−	−	−	−
Chondrocytes	+	−	−	−	−
Endothelial cells	+	−	−	−	−
Skin melanocytes	+	−	−	−	−
Pigment cells	+	−	−	−	−
Sertoli cells of testis	+	−	−	−	−
Ovarian granulosa cells	+	−	−	−	−
Schwann cells	+	−	−	−	−
Neuroblastoma	+	−	−	no data	−
Chinese hamster ovary (CHO)	+	−	−	−	−
BHK–21	+	+	−	−	−
NIL–8	+	−	−	−	−
3T3/SV40–3T3	+	−	−	−	−
Sarcoma 180	+	−	−	−	−
Hamster embryo fibroblasts line 9	+	+	−	−	−
HEF adenovirus transformant 14b	+	−	−	−	−
Melanoma	+	−	−	−	−
Hepatoma	+	−	−	−	−
PNS, CNS neurons	−	−	−	+	−
Glial cells					
Bergman glial fibers	+	−	+	−	−
Astrocytes (white matter)	+	−	+	−	−
Astrocytes (grey matter)	−	−	+	−	−
Astrocytes in culture	+	−	+	−	−
Glial cells in culture	+	−	+	−	−
Glioma cells	+	−	+	−	−
Oligodendrocytes	−	−	+	−	−
Microglia	+	−	+	−	−
Ependymal cells	+	−	+	−	−
Epithelial cells					
All stratified squamous epithelia of ectodermal and endodermal origin	−	−	−	−	+

Table 1 *(Continued)*

Cell type	Vimentin	Desmin	Glial filaments (GFAP)	Neuro-filaments	Keratins
Epithelial cells (continued)					
Epidermal cells (Proliferating and stratum corneum)	−	−	−	−	+
Epidermal appendages	−	−	−	−	+
Ductal and myoepithelial cells of sweat glands	−	−	−	−	+
Hassall's corpuscles of the thymus	−	−	−	−	+
Rest of thymus	+	−	−	−	−
All primary cultured epithelial cells	+	−	−	−	+
Cultured epidermal cells (keratinocytes)	+	−	−	−	+
Tumor cells of epithelial origin					
HeLa (adenocarcinoma)	+	−	−	−	+
D562 (nasopharyngeal carcinoma)	+	−	−	−	+
Squamous cell carcinomas	+	−	−	−	+
Rat kangaroo PtK_1, PtK_2	+	−	−	−	+
liver	−	−	−	−	+
pancreas-pancreatic ducts	−	−	−	−	+
hepatocytes	−	−	−	−	+
submaxillary ducts	−	−	−	−	+
Exceptions					
Epithelia of iris and lens	+	−	−	−	−
Glomerular and tubular cells of kidney	+	−	−	−	−

terns. Amino acid sequencing of some keratin polypeptides and cloning of the keratin genes will greatly advance our knowledge of the structure of these proteins. Finally, since keratin expression changes during epidermal differentiation, the cytoplasmic keratin filaments are likely to be heterogeneous with respect to polypeptide composition at any instant of time during differentiation. It appears that this heterogeneity of filament composition extends to other classes of intermediate filaments (see below).

Desmin Filaments: Integration of Myofibrils at the Z Disc

ADULT MUSCLE While intermediate filaments are present in both the embryonic and adult forms of smooth, skeletal, and cardiac muscles, they are most abundant in adult smooth muscle, which has led to its frequent adoption as a starting point for their purification and study. In adult smooth muscle, these filaments form an interconnected network that links cytoplasmic dense bodies with membrane bound dense plaques (29–31). Morphological studies have shown that actin filaments insert also into these electron-dense structures (32), which by immunocytochemical techniques can be shown to contain α-actinin (33), a protein characteristically found in skeletal myofibril Z discs. Thus, smooth muscle dense bodies appear to be analogous to striated muscle Z lines. Biochemical characterization of the subunit(s) of muscle intermediate filaments was facilitated by the original observation of Cooke (30, 31) that smooth muscle intermediate filaments are insoluble in high ionic strength. Extraction of smooth muscle cells with high-salt buffers renders contractile actin and myosin filaments soluble, leaving behind a cytoskeleton composed predominantly of intermediate filaments still attached to dense bodies. Electrophoretic analysis has shown that this cytoskeleton consists of two major proteins: actin, and a 50,000–55,000-mol wt protein, which has been termed desmin (34) [or skeletin (35)]. Desmin is insoluble in either physiological or high-salt buffers; however, it can be extracted from gizzard smooth muscle with urea (31, 36, 37), or with low pH (e.g. acetic acid) (35, 38), and can be chromatographically purified in the presence of these agents. Upon dialysis against low ionic strength buffers or neutralization, desmin polymerizes into a network of filaments that can then be isolated and resolubilized at low pH or with urea, thereby allowing a cyclic assembly-disassembly purification of this molecule (38). Reconstituted desmin filaments have an average diameter of 100 Å, which strongly suggests that this protein is indeed the major subunit of smooth muscle intermediate filaments (35–38). Immunological and biochemical experiments have shown that desmin is also a component of avian and mammalian skeletal and cardiac muscle cells (34, 39–41). In these three muscle types, the desmins are antigenically related (34, 40) and share the same electrophoretic mobility and isoelectric point, as judged by two-dimensional isoelectric focusing-SDS polyacrylamide gels (2D IEF/SDS PAGE) (39). However, a comparison of avian smooth muscle and mammalian smooth and skeletal muscle desmins has shown that the two desmin types exhibit a small difference in molecular weight (D_m = 52,000 daltons; D_a = 50,000 daltons) as well as a small difference in isoelectric point; mammalian desmin is more acidic than its avian counterpart (40). Furthermore, avian and mammalian desmins can be distinguished immunologically (40), and extensive peptide map and partial amino acid sequence analyses have shown that the two molecules are homologous but not identical (37, 42). Thus, desmin does not appear to exhibit the high degree of evolutionary conservation characteristic of actin. Desmin and vimentin have similar electrophoretic mobility on one-dimensional SDS gels; however, the conclusive identification of a protein as desmin can be made using 2D IEF/SDS PAGE and the immunological crossreaction with desmin-specific antibodies (43, 44). This is especially important after recent observations that certain classes of smooth muscle cells contain vimentin instead of desmin as their major intermediate filament subunit (45, 46; see below). Desmin has been conclusively identified as the major subunit of intermediate filaments in: avian gizzard muscle cells (35, 38); mammalian uterine and stomach smooth muscle cells; most other smooth muscle cells (35, 37); avian and mammalian skeletal muscle (34, 39, 40, 41); avian and mammalian cardiac muscles (34, 39, 47); and the Purkinje conduction fibers of the heart (48; see Table 1). Some aortic cells contain predominantly if not exclusively vimentin and no detectable amounts of desmin; however, other types of vascular smooth muscle cells including some aortic cells contain both desmin and vimentin (45, 46; see below and Table 1). Therefore, with the possible exception of some vascular smooth muscle cells, desmin is an intermediate filament protein that is characteristically expressed in muscle cells. Most nonmuscle cells thus far examined do not express desmin, but certain exceptions exist and are discussed below.

Although intermediate filaments are common in embryonic striated muscle (1), they become progressively less visible as the sarcomeres condense and become laterally registered (49). Intermediate filaments are rarely seen in adult skeletal muscle (50), but substantial amounts of desmin can be detected by electrophoretic and immunological methods (34, 39). Using immunofluorescence with antibodies specific for chicken gizzard desmin on longitudinal sections of myofibrils, it was discovered that desmin is localized in skeletal and cardiac muscle Z lines, in regions of cell-cell juncture, at the site of apposition of the Z line with the plasma membrane, and in cardiac intercalated discs (34, 40, 47). Desmin is thus the third protein (the others being α-actinin and actin), to be localized at the Z line, using immunocytochemical techniques (see Table 2 for a list of the known proteins of the avian myofibril Z disc).

To obtain information on the spatial relationships of these three proteins within the plane of a single Z disc, it was necessary to develop a method that allowed a face-on view of Z discs by light as well as by electron microscopy. Normally when a muscle fiber is sheared, it is cleaved lengthwise, between laterally associated sarcomeres, to produce myofibrils; it was found, however, that if most of the myosin and actin filaments are first extracted from the muscle fiber with buffers containing high salt, the fiber's longitudinal strength is weakened, and subsequent blending shears the fibers transversely between Z planes, which produces honeycomb-like sheets of Z discs (51). Such sheets of Z discs can be visualized clearly by conventional phase contrast microscopy, which reveals a face-on view of the Z disc and

Table 2 Protein composition of the Z disc in chicken muscle cells

Peripheral domain			Central domain	
Protein	Molecular weight		Protein	Molecular weight
Actin	43 K		Actin	43 K
Filamin	250 K		α-Actinin	100 K
Desmin	50 K	Intermediate		
Vimentin	52 K	Filament		
Synemin	230 K	Proteins		

demonstrates that each Z disc is physically linked laterally to neighboring Z discs within a given Z disc plane (51). Electrophoretic analysis of such Z-disc scaffolds has shown that their major components are desmin and actin, which suggests that they are analogous to the high-salt insoluble dense body cytoskeletons obtained from smooth muscle. The ability to view the Z disc by light microscopy side-on in myofibrils (Z line) and face-on in isolated Z-disc sheets allowed the mapping within this structure by immunofluorescence of the various proteins that remain insoluble within this structure after removal of actin and myosin filaments with high salt. These studies have shown that desmin is present at the periphery of each Z disc, and forms an interconnecting network across the myofiber. α-Actinin, which is localized within each disc, gives a face-on fluorescence pattern that is complementary to that of desmin (see Table 2). Actin is present throughout the Z plane, apparently coexisting with both desmin and α-actinin. Thus, the Z disc is composed of two major domains, a peripheral one that contains at least desmin and actin, and a central one that contains at least α-actinin and actin (51). This complementarity in the distribution of α-actinin and desmin extends also into areas of the Z disc that appear to be in the process of subdivision (51–53).

The function of the Z disc is to link all the actin (thin) filaments that emerge from this structure with opposite polarities toward two flanking sarcomeres. The demonstration of the existence of a peripheral domain in the Z disc, composed predominantly of the intermediate filament protein desmin, has led to the hypothesis that the function of this protein in adult muscle is to mechanically integrate all contractile actions of a muscle fiber by linking individual myofibrils laterally at their Z discs and by linking Z discs to the plasma membrane and to other membranous organelles (2, 51). This arrangement will also ensure the maintenance of the proper alignment of cellular structures during the contraction-relaxation cycle of muscle. During the biogenesis and assembly of myofibrils, desmin may play an important role in the generation of the striated appearance of muscle by promoting and stabilizing the lateral registry of adjacent Z discs. This proposed function of desmin has provided the first well-documented cytoskeletal function for intermediate filaments and may hold true for all three muscle types, smooth, skeletal, and cardiac. As mentioned earlier, in adult skeletal muscle fibers, intermediate filaments are rarely seen by conventional electron microscopic techniques. This is in contrast to embryonic skeletal muscle (see below) and adult smooth muscle, where intermediate filaments are clearly identifiable by electron microscopy or by immunofluorescence. However, in certain cases, bundles of intermediate filaments have been observed surrounding muscle Z discs and extending laterally between them (50, 54). The presence of high concentrations of membranous organelles in the intermyofibrillar space may obscure the presence of intermediate filament connections between adjacent Z discs. Alternatively, the polymeric state of desmin may vary depending on the extent of lateral packing of myofibrils. In skeletal muscle, where the lateral proximity of myofibrils is the highest, desmin may not be in the form of elongated filaments that are easily discerned by electron microscopy. In cardiac muscle, where the myofibrils are generally not as tightly packed, bundles of intermediate filaments are more frequently seen, and appear to surround Z discs and to link them to each other and to membranous organelles; occasionally these filaments can be seen to extend from the Z discs to the nuclear and plasma membranes (55–60). Intermediate filaments are especially prominent between Z discs in hypertrophied myocardial cells (61) or in cardiac muscle cells of animals treated with anabolic steroid hormones (58). Recent electron microscopic evidence from canine cardiac conduction fibers, which contain predominantly desmin as their major intermediate filament protein (48), indicate that glycogen particles (composed of glycogen and protein) are attached to intermediate filaments (62). As is the case with the experiments summarized above, in these muscle cells these intermediate filament glycogen particle complexes extend between Z discs and between Z discs and the endoplasmic reticulum of the cells. The biochemical basis of such an association is not known at present.

MYOGENESIS   Intermediate filaments abound in the cytoplasm of embryonic muscle cells; indeed they were first described in skeletal myotubes grown in tissue culture, where they were shown to be distinct from actin filaments (1). Subsequent electrophoretic and immunofluorescent studies showed that desmin is one of the major subunit proteins of these filaments (43, 63–66). The demonstration that the Z disc is composed of two distinct but interconnected domains raised the question of the temporal sequence of events during skeletal myogenesis that culminate in the assembly of these two domains, and the fate of intermediate filaments and desmin during this process. Earlier studies with cardiac muscle cells grown in tissue culture provided the first evidence that desmin exists in the form of cytoplasmic filaments in the less mature cells and in association with Z lines in the more mature cells (47). This observation was in accordance with initial electron microscopic observations, which showed that during skeletal myogenesis cytoplasmic intermediate filaments changed their distribution and became principally disposed around and at right angles to newly assembled myofibrils at the level of their Z bands (49). Thus it became obvious that at some point during muscle differentiation the subunits of intermediate filaments alter their cytoplasmic distribution and begin to associate with the Z disc. Using antibodies to α-actinin as a probe for the newly assembling central domains of the Z discs, and antibodies to desmin as a probe for intermediate filaments and the peripheral domain of the Z disc, it was shown that these two domains assemble sequentially during myogenesis (63). In primary myogenic cultures (consisting predominantly of fibroblasts, myoblasts, and early myotubes), immunofluorescence has shown that the presence of desmin is restricted to a subpopulation of myoblasts as well as all myotubes, which suggests that synthesis of this protein is restricted mainly to postmitotic fusing myoblasts and multinucleate myotubes (63, 65). Biosynthetic experiments have confirmed these immunofluorescent results and have shown that the synthesis of desmin is barely detectable in myoblasts but is increased several fold after the onset of fusion (63, 67). In early myotubes, desmin exists in the form of cytoplasmic filaments, and exhibits no apparent association with the newly assembling Z discs, as assayed by double immunofluorescence with antibodies to α-actinin and desmin (63). During this time the desmin-containing filaments can be induced to aggregate with the antimitotic drug Colcemid (63), a phenomenon characteristic of intermediate filaments at this stage of differentiation (1). However, later in development and several days after the formation of α-actinin containing Z-line striations, desmin begins to associate with and encircle the Z discs (63, 68) with a concomitant reduction in cytoplasmic desmin filaments and resistance to aggregation by Colcemid (63). The transition of desmin from a Colcemid-sensitive to a Colcemid-insensitive state may reflect a molecular change that it undergoes upon its association with the Z disc. The transition of desmin to the Z disc coincides with the lateral registration of Z discs and the appearance of well-developed striations in the muscle fiber (63). Thus, desmin may play not only a passive role in maintaining the striated phenotype of muscle cells, but also an active role in bringing Z discs into lateral registry upon their association with the Z disc. How this process is regulated and what makes desmin and the Z disc competent to associate with each other is presently unknown.

Vimentin Filaments: Their Generalized Occurrence

The majority of cells of mesenchymal or nonmesenchymal origin as well as all cell types grown in tissue culture thus far examined, irrespective of the tissue of origin, contain a class of intermediate filaments whose subunit, vimentin (20), can be distinguished both immunologically and biochemically from the other four classes of intermediate filaments (see Table 1). Antibodies to vimentin react in immunofluorescence with a wavy cytoplasmic filamentous system distinct from microtubules and actin filaments (20, 69). The wide crossreactivity of vimentin antibodies with cells of mammalian, avian, or amphibian origin indicates that this protein is conserved in evolution (20, 64, 65, 70, 71). However, as with desmin, avian and mammalian vimentins exhibit a small difference in molecular weight and isoelectric point, and antibodies to avian vimentin can be produced that will not crossreact with mammalian vimentin (44, 63); avian and mammalian vimentins therefore exhibit both homology and divergence.

Vimentin was first characterized biochemically as a major subunit of intermediate filaments in chick embryo fibroblasts (72). Treatment of these cells, or many other cell types grown in tissue culture, with a nonionic detergent leaves an insoluble cytoskeleton that remains adherent to the substratum. By electron microscopy these cytoskeletons consist predominantly of intermediate filaments, actin filament bundles, and nuclei. The actin filaments can be largely removed by exposing the cells to high-salt buffers (0.6 M KCl or 0.6 M KI), but the intermediate filaments still remain insoluble (43). A prominent 52,000-dalton polypeptide can be extracted, however, from these cytoskeletons in the presence of a strong ionic detergent (Sarkosyl) or by prolonged incubation of the cytoskeletons in a low ionic strength buffer (72). Since the extraction of the 52,000-dalton polypeptide coincided with the disappearance of the intermediate filaments, it was originally concluded that this protein was the major subunit of the intermediate filaments in these cells (72). Subsequent biochemical and immunological studies have shown that vimentin is the major, if not the sole intermediate filament subunit in many normal and transformed cell types grown in tissue culture, including the hamster cell line NIL8 (69), Chinese hamster ovary (CHO) cells (73), all fibroblastic cells (20, 64), macrophages, neuroblastoma (20), Ehrlich ascites tumor cells (74), mouse 3T3 cells (20), and human endothelial cells (70). Additionally, vimentin has been identified ultrastructurally and immunologically in Sertoli cells of the mammalian testis (75), in the histologically analogous granulosa cells of the ovary (76), and in avian red blood cells (94). Mammalian eye lens tissue contains vimentin as the major filament subunit, rather than keratins, even though histologically it is classified as epithelial in origin (26). A characteristic property of vimentin filaments is their aggregation into cytoplasmic or perinuclear filament bundles in cells exposed to Colcemid (69, 77–80). This property of vimentin filaments is shared by desmin filaments (1, 47, 63) prior to their association with the Z disc as well as by glial filaments in glial cells and astrocytes grown in tissue culture (81), but is not shared by keratin filaments or any other class of cytoplasmic filaments. Thus, the Colcemid-induced aggregation of these filaments can be used as a convenient immunocytological assay to determine the association of a protein with vimentin, desmin, or glial filaments. Vimentin filaments can also transiently exist in the form of a perinuclear ring or cap during normal cell adhesion and spreading onto a substrate (82) or in nonmotile tumor cells (83). In later stages of cell spreading (82) or during cell locomotion (83), the filaments are mobilized and become dispersed in the cytoplasm. The birefringent caps, composed almost exclusively of these intermediate filaments, can be isolated rapidly from Colcemid-treated or normally spreading baby hamster kidney cells (BHK-21) by extracting these cells with a combination of high salt and a nonionic detergent (84, 85). However, the isolated filament caps contain two predominant polypeptides in approximately equimolar amounts, one with a molecular weight of 52,000–55,000 and the other with a molecular weight of 50,000–54,000 (84, 85), in contrast to filaments isolated from other cell types grown in tissue culture, which contain predominantly the 52,000–55,000-dalton protein. The BHK-21 52,000–55,000-dalton protein has been named decamin (86), but by a variety of biochemical and immunological criteria, including isoelectric point, molecular weight, peptide mapping, and immunological crossreaction (43), it is indistinguishable from the vimentin polypeptide described in other cell types. For simplicity, it is referred to as vimentin in this review. The 50,000–54,000-dalton polypeptide has been identified as the muscle subunit of intermediate filaments, desmin, by isoelectric point, molecular weight, homology of phosphorylated peptides, and immunological crossreaction with desmin antibodies (43, 87). Vimentin and desmin can be purified from the birefringent caps by repeated cycles of disassembly-reassembly, induced by lowering and raising the ionic strength of the extracting solution (85, 86). Reassembled filaments have an average diameter of 100 Å; these observations provided the first evidence that vimentin and desmin probably have similar polymeric properties, and that in some cell types they could potentially form copolymers by both serving as subunits of one filament (see below for a discussion). However, the presence of either desmin or vimentin is not necessary for the polymerization of the other subunit, and either of them alone will form filaments with an average diameter of 100 Å. Vimentin alone has been purified from a variety of sources and shown to polymerize into filaments with an average diameter of 100 Å (73, 76, 88, 89).

The insolubility properties of vimentin filaments strongly suggest that they serve a structural function in the cytoplasm. As noted earlier, detergent-extracted cytoskeletons of a variety of cells grown in tissue culture retain, in addition to intermediate filaments, the nucleus and the areas of adhesion of the cell to the underlying substrate. Morphological and biochemical evidence have indicated that vimentin filaments are associated with both the nuclear and plasma membranes in cultured fibroblasts (90, 91) and chicken erythrocytes (92–94). Possibly, vimentin filaments, by interacting with the nucleus, provide it with mechanical support and constrain it in a specific place in the cell (93). Mature chicken erythrocyte vimentin filaments do not aggregate when the cells are exposed to Colcemid (94). Perhaps, during erythropoiesis, the structure of vimentin filaments is changed in response to differentiation and they become resistant to aggregation by Colcemid. This may be due to an increased or more stable interaction with the nucleus and the plasma membrane, or a molecular change in the filaments themselves. Since vimentin filaments are such insoluble cytoplasmic structures and they exhibit a strong interaction with the nucleus, what is the fate of vimentin filaments during the cell cycle? Similarly, what is the fate of desmin filaments in cardiac and smooth muscle during proliferation of these cells? The answer to the latter question is unknown. However, in most cell types grown in tissue culture (NIL-8, endothelial cells, HeLa cells), the filaments appear to persist in a polymerized form throughout mitosis (69, 95, 96). Direct isolation of metaphase spindles has demonstrated that they are encased in a network of vimentin filaments that remains tightly associated with the isolated structures (96). All these results in conjunction have indicated that vimentin filaments play a role in maintaining the position of the nucleus and the mitotic spindle during the life cycle of the cell. Whether this role of vimentin filaments is essential and universal is presently undetermined. Nevertheless, the extreme insolubility of vimentin and desmin filaments has suggested that depolymerization of these filaments may not be regulated by disassembly-reassembly but rather by proteolysis of the subunit proteins. Recently a Ca^{2+}-activated neutral protease has been isolated from Ehrlich-ascites tumor cells, which specifically degrades vimentin and desmin but not the glial filament or the neurofilament subunits (74). Since the latter two classes of subunits are also highly susceptible to proteolysis (97), the endogenous protease(s) that degrades them may be distinct from the vimentin protease. The vimentin-specific protease is found associated with the detergent-resistant cytoskeleton of Ehrlich-ascites tumor cells and appears to be resistant to denaturation by urea (74). This would explain why, in total cell extracts or in cytoskeletons dissoved in urea sample buffers in preparation for 2D IEF/SDS-PAGE, vimentin (and desmin) is degraded to a family of progressively smaller and more acidic polypeptides that yield a proteolytically-resistant core(s) 10,000–15,000-daltons less than the parent molecule (43).

Even though intermediate filaments may play a cytoskeletal role and remain insoluble in buffers containing high salt, they are not static structures. On the contrary, intermediate filaments, and in particular vimentin filaments can be induced to reversibly disaggregate. Interference with protein synthesis either with cycloheximide or with toxins that inhibit protein synthesis (e.g. diptheria toxin or exotoxin A) reversibly disaggregates vimentin filaments in some cells (98). Similarly, microinjection of monoclonal antibodies to vimentin (99) or to certain vimentin associated polypeptides (see below) induces a rapid and reversible aggregation of vimentin filaments into cytoplasmic or perinuclear bundles. Under these conditions, the distribution of actin filament bundles and microtubules remains unaltered, which suggests that the effect is specific for intermediate filaments. Thus, the distribution of intermediate filaments can be specifically perturbed, but the meaning of these observations is presently unclear. Similarly, sodium vanadate, a chemical that inhibits the ATPase activity of certain enzymes, will cause the reversible aggregation of vimentin filaments but will not alter the distribution of microtubules in interphase cells (100). Vanadate-treated cells exhibit a dramatic alteration in the positioning of the nucleus and centrioles, which suggests a relationship between these organelles and vimentin filaments, as discussed earlier. Whether vimentin filaments contain an ATPase that is inhibited by vanadate remains an attractive idea, but as yet no enzymatic activity has been found in association with vimentin or desmin filaments. The only exception is the reported association of creatine kinase with cytoplasmic vimentin filaments (101). The biological significance of this observation remains to be established.

Neurofilaments and Glial Filaments

Neurofilaments (average diameter 100 Å) and microtubules (250 Å in diameter) comprise the main structural elements of neuronal axons, dendrites, and perikarya. The neurofilaments can be either randomly or orderly dispersed, and appear to be linked to each other or to microtubules through wispy sidearm appendages. Glial filaments are found randomly dispersed in the cytoplasm of astrocytes and other types of glial cells; in contrast to the neurofilaments, glial filaments are of smaller diameter (80 Å average), are less parallel in their orientation, and lack sidearm appendages (102–104).

A prolonged controversy and confusion has existed in the literature over the subunit composition of neurofilaments and glial filaments and their antigenic relationship. This controversy has ended recently with the clear demonstration by a number of investigators that purified mammalian neurofilaments isolated from spinal cord or brain are composed of three major polypeptides referred to as the neurofilament triplet (106), with approximate molecular weights of 210,000, 160,000, and 68,000 (105). Avian neurofilaments are composed predominantly of two polypeptides with molecular weights of 155,000 and 68,000–70,000; a third polypeptide of 180,000–200,000 mol wt is a minor component (107, 108). The molecular weights of these three polypeptides vary slightly in preparations from different laboratories, probably due to the differences in the electrophoretic systems used; however, it is apparent that the avian and mammalian neurofilament proteins exhibit a reproducible, slight difference in molecular weight. On the other hand, glial filaments are composed of only one major polypeptide with an approximate molecular weight (depending on the electrophoretic system used) of 51,000. The previous difficulty in identifying the subunits of these filaments was due to the lack of any suitable assay that would distinguish between isolated neurofilaments and glial filaments. It is now clear that the two filamentous systems have distinct solubility properites. Mammalian neurofilaments are stable in a filamentous form under isotonic conditions, but disaggregate and become soluble at low ionic strength (109). In contrast to neurofilaments, glial filaments remain mostly polymeric even at low ionic strength (109, 110). Furthermore, the use of antibodies specific for the neurofilament polypeptides or for glial filaments in conjunction with immunofluorescence has aided appreciably in demonstrating that the two classes of filaments are immunologically distinct, with neurofilaments exclusive to neuronal cells, and glial filaments exclusive to glial cells and cells of glial origin.

Using radioactive-labeling techniques in vivo, Hoffman & Lasek were the first to observe that the 210,000-, 160,000-, and 68,000-mol wt polypeptides comigrated as components of slow axoplasmic transport and suggested that this triplet of proteins could be the subunits of neurofilaments (106). Direct evidence that these three polypeptides are indeed the major components of mammalian neurofilaments came from the observation that a similar triplet of proteins was found in filament preparations from peripheral nerves (110, 111); these filaments had an average diameter of 10 nm as judged by electron microscopy, and an antiserum against this peripheral nervous system filament preparation exhibited strong neuronal specificity in immunofluorescence in the central and peripheral nervous systems (CNS and PNS) and bound to the 10 nm filaments in immunoelectron microscopy (112–114). Further evidence that these three proteins are components of neurofilaments comes from the observation that antisera specific for the 210,000-dalton polypeptide will immunoprecipitate all three triplet polypeptides (115, 116). Finally, as discussed in detail later on, in vitro reconstitution experiments, coupled with immunoelectron microscopy have shown that the 68,000-dalton polypeptide makes up the core of the filaments, while the other two higher-molecular-weight polypeptides are peripherally bound to the filaments. The three polypeptides of the triplet are antigenically related. Antibodies raised against any one of the three polypeptides exhibit cross-reactivity with the other two polypeptides but also reveal the existence of unique antigenic determinants on each polypeptide (117, 118). This finding, in conjunction with the observation that the three polypeptides are translated in vitro from distinct mRNA populations (119), strongly indicate that they are not proteolytic derivatives or aggregation products of each other. Rather they suggest the intriguing hypothesis that these three polypeptides share at least small regions of amino acid homology as well as regions of amino acid divergence, which would indicate that they may be evolutionarily related (see discussion of this point later on). Nevertheless, the polypeptide composition of neurofilaments does not appear to be conserved. In contrast to mammalian neurofilaments, squid neurofilaments are composed of two major polypeptides with molecular weights of 200,000 and 60,000 (120–123), while those from the giant axon of the marine worm, *Myxicola infundibulum* are composed of two polypeptides with molecular weights of 160,000 and 150,000 (124, 125). More direct evidence for the conclusion that the 60,000- and 200,000-dalton polypeptides are the major components of squid brain neurofilaments stems from the observation that these two polypeptides copurify by cycles of assembly and disassembly and that when assembled they are in the form of 10-nm filaments (126). In spite of these differences, however, one of the characteristic properties of both vertebrate and invertebrate neurofilaments is their extreme susceptibility to proteolysis in the presence of Ca^{2+} ions (125, 127, 128), as is the case with the other classes of intermediate filament subunits.

The initial isolation of intact intermediate filaments from the mammalian CNS was based on the axonal flotation technique, which depends on the presence of myelin around the axons to float neuronal material away from the rest of the brain tissue (129). The isolated filaments were assumed to represent neurofilaments, since they were derived from myelinated nerve fibers, even though electron microscopy revealed a heterogeneous population of filaments and was unable to differentiate neurofilaments from glial filaments. The isolated filaments are insoluble at low ionic strength and are composed predominantly of a single protein with an approximate molecular weight of 51,000 (130–132). Subsequent immunocytochemical studies have clearly demonstrated that these filaments are predominantly glial filaments and that this protein, the glial fibrillary acidic protein, (GFAP) is their major subunit. Antibodies to this protein reacted with glial, but not neuronal processes in immunofluorescence (114, 115, 130), stained glial filaments using immunoelectron microscopy (133, 134) and cross-reacted with purified GFAP (114, 115, 130, 133–135). In addition, this protein, as isolated by the axonal flotation technique, was found to be identical to the major protein of glial filaments isolated from mammalian CNS tissues (136). From these data it was erroneously concluded that GFAP was identical to the neurofilament protein. The distinction between GFAP and the neurofilament protein was finally clarified with the purification of neurofilaments from the PNS, which is relatively free of glial cells. No protein with a molecular weight of 51,000 was detected in these preparations (110, 111).

The biochemical evidence distinguishing neurofilaments from glial filaments has been supported by immunological studies with antibodies specific for any one of the three subunits of neurofilaments and with antibodies specific for the 51,000-dalton component of glial filaments. Antibodies raised to any one of the three neurofilament polypeptides isolated either from mammalian peripheral nerve (113, 115) or from brain (114) react in immunofluorescence with axons and neurons throughout the mammalian PNS and CNS, but do not react with glial cells. In addition, they react exclusively with neuronal cells in primary cultures from spinal cord (108) and cerebellum (115), but not with cultures of astrocytes. On the other hand, antibodies to GFAP react in immunofluorescence only with glial cells and cells of glial origin both in brain sections and in astrocytes grown in tissue culture, but they exhibit no crossreaction with neurons or their processes (115, 137, 138). Antibodies to GFAP do not react with muscle cells, myofibrils, or fibroblasts, which indicates that this protein is immunologically distinguishable from vimentin and desmin. Thus, these results have demonstrated that any one of the neurofilament triplet polypeptides can be used in immunofluorescence as a specific marker of neuronal cells, while GFAP can be used as a specific marker of glial cells and cells of glial origin. Conclusive demonstration that GFAP is the major component of glial filaments stems from the observation that (a)GFAP is the major component of purified glial filaments (136); (b)GFAP purified from mammalian brain will assemble in vitro into filaments with an average diameter of 100 Å (139); and (c) antibodies to GFAP will react with glial 100-Å filaments in situ using the immunoperoxidase method (133).

While the function of glial filaments is presently undetermined, neurofilaments appear to function as a three-dimensional structural lattice that provides tensile strength to the axons. This idea stems from the observation that the extruded axoplasm from *Myxicola* is a highly structured gel consisting almost exclusively of neurofilaments (124). Exposure of this gel to Ca^{2+} ions results in the degradation of neurofilaments and the dissolution of the gel (125).

Coexistence of Different Intermediate Filament Subunits in the Same Cell

The preceding discussion summarizes evidence establishing the existence of biochemically and immunologically distinct intermediate filament classes, each of which is especially prevalent in a given cell type. This is the case with the keratins in all epithelial cells and cells of epithelial origin (exclusive of the lens epithelium), the neurofilament triplet for neurons, GFAP for glial cells and astrocytes, and with desmin for skeletal, cardiac, and smooth muscle cells, with the possible exception of certain vascular smooth muscle cells. Vimentin, on the other hand, by virtue of its wide cellular and tissue distribution can either be the predominant or exclusive intermediate filament protein; however, in certain cell types, or under certain conditions of cell growth, it can coexist with the other four classes of filament subunits.

COEXISTENCE OF KERATINS AND VIMENTIN   Electron microscopy and immunofluorescence have shown that keratin filaments are abundant in the cytoplasm of the rat kangaroo epithelial cell line PtK_2 and in HeLa cells (15, 140, 141), in accordance with the epithelial origin of these cells. However, these cells contain vimentin, whose presence can be demonstrated immunologically and electrophoretically (15, 140, 141). Normal PtK_2 and HeLa cells react poorly with vimentin antibodies; however, the perinuclear aggregates of intermediate filaments induced by Colcemid stain strongly with antibodies to vimentin, but not with antibodies to keratin, which indicates that these two classes of subunits coexist in the cytoplasm but in distinct filamentous systems. Similarly, a number of established mammalian cell lines of epithelial origin transformed spontaneously or chemically express both keratin- and vimentin-containing filaments (18, 27, 142, 143). However, either normal or chemically transformed primary cultures of epithelial cells express exclusively keratin polypeptides (18, 27, 142). An established epithelial cell line that is transformed with a chemical carcinogen and expresses both keratins and vimentin when injected into nude mice will form tumors that exclusively express keratins (27). These observations have demonstrated that the expression of different intermediate filament subunits varies at different stages of neoplastic progression and in cells maintained in different growth environments. They have also emphasized that the expression of intermediate filament subunits is strictly regulated during normal differentiation in vivo and may be abnormally altered when differentiated cells are established as cell lines in vitro.

COEXISTENCE OF DESMIN AND VIMENTIN   Immunofluorescence, pulse labeling experiments, in vitro mRNA translation experiments, and electrophoretic analysis of intermediate filament–enriched cytoskeletons of fibroblast-free chick embryonic myoblasts and myotubes have demonstrated that vimentin is present throughout myogenesis (43, 63, 65, 71). While the synthesis of desmin increases several fold after the onset of fusion, vimentin is expressed in all myoblasts and persists throughout myogenesis in vitro (63). Both desmin and vimentin continue to be synthesized through at least 20 days in culture. The different rates of synthesis of desmin and vimentin after the onset of fusion have indicated that the synthesis of these two proteins is not coordinately regulated, which results in a dramatic change in the relative composition of intermediate filaments during myogenesis (63); while early in myogenesis vimentin is the predominant filament protein, in late myotubes, vimentin is present in approximately a twofold excess over desmin. Double immunofluorescence has indicated that the distribution of vimentin and desmin is indistinguishable, which suggests that they may both serve as subunits of one filament by forming copolymers. In early myotubes both molecules are localized along cytoplasmic filaments, which can be induced to aggregate into bundles with Colcemid. In late myotubes, both molecules become associated with the Z lines of newly assembled myofibrils, with a concomitant reduction in their association with cytoplasmic filaments, and become resistant to rearrangement by Colcemid (63). Electrophoretic and immunofluorescent analyses of isolated chicken skeletal myofibrils have indicated that both molecules are present as components of the myofibril Z lines (44). Furthermore, examination of the distribution of these two proteins in isolated Z-disc sheets has shown that they exhibit a coincidental distribution at the periphery of the Z disc (44). Similar results are obtained when fresh chicken pectoralis muscle is extracted extensively with solutions of high ionic strength. The predominant insoluble components are desmin and vimentin (along with actin), and extraction of this material at low pH (1 M acetic acid) and subsequent purification of the solubilized proteins by cycles of precipitation and resolubilization result in the copurification of desmin and vimentin (144). When dialyzed against low ionic strength buffers, purified desmin and vimentin will polymerize into filaments with an average diameter of 100 Å (as is the case with BHK-21 vimentin and desmin; see below). Avian muscle vimentin is not confined to pectoralis myofibrils; examination of high-salt insoluble residues from slow muscle (anterior latissimus dorsi) and twitch muscles (posterior latissimus dorsi, peroneus longus, and iliotibialis) by 2D IEF/SDS PAGE and immunofluorescence on isolated myofibrils has revealed the presence of both desmin and vimentin (44, 144). An intriguing observation is that the ratio of desmin to vimentin is variable in these different muscles; it is approximately 1:1 in PLD, 2:1 in iliotibialis, 2.5:1 in pectoralis, and 7:1 in ALD. Vimentin and desmin also are both components of chicken cardiac muscle, where they are present in approximately equal amounts (E. Lazarides, unpublished observations). Electrophoretic analysis of desmin purified from chicken gizzard by cycles of solubilization and precipitation between pH 2.0 and pH 5–7.0 has revealed the presence of vimentin in a concentration approximately 100 fold less than that of desmin (145).

The variable ratio of vimentin to desmin in adult muscle raises an important issue about the functional significance of having both molecules present. Other studies have suggested that the presence of vimentin is not obligatory, since in the mammalian Purkinje fibers (48) and in mammalian fast muscles (41), desmin is the predominant skeletal muscle intermediate filament protein and vimentin is present in very low quantities, if not absent. On the other hand, the presence of desmin is not obligatory for the formation of muscle intermediate filaments, since some mammalian vascular smooth muscle cells express predominantly vimentin and some vascular smooth muscle cells express both desmin and vimentin (45, 46). Most of these latter studies have relied on immunofluorescence for the detection of desmin and vimentin, so that the ratio of the proteins is not evident. Nevertheless, all this evidence emphasizes that the ratio of desmin to vimentin can vary widely in different types of muscle and at different stages of differentiation. In muscles that contain both desmin and vimentin, these molecules exhibit indistinguishable solubility properties, and both molecules copurify in a constant ratio through cycles of solubilization at pH 2.0 and precipitation at pH 5.0–7.0 (144, 145). This is also the case with BHK-21 vimentin and desmin; the two molecules exhibit indistinguishable distributions, both in normal cells and after Colcemid treatment, and copurify at a constant ratio through several cycles of polymerization and depolymerization (85, 86). Desmin is also present in some other types of hamster and chick fibroblastic cells, but in lesser quantities than vimentin (43, 87). In general, virally transformed derivatives of cells that normally express vimentin and small amounts of desmin express only vimentin (87). Variability in the ratios of the two proteins may have important consequences in mediating changes in the overall structure of intermediate filaments in different cell types and, in particular, in different muscle types. Indeed, in vitro polymerization studies have indicated that desmin and vimentin purified to homogeneity will each polymerize into homopolymer 100-Å filaments that exhibit different ionic strength sensitivity to depolymerization. However, when repolymerized together, the intermediate filaments assembled in vitro have solubility properties intermediate between those of the homopolymer intermediate filaments, which indicates that the two subunits had formed copolymer intermediate filaments (146). In polymerization studies where one of the subunits, labeled with ^{35}S-methionine and present in concentrations below its critical concentration for polymerization, was polymerized under conditions that favored polymerization of the other subunit, labeled filaments were obtained, which indicates the formation of copolymers between the two subunits (146). The evidence derived from both in vitro and in vivo systems leads us to conclude that whenever a cell expresses both desmin and vimentin, the two proteins may be capable of forming copolymers, even at widely different ratios of the two subunits.

COEXISTENCE OF GLIAL AND VIMENTIN FILAMENTS   A situation similar to that of the coexpression of desmin and vimentin in muscle cells is also observed with glial filaments and vimentin in glial cells. Human glioma cells (a derivative from a human malignant astrocytoma) (81), normal glial cells grown in tissue culture (108, 115), and astrocytes in frozen sections of brain (115) contain both GFAP, the subunit of intermediate filaments specific for glial cells, and vimentin, as determined by immunofluorescence. The two proteins exhibit indistinguishable cytoplasmic distributions and they coaggregate when cells are exposed to Colcemid. It is intriguing to note that in frozen brain sections, even though all cells of glial origin express GFAP, some glial cells (e.g. Bergmann glial fibers and fibrous astrocytes in white matter) express vimentin, while others (e.g. astrocytes in the granular layer) do not (115); thus, the distributions of the two antigens overlap but do not coincide, which suggests that the expression of vimentin may be required in some glial cells to specifically modify the structure of glial filaments as may be the case with desmin and vimentin. Thus far vimentin has not been detected in neurons (108, 115).

All these observations have indicated that vimentin exhibits the widest distribution of all filament subunits and that in some cases it will coexist with other subunits either in the form of copolymers or in distinct filaments. Furthermore, in some differentiated cells, any one of the five classes of subunits can be the sole subunit expressed.

Structural Similarities and Differences Among Intermediate Filament Subunits

From the observations summarized above, it is evident that the five classes of subunits of intermediate filaments can be distinguished both biochemically and immunologically, which indicates that they exhibit extensive regions of amino acid sequence divergence. However, all five classes of subunits share some unusual properties. For example, they are relatively resistant to extraction in the physiologically relevant pH range (pH 5.5–8.0) and show a marked tendency for filament reassembly under various experimental conditions even with subunits that have been purified under denaturing conditions with very low pH, high concentrations of urea, guanidinium hydrochloride, or SDS. In all five cases, the reassembled filaments have structure and morphology indistinguishable from native intermediate filaments as judged both by electron microscopy and X-ray diffraction (6, 35–38, 73, 89, 146). These results suggest that all subunit classes contain regions with similar sequence arrangements that define common α-helical domains arranged in a coiled-coil conformation and are responsible for the assembly of morphologically similar intermediate-sized filaments (89, 147, 148). The α-helical domains may be flanked by non-α-helical domains that vary in size, configuration, and amino acid sequence in the different subunits and may be responsible for the immunological differences, solubility properties, and structural changes (e.g. Colcemid sensitivity) seen during differentiation (147, 148). Thus intermediate filaments may comprise a complex class of proteins sharing regions of amino acid sequence homology as well as extensive areas of amino acid sequence divergence (148).

Peptide mapping of desmin and vimentin has suggested the existence of peptide homology between these two molecules, especially around some of their phosphorylation sites (43, 149). More recent direct sequencing of vimentin and desmin confirmed these contentions by showing the existence of areas of amino acid homology and areas of amino acid divergence in the carboxyl terminus of these two subunits (150). The existence of structurally homologous areas in them could also permit copolymerization of two or more subunits, as may be the case with desmin and vimentin, GFAP and vimentin, and the various keratins. Thus, by varying the expression of any one filament subunit as well as its ratio to any other subunit, a cell potentially could achieve a wide possibility of polymers with subtle structural differences, tailored to the differentiated state of the cell (148).

Phosphorylation of Intermediate Filament Subunits

All intermediate filament subunits examined to date seem to be phosphorylated in vivo and can be phosphorylated in vitro. For example, phosphorylation of the 200,000- but not the 60,000-mol wt polypeptide from squid axoplasm occurs after intracellular injection of $[\gamma\text{-}^{32}P]ATP$ into giant axons or addition of $[\gamma\text{-}^{32}P]ATP$ to extruded axoplasm (121). In contrast, both *Myxicola* neurofilament polypeptides are phosphorylated in vivo, probably by a soluble kinase that is independent of Ca^{2+}, cAMP, or cGMP for its activity (51, 152). A portion of this activity remains associated with neurofilaments even after extensive purification of the filaments (151, 152). Similarly, all three polypeptides of the mammalian neurofilament triplet are phosphorylated (151, 152).

A nonphosphorylated and several phosphorylated variants of desmin and vimentin are present in assembled structures (e.g. Z discs, dense body attached, and cytoplasmic intermediate filaments) and in muscle and nonmuscle cells (153, 154), while some of the keratin polypeptides are also multiply phosphorylated (7). Thus far the kinase(s) that mediates the phosphorylation reaction of the various intermediate filament subunits has been extensively investigated only for desmin and vimentin. Phosphorylation of both desmin and vimentin is probably mediated by the cAMP-dependent protein kinases both in vivo and in vitro (149, 155, 156). All of the phosphorylation of desmin and most of the vimentin can be shown to be cAMP-dependent. However, a small fraction of the vimentin phosphorylation is cAMP-independent (155). These results have indicated that desmin and vimentin may be regulated similarly through cAMP-dependent phosphorylation reaction, but that vimentin may be capable of being differentially regulated from desmin through phosphorylation by a cAMP-independent kinase. Two-dimensional peptide map analysis has revealed the presence of both homologous and nonhomologous phosphorylation sites on these two molecules, which further indicates the existence of both common and distinct sites, that are potentially regulatory, between desmin and vimentin, as well as some sequence homology between them (149). In vitro RNA translation experiments have shown that RNA isolated from adult chicken gizzard and skeletal muscle (157) or quail myotubes grown in tissue culture (158) direct the translation of the unphosphorylated forms of desmin and vimentin. All this evidence indicates that there is only one major form of desmin and vimentin expressed in the different stages of myogenesis in vitro, and that the multiple, more acidic variants of the two proteins observed in vivo are due to chemical modification by phosphorylation rather than the expression of new isomorphs of these two proteins.

Intermediate Filament-Associated Proteins: The Developmental Regulation of Their Expression and Distribution

As discussed earlier, the composition of intermediate filaments—keratins, desmin, vimentin—is developmentally regulated, with some cells expressing, for example, exclusively vimentin, others exclusively desmin, and still others variable ratios of the two proteins. A second level of developmental regulation in the structure and composition of intermediate filaments is in the expression of their associated proteins as exemplified by that of the desmin- and vimentin-associated proteins, synemin and paranemin, certain desmin- and vimentin-associated proteins defined with monoclonal antibodies, and of some neurofilament-associated proteins.

SYNEMIN   Synemin was originally discovered as a protein that associates with desmin, when this latter molecule was purified from chicken smooth muscle (gizzard) by cycles of solubilization and precipitation between pH

2.0 and 5.0 (145). Synemin has an apparent molecular weight of 230,000, binds tightly to desmin, is highly susceptible to proteolysis, and its quantity in muscle is 50–100 fold less than that of desmin. Immunoautoradiography and immunofluorescence have shown that synemin is present in adult and embryonic skeletal muscle, where it exhibits a distribution and Colcemid sensitivity indistinguishable from that of desmin and vimentin (145). As with desmin, synemin is not present in fibroblastic cells. Synemin is also present in cardiac muscle, which indicates that all three muscle types express synemin, irrespective of the ratio of vimentin to desmin in these cells. More recent studies on the composition of avian erythrocyte intermediate filaments have shown that they are composed of vimentin and synemin, while desmin is absent from these cells (94). Erythrocyte synemin copurifies with vimentin, is present in approximately the same ratio to vimentin as it is to desmin in smooth muscle, and exhibits a cytoplasmic distribution indistinguishable from that of vimentin. Peptide mapping has shown that erythrocyte and smooth muscle synemins are highly homologous proteins and immunologically related (94). These results have indicated that synemin can associate with either desmin- or vimentin-containing intermediate filaments or with filaments containing both desmin and vimentin, irrespective of the ratio of the latter two proteins (94). Thus the expression of synemin does not appear to depend on the expression of either desmin or vimentin. Furthermore, synemin may serve a very specific function in the structure of intermediate filaments in some cells, since it is not expressed in all cells.

PARANEMIN   Paranemin was originally discovered as a high-molecular-weight protein that copurifies with desmin and vimentin filaments when the latter is purified from embryonic chick skeletal muscle (159). Paranemin has an apparent molecular weight of 280,000 and is distinct from synemin by antigenic, isoelectric, and peptide mapping criteria. It is also susceptible to proteolysis, but is present in embryonic muscle in five to ten fold higher quantity than synemin. In skeletal myotubes differentiating in tissue culture, paranemin exhibits a distribution and Colcemid sensitivity indistinguishable from that of vimentin. It is expressed in all presumptive myoblasts, associates with desmin-, vimentin-, and synemin-containing filaments in young myotubes, and transits to the Z line when the other three proteins transit to this structure. However, immunofluorescence and immunoautoradiography of adult chicken skeletal and smooth (gizzard) muscle have shown that paranemin is present in barely detectable quantities. Furthermore, in chick embryonic fibroblasts grown in tissue culture, only a subpopulation of the cells is reactive with antibodies to paranemin, even though these cells all contain vimentin (159). Paranemin is also absent from adult avian erythrocytes (E. A. Repasky, unpublished observations). Thus it appears that paranemin can associate with intermediate filaments containing either vimentin alone or vimentin, desmin, and synemin. Since this protein is not found in all nonmuscle cell types where vimentin is present and since it appears to be removed from muscle cells after the association of desmin, vimentin, and synemin with the Z disc, paranemin appears to play a very specific role in the structure of embryonic filaments; its expression appears to be developmentally regulated and to be independent of the expression of the other three intermediate filament proteins (159). These observations provide further evidence for the earlier postulate that the structure and composition of a desmin-vimentin-containing filament is developmentally regulated and specific for a given cell type and a given stage in the differentiation of a cell (148).

INTERMEDIATE FILAMENT-ASSOCIATED PROTEINS DEFINED WITH MONOCLONAL ANTIBODIES   The observation that intermediate filament components become associated with skeletal myofibril Z lines during differentiation opened the possibility that other intermediate filament proteins might be components of myofibrils in adult muscle. A new appraoch to this question is through the use of monoclonal antibodies. From a bank of monoclonal antibodies raised against whole rat skeletal myofibrils, monoclonals were selected that reacted with intermediate filaments in nonmuscle cells (160). Two such monoclonal antibodies were shown to react with vimentin-containing nonmuscle intermediate filaments and exhibited the characteristic sensitivity to Colcemid. The antigens detected by each of these two monoclonal antibodies were a 95,000- and a 210,000-dalton protein that were shown to be minor components of the intermediate filament system of cultured cells and distinct from desmin and vimentin. Interestingly, however, the antibody recognizing the 210,000-dalton protein reacted with myofibril M lines and on either side of the Z disc, and the antibody recognizing the 95,000-dalton protein reacted only with the M line (160). This raises the possibility that during muscle differentiation intermediate filament components become associated with myofibril structures other than the Z disc. Since lateral connections between myofibrils also exist at the M and N lines of myofibrils in addition to the Z line connections (51), the antigens detected by these monoclonal antibodies may be part of a more extensive intermediate filament system that also functions to link myofibrils laterally at areas other than the Z disc. Microinjections of the anti 95,000-dalton monoclonal antibody into living fibroblasts caused the intermediate filaments to aggregate into tight perinuclear bundles; in contrast, injection of the anti 210,000-dalton monoclonal antibody did not (161). The general morphology, polarity, shape, and ability to spread onto a substratum of the cells injected with the anti 95,000-dalton monoclonal antibody were un-affected, as was the intracellular movement of mitochondria, granules, and vesicles (161). Thus, contrary to earlier suggestions, intermediate filaments are not involved in any of these cellular processes. The use of monoclonal antibodies to define new, intermediate, filament-associated proteins and in microinjection experiments promises to be a useful tool in determining the structure and function of this class of cytoplasmic filaments.

NEUROFILAMENT-ASSOCIATED POLYPEPTIDES   Isolated mammalian neurofilaments are composed of three polypeptides with molecular weights of 200,000 (P200), 160,000 (P160), and 68,000 (P68). Recent in vitro recon-stitution studies with purified proteins indicate that P68 and P160 together will assemble into short filaments but longer filament formations require the presence of P200 (162). Immunoelectron microscopy has indicated that P68 appears to form the "central core" of the filament, whereas P200 composes a structure more loosely and peripherally attached to the central core and periodically arranged along its axis (117). The anti-P200 binding to the filaments has indicated that P200 also appears as a bridge connecting two filaments and is disposed helically, wrapping the central core with a peri-odicity of ~ 1000 Å (117). Antibodies to P160 have also indicated that this polypeptide is peripherally associated with individual neurofilaments. Thus probably P68 is the main subunit protein of neurofilaments and P200 and P160 are peripherally associated polypeptides (117). Furthermore, P200 appears to function to cross-link neurofilaments to each other and possibly to other cytoplasmic organelles. Willard and his colleagues (117) have suggested that the principle of a central core wrapped by a helix could serve as the structural basis for a hypothetical mechanism of energy transduction that would have certain advantages for intracellular transport. Rotation of the helix around the core could result in organelle translocation by a worm-screw mechanism. Such a mechanism would be analogous to the bacterial flagellar motor (163) and provide a molecular basis for axoplasmic flow. The relationship of the neurofilament-associated polypeptides to synemin and paranemin is currently unknown but promises a new avenue of experimen-tation in understanding the structure and function of intermediate fila-ments.

CONCLUSIONS

Over the last few years, intermediate filaments have emerged as a class of proteins composed of chemically heterogeneous subunits with properties unlike those of microtubules and actin filaments. Thus far it is evident that the composition of the filaments changes during differentiation, and each type of differentiated cell exhibits its own intermediate filament subunit composition. In general, the five families of subunits identified thus far are specifically expressed in certain differentiated cell types, but this is a rule with many exceptions. Much remains to be learned about the function of these filaments even though it is clear that during differentiation they assume new distributions and very specific associations. How universal are they and how have they evolved? Answers to these questions will increase our understanding of the relationships among intermediate filaments. As their chemical heterogeneity, structural homology, and variable ratios during differentiation have already contributed different ways of looking at differentiation, intermediate filaments promise to yield still new ways of thinking about cellular differentiation and cytoplasmic morphogenesis. Toward that end, understanding in detail their composition and expression in different cell types as well as their gene structure will aid appreciably in understanding their function and evolution.

Literature Cited

1. Ishikawa, H., Bischoff, R., Holtzer, H. 1968. *J. Cell Biol.* 38:538–55
2. Lazarides, E. 1980. *Nature* 283:249–56
3. Sun, T.-T., Green, H. 1977. *Nature* 269:489–93
4. Steinert, P. M. 1975. *Biochem. J.* 149:39–48
5. Steinert, P. M., Idler, W. W. 1975. *Biochem. J.* 151:603–14
6. Steinert, P. M., Idler, W. W., Zimmerman, S. B. 1976. *J. Mol. Biol.* 108:547–67
7. Sun, T.-T., Green, H. 1978. *J. Biol. Chem.* 253:2053–60
8. Steinert, P. M., Gullino, M. I. 1976. *Biochem. Biophys. Res. Commun.* 70:221–27
9. Fuchs, E., Green, H. 1978. *Cell* 15:887–97
10. Fuchs, E., Green, H. 1979. *Cell* 17:573–82
11. Fuchs, E., Green, H. 1980. *Cell* 19:1033–42
12. Lee, L. D., Kubilus, J., Baden, H. P. 1979. *Biochem. J.* 177:187–96
13. Doran, T. I., Vidrich, A., Sun, T.-T. 1980. *Cell* 22:17–25
14. Sun, T.-T., Green, H. 1977. *Nature* 269:489–93
15. Franke, W. W., Schmid, E., Weber, K., Osborn, M. 1979. *Exp. Cell Res.* 118:95–109
16. Franke, W. W., Weber, K., Osborn, M., Schmid, E., Freudenstein, C. 1978. *Exp. Cell Res.* 116:429–45
17. Wu, Y.-J., Rheinwald, J. G. 1981. *Cell* 25:627–35
18. Sun, T.-T., Green, H. 1978. *Cell* 14:469–76
19. Sun, T.-T., Shih, C., Green, H. 1979. *Proc. Natl. Acad. Sci. USA* 76:2813–17
20. Franke, W. W., Schmid, E., Osborn, M., Weber, K. 1978. *Proc. Natl. Acad. Sci. USA* 75:5034–38
21. Franke, W. W., Appelhans, B., Schmid, E., Freudenstein, C. 1979. *Eur. J. Cell Biol.* 19:255–68
22. Franke, W. W., Schmid, E., Kartenbeck, J., Mayer, D., Hacker, H.-J., Bannash, P., Osborn, M., Weber, K., Denk, H., Wanson, J.-C., Drochmans, P. 1979. *Biol. Cellulaire* 34:99–110
23. Franke, W. W., Denk, H., Kaet, R., Schmid, E. 1981. *Exp. Cell Res.* 131:299–318
24. Freudenstein, C., Franke, W. W., Osborn, M., Weber, K. 1978. *Cell Biol. Int. Rep.* 2:591–99
25. Franke, W. W., Schmid, E., Freudenstein, C., Appelhans, B., Osborn, M., Weber, F., Keenan, T. W. 1980. *J. Cell Biol.* 84:633–54
26. Ramaekers, F. C. S., Osborn, M., Schmid, E., Weber, K., Bloemendal, H., Franke, W. W. 1980. *Exp. Cell Res.* 127:309–27
27. Summerhayes, I. C., Cheng, Y.-S. E., Sun, T.-T., Chen, L. B. 1981. *J. Cell Biol.* 90:63–69
28. Bannasch, P., Zerban, H., Schmid, E., Franke, W. W. 1980. *Proc. Natl. Acad. Sci. USA* 77:4948–52
29. Uehara, Y., Campbell, G. R., Burnstock, G. 1971. *J. Cell Biol.* 50:484–97

30. Cooke, P. H., Chase, R. H. 1971. *Exp. Cell Res.* 66:417–25
31. Cooke, P. H. 1976. *J. Cell Biol.* 68:539–56
32. Somlyo, A. P., Devine, C. E., Somlyo, A. V., Rice, R. V. 1973. *Phil. Trans. R. Soc. London B* 265:223–29
33. Schollmeyer, J. E., Furcht, L. T., Goll, D. E., Robson, R. M., Stromer, M. H. 1976. *Cell Motility*, pp. 361–88. Cold Spring Harbor, NY: Cold Spring Harbor Lab.
34. Lazarides, E., Hubbard, B. D. 1976. *Proc. Natl. Acad. Sci. USA* 73:4344–48
35. Small, J. V., Sobieszek, A. 1977. *J. Cell. Sci.* 23:243–68
36. Huiatt, T. W., Robson, R. M., Arakawa, N., Stromer, M. H. 1980. *J. Biol. Chem.* 255:6981–89
37. Geisler, N., Weber, K. 1980. *Eur. J. Biochem.* 111:425–33
38. Hubbard, B. D., Lazarides, E. 1979. *J. Cell Biol.* 80:166–82
39. Izant, J. G., Lazarides, E. 1977. *Proc. Natl. Acad. Sci. USA* 74:1450–54
40. Lazarides, E., Balzer, D. R. Jr. 1978. *Cell* 14:429–38
41. O'Shea, J. M., Robson, R. M., Hartzer, M. K., Huiatt, T. K., Rathbun, W. E., Stromer, M. H. 1981. *Biochem. J.* 195:345–56
42. O'Shea, J. M., Robson, R. M., Huiatt, T. W., Hartzer, M. K., Stromer, M. H. 1979. *Biochem. Biophys. Res. Commun.* 89:972–80
43. Gard, D. L., Bell, P. B., Lazarides, E. 1979. *Proc. Natl. Acad. Sci. USA* 76:3894–98
44. Granger, B. L., Lazarides, E. 1979. *Cell* 18:1053–63
45. Gabbiani, G., Schmid, E., Winter, S., Chaponnier, C., de Chastonay, C., Vandekerckhove, J., Weber, K., Franke, W. W. 1981. *Proc. Natl. Acad. Sci. USA* 78:298–302
46. Frank, E. D., Warren, L. 1981. *Proc. Natl. Acad. Sci. USA* 78:3020–24
47. Lazarides, E. 1978. *Exp. Cell Res.* 112:265–73
48. Eriksson, A., Thornell, L.-E. 1979. *J. Cell Biol.* 80:231–47
49. Kelly, D. E. 1969. *Anat. Rec.* 163:403–26
50. Page, S. G. 1969. *J. Physiol.* 205:131–45
51. Granger, B. L., Lazarides, E. 1978. *Cell* 15:1253–68
52. Goldspink, G. 1971. *J. Cell Sci.* 9:123–37
53. Shear, C. R. 1978. *J. Cell Sci.* 29:297–312
54. Sandborn, E. B., Cote, M. G., Roberge, J., Bois, P. 1967. *J. Microsc.* 5:169–78
55. Lindner, E., Schaumburg, G. 1968. *Z. Zellforsch. Mikrosk. Anat.* 84:549–62
56. Viragh, S., Challice, C. E. 1969. *J. Ultrastruct. Res.* 28:321–34
57. Oliphant, L. W., Loewen, R. D. 1976. *J. Mol. Cell. Cardiol.* 8:679–88
58. Behrendt, H. 1977. *Cell Tissue Res.* 180:303–15
59. Junker, J., Sommer, J. R. 1977. *35th Ann. Proc. Electron Microsc. Soc. Am.*, pp. 582–83. Boston, Mass.
60. Forbes, M. S., Sperelakis, N. 1980. *Tissue and Cell* 12:467–89
61. Ferrans, V. J., Roberts, W. C. 1973. *J. Mol. Cell. Cardiol.* 5:247–57
62. Rybicka, K. 1981. *J. Histochem. Cytochem.* 29:553–60
63. Gard, D. L., Lazarides, E. 1980. *Cell* 19:263–75
64. Bennett, G. S., Fellini, S. A., Croop, J. M., Otto, J. J., Bryan, J., Holtzer, H. 1978. *Proc. Natl. Acad. Sci. USA* 75:4364–68
65. Bennett, G. S., Fellini, S. A., Holtzer, H. 1978. *Differentiation* 12:71–82
66. Campbell, G. R., Chamley-Campbell, J., Groschel-Stewart, U., Small, J. V., Anderson, P. 1979. *J. Cell Sci.* 37:303–22
67. Devlin, R. B., Emerson, C. P. 1979. *Develop. Biol.* 69:202–16
68. Bennett, G. S., Fellini, S. A., Toyama, Y., Holtzer, H. 1979. *J. Cell Biol.* 82:577–84
69. Hynes, R. O., Destree, A. T. 1978. *Cell* 13:151–63
70. Franke, W. W., Schmid, E., Osborn, M., Weber, K. 1979. *J. Cell Biol.* 81:570–80
71. Fellini, S. A., Bennett, G. S., Toyama, Y., Holtzer, H. 1978. *Differentiation* 12:59–69
72. Brown, S., Levinson, W., Spudich, J. A. 1977. *J. Supramol. Struct.* 5:119–30
73. Cabral, F., Gottesman, M. M., Zimmerman, S. B., Steinert, P. M. 1981. *J. Biol. Chem.* 256:1428–31
74. Nelson, W. J., Traub, P. 1981. *Eur. J. Biochem.* 116:51–57
75. Franke, W. W., Grund, C., Schmid, E. 1979. *Eur. J. Cell Biol.* 19:269–75
76. Albertini, D. F., Kravit, N. G. 1981. *J. Biol. Chem.* 256:2484–92
77. Goldman, R. D. 1971. *J. Cell Biol.* 51:752–62
78. Goldman, R. D., Knipe, D. 1973. *Cold Spring Harbor Symp. Quant. Biol.* 37:523–34
79. Holltrop, M. E., Raisz, L. G., Simmons, H. A. 1974. *J. Cell Biol.* 60:346–55
80. Blose, S. H., Chacko, S. 1976. *J. Cell Biol.* 70:459–66

81. Paetau, A., Virtanen, I., Stenman, S., Kurki, P., Linder, E., Vaheri, A., Westermark, B., Dahl, D., Haltia, M. 1979. *Acta Neuropath.* 47:71–74
82. Goldman, R. D., Follett, E. A. C. 1970. *Science* 169:286–88
83. Felix, H., Strauli, P. 1976. *Nature* 261:604–6
84. Starger, J. M., Goldman, R. D. 1977. *Proc. Natl. Acad. Sci. USA* 74:2422–26
85. Starger, J. M., Brown, W. E., Goldman, A. E., Goldman, R. D. 1978. *J. Cell Biol.* 78:93–109
86. Zackroff, R. V., Goldman, R. D. 1979. *Proc. Natl. Acad. Sci. USA* 76:6226–30
87. Tuszynski, G. P., Frank, E. D., Damsky, C. H., Buck, C. A., Warren, L. 1979. *J. Biol. Chem.* 254:5138–6143
88. Geisler, N., Weber, K. 1981. *FEBS Lett.* 125:253–56
89. Renner, W., Franke, W. W., Schmid, E., Geisler, N., Weber, K., Mandelkow, E. 1981. *J. Mol. Biol.* 149:285–306
90. Lehto, V.-P., Virtanen, I., Kurki, P. 1978. *Nature* 272:175–77
91. Small, J. V., Celis, J. E. 1978. *J. Cell Sci.* 31:393–409
92. Virtanen, I., Kurkinen, M., Lehto, V.-P. 1979. *Cell Biol Int. Rep.* 3:157–62
93. Laurila, P., Virtanen, I., Stenman, S. 1981. *Exp. Cell Res.* 131:41–46
94. Granger, B. L., Repasky, E. A., Lazarides, E. 1982. *J. Cell Biol.* In press
95. Blose, S. H. 1979. *Proc. Natl. Acad. Sci. USA* 76:3372–76
96. Zieve, G. W., Heidemann, R. S., McIntosh, J. R. 1980. *J. Cell Biol.* 87:160–69
97. Day, W. A. 1980. *J. Ultrastruct. Res.* 70:1–7
98. Sharpe, A. H., Chen, L. B., Murphy, J. R., Fields, B. N. 1980. *Proc. Natl. Acad. Sci. USA* 77:7267–71
99. Klymkowsky, M. W. 1981. *Nature* 291:249–51
100. Wang, E., Choppin, P. W. 1981. *Proc. Natl. Acad. Sci. USA* 78:2363–67
101. Eckert, B. S., Koons, S. J., Schantz, A. W., Zobel, C. R. 1980. *J. Cell Biol.* 86:1–5
102. Metuzals, J., Mushynski, W. E. 1974. *J. Cell Biol.* 61:701–22
103. Wuerker, R. B. 1970. *Tissue Cell* 2:1–9
104. Wuerker, R. B., Kirkpatrick, J. B. 1972. *Int. Rev. Cytol.* 33:45–75
105. Mori, H., Kurokawa, M. 1979. *Cell Struct. and Funct.* 4:163–67
106. Hoffman, P. N., Lasek, R. J. 1975. *J. Cell Biol.* 66:351–56
107. Filliatreau, G., Di Giamberardino, L., Belacourte, A., Boutteau, F., Biserte, G. 1981. *Biochimie* 63:369–71

108. Bennett, G. S., Tapscott, S. J., Kleinbart, F. A., Antin, P. B., Holtzer, H. 1981. *Science* 212:567–69
109. Schlaepfer, W. W. 1978. *J. Cell Biol.* 76:50–56
110. Leim, R. K. H., Yen, S.-H., Salomon, G. D., Shelanski, M. L. 1978. *J. Cell Biol.* 79:637–45
111. Schlaepfer, W. W., Freeman, C. A. 1978. *J. Cell Biol.* 78:653–62
112. Schlaepfer, W. W. 1977. *J. Cell Biol.* 74:226–40
113. Schlaepfer, W. W., Lynch, R. G. 1977. *J. Cell Biol.* 74:241–50
114. Anderton, B. H., Thorpe, R., Cohen, J., Selvendrau, S., Woodhams, P. 1980. *J. Neurocytol.* 9:835–44
115. Yen, S.-H., Fields, K. L. 1981. *J. Cell Biol.* 88:115–26
116. Willard, M., Simon, C., Baitinger, C., Levine, J., Skene, P. 1980. *J. Cell Biol.* 85:587–96
117. Willard, M., Simon, C. 1981. *J. Cell Biol.* 89:198–205
118. Dahl, D. 1980. *FEBS Lett.* 111:152–56
119. Czosnek, H., Soifer, D., Wisniewski, H. M. 1980. *J. Cell Biol.* 85:726–34
120. Lasek, R. J., Hoffman, P. N. 1976. *Cell Motility*, pp. 1021–50. Cold Spring Harbor, NY: Cold Spring Harbor Lab.
121. Pant, H. C., Shecket, G., Gainer, H., Lasek, R. J. 1978. *J. Cell Biol.* 78:R23–R27
122. Gilbert, D. 1976. *J. Physiol.* 266:81
123. Roslansky, P. F., Cornell-Bell, A., Rice, R. V., Adelman, W. J. Jr. 1980. *Proc. Natl. Acad. Sci. USA* 77:404–8
124. Gilbert, D. S. 1975. *J. Physiol.* 253:257–301
125. Gilbert, D. S., Newby, B. J., Anderton, B. 1975. *Nature* 256:586–89
126. Zackroff, R. V., Goldman, R. D. 1980. *Science* 208:1152–55
127. Schlaepfer, W. W. 1977. *Brain Res.* 136:1–9
128. Schlaepfer, W. W., Micko, S. 1979. *J. Neurochem.* 32:211–19
129. Shelanski, M. L., Albert, S., DeVries, G. H., Norton, W. T. 1971. *Science* 174:1242–45
130. Yen, S. H., Dahl, D., Schachner, M., Shelanski, M. L. 1976. *Proc. Natl. Acad. Sci. USA* 73:529–33
131. DeVries, G. H., Eng, L. F., Lewis, D. L., Hadfield, M. G. 1976. *Biochim. Biophys. Acta* 439:133–45
132. Schook, W. J., Norton, W. T. 1976. *Brain Res.* 118:517–22
133. Schachner, M., Hedley-White, E. T., Hsu, D. W., Schoonmaker, G., Bignami, A. 1977. *J. Cell Biol.* 75:67–73

134. Eng., L. F., DeVries, G. H., Lewis, D. L., Bigbee, J. 1976. *Fed. Proc.* 35:1766
135. Benitz, W. E., Dahl, D., Williams, K. W., Bignami, A. 1976. *FEBS Lett.* 66:285–89
136. Goldman, J. E., Schaumburg, H. H., Norton, W. T. 1978. *J. Cell Biol.* 78:426–40
137. Bignami, A., Eye, L. F., Dahl, D., Uyeda, C. T. 1972. *Brain Res.* 430: 429–35
138. Bignami, A., Dahl, D. 1977. *J. Histochem. Cytochem.* 25:466–69
139. Rueger, D. C., Huston, J. S., Dahl, D., Bignami, A. 1979. *J. Mol. Biol.* 135:53–68
140. Osborn, M., Franke, W., Weber, K. 1980. *Exptl. Cell Res.* 125:37–46
141. Henderson, D., Weber, K. 1981. *Exptl. Cell Res.* 132:297–311
142. Franke, W. W., Schmid, E., Breitkreutz, D., Luder, M., Boukamp, P., Fusenig, N. E., Osborn, M., Weber, K. 1979. *Differentiation* 14:35–50
143. Borenfreund, E., Schmid, E., Bendich, A., Franke, W. W. 1980. *Exp. Cell Res.* 127:215–35
144. Lazarides, E., Granger, B. L., Gard, D. L., O'Connor, C. M., Breckler, J., Price, M., Danto, S. I. 1982. *Cold Spring Harbor Symp. Quant. Biol.* 46: In press
145. Granger, B. L., Lazarides, E. 1980. *Cell* 22:727–38
146. Steinert, P. M., Idler, W. W., Cabral, F., Gottesman, M. M., Goldman, R. D. 1981. *Proc. Natl. Acad. Sci. USA* 78:3692–96
147. Steinert, P. M., Idler, W. W., Goldman, R. D. 1980. *Proc. Natl. Acad. Sci. USA* 77:4534–38
148. Lazarides, E. 1981. *Cell* 23:649–50
149. O'Connor, C. M., Gard, D. L., Asai, D. J., Lazarides, E. 1981. *Protein Phosphorylation,* Vol. 8, pp. 1157–69. Cold Spring Harbor Conf. Cell Proliferation. Cold Spring Harbor, NY.
150. Geisler, N., Weber, K. 1981. *Proc. Natl. Acad. Sci. USA* 78:4120–23
151. Shecket, G., Lasek, R. J. 1978. *The Cytoskeletal and Contractile Networks of Nonmuscle Cells,* p. 48. Cold Spring Harbor, NY: Cold Spring Harbor Lab.
152. Eagles, P. A. M., Gilbert, D. S. 1979. *J. Physiol.* 287:10P
153. O'Connor, C. M., Balzer, D. R., Lazarides, E. 1979. *Proc. Natl. Acad. Sci. USA* 76:819–23
154. Cabral, F., Gottesman, M. M. 1979. *J. Biol. Chem.* 254:6203–6
155. O'Connor, C. M., Gard, D. L., Lazarides, E. 1981. *Cell* 23:135–43
156. Gard, D. L., Lazarides, E. 1982.
157. O'Connor, C. M., Asai, D. J., Flytzanis, C. N., Lazarides, E. 1981. *Mol. Cell. Biol.* 1:303–9
158. Devlin, R. B., Emerson, C. P. Jr. 1978. *Cell* 13:599–611
159. Breckler, J., Lazarides, E. 1982. *J. Cell Biol.* In press
160. Lin, J. J.-C. 1981. *Proc. Natl. Acad. Sci. USA* 78:2335–39
161. Lin, J. J.-C., Feramisco, J. R. 1981. *Cell* 24:185–93
162. Moon, H. M., Wisniewski, T., Merz, P., De Martini, J., Wisniewski, H. M. 1981. *J. Cell Biol.* 89:560–67
163. Berg, H. C., Manson, M. D., Conly, M. P. 1982. *Soc. Exp. Biol. Symp.* 35: In press

Ann. Rev. Biochem. 1982. 51:251–82

BIOCHEMISTRY OF INTERFERONS AND THEIR ACTIONS[1]

Peter Lengyel[2]

Department of Molecular Biophysics and Biochemistry, Yale University, New Haven, Connecticut 06511

CONTENTS

[1]Abbreviations used are: ds, double-stranded; EAT, Ehrlich ascites tumor; EMC, encephalomyocarditis; IFN, interferon; VS, vesicular stomatitis. (2'–5') (A)$_n$ is also designated as oligoisoadenylate and is sometimes abbreviated as 2–5A. The enzyme that synthesizes (2'–5') (A)$_n$ is designated by different authors as (2'–5')(A)$_n$ synthetase or (2'–5') oligo(A)polymerase or oligo-isoadenylate synthetase E. The endoribonuclease that can be activated by (2'–5')(A)$_n$ is designated as RNase L or RNase F. The protein kinase that can be activated by double-stranded RNA (dsRNA) is also designated as protein kinase PK-i or eIF-2 kinase. The phosphodiesterase that degrades (2'–5')(A)$_n$ is also designated as phosphodiesterase 2'-PDi.

[2]I thank my colleagues for sending me preprints and reprints. Studies in my laboratory are supported by NIH research grants AI12320 and CA16038.

0066-4154/82/0701-0251$2.00

1. Perspectives and Summary

Interferons (IFNs) (1–7) are a family of proteins, many of which share some sequence homology (7–14). They occur in a large variety of vertebrates from fish to man and are biological regulators of cell function. Their activities include the regulation of certain responses to disease. Unless induced, their concentration is below the detectable level in most organs and cells in culture.

Human IFNs are classified into three antigenically distinct types: α (at least eight species), β (at least two species), and γ (number of species unknown). α and β IFNs can be induced in a variety of cells by various agents e.g. certain viruses, certain bacteria, and double-stranded RNA (dsRNA). γ IFNs are induced in lymphoid cells by mitogens and by antigens to which the cells have been sensitized. The first intermediates in the induction are mRNAs that can be translated into IFNs in cell-free systems or Xenopus oocytes. It is this translatability of the mRNAs that has facilitated the isolation of the genes specifying human IFNs. Eight of the α and one of two β genes isolated are free of introns (7, 11, 15–18).

The IFNs that are secreted by the producing cells bind to surface receptors of responsive cells (19, 20) and effect in these numerous, seemingly diverse biological phenomena (2–6).

The first activity of IFNs to be discovered was the interference with the replication of various viruses (21, 22). This is the property for which these agents were named. It also serves as the basis of the most frequent IFN assays (1, 2).

In addition to their antiviral actions, IFNs affect cell motility (23, 24), cell proliferation, and various immunological processes, such as the antibody response, delayed hypersensitivity, graft rejection, expression of several surface antigens, natural killer cell recruitment, and macrophage activation. Many of the IFN effects are transient (1–5, 25–27).

The importance of IFNs as antiviral and antitumor agents has been clearly demonstrated. Mice treated with an antiserum to mouse IFNs were killed by amounts of virus several hundred times less than needed to kill control mice (28, 29). Moreover, IFNs injected could protect animals from some viral diseases (1). The growth of various spontaneous, transplanted, or virus-induced tumors in animals was found to be impaired by treatment with IFNs (1, 27) and to be promoted by treatment with an antiserum to IFNs (1, 30, 31). At the same time IFNs were found to be toxic to newborn mice though not to those older than about 8 days (32).

IFNs are among the most potent biological agents; they impair virus replication in responsive cells at a concentration as low as 3×10^{-14} M (33).

Most of the alterations in cells elicited by IFNs depend on RNA synthesis and protein synthesis, and IFNs were shown to induce, i.e. enhance the level of various mRNAs (34–37) and proteins (38–41). Among the enzymes induced by IFNs is $(2'-5')(A)_n$ synthetase. Upon activation by dsRNA this enzyme generates from ATP $2'-5'$-linked oligoadenylates (42, 43). These in turn activate RNase L, a latent endoribonuclease, that cleaves single-stranded RNAs (e.g. mRNA and ribosomal RNA) at preferred sites (44–51). Another enzyme induced by IFNs is a protein kinase that, if activated by dsRNA, phosphorylates and thereby impairs the activity of a peptide chain initiation factor eIF-2, thus inhibiting protein synthesis (52–54). The activation of these enzymes is reversible.

These enzymes were discovered in experiments with extracts from IFN-treated cells. It is hypothesized that the role that dsRNA plays in vitro is played in vivo, i.e. in virus infected cells, by partially ds intermediates or side products of viral RNA synthesis (48, 55–57). It is in line with the assumed function of the above enzymes in IFN action that viral RNA and protein accumulation are among the processes of virus replication that are impaired in cells treated with IFNs (6). The $(2'-5')(A)_n$ synthetase-RNase L pathway may also be involved in controlling the rate of cell division even in uninfected cells (58, 59).

Other effects of IFNs on cells include an alteration in the methylation of viral (and perhaps host) mRNAs (60–65) and a decrease in the proportion of unsaturated fatty acids in membrane phospholipids, which results in a change in the rigidity of the cell membrane (66, 67). The agents mediating these alterations and the role of the alterations in IFN action remain to be elucidated.

The following are some of the IFN-related biochemical problems that are under investigation: the mechanisms of induction of the different IFNs and of the various proteins induced by IFNs; the mechanism by which the various IFN-induced proteins inhibit virus replication and cell proliferation; and the mechanisms of the modulation of the immune response by IFNs.

There are several recent reviews on various aspects of IFN research (1–7, 25, 26, 68–73). After a short general survey of IFNs this review deals with the biochemistry of IFNs and their actions.

2. Survey of Interferons

a. ASSAY   Many of the IFN assays are based on measuring antiviral activity in cell culture. The procedures usually involve the treatment of cells in culture for 12–24 hr with an IFN preparation at varying dilutions, followed by the exposure of the culture to an appropriate virus (e.g. VS virus) at the suitable dilution. The antiviral effect is then monitored, usually by determining one of the following parameters: protection of the cells from the cytopathic effect of the virus, decrease in virus-specific RNA synthesis, number of viral plaques, or virus yield. The reciprocal value of the highest dilution of the IFN solution giving a 50% decrease in the effect is taken as the titer. The IFN activities are usually expressed in units i.e. in terms of the activity of international reference standard IFN preparations (1, 2).

The isolation of monoclonal antibodies to various IFNs (74–76) allowed the development of radioimmunoassays (77, 78).

b. INDUCTION   IFNs are produced upon induction in a large variety of vertebrates including fish, turtles, birds, and mammals. Within the organism and also in cell culture many, though not all, types of cells can synthesize IFNs. Some cell lines form small amounts of IFNs constitutively.

The inducers of IFNs include: members of most major virus groups; some species of mycoplasmas, bacteria, and protozoa; natural or synthetic dsRNA (e.g. poly(I)·poly(C)); endotoxins; polysaccharides; antigens (in animals that have been sensitized to the particular antigen in advance); mitogens; and some low-molecular weight compounds (1–3, 5, 71). Certain tumor derived or virus transformed cells were also reported to induce IFNs in lymphocytes (79). According to another study this may be due, at least in some cases, to mycoplasma contamination of the tumor cells (80). The same agent that induces IFN formation in certain species of animals or cells may not be an inducer in others (1, 2).

The IFNs synthesized by different animal species are different; moreover one and the same animal or cell species can produce different IFNs. Thus at least three different families of human IFNs have been characterized with little or no immunological cross-reactivity: α, β, and γ. These were designated earlier as leukocyte (Le), fibroblast (F), and type 2 or immune IFNs, respectively (1–5, 81).

The type of IFN produced may depend on the nature of the producing cell and of the inducer. Thus a human fibroblast line produces β type IFNs and only a little if any α type if induced with poly(I)·poly(C). However, upon induction with Newcastle disease virus the same cell line produces about 25% as much α type IFN as β type (82).

c. INTERFERON mRNA AND THE CONTROL OF INTERFERON SYNTHESIS   The first known intermediates in IFN induction are IFN mRNAs. These can be assayed by extracting mRNA (i.e. poly(A)-containing RNA) from induced cells and adding it to a cell-free protein synthesizing system or to a culture of cells (from a species other than that from which the mRNA was extracted) or injecting it into Xenopus oocytes. IFN mRNAs can be translated in each of these systems into biologically active IFNs characteristic of the species from which the IFN mRNA was extracted (3).

The IFN synthesis following the treatment of animals or cells with IFN inducers is shut off after a period whose length depends on the nature of the inducer and of the animal or cell induced (3, 5, 71). As revealed by experiments involving the translation of the mRNA into IFN in oocytes, the shutoff in cultured cells is accompanied (and probably caused) by the disappearance of translatable IFN mRNA (5).

In the case of the induction of human foreskin fibroblasts (FS4) with poly(I)·poly(C), which results in the production of $\beta2$ IFN primarily, the disappearance of translatable IFN mRNA during the shutoff of IFN production was shown to be a selective phenomenon. It is not accompanied by an appreciable decline in the bulk of cellular protein synthesis, protein excretion, the amount of translatable mRNA, or the stability of labeled mRNA (83).

The shutoff of IFN synthesis and the disappearance of translatable IFN mRNA can be delayed by treating the cells at the appropriate time after induction with inhibitors of RNA synthesis (e.g. actinomycin D or 5,6-dichlororibofuranosylbenzimidazole) and/or inhibitors of protein synthesis (e.g. cycloheximide). This procedure results in a greatly increased production of IFNs and is called superinduction (2, 3, 5, 71). The results of studies on superinduction serve as a basis for a hypothesis according to which the shutoff is a consequence of the inactivation of the IFN mRNA by a repressor. The hypothetical repressor is assumed to be induced by poly(I)·poly(C) and/or IFN and to have a short half-life (2, 3, 5, 71). IFNs do induce an endonuclease system [$(2'-5')(A)_n$ synthetase-RNase L system; see Section 9], which can be activated by dsRNA [e.g. poly(I)·poly(C)]. However, the pretreatment of the cells with IFN at a low concentration, which results in a large increase in the level of $(2'-5')(A)_n$ synthetase, has no effect on the induction, shutoff, and superinduction of IFN synthesis. This is consistent with the view that the shutoff of IFN production may not be catalyzed by this inducible endonuclease system (83).

After a round of IFN synthesis or treatment with IFN at higher concentrations, animals or cells in culture may become temporarily refractive to repeated IFN induction (hyporesponsiveness). There are indications that the shutoff of IFN synthesis and hyporesponsiveness are related phenomena. The treatment of certain cell types in culture with IFNs at low concentrations, however, enhances IFN synthesis upon induction (priming). During priming, protein synthesis must take place in the cells. The rate of induced IFN mRNA synthesis can be accelerated in certain cell lines (e.g. Namalva, a human lymphoblastoid line) by various compounds, e.g. bromodeoxy-uridine or butyric acid. The mechanism of this acceleration is unclear (2, 3, 5, 71).

d. MASS PRODUCTION OF HUMAN INTERFERONS   A large portion of the human IFNs used for the small scale clinical trials under way have been produced from leukocytes. These are obtained from human blood by slow speed centrifugation. The process results in a red cell–rich layer at the bottom of the centrifuge tube, a leukocyte-rich layer in the middle, and plasma on top. The leukocyte-rich fraction (buffy coat) is induced to produce IFNs of the α type by infection with Sendai virus (84).

Type β human IFNs are produced in fibroblasts grown in monolayer cultures induced with poly(I)·poly(C) and superinduced with inhibitors of RNA and protein synthesis (85–87). The disadvantage of this system is the limited life span of the fibroblasts: their ability to produce IFNs decreases after less than 50 passages.

Human IFNs can also be produced on a large scale using established cell lines. Lymphoblastoid cells (e.g. Namalva) give rise primarily to the α type (88, 89), and fibroblastoid cells to β type IFNs (90). Such cells may be passaged and induced to form IFNs indefinitely.

A breakthrough in the technology of IFN production occurred at the beginning of 1980. Using recombinant DNA technology human IFN genes were isolated, inserted into *E. coli,* and shown to specify in the bacteria the synthesis of biologically active human IFNs (91–94).

3. Isolation of Human Interferon Genes

The following approach was used to obtain the first bacterial plasmids specifying the production of a human α IFN in *E. coli:* mRNA was isolated from human leukocytes induced by infection with Sendai virus and was fractionated by sucrose gradient centrifugation. The fraction with the highest IFN mRNA activity was identified by the oocyte translation assay. The mRNA in this fraction was used as a template for the synthesis of ds complementary (c) DNA, and the resulting cDNA was inserted into bacterial plasmids (PBR322), which were used to transform *E. coli.* The clones of transformed *E. coli* (about 5000) were divided into pools of about 500. These were screened for DNA segments complementary to IFN mRNA by a hybridization-translation assay. For this purpose mRNA from induced leukocytes was hybridized to plasmid DNA from each of the pools, and the hybridized mRNA was recovered and assayed by translation in oocytes. One of the pools that hybridized IFN mRNA was subdivided into smaller pools, and these were assayed. These and further operations resulted in the isolation of a hybrid plasmid with an insert complementary to a segment of IFN mRNA. The insert (320-base pairs long) was shorter than expected for IFN mRNA. Thus it was used to search among the other plasmids for those with longer inserts. One was found that had an insert of sufficient length (910 base pairs). The *E. coli* transformed with this plasmid synthesized a polypeptide with IFN activity in human cells. Though the plasmid was constructed to allow the synthesis of IFN fused to a part of the β lactamase protein, the translation of the IFN active material apparently started within the inserted IFN-specific region of the mRNA (91).

Essentially similar approaches resulted in the isolation of a human β IFN cDNA clone (13, 95, 96) and its expression in *E. coli* (18, 92, 94, 97).

4. Synthesis of Human Interferons in E. Coli and Yeast

The first of the IFN cDNA–containing plasmids to be constructed, when expressed in *E. coli,* produced little IFN [20,000 IFN units per liter of culture, i.e. ~1–2 IFN molecules per cell (91)]. To increase the IFN yield, the IFN cDNA segments (or IFN gene segments) were subsequently inserted into more efficient "expression plasmids." These usually include a promoter and a ribosome binding site from an efficient bacterial or viral operon (e.g. lac, β lactamase, trp, λ PL). To this segment is linked, at an appropriate distance from the ribosomal binding site, the translation initiator codon (ATG), followed immediately by the protein coding region of the IFN cDNA (18, 93, 94, 97). The insertion of two to three successive promoters in the position preceding the IFN translation initiation site was found to enhance the efficiency of the expression plasmids (94).

The efficient synthesis of several IFNs [including α1 (D) (7), α2 (A) (7, 98), and β1 (94)] from expression plasmids in *E. coli* was reported. In the case of α2 (A), 0.6–1 mg IFN protein (2–2.5×10^8 IFN units) was produced per liter of culture (7, 93).

A human IFN was also produced in yeast. For this purpose an α1 (D) IFN cDNA segment was linked with a DNA segment that included the promoter region of the yeast (*Saccharomyces cerevisiae*) alcohol dehydrogenase 1 gene in a plasmid that is capable of replication in both yeast and

E. coli. Yeast cells transformed by this plasmid synthesized about 2.8 X 10^6 IFN units per liter of culture (99).

Some of the IFNs (though not human α IFNs) are glycoproteins. Thus a possible advantage of yeast over *E. coli* in producing IFNs is that yeast can attach carbohydrate moieties to proteins. Moreover, the attachment of the inner core of carbohydrates occurs in yeast by the same mechanism as is found in animal cells.

5. Structure of Human Interferons and of the Genes Specifying Them

Human IFNs were purified (*a*) from a variety of induced human cells (including leukocytes, i.e. buffy coats); from the blood of healthy donors (100); from leukocytes obtained by leukophoresis from patients with chronic myelogeneous leukemia (101–103); and from lymphoblastoid cells (89, 104), diploid fibroblasts (105–107), and fibroblastoid cells grown in cell culture (108)); and (*b*) from *E. coli* cultures producing human IFNs (7, 109–111).

The purification procedures were: conventional protein fractionation techniques, including high performance liquid chromatography and polyacrylamide gel electrophoresis in sodium dodecyl sulfate; and affinity chromatography with IFN antibodies, lectins, polynucleotides, and some small ligands e.g. Cibachron Blue linked to agarose (1, 2, 4, 5, 106, 112). α2 (A) IFN from *E. coli* has been crystallized (7).

Until recently, the amounts of pure IFNs available for structural studies were sufficient only for microsequencing techniques. They were first used to sequence N-terminal portions (8, 9) and subsequently internal tryptic and chymotryptic peptides (104). The results indicated the existence of sequence heterogeneity first between an α and a β IFN (8, 9) and subsequently also among α IFNs (103). The complete amino acid sequences of α and β type IFNs were first predicted from the nucleotide sequences of cloned IFN cDNAs (12).

The first isolated α cDNA clones used as screening probes allowed the identification of further α clones in cDNA libraries, and restriction endonuclease mapping allowed the division of these into groups. Eight of the α cDNA clones with different sequences are designated as α 1 (D), α 2 (A), α 3 (F), α 5 (G), α 8 (B), Ψ 210 (C), E and H (7, 11, 14). [The α 1 to α 8 and Ψ 210 designations are from C. Weissmann and his colleagues (7). The D, A, F, G, B, and C designations shown in parentheses after the corresponding α designations and the E and H designations are from D. Goeddel and his colleagues (14)]. It should be noted that there is a single amino acid difference between the clones α 1 and D and between the clones α 2 and A (14). These differences and other similarly slight differences in sequence between IFNs may be due to allelic variation. The α IFN cDNA coding sequences specify 23-amino acid long precursor signal peptides. Most IFN clones have sequences specifying 166-amino acid long mature IFN polypeptides, whereas that of the α 2(A) clone (from which codon 44 has been deleted) specifies a 165-amino acid long polypeptide (7, 14). One of the eight clones [Ψ 210 (C)] is from a transcribed IFN pseudogene. This specifies a protein whose sequence from codon 40 on bears no resemblance to those of other IFNs, because of the insertion of a single nucleotide into this codon (14).

In the (six known) signal peptides of α IFN 43% of the amino acids occur in invariant positions, and in the (seven known) mature α IFN proteins 60% occur thus (14). The β 1 IFN has a 21-amino acid long signal peptide and the mature protein is 166 amino acids long (12, 13, 16, 94). There is no homology in signal sequences among β1 IFN and α IFNs, and in the mature IFN sequences only 23% of the amino acids occur in invariant positions. Most of the changes in amino acid sequence are due to single nucleotide substitutions; some are due to the insertion of one nucleotide and a compensatory deletion of a second nucleotide in a nearby position. All the α IFNs code for cysteines in codons 1, 29, 99 or 100 and 139 (14). In α 2 (A) IFN (in which this was determined) cysteine is bonded to 99, and 29 is bonded to 139 (113). There is a pronounced homology between the upstream flanking regions of α 1 IFN and β 1 IFN (114).

The screening of human chromosomal DNA with α IFN cDNA probes, together with restriction and sequence analysis indicate the existence of at least 12 distinct α IFN genes (which are nonallelic), together with several allelic variants. Probably 10 or more α IFN genes are expressed in man (7, 11, 14). In addition, several α IFN pseudogenes were found whose coding sequences contain alterations preventing the expression of full-length IFN-like proteins (7, 14).

It is estimated that the α and β genes must have diverged some five hundred to a thousand million years ago and thus are as old as vertebrates (12). The α gene family seems to have arisen by gene duplication of the ancestral α gene some twenty to eighty million years ago. Further duplications occurred up to recent times. Concurrently α genes were degenerating in consequence of deleterious mutations (7).

Many of the genomic α clones and a genomic β 1 clone direct the synthesis of IFN in *E. coli,* an organism not known to be able to splice (7, 11, 18). This fact, together with the comparison of the sequences of some cDNA clones and genomic clones, indicate that (*a*) the α 1 (D) and β 1 IFN genes lack introns in the entire genes (11, 114, 116) and (*b*) several α genes, other than α 1 (D), are devoid of introns, certainly in their coding sequences (and possibly even in the noncoding segments of the genes) (7,11,16–18, 115). In this respect, these IFN genes resemble the sea urchin and *Drosophila* histone genes, which have no introns either, and differ from all other protein coding genes of higher eukaryotes described so far, which have at least one but frequently several introns.

Most recently, the discovery of even further human α and β IFNs was reported (117–121). This resulted from the following approach. mRNAs from cells exposed to IFN inducers were fractionated according to size by sucrose gradient centrifugation or electrophoresis in agarose CH_3HgOH gels. The mRNA fractions were injected into oocytes for assaying IFN mRNA activity, and the IFNs synthesized were classified as α or β type by neutralization tests with antisera. The approach resulted first in the discovery of a second β type IFN (designated as $\beta2$) from fibroblasts induced with poly(I)·poly(C) (117–119). This was translated from a 1.3-kb mRNA (whereas $\beta1$ was translated from a 0.9-kb mRNA). Poly(I)·poly(C) induces the synthesis of both $\beta1$ and $\beta2$ mRNAs, whereas cyloheximide alone induces the synthesis of $\beta2$ mRNA only (117). The IFN $\beta2$ gene was reported to have introns (18). Furthermore, although, as revealed from cDNA sequencing, there is some homology between $\beta1$ and $\beta2$ cDNAs, the $\beta1$ clone was reported not to cross-hybridize to $\beta2$ mRNA (117, 119).

The same approach resulted in the finding in poly(I)·poly(C)–induced fibroblasts of further β IFN mRNAs ($\beta3$, $\beta4$, and $\beta5$) (120), and in Sendai virus–induced leukocytes of a further population of α IFN mRNAs (1.6–3.5-kb long, designated as α_L mRNAs) in addition to the major population of the earlier discovered α mRNAs [designated in (120) as α_S IFN, 0.7–1.4-kb long] (121). The accumulation of the different IFN mRNAs is under separate control: e.g. 5,6-dichlororibofuranosylbenzimidazole (DRB) inhibits the accumulation of α_S mRNAs, but enhances that of the α_L mRNAs. The α_L IFN mRNAs were reported not to cross-hybridize to cDNA from α_S clones (120, 121). The IFNs specified by α_L, $\beta3$, and $\beta4$ mRNAs have not been characterized as yet.

At least some of the IFN genes are linked in the human genome (7, 115, 122, 123). Thus six of the α IFN genes or pseudogenes can be located in a continuous (36-kb) stretch of genomic DNA (7). Moreover, as revealed by hybridization with labeled cDNA probes to DNA from human-mouse somatic cell hybrids, at least eight α genes and the $\beta1$ gene are located on chromosome 9 (121, 122, 124). One β gene has been assigned to chromosome 5 and two to chromosome 2 (119, 120, 125).

A third type of human IFN, designated as γ IFN, can be produced by lymphocytes from blood or lymphocyte cell lines that are exposed to antigens to which the cells have been sensitized or to staphylococcal enterotoxin or mitogens (e.g. various lectins) (1–5, 126, 127). Certain phorbolesters (e.g.

12-O-tetradecanoyl-phorbol-13 acetate) were found to enhance the induction of γ IFN production by the lectin phytohemagglutinine (128). IFN γ preparations are not neutralized by antisera to α and β IFNs and usually are more acid labile than α or β IFNs (1–5). At least when sized by gel filtration, the γ IFNs appear to be larger (between 40,000 and 70,000 daltons) than human α and β IFNs (128). The isolation, characterization (1–5, 126–132), and cloning of γ IFNs have been made difficult by the scarcity of starting material.

The best-characterized animal IFNs are those from mice. Mouse IFNs were purified to apparent homogeneity (133–135). As revealed by N-terminal sequencing one mouse IFN species (designated as αIFN) shares sequence homology with human α IFNs, and a second mouse IFN species (designated as β IFN) with human β1 IFN (10). Hybridization patterns of a human α cDNA probe with restriction enzyme fragmented mouse DNA indicate the existence of several mouse IFN α genes (122).

Different mouse IFN species differ greatly in their activity on human cells (136, 137) and so do human IFN species in their activity on mouse cells (98, 138, 139).

6. Activity of the Human Interferons Synthesized in E. Coli

Three major, biologically active α IFNs from leukocytes of patients with myelogeneous leukemia lack the ten carboxyl-terminal amino acids specified by the cDNA sequence (102). These and other data indicate that at least some of the IFNs synthesized in mammalian cells undergo postsynthetic modifications (e.g. glycosylation (140) or proteolytic cleavage) and thus are different from the IFNs specified by cDNA clones in E. coli. Nevertheless, the various IFNs from E. coli were found to have all the IFN activities (e.g. antiviral, cell growth inhibitory, and immunomodulatory) for which they were tested (91, 93, 98, 139, 141–143). The different pure IFN species from E. coli differ greatly in biological activities e.g. in antiviral activity against different viruses in homologous cells and in antiviral activity in heterologous cells (98, 138, 139, 144).

α2 IFN synthesized in E. coli has similar antiviral efficiency to buffy coat IFNs in protecting rhesus monkeys from skin lesions caused by vaccinia. Remarkably, however, buffy coat IFNs do cause leukopenia and fever in rhesus monkeys, whereas α2 IFN (even if tested at 20-fold higher amounts) does not. It appears likely that the side effects of the buffy coat IFNs are due to one or more α IFN species (other than α2 IFN) present in the preparation (110).

The existence of common restriction endonuclease cleavage sites in the coding regions of α2 (A) and α1 (D) IFNs allows the in vitro recombination of segments of the two genes to hybrid IFN genes. Inserted into E. coli, these direct the synthesis of hybrid IFNs. The antiviral properties of these are unlike those of the parental IFNs or mixtures of the parental IFNs. Studies on the homologous and heterologous target cell specificities of the parental and hybrid IFNs serve as the basis of a model according to which the IFNs have two regions involved in binding to cellular receptors, one in the carboxyl- and one in the amino-proximal half of the polypeptide (138, 139).

The chemical synthesis of an $\alpha 1$ (D) IFN gene has been accomplished. The nucleotide sequence of the synthetic gene is different from that of the natural gene; wherever possible the former contains codons that may be preferred for high level of expression in bacteria (145).

7. Binding of Interferons to Cells and the Establishment of the Antiviral State

The most studied effect of IFN is the conversion of cells into an "antiviral" state in which they are poor host for virus replication. Much of our knowledge about the early phase of IFN action was gained from studies devoted to this effect.

The IFNs that are secreted from the producing cells have to interact with the surface of cells to be active. Thus, cells in culture that are treated with the IFN inducer, poly(I)·poly(C), together with an antiserum to IFN, are not converted into the antiviral state (146).

Purified mouse α and β IFNs labeled with I^{125} bind specifically to surface receptors of cells of a murine line (L1210S), which is responsive to these IFNs but not to cells of a variant of this line (L1210R), which do not respond to these IFNs (19). The cell bound IFN is apparently not internalized and not degraded; much of it is released from the cells as intact IFN. The number of receptors is about 10^3 per L1210 cell and the K_d is 2 X 10^{-11} M. The binding of IFN is not cooperative and does not result in an alteration in the number of receptors (20). The receptors are inactivated by trypsin treatment, which presumably indicates that they are, at least in part, proteins. α and β IFNs compete for binding, i.e. appear to bind to the same receptors, whereas γ IFNs do not; they seem to bind to different receptors (20, 147).

A human IFN receptor gene is localized on chromosome 21 (148). Cells trisomic for this chromosome require less α or β IFN than normal diploid cells to obtain the same extent of virus growth inhibition (149). Mice immunized by injection with a human-mouse somatic cell hybrid containing mouse chromosomes and the human chromosome 21 produce an antiserum that is presumably directed against cell surface antigens specified by the human chromosome. This antiserum impairs the action of human IFNs on human cells (150).

The earliest reported effect of IFNs on mouse cells is a calcium-dependent increase in the concentration of cGMP. This starts within 5–10 min of the beginning of the exposure of the cells to IFNs and persists for 5–10 min (151, 152).

The conversion of cells into the antiviral state by IFNs requires the presence of the nucleus, RNA synthesis, and protein synthesis (153–156). The conversion is accompanied (and most probably caused at least in part) by a change in the level of several mRNAs and proteins, including enzymes (see Sections 8–13 and 17).

The antiviral state is not fully established in cells treated with certain inhibitors of fatty acid cyclo-oxygenase (e.g. oxyphenylbutazone) (157; see also 158) or superoxidedismutase (e.g. diethyldithiocarbamate) (159). It remains to be seen whether or not this effect of oxypenylbutazone indicates that prostaglandins or their derivatives are involved in IFN action. The proportion of unsaturated fatty acids decreases in the membrane phospholipids of cells after treatment with IFN (66, 67). It is conceivable that this decrease may be related to a reported change in membrane rigidity elicited in cells by IFN treatment (160).

In cells exposed to IFNs in the presence of inhibitors of RNA synthesis, the antiviral state is not established unless the inhibitor is removed within 5–6 hr after the removal of IFNs (161). The antiviral state is transient. It is dissipated within a few days after the exposure of the cells to IFNs ceases (1).

8. Enzymes of Interferon Action

Early attempts to understand the biochemistry of IFN action concerned mechanisms by which virus replication is blocked in cells treated with IFNs. Several observations pointed to viral RNA and protein accumulation as being impaired (162). This prompted a comparison of cell-free extracts from IFN-treated and control cells for their capacity to process, translate, and cleave viral and host mRNAs. The comparison resulted in the discovery of new enzymes whose level is enhanced in cells after treatment with IFNs and that may be involved in the posttranscriptional control of gene expression.

9. The (2'–5') (A)$_n$ Synthetase-RNase L Pathway

This pathway (51) was discovered serendipitously in the course of comparing the rates of cleavage of reovirus mRNAs in extracts from IFN-treated EAT cells with those in extracts from control cells (44). A faster cleavage in the extract from IFN-treated cells turned out to depend on the presence of (genomic) dsRNA from reovirions contaminating the reovirus mRNA preparation: (*a*) reovirus mRNA free of dsRNA was cleaved at the same rate in the two extracts and (*b*) added dsRNA [from reovirions or of poly(I) ·poly(C)] accelerated the cleavage only in the extract from IFN-treated cells (44). Subsequent studies revealed that in addition to dsRNA, ATP was also required for accelerating RNA cleavage, and both of the components were needed only for the activation of an endoribonuclease system, not for its action (163).

The system in the extract from IFN-treated EAT cells was divided into two enzyme fractions. In the presence of dsRNA the first of these catalyzed the production of a small thermostable product from ATP. This in turn activated the second enzyme, a latent endoribonuclease, designated as RNase L (L standing for latent) (45). The small thermostable product was identified as $(2'-5')(A)_n$ (45, 165), a set of compounds discovered originally (42, 43) as inhibitors of protein synthesis that are formed from ATP in extracts from IFN-treated cells if the extracts are supplemented with dsRNA. The addition of $(2'-5')(A)_n$ to a rabbit reticulocyte lysate was found to decrease the translatability of mRNA during incubation (46), and $(2'-5')(A)_n$ added to a HeLa cell extract was shown to accelerate RNA cleavage (47). All these results indicated that $(2'-5')(A)_n$ could inhibit protein synthesis by activating a latent ribonuclease.

a. THE $(2'-5')(A)_n$ SYNTHETASE   The enzyme (70) synthesizing this set of compounds [$(2'-5')(A)_n$ synthetase] was partially purified from mouse L929 cells (166), chicken cells (167), and rabbit reticulocytes (168). An apparently homogeneous enzyme was obtained from IFN-treated mouse EAT cells after a 2,600-fold purification (169). The apparent M_r of the enzyme is 105,000 as determined by gel electrophoresis in sodium dodecyl sulfate, and 85,000 as determined by centrifugation through a sucrose gradient.

In the presence of dsRNA the purified enzyme can convert over 97% of the ATP added to $(2'-5')(A)_n$ (where n extends from 1 to about 15) and pyrophosphate (170). The stoichiometry of the reaction is:

$$(n + 1)\ \text{ATP} \rightarrow (2'-5')\text{pppA(pA)}_n + n \text{ pyrophosphate.}$$

[$(2'-5')$pppA(pA)$_n$ stands for the same set of compounds that are abbreviated otherwise as $(2'-5')(A)_n$. The different abbreviation is used here to emphasize that the 5' terminus of $(2'-5')(A)_n$ is a triphosphate.]

It is notable that the equilibrium of the reaction is strongly in favor of synthesis since it does not require "pulling" by the cleavage of pyrophosphate. This seems to indicate that the free energy of (2'–5')-linked oligoadenylates is different from that of (3'–5')-linked oligoadenylates. It may be in line with this conclusion that, as revealed by nuclear magnetic resonance and circular dichroism studies, (2'–5')-linked oligoadenylates have a stronger tendency to stack than the corresponding (3'–5') compounds (171).

Among the products of the enzyme, usually di-, tri-, and tetra-adenylates are the most abundant. The amount of larger oligoadenylates diminishes with increasing chain lengths (43, 169). The oligoadenylates are extended by the addition of adenylate residues. Diadenylates added to the enzyme are substrates for adenylate addition, but two diadenylates do not become linked to each other (172, 173). (2'–5')(A)$_n$ synthetase was also reported to be capable of adding in (2'–5') linkage: (*a*) one or more adenylate residues or one nucleotide other than adenylate to a variety of compounds with 3'-terminal adenylate moieties e.g. NAD, AppppA; (*b*) one nucleotide other than adenylate to (2'–5')(A)$_n$; and (*c*) at least one adenylate residue to tRNA (174, 175).

The enzyme is maximally active when the concentration of dsRNA (in weight per volume) is about one half of that of the enzyme. In these conditions there is one enzyme molecule for every 80-base pair long segment of the dsRNA (170). Whereas dsRNA shorter than 30 base pairs does not activate the enzyme, dsRNA longer than 65 base pairs causes maximal activation (176).

(2'–5')(A)$_n$ synthetase was also isolated as a homogeneous protein from IFN-treated human (HeLa) cells (177). The HeLa enzyme is somewhat smaller than the enzyme from EAT cells (M_r 100,000 as determined by gel electrophoresis in sodium dodecyl sulfate and 80,000 as determined by sedimentation through a sucrose gradient). The functional characteristics of the two enzymes are, however, similar (177).

Another study of the size distribution of (2'–5')(A)$_n$ synthetase in crude extracts from IFN-treated HeLa and SV80 cells by gel filtration resulted in patterns with two peaks of enzyme activity. The two forms of the enzyme could possibly indicate either that the protein is multimeric or that there is more than one protein with (2'–5')(A)$_n$ synthetase activity. Namalva cells (another human line) appear to have only the small form of (2'–5')(A)$_n$ synthetase (178).

b. RNase L, THE LATENT ENDORIBONUCLEASE ACTIVATED BY (2'–5')(A)$_n$ At present the only well-established biochemical function of (2'–5')(A)$_n$ is the activation of RNase L. The (2'–5')(A)$_n$ has to be the trimer or longer to activate RNase L from EAT, L929, or HeLa cells, and the tetramer or longer to activate RNase L from rabbit reticulocytes (179). Furthermore, to activate any of these RNase L's, (2'–5')(A)$_n$ has to have triphosphate or diphosphate moieties at its 5' terminus (180).

The activation is reversible (181); the removal of (2'–5')(A)$_n$ from the activated enzyme (e.g. by gel filtration) results in the reversion of the enzyme to the latent state. Readdition of $(2'-5')(A)_n$ to the now latent enzyme reactivates it.

The activation appears to involve the binding of $(2'-5')(A)_n$ to the enzyme, whereas free $(2'-5')(A)_n$ passes through a nitrocellulose filter; a partially purified RNase L preparation retains $(2'-5')(A)_n$ on the filter. The agent retaining $(2'-5')(A)_n$ copurifies with RNase L during ion exchange chromatography and gel filtration (181).

The M_r of RNase L is 185,000 as estimated by gel filtration. The activation of the enzyme does not seem to result in a large size change. This and other data make it unlikely that the activation involves the binding or the release of a protein.

RNase L was partially (about 1000-fold) purified from EAT cells (50; see also 182). Studies with extracts from IFN-treated EAT cells indicated that RNase L, if activated by $(2'-5')(A)_n$, cleaves various single-stranded viral RNAs and cellular mRNAs at widely different rates. dsRNAs are, however, not cleaved (48).

From the homopolymers poly(U), poly(A), poly(C), and poly(G), the RNase L purified from EAT cells, if activated, cleaves only poly(U) (50). Other studies with purified RNase L (50), rabbit reticulocyte lysates, extracts from mouse (Krebs ascites) cells, and human (Daudi lymphoblastoid) cells (49) indicate that in natural viral RNAs, activated RNase L cleaves predominantly at the 3' side of UA, UG, and UU sequences and yields 3' phosphate–terminated products. In single-stranded RNA molecules with well-defined secondary structures, RNase L cleavages occur only in single-stranded regions (50). The fact that the enzyme can efficiently cleave single-stranded RNA next to three dinucleotides out of the sixteen possible is consistent with the earlier findings revealing that RNase L is not specific for viral RNA (48).

c. (2'-5') PHOSPHODIESTERASE THAT DEGRADES $(2'-5')(A)_n$ In consequence of its unusual (2'-5') phosphodiester linkages, $(2'-5')(A)_n$ is resistant to many nucleases. It is, however, degraded in extracts from various cells, (e.g. L929, HeLa, and mouse reticulocytes) (183–186). An enzyme designated as (2'-5') phosphodiesterase, which cleaves (2'-5') phosphodiester linkages more efficiently than (3'-5') linkages, was partially purified from L929 cells (185). This degrades $(2'-5')(A)_n$ into ATP and AMP. Phosphate at the 5' terminus of the oligoadenylate does not affect the cleavage, whereas phosphate at the 3' terminus impairs it. This suggests that the enzyme may attack $(2'-5')(A)_n$ starting at the 3' end. (2'-5') phosphodiesterase has an apparent M_r of 40,000. It can catalyze the cleavage of various substances, including that of the CCA termini of tRNAs.

10. Effect of Interferons and other Agents on the Level of the Enzymes of the $(2'-5')(A)_n$ Synthetase-RNase L Pathway

$(2'-5')(A)_n$ synthetase was found in a variety of vertebrates (e.g. mouse, guinea pig, rabbit, dog, and chick) (70). In the mouse it was detected in several tissues including the spleen, lungs, brain, liver, bone marrow, intestinal mucosa, and serum. It is also present in a large variety of cultured human, mouse, monkey, and chick cells. Furthermore it was reported to be packaged into the virions of VS virus and Moloney mouse leukemia virus grown in IFN-treated cells (187).

The level of the enzyme is increased in cells upon treatment with IFNs. The extent of this induction varies greatly: it is several thousand fold in chick cells and only tenfold in HeLa cells. The induction of the enzyme requires RNA synthesis and protein synthesis (70). The amount of the mRNA for $(2'-5')(A)_n$ synthetase is increased in cells treated with IFN, as revealed by translation into active enzyme in Xenopus oocytes (36). The translatability of the mRNA allowed the isolation of a $(2'-5')(A)_n$ synthetase–specific clone from a cDNA library (178).

No difference was detected in the patterns of $(2'-5')(A)_n$ synthetase induction in L929 cells between murine αIFN, βIFN, and a 1:1 mixture of the two IFNs (137). The induction by a γ IFN preparation in HeLa cells was reported to be slower than that by an α IFN preparation (188, 189). α IFN could induce the formation of $(2'-5')(A)_n$ synthetase mRNA even in the presence of an inhibitor of protein synthesis, whereas γ IFN could not. Thus it is conceivable that γ IFN has to induce first intermediary proteins, which in turn may activate the transcription of $(2'-5')(A)_n$ synthetase mRNA (188).

In L929 cells the level of the enzyme starts to increase within 3–5 hr of the beginning of the treatment with α or β IFNs and reaches maximal levels by about 15 hr (137). In some cells (e.g. HeLa), the enzyme induced by IFN seems to be stable (190). In other cells (e.g. quiescent human fibroblasts) the level of $(2'-5')(A)_n$ synthetase increases to the maximum within 24 hr of the beginning of the exposure to IFNs, but thereafter (though the cells remain exposed to IFN) decreases in four days to a level close to that in control cells (191).

The commitment of the cells to the induction of $(2'-5')(A)_n$ synthetase decreases within a few hours of the removal of IFN (188). Thus apparently the effects of the interaction of IFNs with the cell receptors are transient and have to be repeated or maintained for the continual stimulation of the transcription of at least some of the genetic regions activated by IFN.

The level of $(2'-5')(A)_n$ synthetase is affected by agents other than IFNs.

Thus in chick oviducts the enzyme level is low and remains low after treatment with diethylstilbestrol, but increases greatly after withdrawal of diethylstilbestrol (192). In human lymphoblastoid cells (Raji, Daudi), treatment with glucocorticoids boosts the level of $(2'-5')(A)_n$ synthetase up to 30-fold (193). In rabbit reticulocytes and human lymphocytes the level is high (194), and it remains to be seen whether or not this is a consequence of an endogenous exposure of the cells to IFNs.

The level of $(2'-5')(A)_n$ synthetase is high in some cultured cell lines even without treatment with IFNs (195, 196). In at least one case this is a consequence of a constitutive production of IFN (196).

The level of RNase L is affected only slightly if at all by treatment with IFNs (185, 197), but the constitutive level of this enzyme varies greatly in different cells [(198) and C. Baglioni, personal communication].

The level of $(2'-5')$ phosphodiesterase is increased in L cells about three-fold upon treatment with IFNs (185). In other cells, however, the level is unaffected by this treatment. The exposure of mouse spleen lymphocytes to Concanavalin A (ConA) was reported to boost the $(2'-5')$ phosphodiesterase level five to six fold. It is conceivable (see Section 12) that this boost may be relevant to the mitogenic action of ConA (58, 59).

11. The $(2'-5')(A)_n$ Synthetase-RNase L Pathway in Intact Cells

The following results, obtained in experiments with intact cells, are consistent with the assumption that this pathway is functioning in vivo and is mediating at least some of the antiviral activities of IFNs.

(a) In cells treated with IFN in which the reovirus yield is diminished by 80%, the half-life of reovirus mRNA is 5.3 hr, whereas in control cells it is 12.3 hr (199).

(b) Reovirus "subviral particles" (i.e. partially uncoated virions) isolated from cells treated with IFN, when tested in a reaction mixture including ATP, have an endonuclease activity associated with the particles, whereas subviral particles from infected, control cells appear to be free of nuclease activity (200).

(c) The patterns of ribosomal RNA cleavage in intact, IFN-treated (but not in control) cells infected with EMC virus overlap with those obtained on incubation of ribosomes in L cell extracts supplemented with $(2'-5')(A)_n$ (201).

(d) The amount of $(2'-5')(A)_n$ in cells can be assayed by a binding competition assay involving labeled $(2'-5')(A)_n$ and, as the binding agent, a partially purified preparation of RNase L (202, 203). Using this and other assays it was established that dsRNA [poly(I)·poly(C)] added to intact IFN-treated (but not to control) HeLa cells causes an increase in the level of $(2'-5')(A)_n$ and an enhancement of mRNA cleavage and polysome degradation (204).

(e) The level of $(2'-5')(A)_n$ is also substantially increased in intact cells upon treatment with IFN and subsequent infection with EMC virus, whereas IFN treatment alone or virus infection alone causes a much smaller increase (202, 205).

(f) The uptake of $(2'-5')(A)_n$ into intact cells can be facilitated by either permeabilizing the cells by treatment with a hypertonic solution or by using a $CaCl_2$ precipitation method. Using these procedures it was established that $(2'-5')(A)_n$ added to cells accelerates RNA cleavage and inhibits protein synthesis and virus replication in the cells (205–207).

(g) The replication of EMC virus and VS virus is not impaired by IFN in a mouse cell line in which the level of $(2'-5')(A)_n$ synthetase is enhanced by IFNs, but that of RNase L is below the level of detection (198, 208).

All of these observations are in line with the assumption that the $(2'-5')(A)_n$ synthetase-RNase L pathway is one of the mediators of IFN action.

The picture is complicated, however, by observations on various cell lines. Two of these have high levels of $(2'-5')(A)_n$ synthetase (and apparently normal RNase L) activity and still are not in the antiviral state (195, 196). Another line remains low in $(2'-5')(A)_n$ synthetase activity even after treatment with IFN, though the treatment does convert the cells into the antiviral state (209).

The resolution of these anomalies may require, among others, tests of the in vivo $(2'-5')(A)_n$ levels in these cells after treatment with IFN and virus infection. This is because these levels may determine if RNase L becomes activated and because the levels are also affected by agents other than $(2'-5')(A)_n$ synthetase [e.g. by various nucleases, including $(2'-5')$ phosphodiesterase], which degrade $(2'-5')(A)_n$.

Moreover, the $(2'-5')(A)_n$ synthetase-RNase L pathway is apparently only one of several mediators of the antiviral (and other) actions of IFNs (see Sections 13, and 15 to 17). The impairment of the replication of a particular virus might require the functioning of particular mediators. Consequently, in the cells showing an anomalous behavior, IFN mediators other than the $(2'-5')(A)_n$ synthetase-RNase L pathway may also be affected and this may cause the anomalous behavior.

12. Does the $(2'-5')(A)_n$ Synthetase-RNase L Pathway Have Any Functions that are Unrelated to the Antiviral Actions of Interferons or even to Interferon Action?

IFNs can inhibit the growth of cultured cells and mitogenesis induced in lymphocytes by mitogens (25–27). Furthermore, experiments with mouse spleen lymphocytes revealed that not only IFNs but also exogenous $(2'-5')(A)_n$ can block mitogenesis induced by ConA (58, 59). The primary effect of $(2'-5')(A)_n$ in inhibiting mitogenesis seems to be an impairment of the synthesis of various proteins, including histones. In some experiments on mitogenesis, not $(2'-5')(A)_n$ but $(2'-5')(A)_n$ "core" [i.e. the product obtained by removing the 5' triphosphate of $(2'-5')(A)_n$ with alkaline phosphatase] was used. This does not activate RNase L in vitro and is only one tenth as active as $(2'-5')(A)_n$ in inhibiting protein synthesis if added to permeabilized cells (205). It has, however, an advantage over $(2'-5')(A)_n$: it is taken up even by nonpermeabilized cells (58). $(2'-5')(A)_n$ core occurs in vivo, and its level is enhanced in L929 cells upon treatment with IFN and infection with EMC virus (202). The mechanism of action of $(2'-5')(A)_n$ core is unknown. It is conceivable that the compound may be activated in cells by phosphorylation to $(2'-5')(A)_n$ [i.e. $pppA(pA)_n$] or it may increase the level of endogenous $(2'-5')(A)_n$ by blocking $(2'-5')$ phosphodiesterase. It is also conceivable that the compound is affecting agents other than RNase L.

There are several observations to support the possibility that $(2'-5')(A)_n$ may be among the agents controlling the rate of cell division. The treatment of lymphocytes with the mitogen ConA causes a severalfold increase in the level of $(2'-5')$ phosphodiesterase without affecting the level of $(2'-5')(A)_n$ synthetase. The level of $(2'-5')$ phosphodiesterase is threefold higher in dividing cells than in confluent cells of a monkey cell line (AGMK). Serum starvation decreases the level of the enzyme, whereas the addition of serum increases it. Finally, the ratio of $(2'-5')(A)_n$ synthetase to $(2'-5')$ phosphodiesterase, which determines the level of $(2'-5')(A)_n$, is tenfold lower in fast growing (AGMK) cells than in confluent cells (arrested in the G_0 phase by serum starvation) (58, 59).

It remains to be established if the $(2'-5')(A)_n$ synthetase-RNase L pathway or either of the two enzymes involved has any functions unrelated to IFN action. There are several facts consistent with this possibility. The level of RNase L is little, if at all, affected by IFNs. Furthermore, the level of $(2'-5')(A)_n$ synthetase is, as noted earlier, affected by agents other than IFNs (e.g. estrogen withdrawal and glucocorticoids) (192, 193).

Since the pathway, if activated, catalyzes RNA cleavage, it is reasonable to assume that a function possibly unrelated to IFN action may also consist of RNA cleavage, e.g. steps in the processing of heterogeneous nuclear RNA or degradation of mRNA or ribosomal RNA. Heterogeneous nuclear RNA appears to contain essentially double-stranded stretches (210–212); and cellular RNA was reported to serve as a substitute for dsRNA (though a very inefficient one) in activating the $(2'-5')(A)_n$ synthetase-RNase L pathway (213).

Thus, it is conceivable that IFNs are making use of enzymes that, in the absence of IFNs, are utilized for unrelated purposes, and IFN action is based in part on an increased production of such enzymes. It is also possible, however, that the scope of IFN action is wider than imagined at present and that functions seemingly unrelated fall within this scope.

13. dsRNA-Dependent Protein Kinase

The need for dsRNA and ATP for the acceleration of RNA cleavage in extracts from IFN-treated cells, and the knowledge that several proteins are activated or inactivated by phosphorylation (214) prompted tests on the effect of dsRNA on protein phosphorylation in extracts from IFN-treated and control cells.

It was discovered that the addition of dsRNA (from reovirus or of poly(I)·poly(C)) to an extract from IFN-treated EAT cells or L cells (but not, or only to a much lesser extent to an extract from control cells) causes the phosphorylation of at least two proteins: P1, with a molecular weight of 67,000, and P2, with a molecular weight of 37,000 (52–54). Subsequently, a dsRNA activatable protein kinase was purified several thousand fold from L cells and from EAT cells treated with IFN (215, 216). The extent of induction of the kinase by IFN is three to ten fold. IFNs α, β, and γ can all induce the enzyme (137, 217). The purified kinase preparation is essentially free of dsRNA-independent kinase activity. The activation of the partially purified kinase by dsRNA and ATP was reported to be enhanced by an acidic protein (factor A) whose level is not affected by IFN (216). The details of the activation process are under investigation.

The P2 protein, which can be phosphorylated by the activated kinase, is the α subunit of the peptide chain initiation factor eIF-2 (165, 218). The P1 protein copurifies with the dsRNA-dependent protein kinase activity throughout a several thousand fold purification, and at present the most highly purified dsRNA-dependent protein kinase preparation from IFN-treated mouse cells consists of P1 as the major protein and two minor components (219). The activation of the kinase by dsRNA appears to be accompanied (and is probably caused by) the phosphorylation of P1 protein at more than one site. The activated kinase can also phosphorylate some histones (54).

Phosphoprotein phosphatase(s) that can remove the phosphate moieties from phosphorylated eIF-2 and phosphorylated P1 is (are) present in cell extracts (216, 220). The phosphatase action is impaired by dsRNA (221).

The addition of the activated protein kinase preparation to a cell-free protein synthesizing system (from EAT cells or reticulocyte lysates) results in the inhibition of peptide chain initiation. The inhibition can be overcome at least for a short while by the addition of further eIF-2 and appears to be the consequence of the phosphorylation of eIF-2 (165, 222). The mechanisms of the impairment of peptide chain initiation by eIF-2 phosphorylation have been reviewed recently (223).

The fact that dsRNA added to intact IFN-treated cells (but not to control cells) results in the phosphorylation of P1 protein indicates that the activation and action of the protein kinase is not an artifact of the in vitro system (224).

A protein kinase in an extract from IFN-treated and EMC virus–infected L cells (but not in those from control cells, IFN-treated but uninfected cells, or infected but not IFN-treated cells) can phosphorylate the P1 protein as well as added eIF-2, without the need for added dsRNA; i.e. it appears to be preactivated (225). It will have to be tested whether or not the endogenous activating agent is dsRNA in the viral replication complex.

A dsRNA-dependent protein kinase was first discovered in rabbit reticulocyte lysates (222). It is not known whether the presence of this enzyme in the reticulocyte lysate is a consequence of an exposure of the reticulocyte precursor cells to endogenous IFNs. The reticulocyte kinase (M_r 76,000) has been purified and appears to be similar to the kinase from IFN-treated mouse cells (226–228).

14. Involvement of dsRNA in Interferon Induction and Action

As reported above, dsRNA is an inducer of IFN synthesis and also an activator of the two IFN-induced, latent enzymes: the $(2'-5')(A)_n$ synthetase and the dsRNA-dependent protein kinase. It is usually assumed that dsRNA became a modulator of the IFN system because it is an intermediate or side product in the replication of various viruses and thus may serve in the cell as a signal revealing the presence of such replicating viruses.

The validity of this assumption has to be tested for each virus. The dsRNA of viruses with a dsRNA genome (e.g. reovirus) is formed, and also remains, inside a protein coated particle (229). Nevertheless, purified reovirions were found to be able to substitute for dsRNA in activating the $(2'-5')(A)_n$ synthetase-RNase L pathway in an extract from IFN-treated cells. However, as activating agents, the reovirions are only 4% as efficient (on a per milligram of dsRNA basis) as free dsRNA (199).

EMC virus, which has a single-stranded RNA genome, appears to generate during its replication stretches of dsRNA of sufficient length to activate the $(2'-5')(A)_n$ synthetase-RNase L pathway, at least in vitro (56). Moreover, a cross-linking agent (4 aminoethyl-4,5,8-trimethyl psoralen), reported to be specific for dsRNA, cross-links nascent viral RNA strands to template viral RNA strands in EMC virus–infected HeLa cells (57).

It is puzzling why dsRNA, which is involved in the induction of IFNs, is also required for the activation of some IFN-induced enzymes. This complexity can perhaps be rationalized by the following hypothesis: virus infection (perhaps if accompanied by the formation of some dsRNA) may result in the synthesis of IFNs in the infected cells and the secretion and spreading of these IFNs in the organism. In the cells exposed to IFNs the synthesis of $(2'-5')(A)_n$ synthetase and of dsRNA activatable protein kinase are induced. Since both of these enzymes are latent, they do not impair the metabolism of uninfected cells. However, when the cells, previously exposed to IFNs, become infected with virus, this may result in the formation of some dsRNA and thereby the activation of the two enzyme systems. This would in turn impair the metabolism of the virus-infected cells, and thereby inhibit virus replication. Thus, ultimately, this complex control system may contribute to the localization of viral infection (48, 55).

In vitro, the $(2'-5')(A)_n$ synthetase-RNase L system appears to cleave viral and host RNAs without discrimination. In vivo, however, at least in the case of some virus cell systems (e.g. reovirus-L929 cells), viral protein synthesis is inhibited preferentially above host protein synthesis. An intriguing series of experiments in vitro reveals that the involvement of dsRNA in the activation of the $(2'-5')(A)_n$ synthetase-RNase L system might make this system capable of such discrimination (56, 230). For these experiments poly(U) was annealed to the poly(A) segment of a viral RNA (from VS virus), which resulted in a viral RNA in which a single-stranded segment is covalently linked to a double-stranded segment. In an extract from IFN-treated cells supplemented with ATP, this viral RNA, covalently bound to a dsRNA segment, is degraded faster than an identical viral RNA not bound to dsRNA. This result is consistent with the possibility that in IFN-treated cells infected by an RNA virus the RNase L activity may be the highest in a region near to the partially double-stranded replicative intermediate of the virus. Such a localization may be a consequence of (a) the localized synthesis of $(2'-5')(A)_n$ by the $(2'-5')(A)_n$ synthetase bound to the partially double-stranded viral replicative intermediate and (b) the decrease in the concentration of $(2'-5')(A)_n$ away from its site of formation, which is caused by cleavage by $(2'-5')$ phosphodiesterase. It is also conceivable, however, that RNase L may form a complex with $(2'-5')(A)_n$ synthetase and this may be the basis of the localization. It remains to be established if RNase L activity is localized in vivo.

15. Biochemical Manifestations of Interferon Action Not Known to Depend on dsRNA

a. EFFECTS ON mRNA METHYLATION   The 5' termini of eukaryotic mRNAs (including those of most viral mRNAs) are capped and methylated. The translational efficiency of an mRNA that is normally capped and methylated is greatly diminished if it is lacking the methylation (231). The methylation of unmethylated, capped reovirus mRNAs by the cellular methylating enzymes is impaired in extracts from IFN-treated L cells or HeLa cells. The inhibitor has a short half-life in vitro, and thus the impairment of the methylating capacity in the extract disappears during incubation (60, 61).

The methylation of several viral RNAs is diminished in intact IFN-treated cells. In the case of reovirus and vaccinia virus, the 2-O-methylation of ribose moieties is decreased (62, 63); in that of VS virus the 7-methylation of the 5' terminal guanylate residues is decreased. The nonmethylated VS virus mRNA cannot be translated efficiently, since it is located mainly in the nonpolysomal fraction (64). Interestingly, SV40 mRNA synthesized late in infection in IFN-treated cells was found to be overmethylated and at the same time inefficient as a messenger in vivo (65).

b. OTHER CHANGES   The following are only a few of a large variety of reported alterations in the biochemistry of cells after exposure to IFNs (25).

In extracts from IFN-treated cells there is a faster inactivation of some tRNAs than in corresponding extracts from control cells (232, 233). This may be a consequence of an increase in the level of the (2'–5') phosphodiesterase upon IFN treatment, since this enzyme can remove the CCA termini from tRNAs (185). The in vivo relevance of this effect (if any) remains to be seen.

IFNs were reported to accelerate protein degradation (234) and inhibit thymidine uptake (235, 236), protein glycosylation (presumably at the level of the synthesis of N-acetylglucosaminyl pyrophosphoryldolichol) (237), and ornithine transcarbamylase induction by serum growth factors (238). Moreover, IFNs were reported to alter the protein composition of the cell surface by enhancing the expression of certain histocompatibility antigens and of receptors for the Fc fragments of antibodies (239–241).

16. The Inhibition of the Replication of Different Viruses May Be Due to Different Mediators of Interferon Action

In cells of various murine lines (e.g. L929 or EAT) IFNs block the replication of both VS virus and EMC virus. However, in cells of an undifferentiated mouse embryonal carcinoma line, IFN inhibits the replication of

EMC virus, but not that of VS virus (242). In cells of this line the inducing effect of IFNs is only partial: IFN enhances the level of $(2'-5')(A)_n$ synthetase, but not that of the dsRNA-dependent protein kinase (242, 243). (It remains to be seen if the lack of inhibition of VS virus replication by IFN is in any way related to the lack of kinase induction.)

In cells of an NIH 3T3 variant, IFNs impair the replication of neither VS virus nor EMC virus (though they inhibit that of Moloney murine leukemia virus). In cells of this line, IFNs induce the synthesis of both the protein kinase and the $(2'-5')(A)_n$ synthetase. Since, however, the cells are deficient in RNase L, the $(2'-5')(A)_n$ synthetase-RNase L pathway may not function at the appropriate level (208).

Mice (or their macrophages) carrying the dominant Mx allele have an IFN-dependent resistance to influenza virus, whereas Mx-negative mice (or their macrophages) are susceptible to influenza virus. The efficacy of IFNs toward EMC virus and VS virus is, however, independent of Mx (244, 245).

All these examples are in line with the conclusion that the inhibition of the replication of different viruses may be due to different mediators of IFN action.

At the same time, the replication of one and the same virus may be inhibited by IFNs at more than one stage of the replicative cycle, presumably in consequence of the functioning of different mediators of IFN action. Thus, e.g. the replication of Simian vacuolating virus 40 appears to be blocked by IFNs in consequence (*a*) of a decreased accumulation of viral RNA during the early phase of the replicative cycle, and (*b*) an impairment of the translation of viral mRNA during the late phase (65, 73). The nature of the agents mediating these impairments and their modes of action remain to be established.

It should be noted that cells treated with IFNs are rescued from the cytopathic effect of some viruses (e.g. Mengo) but not of that from other viruses (e.g. reovirus) (246, 247). However, in both cases (as long as the infecting virus is sensitive to IFNs) virus replication is impaired by the treatment, and consequently the infection is localized.

17. mRNAs and Proteins Induced by Interferons

The treatment of various human, mouse, and chicken cells with homologous IFNs induces the synthesis of several mRNAs and of the corresponding proteins (34–41). The correspondence was established by translating the mRNAs into the proteins in vitro. The induction was inhibited by actinomycin D, which indicates de novo synthesis (34–37). Studies involving the labeling of proteins with radioactive amino acids in vivo revealed that (*a*) the mRNAs and proteins are induced within a few hours after the start of the exposure of the cells to IFNs; (*b*) even in the continued presence of

IFNs, the synthesis of several induced proteins is only transient; and (c) the time period in which a particular protein is synthesized at the maximal rate is not the same for all of the induced proteins (248).

α IFN and β IFN preparations appear to induce the synthesis of the same set of proteins, whereas that induced by an γ IFN preparation was reported to overlap only partially with those induced by α IFN and β IFN (249). Some of the induced proteins can be retained on a poly(I)·poly(C)-agarose column which indicates their high affinity to dsRNA (39). Among the IFN-induced mRNAs and proteins (up to eight or nine proteins in human cells as revealed by gel electrophoresis and autoradiography) so far only an mRNA specific for (2'-5') (A)$_n$ synthetase has been identified (36).

18. Remaining Problems

The last five years brought much progress in our understanding of the biochemistry of IFNs and IFN action. Nevertheless, many of the basic problems in this field remain unsolved.

Why do so many IFNs exist? Do they differ in target (i.e. cell) specificity, or do they affect different, though perhaps partially overlapping, sets of IFN mediators? How can substances as diverse as IFN inducers all trigger IFN induction, and how selective is the induction for different IFNs? Do IFNs have to enter the cell for action? If they do not have to enter, what are the "second messengers" that transmit the signal into the cell? How do the IFNs control the expression of the various genes specifying the various mediators of IFN action? Which are the IFN mediators impairing the replication of particular viruses and cells, and which are the mediators modulating the various immunological processes affected by IFNs?

It is likely that behind the complicated phenomenology of IFN action there is a complex biochemistry remaining to be elucidated.

Literature Cited

1. Stewart, W. E. II. 1979. *The Interferon System.* New York: Springer. 421 pp.
2. 1977. *Tex. Rep. Biol. Med. Vol.* 35. 573 pp.
3. DeMaeyer, E., DeMaeyer-Guignard, J. 1979. *Compr. Virol.* 15:205–84
4. Vilcek, J., Gresser, I., Merigan, T. C., eds. 1980. *Ann. NY Acad. Sci.* Vol. 350. 641 pp.
5. Sehgal, P. B., Pfeffer, L. M., Tamm, I. 1982. In *Chemotherapy of Viral Infections,* ed. P. E. Came, L. A. Caliguiri. In press
6. Sen, G. C. 1982. *Prog. Nucleic Acids Res.* In press
7. Weissmann, C. 1981. *Interferon* 3: In press

8. Zoon, K. C., Smith, M. E., Bridgen, P. J., Anfinsen, C. B., Hunkapiller, M. W., Hood, L. E. 1980. *Science* 207:527–28
9. Knight, E., Hunkapiller, M. W., Korant, B. D., Hardy, R. W. F., Hood, L. E. 1980. *Science* 207:525–26
10. Taira, H., Broeze, R. J., Jayaram, B. M., Lengyel, P., Hunkapiller, M. W., Hood, L. E. 1980. *Science* 207:528–30
11. Nagata, S., Mantei, N., Weissmann, C. 1980. *Nature* 287:401–8
12. Taniguchi, T., Mantei, N., Schwarzstein, M., Nagata, S., Muramatsu, M., Weissmann, C. 1980. *Nature* 285: 547–49
13. Derynck, R., Content, J., DeClercq, E., Volckaert, G., Tavernier, J., Devos, R., Fiers, W. 1980. *Nature* 285:542–47

14. Goeddel, D. V., Leung, D. W., Dull, T. J., Gross, M., Lawn, R. M., McCandliss, R., Seeburg, P. H., Ullrich, A., Yelverton, E., Gray, P. W. 1981. *Nature* 290:20–26
15. Tavernier, J., Derynck, R., Fiers, W. 1981. *Nucleic Acids Res.* 9:461–71
16. Houghton, M., Jackson, I. J., Porter, A. G., Doel, S. M., Catlin, G. H., Barber, C., Carey, N. H. 1981. *Nucleic Acids Res.* 9:247–66
17. Lawn, R. M., Adelman, J., Franke, A. E., Houck, C. M., Gross, M., Najarian, R., Goeddel, D. V. 1981. *Nucleic Acids Res.* 9:1045–52
18. Mory, Y., Chernajovsky, Y., Feinstein, S. I., Chen, L., Nir, U., Weissenbach, J., Tiollais, P., Marks, D., Ladner, M., Colby, C., Revel, M. 1982. *Eur. J. Biochem.* In press
19. Aguet, M. 1980. *Nature* 284:459–61
20. Aguet, M., Blanchard, B. 1982. *Virology.* In press
21. Nagano, Y., Kojima, Y. 1954. *CR Soc. Biol.* 148:1700–2
22. Isaacs, A., Lindenmann, J. 1957. *Proc. R. Soc. London Ser. B* 147:258–67
23. Pfeffer, L. M., Wang, E., Tamm, I. 1980. *J. Cell Biol.* 85:9–17
24. Brouty-Boye, D., Zetter, B. R. 1980. *Science* 208:516–18
25. Taylor-Papadimitriou, J. 1980. *Interferon* 2:13–46
26. Gresser, I. 1977. *Cell Immunol.* 34:406–15
27. Gresser, I., Tovey, M. G. 1978. *Biochim. Biophys. Acta* 516:231–47
28. Gresser, I., Tovey, M. G., Bandu, M. T., Maury, C., Brouty-Boye, D. 1976. *J. Exp. Med.* 144:1305–15
29. Gresser, I., Tovey, M. G., Maury, C., Bandu, M. T. 1976. *J. Exp. Med.* 144:1316–23
30. Gresser, I., Maury, C., Bandu, M. T., Tovey, M., Maunoury, M. T. 1978. *Int. J. Cancer* 21:72–77
31. Gresser, I., Maury, C., Kress, C., Blangy, D., Maunoury, M. T. 1979. *Int. J. Cancer* 24:178–83
32. Gresser, I., Tovey, M. G., Maury, C., Chouroulinkov, I. 1975. *Nature* 258:76–78
33. Kawakita, M., Cabrer, B., Taira, H., Rebello, M., Slattery, E., Weideli, H., Lengyel, P. 1978. *J. Biol. Chem.* 253:598–602
34. Farrell, P. J., Broeze, R. J., Lengyel, P. 1979. *Nature* 279:523–25
35. Farrell, P. J., Broeze, R. J., Lengyel, P. 1980. *Ann. NY Acad. Sci.* 350:615–16
36. Shulman, L., Revel, M. 1980. *Nature* 288:98–100

37. Colonno, R. J. 1981. *Proc. Natl. Acad. Sci. USA* 78:4763–66
38. Knight, E., Korant, B. D. 1979. *Proc. Natl. Acad. Sci. USA* 76:1824–27
39. Gupta, S. L., Rubin, B. Y., Holmes, S. L. 1979. *Proc. Natl. Acad. Sci. USA* 76:4817–21
40. Rubin, B. Y., Gupta, S. L. 1980. *J. Virol.* 34:446–54
41. Ball, A. L. 1979. *Virology* 94:282–96
42. Hovanessian, A. G., Brown, R. E., Kerr, I. M. 1977. *Nature* 268:537–40
43. Kerr, I. M., Brown, R. E. 1978. *Proc. Natl. Acad. Sci. USA* 75:256–60
44. Brown, G. E., Lebleu, B., Kawakita, M., Shaila, S., Sen, G. C., Lengyel, P. 1976. *Biochem. Biophys. Res. Commun.* 69:114–22
45. Ratner, L., Wiegand, R., Farrell, P., Sen, G. C., Cabrer, B., Lengyel, P. 1978. *Biochem. Biophys. Res. Commun.* 81:947–57
46. Clemens, M. J., Williams, B. R. G. 1978. *Cell* 13:565–72
47. Baglioni, C., Minks, M. A., Maroney, P. A. 1978. *Nature* 273:684–87
48. Ratner, L., Sen, G. C., Brown, G. E., Lebleu, B., Kawakita, M., Cabrer, B., Slattery, E., Lengyel, P. 1977. *Eur. J. Biochem.* 79:565–77
49. Wreschner, D. H., McCauley, J. W., Skehel, J. J., Kerr, I. M. 1981. *Nature* 289:414–17
50. Floyd-Smith, G., Slattery, E., Lengyel, P. 1981. *Science* 212:1030–32
51. Lengyel, P. 1981. *Interferon* 3:77–99
52. Lebleu, B., Sen, G. C., Shaila, S., Cabrer, B., Lengyel, P. 1976. *Proc. Natl. Acad. Sci. USA* 73:3107–11
53. Zilberstein, A., Kimchi, A., Schmidt, A., Revel, M. 1978. *Proc. Natl. Acad. Sci. USA* 75:4734–38
54. Roberts, W. K., Hovanessian, A., Brown, R. E., Clemens, M. J., Kerr, I. M. 1976. *Nature* 264:477–80
55. Kerr, I. M., Brown, R. E., Ball, L. A. 1974. *Nature* 250:57–59
56. Nilsen, T. W., Baglioni, C. 1979. *Proc. Natl. Acad. Sci. USA* 76:2600–04
57. Nilsen, T. W., Wood, D. L., Baglioni, C. 1981. *Virology* 109:82–93
58. Kimchi, A., Shure, H., Revel, M. 1979. *Nature* 282:849–51
59. Kimchi, A., Shure, H., Revel, M. 1981. *Eur. J. Biochem.* 114:5–10
60. Sen, G. C., Lebleu, B., Brown, G. E., Rebello, M. A., Furuichi, Y., Morgan, M., Shatkin, A. J., Lengyel, P. 1975. *Biochem. Biophys. Res. Commun.* 65:427–34
61. Sen, G. C., Shaila, S., Lebleu, B.,

Brown, G. E., Desrosiers, R. C., Lengyel, P. 1977. *J. Virol.* 21:69–83

62. Desrosiers, R. C., Lengyel, P. 1979. *Biochim. Biophys. Acta* 562:471–80
63. Kroath, H., Gross, H. J., Jungwirth, C., Bodo, G. 1978. *Nucleic Acids Res.* 5:2441–54
64. deFerra, F., Baglioni, C. 1981. *Virology* 112:426–35
65. Kahana, C., Yakobson, E., Revel, M., Groner, Y. 1981. *Virology* 112:109–118
66. Chandrabose, K., Cuatrecasas, P., Pottathil, R. 1981. *Biochem. Biophys. Res. Commun.* 98:661–68
67. Apostolov, K., Barker, W. 1981. *FEBS Lett.* 126:261–64
68. Borden, E. C., Ball, L. A. 1982. In *Progress in Hematology*, Vol. XII, ed. E. B. Brown. In press
69. Gordon, J., Minks, M. A. 1981. *Microbiol. Rev.* 45:244–66
70. Ball, L. A. 1982. *The Enzymes.* In press
71. Torrence, P. F., DeClercq, E. 1977. *Pharmacol. Ther. Part A* 2:1–88
72. Baglioni, C. 1979. *Cell* 17:255–64
73. Revel, M. 1979. *Interferon* 1:102–63
74. Secher, D. S., Burke, D. C. 1980. *Nature* 285:446–50
75. Staehelin, T., Durrer, B., Schmidt, J., Takacs, B., Stocker, J., Miggiano, V., Stahli, C., Rubenstein, M., Levy, W. P., Hershberg, R., Pestka, S. 1981. *Proc. Natl. Acad. Sci. USA* 78:1848–52
76. Hochkeppel, H. K., Menge, V., Collins, J. 1981. *Nature* 291:500–1
77. Skurkovich, S. V., Olshansky, A. J., Samoilova, R. S., Eremkina, E. I. 1978. *J. Immunol. Methods* 19:119–24
78. Secher, D. S. 1981. *Nature* 290:501–03
79. Trinchieri, G., Santoli, D., Dee, R. R., Knowles, B. B. 1978. *J. Exp. Med.* 147:1299–313
80. Beck, J., Engler, H., Brunner, H., Kirchner, H. 1980. *J. Immunol. Methods* 38:63–73
81. Stewart, W. E. II. 1980. *Nature* 285:111
82. Havell, E. A., Hayes, T. G., Vilcek, J. 1978. *Virology* 89:330–34
83. Sehgal, P. B., Gupta, S. L. 1980. *Proc. Natl. Acad. Sci. USA* 77:3489–93
84. Cantell, K., Hirvonen, S. 1977. *Tex. Rep. Biol. Med.* 35:138–41
85. Vilcek, J., Kohase, M. 1977. *Tex. Rep. Biol. Med.* 35:57–62
86. Wiranowska-Stewart, M., Chudzio, T., Stewart, W. E. II. 1977. *J. Gen. Virol.* 37:221–23
87. Billiau, A., Van Damme, J., Van Leuven, F., Edy, V. G., DeLey, M., Cassiman, J. J., Van Den Berghe, H., DeSomer, P. 1979. *Antimicrob. Agents Chemother.* 16:49–55

88. Finter, N. B., Bridgen, D. 1978. *Trends in Biochem. Sci.* 3:76–80
89. Zoon, K. C., Smith, M. E., Bridgen, P. J., Zur Nedden, D., Anfinsen, C. B. 1979. *Proc. Natl. Acad. Sci. USA* 76:5601–5
90. Berthold, W., Tan, C., Tan, Y. H. 1978. *J. Biol. Chem.* 253:5206–12
91. Nagata, S., Taira, H., Hall, A., Johnsrud, L., Streuli, M., Ecsodi, J., Boll, W., Cantell, K., Weissmann, C. 1980. *Nature* 284:316–20
92. Derynck, R., Remaut, E., Saman, E., Stanssens, P., DeClercq, E., Content, J., Fiers, W. 1980. *Nature* 287:193–97
93. Goeddel, D. V., Yelverton, E., Ullrich, A., Heyneker, H. L., Miozzari, J., Holmes, W., Seeburg, P. H., Dull, T., May, L., Stebbing, N., Crea, R., Maeda, S., McCandliss, R., Sloma, A., Tabor, J. M., Gross, M., Familletti, P. C., Pestka, S. 1980. *Nature* 287:411–16
94. Goeddel, D. V., Shepard, H. M., Yelverton, E., Leung, D., Crea, R., Sloma, A., Pestka, S. 1980. *Nucleic Acids Res.* 8:4057–74
95. Taniguchi, T., Sakai, M., Fujii-Kuriyama, Y., Muramatsu, M., Kobayashi, S., Sudo, T. 1979. *Proc. Jpn. Acad. Ser. B* 55:464–69
96. Taniguchi, T., Fujii-Kuriyama, Y., Muramatsu, M. 1980. *Proc. Natl. Acad. Sci. USA* 77:4003–6
97. Taniguchi, T., Guarente, L., Roberts, T. M., Kimelmann, P., Douhan, J. III, Ptashne, M. 1980. *Proc. Natl. Acad. Sci. USA* 77:5230–33
98. Yelverton, E., Leung, D., Weck, P., Gray, P. W., Goeddel, D. V. 1981. *Nucleic Acids Res.* 9:731–41
99. Hitzeman, R. A., Hagie, F. E., Levine, H. L., Goeddel, D. V., Ammerer, G., Hall, B. D. 1981. *Nature* 293:717–22
100. Rubinstein, M., Rubinstein, S., Familletti, P. C., Miller, R. S., Waldman, A. A., Pestka, S. 1979. *Proc. Natl. Acad. Sci. USA* 76:640–44
101. Rubinstein, M., Levy, W. P., Moschera, J. A., Lai, C., Hershberg, R. D., Bartlett, R. T., Pestka, S. 1981. *Arch. Biochem. Biophys.* 210:307–18
102. Levy, W. P., Rubinstein, M., Shively, J., Del Valle, U., Lai, C., Moschera, J., Brink, L., Gerber, L., Stein, S., Pestka, S. 1981. *Proc. Natl. Acad. Sci. USA* 78:6186–90
103. Evinger, M., Rubinstein, M., Pestka, S. 1981. *Arch. Biochem. Biophys.* 210:319–29
104. Allen, G., Fantes, K. H. 1980. *Nature* 287:408–11

105. Knight, E. Jr. 1976. *Proc. Natl. Acad. Sci. USA* 73:520–23
106. Knight, E. Jr., Fahey, D. 1981. *J. Biol. Chem.* 256:3609–11
107. Friesen, H., Stein, S., Evinger, M., Familletti, P. C., Moschera, J., Meienhofer, J., Shively, J., Pestka, S. 1981. *Arch. Biochem. Biophys.* 206:432–50
108. Okamura, H., Berthold, W., Hood, L., Hunkapiller, M., Inoue, M., Smith-Johannsen, H., Tan, Y. H. 1980. *Biochemistry* 19:3831–35
109. Stewart, W. E. II, Sarkar, F. H., Taira, H., Hall, A., Nagata, S., Weissmann, C. 1980. *Gene* 11:181–86
110. Schellekens, H., de Reus, A., Bolhuis, R., Fountoulakis, M., Schein, C., Ecsodi, J., Nagata, S., Weissmann, C. 1981. *Nature* 292:775–76
111. Staehelin, T., Kung, H. F., Hobbs, D. S., Pestka, S. 1982. *Methods Enzymol.* 78: In press
112. Pestka, S. ed. 1982. *Methods Enzymol.* Vol. 78. In press
113. Wetzel, R. 1981. *Nature* 289:606–7
114. Degrave, W., Derynck, R., Tavernier, J., Haegeman, G., Fiers, W. 1981. *Gene* 14:137–43
115. Lawn, R. M., Adelman, J., Dull, T. J., Gross, M., Goeddel, D., Ullrich, A. 1981. *Science* 212:1159–62
116. Gross, G., Mayr, U., Bruns, W., Grosveld, F., Dahl, H.-H. M., Collins, J. 1981. *Nucleic Acids Res.* 9:2495–2507
117. Weissenbach, J., Chernajovsky, Y., Zeevi, M., Shulman, L., Soreq, H., Nir, U., Wallach, D., Perricaudet, M., Tiollais, P., Revel, M. 1980. *Proc. Natl. Acad. Sci. USA* 77:7152–56
118. Sagar, A. D., Pickering, L. A., Sussman-Berger, P., Stewart, W. E. II, Sehgal, P. B. 1981. *Nucleic Acids Res.* 9:149–60
119. Sehgal, P. B., Sagar, A. D. 1980. *Nature* 287:95–97
120. Sehgal, P. B., Sagar, A. D., Braude, I. A., Smith, D. 1981. In *The Biology of the Interferon System,* ed. E. De-Maeyer, G. Galasso, H. Schellekens, pp. 43–46. Amsterdam: North Holland-/Elsevier
121. Sagar, A. D., Sehgal, P. B., Braude, I. A. 1981. *Science.* 214:803–5
122. Owerbach, D., Rutter, W. J., Shows, T. B., Gray, P., Goeddel, D., Lawn, R. M. 1981. *Proc. Natl. Acad. Sci. USA* 78:3123–27
123. Nagata, S., Brack, C., Henco, K., Schamböck, A., Weissmann, C. 1981. *J. Interferon Res.* 1:333–36
124. Burke, D. C. 1980. *Interferon* 2:47–64

125. Slate, D. L., Ruddle, F. H. 1979. *Pharmacol. Ther.* 4:221–30
126. Epstein, L. B. 1977. *Tex. Rep. Biol. Med.* 35:42–56
127. Epstein, L. B. 1982. *Interferon* 3: In press
128. Yip, Y. K., Pang, R. H. L., Urban, C., Vilcek, J. 1981. *Proc. Natl. Acad. Sci. USA* 78:1601–5
129. Georgiades, J. A., Langford, M. P., Goldstein, L. D., Blalock, J. E., Johnson, H. M. 1979. In *Interferon: Properties and Clinical Uses,* ed. A. Kan, N. O. Hill, G. L. Dorn, pp. 97–108. Dallas, Tex: Leland Fikes Found. Press of Wadley Inst. Mol. Med.
130. Wiranowska-Stewart, M., Lin, L. S., Braude, I. A., Stewart, W. E. II. 1980. *Mol. Immunol.* 17:625–33
131. Nathan, I., Groopman, J. E., Quan, S. G., Bersch, N., Golde, D. 1981. *Nature* 292:842–44
132. Wallace, D. M., Hitchcock, M. J. M., Reber, S. B., Berger, S. L. 1981. *Biochem. Biophys. Res. Commun.* 100:865–71
133. DeMaeyer-Guignard, J., Tovey, M. G., Gresser, I., DeMaeyer, E. 1978. *Nature* 271:622–25
134. Cabrer, B., Taira, H., Broeze, R. J., Kempe, T. D., Williams, K., Slattery, E., Konigsberg, W. H., Lengyel, P. 1979. *J. Biol. Chem.* 254:3681–84
135. Yonehara, S., Iwakura, Y., Kawade, Y. 1980. *Virology* 100:125–29
136. Stewart, W. E. II, Havell, E. A. 1980. *Virology* 101:315–18
137. Broeze, R. J., Dougherty, J. P., Pichon, J., Jayaram, B. M., Lengyel, P. 1981. *J. Interferon Res.* 1:191–201
138. Streuli, M., Hall, A., Boll, W., Stewart, W. E. II, Nagata, S., Weissmann, C. 1981. *Proc. Natl. Acad. Sci. USA* 78:2848–52
139. Weck, P. K., Apperson, S., Stebbing, N., Gray, P. W., Leung, D., Shepard, M., Goeddel, D. V. 1982. *Nucleic Acid Res.* (Submitted)
140. Fujisawa, J., Kawade, Y. 1981. *Virology* 112:480–87
141. Masucci, M. G., Szigeti, R., Klein, E., Klein, G., Gruest, J., Montagnier, L., Taira, H., Hall, A., Nagata, S., Weissmann, C. 1980. *Science* 209:1431–35
142. Evinger, M., Maeda, S., Pestka, S. 1981. *J. Biol. Chem.* 256:2113–14
143. Herberman, R. B., Ortaldo, J. R., Mantovani, A., Hobbs, D. S., Kung, H., Pestka, S. 1982. *Nature.* In press
144. Streuli, M., Nagata, S., Weissmann, C. 1980. *Science* 209:1343–47

145. Edge, M. D., Greene, A. R., Heathcliffe, G. R., Meacock, P. A., Schuch, W., Scanlon, D. B., Atkinson, T. C., Newton, C. R., Markham, A. F. 1981. *Nature* 292:756–62
146. Pitha, P. M., Vengris, V. E., Reynolds, F. H. 1976. *J. Supramol. Struct.* 4:467–73
147. Ankel, H., Krishnamurthi, C., Besançon, F., Stefanos, S., Falcoff, E. 1980. *Proc. Natl. Acad. Sci. USA* 77:2528–32
148. Tan, Y. H., Tischfield, J., Ruddle, F. H. 1973. *J. Exp. Med.* 137:317–30
149. Tan, Y. H., Schneider, E. L., Tischfield, J., Epstein, C. J., Ruddle, F. H. 1974. *Science* 186:61–63
150. Revel, M., Bash, D., Ruddle, F. H. 1976. *Nature* 260:139–41
151. Tovey, M. G., Rochette-Egly, C., Castagna, M. 1979. *Proc. Natl. Acad. Sci. USA* 76:3890–93
152. Vesely, D. L., Cantell, K. 1980. *Biochem. Biophys. Res. Commun.* 96: 574–79
153. Radke, K. L., Colby, C., Kates, J. R., Krider, H. M., Prescott, D. M. 1974. *J. Virol.* 13:623–30
154. Taylor, J. 1964. *Biochem. Biophys. Res. Commun.* 14:447–51
155. Lockart, R. Z. Jr. 1964. *Biochem. Biophys. Res. Commun.* 15:513–18
156. Friedman, R. M., Sonnabend, J. 1965. *J. Immunol.* 95:696–703
157. Pottathil, R., Chandrabose, K., Cuatrecasas, P., Lang, D. J. 1981. *Proc. Natl. Acad. Sci. USA* 78:3343–47
158. Chandrabose, K., Cuatrecasas, P., Pottathil, R., Lang, D. J. 1981. *Science* 212:329–31
159. Pottathil, R., Chandrabose, K., Cuatrecasas, P., Lang, D. J. 1980. *Proc. Natl. Acad. Sci. USA* 77:5437–40
160. Pfeffer, L. M., Wang, E., Tamm, I. 1980. *J. Exp. Med.* 152:469–74
161. Gordon, I., Stevenson, D. 1980. *Proc. Natl. Acad. Sci. USA* 77:452–56
162. Friedmann, R. M. 1977. *Bacteriol. Rev.* 41:543–67
163. Sen, G. C., Lebleu, B., Brown, G. E., Kawakita, M., Slattery, E., Lengyel, P. 1976. *Nature* 264:370–73
164. Ratner, L., Wiegand, R. C., Farrell, P. J., Sen, G. C., Cabrer, B., Lengyel, P. 1978. *Biochem. Biophys. Res. Commun.* 81:947–54
165. Farrell, P. J., Sen, G. C., Dubois, M. F., Ratner, L., Slattery, E., Lengyel, P. 1978. *Proc. Natl. Acad. Sci. USA* 75:5893–97
166. Hovanessian, A. G., Kerr, I. M. 1979. *Eur. J. Biochem.* 93:515–26
167. Ball, L. A., White, C. N. 1979. In *Regulation of Macromolecular Synthesis by Low Molecular Weight Mediators,* ed. E. Koch, D. Richter, pp. 303–17. New York: Academic
168. Justesen, J., Ferbus, D., Thang, M. N. 1980. *Ann. NY Acad. Sci.* 350:510–21
169. Dougherty, J. P., Samanta, H., Farrell, P. J., Lengyel, P. 1980. *J. Biol. Chem.* 255:3813–16
170. Samanta, H., Dougherty, J. P., Lengyel, P. 1980. *J. Biol. Chem.* 255:9807–13
171. Doornbos, J., Den Hartog, J. A. J., Boom, J. H. van, Altona, C. 1981. *Eur. J. Biochem.* 116:403–12
172. Justesen, J., Ferbus, D., Thang, M. N. 1980. *Proc. Natl. Acad. Sci. USA* 77:9618–22
173. Minks, M. A., Benvin, S., Baglioni, C. 1980. *J. Biol. Chem.* 255:5031–35
174. Ball, L. A., White, C. N. 1978. *Proc. Natl. Acad. Sci. USA* 75:1167–71
175. Ferbus, D., Justesen, J., Besançon, F., Thang, M. N. 1981. *Biochem. Biophys. Res. Commun.* 100:847–56
176. Minks, M. A., West, D. K., Benvin, S., Baglioni, C. 1979. *J. Biol. Chem.* 254:10180–83
177. Yang, K., Samanta, H., Dougherty, J., Jayaram, B., Broeze, R., Lengyel, P. 1981. *J. Biol. Chem.* 256:9324–28
178. Revel, M., Kimchi, A., Friedman, M., Wolf, D., Merlin, G., Panet, A., Rapoport, S., Lapidot, Y. 1982. *Tex. Rep. Biol. Med.* In press
179. Williams, B. R. G., Golgher, R. R., Brown, R. E., Gilbert, C. S., Kerr, I. M. 1979. *Nature* 282:582–86
180. Martin, E. M., Birdsall, N. J. M., Brown, R. E., Kerr, I. M. 1979. *Eur. J. Biochem.* 95:295–307
181. Slattery, E., Ghosh, N., Samanta, H., Lengyel, P. 1979. *Proc. Natl. Acad. Sci. USA* 76:4778–82
182. Chernajovsky, Y., Kimchi, A., Schmidt, A., Zilberstein, A., Revel, M. 1979. *Eur. J. Biochem.* 96:35–41
183. Williams, B. R. G., Kerr, I. M., Gilbert, C. S., White, C. N., Ball, L. A. 1978. *Eur. J. Biochem.* 92:455–62
184. Minks, M. A., Benvin, S., Maroney, P. A., Baglioni, C. 1979. *Nucleic Acids Res.* 6:767–80
185. Schmidt, A., Chernajovsky, Y., Shulman, L., Federman, P., Berissi, H., Revel, M. 1979. *Proc. Natl. Acad. Sci. USA* 76:4788–92
186. Ball, L. A. 1980. *Ann. NY Acad. Sci.* 350:486–96
187. Wallach, D., Revel, M. 1980. *Nature* 287:68–70

188. Baglioni, C., Maroney, P. A. 1980. *J. Biol. Chem.* 255:8390–93
189. Hovanessian, A. G., La Bonnardiere, C., Falcoff, E. 1980. *J. Interferon Res.* 1:125–35
190. Minks, M. A., Benvin, S., Maroney, P. A., Baglioni, C. 1979. *J. Biol. Chem.* 254:5058–64
191. Krishnan, I., Baglioni, C. 1981. *Virology* 111:666–70
192. Stark, G. R., Dower, W. J., Schimke, R. T., Brown, R. E., Kerr, I. M. 1979. *Nature* 278:471–73
193. Krishnan, I., Baglioni, C. 1980. *Proc. Natl. Acad. Sci. USA* 77:6506–10
194. Hovanessian, A. G., Kerr, I. M. 1978. *Eur. J. Biochem.* 84:149–59
195. Verhaegen, M., Divizia, M., Vandenbussche, P., Kuwata, T., Content, J. 1980. *Proc. Natl. Acad. Sci. USA* 77:4479–83
196. Hovanessian, A. G., Meurs, E., Montagnier, L. 1981. *J. Interferon Res.* 1:179–90
197. Ratner, L. 1979. PhD thesis. *Studies on interferon action.* Yale Univ., New Haven, Conn. 248 pp.
198. Czarniecki, C. W., Sreevalsan, T., Friedman, R. M., Panet, A. 1981. *J. Virol.* 37:827–31
199. Lengyel, P., Desrosiers, R., Broeze, R., Slattery, E., Taira, H., Dougherty, J., Samanta, H., Pichon, J., Farrell, P., Ratner, L., Sen, G. 1980. In *Microbiology 1980,* ed. D. Schlessinger, pp. 219–26. Am. Soc. Microbiol. Wash. DC
200. Galster, R. L., Lengyel, P. 1976. *Nucleic Acids Res.* 3:581–98
201. Wreschner, D. H., James, T. C., Silverman, R. H., Kerr, I. M. 1981. *Nucleic Acids Res.* 9:1571–81
202. Knight, M., Cayley, P. J., Silverman, R. H., Wreschner, D. H., Gilbert, C. S., Brown, R. E., Kerr, I. M. 1980. *Nature* 288:189–92
203. Silverman, R. H., Wreschner, D. H., Gilbert, C. S., Kerr, I. M. 1981. *Eur. J. Biochem.* 115:79–85
204. Nilsen, T. W., Maroney, P. A., Baglioni, C. 1981. *J. Biol. Chem.* 256:7806–11
205. Williams, B. R. G., Kerr, I. M. 1978. *Nature* 276:88–90
206. Hovanessian, A. G., Wood, J., Meurs, E., Montagnier, L. 1979. *Proc. Natl. Acad. Sci. USA* 76:3261–65
207. Hovanessian, A. G., Wood, J. N. 1980. *Virology* 101:81–90
208. Epstein, D. A., Czarniecki, C. W., Jacobsen, H., Friedman, R. M., Panet, A. 1981. *Eur. J. Biochem.* 118:9–15
209. Meurs, E., Hovanessian, A. G., Montagnier, L. 1981. *J. Interferon Res.* 1:219–32
210. Jelinek, W., Darnell, J. E. 1972. *Proc. Natl. Acad. Sci. USA* 69:2537–41
211. Robertson, H. D., Dickson, E., Jelinek, W. 1977. *J. Mol. Biol.* 115:571–89
212. Calvet, J. P., Pederson, T. 1979. *Proc. Natl. Acad. Sci. USA* 76:755–59
213. Revel, M., Kimchi, A., Shulman, L., Fradin, A., Shuster, R., Yakobson, E., Chernajovsky, Y., Schmidt, A., Shure, A., Bendori, R. 1980. *Ann. NY Acad. Sci.* 350:459–72
214. Rosen, O. M., Krebs, E. G., eds. 1981. *Protein Phosphorylation.* Cold Spring Harbor, NY: Cold Spring Harbor Lab., 1421 pp.
215. Sen, G. C., Taira, H., Lengyel, P. 1978. *J. Biol. Chem.* 253:5915–21
216. Kimchi, A., Zilberstein, A., Schmidt, A., Shulman, L., Revel, M. 1979. *J. Biol. Chem.* 254:9846–53
217. Hovanessian, A. G., Meurs, E., Aujean, O., Vaquero, C., Stefanos, S., Falcoff, E. 1980. *Virology* 104:195–204
218. Samuel, C. E. 1979. *Proc. Natl. Acad. Sci. USA* 76:600–4
219. Lengyel, P., Samanta, H., Pichon, J., Dougherty, J., Slattery, E., Farrell, P. 1980. *Ann. NY Acad. Sci.* 350:441–47
220. Crouch, D., Safer, B. 1980. *J. Biol. Chem.* 255:7918–24
221. Epstein, D. A., Torrence, P. F., Friedman, R. M. 1980. *Proc. Natl. Acad. Sci. USA* 77:107–11
222. Farrell, P. J., Balkow, K., Hunt, T., Jackson, R. J. 1977. *Cell* 11:187–200
223. Ochoa, S., deHaro, C. 1979. *Ann. Rev. Biochem.* 48:549–80
224. Gupta, S. L. 1979. *J. Virol.* 29:301–11
225. Golgher, R. R., Williams, B. R. G., Gilbert, C. S., Brown, R. E., Kerr, I. M. 1980. *Ann. NY Acad. Sci.* 350:448–58
226. Grosfeld, H., Ochoa, S. 1980. *Proc. Natl. Acad. Sci. USA* 77:6526–30
227. Ranu, R. S. 1980. *Biochem. Biophys. Res. Commun.* 97:252–62
228. Levin, D. H., Petryshyn, R., London, I. M. 1981. *J. Biol. Chem.* 256:7638–41
229. Joklik, W. K. 1974. *Compr. Virol.* 2:231–339
230. Nilsen, T. W., Weissman, S. G., Baglioni, C. 1980. *Biochemistry* 19:5574–79
231. Banerjee, A. K. 1980. *Microbiol. Rev.* 44:175–205
232. Sen, G. C., Gupta, S. L., Brown, G. E., Lebleu, B., Rebello, M. A., Lengyel, P. 1976. *J. Virol.* 17:191–203
233. Falcoff, R., Falcoff, E., Sanceau, J., Lewis, J. A. 1978. *Virology* 86:507–15

234. Panniers, L. R. V., Clemens, M. J. 1980. *Biochem. Soc. Trans.* 8:352–3
235. Brouty-Boye, D., Tovey, M. G. 1978. *Intervirology* 9:243–52
236. Gewert, D. G., Clemens, M. J. 1980. *Biochem. Soc. Trans.* 8:353–54
237. Maheshwari, R. K., Banerjee, D. K., Waechter, C. J., Olden, K., Friedman, R. M. 1980. *Nature* 287:454–56
238. Sreevalsan, T., Rozengurt, E., Taylor-Papadimitriou, J., Burchell, J. 1980. *J. Cell Physiol.* 104:1–9
239. Vignaux, F., Gresser, I. 1977. *J. Immunol.* 118:721–23
240. Fellous, M., Kamoun, M., Gresser, I., Bono, R. 1979. *Eur. J. Immunol.* 9:446–49
241. Fridman, W. H., Gresser, I., Bandu, M. T., Aguet, M., Neauports-Sautes, C. 1980. *J. Immunol.* 124:2436–41
242. Nilsen, T. W., Wood, D. L., Baglioni, C. 1980. *Nature* 286:178–80
243. Wood, J. N., Hovanessian, A. G. 1979. *Nature* 282:74–76
244. Horisberger, M. A., Haller, O., Arnheiter, H. 1980. *J. Gen. Virol.* 50:205–10
245. Haller, O., Arnheiter, H., Gresser, I., Lindenmann, J. 1981. *J. Exp. Med.* 154:199–203
246. Falcoff, R., Sanceau, J. 1979. *Virology* 98:433–38
247. Gupta, S. L., Graziadei, W. D., Weideli, H., Sopori, M. L., Lengyel, P. 1974. *Virology* 57:49–63
248. Gupta, S. L., Rubin, B. Y., Holmes, S. L. 1981. *Virology* 111:331–40
249. Rubin, B. Y., Gupta, S. L. 1980. *Proc. Natl. Acad. Sci. USA* 77:5928–32

ADDED IN PROOF   Clinical studies with the recombinant human α 2 (A) IFN revealed that its effects and side effects are similar to those of partially purified leukocyte IFNs (250). A cDNA sequence coding for a human γ IFN has been obtained. The longest homologous amino acid sequence between this γ IFN and an α IFN covers only four consecutive amino acids (251). Cloned human α and β IFN genes were introduced into mouse cells. The expression of these genes was enhanced after treating the cells with IFN inducers (252, 253). It was reported that in transformed macrophages the stimulation of phagocytosis and the inhibition of growth by IFNs are mediated through an increased level of intracellular cAMP, whereas the antiviral state induced by IFNs is independent of cAMP (254). The ds regions of heterogeneous nuclear RNA from HeLa cells were reported to promote the synthesis of $(2'-5')(A)_n$ by $(2'-5')(A)_n$ synthetase, and labeled heterogeneous nuclear RNA was cleaved more extensively than mRNA by extracts from IFN-treated HeLa cells (255). The activation of the $(2'-5')(A)_n$-dependent RNase L was reported to be inhibited in EAT, L, or HeLa cells after infection with EMC virus. This effect of EMC virus infection was prevented by IFN pretreatment of the cells (256). cDNA segments were cloned, which are complementary to mRNAs induced in EAT cells by IFNs. The increase in the level of IFN induced mRNAs was transient even if the cells were kept continuously in the presence of IFNs (257).

250. Gutterman, J. U. et al. 1982. *Ann. Int. Med.* In press
251. Gray, P. W. et al 1982. *Nature* 295:503–8
252. Ohno, S., Taniguchi, T. 1982. *Nucleic Acids Res.* 10:967–77
253. Mantei, N., Weissmann, C. 1982. In press
254. Schneck, J., Rager-Zisman, B., Rosen, O. M., Bloom, B. R. 1982. *Proc. Natl. Acad. Sci. USA* 79:1879–83
255. Nilsen, T. W., Maroney, P. A., Robertson, H. D., Baglioni, C. 1982. *Mol. Cell. Biol.* 2:154–60
256. Cayley, P. J., Knight, M., Kerr, I. M. 1982. *Biochem. Biophys. Res. Commun.* 104:376–82
257. Lengyel, P., Samanta, H., Dougherty, J. P., Brawner, M. E., Schmidt, H. 1982. *J. Cell. Biochem.* (Suppl.) 6:81

Ann. Rev. Biochem. 1982. 51:283–308
Copyright © 1982 by Annual Reviews Inc. All rights reserved

ENZYMES OF THE RENIN-ANGIOTENSIN SYSTEM AND THEIR INHIBITORS

M. A. Ondetti and D. W. Cushman

The Squibb Institute for Medical Research, Princeton, New Jersey 08540

CONTENTS

PERSPECTIVES AND SUMMARY

The renin-angiotensin system is perhaps the most important of various humoral vasoconstrictor and vasodilator mechanisms implicated in blood pressure regulation; its end product, the octapeptide angiotensin II, is one of the most potent vasoconstrictor agents known.

The constitution of this system as it is presently understood (1), and its interrelationships with the kallikrein-kinin system are depicted in Figure 1. The formation and degradation of angiotensins I, II, and III are controlled by three enzymes or groups of enzymes: renin, angiotensin-converting enzyme, and angiotensinases. Other important actions required for the functioning of the renin-angiotensin system, namely, renin release, and the

0066-4154/82/0701-0283$02.00

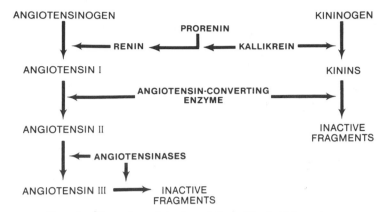

Figure 1   The renin-angiotensin and the kallikrein-kinin systems.

action of angiotensin II on vasculature and on the adrenal gland (aldosterone release), are mediated through agonist/receptor-type interactions and are not reviewed here.

As we discuss below, several lines of evidence indicate that renin may occur as a proenzyme that requires activation in an enzymatic step possibly involving kallikrein, one of the key enzymatic components of the kinin-generating system. The conversion of angiotensin II to angiotensin III and further degradation of this heptapeptide is controlled by a group of enzymes designated as angiotensinases. Space limitations preclude the review here of kallikrein and the angiotensinases, but they are reviewed in the recent literature (2, 3).

Even though the mechanism of the renin-angiotensin system at the molecular level was fairly well understood in the 60s, its relevance for blood pressure regulation was the subject of vigorous controversy. The discovery in the 1970s of specific blockers, in particular, angiotensin-converting enzyme inhibitors, made a significant contribution to the understanding of the role of the renin-angiotensin system in blood pressure homeostasis. The impact of this recently acquired knowledge on the development of new approaches to the therapy of human hypertension has been recently reviewed (4). The biochemistry of the two key enzymes of the renin-angiotensin system, renin and angiotensin-converting enzyme, and their inhibitors are the subject of the present chapter.

RENIN

Renin was first identified as a vasopressor substance in kidney extracts by Tigerstedt and Bergman, but the enzymatic nature of its role within the context of the renin-angiotensin system was not apparent until the work of

Page, Braun-Menendez and collaborators (see 5). In spite of its early identification, progress in the purification and characterization of renin has been slow, mainly because the lack of suitable technology to achieve complete purification led many times to unstable preparations, probably due to contamination with other peptidases. The introduction of affinity chromatography techniques has allowed, in recent years, the isolation of homogeneous renin preparations from several sources (6). It has also been shown that renin occurs in several forms with different molecular weights and with varying degrees of enzymatic activity, from the inactive "pro-renin," to the fully active low-molecular-weight form. The inactive or partially active renins can be activated by different procedures, e.g. acidification, storage at low temperature, enzymatic hydrolysis, or a combination of these.

Assay

Renin is a highly specific peptidase for which the only known substrate is the alpha-globulin angiotensinogen or its N-terminal fragments. Renin splits only one peptide bond from this protein, the Leu-Leu bond between the tenth and eleventh amino acid residues, with the release of the N-terminal decapeptide angiotensin I (Figure 2). Human angiotensinogen is only cleaved by renin from humans or primates, but human renin will cleave angiotensinogen from most mammalian species. The isolation and characterization of angiotensinogen has been recently reviewed (7).

Partial hydrolysis of angiotensinogen with trypsin releases an N-terminal tetradecapeptide that can subsequently yield angiotensin I upon incubation with renin. Utilizing synthetic peptides representing partial sequences of this tetradecapeptide, Skeggs and collaborators were able to define more closely the substrate specificity of renin. The smallest fragment that can be

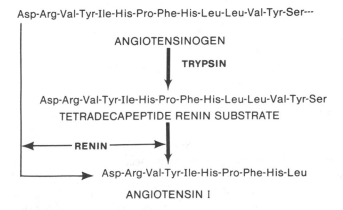

Figure 2 Renin substrates.

hydrolyzed by renin at a significant rate is the octapeptide His[6]----Tyr[13]. Further additions of amino acid residues to the N- or C-terminals of this octapeptide can significantly increase the affinity (lowering the K_m) but without major effects on the V (8).

Since angiotensin I has very high biological activity, due to its ready conversion to angiotensin II in vivo, the first renin assays developed were biological assays (9). The extreme sensitivity of this assay, based on the hypertensive action of angiotensin I, allowed the measurement of the very small concentrations of renin normally present in human plasma. A Gold-blatt Unit (GU), now called the International Reference Unit (IRU), is the quantity of renin which, when injected intravenously into an unanesthe-tized, trained dog, raises the direct mean femoral arterial blood pressure by 30 mm of mercury (10). With the advent of radioimmunoassay techniques, the measurement of very low concentrations of angiotensin by a nonbiologi-cal procedure became feasible, and a renin assay based on the measurement of angiotensin I by radioimmunoassay (9–11) has become the method of choice for determining the values of renin concentration or plasma renin activity (12). In these assays the substrate is usually the endogenous angi-otensinogen, if "plasma renin activity" is to be measured, or plasma from nephrectomized sheep (to ascertain that no exogenous renin is added) if a measure of plasma renin concentration is desired. This second approach is also to be used when measuring renin concentration in sources other than plasma, e.g. during isolation of renin from organs. In order to avoid poten-tial losses of angiotensin I due to degradation by proteolytic enzymes (angi-otensin-converting enzyme or angiotensinases) protease inhibitors such as EDTA, phenylmethanesulfonyl fluoride, dimercaptoprol, and 8-hydroxy-quinoline are added. The angiotensin I formed is measured by radioim-munoassay using antisera generated by immunization with conjugates of angiotensin I and synthetic or natural polymers, and iodinated angiotensin I as radioactive ligand. Kits for angiotensin I radioimmunoassay are now commercially available (12). Iodinated (^{125}I) or carbamylated (^{14}C) angi-otensinogen and ^{14}C labeled tetradecapeptide have been used in renin assays (13–15). To facilitate separation of products, radiolabeled trideca- or hep-tadecapeptide N-terminal fragments of angiotensinogen bound to insoluble polymers were also developed as substrates to measure renin activity (16, 17).

In the nonradioactive assays of Skeggs et al (18) and Galen et al (19) the tetrapeptide product Leu-Val-Tyr-Ser is measured with Lowry reagent or fluorometry.

The fluorometric assay of Reinharz & Roth (20) utilizes a coupled en-zymatic system whereby the product of the renin hydrolysis of the octapep-tide Z-Pro-Phe-His-Leu-Leu-Val-Tyr-Ser-β-naphthylamide, the tetrapeptide Leu-Val-Tyr-Ser-β-naphthylamide, is sequentially degraded by aminopeptidase M to yield the fluorescent β-naphthylamine. With this procedure it is possible to detect renin activity as low as 1 mGU with linear reaction rates observed up to 10–25 mGU.

A somewhat similar procedure was recently reported by Murakami et al (21). The substrate employed is the octapeptide Arg-Pro-Phe-His-Leu-Leu-Val-Tyr-7-amino-4-methyl-coumarin or its N-succinylated derivative. The succinyl derivative was found to be a better substrate even though kinetic constants were not reported. Linear rates of hydrolysis were observed up to 100 mGU, and the minimum detectable amount of renin activity was 5 mGU. Leucine aminopeptidase was employed to degrade the tetrapeptide-coumarin product to 7-amino-4-methylcoumarin, which was then measured by fluorometry. It is important to point out that nonrenin proteases can seriously interfere in all those procedures that are not based on the specific measurement of angiotensin I formation.

Fully Active Low-Molecular-Weight Renin

Kidney, the most physiologically relevant source of renin, was the first organ in which this enzyme was detected. However, it has also been isolated from brain, submaxillary gland, and amniotic fluid. Affinity chromatography is the key step in most of the procedures of renin isolation. Columns with the inhibitor pepstatin as the affinity ligand allow the separation of acidic proteases from other types of enzymes or nonenzymatic protein components (23, 24). The use of casein (22) or hemoglobin (26) as ligands in affinity chromatography columns permitted the separation of renin from other acidic proteases. Finally, cross-reactivity with renin-specific antibodies has served to confirm that the angiotensin I–generating activity isolated from tissues other than kidney, is indeed renin (25, 27, 28). Table 1 summarizes the properties of the renin preparations reported so far.

Activatable Renin

Since the early seventies the literature has reported, on many occasions, the presence of "renins" of molecular size larger than 40,000 with different degrees of enzymatic activity. Boyd (29) described in 1974 the isolation from hog kidney at neutral pH of two different forms of renin, one of molecular weight 60,000 and the other 40,000. Incubation of the high-molecular-weight renin (HMWR) at low pH (2.5) led to the formation of the low-molecular-weight form with 2.5-fold increase in enzymatic activity. Chromatography of the HMWR on DEAE cellulose under specially designed conditions, led to the separation of a low-molecular-weight form and an inactive protein, renin-binding protein (RBP) that could, upon admixture with the low-molecular-weight form, reconstitute the high-molecular-

Table 1 Properties of low-molecular-weight renins

Source	Mol wt × 10^{-3}	pH optimum	Sugar	Iso-enzymes	Specific activity GU/mg	Reference
Submaxillary Gland						
Mouse	36–37	8.0–8.5	—	+	6–6.7[a]	78, 87
Kidney						
Hog	36.4	5.5–7.0	+	+	2,000	23
	36.8		+	+	1,097	24
Dog				+	4,200	79
Rat					134[a]	80
	35–37	6.0–6.5	+	+	500–750	81
Rabbit	40					82
Bull	46			+	16,000[a]	88
Human	40	6.5		+	870	83
	40				400	84
	40	6.0	+	+	950	26
Brain						
Hog	42	6.5–7.5	+	+		85
Amniotic Fluid						
Human	39					86

[a] μmoles AI/h/mg protein.

form. The molecular weight of the RBP was found to be 55,000 (using Sephadex) even though the difference in molecular weight between the high- and low-molecular-weight components was only 20,000. The most recent investigations (30–36) have confirmed these observations and provided some more detail. Renin granules of dogs and rats contain only low-molecular-weight renin (LMWR; molecular weight of 40,000) while the RBP is in the cytosol (30). RBP can be separated from active renin in hog kidney extracts by chromatography on pepstatinylaminohexylagarose (36), but full purification has not yet been achieved. RBP has been described as being nondiffusible, unstable to heat and low pH, and with a molecular weight higher than 60,000. It is not a glycoprotein, and lipids or lipoproteins do not appear to be needed for its action (33).

The interaction of RBP and LMWR produced HMWR. This process can be facilitated by the presence of thiol-blocking agents (26, 30, 32, 35), such as tetrathionate, 5,5'-dithiobis 2-nitrobenzoic acid (DTNB), diimide, N-ethyl maleimide, iodoacetic acid, or iodoacetamide. The effect of these agents was originally interpreted to be the result of the inhibition of a thiol-protease responsible for the conversion of HMWR to LMWR (37). However, more recent studies (32–26) favor the hypothesis that in going from LMWR to HMWR a noncovalent complex is formed between the

LMWR and a portion of RBP and that free SH groups have to be oxidized or derivatized in order to facilitate its formation.

Acidification converts the HMWR to LMWR (30–36). Intermediate HMWR have also been isolated from rat (31) and dog renal extracts by Nakane et al (31). RBP from hog kidney can interact with rat and rabbit kidney renins (36); dog RBP can interact with rat renin and vice-versa (30). However, hog kidney RBP does not interact with human renin, hog brain, or mouse submaxillary renin (36). Therefore, RBP shows some of the species specificity typical of renin. The tissue specificity is somewhat narrower than that of renin, since brain and kidney renin are immunologically related (38, 39).

The HMWRs described above have significant amounts of enzymatic activity (32, 34, 35), and cannot be considered zymogen forms of renin. Completely inactive forms of renin have been characterized in brain, kidney, plasma, and amniotic fluid (40–46). By a combination of hydrophobic and affinity chromatography, an inactive renin precursor can be isolated from kidney extracts. The molecular weight of this renal prorenin has been described as either 43,000 (42) or 53,000 (43) and it has been hypothesized that these two species represent two different steps in the posttranslational mechanism of synthesis of renin (42). Human plasma prorenin can also be separated from active renin by chromatography on affi-gel blue to which the active renin does not bind. Further purification can be achieved with pepstatin-Sepharose (40) or Concanavalin-Sepharose (47). Inactive renin constitutes about 70% of the total renin in plasma, but only 15% of the total renin in kidney extracts (41). Plasma prorenin has a molecular weight of 56,000 (41). Treatment with acids may (46, 48) or may not (40, 43) activate these prorenins, but this ambivalent behavior may be due to the degree of purification of each preparation (40, 49). Trypsin, pepsin, or kallikreins have been used to activate inactive renin with (41, 43) or without (45, 48) change in molecular weight.

The problem of activation of plasma renin has attracted considerable attention in the last few years. Most of these studies have been carried out with whole plasma, a system of such complexity that all interpretations are necessarily tentative. Activation of human plasma renin can be achieved by incubation at low temperature (–4°C) ("cryoactivation"), dialysis at pH 3.3 ("acid activation") or treatment with trypsin (50, 51, 52). Dialysis at pH 3.3 has to be followed by dialysis at pH 7.4 to achieve complete activation (50, 52). Leckie & McGhee have recently shown that complete activation can be achieved by dialysis at pH 3.0 and that the only effect of a second stage of dialysis at pH 7.4 is to make this activation irreversible (53). The most widely accepted interpretation is that these processes destroy protease inhibitors, which leads to the activation of serine-type protease, which act upon the "prorenin" component to yield fully active renin (50, 52). It is possible that acid or cold treatment may be responsible only for a modification of the prorenin molecule, which makes it more susceptible to proteolytic action (54). One other interpretation is that acid treatment dissociates the "inactive" renin into "active" renin and a "renin inhibitor," which is then either hydrolyzed by proteases or bound by them in competition with renin (49, 53).

Chemical, Physical, and Enzymatic Properties

In the above sections we have commented upon some of the properties of purified renin from different tissues, such as molecular weight, pH optimum, carbohydrate content, and specific activity. Amino acid composition has been reported for human, dog, rat renal renin, and mouse submaxillary gland renin. These four enzymes share a remarkable similarity in amino acid composition, although human renin stands out for its relatively higher concentration of polar amino acids. It has not yet been reported whether or not the approximately 350 amino acid residues present in the renin molecule occur in a single chain, but at least four residues of cysteine per mole have been found (6).

In spite of the differences already noted between renins from various sources, there are some strong similarities, not the least important of which is the unique substrate specificity, a characteristic common to very few proteolytic enzymes. Renin has been classified among the acid proteases mainly because it is not inhibited by inhibitors typical of serine, thiol, or metallopeptidases, but is inhibited by pepstatin, an acidic protease inhibitor. Even though the designation of acidic protease is not properly applicable to renin, since its pH optimum is close to neutrality when acting upon its natural substrate (55), closer characterization of its active-site residues has justified its classification in the family of acid proteases like pepsin.

Since mouse submaxillary gland renin can be purified in relatively large amounts, this enzyme has been the first target for extensive studies on mechanism of action. With the aid of active site reagents it has been shown that two acidic amino acid residues, two tyrosine residues, and one arginine residue are present in the active site (86).

On the basis of the somewhat limited structure-activity relationships gathered from the study of substrate analogues as inhibitors (56), and the minimum structural requirements for peptide substrates (8), together with the knowledge of the essential functional groups present at the active site, a tentative model of the renin active site could be depicted as in Figure 3. Subsites S'_2, S'_1, S_1, S_3, and S_4 would be expected to be of hydrophobic nature, since introduction of hydroxyl residues on the side chains of the amino acids interacting with these subsites decreases binding significantly

(56). Since the octapeptide shown in Figure 3 is the minimum sequence to show a significant hydrolysis rate, binding of the substrate to subsites S'_3 to S_5 must be required in order to induce the conformational modification required to properly align the catalytic residues. The residues found to be involved in catalysis at the active site of renin are identical to those shown to occur in other carboxyl proteases, with the possible exception of an arginine residue that appears to be present only in renin. The two carboxyl groups and the tyrosine side chains depicted in the hypothetical model of Figure 3 are assumed to play the role that has been assigned to them in some of the tentative mechanisms of action proposed for pepsin. The protonated carboxyl polarizes the carbonyl of the scissile amide bond to facilitate attack by the carboxylate of the ionized residue, directly or through the intermediacy of a water molecule. Tyrosine protonates the leaving amino group (57). At the present time it is not obvious what might be the mechanistic role of the arginine residue present at the active site of renin.

Inhibitors

The renin inhibitors that have been developed so far can be broadly divided into four groups: (*a*) substrate analogues, (*b*) pepstatin and related analogues, (*c*) phospholipids, and (*d*) renin antibodies.

The studies of Skeggs and co-workers (8) defined the minimal sequence required for an efficient substrate and laid down the basis for the development of specific inhibitors capable of interacting with the active site. The most extensive studies in this area have been carried out by Haber, Burton, and collaborators (56), who prepared a number of analogues of the octapeptide sequence His-Pro-Phe-His-Leu-Leu-Val-Tyr. Full expression of the inhibitory activity of these analogues was hindered by the lack of solubility at neutral pHs, and attempts to increase solubility by replacing hydrophobic

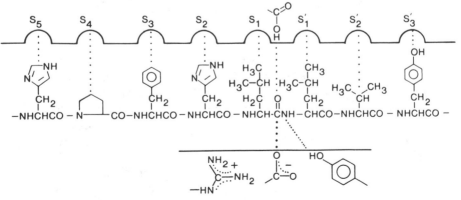

Figure 3   Hypothetical model of the active site of renin.

with hydrophilic residues led to substantial increase in K_i. To separate the two problems they resorted to increasing the lipophilicity of the main sequence to maximize binding, while adding proline residues to the amino end to enhance solubility. The analogue with best in vitro activity was the nonapeptide Pro-His-Pro-Phe-His-Phe-Phe-Val-Tyr which has a K_i of 1 X 10^{-6} M. The inhibitors demonstrate remarkable species specificity, since they interact with human renin much more efficiently than with any other nonprimate renin (58).

The nonapeptide Pro-His-Pro-Phe-His-Phe-Phe-Val-Tyr is very rapidly cleared from the circulation when administered intravenously, and this may explain its lack of in vivo activity as a renin inhibitor. However, this problem has now been overcome with further structural modification. The addition of a C-terminal lysine residue has given the decapeptide Pro-His-Pro-Phe-His-Phe-Phe-Val-Tyr-Lys, which has a K_i of 2 μM and an in vivo half-life more than ten times longer than that of the nonapeptide. When administered intravenously to monkeys (0.2 mg/kg/min) this decapeptide can block the hypertensive response to 0.6 GU of purified human renin without affecting the hypertensive responses to angiotensins I and II (59). It has also shown antihypertensive activity in a dog model of renin-dependent hypertension.

Pepstatin (1 in Scheme 1) is an acyl-pentapeptide-like compound isolated from the culture filtrates of certain species of *Streptomyces* that inhibits a variety of acidic proteases (60), including renin (61–63).

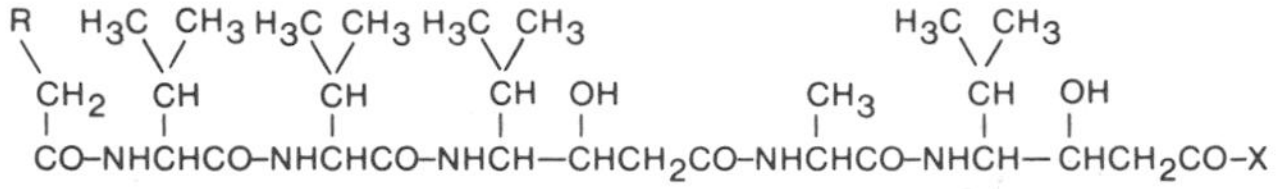

1, R = (CH$_3$)$_2$CH–; X = OH; pepstatin

2, R = H; X = OH; N–acetylpepstatin

3, R = (CH$_3$)$_2$CH–; X = Arg-OMe

Scheme 1

The inhibitory potency of pepstatin (I_{50} or K_i) has been reported to be approximately 10^{-6} M whenever the enzymatic reaction was carried out at pH close to 7 (63–66). When the substrate was polyglutamyl octapeptide (67), which requires a pH of 5.5 for the assay, the K_i was 1.3 X 10^{-10} M. A similar value for the inhibitor constant was obtained with rat spleen pseudorenin, a cathepsin-like enzyme that has an optimal pH of 5 (68), and with pepsin (67).

The type of inhibition has been variously described as being competitive (62, 66) and noncompetitive (64, 67). A similar variability in the analysis of kinetic results has been observed with the inhibition of pepsin by pepstatin, but in this case it appears that careful examination of kinetics supports a competitive mechanism (69), and therefore the conclusion that pepstatin interacts with the active site of pepsin. At this point we can only make the same assumption for renin based on the similarities of their hydrolytic mechanism, and on the observation that pepstatin protects renin against modification by active-site directed reagents (86).

It has been claimed that pepstatin inhibits pepsin by virtue of its interaction as a transition state analogue, since the carbon bearing the hydroxyl group in in the 4-amino-3-hydroxy-6-methylheptanoyl residue in position three, mimics the tetrahedral transition state form of the amide carbonyl carbon during peptide bond hydrolysis. Indeed, this hydroxyl group is very important for the inhibitory activity of pepstatin against pepsin (69). Unfortunately, very few of the analogues of pepstatin have been tested against renin, and no substantial information on structure-activity relationships is yet available. In general pepstatin analogues are two to three orders of magnitude less potent as inhibitors of renin than of pepsin (61, 70).

Pepstatin inhibits the hypertensive action of renin when administered intravenously to rats, and has also antihypertensive action in spontaneously hypertensive rats and rats with renal hypertension (63, 71). All of these effects are of very short duration.

Extensive biological studies with pepstatin have been hindered by the very low solubility of this hydrophobic inhibitor. In order to circumvent this problem, a derivative containing an additional C-terminal arginine methyl ester (3 in Scheme 1) residue has been synthesized (72) that has forty times greater solubility and the same inhibitory potency. Another derivative with increased solubility is N-acetyl pepstatin (2 in Scheme 1), an analogue of pepstatin also isolated from *Streptomyces* (73). Both of these derivatives also have very short duration of action in vivo.

A number of investigators have reported on the renin inhibitory activity of phospholipids, lysophospholipids, and analogues (see 74). These derivatives show a low level of inhibitory activity in vitro, with I_{50}'s or K_i's in the millimolar range (75). The two most studied derivatives are 4 (89) and 5 (75) in Scheme 2.

$$\textbf{4, } CH_3(CH_2)_4 \, (CH=CHCH_2)_4 \, (CH_2)_3O\overset{\overset{O}{\|}}{\underset{\underset{OH}{|}}{P}}CH_2CH_2NH_2$$

$$\textbf{5, } Cl\,C_6H_4OC_6H_4OCH_2CONH(CH_2)_2O\overset{\overset{O}{\|}}{\underset{\underset{OH}{|}}{P}}OCH_2CH_2NH_2 \quad (PE\text{-}104)$$

Scheme 2

In spite of this marginal activity in vitro, 4 lowers blood pressure in the renin-dependent hypertensive rat model (89) at the reasonable dose of 25 mg/kg/day. PE-104 (5 in Scheme 2) has a K_i of 2 mM and is a competitive and reversible inhibitor of dog renal renin. It has no inhibitory effect on pepsin, papain, or trypsin. Infusion of 20 mg/kg/min of PE-104 inhibits the pressor activity of exogenously administered renin and lowers blood pressure in a model of renin-dependent hypertension.

The recent availability of homogeneous or nearly homogeneous preparations of renin has reawakened the interest in developing renin-specific antibodies to be utilized as renin inhibitors. The most recent studies reported by Dzau and collaborators (76) describe the development of specific antibodies to canine renin and demonstrate that these antibodies can inhibit renin activity in vitro and in vivo, and can lower blood pressure in the acute, one kidney, renal hypertensive dog. Similar results can be obtained with the Fab portion of these antibodies, but the duration of the effect is considerably shorter (77).

ANGIOTENSIN-CONVERTING ENZYME

Assay

The obvious biological importance of angiotensin-converting enzyme (ACE) led many early investigators to consider angiotensin I to be its "natural" substrate, and to utilize this decapeptide substrate for various biological, radiochromatographic, colorimetric, or radioimmunological assays of ACE activity (90). Quantitative chromatographic assays employing angiotensin I (91) or [(^{125}I)Tyr^8]-bradykinin (92) have also been described within the last few years. Such assays are often cumbersome, but suffer most seriously from interference by other peptidases that degrade both substrate and products.

The great increase during the last decade in our understanding of the physical, chemical, and enzymatic properties of ACE has been largely a consequence of the development of simple chemical methods for assay of ACE activity on specific amino-substituted tripeptide substrates. Z-Phe-His-Leu has about the same enzyme-binding affinity (K_m) as angiotensin I, and is rapidly cleaved by ACE to form His-Leu, low concentrations of which can be quantitated by a specific fluorometric procedure (93). The similar tripeptide substrate Bz-Gly-His-Leu (Hip-His-Leu) has a high K_m, but can be utilized for an even more specific assay that involves spectrophotometric determination of the extracted hippuric acid (Bz-Gly) product (94), as well as for the fluorometric assay procedure (95, 96).

Sensitive radiochemical assays for ACE activity based on extraction of the labeled N-protected amino acid product have been employed with [1-^{14}C-Gly]-Hip-His-Leu (97), [p-^{3}H-Bz]-Hip-Gly-Gly (98), [p-^{3}H-Bz]-Hip-His-Leu, [p-^{3}H-Bz]-Phe-His-Leu, [p-^{3}H-Bz]-Phe-Ala-Pro, and [p-^{3}H-Bz]-Pro-Phe-Arg (99). A radiochromatographic assay, with fluorescent visualization of the separated product, has been employed with ^{14}C-Dns-Gly-Gly-Gly (100).

The hippuric acid product from Hip-His-Leu or Hip-Gly-Gly can also be quantitated by liquid chromatography (101), or by colorimetric determination of the reaction product with 2,4,6-trichloro-s-triazine (102) or p-dimethylaminobenzaldehyde (103). Dipeptide products may be quantitated fluorometrically after reaction with fluorescamine (104, 105), or spectrophotometrically with an automated ninhydrin method (106).

Direct kinetic spectrophotometric assays of ACE activity were described quite early using Boc-Phe(NO$_2$)-Phe-Gly, Z-Phe(NO$_2$)-Gly-Gly, Z-Phe(NO$_2$)-His-Leu, and Hip(NO$_2$)-Gly-Gly as substrates (107, 108). S-Hippurylthioglycolylglycine is cleaved by ACE to form thioglycolylglycine, which reacts directly in the assay medium with 5,5'-dithiobis(2-nitrobenzoic acid) to produce the yellow 5-thio-2-nitrobenzoic acid (109). In a similar kinetic assay, the rather complex substrate p-nitrobenzyloxycarbonylglycyl-(S-4-nitrobenzo-2-oxa-1,3-diazole)cysteinylglycine (NO$_2$Z-Gly-(S-NBD)Cys-Gly) is cleaved by ACE to form first (S-NBD)Cys-Gly, and then the rearrangement product (N-NBD)Cys-Gly, which reacts with a thiol reagent in the assay medium (110). Two internally quenched fluorescent tripeptide derivatives, 2-aminobenzoyl-Gly-Phe(NO$_2$)-Pro (111) and Z(NO$_2$)-Gly-Trp-Gly (112) may be employed for continuous spectrofluorometric assays of ACE activity, but only at relatively low concentrations with partially purified ACE. Probably the most generally applicable kinetic assay for ACE activity employs furanacryloyl-tripeptides such as FA-Phe-Gly-Gly (113), whose cleavage is attended by a large increase in absorbance at 328 nm.

Localization

Skeggs and his colleagues (114) originally discovered ACE in horse plasma. However, the enzyme was little studied until a decade later when Vane and other investigators demonstrated that the biologically important conversion of blood-borne angiotensin I to angiotensin II was catalyzed by ACE on the luminal surface of small blood vessels in lung, and other tissues (115, 116). Despite the enormous importance of these studies, the bioassay methods usually employed detected only net conversion of circulating angiotensin I to angiotensin II, often obscured by other angiotensin-degrading peptidases, and did not detect nonvascular ACE in various tissues.

The first specific and quantitative measurements of levels of ACE activity in various tissues of the rat utilized the fluorometric assay with Z-Phe-His-Leu as substrate (117). Similar results were obtained with 25 different rat tissues using the spectrophotometric assay employing Hip-His-Leu, with each determination of ACE activity tested for specific activation by chloride ion and specific inhibition by a snake venom peptide (118). Highest ACE activities have been found in lung and male reproductive tissues such as testes, moderate activities in gastrointestinal tissues and in kidneys of all species except the rat, and low activities in brain and liver and in the serum of all species except the guinea pig (117–119, 122). Moderate ACE activity has recently been demonstrated in occular tissues of various species (123); and human semen and prostate gland have been shown to have high levels (121, 124, 125). Highest levels of ACE activity in the brain are usually found in the corpus striatum, substantia nigra, and pituitary, with low levels in cortical tissue (96, 126–128). The highly vascularized area postrema and choroid plexus of rats have exceptionally high levels, with the latter region proposed as the source of angiotensin II found in cerebrospinal fluid (127, 128).

The localization of ACE within a number of tissues has been demonstrated more recently by immunofluorescent techniques (129–132). Most of the ACE activity was found on vascular endothelial cells in lung, intestine, kidney, heart, brain, skeletal muscle, adrenal, pancreas, spleen, and placenta. Electron micrographs of lung slices treated with ACE antibody coupled to microperoxidase indicated a specific localization of the enzyme on the plasma membrane and caveolae of endothelial cells facing the vascular lumen (130). In kidney and intestine, ACE activity is also localized in epithelial cells lining the proximal tubule (129, 131, 132) and intestinal lumen (129), respectively.

Further evidence for the primary vascular localization of ACE in brain (133–135), adrenal cortex (136), and retina (137) is the 6- to 33-fold enrichment of the enzyme in microvessels or capillary segments prepared from these tissues. Cultured vascular endothelial cells also have high levels of ACE activity (130, 138–141). That such cells can release into the culture medium up to forty times their initial ACE content (141) supports the widely held view that the vascular endothelium is the source of plasma ACE. The epithelial localization of ACE in kidney and intestine is corroborated by the 8- to 17-fold enrichment of the enzyme in brush border fractions isolated from these tissues (142, 143). Human urinary ACE is presumed to originate in the proximal tubule (144).

ACE activity can be increased at least 400-fold in human monocytes by growth in adherent monolayers or exposure to dexamethasone (145). Interestingly, cells derived from monocytes are the source of diagnostically important increases in serum ACE activity that occur in certain diseases; these cells include macrophages in leprosy, epithelioid granuloma cells in sarcoidosis, and Gaucher cells in Gaucher's disease (see 145).

Purification

Angiotensin-converting enzyme has been purified to homogeneity or near homogeneity from lung, kidney, blood, and reproductive tissues of various species (Table 2). Activity in all cases was followed during purification by chemical assay methods employing tripeptide substrates, usually Hip-His-Leu. ACE from human or rabbit blood was purified to homogeneity only with the aid of specific immunoabsorbant affinity gels (156, 165).

Partial purification of ACE sufficient for study of certain properties has been achieved from lung of guinea pig and hog (166), human (167), and calf (168); from plasma of hog (107, 169, 170), human (171), and horse (172); from brain of rat (173) and human (174); and from human urine (144).

Chemical, Physical, and Enzymatic Properties

ACE from all sources investigated has been found to be a glycoprotein containing 8–32% carbohydrate by weight (Table 3). The enzymes purified from rabbit lung (148), dog lung (152), and rabbit serum (165) all contain large and roughly equivalent amounts of mannose, galactose, and N-acetyl-glucosamine, with lower amounts of fucose. The serum ACE and the ACE released into the medium from cultured vascular endothelial cells (141)

Table 2 Highly purified preparations of angiotensin-converting enzyme

Tissue	Species	Total purification (– fold)	Specific activity (units/mg)	Substrate	References
Lung	Rabbit	740–1,300	60–90	Hip-His-Leu	95, 119, 120, 146–148
	Hog	1,400–1,500	10–14	Hip-Gly-Gly	100, 149, 150
	Rat	330	18	Hip-His-Leu	151
	Dog	1,800	25	Hip-His-Leu	152
	Bovine	2,200	29	Z-Phe-His-Leu	153
		2,300	14	Hip-His-Leu	154
	Guinea pig	740[a]	17	Hip-His-Leu	155
	Baboon	1,800	10	Hip-His-Leu	156
	Human	6,600–34,000[a]	10–75	Hip-His-Leu	157–159
		—	17	Hip-Gly-Gly	160
Kidney	Hog	—	28–31	Hip-Gly-Gly	160–162
	Bovine	1,600	31	Z-Phe-His-Leu	163
	Human	—	12	Hip-Gly-Gly	160
Plasma	Human	101,000	40	Hip-His-Leu	156–164
Serum	Rabbit	60,000	91	Hip-His-Leu	165
Semen	Human	430	0.7	Z-Phe-His-Leu	124
Prostate	Human	420	19	Hip-His-Leu	125

[a] Purification calculated from specific activity of crude extract reported by others.

have an additional high sialic acid content. Rabbit lung ACE has a min—
or oligosaccharide component (1 mol per mole enzyme) composed of 1
Man: 5 Gal:3 Glc:4 GlcNAc:1 GalNAc: 2 Sia, and a major oligosaccharide component ($\sim$ 12 mol per mole enzyme) composed of 1 Fuc: 4
Man: 4 Gal: 4 GlcNAc: 1 Sia. A partial structure has been proposed
for the major oligosaccharide (176).

Estimates of the molecular weight of ACE from different tissues based
on elution from Sephadex G-200 or G-150 are probably all about two times
too high due to the glycoprotein nature of the enzyme (148, 175; Table 3)
Similarly, molecular weights based on sodium dodecylsulfate (SDS) polyacrylamide gel electrophoresis using gels with less than 7–8% acrylamide are
probably 30% too high (162, 175; Table 3). Molecular weights calculated
from sedimentation equilibrium analysis, from SDS polyacrylamide gel
electrophoresis with more highly cross-linked gels (Table 3), or by sucrose
density gradient centrifugation (166, 170, 171) all range between 130,000
and 160,000. Only the hog lung ACE was reported to be composed of
readily dissociable subunits (150), although human lung and plasma enzymes have been shown to generate lower-molecular-weight forms upon

Table 3 Physical and chemical properties of angiotensin-converting enzyme

Source	Mol wt $\times 10^{-3}$	pI	Zinc content (gram-atom/ mole)	Sugar (%)	$A_{280}^{1\%}$	Ref.
Rabbit lung	180[a], 140[b], 129[c]	—	1.2	16–29	19.5	146, 148, 165, 175
Hog lung	206[a], 139[b], 300[d]	4.3	—	—	—	149, 150, 175
Rat lung	139[b], 270[d]	—	—	8	—	151, 175
Dog lung	140[b]	—	0.9	17	22	152
Bovine lung	126[b], 180[d]	4.5	—	—	—	153, 154
Human lung	165[a], 130[b], 290[d]	4.5	—	—	—	157–159
Hog kidney	195[a], 160[b], 300[d]	5.2	—	8	—	160, 162
Bovine kidney	180[d]	4.5	—	—	—	163
Human plasma	140[b]	4.6	—	—	—	156, 164
Rabbit serum	140[b]	—	—	32	—	165
Horse plasma	113[d]	—	0.8	—	—	172
Human semen	330[d]	4.6–5.0	—	—	—	124
Human prostate	290[d]	4.1	—	—	—	125

[a] By SDS gel electrophoresis with 5% acrylamide gels.
[b] By SDS gel electrophoresis with $\geqslant$ 7% acrylamide gels.
[c] By sedimentation equilibrium analysis.
[d] By gel-permeation chromatography on Sephadex G–200 or G–150.

treatment with trypsin-like enzymes (157, 177), and the hog kidney enzyme upon treatment with neuraminidase (162).

As shown in Table 3, ACE purified from various tissues usually has an isoelectric point around 4.5, although this value, like the pH optimum and the electrophoretic or chromatographic mobilities (141, 161, 162) may be influenced by the sialic acid content of the enzyme. The high extinction coefficient of the enzyme at 280 nm (Table 3) reflects its high content of tryptophan and tyrosine (148). The rabbit lung ACE has an N-terminal threonine residue and a C-terminal alanine residue. However, no N-terminal amino acid residue could be detected in the dog lung enzyme (148). Rabbit and dog lung ACEs and horse plasma ACE have all been shown to contain about one gram-atom of zinc per mole (148, 152, 172). Other catalytically important functional residues at the active site of the rabbit lung or horse plasma enzymes appear to include side chains of arginine, tyrosine, lysine, and glutamic or aspartic acids (172, 178), groups quite similar to most found at the active site of carboxypeptidase A.

The active-sites of ACEs from lung, kidney, brain, serum, and seminal plasma of rabbits were immunologically indistinguishable on the basis of anticatalytic activity of goat antibody to the lung enzyme (120). There were, however, immunologically detectable structural differences in the seminal plasma ACE beyond the active-site region as judged by its lower antibody-binding affinity in a radioimmunoassay employing ^{125}I-labeled rabbit lung ACE. Similar results have been obtained with rat and dog ACEs from different tissues (179, 180). However, the cross-reactivity of ACEs from all but the most closely related species is poor (152, 156, 160).

Although it releases dipeptides rather than amino acids, the general specificity of ACE as an exopeptidase is similar to that of carboxypeptidase A (see 90, 116, 181). The enzyme rapidly hydrolyzes substrates as small as N-protected tripeptides, with greatest activity on those with antepenultimate aromatic amino acids, and little or no activity on those with blocked terminal carboxyl, C-terminal dicarboxylic amino acids, or penultimate proline. Recent studies with high concentrations of purified ACE show that the enzyme can slowly hydrolyze tripeptides with free amino groups and those with C-terminal dicarboxylic amino acids, but not those with penultimate proline residues (167, 182, 183). Ester bonds are cleaved at 1–20% of the corresponding rate for cleavage of a scissile peptide bond (184); and substitution of D-amino acids in terminal, penultimate, or antepenultimate positions on a substrate leads to at least a 500-fold slower rate of cleavage (154, 182).

Several recent studies have attempted to systematically assess the contributions to substrate binding (K_m or K_i) and to rate of hydrolysis

(k_{cat}/K_m) of amino acid side chains in the C-terminal, penultimate, and antepenultimate positions of peptide substrates. These side chains are considered to interact with respective "subsites" S'_2, S'_1, and S_1 at the active site of ACE (Figure 1).

Rate of cleavage (k_{cat}/K_m) of the prototype substrate Bz-Gly-His-Leu by bovine lung ACE was enhanced most effectively by substitution of Phe for the antepenultimate Gly, of Ala or Phe for the penultimate His, or of Ala or Phe for the terminal Leu; and was decreased by substitution of Glu, Ile, or Pro for the antepenultimate Gly, of Pro for the penultimate His, or of Glu or Pro for the terminal Leu (154). In reasonable agreement were other studies on relative rates of cleavage of tripeptide substrates that differed by more than a single amino acid residue: for the rabbit lung ACE, Bz-Gly-Ala-Pro > Bz-Gly-Phe-Arg >> Bz-Gly-His-Leu (147), and Bz-Phe-Phe-Ala > Bz-Gly-Gly-Phe > Bz-Gly-Phe-Ala > Bz-Gly-Phe-Phe (184); for the guinea pig urinary ACE, Bz-Phe-Ala-Pro > Bz-Phe-His-Leu > Bz-Gly-His-Leu > Bz-Pro-Phe-Arg > Bz-Gly-Gly-Gly (99); and for the rabbit lung ACE, FA-Phe-Ala-Phe > FA-Phe-Ala-Gly > FA-Phe-Gly-Gly > FA-Leu-Ala-Gly > FA-Phe-His-Leu > FA-Phe-Leu-Gly > FA-Leu-Leu-Gly > FA-Ala-Leu-Gly > FA-Gly-Gly-Phe > FA-Gly-Leu-Gly > FA-Gly-Leu-Ala (113).

Binding (K_m or K_i) of Bz-Gly-His-Leu to bovine lung ACE was most effectively enhanced by substitution of Arg or Phe for the antepenultimate Gly, of Pro or Arg for the penultimate His, and especially substitution of Arg or Pro for the terminal Leu; it was greatly decreased by substitution of Glu for the terminal Leu (154). Other studies on binding of tripeptide substrates were reasonably consistent: for the rabbit lung ACE, Bz-Gly-Ala-Pro >> Bz-Gly-Phe-Arg > Bz-Gly-His-Leu (147); for the guinea pig urinary ACE, Bz-Phe-Ala-Pro = Bz-Phe-His-Leu >> Bz-Gly-His-Leu > Bz-Pro-Phe-Arg >> Bz-Gly-Gly-Gly (99). However, studies with dipeptide product inhibitors (147) suggested that binding of peptides to ACE would be enhanced most effectively by Val, Ile, or Arg in the penultimate position, and by Trp, Tyr, Pro, or Phe in the terminal position; but that it would be decreased rather than enhanced by a penultimate Pro. The C-terminal sequence Phe-Ala-Pro (6 in Figure 4), analogous to that found naturally in a snake venom peptide inhibitor (see below), is consistently one of the most favorable for binding to subsites S_1, S'_1, and S'_2 at the active site of ACE.

Effects of anions on the activity of ACE are complex, often depending on the substrate, pH, or perhaps even the buffer employed. At alkaline pH, chloride ion decreased the K_m values for Hip-Ala-Pro (147), Hip-Gly-Gly (106), Hip-His-Leu (99, 154), and bradykinin (185); k_{cat} values were increased 2- to 20-fold in most cases (chloride ion activation), but decreased

20-fold for Hip-Ala-Pro (147). ACEs from hog lung and kidney and from human plasma acting on Hip-Gly-Gly or $Z(NO_2)$-Gly-Trp-Gly (106, 112) have been reported to be inhibited by phosphate ion. Hog lung and guinea pig urinary enzymes acting, respectively, on Hip-Gly-Gly and low concentrations of labelled Hip-His-Leu in HEPES buffer were specifically activated by chloride ion, but underwent a further nonspecific activation by increased ionic strength achieved with 1 M NaCl or 0.6 M Na_2SO_4 (99, 106). In the latter case this was shown to be due to decrease in the K_m value for Hip-His-Leu from 3.6 mM to 0.17 mM. Since phosphate inhibition and sodium sulfate activation have always been observed at nonsaturating substrate concentrations, it is possible that these effects, not observed by all investigators, are both due to changes in substrate affinity for ACE. The pH optima for cleavage of Hip-His-Leu, Hip-Phe-Arg, and Hip-Ala-Pro by rabbit lung ACE were all 7.8 in the absence of chloride ion, and shifted to optima of 8.3, 7.5–8.5, and 6.9, respectively, in the presence of the ion (147). Two pH optima at 6.0 and 8.0 were observed for hydrolysis of Hip-Gly-Gly by hog lung ACE at moderate concentrations of chloride ion (106); at high ionic strength, activity at pH 6 was inhibited, and that at pH 8 was increased, to yield a single optimum pH.

Inhibitors

Potent and highly specific inhibitors of ACE have begun to proliferate during the last decade (see 116, 181, 186). These inhibitors fall into two major classes: snake venom oligopeptides and their analogues; and nonpeptidic inhibitors, related in structure to the optimal terminal amino acid sequences of the venom peptides, but designed for multiple effective interactions with ACE, particularly with its important zinc ion.

The most studied venom peptide inhibitors are SQ 20,881 (<Glu-Trp-Pro-Arg-Pro-Gln-Ile-Pro-Pro) and bradykinin-potentiating peptide 5_a or BPP_{5a} (<Glu-Lys-Trp-Ala-Pro). These ACE inhibitors inhibit biological actions of angiotensin I and augment those of bradykinin in vitro and in vivo (181, 186). Structure-activity studies (181, 186) with analogues of SQ 20, 881, and BPP_{5a} indicate that these venom peptides compete with biologically important ACE substrates such as angiotensin I and bradykinin by interaction of their C-terminal tripeptide residues with subsites S_1, S'_1, and S'_2 at the active-site of the enzyme (6 in Figure 4). The optimal C-terminal sequence for binding of these inhibitors to ACE was Trp-Ala-Pro, as found in BPP_{5a}, or the analogous Phe-Ala-Pro (6 in Figure 4). However, the other more distal amino acid residues of SQ 20,881 and BPP_{5a} also contribute to their overall high affinity for ACE by binding to subsites not normally occupied by substrates (181, 186). Replacement of any one of the four proline residues of SQ 20,881 by Δ^3-dehydroproline produced a 20- to

100-fold increase in inhibitory activity (187). SQ 20,881 has a much longer duration of action in vivo than BPP_{5a} (an ACE substrate), and is not significantly metabolized prior to elimination by the kidney (181, 186, 188). Only its lack of oral activity limits its utility as an antihypertensive drug.

The development of a hypothetical working model of the active site of ACE based on the known active site of the similar exopeptidase, carboxypeptidase A, and the use of this model to guide the design of potent inhibitors that led to the orally active antihypertensive drug captopril (7 in Figure 4) have been throughly reviewed in recent years (181, 186, 193). Captopril evolved from a series of dicarboxylic "biproduct analogue inhibitors" designed to bind to the active site of ACE in a manner similar to the binding of the potent competitive inhibitor benzylsuccinic acid to carboxypeptidase A. The terminal amino acid carboxyl of such biproduct analogues was proposed to bind ionically to the positively charged recognition site on the exopeptidase, and the second carboxyl to interact strongly with the catalytically functional zinc ion (189, 190). Such specific proposals for active-site binding interactions allowed thorough testing and refinement of the working model through extensive structure-activity correlations (181, 186, 189, 190), examples of which continue to be reported (191, 192). Replacement of the putative zinc-binding carboxyl group by the more effective sulfhydryl ligand led to captopril (7 in Figure 4), a potent competitive inhibitor of ACE ($K_i = 1.7$ nM) that may be considered a biologically stable analogue of the optimal terminal dipeptide Ala-Pro of substrates or inhibitors, but which also incorporates a strong zinc ligand.

Several recently reported ACE inhibitors are close analogues of captopril. The most thoroughly studied of these, YS-980, substitutes thioproline (4-thiazolidinecarboxylic acid) for the proline of captopril. It is about one third as potent as captopril in vitro ($K_i = 6$ nM), and has similar ACE inhibitory and antihypertensive activity in animals (194, 195). Cyclized lactam analogues of mercaptopropionyl derivatives of glycine (11 in Scheme 3) or other amino acids were all at least 100-times less inhibitory than captopril in vitro or in vivo (196).

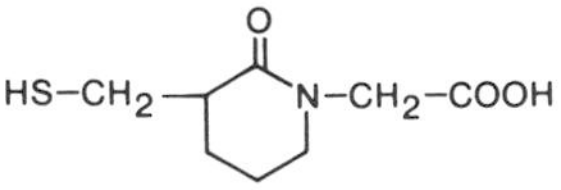

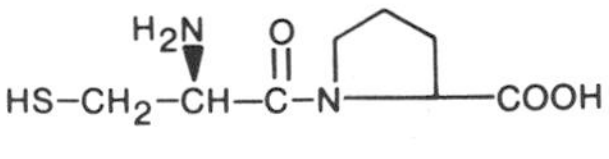

Scheme 3

D-Cysteinyl-L-proline (12 in Scheme 3) which substitutes an amino group for the methyl substituent of captopril, was only 2–3 times less potent than captopril in vitro as an ACE inhibitor (197).

Several newly described inhibitors (8, 9, and 10, Figure 4) bind to the active-site of ACE with somewhat different interactions that are, nevertheless, well within the scope of the hypothetical active-site model used to design captopril (198–201). The close structural relationship between each of these inhibitors and the terminal amino acid sequences of substrates, as well as the improbability of the existence of highly potent noncompetitive inhibitors, suggest that reports (198–200) of noncompetitive inhibition by such compounds under certain conditions are due to a variety of technical problems encountered in the kinetic analysis of tightly binding competitive inhibitors.

N-Phosphoryl-Ala-Pro (8 in Figure 4) is another dipeptide analogue with a strong zinc ligand, the phosphoryl oxygen, that is at least one fourth as potent as captopril in vitro (198). Unlike captopril, this compound maintains an amide nitrogen analogous to that of the scissile peptide bond of a substrate; this nitrogen probably contributes to the overall binding of the inhibitor to the active site of ACE by accepting a hydrogen bond from the catalytically important tyrosine residue thought to be present at the active site (X-H, Figure 4). Substituted O-phenylphosphoryl dipeptides and mercaptoacetyl dipeptides (192) probably bind to ACE in a similar manner, with perhaps both the amide carbonyl and sulfhydryl groups of the mercaptoacetyl compounds interacting with the zinc ion of the enzyme.

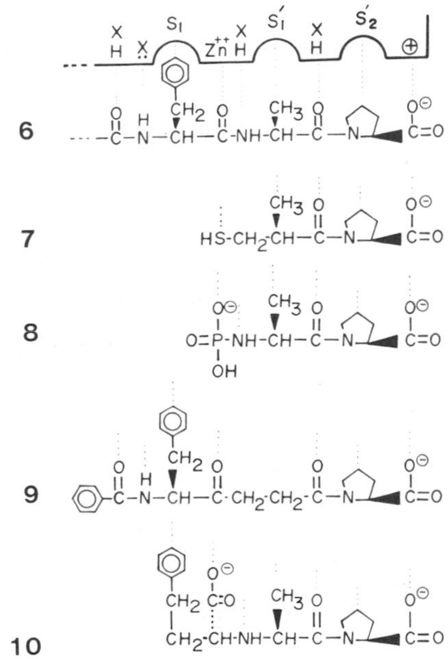

Figure 4 Schematic diagram of proposed binding of a peptide substrate or competitive inhibitor (6) and specific nonpeptide inhibitors (7, 8, 9, and 10) to the active site of angiotensin-converting enzyme.

Compounds 9 and 10 (Figure 4) are both analogues of favorable C-terminal tripeptide sequences of peptide substrates or competitive inhibitors. Compound 9 is an analogue of Bz-Phe-Gly-Pro in which the scissile Phe-Gly peptide bond has been replaced by the nonscissile ketomethylene function. Thus, 9 is a tripeptide analogue with a relatively weak zinc ligand. It is about half as potent as captopril as an ACE inhibitor in vitro (199, 200); but, like N-phosphoryl-Ala-Pro, it does not appear to be very potent in vivo. Much of the inhibitory activity of 9 is probably due to the interaction of its benzyl function with subsite S_1 of ACE, in the same manner as an antepenultimate aromatic amino acid of a substrate. Compound 10 is the most potent of a new series of ACE inhibitors that may be considered to be tripeptide analogues with moderately strong zinc ligands, the carboxyl group (Figure 4). Compound 10 is reported to be approximately ten times more potent than captopril (201), and appears to have most of the active site–binding interactions described for Compounds 6 to 9 in Figure 4, although its developers did not discuss such specific interactions. The ethyl ester of 10 (MK-421) has been shown to be an orally effective ACE inhibitor and an effective antihypertensive agent in animals and man (202, 203).

Literature Cited

1. Peach, M. J. 1977. *Physiol. Rev.* 57:313–55
2. Colman, R. W., Schmaier, A. H., Wong, P. Y. 1981. *Biochemical Regulation of Blood Pressure*, ed. R. L. Soffer, pp. 321–55. New York: Wiley. 456 pp.
3. Khairallah, P. A., Hall, M. M. 1977. *Hypertension*, ed. J. Genest, E. Koiw, O. Kuchel, pp. 179–83. New York: McGraw. 1208 pp.
4. Laragh, J. H. 1981. See Ref. 2, pp. 393–410
5. Fasciolo, J. C. 1977. See Ref. 3, pp. 134–39
6. Inagami, T. 1981. See Ref. 2, pp. 39–71
7. Tewksbury, D. 1981. See Ref. 2, pp. 95–122
8. Skeggs, L. T., Dorer, F. E., Levine, M., Lentz, K. E., Kahn, J. R. 1980. *The Renin-Angiotensin System*, ed. J. A. Johnson, R. R. Anderson, pp. 1–27. New York: Plenum. 307 pp.
9. Gould, A. B., Skeggs, L. T., Kahn, J. R. 1966. *Lab. Invest.* 15:1802–13
10. Gould, A. B., Goodman, S., DeWolf, R., Onesti, G., Swartz, C. 1979. *Anal. Biochem.* 94:125–39
11. Haber, E., Koerner, T., Page, L. B., Kleman, B., Purnode, A. L. 1969. *J. Clin. Endocr.* 29:1349–55
12. Freedlender, A. E., Goodfriend, T. L. 1979. *Methods of Hormone Radioim-*munoassay, ed. B. M. Jaffe, H. R. Behrman, pp. 889–907. New York: Academic. 1040 pp. 2nd ed.
13. Campbell, D. J., Skinner, S. L. 1975. *Clin. Chim. Acta* 65:361–70
14. Lentz, K. E., Skeggs, L. T., Dorer, F. E., Kahn, J. R., Levine, M. 1976. *Anal. Biochem.* 74:1–11
15. Mendelsohn, F. A., Johnston, C. I. 1971. *Biochem. J.* 121:241–44
16. Ontjes, D. A., Majstoravich, J. Jr., Roberts, J. C. 1972. *Anal. Biochem.* 45:374–86
17. Bath, N. M., Gregerman, R. I. 1972. *Biochemistry* 11:2845–53
18. Levine, M., Dorer, F. E., Kahn, J. R., Lentz, K. E., Skeggs, L. T. 1970. *Anal. Biochem.* 34:366–75
19. Galen, F. X., Devaux, C., Grogg, P., Menard, J., Corvol, P. 1978. *Biochim. Biophys. Acta* 523:485–93
20. Reinharz, A., Roth, M. 1969. *Eur. J. Biochem.* 7:334–39
21. Murakami, K., Ohsawa, T., Hirose, S., Takada, K., Sakakibara, S. 1981. *Anal. Biochem.* 110:232–39
22. Hirose, S., Yokosawa, H., Inagami, T. 1978. *Nature* 274:392–93
23. Inagami, T., Murakami, K. 1977. *J. Biol. Chem.* 252:2978–83
24. Corvol, P., Devaux, C., Ito, T., Sicard,

P., Ducloux, J., Menard, J. 1977. *Circ. Res.* 41:616–22
25. Hirose, S., Workman, R. J., Inagami, T. 1979. *Circ. Res.* 45:275–81
26. Yokosawa, H., Holladay, L. A., Inagami, T., Haas, E., Murakami, K. 1980. *J. Biol. Chem.* 255:3498–502
27. Dworshack, R. T., Gregory, T. J., Printz, M. P. 1978. *Fed. Proc.* 37:1395
28. Dzau, V. J., Brenner, A., Emmett, N., Haber, E. 1980. *Clin. Sci.* 59:45s–75s
29. Boyd, G. W. 1974. *Circ. Res.* 35:426–38
30. Ikemoto, F., Kawamura, M., Takaori, K., Yamamoto, K. 1980. *Jpn. Circ. J.* 44:371–74
31. Nakane, H., Nakane, Y., Misumi, J., Sarieta, T., Corvol, P., Menard, J. 1980. *Clin. Sci.* 59:37s–40s
32. Morimoto, J., Matsunaga, M., Hara, A., Pak, C. H., Nagai, H., Hamada, H., Kawai, C. 1980. *Jpn. Heart J.* 21: 205–13
33. Yamamoto, K., Ikemoto, F., Kawamura, M., Takaori, K. 1980. *Clin. Sci.* 59:25s–27s
34. Kawamura, M., Ikemoto, F., Yamamoto, K. 1980. *Clin Sci.* 58:451–56
35. Murakami, K., Takahashi, S., Suzuki, F., Hirose, S., Inagami, T. 1980. *Biomed. Res.* 1:392–99
36. Murakami, K., Chino, S., Hirose, S., Higaki, J. 1980. *Biomed. Res.* 1:476–81
37. Inagami, T., Hirose, S., Murakami, K., Matoba, T. 1977. *J. Biol. Chem.* 252:7733–37
38. Hirose, S., Yokosawa, H., Inagami, T., Workman, R. J. 1980. *Brain Res.* 191:489–99
39. Menard, J., Galen, F. X., Devaux, C., Kopp, N., Auzan, C., Corvol, P. 1980. *Clin. Sci.* 59:41s–44s
40. Yosokawa, N., Takahashi, N., Inagami, T., Page, D. 1979. *Biochim. Biophys. Acta* 569:211–19
41. Atlas, S. A., Laragh, J. H., Sealey, J. E., Hesson, T. E. 1980. *Clin. Sci.* 59:-29s–33s
42. Murakami, K., Takahashi, S., Hirose, S., Takii, Y., Inagami, T. 1980. *Clin. Sci.* 59:21s–24s
43. Takii, Y., Inagami, T. 1980. *Biochim. Biophys. Acta* 94:182–88
44. Inagami, T., Hirose, S. 1979. *Fed. Proc.* 36:636
45. Johnson, R. L., Roisner, A. M., Crist, R. D. 1979. *Biochem. Pharmacol.* 28: 1791–99
46. Hirose, S., Naruse, M., Ohtsuki, K., Inagami, T. 1981. *J. Biol. Chem.* 256:5572–76

47. Atlas, S. A., Sealey, J. E., Laragh, J. H. 1979. *Kidney Int.* 16:791
48. Takii, Y., Takahashi, N., Inagami, T., Yokosawa, N. 1980. *Life Sci.* 26: 347–53
49. Leckie, B. J. 1981. *Clin. Sci.* 60:119–30
50. Atlas, S. A., Laragh, J. H., Sealey, J. E. 1978. *Clin. Sci. Mol. Med.* 55:135s–38s
51. Sealey, J. E. 1980. *Enzymatic Release of Vasoactive Peptides,* ed. F. Gross, G. Vogel, pp. 117–36. New York: Raven
52. Derkx, F. H. M., Bouma, B. N., Tan-Tjiong, H. L., Man in't Veld, A. J., de-Bruyn, J. H. B., Wenting, G. J., Schalekamp, M. A. D. H. 1980. *Clin. Exp. Hypert.* 2(3–4):575–92
53. Leckie, B. J., McGhee, W. K. 1980. *Nature* 288:702–5
54. Hsueh, W. A., Carlson, E. J., O'Connor, D., Warren, S. 1980. *J. Clin. Endocrinol. Metabol.* 51:942–44
55. Inagami, T., Murakami, K., Misono, K., Workman, R. J., Cohen, S., Sketa, Y. 1977. *Acid Proteases, Structure, Function, and Biology,* ed. J. Tang, pp. 225–48. New York: Plenum. 355 pp.
56. Haber, E., Burton, J. 1979. *Fed. Proc.* 38:2768–73
57. Tang, J. 1979. *Molec. Cell Biochem.* 26:93–109
58. Poulsen, D., Haber, E., Burton, J. 1976. *Biochim. Biophys. Acta* 452:533–37
59. Burton, J., Cody, R. J. Jr., Herd, J. A., Haber, E. 1980. *Proc. Natl. Acad. Sci. U.S.A.* 77:5476–79
60. Aoyagi, T., Kunimoto, S., Morishima, H., Takeuchi, T., Umezawa, H. 1971. *J. Antibiot.* 24:687–94
61. Aoyagi, T., Morishima, H., Nishizawa, R., Kunimoto, S., Takeuchi, T., Umezawa, H. 1972. *J. Antibiot.* 25:689–94
62. Gross, F., Lazar, J., Orth, H. 1972. *Science* 175:656
63. Miller, R. P., Poper, C. J., Wilson, C. W., DeVito, E. 1972. *Biochem. Pharmacol.* 21:2941–44
64. Orth, H., Hackenthal, E., Lazar, J., Miksche, V., Gross, F. 1974. *Circ. Res.* 35:52–55
65. Miyazaki, M., Komori, T., Okunishi, H., Toda, H. 1979. *Jpn. Circ. J.* 43:818–23
66. Guyene, T. T., Devaux, C., Menard, J., Corvol, P. 1976. *J. Clin. Endocrinol. Metab.* 43:1301–6
67. McKown, M. M., Workman, R. J., Gregerman, R. I. 1974. *J. Biol. Chem.* 249:7770–74
68. Johnson, R. L., Poisner, A. M. 1977. *Biochem. Pharmacol.* 26:639–41

69. Rich, D. H., Sun, E., Sengh, J. 1977. *Biochem. Biophys. Res. Commun.* 74:762–67
70. Rich, D. H., Sun, E. T. O., Ulm, E. 1980. *J. Med. Chem.* 23:27–33
71. Schölkens, B. A., Jung, W. 1974. *Arch. Int. Pharmacodyn.* 208:24–34
72. Gardes, J., Evin, G., Castro, B., Corvol, P., Menard, J. 1980. *J. Cardiovasc. Pharmacol.,* 2:687–98
73. Miyazaki, M., Okunishi, H., Komori, T., Yamamoto, K. 1978. *Jpn. J. Pharmacol.* 28:171–74
74. Ondetti, M. A., Cushman, D. W. 1978. *Ann. Rep. Med. Chem.* 13:82–91
75. Hosoki, K., Miyazaki, M., Yamamoto, K. 1977. *J. Pharmacol. Exp. Therap.* 203:485–92
76. Dzau, V. J., Kopelman, R. I., Barger, A. C., Haber, E. 1980. *Science* 207:1091–93
77. Haber, E. 1980. *Clin. Sci.* 59:7s–19s
78. Misono, K., Halladay, L. A., Cohen, S., Inagami, T. 1978. *Fed. Proc.* 36:1436
79. Dzau, V. J., Slater, E. E., Haber, E. 1978. *Circulation* 58 (Suppl. II): 249
80. Hackenthal, E., Hackenthal, R., Hilgenfeldt, U. 1978. *Biochim. Biophys. Acta* 522:574–88
81. Matoba, T., Murakami, K., Inagami, T. 1978. *Biochim. Biophys. Acta* 526: 560–71
82. Gross, D. M., Barajas, L. 1978. *J. Med.* 9:53–66
83. Galen, F. X., Devaux, C., Guyenne, T., Menard, J., Corvol, P. 1979. *J. Biol. Chem.* 254:4848–55
84. Slater, E. E., Cohn, R. C., Dzau, V. J., Haber, E. 1978. *Clin. Sci. Mol. Med.* 55:117s–19s
85. Hirose, S., Yokosawa, H., Inagami, T., Workman, R. J. 1980. *Brain Res.* 191:489–99
86. Misono, K. S., Inagami, T. 1980. *Biochemistry* 19:2616–22
87. Cohen, S., Taylor, J. M., Murakami, K., Michelakis, A. M., Inagami, T. 1972. *Biochemistry* 11:4286–93
88. Conio, G., Ghiani, P., Patrone, E., Trefiletti, V., Uva, B., Vallarino, M. 1980. *Biochim. Biophys. Acta* 623:317–28
89. Turcotte, J. G., Yu, C., Lee, H., Pavanaram, S. K. 1975. *J. Med. Chem.* 18:1184–90
90. Bakhle, Y. S. 1974. In *Angiotensin,* ed. I. H. Page, F. M. Bumpus, pp. 41–80. New York: Springer. 591 pp.
91. Spadaro, A. C., Martins, A. R., Berti, J. D., Greene, L. J., 1978. *Anal. Biochem.* 91:410–20
92. Chiu, A. T., Ryan, J. W., Ryan, U. S., Dorer, F. E. 1975. *Biochem. J.* 149:297–300
93. Piquilloud, Y., Reinharz, A., Roth, M. 1970. *Biochim. Biophys. Acta* 206: 136–42
94. Cushman, D. W., Cheung, H. S. 1971. *Biochem. Pharmacol.* 20:1637–48
95. Cheung, H. S., Cushman, D. W. 1973. *Biochim. Biophys. Acta* 293:451–63
96. Yang, H. Y. T., Neff, N. H. 1972. *J. Neurochem.* 19:2443–50
97. Rohrbach, M. S. 1978. *Anal. Biochem.* 84:272–76
98. Ryan, J. W., Chung, A., Ammons, C., Carlton, M. L. 1977. *Biochem. J.* 167:501–4
99. Ryan, J. W., Chung, A., Ryan, U. S. 1980. *Env. Health Perspectives* 35: 165–70
100. Igic, R., Erdös, E. G., Yeh, H. S. J., Sorrells, K., Nakajima, T. 1972. *Circ. Res.* 30:II–51–II–61
101. Chiknas, S. G. 1979. *Clin. Chim. Acta* 25:1259–62
102. Hayakari, M., Kondo, Y., Izumi, H. 1978. *Anal. Biochem.* 84:361–96
103. Filipovic, N., Mijanović, M., Igic, R. 1978. *Clin. Chim. Acta* 88:173–75
104. Conroy, J. M., Lai, C. Y. 1978. *Anal. Biochem.* 87:556–61
105. Hayakari, M., Kondo, Y. 1977. *Tohoku J. Exp. Med.* 122:313–20
106. Dorer, F. E., Kahn, J. R., Lentz, K. E., Levine, M., Skeggs, L. T. 1976. *Biochim. Biophys. Acta* 429:220–28
107. Yang, H. Y. T., Erdös, E. G., Levin, Y. 1971. *J. Pharmacol. Exp. Therap.* 177:291–300
108. Stevens, R. L., Micalizzi, E. R., Fessler, D. C., Pals, D. T. 1972. *Biochemistry* 11:2999–3007
109. Lee, H. J. 1975. *J. Korean Chem. Soc.* 19:246–51
110. Russo, S. F., Persson, A. V., Wilson, I. B. 1978. *Clin. Chem.* 24:1539–42
111. Carmel, A., Yaron, A. 1978. *Eur. J. Biochem.* 87:265–73
112. Persson, A., Wilson, I. B. 1977. *Anal. Biochem.* 83:296–303
113. Holmquist, B., Bünning, P., Riordan, J. 1979. *Anal. Biochem.* 95:540–48
114. Skeggs, L. T., Kahn, J. R., Shumway, N. P. 1956. *J. Exp. Med.* 103:295–99
115. Vane, J. R. 1974. See Ref. 90, pp. 16–40
116. Soffer, R. L. 1976. *Ann. Rev. Biochem.* 45:73–94
117. Roth, M., Weitzman, A. F., Piquilloud, Y. 1969. *Experientia* 25:1247
118. Cushman, D. W., Cheung, H. S. 1971. *Biochim. Biophys. Acta* 250:261–65

119. Cushman, D. W., Cheung, H. S. 1972. In *Hypertension '72*, ed. J. Genest, E. Koiw, pp. 532–41. Berlin: Springer. 617 pp.
120. Das, M., Soffer, R. L. 1976. *Biochemistry* 15:5088–94
121. Depierre, D., Roth, M. 1972. *Experientia* 28:154–55
122. Yang, H. Y. T., Erdös, E. G., Levin, Y. 1971. *J. Pharmacol. Exp. Therap.* 177:291–300
123. Igic, R., Kojović, V. 1980. *Exp. Eye Res.* 30:299–303
124. Depierre, D., Bargetzi, J. P., Roth, M. 1978. *Biochim. Biophys. Acta* 523:469–76
125. Yokoyama, M., Hiwada, K., Kokubo, T., Takaha, M., Takeuchi, M. 1980. *Clin. Chim. Acta* 100:253–58
126. Poth, M. M., Heath, R. G., Ward, M. W. 1975. *J. Neurochem.* 25:83–5
127. Arregui, A., Iversen, L. L. 1978. *Eur. J. Pharmacol.* 52:147–50
128. Arregui, A., Bennett, J. P., Bird, E. D., Yamamura, H. I., Iversen, L. L., Snyder, S. H. 1977. *Ann. Neurol.* 2:294–98
129. Wigger, H. J., Stalcup, S. A. 1978. *Lab. Invest.* 38:581–85
130. Ryan, U. S., Ryan, J. W., Whitaker, C., Chiu, A. 1976. *Tissue & Cell* 8:125–45
131. Caldwell, P. R. B., Seegal, B. C., Hsu, K. C., Das, M., Soffer, R. L. 1976. *Science* 191:1050–51
132. Hall, E. R., Kato, J., Erdös, E. G., Robinson, C. J. G., Oshima, G. 1976. *Life Sci.* 18:1299–1304
133. Gimbrone, M. A. Jr., Majeau, G. R., Atkinson, W. J., Sadler, W., Cruise, S. A. 1977. *Life Sci.* 25:1075–84
134. Brecher, P., Tercyak, A., Gavras, H., Chobanian, A. V. 1978. *Biochim. Biophys. Acta* 526:537–46
135. Brecher, P., Tercyak, A., Chobanian, A. V. 1981. *Hypertension* 3:198–204
136. Del Vecchio, P. J., Ryan, J. W., Chung, A., Ryan, U. S. 1980. *Biochem. J.* 186:605–8
137. Ward, P. E., Stewart, T. A., Hammon, K. J., Reynolds, R. C., Igic, R. P. 1979. *Life Sci.* 24:1419–24
138. Ody, C., Junod, A. F. 1977. *Am. J. Physiol.* 232:C95–C98
139. Johnson, A. R., Schulz, W. W., Noguiera, L. A., Erdös, E. G. 1980. *Clin. Exp. Hypert.* 2:659–74
140. Ryan, U. S., Ryan, J. W. 1980. *Env. Health Perspect.* 35:171–80
141. Ching, S. F., Hayes, L. W., Slakey, L. L. 1981. *Biochim. Biophys. Acta* 657:222–31
142. Ward, P. E., Schultz, W., Reynolds, R.

C., Erdös, E. G. 1977. *Lab. Invest.* 36:599–606
143. Ward, P. E., Sheridan, M. A., Hammon, K. J., Erdös, E. G. 1980. *Biochem. Pharmacol.* 29:1525–29
144. Kokubu, T., Kato, I., Nishimura, K., Hiwada, K., Ueda, E. 1978. *Clin. Chim. Acta* 89:375–79
145. Friedland, J., Setton, C., Silverstein, E. 1978. *Biochem. Biophys. Res. Commun.* 83:843–49
146. Tsai, B. S., Peach, M. J. 1977. *J. Biol. Chem.* 252:4674–81
147. Cheung, H. S., Wang, F. L., Ondetti, M. A., Sabo, E. F., Cushman, D. W. 1980. *J. Biol. Chem.* 255:401–7
148. Das, M., Soffer, R. L. 1975. *J. Biol. Chem.* 250:6762–68
149. Dorer, F. E., Kahn, J. R., Lentz, K. E., Levine, M., Skeggs, L. T. 1972. *Circ. Res.* 31:356–66
150. Nakajima, T., Oshima, G., Yeh, H. S. J., Igic, R., Erdös, E. G. 1973. *Biochim. Biophys. Acta* 315:430–38
151. Lanzillo, J. J., Fanburg, B. L. 1974. *J. Biol. Chem.* 249:2312–18
152. Conroy, J. M., Hartley, J. L., Soffer, R. L. 1978. *Biochim. Biophys. Acta* 524:403–12
153. Elisseeva, Y. E., Orekhovich, V. N., Pavlikhina, L. V. 1976. *Biokhimiya* 41:506–12
154. Rohrbach, M. S., Williams, E. B. Jr., Rolstad, R. A. 1981. *J. Biol. Chem.* 256:225–30
155. Lanzillo, J. J., Fanburg, B. L. 1976. *Biochim. Biophys. Acta* 445:161–68
156. Lanzillo, J. J., Polsky-Cynkin, R., Fanburg, B. L. 1980. *Anal. Biochem.* 103:400–7
157. Nishimura, K., Yoshida, N., Hiwada, K., Ueda, E., Kokubu, T. 1978. *Biochim. Biophys. Acta* 522:229–37
158. Friedland, J., Silverstein, E., Drooker, M., Setton, C. 1981. *J. Clin. Invest.* 67:1151–60
159. Grönhagen-Riska, C., Fyhrquist, F. 1980. *Scand. J. Clin. Lab. Invest.* 40:711–19
160. Oshima, G., Gecse, A., Erdös, E. G. 1974. *Biochim. Biophys. Acta* 350:26–37
161. Oshima, G., Nagasawa, K., Kato, J. 1976. *J. Biochem.* 80:477–83
162. Nagamatsu, A., Inokuchi, J. I., Soeda, S. 1980. *Chem. Pharm. Bull.* 28:459–64
163. Elisseeva, Y. E., Pavlikhina, L. V., Orekhovich, V. N. 1974. *Dokl. Acad. Nauk SSR* 217:953–56
164. Lanzillo, J. J., Fanburg, B. L. 1977. *Biochemistry* 16:5491–95
165. Das, M., Hartley, J. L., Soffer, R. L. 1977. *J. Biol. Chem.* 252:1316–19

166. Lee, H. J., Larue, J. N., Wilson, I. B. 1971. *Biochim. Biophys. Acta* 250: 549–57
167. Beckner, C. F., Caprioli, R. M. 1980. *Biochem. Biophys. Res. Commun.* 93:1290–96
168. Stevens, R. L., Micalizzi, E. R., Fessler, D. C., Pals, D. T. 1972. *Biochemistry* 11:2999–3007
169. Dorer, F. E., Skeggs, L. J., Kahn, J. R., Lentz, K. E., Levine, M. 1970. *Anal. Biochem.* 33:102–13
170. Lee, H. J., Larue, J. N., Wilson, I. B. 1971. *Biochim. Biophys. Acta* 235: 521–28
171. Lee, H. J., Larue, J. N., Wilson, I. B. 1971. *Arch. Biochem. Biophys.* 142: 548–51
172. Fernley, R. J. 1977. *Clin. Exp. Pharmacol. Physiol.* 4:267–81
173. Benuck, M., Marks, N. 1978. *Neurochem.* 30:729–34
174. Arregui, A., Iversen, L. L. 1979. *Biochem. Pharmacol.* 28:2693–96
175. Lanzillo, J. J., Fanburg, B. L. 1976. *Biochim. Biophys. Acta* 439:125–32
176. Hartley, J. L., Soffer, R. L. 1978. *Biochem. Biophys. Res. Commun.* 83: 1545–52
177. Nakahara, M. 1978. *Biochem. Pharmacol.* 27:1651–57
178. Bünning, P., Holmquist, B., Riordan, J. F. 1978. *Biochem. Biophys. Res. Commun.* 83:1442–49
179. Polsky-Cynkin, R., Fanburg, B. L. 1979. *Int. J. Biochem.* 10:669–74
180. Odya, C. E., Hall, E. R., Robinson, C. J. G. 1979. *Biochem. Biophys. Res. Commun.* 86:508–13
181. Cushman, D. W., Ondetti, M. A. 1980. *Prog. Med. Chem.* 17:41–104
182. Oshima, G., Nagasawa, K. 1979. *J. Biochem.* 86:1719–24
183. Krutzch, H. C. 1980. *Biochemistry* 19:5290–96
184. Keung, W. M., Holmquist, B., Riordan, J. P. 1980. *Biochem. Biophys. Res. Commun.* 96:506–13
185. Dorer, F. E., Kahn, J. R., Lentz, K. E., Levine, M., Skeggs, L. T. 1974. *Circ. Res.* 34:824–27
186. Ondetti, M. A., Cushman, D. W. 1981. See Ref. 2, pp. 165–204
187. Fisher, G. H., Ryan, J. W. 1979. *FEBS Lett.* 107:273–76
188. Martin, L. C., Ryan, J. W., Fisher, G. H., Chung, A., Epstein, M., Stewart, J. M. 1979. *Biochem. J.* 184:713–16
189. Ondetti, M. A., Rubin, B., Cushman, D. W. 1977. *Science* 196:441–44
190. Cushman, D. W., Cheung, H. S., Sabo, E. F., Ondetti, M. A. 1977. *Biochemistry* 16:5484–91
191. Oya, M., Matsumoto, J., Takashina, H., Iwao, J. I., Funae, Y. 1981. *Chem. Pharm. Bull.* 29:63–70
192. Holmquist, B., Vallee, B. L. 1979. *Proc. Natl. Acad. Sci.* 76:6216–20
193. Rubin, B., Antonaccio, M. J. 1980. In *Pharmacology of Antihypertensive Drugs*, ed. A. Scriabine, pp. 21–42. New York: Raven. 462 pp.
194. Funae, Y., Komori, T., Sasaka, D., Yamamoto, K. 1980. *Biochem. Pharmacol.* 29:1543–47
195. Unger, T., Rockhold, R. W., Bönner, G., Rascher, W., Schaz, K., Speck, G., Schömig, A., Ganten, D. 1981. *Clin. Exp. Hyp.* 3:121–40
196. Klutchko, S., Hoefle, M. L., Smith, R. D., Essenburg, A. D., Parker, R. B., Nemeth, V. L., Ryan, M. J., Dugan, D. H., Kaplan, H. R. 1981. *J. Med. Chem.* 24:104–9
197. Harris, R. B., Ohlsson, J. T., Wilson, I. B. 1981. *Arch. Biochem. Biophys.* 206:105–12
198. Galardy, R. E. 1980. *Biochem. Biophys. Res. Commun.* 97:94–99
199. Almquist, R. G., Chao, W. R., Ellis, M. E., Johnson, H. L. 1980. *J. Med. Chem.* 23:1392–98
200. Weare, J. A., Stewart, T. A., Gafford, J. T., Erdös, E. G. 1981. *Hypertension* 3:150–153
201. Patchett, A. A. et al. 1980. *Nature* 288:280–3
202. Gross, D. M., Sweet, C. S., Ulm, E. H., Backlund, E. P., Morris, A. A., Weltz, D., Bohn, D. L., Wenger, H. C., Vassil, T. C., Stone, C. A. 1981. *J. Pharmacol. Exp. Therap.* 216:552–57
203. Sweet, C. S., Gross D. M., Arbegast, P. T., Gaul, S. L., Britt, P. M., Ludden, C. T., Weitz, D., Stone, C. A. 1981. *J. Pharmacol. Exp. Therap.* 216:558–66

Ann. Rev. Biochem. 1982. 51:309–33
Copyright © 1982 by Annual Reviews Inc. All rights reserved

PHYTOTOXINS

Gary A. Strobel

Department of Plant Pathology, Montana State University, Bozeman,
Montana 59717

CONTENTS

Perspectives

Worldwide, hunger and malnutrition are two serious problems facing mankind. These conditions result, in part, from man's inability to grow, sustain, and distribute plant life. In addition to insect and weed pests, another major cause of the destruction of plants used for food, fiber, and recreation is disease. Diseases of plants are caused by many of the same classes of agents responsible for diseases of man and animals. However, fungi and bacteria are the most important in terms of distribution, diversity, and total damage to plants in the field as well as in storage. These parasites can, in part, produce phytotoxic compounds that cause disease symptoms.

Many terms have been devised to describe the biology of compounds produced by parasites that are toxic to plants and play some role in symptom expression. For the purposes of this review, however, these compounds are collectively referred to as phytotoxins (1). This term does not include

309

0066-4154/82/0701-0309$02.00

the phytohormones and is to be contrasted with plant toxins, which are compounds produced by plants that adversely affect man or animals, such as Castor bean toxin (ricin), or the death cap mushroom toxin (amanitin).

Until the early 1960s reports on phytotoxins appeared only sporadically. Many of these were focused on the biological activity of phytotoxins rather than their chemistry or mode of action. Recently, new or improved techniques in chromatography and spectroscopy have facilitated the isolation and characterization of phytotoxins. The intense activity in a number of European laboratories in the 1960s and 1970s in isolating, characterizing, and defining the role of fusicoccin, a toxin produced by *Fusicoccum amygdali,* set an excellent example for people working on other toxins. The importance of a plant disease in which toxins are involved was dramatically brought to bear on the US public with the corn blight epidemic in 1970 (2). Since that time a flurry of work has resulted in an expanding literature on phytotoxins. For this reason it seems appropriate that the *Annual Review of Biochemistry* should have its first review on the subject of phytotoxins. This may be a signal that this topic has come of age as a subfield of biochemistry. Furthermore, at a time in history when food and fiber are becoming limited (3), it is appropriate for biochemical studies on phytotoxins to dramatically expand.

Besides their obvious role in the development of symptoms of certain plant diseases, phytotoxins also possess some unusual and intriguing chemical and biological properties. Some of them have extremely high host specificity, that is, they affect only those plant species or cultivars within a species that are similarly affected by the pathogen producing them. On the other hand, some have host specificity that is different from the organism that produces them. At the other extremes are those compounds that have no specificity whatever. Included in this group are some macromolecular compounds that cause wilting in plants more by virtue of physically impairing water movement than by directly interfering with metabolic function or cell integrity.

In the sensitive host, phytotoxins interact with a specific target site, and this ultimately leads to some outward manifestation by the plant. Thus, toxins are useful agents in helping us understand the mechanisms of plant disease development, including pathogen specificity, disease susceptibility, and resistance. Furthermore, phytotoxins have also demonstrated their utility as tools for probing the normal physiological and biochemical activities of plants. Some have even provided a practical means for selecting seeds, tissues, or seedlings for disease resistance (4–6).

While a large number of plant parasites show phytotoxic activity, only a few phytotoxins have been isolated and partially or completely characterized: fewer still have been chemically synthesized, and/or crystallized and subjected to X-ray analysis. A mechanism of action for some phytotoxins has been elucidated. Biosynthetic studies also have been done on a few phytotoxins, but a complete pathway with all intermediates and enzymes is not known for any phytotoxin. Furthermore, studies on the enzymatic degradation of phytotoxins are just beginning.

As a group, phytotoxins have no common structural features. They belong to such diverse classes as peptides (or other derivatives of amino acids), terpenoids, glycosides, phenolics, polyacetate α-pyrone derivatives, and polysaccharides, or a combination of these classes, and several others. Several reviews on phytotoxins and their role in disease have appeared in the past 15 years (1, 7–11).

Methodology

The pathogenic fungi and bacteria of many plants produce one or more phytotoxins. Some of the most notable fungal toxin producers are species of *Alternaria, Helminthosporium, Rhynchosporium, Fusicoccum, Ceratocystis,* and *Stemphyllium.* Toxins have not been isolated and characterized from the fungi that cause rust, smut, and mildew diseases. *Xanthomonas, Pseudomonas, Rhizobium,* and *Corynebacterium* are the most common bacteria associated with plants that produce phytotoxins.

That a toxin is involved in a disease can be determined by examining symptoms developing on the plant. Yellowing, wilting, brightly colored lesions, and necrosis are commonly caused by phytotoxins. If symptoms are produced at sites well removed from the pathogen then the involvement of a phytotoxin in disease development can be strongly suspected.

Phytotoxins are isolated from parasites that are grown under culture conditions. In culture, fungi and bacteria produce from trace amounts up to 2 g of toxin per liter under optimum conditions. Ideally, synthetic media are most desirable as a starting source for extracting toxins, since various plant or animal concoctions in the medium encumber the purification procedures. On the other hand, plant extracts are sometimes necessary for optimum fungal growth and toxin production. Purification of a phytotoxin to homogeneity relies heavily upon the availability of a quantitative bioassay technique. Once the parasite is removed from the culture medium by filtration or centrifugation, substances can be separated on the basis of their molecular size by solvent precipitation or chromatography over molecular sieves. For relatively small phytotoxins, direct solvent extraction with subsequent purification on HPLC, TLC, or other chromatographic techniques can be used. Most often, the biological activity of the molecule is the principal means by which it is followed through the steps of purification until some idea of its chemical nature is obtained, at which time appropriate colorimetric and chemical detection procedures can be employed.

The biological activity of a phytotoxin can be measured in a number of ways depending on the apparent effect it has on the plant (Table 1). The more quantitative assays, and usually the most preferred, are those in which the toxin lesion is measured directly ie. by plasmalemma disruption as determined by electrolytic leakage or membrane depolarization. Many of the assays in which leaf wilt, necrosis, discoloration, or chlorosis are measured provide only semiquantitative estimates of biological activity.

Many times, microorganisms produce toxic substances in culture, but their importance to the disease process is unknown. Some considerations used to implicate a toxin in a given disease are: (a) the comparative symptomology, i. e. do the toxin-induced symptoms resemble those found in the disease, (b) can the toxin be isolated from the diseased plants, (c) is the toxin present in vivo in quantities large enough to account for the symptoms, (d) do toxin-less trains produce toxin-related symptoms, and (e) can genetic variants of the host be found some of which are sensitive and others insensitive to the toxin?

Ideally, a toxin should be isolated and characterized before concerted efforts are made to determine its mechanism of action, or its role in the

Table 1 Examples of assays for phytotoxins

Source of toxin	Disease	Effect on plant	Type of assay	Ref.
Fusicoccum amygdali	Wilt of stone fruits	Wilt	a. Collapse of stems	12
			b. Effects on stomatal opening	13
Ceratocystis ulmi	Dutch elm disease	Wilt, yellowing, death	Water conductivity in stems	14
Helminthosporium maydis	Southern corn leaf blight	Yellowing, necrosis	a. Local lesions on leaves	15
			b. CO_2 fixation	16
			c. Pollen growth inhibition	17
			d. Mitochondrial oxidation	18
Helminthosporium sacchari	Eye-spot disease of sugarcane	Streaks on leaves, necrosis	a. Streaks on leaves	19
			b. Electrolyte leakage	20
Alternaria alternata F. sp. lycopersici	Stem canker of tomato	Cankers on stems and leaf chlorosis	Detached leaf (interveined chlorosis)	21
Helminthosporium victoriae	Victoria blight of oats	Death of plants	a. Root growth inhibition	22, 23
			b. Membrane effects	24
Pseudomonas phaesolicola	Halo blight of beans	Chlorosis	a. Chlorosis on leaves	25
			b. Inhibition of *E. coli*	26
Phyllosticta maydis	Yellow leaf blight of corn	Yellowing, necrosis	Inhibition of root and shoot growth	27

disease process. Unfortunately, because there are literally thousands of plant diseases, and only limited effort toward understanding their chemical basis, there are only a few examples that provide more than a semblance of a complete account of phytotoxin, its biochemistry, and mechanism of action.

Properties of some Selected Phytotoxins

Aspects of chemical and biological properties of compounds that are of microbial origin and possess toxic activity to plants have been discussed in other publications (28–30). Here I deal mainly with those toxins that have been implicated in the plant disease process on which new information has appeared in the past few years. Some examples of phytotoxins that have been recently discovered and have potential for additional biochemical studies are also mentioned. Space limitations preclude a discussion of all phytotoxins or parasites shown to produce phytotoxic activity.

HELMINTHOSPOROSIDE   Eyespot disease of sugarcane is caused by *Helminthosporium sacchari*. The characteristic red runners arising from the eye-shaped lesions on infected leaves may be 1 m long and so numerous as to cause death of the entire leaf. A toxin, given the trivial name helminthosporoside, was isolated from *H. sacchari* (31). Clones of sugarcane susceptible to *H. sacchari* are sensitive to helminthosporoside with a high degree of statistical significance (32). For this reason, helminthosporoside is used as a tool in sugarcane improvement programs to acquire clones resistant to *H. sacchari* (32).

The original structure proposed for helminthosporoside was that of an α-galactoside of cyclopropanediol (31). Recently, helminthosporoside has been subjected to comparative ^{13}C spectroscopy (33) and its nature established as a β 1 → 5–linked oligogalactofuranoside linked to a sesquiterpenoid (34). The general nature of the compound has been confirmed in other laboratories (35, 36), but there has not been universal agreement on the number of galactosyl residues present, or the precise empirical formula of the aglycone (34–36). One group, using the mass spectroscopy technique of fast atom bombardment, has shown that four galactosyl residues are present in the toxin and that the aglycone is $C_{14}H_{24}O_2$ (36), for a total molecular weight of 884, and these observations have been independently confirmed (G. A. Strobel, R. Beier, B. P. Mundy unpublished). It will be interesting to learn whether the aglycone of helminthosporoside bears any structural similarity to helminthosporal, a sesquiterpenoid toxin produced by *Helminthosporium sativum* (37).

At least two reports claim multiple host–specific toxins in cultures of *H. sacchari* (31, 36). The toxins appear to be isomers that differ in the location of the double bond in the aglycone moiety (36). Helmithosporoside described in (31) has the same chromatographic and spectral properties as the toxic substance from *H. sacchari* (34–36).

Helminthosporoside plays a significant role in eyespot disease. The fungus normally remains confined to the small eye-shaped lesions. The red "runner" that forms does not harbor the fungus and has an appearance identical to that caused by the toxin alone (31). Furthermore, the toxin has been isolated from those portions of naturally infected cane leaves showing symptoms of eyespot disease in quantities great enough to account for symptom production (19). The toxin has been noted in isolates of *H. sacchari* obtained from various cane growing areas in the world (38), which indicates that symptom production is probably not related to races of the fungus. Finally, attentuated cultures of *H. sacchari,* producing little or no toxin, are delayed several days in causing symptoms on susceptible sugarcane (39).

The role of the toxin in the initial establishment of the fungus in the host is unknown, but the toxin could be important in causing cell death, which would ultimately allow for colonization of the leaf tissue. It is unlikely that the genetics of the host, as related to its sensitivity to helminthosporoside and production of this toxin by the fungus, will ever be firmly established by classical genetic techniques, since sugarcane is polyploid and difficult to breed, and the sexual or perfect stage of *H. sacchari* has not yet been found.

The selectivity of helminthosporoside for certain clones of sugarcane is apparently determined by the presence of a toxin-binding protein (receptor) (40), which also binds α-galactopyranosides (41). The evidence that the binding protein serves as the recognition site for the toxin is the following: (*a*) Crude membrane preparations of toxin sensitive clones bind ^{14}C-helminthosporoside; similar preparations from toxin insensitive clones do not (40). (*b*) Mutant clones prepared from a toxin sensitive clone by γ irradiation are insensitive to the toxin and lack binding activity (20). (*c*) Toxin sensitive sugarcane leaves are made relatively toxin insensitive by flooding the vascular bundles with 0.1 M solutions of α-galactopyranosides or methyl β-galactofuranoside; β-galactopyranosides do not give this effect [(40); G. A. Strobel and B. P. Mundy unpublished]. (*d*) The binding protein, when purified by affinity chromatography and added to protoplasts of an insensitive sugarcane clone, confers toxin sensitivity on them (42).

The toxin binding protein is associated with the plasmalemma (43, 44). It occurs as a 4-subunit oligomer, has a mol wt of about 48,000, and has K_d's in the order of 10^{-5} for helminthosporoside, melibiose, and raffinose (40). At least two and possibly four toxin binding sites are present on the 4-subunit oligomer (40). The relationship of the binding site(s) for both α-galactopyranosidic and β-galactofuranosidic compounds needs to be further studied. Some lectins, however, do have affinities for both the furano-side and the pyranoside forms of a given sugar (45).

A comparable protein, differing in composition by several amino acid residues, is present in the plasmalemma of resistant sugarcane clones (46). In situ it does not bind the toxin, but when released from the membrane by the action of chaotropic agents or incubated in the presence of deter-gents, toxin binding activity is demonstrable (47).

The toxin binding activity of the purified binding protein from a resistant sugarcane clone is annulled by treatment with submicellar concentrations of digalactosyl diglycerides, which appear to be the major polar lipids in sugarcane plasmalemmas (48). Similar treatment of a purified binding pro-tein from a susceptible clone does not reduce affinity for raffinose and causes some reduction in toxin binding. Thus, although there are intrinsic differ-ences between the proteins from two different sugarcane clones, it appears that their binding activities are regulated by membrane lipids.

The toxin binding protein from resistant or susceptible sugarcane, puri-fied by extraction with chaotropic agents and affinity column chromatogra-phy, behaves anomalously (49). Binding activity at higher protein concentrations (1 μg/ml) is much less than that at low concentrations (0.1 μg/ml). The loss of binding activity at the higher protein concentrations corresponds to the formation of high-molecular-weight multimers. The self-inhibition of binding activity at high protein concentrations arises from a competition between ligand binding by oligomers and self-association of these oligomers into multimeric species that have little or no binding activ-ity. Raffinose and melibiose cause a ligand-dependent increase in binding activity and a corresponding decrease in the relative abundance of multim-ers at any given protein concentration. The association of subunits into multimers may have some bearing on the toxin sensitivity of sugarcane leaves and their relative age. Young leaf tissue of normally toxin sensitive clones is relatively insensitive to the toxin, whereas older tissue is sensitive. The relative susceptibility of these tissues to *H. sacchari* is comparable to that of the toxin. One explanation for this phenomenon is that the binding protein (in a multimeric, relatively inactive state) is the form of this protein incorporated into the membrane early in the life of the cell. Then, as the cell ages, the multimer disassociates into the "active" tetramers and becomes dispersed in the plasmalemma (49).

Some evidence for this hypothesis has come with the use of dimethyl-3,3'-dithiobispropionimidate (a bifunctional cross-linker) in in vivo experiments (50). The cross-linker was added to young and old leaves at low tempera-tures and basic pH in order to covalently stabilize the protein subunits in the form in which they occur in the membrane. In young leaves a multim-eric form of the binding protein was commonly isolated that could only be disassociated into subunits by reducing the –S–S– bond of the cross-linker. Such a stabilized multimer could not be isolated from the older tissue. These data support the idea that toxin insensitivity in young leaves is related to the predominance of the multimeric form of the binding protein, which possesses relatively little affinity for α-galactopyranosides or the toxin.

The role of the binding protein in the healthy plant seems to be that of α-galactopyranoside transport. Thus, sugarcane protoplasts pretreated with sublethal concentrations of helminthosporoside over a short time span fail to show a net uptake of α-galactopyranosides. However, control counterpart protoplasts demonstrate uptake and transport of these sugars. Protoplasts of a resistant clone have initial uptake but not appreciable transport of labeled α-galactopyranosides. α-Galactopyranoside uptake in sugarcane is driven by an electrogenic pump; however, no differences in membrane potential were found between a toxin sensitive and a toxin insensitive clone of sugarcane (51).

α-Galactopyranoside binding proteins have been isolated from membrane preparations of a number of plants including mint, tobacco, and various clones of sugarcane (47, 52) and have been demonstrated in other plants (53). Each of these proteins is oligomeric, has a molecular weight in the range of 100,000 (when isolated by the chaotropic-extraction-affinity column technique) and K_d's of binding such common α-galactosides as raffinose, galactinol, and helminthosporoside in the order of 10^{-5} M. Furthermore, subunits of these proteins from diverse plant species associate in vitro and form functional oligomers. Although mint and tobacco possess functional binding proteins they are insensitive to helminthosporoside and resistant to *H. sacchari*. Thus, binding of the toxin is a necessary first step in the toxin sensitivity of sugarcane, but it is not an exclusive property of sugarcane. Additional steps are obviously involved in causing cellular sensitivity to the toxin and it appears they are in some manner connected with some unique property(s) of the binding protein of susceptible cane and its interaction with membrane lipid.

It seems that helminthosporoside does not exert its toxic effects on sugarcane by primarily acting as an inhibitor of α-galactopyranoside transport. However, its effects do occur at the plasmalemma since membrane depolarization and ion leakage in tissues have been commonly noted (20, 54, 55). Some clue as to the molecular mechanism involved in membrane disruption has come with the observation that normally sensitive clones become totally insensitive to helminthosporoside in the summer months in Hawaii (56). The environmental parameter involved in this phenomenon is temperature, since mild heat treatment of toxin sensitive leaves renders them insensitive to helminthosporoside for at least 24 hours (56). The onset of sensitivity can be precluded by incubating the leaves in N_2 atmosphere or treating the leaves with protein synthesis inhibitors such as cycloheximide (56). Heat treatment does not inactivate the toxin binding protein, but substantially decreases membrane K^+, Mg^{2+} ATPase activity (57). Helminthosporoside activates (30%) membrane ATPase activity in microsomes from toxin sensitive leaves but not in heat-treated (now toxin insensitive) leaves. Once heat-treated leaves again become toxin sensitive, helminthosporoside activation of membrane ATPase activity resumes. Inhibitors of K^+, Mg^{2+} ATPase render normally toxin sensitive leaves resistant to the effects of helminthosporoside (57). Thus, the toxin seems to interfere with the electrochemical gradient at the level of ATPase in plasmalemma, which ultimately results in the demise of the cell. Conclusive evidence for the direct interaction of the binding protein with ATPase will only come when the two proteins are reconstituted into membrane vesicles and then examined for some of the effects observed in vivo.

When certain isolates of *H. sacchari* are successively transferred in synthetic medium attenuated cultures not producing helminthosporoside can result (39). When a "water wash" of leaves is added to the culture medium, some of the attenuated isolates produce toxin. Compounds present on the leaf surface that arise via plant metabolism have been called activators. One such activator was isolated and identified as serinol (2-amino-1,3-propanediol). It activates toxin production in some attenuated cultures at 1 μM. Serinol arises in toxin sensitive sugarcane clone 51NG97 via an enzyme mediated transamination of dihydroxyacetone phosphate; a number of amino acids can serve as amino donors (58). The serinol phosphate formed is hydrolyzed to P_i and serinol. The pathway of serinol formation is not present in resistant clone H50–7209. The traits of toxin binding and production of activators are apparently unrelated. In addition to serinol, another activator(s) is present in sugarcane (39). Stimulation of toxin production in *H. sacchari* by suspension cultures of sugarcane has also been noted (55). Collectively, these observations on activators affecting plant pathogens offer an alternative explanation for the selection theory, which is often used to explain the common phenomenon of attenuation that occurs in cultures of plant pathogens.

HELMINTHOSPORIUM MAYDIS TOXINS (HmT)   *H. maydis* causes Southern corn leaf blight. The yellowing and blighting symptoms associated with this disease are caused by several closely related toxins produced by the fungus. The toxins are host specific and affect corn containing the Texas (T) source of male sterile cytoplasm (59, 60). Thus, sensitivity to the toxins is controlled primarily by cytoplasmic genes. Corn containing other cytoplasms such as C, S, and N are insensitive to the toxins and resistant to *H. maydis* (59, 60). Nevertheless, nuclear genes can have some modulating effect on the level of susceptibility expressed by plants having T-cytoplasm (61).

Toxin preparations with high specific biological activity have been obtained from cultures of *H. maydis* (62). Several proposals concerning the nature of the HmT have been made, ie. that is a pepide (63), a triterpenoid (64), a mannitol-amino acid ester (65) and more recently, a polyketide (62, 66). Most workers agree that there are several closely related toxic components in HmT, that there is no nitrogen in it (62, 64), and that it contains OH and keto functional groups (62, 64).

Relatively crude preparations of HmT have been used in studies of its mode of action (67). This may have led to difficulties in interpretation of data, since nonspecific toxins (ophiobolins) may also be present in such preparations (68). Nevertheless, differential specificity to tissues, membranes, or organelles is characteristically demonstrated in these comparative studies (67).

Even though HmT causes electrolyte leakage and rapid partial depolarization of the plasmalemma electrical potential (69–71), the cytoplasmic inheritance of toxin sensitivity suggests that the mitochondrion or chloroplast could also be involved as a site of toxin action. A chloroplast site seems unlikely, since HmT has no apparent effect on enzyme activities in isolated chloroplast lamellae (72). On the other hand, in the presence of HmT mitochondria in tissues or cell-free preparations from T-cytoplasm corn show swelling, loss of matrix density, and damaged cristae, whereas mitochondria from N (normal) cytoplasm remain unaffected (73, 74). In addition, HmT affects mitochondrial respiration as follows: (*a*) complete inhibition of the oxidation of NAD-linked substrates such as malate (75); (*b*) stimulation of O_2 uptake when NADH is the substrate and; (*c*) variation in succinate metabolism depending upon the assay medium (67). The effects of the toxin on mitochondrial activity are explainable on the basis of induced changes on the permeability of the inner mitochondrial membrane (67). Furthermore, additional characterization of the protease sensitive HmT binding activity that occurs in the cytosol of toxin sensitive and insensitive lines of corn needs to be done and ultimately related to the mode of action of HmT (76).

Restriction endonuclease patterns of mitochondrial DNA of corn containing T and N cytoplasm are distinctly different (77). Gel electrophoresis patterns of mitochondrial proteins show an additional variant peptide in the mitochondria of the T-cytoplasm (78). However, as yet, no functional relationship between the polypeptides or DNA fragments have been directly related to toxin sensitivity.

Southern corn leaf blight may provide a valuable model for the study of the role of toxins in disease and the elucidation of a molecular mechanism for disease resistance, because corn genetics is well understood, the pathogen has a sexual stage that can be used for genetic analysis, and some information is available on the toxin and its probable site of action.

ALTERNARIA ALTERNATA TOXINS   This organism displays a number of pathotypes that affect certain plant species (79). Many of these pathotypes produce host-specific toxins that are primarily responsible for symptom production in the infected host plant. For instance, host-specific toxin activity has been demonstrated from pathotypes of this fungus that cause brown spot of citrus (80), black spot of Japanese pear (81), and black spot of strawberry (82). Two additional *A. alternata* toxins in which some detail is known of the chemistry and mode of action are involved in blotch of apple and stem canker of tomato.

Alternaria alternata F. sp. mali toxins   Alternaria blotch on certain cultivars of apples was noted in the 1950s in Japan. A toxin with host specificity is involved in the blotch disease (83). This toxin is a complex containing at least six components (AM toxin), each of which possesses host specificity (82). The two major toxins, alternariolide and its demethoxy derivative have been characterized (81, 85). Alternariolide is a depsipeptide, cyclo(α-hydroxyisovaleryl-α-amino-*p*-methoxyphenyl-valeryl-α-amino-acryl-alanyllactone) and has been chemically synthesized (Figure 1) (83, 87).

Alternariolide causes interveinal chlorosis on a susceptible cultivar of apples in the nM range, while on resistant apples chlorosis is observed in the 1 μM range (82). It causes an immediate loss of electrolytes from toxin sensitive apple cultivars (82, 83). These findings suggest the involvement of the plasmalemma as the toxin sensitive site, although a toxin receptor and its relative location in apple tissue have not yet been demonstrated.

This susceptibility of apples to *A. mali* is controlled by multiple dominant genes (88), which suggests that they determine the relative sensitivity or insensitivity of apple to the AM toxins (82).

Alternaria alternata F. sp. Lycopersici toxins   This fungus causes stem canker of tomato, and susceptibility to this disease is controlled by a single genetic locus with two alleles (21). Culture filtrates of the fungus yield two

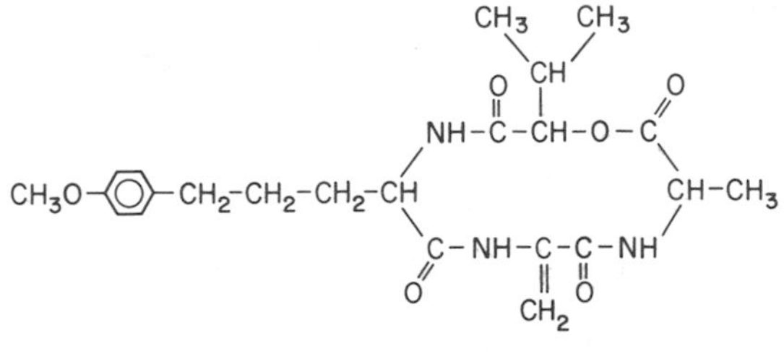

Figure 1   Alternariolide.

ninhydrin-positive fractions, TA and TB, which cause the macroscopic disease symptoms only on those tomato lines that carry the genetic locus that regulates the disease reaction. TA and TB are separable by isoelectric focusing at pH 4–5 or by thin layer chromatography on silica gel using selective solvents (89). Both TA and TB exhibit necrotrophic activity at less than 10 ng/ml, and to date, seem to be produced only by this fungus (90). TA has two components, 2a and 2b, composed (90) of two esters of 1,2,3 propanetricarboxylic acid and a novel aminopentol (Figure 2). The sites of esterification are a terminal carboxyl of the acid and the C_{13} (2a major component) and C_{14} (2b minor component of the aminopentol) (90). The relative toxicities of 2a and 2b remain to be determined (90). TB appears to consist of two components with the same carbon skeletons as 2a and 2b but lacking the C_5 hydroxyl and differing in stereochemistry at one or more chiral centers from C_{11} to C_{15} (90).

Some preliminary studies indicate that the toxins are not preferentially degraded by tomato tissues resistant to them. However, aspartate and certain products of aspartate metabolism protect toxin sensitive tomato plants (91). The effectiveness of orotic acid as a protective agent may indicate that the toxins have an antimetabolite role involving aspartate transcarbamoylase (ATCase) (91). ATCase preparations from hosts of various genotypes (sensitive-rr and insensitive-RR) exhibit sensitivity to toxin inhibition that correlates with their source. The relative activities of the toxins as well as the unique characteristics of the ATCase from toxin sensitive genotypes are areas for further study.

TENTOXIN   Tentoxin is a cyclic tetrapeptide, *cyclo* [-L-leucyl-N-methyl-(Z-dehydrophenylalanyl-glycyl-N-methyl-L-alanyl (92–94) produced by *A. alternata* (95). It has been synthesized (96, 97), and its stereochemical conformation established (Figure 3) (98). Unlike some of the host-specific toxins of *A. alternata,* tentoxin induces chlorisis in many plants including lettuce, potato, cucumber, and spinach, but does not affect certain *Nicotiana* species, tomato, cabbage, or radish (99).

The species selectivity of tentoxin results from its interaction (binding) to chloroplast coupling factor 1(CF$_1$), with consequent inhibition of its normal catalytic function (photophosphorylation) and its Ca^{2+}-dependent ATPase (100). There is one toxin binding site per CF$_1$, associated with the

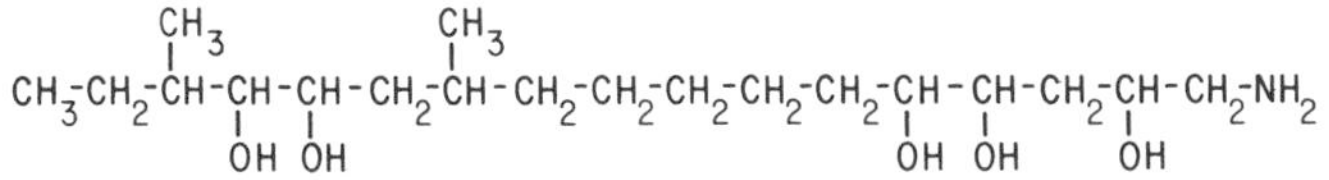

$$CH_3-CH_2-\underset{\underset{OH}{|}}{CH}-\underset{\underset{OH}{|}}{CH}-CH-CH_2-\overset{\overset{CH_3}{|}}{CH}-CH_2-CH_2-CH_2-CH_2-CH_2-\underset{\underset{OH}{|}}{CH}-\underset{\underset{OH}{|}}{CH}-CH_2-\underset{\underset{OH}{|}}{CH}-CH_2-NH_2$$

Figure 2   A novel amino pentol (a portion of the TA component of the host specific toxins of *A. alternata* L. sp. *lycopersici*).

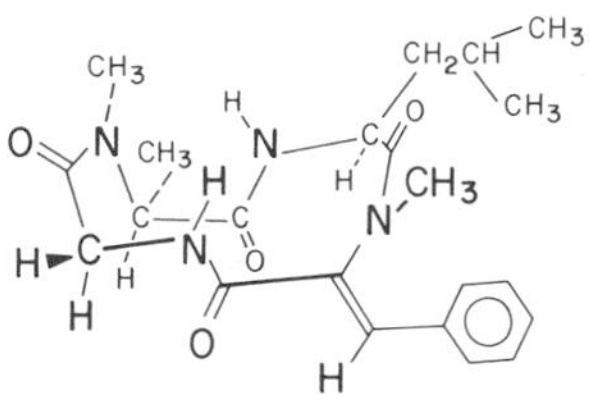

Figure 3 Tentoxin.

α-β subunit complex (101), which is involved in the photophosphorylation activity of CF_1 (102). The inhibition of this reaction by the toxin results in an inhibition of light-driven protein and RNA synthesis in isolated chloroplasts (103); the addition of ATP relieves this inhibition. These effects may explain the chloroplast-specific ultrastructural alterations caused by tentoxin (104) and the development of chlorosis in toxin-treated leaves. The affinity of the toxin for the site in CF_1 is 1.3–20×10^{-7} M. With insensitive species, such as radish, at least 20 times more tentoxin is required for 50% inhibition of photophosphorylation. Thus, the absence of a high-affinity site in the CF_1 from insensitive species would account for their relative inability to develop toxin induced chlorosis.

The kinetics of inhibition suggst that the interaction of tentoxin with CF_1 is at a site that is independent of the one controlling its enzymatic function. Steady-state kinetics best fit an uncompetitive pattern, which suggests that the steps of inhibition follow an irreversible step that occurs after ATP binding (105). Nucleotides, phosphate, and calcium do not compete with tentoxin for its binding site, nor do they affect its affinity for tentoxin (105).

That CF_1 has both a catalytic and structural function is supported by reconstitution experiments (106). Spinach chloroplasts, depleted in CF_1, to which was added CF_1 from toxin insensitive *Nicotiana* species, remain insensitive to tentoxin; however, sensitivity to tentoxin was regained when the CF_1 from sensitive plants was used for reconstitution (106).

A genetic analysis of interspecific hybrids of *Nicotiana* spp., in which only one of the parents was sensitive to tentoxin showed that toxin sensitivity was exclusively transmitted through the female parent (107, 108). Thus, the gene(s) specifying toxin sensitivity, and hence one or both of the α,β subunits in the 5-subunit complex of CF_1 (102) is (are) located in the cytoplasm (107).

RHYNCHOSPOROSIDES Scald disease in barley and other grasses is caused by *Rhynchosporium secalis*. In addition to the scalded lesions that develop on the blades of infected leaves, grey-green lesions may appear on the leaf margins. Toxic metabolites produced by *R. secalis* can cause lesions having the same appearance as those caused by the fungus (109, 110). The glucoside, cellobioside, and cellotrioside linked 1-0-α to 1,2 propanediol are referred to as rhynchosporosides 1,2 and 3, respectively (34, 110) (Figure 4). Each of these glycosides is capable of causing marginal leaf discoloration and necrosis (34, 110). The specificity of toxin is noted from the observation that only certain cultivars of barley and rye are sensitive to the toxin. Sensitivity to the toxin and susceptibility to the fungus are independently inherited traits (110).

Preparations of rhynchosporosides 2 and 3 invariably contain small amounts of other β glucosides (1→6, 1→3, 1→2) and linked 1-O-α to 1,2 propanediol (111). It is unlikely that the contaminating glucosides represent the total biological activity of the toxin preparations, since the synthetic

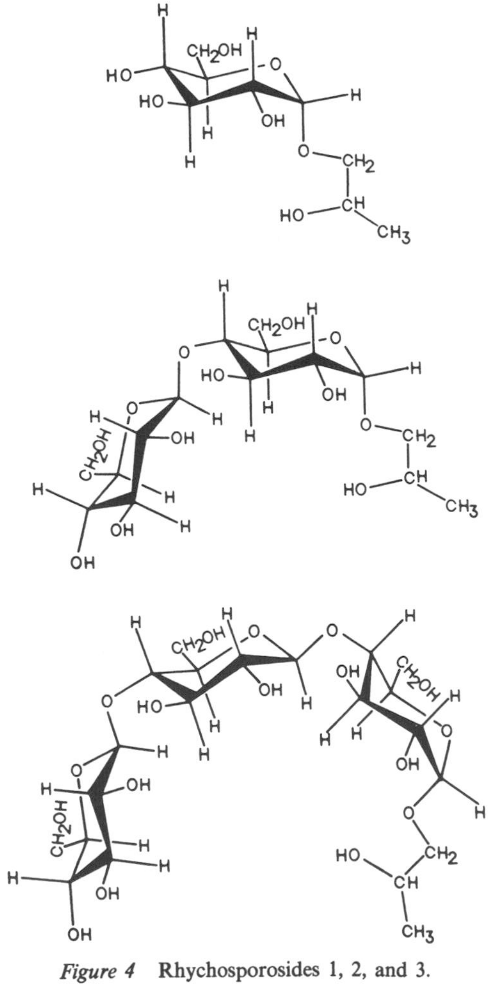

Figure 4   Rhychosporosides 1, 2, and 3.

cellobiosyl derivatives of 1,2 propanediol (1–0–α and 2–0–α linked) possess biological activity (34, 111). Nevertheless, further analytical work is needed to separate these glycosides from toxin preparations, to synthesize them, and to determine if they also possess biological activity.

A toxic oligo-glucosyl derivative of 1,2 propanediol has been isolated from barley plants infected by *R. secalis* in quantities large enough to account for disease symptoms (110). A membrane receptor protein has been isolated from a toxin sensitive cultivar of barley (112). Membrane preparations from a toxin sensitive barley cultivar bind more toxin per unit protein than preparations from toxin insensitive cultivar (112). Conceivably, the pathogenicity of the races of of *R. secalis* (113) may in some way be related to the qualitative and quantitative production of the rhynchosporosides, or related toxic glycosides.

FUSICOCCIN   Fusicoccin is the most extensively investigated phytotoxin. It is the major toxin produced by *Fusicoccum amygdali* and causes many of the symptoms commonly induced by the fungus on almond and peach trees (114, 115). Its structure has been established (116, 117), and a recent review on its chemistry, biological activity, and mode of action has appeared (118). Derivatives of fusicoccin including isofusicoccin, monodeacetyl fusicoccin, and dideacetylfusicoccin have been isolated and their structure-function relationships described (119, 120).

Fusicoccin causes cell enlargement, proton efflux, potassium uptake, and stomatal opening in practically all higher plant species studied to date (118). The promotion of seed germination in antagonism with abscisic acid has been noted in several species. Overall, the toxin exhibits no organ, tissue, or species specificity. However, no effects have been observed on fungi, bacteria, or animals.

The myriad of cellular activities affected by fusicoccin are probably controlled by a single process, namely the conversion of phosphate bond energy into proton electrochemical gradient energy at the plasmalemma level, and the biological effects of the toxin are consequences of this primary effect (118, 121, 122). For instance, proton extrusion would have important consequences on cell wall loosening and cell enlargement, and would eventually affect seed germination. Cytoplasm alkalinization by virtue of toxin induced K^+ uptake results in an increase in osmotic potential that directly affects stomata opening. The increase in respiration and the rate of CO_2 fixation into malate (from PEP), together with a decrease in the C1/C6 ratio in hexose utilization are interpreted to be in agreement with the pH-stat theory (118, 123).

All metabolic consequences of effects on the cell induced by fusicotoxin can be theoretically related to the pH-stat effect. The data also implicate the plasmalemma as the site of the fusicoccin activated H^+ pump and rule out alternative possibilities such as the mitochondrion being the site of H^+ extrusion.

Evidence has been presented that the generation of the electrochemical gradient at the plasmalemma occurs as a result of an interaction between fusicoccin and the K^+, Mg^{2+} activated plasmalemma ATPase (118). For instance, inhibitors of membrane ATPase such as octylguanidine, N,N^1 dicyclohexylcarbodiimide (DCCD), and diethylstilbestrol (DES) produce inhibition of fusicoccin stimulated H^+ extrusion and K^+ uptake (124). Treatments that inhibit intercellular ATP levels also suppress the effects of the toxin (125, 126). ^{3}H-Fusicoccin specifically binds to a component present in plasmalemma-enriched membrane preparation of corn coleoptiles (127). Treatment of oat root membrane vesicles with trypsin or a temperature above 45°C reduces membrane binding of fusicoccin, which suggest the existence of a membrane receptor protein for fusicoccin (128). Solubilization of the membranes with Triton X-100 followed by gel column chromatography demonstrate that fusicoccin binding activity is distinct from membrane ATPase activity.

Nevertheless, it seems likely that the fusicoccin binding protein is associated with the membrane ATPase, but is not one of the subunits of ATPase (128). This toxin binding protein in plants may serve as a transport vehicle or receptor for some common product of plant metabolism, since through washing of root tissues in H_2O increases the binding of fusicoccin severalfold (127). Such a product may be a sugar, or a plant growth substance. Roots may contain one or more water soluble compounds that compete with the fusicoccin receptor (129). Thus, although its activity has been demonstrated, the fusicoccin receptor awaits isolation, biochemical characterization, and an understanding of its normal function in the plant.

The most compelling model for the mechanism of action of fusicoccin is the stimulation of electrogenic H^+ extrusion, caused by the interaction between the toxin and the plasmalemma ATPase (130). To date, the evidence for this model is circumstantial (130), and additional experiments, especially ones employing reconstituted membrane vesicles with the membrane ATPase and the binding protein, should begin.

TOXINS OF PSEUDOMONAS   More phytotoxins have been isolated from this group of plant parasitic bacteria than any other, and some reviews discussing them have appeared (8, 10, 131).

Tabtoxin and 2-serine tabtoxin   *Pseudomonas tabaci,* which causes wildfire disease of tobacco, produces tabtoxin and 2-serine tabtoxin (132, 133) as do other species of Pseudomonas (134). Tabtoxin contains a residue of the β-lactam of tabtoxinine linked to the amino group of L-threonine. 2-Serine tabtoxin possess an L-seryl residue substituted for the L-threonyl residue in the structure.

Tabtoxin induced chlorosis in plants may be related to the inhibition of glutamine synthetase (135), and the hydrolysis of tabtoxins appears to be a vital step in the process. Apparently the β-lactam of tabtoxinine is an inhibitor of glutamine synthetase, whereas the tabtoxins are not (136). Hydrolysis of the tabtoxins occurs in plant tissues by the action of peptidases, which leads to the production of the toxic lactam (136). This represents an interesting case of the metabolic activation of a phytotoxin by the plant system.

Phaseolotoxin *Pseudomonas phaseolicola,* which causes halo blight of beans, produces phaseolotoxin [(N$^\delta$-phosphosulfamyl) ornithylalanyl-homoarginine] (137). Bean tissues treated with *P. phaseolicola* or phaseolotoxin accumulate ornithine (138, 139). This condition results from the inhibition of ornithine carbamoyltransferase by phaseolotoxin (140, 141). However, breakdown products of phaseolotoxin found in the plant soon after the application of the radiolabeled toxin may also account for symptom production (139). Both homoarginine and alanyl residues are removed by plant enzymes, leaving N$^\delta$-phosphosulfamyl-ornithine, which suggests that it acts as a phytotoxin and perhaps is the major residue in the host-parasite interaction (139). Aspects of other toxins produced by *P. phaseolicola* are covered elsewhere (10, 131, 142).

Coronatine Both *Pseudomonas glycinea* and *Pseudomonas coronafaciens* pv. atropurpurea produce coronatine, a chlorosis inducing toxin (143, 144). The structure of coronatine (Figure 5) has been solved by spectroscopic methods and X-ray crystallographic data on coronafacic acid. It causes chlorosis in a variety of plants including bean, soybean, and Italian ryegrass. Studies on its biochemical and biological effects on plants have begun (145).

Syringomycin and Tagetitoxin Some isolates of *Pseudomonas syringae* produce syringomycin, a peptide phytotoxin (146, 147). More recently, a chlorosis inducing toxin has been isolated from *P. syringae* pv tagetis. This toxin (tagetitoxin) is distinct from phaseolotoxin, but both contain sulfur, phosphorus, and an amino functionality (148).

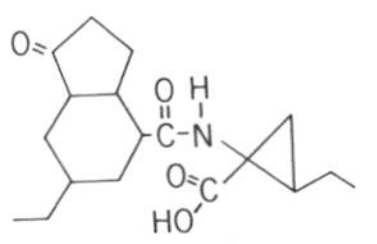

Figure 5 Coronatine.

CERATOCYSTIS ULMI TOXINS   The Dutch elm disease fungus, *C. ulmi*, produces substances in culture that induce wilt symptoms in elm stem cuttings (149, 150) that resemble those observed in the field. One of these substances is a peptidorhamnomannan (151). The polydisperse toxin has an average molecular weight of 130,000 (152) and consists of a 30,000 -mol wt polypeptide backbone to which 2–3 rhamnomannan side chains are linked as O-glycosides to seryl and threonyl residues of the polypeptide (151, 152). Each side chain has a molecular weight of $\sim$ 30,000 and is composed of α 1→6–linked mannosyl residues with a terminal 1→3–linked rhamnosyl residue on alternate mannosyl residues (151).

Small side chains containing 1, 2, 3 or 4 mannosyl residues are also attached to the peptide. The peptidorhamnomannan is antigenic (152, 153) and has been detected in diseased elm tissues by the enzyme-linked immunospecific assay) (ELISA), which implicates it in the disease process (153)

The toxin causes wilting in elm stems by physically interfering with water movement through the xylem (154). The pit membranes of the xylem vessels facilitate lateral water movement to the leaf petioles and ultimately to the leaf. The toxin impedes normal water flow through these pits (154). Thus, if water is lost from the leaf surface at a greater rate than it is supplied by the branch, wilting occurs. If the water deficit is maintained leaves will eventually die.

No host specificity is exhibited by the peptidorhamnomannan of *C. ulmi*. In fact, polysaccharides such as limit dextrans also induce wilting in elm cuttings and other plants (154). This does not eliminate the possibility that molecules of this type, including the small protein toxin (13,000 mol wt) of *C. ulmi*, cerato-ulmin (155), play some role in the development of Dutch elm disease. Some, however, have questioned whether compounds that act by physically impairing water movement should be termed phytotoxins. Nevertheless, there are many examples of plant parasites that produce molecules that appear to act by interfering with water movement in plant tissues (Table 2).

OTHER PHYTOTOXINS   A number of parasitic organisms produce phytotoxins worthy of mention because of the economic importance of the diseases they cause, and the potential they offer for further research.

Alternaria phytotoxins   A wide array of compounds from various species of *Alternaria* are phytotoxic (166). These include zinniol, alternaric acid, tenuazonic acid, alternariol, alternariol methylether (167), radicinin, and radicinol (168–170). None of these compounds are host specific; some have

Table 2 Some examples of macrocolecules produced by parasites that cause wilt symptoms in plants

Pathogen	Disease	Nature of toxin	Estimated mol wt	Ref.
Corynebacterium insidiosum	Alfalfa wilt	Glycopeptide (gluco-galactofucan)	5×10^6	156
Corynebacterium sepedonicum	Ring rot of potato	Glycopeptide (glucomannan)	22,000	157, 158
Erwinia amylovora	Fire blight of apples & pears	Polysaccharide (gluco-glucuronogalactan)	165,000	159, 160
Xanthomonas campestris	Black rot of cabbage	Polysaccharide (glucan and glucomannan)	—	161, 162, 163
Xanthomonas oryzae	Leaf blight of rice	Polysaccharide (manno-glucan) four components	10^4–10^5	164
Phoma tracheiphila	Malsecco of lemon	Glycopeptide	93,000	165

been found in host plants infected by *Alternaria* spp. and their mode of action is not known. The absolute stereochemical configuration obtained by X-ray analysis and NMR is known for some of them.

Helminthosporium carbonum toxin   A peptide toxin containing residues of alanine, proline, and an epoxy-amino acid is known from *H. carbonum,* the cause of Northern corn leaf blight (171, 172). The toxin is host specific (173) and produces gross physiological imbalances in susceptible corn tissue (174).

Helminthosporium victoriae toxin   One of the first host-specific toxins to be studied was that of *H. victoriae,* and earlier work on it has been reviewed (175, 176). It affects only oat cultivars containing the victoria gene. The toxin consists of victoxinine ($C_{17}H_{29}NO$) (177) covalently linked to a small peptide. Its relative instability has precluded more complete chemical characterization. Highly active preparations of the toxin cause plasma membrane disruption of sensitive oat cultivars (178, 179).

Periconia circinata toxins   Host-specific peptide toxins produced by *P. circinata* cause milo disease of sorghum (180, 181). The two toxins are enriched in aspartyl residues and an unidentified polyamine(s) (182). Toxin preparations cause electrolytic loss and plasmalemma damage in susceptible cultivars of sorghum (184).

Stemphylium botryosum toxins   Several phytotoxins, including stemphlin, an aromatic glycoside, have been isolated from this pathogen, and there are probably many more remaining to be discovered from its biotypes (185). Chemical synthesis of the aglycone of stemphylin has proven that its originally proposed structure requires revision (186). Stemphyloxin is the latest *S. botryosum* toxin to be isolated and characterized, and it appears to be a tricyclic β-ketoaldehyde (187). Thus far, none of the toxins isolated from this organism possess host specificity. Necrosis is caused on plants to which these compounds have been administered.

Pyrenophora teres toxins   *P. teres* invades barley and causes the net blotch disease, one of the most important diseases of barley worldwide. The fungus produces at least two toxins in culture: N -(2-amino-2-carboxyethyl) aspartic acid and aspergillomarasmine A (188, 189). The latter compound is known from species of *Colletotrichum* and *Aspergillus* (1). A mixture of the *P. teres* toxins causes a dramatic increase in respiration of susceptible barley cultivars and less so in leaves of more resistant cultivars and nonhosts of *P. teres* (190).

Pyrenochaeta terrestris toxins   Some of the symptoms of the onion pink root disease seem to be attributable to pyrenocine A and B. The structure of crystalline pyrenocines A and B as determined by X-ray analysis are 5-crotonyl-4-methoxy-6-methyl-2-pyrone   and   5-(3-hydroxybutyroyl)-4-methoxy-6-methyl-2-pyrone, respectively (191). The groundwork is now sufficiently developed to permit studies of the mode of action of these interesting compounds.

Practical Applications

Most of the work on phytotoxins is concentrated on those produced by parasites of crop plants; yet another serious threat to croplands are weeds. Parasites of weeds are known, but their phytotoxins are not. Conceivably, studies on the phytotoxins of weed parasites could yield compounds possessing herbicidal activity with the valuable properties of host specificity and biodegradability.

Phytotoxins have proven useful as tools for screening in crop improvement programs (1, 6, 7, 32). The advent of techniques allowing for plant regeneration from single leaf cell protoplasts or tissues of important crop plants may also permit screening at the cellular level with subsequent regeneration of a toxin insensitive plant (6, 192, 193).

Some of the phytotoxins also possess antibiotic activities, such as syringomycin (145–147) and radicinin (170). While none of the phytotoxins are used in modern human or animal medicine, some potential does exist for their application.

Questions

There are many unanswered questions in phytotoxin research. For example: (*a*) What are the pathways of biosynthesis and degradation? (*b*) What are the relationships between races of the parasite and the numbers and kinds of phytotoxins produced? (*c*) What are the mechanisms of toxin action? (*d*) What is the relationship between host specificity of a phytotoxin and the structural entity in the host controlling specificity? (*e*) How can we use phytotoxins in probing normal plant cell function? and (*f*) How might we better apply our knowledge of phytotoxins in solving important problems in food and fiber production?

ACKNOWLEDGMENTS

The author appreciates the help of B. P. Mundy, D. Mathre, R. Durbin, E. Marré, L. Daley, and U. Matern in critically reading portions or all of the manuscript. I also thank Pam Berger for invaluable assistance in preparation of the manuscript. A portion of this work was supported by NSF, Herman Frasch, and Dow Chemical Company grants to the author.

Literature Cited

1. Strobel, G. A. 1974. *Ann. Rev. Plant Physiol.* 25:541–66
2. Tatum, L. A. 1971. *Science* 171:1113–15
3. National Academy of Sciences. 1977. *World Food and Nutrition Study,* Vol. 1. 318 pp.
4. Steiner, G. W., Byther, R. S. 1971. *Phytopathology* 61:691–95
5. Wheeler, H. E., Luke, H. H. 1955. *Science* 122:1229
6. Gegenenbach, B. G., Green, C. E., Donovan, C. M. 1977. *Proc. Natl. Acad. Sci. USA* 74:5113–17
7. Pringle, R. B., Scheffer, R. P. 1964. *Ann. Rev. Phytopathol.* 2:133–56
8. Durbin, R. D. ed. 1981. *Toxins in Plant Disease* New York: Academic. 536 pp.
9. Wheeler, H. 1976. *Specificity in Plant Diseases,* ed. Wood, R. K. S. Graniti, A. pp. 217–35. New York: Plenum. 354 pp.
10. Strobel, G. A. 1977. *Ann. Rev. Microbiol.* 31:205–24
11. Yoder, O. C. 1980. *Ann. Rev. Phytopathol.* 18:103–29
12. Turner, N. C., Graniti, A. 1969. *Nature* 223:1070–71
13. Graniti, A. 1964. *Host-Parasite Relations in Plant Pathology,* ed. Z. Kiraly, G. Ubrizy, pp. 211–17. Budapest, Hungary: Res. Inst. Plant Protection
14. VanAlfen, N. K., Turner, N. C. 1975. *Plant Physiol.* 55:312–16
15. Karr, A. L., Karr, D. B., Strobel, G. A. 1974. *Plant Physiol.* 53:250–57
16. Bhullar, B. S., Daly, J. M., Rehfeld, D. W. 1975. *Plant Physiol.* 56:1–7
17. Laughnan, J. R., Gubay, S. J. 1973. *Crop Sci.* 13:681–84
18. Miller, R. J., Koeppe, D. E. 1971. *Science* 173:67–9
19. Strobel, G. A., Steiner, G. W. 1972. *Physiol. Plant Pathol.* 2:129–32
20. Strobel, G. A., Steiner, G. W., Byther, R. 1975. *Biochem. Genet.* 13:557–65
21. Gilchrist, D. G., Grogan, R. G. 1976. *Phytopathology* 66:165–71
22. Meehan, F., Murphy, H. C. 1947. *Science* 106:270–71
23. Luke, H. H., Wheeler, H. E. 1955. *Phytopathology* 45:453–58
24. Samaddar, K. R., Scheffer, R. P. 1968. *Plant Physiol.* 43:21–8
25. Hoitink, H. A. J., Pelletier, R. L., Coulson, J. G. 1966. *Phytopathology* 56:1062–65

26. Staskawicz, B. J. 1979. *Phytopathology* 69:663-66
27. Yoder, O. C. 1973. *Phytopathology* 63:1361–65
28. Friend, J., Threlfall, D. R., eds. 1976. *Biochemical Aspects of Plant-Parasite Relationships,* New York: Academic. 354 pp.
29. Wood, R. K. S., Ballio, A., Graniti, A. 1972. *Phytotoxins in Plant Diseases,* New York: Academic. 530 pp.
30. Kadis, S., Ciegler, A., Ajl, S. J., eds. 1972. *Microbial Toxins VIII,* New York: Academic. 400 pp.
31. Steiner, G. W., Strobel, G. A. 1971. *J. Biol. Chem.* 246:4350–57
32. Steiner, G. W., Byther, R. S. 1971. *Phytopathology* 61:691–95
33. Beier, R. C., Mundy, B. P., Strobel, G. A. 1980. *Can. J. Chem.* 58:2800–4
34. Beier, R. C. 1980. Carbohydrate chemistry: synthetic and structural investigation of the phytotoxins found in *Helminthosporium sacchari* and *Rhynchosporium secalis.* PhD thesis. Montana State Univ. 334 pp.
35. Livingston, R. S., Scheffer, R. P. 1981. *J. Biol. Chem.* 256:1705–10
36. Macko, U., Goodfriend, K., Wachs, T., Renwick, J. A. A., Acklin, W., Arigoni, D. 1981. *Experientia* 37:923–4
37. DeMayo, P., Williams, R. E., Spencer, E. Y. 1965. *Can. J. Chem.* 43:1357–65
38. Steiner, G. W., Byther, R. S. 1971. *Phytopathology* 61:691–95
39. Pinkerton, F., Strobel, G. A. 1976. *Proc. Natl. Acad. Sci. USA* 73:4007–11
40. Strobel, G. A. 1973. *J. Biol. Chem.* 248:1321–28
41. Strobel, G. A. 1974. *Proc. Natl. Acad. Sci. USA* 71:4231–36
42. Strobel, G. A., Hapner, K. 1975. *Biochem. Biophys. Res. Commun.* 63:1151–56
43. Strobel, G. A., Hess, W. M. 1974. *Proc. Natl. Acad. Sci. USA* 71:1413–17
44. Thom, M., Laetsch, W. M., Maretzki, A. 1975. *Plant Sci. Lett.* 5:245–53
45. Lis, H., Sharon, N. 1973. *Ann. Rev. Biochem.* 42:541–74
46. Strobel, G. A. 1973. *Proc. Natl. Acad. Sci. USA* 70:1693–96
47. Kenfield, D., Strobel, G. A. 1981. *Plant Physiol.* 67:1174–80
48. Kenfield, D., Strobel, G. A. 1981. *Physiol. Plant Pathol.* 19:145–52
49. Kenfield, D., Strobel, G. A. 1980. *Biochim. Biophys. Acta* 600:705–12
50. Kenfield, D., Strobel, G. A. 1981. *Biochem. Int.* 2:249–55
51. Franz, S., Tatar, L. A. 1981. *Plant Physiol.* 67:150–55
52. Kenfield, D. 1978. α-Galactoside binding proteins from plant membranes distribution, function and relation to Helminthosporoside-binding proteins of sugarcane. PhD thesis. Montana State Univ. 126 pp.
53. Kenfield, D., Strobel, G. A. 1977. *ACS Symp. Ser.* 62:35–46
54. Scheffer, R. P., Livingston, R. S. 1980. *Phytopathology* 70:400–4
55. Larkin, P. J., Scowcroft, W. R. 1981. *Plant Physiol.* 67:408–14
56. Byther, R. S., Steiner, G. W. 1975. *Plant Physiol.* 56:415–19
57. Strobel, G. A. 1979. *Biochim. Biophys. Acta* 554:460–68
58. Babczinski, P., Matern, U., Strobel, G. A. 1978. *Plant Physiol.* 61:46–9
59. Hooker, A. L., Smith, D. R., Lim, J., Beckett, B. 1970. *Plant Dis. Rep.* 54:708–12
60. Smith, D. R., Hooker, A. L., Lim, S. M. 1970. *Plant Dis. Rep.* 54:819–22
61. Lim, S. M. 1974. *Plant Dis. Rep.* 58:811–13
62. Kono, Y., Daly, J. M. 1979. *Bioorg. Chem.* 8:391–97
63. Smedegard-Petersen, V., Nelson, R. R. 1969. *Can. J. Bot.* 47:951–57
64. Karr, A., Karr, D., Strobel, G. A. 1974. *Plant Physiol.* 53:250–57
65. Aranda, G., Berville, A., Cassini, R., Fetizon, M., Poiret, B. 1978. *Ann. Phytopathol.* 10:375–79
66. Kono, Y., Takeuchi, S., Kawarada, A., Daly, J. M., Knoche, H. W. 1980. *Tetrahedron Lett.* 21:1537–40
67. Gregory, P., Earle, E. D., Gracen, V. E. 1977. *ACS Symp. Ser.* 62:90–114
68. Tipton, C. L., Paulsen, P. V., Betts, R. E. 1977. *Plant Physiol.* 59:907–10
69. Arntzen, C. J., Koeppe, D. E., Miller, R. J., Peverly, J. H. 1973. *Physiol. Plant Pathol.* 3:79–89
70. Gracen, V. E., Grogan, C. O., Forster, M. J. 1972. *Can. J. Bot.* 50:2167–70
71. Mertz, S. M., Arntzen, C. J. 1973. *Plant Physiol.* 51:16 (Suppl.)
72. Arntzen, C. J., Haugh, M. F., Bobick, S. 1973. *Plant Physiol.* 52:569–74
73. Miller, R. J., Koeppe, D. E. 1971. *Science* 173:67–69
74. Gengenbach, B., Miller, R. J., Koeppe, D. E., Arntzen, C. J. 1973. *Can. J. Bot.* 51:2119–25
75. Flavell, R. 1975. *Physiol. Plant Pathol.* 6:107–16
76. Ireland, C. I., Strobel, G. A. 1977. *Plant Physiol.* 60:26–29
77. Levings, C. S. III, Pring, D. R. 1976. *Science* 193:158–60

78. Forde, B. G., Oliver, J. C., Leaver, C. J. 1978. *Proc. Natl. Acad. Sci. USA* 75:3841–45
79. Nishimura, S. 1980. *Proc. Jpn. Acad.* 56:362–66
80. Kohmoto, K., Scheffer, R. P., Whiteside, J. O. 1979. *Phytopathology* 69:667–71
81. Ueno, T., Nakashima, T., Hayashi, Y., Fukami, H. 1975. *Agric. Biol. Chem.* 39:1115–22
82. Nishimura, S., Kohmoto, K., Otani, H. 1974. *Rev. Plant Prot. Res.* 7:21–32
83. Sawamura, K. 1962. *Bull. Tohoku Natl Agric. Exp. Stn.* Japan 23:163–75
84. Deleted in proof
85. Okuno, T., Ishita, Y., Sawai, K., Matsumoto, T. 1974. *Chem. Lett.* pp. 635–38
86. Deleted in proof
87. Lee, S. 1976. *Tetrahedron Lett.* 11:843–46
88. Tsuchiya, S., Yoshida, Y., Hamiuda, T. 1967. *Bull. Hortic. Res. Stn. Minist. Agric. For. Ser.* C (Morioka) (Engei Shikenjo Hokoku C Morioka). 9–19
89. Bottini, A. T., Gilchrist, D. G. 1981. *Tetrahedron Lett.* 22:2719–22
90. Bottini, A. T., Bowen, J., Gilchrist, D. G. 1981. *Tetrahedron Lett.* 22:2723–26
91. McFarland, B. L., Gilchrist, D. G. 1980. *Am. Phytopath. Soc. Abstr.* p. 125
92. Meyer, W. L., Kuyper, L. F., Lewis, R. B., Templeton, G. E., Woodhead, S. H. 1974. *Biochem. Biophys. Res. Commun.* 56:234–40
93. Meyer, W. L., Kuyper, L. F., Phelps, D. W., Cordes, A. W. 1974. *Chem. Commun.* p. 339
94. Meyer, W. L., Templeton, G. E., Grable, C. I., Jones, R., Kuyper, L. F., Lewis, R. B., Sigel, C. W., Woodhead, S. H. 1975. *J. Am. Chem. Soc.* 97:3802–9
95. Templeton, G. E. 1972. See Ref. 95, pp. 160–92
96. Rich, D. H., Mathiaparanam, P. 1974. *Tetrahedron Lett.* pp. 4037–40
97. Rich, D. H., Bhatnagar, P., Mathiaparanam, P., Grant, J. A., Tam, J. P. 1978. *J. Org. Chem.* 43:296–303
98. Rich, D. H., Bhatnagar, P. K. 1978. *J. Am. Chem. Soc.* 100:2212–18
99. Durbin, R. D., Uchytil, T. F. 1977. *Phytopathology* 67:602–3
100. Steele, J. A., Uchytil, T. F., Durbin, R. D., Bhatnagar, P., Rich, D. H. 1976. *Proc. Natl. Acad. Sci. USA* 73:2245–48
101. Steele, J. A., Uchytil, T. F., Durbin, R. D. 1977. *Biochim. Biophys. Acta* 459:347–50
102. Nelson, N., Deters, D. W., Nelson, H., Racker, E. 1973. *J. Biol. Chem.* 248:2049–55
103. Bennett, J. 1976. *Phytochemistry* 15:263–65
104. Halloin, J. M., deZoeten, G. A., Gaard, D., Walker, J. C. 1970. *Plant Physiol.* 45:310–14
105. Steele, J. A., Durbin, R. D., Uchytil, T. F., Rich, D. H. 1978. *Biochim. Biophys. Acta* 501:72–82
106. Selman, B. R., Durbin, R. D. 1978. *Biochim. Biophys. Acta* 502:29–37
107. Durbin, R. D., Uchytil, T. F. 1977. *Biochem. Genet.* 15:1143–46
108. Burk, L. G., Durbin, R. D. 1978. *J. Hered.* 69:117–20
109. Ayesu-Offei, E. N., Clare, B. G. 1971. *Aust. J. Biol. Sci.* 24:169–74
110. Auriol, P., Strobel, G., Beltran, J. P., Gray, G. 1978. *Proc. Natl. Acad. Sci. USA* 75:4339–43
111. Beltran, J. P., Strobel, G., Beier, R., Mundy, B. 1980. *Plant Physiol.* 65:554–56
112. Beltran, J. P., Strobel, G. A. 1978. *FEBS Lett.* 96:34–36
113. Jackson, L. F., Webster, R. K. 1976. *Phytopathology* 66:726–28
114. Graniti, A. 1964. *Phytopathol. Mediter.* 3:75–86
115. Ballio, A., D'Alessio, V., Randazzo, G., Bottalico, A., Graniti, A., Sparapano, L., Bosnar, B., Casinovi, C., Gribanovski-Sassu, O. 1976. *Physiol. Plant Pathol.* 8:163–69
116. Ballio, A. 1977. *Regulation of Cell Membrane Activities in Plants,* ed. E. Marré, O. Ciferri, pp. 217–23. Amsterdam: North Holland. 332 pp.
117. Barrow, K. D., Barton, D. H. R., Chain, E. B., Ohnsorge, U. F. W., Thomas, R. 1968. *Chem. Commun.* pp. 1198–1200
118. Marré, E. 1979. *Ann. Rev. Plant Physiol.* 30:273–88
119. Ballio, A., Bottalico, A., Framondino, M., Graniti, A., Randazzo, G. 1973. *Phytopathol. Mediter.* 12:22–29
120. Radice, M., Scacchi, A., Pesci, P., Beffagna, N., Marre, M. T. 1981. *Physiol. Plant.* 51:215–21
121. Cleland, R. E., Prins, H. B. A., Harper, J. R., Higinbotham, N. 1977. *Plant Physiol.* 59:395–97
122. Marré, E., Lado, P., Ferroni, A., Denti, A. B. 1974. *Plant Sci. Lett.* 2:257–65
123. Raven, J. A., Smith, F. A. 1974. *Can. J. Bot.* 52:1035–48
124. Marré, E., Lado, P., Rasi-Caldogno, F., Colombo, R. 1973. *Plant Sci. Lett.* 1:185–92

125. Marré, E., Lado, P., Rasi-Caldogno, F., Colombo, R., DeMichelis, M. I. 1974. *Plant Sci. Lett.* 3:365–79
126. Rasi-Caldogno, F., Cerana, R., Pugliarello, M. C. 1980. *Plant Physiol.* 66:1095–98
127. Dohrmann, U., Hertel, R., Pesci, P., Cocucci, S. M., Marré, E., Randazzo, G., Ballio, A. 1977. *Plant Sci. Lett.* 9:291–99
128. Stout, R. G., Cleland, R. E. 1980. *Plant Physiol.* 66:353–59
129. Aducci, P., Crosetti, G., Federico, R., Ballio, A. 1980. *Planta* 148:208–10
130. Marré, E. 1980. *Prog. Phytochem.* 6:253–84
131. Patil, S. S. 1974. *Ann. Rev. Phytopathol.* 12:259–79
132. Stewart, W. W. 1971. *Nature* 229:174–78
133. Taylor, P. A., Schnoes, H. K., Durbin, R. D. 1972. *Biochim. Biophys. Acta* 286:107–17
134. Sinden, S. L., Durbin, R. D. 1970. *Phytopathology* 60:360–64
135. Meister, A., Tate, S. S. 1976. *Ann. Rev. Biochem.* 45:559–604
136. Uchytil, T. F., Durbin, R. D. 1980. *Experientia* 36:301–2
137. Mitchell, R. E. 1978. *Physiol. Plant Pathol.* 13:37–49
138. Ferguson, A. R., Johnston, J. S. 1980. *Physiol. Plant Pathol.* 16:269–75
139. Mitchell, R. E., Bieleski, R. L. 1977. *Plant Physiol.* 60:723–29
140. Tam, L. Q., Patil, S. S. 1972. *Plant Physiol.* 49:808–12
141. Kwok, O. C. A., Ako, H., Patil, S. S. 1979. *Biochem. Biophys. Res. Commun.* 89:1361–68
142. Patil, S. S., Youngblood, P., Christianson, P., Moore, R. E. 1976. *Biochem. Biophys. Res. Commun.* 69:1019–27
143. Mitchell, R. E., Young, H. 1978. *Phytochemistry* 17:2028–29
144. Ichihara, A., Shiraishi, K., Sato, H., Sakamura, S., Nishiyama, K., Sakai, R., Furusaki, A., Matsumoto, T. 1977. *J. Am. Chem. Soc.* 99:636–37
145. Sakai, R., Nishiyama, K., Ichihara, A., Shiraishi, K., Sakamura, S. 1979. *Ann. Phytopath. Soc. Jpn.* 45:645–53
146. Gross, D. C., DeVay, J. E. 1977. *Phytopathology* 67:475–83
147. Gross, D. C., DeVay, J. E., Stadtman, F. H. 1977. *J. Appl. Bacteriol.* 43:453–63
148. Mitchell, R. E., Durbin, R. D. 1981. *Physiol. Plant Pathol.* 18:157–68
149. Zentmeyer, G. A. 1942. *Science* 95:512–13
150. Salemink, C. A., Rebel, H., Kerling, C. P., Tschernoff, V. 1965. *Science* 149:202–3
151. Strobel, G. A., VanAlfen, N., Hapner, K. D., McNeil, M., Albersheim, P. 1978. *Biochim. Biophys. Acta* 538:60–75
152. Nordin, J., Strobel, G. A. 1981. *Plant Physiol.* 67:1208–13
153. Scheffer, R., Elgersma, D. 1981. *Physiol. Plant Pathol.* 18:27–32
154. VanAlfen, N. K., Turner, N. C. 1975. *Plant Physiol.* 55:312–16
155. Stevenson, K. J., Slater, J. A., Takai, S. 1979. *Phytochemistry* 18:235–38
156. Ries, S. M., Strobel, G. A. 1972. *Plant Physiol.* 49:676–84
157. Strobel, G. A. 1970. *J. Biol. Chem.* 245:32–38
158. Strobel, G. A., Talmadge, K. W., Albersheim, P. 1972. *Biochim. Biophys. Acta* 261:365–74
159. Goodman, R. N., Huang, J. S., Huang, P. Y. 1974. *Science* 183:1081–82
160. Eden-Green, S. J., Knee, M. 1974. *J. Gen. Microbiol.* 81:509–12
161. Jansson, P. E., Kenne, L., Lundberg, B. 1975. *Carbohydr. Res.* 45:275–82
162. Holzworth, G. 1976. *Biochemistry* 15:4333–39
163. Sutton, J. C., Williams, P. H. 1970. *Can. J. Bot.* 48:645–51
164. Kuo, T. T., Lim, B. C., Li, C. C. 1970. *Bot. Bull. Acad. Sin.* 11:36–45
165. Nachmias, A., Barash, I., Solel, Z., Strobel, G. A. 1977. *Physiol. Plant Pathol.* 10:147–57
166. Harvan, D. J., Pero, R. W. 1976. *Adv. Chem. Ser.* pp. 344–55
167. Pero, R. W., Main, C. E. 1970. *Phytopathology* 40:1570–73
168. Nukina, M., Marumo, S. 1977. *Tetrahedron Lett.* 37:3271–72
169. Grove, J. F. 1964. *J. Chem. Soc.* pp. 3234–39
170. Robeson, D., Gray, G., Strobel, G. 1982. *Phytochemistry.* In press
171. Pringle, R. 1971. *Plant Physiol.* 48:756–59
172. Weber, D. J., Sweeley, C. C., Liesch, J. M., Staffeld, G. C., Anderson, M., Scheffer, R. P. 1981. *13th Int. Bot. Congr. Proc.* p. 156
173. Scheffer, R. P., Ullstrup, A. J. 1965. *Phytopathology* 55:1037–38
174. Kuo, M., Scheffer, R. P. 1971. *Phytopathology* 60:1391–94
175. Meehan, F. L., Murphy, H. C. 1947. *Science* 106:270–71
176. Scheffer, R. P., Samaddar, K. R. 1970. *Adv. Phytochem.* 3:123–42
177. Dorn, F., Arigoni, D. 1972. *J. Chem. Soc. Commun.* 1342–43

178. Wheeler, H., Black, H. S. 1963. *Am. J. Bot.* 50:686–93
179. Samaddar, K. R., Scheffer, R. P. 1971. *Physiol. Plant Pathol.* 1:319–28
180. Scheffer, R. P., Pringle, R. B. 1961. *Nature* 191:912–13
181. Pringle, R. B., Scheffer, R. P. 1967. *Phytopathology* 57:530–32
182. Wolpert, T. J., Dunkle, L. D. 1980. *Phytopathology* 70:872–76
183. Schertz, K. F., Tai, Y. P. 1969. *Crop Sci.* 9:621–24
184. Gardner, J. M., Mansour, I. S., Scheffer, R. P. 1972. *Physiol. Plant Pathol.* 2:197–206
185. Barash, I., Karr, A., Strobel, G. 1975. *Plant Physiol.* 55:646–51
186. Starratt, A. N., Stoessl, A. 1977. *Can. J. Chem.* 55:2360–62
187. Barash, I., Pupkin, G., Netzer, D., Kashman, Y. 1982. *Plant Physiol.* In press
188. Smedegard-Petersen, V. 1977. *Physiol. Plant Pathol.* 10:203–11
189. Bach, E., Christensen, S., Dalgaard, L., Larsen, P. O., Olsen, C. E., Smedegard-Petersen, V. 1979. *Physiol. Plant Pathol.* 14:41–46
190. Smedegard-Petersen, V. 1977. *Physiol. Plant Pathol.* 10:213–20
191. Sato, H., Konoma, K., Sakamura, S., Furusaki, A., Matsumoto, T., Matsuzaki, T. 1981. *Agric. Biol. Chem.* 45:795–97
192. Matern, U., Strobel, G., Shepard, J. 1978. *Proc. Natl. Acad. Sci. USA* 74:4935–39
193. Shepard, J. F., Bidney, D., Shahin, E. 1980. *Science* 208:17–24

Ann. Rev. Biochem. 1982. 51:335–64

MECHANISMS OF INTRACELLULAR PROTEIN BREAKDOWN

Avram Hershko and Aaron Ciechanover

Unit of Biochemistry, Faculty of Medicine, Technion-Israel Institute of
Technology, Haifa, Israel

CONTENTS

PERSPECTIVES AND SUMMARY

Most cellular proteins are in a dynamic state of constant turnover. Protein turnover is extensive and is highly selective: specific proteins are degraded within cells at widely different rates. This process is involved in basic cellular functions, since the levels of intracellular proteins are determined both by the rates of synthesis and the rates of degradation. An additional role of protein breakdown appears to be the provision of amino acids in times of need, since overall rates of protein degradation are greatly accelerated under conditions of nutritional or hormonal deprivation. Another

335

0066-4154/82/0701-0335$02.00

important function is the disposal of defective proteins: abnormal proteins produced by specific mutations or by various experimental manipulations are selectively recognized and rapidly degraded.

Until recently, studies on protein breakdown have been concerned mainly with the description of its various aspects at the phenomenological level, while the underlying mechanisms were not known. That some unusual mechanisms may carry out this process has long been suspected as it was observed that intracellular protein breakdown in all organisms has an absolute requirement for cellular energy. The energy dependence is not expected on thermodynamic grounds, since proteolysis per se is an exergonic process. A reasonable expectation was that energy is required for control or specificity, and that the mechanisms of intracellular protein breakdown are different from those of known proteases.

Recent information indicates that in mammalian cells, there are separate lysosomal and nonlysosomal mechanisms that may be involved in different aspects of protein breakdown. Strong evidence now indicates that lysosomal autophagy plays a major role, most prominently under conditions of nutritional deprivation. This conclusion is based on extensive correlations between rates of overall protein degradation and alterations in the structural or functional characteristics of the lysosomal system, as well as on selective effects of lysosomal inhibitors on protein breakdown.

Some insight into the nonlysosomal (but energy-requiring) mechanisms of protein breakdown was gained by the establishment of an ATP-dependent cell-free proteolytic system from reticulocytes. The system is composed of several essential components. One of these, a small heat-stable polypeptide, was subsequently identified as ubiquitin, a universally occurring polypeptide of previously unknown function. Ubiquitin is covalently linked to protein substrates in an ATP-requiring reaction, in which several molecules of the polypeptide are conjugated to one molecule of protein in isopeptide linkages. This may be the initial signal event in protein degradation, since the ubiquitin-protein conjugates are degraded rapidly. In addition to the above system, different ATP-dependent proteases have been described in *E. coli.*

The aim of this review is to describe and critically evaluate these recent developments. Because of space limitations, we confine our discussion to data related directly to the mechanisms of lysosomal and ATP-dependent protein breakdown, and we apologize for the omission of much other important recent work in this field. For detailed discussions of the various characteristics and physiological roles of intracellular protein breakdown, the reader is referred to excellent earlier reviews (1–7) and reports of recent symposia (8–11).

INVOLVEMENT OF LYSOSOMES IN PROTEIN BREAKDOWN

The possible participation of lysosomes in intracellular protein breakdown has been considered for a long time, since lysosomes contain a set of acidic proteases and are capable of hydrolyzing endocytozed exogenous proteins (12). Also, a process of autophagic engulfment of cellular constituents has been observed under a variety of pathologic or physiological conditions (12). It was not clear, however, whether apart from the role of autophagocytosis in the disposal of injured cytoplasmic components, it also has a function in the continuous process of intracellular protein turnover. Among the most compelling evidence for the participation of lysosomes in protein turnover is provided by Mortimore and associates with perfused rat liver. They had first observed that the rate of general protein breakdown increases markedly during the perfusion of liver with unsupplemented medium, and that this enhancement can be suppressed by insulin (13). A similar inhibition of enhanced protein breakdown was exerted by supplementation with amino acid mixtures (14). Insulin or amino acids had no influence on the rate of protein synthesis, and it thus seems that in the liver, the regulation of bulk protein metabolism is exerted at the site of protein degradation. Another regulatory agent is glucagon, which markedly accelerates hepatic protein degradation (15); it seems, however, that the effect of this hormone is secondary to the lowering of intracellular amino acid levels (16). Such regulatory effects on protein degradation are not particular to the liver, but occur in most cell types, and probably represent a general regulatory mechanism of this process by nutrients, specific hormones, or growth-promoting agents. Thus, the rate of the degradation of generally labeled cellular proteins is accelerated by serum deprivation in cultured hepatoma cells (17) and fibroblasts (18), and it can be reversed by insulin (18) or a variety of growth-promoting factors (19–21). The realization that this deprivation-induced enhancement of protein breakdown is caused by increased lysosomal autophagy was due, to a large extent, to the extensive work of Mortimore and co-workers and was aided by inhibitor studies of other investigators, as described below.

Correlations with Structural Alterations of the Lysosomal System

The initial experiments indicating that lysosomal autophagy may play a role in enhanced protein breakdown were observations on changes of osmotic sensitivity of lysosomes from livers perfused under conditions affecting protein breakdown (22). When prepared from livers perfused with unsupplemented medium (which enhances protein breakdown), the sensitivity of lysosomes to lysis in hypotonic sucrose was markedly increased. This was prevented by additions of insulin, amino acids, or cycloheximide, which also suppress the acceleration of protein breakdown. Since it was known from previous studies that large autophagic vacuoles are more sensitive to osmotic shock than primary lysosomes (23) it was assumed that the observed osmotic changes reflect increased accumulation of autophagic elements. A striking feature of these lysosomal changes was the rapidity of transition from one state to another: when insulin and amino acids were added after a prior treatment with deficient medium, the osmotic alterations were rapidly reversed with a half-time of about 8 min (22). Assuming that insulin and amino acids prevent the formation of autophagic vacuoles, this would indicate that in the deprived condition, these vacuoles are in a state of rapid turnover of formation and regression. Similar effects of insulin on lysosomal latency have been found in perfused heart (24).

Further evidence supporting the above conclusions showed that the buoyant density of lysosomes is increased by perfusion under conditions of nutritional deprivation (25). This was detected by a significant shift of lysosomal marker enzymes to higher density in sucrose gradients. Again, this physical alteration was correlated to rates of protein breakdown, since it was prevented by the supplementation of insulin and amino acids. The increased density of lysosomes probably reflects increased autophagy, which results in the entrapment of heavy cytoplasmic elements, such as glycogen. Indeed, it was found that the omission of glucose from the perfusate medium, which reduces glycogen content of liver cells, also caused a significant reduction in the shift of lysosomal elements to higher density during amino acid deprivation (26).

These interpretations were fully corroborated by direct morphological examination of liver tissue following treatments that effect protein degradation. Electron micrographs of tissue samples following perfusion with unsupplemented medium showed the appearance of numerous enlarged lysosomal elements or autophagic vacuoles, in contrast to the dense bodies seen in untreated liver (25, 27). Here again, the appearance of autophagic vacuoles was completely prevented by the supplementation of amino acids during perfusion (27).

Relationship to Degradable Protein Entrapped in Lysosomes

In parallel to the above studies correlating physical and morphologic alterations with rates of proteolysis, Mortimore and co-workers used another functional approach. The rationale of these experiments was that though the uptake of proteins by lysosomes cannot be reconstituted in vitro (28), it might be possible to follow the degradation of protein that had already been engulfed by lysosomes in intact cells, by the subsequent incubation of such lysosomes in vitro. Indeed, a good correlation was found between the relative rates of the release of free amino acids from homogenates of livers perfused under various conditions (treatments with unsupplemented medium or with amino acids, insulin, or glucagon) and the rates of protein breakdown in similarly treated intact livers (29). Proteolysis occurred only in incubations of whole homogenates and not in particle-free supernatants. The particulate fraction capable of releasing acid-soluble material from prelabeled endogenous proteins was identified as lysosomal by the comigration of the products of endogenous proteolysis with lysosomal enzymes in density equilibrium centrifugation (30). No acid-soluble label was released upon incubation of unlabeled lysosomes with labeled cytosol (while the opposite mixture did release acid-soluble labeled material), which indicates that the labeled protein substrate was contained within the lysosome prior to homogenization (30, 31). It follows that the amount of endogeous proteolysis in lysosomes incubated in vitro reflects the size of the degradable protein pool entrapped within the lysosomes in the intact tissue. Thus, the observed correlations between protein breakdown in intact livers and incubated homogenates indicate corresponding changes in the steady-state levels of intralysosomal degradable protein in the different physiological situations. This information was used subsequently for the quantitative estimation of lysosomal proteolysis (see below).

Quantitative Determination of Rates of Lysosomal Proteolysis

The above studies strongly indicated that lysosomal autophagy is involved in deprivation-enhanced protein breakdown and in its regulation by amino acids and hormones. Still, the question remained whether lysosomal proteolysis is responsible for all, or only a part of the above phenomena. To answer this problem, the quantitative estimation of lysosomal proteolysis and its comparison to absolute rates of protein breakdown in the perfused liver were required. Rates of lysosomal proteolysis in the intact tissue can be estimated if the rate of the turnover of autophagic particles and the amount of protein entrapped within these particles are known. The turnover rate of autophagic vacuoles has been estimated from the rate of the regression of these vacuoles when amino acids and insulin are added following nutritional deprivation. A half-life of approximately 8 min was estimated by the osmotic sensitivity method (22), and this value was confirmed by quantitative electron microscopic observations (32). From these data the rate constant of the turnover of autophagic vacuole contents can be calculated, assuming that amino acids or insulin prevent the formation of autophagic vacuoles and have no influence on the rate of their regression. A further assumption is that the regression of the autophagic vacuoles is accompanied by the total digestion of entrapped protein.

The quantity of protein entrapped within the lysosomes has been estimated by several methods. The protein content of lysosomes can be determined directly, provided that they are well separated from mitochondria on sucrose gradients. A good separation of lysosomes was achieved following iron loading, which increases the density of lysosomes due to their content of ferritin (33). The increase in the protein content of the heavy lysosomal peak was taken as the measure of the increase in digestible intralysosomal protein pool. The rate of lysosomal proteolysis was computed by multiplying this protein pool by the rate constant of the turnover of autophagic vacuoles. The calculated rate was very close to the observed increase in the rate of protein breakdown by amino acid deprivation in intact perfused liver (33).

Another method used by Mortimore and co-workers was the analysis of electron micrographs to obtain a stereological estimate of the volumes of lysosomal vacuoles. Assuming that the concentration of protein within the vacuoles is similar to that of mean liver protein, the amount of intralysosomal protein can be calculated and hence the rates of lysosomal protein degradation. Again, the calculated rates of lysosomal proteolysis agreed quantititatively with the rates of overall protein degradation over a range of amino acid concentrations (32).

A third method utilized previous information indicating that proteins entrapped within lysosomes in intact cells can be degraded by subsequent incubation of lysosomes in vitro (see above). If proteolysis of endogenous proteins is allowed to go to completion (by prolonged in vitro incubation), and if proteolysis then ceases because of the exhaustion of endogenous substrates, a direct estimate of intralysosomal degradable protein pool can be obtained. The absolute rates of lysosomal proteolysis calculated by this method were in good agreement with rates of overall proteolysis in various treatments, such as perfusion with unsupplemented medium or with glucagon, amino acids, or insulin (34). These, and the above experiments provided impressive evidence that virtually all of deprivation-enhanced degradation of intracellular proteins is carried out by lysosomal autophagy. Furthermore, from the finding that there is an intralysosomal pool of degradable protein, also under basal conditions (in untreated liver or following perfusion with amino acids or insulin), it was concluded that the lysosomal system is involved in basal protein degradation as well [(34) and see below].

Effects of Lysosomal Inhibitors

In addition to the extensive correlations described above, many recent reports have shown that agents that specifically inhibit lysosomal proteolysis also inhibit deprivation-enhanced protein breakdown in animal cells.

Poole and co-workers (35) first showed that chloroquine partially inhibits protein breakdown in cultured embryo fibroblasts. This compound inhibits lysosomal cathepsin B activity (35); in addition (and perhaps more importantly), chloroquine can be regarded as an example of the so-called "lysosomotropic" agents. These are generally weak bases that accumulate within lysosomes and thereby increase intralysosomal pH. It is assumed that within the lysosomes, the base becomes protonated and that the protonated form cannot readily diffuse out through the lysosomal membrane (36). In fact, chloroquine is concentrated about 1000-fold within lysosomes (35, 36) and direct measurements of intralysosomal pH in living macrophages showed a marked increase following exposure to chloroquine (37). This increase (to about pH 6.5) would be sufficient to prevent most of the action of lysosomal acid proteinases. Poole and co-workers also showed that chloroquine mainly inhibits enhanced protein breakdown in the absence of serum, but has much less effect on basal protein breakdown in the presence of serum (36).

A specific inhibitor of lysosomal proteinase was first utilized by Dean (38) to study the involvement of lysosomes in intracellular protein breakdown. The pentapeptide, pepstatin, inhibits carboxyl proteinases such as cathepsin D (39), but does not penetrate cellular membranes. It was therefore introduced into cells within multilammelar liposomes, which are endocytozed and thus enter lysosomes. Addition of liposomes containing pepstatin to the perfused rat liver inhibited protein breakdown by about 50% (38). Since the perfused liver is in a state of enhanced protein catabolism (unless insulin or high concentrations of amino acids are supplemented), it cannot be concluded from this study which type of protein breakdown was affected by pepstatin.

Other workers have subsequently tested the effects of a variety of proteinase inhibitory peptides, isolated from Actinomycetes by Umezawa and co-workers (39). Knowles & Ballard (40) showed that leupeptin [an inhibitor of cathepsin B and some other serine proteases (41)] and antipain [an inhibitor of cathepsins A, B, papain, and trypsin (41)] inhibit the degradation of normal proteins, but not canavanine-containing abnormal proteins in Reuber hepatoma cells. Libby & Goldberg (42) reported that leupeptin inhibits protein breakdown in skeletal and heart muscle, which may account for the observations of Stracher and co-workers (43, 44) that leupeptin, antipain, and pepstatin delay the degeneration of dystropic muscle cells. A further study has shown (45) that leupeptin, antipain, and chymostatin inhibit the degradation of long-lived proteins in cultured hepatocytes, but not of short-lived proteins or abnormal protein containing amino acid analogues (for definition of these protein classes, see below). Quite similar effects were observed in hepatocytes by others (46, 47). Since most of the peptide inhibitors may inhibit other intracellular proteinases in addition to lysosomal proteinases (with the exception of pepstatin, which appears to inhibit only cathepsin D), the above studies do not provide rigorous proof for the participation of lysosomes in protein breakdown. They do support other evidence for the existence of multiple pathways in intracellular protein breakdown (see below).

EVIDENCE FOR MULTIPLE PATHWAYS IN PROTEIN DEGRADATION

That several distinct mechanisms operate in intracellular protein breakdown has been pointed out by several authors (4, 36); here we mainly discuss additional recent work. For this discussion, we define several classes of cellular proteins with distinct features of degradation: (a) *Long-lived proteins* constitute the bulk of cellular proteins with relatively slow turnover rates. The degradation of this class can be studied by a prolonged exposure of cells to labeled amino acid (which is sufficient to label most cell proteins), followed by a "chase" with unlabeled amino acid to allow the breakdown of the rapidly degrading class. (b) *Short-lived proteins* are a class of normal cellular proteins of limited size, but of exceptionally high turnover rate. This protein population can be preferentially labeled by a short pulse with a radioactive precursor. (c) *Abnormal proteins* are produced with amino acid analogues or puromycin, or by specific mutations (3); they are broken down even more rapidly than short-lived proteins. The above classification is entirely operational and quite arbitrary. For example, when short-lived proteins are preferentially labeled by a short pulse, a sizable portion of long-lived proteins also becomes labeled (19, 48), and each "class" consists of many proteins with heterogeneous degradation rates. Nevertheless, this rough classification allows us to discuss evidence for the multiplicity of proteolytic pathways.

Most available evidence indicates that degradation of most short-lived proteins or of abnormal proteins is probably nonlysosomal in nature. The degradation of these classes is not much influenced by nutritional deprivation, hormones, or inhibitors of protein synthesis (19, 40, 48), as opposed to the breakdown of long-lived proteins. More recently, it was shown that inhibitors of lysosomal proteases have no influence on the breakdown of abnormal or of short-lived normal cellular proteins [(40, 45) and see above]. Decreased temperature affects much more the degradation of long-lived proteins than that of short-lived normal or analogue containing proteins (45), which indicates a difference in the activation energy of the rate-limiting steps for these processes. Though puromycyl peptides are degraded more rapidly in the normal than in the regenerating liver (49) (as are long-lived proteins), the interpretation of these data has been challenged by

Hendil (50) who found that nongrowing fibroblasts are more sensitive to the action of puromycin than are growing cells, and thus may produce a higher proportion of shorter and more rapidly degradable puromycyl peptides.

With regard to the degradation of long-lived proteins, a distinction can be made between "enhanced" (under conditions of nutritional deprivation) and "basal" (fully supplemented with nutrients, insulin, or growth-promoting factors) states. Little doubt remains that most or all of enhanced protein breakdown in mammalian cells is carried out by the lysosomal system, but there are conflicting reports concerning the involvement of the lysosomal pathway in the basal degradation of long-lived proteins. Strong evidence exists indicating that basal protein breakdown is insensitive to agents that inhibit lysosomal proteolysis. As noted above, chloroquine and a variety of lysosomotropic agents inhibit enhanced protein breakdown (in the absence of serum) to a much greater degree than basal degradation (in the presence of serum) in cultured rat embryo fibroblasts (36). The effects of chloroquine on the degradation of endocytozed exogenous proteins and of cellular endogenous proteins were compared in cultured mouse peritoneal macrophages under basal conditions (36). Since macrophages readily endocytose a variety of proteins, it was possible to compare endogenous and exogenous protein breakdown in the same cells. It was found that chloroquine inhibits the degradation of exogenous proteins much more than it does that of endogenous proteins. This difference persisted when the source of exogenous material was labeled macrophage proteins, so that the breakdown of similar protein populations could be compared. It was concluded that the degradation of endocytosed proteins, or that of cellular proteins under conditions of nutritional deprivation, occurs at a cell compartment different from that involved in the degradation of cellular proteins in the process of normal protein turnover. The authors cautioned, though, that these may represent different subsets of lysosomes (36).

In agreement with the above results Ballard and co-workers found that in the yolk sac, the degradation of exogenous internalized protein is much more inhibited by lysosomotropic agents (51) or microbial proteinase inhibitors (52) than is endogenous protein breakdown. Similarly, lysosomal breakdown of membrane acetylcholine receptor in cultured myotubes is drastically inhibited by the above agents (53), which however have little if any effect on endogenous protein breakdown in the same cells (54).

Amenta et al (55) showed that microtubular inhibitors such as vinblastine, vincristine, or colchicine inhibit enhanced proteolysis in cultured rat embryo fibroblasts incubated in serum-free medium, but not basal degradation occurring in the complete medium. The microtubular poisons were known to inhibit the function of the lysosomal-vacuolar system and to block the digestion of exogenous proteins. In both hepatocytes (see below), and cultured fibroblasts, ammonia inhibits only deprivation-enhanced, but not basal proteolysis (56). Ammonia is a potent inhibitor of protein breakdown (57) with a lysosomotropic mode of action (58), and can be used to distinguish between lysosomal and nonlysosomal pathways of protein breakdown in isolated hepatocytes. Isolated hepatocytes exhibit a marked protein catabolic state, and thus a large proportion of protein degradation resembles that induced by nutritional deprivation. At maximally effective concentrations, NH_4Cl inhibits about 75% of the degradation of long-lived proteins (59). Other lysosomotropic agents, such as methylamine or chloroquine (59), as well as a great number of other amines (60) inhibit protein breakdown to a similar maximal extent, but do not suppress further the residual protein breakdown in the presence of ammonia (47, 59). Leupeptin and antipain inhibit breakdown of the ammonia-resistant component less effectively than the ammonia-sensitive fraction (47), which was taken as evidence for a nonlysosomal pathway in basal protein breakdown.

As opposed to the above findings, Dean (61) has reported that protein breakdown in cultured macrophages is inhibited by pepstatin equally in the presence or absence of serum. Similarly, in perfused liver and heart, the joint addition of leupeptin and pepstatin inhibit deprivation-enhanced and basal protein breakdown to a similar degree (62), while in cultured hepatocytes, leupeptin and chymostatin inhibit protein breakdown similarly in the presence or absence of serum (45). It was concluded from these results that the lysosomal system is involved in both basal and enhanced types of protein breakdown.

It is difficult to explain satisfactorily these apparently conflicting observations in different experimental systems. Most of the evidence for basal protein breakdown resistant to lysosomal inhibitors was obtained from growing cultured cells where serum deprivation is accompanied by the cessation of cell multiplication (18, 20). It is possible (45, 54) that the proportion of the lysosomal pathway is greater in nongrowing tissues, such as the normal liver or muscle. In addition, the definition of the basal state is largely experimental, and depends on the conditions employed in a particular study.

Both types of the above mentioned studies are based on the effects of inhibitors, and inhibitor studies are always subject to uncertainties concerning the specificity of the agent or the completeness of its action. However, studies in which inhibitors were not used also yielded conflicting conclusions in different experimental systems. As mentioned earlier, there is evidence for the existence of a degradable pool of intralysosomal proteins in the liver under basal conditions (29, 34). On the other hand, proteins microinjected into cultured cells are degraded mainly in the cytosol (63).

Red cell–mediated microinjection appears to be an excellent tool for following the degradation of specific cytosolic proteins, since microinjected proteins are degraded at specific and heterogenous rates (64) and show the normal correlations between protein half-lives and molecular size or charge (65); also the degradation of microinjected glutamine synthetase is subject to regulation by glutamine (66) as is the case with the endogenous enzyme (67).

A conceptual difficulty in the assumption that all basal protein breakdown is carried out by the lysosomal pathway is how to account for the heterogeneity of the turnover rates of specific proteins under normal steady-state conditions (3, 68). Although the possibility of a selective adsorption of proteins to lysosomal membranes has been raised (69–72), there is little evidence to support this notion except for the correlation between hydrophobicity and protein half-lives (70–72). On the contrary, when protein breakdown is accelerated by severe diabetes or starvation, the normal correlations between protein half-lives and subunit size or protein isoelectric points tend to disappear (73, 74). The simplest explanation of these results would be that the enhanced lysosomal autophagy that is associated with these conditions is essentially nonselective. It is possible, though, that under basal conditions a different type of vacuolar sequestration takes place, which may have other mechanisms for selectivity.

In spite of these uncertainties, the general notion that emerges from the above studies does indicate a multiplicity of proteolytic pathways. Since all types of intracellular protein breakdown require energy [(3) and see below], a search for energy-dependent nonlysosomal systems appears to be justified.

ATP-DEPENDENT PROTEOLYTIC SYSTEMS

Energy Dependence of Intracellular Protein Breakdown

The requirement of intracellular protein breakdown for a supply of metabolic energy was discovered by Simpson in 1953, who observed that release of amino acids from labeled proteins in rat liver slices is inhibited by anaerobic conditions or by inhibitors of cellular energy production (75). Similar results were later reported by Steinberg & Vaughan (76). Similarly, Rapoport and co-workers found that the breakdown of proteins to "nonprotein nitrogen" in reticulocytes is blocked by the uncoupling or inhibition of oxidative phosphorylation (77). Numerous studies have since shown that the energy dependence of intracellular protein breakdown is a universal feature of this process in a variety of biological systems. For example, protein breakdown is suppressed by energy inhibitors in bacteria (78–80), yeasts (81, 82), and cultured mammalian cells (17, 19, 48), and energy is required for the degradation of both abnormal proteins (83–85) and normal cellular enzymes (17, 86, 87). For earlier literature on this topic see (3); we limit our discussion here to a few main points.

Although the effect of energy inhibitors in intact cells may reflect a direct energy requirement for protein breakdown, it may also result indirectly. For example, protein breakdown may be linked to protein synthesis through some unspecified mechanism, and thus inhibition of energy production may affect it indirectly (76). However, Hershko & Tomkins (17) used nutritional deprivation to show that inhibitors of protein synthesis block only the enhancement of protein breakdown, while inhibitors of energy production also suppress basal protein degradation. In addition, energy depletion in intact cells could interfere with several separate processes which carry out different types of intracellular protein breakdown. It appears reasonable to assume that the lysosomal pathway may require cellular energy, either directly or indirectly. The initial events in autophagy, although still obscure, may involve processes of membrane assembly or rearrangement, which may well require energy at some or several sites. In addition, energy may be required for the maintenance of acidic intralysosomal pH, necessary for the action of acidic proteases. Several investigators have provided evidence for an ATP-requiring lysosomal proton pump (37, 88–92), although this has been questioned (93), and at least part of intralysosomal acidity is due to a proton Donnan equilibrium (93, 94).

Energy dependence of nonlysosomal intracellular protein breakdown may not be as surprising as it seems at first glance. Though exergonic, it is a highly selective and specific process, and energy may be required to attain this high degree of selectivity. Although the biochemical mechanisms involved have resisted elucidation for a long time, several cell-free ATP-dependent proteolytic systems have been recently established. One of the most extensively studied is the ATP-dependent proteolytic system from reticulocytes, which is described in some detail below.

ATP-Dependent Proteolytic System from Reticulocytes: The Ubiquitin Pathway

PROTEIN BREAKDOWN IN INTACT RETICULOCYTES   The reticulocyte is a highly specialized cell that synthesizes predominantly one protein, hemoglobin (95). It has a relatively simple cellular structure, lacking nucleus, endoplasmic reticulum, and Golgi membranes (96). When reticulocytes were incubated in physiological saline, Rapoport and co-workers (77) observed a massive release of amino acids. No comparable protein breakdown occurred in mature erythrocytes. This breakdown of reticulocyte proteins is energy dependent, since it was blocked by anaerobiosis or 2,4-dinitrophenol (77). When reticulocytes are labeled with radioactive amino acids, very little degradation of labeled protein takes place, unless abnormal proteins have been synthesized during labeling (see below). This indicates that under normal conditions, most of the proteins that are degraded had been synthesized at a much earlier stage of erythroid cell differentiation. Rapoport and co-workers suggest that the major endogenous proteins that are degraded in the reticulocytes are "stromal" (mainly mitochondrial) proteins, and that such degradation plays a role in reticulocyte maturation [(97) and see also below].

The reticulocyte proteolytic system is highly active in the degradation of abnormal globin chains, whereas normal hemoglobin is stable. Rabinowitz & Fisher (98, 99), first showed that globin chains that incorporate the lysine analogue S-(β-aminoethyl)-cysteine (4-thialysine) or the valine analogue, t-α-amino-β-chlorobutyric acid are degraded with extreme rapidity. Rapid degradation of puromycyl peptides also occurs in rabbit reticulocytes (100). In addition, the degradation of certain mutant hemoglobin variants (101–103) or of excess of normal α-chains in β-thalassemic patients (104) has been noted (reviewed in 3).

Except for its higher activity, the general characteristics of abnormal protein degradation in reticulocytes resemble those observed in other animal cells or even in bacteria. The degradation of amino acid analogue-containing globin chains is blocked by inhibitors of ATP formation (84, 85), but not by inhibitors of protein synthesis (85, 105). As in other cases of massive formation of abnormal proteins (106, 107), abnormal globin chains tend to associate in large aggregates or in a sedimentable form (105, 108). It is not clear whether this aggregation has any significance other than reflecting the hydrophobicity of denatured proteins.

The physiological function of this active proteolytic system in reticulocytes is not known with certainty. It may serve to remove globin molecules containing biosynthetic errors, or excessive amounts of one species of globin that is not assembled into hemoglobin (though the synthesis of α and β chains of globin is normally well balanced (95)). In addition, the proteolytic system may have a major role in maturation of the reticulocyte, which is characterized by the disappearance of mitochondria, ribosomes, and many cellular proteins and enzymes that have no further functions in the mature erythrocyte (109). The suggestion that mitochondrial proteins are the main substrate for the energy-dependent degradative system, first proposed by Rapoport and co-workers more than 20 years ago (77), was corroborated by recent work from the same laboratory (97, 110). Absolute rates of protein breakdown in incubated reticulocytes, measured by a technique based on the isotopic dilution of lysine, agreed quantitatively with the amounts of protein lost from the mitochondrial fraction during incubation (97). Furthermore, by the use of the ATP-dependent cell-free system from reticulocyte cytosol (see below), these investigators could show directly that mitochondria, but not ribosomes or the cytosol, contain the bulk of endogenous substrates for proteolysis (97). According to these authors, the initial event in the breakdown of mitochondria is the peroxydation of lipids of the mitochondrial membrane by a specific lipoxygenase, which leads to their lysis (111). In agreement with the above hypothesis, it was shown that salicylhydroxamic acid, which inhibits the lipoxygenase, also inhibits markedly the degradation of mitochondrial proteins (110). An interesting question is what mechanisms determine the specific recognition of a protein by the degradative system as being of mitochondrial origin. Activity of the energy-dependent proteolytic system decreases sharply during maturation of the reticulocytes (85, 112), and is completely absent in mature red cells (77, 84, 85). One wonders whether the proteolytic system is "self-terminating" during maturation, or whether some other factors play a role in its inactivation.

We became interested in reticulocytes as a model system because of the possibility of following the fate of one predominantly synthesized protein (analogue-containing globin) in a highly active degradative system. Analysis of the size distribution of labeled analogue-containing polypeptides on high cross-linkage SDS-polyacrylamide gels showed most of the label in complete globin chains, and no significant amounts of cleavage fragments smaller than globin could be detected in the course of a "pulse-chase" incubation. In addition, there was no accumulation of cleavage fragments when degradation was blocked by energy deprivation (85). We concluded that the initial reaction(s) in the degradation of globin must be strongly rate limiting relative to the subsequent rapid proteolysis of intermediary cleavage fragments, and that energy is required at or before the initial cleavage reactions, otherwise such fragments would accumulate under conditions of energy deprivation (85).

CHARACTERISTICS OF THE CELL-FREE SYSTEM   A cell-free ATP-dependent proteolytic system from reticulocyte lysates was first established by Etlinger & Goldberg (84). The system did not seem to be of lysosomal origin, since it was soluble and had a slightly alkaline pH optimum. Furthermore, the system is not inhibited by peptides that inhibit lysosomal proteases, such as leupeptin, pepstatin, or chymostatin (105). ATP was specifically required in the presence of Mg^{+2} ions. The following evidence indicates that the cell-free system is similar to that responsible for the degradation of abnormal proteins: (a) the system requires an energy-rich compound, as does protein degradation in intact cells; (b) the cell-free system degrades abnormal protein but not normal hemoglobin, as is the case in intact reticulocytes; and (c) the cell-free proteolytic system is inhibited by the same agents that inhibit protein degradation in intact reticulocytes, such as sulfhydryl blocking agents, chloromethyl ketones, metal chelators (84), or hemin (113). The comparison of the effects of some of these inhibitors in intact cells and in cell-free lysates is not conclusive, since sulfhydryl reagents inhibit protein breakdown in intact reticulocytes only at concentrations that drastically reduce cellular ATP levels (85). However, additional evidence strongly indicates that the cell-free system is very similar to that carrying out protein breakdown in intact reticulocytes. Botbol & Scornik (114) have used bestatin, an inhibitor of aminopeptidases, to cause the accumulation of intermediates in the degradation of abnormal proteins in mouse reticulocytes. Under usual conditions, the only detectable products are free amino acids, but in the presence of bestatin a marked accumulation of di- and tripeptides was observed. A similar accumulation of peptides by bestatin was found in the cell-free system in the presence of ATP (115). Most significantly, the fingerprint pattern of peptides accumulated in cell-free extracts appeared strikingly similar to that in intact reticulocytes, which indicates that the cell-free system represents proteolytic events identical to those occurring in intact cells (115).

RESOLUTION OF THE COMPONENTS OF THE SYSTEM   To elucidate the role of ATP and the intermediary reactions of the ATP-dependent proteolytic system, the resolution and purification of its components was necessary. Several possibilities for the role of ATP in protein breakdown were considered: (*a*) ATP may be required for the covalent modification of the substrate protein, which would be a signal for its degradation; (*b*) covalent modification of a proteolytic enzyme may be required for its activation; (*c*) ATP hydrolysis (without protein modification) may be required for the action of an unidentified protease; (*d*) there may be allosteric activation by ATP of an unknown proteolytic enzyme. In cases *a* and *b,* there should be at least two necessary components, i.e. the modifying enzyme and the protease. Therefore, our initial approach was to separate the lysate into crude fractions and to search for complementation of activities between the fractions. In fact, the reticulocyte system has been resolved into several components necessary for its activity (116, 117). Initially, lysates from ATP-depleted reticulocytes were separated on DEAE-cellulose into two crude fractions: unadsorbed material (Fraction I), which contains hemoglobin and a few basic or neutral proteins, and a high-salt elute (Fraction II), which consists of most of the nonhemoglobin proteins that bind to the resin. Neither fraction had appreciable ATP-dependent proteolytic activity by itself, but activity was restored upon the combination of the two fractions (116).

The active component in Fraction I showed rather unusual features. It was remarkably stable to heating at 90°C, but appeared to be of polypeptide nature by the following criteria: it was nondialysable, it was precipitable by ammonium-sulfate, and its activity was destroyed by treatment with proteo-lytic enzymes such as chymotrypsin or pronase. By gel filtration on Se-phadex G-75, molecular weight of approximately 9,000 was determined (116). The heat-stable polypeptide, designated as APF-1 (ATP-dependent Proteolysis Factor I) was purified to apparent homogeneity and character-ized (118). It was found to be present in several tissues of the rat in significant amounts (118). Subsequently, it was found to be closely similar to ubiquitin, a widely distributed polypeptide of previously unknown func-tion [(119) and see below].

Fraction II also contains several separable essential components of the ATP-dependent proteolytic system. Ammonium sulfate fractionation yielded two mutually required subfractions, fractions IIA and IIB. Fraction IIA contains a high-molecular-weight component of the proteolytic system that is extremely heat labile, but is remarkably stabilized by ATP (117). However, the stimulatory effect of ATP on proteolysis is not due merely to the stabilization of this factor, since it was also stabilized by ADP or ATP analogues that cannot replace ATP in the stimulation of proteolysis (117).

CONJUGATION OF THE HEAT-STABLE POLYPEPTIDE TO PROTEINS
The results cited above showed that ATP-dependent protein breakdown in reticulocytes is carried out by a complex, multicomponent system. To gain an insight into the roles of the different factors and of ATP in this system, we next examined whether an ATP-dependent association of the heat-stable polypeptide (APF-1) with other cellular proteins takes place. Following incubation of purified, radiolabeled APF-1 with fraction II in the presence of ATP, binding of the polypeptide to high-molecular-weight material was observed, as analyzed by gel filtration on Sephadex-G-75 columns (120). The reaction specifically required ATP (in the presence of Mg^{+2}) and no significant binding was observed with other nucleoside triphosphates or ADP. The binding required Fraction II at concentrations similar to those required for protein breakdown, and was blocked by N-ethyl-maleimide or EDTA. All these characteristics of the binding reaction were similar to those of ATP-dependent protein breakdown, and therefore it seemed rea-sonable to assume that the binding of the heat-stable polypeptide to proteins is relevant to the proteolytic process. It was surprising to find, however, that the binding is of covalent nature: bound polypeptide was not released by treatments with acid, mild alkali, or heating at 100°C in the presence of high concentrations of sodium dodecyl sulfate and 2-mercaptoethanol (120).

Analysis of the reaction products by SDS-polyacrylamide gel electro-phoresis indicated the association of the labeled polypeptide with numerous high-molecular-weight proteins (120). It did not seem likely, therefore, that the polypeptide is bound to some specific enzyme(s) involved in the degradative process, but rather that substrates of the proteolytic system, present endogenously in crude Fraction II, are the target of the binding reaction. Indeed, it was found that proteins that are good substrates for ATP-dependent proteolysis (such as lysozyme, globin, or α-lactalbumin), form multiple conjugates with the polypeptide in the ATP-requiring reaction (121). Analysis of the ratio of APF-1 to lysozyme radioactivities in the various APF-1-lysozyme adducts and of the apparent molecular weights of these compounds (by SDS gel electrophoresis) indicated that they consist of increasing numbers of APF-1 bound to one molecule of lysozyme (121). Preliminary examination of the nature of the linkage between the heat-stable polypeptide and proteins indicated that it is probably an amide bond, since it is stable to prolonged incubation in 0.1 N NaOH or with 1 M hydroxylamine at pH 9 (121). Since poly–L–lysine also could serve as a substrate for conjugation with APF-1, it appears likely that an isopeptide linkage is formed between ϵ-NH$_2$ groups of lysine residues of the protein substrates and a carboxyl residue of the heat-stable polypeptide (121).

The conjugates of the heat-stable polypeptide with proteins, formed by reticulocyte Fraction II with ATP, appeared to be in a dynamic state of continuous synthesis and breakdown (120, 121). When ATP was removed with hexokinase and glucose, rapid degradation (though with heterogenous kinetics) of the conjugates was observed (120). This was accompanied by the release of free and reusable APF-1, as shown by the finding that the released polypeptide could be quantitatively converted back to conjugates when incubated again with Fraction II and ATP (121). Based on the above results, a model was proposed for ATP-dependent protein breakdown, according to which the covalent linkage of the heat-stable polypeptide to the substrate is followed by the proteolytic breakdown of the modified substrate [(121) and see below].

SIMILARITY OF THE HEAT-STABLE POLYPEPTIDE TO UBIQUITIN
Following the above studies of Hershko, Rose, and co-workers Wilkinson and associates (119) noted a marked resemblance of these conjugates to the previously characterized structure of chromosomal protein A24, which consists of ubiquitin conjugated to histone 2A by an isopeptide linkage (see below). This is an interesting case where research from three apparently unrelated areas (i.e. protein breakdown, the structure and function of chromosomal proteins, and the isolation of thymic hormones) has converged, and therefore we describe it briefly. Protein A24 was first discovered by Busch and co-workers as one of many nucleolar proteins resolved by two-dimensional electrophoresis (122). Attention of this group had been focused on this particular protein because the levels of protein A24 in rat liver nucleoli were markedly decreased following thioacetamide treatment

or in regeneration following partial hepatectomy (123–125), which suggests some relationship to the activity of gene expression. Progress achieved prior to 1978 was summarized in an excellent review by Goldknopf & Busch (126). Briefly, protein A24 was purified to homogeneity from calf thymus chromatin (127) and its structure was characterized. It was found to have a branched structure (128), in which the COOH-terminal-gly-gly of ubiquitin is bound to the ϵ-NH$_2$ group of lysine 119 of histone 2A by an isopeptide linkage (128–131). Recently it was reported that ubiquitin is also conjugated to histones 2B (132) and to all subtypes of histone 2A (133).

Ubiquitin is a small polypeptide ($M_r = 8500$) initially isolated during a search for thymic hormones by Goldstein and co-workers, from bovine thymus. It was subsequently found to be present in all tissues, and hence its name (134). It has been detected by radioimmunoassay in all living organisms examined, including animal cells, plants, yeasts, and bacteria (134). Its amino acid sequence shows a high degree of evolutionary conservation, the sequence being identical in ubiquitin of bovine (135) and human (136) origin, and apparently similar to that of trout ubiquitin, with an uncertainty (apparently technical) only in residues 68–71 (137). This indicated a basic and universally occurring cellular function, the identity of which was, however, unknown. Initial reports that ubiquitin induces the differentiation of lymphocytes (134) and stimulates adenylcyclase (138) could not be reproduced by other workers, and were claimed to be due to some contamination in the preparation (139, 140). A single clue for a physiological function of ubiquitin was the observation of its conjugation to histone, described earlier. In the chromatin, ubiquitin exists not only as a histone conjugate, but also in the free form (137, 141). The conjugated protein A24 was found to be tightly associated with the nucleosomes (142–144), along with the histones normally present in the nucleosome structure. On the other hand, free ubiquitin is apparently located at the "linker" regions of the chromatin (i.e. between nucleosomes), since it was released by limited digestion of DNA by micrococcal nuclease (137). Goldknopf et al (145) observed that free ubiquitin is solubilized from rat liver chromatin by DNase II digestion, while protein A24 is not; this was interpreted by the authors as indicating that free ubiquitin is located at actively transcribed regions of DNA.

The functions of chromatin-associated ubiquitin or of its conjugates with histones are not known. There is no simple correlation between transcriptional activity and the levels of protein A24. As noted originally by Busch and co-workers (123–125), protein A24 levels in liver nucleoli are decreased under conditions associated with high rates of synthesis of ribosomal RNA. On the other hand, a more recent study showed that the levels of protein A24 and ubiquitin are actually increased in actively transcribing avian erythroid cells, as compared to inactive mature chicken erythrocytes (146). It is also possible that histone-ubiquitin conjugates have a structural role, such as in the supercoiling of chromatin (142). It has been reported that protein A24 disappears during mitosis in synchronized cultured cells, as opposed to its high level in interphase cells (147). In addition, the rates of the synthesis of ubiquitin and of its conjugation with histone 2A are also lower during cell division than in the G_1 phase of the cell cycle (148). While these phenomena may be related to the marked alterations in the structure and organization of chromatin that occur in mitosis, a possible relationship to any of the multitude of other changes that occur in mitotic nuclei cannot be ruled out.

The similarity of ubiquitin to the heat-stable polypeptide of the proteolytic system was noted (119) because of the following common features: (a) the conjugation of APF-1 with proteins in an isopeptide linkage (121) resembles the similar bond between ubiquitin and histone (128); (b) both APF-1 and ubiquitin are heat-stable proteins (116, 134) of similar molecular size (116, 134) and neutral isoelectric point (118, 139); (c) the amino acid composition of APF–1, analyzed by Ciechanover et al (118), is very similar to the known composition of ubiquitin (135). A single, notable difference (see also below) was the presence of six glycine residues in APF-1 (118) versus four in ubiquitin (135). Ubiquitin from calf thymus migrated identically with rabbit reticulocyte APF-1 in five different polyacrylamide gel electrophoresis systems, and calf thymus ubiquitin stimulated protein breakdown and formed conjugates with proteins in the presence of rabbit reticulocyte fraction II and ATP (119). In both of the above processes, however, thymus ubiquitin was only about 65% as active as APF-1.

While ubiquitin and APF-1 are obviously similar, an important point remained unsettled. The carboxy end of ubiquitin is conjugated to histone through a gly-gly residue (128), but arginine-74 has been reported as the COOH-terminal amino acid of thymus ubiquitin (135, 136, 140). Dixon and co-workers (137) noted a heterogeneity at the COOH-terminus of trout testis ubiquitin; about half of the molecules are terminated by arginine, and the other half by glycine. On the other hand, our subsequent work has shown that a COOH-terminal glycine of APF-1 is the single amino acid residue that is activated by the reticulocyte ubiquitin-activating enzyme [(149) and see below]. These discrepancies now appear to be resolved by the findings that APF-1 from human (150) or rat (151) erythrocytes contains a COOH-terminal -gly-gly residue, and that only the species having the -gly-gly terminus is active in the stimulation of ATP-dependent protein breakdown (150). It is not yet clear whether ubiquitin that lacks the C-terminal -gly-gly is a proteolytic artefact, or whether it has some as yet unknown physiological function.

Since APF-1 is apparently identical with ubiquitin -gly-gly, the question arises whether chromosomal protein A24 is an intermediate in the degradation of the histone. This cannot be answered at present. The turnover of histones is slow, while that of the ubiquitin moiety of the conjugated form is quite rapid (152). Moreover, cleavage of the isopeptide linkage in protein A24 by an enzyme from nucleoli of thioacetamide-treated rat liver has been described (151, 153). This enzyme releases free ubiquitin and histone, without proteolytic breakdown of the latter. Thus, conjugation of ubiquitin with histone may represent a special type of protein modification, as proposed originally by Busch and co-workers (126). It is not unlikely that "ubiquitination" of proteins has evolved to fulfill several distinct biological functions, such as protein breakdown or protein modification.

INTERMEDIARY REACTIONS IN THE FORMATION OF UBIQUITIN-PROTEIN CONJUGATES   Information concerning the enzymatic steps involved in the conjugation of ubiquitin to proteins is still fragmentary. We first concentrated on the activation reaction necessary for conjugation. An enzyme was partially purified from reticulocyte Fraction II that catalyzes ubiquitin-dependent PP_i : ATP and AMP : ATP exchange reactions (154). A likely explanation of these observations appears to be the following two-step mechanism:

$$\text{ubiquitin} + \text{ATP} \leftrightarrows \text{ubiquitin} \sim \text{AMP} + PP_i$$

$$\text{ubiquitin} \sim \text{AMP} + \text{E-SH} \leftrightarrows \text{E-S} \sim \text{ubiquitin} + \text{AMP}$$

In the first step, to account for the PP_i : ATP exchange, an adenylate of the polypeptide is assumed to be formed, with release of pyrophosphate. The ubiquitin-dependent AMP : ATP exchange indicates that the activated group is transferred to a secondary acceptor, with the liberation of AMP. The acceptor appears to be a sulfhydryl group of the enzyme, since labeled ubiquitin becomes associated with the enzyme by a linkage that has the characteristics of a thiolester bond (154). The enzyme is specific for ATP (in the presence of Mg^{2+}), and ubiquitin is effective at low concentrations similar to those required for protein degradation. Though the reaction is unusual for a polypeptide, it is analogous to processes such as the activation of amino acids for protein synthesis, and the nonribosomal biosynthesis of peptide antibiotics. In the latter case the aminoacyl adenylates are transferred to thiolester intermediates on the same enzyme (155), which preserves the activated state of the acyl group and functions in the formation of the final amide bond (155). It appears likely that ubiquitin bound to the activating enzyme by the energy-rich thiolester linkage represents a similar reactive intermediate.

We next attempted to identify the functional amino acid residue of ubiquitin by reductive cleavage of the activated intermediate with [^{3}H]-borohydride (149). Following incubation of ubiquitin with the activating enzyme and ATP, NaB^{3}H$_4$ label became incorporated into ubiquitin. The reductively labeled polypeptide was isolated and hydrolyzed, and the labeled amino alcohol derivative identified as ethanolamine. Since ethanolamine is the reduction product of glycine, it was concluded that the activated amino acid residue of the polypeptide is a COOH-terminal glycine (149). This conclusion was subsequently corroborated by the finding (see above) that only the species of ubiquitin that contains the C-terminal -gly-gly residue is active in ATP-dependent proteolysis (150).

The ubiquitin-activating enzyme has been recently purified by covalent affinity chromatography on ubiquitin-Sepharose (156, 157). It does not catalyze the formation of conjugates by itself, but it is required for conjugate formation in the presence of two other enzymes isolated from reticulocyte Fraction II (A. Ciechanover, S. Elias, N. Heller, and A. Hershko, in preparation). Both of the above enzymes are also required for ATP-dependent proteolysis in the presence of ubiquitin and other components of the system from Fraction II.

PROPOSED PATHWAY FOR ATP-UBIQUITIN-DEPENDENT PROTEIN BREAKDOWN   These relationships between conjugation of ubiquitin to proteins and its participation in protein breakdown have led to the proposal that conjugation with the heat-stable polypeptide is an intermediary step in the degradation of the substrate protein (121). It appeared reasonable to assume that conjugation serves as a recognition signal, providing a mechanism for the specificity and selectivity of the degradation of intracellular proteins, and that specific proteolytic enzymes exist that selectively recognize and degrade ubiquitin-protein conjugates. It appeared likely that the binding of several molecules of the heat-stable polypeptide to the protein produces a marked alteration in the conformation of the substrate protein and thus renders it highly susceptible to attack by such proteases. The proposed sequence of events in this process is shown in Figure 1. Four main stages are visualized: 1. The initial activation of the glycine carboxyl terminus of ubiquitin as an acyl adenylate (149, 154), with transfer of the activated polypeptide to an –SH group of the enzyme to give a thiolester linkage. 2. Transfer of the activated COOH-terminus of ubiquitin to ε-amino groups of lysine residues in the substrate protein. This reaction is either precessive or has a strong preference for free proteins conjugated to ubiquitin, since conjugates containing multiple molecules of ubiquitin appear under conditions of excess substrate (121). 3. Breakdown of the conjugates by an endoproteolytic attack on the protein substrate moiety, which would produce short peptides still attached to ubiquitin by the isopeptide linkage. 4. In the last phase, small peptides and free (and reutilizable) ubiquitin may be released by the action of an amidase that cleaves the isopeptide bond. The assumption that short peptides are the products of the ATP-dependent proteolytic system is based on the observations of Botbol & Scornik (114, 115) that such peptides accumulate when aminopeptidase action is blocked by bestatin. Steps 3 and 4 are entirely hypothetical at present. It is known that ubiquitin-protein conjugates are broken down (120) with the release of free and reusable ubiquitin (121), but the intermediary reactions of the breakdown process have not yet been investigated. A valuable tool for the investigation of this step appears to be hemin, which blocks the breakdown of ubiquitin-protein conjugates (158a) and inhibits proteolysis (113). In addition, a "correction enzyme" (121) (i.e. an isopeptidase that would release ubiquitin conjugated to proteins not destined for degradation) may exist, similar to the nucleolar isopeptidase that cleaves protein A24 (151, 153). While the proposed model is consistent with all the presently available findings, it obviously requires much further investigation. For example, the high-molecular-weight component of the system that is stabilized by ATP and some other adenine nucleotides (117) is not required for conjugate formation, but appears to co-purify with an activity that breaks down ubiquitin-protein conjugates [(159); E. Leshinsky, A. Ciechanover, A. Heller, and A. Hershko, unpublished.] It remains to be seen whether ATP has a second role in this pathway.

The physiological roles of this system in cell types other than reticulocytes presents another problem. Conjugation of ubiquitin with proteins has been observed in all rat tissues examined [(160); S. Ferber and A. Hershko, unpublished.] but not in *E. coli* [(161); D. Rafaeli-Eshkol and A. Hershko, unpublished.] It may be that prokaryotes have a different polypeptide of a

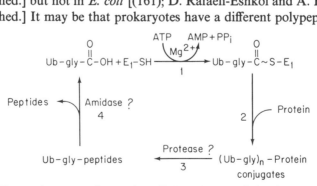

Figure 1 Proposed sequence of events in ATP-dependent protein breakdown. See the text for details. 1. Activation of the carboxyl terminal glycine of ubiquitin and its intermediary binding to the activating enzyme (E_1). 2. Conjugation of several molecules of activated ubiquitin to ϵ-NH_2 lysine groups of the substrate protein. 3. Breakdown of the conjugates by an endoprotease acting on the protein substrate moiety. 4. Release of short peptides and free ubiquitin by a specific amidase that cleaves the isopeptide linkage.

similar function, and thus the bacterial system would not recognize mammalian ubiquitin. It is also possible that in *E. coli,* at least some types of protein breakdown are carried out by other proteolytic systems, since different ATP-dependent proteases have been recently described in bacteria (see below).

Occurrence of Different ATP-Dependent Proteolytic Systems

In addition to the ATP-ubiquitin proteolytic system described in reticulocytes, other, apparently unrelated ATP-dependent proteolytic processes have been found in other biological systems, notably in *E. coli.* Among the best characterized of these is the proteolytic inactivation of the repressor of bacteriophage lambda, studied mainly by Roberts and co-workers. This repressor, which prevents the expression of viral genes required for lytic growth, was known to be inactivated by treatments affecting DNA, such as mitomycin C or irradiation with ultraviolet light. Roberts et al first found that in cells treated with mitomycin C, the repressor is cleaved to smaller fragments (162). This proteolytic mechanism for repressor inactivation differs fundamentally from the classical case in which induction occurs by the reversible binding of a small molecule to a repressor protein. Its relevance to inactivation of the repressor was shown by the findings that mutations that prevent the induction of the phage (bacterial *recA*⁻ mutation or phage *ind* mutation) also prevent the cleavage of the repressor (162).

Next, Roberts and co-workers reproduced this specific cleavage reaction in vitro. They found that extracts of a strain of *E. coli* in which the repressor is constitutively inactivated are able to carry out its cleavage in vitro in the presence of ATP and Mg^{+2} (163). ATP could not be replaced by other nucleotides such as GTP or ADP (163). The enzyme responsible for this reaction was purified and identified as the product of the *recA* gene (164). The purified *recA* protein cleaved the purified repressor at a reduced rate, but its activity could be restored by addition of single-stranded DNA or a variety of polynucleotides (165, 166). It was suggested that formation of a ternary complex between *recA* protein, a polynucleotide, and ATP is necessary for activation of the proteolytic function of the enzyme (166–168). The *recA* gene product of *E. coli* is a multifunctional protein of considerable regulatory importance. It is essential for genetic recombination and also for the expression of the so-called SOS functions, which are a set of responses believed to aid cell survival in response to agents that damage DNA, including inducible DNA repair, induced mutagenesis, inhibition of cell division, and prophage induction (169). In addition to its action as a highly specific protease, purified *recA* protein also promotes the homologous pairing of DNA molecules in vitro (170, 171). The protease function appears to be involved in the induction of some, if not all of the SOS responses, since purified *recA* protein cleaves not only the repressor of phage λ and of *S. typhimurium* phage P22 (165, 172), but it also cleaves *lexA* protein, which appears to be a repressor of *recA* protein formation (173). Its action in catalyzing pairing of single-stranded DNA to homologous region of double-stranded DNA, probably explains the role of the *recA* protein in genetic recombination. These distinct biochemical activities of *recA* protein may have a common initial activation mechanism, since the ATP analogue, adenosine-5'-[γ-thio]triphosphate (ATP-γ-S), which is not hydrolyzed by *recA* protein (166), can effectively promote both repressor cleavage (166) and the binding of *recA* protein to single-stranded DNA [which also appears to be the initial step in DNA recombination (174, 175)]. The activation of the proteolytic function of *recA* protein by single-stranded DNA in vitro may represent the physiological mechanism that triggers the expression of SOS functions: it is possible that damage to DNA may expose regions of single-stranded DNA, or liberate DNA fragments, which may then activate the enzyme (166, 168). The exact role of ATP in promoting this association, and the putative conformational change in the enzyme that activates its proteolytic function, remain to be elucidated.

Another recently discovered ATP-dependent protease in *E. coli* may be involved in the degradation of abnormal proteins. Goldberg and co-workers showed that in soluble extracts of *E. coli,* ATP stimulates the breakdown of a nonsense fragment of β-galactosidase and the degradation of [^{14}C]methyl-globin to acid-soluble material (176). Other nucleotides were less effective, though the enzyme was stimulated significantly by CTP and GTP. When the bacteria are broken by more gentle methods, most of the ATP-stimulated protease activity is associated with the cell membrane (177). It is not clear, however, whether the soluble enzyme is released from the membrane, or whether these represent two different enzyme systems. The soluble ATP-dependent enzyme has been separated from a variety of other ATP-independent proteases by chromatography on DEAE-cellulose (178). The partially purified enzyme has an essentially complete requirement for ATP and Mg^{+2}; it has been designated protease La (178). More recently, it was shown that protease La appears to be similar to the product of the *lon* gene in *E. coli* (179, 180). The relationship of the *lon* (*capR, deg*) gene function to the breakdown of abnormal proteins was originally discovered by Zipser and co-workers, who first isolated mutants defective in the degradation of nonsense fragments of β-galactosidase (181) and then found that the mutation is identical to the previously known *lon* mutation (182, 183). Mutations in the *lon* gene have no effect on the degradation of most normal *E. coli* proteins (3), but reduce the rate of degradation of nonsense polypeptides (83, 181), missense proteins (183), puromycyl-peptides (2, 184), or canavanine-containing abnormal proteins (184). However, in none of these cases was the process completely inhibited by any of the *lon*⁻ mutations, but rather a partial inhibition of the breakdown of abnormal protein is usually observed. Gottesman et al (185) recently examined the influence of the *lon* mutation on the breakdown of physiologically unstable regulatory proteins of bacteriophage λ, and found that while the degradation of N protein was slowed down, and of O protein was not affected, while the breakdown of protein cII was actually increased. It was suggested that multiple proteolytic pathways may exist for these different highly unstable lambda proteins (185). Mutation in the *lon* gene also produces a variety of other pleiotropic effects including overproduction of capsular polysaccharides, increased sensitivity to irradiation, and abnormal cell division (186). Could these multiple effects be secondary to a primary lesion in the degradation of some regulatory proteins; or conversely, could it be that protein breakdown is inhibited as a result of some other alteration that interferes with a variety of processes? Significant progress in the elucidation of the mode of action of the *lon* gene product was achieved by Markovitz and co-workers who identified the polypeptide specified by the use of cloning techniques (187, 188) and then purified the gene product from bacteria containing cloned *lon* genes (189). The purified *lon* gene product, has an ATP-dependent protease activity (180). Comparison of its properties with protease La showed similar features with regard to nucleotide and substrate specificities, susceptibility to inhibitors, heat inactivation, and apparent molecular weight (179). It cannot be ruled out, however, that the *lon* gene product may be a cofactor of a highly active protease that is present in small amount in the purified preparation. At any rate, the relationship between *lon* gene function and at least a part of the degradation of abnormal proteins in intact bacteria indicates the relevance of this ATP-dependent protease to the degradative process. Though the *lon* gene product is obviously different from the *recA* protein (which carries out limited proteolysis of specific proteins) they do have some common features. It is remarkable that both the *lon* gene product and *recA* protein bind to DNA (189), and that both have an ATPase activity that is stimulated by DNA and other polynucleotides [165; A. Markovitz, personal communication]. It may well be that the *lon* gene product is also a multifunctional enzyme that is involved in DNA metabolism as well as in protein breakdown. A notable difference is that while ATP hydrolysis does not appear to be essential for the proteolytic action of *recA* protein (166), it is required for the *lon* protease (179, 180), and stimulation of the ATPase activity of the latter by the protein substrate has been observed [(180, 191); A. L. Goldberg, personal communication.]

The above *E. coli* enzymes are obviously different from the multicomponent ATP-ubiquitin proteolytic pathway described in reticulocytes. At present, it is not known whether such ATP-dependent proteases occur in other biological systems. DeMartino & Goldberg (192) and Rose et al (193) have independently described a high-molecular-weight protease from liver cytosol that appeared to be stimulated by ATP. However, the enzyme is also stimulated by other polyanions such as pyrophosphate (192), creatine phosphate, or citrate (193) and does not require Mg^{+2} (192). The effect of the polyanions appears to be due to the stabilization of the enzyme during assay at 37°C; at lower temperatures (in which the enzyme is stable), no effects of ATP or other polyanions can be observed (193). Furthermore, the apparent stimulation by ATP is only partial even following several steps of purification (192, 193). This is in contrast to the absolute ATP dependence of the ubiquitin proteolytic system and the *recA* or *lon* proteases, as well as to the complete energy requirement of protein breakdown in intact cells. Thus, the relationship of this protease to intracellular protein breakdown is doubtful. A similar protease was found in various tissues (193) and in reticulocytes (194). [^{14}C]-Methylglobin, which was mainly used for the assay of the reticulocyte protease (194), would not be a substrate for the ATP-ubiquitin system, since the ϵ-NH_2 lysine residues are blocked by methylation.

CONCLUDING REMARKS

From the above discussion it seems clear that though significant progress has been made recently, it provides only the first steps toward elucidation of the mechanisms of intracellular protein breakdown. For the lysosomal system, major unsolved problems are the mechanisms that regulate the formation of autophagic vacuoles, and whether this organelle is capable of carrying out selective protein breakdown under steady-state conditions. Concerning the soluble ATP-dependent proteolytic systems, apart from obvious questions of the enzymatic steps involved, their physiological roles in various biological systems remain to be established. Most importantly, we are still ignorant of the mechanisms that determine the high selectivity of intracellular protein breakdown. In the case of bacterial ATP-dependent proteases, it is not clear how ATP can enhance selectivity. On the other hand, the ubiquitin-conjugation pathway may provide an elaborate mechanism for the recognition of protein conformations, but this remains to be investigated.

Acknowledgments

We wish to thank Dr. Irwin A. Rose for helpful comments on the manuscript. We also thank many of our colleagues for sending us preprints of publications. Our own studies have been supported by United States Public Health Service Grant AM-2561401 and a grant from the United States-Israel Binational Science Foundation.

Literature Cited

1. Schimke, R. T. 1973. *Adv. Enzymol.* 37:135–87
2. Goldberg, A. L., Dice, F. J. 1974. *Ann. Rev. Biochem.* 43:835–69
3. Goldberg, A. L., St. John, A. C. 1976. *Ann. Rev. Biochem.* 45:747–803
4. Ballard, F. J. 1977. *Essays Biochem.* 13:1–37
5. Waterlow, J. C., Garlick, P. J., Millward, D. J. 1978. *Protein Turnover in Mammalian Tissues and in the Whole Body.* Amsterdam: North-Holland. 804 pp.
6. Maurizi, M. R., Switzer, R. L. 1980. *Curr. Top. Cell. Regul.* 16:163–224
7. Holzer, H., Heinrich, P. C. 1980. *Ann. Rev. Biochem.* 49:63–91
8. Segal, H. L., Doyle, D. J., eds. 1978. *Protein Turnover and Lysosome Function.* New York: Academic. 790 pp.
9. Cohen, G. N., Holzer, H., eds. 1979. *Limited Proteolysis in Microorganisms,* DHEW Publ. No. 79–1591 (NIH). Washington DC: USGPO
10. Morgan, H., Wildenthal, K., eds. 1980. *Fed. Proc.* 39:7–52
11. Evered, D., Whelan, J., eds. 1980. *CIBA Found. Symp.* 75:1–417
12. De Duve, C., Wattiaux, R. 1966. *Ann. Rev. Physiol.* 28:435–92
13. Mortimore, G. E., Mondon, C. E. 1970. *J. Biol. Chem.* 245:2375–83
14. Woodside, K. H., Mortimore, G. E. 1972. *J. Biol. Chem.* 247:6474–81
15. Woodside, K. H., Ward, W. F., Mortimore, G. E. 1974. *J. Biol. Chem.* 249:5458–63
16. Schworer, C. M., Mortimore, G. E. 1979. *Proc. Natl. Acad. Sci. USA* 76:3169–73
17. Hershko, A., Tomkins, G. M. 1971. *J. Biol. Chem.* 246:710–14
18. Hershko, A., Mamont, P., Shields, R., Tomkins, G. M. 1971. *Nature New Biol.* 232:206–11
19. Poole, B., Wibo, M. 1973. *J. Biol. Chem.* 248:6221–26
20. Warburton, M. J., Poole, B. 1977. *Proc. Natl. Acad. Sci. USA* 74:2427–31
21. Ballard, F. J., Knowles, S. E., Wong, S. S. C., Bodner, J. B., Wood, C. M., Gunn, J. M. 1980. *FEBS Lett.* 114:209–12
22. Neely, A. N., Nelson, P. B., Mortimore, G. E. 1974. *Biochim. Biophys. Acta* 338:458–72
23. Deter, R. L., De Duve, C. 1967. *J. Cell. Biol.* 33:437–49
24. Rannels, E. D., Kao, R., Morgan, H. E. 1975. *J. Biol. Chem.* 250:1694–701
25. Neely, A. N., Cox, J. R., Fortney, J. A., Schworer, C. M., Mortimore, G. E. 1977. *J. Biol. Chem.* 252:6948–54
26. Schworer, C. M., Cox, J. R., Mortimore, G. E. 1979. *Biochem. Biophys. Res. Commun.* 87:163–70
27. Mortimore, G. E., Schworer, C. M. 1977. *Nature* 270:174–76
28. Huisman, W., Lanting, L., Bouma, J. M. W., Gruber, M. 1974. *FEBS Lett.* 45:129–31
29. Mortimore, G. E., Neely, A. N., Cox, J. R., Guinivan, R. A. 1973. *Biochem. Biophys. Res. Commun.* 54:89–95
30. Neely, A. N., Mortimore, G. E. 1974. *Biochem. Biophys. Res. Commun.* 59:680–87
31. Mortimore, G. E., Ward, W. F., Schworer, C. M. 1978. See Ref. 8, pp. 67–87
32. Schworer, C. M., Shiffer, K. A., Mortimore, G. E. 1981. *J. Biol. Chem.* 256:7652–58
33. Ward, W. F., Cox, J. R., Mortimore, G. E. 1977. *J. Biol. Chem.* 252:6955–61
34. Mortimore, G. E., Ward, W. F. 1981. *J. Biol. Chem.* 256:7659–65
35. Wibo, M., Poole, B. 1974. *J. Cell. Biol.* 63:430–40
36. Poole, B., Ohkuma, S., Warburton, M. 1978. See Ref. 8, pp. 43–58
37. Ohkuma, S., Poole, B. 1978. *Proc. Natl. Acad. Sci. USA* 75:3327–31
38. Dean, R. T. 1975. *Nature* 257:414–16
39. Aoyagi, T., Umezawa, H. 1975. In *Proteases and Biological Control,* ed. E. Reich, D. B. Rifkin, E. Shaw, pp. 429–54. Cold Spring Harbor, New York: Cold Spring Harbor Lab.
40. Knowles, S. E., Ballard, F. J. 1976. *Biochem. J.* 156:609–17
41. Umezawa, H., Aoyagi, T. 1977. In *Proteinases in Mammalian Cells and Tissues,* ed. A. J. Barrett, pp. 637–62. Amsterdam: North-Holland
42. Libby, P., Goldberg, A. L. 1978. *Science* 199:534–36
43. McGowan, E. B., Shafiq, S. A., Stracher, A. 1976. *Exp. Neurol.* 50:649–57
44. Stracher, A., McGowan, E. B., Shafiq, S. A. 1978. *Science* 200:50–51
45. Neff, N. T., DeMartino, G. N., Goldberg, A. L. 1979. *J. Cell. Physiol.* 101:439–58
46. Hopgood, M. F., Clark, M. G., Ballard, F. J. 1977. *Biochem. J.* 164:399–407
47. Grinde, B., Seglen, P. O. 1980. *Biochim. Biophys. Acta* 632:73–86
48. Epstein, D., Elias-Bishko, S., Hershko, A. 1975. *Biochemistry* 14:5199–5204

49. Amils, R., Conde, R. D., Scornik, O. A. 1977. *Biochem. J.* 164:363–69
50. Hendil, K. B. 1981. *FEBS Lett.* 129:77–79
51. Livesey, G., Williams, K. E., Knowles, S. E., Ballard, F. J. 1980. *Biochem. J.* 188:895–903
52. Knowles, S. E., Ballard, F. J., Livesey, G., Williams, K. E. 1981. *Biochem. J.* 196:41–48
53. Libby, P., Bursztajn, S., Goldberg, A. L. 1980. *Cell* 19:481–91
54. Libby, P., Goldberg, A. L. 1981. *J. Cell. Physiol.* 107:185–94
55. Amenta, J. S., Sargus, M. J., Baccino, F. M. 1977. *Biochem. J.* 168:223–27
56. Amenta, J. S., Hlivko, T. J., McBee, A. G., Shinozuka, H., Brocher, S. 1978. *Exp. Cell Res.* 115:357–66
57. Seglen, P. O. 1975. *Biochem. Biophys. Res. Commun.* 66:44–52
58. Seglen, P. O., Reith, A. 1976. *Exp. Cell Res.* 100:276–80
59. Seglen, P. O., Grinde, B., Solheim, A. E. 1979. *Eur. J. Biochem.* 95:215–25
60. Seglen, P. O., Gordon, P. B. 1980. *Mol. Pharmacol.* 18:468–75
61. Dean, R. T. 1979. *Biochem. J.* 180:339–45
62. Ward, W. F., Chua, B. L., Li, J. B., Morgan, H. E., Mortimore, G. E. 1979. *Biochem. Biophys. Res. Commun.* 87:92–98
63. Bigelow, S., Hough, R., Rechsteiner, M. 1981. *Cell* 25:83–91
64. Zavortink, M., Thacher, T., Rechsteiner, M. 1979. *J. Cell. Physiol.* 100:175–86
65. Neff, N. T., Bourret, L., Miao, P., Dice, J. F. 1981. *J. Cell. Biol.* 91:184–94
66. Freikopf-Cassel, A., Kulka, R. G. 1981. *FEBS Lett.* 128:63–66
67. Arad, G., Freikopf, A., Kulka, R. G. 1976. *Cell* 8:95–101
68. Schimke, R. T., Doyle, D. 1970. *Ann. Rev. Biochem.* 39:929–76
69. Bohley, P. 1968. *Naturwissenschaften* 55:211–17
70. Bohley, P., Riemann, S. 1977. *Acta Biol. Med. Germ.* 36:1823–27
71. Segal, H. L., Rothstein, D. M., Winkler, J. R. 1976. *Biochem. Biophys. Res. Commun.* 73:79–84
72. Dean, R. T. 1975. *Biochem. Biophys. Res. Commun.* 67:604–9
73. Dice, J. F., Walker, C. D., Byrne, B., Cardiel, A. 1978. *Proc. Natl. Acad. Sci. USA* 75:2093–97
74. Dice, J. F., Walker, C. D. 1980. *Ciba Found. Symp.* 75:331–50
75. Simpson, M. V. 1953. *J. Biol. Chem.* 201:143–54
76. Steinberg, D., Vaughan, M. 1956. *Arch. Biochem. Biophys.* 65:93–105
77. Schweiger, H. G., Rapoport, S., Schölzel, E. 1956. *Nature* 178:141–42
78. Mandelstam, J. 1960. *Bacteriol. Rev.* 24:289–308
79. Shechter, Y., Rafaeli-Eshkol, D., Hershko, A. 1973. *Biochem. Biophys. Res. Commun.* 54:1518–24
80. Olden, K., Goldberg, A. L. 1978. *Biochim. Biophys. Acta* 542:385–98
81. Halvorson, H. 1958. *Biochim. Biophys. Acta* 27:255–66
82. Martegani, E., Alberghina, L. 1979. *J. Biol. Chem.* 254:7047–54
83. Kowit, J. D., Goldberg, A. L. 1977. *J. Biol. Chem.* 252:8350–57
84. Etlinger, J. D., Goldberg, A. L. 1977. *Proc. Natl. Acad. Sci. USA* 74:54–58
85. Hershko, A., Heller, H., Ganoth, D., Ciechanover, A. 1978. See Ref. 8, pp. 149–69
86. Kulka, R. G., Cohen, H. 1973. *J. Biol. Chem.* 248:6738–43
87. Waindle, L. M., Switzer, R. L. 1973. *J. Bacteriol.* 114:517–27
88. Mego, J. L., Farb, R. M., Barnes, J. 1972. *Biochem. J.* 128:763–69
89. Mego, J. L. 1979. *FEBS Lett.* 107:113–16
90. Schneider, D. L., Cornell, E. 1978. See Ref. 8, pp. 59–66
91. Schneider, D. L. 1981. *J. Biol. Chem.* 256:3858–64
92. Reeves, J. P., Reames, T. 1981. *J. Biol. Chem.* 256:6047–53
93. Hollemans, M., Reijngoud, D.-J., Tager, J. M. 1979. *Biochim. Biophys. Acta* 551:55–66
94. Reijngoud, D.-J., Tager, J. M. 1977. *Biochim. Biophys. Acta* 472:419–49
95. Lodish, H. F., Desalu, O. 1973. *J. Biol. Chem.* 248:3520–27
96. Bessis, M. 1973. *Living Blood Cells and their Ultrastructure.* pp. 128–38. Berlin: Springer
97. Müller, M., Dubiel, W., Rothmann, J., Rapoport, S. 1980. *Eur. J. Biochem.* 109:405–10
98. Rabinovitz, M., Fisher, J. M. 1962. *Biochem. Biophys. Res. Commun.* 6:449–51
99. Rabinovitz, M., Fisher, J. M. 1964. *Biochim. Biophys. Acta* 91:313–22
100. McIlhinney, A., Hogan, B. L. M. 1974. *FEBS Lett.* 40:297–301
101. Huehns, E. R. 1970. *Bull. Soc. Chim. Biol.* 52:1131–46
102. Shaeffer, J. R., Kleve, L., DeSimone, J. 1976. *Br. J. Haemotol.* 32:365–72
103. Rieder, R. F., Wolf, D. J., Clegg, J. B., Lee, S. L. 1975. *Nature* 254:725–27

104. Chalevelakis, G., Clegg, J. B., Weatherall, D. J. 1975. *Proc. Natl. Acad. Sci. USA* 72:3853–57
105. Goldberg, A. L., Kowit, J., Etlinger, J., Klemes, Y. 1978. See Ref. 8, pp. 171–96
106. Prouty, W. F., Goldberg, A. L. 1972. *Nature New Biol.* 240:147–50
107. Prouty, W. F., Karnovsky, M. J., Goldberg, A. L. 1975. *J. Biol. Chem.* 250:1112–22
108. Daniels, R. S., Worthington, V. C., Atkinson, E. M., Hipkiss, A. R. 1980. *FEBS Lett.* 113:245–48
109. Rapoport, S. M., Rosenthal, S., Schewe, T., Schultze, M., Müller, M. 1974. In *Cellular and Molecular Biology of Erythrocytes*, ed. H. Yoshikawa, S. M. Rapoport, pp. 93–141. Tokyo: Univ. Tokyo
110. Dubiel, W., Müller, M., Rothmann, J., Hiebsch, C., Rapoport, S. M. 1982. *Acta Biol. Med. Germ.* In press
111. Rapoport, S. M., Schewe, T., Wiesner, R., Halangk, W., Ludwig, P., Janicke-Höhne, M., Tannat, C., Hiebsch, C., Klatt, D. 1979. *Eur. J. Biochem.* 96:545–61
112. McKay, M. J., Daniels, R. S., Hipkiss, A. R. 1980. *Biochem. J.* 188:279–83
113. Etlinger, J. D., Goldberg, A. L. 1980. *J. Biol. Chem.* 255:4563–68
114. Botbol, V., Scornik, O. A. 1979. *Proc. Natl. Acad. Sci. USA* 76:710–13
115. Botbol, V., Scornik, O. A. 1979. *J. Biol. Chem.* 254:11254–57
116. Ciechanover, A., Hod, Y., Hershko, A. 1978. *Biochem. Biophys. Res. Commun.* 81:1100–5
117. Hershko, A., Ciechanover, A., Rose, I. A. 1979. *Proc. Natl. Acad. Sci. USA* 76:3107–10
118. Ciechanover, A., Elias, S., Heller, H., Ferber, S., Hershko, A. 1980. *J. Biol. Chem.* 255:7525–28
119. Wilkinson, K. D., Urban, M. K., Haas, A. L. 1980. *J. Biol. Chem.* 255:7529–32
120. Ciechanover, A., Heller, H., Elias, S., Haas, A. L., Hershko, A. 1980. *Proc. Natl. Acad. Sci. USA* 77:1365–68
121. Hershko, A., Ciechanover, A., Heller, H., Haas, A. L., Rose, I. A. 1980. *Proc. Natl. Acad. Sci. USA* 77:1783–86
122. Orrick, L. R., Olson, M. O. J., Busch, H. 1973. *Proc. Natl. Acad. Sci. USA* 70:1316–20
123. Ballal, N. R., Busch, H. 1973. *Cancer Res.* 33:2737–43
124. Ballal, N. R., Goldknopf, I. L., Goldberg, D. A., Busch, H. 1974. *Life Sci.* 14:1835–45
125. Ballal, N. R., Kang, Y.-J., Olson, M. O. J., Busch, H. 1975. *J. Biol. Chem.* 250:5921–25
126. Goldknopf, I. L., Busch, H. 1978. *Cell Nucl.* 6:149–80
127. Goldknopf, I. L., Taylor, C. W., Baum, R. M., Yeoman, L. C., Olson, M. O. J., Prestayko, A. W., Busch, H. 1975. *J. Biol. Chem.* 250:7182–87
128. Goldknopf, I. L., Busch, H. 1977. *Proc. Natl. Acad. Sci. USA* 74:864–68
129. Olson, M. O. J., Goldknopf, I. L., Guetzow, K. A., James, G. T., Hawkins, C. T., Mays-Rothberg, C. J., Busch, H. 1976. *J. Biol. Chem.* 251:5901–3
130. Hunt, L. T., Dayhoff, M. O. 1977. *Biochem. Biophys. Res. Commun.* 74:650–55
131. Goldknopf, I. L., Busch, H. 1980. *Biochem. Biophys. Res. Commun.* 96:1724–31
132. West, M. H. P., Bonner, W. M. 1980. *Nucleic Acids Res.* 8:4671–80
133. West, M. H. P., Bonner, W. M. 1980. *Biochemistry* 19:3238–45
134. Goldstein, G., Scheid, M., Hammerling, U., Boyse, E. A., Schlesinger, D. H., Niall, H. D. 1975. *Proc. Natl. Acad. Sci. USA* 72:11–15
135. Schlesinger, D. H., Goldstein, G., Niall, H. D. 1975. *Biochemistry* 14:2214–18
136. Schlesinger, D. H., Goldstein, G. 1975. *Nature* 255:423–24
137. Watson, D. C., Levy, W. B., Dixon, G. H. 1978. *Nature* 276:196–98
138. Scheid, M. P., Goldstein, G., Hammerling, U., Boyse, E. A. 1975. *Ann. NY Acad. Sci.* 249:531–40
139. Low, T. L. K., Thurman, G. B., McAdoo, M., McClure, J., Rossio, J. L., Naylor, P. H., Goldstein, A. L. 1979. *J. Biol. Chem.* 254:981–86
140. Low, T. L. K., Goldstein, A. L. 1979. *J. Biol. Chem.* 254:987–95
141. Walker, J. M., Goodwin, G. H., Johns, E. W. 1978. *FEBS Lett.* 90:327–30
142. Goldknopf, I. L., French, M. F., Musso, R., Busch, H. 1977. *Proc. Natl. Acad. Sci. USA* 74:5492–95
143. Albright, S. C., Nelson, P. P., Garrard, W. T. 1979. *J. Biol. Chem.* 254:1065–73
144. Martinson, H. G., True, R., Burch, J. B. E., Kunkel, G. 1979. *Proc. Natl. Acad. Sci. USA* 76:1030–34
145. Goldknopf, I. L., French, M. F., Daskal, Y., Busch, H. 1978. *Biochem. Biophys. Res. Commun.* 84:786–93
146. Goldknopf, I. L., Wilson, G., Ballal, N. R., Busch, H. 1980. *J. Biol. Chem.* 255:10555–58
147. Matsui, S. I., Seon, B. K., Sandberg, A. A. 1979. *Proc. Natl. Acad. Sci. USA* 76:6386–90

148. Goldknopf, I. L., Sudhakar, S., Rosenbaum, F., Busch, H. 1980. *Biochem. Biophys. Res. Commun* 95:1253–60
149. Hershko, A., Ciechanover, A., Rose, I. A. 1981. *J. Biol. Chem.* 256:1525–28
150. Wilkinson, K. D., Audhya, T. K. 1981. *J. Biol. Chem.* 256:9235–41
151. Andersen, M. W., Goldknopf, I. L., Busch, H. 1981. *FEBS Lett.* 132:210–14
152. Wu, R. S., Kohn, K. W., Bonner, W. M. 1981. *J. Biol. Chem.* 256:5916–20
153. Andersen, M. W., Ballal, N. R., Goldknopf, I. L., Busch, H. 1981. *Biochemistry* 20:1100–4
154. Ciechanover, A., Heller, H., Katz-Etzion, R., Hershko, A. 1981. *Proc. Natl. Acad. Sci. USA* 78:761–65
155. Lipmann, F. 1971. *Science* 173:875–84
156. Ciechanover, A., Elias, S., Heller, H., Hershko, A. 1982. *J. Biol. Chem.* In press
157. Haas, A. L., Rose, I. A., Hershko, A. 1981. *Fed. Proc.* 40:1691
158. Deleted in proof
158a. Haas, A. L., Rose, I. A. 1981. *Proc. Natl. Acad. Sci. USA* 78:6845–48
159. Deleted in proof
160. Deleted in proof
161. Deleted in proof
162. Roberts, J. W., Roberts, C. W. 1975. *Proc. Natl. Acad. Sci. USA* 72:147–51
163. Roberts, J. W., Roberts, C. W., Mount, D. W. 1977. *Proc. Natl. Acad. Sci. USA* 74:2283–87
164. Roberts, J. W., Roberts, C. W., Craig, N. L. 1978. *Proc. Natl. Acad. Sci. USA* 75:4714–18
165. Roberts, J. W.,Roberts, C. W., Craig, N. L., Phizicky, E. M. 1978. *Cold Spring Harbor Symp. Quant. Biol.* 43:917–20
166. Craig, N. L., Roberts, J. W. 1980. *Nature* 283:26–30
167. Roberts, J. W., Roberts, C. W. 1981. *Nature* 290:422–24
168. Phizicky, E. M., Roberts, J. W. 1981. *Cell* 25:259–67
169. Witkin, E. M. 1976. *Bacteriol. Rev.* 40:869–907
170. Weinstock, G. M., McEntee, K., Lehman, I. R. 1979. *Proc. Natl. Acad. Sci. USA* 76:126–30
171. Shibata, T., DasGupta, C., Cunningham, R. P., Radding, C. M. 1979. *Proc.*

Natl. Acad. Sci. USA 76:1638–42
172. Phizicky, E. M., Roberts, J. W. 1980. *J. Mol. Biol.* 139:319–28
173. Little, J. W., Edmiston, S. H., Pacelli, L. Z., Mount, D. W. 1980. *Proc. Natl. Acad. Sci. USA* 77:3225–29
174. Cunningham, R. P., Shibata, T., Dasgupta, C., Radding, C. M. 1979. *Nature* 281:191–95
175. McEntee, K., Weinstock, G. M., Lehman, I. R. 1979. *Proc. Natl. Acad. Sci. USA* 76:2615–19
176. Murakami, K., Voellmy, R., Goldberg, A. L. 1979. *J. Biol. Chem.* 254:8194–200
177. Voellmy, R. W., Goldberg, A. L. 1981. *Nature* 290:419–21
178. Swamy, K. H. S., Goldberg, A. L. 1981. *Nature* 292:652–54
179. Chung, C. H., Goldberg, A. L. 1981. *Proc. Natl. Acad. Sci. USA* 78:4931–35
180. Charrette, M. F., Henderson, G. W., Markovitz, A. 1981. *Proc. Natl. Acad. Sci. USA* 78:4728–32
181. Bukhari, A. I., Zipser, D. 1973. *Nature New Biol.* 243:238–41
182. Shineberg, B., Zipser, D. 1973. *J. Bacteriol.* 116:1469–71
183. Gottesman, S., Zipser, D. 1978. *J. Bacteriol.* 133:844–51
184. Simon, L. D., Gottesman, M., Tomczak, K., Gottesman, S. 1979. *Proc. Natl. Acad. Sci. USA* 76:1623–27
185. Gottesman, S., Gottesman, M., Shaw, J. E., Pearson, M. L. 1981. *Cell* 24:225–33
186. Markovitz, A. 1977. In *Surface Carbohydrates of the Prokaryotic Cell*, ed. I. W. Sutherland, pp. 415–62. New York: Academic
187. Zehnbauer, B. A., Markovitz, A. 1980. *J. Bacteriol.* 143:852–63
188. Schoemaker, J. M., Markovitz, A. 1981. *J. Bacteriol.* 147:46–56
189. Zehnbauer, B. A., Foley, E. C., Henderson, G. W., Markovitz, A. 1981. *Proc. Natl. Acad. Sci. USA* 78:2043–47
190. Deleted in proof
191. Deleted in proof
192. DeMartino, G. N., Goldberg, A. L. 1979. *J. Biol. Chem.* 254:3712–15
193. Rose, I. A., Warms, J. V. B., Hershko, A. 1979. *J. Biol. Chem.* 254:8135–38
194. Boches, F. S., Klemes, Y., Goldberg, A. L. 1980. *Fed. Proc.* 39:1682

Ann. Rev. Biochem. 1982. 51:365–94

MAGNETIC RESONANCE STUDIES OF ACTIVE SITES IN ENZYMIC COMPLEXES

Mildred Cohn and George H. Reed[1]

Department of Biochemistry and Biophysics, University of Pennsylvania, Philadelphia, Pennsylvania 19104

CONTENTS

Summary and Perspectives

Prior to any attempt to elucidate the mechanism of enzyme action in complete molecular detail, it is necessary to know the geometric and electronic structure of the reacting species, i.e. the active enzyme-substrate(s) complex and the enzyme-product(s) complex. The impressive successes of X-ray crystallography in determining three-dimensional structures have been limited to enzymes alone or enzyme-inhibitor complexes. For obvious reasons, rapidly interconverting active enzyme-substrate and enzyme-product structures cannot be accommodated by X-ray crystallographic analysis, although recent work at low temperatures is promising in this

---

[1]The authors wish to thank Drs. Jardetzky, Roberts, and Coleman for making their manuscripts available before publication.

365

0066-4154/82/0701-0365$02.00

regard. Of the various spectroscopic techniques available, magnetic resonance spectroscopy is the choice in solution for delineating the active site of an enzyme, including the specific interaction of groups in the substrates with specific protein residues, specification of metal ligands, and the disposition of substrates within the active site. The observable parameters are those of individual atoms, and many atoms of interest have net nuclear magnetic moments (^{1}H, ^{13}C, ^{15}N, ^{17}O, ^{19}F, ^{31}P, and ^{113}Cd) and a more limited number have net electron magnetic moments [Cu(II), Mn(II), Co(II), Mo(V), Fe-(III), and lanthanides] in addition to organic free radicals. In magnetic resonance, dynamic phenomena, e.g. interconverting protein conformations or interconverting substrate and product complexes, do not vitiate the usefulness of this type of spectroscopy. On the contrary, such phenomena can lead to spectral effects that are interpretable in terms of dynamic as well as static parameters of the structures. Perturbations of the structure at the active site, of utmost importance to enzyme function, which arise from substrate binding, are often reflected in NMR and EPR parameters, but may induce distance changes that are too small to detect by any available method. Magnetic resonance studies of active sites have strongly reinforced the concept that the active site of an enzyme is conformationally flexible and that binding at one subsite frequently modifies the affinity, specificity, and mode of binding at a second subsite. The implications of the static and dynamic structural features of the active site for mechanisms of enzyme action are constantly evolving and being revised as more precise structural and kinetic data become available. A discussion of such implications is beyond the scope of this review.

NMR and EPR measurements involve stimulation of resonant transitions between energy levels of nuclear and electronic spin states, respectively. These energy levels are influenced by the electronic and magnetic environments of the nucleus or paramagnetic center under observation and the spectra therefore reflect the structure of the site. Chemical shifts, nuclear spin-nuclear spin interactions, and electron spin-nuclear spin interactions, in the case of NMR; and ligand fields, spin-spin coupling with surrounding magnetic nuclei, and spin-spin interactions with nearby paramagnetic species, in the case of EPR, contribute to the spectral features of a given species. Hence, NMR and EPR spectroscopy can provide detailed information concerning the identity and arrangement of ligand groups and proximity to other magnetic species, although the procedures for extracting such information from the NMR and EPR spectra may require considerable interpretive effort and are fraught with pitfalls.

The use of NMR spectroscopy for the study of enzymic complexes has burgeoned in recent years and will continue to do so with the increased availability of high frequency instruments. The improved resolution plus the sophisticated technological advances, which make resolution enhancement techniques and double irradiation techniques simple to implement, will aid in assignment of amino acid residues in the protein and of nuclei in the substrate. Also the increased sensitivity will make experiments with ^{13}C and other nonproton nuclei more attractive. In EPR spectroscopy, multiple irradiation techniques, pulsed-spin echo methods, and isotopic substitution promise to provide new structural details.

NMR Studies of Enzymic Complexes

A number of recent reviews on the application of NMR to studies of enzyme structure and mechanism are available, including the broad coverage in several chapters of Jardetzky & Roberts (1) and more specialized reviews on the geometry of enzyme-bound substrates (2), ^{19}F studies (3, 4), ^{31}P studies (5), catalytic groups of serine proteases (6), alkaline phosphatase (7), and some ATP reactions (8, 9). Qualitative and quantitative structural information obtainable is illustrated in this review by recent multifaceted NMR studies of several representative enzyme systems, catalyzing hydrolysis (*Escherichia coli* alkaline phosphatase), hydrogen transfer (dihydrofolate reductase), and phosphoryl transfer (pyruvate kinase). For large proteins, in general, proton NMR of native proteins currently has insufficient resolution to be useful, and the same difficulty holds even for some small proteins such as dihydrofolate reductase (mol wt 18,000). Simplification of the proton spectra has been achieved in some cases by selective deuteration, by selecting on the basis of relaxation rates, or by photo-CIDNP. More generally the problem has been avoided for bacterial enzymes by using ^{13}C NMR of specifically enriched ^{13}C proteins and ^{19}F spectra of specifically labeled ^{19}F enzymes. Experiments with large nonbacterial enzymes have been limited primarily to spectroscopy of substrates and cofactors.

The most frequently measured NMR parameter is the chemical shift, which essentially reflects the electronic environment of the nucleus and is perturbed by temperature, ionic strength, state of protonation, and ligand binding. Absolute values of chemical shifts are difficult to interpret definitively, but changes are most useful and often interpretable. From changes in chemical shifts, pK values of individual amino acid residues in proteins and of substrates in the free and enzyme-bound state may be determined, and the state of protonation of these components in the enzymic complexes established. Reciprocal effects on chemical shifts of nuclei in proteins and those in substrates, inhibitors, and metal ligands may be observed. In favorable cases, it is possible to differentiate direct effects due to contact between ligand and enzyme and indirect effects due to conformational change. Observation of nuclear spin-spin coupling mediated through chemical bonds gives unequivocal evidence for the existence of a given structure, e.g. ^{113}Cd chelation, and the magnitude of the coupling constant may yield conformational information, e.g. ^1H-^{31}P coupling constants of nucleotides.

There are a number of NMR parameters that arise from spin-spin interaction and are a function of distance. One is a nuclear spin-nuclear spin interaction, the intra- or intermolecular nuclear Overhauser effect (NOE), where, upon saturation by strong irradiation of one nucleus, a change in the intensity of the resonance signal of another nucleus results from cross-relaxation between the two nuclei (10). Where distances between nuclei are known, NOE can be used for assignment of nearby nuclei in complex spectra. Relative distances and consequently conformations can be determined from comparisons of the magnitude of NOEs produced on the observed signal. Irradiating either of two other nuclei will indicate which one is closer to the observed one. Observation of intermolecular NOE between protein and ligand nuclei in an enzymic complex can identify the amino acid residue in proximity to the ligand nucleus observed.

Since magnetic dipole relaxation falls off as the sixth power of the distance between the two magnetic species involved, measurement of relaxation rates would provide structural information if the system under observation consisted of a single pure species and if simple dipole-dipole interaction dominated the spin-spin interaction. Alas, the system is "never pure and rarely simple."[2] Even for the longitudinal relaxation rate, $1/T_1$, of ^{13}C in diamagnetic systems, where the primary contribution to T_1 does arise from dipole-dipole interaction, the interpretation of an increase in T_1 upon binding a ^{13}C ligand to enzyme is ambiguous; either immobilization has occurred or decreased proximity to other magnetic nuclei, particularly strong relaxers such as protons, or both factors contribute. With T_1 relaxation rate changes of nuclei induced by paramagnetic ions, the dipole-dipole interaction (proportional to r^6) dominates T_1, and in principle should be an excellent parameter for determining distances between the observed nucleus and the paramagnetic ion (2). However, since only a small fraction of the nuclei under observation are in complexes containing the paramagnetic ion (otherwise the resonances would become too broad to detect), the rate of exchange ($1/\tau_M$) between the coordination sphere of the paramagnetic ion and the diamagnetic species must be sufficiently fast so that the observed T_1 is a true average, i.e. $\tau_M \ll T_{1M}$, a condition that obtains only in a limited number of enzymic complexes, as is discussed in a later section. Other stringent criteria that must be met for valid distance determinations have been noted (2, 12).

Dynamic parameters, including rates of substrate binding, of conforma-

---

[2]*The Importance of Being Earnest,* Act I, by Oscar Wilde.

tional interconversions, and of interconversions of active complexes may be determined either from line widths (T_2 relaxation) or transfer of saturation (double irradiation) between relevant nuclei. All of these methods have been used and the information derived from them will be illustrated.

E. Coli Alkaline Phosphatase

Alkaline phosphatase, a nonspecific phosphomonoesterase, is a dimeric zinc metalloenzyme of molecular weight 94,000, and is by far the largest of the hydrolytic enzymes that has been investigated extensively by NMR. The amino acid sequence (13) and the crystal structure (14) have recently been solved. However, the NMR studies described below were done prior to the availability of the sequence and X-ray data. Nuclei other than protons, $^{13}C, ^{19}F, ^{31}P$, and ^{113}Cd, have been used profitably to delineate the active site. γ–^{13}C His, 3-^{19}F Tyr, and 4-^{19}F Phe have been incorporated biosynthetically. The symmetrical disposition of the two subunits follows from the observation of only 11 ^{19}F (F-Tyr) (15) and 10 ^{13}C ($C\gamma$-His) (16) resonances, the known number of Tyr and His residues in each subunit. In addition, ^{113}Cd (spin ½) may replace the magnetically inert Zn in the enzyme with the retention of ~1% activity. NMR studies through 1978 have been thoroughly reviewed (6) and the more recent results are emphasized in this review.

The enzyme is known to catalyze hydrolysis with the formation of a phosphoseryl intermediate (see 102) by the following sequence of reactions:

$$E + ROP \; \underset{\leftarrow}{\overset{\rightarrow}{\rightleftarrows} } \; E \cdot ROP \; \underset{\leftarrow}{\overset{RO^-}{\rightleftarrows} } \; E-P \; \underset{\leftarrow}{\overset{H_2O}{\rightleftarrows} } \; E \cdot P \; \underset{\leftarrow}{\overset{\rightarrow}{\rightleftarrows} } \; E + P_i \qquad 1.$$

E–P may be formed from either direction. For the Zn enzyme, the equilibrium concentration of E–P is high at low pH but low at alkaline pH (6). For the Cd enzyme, the equilibrium concentration of E–P is high even at alkaline pH.

^{31}P NMR   In view of the well-established existence of an E–P intermediate, it is not surprising that ^{31}P was the first nucleus to be investigated in this system by NMR (17). It was found that the chemical shift of Zn E–P was unusually low, +8.5 ppm, and could readily be distinguished from that of E·P_i at ~+3.5 ppm (17–21). The resonances were assigned on the basis of pH dependence of the relative intensities from the known equilibria (see Equation 1) and were confirmed with the alkaline stable Cd E–P (17, 18), which has its resonance at +8.1 ppm. No multiplet structure is observed in E–P, but proton decoupling narrows the line for both Zn E–P (17) and

Cd E–P (18) as anticipated for serylphosphate. The unusual shift of the ^{31}P resonance of E–P ($\sim$8 ppm downfield from that of phosphoserine) has been attributed to a 2° narrowing of the O–P–O angle (20). It has been suggested (17) that strain might arise from a hydrogen bond acting in concert with a covalent bond to serine to mimic a strained cyclic ester. The shift appears not to be due to metal chelation. Even in the apoenzyme (no metal ion) the covalently bound ^{31}P is observed at +6.5 ppm (17, 18), and in the ^{113}Cd enzyme, ^{113}Cd-^{31}P spin-spin coupling is observed only for the noncovalently bound P_i in E·P_i, but not for the covalently bound ^{31}P in E–P (22).

A great deal of confusion has existed over the years concerning the role and stoichiometry of the Zn (2–6 per dimer). The latter appears to depend on the method of preparation, the pH, buffer conditions, and the interaction with both Mg(II) and phosphate binding. The stoichiometry of phosphate binding has also been controversial. It has now been established (23) that the apoalkaline phosphatase dimer can bind a pair of metal ions at each of three sites (A, B, and C) in each monomer. The metal at site A with three His ligands is separated at the active site by $\sim$5.8 Å from the metal at site B, which has carboxyl ligands. Full activity is attained when sites A and B are fully occupied either by Zn at both sites or Zn at A site and Mg at B site; occupation of C site does not affect activity. ^{113}Cd (spin 1/2) has been most useful as a probe, particularly in conjunction with ^{13}C and ^{31}P NMR studies. The Cd(II) E analogue exhibits properties quantitatively different from the native enzyme, namely Cd(II) is bound less tightly to both A and B sites, by two orders of magnitude at the A site, and Cd(II) self-exchanges more rapidly; the catalytic activity is reduced to 1%, and the pK_a controlling the E·P $\leftrightharpoons$ E–P equilibrium is shifted $\sim$3 pH units more alkaline. Nevertheless, Cd(II) E forms the same phosphoenzyme intermediates and Cd(II) qualitatively exhibits the same binding pattern as a function of pH, phosphorylated state of enzyme, and cooperative effects between sites. The Co(II) enzyme has also been shown to differ from the native Zn enzyme in kinetic partition patterns of the intermediate E·P_i to E–P and E + P_i, respectively (See Equation 1). E–P formation is favored for the Cd enzyme relative to the Zn enzyme as demonstrated from the time course of the distribution of ^{31}P–^{18}O$_n$ isotopic species during the exchange of P_i (^{18}O$_4$) and H$_2^{16}$O (24).

The ligands to the metal ions have been unequivocally assigned by a combination of ^{113}Cd with ^{31}P and ^{13}C. Figure 1, A and B show the effect of ^{112}Cd and ^{113}Cd on the ^{31}P NMR spectra of E· and E–P compared to their Zn counterparts, Figure 1C (22). The change in chemical shift for E–P from ZnE to CdE (Figure 1, A and B) is not large; the shift is 8.70 ppm in Zn$_4$Mg$_2$E and Zn$_4$E compared to 9.05 ppm in Cd$_6$E and 8.27 ppm in Cd$_2$E. On the other hand, a marked change in chemical shift is observed between the Cd and Zn species of the noncovalent E·P complex, 4 ppm in Zn_4Mg_2E compared to 13 ppm in Cd_2E and Cd_6E (Figure 1, A and B). The large difference in the E·P species suggests direct coordination of phosphate to the metal, which is proven by the ^{113}Cd–^{31}P coupling (33 Hz) observed in Figure 1C (22).

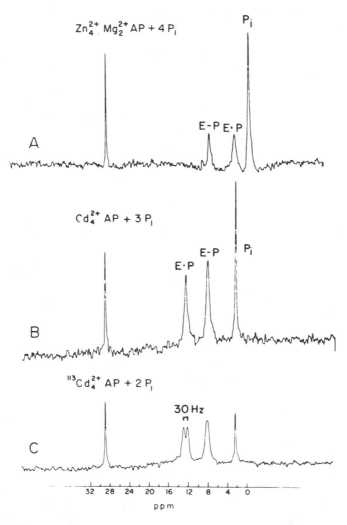

Figure 1 ^{31}P NMR spectra at 36.4 MHz of Zn(II) and Cd(II) alkaline phosphatases (23). The metal ion and phosphate components are indicated in the figure: A, 2.4 mM enzyme; B, 1.75 mM enzyme, pH 8.9; and C, 2.9 mM enzyme, pH 9.1. Resonance at 29.3 ppm arises from an external methylphosphonate standard.

^{13}C NMR   The quaternary nonprotonated γ-carbon of histidine is an excellent probe of the active site structure of alkaline phosphatase in which $\beta,\beta[\gamma$-^{13}C]dideuteriohistidine (deuterium in the β-position reduces the ^{13}C linewidths by 30%) has been incorporated biosynthetically (16). The chemical shift (δ) of γ-^{13}C is sensitive to (a) ionization state, (b) tautomeric forms of the imidazole side chain, (c) metal coordination, and (d) protein environment. Coupling to [113]Cd is observable upon coordination to N$^\tau$ (3 bond), but not upon coordination to N$^\tau$ (2 bond) (25). Figure 2B shows the ^{13}C spectrum of the ten His residues of the diphosphorylated form of the Cd$_6$ enzyme with ten resolved peaks spread over 14 ppm (the very similar spectrum of the native Zn enzyme is somewhat less resolved); [γ-^{13}C] shift ranges of some model compounds are indicated below the spectrum. Peaks 1, 2, and 9 of the Zn or Cd enzyme fall outside the normal range of nonmetalloproteins, but are shifted within the normal range in the apoenzyme (upfield >3 ppm for 1 and 2, downfield >2 ppm for 9). The downfield shifts of His-1 and -2 upon addition of two metal ions to apoenzyme implicate these His residues as ligands to Zn at A sites coordinated via N$^\tau$ (6, 25). The spin-spin splitting due to [113]Cd in the Cd$_6$ enzyme is clearly observable for His-7, -8, and -9 and establishes these His residues as Zn ligands coordinated via N$^\pi$ (Figure 2, A and B). The exchange of [113]Cd by [112]Cd (Figure 2C) in Cd coordinated to His-7 and -8 but not to His-9, taken together with data on [113]Cd spectra (26), permits assignment of four His residues, 7 and 8 as well as 1 and 2, as metal ligands at site A, and His-9 as ligand to metal at site B. Additional evidence that supports these structural identifications are broadening by Mn(II) and absence of normal pH titration (25).

[113]Cd NMR   The results of titrations of the apoenzyme with [113]Cd, particularly in the range between two Cd atoms/dimer and six Cd atoms/dimer and of the nonphosphorylated and phosphorylated forms of the enzyme have been as difficult to interpret as the welter of contradictory stoichiometries found in the literature. The complexity appears to arise from the dependence of the binding affinities and exchange rates at the three metal binding sites A, B, and C on pH, phosphate, and magnesium concentrations and upon occupancy of other sites, i.e. cooperativity of binding (23, 26). Independent determinations of affinities and rates have been helpful (23).

If two Cd atoms are added per dimer of apoenzyme at pH 6.5, the result (19, 23, 26) is clear-cut: one resonance is observed at 120–180 ppm depending on halide concentration, and the resonance is justifiably ascribed to identical binding at each A site monomer. However, upon covalent phosphorylation at one active site of Cd$_2$E, i.e. in the species Cd$_2$E–P$_1$, two

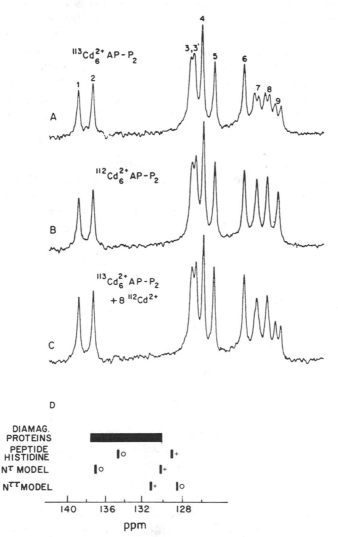

Figure 2 ^{13}C NMR spectra at 50.3 MHz, pH 6.2, of ^{112}Cd(II)$_6$ and ^{113}Cd(II)$_6$ diphosphorylated alkaline phosphatase enriched with ^{13}C in β, $\beta[\gamma$-^{13}C] dideuteriohistidine (24). The Cd(II) components are indicated in the figure: A and B, 1.3 mM enzyme; C, sample A after addition of eight equivalents of ^{112}Cd(II), equilibration for 72 h, and removed of excess Cd(II) (Relevant chemical shifts of model compounds (16) are indicated below the spectra); I, range of His C$_\gamma$ in proteins; II, neutral (0) and protonated (+) His C$_\gamma$ in small peptides; III, 0 and + forms of N$^\tau$ methylimidazole peptide; and IV, 0 and + forms of N$^\tau$ methylimidazole peptide.

resonances appear at 141 and 56 ppm, both upfield from the original resonance at 169 ppm (19). The initial explanation (19) ascribed the resonances to A sites of each monomer, one phosphorylated (subunit 1), the other nonphosphorylated (subunit 2). With the later measurement of the ^{113}Cd resonances of the diphosphorylated $Cd_4Mg_2EP_2$ (26), which shows two peaks with approximately the same δ values as Cd_2E-P_1, and with the realization that phosphorylation increases the binding affinity of both A and B sites, the explanation of the two peaks in Cd_2E-P_1 was modified and the two peaks were ascribed to a species with ^{113}Cd bound to one A site and one B site on the phosphorylated monomer and no Cd bound to the nonphosphorylated monomer (26). These assignments are reinforced by the finding (23) that even in the nonphosphorylated Cd_2E the symmetry of the two A sites reflected in the single observable resonance at 168 ppm is destroyed merely by raising the pH from 6.5 to 7.5, as evidenced by the appearance of two peaks, 144 and 52 ppm (23). The increase in pH has so increased the affinity of binding at the B site that it apparently makes occupation of A and B sites on one monomer and none on the other, the more favorable configuration. Other puzzling phenomena observed upon further addition of Cd to form Cd_4E, namely apparent loss of signal, is explicable in terms of exchange broadening, which is consistent with exchange rates measured by radioisotopes; the rates are pH dependent (23). In principle, an EPR investigation with $Mn(II)_2·E$ should give direct evidence for the migration of the metal ion upon raising the pH or upon monophosphorylation, since two Mn(II) atoms separated by only $\sim$5.8 Å would be expected to exhibit spin-spin exchange detectable in Mn(II) EPR spectra.

Dihydrofolate Reductase

Functionally, dihydrofolate reductase is a representative of the very large class of hydrogen transfer catalysts linked to the oxidation-reduction of pyridine nucleotide coenzymes; the hydrogen acceptor in this case is dihydrofolate, which is converted to tetrahydrofolate. The enzyme has been the focus of intensive investigation by X-ray diffraction (27, 28) and NMR not only because it is the smallest of this class of enzymes (mol wt 18,000) but because it is the site of action of antitumor and antimicrobial drugs. The structural basis of the 1000-fold increase in binding constant of the reduced coenzyme NADPH relative to NADP and of the 10,000-fold increase in binding constant of the inhibitor methotrexate relative to the substrate has been sought (the essential difference in structure between methotrexate and substrate is the substitution of the 4-oxo of the substrate by 4-NH_2 on the pteridine ring). Toward this end, changes in both protein resonances and ligand resonances in binary and ternary complexes have been monitored.

PROTEIN SPECTRA   Unlike other small proteins such as ribonuclease and lysozyme, dihydrofolate reductase yields poorly resolved ^{1}H and ^{13}C spectra even at high magnetic fields. Thus to follow conformational changes in the protein specifically labeled ^{19}F (29, 30) and ^{13}C (31, 32, 33), amino acids have been introduced biosynthetically; selective deuteration (34, 35) and photo-CIDNP (36) have been used to simplify the proton spectra of the aromatic region. The early work on substrate and inhibitor binding to ^{19}F labeled protein (6-F-Trp, 3-F-Tyr, m-F-Phe) and also proton NMR of binary complexes of selectively deuterated dihydrofolate reductase from *Lactobacillus casei* has been reviewed by Roberts et al (37). The most striking findings were (*a*) in the respective enzymic complexes of folate and methotrexate, the orientation of the pteridine ring differs by about 180° and (*b*) sharpening of the ^{19}F and ^{1}H resonances of Tyr and of Trp, upon binding methotrexate, resulted from locking the many interconverting conformations of the naked enzyme into one form of enzyme·inhibitor complex.

The most cogent evidence for the existence of conformational equilibria of the free enzyme was obtained subsequently from the studies of [8-^{13}C-Trp]dihydrofolate reductase from *Streptococcus faecium* (33). One of the four 8-^{13}C-Trp resonances is split into two resonances and another is broad; upon addition of methotrexate, the split resonance coalesces and the broad resonance sharpens dramatically. The most likely explanation is that the complex has been locked into one conformation, but one cannot exclude the possibilities that either exchange between conformers of the complex is much slower with one resonance so broad that it is undetectable or that the chemical shift difference of the two liganded conformations is now sufficiently small that the system is consequently in the fast exchange region. Two distinct conformational forms of a ternary inhibitor complex, E·NADP· trimethoprim, also exist, as evidenced by the presence of split C2 H resonances for two of the histidines at low temperatures (38). As the temperature is raised, the rate of interconversion increases, as reflected in the behavior of two species of bound NADP in ^{1}H and ^{31}P spectra as described below.

Once the three-dimensional structure of the enzyme-methotrexate-NADPH complex in the crystal was established (27, 28), which indicated that His-28 forms an ion pair with the γ-carboxyl group of the glutamate moiety of methotrexate, and His-64 binds to the 2'-phosphate of NADPH, the C2 proton resonances of those two histidines could be assigned (39). The specific line broadening of one His-C2 peak (His-64) due to the paramagnetic [CR(CN)$_6^{3}$] anion, which binds competitively with etheno-NADP, and the disappearance of another His-C2 peak upon binding methotrexate alone led to the assignment of His-28. The only other amino acid residue at the active site unequivocally assigned is Trp-21, based on N-bromosuccinimide modification (35).

Several features of the disposition of the substrates at the active site may be deduced from the protein spectral changes. For example, the C2 proton peak of Trp-21 is unaffected by NADPH binding (35) but the 6 position in ^{19}F-substituted Trp-21 is shifted by 2.7 ppm (29, 30), thus establishing that the carboxamide group of NADPH is close to the 6 position but distant from the 2 position. This result agrees with the orientation observed in the crystal (28).

A novel approach, photo-CIDNP (36), has aided in identifying the His, Tyr, and Trp residues involved in ligand binding, because of changes in accessibility of the aromatic residues of the protein in enzymic complexes to a photo excited flavin dye in solution. The spectrum is simplified, since only 7 or 8 of the 25 aromatic residues are observed in the light-dark difference spectrum (See Table 1), and also the most likely ligand binding sites, the accessible residues, have been selected. Decreased accessibility upon ligand binding indicates either direct binding or a conformational change, but increased accessibility is uniquely ascribable to a conformational change. Table 1 (36) summarizes the effects of ligand binding on the accessible His and Trp residues. The accessibility of complexes observed in solution is consistent with the anticipated accessibility from the crystal structure of E·NADPH·methotrexate. The inaccessibility of His-C (His-64) in complexes containing NADP (see Table 1) is in agreement with the effect of NADP binding observed on the C-2 proton resonance of His-64 (38), on the adenine proton resonances, and on the ^{31}P of the 2'-P of NADP (38, 40). The loss of His-F (His-28) accessibility upon binding p-aminobenzoyl glutamate but not trimethoprim is in agreement with previous results showing interaction of His-28 only with those ligands containing a p-aminobenzoyl glutamate moiety (39). The *increased* accessibility of His-A and Trp-2 upon ligand binding is most convincing evidence for a conformational change.

^{13}C spectra of *S. faecium* enzyme labeled with methyl-^{13}C methionine (31) indicate differences between the binary NADP and NADPH complexes. Binding of methotrexate or other inhibitors produces shifts of the ^{13}C methyl in the opposite direction from those of substrates, which gives further evidence of the different modes of binding for substrates and inhibitors. In similar experiments with [guanidino-^{13}C]arginine (32) where five of the eight arg residues are resolved, the authors conclude that the shifts resulting from NADPH and inhibitor binding monitor conformational changes, since free enzyme and all binary and ternary complexes have the same temperature coefficient of the chemical shift.

In summary, with the use of many approaches it has been possible (*a*)

Table 1 Accessibility of aromatic residues from ^{1}H photo-CIDNP spectra of dihydrofolate reductase (36)

Complex	Histidine residues					Trp	
	A	B	C	E	F	1	2
E alone	+[a]	+	+	+	+	+	0[a]
E.trimethoprim	+	+	+	+	+	+	+
E.folate	0	+	+	0	−[a]	+	+
E.P-aminobenzoyl glutamate	+	+	+	0	0	+	+
E.methotrexate	+	+	+	±[a]	−	+	+
E.aminopterin	+	+	+	±	−	+	+
E.NADP	±	0	±	+	+	+	+
E.NADP.trimethoprim	+	0	±	+	+	+	+
E.NADP.folate	+	0	±	±	−	+	+
E.NADP.methotrexate	+	0	±	±	−	+	+
E.NADP.aminopterin	+	0	±	±	−	+	+

[a] The symbol + signifies accessible; ±, slightly accessible; 0, inaccessible; and −, ambiguous.

to assign two His and one Trp resonance, (b) to demonstrate unequivocally conformational equilibria of the enzyme, (c) to differentiate in some cases which observed changes in individual amino acid residues upon ligand binding are due to direct binding and which to conformational changes, and (d) to obtain evidence with several nuclei that the folate substrate and inhibitors bind differently to the enzyme.

LIGAND SPECTRA Since the coenzyme has three ^{31}P atoms, simple spectra of the reduced and oxidized forms and fragments in the free and bound species are easily observable. Questions have been directed toward delineating the groups on the coenzyme important to binding and distinguishing the modes of binding between NADPH and NADP to explain the difference of 10^3 in binding constants. When NADPH is bound to the reductase from L. casei (41), three ^{31}P peaks appear (See Figure 3B): one arises from the 2'P and two, coupled to each other, from the pyrophosphate moiety; the latter two are equivalent in the free coenzyme (See Figure 3A). The large downfield shift of the 2'P is pH independent and has been partially ascribed to preferential binding of the dianionic form (42) by analogy with the ^{31}P shift of the bound form of 2'AMP. The latter is pH dependent and the dianionic species is preferred by a factor of 16, since its pK as determined from ^{31}P NMR titration is lowered by 1.2 pH units in the enzymic complex. The 2'P shift is much smaller in the E. coli enzymic complex (34) in which His-64 of the L. casei enzyme has been replaced by serine. A comparison of B and C in Figure 3, with and without ^{1}H decoupling, reveals the difference in H-5'–^{31}P coupling constants of the two phosphorus atoms of the pyrophosphate moiety; the coupling constant is determined by the conformation of the 5'–O–P bond. The very small unobservable coupling on the highest field peak (Figure 3C) implies a gauche-gauche conformation about the 5'–O bond as in the free coenzyme. The other ^{31}P peak of the pyrophosphate moiety has a much larger coupling constant (See Figure 3C), and simulation (42) is consistent with a C4'–C5'–O–P dihedral angle that differs from the free NADPH by ~50°. The X-ray data obtained subsequently (28) agree with this result. Recent higher resolution X-ray data of the E·NADP·methotrexate complex indicate that the lower field ^{31}P resonance with the large coupling constant (See Figure 3) is that on the nicotinamide (44). The coupling constant for the lower field ^{31}P signal in the *E. coli* enzymic complex (43) is much smaller, which indicates that the conformation of NADPH differs in the two enzymic complexes. As in the case of NADPH complexes, it is concluded from ^{31}P chemical shifts that the variations in mode of binding NADP and its analogues reside in the nicotinamide moiety of the coenzyme (44).

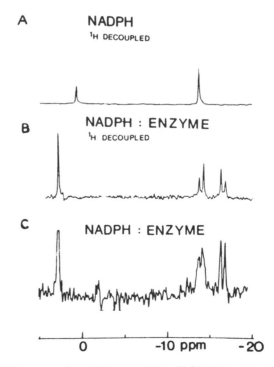

Figure 3 ^{31}P NMR spectra of NADPH at 40.5 MHz, pH 6.9 (41). A, 10 mM NADPH proton decoupled; B, 1.3 mM NADPH plus 1.3 mM dihydrofolate reductase, proton decoupled; and C, same as B without proton decoupling.

Because of slow exchange and the consequent inability to observe averaged shifts in the course of titration with ligand, it is difficult to establish a 1:1 correspondence between the resonances of individual coenzyme protons in their free and enzyme-bound forms. This difficulty of assignment of bound NADP was overcome by saturation transfer experiments (40) where the intensities of the known resonances in excess free ligand were monitored during systematic irradiation over the aromatic proton spectral range. Structural modifications in NADP produce marked changes in the chemical shifts of the nicotinamide moiety, but not of the adenine in the enzymic complexes. Again, the chemical shifts of the nicotinamide moiety and not of the adenine respond strongly to occupation of the second subsite by inhibitor or substrate, as illustrated in Table 2.

The two conformations of the enzyme·NADP·trimethroprim complex are reflected not only in the protein spectrum but also in the two signals ($\Delta\delta = 1$ ppm) arising from each proton of the nicotinamide of bound NADP (45). Two sets of signals are also clearly separated in the ^{31}P spectra of the two pyrophosphate ^{31}P atoms, one corresponding to the E·NADP·methotrexate complex and the other to the E·NADP· trimethoprim complex. In the two conformations of the ternary complex, the ^1H–^{31}P and ^{31}P–^{31}P coupling constants differ, thus pinpointing the conformational differences at the 5'C–O and the P–O–P bond.

Chemical shifts of enzyme-bound inhibitor and/or substrate have been determined in ^1H NMR by difference spectroscopy between trimethoprim and selectively deuterated trimethoprim (46) and in ^{13}C NMR for folate, methotrexate, and aminopterin with all three compounds enriched to the extent of 90% ^{13}C in the C2 position (47). The postulated structure based on chemical shifts predicts that the pyrimidine H6 and benzyl H_2' are very close, and the observation of an NOE effect between the two protons is consistent with the model.

Table 2 Chemical shift changes on ligand binding to dihydrofolate reductase (40) in ppm

			E.NADP	
Proton[a]	E.NADPH	E.NADP	folate	methotrexate
A2	−0.93	−0.86	−0.83	−0.86
A8	−0.28	−0.39	−0.47	−0.50
N2	0.15	0.61	0.56	0.53
N4	0.64	1.36	0.14	0.92
N5	—	0.97	1.41	0.79
N6	0.47	0.73	0.69	0.73
A1'	0.47	0.56	0.53	0.48
N1'	—	−0.42	0.43	−0.45

[a] A = adenosine; N = nicotinamide.

Evidence for the binding of methotrexate in its protonated form had accumulated from many types of spectroscopic measurements, and the increased affinity for the reductase relative to that of the substrate had been ascribed to the increased basicity of methotrexate. The ^{13}C NMR data on the pKa's of the free and bound ligands provide the most convincing support for this view (47). Changes in chemical shift of [13]C2 of free ligands upon protonation ranged from 2 to almost 10 ppm. In the binary methotrexate complex, the chemical shift of [13]C2 is invariant with pH between pH 6 and 10, and is 0.93 ppm downfield from the free protonated form and 5.44 ppm upfield from the unprotonated form. It was concluded that the pKa of methotrexate or aminopterin, when bound to dihydrofolate reductase from *S. faecium,* increased from 5.7 (N1) to ~10, probably due to interaction with Asp 26. The same result was obtained in ternary complexes with NADPH$_4$, a competitive inhibitor of NADPH. On the other hand, the substrate, folate, had the same pKa, 2.40 (N1), in the free state and in the binary and ternary complexes.

The binding of coenzyme, substrates, and inhibitors to a number of other dehydrogenases has been investigated by NMR, although not so systematically as for dihydrofolate reductase. Probes other than those used for the latter are briefly noted. With the alcohol dehydrogenases, a zinc metalloenzyme, the Zn has been replaced by a paramagnetic probe, Co(II) (48, 49), and from relaxation effects on ethanol and isobutyramide both groups of investigators have concluded that the metal-substrate distance excludes direct coordination to metal. The contradiction between these NMR results and the X-ray crystallographic results (50a) remains unresolved. When the Zn was replaced with [113]Cd (50b), it was concluded, from the [113]Cd chemical shift data, that imidazole is in the first coordination sphere of the metal ion, in agreement with crystallographic data. [19]F in the form of trifluoroacetonylated Cys-149 has been used (51) to label glyceraldehyde-3-P dehydrogenase to demonstrate the existence of two pairs of nonequivalent subunits subsequently observed crystallographically. With glutamate dehydrogenase (52) measurements were made of ^{1}H and ^{31}P line broadening of 2-oxoglutamate and NADP in complexes with nitroxide spin labels linked to the β-P of ADP, an activator, or to the protein near the active site. Relative distances of the components from the spin labels could be estimated and a model consistent with all of the data was constructed.

Pyruvate Kinase

Pyruvate kinase is a key enzyme in glycolysis, which catalyzes the formation of ATP and pyruvate from P-enolpyruvate and ADP. The crystal structure of the cat muscle enzyme has been determined to a resolution of 2.6 Å (53). The tetrameric enzyme from rabbit muscle is too large (mol wt

~237,000) to attempt structural analysis of the active site by ^1H or ^{13}C NMR of the protein, although conformational changes upon binding Mg(II) have been followed by difference spectroscopy (54), and the C2 protons of 6 of the 14 His residues per subunit are sufficiently resolved so that their pKa's could be determined (55). NMR studies with ^{31}P, ^1H, ^{13}C, and ^{19}F (55–62) have focused on the characterization of enzymic complexes. Two divalent cations and one monovalent cation are required for activity, and paramagnetic ions including Mn(II), Co(II), and Cr(III) have been used as relaxation probes to distances from protein-bound metal or nucleotide-bound metal to components of various enzymic complexes. A large number of monovalent ions with net nuclear spin (63–65) have been used as probes to determine distances from the paramagnetic divalent ion site. Advantage has been taken of diverse reactions catalyzed by the enzyme including the enolization of pyruvate, and the phosphoryl transfer from ATP to fluoride ion and to glycolate.

^{31}P NMR   The chemical shifts of ^{31}P in various binary, ternary, and quaternary enzymic complexes of ATP, ADP, and P-enolpyruvate are listed in Table 3 (56). Although there are no significant changes in shifts for the nucleotides due to binding, irrespective of other components in the system, the ^{31}P shift and line shape (56) of P-enolpyruvate (PEP) is very sensitive to other components in the system and shifts downfield $\Delta\delta = +1.1$ ppm in E·PEP and 3.2 ppm in E·Mg·PEP and finally, in an equilibrium mixture,

Table 3 ^{31}P chemical shifts for free and enzyme-bound complexes of pyruvate kinase substrates

	ATP			ADP		P-enol-pyruvate	P_i + AMP
Complex	α	β	γ	α	β		
Free substrates	–10.8	–21.4	–5.8	–10.4	–6.2	0.9	
Free substrates + Mg^{2+}	–10.6	–19.2	–5.5	–10.1	–5.9	—	
Bound substrates	–11.1	–21.7	–6.4	–10.7	–6.0	0.2	
Bound substrates + Mg^{2+a}	–10.6	–19.7	–5.8	–10.4	–5.9	—	
Bound substrates + Mg^{2+b}	–10.5	–19.0	–5.8	–10.0	–5.7	2.3	
Bound substrates + Mg^{2+} + oxalate	–10.7	–19.3	–5.7	–11.5	–4.1	—	
E.P-enolpyruvate $(-K^+)$	—	—	—	—	—	0.1	
Equilibrium mixture	–10.7	–19.2	–5.8	–10.7	–5.8	3.6	3.6
Equilibrium mixture + EDTA	–10.9	–21.6	–6.2	–10.9	–6.2	3.1	3.7

$^a[Mg^{2+}] = [E]$.
$^b[Mg^{2+}]$ is ~35% in excess of the sum of enzyme and nucleotide concentrations.

i.e. in $Mg \cdot E \cdot MgADP \cdot PEP$, $\Delta\delta = 4.5$ ppm. The equilibrium constant for the interconversions of enzyme-bound substrates and products for both the pyruvate + ATP reaction and the glycolate + ATP reaction are close to 1, although the equilibrium constant for the overall reaction with catalytic amounts of enzyme are 3×10^{-4} and >50 respectively. The rates of interconversion of the ternary enzymic complexes could be estimated from the line broadening of the β-P of MgATP in the equilibrium mixture. The difference in chemical shifts between the β-P resonances of ATP and ADP are large compared to the reciprocal lifetimes in either state, thus fulfilling the slow exchange condition.

PARAMAGNETIC PROBES   The early work on the effect of Mn(II) on the proton relaxation rate of water of the $E \cdot Mn(II)$ complex has recently been reanalyzed (66) and it was confirmed that the number, q, of rapidly exchanging water molecules in the first coordination sphere of Mn(II) is between 2 and 3. For the ternary $E \cdot Mn \cdot P$-enolpyruvate complex, q equals 0.4 (67). The decrease in q is caused either by a displacement of water ligands upon addition of P-enolpyruvate or to a decrease in the rate of exchange of the coordinated water protons with solvent; the two possibilities cannot be distinguished by water proton relaxation rate measurements. However, in the case of a transition state analogue complex of creatine kinase with a value of $q < 1$, it has been demonstrated by the superhyperfine effect of $H_2^{17}O$ on the Mn(II) EPR spectrum (68) that three coordinated water molecules remain, and consequently in that case it is the insufficiently rapid rate of exchange that is responsible for the low value of q.

The distance between the divalent and monovalent ions, r, bound to pyruvate kinase was calculated to be 8.2 Å and shortened to 4.9 Å upon binding P-enolpyruvate in enzymic complexes of Mn(II) and $^{205}Tl(\pm)$ from relaxation rate measurements of ^{205}Tl (63). The effect of Mn(II) on the relaxation rates of a whole series of monovalent ions, ^6Li, ^7Li, $^{14}NH_4$, $^{15}NH_4$, ^{133}Cs, ^{23}Na, ^{39}K, and ^{87}Rb, has been measured (64) and the correlation time determined when possible from a comparison of pairs of isotopic nuclei. The dramatic shortening of r upon binding P-enolpyruvate was confirmed with all ions. Furthermore, an inverse relationship was found between the kinetic activity of the monovalent ion and the magnitude of r in the $E \cdot Mn \cdot P$-enolpyruvate complex, e.g. the percent activity, relative to K^+, NH_4^+, Tl^+, Li^+ is 81, 61, and 2, respectively and the corresponding r values are 4.4, 4.8, and 5.7 Å (64). From measurements of proton NMR of the methyl group of $CH_3NH_3^+$ (65), it was found that the r value in the binary $E \cdot Mn$ complex, 8.5 Å, was not altered by nucleotide binding, which complements the earlier result that the ^{31}P chemical shift of enzyme-bound nucleotides was not altered by binding of other ligands. In the ternary complex with P-enolpyruvate, r becomes 6.3 Å, with pyruvate $r = 6.9$ Å, and in the quaternary complex of pyruvate and ATP, r is reduced to 6.1 Å. Thus the monovalent ion is close to the active site in the binary E·Mn complex and even closer in the ternary E·Mn·P-enolpyruvate complex.

It was concluded from the distances calculated on the basis of the magnitude of the Mn(II) effect on the relaxation rate of ^{13}C in the carboxyl and carbonyl groups of pyruvate (60) and the methyl protons (58) in the E·Mn·pyruvate complex that Mn(II) is not directly coordinated to the pyruvate carboxylate group. More recent work with the paramagnetic probe Cr(III)ATP (61) yielded distances of 6.1 Å to each carbon atom of pyruvate and 7.9 Å to the methyl protons. Similar experiments could not be done with P-enolpyruvate, since its exchange rate on and off the enzyme is too slow compared to $1/T_1$ in the Mn(II) enzyme complex. From experiments substituting Co(II) (59), with its considerably decreased paramagnetic relaxation effect compared to Mn(II), to avoid the exchange rate limitation, it was again concluded that no direct coordination exists between metal ion and P-enolpyruvate. Difficulties involved in calculations from Co(II) data are discussed on page 382 of (1). The finding that two divalent metal ions (69) are required for activity, one bound to enzyme and the other to nucleotide, requires reassessment of the interpretation of earlier work on the distance between Mn(II) and the phosphorus atoms of ATP based on one Mn(II) in the enzyme·nucleotide complex.

The various distances between different paramagnetic probes and various substrate nuclei have been used to construct a model (70) of the three-dimensional arrangement of substrates in the active site. In this model, the substrate phosphoryl groups are not directly coordinated to the enzyme-bound divalent cation; a water molecule intervenes.

It should be pointed out that for the monovalent ions and analogues of P-enolpyruvate, which bind more weakly to pyruvate kinase and probably dissociate more rapidly from the enzyme, the exchange rate limitation is not likely to be so severe. In fact the Mn-proton distances in complexes with P-lactate and P-glycolate (71) and the methylene analogue of P-enolpyruvate (58) are consistent with direct coordination to phosphate. On the other hand, for normal substrates that bind tightly in all enzyme systems thus far investigated, the distance between metal ions and substrates appears to be too large for first sphere coordination. The exchange rate limitations are particularly treacherous in tightly bound, directly coordinated ligands, because with a given rate of exchange, $1/\tau_M$, the criterion that $T_{1M} \gg \tau_M$ becomes more difficult to meet the smaller the distance, r, between the metal and the nucleus under observation, because $T_{1M} \propto r^6$. As pointed out previously, in the Mn(II) transition state analogue complex of creatine kinase, even the rate of exchange of a water ligand is so slow that T_{1M} is not greater than τ_M (68). It is of some concern that with a tightly bound ligand, the calculated distance is never small enough for direct coordination.

Enzymic complexes of kinases other than pyruvate kinase have been widely investigated. ^{31}P experiments yielding chemical shifts of various enzymic complexes, equilibrium constants, and rates of interconversion for enzyme-bound substrates and products have been summarized (72). A conformation of the peptide substrate bound to protein kinase has been suggested on the basis of proton relaxation rates induced by Mn(II) and Cr(III) probes (73). The proton NMR spectra reported for the smallest kinase enzyme, adenylate kinase, and its substrate complexes (74) look sufficiently promising to warrant further investigation. Rates of interconversion of the central complexes were determined for adenylate kinase by transfer of saturation in ^{31}P NMR (75). For the enzyme creatine kinase, an intermolecular NOE effect (76) has been observed between the anion formate, a transition state analogue of the γ-phosphoryl group of ATP and a lysyl group on the enzyme. The finding of an intermolecular NOE effect in ADP-creatine kinase complexes between the protein peaks assigned to Arg and Lys residues (77) was extended in a study using "truncated driven nuclear Overhauser effects" (78). Interaction of the adenine protons with additional aromatic amino acid residues were observed in the latter investigation.

EPR Studies of Enzymic Complexes

Various redox enzymes, electron transfer proteins, and oxygen transport proteins, which naturally possess paramagnetic ions in one or more of their oxidation states have been studied extensively by EPR methods, and many aspects have been reviewed (79–87). Ions of Fe, Cu, Mo, and Mn are native constituents of many of the proteins in this class, and progress in EPR investigations of Cu, Mo, and Fe-containing proteins has been recently reviewed (81–83). EPR investigations of redox enzymes for which free radicals participate in catalysis have also been reviewed (79). Another class of EPR studies involves enzymes that naturally function with diamagnetic metal ions, such as Zn(II), Mg(II), or Ca(II) for which paramagnetic ions such as Mn(II), Co(II), or Gd(III) may be substituted. Some recent contributions in this category are discussed subsequently. Finally, another category of investigation involves use of extrinsic free radical probes such as nitroxide-containing compounds, which may be covalently linked to the protein or substrates for EPR studies (88).

Information regarding electron spin-nuclear spin coupling in enzymic complexes has the potential of revealing not only ligand groups to the metal ion but other nuclei in close proximity (i.e. ≤ 6 Å) to the paramagnetic center. Unfortunately, such information is often difficult to obtain because the electron spin-nuclear spin couplings may be smaller than the intrinsic linewidths of the EPR signals. Electron nuclear double resonance (ENDOR) spectroscopy, wherein the electron spin and nuclear spins are irradiated simultaneously, was developed to overcome such difficulties (89), and there have been several applications of ENDOR spectroscopy to biological problems (90, 91). However, the somewhat stringent requirements of ENDOR on relaxation properties of the spin system and the modest sensitivity of the method have thus far limited applications to enzymic complexes. Relatively recent variations on the basic ENDOR experiment involving simultaneous irradiation at two nuclear resonance frequencies (ENDOR TRIPLE) (92) ameliorate some of the sensitivity problems.

Another recent method for extracting otherwise hidden nuclear spin coupling information is an electron spin echo technique through which coupled nuclear spins are identified by their characteristic modulation of the amplitudes of electron spin echoes (84, 85). There is also a double resonance version of this spin echo technique that is reported to be more sensitive than conventional ENDOR (84). Improving circumstances with respect to the availability of sources of stable magnetically active isotopes such as ^{17}O ($I = 5/2$) have stimulated activity in EPR investigations of electron spin-nuclear spin coupling in enzymic complexes. Some recent EPR studies involving ^{17}O labeled ligands are discussed below.

Analysis of spin-spin coupling between two or more paramagnetic centers within a single enzyme complex or involving centers on different proteins has been actively pursued. In principle, interpretation of these phenomena can provide information on the proximity of the paramagnetic centers and on the relative magnetic geometry of the coupled spin systems (93–100). Magnetically coupled clusters of paramagnetic ions such as the FeS clusters are common in electron transfer proteins, and EPR has long been used to characterize these centers (101, 102). There are several enzymes that require more than a single equivalent of metal ion at the active site, and it is likely that such requirements may be more common than was previously recognized. Electron spin-electron spin interactions should be significant at the active sites of this class of enzymes. Some recent findings in this category are discussed in more detail below. The spin-spin interaction between paramagnetic ions and free radical spin labels (93, 103) has been the basis of a popular method for mapping intersite distances on enzymes. The original treatment of this problem (93) dealt with the through-space, dipole-dipole interaction between two spins affixed to a rigid lattice, where the lifetime of one spin was much shorter than that of the other. There have been numerous applications of this methodology, and a recent review article has covered contributions up to 1978 (103). A potential complication arises at distances $\leqslant 14$ Å, due to the electron spin exchange interaction (98). The latter interaction can have both isotropic and anisotropic components, and in some instances it may be difficult, if not impossible, to separate the electron spin exchange and dipole-dipole contributions (98). Recent studies with well-defined model complexes that contain a paramagnetic metal ion and a nitroxyl group clearly demonstrate the presence of the electron spin exchange interaction (104–109).

Cytochrome c Oxidase

Cytochrome c oxidase catalyzes the four electron reduction of O_2 to H_2O. The central importance of this enzyme in respiration and its intriguing network of redox sites (hemes a and a_3 and two Cu's) have stimulated much effort in biophysical and biochemical characterization of the enzyme (110–112). EPR studies have been directed toward understanding the spin states, oxidation states, and liganding of Fe in the two heme groups, and the environments and redox properties of the two Cu's. Magnetic interactions among the four centers have also been of considerable interest, since EPR signals for only a single Cu^{2+} and one heme (heme a) have been observed in the fully oxidized enzyme (113). Recent EPR studies of cytochrome c oxidase have provided new insight into two long-standing questions about the Cu sites in this enzyme, and these studies illustrate several old and new strategies for EPR characterization of redox enzymes.

EPR spectra for the fully oxidized enzyme show a signal characteristic of one low spin Fe^{3+} heme (heme a) and an atypical signal for one Cu^{2+} (110). The atypical g values and absence of Cu nuclear hyperfine structure for the latter signal have raised some doubt as to whether this signal is from Cu^{2+} or from another species such as a sulfur radical (114). Several recent studies have been directed at this question (115–117). Froncisz et al (115) measured the EPR spectra for the fully oxidized enzyme at 2.6 GHz and 3.8 GHz and found hyperfine structure that had not been resolved previously at higher observation frequencies, apparently because of inhomogeneous broadening resulting from "g-strain" (i.e. a distribution of g values throughout the population of sites in the sample). The analysis of the hyperfine structure was tentative but the magnitudes of the splittings were consistent with a nuclear hyperfine interaction with Cu (115). Hoffman et al (116) subsequently used ENDOR to measure nuclear hyperfine couplings associated with this signal. Small couplings, assigned to ^1H and ^{14}N, were observed, together with a much larger coupling that was assigned to $^{63,65}Cu$ (116). The results were discussed in terms of the common models for this redox site. Although the magnitude of the coupling constants for Cu were only about half as large as those observed for other Cu(II) protein complexes, the authors indicated that the proton hyperfine coupling constant and g tensor were inconsistent with a thiyl radical coordinated to a

Cu(I). The center was described as one in which Cu(II) was bound with considerable covalency in the metal-ligand bonding. Mims et al (117) investigated the linear electric field shifts of the g values of the signal and the nuclear modulation patterns in a pulsed EPR study of oxidized cytochrome c oxidase. The linear electric field effects and nuclear modulation data were both unlike those of other Cu(II) protein complexes, and the authors suggested that the results might be explained by a Cu(II) center that receives an electron from an RS^- ligand (117). Thus, while the detailed structure of this site on cytochrome oxidase remains somewhat controversial, the association of Cu with this center seems unequivocal (116), and it is also likely that there are one or two sulfur ligands to the Cu.

The other Cu center of cytochrome c oxidase, Cu_{a_3}, is EPR silent in the oxidized, resting enzyme, and this situation has been attributed to an antiferromagnetic coupling between Cu(II) and the Fe^{3+} of heme a_3 (113). This site on the enzyme is the one that interacts with exogenous ligands including O_2, and a close proximity of this Cu site and cytochrome a_3 has been suggested from several other types of experiments (118–120). Recently, a new EPR signal from Cu(II) has been observed upon reoxidation of reduced cytochrome c oxidase (120–122). The new Cu(II) signal is seen in addition to the standard Cu_a signal discussed above, and Cu hyperfine structure is clearly visible in both 9 GHz and 35 GHz spectra (121). Moreover, the total intensity in the two Cu signals corresponds to 1.3 of the 2 Cu's present in the functional unit of the enzyme (122), so it is clear that both types of Cu contribute EPR signals in this state. The spectral properties of the new Cu signal for cytochrome c oxidase were similar to those stemming from Cu(II) in metal pairs of two laccase enzymes (121). These particular Cu's in laccase are also antiferromagnetically coupled to an EPR silent state in the fully oxidized enzyme (122). The new Cu(II) signal in cytochrome c oxidase has been assigned to the Cu that is closely associated with heme a_3 (122). The state of the enzyme in which the signal for the second Cu is visible is suggested to be

$$a^{3+} \; Cu_a^{2+} \; Cu_{a_3}{}^{2+} \; a_3^{2+}\text{-}O_2$$

or

$$a^{3+} \; Cu_a^{2+} \; Cu_{a_3}{}^{2+} \; a_3^{3+} \; \text{-}O_2^-$$

where the cytochrome a_3-O_2 complex has magnetic properties analogous to that of oxyhemoglobin. This model for the a_3-O_2 complex provides an explanation for breaking the antiferromagnetic coupling between the Fe^{3+} of cytochrome a_3 and the $Cu_{a_3}{}^{2+}$. The observation that partial reduction in the absence of oxygen does not give this state, but one in which Cu_{a_3} is reduced and heme a_3 is oxidized is also consistent with this model (122).

Other recent studies of cytochrome c oxidase have focused on characterizing transient intermediates in reoxidation of the reduced enzyme (123). The nitric oxide complexes at the heme a_3-Cu_{a_3} center have also received attention (120, 124, 125). Brudvig et al (125) have characterized several redox reactions involving nitric oxide and cytochrome c oxidase. Binding of nitric oxide to the oxidized enzyme leads to the appearance of a high-spin cytochrome a_3 EPR signal (120, 125). In this case it appears that nitric oxide binds to Cu_{a_3} and thereby breaks the antiferromagnetic coupling between Cu_{a_3} and cytochrome a_3 (120)—the converse of the effect with O_2 binding discussed above (122).

Superhyperfine Coupling

Observation of nuclear spin superhyperfine interactions in EPR spectra for enzymic complexes enhances the possibilities for detailed structural interpretations. Oxygen is a common ligand atom in enzymic complexes of metal ions. However, the magnetically active isotope, ^{17}O (I - 5/2), has a low natural abundance ($3.7 \times 10^{-2}\%$), and until recently highly enriched sources of this isotope were not generally available. This situation has recently improved, and there is considerable activity in biological EPR studies with ligands enriched in ^{17}O (68, 126–132). In some cases superhyperfine splittings from ^{17}O are resolved in the EPR spectra, and in other cases, where the splitting is smaller than the EPR linewidths, the ^{17}O coupling is inferred from a characteristic inhomogeneous broadening of the EPR signals.

The Mo enzymes have provided an especially fruitful class for EPR investigations with ^{17}O labeling (129–132). The Mo(V) state ($S = \frac{1}{2}$) is EPR active, but this state is frequently a transient intermediate that requires special trapping techniques for observation (82). Cramer et al (129) observed ^{17}O coupling on the Mo(V) EPR signals from sulfite oxidase and xanthine dehydrogenase upon reduction of these enzymes in $H_2^{17}O$. These results established the presence of an exchangeable oxygen ligand in the coordination sphere of Mo (129). Gutteridge et al (130) observed ^{17}O coupling in the intermediate Mo(V) EPR signal known as "Rapid" in xanthine oxidase that was generated in $H_2^{17}O$. Subsequently, Gutteridge & Bray (131) characterized an essentially isotropic ^{17}O coupling in the signal for an earlier intermediate, "Very Rapid," in xanthine oxidase reduced in $H_2^{17}O$. For both the Rapid and Very Rapid species the coupling patterns were those expected for a single ^{17}O ligand, and the authors discuss the possibility that ^{17}O is present as an Mo–OH group in the Rapid species and as an Mo–O–C group in the Very Rapid intermediate (130, 131).

Gutteridge et al (132) have also investigated the mechanism for phosphate inhibition of sulfite oxidase using ^{17}O labeling. When reduction of the enzyme by sulfite was carried out in $H_2^{17}O$ with unlabeled phosphate present, the ^{17}O effects on the Mo(V) signal (129) were not observed (132). On the other hand, reduction by sulfite in normal water but with ^{17}O labeled phosphate present revealed Mo(V)-^{17}O-phosphate superhyperfine coupling (132). These results clearly show that phosphate inhibits the enzyme by binding directly to Mo at the active site. Since phosphate is structurally analogous to sulfate, the product of the reaction, it is likely that sulfate may be directly bound to the Mo. Moreover, the oxygen that is added to sulfite upon oxidation by the enzyme may be transferred directly from a Mo oxygen ligand (132).

Gupta et al (128) have observed inhomogeneous broadening of the ferric EPR signals of horseraddish peroxidase in $H_2^{17}O$ when benzohydroxamic acid was present. The interpretation was that the benzohydroxamic acid permits a water molecule to bind to the Fe^{3+} site (128). Inhomogeneous broadening of Mn(II) EPR signals due to ^{17}O superhyperfine coupling was used to identify all six ligands to Mn(II) in transition state analogue complexes of creatine kinase (68).

The Cu(II) site of galactose oxidase has been investigated by pulsed EPR methods (133). The ^{14}N modulation patterns observed in the spin echo envelopes show that Cu(II) is coordinated to at least one and possibly two histidine imidazole groups (133). A comparison of nitrogen coupling from histidine groups to the Cu(II) site of superoxide dismutase in the presence and absence of zinc has also been investigated by the pulsed EPR method (134). ENDOR studies of cytochrome P-450 complexes revealed a strongly coupled proton in the low spin Fe^{3+} complex in the absence of a substrate (135).

Electron Spin-Electron Spin Coupling

Magnetic interactions between closely spaced paramagnetic centers can dominate other interactions, and this is the underlying reason for seeking conditions of magnetic dilution in most EPR experiments (136). However, in some circumstances, including several biologically interesting complexes, paramagnetic centers occur in pairs or clusters, and the spin-spin interactions can provide considerable information about the geometrical arrangement of the sites and about the sign and magnitude of the exchange interaction (93, 98, 137). The antiferromagnetic coupling mentioned above is an example of a strong exchange interaction with a positive sign. There are two major components in the electron spin-electron spin interaction: a through space or dipole-dipole coupling, which is anisotropic, and an exchange interaction that can have both isotropic and anisotropic contributions (136, 137). The exchange interaction can occur as a result of direct overlap of orbital wavefunctions for the two paramagnetic species or can be mediated by intervening atoms such as bridging ligand groups (136). The latter mechanism is known as superexchange. Anisotropies in the exchange interaction are anticipated for paramagnetic species with appreciable orbital contributions to the magnetic moment (e.g. anisotropic g values or fine structure in ions with $S > \frac{1}{2}$) (98, 136). The dipole-dipole interaction follows a $1/r^3$ dependence (where r is the distance between the two centers) and vanishes at specific orientations of the magnetic field with respect to the coordinate system of the two spins (93, 98). In the absence of an exchange interaction, analysis of the dipole-dipole coupling can be used to estimate the distance between the paramagnetic centers (93, 98, 137). However, the significance of exchange contributions in well-defined model complexes (103–109) indicates that interpretation of spin-spin interactions in terms of pure dipolar contributions may not be sound for paramagnetic centers that are separated by less than 14 Å (98). A warning of the potential difficulties in interpretation of this currently popular approach is timely.

In addition to the experiments with paramagnetic metal ions and nitroxide radical species (103), EPR has recently been used to investigate electron spin-electron spin coupling in vitamin B_{12} enzyme complexes (94, 97, 99) and coupling between two paramagnetic metal ions at the active site of glutamine synthetase (138) and S-adenosylmethionine synthetase (139). Glutamine synthetase from *E. coli* requires two equivalents of divalent cation for activity, one of which is associated with the nucleotide substrate, ATP. Balakrishnan & Villafranca used the exchange inert Cr(III) nucleotides, Cr(III)ADP and Cr(III)ATP, for the metal nucleotide requirement, and Mn(II) for the second site (138). The EPR signals for the enzyme-bound Mn(II) were followed as the paramagnetic Cr(III) complexes were added. The Mn(II) EPR spectrum for the complex of the enzyme with methionine sulfoximine exhibits very narrow signals, and these signals were virtually eliminated when the metal nucleotide site was saturated with Cr(III)ATP. The results were analyzed assuming that the interaction between Cr(III) and Mn(II) was dipolar, and a distance of 6.8 Å was calculated (138). The possibility for an exchange interaction was not considered; however, the qualitative conclusion that the two metal ions are bound closely to each other at the active site is clear from these results (138).

Markham (139) has shown that two equivalents of divalent cation and one equivalent of monovalent cation (140) are required for maximal activity of S-adenosylmethionine synthetase from *E. coli*. Binding studies showed that one equivalent of Mn(II) was bound to the enzyme in the absence of substrate and that a second equivalent was bound in the presence of an ATP analogue or of pyrophosphate (139). A spin exchange interaction between two Mn(II) ions at the active site was apparent in the EPR spectra for complexes with S-adenosylmethionine, a polyphosphate, and K^+ present.

For other complexes, which binding studies had shown to contain two Mn(II) ions, the EPR spectra did not show clear evidence of an exchange interaction. EPR spectra for the exchange coupled Mn(II) ions at 9 GHz showed a symmetrical signal centered at $g = 2$ with ^{55}Mn hyperfine structure consisting of more than sixteen lines spaced at ½ the normal ^{55}Mn hyperfine coupling constant. The spacing is diagnostic for exchange coupled pairs of Mn(II), and the number of lines is indicative of an exchange coupling that is comparable to the ^{55}Mn hyperfine coupling constant (139). However, the theoretical difficulties for this complicated manifold of spin states precluded the possibility for a more detailed analysis of the spectra. The results were discussed in terms of a superexchange coupling mechanism wherein the interaction could be mediated by a ligand group common to both metal ions.

Conclusions

The reader should recognize that the limited number of investigations that have been discussed in this review represent only a small fraction of the recent enzymological studies involving magnetic resonance spectroscopy. There has been an evolution in the sophistication of the questions challenging magnetic resonance spectroscopy that has in turn stimulated the development of new methodologies and improved instrumentation. In some cases, the structural conclusions from studies of macromolecular complexes have approached the level that was feasible only for small molecules a decade ago. Magnetic resonance experiments with enzymic complexes still require large amounts of homogeneous protein and this requirement has long been an impediment to investigations of less abundant enzymes. The recent developments in cloning technology promise to obviate this difficulty and permit sufficient production of native and appropriately modified enzymes to extend magnetic resonance studies to systems whose feasibility of investigation currently exists only as a figment of the imagination in the minds of magnetic resonance practitioners.

Literature Cited

1. Jardetzky, O., Roberts, G. C. K. 1981. *NMR in Molecular Biology,* Chaps. 9–11 New York: Academic. 640 pp.
2. Mildvan, A. S., Gupta, R. K. 1978. *Methods Enzymol.* 49:322–59
3. Gerig, J. R. 1978. In *Biological Magnetic Resonance,* ed. L. J. Berliner, J. Reuben, pp. 139–203. New York: Plenum. 345 pp.
4. Cohn, M., Nageswara Rao, B. D. 1979. *Bull. Magn. Reson.* 1:38–60
5. Markley, J. L. 1979. In *Biological Applications of Magnetic Resonance,* ed. R. G. Shulman, pp. 397–462. New York: Academic. 595 pp.
6. Coleman, J. E., Armitage, I. M., Chlebowski, J. F., Otvos, J. D., Schoot Uiterkamp, A. J. M. 1979. See Ref. 5, pp. 345–95
7. Mildvan, A. S. 1979. *Adv. Enzymol.* 49:103–26
8. Villafranca, J. J., Raushel, F. M. 1980. *Ann. Rev. Biophys. Bioeng.* 9:363–92
9. Sykes, B. D. 1980. In *Magnetic Resonance in Biology,* ed. J. S. Cohen, 1:171–96. New York: Wiley. 309 pp.
10. Bothner-By, A. A. 1979. See Ref. 5, pp. 177–219
11. Deleted in proof

12. McLaughlin, A., Leigh, J. S., Cohn, M. 1976. *J. Biol. Chem.* 251:2777–87
13. Bradshaw, R. A., Cancedda, F., Ericsson, L. H., Neumann, P. A., Piccoli, S. P., Schlesinger, M. J., Shriefer, K., Walsh, K. A. 1981. *Proc. Natl. Acad. Sci. USA* 78:3473–77
14. Sowadski, J. M., Foster, B. A., Wyckoff, H. W. 1981. *J. Mol. Biol.* 150:245–72
15. Hull, W. E., Sykes, B. D. 1974. *Biochemistry* 13:3431–37
16. Otvos, J. D., Browne, D. T. 1980. *Biochemistry* 19:4011–20
17. Bock, J. L., Sheard, B. 1975. *Biochem. Biophys. Res. Comun.* 66:24–30
18. Chlebowski, J. F., Armitage, I. M., Tusa, P. P., Coleman, J. E. 1976. *J. Biol. Chem.* 251:1207–16
19. Chlebowski, J. F., Armitage, I. M., Coleman, J. E. 1977. *J. Biol. Chem.* 252:7053–61
20. Hull, W. E., Halford, S. E., Gutfreund, H., Sykes, B. D. 1976. *Biochemistry* 15:1547–61
21. Bock, J. L., Kowalsky, A. 1978. *Biochim. Biophys. Acta* 526:135–46
22. Otvos, J. D., Alger, J. R., Coleman, J. E., Armitage, I. M. 1979. *J. Biol. Chem.* 254:1778–80
23. Gettins, P., Coleman, J. E. 1982. *Fed. Proc.* In press
24. Bock, J. L., Cohn, M. 1978. *J. Biol. Chem.* 253:4082–85
25. Otvos, J. D., Armitage, I. M. 1980. *Biochemistry* 19:4021–30
26. Otvos, J. D., Armitage, I. M. 1980. *Biochemistry* 19:4031–43
27. Matthews, D. A., Alden, R. A., Bolin, J. T., Filman, D. J., Freer, S. T., Hamlin, R., Hol, W. G., Kisliuk, R. L., Pastore, E. J., Plante, L. T., Xuong, N., Kraut, J. 1978. *J. Biol. Chem.* 253:6946–54
28. Matthews, D. A., Alden, R. A., Freer, S. T., Xuong, N., Kraut, J. 1979. *J. Biol. Chem.* 249:4144–51
29. Kimber, B. J., Griffiths, D. V., Birdsall, B., King, R. W., Scudder, P., Feeney, J., Roberts, G. C. K., Burgen, A. S. V. 1977. *Biochemistry* 16:3492–500
30. Kimber, B. J., Feeney, J., Roberts, G. C. K., Birdsall, B., Griffiths, D. V., Burgen, A. S. V., Sykes, B. D. 1978. *Nature* 271:184–85
31. Blakley, R. L., Cocco, L., London, R. E., Walker, T. E., Matwiyoff, N. A. 1978. *Biochemistry* 17:2284–92
32. Cocco, L., Blakley, R. L., Walker, T. E., London, R. E., Matwiyoff, N. A. 1978. *Biochemistry* 17:4285–90

33. London, R. E., Groff, J. P., Blakley, R. L. 1979. *Biochem. Biophys. Res. Commun.* 86:779–86
34. Feeney, J., Roberts, G. C. K., Birdsall, B., Griffiths, D. V., King, R. W., Scudder, P., Burgen, A. S. V. 1977. *Proc. R. Soc. London Ser. B* 196:267–90
35. Feeney, J., Roberts, G. C. K., Thomson, J. W., King, R. W., Griffiths, D. V., Burgen, A. S. V. 1980. *Biochemistry* 19:2316–21
36. Feeney, J., Roberts, G. C. K., Kaptein, R., Birdsall, B., Gronenborn, A., Burgen, A. S. V. 1980. *Biochemistry* 19:2466–72
37. Roberts, G. C. K., Feeney, J., Birdsall, B., Kimber, B., Griffiths, D. V., King, R. W., Burgen, A. S. V. 1977. In *NMR in Biology*, ed. R. A. Dwek, I. D. Campbell, R. E. Richards, R. J. P. Williams, 16:95–108. New York: Academic. 381 pp.
38. Wyeth, P., Gronenborn, A., Birdsall, B., Roberts, G. C. K., Feeney, J., Burgen, A. S. V. 1980. *Biochemistry* 19:2608–15
39. Gronenborn, A., Birdsall, B., Hyde, E. I., Roberts, G. C. K., Feeney, J., Burgen, A. S. V. 1981. *Biochemistry* 20:1717–22
40. Hyde, E. I., Birdsall, B., Roberts, G. C. K., Feeney, J., Burgen, A. S. V. 1980. *Biochemistry* 19:3738–46
41. Feeney, J., Birdsall, B., Roberts, G. C. K., Burgen, A. S. V. 1975. *Nature* 257:564–66
42. Birdsall, B., Roberts, G. C. K., Feeney, J., Burgen, A. S. V. 1977. *FEBS Lett.* 80:313–16
43. Cayley, P. J., Feeney, J., Kimber, B. J. 1980. *Int. J. Biol. Macromol.* 2:251–55
44. Hyde, E. I., Birdsall, B., Roberts, G. C. K., Feeney, J., Burgen, A. S. V. 1980. *Biochemistry* 19:3746–54
45. Gronenborn, A., Birdsall, B., Hyde, E. I., Roberts, G. C. K., Feeney, J., Burgen, A. S. V. 1982. *Mol. Pharmacol.* In press
46. Cayley, P. J., Albrand, J. P., Feeney, J., Roberts, G. C. K., Piper, E. A., Burgen, A. S. V. 1979. *Biochemistry* 18:3886–95
47. Cocco, L., Groff, J. P., Temple, C. Jr., Montgomery, J. A., London, R. E., Matwiyoff, N. A., Blakley, R. L. 1981. *Biochemistry* 20:3972–78
48. Sloan, D. L., Young, J. M., Mildvan, A. S. 1975. *Biochemistry* 14:1998–2008
49. Drysdale, B., Hollis, D. P. 1980. *Arch. Biochem. Biophys.* 205:267–79
50a. Plapp, B. V., Eklund, H., Branden, C. I. 1978. *J. Mol. Biol.* 122:23–32

50b. Bobsein, B. R., Myers, R. J. 1980. *J. Am. Chem. Soc.* 102:2454–5
51. Bode, J., Blumenstein, M., Raftery, M. A. 1975. *Biochemistry* 14:1153–60
52. Zantema, A., DeSmet, M. J., Robillard, G. T. 1979. *Eur. J. Biochem.* 96:465
53. Stuart, D. I., Levine, M., Muirhead, H., Stammers, D. K., 1979. *J. Mol. Biol.* 134:109–42
54. Cohn, M., Leigh, J. S. Jr., Reed, G. H. 1971. *Cold Spring Harbor Symp. Quant. Biol.* 26:533–40
55. Meshitsuka, S., Smith, G. M., Mildvan, A. S. 1981. *J. Biol. Chem.* 256:4460–65
56. Nageswara Rao, B. D., Kayne, F. J., Cohn, M. 1979. *J. Biol. Chem.* 254:2689–96
57. sloan, D. L., Mildvan, A. S. 1976. *J. Biol. Chem.* 251:2412–20
58. James, T. L., Cohn, M. 1974. *J. Biol. Chem.* 245:3519–26
59. Melamud, E., Mildvan, A. S. 1975. *J. Biol. Chem.* 250:8193–201
60. Fung, C. H., Mildvan, A. S., Allerhand, A., Komoroski, R., Scrutton, M. C. 1973. *Biochemistry* 12:620–29
61. Gupta, R. K., Oesterling, R. M., Mildvan, A. S. 1976. *Biochemistry* 15:2881–87
62. Nowak, T. 1978. *Arch. Biochem. Biophys.* 186:343–50
63. Reuben, J., Kayne, F. J. 1971. *J. Biol. Chem.* 246:6227–34
64. Raushel, F. M., Villafranca, J. J. 1980. *Biochemistry* 19:5481–85
65. Nowak, T. 1978. *J. Biol. Chem.* 253:1998–2004
66. Burton, D. R., Forsen, S., Karlstrom, G., Dwek, R. A. 1979. *Prog. NMR Spectrosc* 13:1–45
67. James, T. L., Reuben, J., Cohn, M. 1973. *J. Biol. Chem.* 248:6443–49
68. Reed, G. H., Leyh, T. 1980. *Biochemistry* 19:5472–80
69. Gupta, R. K., Fung, C. H., Mildvan, A. S. 1976. *J. Biol. Chem.* 251:2421–30
70. Mildvan, A. S., Sloan, D. L., Fung, C. H., Gupta, R. K., Melamud, E. 1976. *J. Biol. Chem.* 251:2431–34
71. Nowak, T., Mildvan, A. S. 1972. *Biochemistry* 11:2819–28
72. Nageswara Rao, B. D. 1979. In *NMR and Biochemistry,* ed. S. J. Opella, P. Lu, pp. 371–88. New York: Dekker, 434 pp.
73. Granot, J., Mildvan, A. S., Bramson, H. N., Thomas, N., Kaiser, E. T. 1981. *Biochemistry* 20:602–10
74. McDonald, G. G., Cohn, M., Noda, L. 1975. *J. Biol. Chem.* 250:6947–54
75. Brown, T. R., Ogawa, S. 1977. *Proc. Natl. Acad. Sci. USA* 74:3627–31

76. James, T. L., Cohn, M. 1974. *J. Biol. Chem.* 249:2599–2603
77. James, T. L. 1976. *Biochemistry* 15:4724–30
78. Vasak, M., Nagayama, K., Wuthrich, K., Mertens, M. L., Kagi, J. H. R. 1979. *Biochemistry* 18:5050–55
79. Edmondson, D. E. 1978. See Ref. 3, pp. 205–38
80. Warden, J. T. 1978. See Ref. 3, pp. 239–76
81. Boas, J. F., Pilbrow, J. R., Smith, T. D. 1978. See Ref. 3, pp. 277–342
82. Bray, R. C. 1980. In *Biological Magnetic Resonance,* ed. L. J. Berliner, J. Reuben, 2:45–84. New York: Plenum. 351 pp.
83. Smith, T. D., Pilbrow, J. R. 1980. See Ref. 82, pp. 85–168
84. Norris, J. R., Thurnauer, M. C., Bowman, M. K. 1980. *Adv. Biol. Med. Phys.* 17:365–416
85. Mims, W. B., Peisach, J. 1979. See Ref. 5, pp. 221–69
86. Orme-Johnson, W. H., Sands, R. H. 1973. In *Iron-Sulfur Proteins,* ed. W. Lovenberg, 2:195–238. New York: Academic Press, 343 pp.
87. Palmer, G. 1975. *The Enzymes* 12:1–56
88. Morrisett, J. D. 1976. In *Spin Labeling Theory and Applications,* ed. L. J. Berliner, pp. 339–72. New York: Academic. 592 pp.
89. Feher, G. 1956. *Phys. Rev.* 103:834–35
90. Scholes, C. P. 1979. In *Multiple Electron Resonance Spectroscopy,* ed. M. M. Dorio, J. H. Freed, pp. 297–329. New York: Plenum. 512 pp.
91. Sands, R. H. 1979. See Ref. 88, pp. 331–74
92. Möbius, K., Biehl, R. 1979. See Ref. 88, pp. 475–507
93. Leigh, J. S. Jr. 1970. *J. Chem. Phys.* 52:2608–12
94. Buettner, G. R., Coffman, R. E. 1977. *Biochim. Biophys. Acta* 480:495–505
95. Carr, S. G., Smith, T. D., Pilbrow, J. R. 1974. *J. Chem. Soc. Faraday Trans. 1* 70:497–511
96. Boyd, P. D. W., Toy, A. D., Smith, T. D., Pilbrow, J. R. 1973. *J. Chem. Soc. Dalton Trans.,* pp. 1549–63
97. Boas, J. F., Hicks, P. R., Pilbrow, J. R., Smith, T. D. 1978. *J. Chem. Soc. Faraday Trans. 2* 74:417–31
98. Coffman, R. E., Buettner, G. R. 1979. *J. Phys. Chem.* 83:2392–2400
99. Schlepler, K. L., Dunham, W. R., Sands, R. H., Fee, J. A., Abeles, R. H. 1975. *Biochim. Biophys. Acta* 397:510–18

100. Lowe, D. J., Bray, R. C. 1978. *Biochem. J.* 169:471–79
101. Sands, R. H., Dunham, W. R. 1974. *Quart. Rev. Biophys.* 7:443–504
102. Ohnishi, T. 1979. In *Membrane Proteins in Energy Transduction,* ed. R. A. Capaldi, pp. 1–87. New York: Dekker
103. Eaton, S. S., Eaton, G. R. 1978. *Coord. Chem. Rev.* 26:207–62
104. DuBois, D. L., Eaton, G. R., Eaton, S. S. 1979. *J. Am. Chem. Soc.* 101:2624–27
105. Branden, G. A., Trevor, K. T., Neri, J. M., Greenslade, D. J., Eaton, G. R., Eaton, S. S. 1977. *J. Am. Chem. Soc.* 99:4854–55
106. DuBois, D. L., Eaton, G. R., Eaton, S. S. 1979. *Inorg. Chem.* 18:75–79
107. Boymel, P. M., Eaton, G. R., Eaton, S. S. 1980. *Inorg. Chem.* 19:727–35
108. Moore, K. M., Eaton, G. R., Eaton, S. S. 1979. *Inorg. Chem.* 18:2492–96
109. Boymel, P. M., Braden, G. A., Eaton, G. R., Eaton, S. S. 1980. *Inorg. Chem.* 19:735–39
110. Malmström, B. G. 1979. *Biochim. Biophys. Acta* 549:281–303
111. Beinert, H. 1978. *Methods Enzymol.* 54:133–50
112. Caughey, W. S., Wallace, W. J., Volpe, J. A., Yoshikawa, S. 1976. In *The Enzymes* 13:299–344
113. Van Gelder, B. F., Beinert, H. 1969. *Biochim. Biophys. Acta* 189:1–24
114. Peisach, J., Blumberg, W. E. 1974. *Arch. Biochem. Biophys.* 165:691–708
115. Froncisz, W., Scholes, C. P., Hyde, J. S., Wei, Y.-H., King, T. E., Shaw, R. W., Beinert, H. 1979. *J. Biol. Chem.* 254:7482–84
116. Hoffman, B. M., Roberts, J. E., Swanson, M., Speck, S. H., Margoliash, E. 1980. *Proc. Natl. Acad. Sci. USA* 77:1452–56
117. Mims, W. B., Peisach, J., Shaw, R. W., Beinert, H. 1980. *J. Biol. Chem.* 255:6843–46
118. Alben, J. O., Moh, P. P., Fiamingo, F. G., Altschuld, R. A. 1981. *Proc. Natl. Acad. Sci. USA* 78:234–37
119. Powers, L., Chance, B., Ching, Y., Angiolillo, P. 1981. *Biophys. J.* 34:465–98
120. Stevens, T. H., Brudvig, G. W., Bo-
crian, D. F., Chan, S. I. 1979. *Proc. Natl. Acad. Sci. USA* 76:3320–24
121. Reinhammar, B., Malkin, R., Jensen, P., Karlsson, B., Andreasson, L.-E., Aasa, R., Vänngärd, T., Malström, B. G. 1980. *J. Biol. Chem.* 255:5000–3
122. Karlsson, B., Andreasson, L.-E. 1981. *Biochim. Biophys. Acta* 635:73–80
123. Shaw, R. W., Hansen, R. E., Beinert, H. 1979. *Biochim. Biophys. Acta* 548:386–96
124. Morse, R. H., Chan, S. I. 1980. *J. Biol. Chem.* 255:7876–82
125. Brudvig, G. W., Stevens, T. H., Chan, S. I. 1980. *Biochemistry* 19:5275–85
126. Gupta, R. K., Mildvan, A. S., Yonetani, T., Srivastava, T. S. 1975. *Biochem. Biophys. Res. Commun.* 67:1005–12
127. Deinum, J. S. E., Vänngård, T. 1975. *FEBS Lett.* 58:62–65
128. Gupta, R. K., Mildvan, A. S., Schonbaum, G. R. 1979. *Biochem. Biophys. Res. Commun.* 89:1334–40
129. Cramer, S. P., Johnson, J. L., Rajagopalan, K. V., Sorrell, T. N. 1979. *Biochem. Biophys. Res. Commun.* 91:434–39
130. Gutteridge, S., Malthouse, J. P. G., Bray, R. C. 1979. *J. Inorg. Biochem.* 11:355–60
131. Gutteridge, S., Bray, R. C. 1980. *Biochem. J.* 189:615–23
132. Gutteridge, S., Lamy, M. T., Bray, R. C. 1980. *Biochem. J.* 191:285–88
133. Kosman, D. J., Peisach, J., Mims, W. B. 1980. *Biochemistry* 19:1304–8
134. Fee, J. A., Peisach, J., Mims, W. B. 1981. *J. Biol. Chem.* 256:1910–14
135. LoBrutto, R., Scholes, C. P., Wagner, G. C., Gunsalus, I. C., Debrunner, P. G. 1980. *J. Am. Chem. Soc.* 102:1167–70
136. Abragam, A., Bleaney, B. 1970. In *Electron Paramagnetic Resonance of Transition Ions,* pp. 491–540. Oxford: Clarendon Press. 911 pp.
137. Luckhurst, G. R. 1976. See Ref. 88, pp. 133–81
138. Balakrishnan, M. S., Villafranca, J. J. 1978. *Biochemistry* 17:3531–38
139. Markham, G. D. 1981. *J. Biol. Chem.* 256:1903–09
140. Markham, G. D., Hafner, E. W., Tabor, C. W., Tabor, H. 1980. *J. Biol. Chem.* 255:9082–92

Ann. Rev. Biochem. 1982.51:395–427

THE THREE-DIMENSIONAL STRUCTURE OF DNA[1]

Steven B. Zimmerman

Laboratory of Molecular Biology, National Institute of Arthritis, Diabetes, and Digestive and Kidney Diseases, National Institutes of Health, Bethesda, Maryland 20205

CONTENTS

Perspectives and Summary

This is a particularly interesting time to review the three-dimensional structures of DNA. Among recent dramatic events are the demonstration of a radically new type of DNA structure—the left-handed Z-helix—as well as the determination of the detailed structure of a segment of B form DNA. The latter is now especially relevant, as an increasing number of heretical candidates are being proposed to replace several familiar models for DNA, including the Watson-Crick model for the B form. Accordingly, the emphasis of this review is on the regular structures that have been proposed for double-stranded DNA.

The review starts by considering two of the major conformations (or "forms") that have been proposed for DNA, namely the B and Z forms,

[1]The US Government has the right to retain a nonexclusive, royalty-free license in and to any copyright covering this paper.

with a brief mention of several others, including the A, C, and D forms. Although this survey contains some recent information on the interconversions between certain of these forms and a mention of studies on their dynamic behavior, no attempt is made to be comprehensive in these areas. The status of the structure of DNA-RNA hybrids is briefly reviewed. Obviously, a detailed review of all of these areas is impossible and considerable selection has been exercised. Studies of transfer RNA or of oligonucleotide or polynucleotide complexes with drugs or proteins are only discussed as they seem directly pertinent to DNA structure per se. Theoretical studies are deemphasized relative to experimental studies. The literature covered by this review extends through the middle of 1981.

DNA structure was last reviewed in this series by Jovin (1). A number of other reviews have appeared on nucleic acid structure (2–8) and physical properties (9) as well as on the properties of DNA-protein complexes (10, 11), chromatin (12), and topoisomerases (13–15), and complexes of nucleic acids or their components with metals (16, 17), with drugs (18), or with water (19). Related reviews have centered upon the data obtained from optical techniques (20) or NMR (21–25) (24 reviews nucleosides and nucleotides). Superhelical DNA has been also been reviewed (26–29).

Structural Models for DNA

ORIGINS OF POLYNUCLEOTIDE MODELS   With the exception of the recently discovered Z form, the various conformations of DNA were all originally distinguished and defined by their X-ray fiber diffraction patterns. In all cases, this type of diffraction record has provided the richest source of data against which detailed structural proposals for polymeric DNA can be tested. Such fiber diffraction patterns provide two major types of information (30): the helical parameters (pitch, residue repeat distance, and residues/turn), which are often readily obtainable from spacings of the reflections in the X-ray patterns, and the intensity distribution pattern itself. The helical parameters are used to set limits on the type of detailed structural proposals ("models") that need to be considered, and trial structures are built within these constraints. Such provisional models are further subjected to the laws of structural chemistry: covalent bond angles or distances must be within the ranges determined by accurate structure determinations on small molecules. The coordinates of stereochemically acceptable models are used to calculate predicted diffraction patterns, which are then compared with the intensity distribution in the observed diffraction pattern. The model typically goes through a number of cycles of adjustment to improve the fit of its calculated diffraction pattern to the one observed.

It is in the nature of this train of argument that the best model so obtained is not necessarily unique. The coordinates are presented at atomic resolution because they must be so specified for testing of the model, not because they are necessarily a unique solution dictated by the diffraction data. Because of the shortage of diffraction data, assumptions are incorporated into the model building. For example, base-pairing schemes are commonly assumed from physicochemical or other sources. Sugar puckers or backbone dihedral angles may be fixed to certain values or constrained to be within certain ranges. Indeed, even the hand of the basic helix must be assumed, at least initially, although subsequently right- vs left-handed models can be tested against each other.

These limitations on DNA models derived from fiber data have stimulated major reinterpretations and have led to proposals for left-handed and "side-by-side" models. These proposals are discussed in some detail. In the case of Z DNA, the original structure was obtained from single crystal X-ray diffraction analysis of short oligomers. This approach does yield a unique structure; however, it must then be shown that polymers can assume the same structure as did the oligomers. This stage of the demonstration once again has rested heavily on fiber diffraction studies, in conjunction with other physical chemical evidence. Several models are compared in Figure 1 and their parameters are summarized in Table 1. Finally, it should be noted that the original structural proposals for the various forms of DNA stem from data on solid samples, i.e. fibers or crystals. A number of recent studies have concluded that the original models must be modified if they are to portray realistically the structures present in solution.

TERMINOLOGY AND CONVENTIONS   For those interested in the definitions of specific conformational features such as the various sugar puckers and the ranges of dihedral angles the descriptions of Saenger (47) or of Arnott (48) are suggested.

A designation such as "the B form" of DNA is commonly used in the literature to refer to both the actual structure that gives rise to the characteristic diffraction pattern and to the hypothetical structural model that is proposed to rationalize the diffraction pattern. In this review, the former meaning is employed. Specific structural proposals such as types of sugar puckers and sets of atomic coordinates are considered to be attributes of models for the actual structure.

B FORM DNA   The Watson-Crick proposal (49) provided the basis for the familiar model of B form DNA. Elegant studies (50) led to a stereochemically acceptable version of this model that was consistent with the observed

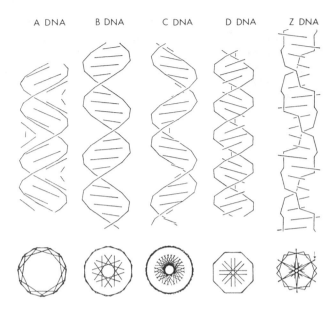

Figure 1 Models for various conformations of DNA. Segments containing 20 base pairs are shown for right-handed models of A (31), B (31), C (32) or D DNA (33) and for the left-handed Z_1 form of DNA (34). The upper views are perpendicular to the helical axes and the lower views look along the helical axes. The continuous helical lines are formed by linking the phosphorus atoms along each strand. The line segments indicate the positions of the base pairs and are formed by joining the C1' atoms of each base pair. This simplified mode of representation emphasizes the differences in helical parameters and in positions of the base pairs among these models. The arrow in the lower view of Z DNA indicates the C1' of the deoxyguanosine residue, the base of which is relatively exposed in this structure (see text); the deoxycytidine residue is more centrally located.

diffraction pattern. Arnott & Hukins (31) refined this structure and provided a set of atomic coordinates that have been widely used in discussions of B DNA. Certain aspects of the B form are not presently controversial: all current models assume a double helix with antiparallel strands and with Watson-Crick base pairs oriented roughly at right angles to the helix axis (Figure 1). Controversy exists as to whether the helix is right-or left-handed or both, as to the exact number of base pairs/turn, and as to the disposition of the bases (coplanarity of bases within the pairs and their orientation with respect to the helix axis).

Most DNA can adopt the B form as defined by its characteristic X-ray fiber diffraction pattern: The bulk of the sequences of natural DNA over a wide range of base compositions (51, 52) as well as synthetic DNA of several simple base sequences (35) yields the B pattern under appropriate

DNA conformation	Occurrence	Axial rise per base pair (Å)	Base pairs per turn	Base pairs per repeating unit	Ref
A	Most natural and synthetic DNA	2.6	11.0	1	31, 35, 36
B	Most natural and synthetic DNA	3.4	10.0[a]	1	31, 35, 37
Alternating B	Several alternating purine-pyrimidine DNA	3.4	10.0	2	38
C	Most natural and synthetic DNA	3.3	7.9–9.6	1	33, 35, 39, 40
D	Several synthetic DNA	3.0	8.0	1	33, 35, 41
T	Glucosylated bacteriophage DNA	3.3–3.4	8.0–8.4	1	42, 43
Z	Several alternating purine-pyrimidine DNA	3.6–3.8	12.0	2	34, 44–46

[a] Alternative models with different parameters are discussed in text.

conditions. Further, the patterns are similar to each other in detail, which suggests a narrow range of variability in B form structure. This apparent homogeneity is in contrast to recent suggestions from nondiffraction techniques and from single-crystal diffraction studies on oligonucleotides, which indicate either static or dynamic structural heterogeneity (see below). Early indications of deviant structures in DNA of high AT-content have been correlated with the presence of non-DNA material (53). There are a few species of synthetic DNA, namely poly(dA)·poly(dT), poly(dI)·poly(dC), and poly(dA-dI)·poly(dC-dT), that have a B form that is significantly different from that of the other complementary deoxypolymers. These polymers have different intermolecular packing arrangements and a slightly changed value for the rise per residue (35, 54). A striking indication of altered structure or structures for the conformations of poly(dA)·poly(dT) and poly(dI)·poly(dC) is their refusal to undergo a transition from the B form to the A form (35, 54–56). These unusual properties assume particular interest given the existence of relatively long dA·dT sequences in vivo (57).

B form adopted by self-complementary oligodeoxynucleotides   The structure of a long self-complementary oligodeoxynucleotide, d(CGCGAAT-TCGCG), has recently been solved by single-crystal diffraction techniques (58–61). The dodecanucleotide forms more than a full turn of a helix whose overall structure is notably like that proposed for the canonical model for B DNA. This study therefore allows an unambiguous examination at high resolution of a prototype for the B form of DNA, and is certainly a most important addition to our knowledge.

The structure adopted by the dodecanucleotide is a right-handed antiparallel double-stranded helix with the bases essentially perpendicular to the helix axis (58). While these major attributes of the B form model are present, the detailed structure of the oligomer departs from that of the familiar model in a number of significant ways.

The well-defined molecular axis is not straight; rather the axis traces a smooth bend of significant curvature (a total of 19° bend distributed over the entire length, which corresponds to a radius of curvature of 112 Å; for reference, the diameter of the B helix itself is ~ 21 Å.) A clear basis for the bend is seen in the intermolecular contacts within the crystal (58). The terminal three G-C base pairs of each duplex form five H bonds with atoms of the next duplex "above" themselves in a manner that requires a bend in the helix axis. There are two formal possibilities: either the DNA is intrinsically bent and the crystal packing simply accommodates this innate tendency or, alternatively, the packing causes the bending. The authors provide several reasonable arguments for the latter interpretation, but ultimately we must await the results of similar studies with other oligomers. It is the terminal alternating G-C sequences that are involved in the canted interactions between duplexes, so that it will be important to see what structures form in oligomers lacking these terminal sequences or, alternatively, containing longer internal sequences. The bend in the dodecamer clearly poses a problem if we wish to know the structure of unbent DNA, since deconvoluting the bent structure requires some arbitrary decisions. Notwithstanding, it is most useful to examine this remarkable structure.

The crystal structure differs from the usual models derived from fiber diffraction in a fundamental way: The dodecamer duplex is not simply a set of 24-nucleotidyl residues of essentially identical conformation joined monotonously into 12 base pairs of essentially identical structure. Rather there are enormous variations in the helical relationships between successive base pairs, and often significant conformational differences between the residues (59, 60). The helical variations span the full range observed by fiber diffraction for structures as disparate as those of the A, B, and D forms of DNA. For example, the eleven local helical steps correspond to 9.4, 9.1, 10.8, 9.6, 9.6, 11.2, 10.0, 8.7, 11.1, 8.0, and 9.7 residues/turn (59). These values may be compared to the values from fiber diffraction of ~ 11, 10, and 8 residues/turn for the A, B, and D helices, respectively (Table 1). Hence, although the average (local) helical parameters in the dodecanucleotide (9.65 residues/turn and 3.33 Å rise/residue) are within a few percent of those inferred from the B form diffraction, the structure has tremendous local variation. Indeed, while the base pair tilt and position relative to the helix axis for the central six base pairs is similar to that in the B DNA models from fiber diffraction studies (31), the adjacent base pair on either side has a distinctly A DNA conformation, and another peripheral base pair is like that of the D form. The other backbone conformational angles are generally similar to those inferred from earlier fiber diffraction or theoretical studies, with a few values in unusual ranges.

The bases of a given base pair are not coplanar in the dodecanucleotide (59, 60). The average propellor twist (i.e. total dihedral angle between base planes in one H-bonded base pair) is $17.3° \pm 0.4°$ for the central four A-T residues and $11.5° \pm 5.1°$ for the G-C pairs at the ends. [In contrast, an earlier model for B DNA from fiber diffraction studies (31) has an almost coplanar arrangement of the bases within a base pair (twist = 4°).]

The variation in sugar conformations within the dodecamer has provided several insights (59, 59a). First, the conformations span essentially the whole range possible for deoxyfuranose rings. They show a correlation between the glycosidic torsion angle and the sugar ring conformation. The purine residues tend toward the C2'-*endo* sugar pucker and a high value for their glycosidic angle, while the pyrimidine residues tend toward a lower glycosidic angle and their sugars are more like C3'-*endo* puckered sugars. This behavior was rationalized in terms of steric contacts between the sugars and the 02 of the pyrimidines. The authors note a further striking fact, which they formalize as the "principle of anticorrelation:" the two sugars of a given base pair tend to have values of their internal torsion angles (specifically about C4'–C3') that are equidistant from that of a central value corresponding to the C1'-*exo* conformation. In other words, if one sugar in a given base pair is C3'-*endo,* the other sugar in that pair tends to have a C2'-*endo* conformation. Several deoxyribosyl structures and a recent DNA-RNA hybrid structure seem to be consistent with this principle, whereas the sugars in yeast phenylalanine tRNA are not.

In addition to the striking static heterogeneity in local structure, the dodecanucleotide provides evidence of dynamic heterogeneity. The thermal vibrations of individual atoms were inferred from their temperature factors as obtained in the X-ray structure determination. Drew et al (59) indicate that the relatively larger vibrations of peripheral atoms are consistent with the rapid intramolecular motions ascribed to DNA from NMR and other measurements (see below)

DNA in solution is often presumed to be in the B form (see section on the helical repeat of DNA). While modeling studies of solvent accessibility to the surfaces of the B form and of other polynucleotide structures have appeared (62, 63), experimental knowledge of the organization of water about duplex DNA has been sparse. [The single-crystal structure of the proflavine complex with dCpdG is a notable exception (64).] The dodecanucleotide structure provides a unique opportunity to visualize water structure in a sizeable unit of B form DNA (61). There are 72 ordered molecules of water per duplex. Of these, 50 are either in the grooves or closely associated with the phosphate groups. The most striking arrangement of water molecules is in the region of the minor groove associated with the central AATT sequence. Solvent is organized in layers up to three or four molecules deep. The innermost layer forms a regular "spine" of water bridging the hydrophilic groups of alternate bases. This backbone of water is compatible with any sequence of A-T or I-C base pairs, but *not* with G-C residues, due to the disrupting influence of the 2-amino group of guanine. This pattern suggests that guanine residues will destabilize the B form, an inference that is in general concurrence with the fiber diffraction survey of Leslie et al (35).

Finally, we note a study that suggests that the dodecanucleotide structure may represent a reasonable model for B DNA in solution. Lomonosoff et al [(65) and further discussion in (60)] have found a good correlation between the rates of nuclease cleavage of internucleotide bonds of the dodecamer in solution and their "exposure" (magnitude of the local helical rotation between base pairs) in the crystal.

A complex of a second sizable oligodeoxynucleotide, d(CGTACG), with daunomycin has been shown by single-crystal diffraction analysis to contain a short segment that adopts a structure similar to that of the canonical B form. This complex forms a self-complementary right-handed anti parallel mini-DNA helix, with a molecule of drug intercalated between the C-G base pairs at each end (66). The structure of the ends of the molecules is modified for intercalation, but the central A and T residues share many characteristics with the familiar B DNA model. The sugars of the central A and T residues conform to the principle of anticorrelation above (59). The authors indicate however that the backbone conformations of the central base pairs of the hexanucleotide have significant departures from those in the usual A or B form models.

B form model for alternating copolymers   A single-crystal X-ray structure determination (67, 67a) on the tetranucleotide d(ATAT) has led to interesting proposals, (38, 67a) for a modified B conformation suitable for a regularly alternating copolymer. In the crystal, the two base pairs at each end of the tetranucleotide form H bonds to two different adjacent tetranucleotide molecules, which yield short segments of right-handed antiparallel double helix. The most striking feature of the crystal structure is the regular alternation in sugar pucker and glycosidic dihedral angle, χ. The adenosine residues both have a C3'-*endo* sugar conformation, while the thymidine residues have a C2'-*endo* pucker. This alternation has been incorporated into a model for the polymer poly(dA-dT)·poly(dA-dT), which is generally rather similar to the canonical B form model [cf Figure 10 in (68)]. This model is proposed as one of a family of basically similar alternating structures that could be built to maximize base-stacking interactions (38; see also 67a, 69). Such alternating B models are consistent with a number of the properties of poly(dA-dT)·poly(dA-dT) previously observed in solution [see (38) for references]. Shindo et al (70) have independently suggested a model with an alternating backbone conformation for this polymer in solution based on two resonance peaks in the ^{31}P NMR spectrum. Recent ^{31}P NMR studies have also detected the two resonance peaks expected for the alternating B model in fibers of poly(dA-dT)·poly(dA-dT) (71), as well as in solutions of poly(dA-dbr^{5}U)·poly(dA-dbr^{5}U) and poly(dI-dC)·poly(dI-dC) (68). The tetramethylammonium ion has been suggested to favor an alternating conformation for poly(dA-dT)·poly(dA-dT) (72).

Lomonosoff et al (65) have provided evidence that poly(dG-dC)·poly(dG-dC) in low-salt solutions can also adopt an alternating type of conformation, presumably of the B form, based upon pancreatic DNase I digestion patterns of the polymer. [An alternating B model was proposed (73) for the high-salt form of (dG-dC)-oligomers in an earlier study; in hindsight, it seems as likely that the dinucleotide repeat inferred is related to the more recently delineated Z form (see below).] Less detailed models for alternating structures have been suggested from theoretical studies (74, 75).

Bent models for B DNA   Structural models for DNA have generally been built as unbent double helices. There have, however, been several suggestions of smoothly bent or discontinuously bent ("kinked") models. Smoothly bent DNA has actually been observed, in the form of the dodecanucleotide duplex (58) discussed in a previous section. There have been a number of theoretical studies of smoothly bent models for DNA wound around the histone cores of chromatin (76–79).

Various kinked models feature straight segments of B form DNA interspersed at regular intervals with residues having alterations in certain torsional angles of their sugar-phosphate backbones (80–83). A variety of superhelical arrangements can result, depending upon both the frequency of kinking and its detailed mechanism. Such kinked models form a conceptual framework for observations as diverse as the compaction of chromatin and its regular patterns of nuclease digestion, the "breathing" of DNA, or the mechanism of drug intercalation. Solvent bombardment or thermal sources have been suggested as possible origins for kinks and other types of localized or traveling fluctuations in DNA structure (84–86). There is no direct evidence for kinking at present. ^{31}P NMR studies of DNA in either chromatin or free in solution have not indicated the presence of species with an altered chemical shift as might be expected if kinking occurred and if the phosphodiester geometry was markedly changed at the kinked site (87–90).

Left-handed models for B DNA There has in the past generally been a consensus that B form DNA is a right-handed helix. As particularly emphasized by Sasisekharan and his collaborators, this assumption does not have a secure experimental basis. Let us review the arguments for left-handed DNA, reserving for the moment a discussion of mixtures of left- and right-handed DNA in the form of the side-by-side DNA model. It has long been known that DNA in the solid state can rapidly and reversibly be interconverted between the A, B, and C forms, simply by varying the relative humidity and salts present (37). The ease of these transitions in fibers was interpreted to mean that all of these forms are of the same hand. Since the original model building for the A form appeared to rule out a left-handed structure (cf section on A DNA below), the A, B, and C forms were all inferred to be right handed. This reasoning is subject to dispute at several points: First, it is only an intuitive conclusion at present that transitions in fibers can not change the hand of the structures involved. Second, even if this is accepted as a working hypothesis, several groups have shown that the B form of DNA interconverts in fibers with the Z form (46, 91), a presumably left-handed structure; hence, if anything, this line of argument suggests that the B form could be left handed. Third, satisfactory left-handed models apparently can be built for the A form (and also for the B and D forms) (69, 92). Detailed sets of coordinates have been supplied by Gupta et al (69) that are stated to fulfill the usual stereochemical criteria. Several groups have noted that, surprisingly enough, only minor differences in the values of backbone dihedral angles need occur between left- and right-handed conformations, although the orientation of the base undergoes a significant shift (2, 7, 75, 92–94). Energy calculations have indicated to some a preference for right-handed conformations (2, 7, 94), although base stacking energies seem to be similar (95). A major criterion for an acceptable model is that it yield a predicted diffraction pattern consistent with the observed diffraction pattern. The left-handed B DNA model appears satisfactory in this respect (69). Hence, from the results of the fiber diffraction approach, there is no clear basis for preferring left- to right-handed DNA.

It may be noted that there are several unambiguous observations of right-handed helices in oligodeoxynucleotides of mixed base sequence in both the A and B forms and in RNA [the short A form–like helices of yeast phenylalanine tRNA (96)]. As yet, the only examples of a left-handed helix are seen in certain oligodeoxynucleotides and polydeoxynucleotides, which have a regularly alternating purine-pyrimidine sequence and a structure that is presumably not of general occurrence (see section on Z form DNA), or in short oligoribonucleotides that are constrained by a second covalent link between base and sugar (97). In sum, while there are indications that natural DNA is right handed, there seems no obvious reason why it should not be able to adopt a left-handed conformation.

Side-by-side models for DNA   Two groups have independently proposed a new class of model for the B form of DNA (98–101). These so-called side-by-side or SBS models are basically different from either the uniformly right- or uniformly left-handed structures discussed above. All of the SBS models feature a regular alternation of short segments of left- and right-handed double-stranded DNA. The versions proposed by Sasisekharan and collaborators (98, 100) have alternating segments of five base pairs, so that the two strands do not undergo a net winding around themselves. The most recent SBS model of Millane & Rodley (101) has a slight right-handed bias so that the strands wind around each other, i.e. they are "linked" about once for every 77 base pairs. In contrast, the uniform left- or right-handed models previously discussed are linked once for every pitch length of approximately ten base pairs. A debate has developed in the literature between the proponents of the double helix and those favoring the SBS conformation. We first consider the arguments based on model building and X-ray diffraction studies and then consider topological and other types of evidence.

Models of SBS DNA have many more degrees of freedom available than do models of uniform helices, since the repeating unit of SBS DNA is long (ten base pairs rather than one or two base pairs) and heterogeneous (left-handed, right-handed, and junction regions). Consequently, while modeling of such a large unit is a technically difficult problem, there seems little doubt that stereochemically acceptable SBS models can be built (101, 102). Detailed studies of the conformation and base stacking have appeared (95). It may be surprising to the reader that the expected diffraction patterns for such models are not totally different from those of the regularly helical models. Similarities arise because the fiber diffraction patterns are dominated by two features that appear similarly in SBS and uniform models. The first feature is the apparent helical repeat: canonical models have a pitch of ~ 34 Å, consistent with a tenfold helix and implying diffraction layer lines at spacings of $N/34$ Å (where N is an integer). Similarly spaced layer lines may be generated by SBS models with an exact or approximate structural repeat distance of 34 Å, which corresponds to a unit of five left-handed and five right-handed base pairs. The second common feature of SBS and double-helical models is the centrally located Watson-Crick base pairs, which are situated approximately at right angles to the helix axis and spaced at an average of ~ 3.4 Å. This 3.4-Å average spacing between base pairs results in the strong meridional reflection that is observed at $N/3.4$ Å and that also dominates the form of an interatomic scattering function called the axial Patterson function, which has been applied to this problem (103). The general similarities in this function for the SBS and canonical B DNA models therefore do not form a basis for distinguishing between these structures as candidates for the B form. A more incisive test is to calculate the actual pattern expected from the atomic coordinates of the models in question. This has been done by two groups for several versions of SBS models (102, 104). In both studies, the authors concluded that the predicted diffraction patterns of the particular SBS models they examined do not fit the observed diffraction data nearly as well as do the predicted patterns of the best double-helical structures. In particular, SBS models predict several meridional reflections that are not observed. These studies have dealt with diffraction from fibers of noncrystalline DNA, the so-called continuous transform. There is also considerable diffraction data from fibers of semi-crystalline B DNA. Arnott (105) notes that the intensity distribution of the crystalline data agrees better with the double-helical rather than the SBS model. Further, a point originally made by Dover (106) is applied (105). The crystal lattice adopted by the lithium salt of DNA generates precisely regular intermolecular contacts between neighboring molecules for base pairs related by steps of $36.15° \pm 0.25°$ of helical rotation, i.e. for a helix with precisely ten base pairs per turn of the helix. This relationship makes immediate sense in that it optimizes favorable lattice interactions for each base pair if the structure involved is that of a regular tenfold helix; however, it would be essentially a fortuitous result for an SBS type of structure, since SBS structures do not have a regular repeat at 36° intervals about the molecular axis.

As mentioned, the two strands of SBS DNA are relatively unlinked because of the alternation between left- and right-handed segments. A possible consequence of this feature, as noted a number of years ago (107), and presumably one of the motive forces behind SBS DNA, is that the relative lack of linking of the strands might facilitate the separation of the two strands during replication or other processes. However, the cell has developed elegant enzymatic mechanisms for winding and unwinding DNA

(13–15), so that the ease of strand separation in SBS models is not a compelling argument in their favor (cf 108). Further, there is topological evidence that the two strands of closed-circular DNA are indeed linked by intertwining about each other once every 10 base pairs (109). In contrast, under special circumstances, several studies have observed coiling of DNA strands without linking, specifically when closed-circular single-stranded DNA interacts with closed-circular single- or double-stranded DNA of complementary sequence (110, 111). The regularity and conformations involved in these interactions are unknown, but such situations are possible candidates for mixtures of left- and right-handed helical segments. Finally, Greenall et al (104) raise the question of the conformations involved in the A ⇌ B interconversion if the B form is indeed in SBS structure, since an SBS version of the A form has apparently not yet been described.

In sum, there seems to be little experimental basis for SBS DNA, while there is evidence for uniform double helices. Until some more convincing evidence is brought forth, SBS DNA seems to be an unlikely possibility.

Z FORM DNA   Some years ago, Pohl & Jovin observed that the circular dichroism (CD) spectrum of the alternating copolymer, (poly(dG-dC)·poly(dG-dC), underwent a novel inversion when the polymer was exposed to high salt or ethanol concentrations (112, 113). Recent demonstrations (34, 44, 45, 114) that oligomers of this material can assume a left-handed conformation, the Z form, have suggested that this structure is the basis for the inverted CD spectrum (see below).

Z Form in crystalline oligonucleotides of d(C-G)·d(C-G)   The structures of several oligomers of alternating d(C-G)·d(C-G) have been determined within the last few years using single-crystal diffraction techniques. Given the large number of departures of these oligomer structures from those of canonical polynucleotides, it is fortunate that this methodology yields structures that are free of the kinds of assumptions that typically are made in fiber diffraction analysis. The initial structure in this series was derived from crystals of the hexamer, d(CGCGCG) (44), and was followed shortly by determinations on the tetramer, d(CGCG) (45, 114). There are now a number of examples of these oligomers crystallized from a variety of media. The major features of their structures are similar. I first discuss the hexamer (44) as a prototype, and then summarize the differences among the various oligomer structures.

In the crystals of the hexamer, two molecules of d(CGCGCG) join in an antiparallel fashion to form a left-handed minihelix. The C and G residues on one strand form Watson-Crick hydrogen bonds to the G and C residues on the other strand. The six base pairs so formed comprise one half of a left-handed helical turn.

The conformations of the residues also alternate in this structure of alternating base sequence. In general, the dC residues are similar to the residues in the canonical models for B DNA, while the dG residues are very different. For example, the sugars of the dC and dG residues display a regular alternation between the C2'-*endo* and C3'-*endo* conformations, respectively. Further, the dihedral angle around the C4'–C5' bond alternates markedly (between a *gauche-gauche* conformation in dC residues, as in B DNA, and a *gauche-trans* conformation in dG residues). Since the C and G residues have different conformations, the repeating unit on a given strand is a *dinucleotide*. These dinucleotide units are very regularly arrayed in the minihelix, although the symmetry of the lattice does not demand this regularity. The regularity even extends across the gap where duplexes stack upon each other except, of course, for the missing phosphate groups. The result is that the crystal structure approximates that of a continuous polymer, with 6 dinucleotides or 12 bases for each helical turn of a given strand. These alternating conformations produce an irregular zig-zag course of the sugar-phosphate backbone, hence the Z DNA designation for this structure (Figure 1). The residues also alternate in the relationship of the base to its sugar. The dC residue has an *anti* relationship, which seems to be generally needed to relieve steric interactions between pyrimidine bases and their sugars (115). The guanine residues in contrast assume a *syn* conformation wherein the base closely approaches its sugar. This is the first example of a residue of a natural nucleotide that adopts the *syn* position in an oligomeric or larger system, although the bases in polynucleotides can be constrained toward this position by a bulky base substituent (116) or by a second covalent linkage between base and sugar (97).

As might be expected from the unique alternation of backbone and base-sugar conformations, the positioning of base pairs within the helix and the base-stacking relationships are remarkable. The base pairs are relatively peripheral in location in Z DNA. Because of the *syn* conformation of the guanosine, the reactive N7 and C8 positions of guanine are located at the margins of the structure and thereby made accessible to environmental insults (see below). The base pairs at the d(CpG) sequences are displaced laterally by 7 Å relative to each other, so that ordinary base overlaps do not occur. Rather, the cytosine residue of one base pair stacks with a neighboring cytosine residue on the opposite strand. The guanine residues at this sequence do not stack with the adjacent bases, but rather overlap the furanose ring oxygens of the adjacent sugars. In contrast the stacks at the d(GpC) sequences are relatively similar to those of B DNA with overlap of the bases that adjoin on the same strand. The Z form has a single very deep helical "groove" that corresponds in location to the minor groove (i.e. the groove between the sugar-phosphate backbones) in the A or B forms.

Limited but significant variation has been observed in the structures solved to date for d(CGCG) or d(CGCGCG) crystallized under various conditions (34, 44, 45, 114, 117). The principal variation occurs in the orientation of the phosphate group of the GpC sequence. In the various crystals that have been solved, one or more of the phosphate residues has rotated from its position in the predominant conformation so that it lies further outwards and away from the minor groove (34, 45, 114, 117). Models for continuous polymers based upon each of these two phosphate conformations, labelled Z_I and Z_{II}, have been described (see below). A related change in the conformation of the phosphate group occurs in the crystalline tetramer solved by Drew et al (45). In this structure, labeled Z', the phosphate group of the GpC sequence again rotates outward, as in going from Z_I to Z_{II}. The reasons for the rotations in Z' and Z_{II} are apparently different, being correlated with repulsion by a neighboring chloride ion for Z' (45) or, in some cases at least, with binding to a nearby hydrated magnesium ion for Z_{II} (34, 114). The altered phosphate positioning in Z' is correlated with a change of sugar pucker of the internal deoxyguanosine residues, so that those sugars adopt a conformation (Cl'-exo) similar to that in the deoxycytidine residues. The result is that the Z' backbone has an almost uniform sugar pucker, which demonstrates that the characteristic dinucleotide repeat of Z DNA does not necessarily need to extend to a major involvement of the sugar conformation (45).

Z form in fibers of alternating polymers   The demonstration of the Z structure in crystals of oligomers has led to attempts to identify such a structure in fibers or in solutions of polymeric DNA. X-ray fiber diffraction has provided relatively unambiguous evidence in the case of fibers. Arguments for the Z form in solution are summarized in the next section.

Fibers of certain alternating deoxypolymers yield a distinctive X-ray diffraction pattern identified with the Z form [which has also been called the S form in polymers (35)]. The original observation of Z form diffraction *from fibers* was made with poly(dG-dC)·poly(dG-dC) (46). Basically similar patterns have been collected from fibers of poly(dA-dC)·poly(dG-dT) (35) and poly(dG-dm⁵C)·poly(dG-dm⁵C) (118). A related pattern from poly(dA-ds⁴T)·poly(dA-ds⁴T) (119) has been reinterpreted in terms of the Z structure (46).

Models play an important role in fiber diffraction in validating proposed structures, as outlined earlier. In the case of polymeric DNA, the models for the Z form have followed the structures that were determined by single-crystal diffraction of the oligomers. The initial model for poly(dG-dC)·poly(dG-dC) (46) differs significantly in backbone position and base-stacking arrangements from the oligomer structures; a more recent model (120) approaches the oligomer structure more closely. In neither case were predicted diffraction patterns presented for the models. The discussion of Wang et al (34) contains the most detailed polymeric models based on the oligomers. These authors have generated full sets of atomic coordinates for each of the two variations on the Z form described above, i.e. Z_I and Z_{II}, which they observed in their oligomer studies. The predicted diffraction pattern for their model of the predominant conformation (Z_I) (34) fits the observed pattern (46), which provides strong evidence for the Z conformation in fibers of this polymer.

Is an RNA backbone compatible with the Z form? In the initial description of the Z form, the authors (44) indicated that the 2'-hydroxyl position of the sugar would point outwards from the helix and so would not necessarily be sterically encumbered. There are, however, no spectral indications that the ribopolymer, poly(rG-rC)·poly(rG-rC), enters a high-salt form under conditions where the deoxypolymer does (113). The relationship between such high-salt conformations and the Z form is considered next.

The high-salt form of alternating polymers and its relationship to the Z form
The studies of Pohl & Jovin on the high-salt form of poly(dG-dC)·poly(dG-dC) were clearly an important factor in the choice of d(CGCG) and d(CGCGCG) for crystallization, which led to the elucidation of the Z structure. The original studies, which showed a requirement for high-salt concentrations (113) or intermediate levels of ethanol (112), have been extended to include a variety of cations that are able to elicit this form (based on spectral criteria) at much lower concentrations (121). For example, poly(dG-dm^{5}C)·poly(dG-dm^{5}C) undergoes an inversion in its circular dichroism spectrum in the presence of 5 μM hexamine cobalt or 2 μM spermine. This study also demonstrates a dramatic influence of polymer composition: the methylated polymer undergoes the spectral transition at $\sim$ 1 mM Mg^{2+}, a concentration about 1000-fold lower than that needed for the unmethylated polymer (121). Although I continue to use the high-salt designation for the conformation with inverted CD spectrum described by Pohl & Jovin (113), these results clearly show it can be induced under specific conditions at low ionic strength. As is described in the next section, binding of certain substituents also favors the occurrence of the high-salt form.

Other properties besides circular dichroism and ultraviolet absorbance (112, 113) have been used to measure the transition. The ^{3}H-exchange properties of the high-salt form are markedly different from those of the low-salt form. Two protons have half-times at least 50-fold longer than those in more usual polynucleotide conformations. Ramstein & Leng (122) note their potential use to assay for the high-salt conformation in natural DNA. The transition is also accompanied by marked changes in Raman scattering spectra (123).

There is considerable evidence that the high-salt form corresponds to the Z form. First, the laser Raman spectrum of crystalline d(CGCGCG) is similar that of poly(dG-dC)·poly(dG-dC) in high-salt solution and different from that of the polymer in low-salt solution (123a). Second, the increase in molecular length going through the transition at intermediate ethanol concentrations corresponds to that expected for the conversion from the B to the Z form (123b). Third, the high-salt form (as evidenced by its CD spectrum) is observed for polymers with regularly alternating purine-pyrimidine sequences, i.e. polymers of alternating (dG-dC) (113), (dG-dm^{5}C) (121), or (dI-dbr^{5}C) (73), but not for the nonalternating polymer, poly(dG)·poly(dC) (113). This specificity is consistent with the basic structure of the Z form, which requires alternation of purines and pyrimidines. Incidentally, the inversion of the CD spectrum might seem to be evidence for the presumed conversion from a right-handed B form to a left-handed Z form. However, theoretical considerations indicate that inversion does not necessarily correspond to a change in helical sense (124). In addition, there are instances of similar inverted CD spectra for which there are viable alternative explanations. For example, the product of the annealing of complementary closed-circular DNA, Form V DNA, has a related spectrum (111), although its irregular base sequence seems incompatible with a regular Z conformation. A second inverted spectrum is that of poly(dI-dC)·poly(dI-dC) in low-salt media. As described in the section on D form DNA, it has been variously interpreted by different groups as due to left- or right-handed helices quite different from the Z form.

Further evidence for correspondence between the high-salt and Z forms comes from NMR experiments. These studies suggest an alternating conformation for the high-salt form of alternating (dG-dC). The ^{1}H NMR spectrum of (dG-dC)$_8$ (73) indicates that one but not both glycosidic torsion angles and one but not both sugar puckers change in the Pohl-Jovin transition. The ^{31}P NMR spectrum of this material (73) as well as that of a 145 base-pair length of alternating (dC-dG) (68, 125) splits into two resonances of approximately equal intensity, which indicates an alternating conformation about the phosphodiester linkage. Magnetic shielding constants calculated for poly(dG-dC)·poly(dG-dC) in high salt were in agreement with the presence of the Z form but not of the B form (126). These various NMR results are consistent with but not uniquely diagnostic for the Z form. Adoption of the Z structure is also suggested by studies of derivatives of alternating sequence polymers (see next section).

There are two studies that raise the possibility that the high-salt form may not be the Z form. First, determination of the helical periodicity of either poly(dG-dC)·poly(dG-dC) or poly(dG-dm^{5}C)·poly(dG-dm^{5}C) in the high-salt form by nuclease digestion while the polymers are bound to absorbents (see section on the helical repeat of DNA) gives $\geq$ 13 base pairs/turn rather than the value of 12 expected for the Z structure. It is not clear whether the discrepancy reflects the effects of adsorption or whether a different structure is present. Second, the calculated CD spectrum for the Z form does not agree with that observed for the high-salt form of poly(dG-dC)·poly(dG-dC) (127). Whether the disagreement arises from assumptions in the calculations or in the choice of structure has not been resolved.

The structural transition of alternating (dG-dC) in solution generally occurs at a relatively high-salt concentration (113). Crawford et al (114) note that, despite this, the various hexamer and tetramer crystals containing segments of Z DNA were generally obtained from low-salt solutions. They therefore suggest that there is an equilibrium between left- and right-handed conformations, which shifts toward the Z form as crystallization proceeds. The crystals themselves nominally qualify as high-salt environments, having concentrations of charged groups of the order of several molar (44, 45). A form of d(CGCG) crystallized at lower salt concentrations has been described but not solved in detail; this form undergoes a reversible salt-dependent transition to the Z' structure while still in crystalline form (128).

Reaction of alternating (dG-dC) with ligands Several ligands favor the transition of poly(dG-dC)·poly(dG-dC) to a form with an inverted CD spectrum similar to that induced by high-salt levels. Mitomycin C induces such a change with a polymer specifity like that for the salt-induced conversion: alternating deoxypolymers of (dG-dC) are affected, but not the corresponding ribopolymer or the nonalternating deoxypolymer. The drug also can cause a qualitatively similar but much less extensive change in the CD spectrum of natural DNA (129).

Extensive reaction of the carcinogen, N-acetoxy-N-acetyl-2-aminofluorene, can cause poly(dG-dC)·poly(dG-dC) to undergo an inversion in its CD spectrum even at low ionic strengths (130–132). At relatively low levels of derivatization, the polymer remains more prone to undergo the transition, as judged by reduced levels of ethanol needed for the reaction. The major site of covalent attachment of the bulky fluorene derivative is at the

C8 position of guanine, which leads to very interesting speculation in terms of the Z structure. (The usual caveat applies: we do not know that the inverted spectrum of the fluorene derivative corresponds to that of the Z form, although it is certainly worth making this working hypothesis to entertain the implications.) Reactive sites on guanine (N7 and C8) are particularly exposed to the medium in the Z structure, and the potential importance of the Z form in reactions with chemicals in the environment was noted by Wang et al (44). Several consequences have been suggested. First, the guanine residues may be more reactive in the Z form. While this suggestion was difficult to test in high-salt or ethanolic media (131), it may be approachable in the aqueous media recently described (121). Second, once reacted, the altered residues may lock the sequence into the Z form.

Related studies have been performed with a second carcinogen, N-hydroxy-N-2-aminofluorene (130) or with dimethyl sulfate (132a). Also the intercalating dye, ethidium bromide, was early noted to be an effective inhibitor of the transition to the high-salt form, presumably due to a preference for binding to the low-salt form (133).

Z form in natural DNA?   The unusual properties associated with Z DNA and with the high-salt form of DNA of alternating purine-pyrimidine sequence have prompted experiments that seek evidence for the occurrence of either or both states in DNA of heterogeneous sequence. As just outlined, ligands may induce limited amounts of the high-salt form in natural DNA. Several studies that do not depend on such ligands are available at the moment; more are expected.

Supercoiled phage PM2 DNA, a closed-circular DNA of heterogeneous sequence, undergoes a salt-induced cooperative transition to a form adsorbed by nitrocellulose filters (134). The basis of the adsorption is unknown; formation of a Z structure is proposed, with either the Z form itself or single-stranded regions between B and Z regions binding to the filter. (The degree of retention of poly(dG-dC)·poly(dG-dC) under these conditions would be of interest.) The NaCl concentration required for the transition is similar to that for the high-salt transition of alternating d(G-C). A higher salt level is required for *linear* PM2 DNA, in accord with earlier suggestions (135, 136) that the excess energy of negative supercoiling would aid the transition from a right-handed to a left-handed structure. Prior treatment of the DNA with N-acetoxy-N-acetyl-2-aminofluorene (see preceding section) or with ultraviolet irradiation lowered the salt concentration needed to induce the transition. The authors speculate that an increased tendency to enter the proposed Z conformation as a result of DNA damage may provide a signal to DNA repair systems.

Klysik et al (137) have cloned plasmids containing inserts of various lengths of alternating d(G-C). Inserts of longer than 40 base pairs of d(G-C) were unstable and suffered deletions. The authors generated a short segment (138–176 base pairs) of heterogeneous sequence DNA containing near its center about a third of its length as alternating d(G-C) residues [which were in turn interrupted by a single d(GATC) sequence]. This DNA segment underwent a salt-induced transformation to a form with a partially inverted CD spectrum. The high-salt spectrum was equivalent to that of a mixture of heterogeneous sequence DNA with an amount of alternating d(G-C) lower than that which was actually present, which lead the authors to suggest that d(G-C) residues near the edges of the insert might not be in a Z type of conformation. As they note, this argument makes several assumptions, including a lack of signal from the proposed junction regions. In contrast, the ^{31}P NMR spectrum in high-salt media showed the appearance of a second resonance in about the amount expected for total conversion of the insert. Klysik et al further applied an important test of the proposed conversion from right- to left-handed regions. A DNA segment with alternating d(G-C) ends was inserted into a plasmid; the linking number of the DNA was estimated from the band pattern in gels run at salt concentrations that spanned the transition. A high-salt conversion clearly occurred that was formally equivalent to the unwinding of about half of the expected number of supercoils for a B to Z transition. The authors suggest that the relatively small change reflects a partial conversion. An alternative interpretation may be considered: Because of their self-complementary sequences, the d(G-C) inserts could "loop out" to make a cruciform structure, with an expected change in linking number about equal to that actually observed. This ambiguity can be avoided in principle by using inserts of a sequence that is not self-complementary but that forms the Z structure. Poly(dA-dC)·poly(dG-dT) appears to fit these requirements (35). It may be noted that a stretch of 62 base pairs of this latter sequence has been found in vivo (138), while comparably long sequences of alternating d(G-C) have apparently not been described. Dickerson & Drew (60) suggest that the tendency of alternating (dC-dG) sequences to adopt the Z form may indeed be small, given the absence of Z structure in their crystals of d(CGCGAATTCGCG). Those crystals were comparable in salt level to those of d(CGCG) or d(CGCGCG) containing the Z form.

Antibodies specific for the high-salt form have been elicited in several animals in response to injections of either brominated or unbrominated poly (dG-dC)·poly(dG-dC) (140a). The fluorescence staining patterns of these antibodies on the polytene chromosomes of Drosophila have provided the first evidence for the Z form in natural DNA (140b). A regulatory role has been suggested for such Z form regions (121, 140b).

Transition between B and Z forms   Arnott et al (35, 46) observed the Z form in fibers of alternating d(G-C) that earlier had given the B form diffraction pattern. Sasisekharan & Brahmachari (91) showed a relatively rapid conversion between these forms that was controlled by changes in the ambient relative humidity. The two groups reached opposite conclusions: one argued that B and Z forms are of different hand (46), while the other concluded that a ready transition under mild conditions ruled out a change in handedness of the helix (91). Whatever the outcome in fibers, Simpson & Shindo (125) note that the transition in solution between the high- and low-salt forms of the 145 base pair pieces of alternating d(G-C) can occur without total strand dissociation; it is, of course, not clear that the same forms are involved as in fibers.

The Z form (74, 139, 140) and the B $\leftrightarrows$ Z transition (136) have also been considered from a theoretical point of view. Possible model structures for the interface between B and Z helices have been discussed, both with unstacked (44) or completely stacked junctions (141).

OTHER FORMS OF DNA   Studies of polymers of simple repeated sequence have been crucial in delimiting the variation expected from base sequences. A recent survey of many complementary deoxypolymers of defined repeating sequences, by Leslie et al (35), based on X-ray fiber diffraction methods should be noted.

A form DNA   Most DNA will enter the A form. Notable exceptions are poly(dA)·poly(dT) (see above), poly(dA-dG)·poly(dC-dT) (35), and the glucosylated DNA from bacteriophage T2 (142). The A form of poly(dA-dT)·poly(dA-dT) was originally reported to be unstable relative to the D form (described in the next section) (41), although exceptions to this behavior have been noted (35, 143). Fibers of different materials in the A form tend to be quite crystalline and to share the same space group and the same lattice parameters (35, 36), which suggests that the A conformation may be favored by specific packing interactions. The A form of DNA can apparently, however, occur in solution in ethanolic solvents (144, 145).

The original model for A DNA (36) and a refined model (31) are right-handed helices with 11 base pairs per turn (Figure 1). The structures of two self-complementary oligodeoxynucleotides, d(GGTATACC)(145a) and d(iodo-CCGG) (145b), have recently been determined by single-crystal methods to be of the A form. It is very significant that both form right-handed helices. Left-handed models for polymeric A form DNA which are stated to be stereochemically acceptable have been reported, but the details are not yet available (69, 92). Given the differences in the molecular transforms of left- and right-handed versions (69), a quantitative comparison with the semi-crystalline fiber diffraction data must be awaited. NMR results indicate that the A and B conformations can exist in apposition in a single RNA-DNA duplex (146). A model-building study of this duplex has appeared (147).

DNA in solution can be induced to undergo a reversible cooperative transition in secondary structure over a small range of ethanol concentrations. The species involved have distinctive CD spectra that were early correlated with the presence of the B form at low ethanol levels and the A form at high ethanol levels (148). This assignment has been corroborated by a direct demonstration of the species involved in the ethanol-induced transition by diffraction techniques (149, 150).

C form DNA   The characteristic C form X-ray diffraction pattern can be obtained from fibers of a wide variety of natural and synthetic DNAs held under relatively dehydrated conditions (35, 39, 40). The original model for the C form (40) as well as a subsequent refinement (32) correspond to a right-handed structure with 9.3 base pairs/turn (Figure 1). There is a widespread tendency in the literature to equate the C form with these models. However, unlike the forms so far discussed, the C form diffraction pattern is obtained from a family of structures possessing a wide range of helical parameters [7.9 to 9.6 residues/turn (39, 40), a range wider than that which separates the A and B forms].

DNA in concentrated salt solutions or in certain organic solvent-water mixtures adopts a distinctive CD spectrum. A similar spectrum is assumed by DNA in chromatin and in certain viruses. This spectrum had been correlated with the presence of the C form (148). However, a number of studies indicate that the conformation actually present is close to that of the B form (39, 151–157), which leaves the CD spectrum of the C form presently undefined.

D form DNA   The characteristic X-ray diffraction pattern of the D form of DNA was originally obtained by Davies & Baldwin from fibers of poly (dA-dT)·poly(dA-dT) (41). The closely related pattern from poly(dI-dC) ·poly(dI-dC) (158) corresponded to an eightfold helix. In addition to those just mentioned, several other nonguanine-containing deoxypolymers have been found to adopt the D form (35, cf 33): poly(dA-dA-dT)·poly(dA-dT-dT), poly(dA-dI-dT)·poly(dA-dC-dT), poly(dA-dI-dC)·poly(dI-dC-dT), and poly(dA-dC)·poly(dI-dT). In general, the D form seems to occur in fibers under relatively dehydrated conditions, more like those favoring the A or C forms than those yielding the B form.

Mitsui et al (158) were unable to build a fully satisfactory *right-handed* model consistent with the D form pattern from poly(dI-dC)·poly(dI-dC). They made the then startling suggestion that a left-handed model (with an unusual sugar conformation) was at least as likely as a right-handed model. Details of the models and their predicted diffraction patterns were not presented. The inverted CD spectrum of the polymer (which occurs at low ionic strengths) was taken as encouragement for the left-handed model. Whatever the handedness of the structure in solution, it is a substrate for a variety of enzymes that act on natural DNA (159). Such enzymatic action could be used to argue for a similar helical sense in solution for natural DNA and poly(dI-dC)·poly(dI-dC) if we knew more about the binding sites of the enzymes. Arnott et al (33) subsequently rejected the left-handed model of Mitsui et al on the grounds of its unusual nature and sugar conformation, and suggested a more usual right-handed model (Figure 1) to explain the diffraction pattern. Very recently, several residues of the dodecamer structure of Drew et al (59) (see section on B form DNA) have provided instances of the O'-endo sugar pucker suggested by Mitsui et al, and certainly the left-handedness of their model cannot currently be used as a basis to reject their structure. What does prevent serious consideration of their proposal for D DNA is the lack of a detailed model to provide a basis for judging stereochemical acceptability and fit to the diffraction pattern. However, as part of their survey of left-handed structures, Sasisekharan and collaborators (69) have built a left-handed model for D DNA with an unexceptional sugar conformation (C2'-endo), as well as a right-handed version. They find their left-handed model agrees with the crystalline diffraction data about as well as does the previous right-handed model (33).

Extended DNA   Arnott et al (160) have recently interpreted the fiber diffraction pattern from a complex of DNA with a platinum-containing intercalator in terms of an unusual linear (i.e. nonhelical) model, which they call L DNA. The proposed structure is stabilized by intercalation between every second base pair. In general terms, the model is obtained by unwinding the helical turns of a double helix without unpairing the base pairs. The result is a ladder-like structure with the base pairs as rungs, generally much like the "*cis*-ladder" proposed by Cyriax & Gäth (161) from a modeling study of nonintercalated DNA. These developments recall the venerable observation of Wilkins et al (162) of a reversibly altered phase of DNA formed by the mechanical extension of DNA fibers. The extension of the fibers was tentatively suggested to result from the actual extension of the molecules.

Miscellaneous models   Two novel types of models have been proposed on theoretical grounds. In the "vertical helix" a relatively open duplex with its bases oriented parallel to the helical axis is generated by using the high *anti* range of the glycosidic torsion angle (163). In another series of models, a reversed backbone polarity has been suggested, which leads to the *syn* conformation for right-handed duplexes or the *anti* conformation for left-handed duplexes (164).

RNA-DNA hybrids   The early fiber diffraction studies of RNA-DNA hybrids (165–167) yielded RNA-like patterns with 11 or 12 base pairs/turn. These results are often cited to show that hybrids adopt only such RNA-like conformations. The NMR spectrum of a model system containing hybrid sequences [(rC)$_{11}$-(dC)$_{16}$ annealed to poly(dG)] is also consistent with an A form for the hybrid sequences (146).

A number of observations, however, indicate that at least some hybrids may rival DNA-DNA duplexes in structural capabilities. For example, Gray & Ratliff (168) have shown that certain synthetic hybrids undergo an ethanol-induced transition that is similar to the transition between the A and B forms of natural DNA (see section on A form DNA). Further, a diffraction study of highly solvated fibers of poly(rA)·poly(dT) yielded a pattern similar to that of B DNA (169). A structural model has been proposed that has major similarities to the canonical B form model except that the backbone conformations of the two strands differ from each other and from those in the usual B form model (31, 50). Poly(rA)·poly(dT) can also adopt A- or A'-like forms (6, 169) with 11 or 12 residues/turn, respectively, which emphasizes that at least some hybrid sequences are polymorphic. The hybrid poly(dI)·poly(rC) also forms a tenfold helix in fibers (6); its structure has been suggested to be similar to that of the original Watson-Crick model for B DNA (49), which had the sugar pucker (C3'-endo) most often associated with polyribonucleotide models.

Helical Repeat of DNA

The X-ray techniques that have had such a dominant role in deriving structural models for DNA require samples with considerable local order. The ordering may be as low as that in a rudimentary fiber where elongated molecules tend to pack with their axes in the same direction, or the ordering may be as high as that in a crystal with its precisely repeated units. In either case, order is usually obtained by repetitive intermolecular interactions in solid samples. Such interactions can influence the structure. Hence, studies in solution assume a double importance, being useful not only in themselves but also to help in evaluation of the effect of the solid nature of the samples on the characteristics of the structure. Recently, two new techniques and a variation on an old technique have been used to determine one of the most basic characteristics of a model, its helical repeat, under conditions relatively free from intermolecular interactions (Table 2).

Wang and his collaborators developed an elegant technique employing dilute aqueous solutions of DNA. Sequences of known length were inserted into closed circular DNA. The resulting changes in linking number were evaluated from gel electrophoretic patterns and used to deduce the number of base-pairs/turn. Insets of heterogeneous sequences had a repeat of 10.6 ± 0.1 residues/turn, while inserts of the homopolymer poly(dA)·poly(dT) were distinctly different (10.1 ± 0.1 residues/turn) (171, 171a). Values for inserts of other sequences are summarized in Table 2.

A second approach is based on the observation of Liu & Wang (175) that when DNA that has been adsorbed to crystallites of calcium phosphate is digested with pancreatic DNase I, a regular pattern of single-strand cuts is found, presumably due to limited steric accessibility. The periodicity in lengths of the DNase products is equated by the authors with the helical repeat. Rhodes & Klug (172, 173) have extensively characterized this technique and shown the periodicity to be independent of the particular choice of adsorbent and nuclease. They argue that the helical repeats so obtained are indicative of solution values based on the indifference of the repeat to details of the adsorption or digestion conditions. Although the technique is likely to be free of DNA-DNA intermolecular interactions, the possibility of changes induced by surface adsorption must be kept in mind. There is good correspondence in the results of this and the preceding technique for the materials so far tested by both approaches. The adsorption technique gave helical repeats of 10.6 ± 0.1 and 10.0 ± 0.1 for DNA of heterogeneous sequence and for poly(dA)·poly(dT), respectively, in agreement with the insertion technique (Table 2). Values for several alternating copolymers, namely poly(dA-dT)·poly(dA-dT), poly(dG-dC)·poly(dG-dC) and poly(dG-dm^{5}C)·poly(dG-dm^{5}C) are summarized in Table 2.

In comparing the helical repeats obtained by these techniques to those from fiber data, we must decide which fiber conformation is most appropriate. The canonical B form is an obvious, but not necessarily correct, choice to correspond to DNA in solution. The poly(dA)·poly(dT) periodicity does indeed agree exactly with that estimated from fiber data for the B form, perhaps reflecting the previous discussed reluctance of this polymer to leave the B form. In contrast, poly(dA-dT)·poly(dA-dT), poly(dG)·poly(dC), and DNA of heterogeneous sequence all had helical repeats between those of the classical A and B forms, i.e. between 11 and 10 residues/turn,

Table 2 Experimental estimates of the residues per turn of DNA duplexes

DNA sample	Technique	Base pairs per helical turn	Ref
Natural DNA			
B form fibers (high humidity)	X-ray	10.0 ± 0.15	170
Wetted fibers	X-ray	9.9 ± 0.14	170
Solution	Linking changes	10.6 ± 0.1	171, 171a
Adsorbed on surface	Nuclease digestion	10.6 ± 0.1	172, 173
d (CGCGAATTCGCG)			
Crystal	X-ray	9.65[a], 10.1[b]	60
Poly (dA) · poly (dT)			
B form fibers	X-ray	10.0	54
Solution	Linking changes	10.1 ± 0.1	171, 171a
Adsorbed to surface	Nuclease digestion	10.0 ± 0.1	173
Poly (dA–dT) · poly (dA–dT)			
B form fibers	X-ray	10.0	41
Solution	Linking changes	10.7 ± 0.1	171a
Adsorbed to surface	Nuclease digestion	10.5 ± < 0.1	173
Poly (dG) · poly (dC)			
B form fibers	X-ray	10.0	174
Solution	Linking changes	10.7 ± 0.1	171
Poly (dG–dC) · poly (dG–dC)			
B form fibers	X-ray	10.0	35
Adsorbed to surface[c]	Nuclease digestion	10 ± 1	118
Z form fibers	X-ray	12.0	46
Adsorbed to surface[d]	Nuclease digestion	13 ± 1	118
Poly (dG–dm⁵C) · poly (dG–dm⁵C)			
Adsorbed to surface[c]	Nuclease digestion	10.4 ± 0.4	118
Z form fibers	X-ray	12.0	118
Adsorbed to surface[d]	Nuclease digestion	13.6 ± 0.4	118

[a] Average value using local helical axes.
[b] Using a single helical axis for all residues.
[c] Adsorbed under conditions favoring low-salt form.
[d] Adsorbed under conditions favoring high-salt form.

respectively, while the repeats of the alternating d(G-C) and d(G-m⁵C) polymers were significantly greater than the fiber repeats for any duplex DNA, including Z DNA. Generally, then, the surface adsorption technique and the insertion technique give values for the residues per turn that are greater than those expected from the fiber or crystal structures. A model for DNA of heterogeneous sequence with 10.6 residues/turn has also been proposed based on energy calculations for DNA in the absence of water (76).

An alternative approach (170) attempted to bridge the gap between fiber and solution structures. Fibers were swollen with up to three volumes of water, and their X-ray fiber diffraction patterns were collected. Molecular orientation was sufficiently retained in this range to obtain helical parameters. The diffraction pattern remained of the B form, and the helical parameters were not significantly altered from those in fibers. There is therefore disagreement between the helical repeat of wetted DNA fibers (9.9 ± 0.14 residues/turn) and that from the insertion or adsorption techniques (10.6 ± 0.1 residues/turn). The behavior of the wetted fibers has been suggested to be influenced by remaining intermolecular interactions which either directly (172) or through multivalent cations (172a) might constrain the helical repeat to stay near 10.0 residues/turn as in the fibers. Such interactions would have to extend through water layers which average at least several molecules in thickness. There is no evidence that multivalent ions occur in significant amounts in the wetted fiber samples.

Heterogeneity in DNA Structure

STATIC HETEROGENEITY   DNA is highly polymorphic. Both natural DNA of heterogenous sequence and a variety of synthetic DNAs of simple repetitive sequence are capable of adopting more than one conformation. Base sequence rather than overall base composition often clearly influences the preferred conformation under a given set of conditions. Among many examples studied by fiber diffraction are the ability of poly(dA-dt)·poly(dA-dT) to adopt the A,B,C, and D forms (41), while poly(dA)·poly(dT) has only been observed in a B form (54), or the ability of poly(dG-dC)·poly(dG-dC) to adopt the A,B, or Z forms (46), whereas poly(dG)·poly(dC) is apparently restricted to the A and B forms (174). In addition, there are numerous instances of variation in physical, chemical, or other properties depending on base sequence, which undoubtedly involve conformational changes (3).

Examples of restricted variation in local nucleotide conformation are those observed in Z DNA or suggested for the alternating B DNA model; in each case two different nucleotidyl conformations alternate regularly. At a more complex level, the dodecanucleotide structure from Dickerson's group is a marvelous example of local structural variation, as discussed in the section on B form DNA. Considerable heterogeneity in backbone conformation is also suggested by the ^{31}P NMR spectra of an oligodeoxynucleotide (176) and of fibers of natural DNA in the B form (177); such heterogeneity is not ascribed to the A form of natural DNA (177, 178).

Natural DNA sequences are not random. Analyses of the deviations from randomness have revealed patterns that raise the possibility of more subtle influences of sequence upon structure (179–182).

DYNAMIC HETEROGENEITY   The heterogeneity so far discussed is static in nature, i.e. under a given set of conditions a particular conformation is assumed to accompany a particular base sequence. An alternative source of heterogeneity is a dynamic variation of structure, not necessarily associated with particular base sequences. DNA in solution or in B form DNA fibers [but not in the A form (177, 178)] has often been suggested to undergo internal motions over a wide range of time constants and with significant amplitudes [however, note (183)]. A detailed review of dynamic structural heterogeneity is beyond the scope of this review, but a few references are cited in this rapidly expanding field. Evidence concerning dynamic heterogeneity in DNA or in DNA-ligand complexes has been obtained by fluorescence depolarization (184, 185), by electron paramagnetic resonance (186), by NMR (87, 177, 183, 187–192); for review see 21), and, as mentioned in the section on B form DNA, by crystallography (59).

Concluding Remarks

A number of detailed models for DNA have been discussed. To the extent to which the models conform to reality, they provide snapshots of likely conformations for biological contexts. However, much is missing from the pictures, particularly the selection and distortion enforced by interaction with proteins or other molecules. Only recently have detailed structures for the classes of protein of interest begun to become available from single-crystal diffraction results. The structures of two regulatory proteins that interact with DNA have been solved at high resolution (193). Neither crystal structure includes oligodeoxynucleotides or other representatives of DNA, so that the interactions of the proteins with DNA had to be inferred from modeling studies. In both cases, it was proposed that exposed α-helices protruding from the protein surface interact with the major grooves of a basically B form model for DNA. However, there is a dramatic difference between the two models: the catabolite gene activator protein of *E. coli* (194) is proposed to interact with *left*-handed B form DNA, while a *right*-handed B form DNA gives a neat fit for the cro repressor of bacteriophage λ (195). These models of the DNA-protein complexes must be regarded as hypothetical, and may not prove to be representative. Nevertheless, these studies suggest the beginning of an exciting period wherein the functional importance of a variety of DNA structures can be judged not only by detailed studies of uncomplexed DNA but also by detailed studies of complexes with biologically important proteins or other molecules.

ACKNOWLEDGMENTS

The many editorial and logistic contributions of Barbara H. Pheiffer have been invaluable and are acknowledged with appreciation. I also thank the authors who sent preprints of unpublished work. Critical comments from my colleagues at the National Institutes of Health have been most valuable.

Literature Cited

1. Jovin, T. M. 1976. *Ann. Rev. Biochem.* 45:889–920
2. Sundaralingam, M. 1979. In *Symposium on Biomolecular Structure, Conformation, Function and Evolution,* ed. R. Srinivasan, pp. 259–83. New York: Pergamon
3. Wells, R. D., Goodman, T. C., Hillen, W., Horn, G. T., Klein, R. D., Larson, J. E., Müller, U. R., Neuendorf, S. K., Panayotatos, N., Stirdivant, S. M. 1980. *Prog. Nucleic Acid Res. Mol. Biol.* 24:167–267
4. Wells, R. D., Blakesley, R. W., Hardies, S. C., Horn, G. T., Larson, J. E., Selsing, E., Burd, J. F., Chan, H. W., Dodgson, J. B., Jensen, K. F., Nes, I. F., Wartell, R. M. 1977. *Crit. Rev. Biochem.* 4:305–40
5. Arnott, S., Chandrasekaran, R., Bond, P. J., Birdsall, D. L., Leslie, A. G. W., Puigjaner, L. C. 1980. *7th Ann. Katzir Katchalsky Conf. Struct. Aspects Recognition Assem. Biolog. Macromol.* Glenside, Pa: Int. Sci. Serv.
6. Chandrasekaran, R., Arnott, S., Banerjee, A., Campbell-Smith, S., Leslie, A. G. W., Puigjaner, L. 1980. *ACS Symp. Ser.* 141:483–502
7. Sundaralingam, M., Westhof, E. 1979. *Int. J. Quantum Chem. Symp.* 6:115–30
8. Jack, A. 1979. *Int. Rev. Biochem: Chem. Macromolecules II A* 24:211–56
9. Record, M. T. Jr., Mazur, S. J., Melançon, P., Roe, J.-H., Shaner, S. L., Unger, L. 1981. *Ann. Rev. Biochem.* 50:997–1024
10. Champoux, J. J. 1978. *Ann. Rev. Biochem.* 47:449–79
11. von Hippel, P. 1979. *Biol. Regul. Develop.* 1:279–347
12. McGhee, J. D., Felsenfeld, G. 1980. *Ann. Rev. Biochem.* 49:1115–56
13. Cozzarelli, N. R. 1980. *Cell* 22:327–28
14. Cozzarelli, N. R. 1980. *Science* 207:953–60
15. Gellert, M. 1981. *Ann. Rev. Biochem.* 50:879–910
16. Swaminathan, V., Sundaralingam, M. 1979. *Crit. Rev. Biochem.* 6:245–336
17. Lippard, S. J. 1978. *Acc. Chem. Res.* 11:211–17
18. Patel, D. J. 1979. *Acc. Chem. Res.* 12:118–25
19. Texter, J. 1978. *Prog. Biophys. Mol. Biol.* 33:83–97
20. Tinoco, I. Jr., Bustamante, C., Maestre, M. F. 1980. *Ann. Rev. Biophys. Bioeng.* 9:107–41
21. Shindo, H. 1982. In *Magnetic Resonance in Biology,* Vol. 2, ed. J. Cohen, New York: Wiley. In press
22. Kearns, D. R. 1977. *Ann. Rev. Biophys. Bioeng.* 6:477–523
23. Cohen, J. S. 1980. *Trends Biochem. Sci.* 5:58–60
24. Davies, D. B. 1978. *Prog. NMR Spectrosc.* 12:135–225
25. Schweizer, M. P. 1980. In *Magnetic Resonance in Biology,* ed. J. Cohen, 1:259–302. New York: Wiley
26. Bauer, W. R. 1978. *Ann. Rev. Biophys. Bioeng.* 7:287–313
27. Crick, F. H. C. 1976. *Proc. Natl. Acad. Sci. USA* 73:2639–43
28. Wang, J. C. 1980. *Trends Biol. Sci.* 5:219–21
29. Bauer, W. R., Crick, F. H. C., White, J. H. 1980. *Sci. Am.* 243:118–33
30. Wilson, H. R. 1966. *Diffraction of X-rays by Proteins, Nucleic Acids and Viruses.* London: Arnold. 137 pp.
31. Arnott, S., Hukins, D. W. L. 1972. *Biochem. Biophys. Res. Commun.* 47:1504–9
32. Arnott, S., Selsing, E. 1975. *J. Mol. Biol.* 98:265–69
33. Arnott, S., Chandrasekaran, R., Hukins, D. W. L., Smith, P. J. C., Watts, L. 1974. *J. Mol. Biol.* 88:523–33
34. Wang, A. H.-J., Quigley, G. J., Kolpak, F. J., van der Marel, G., van Boom, J. H., Rich, A. 1981. *Science* 211:171–76
35. Leslie, A. G. W., Arnott, S., Chandrasekaran, R., Ratliff, R. L. 1980. *J. Mol. Biol.* 143:49–72
36. Fuller, W., Wilkins, M. H. F., Wilson, H. R., Hamilton, L. D., Arnott, S. 1965. *J. Mol. Biol.* 12:60–80
37. Langridge, R., Wilson, H. R., Hooper,

C. W., Wilkins, M. H. F., Hamilton, L. D. 1960. *J. Mol. Biol.* 2:19–37
38. Klug, A., Jack, A., Viswamitra, M. A., Kennard, O., Shakked, Z., Steitz, T. A. 1979. *J. Mol. Biol.* 131:669–80
39. Zimmerman, S. B., Pheiffer, B. H. 1980. *J. Mol. Biol.* 142:315–30
40. Marvin, D. A., Spencer, M., Wilkins, M. H. F., Hamilton, L. D. 1961. *J. Mol. Biol.* 3:547–65
41. Davies, D. R., Baldwin, R. L. 1963. *J. Mol. Biol.* 6:251–55
42. Mokul'skii, M. A., Kapitonova, K. A., Mokul'skaya, T. D. 1972. *Mol. Biol.* 6:883–901
43. Mokul'skaya, T. D., Smetanina, E. P., Myshko, G. E., Mokul'skii, M. A. 1975. *Mol. Biol.* 9:552–55
44. Wang, A. H.-J., Quigley, G. J., Kolpak, F. J., Crawford, J. L., van Boom, J. H., van der Marel, G., Rich, A. 1979. *Nature* 282:680–86
45. Drew, H., Takano, T., Tanaka, S., Itakura, K., Dickerson, R. E. 1980. *Nature* 286:567–73
46. Arnott, S., Chandrasekaran, R., Birdsall, D. L., Leslie, A. G. W., Ratliff, R. L. 1980. *Nature* 283:743–45
47. Saenger, W. 1973. *Angew. Chem.* 12:591–601
48. Arnott, S. 1970. *Prog. Biophys. Molec. Biol.* 21:265–319
49. Crick, F. H. C., Watson, J. D. 1954. *Proc. R. Soc.* 223A:80–96
50. Langridge, R., Marvin, D. A., Seeds, W. E., Wilson, H. R., Hooper, C. W., Wilkins, M. H. F., Hamilton, L. D. 1960. *J. Mol. Biol.* 2:38–64
51. Hamilton, L. D., Barclay, R. K., Wilkins, M. H. F., Brown, G. L., Wilson, H. R., Marvin, D. A., Ephrussi-Taylor, H., Simmons, N. S. 1959. *J. Biophys. Biochem. Cytol.* 5:397–403
52. Prémilat, S., Albiser, G. 1975. *J. Mol. Biol.* 99:27–36
53. Selsing, E., Arnott, S. 1976. *Nucleic Acids Res.* 3:2443–50
54. Arnott, S., Selsing, E. 1974. *J. Mol. Biol.* 88:509–21
55. Pilet, J., Blicharski, J., Brahms, J. 1975. *Biochemistry* 14:1869–76
56. Langridge, R. 1969. *J. Cell Physiol.* 74: Suppl. 1, pp. 1–20
57. Baralle, F. E., Shoulders, C. C., Goodbourn, S., Jeffreys, A., Proudfoot, N. J. 1980. *Nucleic Acids Res.* 8:4393–404
58. Wing, R., Drew, H., Takano, T., Broka, C., Tanaka, S., Itakura, K., Dickerson, R. E. 1980. *Nature* 287:755–58
59. Drew, H. R., Wing, R. M., Takano, T., Broka, C., Tanaka, S., Itakura, K.,

Dickerson, R. E. 1981. *Proc. Natl. Acad. Sci. USA* 78:2179–83
59a. Dickerson, R. E., Drew, H. R. 1981. *Proc. Natl. Acad. Sci. USA* 78:7318–22
60. Dickerson, R. E., Drew, H. R. 1981. *J. Mol. Biol.* 149:761–86
61. Drew, H. R., Dickerson, R. E. 1981. *J. Mol. Biol.* 151:535–56
62. Alden, C. J., Kim, S.-H. 1979. *J. Mol. Biol.* 132:411–34
63. Thiyagarajan, P., Ponnuswamy, P. K. 1979. *Biopolymers* 18:2233–47
64. Neidle, S., Berman, H. M., Shieh, H. S. 1980. *Nature* 288:129–33
65. Lomonossoff, G. P., Butler, P. J. G., Klug, A. 1981. *J. Mol. Biol.* 149:745–60
66. Quigley, G. J., Wang, A. H.-J., Ughetto, G., van der Marel, G., van Boom, J. H., Rich, A. 1980. *Proc. Natl. Acad. Sci. USA* 77:7204–8
67. Viswamitra, M. A., Kennard, O., Jones, P. G., Sheldrick, G. M., Salisbury, S., Falvello, L. 1978. *Nature* 273:687–88
67a. Viswamitra, M. A., Shakked, Z., Jones, P. G., Sheldrick, G. M., Salisbury, S. A., Kennard, O. 1981. *Biopolymers.* In press
68. Cohen, J. S., Wooten, J. B., Chatterjee, C. L. 1981. *Biochemistry* 20:3049–55
69. Gupta, G., Bansal, M., Sasisekharan, V. 1980. *Int. J. Biol. Macromol.* 2:368–80
70. Shindo, H., Simpson, R. T., Cohen, J. S. 1979. *J. Biol. Chem.* 254:8125–28
71. Shindo, H., Zimmerman, S. B. 1980. *Nature* 283:690–91
72. Marky, L. A., Patel, D., Breslauer, K. J. 1981. *Biochemistry* 20:1427–31
73. Patel, D. J., Canuel, L. L., Pohl, F. M. 1979. *Proc. Natl. Acad. Sci. USA* 76:2508–11
74. Sasisekharan, V., Gupta, G., Bansal, M. 1981. *Int. J. Biol. Macromol.* 3:2–8
75. Yathindra, N., Jayaraman, S. 1980. *Current Sci.* 49:167–71
76. Levitt, M. 1978. *Proc. Natl. Acad. Sci. USA* 75:640–44
77. Sussman, J. L., Trifonov, E. N. 1978. *Proc. Natl. Acad. Sci. USA* 75:103–7
78. Camerini-Otero, R. D., Felsenfeld, G. 1978. *Proc. Natl. Acad. Sci. USA* 75:1708–12
79. Olson, W. K. 1979. *Biopolymers* 18: 1213–33
80. Sobell, H. M., Tsai, C.-C., Gilbert, S. G., Jain, S. C., Sakore, T. D. 1976. *Proc. Natl. Acad. Sci. USA* 73:3068–72
81. Sobell, H. M., Tsai, C.-C., Jain, S. C., Sakore, T. D. 1978. *Philos. Trans. R. Soc. London Ser. B* 283:295–98
82. Crick, F. H. C., Klug, A. 1975. *Nature* 255:530–33

83. Yathindra, N. 1979. *Current Sci.* 48:753–56
84. Sobell, H. M., Lozansky, E. D., Lessen, M. 1978. *Cold Spring Harbor Symp. Quant. Biol.* 43:11–19
85. Lozansky, E. D., Sobell, H. M., Lessen, M. 1979. In *Stereodynamics of Molecular Systems*, ed. R. Sarma, pp. 265–82. New York: Pergamon
86. Englander, S. W., Kallenbach, N. R., Heeger, A. J., Krumhansl, J. A., Litwin, S. 1980. *Proc. Natl. Acad. Sci. USA* 77:7222–26
87. Klevan, L., Armitage, I. M., Crothers, D. M. 1979. *Nucleic Acids Res.* 6:1607–16
88. Simpson, R. T., Shindo, H. 1979. *Nucleic Acids Res.* 7:481–92
89. Shindo, H., McGhee, J. D., Cohen, J. S. 1980. *Biopolymers* 19:523–37
90. Kallenbach, N. R., Appleby, D. W., Bradley, C. H. 1978. *Nature* 272:134–38
91. Sasisekharan, V., Brahmachari, S. K. 1981. *Curr. Sci.* 50:10–13
92. Gupta, G., Bansal, M., Sasisekharan, V. 1980. *Proc. Natl. Acad. Sci. USA* 77:6486–90
93. Sasisekharan, V., Pattabiraman, N. 1978. *Nature* 275:159–62
94. Yathindra, N., Sundaralingam, M. 1976. *Nucleic Acids Res.* 3:729–47
95. Gupta, G., Sasisekharan, V. 1978. *Nucleic Acids Res.* 5:1655–73
96. Rich, A., RajBhandary, U. L. 1976. *Ann. Rev. Biochem.* 45:805–60
97. Ikehara, M., Uesugi, S., Yano, J. 1972. *Nature New Biol.* 240:16–17
98. Sasisekharan, V., Pattabiraman, N., Gupta, G. 1978. *Proc. Natl. Acad. Sci. USA* 75:4092–96
99. Rodley, G. A., Scobie, R. S., Bates, R. H. T., Lewitt, R. M. 1976. *Proc. Natl. Acad. Sci. USA* 73:2959–63
100. Sasisekharan, V. 1981. *Curr. Sci.* 50:107–11
101. Millane, R. P., Rodley, G. A. 1981. *Nucleic Acids Res.* 9:1765–73
102. Albiser, G., Prémilat, S. 1980. *Biochem. Biophys. Res. Commun.* 95:1231–37
103. Bates, R. H. T., McKinnon, G. C., Millane, R. P., Rodley, G. A. 1980. *Pramaña* 14:233–52
104. Greenall, R. J., Pigram, W. J., Fuller, W. 1979. *Nature* 282:880–82
105. Arnott, S. 1979. *Nature* 278:780–81
106. Dover, S. D. 1977. *J. Mol. Biol.* 110:699–700
107. Pohl, F. M. 1967. *Naturwissenschaften* 54:616
108. Pohl, W. F., Roberts, G. W. 1978. *J. Math. Biol.* 6:383–402
109. Crick, F. H. C., Wang, J. C., Bauer, W. R. 1979. *J. Mol. Biol.* 129:449–61
110. DasGupta, C., Takehiko, S., Cunningham, R. P., Radding, C. M. 1980. *Cell* 22:437–46
111. Stettler, U. H., Weber, H., Koller, T., Weissmann, C. 1979. *J. Mol. Biol.* 131:21–40
112. Pohl, F. M. 1976. *Nature* 260:365–66
113. Pohl, F. M., Jovin, T. M. 1972. *J. Mol. Biol.* 67:375–96
114. Crawford, J. L., Kolpak, F. J., Wang, A. H.-J., Quigley, G. J., van Boom, J. H., van der Marel, G., Rich, A. 1980. *Proc. Natl. Acad. Sci. USA* 77:4016–20
115. Haschemeyer, A. E. V., Rich, A. 1967. *J. Mol. Biol.* 27:369–84
116. Govil, G., Fisk, C. L., Howard, F. B., Miles, H. T. 1981. *Biopolymers* 20:573–603
117. Drew, H. R., Dickerson, R. E. 1981. *J. Mol. Biol.* 152:723–36
118. Behe, M., Zimmerman, S., Felsenfeld, G. 1981. *Nature* 293:233–35
119. Saenger, W., Landmann, H., Lezius, A. G. 1973. *Jerusalem Symp. Quantum Chem. Biochem.* 5:457–66
120. Gupta, G., Bansal, M., Sasisekharan, V. 1980. *Biochem. Biophys. Res. Commun.* 95:728–33
121. Behe, M., Felsenfeld, G. 1981. *Proc. Natl. Acad. Sci. USA* 78:1619–23
122. Ramstein, J., Leng, M. 1980. *Nature* 288:413–14
123. Pohl, F. M., Ranade, A., Stockburger, M. 1973. *Biochim. Biophys. Acta* 335:85–92
123a. Thamann, T. J., Lord, R. C., Wang, A. H.-J., Rich, A. 1981. *Nucleic Acids Res.* 9:5443–57
123b. Wu, H. M., Dattagupta, N., Crothers, D. M. 1981. *Proc. Natl. Acad. Sci. USA* 78:6808–11
124. Bayley, P. M. 1973. *Prog. Biophys.* 27:1–76
125. Simpson, R. T., Shindo, H. 1980. *Nucleic Acids Res.* 8:2093–2103
126. Mitra, C. K., Sarma, M. H., Sarma, R. H. 1981. *Biochemistry* 20:2036–41
127. Vasmel, H., Greve, J. 1981. *Biopolymers* 20:1329–32
128. Drew, H. R., Dickerson, R. E., Itakura, K. 1978. *J. Mol. Biol.* 125:535–543
129. Mercado, C. M., Tomasz, M. 1977. *Biochemistry* 16:2040–46
130. Sage, E., Leng, M. 1980. *Proc. Natl. Acad. Sci. USA* 77:4597–601
131. Santella, R. M., Grunberger, D., Weinstein, I. B., Rich, A. 1981. *Proc. Natl. Acad. Sci. USA* 78:1451–55
132. Sage, E., Leng, M. 1981. *Nucleic Acids Res.* 9:1241–50

132a. Möller, A., Nordheim, A., Nichols, S. R., Rich, A. 1981. *Proc. Natl. Acad. Sci. USA* 78:4777–81

133. Pohl, F. M., Jovin, T. M., Baehr, W., Holbrook, J. J. 1972. *Proc. Natl. Acad. Sci. USA* 69:3805–9

134. Kuhnlein, U., Tsang, S. S., Edwards, J. 1980. *Nature* 287:363–64

135. Gellert, M., Mizuuchi, K. 1980. As cited in Davies, D. R., Zimmerman, S. B. *Nature* 283:11–12

136. Benham, C. J. 1980. *Nature* 286:637–38

137. Klysik, J., Stirdivant, S. M., Larson, J. E., Hart, P. A., Wells, R. D. 1981. *Nature* 290:672–77

138. Nishioka, Y., Leder, P. 1980. *J. Biol. Chem.* 255:3691–94

139. Zakrzewska, K., Lavery, R., Pullman, A., Pullman, B. 1980. *Nucleic Acids Res.* 8:3917–32

140. Jayaraman, S., Yathindra, N. 1980. *Biochem. Biophys. Res. Commun.* 97:1407–19

140a. Lafer, E. M., Möller, A., Nordheim, A., Stollar, B. D., Rich, A. 1981. *Proc. Natl. Acad. Sci. USA* 78:3546–50

140b. Nordheim, A., Pardue, M. L., Lafer, E. M., Möller, A., Stollar, B. D., Rich, A. 1981. *Nature* 294:417–22

141. Gupta, G., Bansal, M., Sasisekharan, V. 1980. *Biochem. Biophys. Res. Commun.* 97:1258–67

142. Skuratovskii, I. Ya., Bartenev, V. N. 1978. *Mol. Biol.* 12:1359–76

143. Shindo, H., Wooten, J. B., Zimmerman, S. B. 1981. *Biochemistry* 20:745–50

144. Potaman, V. N., Bannikov, Yu. A., Shlyachtenko, L. S. 1980. *Nucleic Acids Res.* 8:635–42

145. Zavriev, S. K., Minchenkova, L. E., Frank-Kamenetskii, M. D., Ivanov, V. I. 1978. *Nucleic Acids Res.* 5:2657–63

145a. Shakked, Z., Rabinovich, D., Cruse, W. B. T., Egert, E., Kennard, O., Sala, G., Salisbury, S. A., Viswamitra, M. A. 1981. *Proc. R. Soc. Ser. B.* 213:479–87

145b. Conner, B. N., Takano, T., Tanaka, S., Itakura, K., Dickerson, R. E. 1981. *Nature.* 295:294–99

146. Selsing, E., Wells, R. D., Early, T. A., Kearns, D. R. 1978. *Nature* 275:249–50

147. Selsing, E., Wells, R. D., Alden, C. J., Arnott, S. 1979. *J. Biol. Chem.* 254:5417–22

148. Tunis-Schneider, M. J. B., Maestre, M. F. 1970. *J. Mol. Biol.* 52:521–41

149. Zimmerman, S. B., Pheiffer, B. H. 1979. *J. Mol. Biol.* 135:1023–27

150. Gray, D. M., Edmondson, S. P., Lang, D., Vaughan, M. 1979. *Nucleic Acids Res.* 6:2089–2107

151. Wang, J. C. 1969. *J. Mol. Biol.* 43:25–39

152. Maniatis, T., Venable, J. H. Jr., Lerman, L. S. 1974. *J. Mol. Biol.* 84:37–64

153. Goodwin, D. C., Brahms, J. 1978. *Nucleic Acids Res.* 5:835–50

154. Gray, D. M., Taylor, T. N., Lang, D. 1978. *Biopolymers* 17:145–57

155. Baase, W. A., Johnson, W. C. Jr. 1979. *Nucleic Acids Res.* 6:797–814

156. Sprecher, C. A., Baase, W. A., Johnson, W. C. Jr. 1979. *Biopolymers* 18:1009–19

157. Lee, C.-H., Mizusawa, H., Kakefuda, T. 1981. *Proc. Natl. Acad. Sci. USA* 78:2838–42

158. Mitsui, Y., Langridge, R., Shortle, B. E., Cantor, C. R., Grant, R. C., Kodama, M., Wells, R. D. 1970. *Nature* 228:1166–69

159. Grant, R. C., Kodama, M., Wells, R. D. 1972. *Biochemistry* 11:805–15

160. Arnott, S., Bond, P. J., Chandrasekaran, R. 1980. *Nature* 287:561–63

161. Cyriax, B., Gäth, R. 1978. *Naturwissenschaften* 65:106–8

162. Wilkins, M. H. F., Gosling, R. G., Seeds, W. E. 1951. *Nature* 167:759–60

163. Olson, W. K. 1977. *Proc. Natl. Acad. Sci. USA* 74:1775–79

164. Hopkins, R. C. 1981. *Science* 211:289–91

165. Milman, G., Langridge, R., Chamberlin, M. J. 1967. *Proc. Natl. Acad. Sci. USA* 57:1804–10

166. O'Brien, E. J., MacEwan, A. W. 1970. *J. Mol. Biol.* 48:243–61

167. Higuchi, S., Tsuboi, M., Iitaka, Y. 1969. *Biopolymers* 7:909–16

168. Gray, D. M., Ratliff, R. L. 1975. *Biopolymers* 14:487–98

169. Zimmerman, S. B., Pheiffer, B. H. 1981. *Proc. Natl. Acad. Sci. USA* 78:78–82

170. Zimmerman, S. B., Pheiffer, B. H. 1979. *Proc. Natl. Acad. Sci. USA* 76:2703–7

171. Peck, L. J., Wang, J. C. 1981. *Nature* 292:375–78

171a. Strauss, F., Gaillard, C., Prunell, A. 1981. *Eur. J. Biochem.* 118:215–22

172. Rhodes, D., Klug, A. 1980. *Nature* 286:573–78

172a. Mandelkern, M., Dattagupta, N., Crothers, D. M. 1981. *Proc. Natl. Acad. Sci. USA* 78:4294–98

173. Rhodes, D., Klug, A. 1981. *Nature* 292:378–80

174. Arnott, S., Selsing, E. 1974. *J. Mol. Biol.* 88:551–52

175. Liu, L. F., Wang, J. C. 1978. *Cell* 15:979–84

176. Patel, D. J., Canuel, L. L. 1979. *Eur. J. Biochem.* 96:267–76

177. Shindo, H., Wooten, J. B., Pheiffer, B.
H., Zimmerman, S. B. 1980. *Biochemistry* 19:518–26
178. Nall, B. T., Rothwell, W. P., Waugh, J.
S., Rupprecht, A. 1981. *Biochemistry* 20:1881–87
179. Nussinov, R. 1980. *Nucleic Acids Res.* 8:4545–62
180. Zhurkin, V. B. 1981. *Nucleic Acids Res.* 9:1963–71
181. Trifonov, E. N. 1980. *Nucleic Acids Res.* 8:4041–53
182. Grantham, R. 1980. *FEBS Lett.* 121:193–99
183. DiVerdi, J. A., Opella, S. J. 1981. *J. Mol. Biol.* 149:307–11
184. Wahl, Ph., Paoletti, J., LePecq, J.-B. 1970. *Proc. Natl. Acad. Sci. USA* 65: 417–21
185. Barkley, M. D., Zimm, B. H. 1979. *J. Chem. Phys.* 70:2991–3007
186. Robinson, B. H., Lerman, L. S., Beth,
A. H., Frisch, H. L., Dalton, L. R., Auer, C. 1980. *J. Mol. Biol.* 139:19–44
187. Hogan, M. E., Jardetzky, O. 1980. *Biochemistry* 19:2079–85
188. Hogan, M. E., Jardetzky, O. 1980. *Biochemistry* 19:3460–68
189. Bolton, P. H., James, T. L. 1980. *J. Am. Chem. Soc.* 102:25–31
190. Rill, R. L., Hilliard, P. R., Jr., Bailey, J. T., Levy, G. C. 1980. *J. Am. Chem. Soc.* 102:418–20
191. Hogan, M. E., Jardetzky, O. 1979. *Proc. Natl. Acad. Sci. USA* 76:6341–45
192. Early, T. A., Kearns, D. R. 1979. *Proc. Natl. Acad. Sci. USA* 76:4165–69
193. Davies, D. R. 1981. *Nature* 290:736–37
194. McKay, D. B., Steitz, T. A. 1981. *Nature* 290:744–49
195. Anderson, W. F., Ohlendorf, D. H., Takeda, Y., Matthews, B. W. 1981. *Nature* 290:754–58

Ann. Rev. Biochem. 1982. 52.429–57
Copyright © 1982 by Annual Reviews Inc. All rights reserved

FIDELITY OF DNA SYNTHESIS[1]

Lawrence A. Loeb and Thomas A. Kunkel

Joseph Gottstein Memorial Cancer Research Laboratory, Department of
Pathology, University of Washington, School of Medicine, Seattle, Washington
98195

CONTENTS

Perspectives and Summary

A determination of cellular mechanisms for achieving accuracy in the
copying of DNA, i.e. the fidelity of DNA synthesis, is pivotal to our
understanding of fundamental biological processes. Organisms must replicate and repair their DNA with high accuracy so as to maintain their
genetic identity, yet a few errors must be permitted for species evolution.
An understanding of the fidelity of DNA synthesis is necessary for elucidating the molecular basis of mutagenesis. Mutagenesis is not only an important process within itself, but it appears to be central to pathological
processes including aging, cancer, arteriosclerosis, and other diseases.

[1]This work was supported by Grants CA 24845 and CA 24498 from the National Cancer
Institute and Grant PCM 76-80439 from the National Science Foundation.

429

Measurements of spontaneous mutation frequencies suggest that the average frequency of base pair substitutions is in the range of 10^{-7} to 10^{-11} misincorporations per base pair replicated (1–3). Three disciplines have focused on how such high accuracy can be achieved. Theoreticians have proposed multistep mathematical models (4–8), which focus on the cost in energy to achieve accuracy. Geneticists have identified multiple loci in prokaryotes, which affect the overall accuracy of DNA replication (9–16). The proteins coded by these genes are being rapidly identified (17–20). Biochemists have measured the frequency of base substitution errors during copying of synthetic polynucleotides by purified DNA polymerases (21–23). More recently, these studies are being extended to biologically active DNA (24–30) and to measuring accuracy with multiprotein systems (31–36). The overall conclusions from these divergent approaches indicate that highly accurate DNA replication is a multicomponent process involving sequential steps (37). In this review, we discuss each of these steps in light of recent experimental studies that measure base substitituon error frequencies during in vitro DNA synthesis with defined components. Emphasis is on purified DNA polymerases and other proteins that may also function in fidelity. In addition we briefly summarize studies on the effects of perturbations in reaction conditions, which result in decreased fidelity and mutagenesis.

In our view, the final accuracy of DNA synthesis results from three discrimination steps that reduce errors. The first occurs as a base is actually being inserted at the growing point of the DNA chain. Here the replication machinery can reduce errors by discriminating against insertion of an incorrect nucleotide. This first discrimination step is thought to result from a difference in free energy (ΔG) between correct and incorrect base pairings, a difference that is in part a property of the base pairs themselves, but that may be amplified by DNA polymerase and other proteins (error prevention or base selection). Once an incorrect nucleotide is inserted it may be excised. At this second step, correction during replication occurs at the growing point, removing mistakes at the time of or immediately after incorporation (proofreading). We review recent experiments with prokaryotic DNA polymerases in vitro that demonstrate the existence and importance of proofreading to fidelity. A role for proofreading in eukaryotic cells remains to be rigorously determined. Both steps in discrimination during DNA synthesis exhibit specificity, presumably resulting from the replication proteins, the nature of the mismatch, and the surrounding nucleotide sequence. Lastly, there is evidence for a third opportunity to reduce errors, a postsynthetic correction of mismatches, which is discussed only briefly.

Methodology

As a result of the high accuracy of DNA synthesis, assays to quantitate errors need to be exceptionally sensitive. Methods to measure the frequency of misincorporation by DNA polymerases or DNA replicating complexes can be grouped into two categories based on the nature of the template.

ASSAYS WITH SYNTHETIC POLYNUCLEOTIDES   Until recently most studies on the fidelity of DNA synthesis utilized synthetic polynucleotides as templates. These templates are frequently of defined size and contain only one or two bases. An error is defined as incorporation of any base that is noncomplementary to the base(s) in the template, the error rate being the ratio of noncomplementary to total nucleotide incorporation (23). Initial experiments involved parallel reactions containing labeled complementary and noncomplementary nucleotides (21). Increased precision can be achieved with a double label assay (23), using for example an $[\alpha\text{-}^{32}P]$-labeled correct deoxynucleoside triphosphate substrate of low specific activity and an $[^3H]$-labeled incorrect deoxynucleoside triphosphate substrate of high specific activity, both in the same synthetic reaction. The error rate is then inferred from the ratio of $[^3H]$ to $[^{32}P]$.

The major limiting technical factor is sensitivity, which is limited by cost, purity, and specific activity of the deoxynucleotide substrates as well as the purity of the template. Most homopolymer templates are a product of the random polymerization of deoxynucleoside triphosphates with terminal transferase (38, 39); as a result, the purity of the templates is proportional to the initial purity of the substrates. However, greater template purity can be obtained by synthesizing multiple copies of pure template with a faithful polymerase (40). With respect to highly accurate synthetic reactions, even a minute amount of contaminating DNA in the enzyme preparation would invalidate the assay. The nucleotide noncomplementary to the synthetic template is complementary to the DNA contaminant. Methods devised to overcome this difficulty involve selective suppression of the template activity of the contaminating DNA and its removal by hydrolysis with restriction enzymes that fail to cleave the added synthetic template or the product of the reaction (41). These questions of purity have necessitated large numbers of controls to validate that the noncomplementary nucleotide is indeed incorporated into phosphodiester linkage and is present as a single base substitution (21, 41, 42).

As a consequence of their repeating nature, synthetic polynucleotide templates may allow more errors than natural DNA. With synthetic polynucleotides, there is slippage of the primer along the template during synthesis (43) and mismatched bases can more easily loop out of the DNA helix (44, 45). Both secondary structure and stacking interactions between neighboring nucleotides (46–51) could potentially influence base substitution frequencies, and these parameters are clearly different between natural DNA and synthetic templates. The excision of incorrect nucleotides during polymerization may be more active on natural DNA (29, 30) than on synthetic templates (52). Finally, protein subunits of DNA polymerases, which could conceivably affect accuracy, may not be as active with synthetic templates as with natural DNA. For example. *E. coli* DNA polymerase III core (three subunits) and holoenzyme (seven subunits) exhibit very similar properties with synthetic templates, yet are distinctly different in copying natural DNA (53, 54). Despite these limitations, assays using synthetic polynucleotides have been very informative. Their defined composition allows clear distinction between correct and incorrect substrates. Measurements can be made of both stable misincorporation and nucleoside monophosphate generation (turnover) with either normal deoxynucleotide substrates (55–62) or with base analogues (6, 49, 63–66). Since large numbers of assays can be performed rapidly, synthetic polynucleotides are advantageous in screening for agents that perturb fidelity (67, 68).

ASSAYS WITH NATURAL DNA TEMPLATES   Since natural DNA contains all four bases, errors cannot be defined by simple direct measurements of misincorporation. Rather, a genetic approach has been used, in which a biologically active DNA template, containing a defined single base substitution mutation, is copied in vitro and then analyzed in vivo (24). Errors are detected by determining the frequency of reversion of the mutation to wild type. To date, this approach has been successfully applied only to amber mutations in ϕX174 DNA (24–36). The phage DNA is copied in vitro and transfected into *E. coli* spheroplasts. Subsequent plating on permissive and nonpermissive indicator bacteria yields a reversion frequency that can be converted into an error rate for in vitro synthesis. The sensitivity of the assay is limited by the spontaneous reversion frequency of the mutant DNAs (10^{-4}–10^{-7} for different amber mutations) and can be two to three orders of magnitude more sensitive than assays utilizing synthetic polynucleotides.

The three methods developed thus far differ primarily in the components used for in vitro DNA synthesis (Figure 1). The simplest method (Figure 1A) developed in our laboratory (24), used ϕX174 single-stranded viral (+) DNA primed with a single restriction endonuclease fragment. DNA synthesis can be carried out under a variety of conditions using DNA polymerase from any source (24, 25, 27, 28, 69), alone or in conjunction with added replicating proteins (31). The product of the reaction contains a partial minus strand whose phenotype is expressed in spheroplasts with measurable efficiency. The critical requirement is that the polymerase copies past the amber mutation in vitro. Because of the possibility of unspecified heteroduplex repair in spheroplasts, a number of controls are required to quantitate the frequency of expression of the minus strand (24, 27). In principle, this method can be used with crude extracts, including those from eukaryotic cells (J. Silber, unpublished results).

The second method (Figure 1B) developed by Hibner, Alberts, and co-workers (32, 34, 35), utilizes randomly nicked double-stranded ϕX174 RFII DNA as a template. Synthesis is catalyzed by the seven-protein T4 replication complex and produces—by a rolling circle mechanism—25–30 product molecules per input RFII molecule. These linear double-stranded molecules are cut with restriction endonuclease *Pst* I and then ligated together to form infective double-stranded circles, whose reversion frequency is determined by transfection. Only those molecules containing mistakes made during lagging strand synthesis are subject to heteroduplex repair. The requirement for the entire T4-replicating complex makes this assay closer to the realities of in vivo replication. However, this requirement

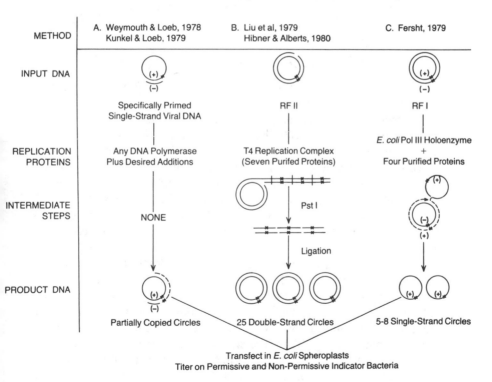

Figure 1. Fidelity assays with ϕX174 DNA.

prevents one from easily assigning functions to each of the components, since one cannot delete individual factors. Also, rigorous interpretations on neighboring nucleotide effects, proofreading contributions, specificity, and frequency of mismatches are more difficult when copying two complementary strands rather than only one.

The third method (Figure 1C), developed by Fersht (33, 36), utilizes RFI ϕX174 DNA as a template. Synthesis is catalyzed by a purified *E. coli* synthetic apparatus, consisting of the holoenzyme complex [containing at least seven established proteins (53)] and four other added proteins. This method produces, per input RFI molecule, five to eight covalently closed circles of plus orientation exclusively from the minus strand. As in the first method (Figure 1A), synthesis is directed by only one strand. This allows a more direct analysis of error rates extrapolated from reactions carried out with varying concentrations of differenct nucleotide substrates (pool bias). Since the product is a single-stranded molecule, there is no difficulty in interpretation due to heteroduplex repair. However, this method is limited to the normal ϕX174 RF to SS DNA replication proteins, and one cannot substitute DNA polymerases and replication proteins from other sources.

Nucleotide-Polynucleotide Interactions

Before considering the role of DNA polymerases in fidelity, it is important to summarize some experiments on the specificity of nucleotide-polynucleotide interactions. If replication proteins do not enhance accuracy, then the difference in free energy (ΔG) between complementary and noncomplementary base pairs would be the determinant of discrimination at the level of insertion. In this hypothetical model, the replication machinery would act only as a zipper to polymerize substrates prealigned with the template. The difference in free energy between correct and incorrect Watson-Crick base pairing at equilibrium is not precisely known. Based on measurements of nucleotide interactions (70, 71), and the stability of polynucleotide helices containing varying numbers of noncomplementary base pairs (44, 46, 72), ΔG is usually estimated to be one to three kcal/mol. The frequency of misincorporation is given by the equation: $-\Delta G = RT\ln$ Incorrect/Correct. One error in 10^n base pairs requires a free energy difference of $1.4(n)$ kcal/mol between a complementary and noncomplementary base pair (73). Thus a 1 to 3 kcal/mol ΔG would predict an error frequency of 1 mispaired nucleotide out of every 10 to 100 nucleotides incorporated. The recent observation that the nonenzymatic formation of synthetic oligonucleotides occurs with an error rate of 10^{-1} to 10^{-2} (74, 75) supports these conclusions. Reactions catalyzed by DNA polymerase are much more accurate.

Enchanced Discrimination by Polymerase Without Excision

Initial evidence for the participation of the DNA polymerases themselves in reducing misincorporation came from the studies of bacteriophage T4 (9–12). Mutants of T4 gene 43 (DNA polymerase) exhibit significant increases or decreases in frequency of spontaneous and base analogue–induced mutations when compared to wild-type T4. In vitro studies with certain of these mutant polymerases have demonstrated differential discrimination at the level of insertion (57–59, 62). A compilation of error rates observed with purified DNA polymerases using synthetic polynucleotides (Table 1) and ϕX174 DNA (Table 2) indicate that these enzymes synthesize DNA with a high degree of discrimination. In this tabulation, Mg^{2+} was the metal activator. In certain reports, error rates were corrected in accord with differences in concentrations between complementary and noncomplementary nucleotides. In general, prokaryotic DNA polymerases are more accurate than DNA polymerases from mammalian sources or RNA tumor viruses. This difference is most pronounced with the ϕX174 assay, and results in part from proofreading by prokaryotic DNA polymerases (see below). In nearly all reports, fidelity is greater than that indicated by a one to three kcal/mol free energy difference between correct and incorrect base pairs measured in the absence of polymerase. This enzyme mediated enhancement is observed with polymerases that lack a proofreading exonuclease activity; i.e. those from RNA tumor viruses or eukaryotic sources, as well as with those polymerases that have their potential proofreading activity reduced or eliminated (e.g. Pol I and T4; see below). It should also be noted that the frequency of misinsertions varies with different DNA polymerases. Under identical reaction conditions, misinsertion at position 587 in ϕX174 DNA is tenfold more frequent with Pol-β than with Pol-α (Table 2). It should also be noted that the misinsertion frequency with DNA polymerase-α is base specific. An A or a G is misinserted (in place of T opposite a template A) much more frequently than is C at position 587 (Figure 2A). While the reasons for these specificities are not yet clear, they are not easily explained on the basis of the frequency of different tautomers (see below).

The actual mechanism by which the polymerase mediates enhanced discrimination at the level of insertion is not known. The simplest model for discrimination without proofreading involves an enzyme mediated increase in ΔG; that is to say, the difference in free energy between correct and incorrect base pairs is greater in the presence of polymerase than that suggested from analyses of base pairings in the absence of polymerase. This concept is implied in the passive polymerase model of Goodman & Branscomb (87). In this model, it is assumed that the "on rates" for binding at

Table 1 Fidelity of purified DNA polymerases with synthetic polynucleotide template[a]

Enzyme source	Template	Incorrect substrate	Error rate	Ref
Bacteria and bacteriophage				
E. coli Pol I	poly (dA–dT)	G	1/80,000	40
	poly (dA–dT)	C	1/8,000	40
	poly (dA) · oligo (dT)	C	1/3,200[b]	76
E. coli Pol II	poly (dA) · oligo (dT)	C	1/13,300[b]	76
E. coli Pol III (core)	poly (dA) · oligo (dT)	C	1/16,000[b]	76
E. coli Pol III*	poly (dA) · oligo (dT)	C	1/4,050	77
Bacteriophage T4 (wild-type)	poly (dA–dT)	G	1/42,300	31
	poly (dC) · oligo (dG)	T	1/96,600[b]	21
Bacteriophage T4, mutator L56	poly (dC) · oligo (dG)	T	1/27,700[b]	21
Eukaryotic cells				
DNA polymerase –α				
Regenerating rat liver	poly (dA–dT)	G	1/100,000[c]	78
Normal rat liver	poly (dA–dT)	G	1/16,500	79
Rat hepatoma	poly (dA–dT)	G	1/3,360	80
Human placenta	poly (dA–dT)	G	1/12,000	81
Calf thymus	poly (dA–dT)	G	1/4,500	31
Human lymphocytes	poly (dA–dT)	G	1/5,330	80
HeLa cell	poly (dA–dT)	G	1/4,830	80
DNA polymerase –β				
Calf thymus	poly (dC) · oligo (dG)	T	1/8,000	22
		C	1/8,000	22
		A	1/1,400	22
	poly (dA) · oligo (dT)	C	1/180,000[d]	22
	poly (dA–dT)	G	1/11,800	80
Calf liver	poly (dA–dT)	G	1/15,200	80
Rat hepatoma	poly (dA–dT)	G	1/4,500	31
Rat liver	poly (dA–dT)	G	1/18,900	79
Guinea pig liver	poly (dA–dT)	G	1/44,200	80
Human placenta	poly (dA–dT)	G	1/20,000	81
DNA polymerase –γ				
Human placenta	poly (dI) · poly (dC)	T	1/1,440[b]	82
Fetal calf liver	poly (dA–dT)	G	1/7,250	(T. A. Kunkel, unpublished)
HeLa cell	poly (dA–dT)	G	1/3,120	(T. A. Kunkel, unpublished)
Animal tumor viruses				
Avian myeloblastosis virus	poly (rA) · oligo (dT)	C	1/700	23
	poly (dA–dT)	G	1/1,400	83
	poly (dG) · poly (dC)	A	1/670	23
Avian leukosis virus	poly (dA) · oligo (dT)	C	1/235	84
Rauscher leukemic virus	poly (dA) · oligo (dT)	C	1/525	85
Spleen necrosis virus	poly (dA) · oligo (dT)	G	~1/300[e]	86

[a] The data in this table are from Mg^{2+}–activated reactions using purified DNA polymerases. The table is not intended to be a comprehensive compilation of literature values; "less than" values have been omitted and many other values are available under many different copying conditions. The designation poly (dA–dT) refers to alternating copolymer containing an A and T residue in each strand.

[b] Extrapolated from pool biased experiments.

[c] Higher value shown, assuming equal concentrations of complementary and noncomplementary substrates.

[d] Determined by subtraction of misincorporation due to terminal addition.

[e] Error rate of 1 in 50 corrected six-fold for pool bias.

Table 2 DNA polymerase fidelity with ϕX174 *am*3 DNA[a]

DNA polymerase	Incorrect deoxynucleotide(s)	Error rate	Ref
E. coli Pol I	C	1/680,000 → 1/2,000,000	27, 29, 30
	CαS	1/100,000	30
	A	⩽ 1/6,300,000	27
	G	1/1,080,000	29
E. coli Pol III	C	< 1/12,200,000	69
	A	< 1/12,200,000	69
Bacteriophage T4 (wild-type)	C	< 1/10^7	29, 30
	A	< 1/10^7	29
	CαS	1/20,000	30
Avian myeloblastosis virus (AMV)	C, A, G	1/329 → 1/17,000	25, 29, 30
Calf thymus Pol-α	C, A, G	1/23,800 → 1/30,500	28
Mouse myeloma Pol-α	C, A, G	1/47,500	28
Rat hepatoma Pol-β	C, A, G	1/6,600	28
Mouse myeloma Pol-β	C, A, G	1/2,930 → 1/4,660	28
HeLa cell Pol-γ	C, A, G	1/6,660 → 1/8,070	28

[a] The measurements were performed as described in Figure 1A. Error rates for individual nucleotides were extrapolated from pool bias experiments as described in the references. Each value was determined from only copied molecules, except for DNA polymerase –α from mouse myeloma; the value of the latter has been corrected for the presence of uncopied DNA.

the active site are diffusion limited and the same for correct and incorrect substrates. Discrimination is based on differences in "off rates," which result from difference in hydrogen bonding and stacking energies. One prediction is that the error rate for a nonproofreading polymerase is proportional to differences in the K_m between correct and incorrect nucleotides. With DNA polymerase-α, Watanabe & Goodman (88) observed that the K_m for insertion of C (incorrect) opposite template 2-AP was 20- to 40-fold higher than that of T (correct) opposite template 2-AP. Similarly, there was a sixfold increase in the K_m of 2-AP opposite template T than A opposite template T (65). In both cases, these differences are proportional to their misincorporation rate. The significance of these observations would be considerably strengthened by detailed studies using normal substrates; the predicted difference in K_m would be three to five orders of magnitude (the reciprocal of the misinsertion frequency). Two such studies have been performed on polynucleotides in which proofreading might be minimal (Table 1). Using poly(dA)·poly(dT) and purified T4 DNA polymerase, Gillin & Nossal (89) reported a 10- to 300-fold higher apparent K_m for the total utilization of noncomplementary dGTP and dCTP than for complementary dATP and dTTP. A kinetic study of competition between incorrect and correct normal substrates with Pol I and polynucleotides was reported by Travaglini et al (90). The ratio of the K_m of the incorrect substrate to the K_m of the correct substrate was in most comparisons only 10^2–10^3. If one assumes that proofreading by Pol I with homopolymers is minimal, then neither study supports the passive polymerase model, since error rates with these enzymes are 10^{-4} or lower. More definitive studies are needed with natural DNA and polymerases lacking proofreading ability.

A structural basis for enzyme mediated enhancement in accuracy is suggested by the data of Sloan et al (91), which demonstrates that *E. coli* Pol I in the absence of any template changed the conformation of the bound nucleotide at the active site to one that would fit more precisely into double helical B-DNA. Also, fluorescent polarization studies indicate a decreased mobility of purine rings when the nucleotide is bound to the polymerase (92, 93). Such reorientation and immobilization of the nucleotide by the enzyme would increase the ΔG of discrimination if the template is also immobilized. This structural mechanism would be likely to result in differences in binding between correct and incorrect nucleotides and does not require a change in enzyme conformation in response to each template base.

Other mechanisms proposed to explain enhanced discrimination have implied a conformation change repeated at each nucleotide selection step. These include a template change in conformation, so that the enzyme will accommodate only the correct substrate (9, 11, 94), or the correct base pair (10, 95); base-specific nucleotide binding subsites (59, 96) and tightening of enzyme-template binding in the presence of the correct nucleotide (59). Conformational discrimination at the level of insertion should involve a slower rate of incorporation of an incorrect nucleotide than of a correct one (a lower V). While such a difference was not observed with DNA polymerase-α for 2-AP mispairing (88), this possibility has not been extensively examined. This is a critical test of the passive polymerase model (87). However, it is very difficult to design assays to measure the rate of incorporation of a single incorrect nucleotide.

The chemical structure of the bases may also provide a lower limit for each step in fidelity. In fact, Watson & Crick (97, 98) suggest in their initial papers that transition errors in DNA synthesis arise when incorrect (non-Watson-Crick) nucleotide bases assume rare imino and enol tautomeric forms, which allow inappropriate hydrogen bonds. Recently, Topal & Fresco (99) have invoked rare imino and enol tautomers, protonated bases, and syn isomers to extend possible "correct" pairings to account for transversions. Their model predicts that transversions occur predominantly by purine-purine and not pyrimidine-pyrimidine mispairs. Since some studies have indicated that rare tautomeric forms exist at a level of 10^{-4}–10^{-5} (92, 99–103), it has been postulated that this is the upper limit of discrimination in a one step model. A second discrimination step (e.g. proofreading) with comparable fidelity could then achieve an error rate of 10^{-8}–10^{-10}. Support for this concept is the data on relative rates of transitions and tranversions in vivo (3, 104, 105). The degree and specificity of 5-bromodeoxyuracil has traditionally been invoked in support of the importance of tautomers in mutagenesis (99, 106). However, recent studies have suggested the contrary; mutagenesis by 2-aminopurine results primarily from the hydrogen bonding properties of the analogue in its favored tautomeric form (87). Also, the mutagenic effects of 5-bromodeoxyuridine may be partly explained in the basis of alterations in cellular nucleotide pools (107; reviewed in 108). The fact that tautomerization may not completely account for the frequency and specificity of base analogue-induced mutagenesis does not preclude an important role for tautomers in spontaneous mutagenesis with normal substrates. Conversely, tautomer models do not provide information about the mechanisms of avoiding or correcting errors when the bases are in their favored tautomeric states.

Most of the studies performed to date on base selection have been conducted with purified DNA polymerases. However, these enzymes clearly function in vivo in concert with other proteins, which often affect binding constants, polymerization rate, processivity, and stacking interactions. Since each of these parameters is expected to affect accuracy, the search for such accessory proteins and an analysis of their effects on fidelity should prove useful.

Proofreading

All DNA polymerases from bacteria and bacteriophage contain a 3'→5' exonuclease activity. In studies with E. coli DNA polymerase I, Brutlag & Kornberg (109), demonstrated preferential excision of noncomplementary nucleotides at the 3'-OH primer terminus and suggested that the 3'→5' exonuclease functions in proofreading mistakes during polymerization. Using mutant T4 DNA polymerases, Muzyczka et al (55), extended these studies and showed a correlation between the in vivo mutator phenotype and the ratio of exonuclease to polymerase activity in vitro. They proposed that the important considerations in determining fidelity were the ratios of the rates of exonuclease and polymerase activity. A higher rate of deoxynucleoside monophosphate generation relative to polymerization results in fewer stable misincorporations. Strong support for this concept comes from the numerous subsequent experiments measuring turnover of normal correct and incorrect substrates (55–62, 89) and 2-aminopurine triphosphate vs deoxyadenosine triphosphate (6, 63–65) by mutant and wild-type T4 DNA polymerases.

These turnover data qualitatively support proofreading, but only to the extent that the measured dNMP is actually formed at the growing primer terminus during ongoing polymerization. A recent study by Das &

Fujimura (110) provides evidence that generation of nucleoside monophosphates occurs at the growing point. In measuring dTTP utilization on a poly(dA)·oligo(dT) template with T5 DNA polymerase, the ratio of free dTMP generated to total dTMP utilized was constant with time under conditions in which the polymerase did not dissociate from the primer template. In addition, it was shown that the enzyme could switch its direction from 3'→5' hydrolysis to 5'→3' polymerization without leaving the template.

The extent of proofreading by a given polymerase may depend on the DNA template. A recent study of incorrect nucleoside monophosphate generation during Mg^{2+}-activated copying of poly(dA-dT) by Pol I indicates that proofreading can increase the accuracy of this enzyme on poly (dA-dT) by at most two to three fold (52). Also, Hall & Lehman have shown only a fourfold average difference in misincorporation of dTTP between mutator L56 and wild-type T4 DNA polymerase when copying poly(dC) (21). These data indicate that proofreading may not be highly active with synthetic polynucleotide templates, a conclusion consistent with a comparison of the error rates in Tables 1 and 2.

While it is clear that proofreading makes a sizeable contribution to the accuracy of catalysis by T4 DNA polymerase, its contribution in other systems has not been firmly established. T4 may be atypical in its use of proofreading, since the exonuclease to polymerase ratio of T4 DNA polymerase (55–65) is 10- to 100-fold greater than that of E. coli Pol I (17). Also, the relative importance of proofreading must be interpreted with regard to its specificity. The genetic studies (9–12, 111–113) of T4 mutants demonstrate large differential mutator and antimutator effects, depending on the particular substitution mutation being measured, the mutagen used, and the method of measurement (for review see 114). Thus, depending on the site being examined, an antimutator can in fact have mutator chracteristics. Several studies (56–58) have shown differential nucleoside monophosphate generation for each of the four dNTP substrates. Muzyczka et al (55) have shown that excision of a terminal mismatched nucleotide in the absence of polymerization depends on the nature of the mismatch and the DNA polymerase. Thus the definition of a polymerase as a mutator or antimutator depends on experimental conditions.

PROOFREADING WITH ΦX174 DNA   The error rates of eukaryotic DNA polymerases are similar when comparing results with synthetic polynucleotides and natural DNA templates (Tables 1 and 2). In contrast, both E. coli Pol I and T4 DNA polymerase are much more accurate with ΦX174 DNA. This suggests enhanced proofreading with natural DNA. However, greater accuracy could also be the result of decreased misinsertion. Studies with

ϕX174 DNA to assess the relative contributions of error prevention and proofreading to accuracy have been designed to diminish or eliminate proofreading and observe the effect on mutagenesis. The following two approaches have been utilized.

Next Nucleotide Effect   Considering that proofreading involves the selective hydrolysis of a terminally incorrect base, one can intuitively reason that decreased fidelity would result from more rapid synthesis. The concept, reviewed by Bernardi & Ninio (115) and embodied in several mathematical models of proofreading (5–7), involves competition between excision and polymerization. When a mistake is inserted, it can be removed by excision or locked into the growing DNA chain by the incorporation of the next nucleotide. This concept is supported by 2-aminopurine incorporation data with T4 DNA polymerase (6, 65). This logic has been tested in all three of the assays outlined in Figure 1. First, Fersht (33) copied the minus strand of ϕX174 *am*3 DNA with *E. coli* replication proteins (Figure 1, assay C). The relevant DNA sequence in the minus strand is 3'-ATC-5' (complementary to the 5'-TAG-3' amber codon in the plus strand). True wild-type revertants are produced by misincorporation of G instead of A at position 587 (opposite the template T in the minus strand). Since the next correct nucleotide is also a G (opposite a template C), then increasing the concentration of dGTP, relative to correct dATP during polymerization, should have two effects: more errors should be misinserted due to an increased ratio of G to A; and these should be less frequently excised due to more rapid insertion of G as the next correct nucleotide. The observation that the frequency of production of wild-type revertants in this system was proportional to the $[dGTP]^2/dATP$ ratio is consistent with the hypothesis and provides evidence that proofreading does occur with these replicating proteins.

In order to measure the relative importance of error prevention vs proofreading in determining the final accuracy of purified DNA polymerases, we used the assay described in Figure 1A (29). In this system, the plus strand is copied in the direction 3'-GAT-5'. As shown in Figure 2B, a dATP pool bias during polymerization has only a negligible effect on reversion frequency. However, both dCTP and dGTP pool biases significantly increase the reversion frequency. When sequenced, the revertants contain the specifically biased nucleotide, always at position 587 (opposite a template A), in place of the normally correct T (27, 29). We examined the effect of increasing dATP during polymerization under conditions in which either C or G mistakes could be easily detected (Figure 2C), making use of the fact that A misincorporation is very infrequent at 587. At a constant dCTP (incorrect) to dTTP (correct) ratio of 50:1, increasing dATP 50-fold (to 250 μM) increased the reversion frequency 25-fold. Seven of eight revertants contained at C at 587. This enhancement of dCMP misincorporation by increasing dATP was not observed with AMV DNA polymerase, which lacks an associated proofreading exonuclease. With Pol I, the enhancement in mutagenesis appears to reach a maximum at 250 μM dATP, which suggests that the contribution of proofreading is at the most 25-fold. The residual fidelity ($1/10^5$; Table 2) is considered to result from discrimination at the level of insertion. A similar experiment performed with a dGTP to dTTP ratio of 50:1 gave only a two to three fold response, even when the dATP concentration was increased to 2500 μM. This suggests that a G:A mismatch at 587 is not excised as efficiently by Pol I as a C:A mismatch. An extension of this approach to other amber mutations should permit the quantitation of proofreading as a function of the type of mismatch and the adjacent sequence.

A specificity for proofreading may further be indicated by recent pool bias experiments of Fersht (36) with amber 16 ϕX174 DNA. With the *E. coli* Pol III holoenzyme system (Figure 1C), the error frequency for substitutions at the first position (residue number 5276) in the amber 16 codon was independent of next nucleotide effect, while those viable substitutions at the middle position (5277) exhibit the response indicative of proofreading. Such differential effects could reflect differences in the rate of excision of mismatched bases resulting from the nature of the mismatch, stacking interactions with adjacent nucleotides, or secondary structural effects. Also, the next nucleotide effect may be diminished by varying experimental conditions. Hibner & Alberts (34) and Sinha & Haimes (35) have shown with the T4 replication complex (Figure 1B) that errors at the middle position of the amber 3 codon show no next nucleotide effect. However, their reactions were intentionally performed at relatively high dNTP concentrations. An effect can only be observed if the concentration of the next nucleotide is rate limiting for locking in mistakes. Thus, the difference in the next nucleotide effect exhibited by *E. coli* and T4 replication proteins may indicate a basic difference in proofreading or a difference in saturation levels, or may simply reflect different experimental conditions.

Deoxynucleoside thiotriphosphates abolish proofreading   Recently, we have used a second, independent approach to quantitate proofreading. Deoxynucleoside thiotriphosphates contain a sulfur atom in place of an oxygen on the α phosphorus (116). They are incorporated normally into DNA (116, 117), and once incorporated are not excised (117, 118). ϕX174 DNA containing normal dTMP and αS-substituted dAMP, dCMP, and dGMP was insensitive to hydrolysis by the exonucleases associated with Pol I and T4 DNA polymerase (30), which suggests that their exonucleolytic activities are unable to hydrolyze the phosphorothioate-diester bonds. When (incorrect) dCTPαS is substituted for (incorrect) normal dCTP during

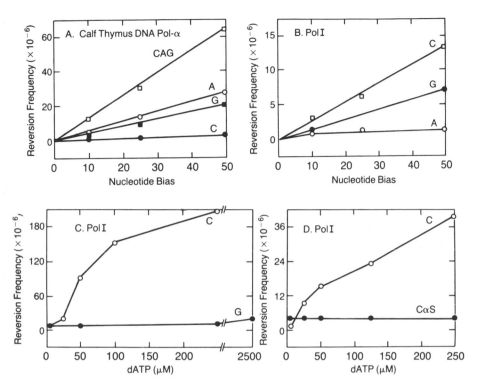

Figure 2. The effect of substrate pool biases on fidelity. φX174 *am* 3 DNA was copied using reaction conditions as described in (27–30). Reversion frequencies were then measured, as illustrated in Figure 1A. *A.* Individual substrate biases (C or A or G) were performed with three substrates at 100 μM, while the biased substrate was increased from 100 μM to 5000 μM. The combination bias (C plus A plus G) was performed with dCTP, dATP, and dGTP at 1000 μM, while dTTP was lowered from 1000 μM to 20 μM. *B.* Three substrates at 5 μM; biased substrate increased from 5 μM to 250 μM. *C.* ●——●, 5 μM dCTP, dTTP, 250 μM dGTP, increasing dATP; ○——○, 5 μM dGTP, dTTP, 250 μM dCTP, increasing dATP. *D.* ●——●, 5μM dGTP, dTTP, 25 μM dCTPαS, increasing dATP; ○——○, 5 μM dGTP, dTTP, 250 μM dCTP, increasing dATP.

polymerization by Pol I or T4 DNA polymerase, the fidelity of synthesis is diminished (Table 2). The accuracy of Pol I decreases 20-fold (similar to that observed with the next nucleotide effect), while that of T4 DNA polymerase decreases > 500 fold. No decrease in accuracy of AMV DNA polymerase or DNA polymerase-β is observed, consistent with the fact that these enzymes lack associated exonuclease activities. To the extent that the mistake is not hydrolyzed by the proofreading activities of Pol I or T4, this difference between normal and αS substrates represents the contributions of proofreading by these enzymes to removing C mistakes at position 587 in φX174 DNA. Strong support for this interpretation is obtained from a next nucleotide experiment with an incorrect αS substrate (Figure 2D). The increase in reversion frequency observed with normal dCTP and increasing dATP is eliminated when dCTPαS is substituted for dCTP.

PROOFREADING IN EUKARYOTIC SYSTEMS   The importance of proofreading to the fidelity of eukaryotic DNA replication is not known. Based on incorporation of 2-aminopurine triphosphate vs dATP, Wang et al (119) suggest that proofreading is not occurring in HeLa nuclei. Furthermore, many highly purified DNA polymerases from eukaryotes and animal viruses lack associated exonuclease activity (120–132). These enzymes are inaccurate relative to in vivo spontaneous mutation frequencies with synthetic templates (Table 1) and ϕX174 (Table 2). Several possibilities could explain this difference: 1. DNA replication in eukaryotes may in fact be quite inaccurate and depend heavily on postreplication repair; 2. Misinsertion frequencies may be substantially reduced by accessory proteins in the putative replication complexes; 3. A proofreading activity may be removed during purification, either due to dissociation of a subunit or accessory protein or due to proteolysis. Reports now exist for the presence of exonuclease in preparations of certain purified DNA polymerases from *Saccharomyces cerevisiae* (133–136), *Tetrahymena pyriformis* (137), *Euglena gracilis* (138), *Ustilago maydis* (139–141), *Cylindrotheca fusiformis* (142), *Chlamydomonas reinhardii* (143), calf thymus and rabbit bone marrow DNA polymerase-δ (144–146), and mouse myeloma DNA polymerase (147). Mosbaugh and Meyer describe rat ascites hepatoma DNAse V, having both 3'→5' and 5'→3' exonuclease activity (148), which is specifically associated (149) with homogeneous DNA polymerase-β from the same source (123). The exonucleases associated with the DNA polymerase from yeast (135) and *Ustilago* (141) are capable of excising primer terminal mismatched bases. Furthermore, with the *Ustilago* polymerase, the terminal mismatch is hydrolyzed before synthesis begins, which suggests a role for the exonuclease in proofreading. However, the effect of the associated exonuclease activity on fidelity has not been measured with any of these eukaryotic polymerases. Circumstantial evidence that a proofreading activity may be removed during purification has been reported by Radman, et al (150). They correlated a decrease in accuracy for DNA polymerase-α with a decrease in the activity of a nuclease that hydrolyzes a terminal mismatch, through several purification steps. A hint that proofreading may occur in eukaryotes can be found in recent data measuring the fidelity of crude preparations of cytoplasmic DNA polymerase from human fibroblasts (151). In pool bias experiments with poly(dA-dT) in which the concentration of the correct nucleotides was varied over a 20-fold range, a 2- to 3-fold decrease in discrimination was observed. This next nucleotide effect, a hallmark of proofreading, is not observed with AMV DNA polymerase (23, 115), or in the copying of poly(dA-dT) with a relatively "weak" proofreading enzyme like Pol I (151).

Formalized Models for Kinetic Amplification

Experiments on the mechanism of fidelity have been greatly stimulated by mathematical descriptions of kinetic amplification. In these models, enzymes can achieve accuracy greater than that obtained from differences in free energies between base pairings by utilizing a branched catalytic pathway. In Hopfield's model (4), the nonspecific driving force of pyrophosphate release upon incorporation is coupled with the preferential hydrolysis of the incorrect nucleotide to yield an accuracy approaching the square of the error rate obtained by differences in basepairing at the insertion step. In the model of Ninio (5), the polymerase nucleotide intermediate is subject to a time delay in which the nucleoside monophosphate can be excised, but may not be incorporated. The second discrimination step results from differences in the probability that a correct or an incorrect nucleotide-polymerase intermediate will escape hydrolysis. Even though the schemes of Hopfield and Ninio differ considerably in topology, the end results are nearly identical (52); and, in the case of DNA synthesis, increased accuracy is at the cost of hydrolysis of deoxynucleoside triphosphates to monophosphates. Analysis of these models has permitted quantitative predictions on the effects of added deoxynucleoside monophosphates and pyrophosphates to DNA polymerization reactions catalyzed by prokaryotic DNA polymerases (52, 152, 153.) In this regard, So and co-workers (145, 154) and Granger et al (155) reported that a variety of nucleoside monophosphates inhibited the exonucleolytic activity of *E. coli* DNA polymerase I, while having no effect on incorporation. In the poly(dA-dT) system, we observed no effects of monophosphates on fidelity (52) consistent with our argument that proofreading by Pol I is minimal with synthetic polynucleotides. However, deoxynucleoside monophosphates (29) and pyrophosphate (T. A. Kunkel, unpublished results) increase errors in the ϕX system. An interesting feature of enhanced infidelity by monophosphates is the lack of specificity; i.e. all four dNMPs are mutagenic. This lack of specificity of monophosphate inhibition combined with the preferential inhibition of exonuclease compared to polymerase activity (29, 145, 154) is consistent with models in which the terminal residue has to leave the template so as to occupy the catalytic site of the 3'→5' exonuclease. The frayed-unfrayed model of Brutlag & Kornberg (109) suggests that excision by 3'→5' exonuclease is dependent on the degree to which the terminal residue is frayed or melted out. Proofreading would be expected to exhibit mismatch specificity and position effects, which are dependent on hydrogen bonding and stacking interactions, but not necessarily specificity with respect to monophosphates.

An entirely different scheme for achieving accuracy without utilizing excess substrates has recently been designed by Hopfield (8). This nonequilibrium mechanism, referred to as "energy relay," utilizes the energy of the previous addition step to enhance fidelity. There is no requirement for a $3' \rightarrow 5'$ exonuclease or for the hydrolysis of substrates in excess of incorporation. In this mechanism, the first nucleotide inserted in each processive replication segment is inserted at a high error rate. Thus, decreased processivity would be mutagenic. Processivity differs substantially among purified DNA polymerases (17, 54, 122, 128, 156, 164) and is affected by other replication proteins (54, 162–165). For example, DNA polymerase-β is distributive (122, 128, 156); i.e. it dissociates from the template after every nucleotide addition step. In contrast, DNA polymerase-α is processive, adding about ten nucleotides per association event (158, 160–164). These data correlate with an observed tenfold difference in accuracy for these enzymes with ϕX174 DNA (28), consistent with the energy relay model. In cases where processivity is affected by the rate of DNA synthesis, a decrease in rate would be mutagenic. This is just the opposite of kinetic proofreading, which predicts that accuracy would increase from a decrease in the rate of synthesis. Whether or not the energy relay mechanism for accuracy is biologically operative in any aspect of DNA replication remains to be determined.

Mismatch Repair

Perhaps the simplest mechanism to increase accuracy would involve excision of noncomplementary base pairs postsynthetically by mismatch specific endonucleases. In bacteria, evidence indicates that this pathway is operative (reviewed in 166–169), and recognition of the newly synthesized strand is determined by a delay in methylation of specific DNA sequences; strand recognition is dependent on the presence of GATC sequences containing N^6-methyladenine (167–173). When the excised bases are resynthesized by a DNA polymerase, the final fidelity approaches the error rate of replication, multiplied by the error rate of repair, multiplied by the number of nucleotides polymerized in replacing the mismatch. Repetitive cycles of mismatch repair could provide exceptionally high accuracy. The presence of 6-methyladenine (6-MeA) in eukaryotes has not been reported, which implies that if mismatch correction occurs, it is in response to different signals to ensure that excision occurs only in the daughter strand. In eukaryotes, this signal could be provided by a subset of methylcytidines or by the association of histones or other proteins with parental DNA.

Fidelity of Multicomponent Prokaryotic Systems

While DNA replication in vivo involves the concerted action of several proteins, most studies on parameters that determine normal fidelity have been limited to purified DNA polymerases. Recently however, several laboratories have begun to use the multiprotein replication complexes purified from *E. coli* (33, 36) and T4-infected *E. coli* (32, 34, 35). Fidelity is measured by reversion of amber mutations in ϕX174 DNA synthesized in vitro, as described in Figure 1 B and C. The experimental approach is similar in both systems; when DNA is synthesized with dNTP substrates at equimolar concentrations no revertants above the background reversion frequency are observed. To observe revertants, large substrate pool imbalances are employed in order to favor potentially incorrect substitutions. An important conclusion from these substrate pool bias experiments is that these replication complexes can synthesize ϕX174 DNA in vitro with high accuracy, approaching at some sites that estimated to occur in vivo (1–3). A careful analysis of the interactions between these components should produce better understanding of the mechanisms used to achieve high fidelity in prokaryotes. The frequency of stable misincorporations by these systems using different ϕX174 amber mutations is summarized in Table 3. We emphasize that these are extrapolated values, based on several assumptions, only some of which have been experimentally verified. For example, in only one instance has the substitution inferred by pool imbalances been confirmed by DNA sequence analysis (35). Despite such uncertainties, the data support the conclusion that reversion frequencies are dependent on the position being analyzed, both with respect to the type of mismatch and the adjacent bases. The conclusion that transversions occur most frequently via purine-purine mispairs (G-A and A-A, See Table 3) rather than pyrimidine-pyrimidine mispairs (C-T, See Table 3) is in accord with the tautomer model of Topal and Fresco (99). However, an alternate explanation is that proofreading could be more efficient with pyrimidine-pyrimidine mispairs.

Table 3   Fidelity of multicomponent prokaryotic systems[a]

Mispair	Estimated reversion frequency at various sites				
	*am*3 (587)	*am*16 (5276)	*am*18 (23)	*am*86 (4116)	*am*9 (?)
E. coli replication proteins (33, 36; Figure 1C)					
G–T	5×10^{-7}	$\sim 2 \times 10^{-7}$ (5277)	—	—	—
C–T	—	$\sim 5 \times 10^{-9}$ (5277)	—	—	—
G–A	—	5×10^{-7}	—	—	—
C–A	—	4×10^{-7}	—	—	—
A–A	—	4×10^{-7}	—	—	—
T4 replication proteins (32, 34, 35; Figure 1B)					
G–T[b]	$2.2 - 6.2 \times 10^{-6}$	1.6×10^{-7}	9.1×10^{-6}	6.9×10^{-7}	—
C–T	$<5 \times 10^{-8}$	$<3.8 \times 10^{-9}$	—	$<1.6 \times 10^{-7}$	$<2.5 \times 10^{-9}$
G–A	4×10^{-7}	1.2×10^{-7}	—	$<3.0 \times 10^{-7}$	3.9×10^{-8}
C–A	$0.83 - 3.3 \times 10^{-7}$	$<2.1 \times 10^{-8}$	2.0×10^{-6}	$<6.6 \times 10^{-8}$	—

[a] The reversion frequencies are extrapolated from pool bias experiments. The numbers in parentheses indicate the nucleotide position in ϕX174 DNA that is presumably being mutagenized by the indicated mispair.

[b] The G–T mismatch at position 587 has been confirmed by DNA sequence in two revertants.

From a comparison of error rates between Table 2 (purified DNA polymerases) and Table 3 (multicomponent systems) it seems likely that DNA polymerase itself is the major contributor to the fidelity of DNA synthesis in prokaryotes. The known mutator effects of genes that code for other replicating proteins (13–16) are small compared to the effects of mutations in the DNA polymerase gene itself. Thus, even though other proteins function in maintaining accuracy, their relative contributions may not be large. Several possibilities exist for the role of other proteins in enhancing discrimination. These include direct effects on either insertion or proofreading due to altering polymerization rate, increasing processivity, altering stacking interactions, modifying secondary structure, or changing the conformation of the polymerase. Indirect effects are also possible; e.g. altered substrate concentrations at the growing point.

Evidence for the direct effect of a protein other than the polymerase on discrimination was obtained by Gillin & Nossal (59), who showed that the addition of T4 gene 32 protein, a single-strand DNA-binding protein, causes a two to four fold decrease in the ratio of incorrect to correct nucleotides utilized by wild-type and ts L88 mutator T4 DNA polymerases in copying poly(dA-dT), thus implying an increased fidelity. Studies in our laboratory indicate *E. coli* single-strand binding protein increases by two to ten fold the fidelity of copying poly(dA-dT) and ϕX174 DNA using different DNA polymerases (31), including several lacking exonucleolytic activity. The data are consistent with a model in which both the template and substrate were immobilized by single-stranded binding protein and polymerase, respectively (69). Enhanced fidelity would result from rejection of incorrect bases due to increased structural constraints.

Parameters that Diminish Fidelity

A large effort has been made to catalogue factors that diminish fidelity. The guiding assumption is that misincorporations made during DNA replication are not always corrected. Thus, the fidelity of DNA polymerization may be a rate-limiting step in mutagenesis. Agents that diminish fidelity in vitro may be mutagenic by this mechanism in vivo. Paradoxically, one of the most useful approaches in understanding the normal mechanism of fidelity is to perturb it and measure the consequences. We consider representative studies on the effects of altering polymerases, templates, and substrates, as well as the effect of various metals on fidelity.

ALTERATIONS IN DNA POLYMERASES   Multiple forms of DNA polymerase are present in both prokaryotes and eukaryotes (for review see 17, 130–132). These enzymes appear to have intrinsic differences in fidelity. For example, with amber 3 ϕX174 DNA, DNA Polymerase-α is five to ten times more accurate than DNA polymerase-β (28). Severalfold differences were noted in the accuracy by which various forms of human placental DNA polymerases-α and -β copy synthetic polynucleotides (82), which implies that subunit composition may influence accuracy. It is interesting to speculate that differences in fidelity between polymerases may reflect their function in cells. Eigen et al (174, 175) have argued that the organisms with small genomes have evolved DNA-replicating systems with inherently higher error rates. By analogy, replicative DNA polymerases, synthesizing the bulk of the genome, may be more accurate than repair DNA polymerases that synthesize, by gap filling, only short stretches of DNA. Also, in bacteria there is evidence that one polymerase can substitute for another (176, 177), and we have suggested that error prone DNA polymerases from tumor viruses may copy host cell DNA and cause increased numbers of mutations (178, 179).

The effect of modification of DNA polymerase has not been studied extensively with respect to fidelity. Modification of *E. coli* DNA polymerase I by methylnitrosourea or ionizing radiation diminished fidelity in copying synthetic polynucleotides (180). In contrast, treatment of rat liver DNA polymerases-α and -β with methylnitrosourea or N-acetoxy-N-2-acetylaminofluorene treatment decreased activity but did not decrease fidelity (181). Decreased activity without a change in fidelity also resulted from nitrosoguanidine treatment of *E. coli* DNA polymerase III* (77). Most importantly, the effects of potential physiological modifications (phosphorylation, acetylation, glycosylation, etc) on fidelity have not been reported.

Tumor progression   The presence of error prone DNA polymerases in malignant cells was first indicated by Springgate & Loeb (182, 183) in studies on normal vs leukemic human lympohocytes. The use of crude enzyme preparation in these studies precluded rigorous interpretation. A more extensive study suggesting the involvement of an error prone polymerase during tumor progression has been reported by Chan & Becker (79). They purified an error prone cytoplasmic DNA polymerase-α from livers of rats exposed to a programmed treatment with the hepatocarcinogen, N-2-fluorenylacetamide. The error prone fraction, representing about 5% of the total polymerase activity was purified 10,000-fold and shown to be severalfold less accurate in copying synthetic polynucleotides than DNA polymerase-α purified from controls. The guiding concept in such studies is that tumor progression proceeds by a cascading of genetic errors, which allows for selection of clones with more malignant phenotypes (179).

Aging   In an attempt to test the "error catastrophe" theory of aging, Linn, et al (184) and Murray & Holliday (151) have examined the accuracy of DNA polymerases obtained from early and late passage cultured human fibroblasts. With partially purified enzyme preparations containing either

DNA polymerases-α or -γ they observed a severalfold decrease in accuracy in copying synthetic polynucleotides from late compared to early passage cells. In contrast, Fry & Weisman-Shomer (185) could not detect differences in fidelity using DNA polymerase-α purified from chicken fibroblasts of varying age. No diminution in fidelity was noted using partially purified DNA polymerase-α obtained from phytohemagglutinin-stimulated lymphocytes from young and old individuals (186). Studies by Fry et al (187) with DNA polymerase-β in isolated chromatin from rat liver failed to show any differences in fidelity with respect to age of mice or species longevity. It should be noted that if misincorporation is due to altered DNA polymerases that are thermal sensitive, misincorporation would be decreased by preincubating the extract at elevated temperatures; any altered enzymes, presumably more thermal labile, would be preferentially inactivated. So far, all fidelity studies on aging have been limited to synthetic polynucleotides. The need is for detailed studies with DNA polymerases using natural DNA templates.

TEMPLATE ALTERATIONS   Since the key step in discrimination between "correct" and "incorrect" substrates resides in the template, the effect of template differences on fidelity can be large. Among synthetic polynucleotides, alternating copolymers typically are copied with greater accuracy than homopolymers (Table 1 and references therein).

While the effects of base modifications have been most extensively examined with synthetic polyribonucleotides and RNA polymerase (B. Singer & J. T. Kuśmierek, see article in this volume), there are a few studies with DNA polymerases. The misincorporation induced by aminopurine and bromodeoxyuridine substitutions in both synthetic and natural DNA templates have been extensively investigated. From these and a number of studies on alkylation of polynucleotides (188–191), the following generalizations seem reasonable. Many modifications, particularly those at a distance on the nucleoside from the H-bonding groups, have no discernible effect on base pairings. Alkylations of positions involved in Watson-Crick bonding frequently result in random misincorporations when the altered polynucleotides are copied by polymerases. This has been referred to as ambiguity by Singer & Kröger (192). Only a few modifications have been shown to result in chemically predicted mispairings (193). The covalent attachment of large aromatic rings invariably terminates DNA synthesis at or next to the site of modification (194–196). These generalizations are in accord with studies using RNA polymerase (B. Singer and J. T. Kuśmierek; see article in this volume).

The question of what happens when a DNA polymerase encounters apurinic sites in DNA has been studied in detail (197–201). Apurinic sites can arise spontaneously because of the inherent chemical instability of the glycosylic bond (202) or by action of glycosylases that remove altered bases (203). Furthermore, alkylation of purines at the N-3 or N-7 positions enhance rates of depurination several orders of magnitude (204–207). Depurination of synthetic polynucleotides and ϕX174 DNA has been shown to enhance misincorporation when these templates are copied by DNA polymerases from a variety of sources (197–199). So far, apurinic sites have been correlated with mutagenesis in vivo only when error prone DNA synthesis is induced (200). A common mechanism has been suggested (201) by which mutagenesis could result from different modifications of bases, particularly purines, by bulky chemical carcinogens. Conceivably, a bulky adduct could stall DNA replication (194–196) and induce an alteration in the DNA replication complex. The altered replicating complex might proceed most easily after removal of the adduct by spontaneous or glycosylase-catalyzed depurination. Since the replication complex is poised at the site, replication would proceed before the apurinic site is repaired.

NUCLEOTIDE SUBSTRATE VARIATIONS   Variations in the relative ratios of the deoxynucleotide substrates during polymerization reactions in vitro (pool bias) have provided valuable information on both misinsertion frequencies and proofreading capabilities. The effects of relative substrate concentrations on fidelity are particularly important, since replication complexes contain proteins that are involved in the control of nucleotide levels (208–215). Intuitively, one might expect cells to stringently regulate nucleoside triphosphate concentrations, since imbalances can be mutagenic. However, if such controls do exist, they are not absolute, since the addition of nucleosides to cell cultures alters intracellular nucleotide pools (216–218). Alterations in nucleotide pools have been invoked to explain a number of biological observations, including bromodeoxyuridine mutagenesis (107), deoxynucleoside toxicity and mutagenesis (219–227), potentiation of mutagenesis by alkylating agents (228), and the enhanced mutagenesis observed with mutant cultured cell lines (229–231). It will also be of interest to study mutagenesis in those diseases characterized by altered nucleotide metabolism (232).

DNA polymerases can incorporate certain modified deoxynucleoside triphosphates. Studies with nucleoside triphosphate analogues, including those of aminopurine, bromodeoxyuridine, 6-methyladenine, as well as phosphorothioates, have been instructive in studies of fidelity and mutagenesis. Recently, O^4-methyl-[6-^{3}H]-thymidine has been shown to be incorporated in vivo into the newly synthesized DNA of cultured Chinese hamster V79A cells (233). In vitro studies with 5-alkyl-2'-deoxyuridine triphosphate (234) and methylnitrosourea-modified dATP (235) have shown that altered nucleotide substrates can be incorporated by purified DNA polymerases. In the latter report (235), ambiguous base pairing specificities were demonstrated, although the nature of the base modification was not determined.

EFFECT OF METALS ON FIDELITY   All DNA polymerases require a divalent metal ion for activity (93). Mg^{2+} is considered the physiological activator because of its presence in cells; however, Ni^{2+}, Mn^{2+}, Co^{2+}, and Zn^{2+} can substitute for Mg^{2+}. In each case, substitution by these metals reduced fidelity (67). The simplest mechanism would involve substitution at the active site, though this need not be the case, since there are multiple metal-binding sites on DNA polymerases (93). A number of nonactivating metals that are mutagenic and/or carcinogenic in vivo reduce the fidelity of DNA synthesis in vitro (67). One likely explanation is that the metal ions bind to bases on the DNA templates and change their hydrogen-bonding properties. Even though there is ample basis for invoking metal template interactions, (236), this hypothesis remains to be proven. Furthermore, the reduction of fidelity during DNA polymerization has not yet been shown to be the cause of metal mutagenesis.

Future Directions

Knowledge of how agents alter the fidelity of DNA synthesis is necessary for elucidating the molecular basis of mutagenesis. A catalogue of agents has been shown to increase misincorporation by DNA polymerases. While many of these agents alter the DNA templates, in no instance has it been shown that the misincorporation is opposite an altered base and is the result of a specific alteration at a particular site. Most importantly, the fact that a mutagen causes infidelity in vitro does not necessarily indicate that this is the mechanism by which the agent causes mutation in vivo.

In prokaryotic systems, considerable progress has been made in establishing how cells can copy DNA with an accuracy estimated at 10^{-7} to 10^{-11}. The error rates of DNA polymerases are being established, and systems have been isolated that can copy DNA with an accuracy approaching that inferred from measurements of rates of spontaneous mutations. We currently lack sufficient information on the specificity of misinsertion and proofreading, which is needed to define mechanisms of discrimination and explain marked variations in mutation frequency at different loci. It should be noted that present in vitro measurements are limited to frequencies of base substitutions, and there are as yet no sensitive in vitro systems to measure other mutagenic processes such as frameshifts. Also, the molecular details of how polymerases enhance fidelity await studies on the three-dimensional structure of these enzymes as well as new ideas on how to measure the kinetics of rare events.

There is a major gap in our understanding of the fidelity of DNA replication in eukaryotes. The development of eukaryotic systems that copy DNA with high fidelity will be of particular importance. The polymerases themselves exhibit error rates of 10^{-3} to 10^{-5}. If anything, mutations per base pair may be less frequent than in prokaryotes. So far, there is no rigorous evidence for what seems the most likely possibility, i.e. eukaryotes use the same mechanism as prokaryotes (proofreading and mismatch repair) to achieve accuracy. In fact, initial evidence suggests that these mechanisms may not be major contributors to fidelity in eukaryotes. Thus, new mechanisms and unanticipated results may be forthcoming.

Literature Cited

1. Drake, J. W. 1969. *Nature* 221:1132
2. Drake, J. W. 1970. *The Molecular Basis of Mutation,* San Francisco, Calif: Holden Dag
3. Cox, E. C. 1976. *Ann. Rev. Genet.* 10: 135–56
4. Hopfield, J. J. 1974. *Proc. Natl. Acad. Sci. USA* 71:4135–39
5. Ninio, J. 1975. *Biochimie* 57:587–95
6. Goodman, M. F., Gore, W. C., Muzyczka, N., Bessman, M. J. 1974. *J. Mol. Biol.* 88:423–35
7. Galas, D. J., Branscomb, E. W. 1978. *J. Mol. Biol.* 124:653–87
8. Hopfield, J. J. 1980. *Proc. Natl. Acad. Sci. USA* 77:5248–52
9. Speyer, J. F. 1965. *Biochem. Biophys. Res. Commun.* 21:6–8
10. Freese, E. B., Freese, E. 1967. *Proc. Natl. Acad. Sci. USA* 57:650–57
11. Drake, J. W., Allen, E. F. 1968. *Cold Spring Harbor Symp. Quant. Biol* 33: 339–44
12. Drake, J. W., Allen, E. F., Forsberg, S. A., Preparata, R. M., Greening, E. O. 1969. *Nature* 221:1128–32
13. Bernstein, C., Bernstein, H., Mufti, S., Strom, B. 1972. *Mutat. Res.* 16:113–19
14. Watanabe, S. M., Goodman, M. F. 1978. *J. Viol.* 25:73–77
15. Mufti, S. 1979. *Virology* 94:1–9
16. Drake, J. W. 1973. *Genet. Suppl.* 23: 45–64
17. Kornberg, A. 1980. *DNA Replication,* San Francisco, Calif: Freeman
18. Ogawa, T., Okazaki, T. 1980. *Ann. Rev. Biochem.* 49:421–57
19. Alberts, B., Morris, C. F., Mace, D., Sinha, N., Bittner, M., Moran, L. 1975. *DNA Synthesis and Its Regulation,* ICN-UCLA Symp. Mol. Biol., ed. M. Goulian, P. Hanawalt, C. F. Fox, Menlo Park, Calif: Benjamin 3:241–69
20. Nossal, N. G. 1979. *J. Biol. Chem.* 254:6026–31
21. Hall, Z. W., Lehman, I. R. 1968. *J. Mol. Biol.* 36:321–33
22. Chang, L. M. S. 1973. *J. Biol. Chem.* 248:6983–92
23. Battula, N., Loeb, L. A. 1974. *J. Biol. Chem.* 249:4086–93
24. Weymouth, L. A., Loeb, L. A. 1978. *Proc. Natl. Acad. Sci. USA* 75:1924–28
25. Gopinathan, K. P., Weymouth, L. A., Kunkel, T. A., Loeb, L. A. 1979. *Nature* 278:857–59
26. Kunkel, T. A., Loeb, L. A. 1979. *J. Biol. Chem.* 254:5718–25
27. Kunkel, T. A., Loeb, L. A. 1980. *J. Biol. Chem.* 255:9961–66
28. Kunkel, T. A., Loeb, L. A. 1981. *Science* 213:765–67
29. Kunkel, T. A., Schaaper, R. M., Beckman, R., Loeb, L. A. 1981. *J. Biol. Chem.* 256:9883–89
30. Kunkel, T. A., Eckstein, F., Mildvan, A. S., Koplitz, R. M., Loeb, L. A. 1981. *Proc. Natl. Acad. Sci. USA* 78:6734–38
31. Kunkel, T. A., Meyer, R. R., Loeb, L. A. 1979. *Proc. Natl. Acad. Sci. USA* 76:6331–35
32. Liu, C. C., Burke, R. L., Hibner, U., Barry, J., Alberts, B. M. 1979. *Cold Spring Harbor Symp. Quant. Biol.* 43: 469–87
33. Fersht, A. R. 1979. *Proc. Natl. Acad. Sci. USA* 76:4946–50
34. Hibner, U., Alberts, B. M. 1980. *Nature* 285:300–5
35. Sinha, N. K., Haimes, M. D. 1980. In *Mechanistic Studies of DNA Replication and Genetic Recombination,* ed. B. M. Alberts, C. F. Fox, ICN-UCLA Symp. Mol. Cell. Biol. 19:707–23. New York: Academic
36. Fersht, A. R., Knill-Jones, J. W. 1981. *Proc. Natl. Acad. Sci. USA* 78:4251–55
37. Loeb, L. A., Weymouth, L. A., Kunkel, T. A., Gopinathan, K. P., Beckman, R. A., Dube, D. K. 1979. *Cold Spring Harbor Symposium on Quantitative Biology* 43:921–27
38. Bollum, F. J. 1974. *The Enzymes* 10: 145–71
39. Bollum, F. J. 1978. *Adv. Enzymol.* 47: 347–74
40. Agarwal, S. S., Dube, D. K., Loeb, L. A. 1979. *J. Biol. Chem.* 254:101–6

41. Fry, M., Shearman, C. W., Martin, G. M., Loeb, L. A. 1980. *Biochemistry* 19:5939–46
42. Battula, N., Loeb, L. A. 1975. *J. Biol. Chem.* 250:4405–9
43. Change, L. M. S., Cassani, G. R., Bollum, F. J. 1972. *J. Biol. Chem.* 247: 7718–23
44. Fresco, J. R., Alberts, B. M. 1960. *Proc. Natl. Acad. Sci. USA* 46:311–21
45. Wang, A. C., Kallenbach, N. R. 1971. *J. Mol. Biol.* 62:591–611
46. Fresco, J. R., Broitman, S., Lane, A. E. 1980. See Ref. 35 pp. 753–68
47. Topal, M. D., Warshaw, M. M. 1976. *Biopolymers* 15:1775–93
48. Topal, M. D., DiGuiseppi, S. R., Sinha, N. K. 1980. *J. Biol. Chem.* 255:11717–724
49. Watanabe, S. M., Goodman, M. F. 1981. *Proc. Natl. Acad. Sci. USA* 78: 2864–68
50. Tinoco, I. Jr., Borre, P. N., Dengler, B., Levine, M. D., Uhlenbeck, O. C., Crothers, D. M., Gralla, J. 1973. *Nature New Biol.* 246:40–41
51. Borer, P. N., Dengler, B., Tinoco, I. Jr., Uhlenbeck, O. C. 1974. *J. Mol. Biol.* 86:843–53
52. Loeb, L. A., Dube, D. K., Beckman, R. A., Gopinathan, K. P. 1981. *J. Biol. Chem.* 256:3978–87
53. McHenry, C., Kornberg, A. 1982. *The Enzymes* 14: In press
54. Fay, P. J., Johanson, K. O., McHenry, C. S., Bambara, R. A. 1981. *J. Biol. Chem.* 256:976–83
55. Muzyczka, N., Roland, R. L., Bessman, M. J. 1972. *J. Biol. Chem.* 247:7116–22
56. Hershfield, M. S., Nossal, N. G. 1972. *J. Biol. Chem.* 247:3393–3404
57. Hershfield, M. S. 1973. *J. Biol. Chem.* 248:1417–23
58. Gillin, F. D., Nossal, N. G. 1976. *J. Biol. Chem.* 251:5219–24
59. Gillin, F. D., Nossal, N. G. 1976. *J. Biol. Chem.* 251:5225–32
60. Lo, K. Y., Bessman, M. J. 1976. *J. Biol. Chem.* 251:2475–79
61. Reha-Krantz, L. J., Bessman, M. J. 1977. *J. Mol. Biol.* 116:99–113
62. Reha-Krantz, L. J., Bessman, M. J. 1981. *J. Mol. Biol* 145:677–95
63. Bessman, M. J., Muzyczka, N., Goodman, M. F., Schnaar, R. L. 1974. *J. Mol. Biol.* 88:409–21
64. Bessman, M. J., Reha-Krantz, L. J. 1977. *J. Mol. Biol.* 116:115–23
65. Clayton, L. K., Goodman, M. F., Branscomb, E. W., Galas, D. J. 1979. *J. Biol. Chem.* 254:1902–12
66. Engel, J. D., von Hippel, P. H. 1978. *J. Biol. Chem.* 253:935–39
67. Sirover, M. A., Loeb, L. A. 1976. *Science* 194:1434–36
68. Sirover, M. A., Loeb, L. A. 1980. *Chem. Biol. Interact.* 30:1–8
69. Loeb, L. A., Kunkel, T. A., Schaaper, R. M. 1980. See Ref. 35, pp. 735–51
70. Huang, W. M., Tso, P. O. P. 1966. *J. Mol. Biol.* 16:523–43
71. Pitha, P. M., Huang, W. M., Tso, P. O. P. 1968. *Proc. Natl. Acad. Sci. USA* 61:332–39
72. Fink, T. R., Crothers, D. M. 1972. *J. Mol. Biol.* 66:1–12
73. Mildvan, A. S. 1974. *Ann. Rev. Biochem.* 43:357–99
74. Orgel, L. E., Lohrmann, R. 1974. *Acc. Chem. Res.* 1:368–77
75. Ninio, J. 1978. *Approaches Moléculaires de l'Evolution.* Paris: Masson
76. Miyaki, M., Murata, I., Osabe, M., Ono, T. 1977. *Biochem. Biophys. Res. Commun.* 77:854–57
77. Jimenez-Sanchez, A. 1976. *Mol. Gen. Genet.* 145:113–17
78. Salisbury, J. G., O'Connor, P. J., Saffhill, R. 1978. *Biochim. Biophys. Acta* 517:181–85
79. Chan, J. Y. H., Becker, F. F. 1979. *Proc. Natl. Acad. Sci. USA* 76:814–18
80. Dube, D. K., Kunkel, T. A., Seal, G., Loeb, L. A. 1979. *Biochim. Biophys. Acta* 561:369–82
81. Seal, G., Shearman, C. W., Loeb, L. A. 1979. *J. Biol. Chem.* 254:5229–37
82. Krauss, S. W., Linn, S. 1980. *Biochemistry* 19:220–28
83. Sirover, M. A., Loeb, L. A. 1977. *J. Biol. Chem.* 252:3605–10
84. Weymouth, L. A., Loeb, L. A. 1977. *Biochim. Biophys. Acta* 478:305–15
85. Sirover, M. A., Loeb, L. A. 1974. *Biochem. Biophys. Res. Commun.* 61: 410–14
86. Mizutani, S., Temin, H. M. 1976. *Biochemistry* 15:1510–16
87. Goodman, M. F., Branscomb, E. W. 1981. *Accuracy in Molecular Biology.* ed. D. J. Falas. New York: Dekker. In press
88. Watanabe, S. M., Goodman, M. F. 1982. *Molecular and Cellular Mechanisms of Mutagenesis,* ed. J. F. Lemontt, W. M. Generoso. New York: Plenum. In press
89. Gillin, F. D., Nossal. N. G. 1975. *Biochem. Biophys. Res. Commun.* 64: 457–64
90. Travaglini, E. C., Mildvan, A. S., Loeb, L. A. 1975. *J. Biol. Chem.* 250:8647–56

91. Sloan, D. L., Loeb, L. A., Mildvan, A. S., Feldman, R. J. 1975. *J. Biol. Chem.* 250:8913–20
92. Mildvan, A. S., Stein, P. J., Abboud, M. M., Koren, R., Bean, B. L. 1978. *Protons and Ions Involved in Fast Dynamic Phenomena*, ed. P. Laszlo, Amsterdam: Elsevier: pp. 413–33.
93. Mildvan, A. S., Loeb, L. A. 1979. *Crit. Rev. Biochem.* 6:219–44
94. Loftfield, R. B. 1963. *Biochem. J.* 89: 82–92
95. Kornberg, A. 1969. *Science* 163: 1410–18
96. Battula, N., Dube, D. K., Loeb, L. A. 1975. *J. Biol. Chem.* 250:8404–8
97. Watson, J. D., Crick, F. H. C. 1953. *Cold Spring Harbor Symp. Quant. Biol.* 18:123–31
98. Watson, J. D., Crick, F. H. C. 1953. *Nature* 171:964–67
99. Topal, M. D., Fresco, J. R. 1976. *Nature* 263:285–89
100. Tucker, G., Irvin, J. 1951. *J. Am. Chem. Soc.* 73:1923–29
101. Kenner, G. W., Reese, C. B., Todd, A. R. 1955. *J. Chem. Soc.* pp. 855–59
102. Katritzky, A. R., Waring, A. J. 1962. *J. Chem. Soc.* pp. 1540–44
103. Wolfenden, R. V. 1969. *J. Mol. Biol.* 40:307–10
104. Fowler, R. G., Degnen, G. E., Cox, E. C. 1974. *Molec. Gen. Genet.* 133:179–91
105. Coulondre, C., Miller, J. H. 1977. *J. Mol. Biol.* 117:577–606
106. Freese, E. 1959. *J. Mol. Biol.* 1:87–105
107. Kaufman, E. R., Davidson, R. L. 1978. *Proc. Natl. Acad. Sci. USA* 75:4982–86
108. Hopkins, R. L., Goodman, M. F. 1980. *Proc. Natl. Acad. Sci. USA* 77:1801–5
109. Brutlag, D., Kornberg, A. 1972. *J. Biol. Chem.* 247:241–48
110. Das, S. K., Fujimura, R. K. 1980. *J. Biol. Chem.* 255:7149–54
111. Speyer, J. F., Karam, J. D., Lenny, A. B. 1966. *Cold Spring Harbor Symp. Quant. Biol.* 31:693–97
112. Ripley, L. S. 1975. *Molec. Gen. Genet.* 141:23–40
113. de Vries, F. A. J., Swart-Idenberg, C. J. H., deWaard, A. 1972. *Molec. Gen. Genet.* 117:60–71
114. Ripley, L. S. 1981. In *Induced Mutagenesis: Molecular Mechanisms and their Implications for Environmental Protection,* ed. C. W. Lawrence, L. Prakash, F. Sherman. New York: Plenum. In press
115. Bernardi, F., Ninio, J. 1978. *Biochimie* 60:762–78
116. Vosberg, H. P., Eckstein, F. 1977. *Biochemistry* 16:3633–40
117. Burgers, P. M. J., Eckstein, F. 1979. *J. Biol. Chem.* 254:6889–93
118. Brody, R. S., Frey, P. A. 1981. *Biochemistry* 20:2145–52
119. Wang, M. L. J., Stellwagen, R. H., Goodman, M. F. 1981. *J. Biol. Chem.* 256:7097–7100
120. Chang, L. M. S., Bollum, F. J. 1973. *J. Biol. Chem.* 248:3398–404
121. Battula, N., Loeb, L. A. 1976. *J. Biol. Chem.* 251:982–86
122. Wang, T. S. F., Sedwick, W. D., Korn, D. 1974. *J. Biol. Chem.* 249:841–850
123. Stalker, D. M., Mosbaugh, D. W., Meyer, R. R. 1976. *Biochemistry* 15: 3114–21
124. Sedwick, W. D., Wang, T. S., Korn, D. 1975. *J. Biol. Chem.* 250:7045–56
125. Wang, T. S. F., Eichler, D. C., Korn, D. 1977. *Biochemistry* 16:4927–33
126. Mechali, M., Abadiedebat, J., de Recondo, A. M. 1980. *J. Biol. Chem.* 155:2114–22
127. Kunkel, T. A., Tcheng, J. E., Meyer, R. R. 1978. *Biochim. Biophys. Acta* 520: 302–16
128. Tanabe, K., Bohn, E. W., Wilson, S. H. 1979. *Biochemistry* 18:3401–3406
129. Fansler, B. S., Loeb, L. A. 1974. *Methods Enzymol.* 29:53–70
130. Bollum, F. J. 1975. *Prog. Nucleic Acid Res. Mol. Biol.* 15:109–44
131. Weissbach, A. 1977. *Ann. Rev. Biochem.* 46:25–47
132. Loeb, L. A. 1974. *The Enzymes* 10:173–209
133. Helfman, W. B. 1973. *Eur. J. Biochem.* 32:42–50
134. Wintersberger, E. 1974. *Eur. J. Biochem.* 50:41–47
135. Chang, L. M. S. 1977. *J. Biol. Chem.* 252:1873–80
136. Wintersberger, E. 1978. *Eur. J. Biochem.* 84:167–72
137. Crerar, M., Pearlman, R. E. 1974. *J. Biol. Chem.* 249:3123–31
138. McIennan, A. G., Kier, H. M. 1975. *Biochem. J.* 151:239–47
139. Banks, G. R., Holloman, W. K., Kairis, M. V., Yarranton, G. T., Spanos, A. 1976. *Eur. J. Biochem.* 62:131–42
140. Banks, G. R., Yarranton, G. T. 1976. *Eur. J. Biochem.* 62:143–50
141. Yarranton, G. T., Banks, G. R. 1977. *Eur. J. Biochem.* 77:521–27
142. Okita, T. W., Volcani, B. E. 1977. *Biochem. J.* 167:611–19
143. Ross, C. A., Harris, W. J. 1978. *Biochem. J.* 171:241–49
144. Byrnes, J. J., Downey, K. M., Black, V. L., So, A. G. 1976. *Biochemistry* 15: 2817–23

145. Byrnes, J. J., Downey, K. M., Que, B. G., Lee, M. Y. W., Black, V. L., So, A. G. 1977. *Biochemistry* 16:3740–46
146. Lee, M. Y. W. T., Tan, C.-K., So, A. G., Downey, K. M. 1980. *Biochemistry* 19:2096–2101
147. Chen, Y.-C., Bohn, E. W., Planck, S. R., Wilson, S. H. 1979. *J. Biol. Chem.* 254:11678–87
148. Mosbaugh, D. W., Meyer, R. R. 1980. *J. Biol. Chem.* 255:10239–47
149. Mosbaugh, D. W., Stalker, D. M., Probst, G. S., Meyer, R. R. 1977. *Biochemistry* 16:1512–18
150. Radman, M., Spadari, S., Villani, G. 1978. *Natl. Cancer Inst. Monogr.* 50: 121–27
151. Murray, V., Holliday, R. 1981. *J. Mol. Biol.* 146:55–76
152. Bernardi, F., Saghi, M., Dorizzi, M., Ninio, J. 1979. *J. Mol. Biol.* 129:93–112
153. Herbomel, P., Ninio, J. 1980. *CR Acad. Sci.* 291:881–84
154. Que, B. G., Downey, K. M., So, A. G. 1978. *Biochemistry* 17:1603–1607
155. Granger, M., Toulmé, F., Hélène, C. 1980. *CR Acad. Sci. P.* 291:203–6
156. Chang, L. M. S. 1975. *J. Mol. Biol.* 93:219–35
157. Uyemura, D., Bambara, R., Lehman, I. R. 1975. *J. Biol. Chem.* 250:8577–84
158. Wilson, S. H., Matsukage, A., Bohn, E. W., Chen, Y.-C., Sivarajan, M. 1977. *Nucleic Acids Res.* 4:3981–96
159. Bambara, R. A., Uyemura, D., Choi, T. 1978. *J. Biol. Chem.* 253:413–23
160. Das, S. K., Fujimura, R. K. 1979. *J. Biol. Chem.* 254:1227–32
161. Fisher, P. A., Wang, T. S.-F., Korn, D. 1979. *J. Biol Chem.* 254:6128–37
162. Hockensmith, J. W., Bambara, R. A. 1981. *Biochemistry* 20:227–32
163. Detera, S. D., Becerra, S. P., Swack, J. A., Wilson, S. H. 1981. *J. Biol. Chem.* 256:6933–43
164. Villani, G., Fay, P. J., Bambara, R. A., Lehman, I. R. 1981. *J. Biol. Chem.* 256:8202–7
165. Huang, C.-C., Hearst, J. E., Alberts, B. M. 1981. *J. Biol. Chem.* 256:4087–94
166. Radding, C. M. 1978. *Ann. Rev. Biochem.* 47:847–80
167. Glickman, B. W., Radman, M. 1980. *Proc. Natl. Acad. Sci. USA* 77:1063–67
168. Glickman, B. W. 1982. *Molecular and Cellular Mechanisms of Mutagenesis,* ed. J. F. Lemontt, W. M. Generoso, New York: Plenum. In press
169. Radman, M., Wagner, R. E. Jr., Glickman, B. W., Meselson, M. 1980. *Progress in Environmental Mutagenesis,*

ed. M. Alacevic, Amsterdam: Elsevier/North Holland, Biomed.
170. Marinus, M. G., Morris, N. R. 1975. *Mutat. Res.* 28:15–26
171. Wagner, R. Jr., Meselson, M. 1976. *Proc. Natl. Acad. Sci. USA* 73:4135–39
172. Lacks, S., Greenberg, B. 1977. *J. Mol. Biol.* 114:153–68
173. Glickman, B., van den Elsen, P., Radman, M. 1978. *Molec. Gen. Genet.* 163:307–12
174. Eigen, M., Schuster, P. 1977. *Naturwissenschaften* 64:541–65
175. Eigen, M., Gardiner, W., Schuster, P., Winkler-Oswatitsch, R. 1981. *Sci. Am.* 244:88–118
176. Niwa, O., Bryan, S. K., Moses, R. E. 1979. *Proc. Natl. Acad. Sci. USA* 76: 5572–76
177. Niwa, O., Bryan, S. K., Moses, R. E. 1980. *Mechanistic Studies of DNA Replication and Genetic Recombination,* ed. B. M. Alberts, ICN-UCLA Symp. Mol. Cell. Biol. 19:597–604. New York: Academic
178. Loeb, L. A., Battula, N., Springgate, C. F., Seal, G. 1975. *Fundamental Aspects of Neoplasia,* ed. A. A. Gottlieb, O. S. Plescia, D. H. L. Bishop, New York: Springer
179. Loeb, L. A., Springgate, C. F., Battula, N. 1974. *Cancer Res.* 34:2311–21
180. Saffhill, R. 1974. *Biochem. Biophys. Res. Commun.* 61:752–58
181. Chan, J. Y. H., Becker, F. F. 1981. *J. Supramolec. Struct.* 5:209 (Suppl.)
182. Springgate, C. F., Loeb, L. A. 1972. *Res. Commun. Chem. Path. Pharm* 4: 651–5
183. Springgate, C. F., Loeb, L. A. 1973. *Proc. Natl. Acad. Sci. USA* 70:245–49
184. Linn, S., Kairis, M., Holliday, R. 1976. *Proc. Natl. Acad. Sci. USA* 73:2818–22
185. Fry, M., Weisman-Shomer, P. 1976. *Biochemistry* 15:4319–29
186. Agarwal, S. S., Tuffner, M., Loeb, L. A. 1978. *J. Cell. Physiol.* 96:235–43
187. Fry, M., Loeb, L. A., Martin, G. M. 1981. *J. Cell. Physiol.* 106:435–44
188. Abbott, P. J., Saffhill, R. 1977. *Nucleic Acids Res.* 4:761–69
189. Saffhill, R., Abbott, P. J. 1978. *Nucleic Acids Res.* 5:1971–78
190. Abbott, P. J., Saffhill, R. 1979. *Biochim. Biophys. Acta* 562:51–61
191. Hall, J. A., Saffhill, R. 1982. *Carcinogenesis.* In press
192. Singer, B., Kröger, M. 1979. *Prog. Nucleic Acid Res. Mol. Biol.* 23:151–94
193. Singer, B., Spangler, S. 1981. *Biochemistry* 20:1127–32

194. Moore, P., Strauss, B. S. 1979. *Nature* 278:664–66
195. Moore, P. D., Rabkin, S. D., Strauss, B. S. 1980. *Nucleic Acids Res.* 8:4473–84
196. Moore, P. D., Bose, K. K., Rabkin, S. D., Strauss, B. S. 1981. *Proc. Natl. Acad. Sci. USA* 78:110–14
197. Shearman, C. W., Loeb, L. A. 1977. *Nature* 270:537–38
198. Shearman, C. W., Loeb, L. A. 1979. *J. Mol. Biol.* 128:197–218
199. Kunkel, T. A., Shearman, C. W., Loeb, L. A. 1981. *Nature* 291:349–51
200. Schaaper, R. M., Loeb, L. A. 1981. *Proc. Natl. Acad. Sci. USA* 78:1773–77
201. Kunkel, T. A., Schaaper, R. M., James, E., Loeb, L. A. 1981. See Ref. 114. In press
202. Lindahl, T., Nyberg, B. 1972. *Biochemistry* 11:3610–18
203. Lindahl, T. 1979. *Prog. Nucleic Acid Res. Mol. Biol.* 22:135–92
204. Lawley, P. D., Brookes, P. 1963. *Biochem. J.* 89:127–38
205. Margison, G. P., O'Connor, P. J. 1973. *Biochim. Biophys. Acta* 331:349–56
206. Margison, G. P., Capps, M. J., O'Connor, P. J., Craig, A. W. 1973. *Chem. Biol. Interact.* 6:119–24
207. Strauss, B., Scudiero, D., Henderson, E. 1975. *Molecular Mechanisms for Repair of DNA*, ed. P. C. Hanawalt, R. B. Setlow, New York: Plenum
208. Wovcha, M. G., Tomich, P. K., Chiu, C.-S., Greenberg, G. R. 1973. *Proc. Natl. Acad. Sci. USA* 70:2196–2220
209. Tomich, P. K., Chiu, C.-S., Wovcha, M. G., Greenberg, G. R. 1974. *J. Biol. Chem.* 249:7613–22
210. Reddy, G. P. V., Singh, A., Stafford, M. E., Mathews, C. K. 1977. *Proc. Natl. Acad. Sci. USA* 74:3152–56
211. Mathews, C. K., North, T. W., Reddy, G. P. V. 1979. *Adv. Enzyme Regul.* 17:133–56
212. Reddy, G. P. V., Pardee, A. B. 1980. *Proc. Natl. Acad. Sci. USA* 77:3312–16
213. Allen, J. R., Reddy, G. P. V., Lasser, G. W., Mathews, C. K. 1980. *J. Biol. Chem.* 255:7583–88
214. Reddy, G. P. V., Mathews, C. K. 1978. *J. Biol. Chem.* 253:3461–67
215. Flanegan, J. B., Greenberg, G. R. 1977. *J. Biol. Chem.* 252:3019–27
216. Tattersall, M. H. N., Jackson, R. C., Jackson, S. T. M., Harrap, K. R. 1974. *Eur. J. Cancer* 10:819–26
217. Tattersall, M. H. N., Brown, B., Frei, E. 1975. *Nature* 253:198–200
218. Tattersall, M. H. N., Ganeshaguru, K., Hoffbrand, A. V. 1975. *Biochem. Pharmacol.* 24:1495–98
219. Bjursell, G., Reichard, P. 1973. *J. Biol. Chem.* 248:3904–9
220. Carson, D. A., Kaye, J., Matsumoto, S., Seegmiller, J. E., Thompson, L. 1979. *Proc. Natl. Acad. Sci. USA* 76:2430–33
221. Gelfand, E. W., Lee, J. J., Dosch, H.-M. 1979. *Proc. Natl. Acad. Sci. USA* 76:2998–3002
222. Wortmann, R. L., Mitchell, B. S., Edwards, N. L., Fox, I. H. 1979. *Proc. Natl. Acad. Sci. USA* 76:2434–37
223. Bradley, M. O., Sharkey, N. A. 1978. *Nature* 274:607–8
224. Bresler, S. E., Mosevitsky, M. I., Vyacheslovov, L. G. 1973. *Mutat. Res.* 19:281–93
225. Smith, M. D., Green, R. R., Ripley, L. S., Drake, J. W. 1973. *Genetics* 74:393–403
226. Barclay, B. J., Little, J. G. 1978. *Molec. Gen. Genet.* 160:33–40
227. Anderson, D., Richardson, C. R., Davies, P. J. 1981. *Mutat. Res.* 91:265–72
228. Peterson, A. R., Landolph, J. R., Peterson, H., Heidelberger, C. 1978. *Nature* 276:508–10
229. Meuth, M., Trudel, M., Siminovitch, L. 1979. *Somatic Cell Genet.* 5:303–18
230. Meuth, M., L'Heureux-Huard, N., Trudel, M. 1979. *Proc. Natl. Acad. Sci. USA* 76:6505–9
231. Weinberg, G., Ullman, B., Martin, D. W. Jr. 1981. *Proc. Natl. Acad. Sci. USA* 78:2447–51
232. Giblett, E. R., Anderson, J. E., Cohen, F., Pollara, B., Meuwissen, H. J. 1972. *Lancet* 2:1067–69
233. Saffhill, R., Fox, M. 1980. *Carcinogenesis* 1:487–93
234. Sági, J., Nowak, R., Zmudzka, B., Szemző, A., Ötvös, L. 1980. *Biochim. Biophys. Acta* 606:196–201
235. Baker, M. S., Topal, M. D. 1981. *J. Supramolec. Struct.* 5:170 (Suppl.)
236. Marzilli, L. G., Kistenmacher, T. J., Eichhorn, G. L. 1980. *Nucleic Acid-Metal Ion Interactions*, ed. T. G. Spiro, New York: Wiley

Ann. Rev. Biochem. 1982. 51:459–89

SPECIFIC INTERMEDIATES IN THE FOLDING REACTIONS OF SMALL PROTEINS AND THE MECHANISM OF PROTEIN FOLDING[1]

Peter S. Kim and Robert L. Baldwin

Department of Biochemistry, Stanford University School of Medicine, Stanford, California 94305

CONTENTS

[1]Abbreviations used for proteins are: BPTI, bovine pancreatic trypsin inhibitor; cyt c, horse heart cytochrome c; RNase A, bovine pancreatic ribonuclease A; RNase S, a derivative of RNase A cleaved at the peptide bond between residues 20 and 21; S peptide, residues 1–20 of RNase S; S protein, residues 21–124 of RNase S. Other abbreviations are: N, native; U, unfolded; I, intermediate; I_N, quasinative intermediate; U_F and U_S, fast- and slow-folding species, respectively; k, rate constant; GuHCl, guanidinium chloride; t_m, temperature at the midpoint of an unfolding transition; ΔC_P, change in heat capacity at constant pressure; θ, mean residue ellipticity. Lysozyme refers to hen egg white lysozyme and myoglobin to sperm whale myoglobin.

459

0066-4154/82/0701-0459$02.00

PERSPECTIVES AND SUMMARY

The stage is set for determining the pathway of folding of representative small proteins by characterizing the structures of well-populated folding intermediates. Structural intermediates accumulate in kinetic experiments under conditions (especially low temperatures) where the intermediates are stable relative to the unfolded form. The major problem appears to be in tracing the folding pathway for a single unfolded form, because the *cis-trans* isomerization of proline residues about X-Pro peptide bonds often gives multiple unfolded forms, with different rates of refolding. The role of proline isomerization in refolding is beginning to be understood. Covalent intermediates have been trapped and characterized in the refolding process, which accompanies reoxidation of the disulfide bonds for two small proteins. Equilibrium intermediates have been found and characterized for some unusual small proteins, and it appears that the unfolding reactions induced by certain salts or methanol yield equilibrium intermediates even for proteins that normally show highly cooperative folding.

The pathway of folding should reveal the mechanism of folding and help in determining the code by which the amino acid sequence of a protein specifies its tertiary structure. This will aid in engineering changes in protein structure using recombinant DNA techniques. Knowledge of the pathway can show at which stage an amino acid substitution causes a change in folding.

Earlier work searching for equilibrium intermediates with small, single-domain proteins such as RNase A and lysozyme gave chiefly negative results and led to the use of the two-state approximation (N $\rightleftharpoons$ U, N = native, U = unfolded), in which folding intermediates are neglected. Populated intermediates were thought to be ruled out by the success of the two-state approximation. However, such highly cooperative folding is found inside the folding transition zone where intermediates are only marginally stable. Folding can be measured kinetically in conditions where intermediates are more stable.

The evidence is now essentially complete that multiple forms of an unfolded protein are produced by the *cis-trans* isomerization of proline residues about peptide bonds after unfolding. Slow-folding (U_S) and fast-folding (U_F) forms of an unfolded protein can be recognized in refolding experiments by the fact that native protein is formed in separate slow ($U_S \rightarrow N$) and fast ($U_F \rightarrow N$) refolding reactions. They also can be recognized in unfolding experiments by refolding assays made during the fast (N $\rightarrow U_F$) and slow ($U_F \rightleftharpoons U_S$) phases of unfolding. The kinetic properties of the $U_F \rightleftharpoons U_S$ reaction, measured during unfolding, match those of proline isomerization in such specific aspects as catalysis by strong acid and cleavage of X–Pro bonds by an enzyme specific for the *trans* proline isomer. Folding in strongly native conditions occurs before proline isomerization and can go almost to completion. An enzymatically active intermediate that still contains a wrong proline isomer is found when RNase A folds at 0°–10°C. Partial folding increases the rate of proline isomerization, possibly because strain is produced in folding intermediates. The unexpectedly low proportion of U_S species in several unfolded proteins suggests that not all proline residues produce slow-folding species.

Information about the folding pathway is preliminary in all cases but, surprisingly, S–S bond formation in BPTI proceeds via obligatory two-disulfide intermediates each having a nonnative S–S bond. The major folding reaction occurs in a single S–S rearrangement. Both kinetic and equilibrium results for folding with S–S bonds intact are consistent with a framework model in which the H-bonded secondary structure is formed early in folding. The tertiary structures of α-lactalbumin, penicillinase, and carbonic anhydrase are disrupted before their secondary structures unfold, in denaturant-induced unfolding. Kinetically, secondary structure is formed early in the folding of RNase A and RNase S, as judged by stopped-flow CD studies and protection of NH protons against exchange with solvent. The kinetic mechanism of folding appears to be sequential folding with defined intermediates. Folding probably proceeds along a pathway determined by the most stable intermediates. Independently folding domains have been demonstrated via fragment isolation for the small proteins ovomucoid and elastase, the α subunit of tryptophan synthase, and for larger proteins. For multidomain proteins, the pathway of folding involves independent folding of individual domains followed by domain interaction. Mutants blocked kinetically in the folding and assembly of a trimeric protein, the tail spike protein of phage P22, have been demonstrated.

INTRODUCTION

Statement of the Problem

Small, monomeric proteins fold to thermodynamically stable structures, as judged by reversibility of folding. Nevertheless, folding is very rapid in most cases (seconds or less). The number of possible conformations is astronomical. If the pathway is under thermodynamic control, how does the protein find the most stable structure so quickly? If the pathway is under kinetic control, how are incorrectly folded structures avoided?

The amino acid sequence codes for the folding of a protein, but amino acid substitutions (changes in the code) are allowed at almost all residue positions without drastic changes in the folding pattern. X-ray structures have been determined for globins that are related only distantly through evolution and have only a few amino acids in common; yet the "globin fold" is strikingly similar in each case. The qualitative features of folding appear to be the same in horse cyt c and yeast cytochrome c, even though they differ by 46% in amino acid sequence. The code for folding is not a simple code like the mRNA triplet code.

The complexity of the code probably reflects the complexity of the folding process. Determination of the pathway of folding may be the chief means of solving the code, because the specific interactions can be measured at different stages in folding. This can be illustrated by a current model for folding, according to which α helices and β sheets form at their correct locations in the otherwise unfolded chain. Amino acid substitutions may be allowed because the choice between helix, sheet, or no folding is averaged over several residues, so that one substitution need not tip the balance, or because only a few specific interactions determine the locations of the α helices and β sheets. At the second stage in folding, the model predicts that α helices and β sheets interact via special pairing sites,[2] which are coded by only a few residues, thus allowing substitutions at other residue positions. A decade ago it was difficult to understand how an amino acid substitution could be tolerated in the interior of a protein because the side chains are closely packed and the protein structure was thought to be rigid. Today it is known, especially from NMR studies of tyrosine and phenylalanine ring flips, that the interior of a protein is flexible.

---

[2]Alternatively, the pairing between α helices and/or β strands may be determined by their relative positions in the chain (1a).

The practical problems in determining the folding pathway are severe. The methods that give structural information about protein folding (e.g. X-ray, NMR, and CD) are intrinsically slow, so that equilibrium intermediates are needed. Moreover, the intermediates must be well-populated, since these methods require reasonably pure materials. The pathway of folding should first be determined for the simplest case, that of small, "single-domain" proteins like BPTI, RNase A, myoglobin, or staph nuclease. But folding of these small proteins is highly cooperative in most cases, and equilibrium intermediates are not populated. On the other hand, although kinetic intermediates may be well populated, steps in folding are often fast (1–100 msec). It is necessary to find methods of trapping intermediates in a stable form, so that they can be studied at leisure, of slowing down folding, or of adapting spectroscopic methods so that they will give structural information rapidly. Some solutions to these problems have been found and are discussed here.

Other Reviews

An excellent summary of work on protein folding, as of September 1979, is contained in a set of symposium papers collected by Jaenicke (1) into a book, *Protein Folding*. It includes a separate review of recent experimental work. A new monograph on protein folding is being prepared by Ghélis & Yon (2). The study of folding intermediates was reviewed in 1978 by Creighton (3) and in 1975 by Baldwin (4). Wetlaufer (5) has just reviewed the folding of protein fragments. Structural studies of folding, and the possible relationships between the final structure and the pathway of folding have been reviewed recently by Richardson (6), Ptitsyn & Finkelstein (7), Thomas & Shechter (8), and Rossmann & Argos (9). Thermodynamic data on the energetics of folding, obtained by calorimetric studies of folding transitions, have been reviewed by Privalov in 1979 (10). Nemethy & Scheraga consider both theoretical and experimental aspects in a 1977 review (11), which stresses the possible stereochemical determinants of folding. The use of packing principles and surface areas in analyzing the folding process was reviewed in 1977 by Richards (12). Ikegami (13) has just reviewed work on interpreting folding transitions by a cluster model. Earlier general reviews of folding have been given by Anfinsen & Scheraga in 1975 (14) and by Wetlaufer & Ristow in 1973 (15). The nature of protein folding transitions and of the unfolded state was discussed by Tanford (16, 17) in a pair of classic reviews.

MODELS FOR THE FOLDING PATHWAY

These models are of two kinds: kinetic and structural. Since structural data on folding intermediates are only now starting to be available, both classes of models still consist of guesses about the folding process. Present structural models are based chiefly on reflection about the X-ray structures of native proteins. The aim of a structural model is to give the actual structures of intermediates as well as their order on the pathway, without being too specific about the factors that control the rate of folding. The aim of a kinetic model is to indicate the dominant intermediates and give the factors that control the rate of folding without being too specific about the structures of the intermediates.

Kinetic Models

1. BIASED RANDOM SEARCH   The possibility that proteins might fold by a purely random search of all possible conformations was considered by Levinthal (18) and then dismissed, because the time required for folding would be impossibly long [10^{50} years for a chain of 100 residues (19)]. However, it is possible that a biased random search could occur in a reasonable time. By means of a computer-simulated folding for a lattice model, it is possible to show that the number of possible chain conformations is drastically reduced if only self-avoiding (sterically possible) conformations are allowed (M. Levitt, personal communication, 1978). Levitt also found that a significant fraction of the self-avoiding conformations are fairly compact, and he suggested that rapid folding might begin whenever the unfolded chain assumes a backbone conformation sufficiently like that of the native protein, because the major free energy barrier to folding (the necessary reduction in entropy of the polypeptide chain) has been overcome [compare (20)]. This result, taken together with the possibility of a rapid, nonspecific collapse when refolding is initiated, [compare (21)] suggests that a biased random search could play an important role in early stages of folding. However, it may not be easy to test the prediction that a nonspecific collapse precedes specific folding, because specific structures can be formed very rapidly: α helix formation in model systems occurs in 10^{-5}–10^{-7} sec (22–24).

2. NUCLEATION-GROWTH   The term nucleation has been used with two quite different meanings to describe models for protein folding. In both cases the nucleus is the structure formed at the beginning of folding, which guides subsequent steps. With the first meaning of nucleation, folding is sequential, and early folding intermediates may be populated: the nucleated molecule may be directly observable in kinetic folding experiments. With the second meaning, the folding reaction proceeds rapidly as soon as a nucleus is provided (as in crystallization of a supercooled liquid after seeding with a crystal, or as in α helix formation) and the nucleated molecule is not observable as a populated species either because folding occurs rap-

idly after nucleation or because the nucleus is unstable by itself and breaks down if it is not stabilized by further folding. The second meaning of nucleation is the classical one in chemistry, and the first meaning is a special usage that has developed gradually in protein folding work. The term was used originally in its correct (second) sense, but when it became apparent that the "nucleated" molecules might nevertheless be observable in kinetic folding experiments, the term was still retained by several workers. We suggest that the term nucleation now be dropped in protein folding studies unless it is used with its classical meaning, and that the other type of folding be referred to as *sequential folding* in which the first structure formed is the *kernel* (see Model 4).

In the nucleation-growth model, folding cannot start until an initial reaction occurs (nucleation), and subsequent folding takes place rapidly compared to the observed folding reaction. Folding intermediates are not populated, because folding is too fast once it starts, and the rate of folding is determined by the nucleation reaction. Therefore, demonstration of a populated intermediate in folding rules out the nucleation-growth model. Populated kinetic intermediates have been demonstrated in the folding of several proteins (see section on kinetic intermediates), so it appears that protein folding is not a nucleation-limited reaction.

3. DIFFUSION-COLLISION-ADHESION   In this *microdomain coalescence* model, short segments of the unfolded chain fold independently into microdomains. These are unstable, but they diffuse, collide, coalesce, and become stable (19, 25). Karplus & Weaver (19, 26) calculated the folding rate for a diffusion-collision model in which the diffusional collision is rate limiting, and computer simulations for the folding of apoMb have been made on the assumption that diffusion is rate limiting (27). However, adhesion or coalescence of the two microdomains may be the rate-limiting step, in this model. The two steps can be written as:

$$\underline{A \quad B} \quad \overset{k_{12}}{\underset{k_{21}}{\rightleftharpoons}} \quad A{\cdot}B \quad \overset{k_{23}}{\underset{k_{32}}{\rightleftharpoons}} \quad C$$

where A__B are the two microdomains, linked by the polypeptide chain, which diffuse together to form the encounter complex A·B, and C is the product formed by adhesion. There are two limiting cases: (*a*) If k_{21}, the rate of dissociation of A·B, is large compared to k_{23}, the rate of adhesion, then $v = (k_{12}/k_{21})k_{23}$, where v is the overall rate of forming C from A__B. In this case, it is the equilibrium ratio of A·B to A__B and not the diffusion-controlled rate of forming A·B that enters into the overall rate expression. This situation is found commonly for reactions in solution (28) because k_{21} is always large (of the order of 10^{10} s^{-1}). (*b*) In special cases, such as proton transfer reactions, k_{23} can be very large (10^{12} s^{-1} for proton transfer) so that $k_{23} \gg k_{21}$ and $v = k_{12}$. This is the situation envisaged in the original diffusion-collision model.

The diffusion-collision model (19, 26) has been tested (29, 30) by asking if the folding rate of RNase A depends on solvent viscosity in the direct $(U_F \rightarrow N)$ folding reaction. Model compound studies have shown that diffusion of one segment of a chain molecule relative to another is dependent on solvent viscosity (31). The overall folding rate of RNase A was found to be independent of solvent viscosity (29) when either glycerol or sucrose was added, which demonstrates that diffusion is not rate limiting. Recently, a fast reaction (msec) of RNase A has been found (32) that is strongly affected by solvent additives that change the viscosity. Since it can be measured at temperatures far below the transition zone for unfolding, its relation to the folding process is not yet clear.

4. SEQUENTIAL FOLDING   Folding occurs in a unique and definite sequence of steps, analogous to a metabolic pathway. Intermediates may be populated in suitable conditions of folding. To demonstrate sequential folding, it is necessary to show that there are specific, well-populated intermediates. This test has been satisfied for the folding at low temperatures $(0°-10°C)$ of the major slow-folding species of RNase A (33–38) and also of RNase S (30, 39, 40). Proline isomerization is one step in these folding reactions, but folding can proceed to an enzymatically active, native-like form of RNase A before proline isomerization occurs (33, 36–38). Therefore, the folding process is probably similar (i.e. sequential) for both the fast and slow folding species.

Structural Models

The increasing number of X-ray crystal structures has stimulated the proposal of many structural models for protein folding. The possible relationship between the folding pathway and the final structure of a protein has been the subject of several recent reviews (6–9, 12, 25, 41–46) and we do not review these models here.

Some models for the folding of an entire class of proteins postulate that folding begins by forming a "primitive" H-bonded structure that breaks down to generate the observed structure. The primitive postulated by Ptitsyn & Finkelstein (7) for all β proteins is a long two-stranded antiparallel β structure with a central hairpin loop. The Greek key pattern of connections between β strands (47) then results from breaking this hairpin helix into shorter segments by opening unpaired loops. The particular "swirl" of the Greek key (only one is found, and not its isomer) results from the right-handed twist of the β sheet. An α-helical folding primitive has been postulated by Lim (45).

WORKING MODELS   In comparing experimental results with models, experimentalists have two choices: either to take an existing model and to test it against their results, or else to extract a working model from the experimental data. Since present structural information is "low resolution," such working models are necessarily low resolution. Two generalized working models are being tested currently. The first is the *framework model* in which the H-bonded secondary structure is formed early in folding. The second is *modular assembly,* in which essentially complete folding of any part of a protein occurs at one time, although different parts of the protein fold at different times (folding by parts). Note that the folding process may combine features of both models: formation of H-bonded secondary structure may precede tertiary interactions, as in the framework model, while separate subdomains (each capable of forming its own secondary structure) may fold at different times, as in modular assembly.

PROLINE ISOMERIZATION AND SLOW-FOLDING SPECIES

Formation of Slow-Folding Species

Proline isomerization as a slow step in protein folding was suggested by Brandts and co-workers (48) as a possible explanation for the two unfolded forms of RNase A (49). The folding of RNase A shows biphasic kinetics: a fast phase (50 msec at 25°C) precedes a major (80%) slow phase (20 sec at 25°C); both the fast- and the slow-folding reactions produce native enzyme (49). In unfolding experiments, the fast-folding species is formed rapidly, and at least two slow-folding species (U_S^{II}, U_S^{I}) are formed slowly (38, 48, 50–52). A quantitative study of the unfolding and refolding kinetics of RNase A (53) has demonstrated that the 3-species mechanism

$$N \longleftrightarrow U_F \longleftrightarrow U_S$$

explains the unfolding and refolding kinetics in the folding transition zone, where N is only marginally stable, and folding intermediates are not well populated. The fast-folding species (U_F) of RNase A is not a partly folded or nucleated molecule, since its concentration (20% of the unfolded molecules) is not affected by high temperature or strong denaturants such as 6 M GuHCl or 8.5 M urea (49, 54). Recently, the same tests used for RNase A have demonstrated the existence of both fast- and slow-folding molecules in hen lysozyme (55, 56) and in horse cyt c (57). Urea-gradient electrophoresis experiments at 2°C show that unfolded chymotrypsinogen and α-chymotrypsin also contain both slow- and fast-folding molecules

(58). Unfolding occurs in two stages, consistent with a $N \to U_F \rightleftharpoons U_S$ unfolding mechanism for a BPTI derivative (59), yeast isocyt c (60), and pepsinogen (61).

The first good evidence that the $U_F \rightleftharpoons U_S$ reaction of RNase A is proline isomerization was based on a comparison of the kinetics of the $U_F \rightleftharpoons U_S$ reaction in the unfolded protein (51) with the $cis \rightleftharpoons trans$ isomerization of prolyl residues in model compounds (48, 50, 62–65), and it made use of the conclusion that both U_F and U_S are completely unfolded (49, 54). The most striking characteristics, which are common to both reactions, are (51): (a) a high activation enthalpy ($\sim$ 20 kcal/mol); (b) catalysis by strong acids; and (c) kinetics that are independent of GuHCl concentration, which confirm that the interconversion of U_S and U_F does not involve residual structure (223). RNase A has three nitratable tyrosine groups including Tyr 115, which follows Pro 114; the kinetics of the $U_F \rightleftharpoons U_S$ reaction for nitrotyrosyl RNase A have been correlated with the pK changes during unfolding (66), and it has been suggested that NO_2-Tyr-115 provides an optical probe monitoring isomerization of Pro 114.

In recent work (L. -N. Lin and J. F. Brandts, personal communication, 1981), the appearance of a specific wrong proline isomer during unfolding has been correlated directly with the formation of a slow-folding species. Enzymatic cleavage specific for the $trans$ X-Pro bond has been used to break the peptide bond between Tyr 92 and Pro 93 during the unfolding of RNase A, and the results can be correlated with the formation of U_S^{II}. Pro 93 is cis in native Rnase A and $trans$ in U_S^{II} (the major unfolded species).

Proline Isomerization During Folding

Although the $U_F \rightleftharpoons U_S$ reaction of RNase A is almost certainly proline isomerization, the refolding of U_S^{II} has kinetic properties very different from proline isomerization. The activation enthalpy for the $U_S \to N$ reaction is small ($<$ 5 kcal/mol at pH 6 and 20–40°C) as compared with 20 kcal/mol for proline isomerization (50). Furthermore, the rate of the $U_S \to N$ reaction is strongly dependent on the GuHCl concentration, unlike proline isomerization (50). Thus, proline isomerization is not the initial and rate-limiting step in the folding of the major U_S species of RNase A, in contrast to the original proposal (48).

At low temperatures (0°–10°C), a native-like intermediate (I_N) is formed in the folding of U_S^{II}. I_N has nearly the same tyrosine absorbance and enzymatic activity as the native protein (33, 38), but it differs from N in having a wrong proline isomer. I_N unfolds to give U_S, whereas N unfolds to give U_F (33). Proline isomerization appears to be the final or nearly final step in folding ($I_N \to N$). Proline isomerization can be 20 to 40 times faster in I_N than in the unfolded protein (33), perhaps because the non-native proline residues are under strain in I_N. The rate of proline isomerization in a cyclic pentapeptide is six times faster than in the corresponding blocked linear peptide, probably because of strain (67). In the case of U_S^{II}, an incorrect proline isomer does not block folding in moderate or strongly native folding conditions, but rather slows down the folding process: the probable explanation is that folding intermediates are less stable with a non-native proline isomer.

The kinetics of folding for three different carp parvalbumins provide further evidence for the role of proline isomerization in protein folding (68, 69). They have similar amino acid sequences and spectroscopic properties, but one of the parvalbumins contains a proline residue and the other two do not (69). All three proteins show complex folding kinetics; however, the parvalbumin that contains a proline residue shows an additional slower phase not seen in the other two proline-free proteins, and this phase has some kinetic properties like those of proline isomerization (69). However, a similar comparison of cytochrome c molecules from two different species has not given comparable results (70). The probable existence of nonessential proline residues complicates this kind of comparison. The role of proline isomerization has been studied in the refolding kinetics of a specific fragment of procollagen (71).

Nonessential Proline Residues

Not all prolines in a protein may affect the kinetics of folding: there may be "essential" and "nonessential" proline residues (51). The X-ray crystal structure of RNase S suggests that Pro 114 may be accommodated in either the *cis* or *trans* configuration [H. W. Wyckoff, quoted in (51)]. Levitt (72), used conformational energy calculations to study the effect of non-native proline isomers in BPTI, which has four *trans* proline residues. Proline residues can be classified into three groups, based on the energy difference between the native protein and the minimum energy structure with a wrong proline isomer (72). He suggested that these types of proline residues should produce different types of folding reactions (59, 72). Type I (small energy difference) should not affect the rate of folding, type II (intermediate energy difference) should slow down but not block folding, and type III (large energy difference) should block folding in the manner originally suggested by Brandts and co-workers (48).

So far, only type II prolines have been characterized (59). The "type" of folding reaction will depend on the folding conditions. In the folding of RNase A at 25°C, the activation enthalpy changes from 3 kcal/mol (type

II folding) in 0.1 M GuHCl to 18 kcal/mol (type III folding) in 2 M GuHCl (37, 50).

Wüthrich and co-workers have studied proline-containing linear oligopeptides and shown that the *cis/trans* ratio and the isomerization rate of the X-Pro bond depend on the charge and nature of the amino acid preceding the proline residue (62, 63, 67, 73, 74).

Nonproline peptide bond isomerization may be an important factor in the folding of some proteins. For example, the crystal structure of carboxypeptidase A shows three *cis* peptide bonds that are not N-terminal to prolyl residues (75).

KINETIC INTERMEDIATES

Multiple Unfolded Forms

The existence of multiple unfolded forms of a protein, arising from proline isomerization, is discussed above. It presents a serious problem in working out the kinetic pathway of folding, since the pathway must be studied separately for each unfolded form. In the case of RNase A, it has been possible to study the major slow-folding species U_S^{II} [(80% of the total slow folding species (38)]. In the case of hen lysozyme, it is possible to study the direct folding reaction ($U_F \rightarrow N$) of the species with correct essential proline isomers (55, 56).

The standard test for the presence of a kinetic intermediate is the existence of two kinetic phases: whenever more than one phase is observed, at least three species must be involved. However, the three species could be U_F, U_S, and N, and the two phases could be $U_F \rightarrow N$ and $U_S \rightarrow N$, so that a structural intermediate need not be present.

Tests for Structural Intermediates

The standard test for a structural intermediate is the *kinetic ratio test*, whose application to the problem of protein folding has been discussed (39). If the folding reaction shows different kinetics when measured by two different probes, then a structural intermediate must be present, provided that all unfolded species appear alike when measured by each probe. If two (or more) well-resolved kinetic phases are found, and the two probes change differently in different phases, then there is at least one structural intermediate that can be studied readily. Three probes that are particularly informative are: (*a*) stopped-flow CD (76, 77); (*b*) enzyme activity, measured by combination with specific ligands (38, 39, 56, 78) or measured directly (38, 49); (*c*) protection of NH protons against exchange with solvent (34, 35). The test of *specific combination* between fragments has been applied to the folding of RNase S (39). The principle is that, if combination between S peptide (residues 1–20) and S protein (residues 21–124) occurs early in the folding of S protein, then a structural intermediate must be present to provide the combining site.

RNase A and RNase S

The present minimal mechanism for the folding of the U_S^{II} species of RNase A at 0°–10°C is:

$$U_S \rightleftharpoons I_1 \rightleftharpoons I_N \rightleftharpoons N$$

but it is probable that additional intermediates are populated between I_1 and I_N. I_1 has been observed by protection of NH protons against exchange with solvent (34, 35). The protected protons are trapped early and remain trapped throughout folding. The average degree of protection in I_1 is at least 100 (pH 7.5, 10°C) (35) for the 50 most protected NH protons of native RNase A. In native proteins, some NH protons are protected by as much as 10^8 (79, 80). I_N is highly folded as measured by tyrosine absorbance or binding of the specific inhibitor 2'-CMP (33, 38), and I_N is enzymatically active (38). Nevertheless, I_N still contains a wrong proline isomer (33) and the $I_N \rightleftharpoons$ reaction can also be followed by a fluorescence change (36, 37).

The unfolding pathway of RNase A contains an additional intermediate I_U (53):

$$N \rightleftharpoons I_U \rightleftharpoons U_F \rightleftharpoons U_S$$

The refolding kinetics of I_U have been measured, using a sequential stopped-flow apparatus (52), and it is known that proline isomerization does not occur freely in I_U (81). Thus far I_U has been studied only in unfolding conditions, and it is not known whether I_U is also populated in refolding experiments.

The refolding kinetics of RNase S are more complex (39, 40, 82) than those of RNase A, but they contain additional information about the role of the S-peptide moiety (residues 1–20) in folding. Recent stopped-flow CD measurements on the folding of RNase S (U_S) show that sequential steps in folding can be resolved: β-sheet formation precedes the S-peptide α-helix formation (A. M. Labhardt, personal communication, 1981). Enzymatic activity, as measured by the ability to bind 2'-CMP, is regained together with the α-helix formation. It is not yet known when proline isomerization occurs.

Lysozyme

The refolding kinetics of hen egg white lysozyme (83) have recently been reinvestigated by Utiyama and co-workers (55, 56). Their results demonstrate that unfolded lysozyme also exists in a mixture of fast and slow folding species ($U_F:U_S$ ratio of 90:10) [cf Hagerman (84)]. The refolding kinetics outside the transition zone are biphasic, and both phases produce native enzyme (56). The unfolding kinetics are monophasic, and the formation of U_F precedes U_S (56). Evidence for an early intermediate in folding is based on an absorbance change that occurs in the dead time of the stopped-flow instrument (20 msec), after correction for solvent effects (55). The activation enthalpy for the folding of U_S is only 11 kcal/mol (vs 20 kcal/mol for proline isomerization), which suggests that some folding occurs before proline isomerization in U_S (56).

Cytochrome c

The refolding kinetics of horse Fe(III) cyt c (85) have been reinvestigated recently (57). As in RNase A and lysozyme, unfolded cyt c consists of an equilibrium mixture of U_F and U_S (in a 78:22 ratio), which gives rise to biphasic refolding kinetics, with native enzyme formed in both phases (57). An earlier suggestion that the fast refolding reaction is the formation of an abortive intermediate (85) has been ruled out. Sequential unfolding/refolding ("double jump") experiments demonstrate that the unfolding of cyt c produces U_F, which then isomerizes to U_S (57). Similar results have been obtained with yeast iso-2 cyt c (60). An intermediate has been identified in the $U_F \rightarrow N$ reaction of cyt c; the Soret absorbance change precedes the recovery of the native 695 nm band spectrum. In the $U_S \rightarrow N$ reaction, an ascorbate-reducible intermediate is formed before native enzyme is produced (57). Several kinetic intermediates have been found in unfolding by monitoring heme absorbance (86).

Carbonic Anhydrase

This protein (mol wt 29,000) is about twice as large as the other proteins discussed above. The folding of bovine carbonic anhydrase is much slower than the folding of RNase A, lysozyme, or cyt c. Both refolding and unfolding kinetics show multiple phases (78, 87–89), but their relationship to possible multiple forms of the unfolded protein has not yet been investigated (carbonic anhydrase has 20 prolines). Carbonic anhydrase contains Zn^{2+} and the presence or absence of Zn^{2+} strongly affects the refolding kinetics (90, 91). There is evidence for early formation of the H-bonded framework: changes in θ_{222} (secondary structure) precede changes in θ_{270} (tertiary structure) (88). A spin-label study of carbonic anhydrase shows that an intermediate is formed within 0.1 sec after the start of refolding (89). A late intermediate can bind a specific inhibitor, but does not have enzymatic activity (78). Interestingly, the fastest observed phase in the refolding of carbonic anhydrase has a rate that increases with increasing GuHCl concentration, which suggests that there is an early abortive intermediate in a rapid preequilibrium with the unfolded state (88). However, carbonic anhydrase is known to precipitate easily during attempts at renaturation (92–94), and it is possible that this is aggregation dependent.

Other Proteins

Creighton (58) has recently introduced urea gradient electrophoresis as a method of studying folding intermediates. A linear gradient of urea perpendicular to the direction of migration is used, and the migration pattern is observed as a function of time; the patterns obtained with native and unfolded proteins are compared (58). The temperature of electrophoresis is low (2°C), to decrease the rates of folding and of proline isomerization. Slow-folding (U_S) forms of the unfolded protein have been demonstrated with RNase A, chymotrypsinogen, and α-chymotrypsin (58). Slow-folding species have not been detected in several other small proteins, including lysozyme and cyt c. However, this is not surprising, since reactions with half times less than 8 min are too fast to measure by this method, and the fraction of U_S molecules is small in these proteins (55–57). Several examples of compact kinetic intermediates in folding have been demonstrated with this method (58).

Kinetic and urea gradient electrophoresis experiments also demonstrate the existence of at least two unfolded forms in the α subunit of tryptophan synthase, in addition to two rapidly formed intermediates (95, 96). Moreover, binding of a substrate analogue during refolding displays biphasic kinetics; the rates for the two phases are identical with those observed for folding in the absence of the analogue (96). The α subunit can be cleaved proteolytically into two fragments, each of which can fold by itself, but neither one alone can bind the substrate analogue (97). The relationship between these fragments and the two kinetic intermediates of the intact α subunit is not yet known.

The unfolding of apomyoglobin has been studied by stopped-flow CD and by fluorescence (77) as has the folding of the β chain of hemoglobin (98). Both studies suggest the existence of specific interactions apart from helix formation. The helices of apomyoglobin break down more rapidly in unfolding, as judged by CD, than interactions detected by a fluorescence probe (77). In refolding, the β chain interacts rapidly and specifically with the heme, followed by slower helix formation (98).

EQUILIBRIUM INTERMEDIATES

Tests for Intermediates

The equilibrium unfolding transitions of most small, globular proteins are highly cooperative, and the two-state approximation (N $\rightleftharpoons$ U) is usually a good working model for equilibrium studies. However, in the past few years, several examples of proteins with populated equilibrium intermediates have been reported. There are two tests for an equilibrium intermediate based on the use of probes (16) (Figure 1): (*a*) a biphasic transition as measured by a single probe and (*b*) noncoincident transitions as measured by different probes. Either one of these observations is sufficient evidence for an equilibrium intermediate, and they are not mutually exclusive. There is also a calorimetric test for intermediates (99, 100): $\Delta H_{vH} < \Delta H_{cal}$, where ΔH_{cal} is the calorimetrically determined enthalpy of unfolding and ΔH_{vH} is the apparent enthalpy computed from the temperature dependence of the $N \rightleftharpoons U$ equilibrium constant, by the van't Hoff relation. Equilibrium measurements cannot demonstrate that an intermediate is actually on the pathway of folding: it may be an abortive intermediate or an alternative native form. Also, aggregation of an unfolded protein is known to occur inside the unfolding transition zone in some cases.

Modular Assembly Versus Framework Formation

A biphasic transition (Figure 1 a) is evidence for modular assembly, or folding by parts. To use this as evidence for the mechanism of folding, it is necessary to know whether or not the molecule contains two or more stable domains. If so, the folding must be judged complex: the first goal is to understand the mechanism of folding for small "single-domain" proteins. Noncoincident transition curves (Figure 1 b), which show that the secondary structure is more stable than the tertiary structure, provide evidence for the framework model. Far-UV CD (210–240 nm) has been used as a probe of secondary structure, and either enzyme activity, specific ligand binding, or spectroscopic bands of aromatic residues (270–300 nm) can be used as probes of tertiary structure.

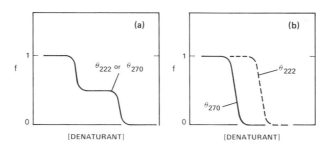

Figure 1   (*a*) Different probes of secondary and tertiary structure each give similar biphasic transition curves, which is interpreted as evidence for modular assembly or domain folding. (*b*) The secondary structure is more resistant than the tertiary structure to denaturant-induced unfolding, as measured by CD, which is interpreted as evidence for a framework model of folding.

a-Lactalbumin

The three-state (N $\rightleftharpoons$ I $\rightleftharpoons$ U) equilibrium unfolding transition of α-lactalbumin has been well characterized by Kuwajima, Sugai, and co-workers (101–108) and recently also by Ptitsyn and co-workers (224). The GuHCl unfolding transition shows non-coincident changes in CD at different wavelengths; aromatic signals (θ_{270} and θ_{296}) show an unfolding transition at a lower GuHCl concentration than the unfolding of the secondary structure (θ_{222}) (102, 104). Three important properties of I are: (a) I is in rapid equilibrium with U (a time constant of less than 1 msec), (b) I is present even when the protein's disulfide bonds are reduced, and (c) I has a far-UV CD spectrum close to that of N (102, 104). The fast interconversion (U $\rightleftharpoons$ I) may be compared to the fast helix-coil transition of synthetic polypeptides (22–24). In contrast, the I $\rightleftharpoons$ N reaction can be measured in seconds (101, 103).

Similar intermediates (as judged by CD) are formed in 2 M GuHCl or 3 M NaClO$_4$ (105), and in unfolding by acid (102, 104) or base (108). Unfolding by acid or base involves the titration of ionizable groups that have abnormal pKs in N but normal pKs in I. These groups with abnormal pKs are responsible for a 10^4-fold increase in the unfolding rate (N $\rightarrow$ I) at low and high pH (103, 108).

α-lactalbumin is not very stable (102). This may explain why an equilibrium intermediate is populated. The secondary structure may be intrinsically stable, while the tertiary structure is weak and easily disrupted by extremes of pH or by moderate concentrations of GuHCl or NaClO$_4$. It is interesting to compare α-lactalbumin with lysozyme, since these two proteins are believed to be structurally homologous (109). Lysozyme is a more stable protein, with a t_m approaching 80° at pH 5 (110). The high stability of its tertiary structure may prevent a comparable intermediate from being observed for lysozyme, because more drastic conditions are needed to unfold lysozyme than α-lactalbumin. Recently α-lactalbumin has been found to bind Ca^{2+} (107), and the stability of its tertiary structure is markedly increased by Ca^{2+} binding.

Penicillinase

Penicillinase is a single-chain protein (mol wt 29,000) with no disulfide bonds. Studies of the GuHCl-induced unfolding of penicillinase by Pain, Robson and co-workers (111–116) demonstrate at least one well-populated equilibrium intermediate and were initially interpreted as providing evidence for a framework model (111). The equilibrium transition curves fall into two categories (111, 112): (a) UV absorbance, aromatic CD, enzymatic activity, and viscosity measurements all show the same transition whereas

(b) far-UV CD and ORD changes occur at much higher GuHCl concentrations. The intermediate is a monomer, has significant secondary structure as judged by CD, and is expanded (112, 114). The existence of at least one equilibrium intermediate has been verified by NMR (116) and another intermediate has been found by urea-gradient electrophoresis (115). There are several similarities with results found for α-lactalbumin. (a) The intermediate has little tertiary structure (as measured with spectroscopic probes), but has a far-UV CD spectrum like that of N. (b) N is not very stable. (c) The intermediate is formed too rapidly to be measured with manual methods (i.e. within 30 sec), whereas the formation of N is slow (several minutes) (112, 116). Unlike penicillinase, the intermediate of α-lactalbumin is compact (224).

There is evidence for folding by domains (or by subdomains) in penicillinase. Digestion of the protein with CNBr results in three large fragments (mol wts 10,500, 9,500, and 8,500) that have been isolated. They combine to form a compact globular complex whose far-UV CD spectrum is indistinguishable from that of native penicillinase (113, 116). The reassembled complex has a different aromatic CD, and does not have enzymatic activity. The isolated fragments combine specifically with antibodies directed against native penicillinase, but it is not known whether they constitute stable domains by the test of showing a folding transition. The folding of penicillinase probably involves both framework formation and subdomain assembly.

Ovomucoid

Ovomucoid is a small (186-residue) protease inhibitor that is clearly a three-domain protein. The equilibrium unfolding transition is complex when measured by a single probe (117–121). Sequencing studies first indicated the presence of three homologous domains (121, 122). This explained earlier observations on the inhibitory properties of different avian ovomucoids; some have one site for trypsin inhibition, others have two sites, one each for trypsin and chymotrypsin; and still others have three sites, two for trypsin and one for chymotrypsin (123–125).

Domains can be isolated by digesting ovomucoid at interdomain sites (121, 126, 127). The isolated domains can refold independently into their native conformations. The isolated fragments: (a) have significant secondary and tertiary structure as judged by CD (128), (b) show both acid and thermal unfolding transitions (119, 121), (c) react with antisera to the native protein (126), (d) retain their inhibitory activity (121, 122), and (e) reform the correct disulfide bonds after reduction (120). One of the domains has recently been crystallized and the structure refined to 2.5 Å resolution (129). Recent analyses of the DNA and mRNA for ovomucoid demonstrate that the gene segments coding for each domain are separated from each other by intervening sequences (130). The DNA sequence of the ovomucoid gene suggests that it evolved from a primordial ovomucoid gene by two separate intragenic duplications (130).

The unfolding transition of intact ovomucoid induced by GuHCl or urea is biphasic; both phases of the transition can be monitored by viscosity or tyrosine absorbance (117, 118). Since these probes usually monitor tertiary structure, it is not yet clear whether the two phases of the transition arise from independent unfolding of domains or from a two-step unfolding of the entire molecule (e.g. step 1: disruption of domain/domain contacts; step 2: unfolding of the individual domains).

Other Proteins

The GuHCl unfolding transition of growth hormone (mol wt 22,000) follows a framework model; changes in tertiary structure (A_{290}) occur at significantly lower GuHCl concentrations than changes in secondary structure (θ_{222}) (131). Moreover, as with α-lactalbumin, the intermediate is very similar to acid-denatured growth hormone (131, 132). Equilibrium studies on the GuHCl unfolding of carbonic anhydrase (mol wt of $\sim$ 29,000) in the absence of Zn^{2+} also support a framework model (133), as do kinetic studies on this protein (78, 88). The equilibrium unfolding transition of cyt c shows spectrally measurable intermediates (134–139); loosening of the polypeptide chain around the heme precedes the unfolding of the remainder of the molecule. However, the calorimetric criterion for a two-state transition is satisfied with cyt c (99). The thermal unfolding transition of the lac repressor headpiece has been monitored by NMR; the changes in chemical shifts show that it is clearly not a two-state transition (140).

Proteins that have equilibrium properties suggesting domain assembly include: the α subunit of tryptophan synthase (141, 142), the β subunit of tryptophan synthase (143, 144), phosphorylase b (145), Bence Jones proteins (146), phosphoglycerate kinase (116), paramyosin (147), and myosin (148). In addition, papain, which has two structural domains separated by a deep cleft, gives a ratio of ΔH measured calorimetrically to the van't Hoff value ($\Delta H_{cal}/\Delta H_{vH}$) of 1.8, which suggests that the two domains fold almost independently (149).

Salt- and Methanol-Induced Unfolding

The role of neutral salts in stabilizing or denaturing folded proteins is complex (150–154). The effects of salts have been broken down into effects on the peptide group and effects on the nonpolar side chains (151, 153). Certain salts (e.g. LiCl, $LiClO_4$, and $CaCl_2$) induce unfolding of proteins at high concentrations. These unfolding transitions appear to be incomplete, based on comparisons of the physical properties of the salt-denatured and GuHCl-denatured protein (16, 155, 156). Recently, it has been found that the addition of urea to salt-denatured RNase A produces a second transition (156).

Alcohols can also induce the unfolding of proteins (16). In particular, adding methanol or ethanol lowers the t_m of RNase A, and decreases the cooperativity of the transition (157, 158). Recently, proton NMR measurements at low temperatures in MeOH-H_2O mixtures have shown that equilibrium intermediates are well populated (158); the results have been interpreted by a framework model.

ENERGETICS OF FOLDING

These questions about the energetics of folding are of particular importance for understanding the mechanism of folding. (*a*) Which is the most stable structure of those proposed for the initial stage in folding? (*b*) Can the stabilities of possible folding intermediates be correlated with some property, such as water-accessible surface area, that can be computed directly from the structure (12, 41, 159)? (*c*) Are there specific structural interactions (e.g. H bonds or salt bridges) that are important energetically and that guide the formation of structure? Definitive answers will come from the structures of actual folding intermediates. Meanwhile, some specific questions are being answered from studies of model compounds and protein fragments. Also, accurate data are being obtained from studies of intact proteins on the factors that affect protein stability.

Model Compound Studies

In the last decade Scheraga and co-workers have determined the stability constant (s) and nucleation constant (σ) for α helix formation for most of the amino acid residues. A given residue is incorporated randomly as a "guest" in a water-soluble, helix-forming polypeptide "host" (derivatives of poly-L-glutamine). The results give the helix-forming propensies of the different amino acid residues as a function of temperature. A short, informative review of the method and results has been given (160).

The striking fact that emerges is that short α helices (10–20 residues) are intrinsically unstable in water. The ratio of helix to random coil is given approximately by: (helix)/(coil) $= \sigma s^{n-1}/(s-1)^2$ for short helices (175), where n is the number of amino acid residues. The largest value of s measured for any residue is 1.3 (Met at 0°C) (160). With $\sigma \leqslant 10^{-4}$, it is clear that a short helix of any composition should be unstable in water.

This conclusion had been predicted from studies of the stability of the amide (–NH...O–C–) H bond in aqueous solution (161, 162). Using an infrared technique, H-bonded dimers and higher oligomers could barely be detected in aqueous solutions of N-methylacetamide (161). A lactam (δ-valerolactam), which forms two amide H bonds per dimer, forms a stronger dimer in water (163, 164), with an association constant at 25°C of 0.01 M⁻¹ (164). This reaction of dimer formation shows a substantial, favorable enthalpy change: −3 kcal/mol H bond (164).[3]

A stronger H bond, which is easily measured in water by an NMR technique, is the charge stabilized H bond (–COO⁻...HN–), formed when the side chain of a glutamic acid residue bends back to bond with its own peptide NH (165). The bond is disrupted by protonation of the carboxylate anion.

The free energy change for burying an amino acid side chain inside a protein as the protein folds up has been estimated from the free energy of transfer of the side chain from water to organic solvents (166). This transfer free energy has been correlated with the water-accessible surface area of hydrophobic side chains (12, 167, 168). Dispersion forces may make an important contribution to binding of hydrophobic amino acids to tRNA synthetases (169) and, similarly, they may be important in stabilizing protein folding. A theoretical study of the hydrophobic effect (170, 171) indicates that the correlation between reduction in water-accessible surface area and transfer free energy is not general and should not be extended to folding intermediates (171) without further justification. Transfer free energy data for the peptide group (151, 162) indicate that it strongly prefers to remain in water, and this should be considered in estimating the stability of a possible folding intermediate.

Studies with small molecules show that it is possible to demonstrate ion pairs in aqueous solution (e.g. guanidinium⁺··· carboxylate⁻ pairs) in a bimolecular reaction, but the strength of the interaction is not large (172, 173).

Protein Fragments

A decade ago several workers tested the possibility of using protein fragments as models for folding intermediates. If a protein folds first into microdomains, and these then coalesce into subdomains, and so on, then appropriately chosen fragments should fold at least partially, and their structures should give clues about the folding process. It is well documented

---

The free energy change for forming the peptide H bond in water has been estimated from the s and σ values for α helix formation (225). Since three residues are fixed in a helical conformation in the nucleation reaction without forming H bonds, we may take the free energy of the nucleation reaction (+ 5.6 kcal/mole if σ = 10⁻⁴) and divide by 3 to get the free energy change per residue (+2 kcal/mole) when it adopts the helical conformation without forming an H bond. Since the s values obtained by the host-guest technique are close to 1, the overall free energy change including the H bond is close to 0, and the free energy change per H bond is −2 kcal/mol. This estimate does not distinguish between the entropic and enthalpic contributions to H bond formation.

that separate domains of larger proteins fold independently or nearly so [see below and also (5)]. However, most studies of subdomain fragments have given the disappointing result that the fragment is predominantly unfolded in aqueous solution.

Nevertheless, an intriguing example has been reported of a "microdomain" that shows partial helix formation in water (174, 175). The C peptide of RNase A (residues 1–13) is essentially unfolded in aqueous solution at 25°C but not at 1°C (174). Aggregation occurs above 2 mg/ml but the helix forms intramolecularly, and helix formation is observed by CD at concentrations as low as 40 μg/ml. NMR data indicate that all, or nearly all, residues participate in helix formation (175). Up to $\sim 30\%$ helix content can be observed in water (175), which is 1000 times greater than the helix content predicted by the host-guest studies. pH titration shows that protonation of His 12, and deprotonation of either Glu 2 or Glu 9 or of both are required for significant helix formation (175). The results suggest that specific salt bridge(s) (e.g. His 12^+...Glu 9^-) nucleate the helix by stabilizing the first turn.

Intact Proteins

Salt bridges have been suspected of being important for folding since their discovery in α-chymotrypsin (176) and hemoglobin (177–179). Recently, salt bridges have been demonstrated in several proteins, including phosphorylase (180) and BPTI (181). Estimates of the strength of individual salt bridges range from –1 to –3 kcal/mol (179, 181, 182).

An advance has been made in treating the electrostatic properties of proteins. A simple discrete charge model, which makes use of X-ray structures to give the locations and solvent accessibilities of the ionizing groups, predicts rather well the pKs of individual groups observed by NMR [(183, 184), see however (185)]. In doing this the model also predicts the electrostatic contribution to the free energy of folding.

The overall thermodynamics of folding are now known accurately from calorimetry for several proteins (10). The data confirm that the folded structures of globular proteins are only marginally stable ($\Delta G = -5$ to -15 kcal/mol), and demonstrate that the thermodynamics of folding as a function of temperature are dominated by the large and approximately constant value of ΔC_P: both ΔH and ΔS are strongly temperature dependent (10, 186).

The striking conclusion that emerges from these studies is that there is a large and favorable contribution to the enthalpy of unfolding that cannot come from hydrophobic interactions (10). Privalov argues that this nonhydrophobic contribution to the unfolding enthalpy is nearly temperature independent and that it arises from H bonds and dispersion forces (10). If one assumes that it arises entirely from H bond enthalpy, then its magnitude corresponds to about -1.7 kcal/mol H bond for several proteins (10), which can be compared with the -3 kcal/mol H bond estimated for the dimerization of δ-valerolactam (163, 164). These results suggest that peptide H bonds may make a major contribution to the stabilization of the native structure.

RELATED TOPICS

The following topics are closely related to our review but limitations of space prevent their review here. We give references to reviews and to some recent papers, and comment on the relationships to work reviewed here.

Disulfide Intermediates

The best evidence for specific intermediates in protein folding experiments comes from the studies by Creighton of trapped disulfide intermediates in BPTI. The work has been reviewed recently (3, 187). Some basic properties of the system are as follows. (a) There are multiple intermediates and multiple pathways: however, there is a "most-favored" pathway and obligatory intermediates. (b) The obligatory two-disulfide intermediates each have one non-native S–S bond. Moreover, an abortive, or dead-end, intermediate has two native S–S bonds. Theorists working with computer simulation of folding have sought to explain these surprising facts. (c) The spectrum of intermediates narrows down toward the most-favored pathway in conditions favoring folding (188) (i.e. low temperatures, or the presence of stabilizing Hofmeister anions such as SO_4^{2-}). This increases the overall rate of the refolding/reoxidation process. Thus, the most-favored pathway proceeds via the most stable intermediates, and the rate of the overall process depends on how well these intermediates are populated. (d) The one-disulfide intermediates equilibrate with each other before the second S–S bond is formed and so do the two-disulfide intermediates, albeit more slowly [compare the recent study of RNase A by Konishi et al (189)]. To a first approximation, there is "thermodynamic control" of the refolding/reoxidation pathways. (e) The major folding process occurs in a single S–S rearrangement, near the end of the pathway. The study of earlier trapped intermediates has shown chiefly that it is difficult to detect and characterize any specific structure (190). Nevertheless, the importance of specific interactions is shown by the close correlation between a narrow or broad spectrum of these intermediates and whether the folding conditions are strongly native or marginally so (188). (f) Unfolding/reduction and refolding/reoxidation can be studied in the same conditions by varying the ratio of oxidant to reductant. Since the conditions are the same and the process is reversible, the pathways of unfolding and refolding are the same.

Domain Folding and Exons

Stable domains that show folding transitions have been isolated after limited proteolysis from numerous large proteins, and the roles of these domains in folding have been investigated in a few cases, including the β_2 subunit of tryptophan synthase (143, 144, 191), the "double-headed" enzyme aspartokinase-homoserine dehydrogenase I of *E. coli,* which has both enzyme activities joined in a single chain (192–194), immunoglobulins, including Bence-Jones proteins (146, 195, 196), elastase (197), and the λ repressor (198). A generalization from these studies is that domains often fold independently in kinetic terms, and are thermodynamically stable, but that a subsequent slow rearrangement, which involves interactions between domains, is commonly required for full activity. We have already discussed domain folding in the case of the small proteins ovomucoid, penicillinase, and the α subunit of tryptophan synthase.

Gilbert (199, 200) has proposed that exons code for protein domains and that genetic recombination within introns speeds up the evolution of new proteins. Introns that separate domain coding regions have been demonstrated for immunoglobulins (201, 202) and for ovomucoid (130). A large intron occurs inside the coding region for the C peptide, thus separating the A and B chains of rat insulin (203). Introns occur within the coding regions for α helices in the globin genes (204): however, the polypeptide fragment coded by the central exon does bind heme specifically and tightly (205).

Oligomeric Proteins

A systematic study of the folding and assembly of several oligomeric enzymes has been made by Jaenicke and co-workers; this work has been reviewed recently (206). Some central points are as follows. (*a*) Only in rare cases is thermodynamic equilibrium ever reached between unfolded monomer and folded oligomer. Kinetic studies give information about a folding pathway that is not readily reversible in most cases. Nevertheless, refolding can give native enzyme in almost quantitative yield in special conditions. (*b*) Aggregation of partly folded chains is the major technical problem. It can be minimized by special procedures, including folding conditions that stabilize folded monomers. (*c*) In general, folded monomers are inactive; rabbit muscle aldolase is an exception. (*d*) Specificity of association is high: mixtures of closely related enzymes (e.g. lactate and malate dehydrogenase) do not refold to give "chimeric" species (207).

The pathways of assembly of aspartate transcarbamylase from catalytic and regulatory subunits have been studied by Schachman & co-workers (208–210). Pulse-chase experiments with radioactively labeled subunits, followed by separation of intermediates in electrophoresis, have been used to give a model for the assembly process.

Temperature-sensitive mutants in the assembly or folding of the phage P22 tail spike protein have been studied by King and co-workers (211–213). These mutants appear to be kinetically blocked in folding or assembly at restrictive temperatures. When synthesized at the permissive temperature, the mutants are as stable as the wild-type protein, as measured by the kinetics of irreversible thermal denaturation. When synthesized at the restrictive temperature, many of the mutant proteins can be reactivated in the absence of new protein synthesis by a shift to the permissive temperature, which indicates that the mutant chains that accumulate at the restrictive temperature are capable of folding and assembly (213).

Fragment Exchange and Local Unfolding Reactions

Certain pairs of polypeptide fragments, which individually are unfolded, can combine to form a native-like, enzymatically active, complex. This type of complementation has been studied extensively by Taniuchi and co-workers for two small proteins, staph nuclease (20) and cyt c (214). The probability of successful complementation increases if the fragments are overlapping. Even three fragments can combine to form a complex (215). The dissociation of these complexes is of particular interest as a model system for studying local unfolding reactions. Local unfolding has often been proposed as a chief means of allowing amide proton exchange in native proteins (216), and local unfolding reactions could be part of the overall unfolding pathway.

Current views of protein flexibility indicate that water can penetrate readily into the interior of a globular protein, and it has been suggested that minor perturbations of the folded structure may permit amide proton exchange (217–219). To decide between "deep breathing" and highly local breathing as dominant mechanisms of exchange, it is necessary to measure the kinetics and equilibria of the breathing reactions by independent methods. Fragment exchange studies can yield both kinetic and equilibrium data and, since the contact regions between fragments are extensive, it is fairly certain that deep breathing reactions are required to break the contacts. A major conclusion from the fragment exchange studies is that the rates and equilibrium constants are in a range where they can be expected to contribute significantly to amide proton exchange in native proteins. This has been demonstrated directly for the dissociation of S peptide from RNase S (220, 221) where amide proton exchange of ^{3}H-labeled S peptide combined with S protein is concentration dependent, and therefore occurs partly by dissociation to free S peptide, even at 0°C, pH 7. A local unfolding reaction in staph nuclease has been demonstrated directly by using antibodies directed against the unfolded protein (222).

CONCLUDING REMARKS

The kinetic mechanism of protein folding proves to be *sequential folding* with defined intermediates in two cases: the major folding reaction of RNase A ($U_S^{II} \rightarrow N$) and the refolding/reoxidation of reduced BPTI or of reduced RNase A. It is likely that sequential folding is a general mechanism. The main objection to it has been the high cooperativity of folding measured by equilibrium experiments inside the folding transition zone, which implies that all folding intermediates are unstable. This objection does not apply if the kinetics of folding are studied outside the transition zone, where intermediates can be stable relative to the unfolded protein. Finding small proteins that do show equilibrium intermediates (especially α-lactalbumin, which is probably a single-domain protein) adds to the evidence that the cooperativity of folding is marginal, not absolute.

Understanding the role of *proline isomerization* in the kinetics of folding has been an essential part of the recent progress. In order to study the kinetic pathway of folding, it is necessary to identify the different unfolded species of a protein and to study the folding of each one separately.

Most present results support a *framework model* of folding in which the H-bonded secondary structure is formed at an early stage, and they are not consistent with strict *modular assembly* of small proteins, or folding by parts, in which both the secondary and tertiary structures of any part of a protein are formed at the same time. However, the H-bonded secondary structure even of a small protein may itself be formed in distinct stages. Modular assembly may apply to the domain folding of larger, multidomain proteins; however, present evidence suggests that a structural rearrangement occurs after the initial folding of separate domains and before the full activities of the native protein are regained. The framework model was first suggested by finding that the secondary structures of some unusual small proteins (e.g. penicillinase, carbonic anhydrase, and α-lactalbumin) are more resistant to unfolding by denaturants than are their tertiary structures. Kinetic intermediates consistent with a framework model were found by ^{3}H trapping experiments with RNase A, which showed that many NH protons are protected from exchange with solvent early in folding and then throughout the folding process, as expected if H-bonded secondary structure is formed early in folding. Stopped-flow CD measurements appear capable of resolving stages in the formation of α helices and β sheets, during the $U_S \rightarrow N$ folding reaction of RNase S.

Well-populated intermediates in the folding of small proteins are now a reality. Characterization of these intermediates will provide a benchmark for theorists working on prediction of folding from sequence and the elucidation of the folding pathway.

ACKNOWLEDGMENTS

We thank Drs. K. Kuwajima and O. B. Ptitsyn for discussion of this review, and many colleagues for sending manuscripts not yet in print. P. S. Kim is a predoctoral fellow of the Medical Scientist Training Program supported by the National Institutes of Health (GM 07365). This work has been supported by grants to R. L. Baldwin from the National Science Foundation (PCM 77-16834) and the National Institutes of Health (2 RO1 GM 19988-21). Use of the Stanford Magnetic Resonance Facility (supported by NSF Grant GP 23633 and NIH Grant RR 00711) is gratefully acknowledged.

Literature Cited

1. Jaenicke, R., ed. 1980. *Protein Folding, Proc. 28th Conf. German Biochem. Soc.* Amsterdam: Elsevier/North-Holland Biomed. Press. 587 pp.
1a. Ptitsyn, O. B. 1981. *FEBS Lett.* 131: 197–202
2. Ghélis, C., Yon, J. 1982. *Protein Folding.* New York: Academic. In press
3. Creighton, T. E. 1978. *Prog. Biophys. Mol. Biol.* 33:231–97
4. Baldwin, R. L. 1975. *Ann. Rev. Biochem.* 44:453–75
5. Wetlaufer, D. B. 1981. *Adv. Protein Chem.* 34:61–92
6. Richardson, J. S. 1981. *Adv. Protein Chem.* 34:167–339
7. Ptitsyn, O. B., Finkelstein, A. V. 1980. *Q. Rev. Biophys.* 13:339–86
8. Thomas, K. A., Schechter, A. N. 1980. In *Biological Regulation and Development,* ed. R. F. Goldberger, 2:43–100. New York: Plenum
9. Rossmann, M. G., Argos, P. 1981. *Ann. Rev. Biochem.* 50:497–532
10. Privalov, P. L. 1979. *Adv. Protein Chem.* 33:167–241
11. Némethy, G., Scheraga, H. A. 1977. *Q. Rev. Biophys.* 10:239–352
12. Richards, F. M. 1977. *Ann. Rev. Biophys. Bioeng.* 6:151–76
13. Ikegami, A. 1981. *Adv. Chem. Phys.* 46:363–413
14. Anfinsen, C. B., Scheraga, H. A. 1975. *Adv. Protein Chem.* 29:205–300
15. Wetlaufer, D. B., Ristow, S. 1973. *Ann. Rev. Biochem.* 42:135–58
16. Tanford, C. 1968. *Adv. Protein Chem.* 23:121–282
17. Tanford, C. 1970. *Adv. Protein Chem.* 24:1–95
18. Levinthal, C. 1968. *J. Chim. Phys.* 65: 44–45
19. Karplus, M., Weaver, D. L. 1976. *Nature* 260:404–6
20. Taniuchi, H. 1973. *J. Biol. Chem.* 248: 5164–74
21. Levitt, M., Warshel, A. 1975. *Nature* 253:694–98
22. Hammes, G. G., Roberts, P. B. 1969. *J. Am. Chem. Soc.* 91:1812–16
23. Barksdale, A. D., Stuehr, J. E. 1972. *J. Am. Chem. Soc.* 94:3334–38
24. Gruenewald, B., Nicola, C. U., Lustig, A., Schwarz, G., Klump, H. 1979. *Biophys. Chem.* 9:137–47
25. Ptitsyn, O. B., Rashin, A. A. 1975. *Biophys. Chem.* 3:1–20
26. Karplus, M., Weaver, D. L. 1979. *Biopolymers* 18:1421–37
27. Cohen, F. E., Sternberg, M. J. E., Phillips, D. C., Kuntz, I. D., Kollman, P. A. 1980. *Nature* 286:632–34
28. Hammes, G. G. 1978. *Principles of Chemical Kinetics.* New York: Academic. 268 pp.
29. Tsong, T. Y., Baldwin, R. L. 1978. *Biopolymers* 17:1669–78
30. Baldwin, R. L. 1980. See Ref. 1, pp. 369–85
31. Haas, E., Katchalski-Katzir, E., Steinberg, I. Z. 1978. *Biopolymers* 17:11–31
32. Tsong, T. Y. 1982. *Biochemistry.* In press
33. Cook, K. H., Schmid, F. X., Baldwin, R. L. 1979. *Proc. Natl. Acad. Sci. USA* 76:6157–61
34. Schmid, F. X., Baldwin, R. L. 1979. *J. Mol. Biol.* 135:199–215
35. Kim, P. S., Baldwin, R. L. 1980. *Biochemistry* 19:6124–29
36. Schmid, F. X. 1980. See Ref. 1, pp. 387–400
37. Schmid, F. X. 1981. *Eur. J. Biochem.* 114:105–9

38. Schmid, F. X., Blaschek, H. 1981. *Eur. J. Biochem.* 114:111–17
39. Labhardt, A. M., Baldwin, R. L. 1979. *J. Mol. Biol.* 135:231–44
40. Labhardt, A. M. 1980. See Ref. 1, pp. 401–25
41. Rashin, A. A. 1979. *Stud. Biophys.* 77:177–84
42. Wako, H., Saitô, H. 1978. *J. Phys. Soc. Jpn.* 44:1931–38
43. Abe, H., Gō, N. 1981. *Biopolymers* 20: 1013–31
44. Gō, N., Abe, H. 1981. *Biopolymers* 20: 991–1011
45. Lim, V. 1980. See Ref. 1, pp. 149–66
46. Lesk, A. M., Rose, G. D. 1981. *Proc. Natl. Acad. Sci. USA* 78:4304–8
47. Richardson, J. S. 1977. *Nature* 268: 495–500
48. Brandts, J. F., Halvorson, H. R., Brennan, M. 1975. *Biochemistry* 14:4953–63
49. Garel, J.-R., Baldwin, R. L. 1973. *Proc. Natl. Acad. Sci. USA* 70:3347–51
50. Nall, B. T., Garel, J.-R., Baldwin, R. L. 1978. *J. Mol. Biol.* 118:317–30
51. Schmid, F. X., Baldwin, R. L. 1978. *Proc. Natl. Acad. Sci. USA* 75:4764–68
52. Hagerman, P. J., Schmid, F. X., Baldwin, R. L. 1979. *Biochemistry* 18: 293–97
53. Hagerman, P. J., Baldwin, R. L. 1976. *Biochemistry* 15:1462–73
54. Garel, J.-R., Nall, B. T., Baldwin, R. L. 1976. *Proc. Natl. Acad. Sci. USA* 73: 1853–57
55. Kato, S., Okamura, M., Shimamoto, N., Utiyama, H. 1981. *Biochemistry* 20:1080–85
56. Kato, S., Shimamoto, N., Utiyama, H. 1982. *Biochemistry.* 21:38–43
57. Ridge, J. A., Baldwin, R. L., Labhardt, A. M. 1981. *Biochemistry* 20:1622–30
58. Creighton, T. E. 1980. *J. Mol. Biol.* 137:61–80
59. Jullien, M., Baldwin, R. L. 1981. *J. Mol. Biol.* 145:265–80
60. Nall, B. T., Landers, T. A. 1981. *Biochemistry* 20:5403–11
61. McPhie, P. 1982. *J. Biol. Chem.* 257:689–93
62. Grathwohl, C., Wüthrich, K. 1976. *Biopolymers* 15:2025–41
63. Grathwohl, C., Wüthrich, K. 1976. *Biopolymers* 15:2043–57
64. Cheng, H. N., Bovey, F. A. 1977. *Biopolymers* 16:1465–72
65. Steinberg, I. Z., Harrington, W. F., Berger, A., Sela, M., Katchalski, E. 1960. *J. Am. Chem. Soc.* 82:5263–79
66. Garel, J.-R. 1980. *Proc. Natl. Acad. Sci. USA* 77:795–98

67. Grathwohl, C., Wüthrich, K. 1981. *Biopolymers* 20:2623–33
68. Brandts, J. F., Brennan, M., Lin, L.-N. 1977. *Proc. Natl. Acad. Sci. USA* 74: 4178–81
69. Lin, L.-N., Brandts, J. F. 1978. *Biochemistry* 17:4102–10
70. Babul, J., Nakagawa, A., Stellwagen, E. 1978. *J. Mol. Biol.* 126:117–21
71. Bruckner, P., Bächinger, H. P., Timpl, R., Engel, J. 1978. *Eur. J. Biochem. 90*, 595–604
72. Levitt, M. 1981. *J. Mol. Biol.* 145: 251–63
73. Wüthrich, K., Grathwohl, C. 1974. *FEBS Lett.* 43:337–40
74. Hetzel, R., Wüthrich, K. 1979. *Biopolymers* 18:2589–606
75. Rees, D. C., Lewis, M., Honzatko, R. B., Lipscomb, W. N., Hardman, K. D. 1981. *Proc. Natl. Acad. Sci. USA* 78: 3408–12
76. Luchins, J., Beychok, S. 1978. *Science* 199:425–26
77. Kihara, H., Takahashi, E., Yamamura, K., Tabushi, I. 1980. *Biochem. Biophys. Res. Commun.* 95:1687–94
78. Ko, B. P. N., Yazgan, A., Yeagle, P. L., Lottich, S. C., Henkens, R. W. 1977. *Biochemistry* 16:1720–25
79. Willumsen, L. 1971. *CR Trav. Lab. Carlsberg* 38:223–95
80. Karplus, S., Snyder, G. H., Sykes, B. D. 1973. *Biochemistry* 12:1323–29
81. Rehage, A., Schmid, F. X. 1982. *Biochemistry.* In press
82. Labhardt, A. M., Baldwin, R. L. 1979. *J. Mol. Biol.* 135:245–54
83. Tanford, C., Aune, K. C., Ikai, A. 1973. *J. Mol. Biol.* 73:185–97
84. Hagerman, P. J. 1977. *Biopolymers* 16:731–47
85. Ikai, A., Fish, W. F., Tanford, C. 1973. *J. Mol. Biol.* 73:165–84
86. Tsong, T. Y. 1973. *Biochemistry* 12: 2209–14
87. Stein, P. J., Henkens, R. W. 1978. *J. Biol. Chem.* 253:8016–18
88. McCoy, L. F., Rowe, E. S., Wong, K.-P. 1980. *Biochemistry* 19:4738–43
89. Carlsson, U., Aasa, R., Henderson, L. E., Jonsson, B.-H., Lindskog, S. 1975. *Eur. J. Biochem.* 52:25–36
90. Yazgan, A., Henkens, R. W. 1972. *Biochemistry* 11:1314–18
91. Wong, K.-P., Hamlin, L. M. 1975. *Arch. Biochem. Biophys.* 170:12–22
92. Nilsson, A., Lindskog, S. 1967. *Eur. J. Biochem.* 2:309–17
93. Wong, K.-P., Hamlin, L. M. 1974. *Biochemistry* 13:2678–83

94. Ikai, A., Tanaka, S., Noda, H. 1978. *Arch. Biochem. Biophys.* 190:39–45
95. Matthews, C. R., Crisanti, M. M. 1981. *Biochemistry* 20:784–92
96. Crisanti, M. M., Matthews, C. R. 1981. *Biochemistry* 20:2700–6
97. Higgins, W., Fairwell, T., Miles, E. W. 1979. *Biochemistry* 18:4827–35
98. Leutzinger, Y., Beychok, S. 1981. *Proc. Natl. Acad. Sci. USA* 78:780–84
99. Privalov, P. L., Khechinashvili, N. N. 1974. *J. Mol. Biol.* 86:665–84
100. Privalov, P. L. 1974. *FEBS Lett.* 40: (Suppl) S140–53
101. Kuwajima, K., Nitta, K., Sugai, S. 1975. *J. Biochem. Tokyo* 78:205–11
102. Kuwajima, K., Nitta, K., Yoneyama, M., Sugai, S. 1976. *J. Mol. Biol.* 106: 359–73
103. Kita, N., Kuwajima, K., Nitta, K., Sugai, S. 1976. *Biochim. Biophys. Acta* 427:350–58
104. Kuwajima, K. 1977. *J. Mol. Biol.* 114: 241–58
105. Maruyama, S., Kuwajima, K., Nitta, K., Sugai, S. 1977. *Biochem. Biophys. Acta* 494:343–53
106. Nozaka, M., Kuwajima, K., Nitta, K., Sugai, S. 1978. *Biochemistry* 17: 3753–58
107. Hiraoka, Y., Segawa, T., Kuwajima, K., Sugai, S., Murai, N. 1980. *Biochem. Biophys. Res. Commun.* 95:1098–1104
108. Kuwajima, K., Ogawa, Y., Sugai, S. 1981. *J. Biochem. Tokyo* 89:759–70
109. Browne, W. J., North, A. C. T., Phillips, D. C., Brew, K., Vanaman, T. C., Hill, R. L. 1969. *J. Mol. Biol.* 42:65–86
110. Khechinashvili, N. N., Privalov, P. L., Tiktopulo, E. I. 1973. *FEBS Lett.* 30:57–60
111. Robson, B., Pain, R. H. 1973. In *Conformation of Biological Molecules Polymers, 5th Jerusalem Symp.* ed. E. D. Bergmann, B. Pullman, pp. 161–72. New York: Academic
112. Robson, B., Pain, R. H. 1976. *Biochem. J.* 155:331–44
113. Carrey, E. A., Pain, R. H. 1977. *Biochem. Soc. Trans.* 5:689–92
114. Carrey, E. A., Pain, R. H. 1978. *Biochim. Biophys. Acta* 533:12–22
115. Creighton, T. E., Pain, R. H. 1980. *J. Mol. Biol.* 137:431–36
116. Adams, B., Burgess, R. J., Carrey, E. A., Mackintosh, I. R., Mitchinson, C., Thomas, R. M., Pain, R. H. 1980. See Ref. 1, pp. 447–67
117. Waheed, A., Qasim, M. A., Salahuddin, A. 1977. *Eur. J. Biochem.* 76:383–90
118. Baig, M. A., Salahuddin, A. 1978. *Biochem. J.* 171:89–97
119. Matsuda, T., Watanabe, K., Sato, Y. 1981. *Biochem. Biophys. Acta* 669: 109–12
120. Matsuda, T., Watanabe, K., Sato, Y. 1981. *FEBS Lett.* 124:185–88
121. Kato, I., Kohr, W. J., Laskowski, M. 1978. *FEBS Meet.* 47:197–206
122. Kato, I., Schrode, J., Wilson, K. A., Laskowski, M. 1976. In *Protides of the Biological Fluids, 23rd,* ed. H. Peeters, pp. 235–43. Oxford: Pergamon. 699 pp.
123. Rhodes, M. B., Bennett, N., Feeney, R. E. 1960. *J. Biol. Chem.* 235:1686–93
124. Osuga, D. T., Bigler, J. C., Uy, R. L., Sjøberg, L., Feeney, R. E. 1974. *Comp. Biochem. Physiol. B* 48:519–33
125. Laskowski, M., Kato, I. 1980. *Ann. Rev. Biochem.* 49:593–626
126. Beeley, J. G. 1976. *Biochem. J.* 155:345–51
127. Bogard, W. C., Kato, I., Laskowski, M. 1980. *J. Biol. Chem.* 255:6569–74
128. Watanabe, K., Matsuda, T., Sato, Y. 1981. *Biochem. Biophys. Acta* 667: 242–50
129. Weber, E., Papamokos, E., Bode, W., Huber, R., Kato, I., Laskowski, M. 1981. *J. Mol. Biol.* 149:109–23
130. Stein, J. P., Catterall, J. F., Kristo, P., Means, A. R., O'Malley, B. W. 1980. *Cell* 21:681–87
131. Holladay, L. A., Hammonds, R. G., Puett, D. 1974. *Biochemistry* 13: 1653–61
132. Burger, H. G., Edelhoch, H., Condliffe, P. G. 1966. *J. Biol. Chem.* 241:449–57
133. Wong, K.-P., Tanford, C. 1973. *J. Biol. Chem.* 248:8518–23
134. Schejter, A., George, P. 1964. *Biochemistry* 3:1045–49
135. Myer, Y. P. 1968. *Biochemistry* 7: 765–76
136. Stellwagen, E. 1968. *Biochemistry* 7: 2893–98
137. Kaminsky, L. S., Miller, V. J., Davison, A. J. 1973. *Biochemistry* 12:2215–21
138. Drew, H. R., Dickerson, R. E. 1978. *J. Biol. Chem.* 253:8420–27
139. Myer, Y. P., MacDonald, L. H., Verma, B. C., Pande, A. 1980. *Biochemistry* 19:199–207
140. Wemmer, D., Ribeiro, A. A., Bray, R. P., Wade-Jardetzky, N. G., Jardetzky, O. 1981. *Biochemistry* 20:829–33
141. Yutani, K., Ogasahara, K., Suzuki, M., Sugino, Y. 1979. *J. Biochem. Tokyo* 85:915–21
142. Yutani, K., Ogasahara, K., Sugino, Y. 1980. *J. Mol. Biol.* 144:455–65
143. Goldberg, M. E., Zetina, C. R. 1980. See Ref. 1, pp. 469–84

144. Zetina, C. R., Goldberg, M. E. 1980. *J. Mol. Biol.* 137:401–14
145. Chignell, D. A., Azhir, A., Gratzer, W. B. 1972. *Eur. J. Biochem.* 26:37–42
146. Azuma, T., Hamaguchi, K., Migita, S. 1972. *J. Biochem. Tokyo* 72:1457–67
147. Riddiford, L. M. 1966. *J. Biol. Chem.* 241:2792–802
148. Tsong, T. Y., Karr, T., Harrington, W. F. 1979. *Proc. Natl. Acad. Sci. USA* 76:1109–13
149. Tiktopulo, E. I., Privalov, P. L. 1978. *FEBS Lett.* 91:57–58
150. von Hippel, P. H., Wong, K. Y. 1965. *J. Biol. Chem.* 240:3909–23
151. Nandi, P. K., Robinson, D. R. 1972. *J. Am. Chem. Soc.* 94:1299–1315
152. Schrier, E. E., Schrier, E. B. 1967. *J. Phys. Chem.* 71:1851–60
153. von Hippel, P. H., Peticolas, V., Schack, L., Karlson, L. 1973. *Biochemistry* 12:1256–64
154. Melander, W., Horváth, C. 1977. *Arch. Biochem. Biophys.* 183:200–15
155. Sharma, R. N., Bigelow, C. C. 1974. *J. Mol. Biol.* 88:247–57
156. Ahmad, F., Bigelow, C. C. 1979. *J. Mol. Biol.* 131:607–17
157. Fink, A. L., Grey, B. L. 1978. In *Biomolecular Structure and Function* eds. P. F. Agris, R. N. Loeppky, B. D. Sykes, pp. 471–77. New York: Academic. 614 pp.
158. Biringer, R. G., Fink, A. L. 1982. *J. Mol. Biol.* In press
159. Lesk, A. M., Chothia, C. 1980. In *Biophysical Discussions, Proteins and Nucleoproteins, Structure, Dynamics, and Assembly,* ed. V. A. Parsegian. pp. 35–47. New York: Rockefeller Univ. Press. 676 pp.
160. Scheraga, H. A. 1978. *Pure Appl. Chem.* 50:315–24
161. Klotz, I. M., Franzen, J. S. 1962. *J. Am. Chem. Soc.* 84:3461–66
162. Klotz, I. M., Farnham, S. B. 1968. *Biochemistry* 7:3879–82
163. Susi, H., Timasheff, S. N., Ard, J. S. 1964. *J. Biol. Chem.* 239:3051–54
164. Susi, H. 1969. In *Structure and Stability of Biological Macromolecules,* eds. S. N. Timasheff, G. D. Fasman, pp. 2:575–663. New York: Dekker. 694 pp.
165. Bundi, A., Wüthrich, K. 1979. *Biopolymers* 18:299–311
166. Nozaki, Y., Tanford, C. 1971. *J. Biol. Chem.* 246:2211–17
167. Chothia, C. 1975. *Nature* 254:304–8
168. Reynolds, J. A., Gilbert, D. B., Tanford, C. 1974. *Proc. Natl. Acad. Sci. USA* 71:2925–27

169. Fersht, A. R., Shindler, J. S., Tsui, W.-C. 1980. *Biochemistry* 19:5520–24
170. Pratt, L. R., Chandler, D. 1977. *J. Chem. Phys.* 67:3683–704
171. Karplus, M. 1980. See Ref. 159, pp. 45–46
172. Tanford, C. 1954. *J. Am. Chem. Soc.* 76:945–46
173. Springs, B., Haake, P. 1977. *Bioorg. Chem.* 6:181–90
174. Brown, J. E., Klee, W. A. 1971. *Biochemistry* 10:470–76
175. Bierzynski, A., Kim, P. S., Baldwin, R. L. 1982. *Proc. Natl. Acad. Sci. USA.* In press
176. Birktoft, J. J., Blow, D. M. 1972. *J. Mol. Biol.* 68:187–240
177. Perutz, M. F., TenEyck, L. F. 1972. *Cold Spring Harbor Symp. Quant. Biol.* 36:295–310
178. Kilmartin, J. V., Hewitt, J. A. 1972. *Cold Spring Harbor Symp. Quant. Biol.* 36:311–14
179. Perutz, M. F. 1978. *Science* 201:1187–91
180. Sprang, S., Fletterick, R. J. 1980. See Ref. 159, pp. 175–92
181. Brown, L. R., DeMarco, A., Richarz, R., Wagner, G., Wüthrich, K. 1978. *Eur. J. Biochem.* 88:87–95
182. Fersht, A. R. 1972. *Cold Spring Harbor Symp. Quant. Biol.* 36:71–73
183. Shire, S. J., Hanania, G. I. H., Gurd, F. R. N. 1975. *Biochemistry* 14:1352–58
184. Matthew, J. B., Hanania, G. I. H., Gurd, F. R. N. 1979. *Biochemistry* 18:1919–36
185. Kilmartin, J. V., Fogg, J. H., Perutz, M. F. 1980. *Biochemistry* 19:3189–93
186. Schellman, J. A., Hawkes, R. B. 1980. See Ref. 1, pp. 331–43
187. Creighton, T. E. 1980. See Ref. 1, pp. 427–46
188. Creighton, T. E. 1980. *J. Mol. Biol.* 144:521–50
189. Konishi, Y., Ooi, T., Scheraga, H. A. 1981. *Biochemistry* 20:3945–55
190. Kosen, P. A., Creighton, T. E., Blout, E. R. 1980. *Biochemistry* 19:4936–44
191. Högberg-Raibaud, A., Goldberg, M. E. 1977. *Proc. Natl. Acad. Sci. USA* 74:442–46
192. Garel, J.-R., Dautry-Varsat, A. 1980. See Ref. 1, pp. 485–99
193. Garel, J.-R., Dautry-Varsat, A. 1980. *Proc. Natl. Acad. Sci. USA* 77:3379–83
194. Dautry-Varsat, A., Garel, J.-R. 1981. *Biochemistry* 20:1396–401
195. Isenman, D. E., Lancet, D., Pecht, I. 1979. *Biochemistry* 18:3327–36
196. Goto, Y., Hamaguchi, K. 1981. *J. Mol. Biol.* 146:321–40

197. Ghélis, C., Tempete-Gaillourdet, M., Yon, J. M. 1978. *Biochem. Biophys. Res. Commun.* 84:31–36
198. Pabo, C. O., Sauer, R. T., Sturtevant, J. M., Ptashne, M. 1979. *Proc. Natl. Acad. Sci. USA* 76:1608–12
199. Gilbert, W. 1978. *Nature* 271:501
200. Gilbert, W. 1980. *J. Supramolec. Structure* Suppl. 4, p. 115. (Abstr. 9th ICN-UCLA Symp.)
201. Brack, C., Tonegawa, S. 1977. *Proc. Natl. Acad. Sci. USA* 74:5652–56
202. Brack, C., Hirama, M., Lenhard-Schuller, R., Tonegawa, S. 1978. *Cell* 15:1–14
203. Lomedico, P., Rosenthal, N., Efstratiadis, A., Gilbert, W., Kolodner, R., Tizard, R. 1979. *Cell* 18:545–58
204. Lawn, R. M., Fritsch, E. F., Parker, R. C., Blake, G., Maniatis, T. 1978. *Cell* 15:1157–74
205. Craik, C. S., Buchman, S. R., Beychok, S. 1980. *Proc. Natl. Acad. Sci. USA* 77:1384–88
206. Jaenicke, R., Rudolph, R. 1980. See Ref. 1, pp. 525–48
207. Jaenicke, R., Rudolph, R., Heider, I. 1981. *Biochem. Int.* 2:23–31
208. Bothwell, M., Schachman, H. K. 1974. *Proc. Natl. Acad. Sci. USA* 71:3221–25
209. Bothwell, M. A., Schachman, H. K. 1980. *J. Biol. Chem.* 255:1962–70
210. Bothwell, M. A., Schachman, H. K. 1980. *J. Biol. Chem.* 255:1971–77
211. Smith, D. H., Berget, P. B., King, J. 1980. *Genetics* 96:331–52
212. Goldenberg, D. P., King, J. 1981. *J. Mol. Biol.* 145:633–51
213. Smith, D. H., King, J. 1981. *J. Mol. Biol.* 145:653–76
214. Hantgan, R. R., Taniuchi, H. 1978. *J. Biol. Chem.* 253:5373–80
215. Andria, G., Taniuchi, H., Cone, J. L. 1971. *J. Biol. Chem.* 246:7421–28
216. Englander, S. W., Downer, N. W., Teitelbaum, H. 1972. *Ann. Rev. Biochem.* 41:903–24
217. Woodward, C. K., Hilton, B. D. 1979. *Ann. Rev. Biophys. Bioeng.* 8:99–127
218. Richards, F. M. 1979. *Carlsberg Res. Commun.* 44:47–63
219. Wagner, G., Wüthrich, K. 1979. *J. Mol. Biol.* 134:75–94
220. Schreier, A. A., Baldwin, R. L. 1976. *J. Mol. Biol.* 105:409–26
221. Schreier, A. A., Baldwin, R. L. 1977. *Biochemistry* 16:4203–9
222. Furie, B., Schechter, A. N., Sachs, D. H., Anfinsen, C. B. 1975. *J. Mol. Biol.* 92:497–506
223. Schmid, F. X., Baldwin, R. L. 1979. *J. Mol. Biol.* 133:285–87
224. Dolgikh, D. A., Gilmanshin, R. I., Brazhnikov, E. V., Bychkova, V. E., Semisotnov, G. V., Venyaminov, S. Y., Ptitsyn, O. B. 1981. *FEBS Lett.* 136:311–15
225. Ptitsyn, O. B. 1972. *Pure Appl. Chem.* 31:227–44

Ann. Rev. Biochem. 1982. 51:491–530

THE NICOTINIC CHOLINERGIC RECEPTOR: Correlation of Molecular Structure with Functional Properties

B. M. Conti-Tronconi

Department of Pharmacology, School of Medicine, Universita' degli Studi di Milano, Via Vanvitelli 32, Milano, Italy

M. A. Raftery

Church Laboratory of Chemical Biology, Division of Chemistry and Chemical Engineering, California Institute of Technology, Pasadena, California 91125

CONTENTS

491

0066-4154/82/0701-0491$02.00

SUMMARY AND PERSPECTIVES

The nicotinic acetylcholine receptor (AcChR) is the first neurotransmitter receptor to be identified as a molecular entity, to be isolated and purified in an active form, and to be reconstituted in artificial membrane systems with quantitative retention of physiological properties. It is possible to purify substantial amounts of the AcChR protein and its constituent subunits and it is expected that the complete primary structure will soon be elucidated. Such information can be used to identify and locate agonist binding sites responsible for AcChR activation and desensitization and the channel for cation flux through the membrane. Conformational changes associated with ligand binding and consequent channel activation can then be identified, using spectroscopic approaches, so that a complete, formal mechanism for AcChR activation and the cation gating property will be eludicated. The structure and regulation of the genes coding for this molecule will be investigated in the near future as well as the evolution of the receptor and its expression in different tissues of the same organism.

INTRODUCTION

The AcChR is a membrane protein that occurs in many neuronal systems and transduces a chemical signal into an electrical event via a local depolarization of the membrane in which the receptor resides. It is now possible to characterize the AcChR at the molecular level and to study its component parts using biochemical and biophysical methods. The identification, isolation, characterization, and purification of electric organ AcChR and subsequently of muscle membrane AcChR has been greatly facilitated by the use of small protein toxins (α-neurotoxins) from the venom of some Elapid snakes (*Bungarus multicinctus* and various *Naja* species).

Here we discuss the molecular properties of the AcChR with emphasis on its structure, its interactions with cholinergic ligands, and the characterization of the ion channel associated with the receptor.

THE ELECTROPLAQUE AcChR MOLECULE

AcChR has been purified from detergent extracts of the electric organs of *Torpedo, Narcine,* and *Electrophorus* and of mammalian muscle, using affinity chromatography that employs either cholinergic ligands (30, 34) or α-neurotoxins (2, 18; Tables 1 and 6). A broad range of specific activities for α-neurotoxin binding to purified AcChR has been reported (6–12 nmol of α-neurotoxin bound/mg of receptor protein; Tables 1 and 6). Since the molecular weight of the monomeric form of the AcChR is 255×10^3, and two α-neurotoxin molecules bind to this unit, the specific activity should be 8 nmol/mg protein. The variable values reported are presumably due to uncertainties in protein determination and radiolabeled toxin calibration, different assay procedures, and proteolytic degradation.

A major controversy regarding the subunit composition of electric organ AcChR arose from conflicting subunit patterns consistently found in different laboratories. Some reports described preparations of purified AcChR composed of four different subunits with molecular weights of 40,000, 50,000, 60,000, and 65,000[1] while other studies reported simpler subunit patterns consisting of three, two, or even only one subunit (see Table 1). The observation (6, 7, 8, 11, 31) that addition of potent inhibitors of proteases activated by divalent cations tended to consistently yield preparations containing four subunits (see Table 1) favored the possibility that the native AcChR molecule composed of subunits of four different types was partially degraded during the purification, with retention of the integrity of only the most resistant parts. The inherent susceptibility of the AcChR to proteolytic attack is evident from the effect of exogenous proteases on the subunit pattern. However, such proteolysis does not obviously affect many physical and functional properties of the AcChR molecule. Extensive degradation of *T. Californica* AcChR by trypsin or papain was shown to drastically alter the subunit pattern in SDS gel electrophoresis (13, 85, 142, 143) without significantly affecting the binding capacity for α-neurotoxins, antibodies, and cholinergic ligands (13, 85). The sedimentation behavior and morphology were also only slightly affected, which indicates that extensive proteolytic cleavage of peptide bonds resulted in minimal fragmentation until the protein was denatured in SDS (13, 85, 142, 143). In the case of solubilized, purified AcChR, the morphology appears to be the same in different preparations, irrespective of the subunit patterns observed in SDS gel profiles (23, 29, 31, 41, 100, 142).

Purified electric eel AcChR was reported to be composed of only three subunits with molecular weights comparable to the three lighter subunits of *Torpedo* AcChR (7, 26), even after purification in the presence of EDTA. The likely occurrence of sulfhydryl-containing proteases was indicated by the demonstration (43) that inclusion of iodoacetamide in the isolation medium allowed the identification of a fourth subunit with a molecular weight of $\sim$ 64,000 in *Electrophorus*.

---

[1] These and similar subunits are referred to in text as the 40K, 50K, etc subunits.

Table 1 Subunit composition and specific activity of purified AcChR fish electric organs

Source	Subunits (Daltons × 10⁻³)				Specific activity (nmol mg⁻¹)	Authors	Year	Ref
Torpedo californica	(26→35) 42				6.3	Schmidt & Raftery	1973	1
	(35→45) and higher				—	Raftery	1973	2
	40	49	60	67	6.0	Raftery et al	1974	3
	39	48	58	64	—	Weill et al	1974	4
	40	49	60	67	—	Reed et al	1975	5
	41	57	60	64	10.0	Raftery et al	1976	6
	39	48	58	64	8.0	Karlin et al	1976	7
	40	50	60	65	9–10	Vandlen et al	1976	8
	40	48	59	67	10–12	Chang & Bock	1977	9
	44	53	58	64	—	Froehner et al	1977	10
	38	49.5	57	64	—	Lindstrom et al	1978	11
	40	50	60	65	8–9	Vandlen et al	1979	12
	40	48	55	64	10.6	Bartfeld & Fuchs	1979	13
	38	47	57	68	—	Sumikawa & O'Brien	1979	14
	43	52	58	63	5.6–12.2	Froehner & Rafto	1979	15
	44	53	60	65	—	Nathanson & Hall	1979	16
	44	52	58	66	5.9	Lennon et al	1980	17
Torpedo marmorata	—	—	—	—	3.3–7.1	Karlsson et al	1972	18
	—	—	—	—	8.9–9.12	Eldefrawi & Eldefrawi	1973	19
	46 and higher				—	Carroll et al	1973	20
	45 →	50			1.71 (74, 93, 148)	Heilbronn & Mattson	1974	21
	37	49			5.0	Gordon et al	1974	22

Species							Author	Year	Ref
	40					9.0	Sobel et al	1977	23
	41	51	59	64		—	Claudio & Raftery	1977	24
	—	—	—	—		10.0	Schorr et al	1978	25
	41	51	58	63		5.67	Deutsch & Raftery	1979	26
	40	50	57	66		—	Saitoh et al	1980	27
Torpedo nobiliana	33 35 38 43					12.5	Ong & Brady	1974	28
	40	48	59	64		—	Claudio & Raftery	1977	24
	40	50	60	65		4.6	Deutsch & Raftery	1979	26
Torpedo oscellata	40	50	61	(81)		—	Eldefrawi et al	1978	29
Narcine entemedor	(28) 38	45				—	Schmidt & Raftery	1972	30
Narcine braziliensis	44	48	58	65		9–10	Chang et al	1977	31
	43	52	59	64		—	Claudio & Raftery	1977	24
	42	51	58	64		5.94	Deutsch & Raftery	1979	26
	42	50	58	66 and higher		—	Kemp et al	1980	32
Narke japonica	33 43 (38→51)					2.17	Ishikawa et al	1980	33
Electrophorus electricus	44	(50)				—	Olsen et al	1972	34
electricus	—	—			(95→100)	4.5	Biesecker	1973	35
	—	—				11.0	Klett et al	1973	36
	45	54			(90)	5.4	Hucho & Changeux	1973	37
	41	47 53			(110)	—	Karlin & Cowburn	1973	38
	(37) 42	49			(102 and higher)	5–6.5	Chang	1974	39
	42	54			(and higher)	—	Lindstrom & Patrick	1974	40
	43	48			(90)	6.7	Meunier et al	1974	41
		48 53	60		(and higher)	3.6	Patrick et al	1975	42
	39	48				—	Sumikawa & O'Brien	1979	14
	40	49	56			3.8	Deutsch & Raftery	1979	26
	41	50	55	64		—	Lindstrom et al	1980	43

Evidence for Four Subunits

The four AcChR polypeptides appear to interact strongly, since they remain tightly associated in the presence of nondenaturing detergents, and only the strong denaturant SDS has been shown to be effective in dissociating them (Table 1). The question arose as to whether these four polypeptides are part of the AcChR molecule or whether they only copurify with the AcChR *sensu stricto*. The isolation of membrane protein complexes in detergent solution does not necessarily mean that these same proteins are part of a discrete complex in the membrane (8).

Using dimethyl suberimidate to cross-link detergent solubilized preparations of *Narcine* and *Torpedo californica* AcChR, products were formed that contained all four peptides (31, 59). Antisera selective for each subunit type of *T. californica* AcChR could precipitate to the same extent the intact receptor, which indicates physical association of the polypeptides in nondenaturing detergent solution (11, 24, 144). In view of possible differences in interactions of membrane proteins in detergent solution and in membranes, the subunits were studied to determine which could be cross-linked to photolabile α-neurotoxin derivatives in membranes. Either two (31, 129, 145), or all four (147) of the AcChR subunits were cross-linked to the toxin upon photolysis, which supports the notion that all four peptides present in the membrane preparation are associated with each other.

It was difficult to imagine how the AcChR complex might be formed from four different polypeptides, only two of which, i.e. the 40K subunits, were structurally related. Since complex protein systems (148) seem to be generally constructed from identical or related subunits, the suggestion that the AcChR was a hexamer of 40K subunits (93, 149–151) held a certain measure of appeal. This dilemma has been resolved by the demonstration of extensive amino acid sequence homology in *T. californica* AcChR by amino acid sequence analysis of the four subunits with molecular weights of 40,000, 50,000, 60,000, and 65,000 (152). All four subunits are identical in 11 out of the first 54 positions; in many other positions either 2 or 3 of the residues are identical or related and in only 4 of the 54 positions do the 4 polypeptides contain different and unrelated amino acids. The structural similarity of the four AcChR subunits allows the generation of a pseudo-symmetric oligomer in a manner similar to many complex proteins (148). The identity between pairs of subunits ranges from 35% to 50%. In addition, many conservative substitutions tend to further relate the four polypeptides.

Interestingly, *T. californica* AcChR subunit structure had been investigated earlier by peptide mapping of the separated polypeptides (15, 16, 153). It was concluded that the subunits were not derivatives of each other (15), that there was no extensive homology between the chains (153), and that the four polypeptides appeared to be structurally unrelated (16). It is clear that peptide mapping is not a high resolution method for detection of homology.

The high degree of homology between the AcChR subunits explains why antibodies raised against purified subunits cross-reacted to a lesser extent with one or more of the other subunits. This is particularly true for the pair of polypeptides with molecular weights of 60,000 and 65,000. In fact most attempts to raise antisera or monoclonal antibodies specific for one of these subunits have resulted in cross-reacting preparations (24, 43, 144, 154, 155). Antibodies raised against individual subunits of *T. californica* AcChR cross-react with subunits of corresponding molecular weight from other electric rays (24), and from electric eel AcChR (43, 144). In addition the "subunit specific" antibodies cross-reacting with both the 60K and 65K subunits of *Torpedo* AcChR also cross-react with both corresponding subunits of eel AcChR (43, 144), and *Narcine braziliensis* (24). This lack of specificity is not surprising, since it has been recently demonstrated that *Electrophorus* AcChR subunits are highly homologous with each other and with the corresponding polypeptides from *T. californica* (156, 157).

Comparison of the amino-terminal sequences of the 40K subunit from *Torpedo marmorata* (158) and *T. californica* (159) reveals a sequence identity for the first 20 amino acids.

Determination of the subunit stoichiometry in *T. californica* AcChR has been achieved by two different experimental approaches. Amino acid analysis of subunits extracted from analytical SDS gels and normalization using the subunit molecular weight values suggested that the ratio of the 40K, 50K, 60K, and 65K subunits is approximately $2:1:1:1$ (153). A different approach used the known amino-terminal sequences of the individual polypeptide chains. With *Torpedo* AcChR the four amino-terminal sequences were determined simultaneously, and from the ratio of the four different amino acids present at three known positions the stoichiometry was determined (62, 152) to be $2:1:1:1$. The definitive stoichiometry of *Electrophorus* AcChR subunits is not yet known. However, the molecular weights of complexes formed between eel AcChR and different amounts of monoclonal antibodies directed against the 40K chain indicate that in this AcChR molecule two copies of the 40K chain are also present (160).

Receptor Composition

The amino acid composition of *Torpedo, Narcine,* and *Electrophorus* AcChR are remarkably similar (12), and little difference is observed in the composition of *T. californica* AcChR subunits (12, 153). The AcChR from both *Torpedo* species and *Electrophorus electricus* contains carbohydrates

(12, 26, 41, 161), which are present, at least in *Torpedo,* on all the four subunits (7, 153, 161–163).

Electrophorus and muscle AcChRs exist as two isoelectric forms (73, 52) possibly differing in net charge as a result of phosphorylation (73). Kinase and phosphatase activities have been demonstrated in membrane fragments from *Torpedo* and *Electrophorus* electroplax (164–167). A 65K phosphorylated component for *T. californica* AcChR (164, 165) and a 48–50K component for *Electrophorus* AcChR (145, 146) were identified. In *T. marmorata* membranes, polypeptides with molecular weights of 60,000, 50,000 and 43,000 were labeled by γ-^{32}P-ATP (169).

It is difficult to assess the extent of amino acid phosphorylation on a molar basis from these studies. However, the occurrence of phosphoserine (7 mol/mol) in *T. californica* AcChR preparations has been demonstrated by amino acid analysis (12). The specificity of kinase-mediated phosphorylation in vitro is open to question. In initial studies (164, 165, 167), identification of the phosphorylated subunit(s) relied on cross-immunoelectrophoresis performed with an antiserum prepared against native AcChR, which cross-reacts with SDS denatured subunits with a much lower affinity, if at all, and it is therefore possible that other labeled subunits were not detected. Indeed, apparent labeling of a 50K subunit in addition to that of a 66K subunit was recently observed for *T. marmorata* AcChR (169). This apparent labeling of a 50K peptide, as well as another 43K chain is open to question, since the same authors (27) have recently shown that 50K and 45K products are readily generated from the 66K and 60K receptor subunits by the action of endogeneous proteases. This can be true also for the labeling of a 48–50K component(s) of *Electrophorus* AcChR by endogeneous kinases (167, 168) since until recently (43) all AcChR preparations from *Electrophorus* were partially degraded, with the 64K subunit being consistently absent (see Table 1). No relationship to physiological function has yet been ascribed to phosphorylation.

Physical Properties

The AcChR purified from *Torpedo* species has been shown to occur as monomeric and dimeric forms, with sedimentation coefficients of 9 S and 13 S respectively (see Table 2). The monomer is composed of the 40K, 50K, 60K, and 65K subunits in the ratio 2:1:1:1. The dimer is composed of two monomers held together by a disulfide bridge(s) between the 65K subunits (9, 170–172). In the membrane bound state in *T. californica,* the dimer is the predominant if not the exclusive form.

For *Electrophorus* AcChR only 9 S forms have been described (40, 41, 50). It is difficult to assess the possible occurrence of dimers, since until recently a 65K subunit for *Electrophorus* receptor was unknown, and there

Table 2 Sedimentation values of purified and/or solubilized AcChR

Sources	S values	Authors	Year	Ref
Torpedo californica	9, 13.7	Raftery et al	1972	44
	9, 13.3	Gibson et al	1976	45
	9, 13	Karlin et al	1976	7
	9, 13	Chang & Bock	1977	9
	9	Kalderon & Silman	1977	46
	8.6, 12.8	Reynolds & Karlin	1978	47
Torpedo marmorata	8.8, 18, 26	Carroll et al	1973	20
	8.6, 12.5	Gibson et al	1976	45
	9	Sobel et al	1977	23
	9.2, 14.4	Doster et al	1980	48
Electrophorus	11–12.5	Meunier et al	1972	49
electricus	9.5–10	Meunier et al	1972	50
	9.5	Lindstrom & Patrick	1974	40
	9	Meunier et al	1974	41
Torpedo ocellata	8.6, 12.5	Gibson et al	1976	45
Muscle				
rat	9	Berg et al	1972	51
rat	9	Brockes & Hall	1975	52
mouse	9.5	Boulter & Patrick	1977	53
rat	9, 13	Froehner et al	1977	10
cat	9.5	Merlie et al	1978	54
cat	9, 13	Barnard et al	1979	55
rat	9, 13	Barnard et al	1979	55
cat	4, 9	Shorr et al	1981	56

is evidence that a loss of antigenic determinants occurred when the 65K subunit was not present [Figure 2 in (43)], which indicates a physical loss of mass in degraded forms of eel AcChR lacking the 65K subunit.

The molecular weight of the electroplax AcChR monomer [estimated using a variety of physical methods (Table 3)] appears to be about 250,000. Gel filtration yielded values that are higher than the true molecular weight (Tables 3 and 4). The most reliable methods for determining the molecular weight of this membrane protein include: chemical cross-linking (35, 37, 58); physical approaches such as membrane osmometry (57); sedimentation equilibrium and sedimentation velocity (20, 45, 47, 49, 58); X-ray diffraction (60); laser light scattering combined with sedimentation velocity (48); low angle neutron scattering (61); and subunit stoichiometry determination (62, 152, 153). The last method requires that the molecular weight values of 40K, 50K, 60K, and 65K for the four subunits be accurate.

The isoelectric points of AcChR from *Torpedo, Narcine, Electrophorus,* and mammalian muscle are similar, with values in the slightly acidic range (see Table 5). For muscle (16, 52), and *Electrophorus* AcChR (73) two

Table 3 Molecular weight of AcChR

Source	Method	Mol wt ($\times 10^{-3}$)	Authors	Year	Ref
Torpedo californica electric organ	Membrane osmo-metry	270 ± 30	Martinez-Carrion et al	1975	57
	Sedimentation	330 and 660	Edelstein et al	1975	58
	Sedimentation	190 and 330	Gibson et al	1976	45
	Cross-linking	270	Hucho et al	1978	59
	Sedimentation	250 and 500	Reynolds & Karlin	1978	47
	X-ray diffraction	250–310	Klymkowski & Stroud	1979	60
	Low angle neutron scattering	240 ± 40	Wise et al	1979	61
	NH_2-terminal sequence analysis	255	Strader et al	1980	62
Torpedo marmorata electric organ	Gel filtration	400	Molinoff & Potter	1972	63
	Sedimentation	500 and 1000	Carroll et al	1973	20
	Sedimentation	330 and 1300	Edelstein et al	1975	58
	Laser scattering/sedimentation velocity	298	Doster et al	1980	48
Electrophorus electricus electric organ	Sedimentation	323 and 470	Meunier et al	1972	49
	Cross-linking	230 ± 15	Hucho & Changeux	1973	37
	Cross-linking	260	Biesecker	1973	35
	Cross-linking	250	Lindstrom et al	1976	64
Narcine braziliensis electric organ	Cross-linking	400	Chang et al	1977	31
	Gel filtration	530	Chang et al	1977	31
Mammalian muscle: rat	Gel filtration	550 (and higher)	Chiu et al	1973	65
cat	Gel filtration	430	Dolly & Barnard	1975	66
rat	Gel filtration	300	Colquhoun & Rang	1976	67
rabbit	Gel filtration	390	Bradley et al	1976	68
rat and cat	Gel filtration	370	Dolly & Barnard	1977	69
rat	Gel filtration	390 (plus traces of dimers)	Kemp et al	1980	32

isoelectric forms have been observed that differ by approximately 0.2 pH units. It has been suggested that they represent synaptic and extrasynaptic receptors, with the more acidic form originating from the synaptic membrane (52).

THE MUSCLE AcChR

Purification of the AcChR from mammalian muscle posed difficulties in comparison with the fractionation of solubilized preparations from electric ray. Isolation of highly enriched postsynaptic membrane fragments is not feasible because of the very low content of AcChR in muscle sarcolemma (173). In normal, innervated mammalian muscle, the AcChR content ranges from 0.1 to about 0.3% of that of *Torpedo* electric organ (70), i.e. 0.002% of the total protein of the tissue (147). In denervated or fetal muscle (51) and in cultured muscle cells (53) the AcChR content is considerably higher (from six to twenty fold increases have been recorded after denerva-

Table 4 Stokes radius of AcChR

Sources	Stokes radius	Authors	Year	Ref
Torpedo californica	125Å ± 10	Kalderon & Silman	1977	46
	73Å (monomer) 95Å (dimer)	Reynolds & Karlin	1978	47
Torpedo marmorata	42Å (degraded) 69Å (monomer) 85Å (dimer)	Potter	1973	70
Electrophorus electricus	73Å	Meunier et al	1973	71

tion) (51, 70, 75). The specific activities of purified receptor preparations from muscle (10, 16, 25, 32, 53–55, 66, 69, 75, 76) are comparable to those reported for electric tissue AcChR (see Table 6). These preparations contained either a single 41–43K subunit (25, 76) or several subunits with different molecular weights (see Table 6). There are indications that muscle AcChR is composed of more than one type of subunit (10, 16, 32, 53), with molecular weights roughly comparable to those reported for electric tissue AcChR (10, 16, 53).

Table 5 Isoelectric points of AcChR

Sources	pI	Authors	Year	Ref
Torpedo californica	RT[a] 4.9	Raftery et al	1972	44
Torpedo marmorata	R[b] 5.3 (> 2 M urea → 5.6)	Sobel et al	1977	23
Narcine brazielensis	RT 5.0	Chang et al	1977	31
Electrophorus electricus	RT 5.2	Raftery et al	1971	72
	R 4.7 RT 5.2	Biesecker	1973	35
	R 4.57 R 4.85	Teichberg & Changeux	1976	73
Mammalian muscle	RT 5.1 J[c] RT 5.3 EJ[d]	Brockes & Hall	1975	52
	RT 5.3	Dolly & Barnard	1977	69
	R 5.0	Barnard et al	1978	55
	RT 5.2 J RT 5.47 EJ	Nathanson & Hall	1979	16
	RT 5.2 R 5.6	Stephenson et al	1979	74
	R 5.0	Shorr et al	1981	56

[a] RT, AChR-toxin complex   [c] Junctional
[b] R, AChR   [d] Extrajunctional

Table 6 Subunits of purified muscle AcChR

Source	Mol wt of subunits ($\times\ 10^{-3}$)				Specific activity	Authors	Year	Ref
Muscle								
rat junctional	—	—	—	—	0.19	Brockes & Hall	1975	75
rat extrajunc- tional	—	—	—	—	0.5			
cat denervated	—	—	—	—	6.0	Dolly & Barnard	1975	66
rat denervated	45	(49)51	(56, 62)	(110)	8–10	Froehner et al	1977	69
cat denervated	—	—	—	—	6.0	Dolly & Barnard	1977	69
cell line	44	53		65 (72)	2.6	Boulter & Patrick	1977	53
rat denervated	—	—	—	—	10.5	Barnard et al	1978	55
calf fetal muscle	41	(46)	—	—	—	Merlie et al	1978	54
calf denervated cultured	41	—	—	—	10–11	Schorr et al	1978	25
rat junctional	45	49, 51	56, 62		—	Nathanson & Hall	1979	16
cat denervated	43	—	—	—	—	Lyddiatt et al	1979	76
human	42	—	—	66	—	Stephenson et al	1979	74
cat denervated	43	—	—	—	10–11.5	Schorr et al	1981	56
mouse cultured	42	46, 48	60		—	Merlie	1981	77

Two findings support the notion that mammalian muscle AcChR contains structural domains related to all the *Torpedo* subunits. First, monoclonal antibodies or antisera induced against each individual *Torpedo* AcChR subunit can cross-react with mammalian muscle AcChR (11, 43, 144, 155), and second it is possible to induce in rats an immune response against their AcChR using any of the four *Torpedo* subunits (11).

Affinity labeling of dithiothreitol-treated AcChR from denervated muscle by the cholinergic agonist [^{3}H]bromoacetylcholine resulted in labeling of a 43K peptide (76). Since this was the only subunit type present in this preparation it is not possible to compare these results with those obtained using [^{3}H]-MBTA (maleimido-benzyl trimethylammonium chloride), which labeled two muscle AcChR subunits (174) with molecular weights of 45,000 and 49,000. This dual labeling could be due to a difference in subunit glycosylation. Selective staining of SDS gels revealed that the 41K subunit of the muscle receptor, like the *Torpedo* subunits, contains some carbohydrate (25).

As was found for eel AcChR, results obtained using monoclonal antibodies against the 40K subunit suggest the presence of two such subunits in the muscle AcChR molecule (160).

The molecular weight of solubilized or purified muscle AcChR has been estimated only by gel filtration or sedimentation in sucrose gradients (see Tables 2 and 3) and the values obtained ranged from 370,000 to 550,000 (see Table 3). Sedimentation coefficients ranging from 9 S to 9.5 S have been observed for most preparations, irrespective of their subunit patterns, upon SDS gel electrophoresis. Interestingly, muscle AcChR from two species (rat and cat) occurs as both 9S and 13 S species (10, 55) composed of the same subunit types (10) (see Table 2). The finding that this structural feature is conserved in species as phylogenetically distinct as *Torpedo*, rat, and cat raises the possibility that dimers are a common feature of AcChR in most or all creatures. The morphology of AcChR purified from calf fetal muscle has been studied by negative staining (175) and appears to have the same shape and dimensions as *Torpedo* and *Electrophorus* AcChR (see later section).

The structural similarity between *Torpedo* and mammalian AcChR has ramifications with regard to human pathology, since it is known that a paralytic syndrome, myasthenia gravis, results from an autoimmune reaction against neuromuscular AcChR (176). In myasthenia gravis the destruction of AcChR is achieved by specific autoantibodies, mainly through antigenic modulation of the AcChR, i.e. acceleration of catabolism of AcChR molecules cross-linked by antibodies. If mammalian AcChR has the same subunit structure and stoichiometry as *Torpedo* AcChR, then repetitive homologous antigenic determinants will be present on its surface. As a consequence, it is possible that antibodies can form intramolecular cross-links between homologous antigenic determinants, and they would be inefficient in causing antigenic modulation. This possibility is exemplified by the demonstration (160) that a high proportion of the monoclonal antibodies specific for the 40K subunit of *Torpedo* AcChR cannot cross-link two monomeric molecules of AcChR, because of intramolecular complex formation.

THE AcChR IN POSTSYNAPTIC MEMBRANES

The isolation from *Torpedo* species of membrane fragments highly enriched in AcChR was an important technical development for in vitro studies and was based on the high packing density of the AcChR (78). The first reports of such preparations were from *T. marmorata* (86) and *T. californica* (78).

In addition to the four AcChR subunits, the membrane preparations contained proteins of 105K and 43K daltons (see Table 7). Preparations lacking these polypeptides can be obtained by treatment of the membranes at pH 11 in low ionic strength (82, 83, 91), by extraction with lithium diiodosalicylate (83), or by affinity partitioning methods (84).

AcChR Mobility in Postsynaptic Membranes
The AcChR is packed at high density in postsynaptic regions both in electric organ and in muscle (see later section and Table 8), with the exclusion of other intrinsic membrane proteins. The receptor molecules do not undergo lateral diffusion in or out of this area (177–179), and in *Torpedo*

Table 7 Polypeptide composition of AcChR enriched membranes

Source (subunits in membranes)	Mol wt of subunits (× 10⁻³)					Specific activity	Authors	Year	Ref
	40	50	60	65					
Torpedo californica	40 (and higher)				105 [a]	—	Duguid & Raftery	1973	78
	40	49	60	67		—	Raftery et al	1974	3
	40	49	60	67	105	—	Reed et al	1975	5
	40	49	60	67	105	2.9–4.6	Flanagan et al	1976	79
	40	48	62	66	100	4.2	Hucho et al	1976	80
	42	48	62	68		3.3	Hucho et al	1978	59
	40	49	60	67	105	2–4.5	Weiland et al	1979	81
	40(43)	50	60	65		—	Elliott et al	1979	82
	40(43)	50	60	65		—	Elliott et al	1980	83
	40(43)	50	60	65		7.8	Johansson et al	1980	84
	~40	~50	~60	~65	~100	—	Klymkowsky et al	1980	85
Torpedo marmorata	—	—	—	—		2.3	Cohen et al	1972	86
	40 43		(50)	(66)		4.0	Sobel et al	1977	23
	40	50	57	66		—	Saitoh et al	1980	27
	40(43)	50	(60)	65	~100	—	Barrantes et al	1980	87
	40	50	58	66		—	Lo et al	1980	88
	40(43)	(45)50(54)(55)		66	95	—	Oswald et al	1980	89
Torpedo nobiliana	40(43)	50		65		—	Neubig et al	1979	91
Narcine braziliensis	44	48	58	65		7.0	Allen & Potter	1979	90

[a] Figures in parentheses indicate polypeptides present in trace amounts.

they have been shown to have a rotational correlation time greater than 10^{-3} sec (88, 180). It seems likely that synaptic AcChR immobility is at least partially due to receptor-receptor interactions, such as dimer formation in *Torpedo* and *Narcine* electric organ and possibly in mammalian muscle (7, 9, 10, 44, 45, 47, 48, 55). Higher order oligomers of AcChR molecules appear in freeze-etching and freeze-fracture images of *Torpedo* (101) and muscle (122) postsynaptic membranes. Solubilization with detergents of AcChR from these membranes leads to a nearly complete loss of these supramolecular aggregates. The possibility exists that lack of mobility is due to protein-protein interaction between the AcChR molecules and other membrane protein(s), such as a 43K extrinsic membrane protein, which is consistently present in AcChR-rich membrane preparations. Removal of this component by base extraction (82, 91), results in enhanced thermal lability of the AcChR (181). Alkali extraction of *T. marmorata* membranes also results in an increase of AcChR rotational mobility by a factor of 50–100 (88), and in a more uniform distribution of AcChR molecules in the membrane (87). None of these studies demonstrated a direct interaction between the AcChR and the 43K protein. A simple alternative is that the latter interacts with phospholipids and indirectly affects receptor-receptor interactions.

AcChR as a Transmembrane Protein

The first attempts to demonstrate that the AcChR is a transmembrane protein utilized anti-AcChR antibodies visualized by electron microscopy.

Using ferritin-labeled antibodies bound to AcChR-rich vesicles (182, 183) it was shown that the antibodies consistently bound to their outer surfaces. The inner surface was labeled in one study (182), unlabeled in the other one (183). The discrepancy could be due to lack of access of the large antibody ferritin conjugate to the inside of sealed vesicles. Another approach used intact or sonicated vesicles attached to coverslips; the sonicated vesicles had their cytoplasmic surfaces exposed (184). Both surfaces were consistently labeled by an anti-AcChR antiserum.

The transmembrane exposure of the AcChR subunits in *T. marmorata*-enriched membranes has been investigated by observing the effects of proteases on membrane bound and solubilized AcChR (163). Enzymes added to the outside of sealed vesicles caused no degradation of AcChR subunits. When the enzymes were present internally, all three AcChR subunits, (the preparations lacked a 60K subunit) were degraded. With *T. californica* membranes containing all four AcChR polypeptides (143, 185), all the AcChR subunits were susceptible to tryptic degradation both from the outside and the inside. From this result and from the different time courses of the degradation of the single subunits from inside and outside it was concluded that all the AcChR sub-units cross the membrane, protrude outside to the same extent, and inside to an extent proportional to their molecular weight.

AcChR-Lipid Interactions

For *Torpedo* species the lipid composition of both unfractionated (186–188) or fractionated (189, 190) electroplax membranes has been described. The protein to lipid ratio ranged from 1.25 to 2.3 (98, 189, 196). In lipid monolayer studies it was shown that the AcChR protein interacted preferentially with cholesterol and long-chain phosphatidylcholines (190). Alteration of fatty acyl groups in cultured myotube membranes, however, had only small effects on AcChR mobility (191).

Spin-label studies of fatty acid derivatives that inhibit binding of acetylcholine showed that fluid lipids exist close to the AcChR protein (192). The purity of the membranes used (2.3 nmol α-toxin/mg protein; see Table 7) and the specificity of the probes lend credence to the results. A different spin-label approach to monitor the fluidity of *T. marmorata* AcChR–enriched membranes used stearic acid and steroid probes (193). The conclusions were that the proportion of immobilized lipids corresponded to the total lipid. This calculation was, however, based on the area of negative contrast observed in electron micrographs of the membranes (fixed and unfixed) and on the assumption that the protein to lipid ratio was approximately 4:1. In view of the above quoted experimental determination of this ratio (98, 190), this calculation is likely to be in error, and the conclusion that the total lipid is immobile seems unwarranted. In addition, the discon-

Table 8 Density and distribution of AcChR in postsynaptic membranes

System	Method	Particle density (× 10⁻³)	Size (Å)	Lattice	Localization and density (× 10⁻³)	Authors	Year	Ref
Torpedo	FF[a]	5-8	50-100	—	—	Waser	1970	92
	NS[b]	10-15	60-70	+, —	—	Nickel & Potter	1973	93
	NS/FE[c]	—	80-90	+	—	Cartaud et al	1973	94
	FE	—	60, 85	+	—	Orci et al	1974	95
	FF	—	80-90	—	—	Clementi et al	1975	96
	TS/FF	5	70 (TS) / 115 (FF)	—	All over	Rosenbluth	1975	97
	NS/FE	12-15	80-90	+	—	Changeux et al	1976	98
	NS/Xr[d]	—	110 (X-r)	P_1	—	Ross et al	1977	99
	FE/FF/NS	8-10 (FF) / 10-12 (FE)	60-80 (FF) / 70 (NS)	+	—	Cartaud et al	1978	100
	FF/FE	10 (FE) / 5 (FF)	~85	Rows	—	Heuser & Salpeter	1979	101
	NS	9.5/5-6[e]	—	—	—	Barrantes et al	1980	87
	NS	6.6/9.4[f]	~70	—	—	Lo et al	1980	88
Narcine	TS/FF	6.1 (TS) / 5.7 (FF)	60/155	Semiregular array	All over	Potter & Smith	1977	102
	FF	5.7	90-125 (125-150)	Semiregular	All over	Allen et al	1977	103
Electrophorus	AR[g]	33	—	—	—	Bourgeois et al	1972	104
	AR	49.6	—	—	—	Bourgeois et al	1978	105
Muscle								
mouse	AR	12	—	—	—	Barnard et al	1971	106
rat, frog	AR	10	—	—	—	Miledi & Potter	1971	107

mouse	AR	13.7	—	—	—	Potter et al	1973	108
mouse/rat	AR	5–10	—	—	—	Salpeter & Eldefrawi	1973	109
mouse	AR	7 (average)	—	—	Crest (28)	Fertuck & Salpeter	1974	110
toad/frog	TS	10	60–120	+	Crest	Rosenbluth	1974	111
frog	FF	7.5	—	—	Crest	Peper et al	1974	112
rat	TS/FF	1.8	110–140	Rows	—	Rash & Ellisman	1974	113
frog	FF	6	80–120 60–70	—	Crest	Heuser et al	1974	114
mouse	AR	8.8	—	—	Crest (18)	Albuquerque et al	1974	115
bat	AR	8.8	—	—	Crest (20–25)	Porter & Barnard	1975	116
mouse	AR	8.5	—	—	Crest (22)	Porter & Barnard	1975	117
mouse	AR	8.7	—	—	Crest (20–25)	Porter & Barnard	1976	118
rat	FF	3 ± 1	100	Rows	—	Ellisman et al	1976	119
mouse	AR	30.5 ± 27%	—	—	Crest (30.5)	Fertuck & Salpeter	1976	120
rat	FF	2	110–140	Linear arrays	—	Ellisman & Rash	1977	121
worm	FF	4	70–150	Paracrystalline	—	Rosenbluth	1978	122
rat	FF	~3	100–140	Rows	Crest	Rash et al	1978	123
frog	AR	—	—	—	Crest (26 ± 6)	Matthew-Bellinger & Salpeter	1978	124
Muscle cells								
chick	AR	9	—	—	Cluster	Sytkowski et al	1973	125
rat myotubes	AR	3–4	—	—	Cluster	Land et al	1977	126
chick myotubes	FF	3.5–4	>100	—	Cluster	Cohen & Pumplin	1980	127

[a] FF, freeze fracture
[b] NS, negative stain
[c] FE, freeze etching
[d] X-r, X-ray diffraction
[e] 9.5 before and 5–6 after base extraction
[f] 6.6 before and 9.4 after base extraction
[g] AR, autoradiography

tinuous relationship between AcChR channel conductance and temperature (194) emphasizes the necessity of a fluid environment for a normal gating function of AcChR.

Identification of the receptor subunits in contact with the lipids of the postsynaptic membrane has been attempted using lipid-soluble photolabile azido compounds. [^{3}H]pyrenesulfonylazide (195, 196) was interpreted to label 48K and 55K polypeptides in *T. californica* postsynaptic membranes. However, this seemed preferential rather than specific, since a significant degree of incorporation occurred on 40K and 68K polypeptides. A related reagent, 5-[^{125}I]iodonaphthyl-1-azide (197) was found to selectively label the 40K subunit in membranes. The variations could be due to preferential reactivity of the photogenerated nitrenes with particular subunits. In fact, given the pseudosymmetrical structure of AcChR (152), it is likely that all the subunits interact with the surrounding membrane environment in a related fashion.

The Shape of the Electric Organ AcChR Molecule

The shape of AcChR and the extent to which it protrudes from the synaptic membrane were investigated with physical and morphological approaches. AcChR protrudes into the synaptic cleft 55–70 Å. This was first suggested by the observation that in thin sections of *Torpedo* (97) and *Narcine* (102) electrocytes, the outer lamina of the postsynaptic membrane is unusually thick and formed by juxtaposed granules 65–70 Å in diameter. Labeling of the external surface of *T. californica* membrane fragments with anti-AcChR antibodies attached to ferritin demonstrated that antibodies appeared to bind to AcChR-rich membranes with a gap of 20–50 Å (182). In negatively stained preparations from *Torpedo,* the edges of some vesicles have grommet-like structures that protrude approximately 50 Å from the plane of the postsynaptic membrane, which may represent the external portion of the AcChR molecule (60). In freeze-fracture replicas of purified AcChR-rich membranes, particles whose distribution and size were compatible with known dimensions and density of AcChR molecules [(100, 101) and later section] were present on both faces of the membrane bilayer.

Low angle X-ray diffraction of membrane fragments enriched in AcChR from *T. marmorata* (198) and *T. californica* (199) gave an electron density profile indicating that the AcChR extends through the membrane and protrudes on both sides (60, 99). The overall length of the receptor was estimated to be 110 Å normal to the plane of the membrane, and to extend 55 ± 5 Å on one side and 15 ± 5 Å on the other. An analysis of the shape of *T. californica* AcChR solubilized with the detergent Triton X-100 has been attempted using low-angle neutron scattering (61), and yielded a radius of gyration of 46 ± 1 Å for monomeric AcChR. This parameter, together with the calculated molecular volume (350,000 Å³) appeared to be inconsistent with a compact spherical shape and with the compact shapes proposed by Klymkowsky & Stroud (60), but is compatible with many different simple shapes, such as oblate or prolate cylinders and also with more complex models. This approach to molecular structure places vague limits on the variety of shapes that we can assume for the AcChR.

The shape of the external part of the AcChR molecule was studied by freeze-etching of the postsynaptic membranes in situ or in isolated fragments, and by negative staining of AcChR-rich membrane fragments. In deep-etched images, the AcChR molecules appeared as densely packed dimpled particles 90 ± 10 Å in diameter (94, 98, 101). By negative staining, postsynaptic membrane fragments appeared to be covered by rosette-like structures 90 ± 10 Å in diameter (93–95, 199), with a central pit 15–20 Å in diameter (93, 94). These same structures were observed with solubilized, purified AcChR from *Torpedo, Narcine,* and *Electrophorus* (29, 31, 41, 71, 94, 100, 142), thus validating their identification as AcChR molecules. It appears that each rosette corresponds to a monomer; the dimers are formed by two side-by-side rosettes (142, 200, 201). Freeze-fracture replicas of postsynaptic membranes had also shown the frequent presence of coupled particles, which were interpreted as dimers (122). A confirmation that these rosette-like structures are AcChR molecules was obtained by studying the binding of colloidal gold-anti-AcChR antibody complexes to the AcChR-enriched membranes (60). Because of the known orientation of the membrane vesicles (>95% outside out) (202) it is most likely that the negatively stained images represent the external portion of AcChR molecules.

The rosette-like structures seen with negative staining were interpreted in various instances to be composed of several subunits; the most frequent numbers quoted were 5 or 6 (41, 71, 93, 94, 100). The identification of such "subunits" was based on subjective judgment related to areas of higher and lower electron density potentially due to stain granularity, enhanced by electron irradiation. An attempt to obtain a more objective evaluation of areas of major and minor electron density was done by noncrystallographic averaging of electron micrographs (203). The receptor appeared to consist of an annular asymmetric structure containing three major areas of electron density.

In summary, these physical and morphological approaches allowed the following conclusions to be drawn with respect to the size and membrane disposition of *Torpedo* AcChR. The receptor is a transmembrane protein approximately 100 Å in length normal to the plane of the membrane. On the synaptic face it extends approximately 50 Å above the surface. This external AcChR structure is annular shaped, ~90 Å in diameter, and contains a central hole 15–25 Å in diameter that extends to an unknown depth within the molecule. The receptor extends 10–15 Å beyond the cytoplasmic surface of the membrane, and the diameter of this internal projection is not known.

The Organization of AcChR Molecules in the Postsynaptic Membranes

Functioning of the nicotinic synapse requires not only the intrinsic ion-gating properties of AcChR molecules but also a critical density of AcChR molecules. For full function, a massive and rapid flux of cations must occur (204) to achieve the depolarization necessary for activation of surrounding electrically excitable membranes. Katz & Miledi (205) first suggested that the nicotinic synapse contains regions of high AcChR density. For iontophoretically applied AcCh to end terminals of known geometry (206), the estimated conductance/μm^2, together with the known single channel conductance (207, 208), allowed the authors to calculate a density of 6,500 ion channels/μm^2. This is probably a lower limit, since not all the AcChRs present are necessarily activated for the physiological response (209).

The results obtained using various morphological techniques show that the AcChR density is very high (Table 8). In the case of *Torpedo*, *Narcine*, and *Electrophorus* electric organ synapses, this high density of AcChR is present over the whole postsynaptic membrane (97, 102, 103, 105) and the density remains unchanged after denervation (96, 105, 178). At the neuromuscular junction (110–112, 114–117, 120, 124) the AcChR is concentrated mostly at the top of the junctional fold. In cultured muscle cells "hot spots" of densely packed AcChR have been described (Table 8) even in the absence of neuronal cells (125–127). Thus the assembly of AcChR molecules in densely packed areas seems to be an intrinsic property both of muscle and electroplax membrane systems.

Despite the general agreement about the high synaptic density of AcChR, the numbers obtained in different systems are variable (Table 8). Negative staining of isolated AcChR-rich membranes of *Torpedo* (60, 87, 88, 93, 98, 100) consistently showed (see Table 8) 10–15 $\times$ 10^3 annular structures/μm^2 in the areas of maximal density, and the results obtained for this same tissue using freeze-etching procedures agree quite well with these estimates (10–15 $\times$ 10^3 particles/μm^2; Table 8). Less agreement exists about the number of particles observed in freeze-fracture replicas of postsynaptic membranes from other sources (Table 8). The particles were present in different proportions on the P and E faces (100, 101), and their overall number varied over a wide range in muscle postsynaptic membranes (from 1,800 to 7,500/μm^2) and to a lesser extent (from 5,000 to 8,000/μm^2) in electric ray electroplax membranes (see Table 8).

In *Electrophorus* electroplax (104, 105, 178) and in muscle (106–110, 115–118, 120, 124) the localization and density of AcChRs have been studied by electron microscopy autoradiography using radiolabeled α-neurotoxin (Table 8). The average value of AcChR density obtained by this approach in *Electrophorus* postsynaptic membrane was approximately 42,000 ± 22,000/μm^2. In early studies the values obtained for muscle postsynaptic membrane were averages of the densities for the whole postsynaptic membrane (106–108). Following recognition that the AcChR was concentrated at the crest of postsynaptic folds (110, 111) the densities at the crests ranged from 18,000 to 30,000 α-neurotoxin binding sites/μm^2. Since the AcChR monomer contains two α-bungarotoxin (α-BuTx) binding sites, the density in muscle agrees quite well with that observed in *Torpedo* tissue by negative staining and freeze-etching procedures. The values for *Electrophorus* electroplax seem to differ by approximately a factor or two but, given the large error involved, these values could also fit in the general picture of the density of postsynaptic AcChR in *Torpedo,* following normalization for the number of toxin binding sites. Thin sections of muscle postsynaptic membrane revealed granular structures, interpreted as AcChR molecules, occurring at a density of 10,000/μm^2 (111), which is reasonably consistent with the data yielded by autoradiography. The data obtained in different morphological studies (Table 8) present a coherent picture regarding the postsynaptic density of AcChR molecules. Assuming that the annular structures observed in negative staining or the dimpled particles seen in freeze-etching replicas in *Torpedo* electroplax both represent 255K AcChR monomers, these two approaches are in complete agreement. The results obtained with thin sectioning in *Narcine* electric organ and in muscle are lower by no more than a factor of two. This difference could easily be accounted for, since the receptor exists both as monomeric and dimeric forms in these tissues, and limitation in resolution does not always allow distinction between the two forms.

The main discrepancy in the numerology reported for AcChR density in a variety of tissues arises from the results obtained by freeze-fracture methods, which were consistently lower than the estimates discussed above. In the case of *Torpedo* (97, 101, 100) and *Narcine* (103) preparations, the freeze-fracture estimates are approximately twofold lower. This could be reconciled if AcChR dimers accounted for the preponderance of particles in the freeze-fracture images. In the case of muscle, however, the discrepancy is as large as a factor of three to five (113, 114, 119, 122, 123). To explain these low estimates it has been suggested (101) that a significant proportion of receptor molecules may be cleaved in correspondence to the fracture plane that separates the inner and outer leaflets of the lipid bilayer and are normally not visualized in shadowed replicas. This interpretation is supported by the fact that low angle rotary shadowing reveals, in addition to the usual tall protruding particles, other structures with a lower degree of relief with respect to the plane of the membrane.

In summary, a battery of morphological methods adequately confirm that the AcChR occurs at high density in the postsynaptic membrane of all species investigated. AcChR density values probably range between 10,000 and 15,000 AcChR molecules/μm^2 (Table 8); this is only two to three fold greater than those estimated by electrophysiological methods (206).

Other Proteins Associated with AcChR Membrane Preparations

When AcChR-enriched membrane fragments from *Torpedo* electric organ are analyzed by SDS gel electrophoresis, other protein components in addition to the four AcChR subunits are consistently associated with the preparations (Table 7). One component that could possibly interact with the AcChR has a molecular weight of 43,000, since other polypeptides with molecular weights of $\sim$ 45,000 [*Torpedo* actin (210)], and 90,000–150,000 (generally considered, although not proven, to be a component of *Torpedo* Na$^+$,K$^+$-ATPase) are absent from membrane fractions purified by affinity partitioning (84).

Extraction with base is an effective procedure for removing peripheral membrane proteins and results in removal of all components except the AcChR from the membranes (82, 91). Since the 43K protein is removed by this treatment, it must be a peripheral protein. The suggestion, based on the effect of proteolysis on membrane vesicles, that the 43K protein occurs exclusively on the cytoplasmic surface of the postsynaptic membranes (163) has not been confirmed (143).

It has been suggested that the 43K protein is associated with the AcChR, since after treatments effective in removing this protein (i.e. base extraction), the AcChR was less resistent to heat inactivation, more uniformly distributed in the membrane fragments, and had a greater rotational mobility (87, 88, 181). However, none of these studies demonstrated a direct interaction between the AcChR and this 43K protein.

LIGAND BINDING TO THE AcChR

Rapid Kinetic Studies of Cholinergic Ligand Binding

Due to desensitization of the AcChR it is necessary to study ligand-AcChR association on a time scale faster than seconds. Accordingly, four experimental approaches have been used in stopped-flow kinetic studies: (*a*) use of noncovalent extrinsic fluorescent probes (211–214); (*b*) use of fluorescent ligands that bind to the AcChR (215–217); (c) monitoring of the intrinsic flourescence of AcChR tryptophan residues (128, 129, 218); and (d) monitoring of the fluorescence of a fluorophore covalently attached to the receptor protein (134). The results of these studies are summarized in Table 9. Conformational transitions in the subsecond time domain were observed in all studies, but the rate constants obtained were in all cases too slow to account for activation of receptor-associated channels on the millisecond time scale. In one case (130, 131) this claim was made for a step (k_3 in Mechanism 2, Table 9) with a rate constant of 60 sec^{-1}, but it seems unwarranted by the data. Other mechanisms (e.g 3, 4, and 5 of Table 9) involve more than a single ligand's associating with the AcChR and therefore are in agreement with physiological data (219–222), but the rates are too slow to account for channel activation.

Limited information has been obtained regarding mechanisms of cholinergic antagonist binding to the AcChR (133). Kinetic analyses of cholinergic ligand association with detergent solubilized and purified AcChR have been performed. Rate constants for acetylcholine (AcCh) binding (2.4 ± 0.5 X 10^7 M^{-1} sec^{-1}) to T. californica AcChR (223) and for nitrobenz-2-oxa-1,3-diazole (NBD)-ω-amino-hexanoic acid binding (1 X 10^8 M^{-1} sec^{-1}) to E. electricus AcChR (224) are in reasonable agreement with the value (9.5 ± 3.5 X 10^7 M^{-1} sec^{-1}) obtained (135) for C$_6$-dansylcholine binding to T. marmorata membrane bound AcChR.

In summary, cholinergic ligand binding can be studied by rapid kinetic methods using membrane preparations highly enriched in AcChR. Conformational transitions have been consistently observed, but the rates of these processes are too slow to account for activation of the receptor channel. Either the proper probes have not been used (i.e. the transitions of interest are mute) or channel activation results in signals of too low amplitude. It is possible that the slow transitions observed are related to the desensitization process or to other inactivation mechanisms.

Binding of Local Anesthetics, Histrionicotoxin, and Related Compounds

Analysis of the effects of local anesthetics (LA) on single channel conductances led to the conclusion that they bind to the AcChR channel in the open form, causing a decrease in conductance. LA may also increase the rate of conversion of the AcChR to a nonconducting desensitized state (225, 226). Histrionicotoxin (HTX) can also reversibly block the AcCh-mediated ion conductance at the neuromuscular junction, possibly via a direct block of the AcChR ion channel (227, 228). HTX blocks the depolarization response to Carb in Electrophorus electroplaque and potentiates, rather than inhibits, [^{3}H]-AcCh binding to Torpedo membrane fragments (229).

Table 9 AcChR ligand binding mechanisms

Mechanism	Signal	Rate constants sec^{-1}	Equilibrium constants	Remarks	References
1. $L+R \underset{k_2}{\overset{k_1}{\rightleftharpoons}} LR \underset{k_4}{\overset{k_3}{\rightleftharpoons}} LR''$	Intrinsic Fluorescence (Tryptophan)	$k_3 = 0.37$ Carb $k_4 = 0.01 - 0.1$	$\frac{k_1}{k_L}$ 66 µM $\frac{k_3}{k_4}$ 37 µM		128,129
2. $L+R \underset{k_2}{\overset{k_1}{\rightleftharpoons}} RL \underset{k_4}{\overset{k_3}{\rightleftharpoons}} R'L \rightleftharpoons R'L$	Extrinsic Probe (Quinacrine)	$k_3 = 10 \pm 5$ Carb $k_4 = 0.2 - 1.5$		k_3 considered related to activation	130,131
3. $2L+R \rightleftharpoons LR \rightleftharpoons L_2R$; C_1, C_2	Extrinsic Probe (Ethidium)	$k'_1 = 0.03$ Carb $k'_{-1} = 20 \times 10^{-3}$ $k'_2 = 2.3$ $k'_{-2} = 0.1$	$K_1 = 1.3$ µM	C_1 & C_2 do not inter-convert	132,133
4. $R+L \underset{k_{-2}}{\overset{K_1, k_2}{\rightleftharpoons}} RL_2 \cdots R^*L \rightleftharpoons R^*L_2$; C_1, C_2	Covalent Probe (Acetamido salicylate)	$k_2 = 8.1$ Carb $k_{-2} = 0.9$ $k_3 = 0.42$ $k_{-3} = 0.008$ $k_5 = 2.7$ $k_{-5} = 0.04$	$K_1 = 50.8$ µM $K_4 = 50.8$ µM		134
5. $R \overset{k_5}{\rightleftharpoons} R'$; $RL \overset{k_7}{\underset{k_8}{\rightleftharpoons}} R'L$	Fluorescent Agonist (C_6-DNS Chol)	$k_7 = 0.18 \pm 0.04$ $k_8 = 2.7 \pm 3 \times 10^{-3}$			135
5a. $L+R \underset{k_{-1}}{\overset{2k_1}{\rightleftharpoons}} RL \underset{2k_{-1}}{\overset{k_1}{\rightleftharpoons}} RL_2 \underset{k_{-2}}{\overset{k_2}{\rightleftharpoons}} IL_2$	Fluorescent Agonist (C_6-DNS Chol)	$k_2 = 180 \pm 20$ $k_{-2} = 1.2 \pm 0.3$	$\frac{k_{-1}}{k_1} = 119$ µM		136

A toxic compound from the venom of *Bungarus caeruleus* was suggested to be a specific ligand for the AcChR ion channel (230); however, it was subsequently shown to be a phospholipase (231). Two other compounds, phencyclidine and amantidine, are believed to interact with the AcChR's channel directly or indirectly in a manner reminiscent of LAs or HTX (232–237).

Evidence for multiple association of LAs with the AcChR was obtained from the inhibitory effects of these compounds on the kinetics of [H^3]-α-neurotoxin-AcChR complex formation (238) and from the effects of different LAs on the fluorescence of ethidium bromide bound to AcChR-enriched membranes (213, 214). This indicates that the LAs may interact with the AcChR at multiple sites, possibly associated with some or all of the homologous subunits.

Direct interaction of LAs with the AcChR has been monitored using radiolabeled quaternary methyl derivatives of proadifen and dimethisoquin (239). In the presence of carbamyl choline (Carb), [^{14}C]meproadifen bound with a K_d of 0.5 μM; in the absence of Carb it bound with a K_d of $\sim$5 μM, which emphasizes synergistic binding. This confirms earlier studies (240) that showed that the LA prilocaine slightly increased the affinity of *T. marmorata* membranes for [^{3}H]-AcCh. Localization of LA binding sites on individual polypeptide chains was attempted by photoaffinity labeling. A peripheral membrane protein with a molecular weight of 43,000 daltons and another 90K component were labeled by *p*-azidoprocainamide (241); in addition minor labeling was seen of the AcChR polypeptides. Interestingly, labeling of several subunits was also observed with the local anesthetic 5-azido[^{3}H]trimethisoquin (5-[^{3}H]-AT) (242). In the case of 5-[^{3}H]-AT, the 50K photolabeled product has been shown to be derived by proteolysis from the 66K subunit (27, 243). Whether a binding site(s) for LA is within the channel is not answered by his approach because in the presence of an agonist at equilibrum, the AcChR is desensitized.

The interaction of HTX with receptor-enriched preparations was first studied using [^{3}H]-H_{12}-HTX (244, 245). In one study (244) failure to recognize nonspecific effects led to the claim that separation of [^{3}H]-ACh and [^{3}H]-HTX binding activities was achieved and that the ACh binding protein and the "ion conductance modulator" were different proteins. In other studies, which took nonspecific binding into account, a dissociation constant of 0.4 μM was obtained, which decreased to 0.29 μM in the presence of Carb (245, 246). This was subsequently confirmed (247). Similarly the interaction of a 43K component with both quinacrine and HTX was mistakenly considered to be specific and the wrong conclusion was drawn that this polypeptide was an ionophore associated with the "AcChR regulator" protein (248).

The question of whether HTX binds to the AcChR *sensu stricto* or to a different protein has been answered by showing that a cholate extract of *T. californica* membranes lost the capability of binding both α-BuTx and [³H]-HTX after specific removal of the AcChR (246) and been confirmed by showing that membranes containing only the AcChR bind local anesthetics and HTX (82, 91). For [³H]-H_{12}-HTX, binding ratios with respect to α-toxin binding have been reported ranging from 2:1 (244, 247) to 0.26:1 (245). For local anesthetics ratios ranging from $0.3 \pm 0.1:1$ (91) to 0.5:1 (239) have been reported.

LAs increased the rate of association of the fluorophore DNS-C_6-Cho with the AcChR (135). LA association with *T. californica* AcChR labeled with a fluorescent probe (IAS) has also been studied. Preincubation with saturating concentrations of several LAs had no significant effect on the equilibrium constant for Carb binding (249); however, the mechanism of Carb binding (see Table 9, Mechanism 4) was altered and could be described as:

$$R + L \underset{k_{-1}}{\overset{k_1}{\rightleftharpoons}} RL \underset{k_{-2}}{\overset{k_2}{\rightleftharpoons}} R*L$$

in which initial complex formation was followed by an isomerization. This differs from the mechanism in the absence of local anesthetics in that there is no evidence for the association of a second Carb molecule (see Table 9). The final (equilibrium) receptor conformation, R*L is reached only in the presence of bound cholinergic ligand, which indicates that prior binding of local anesthetic does not induce a receptor state having an intrinsic high affinity for agonists, as previously suggested (135).

Using ethidium as a probe (133) a binding mechanism for HTX was deduced that involved formation of an initial complex followed by a slow isomerization (250):

$$R + T \underset{}{\overset{K}{\rightleftharpoons}} RT \underset{k_{-1}}{\overset{k_1}{\rightleftharpoons}} R'T$$

where the signal arose from the isomerization and $k_{-1} = 2 \times 10^{-2} \text{ sec}^{-1}$. A possible explanation of the origin of this slow process arose from a study of the binding of a homologous series of alkylguanidines. The efficacy of these compounds in displacing HTX increased with the length of the alkyl chain and could be correlated with the partition coefficient between the aqueous phase and the membranes (251). This suggests that the first step in the binding to AcChR of the alkyl guanidines and possibly other amphiphiles, such as HTX or LA, is their dissolution in the lipid phase, so that they approach their binding site(s) by diffusion from this phase. The slow fluorescence change described for HTX in the presence of ethidium (250) could represent rate-limiting diffusion of the ligand.

In summary, LAs, HTX, or other agents such as alkylguanidines, which all block the physiological response to Ach, may achieve this by one or more mechanisms: (a) they may cause a physical block of the activated channel, (b) they may bind to cholinergic ligand binding sites on some or all of the related subunits of the AcChR and affect channel properties in an indirect manner, and (c) they may modify the properties of the membrane and by so doing, at least in some cases, enhance the rate of desensitization.

Affinity Labeling of Cholinergic Ligand Binding Sites

After reduction, the AcChR is specifically labeled by quaternary ammonium derivates such as MBTA (252). An ~40K polypeptide was specifically labeled in the intact electroplax of *Electrophorus* (252) and in purified AcChR from *Torpedo* (4, 7) and *Electrophorus* (7, 38). In mammalian muscle (174) [^{3}H]-MBTA labeled two subunits with apparent molecular weights of 45,000 and 49,000 that must be closely related or identical, since they give identical peptide maps (16). Using a similar approach, the agonist bromoacetylcholine (BrAcCh) was covalently attached to purified AcChR from *N. braziliensis* (31), *T. californica* (253, 254), *T. marmorata* (76, 255), and cat denervated muscle (76, 255). In all cases a single 40–44K polypeptide was labeled. Results obtained with cholinergic affinity labels that do not depend on reduction of disulfides suggest that two or three of the other subunits may contain cholinergic ligand binding sites; 4-azido-2-nitrobenzyltrimethylammonium fluoroborate (80) labeled all four subunits, and bis-(3-azidopyridinium)-1,10-decane diiodide (DAPA) labeled three of the four subunits following photolysis (256). In contrast, [^{3}H]-*p*-(trimethylammonium)benzendiazonium fluoroborate ([^{3}H]-TDF) specifically labeled only the 40K subunit in *T. californica* membranes (51). α-Neurotoxins have also been used for affinity labeling of the AcChR. Early studies (35) of purified *Electrophorus* AcChR demonstrated that *Naja naja atra*-α-toxin could be cross-linked to a 40K polypeptide; the preparation, however, lacked intact subunits with molecular weights > 40,000. Cross-linking of *Narcine* AcChR with α-BuTx resulted in covalent attachment to two subunits of M_r ~44 and 60K (31), and photolabile derivatives of α-BuTx were cross-linked to at least two subunits with molecular weights of 40,000 and 65,000 in *T. californica* AcChR (145, 146). Different subunits of *T. marmorata* AcChR were specifically labeled by α-BuTx and *Naja naja siamensis* toxin (257), but unambiguous identification of polypeptides was not achieved. In situ (147) labeling with a different α-BuTx derivative demonstrated that all four *T. californica* AcChR subunits were labeled, and five polypeptides of purified muscle AcChR were labeled by the same derivatives. These studies emphasize the difficulties of unambiguously assigning the α-neurotoxin binding sites to a particular subunit.

Under equilibrium conditions, using *Torpedo* membranes, the ratio between the numbers of binding sites for cholinergic ligands and for α-BuTx range from 1 to 0.5 (258, 259). MBTA labels 50% of the α-BuTx binding sites in *Electrophorus* (7) and *Torpedo* (5,260), but 100% of such sites in mammalian muscle AcChR preparations (174). Similarly, BrAcCh has been reported to label only half of the α-BuTx binding sites in *T. californica* (253, 254) and in *T. marmorata* AcChR (255) at 4°C; at room temperature the additional 50% of toxin binding sites was labeled (255), which implies the existence of nonequivalent sites on the two 40K subunits. Similar results have been found in mammalian muscle (255). In equilibrium binding experiments using *T. californica* membranes, BrAcACh bound to half the number of α-BuTx sites (254), and *d*-tubocurarine (*d*-Tc) bound to a number of sites equal to those for α-toxins (261), with two populations of nonequivalent sites. Several studies of acetylcholine and carbamylcholine binding to membranes have yielded values ranging from 0.5 to 1 with respect to the number of α-toxin sites. Further equilibrium studies on *Electrophorus* membranes of the binding of decamethonium, Carb, *d*-Tc, and α-BuTx led to the suggestion that although the stoichiometry of binding for all these ligands was 1:1, there was only partial overlap of binding sites for agonists and antagonists (262).

Ligand Affinity Changes Related to Desensitization

Desensitization of nicotinic AcChR was first demonstrated electrophysiologically (225, 263–266), and it was suggested that the desensitized state has a higher affinity for agonists than for the resting state (263). A biochemical parallel has been observed by study of (*a*) kinetics of α-neurotoxin-AcChR complex formation (98, 267–269), (*b*) estimation of Ach binding by rapid mixing (128), (*c*) immobilization of a spin-label derivative of decamethonium (268, 271), and (*d*) agonist-induced alteration in fluorophore properties (82, 128, 129, 131, 133, 135, 218). The electrophysiological and biochemical effects occur on the same time scale (subsecond to second), are dependent on ligand concentration, and are accelerated by Ca^{2+} and by some local anesthetics. Other local anesthetics, however, have no effect or even decrease the rate of the change in affinity (272). Whereas the affinity for agonists increases by factors ranging from ~3 (270) to 300 (273), the affinity for antagonists either does not change (273) or is increased approximately twofold for the form of high affinity (270). The rate of recovery to a state of low affinity for agonists is independent of the ligand used to cause the initial conversion (267, 270). Slow transitions of liganded forms of *Torpedo* sp. AcChR have also been studied by stopped-flow methods (135, 274), and rate constants for onset of and recovery from states of high affinity were determined.

The interconversion of affinity states induced by agonists is lost upon dissolution of the membranes by neutral detergents or in sodium cholate solution (275), unless exogenous lipids are present (276). Furthermore, apparent disruption of AcChR-membrane lipid interactions by general anesthetics such as halothane, chloroform, or diethyl ether results in an enhanced rate of conversion of the AcChR to a state(s) of high affinity for carbamylcholine (277).

CATION GATING FUNCTION OF THE AcChR

Electrophysiology of a Single Receptor Channel

The kinetic characteristics elucidated by electrophysiological methods represent basic quantitative criteria by which the purified AcChR must be evaluated. Among these is determination of the number of monovalent cations that upon activation flux through a single receptor molecule in unit time. This has been determined in vivo by voltage noise analysis in the muscle postsynaptic membrane (207, 208, 278–280), which led to estimates of single channel conductance corresponding to a flux of $\sim10^7$ cations (281) transported per second per channel. It has been suggested that either one or two Ach molecules can activate the AcChR channel, but that binding of two agonist molecules is more effective (219–222, 282). Another important feature of the AcChR concerns its selectivity for cations. Mono- or divalent inorganic cations (283) as well as many small organic cations can be transported (284, 285) and the channel is depicted as a large aqueous pore, $\sim6.5 \times 6.5$ Å at its narrowest point, that barely discriminates between small cations and does not allow anions to permeate.

Cation Flux in Vesicles Containing AcChR

Cation flux has been extensively studied in vitro using radiotracers such as $^{22}Na^+$ (286). The temporal resolution of the method is poor (the time scale is from seconds to hours) and the $t_{1/2}$ of Carb-mediated $^{22}Na^+$ flux for *Electrophorus* vesicles is ~7.5 min. This discrepancy of a factor of $\sim10^6$ compared to the physiological time scale (1–2 msec) raised serious questions regarding the integrity of the AcChR in isolated membrane fractions (287). The slow kinetics are surprising, since it was demonstrated that in intact electroplax under comparable experimental conditions desensitization followed an exponential time course with a $t_{1/2}$ of 46 sec (288). It has been suggested that desensitized AcChR from *Electrophorus* permits a slow agonist-mediated transport of cations (289). The maximum rate of cation transport in *Electrophorus* membrane vesicles (289, 290) corresponded to 1.6×10^4 cations sec^{-1} $AcChR^{-1}$, which differs from the physiological transport number by a factor of 10^3.

With *Torpedo* membrane vesicles the filtration assay (291–294) showed slow kinetics over seconds to minutes or no kinetics, only amplitude changes (295–297) that exhibited the expected pharmacology, such as block by α-BuTx, HTX, antagonists, and desensitization.

Using a flow-quench method (298) it has been shown that $^{86}Rb^+$ is translocated into *Electrophorus* vesicles rapidly (over a few hundred milliseconds) upon mixing with Carb (299). A complex model has appeared in several papers (140, 141, 299–301) that involves two cholinergic agonists associating with the AcChR to form five different complexes with eight variable parameters, and data from activation and inactivation studies were found to conform to the model. A flow-quench approach has also been reported for studies of $^{22}Na^+$-mediated transport in *T. californica* and *T. nobiliana* membrane vesicles with time resolution of 25 msec (139). The maximum rate corresponded to 0.1% of that expected from the known single channel conductance (Table 10). In a different study (302) that assayed Na^+ transport in cultured muscle cells, a transport number ~3% of the physiological one was obtained.

A rapid stopped-flow spectroscopic method with a 2 msec resolution, for quantitation of monovalent cation transport into sealed *Torpedo* postsynaptic membrane vesicles, is based on Tl^+ quenching of a water soluble fluorophore trapped within the vesicles (138). The transport number determined

Table 10 Kinetic studies of cation flux in vitro

Tissue	Method	Agonist	Rate constant (sec^{-1})	$K_{d\ app}$ (mM)	Transport (number of M^+ $AcChR^{-1}$ sec^{-1})	Authors	Year	Ref
Muscle	Noise analysis (review)				1×10^7	Adams	1981	137
Torpedo vesicles	Spectroscopic Tl^+ flux	Carb[a]	1100 or 1500[b]	1.0 or 5.0[b]	7×10^6	Moore & Raftery	1980	138
Torpedo vesicles	Flow-quench ^{22}Na flux	Carb PTA[c]	65 0.8	0.6 0.2	1.75×10^3	Neubig & Cohen	1980	139
Electrophorus vesicles	Flow-quench ^{86}Rb	Carb AcCh[d]	37 37	1.9 0.08	6×10^6 6×10^6	Hess et al Cash et al	1981 1981	140 141

[a] Carbamyl choline.
[b] Values dependent on whether one or two ligands are considered necessary to effect transport.
[c] Phenyl trimethyl ammonium.
[d] Acetylcholine.

for Carb-activated *Torpedo* AcChR was 7×10^6 cations sec^{-1} AcChR^{-1} in *Torpedo* Ringer solution (Table 10). This transport number determined for *Torpedo* membrane vesicles in vitro is sufficiently close to the value deduced from single-channel conductance measurements in vivo to warrant the statement that the isolated membranes contain fully functional AcChR molecules. A similar value for AcChR transport was subsequently determined for *Electrophorus* AcChR membrane vesicles (140).

Agonist Binding and Cation Flux

Although the transport numbers determined for *Torpedo* preparations and the neuromuscular junction using various methods vary from 7×10^3 (139) to 7×10^6 (138) or 10^7 (137), the apparent K_d values for Carb activation obtained from dose response curves are comparable (0.6–1.9 nM; Table 10). Using the slow filtration assay for $^{22}\text{Na}^+$ flux, misleading dose-response curves for agonists were obtained (291, 295, 296) for *Torpedo* membrane vesicles; e.g. for Carb the midpoint of the dose-response curve was approximately 20 μM.

The K_ds determined for cation flux induced by agonists are different from those obtained from binding studies (270, 271, 274); these yielded values of 20–40 μM and 25–50 nM for Carb binding to the resting state and the high affinity desensitized state of *Torpedo* AcChR. It is generally assumed that these latter values pertain to association with the binding site on the 40K subunit(s) labeled by MBTA and BrAcCh (see earlier section). Based on the discrepancies in K_d values and the knowledge that the AcChR is composed of homologous polypeptides it is probable that binding sites exist on the 50, 60, and 65K subunits and are possibly responsible for activation. This suggests that AcChR functions such as activation and desensitization might involve parallel pathways rather than sequential transitions (98, 137, 139–141, 299–301).

Electrophysiological data suggest that two bound agonists are effective in causing AcChR-induced membrane depolarization. This is consistent with recent results from ^{22}Na flux studies of *Torpedo* vesicles (139) and BC3H-1 clonal muscle cells (303, 304), and from Tl^+ flux studies of *Torpedo* vesicles (138).

Many variants of formal mechanisms have been presented based on physiological data (reviewed in 137), as well as on biochemical data (98, 139, 299–301). Detailed electrophysiological studies of agonist concentration jumps (305, 306), voltage jump relaxations (220, 219, 137), and noise analysis (137) provide complementary data on channel opening and closing parameters. No comprehensive mechanism that also takes conformational transitions and binding rates into account is presently available.

Protein Components Participating in AcChR Function

Whether the AcChR complex formed by the four subunits is able to carry out all receptor function is a question of paramount importance. $^{22}Na^+$ flux in membrane vesicles that contained either the AcChR alone or other protein species as well was found to be the same, which indicates that only the AcChR polypeptides were necessary for cation transport (297). In addition, quantitation of Tl^+ transport (138) showed that the kinetics of cation transport in response to Carb were comparable in these two types of *Torpedo* membrane vesicle preparations. It was shown that the transport number of the membranes containing only the AcChR is within a factor of two of that predicted from single-channel conductance studies, thus demonstrating that the only protein necessary for fully active cation transport is the AcChR. *Torpedo* vesicles are the only system where, due to homogeneity of components, correlation between AcChR structure and function (138) can be made. In *Electrophorus* vesicles (140, 300), cultured muscle cells (302), or intact tissue, AcChR molecules represent too small a fraction of the protein components present to allow definitive conclusions to be drawn. By analogy only (cf subunit composition, Tables 1 and 6) is it possible to infer similarity.

The second approach used to define essential AcChR components has been the reconstitution of AcChR preparations into planar or vesicular lipid bilayers. AcChR-rich membranes from *T. marmorata* were reconstituted into planar bilayers (307) and single-channel fluctuations of 20–25 pS conductance, and lifetimes of 1.3 msec were observed in the presence of Carb, which were blocked by *d*-Tc and α-BuTx; the reconstituted system underwent desensitization. Solubilized, purified AcChR from *T. californica* was also reconstituted into planar bilayers (308) and responded to Carb with an increase in membrane conductance (16 ± 3 pS per single channel with a mean open time of 35 ± 5 msec) and became desensitized. The origin of the unusually long mean open time in this system is not clear. Unfortunately, the polypeptide composition of the preparations used for these experiments (307, 308) was not recorded. A recent report of reconstitution studies of *T. marmorata* AcChR into planar bilayers (309) has confirmed the earlier conclusions regarding the complete protein components of the AcChR (138) using the Tl^+ flux assay into *T. californica* membrane vesicles.

Reconstitution of AcChR into membrane vesicles was first attempted using AcChR containing membranes from *T. marmorata* electroplax dissolved in sodium cholate and mixed with crude lipid fractions from the same source (310). Removal of detergent by dialysis resulted in formation of lipid vesicles containing reassociated AcChR that responded to Carb with an increased $^{22}Na^+$ efflux. These studies demonstrated that reconstitution was possible in principle and that definition of necessary protein components could be investigated.

The first attempt to reconstitute a solubilized, purified AcChR from *T. californica* (186) yielded vesicle preparations responding to Carb with an increased $^{22}Na^+$ permeability that could be blocked by α-BuTx. However, the response to Carb was erratic. At the same time other reports appeared (311–313) describing failures to reconstitute Carb-induced ^{22}Na flux of purified AcChR. Limited success in reconstitution of agonist-mediated cation transport was later reported for *T. californica* AcChR (189). Lipid-AcChR interaction has a crucial role for maintenance of AcChR viability. A critical ratio between lipid and detergent concentrations throughout all steps of the purification and reconstitution procedure is necessary in order to consistently reproduce Carb-induced cation transport (314–319). The expected pharmacological effects of antagonists and desensitization were demonstrated in these studies. Assuming a Poisson distribution of AcChR molecules in vesicles containing an average of one AcChR molecule, it was calculated that essentially all the AcChR molecules were still active in translocating cations (320).

The question remained as to whether the transport efficiency of the reconstituted AcChR was comparable with the single-channel conductance in neuromuscular junction (281) or in intact membranes (138). This was determined from a study of Carb-induced Tl^+ flux in reconstituted vesicles that contained an average of one AcChR molecule per vesicle. The transport number (322) was found to be approximately 3×10^6 M^+ ions sec^{-1} AcChR^{-1}, remarkably close to those previously determined for *Torpedo* membrane preparations and calculated from single-channel conductance measurements.

It is clear from these studies that even in the absence of a membrane potential a rapid and highly efficient transport of cations can be accomplished by reconstituted AcChR molecules in response to agonist binding. Application of quantitative approaches of the type described are mandatory for reaching conclusions regarding the physiological properties, i.e. the kinetic constants and ion transport number, and definition of the molecular components necessary for the full physiological function of AcChR both in the native and in the reconstituted state. These considerations are reinforced by comparison with the conclusions reported for reconstituted AcChR preparations whose properties were investigated using the $^{22}Na^+$ filtration assay (319, 321). Using this qualitative approach the authors tried to draw quantitative conclusions and obtained transport numbers ranging from 64 to 70 Na$^+$ ions per AcChR molecule per minute (incubation time), i.e. about six orders of magnitude lower than expected. There are grave dangers inherent in drawing conclusions regarding the molecular components essential for AcChR function from data obtained with this qualitative assay.

Literature Cited

1. Schmidt, J., Raftery, M. A. 1973. *Biochemistry* 12:852–56
2. Raftery, M. A., 1973. *Arch. Biochem. Biophys.* 154:270–76
3. Raftery, M. A., Vandlen, R. Michaelson, D., Bode, J., Moody, T., Chao, Y., Reed, K., Deutsch, J., Duguid, J. 1974. *J. Supramol. Struct.* 2:582–92
4. Weill, C. L., McNamee, M. G., Karlin, A. 1974. *Biochem. Biophys. Res. Commun.* 61:997–1003
5. Reed, K., Vandlen, R., Bode, J., Duguid, J., Raftery, M. A. 1975. *Arch. Biochem. Biophys.* 167:138–44
6. Raftery, M. A., Vandlen, R. L., Reed, K. L., Lee, T. 1976. *Cold Spring Harbor Symp. Quant. Biol.* 40:193–202
7. Karlin, A., Weill, C. L., McNamee, M. G., Valderrama, R. 1976. *Cold Spring Harbor Symp. Quant. Biol.* 40:203–10
8. Vandlen, R. L., Schmidt, J., Raftery, M. A. 1976. *J. Macromol. Sci. Chem.* A10 (1 & 2):73–109
9. Chang, H. W., Bock, E. 1977. *Biochemistry* 16:4513–20
10. Froehner, S. C., Reiness, C. G., Hall, Z. W. 1977. *J. Biol. Chem.* 252:8589–96
11. Lindstrom, J., Einarson, B., Merlie, J. 1978. *Proc. Natl. Acad. Sci. USA* 75:769–73
12. Vandlen, R. L., Wu, W. C.-S., Eisenach, J. C., Raftery, M. A. 1979. *Biochemistry* 10:1845–56
13. Bartfeld, D., Fuchs, S. 1979. *Biochem. Biophys. Res. Cummun.* 89:512–19
14. Sumikawa, K., O'Brien, R. D. 1979. *FEBS Lett.* 101:395–98
15. Froehner, S. C., Rafto, S. 1979. *Biochemistry* 18:301–7
16. Nathanson, N. M., Hall, Z. W. 1979. *Biochemistry* 18:3401–6
17. Lennon, V. A., Thompson, M., Chen, J. 1980. *J. Biol. Chem.* 255:4395–98
18. Karlsson, E., Heilbronn, E., Widlund, L. 1972. *FEBS Lett.* 28:107–11
19. Eldefrawi, M. E., Eldefrawi, A. T. 1973. *Arch. Biochem. Biophys.* 159:362–73
20. Carroll, R. C., Eldefrawi, M. E., Edelstein, S. J. 1973. *Biochem. Biophys. Res. Commun.* 55:864–72
21. Heilbronn, E., Mattson, C. 1974. *J. Neurochem.* 22:315–17
22. Gordon, A., Bandini, G., Hucho, F. 1974. *FEBS Lett.* 47:204–8
23. Sobel, A., Weber, M., Changeux, J.-P. 1977. *Eur. J. Biochem.* 80:215–24
24. Claudio, T., Raftery, M. A. 1977. *Arch. Biochem. Biophys.* 181:484–89
25. Shorr, R. G., Dolly, J. O., Barnard, E. A. 1978. *Nature* 274:283–84
26. Deutsch, J. W., Raftery, M. A. 1971. *Arch. Biochem. Biophys.* 197:503–15
27. Saitoh, T., Oswald, R., Wennogle, L. P., Changeux, J.-P. 1980. *FEBS Lett.* 116:30–35
28. Ong, D. E., Brady, R. N. 1974. *Biochemistry* 13:2822–27
29. Eldefrawi, A. T., Eldefrawi, M. E., Mansour, N. A. 1978. *Pesticides and Venom Neurotoxicity,* ed. D. L. Shankland, R. M. Hollingworth, T. Smyth, Jr., pp. 27–42, New York: Plenum
30. Schmidt, J., Raftery, M. A. 1972. *Biochem. Biophys. Res. Commun.* 49:572–78
31. Chang, R. S. L., Potter, L. T., Smith, D. S. 1977. *Tissue & Cell* 9:623–44
32. Kemp, G., Morley, B., Dwyer, D., Bradley, R. J. 1980. *Membr. Biochem.* 3:229–56
33. Ishikawa, Y., Yoshida, H., Tamiya, N. 1980. *J. Biochem.* 87:313–21
34. Olsen, R. W., Meunier, J.-C., Changeux, J.-P. 1972. *FEBS Lett.* 28:96–100
35. Biesecker, G. 1973. *Biochemistry* 12:4403–9
36. Klett, R. P., Fulpius, B. W., Cooper, D., Smith, M., Reich, E., Possani, L. D. 1973. *J. Biol. Chem.* 248:6841–53
37. Hucho, F., Changeux, J.-P. 1973. *FEBS Lett.* 38:11–15
38. Karlin, A., Cowburn, D. 1973. *Proc. Natl. Acad. Sci. USA* 70:3636–40
39. Chang, H. W. 1974. *Proc. Natl. Acad. Sci. USA* 71:2113–17
40. Lindstrom, J., Patrick, J. 1974. *Synaptic Transmission and Neuronal Interaction,* ed. M. V. L. Bennett, pp. 191–216. New York: Raven. 388 pp.
41. Meunier, J.-C., Sealock, R., Olsen, R., Changeux, J.-P. 1974. *Eur. J. Biochem.* 45:371–394
42. Patrick, J., Boulter, J., O'Brien, J. C. 1975. *Biochem. Biophys. Res. Commun.* 64:219–225
43. Lindstrom, J., Cooper, J., Tzartos, S. 1980. *Biochemistry* 19:1454–58
44. Raftery, M. A., Schmidt, J., Clark, D. G. 1972. *Arch. Biochem. Biophys.* 152:882–86
45. Gibson, R. E., O'Brien, R. D., Edelstein, S. J., Thompson, W. R. 1976. *Biochemistry* 15:2377–83
46. Kalderon, N., Silman, I. 1977. *Biochim. Biophys. Acta* 465:331–40
47. Reynolds, J. A., Karlin, A. 1978. *Biochemistry* 17:2035–38
48. Doster, W., Hess, B., Watters, D., Maelicke, A. 1980. *FEBS Lett.* 113:312–15

49. Meunier, J. C., Olsen, R. W., Changeux, J.-P. 1972. *FEBS Lett.* 24:63–68
50. Meunier, J.-C., Olsen, R. W., Menez, A., Fromageot, P., Boquet, P., Changeux, J.-P. 1972. *Biochemistry* 11:1200–10
51. Berg, D. K., Kelley, R. B., Sargent, P. B., Williamson, P., Hall, Z. W. 1972. *Proc. Natl. Acad. Sci. USA* 69:147–151
52. Brockes, J. P., Hall, Z. W., 1975. *Biochemistry* 14:2100–6
53. Boulter, J., Patrick, J. 1977. *Biochemistry* 16:4900–8
54. Merlie, J. P., Changeux, J.-P., Gros, F. 1978. *J. Biol. Chem.* 253:2882–91
55. Barnard, E. A., Dolly, J. O., Lo, M., Mantle, T. 1978. *Biochem. Soc. Trans.* 6:649–51
56. Shorr, R. G., Lyddiatt, A., Lo, M. M. S., Dolly, J. O., Barnard, E. A. 1981. *Eur. J. Biochem.* 116:143–53
57. Martinez-Carrion, M., Sator, V., Raftery, M. A. 1975. *Biochem. Biophys. Res. Commun.* 65:129–37
58. Edelstein, S. J., Beyer, W. B., Eldefrawi, A. T., Eldefrawi, M. E. 1975. *J. Biol. Chem.* 250:6101–6
59. Hucho, F., Bandini, G., Suarez-Isla, B. A. 1978. *Eur. J. Biochem.* 83:335–40
60. Klymkowsky, M. W., Stroud, R. M. 1979. *J. Mol. Biol.* 128:319–34
61. Wise, D. S., Karlin, A., Schoenborn, B. P. 1979. *Biophys. J.* 28:473–96
62. Strader, C. D., Hunkapiller, M. W., Hood, L. E., Raftery, M. A. 1980. In *Psychopharmacology and Biochemistry of Neurotransmitter Receptors*, ed. H. Yamamura, R. Olsen, E. Usdin, pp. 35–46. Amsterdam: Elsevier/North Holland
63. Molinoff, P. B., Potter, L. T. 1972. *Adv. Biochem. Psychopharmacol.* 6:111–34
64. Lindstrom, J. M., Lennon, V. A., Seybold, M. E., Whittingham, S. 1976. *Ann. NY Acad. Sci.* 274:254–74
65. Chiu, T. H., Dolly, J. O., Barnard, E. A. 1973. *Biochem. Biophys. Res. Commun.* 51:205–13
66. Dolly, J. O., Barnard, E. A. 1975. *FEBS Lett.* 57:267–71
67. Colquhoun, D., Rang, H. P. 1976. *Molec. Pharmacol.* 12:519–35
68. Bradley, R. J., Howell, J. H., Romine, W. O., Carl, G. F., Kemp, G. E. 1976. *Biochem. Biophys. Res. Commun.* 68:577–84
69. Dolly, J. O., Barnard, E. A. 1977. *Biochemistry* 16:5053–60
70. Potter, L. T. 1973. *Drug Receptors,* ed. H. P. Rang, pp. 295–312. London: Macmillan. 321 pp.
71. Meunier, J.-C., Sugiyama, H., Cartaud, J., Sealock, R., Changeux, J.-P. 1973. *Brain Res.* 62:307–15
72. Raftery, M. A., Schmidt, J., Clark, D. G., Wolcott, R. G. 1971. *Biochem. Biophys. Res. Commun.* 45:1622–9
73. Teichberg, V. I., Changeux, J.-P. 1976. *FEBS Lett.* 67:264–7
74. Stephenson, F. A., Harrison, R., Lunt, G. G. 1979. *Biochem. Soc. Transactions* 7:971–2
75. Brockes, J. P., Hall, Z. W. 1975. *Biochemistry* 14:2092–9
76. Lyddiatt, A., Sumikawa, K., Wolosin, J. M., Dolly, J. O., Barnard, E. A. 1979. *FEBS Lett.* 108:20–24
77. Merlie, J.-P., Sebbane, R. 1981. *J. Biol. Chem.* 256:3605–8
78. Duguid, J. R., Raftery, M. A. 1973. *Biochemistry* 12:3593–7
79. Flanagan, S. D., Barondes, S. H., Taylor, P. 1975. *J. Biol. Chem.* 251:858–65
80. Hucho, F., Layer, P., Kiefer, H. R., Bandini, G. 1976. *Proc. Natl. Acad. Sci. USA* 73:2624–8
81. Weiland, G., Frisman, D., Taylor, P. 1979. *Mol. Pharmacol.* 15:213–26
82. Elliott, J., Dunn, S. M. J., Blanchard, S. G., Raftery, M. A. 1979. *Proc. Natl. Acad. Sci. USA* 76:2576–9
83. Elliott, J., Blanchard, S. G., Wu, W., Miller, J., Strader, C. D., Hartig, P., Moore, H.-P., Racs, J., Raftery, M. A. 1980. *Biochem. J.* 185:667–77
84. Johansson, G., Gysin, R., Flanagan, S. D. 1980. *J. Biol. Chem.* 256:9126–35
85. Klymkowsky, M. W., Heuser, J. E., Stroud, R. M. 1980. *J. Cell Biology* 88:823–38
86. Cohen, J. B., Weber, M., Huchet, M., Changeux, J.-P. 1972. *FEBS Lett.* 26:43–47
87. Barrantes, F. J., Neugebauer, D.-Ch., Zingsheim, H. P. 1980. *FEBS Lett.* 112:73–78
88. Lo, M. M. S., Garland, P. B., Lamprecht, J., Barnard, E. A. 1980. *FEBS Lett.* 111:407–12
89. Oswald, R., Sobel, A., Waksman, G., Roques, B., Changeux, J.-P. 1980. *FEBS Lett.* 111:29–34
90. Allen, T., Potter, L. T. 1977. *Tissue & Cell* 9:609–22
91. Neubig, R. R., Krodel, E. K., Boyd, N. D., Cohen, J. B. 1979. *Proc. Natl. Acad. Sci. USA* 76:690–4
92. Waser, P. G. 1970. *Ciba Found. Symp. Mol. Biol.* 32:59–75
93. Nickel, E., Potter, L. T. 1973. *Brain Res.* 57:508–17
94. Cartaud, J., Benedetti, E. L., Cohen, J.

B., Meunier, J.-C., Changeux, J.-P. 1973. *FEBS Lett.* 33:109–13
95. Orci, L., Perrelet, A., Dunant, Y. 1974. *Proc. Natl. Acad. Sci. USA* 71:307–10
96. Clementi, F., Conti-Tronconi, B., Peluchetti, D., Morgutti, M. 1975. *Brain Res.* 90:133–8
97. Rosenbluth, J. 1975. *J. Neurocytology* 4:697–712
98. Changeux, J.-P., Benedetti, L., Bourgeois, J.-P., Brisson, A., Cartaud, J., Devaux, P., Grunhagen, H., Moreau, M., Popot, J.-L., Sobel, A., Weber, M. 1976. *Cold Spring Harbor Symp. Quant. Biol.* 40:211–30
99. Ross, M. J., Klymkowsky, M. W., Agard, D. A., Stroud, R. M. 1977. *J. Mol. Biol.* 116:635–59
100. Cartaud, J., Benedetti, E. L., Sobel, A., Changeux, J.-P. 1978. *J. Cell. Sci.* 29:313–37
101. Heuser, J. E., Salpeter, S. R. 1979. *J. Cell. Biol.* 82:150–73
102. Potter, L. T., Smith, D. S. 1977. *Tissue & Cell* 9:585–94
103. Allen, T., Baerwald, R. Potter, L. T. 1977. *Tissue & Cell* 9:595–608
104. Bourgeois, J.-P., Ryter, A., Menez, A., Fromageot, P., Boquet, P., Changeux, J.-P. 1972. *FEBS Lett.* 25:127–33
105. Bourgeois, J.-P., Popot, J.-L., Ryter, A., Changeux, J.-P. 1978. *J. Cell Biol.* 79:200–16
106. Barnard, E. A., Wieckowski, J., Chiu, T. H. 1971. *Nature* 234:205–9
107. Miledi, R., Potter, L. T. 1971. *Nature* 233:599–603
108. Porter, C. W., Chiu, T. H., Wieckowski, J., Barnard, E. A. 1973. *Nature New Biol.* 241:3–7
109. Salpeter, M. M., Eldefrawi, M. E. 1973. *J. Histochem. Cytochem.* 21:769–78
110. Fertuck, H. C., Salpeter, M. M. 1974. *Proc. Natl. Acad. Sci. USA* 71:1376–8
111. Rosenbluth, J. 1974. *J. Cell Biol.* 62:755–66
112. Peper, K., Dreyer, F., Sandri, C., Akert, K., Moor, H. 1974. *Cell Tissue Res.* 149:437–55
113. Rash, J. E. Ellisman, M. H. 1974. *J. Cell Biol.* 63:567–86
114. Heuser, J. E., Reese, T. S., Landis, D. M. D. 1974. *J. Neurocytology* 3:109–31
115. Albuquerque, E. X., Barnard, E. A., Porter, C. W., Warnick, J. E. 1974. *Proc. Natl Acad. Sci. USA* 71:2818–22
116. Porter, C. W., Barnard, E. A. 1975. *J. Membr. Biol.* 20:31–49
117. Porter, C. W., Barnard, E. A. 1975. *Exp. Neurol.* 48:542–56
118. Porter, C. W., Barnard, E. A. 1976. *Ann. NY Acad. Sci.* 274:85–107

119. Ellisman, M. H. Rash, J. E., Staehelin, L. A., Porter, K. R. 1976. *J. Cell Biol.* 68:752–74
120. Fertuck, H. C., Salpeter, M. M. 1976. *J. Cell Biol.* 69:144–58
121. Ellisman, M. H., Rash, J. E. 1977. *Brain Res.* 137:197–206
122. Rosenbluth, J. 1978. *J. Cell Biol.* 76:76–86
123. Rash, J. E., Hudson, C. S., Ellisman, M. H. 1978. *Cell Membrane Receptors for Drugs and Hormones: A Multidisciplinary Approach,* ed. R. W. Straub, L. Bolis, pp. 47–68. New York: Raven 356 pp.
124. Matthews-Bellinger, J., Salpeter, M. M. 1978. *J. Physiol.* 279:197–213
125. Sytkowski, A. J., Vogel, Z., Nirenberg, M. W. 1973. *Proc. Natl. Acad. Sci. USA* 70:270–4
126. Land, B. R., Podleski, T. R., Salpeter, E. E., Salpeter, M. M. 1977. *J. Physiol.* 269:115–76
127. Cohen, S. A., Pumplin, D. W. 1979. *J. Cell Biol.* 82:494–516
128. Bonner, R., Barrantes, F. J., Jovin, T. 1976. *Nature* 263:429–31
129. Barrantes, F. J. 1976. *Biochem. Biophys. Res. Commun.* 72:479–88
130. Grunhagen, H. H., Iwatsubo, M., Changeux, J.-P. 1976. *CR Acad. Sci. Ser. B Paris* 283:1105–8
131. Grunhagen, H. H., Iwatsubo, M., Changeux, J.-P. 1977. *Eur. J. Biochem.* 80:225–42
132. Quast, U., Schimerlik, M., Raftery, M. A. 1978. *Biochem. Biophys. Res. Commun.* 81:955–64
133. Quast, U., Schimerlik, M. I., Raftery, M. A. 1979. *Biochemistry* 18:1891–901
134. Dunn, S. M. J., Blanchard, S. G., Raftery, M. A. 1980. *Biochemistry* 19: 5645–52
135. Heidmann, T., Changeux, J.-P. 1979. *Eur. J. Biochem.* 94:255–79
136. Heidmann, T., Changeux, J.-P. 1980. *Biochem. Biophys. Res. Commun.* 97: 889–96
137. Adams, P. R. 1981. *J. Membr. Biol.* 58:161–74
138. Moore, H.-P. H., Raftery, M. A. 1980. *Proc. Natl. Acad. Sci. USA* 77:4509–13
139. Neubig, R. R., Cohen, J. B. 1980. *Biochemistry* 19:2770–9
140. Hess, G. P., Aoshima, H., Cash, D. J., Lenchitz, B. 1981. *Proc. Natl. Acad. Sci. USA* 78:1361–5
141. Cash, D. J., Aoshima, H., Hess, G. P. 1981. *Proc. Natl. Acad. Sci. USA* 78:3318–22
142. Lindstrom, J., Gullick, W., Conti-Tron-

coni, B., Ellisman, M. 1980. *Biochemistry* 19:4791–5
143. Conti-Tronconi, B. M., Dunn, S. M. J., Raftery, M. A. 1982. *Biochemistry.* 21:893–99 In press
144. Lindstrom, J., Walter (Nave), B., Einarson, B. 1979. *Biochemistry* 18: 4470–80
145. Witzemann, V., Raftery, M. A. 1978. *Biochem. Biophys. Res. Commun.* 85:623–31
146. Witzemann, V., Muchmore, D., Raftery, M. A. 1979. *Biochemistry* 18: 5515–8
147. Nathanson, N. M., Hall, Z. W. 1980. *J. Biol. Chem.* 255:1698–1703
148. Matthews, B. W., Barnard, S. A. 1973. *Ann. Rev. Biophys. Bioeng.* 4:257–317
149. Sobel, A., Hofler, J., Heidmann, T., Changeux, J.-P. 1979. *Adv. Inflammation Res.* 1:33–42
150. Sobel, A., Hofler, J., Heidmann, T., Changeux, J.-P. 1979. *Adv. Cytopharmacol.* 3:191–6
151. Hider, R. C. 1979. *Nature* 281:340–41
152. Raftery, M. A., Hunkapiller, M. W., Strader, C. D., Hood, L. E. 1980. *Science* 208:1454–7
153. Lindstrom, J., Merlie, J., Yogeeswaran, G. 1979. *Biochemistry* 18:4465–70
154. Claudio, T. R., Raftery, M. A. 1980. *J. Immunol.* 124:1130–40
155. Tzartos, S. J., Lindstrom, J. M. 1980. *Proc. Natl. Acad. Sci. USA* 77:755–9
156. Conti-Tronconi, B. M., Hunkapiller, M. W., Lindstrom, J. M., Raftery, M. A. 1982. *Biochem. Biophys. Res. Commun.* In press
157. Conti-Tronconi, B. M., Hunkapiller, M. W., Lindstrom, J. M., Raftery, M. A. 1982. *Proc. Natl. Acad. Sci. USA* In press
158. Devillers-Thiery, A., Changeux, J.-P., Paroutaud, P., Strosberg, A. D. 1979. *FEBS Lett.* 104:99–105
159. Hunkapiller, M. W., Strader, C. D., Hood, L., Raftery, M. A. 1979. *Biochem. Biophys. Res. Commun.* 91: 164–9
160. Conti-Tronconi, B., Tzartos, S., Lindstrom, J. 1981. *Biochemistry* 20: 2181–91
161. Mattsson, C., Heilbronn, E. 1975. *J. Neurochem.* 25:899–901
162. Raftery, M. A., Schmidt, J., Vandlen, R., Moody, T. 1974. *Neurochemistry of Cholinergic Receptors,* ed. E. de Robertis, J. Schacht, pp. 5–18. New York: Raven 146 pp.
163. Wennogle, L. P., Changeux, J.-P. 1980. *Eur. J. Biochem.* 106:381–93

164. Gordon, A. S., Davis, C. G., Diamond, I. 1977. *Proc. Natl. Acad. Sci. USA* 74:263–7
165. Gordon, A., Davis, C., Milfay, D., Diamond, I. 1977. *Nature* 267:539–40
166. Gordon, A. S., Milfay, D., Davis, C. G., Diamond, I. 1979. *Biochem. Biophys. Res. Commun.* 87:876–83
167. Teichberg, V., Sobel, S., Changeux, J.-P. 1977. *Nature* 267:540–2
168. Teichberg, V. I., Changeux, J.-P. 1977. *FEBS Lett.* 74:71–6
169. Saitoh, T., Changeux, J.-P. 1980. *Eur. J. Biochem.* 105:51–62
170. Suarez-Isla, B. A., Hucho, F. 1977. *FEBS Lett.* 75:65–9
171. Hamilton, S. L., McLaughlin, M., Karlin, A. 1979. *Biochemistry* 18:115–63
172. Witzemann, V., Raftery, M. A. 1978. *Biochem. Biophys Res. Commun.* 81:1025–31
173. Wallis, I., Koenig, E., Rose, S. 1980. *Biochem. Biophys. Acta* 599:505–17
174. Froehner, S. C., Karlin, A., Hall, Z. W. 1977. *Proc. Natl. Acad. Sci. USA* 74:4685–8
175. Einarson, B., Gullick, W., Conti-Tronconi, B. M., Ellisman, M., Lindstrom, J. 1982. *Biochemistry.* In press
176. Drachman, D. B. 1978. *New Engl. J. Med.* 298:136–42, 186–93
177. Axelrod, D., Ravdin, P., Koppel, D. E., Schlessinger, J., Webb, W. W., Elson, E. L., Podleski, T. R. 1976. *Proc. Natl. Acad. Sci. USA* 73:4594–8
178. Bourgeois, J.-P., Popot, J. L., Ryter, A., Changeux, J.-P. 1973. *Brain Res.* 62: 557–63
179. Weinberg, C. B., Reiness, C. G., Hall, Z. W. 1981. *J. Cell Biol.* 88:215–8
180. Rousselet, A., Devaux, P. F. 1977. *Biochem. Biophys. Res. Commun.* 78: 448–54
181. Saitoh, T., Wennogle, L. P., Changeux, J.-P. 1979. *FEBS Lett.* 108:489–94
182. Karlin, A., Holtzman, E., Valderrama, R., Damle, V., Hsu, K., Reyes, F. 1978. *J. Cell. Biol.* 76:577–592
183. Tarrab-Hazdai, R., Geiger, B., Fuchs, S., Amsterdam, A. 1978. *Proc. Natl. Acad. Sci. USA* 75:2497–501
184. Strader, C. B. D., Revel, J.-P., Raftery, M. A. 1979. *J. Cell Biol.* 83:499–510
185. Strader, C. D., Raftery, M. A. 1974. *Proc. Natl. Acad. Sci. USA* 77:5807–11
186. Michaelson, D. M., Raftery, M. A. 1974. *Proc. Natl. Acad. Sci. USA* 71:4768–72
187. Kreps, E. M., Krasilnikova, V. I., Poniazanskaya, L. F., Pravdina, N. I., Smirnov, A. A., Chirkovskaya, E. V.

1973. *Zh. Evol. Biokhim. Fiziol.* 9: 20–29
188. Bleadsdale, J. E., Hawthorne, J. N., Widlund, L., Heilbronn, E. 1976. *Biochem. J.* 158:557–65
189. Schiebler, W., Hucho, F. 1978. *Eur. J. Biochem.* 85:55–63
190. Popot, J.-L., Demel, R. A., Sobel, A., Van Deenen, L. L. M., Changeux, J.-P. 1978. *Eur. J. Biochem.* 85:27–42
191. Axelrod, D., Wight, A., Webb, W., Horwitz, A. 1978. *Biochemistry* 17: 3604–9
192. Bienvenue, A., Rousselet, A., Kato, G., Devaux, P. F. 1977. *Biochemistry* 16: 841–8
193. Marsh, D., Barrantes, F. J. 1978. *Proc. Natl. Acad. Sci. USA* 75:4329–33
194. Lass, Y., Fischbach, G. 1976. *Nature* 263:150–1
195. Sator, V., Gonzalez-Ros, J. M., Calvo-Fernandez, P., Sator, V., Martinez-Carrion, M. 1979. *Biochemistry* 18:1200–6
196. Gonzalez-Ros, J. M., Calvo-Fernandez, P., Sator, V., Martinez-Carrion, M. 1979. *J. Supramol. Struct.* 11:327–38
197. Tarrab-Hazdai, R., Bercovici, T., Goldfarb, B., Gitler, C. 1980. *J. Biol. Chem.* 255:1204–9
198. Dupont, Y., Cohen, J. B., Changeux, J.-P. 1974. *FEBS Lett.* 40:130–3
199. Raftery, M. A., Bode, J., Vandlen, R., Michaelson, D., Deutsch, J., Moody, T., Ross, M. J., Stroud, R. M. 1975. *Protein-Ligand Interactions,* ed. H. Sund, G. Blauer, pp. 328–55. Berlin: de Gruyter. 486 pp.
200. Cartaud, J., Popot, J.-L., Changeux, J.-P. 1980. *FEBS Lett.* 121:327
201. Wise, D. S., Schoenborn, B. P., Karlin, A. 1981. *The J. Biol. Chem.* 256:4124–6
202. Hartig, P. R., Raftery, M. A. 1979. *Biochemistry* 18:1146–50
203. Zingsheim, H. P., Neugebauer, D.-Ch., Barrantes, F. J., Frank, J. 1980. *Proc. Natl. Acad. Sci. USA* 77:952–6
204. Fatt, P., Katz, B. 1951. *J. Physiol.* 115:320–70
205. Katz, B., Miledi, R. 1973. *J. Physiol.* 231:549–74
206. Dreyer, F., Peper, K. 1975. *Nature* 253:641–3
207. Katz, B., Miledi, R. 1972. *J. Physiol.* 224:665–99
208. Anderson, C. R., Stevens, C. F. 1973. *J. Physiol.* 235:655–91
209. Albuquerque, E. X., Barnard, E. A., Jansson, S.-E., Wieckowski, J. 1973. *Life Sci.* 12:542–52
210. Strader, C. D., Lazarides, E., Raftery, M. A. 1980. *Biochem. Biophys. Res. Commun.* 92:365–73

211. Grunhagen, H., Changeux, J.-P. 1976. *J. Mol. Biol.* 106:497–516
212. Grunhagen, H., Changeux, J.-P. 1976. *J. Mol. Biol.* 106:517–35
213. Schimerlik, M., Raftery, M. A. 1976. *Biochem. Biophys. Res. Commun.* 73: 607–13
214. Schimerlik, M., Quast, U., Raftery, M. A. 1979. *Biochemistry* 10:1884–90
215. Cohen, J. B., Changeux, J.-P. 1973. *Biochemistry* 12:4855–64
216. Martinez-Carrion, M., Raftery, M. A. 1973. *Biochem. Biophys. Res. Commun.* 55:1156–64
217. Waksman, G., Fournie-Zaluski, M. C., Roques, B., Heidmann, T., Grunhagen, H., Changeux, J.-P. 1976. *FEBS Lett.* 67:335–42
218. Barrantes, F. J. 1978. *J. Mol. Biol.* 124:1–26
219. Sheridan, R. E., Lester, H. A. 1977. *J. Gen. Physiol.* 70:187–219
220. Dionne, V. E., Stevens, C. F. 1975. *J. Physiol.* 251:245–70
221. Adams, P. R. 1975. *Pfleugers Arch.* 360:145–53
222. Adams, P. R. 1977. *J. Physiol.* 268: 271–89
223. Neumann, E., Chang, H. W. 1976. *Proc. Natl. Acad. Sci. USA* 73:3994–8
224. Jurss, R., Prinz, H., Maelicke, A. 1979. *Proc. Natl. Acad. Sci. USA* 76:1064–8
225. Magazanik, L. G., Vyskocil, F. 1973. *Drug Receptors,* ed. H. P. Rang, pp. 105–19. London: Macmillan. 321 pp.
226. Magazanik, L. G. 1976. *Ann. Rev. Pharmacol.* 16:161–75
227. Albuquerque, E. X., Barnard, E. A., Chiu, T. H., Lapa, A. J., Dolly, J. O., Jansson, S., Daly, J., Witkop, B. 1973. *Proc. Natl. Acad. Sci. USA* 70:949–53
228. Dolly, J. O., Albuquerque, E. X., Sarvey, J. M., Mallick, B., Barnard, E. A. 1977. *Mol. Pharmacol.* 13:1–14
229. Kato, G., Changeux, J.-P. 1976. *Mol. Pharmacol.* 12:92–100
230. Bon, C., Changeux, J.-P. 1977. *Eur. J. Biochem.* 74:43–51
231. Moody, T. W., Raftery, M. A. 1978. *Arch. Biochem. Biophys.* 189:115–21
232. Albuquerque, E. X., Oliveira, A. C. 1979. In *Advances in Cytopharmacology,* ed. B. Ceccarelli, F. Clementi, 3:197–211. New York: Raven. 475 pp.
233. Albuquerque, E. X., Eldefrawi, A. T., Eldefrawi, M. E., Mansour, N. A., Tsai, M.-C. 1978. *Science* 199:788–90
234. Albuquerque, E. X., Tsai, M.-C., Aronstam, R. S., Witkop, B., Eldefrawi, A. T., Eldefrawi, M. E. 1980. *Proc. Natl. Acad. Sci. USA* 77:1224–8

235. Tiedt, T. N., Albuquerque, E. X., Bakry, N. M., Eldefrawi, M. E., Eldefrawi, A. T. 1979. *Mol. Pharmacol.* 16:909–21

236. Tsai, M.-C., Albuquerque, E. X., Aronstam, R. S., Eldefrawi, A. T., Eldefrawi, M. E., Triggle, D. J. 1980. *Mol. Pharmacol.* 18:159–66

237. Albuquerque, E. X., Tsai, M.-C., Aronstam, R. S., Eldefrawi, A. T., Eldefrawi, M. E. 1980. *Mol. Pharmacol.* 18:167–78

238. Weber, M., Changeux, J.-P. 1974. *Mol. Pharmacol.* 10:35–40

239. Krodel, E. K., Beckman, R. A., Cohen, J. B. 1979. *Mol. Pharmacol.* 15:294–312

240. Cohen, J. B., Weber, M., Changeux, J.-P. 1974. *Mol. Pharmacol.* 10:904–32

241. Blanchard, S. G., Raftery, M. A. 1979. *Proc. Natl. Acad. Sci. USA* 76:81–85

242. Witzemann, V., Raftery, M. A. 1978. *Biochemistry* 17:3598–604

243. Wennogle, L. P., Oswald, R., Saitoh, T., Changeux, J.-P. 1981. *Biochemistry* 20:2492–7

244. Eldefrawi, A. T., Eldefrawi, M. E., Albuquerque, E. X., Oliveira, A. C., Mansour, N., Adler, M., Daly, J. W., Brown, G. B., Burgermeister, W., Witkop, B. 1977. *Proc. Natl. Acad. Sci. USA* 74:2172–6

245. Elliott, J., Raftery, M. A. 1977. *Biochem. Biophys. Res. Commun.* 77: 1347–53

246. Elliott, J., Raftery, M. A. 1979. *Biochemistry* 10:1868–74

247. Eldefrawi, M. E., Eldefrawi, A. T., Mansour, N. A., Daly, J. W., Witkop, B., Albuquerque, E. X. 1978. *Biochemistry* 17:5474–84

248. Sobel, A., Heidmann, T., Hofler, J., Changeux, J.-P. 1978. *Proc. Natl. Acad. Sci. USA* 75:510–4

249. Dunn, S. M. J., Blanchard, S. G., Raftery, M. A. 1981. *Biochemistry* 20: 5617–24

250. Schimerlik, M. I., Quast, U., Raftery, M. A. 1979. *Biochemistry* 10:1902–5

251. Miller, J. 1980. *Studies on ligand binding to the histrionicotoxin and the agonist binding sites of membrane bound acetylcholine receptor from "Torpedo californica".* PhD thesis. Calif. Inst. Technol., Pasadena, Calif. 181 pp.

252. Reiter, M. J., Cowburn, D. A., Prives, J. M., Karlin, A. 1972. *Proc. Natl. Acad. Sci. USA* 69:1168–72

253. Damle, V. N., McLaughlin, M., Karlin, A. 1978. *Biochem. Biophys. Res. Commun.* 84:845–51

254. Moore, H.-P., Raftery, M. A. 1979. *Biochemistry* 10:1862–7

255. Wolosin, J. M., Lyddiatt, A., Dolly, J. O., Barnard, E. A. 1980. *Eur. J. Biochem.* 109:495–505

256. Witzemann, V., Raftery, M. A. 1977. *Biochemistry* 16:5862–8

257. Hucho, F. 1979. *FEBS Lett.* 103:27–32

258. Heidmann, T., Changeux, J.-P. 1978. *Ann. Rev. Biochem.* 47:317–57

259. Karlin, A. 1980. *Cell Surface Rev.* 6:191–260

260. Damle, V. N., Karlin, A. 1978. *Biochemistry* 17:2039–45

261. Neubig, R. R., Cohen, J. B. 1979. *Biochemistry* 18:5464–75

262. Bulger, J. E., Fu, J.-J. L., Hindy, E. F., Silberstein, R. L., Hess, G. P. 1977. *Biochemistry* 16:684–92

263. Katz, B., Thesleff, S. 1957. *J. Physiol.* 138:63–80

264. Rang, H. P., Ritter, J. M. 1970. *Mol. Pharmacol.* 6:383–90

265. Rang, H. P., Ritter, J. M. 1970. *Mol. Pharmacol.* 6:357–82

266. Rang, H. P., Ritter, J. M. 1969. *Mol. Pharmacol.* 5:394–411

267. Weber, M., David-Pfeuty, T., Changeux, J.-P. 1975. *Proc. Natl. Acad. Sci. USA* 72:3443–7

268. Weiland, G., Georgia, B., Wee, V. T., Chignell, C. F., Taylor, P. 1976. *Mol. Pharmacol.* 12:1091–105

269. Lee, T., Witzemann, V., Schimerlik, M., Raftery, M. A. 1977. *Arch. Biochem. Biophys.* 183:57–63

270. Quast, U., Schimerlik, M., Lee, T., Witzemann, V., Blanchard, S., Raftery, M. A. 1978. *Biochemistry* 17:2405–14

271. Weiland, G., Georgia, B., Lappi, S., Chignell, C. F., Taylor, P. 1977. *J. Biol. Chem.* 252:7648–56

272. Blanchard, S. G., Elliott, J., Raftery, M. A. 1979. *Biochemistry* 18:5880–5

273. Weiland, G., Taylor, P. 1979. *Mol. Pharmacol.* 15:197–212

274. Boyd, N. D., Cohen, J. B. 1980. *Biochemistry* 19:5344–53

275. Sugiyama, H., Changeux, J.-P. 1975. *Eur. J. Biochem.* 55:505–15

276. Heidmann, T., Sobel, A., Popot, J.-L., Changeux, J.-P. 1980. *Eur. J. Biochem.* 110:35–55

277. Young, A. P., Brown, F. F., Halsey, M. J., Sigman, D. S. 1978. *Proc. Natl. Acad. Sci. USA* 75:4563–7

278. Katz, B., Miledi, R. 1970. *Nature* 226:962–3

279. Sachs, F., Lecar, H. 1973. *Nature New Biol.* 246:214–216

280. Neher, E., Sakmann, B. 1976. *Nature* 260:799–802

281. Lester, H. A. 1977. *Sci. Am.* 236: 106–18

282. Dionne, V. E., Steinbach, J. H., Stevens, C. F. 1978. *J. Physiol.* 281:421–44
283. Adams, D. J., Dwyer, T. M., Hille, B. 1980. *J. Gen. Physiol.* 75:493–510
284. Maeno, T., Edwards, C., Anraku, M. 1977. *J. Neurobiol.* 8:173–84
285. Dwyer, T. M., Adams, D. J., Hille, B. 1980. *J. Gen. Physiol.* 75:469–92
286. Kasai, M., Changeux, J.-P. 1971. *J. Membr. Biol.* 6:1–80
287. Cohen, J. B., Changeux, J.-P. 1975. *Ann. Rev. Pharmacol.* 15:83–103
288. Lester, H. A., Changeux, J.-P., Sheridan, R. E. 1975. *J. Gen. Physiol.* 65:797–816
289. Hess, G. P., Lipkowitz, S., Struve, G. E. 1978. *Proc. Natl. Acad. Sci. USA* 75:1703–7
290. Hess, G. P., Andrews, J. P. 1977. *Proc. Natl. Acad. Sci. USA* 74:482–6
291. Popot, J.-L., Sugiyama, H., Changeux, J.-P. 1976. *J. Mol. Biol.* 106:469–83
292. Sugiyama, H., Popot, J. L., Changeux, J.-P. 1976. *J. Mol. Biol.* 106:485–96
293. Popot, J.-L., Sugiyama, H., Changeux, J.-P. 1974. *CR Acad. Sci. Ser. B* 279:1721–4
294. Bernhardt, J., Neumann, E. 1978. *Proc. Natl. Acad. Sci. USA* 75:3756–60
295. Miller, D. L., Moore, H.-P. H., Hartig, P. R., Raftery, M. A. 1978. *Biochem. Biophys. Res. Commun.* 85:632–40
296. Moore, H.-P. H., Hartig, P. R., Wu, W. C.-S., Raftery, M. A. 1979. *Biochem. Biophys. Res. Commun.* 88:735–43
297. Moore, H.-P. H., Hartig, P. R., Raftery, M. A. 1979. *Proc. Natl. Acad. Sci. USA* 76:6265–9
298. Fersht, A. R., Jakes, R. 1975. *Biochemistry* 14:3350–6
299. Hess, G. P., Cash, D. J., Aoshima, H. 1979. *Nature* 282:329–32
300. Cash, D. J., Hess, G. P. 1980. *Proc. Natl. Acad. Sci. USA* 77:842–6
301. Aoshima, H., Cash, D. J., Hess, G. P. 1980. *Biochem. Biophys. Res. Commun.* 92:896–904
302. Catterall, W. A. 1975. *J. Biol. Chem.* 250:1776–81
303. Sine, S. M., Taylor, P. 1980. *J. Biol. Chem.* 255:10144–56
304. Sine, S. M., Taylor, P. 1981. *J. Biol. Chem.* 256:6692–9
305. Lester, H. A., Chang, H. W. 1977. *Nature* 266:373–4
306. Nass, M. M., Lester, H. A., Krouse, M. E. 1978. *Biophys. J.* 24:135–60
307. Schindler, H., Quast, U. 1980. *Proc. Natl. Acad. Sci. USA* 77:3052–6
308. Nelson, N., Anholt, R., Lindstrom, J., Montal, M. 1980. *Proc. Natl. Acad. Sci. USA* 77:3057–61
309. Boheim, G., Hanke, W., Barrantes, F. J., Eibl, H., Sakmann, B., Fels, G., Maelicke, A. 1981. *Proc. Natl. Acad. Sci. USA* 78:3586–90
310. Hazelbauer, G. L., Changeux, J.-P. 1974. *Proc. Natl. Acad. Sci. USA* 71:1479–83
311. McNamee, M. G., Weill, C. L., Karlin, A. 1975. See Ref. 199:317–27
312. McNamee, M. G., Weill, C. L., Karlin, A. 1975. *Ann. N.Y. Acad. Sci.* 264:175–82
313. Howell, J., Kemp, G., Eldefrawi, M. E. 1978. In *Membrane Transport Processes,* ed. D. C. Tosteson, Y. A. Ovchinnikov, R. Latorre, 2:207–15. New York: Raven. 452 pp.
314. Epstein, M., Racker, E. 1978. *J. Biol. Chem.* 253:6660–2
315. Wu, W. C.-S., Raftery, M. A. 1979. *Biochem. Biophys. Res. Commun.* 89:26–35
316. Huganir, R. L., Schell, M. A., Racker, E. 1979. *FEBS Lett.* 108:155–60
317. Changeux, J.-P., Heidmann, T., Popot, J.-L., Sobel, A. 1979. *FEBS Lett.* 105:181–7
318. Gonzalez-Ros, J. M., Paraschos, A., Martinez-Carrion, M. 1980. *Proc. Natl. Acad. Sci. USA* 77:1796–1800
319. Lindstrom, J., Anholt, R., Einarson, B., Engel, A., Osame, M., Montal, M. 1980. *J. Biol. Chem.* 255:8340–50
320. Wu, W. C.-S., Raftery, M. A. 1981. *Biochemistry* 20:694–701
321. Anholt, R., Lindstrom, J., Montal, M. 1980. *Eur. J. Biochem.* 109:481–7
322. Wu, W., Moore, H-P., Raftery, M. A. 1981. *Proc. Natl. Acad. Sci. USA* 78:775–9

Ann. Rev. Biochem. 1982. 51:531–54

CARBOHYDRATE-SPECIFIC RECEPTORS OF THE LIVER[1]

Gilbert Ashwell and Joe Harford[2]

National Institute of Arthritis, Diabetes, Digestive and Kidney Diseases, National Institutes of Health, Bethesda, Maryland 20205

CONTENTS

PERSPECTIVES AND SUMMARY

Our current appreciation of the role played by carbohydrates in cellular recognition has its roots in the classical investigations of the Australian school (1). Their early studies demonstrated that the presence of sialic acid residues on the surface of the erythrocyte membrane is an absolute requirement for the attachment of influenza virus; prior treatment with neuraminidase abolished the binding reaction. In the course of the next decade, it became increasingly apparent that cell surface carbohydrates are intimately involved in such diverse phenomena as lymphocyte "homing," tumor invasiveness, trophoblast implantation, intercellular adhesion, etc (2). However, despite a growing elaboration of complex, carbohydrate-related phenome-

---

[1]The US Government has the right to retain a nonexclusive royalty-free licence in and to any copyright covering this paper.

[2]Supported during preparation of this chapter by the American Cancer Society and the US Public Health Service.

nology, the underlying unity of the informational content of sugar residues has remained obscure.

This review emphasizes the emergence and development of carbohydrate-specific binding reactions in the liver. The initial description of an hepatic receptor responsible for the clearance of desialylated glycoproteins from the serum has proved to be a paradigm for the successful characterization of several alternative hepatic receptors capable of recognizing and binding specific nonreducing carbohydrate termini. As described here, at least five such systems are presently known to be operative in the liver. By far the most widely studied is the asialoglycoprotein receptor. The physical properties of this system are eminently suitable for studies on the mechanism of receptor-mediated endocytosis. It is in this area, particularly, that recent new insights have shaped, and are continually reshaping our current understanding of the cellular processes accompanying ligand internalization. Whether, or to what extent, any of the known binding proteins participate in the subsequent determination of subcellular events leading to lysosomal degradation, or to plasma membrane receptor recycling, is unknown, although it is exactly to such questions that extensive efforts are being directed.

The last few years have also witnessed the emergence of at least two areas of promise for clinical application. This material has been covered in more detail in a current review (3) and is merely indicated here. The first relates to the improved ability to target specific therapeutic agents to the parenchymal cells of the liver. For this purpose, a variety of biochemical techniques have been employed whereby biologically active molecules have been covalently linked with galactose-terminated saccharides, or incorporated into liposomes, as a means of guiding them effectively to the desired tissues. Clearly, the design of techniques for the delivery of corrective agents to metabolically impaired cells is being markedly facilitated by the appropriate exploitation of uniquely located cell surface receptors.

The second stems from an awareness of the existence of a pool of circulating asialoglycoproteins with potential value as a diagnostic tool in patients with diseases of the liver. The availability of sensitive analytic techniques for monitoring desialylated proteins in the serum has permitted the assessment of hepatocellular damage in patients with cirrhosis, hepatitis, and primary cancer of the liver. In the latter case, a direct correlation between the size of the tumor and the ratio of intact to desialylated transferrin has been established (4).

Finally, the physiological role of the various receptor systems must be considered. Based on the consistent correlation between the level of serum asialoglycoproteins and the functional ability of the liver, it is generally presumed that the asialoglycoprotein receptor participates in the regulation of serum glycoprotein homeostasis. However attractive this presumption may be, unequivocal evidence is still lacking. In view of the major intracellular locus of the asialoglycoprotein receptor, an alternate hypothesis has been advanced whereby this receptor is viewed as a participant in the subcellular recognition events that determine the migration of endocytosed vesicles to the lysosomes. A more recent hypothesis has surfaced that envisages the clearance of biologically reactive glycoproteins after they have served a particular physiological function. Indeed, evidence has been adduced to implicate the clearance of antigen-antibody complexes formed with both IgG and IgM (5, 6). Other potential candidates for such a mechanism may well include complexes of haptoglobin-hemoglobin or a_2-macroglobulin-proteases, etc. For the remaining receptor systems, the evidence for a specific physiological function is, at best, equally tenuous.

THE RECEPTOR FOR ASIALOGLYCOPROTEINS

Introduction

Desialylated serum glycoproteins have drastically reduced survival times in the circulation compared to native forms of the same proteins. A specific receptor on hepatocytes mediates their clearance by recognition of galactose residues made terminal by removal of sialic acid. This phenomenon was first demonstrated using desialylated ceruloplasmin (7). The newly exposed galactose moieties were implicated as recognition determinants by the findings that treatment of asialoceruloplasmin with galactose oxidase or β-galactosidase (7) or enzymatic replacement of the missing sialic acid residues (8) markedly prolonged survival in circulation. Subsequently, the generality of the original observation has been established by extension to alternate serum glycoproteins (7) and by the growing list of known glycoproteins susceptible to enhanced clearance due to desialylation (Table 1).

Upon the foundation of these initial studies [summarized in an earlier review, see (9)], our current understanding of this process has been built and continues to grow. Two broad and overlapping categories define the studies that have contributed to this understanding. In some, the nature of the receptor molecule and its interaction with ligands have been examined. In others, the relationship between the receptor molecule and the cell in which it resides has been probed. Here, we discuss the results of both of these approaches.

Physical and Chemical Characterization of the Receptor for Asialoglycoproteins

The activity for binding asialoglycoproteins was recovered in membrane fractions (10) and subsequently in Triton X-100 extracts of rabbit liver (11). Hudgin et al (12) employed affinity chromatography on asialo-orosomucoid

Table 1 Proteins known to be susceptible to clearance via the asialoglycoprotein receptor

Modulators of cellular responses	Protein complexers
Erythropoietin (128)[a]	α_1-Antitrypsin (138, 139)
Follicle stimulating hormone (13)	Haptoglobin (13)
Human chorionic gonadotropin (13)	α_2-Macroglobulin (13)
Interferon (129, 130)	**Molecules of immunological importance**
Carrier proteins for smaller molecules	Immunoglobulin G (98, 140)
	IgG-antigen complex (5, 141)
Ceruloplasmin (7)	**Other**
Hemopexin (131)	
Thyroglobulin (132, 133)	Carcino-embryonic antigen (142)
Thyroxin-binding globulin (134)	Fetuin (13)
Transcobalamine (70)	Glycophorin (143)
Transcortin (135, 136)	Orosomucoid (13)
Transferrin (137)	Prothrombin (144)

[a] The numbers in parentheses refer to references.

linked to Sepharose to purify a water-soluble, lipid-free material possessing asialoglycoprotein binding activity. The ligand specificity, pH optimum, and an absolute requirement for calcium of this preparation were similar to those observed in clearance studies (13) and for binding to plasma membranes (10).

The affinity-purified receptor was found to be 10% carbohydrate, consisting of sugars commonly found in asparagine-linked carbohydrate chains. The presence of both "complex" and "polymannose" type glycans [for definition see (14)] was deduced by analyses of glycopeptides generated from the isolated receptor for rabbit liver (15). The integrity of the receptor's carbohydrate units appeared to be essential for functional activity, since treatment of plasma membranes (10) or the purified receptor (12) with neuraminidase resulted in loss of binding activity. It was later reported that activity of neuraminidase-inactivated receptor could be restored by exposure to galactose oxidase or β-galactosidase (16), or by enzymatic replacement of sialic acid residues (17). Both findings supported the interpretation that the loss of activity was a consequence of binding by the receptor preparation of its own galactose residues exposed by the neuraminidase treatment.

The isolated rabbit-binding protein exhibited a high degree of aggregation when subjected to physical studies. Addition of Triton X-100 yielded a species with a molecular weight of 260,000 from which the aggregates were formed (18). Polyacrylamide gel electrophoresis after sodium dodecyl sulfate (SDS) treatment allowed separation of polypeptides with apparent molecular weights of 48,000 and 40,000. It was concluded (15) that each 48,000 subunit contained three "complex" glycan units and one "polymannose" structure, while the 40,000 subunit possessed two of the complex structures. Two of the larger polypeptides and four of the smaller were presumed to define the 260,000-dalton species produced by disaggregation in Triton X-100. Differential scanning calorimetry has been used to examine the thermal denaturation of the rabbit receptor (19). The results of this study indicated that calcium binding stabilizes the receptor toward heat denaturation. An additional but smaller degree of stabilization occurred upon binding of specific ligands.

An effect of calcium binding has also been observed (20) in the analogous receptor from rat liver. Calcium binding to this protein resulted in aggregation leading to a decrease in the protein's intrinsic fluorescence. A dissociation constant for the protein-calcium complex in the range of 1–2 mM was calculated. Circular dichroism indicated that no major changes in the secondary structure accompanied the calcium-induced aggregation.

The rat liver receptor was originally isolated using affinity chromatography (21). Polyacrylamide gel electrophoresis in SDS revealed a major band at 47,000 daltons and two relatively minor bands. The proportion and sizes of these electrophoretically separable species vary somewhat in reports from different laboratories. Similarities in peptide fragments have been interpreted to indicate areas of homology or identity in the primary structures of the three observed bands (22, 23). Immunoprecipitates of a receptor preparation formed using monoclonal antibodies to the rat receptor have been reported to contain all of the polypeptides observed in the preparation (22). Sawamura et al (24) have similarly isolated the rat liver receptor and report a single polypeptide with a molecular weight of 52,000. A molecular weight of 105,000 has been assigned to the receptor function based on target analysis of data from radiation inactivation of isolated rat liver plasma membranes (25). The human liver asialoglycoprotein-binding protein, isolated and characterized by Baenziger & Maynard (26), was found to be composed of a single subunit with an apparent molecular weight of 41,000. Amino acid and carbohydrate compositions of the rat liver receptor (24) and the human receptor (26) are quite similar to those of the rabbit liver preparation (15).

The Binding of Asialoglycoproteins

The binding site of the rabbit receptor for asialoglycoproteins has been examined using the binding protein immobilized on Sepharose (27) and revealed to be relatively small, since only the nonreducing sugar and a portion of the penultimate sugar influenced the ability of saccharides to inhibit ^{125}I-asialo-orosomucoid binding. Methylglycosides of N-acetyl-galactosamine were more potent than those of galactose. This preference had been inferred from an earlier study (28) on the blood type–specific agglutination of red blood cells by the isolated asialoglycoprotein receptor. The analogous human lectin is also most potently inhibited by N-acetyl-galactosamine (26). Moreover, the displacement of hepatocyte-bound ^{125}I-asialo-orosomucoid by the acetylated amino sugar occurs more rapidly than with galactose (29).

Sepharose-immobilized receptor has also been probed using proteins to which carbohydrates have been chemically attached. The synthesis of these neoglycoproteins has been reviewed (30). Bovine serum albumin (BSA) to which D-galactosides were attached was specifically bound, whereas analogous neoglycoproteins containing mannosides or N-acetylglucosaminides were not (31). Surprisingly, these studies indicated that the binding protein could not discrimate between D-galacto and D-gluco configurations. Indeed, the rabbit receptor has been isolated by affinity chromatography on columns of glucosyl-BSA (32). However, further investigations have indicated that only when the glucose is attached to the albumin via an amidino group is the resultant neoglucoprotein taken up by isolated hepatocytes (33). The positive charge near the sugar appears to be responsible for markedly enhancing the affinity of the glucosyl-BSA. The galactose-containing ligands were taken up independent of the means of sugar attachment. These results have led to the speculation that an anionic group exists in the proximity of the receptor's binding site. Additional structural parameters of the glycose determinants capable of influencing binding have been defined using the neoglycoproteins approach (34).

Kawaguchi et al (35) attached 1, 2, or 3 galactose residues to Tris and covalently attached these "cluster" glycosides to albumin. On an attachment sites per albumin basis, the neoglycoproteins containing three sugars per site were more effective inhibitors of receptor-ligand interaction, than those with two per site, which were in turn better than those with one galactose per attachment site. However, when compared on a sugar per protein basis, at low levels of derivatization, all were equally as effective. A "positioning effect" had been postulated earlier (9) to explain in part the variation in the affinities of different ligands for the receptor. The spectrum of avidities for binding of serum proteins is bounded by asialo-orosomucoid (high affinity) and asialotransferrin (low affinity). In fact, asialotransferrin was thought to be unrecognized by the receptor (13) until, by extending the period of observation, Regoeczi et al (36) showed that its serum survival time is shorter than native transferrin. Inspection of the carbohydrate content and architecture of these ligands reveals that orosomucoid contains five branched chain carbohydrate units that are assymmetrically arranged on the polypeptide near the amino terminus (37) and that transferrin, for the most part, contains only two biantennary type glycose chains (38). Arrangement of glycose residues, as well as number, has been invoked by Baenziger & Maynard (26) to explain the avidities of glycoproteins, glycopeptides, and oligosaccharides for the human lectin. It was suggested that simultaneous binding at two receptor sites 25–30 Å apart rather than a statistical effect on binding at a single site accounts for observed differences. This general picture is supported by the finding that as increasing amounts of galactose are added to synthetic substrata, a threshold for binding rat hepatocytes is observed (39). This can best be explained in terms of a critical density being required for appropriate spacing of sugars.

Endocytosis and the Fate of Asialoglycoproteins

The initial interaction of ligand with receptor in the course of endocytosis is at the hepatocyte plasma membrane. The binding characteristics of plasma membranes (10) and of intact hepatocytes isolated by collagenase perfusion (29, 40–44) are similar to those discussed above for the solubilized receptor, although certain differences have been noted (44). Rate constants for surface binding and dissociation from hepatocytes have been determined at reduced temperatures where internalization is minimal (43, 45). Although the values obtained in the two laboratories differ, both concluded that the dissociation rate is extremely slow. This may contribute to the diversity of reported values for binding constants based on Scatchard analysis of asialo-orosomucoid binding to isolated rat hepatocytes (23, 29, 43, 44, 46, 47). Since attaining equilibrium during the experiment is an assumption underlying this calculation, the wide variation (200-fold differences) clearly indicates that this, and perhaps other experimental difficulties, influence such determinations.

The interaction of ligands with the asialoglycoprotein receptor occurs at least in part in coated pits (48–50). These specialized regions of the plasma membrane are so named because of the fuzzy coat that decorates their cytoplasmic surface. These structures have been implicated in adsorptive endocytosis of several ligands in a variety of cell types (51). It has been suggested (50) that association of colloidal gold-tagged ligand with coated pits on isolated hepatocytes is preceded by binding at diffusely distributed sites along the former sinuisoidal surface and local microaggregation of receptor-ligand complexes. Data also indicate that the microaggregation and the presumed migration into coated pits could proceed at 4°C. Lateral mobility of the asialoglycoprotein receptor has also been inferred from the finding that adhesion of hepatocytes to a ligand-bearing substratum results in patching of the surface receptor population (52).

While binding of ligand to surface receptors is clearly central to the process of adsorptive endocytosis, the complexity of this process goes beyond these considerations. Injection of small amounts ($<1\,\mu g/100g$ body weight) of asialotransferrin has been reported to result in an equilibrium between the circulation and the receptor on the hepatocyte plasma membrane, with no significant internalization (53). Thus binding to the receptor is necessary but may not be sufficient to elicit endocytosis. Subsequently it was reported (54) that transferrin molecules can act synergistically to induce mutual endocytosis. In this context, Baenziger & Fiete (44) have reported that glycopeptides differing in affinities in the isolated receptor by as much as 750-fold, display similar kinetics of endocytosis in isolated hepatocytes. The interaction of the receptor with other components of the cell that may be responsible for this observation remains to be elucidated. These authors have proposed that multivalent ligands with appropriate determinant spacing induce conformational changes in the receptor molecule, which triggers endocytosis.

The portion of the receptor responsible for interaction with other molecules leading to endocytosis may be located on the cytoplasmic side of the plasma membrane. Using antibodies raised to the soluble receptor, evidence has been obtained indicating that the receptor spans the lipid bilayer of the hepatocyte plasma membrane (55). The transmembrane nature of the receptor has been supported by the suggestion that the binding protein is synthesized on membrane-bound ribosomes with its amino-terminal segment and the carbohydrate exposed to the luminal side and the carboxyl terminus on the cytoplasmic side (56). Conceivably, that portion of the receptor exposed on the cytoplasmic surface may mediate receptor-coated pit interaction, since clathrin, the major coated pit protein (57) is on the internal side of the plasma membrane. An analogous interaction has been postulated for the receptor for low density lipoprotein (58).

Internalization of receptor-bound ligand is rapid. A first order rate constant at 37°C has been calculated to be $3.7 \times 10^{-3}\ \text{sec}^{-1}$ by Tolleshang (45), a number in close agreement with the $0.2\ \text{min}^{-1}$ rate coefficient generated by computer modeling of internalization data (59). This corresponds to a half-life of approximately 3 min for the occupied receptor on the cell surface. A similar residency has been estimated from the amount of ligand internalized and degraded by hepatocytes and the number of surface receptors involved (29). In keeping with rapid internalization, sections of livers fixed as early as 30 s after injection of asialoglycoproteins linked to horse radish peroxidase show the enzyme reaction product in small (<1000Å) vesicular profiles and large, more irregular structures in the peripheral cytoplasm near the sinusoids (48). A portion ($\sim 45\%$) of the apparently vesicular structures were open to the cell's exterior. Stockert et al (49) have described "elongated pinocytic channels or tubules and pleomorphic tubular structures" containing ligand. It has been suggested elsewhere (60) that structures appearing in micrographs as coated vesicles are, in fact, sections through coated pits, and that free coated vesicles are not an intermediate in the internalization process.

It does appear clear that larger uncoated vesicles are involved in intracellular translocation of ligand. The term "receptosome" has been coined (60) to reflect the involvement of these vesicles in receptor-mediated endocytosis. Interestingly, Dunn et al (61) observed that degradation of ^{125}I-asialofetuin by perfused livers did not occur below 20°C. Internalization continued and ligand appeared to accumulate in the larger intermediate compartment in the peribiliary region of the hepatocytes. It was concluded that transfer from these structures to lysosomes was blocked by the temperature reduction. A related study (62) concluded that a different process is rate limiting for internalization below 20°C than for that at 37°C. Additional information was gained by the use of ferritin to which lactose was attached as ligand (48). Higher numbers of ligand particles were seen in cross sections of the larger vesicles than in individual coated pits and coated vesicular profiles, which suggests that the former structures may arise by fusion of several smaller vesicles. It was also noted that the ferritin particles appeared to be no longer closely apposed to the limiting membrane and it was suggested that ligand was not receptor bound at this stage in endocytosis.

The intracellular separation of ligand from the receptor that mediated its endocytosis has also been indicated by a more biochemical approach (59). Here advantage was taken of the ability to distinguish receptor-bound and -unbound ^{125}I-asialo-orosomucoid after hepatocyte solubilization in nonionic detergent. After occupation of hepatocyte surface receptors at 4°C, washed cells were warmed to 37°C, where internalization was rapid and ligand was ultimately degraded. Ligand appeared to enter the cell bound to receptor but intracellular separation preceded ligand destruction. From a model that fits well the experimental data, compartmental residency times and rate coefficients for transfer of ligand were computed. Although low pH or removal of calcium are known to cause release of ligands from the receptor for asialoglycoproteins (10), the intracellular conditions responsible for ligand dissociation, as well as for where it occurs, remain to be determined. The implications of this separation for receptor reutilization are discussed below. The participation of microtubules and microfilaments in the movement of asialoglycoproteins has been probed (63). Both endocytosis and degradation of ligand were impaired by colchicine, an inhibitor affecting microtubule function. In contrast, cytochalasin B, which interferes with microfilament-related processes, inhibits only degradation. Lack of additivity of these effects argues for inhibition at different points in the process.

Initial evidence implicating lysosomes as the site of asialoglycoprotein catabolism was provided by Gregoriadis et al (64). Asialoceruloplasmin labeled with ^{3}H-galactose and protein-bound ^{64}Cu migrated with lysosomal markers in sucrose density gradients. Kinetic analyses by LaBadie et al (65) indicated that between 5 min and 13 min after injection, ^{125}I-asialofetuin shifted from migration with plasma membrane markers to comigration with lysosomes. Consistent with lysosomes as being the site of catabolism are the observed inhibition of ligand degradation by leupeptin (66) and NH$_4$Cl or chloroquine (40, 67), agents that affect proteolysis in the lysosomal compartment. Morphological studies are in similar accord with this concept (48, 68, 69).

However, endocytosis mediated by the asialoglycoprotein receptor does not obligatorily result in ligand degradation in lysosomes. A small portion of certain injected proteins that are removed from circulation by this system escape degradation and are secreted intact into the bile (70, 71). This may be qualitatively related to the more recent findings of Tolleshaug et al (72) who showed that subsequent to endocytosis of low amounts of asialotransferrin, isolated hepatocytes exocytose the preponderance of the ligand. The intracellular half-life was estimated to be 20 min. The authors suggested that the endocytic vesicles lacked some element responsible for their being "homed to lysosomes." As in earlier in vivo studies (54), the relative extent of asialotransferrin catabolism could be increased by elevating the extracellular ligand concentration. These intriguing results suggest that ligand participates, in some fashion, in intracellular targeting. The mechanism responsible for this phenomenon remains unclear.

Subcellular Distribution and Reutilization of Receptor

In addition to being present in isolated plasma membrane fractions (10), the receptor for asialoglycoproteins was found in other subcellular fractions (73). In fact, the preponderance of receptor activity appeared to be intracellular. Certain observations indirectly support the contention that not all the receptor molecules of the cell participate in ligand catabolism. For example, treatment of isolated hepatocytes with neuraminidase abolishes binding of asialo-orosomucoid, but the cells continued to catabolize desialylated ovine submaximally mucin (74), a ligand of higher affinity for the receptor (16). Although the internal receptor pool was unaffected by this treatment, the altered surface specificity of the enzyme-treated cells was maintained throughout continuing internalization of mucin in excess of surface-binding capacity. Similarly, infusion of a single passage of antireceptor IgG was found to block ligand binding by a perfused liver for at least 90 min (75). Since homogenates of the liver retained greater than 85% of the control-binding activity, these results were interpreted to indicate that intracellular receptors were not recruited to the cell surface. This would imply that receptors do not cycle to and from the cell surface in the absence of ligand. Clearly a caveat should be considered regarding the possibility that antibody bound to surface receptors or neuraminidase treatment may disrupt normal cycling and/or affect the ability to detect it. An opposing view has been presented by Tolleshaug & Berg (76) who have presented Scatchard plots indicating chloroquine-induced reduction in surface receptor sites. These authors suggested that the drug impeded return of receptor to the surface and, since added ligand was absent, constitutive cycling of unoccupied receptor was implied. Oka & Weigel (77) have also suggested that occupied and unoccupied receptors are internalized at similar rates.

Other results indicate that surface receptor is sufficient to mediate catabolism of asialoglycoproteins. Mouse L-929 cells that are impotent in this regard have been fused with rat hepatocyte membrane vesicles (78) or reconstituted vesicles containing the receptor (79). This fusion resulted in cells capable of endocytosis and degradation of asialo-orosomucoid under conditions where the introduced receptor apparently did not redistribute to produce an internal pool analogous to that of hepatocytes. While it was not conclusively shown that the chimeric cells processed ligand in excess of their receptor content, these provocative findings suggest that a large intracellular pool(s) of receptor is not required for ligand catabolism. These data have been summarized in more detail elsewhere (80).

The binding proteins isolated independently from various subcellular fractions appeared to be chemically and immunologically indistinguishable, but the activities of the several fractions differed markedly in response to added nonionic detergent (73). On the assumption that the addition of Triton X-100 made accessible cryptic receptor sites, it was concluded that the receptor binding sites in Golgi- and smooth microsome–enriched fractions were largely oriented toward the lumen of vesicles contained in these fractions. This finding was supported by Tanabe et al (21) who assessed the amount of antireceptor antibody needed to neutralize binding activity in subcellular fractions and membranes derived from these fractions. Finding the receptor in microsomes and Golgi-enriched fractions was rationalized on the basis of the presumed biosynthetic pathway of the receptor. This has been supported by results of Nakada et al (56) in their study of binding protein synthesis. In contrast to Golgi- and microsome-enriched fractions, none of the binding activity observed in lysosome-enriched fractions appeared to be cryptic (21). An orientation of binding site toward the cytosol would not be that anticipated if receptor-ligand complex were enclosed within an endocytic vesicle that fused with the lysosomal compartment. Rather, the receptor binding site might be expected to face the lysosome's interior and suffer the same fate as its ligand, i.e. degradation.

The asialoglycoprotein receptor apparently outlives the ligand whose endocytosis it mediates. An in vivo survival half-time of 88 hr has been calculated for the binding protein, and injection of ligand does not alter this value (21). A somewhat shorter half-life of 20 hr has been estimated for the receptor in cultured hepatocytes (23). In either case, these findings suggest a divergence in the metabolic pathways for receptor and its ligand that allows sparing of receptor and destruction of ligand. Regoeczi et al (81) also reached the conclusion that receptors are reutilized, based on the amounts of asialoglycoproteins removed from the serum by the liver. This is further supported by studies showing that isolated hepatocytes internalize and degrade ligand in excess of their total content of receptor, even when de novo receptor synthesis is blocked by inhibitors (29, 45). It has been estimated that each receptor of the cultured rat hepatocyte delivers 1000 molecules of ligand for degradation (23). A calculation of this type of course is based on estimation of the number of functional receptors involved. Caution should be exercised because this quantification can be dramatically altered by conditions used in the isolation of cells (47).

Based on the apparent durability of the receptor and the enigmatic topology of the activity in lysosome-enriched fractions, it was postulated that the receptor may reorient in the lysosomal membrane to avoid proteolysis (21). The feasibility of this has now been addressed in studies of the insertion of soluble rabbit receptor into model membranes. The binding protein was found to associate spontaneously via a hydrophobic interaction with phospholipid vesicles (82) and black lipid membranes (83). Furthermore, receptor was found to undergo an apparent translocation from one side of the black lipid membrane to the other, thereby orienting its binding site toward the positive pole of an imposed electrical potential. Assuming that this could occur in biological membranes, and based on the inside-negative state of hepatocytes (84) and lysosomes (85, 86), these results predict a binding site orientation in the plasma membrane toward the cell's exterior and that any receptor present in the lysosomal membrane would be oriented toward the cytosol. When galactosyl-BSA linked to colloidal gold was allowed to diffuse into broken hepatocytes, specific labeling of the cytosolic surface of "Golgi-derived vesicles" was observed (50). This may represent the asialoglycoprotein receptor activity observed in lysosome-enriched fractions (21, 73).

An alternative means by which receptor could escape degradation would be for the binding protein to avoid the lysosomal compartment altogether. As summarized above, separation of ligand and receptor before ligand reaches morphologically identifiable lysosomes (48), and prior to ligand degradation (59), has been suggested. Since virtually all studies utilize tagged ligand of one type or another, the intracellular traffic in receptor itself cannot be followed after such a separation. Indeed, it has even been speculated that receptor functions without leaving the plasma membrane (74). However, based on computed residency times, it was concluded (59) that ligand-receptor complex exists in a state inaccessible to dissociation by EGTA and is therefore presumably inside the cell. At the same time no reduction in cell surface receptors occurred and this apparently was due to a replacement from a previously cryptic (presumably intracellular) pool of unoccupied receptors. Weigel (87) has also concluded that receptors enter the cell carrying ligands and are subsequently replaced, although the rate of replacement appears to be significantly lower than the rate of endocytosis of ligand. Clearly, the mechanism for making receptors available for reutilization remains one of the more interesting facets of receptor-mediated endocytosis.

HEPATIC RECEPTORS IN NONMAMMALIAN SPECIES

In 1975, Regoeczi et al (88) isolated α-acid glycoprotein from chicken serum which, upon injection into rabbits, was promptly cleared from the circulation and accumulated in the liver. Analysis of this preparation revealed it to be undersialylated in that it contained 2.6 mol of terminal galactose residues per mole of protein as compared with a corresponding value of 0.97 for human α-acid glycoprotein. As a result of this observation, Lunney (89) looked for, and found, levels of circulating desialylated glycoproteins in the serum of both avian and reptilian species that were approximately an order of magnitude greater than the corresponding values in mammals.

On the premise that such levels were inconsistent with the presence of a functioning asialoglycoprotein receptor in chicken liver, further studies were undertaken. Independent and concurrent reports by Regoeczi and Hatton (90) and Lunney and Ashwell (91) established the absence of this receptor in chickens, the former on the basis of clearance studies, the latter on the basis of a binding assay in detergent extracts of chicken liver. The correlation between elevated levels of incompletely sialylated serum glycoproteins and the absence of the asialoglycoprotein receptor in chickens was held to be consistent with the presumption, albeit unproven, that in mammals this binding protein functions in the regulation of glycoprotein homeostasis.

Further indirect support for this presumption came from the identification and isolation of a closely related avian hepatic receptor which, although inert toward galactose, reacted strongly with terminal, nonreducing N-acetylglucosamine residues (91). The chicken protein was purified to apparent homogeneity by affinity chromatography on a column of N-acetylglucosamine-terminated orosomucoid covalently linked to Sepharose (92).

The receptor was isolated as a water-soluble glycoprotein in which sialic acid, galactose, mannose, and N-acetylglucosamine comprised 8% of the total weight.

Comparison of the avian with the mammalian binding protein isolated from rabbit liver revealed many similarities. In both systems, the major locus of binding was restricted to the liver from which the receptors were solubilized, purified, and assayed by closely analogous techniques. In both cases, calcium was an absolute requirement for binding. Upon purification, both preparations proved to be glycoproteins containing the same set of carbohydrate constituents. In aqueous solution, the two proteins were recovered in aggregated states, which reversibly converted to single molecular weight components in the presence of detergent.

The major differences noted, aside from their ligand specificites, were confined to the physical and kinetic properties of the two proteins. In contrast to the two different subunits of 48,000 and 40,000 daltons reported for the rabbit protein (18), SDS gel electrophoresis of the avian receptor preparation revealed a single subunit with an estimated molecular weight of 26,000. A second major point of differentiation was the ready reversibility of the avian protein-ligand complex for which a rate constant for dissociation of 1.3×10^{-3} sec^{-1} was calculated. Under similar conditions, the rabbit-ligand complex was only minimally reversible. The ready attainment of equilibrium in the avian system permitted the demonstration of a single, high affinity binding site with a dissociation constant of 1.4×10^{-9} M as determined by a Scatchard plot. Finally, whereas the binding activity of the rabbit protein was destroyed by exposure to neuraminidase, the chicken protein was unaffected. The activity of the latter preparation was abolished only after β-galactosidase was added to the incubation mixture. This initially puzzling behaviour has since been clarified by the realization that the exposure of terminal galactose residues on the rabbit binding protein permits this preparation to recognize and bind to itself, or to adjacent molecules, thereby rendering it incapable of binding added ligand (16, 17). Presumably the avian receptor undergoes a similar reaction toward N-acetylglucosamine residues exposed by the action of β-galactosidase.

As originally isolated and assayed, it was thought that the avian binding protein was specific for N-acetylglucosamine-terminated ligands. More recently, it has been reported that neoglycoproteins containing mannose or glucose covalently bound to BSA were equally good ligands (93). The significance of the observation that glucosyl-BSA is recognized is uncertain, since glucose is rarely encountered as a constituent of circulating glycoproteins. However, the finding that the chicken binding protein reacts to both N-acetylglucosamine and mannose raises the question of its possible relationship to an alternate mammalian hepatic receptor with similar ligand specificity (see the section on the mannose/N-acetylglucosamine receptor).

In a recent publication, Drickamer (94) has reported the complete amino acid sequence of the chicken receptor. The single subunit of 26,000 daltons was shown to be comprised of 207 amino acids in which the blocked amino terminus was identified as N-acetylmethionine. The sequence of residues from number 24 to 48 contains no amino acids with charged side chains, and this predominantly hydrophobic region is thought likely to be an intramembranous segment. A single carbohydrate chain was found to be linked to asparagine at residue 67, a location presumably external to the cell. On the assumption that the receptor is a transmembrane protein, comparable to that of the analogous rat liver receptor (55) and based on the report that many cytoplasmic proteins are acetylated at the amino terminus (95), it was speculated that the N-terminal 23 residues may be situated in the cell cytoplasm.

In addition to the well-characterized receptor described above, a large and growing literature has emerged devoted to the study of developmentally regulated lectins. The prototype of these saccharide-binding proteins, originally named "electrolectin" by Teichberg et al in 1975 (96), was a soluble β-D-galactoside binding protein isolated from the electric organ of the eel and shown to be present in the tissues of several animal species including cultures of skeletal muscle and neuroblastoma cells. Since that time, at least three distinct vertebrate lectins have been identified, each of which may become especially prominent at a specific stage in the development of individual tissues. Thus, a lectin purified from adult chicken liver by affinity chromatography appears indistinguishable from that purified from embryonic skeletal muscle. This lectin, although detectable in many chicken tissues, is distinct from a separate lactose-inhibitable lectin (designated chicken lactose lectin-I). The alternate binding protein, chicken lactose lectin-II, is scarce in embryonic muscle but prominent in embryonic liver where its content falls to much lower levels in the mature chicken (97). Although no specific cellular function can as yet be assigned to these saccharide-binding proteins, they are presumed to play roles in differentiation, and may be required in diminished or augmented amounts as the tissue matures. This potentially rewarding area of investigation has been extensively reviewed by Barondes in the preceding volume of this series (97)

THE MANNOSE/N-ACETYLGLUCOSAMINE RECEPTOR

In a carefully executed study on the effect of glycosidases on the serum survival time of homologous antibodies, Winkelhake & Nicolson (98) showed that desialylated antibodies were rapidly cleared from the circulation by the hepatic mechanism described earlier for asialoglycoproteins (9). The subsequent removal of galactose restored the serum survival time and the organ distribution to near normal values, whereas further removal of the antepenultimate N-acetylglucosamine residues resulted in increased kidney localization. Shortly thereafter, in examining the fate of desialylated orosomucoid in the rat, Stockert et al (99) found that removal of galactose, with the resultant exposure of terminal N-acetylglucosamine residues, was again accompanied by rapid hepatic uptake. Significantly, the clearance of this derivative was not inhibited by the simultaneous injection of asialo-orosomucoid and a second recognition system, specific for N-acetylglucosamine, was postulated.

At about this time, considerable interest had been aroused in the metabolic fate of lysosomal glycosidases, based, in large part, on the work of Neufeld and her co-workers, which had implicated carbohydrates in the specificity of binding and uptake of hydrolases by fibroblasts (100). Upon the basis of these findings, Stahl et al (101) examined clearance rates in rats of several purified lysosomal glycosidases and found them to be markedly prolonged by the simultaneous infusion of N-acetylglucosamine-terminated orosomucoid. The corresponding galactose-terminated derivative was totally unaffective. This finding, supportive of a role for an N-acetyl-glucosamine-specific recognition site, was quickly modified by the surprising observation of Achord et al (102) that the clearance of agalacto-orosomucoid (N-acetylglucosamine-terminated) from rat plasma was inhibited by infused mannans or by mannose alone. Conversely, mannan uptake was similarly blocked by a galacto-orosomucoid and it was concluded that the presumed receptor was capable of recognizing and binding both sugars; galactose-terminated ligands were inert in this system.

Extension of these studies with RNase B, a naturally occuring mannose-terminated glycoprotein (103) and with RNase A, a nonglycosylated analogue to which mannose residues were covalently attached (104), provided further confirmation of the participation of mannose in this system. Concurrent studies in several laboratories demonstrated convincingly, by histological examination (105), by isolation and direct binding studies (106), and by electron microscopic autoradiography (107) that the reticuloendothelial system was the principle locus of uptake. Thus, cell type localization is a second parameter of differentiation between this uptake system and that for the uptake of asialoglycoproteins. It should be noted, however, that Kupffer cells as well as hepatocytes possess the ability to form rosettes with desialylated erythrocytes and lymphocytes and that this interaction appears to be galactose specific (108, 109).

The conclusion that elements of the reticuloendothelial system were involved in mannose/N-acetylglucosamine-mediated clearance led Stahl et al (110) to investigate isolated alveolar macrophages. The specificity of binding by these cells mimicked the uptake seen in the intact rat. Mannose-

or N-acetylglucosamine-terminated glycoproteins as diverse as RNase B, ovalbumin, β-glucuronidase, or agalacto-orosomucoid were bound specifically. In no case was binding, or inhibition of binding, seen with glycoproteins bearing terminal galactose residues.

Most of these early studies were performed in vivo by injecting into rats an appropriately labeled ligand and monitoring its disappearance from the circulation and accumulation in the liver. Furthermore, at that time, there was no clear appreciation of the specific nature of the recognition marker for the lysosomal glycosidases, which was only subsequently identified as mannose-6-phosphate (see the section on the phosphomannosyl receptor). Consequently, considerable confusion existed as to whether all glycoproteins bearing terminal mannose or N-acetyl-glucosamine residues, including the lysosomal hydrolases, were cleared by a single receptor.

The first definitive indication that at least two separate recognition systems were involved came from the work of Kawasaki et al (111) who described the isolation and purification of a mannan-binding protein from rat liver. Acetone powders of rat liver were extracted with Triton X-100, and the binding protein was isolated by affinity chromatography using Sepharose to which mannan had been covalently linked. Upon SDS gel electrophoresis, a single major band was seen with an estimated molecular weight of 31,000. Binding studies confirmed the specificity for mannose and N-acetylglucosamine with the interesting exception that N-acetylmannosamine, which appears to have not been tested previously, was the most potent monosaccharide inhibitor of mannan binding. In agreement with the in vivo studies of Achord et al (102) and the in vitro macrophage binding studies of Stahl et al (110), those mannose oligosaccharides linked α 1–6 were better inhibitors than similarly sized compounds in which the linkage was α 1–2 or α 1–4. Mannose-6-phosphate was without inhibitory effect. These preliminary studies have since been extended to provide more detailed information on the physical properties of the purified receptor (112). In addition, an alternate method of preparation has been discribed by Townsend & Stahl (113). In the latter study, it was noted that the binding of ^{125}I-mannan to the purified receptor was strongly inhibited by the neoglycoprotein, fucosyl-BSA, as well as by the analogous mannose- and N-acetylglucosamine-terminated proteins. Glucosyl-BSA was about two orders of magnitude less effective and galactosyl-BSA was inactive in this system (see the section on the fucose receptor).

Another area of controversy has now arisen as to the exact locus of the mannose/N-acetylglucosamine receptor in the liver. As indicated above, all of the available data are in agreement that the uptake of these ligands is restricted to the Kupffer, or sinusoidal, cells of the liver, and it has been presumed that the hepatocytes are devoid of this activity. However, in a preliminary report, Maynard & Baenziger (114) have described the purification of this lectin from isolated hepatocytes and shown that as much as 70% can be released from an acetone powder with 0.2 M NaCl; the remainder requires 2% Triton X-100 for solubilization. Both forms appeared to be similar, if not identical, to the receptor mediating endocytosis by Kupffer cells. It was not possible, however, to detect this lectin on the surface of isolated hepatocytes. In an attempt to reconcile these apparently contradictory findings, it has been speculated that the material present in the hepatocytes may represent a cytosolic binding protein rather than a cell surface receptor. At present, the function of the cytosolic lectin for mannose/N-acetylglucosamine and its relationship to the surface receptor on Kupffer cells is unknown.

A RECEPTOR SPECIFIC FOR FUCOSE

Evidence has been presented that glycoproteins bearing fucose residues bound in α 1–3 linkage to N-acetylglucosamine are rapidly cleared from the serum of mice with over 90% of the injected material recoverable in hepatocytes (115). Uptake was not inhibited by compounds in which the fucose was bound in either α 1–2 or α 1–6 linkage. The original studies compared the characteristic survival behavior of two closely related proteins, transferrin from human serum and lactoferrin from human milk. Both proteins contain two iron-binding sites, an homologous amino acid sequence, and two biantennary oligosaccharide chains with identical core carbohydrate structures. However, in addition to the three terminal sugars of transferrin, (sialic acid, galactose, and N-acetylglucosamine) lactoferrin contains additional fucose residues in α 1–3 linkage to the N-acetylglucosamine adjacent to galactose; those chains containing fucose are devoid of sialic acid.

Whereas lactoferrin promptly disappeared from the circulation after injection into mice, the survival times of both transferrin and asialotransferrin were prolonged. Lactoferrin uptake was shown to be mediated through its carbohydrate groups, since clearance was delayed after periodate oxidation or after extensive cleavage of the oligosaccharide chains by exposure to glycosidases. Additionally, lactoferrin glycopeptides significantly inhibited uptake of the intact protein.

More specifically, when fucose was added enzymatically to asialotransferrin in an α 1–3 linkage to N-acetylglucosamine, the resulting fucosylated derivative was rapidly cleared. Further confirmation of the involvement of fucose was obtained by the demonstration that an almond fucosidase, specific for α 1–3 fucosides and free from other glycosidases, quantitatively removed the added fucose, and the resulting asialotransferrin was only poorly cleared (116). Neither mannan nor derivatives of orosomucoid that are cleared by binding receptors for galactose, N-acetylglucosamine, or mannose inhibited clearance of fucosylated asialotransferrin. Fucoidin, a compound containing a 1–3 linked fucose, was inhibitory. Evidence for the participation of fucose in the hepatic clearance and uptake of glucocerebrosidase, a lysosomal hydrolase deficient in patients with Gaucher's disease, has been published by Furbish et al (117).

Subsequent studies directed toward the isolation and characterization of a specific hepatic protein responsible for the observed clearance of fucose-terminated macromolecules has been reported by Lehrman and Hill (118). Rat liver homogenates were extracted with 2% Triton X-100 and the solubilized material precipitated with 6% polyethyleneglycol. The precipitate, solubilized in buffered 0.5% Triton X-100 containing 20 mM calcium, was applied to a column of Sepharose to which the neoglycoprotein, fucosyl-BSA, was covalently linked. Both the fucose- and the mannose/N-acetylglucosamine-binding proteins adsorbed to the column; the galactose-binding protein was unretarded. Both activities were eluted with EDTA and, after addition of calcium, reapplied to the same column in the presence of 50 mM N-acetylglucosamine. The mannose/N-acetylglucosamine-binding activity emerged unretarded; the fucose-binding activity remained on the column and was eluted with EDTA. The isolated binding protein, approximately 100,000-fold purified, was recovered in 20% yield. Binding activity was calcium dependent and readily inactivated by dithiothreitol.

Gel electrophoresis in SDS of the purified receptor revealed two major bands of 67,000 and 73,000 daltons. In contrast, a value of 32,000 daltons was obtained for the purified mannose/N-acetylglucosamine protein. Antibody to the fucose-binding protein, prepared in rabbits, failed to cross-react with the mannose/N-acetylglucosamine receptor preparation. On the basis of these data, it was concluded that these two receptors are separate entities. (see the section on the mannose/N-acetylglucosamine receptor).

THE PHOSPHOMANNOSYL RECEPTOR

In contrast to the hepatic locus of the preceding carbohydrate recognition systems, the role of mannose-6-phosphate in the uptake of lysosomal hydrolases has emerged largely from studies on human fibroblasts in tissue culture. These studies, originating in the laboratory of Neufeld, demonstrated that normal fibroblasts secrete "corrective factors" active in reducing the abnormally high levels of mucopolysaccharides that accumulate in the cells of patients exhibiting various forms of mucopolysaccharidosis (119). It was subsequently shown that the corrective factors were the lysosomal enzymes missing in the mutant cells. These provocative findings indicate that fibroblasts possess the capacity to actively secrete lysosomal enzymes as well as the ability to internalize and utilize the exogenous hydrolases.

Further studies (120) revealed the presence of at least two forms of the catalytically active lysosomal glycosidases, one of which was readily internalized by fibroblasts (high-uptake form) and one of which was poorly internalized (low-uptake form). The identification of two active forms suggested the existence of a specific structure, or recognition marker, needed for cellular uptake, but distinct from the catalytic site. Initial evidence that the recognition marker is carbohydrate in nature arose from the observation that mild periodate oxidation converted the high-uptake to a low-uptake form (100). From the perceptive studies of Kaplan et al (121) concerning the inhibitory effects of polymeric mannans on cellular uptake, it was inferred that mannose-6-phosphate is a structural analogue of the recognition marker. Ensuing studies by several independent investigators culminated in the definitive identification of this structure, and have provided new insights into the role of this marker in intracellular enzyme transport. Concise and elegant summaries of this work are available in recent reviews by Neufeld (122) and Sly et al (123).

Extension of the earlier studies on fibroblasts to hepatic tissue was reported by Ullrich et al (124). Utilizing a cultured epithelial rat liver cell line, it was shown that the receptor-mediated endocytosis of α-N-acetylglucosaminidase was effectively inhibited by mannose-6-phosphate. Furthermore, the ability of these cells to recognize and take up this substrate was abolished by prior treatment of the enzyme with alkaline phosphatase.

These findings were confirmed and extended by Fischer et al (125) who examined the binding of β-hexosaminidase B to rat liver subcellular membranes. It was found that approximately 90% of the phosphomannosylenzyme receptors were intracellularly located in the membranes of the endoplasmic reticulum, Golgi apparatus, and the lysosomes; only about 10% of the receptors were identified on the plasma membrane. These findings, together with the gradient of receptor occupancy from endoplasmic reticulum to lysosomes, were interpreted as supporting the role of phosphomannosyl receptors as being the intracellular carriers of newly synthesized acid hydrolases to the lysosomes. In addition to the liver, comparable receptor activity was found in various other rat organs including testes, spleen, lung, and kidney.

A description of the successful isolation and initial characterization of the hepatic receptor for mannose-6-phosphate is now available from the work of Sahagian et al (126) who have reported the purification of a receptor that specifially binds the phosphomannosyl recognition marker of bovine testicular β-galactosidase. A crude plasma membrane preparation from bovine liver was extracted with Triton X-100 and the soluble β-galactosidase-receptor complex immunoprecipitated with anti-β-galactosidase. The receptor, dissociated from the precipitate with mannose-6-phosphate, was purified by affinity chromatography on β-galactosidase-Sepharose 4B. The apparently homogeneous protein was characterized as having a subunit molecular weight of 215,000 and a dissociation constant of $2 \times 10^{-8}M$ for the high-uptake form of the enzyme. In contrast to the metal requirement for the hepatic binding proteins described earlier in this chapter, no divalent cation was required. In a subsequent study on Chinese hamster ovary cells (127), the lysosomal enzyme, β-galactosidase, was found to be bound exclusively in coated pits, and the pathway for internalization, via the mannose-6-phosphate receptor, was shown to be similar to that established for other ligands, such as low density lipoprotein (51, 58) and asialoglycoproteins (48–50).

CONCLUDING REMARKS

It is, we believe, significant that the origin and development of the several recognition systems summarized in this chapter encompass the relatively brief span of approximately one decade. Inevitably, the emergence of such newly acquired insights gives rise to a new round of questions; with each cycle the level of sophistication increases. Such may be anticipated here. The isolation and characterization of specific receptors provides merely the initial steps in developing tools, both necessary and appropriate, for probing the more complicated mechanisms of cellular endocytosis. Indeed, encouraging steps in this direction have already been taken by exploiting the unique properties of the asialoglycoprotein receptor. It is from such studies, together with the growing realization that unrecognized receptor systems are yet to be identified, that a more comprehensive picture of the function and participation of these molecules in the totality of cellular processes may become a reality.

Literature Cited

1. Burnet, F. M. 1951. *Physiol. Rev.* 31:131–50
2. Ashwell, G. 1977. In *Mammalian Cell Membranes,* Vol. 4, pp. 57–71. London: Butterworth
3. Harford, J., Ashwell, G. 1982. In *The Glycoconjugates,* Vol. 4, ed. M. Horowitz, pp. 27–55. New York: Academic
4. Sobue, G., Kosaka, A. 1980. *Hepato-Gastroenterology* 27:200–3
5. Thornburg, R. W., Day, J. F., Baynes, J. W., Thorpe, S. R. 1980. *J. Biol. Chem.* 255:6820–25
6. Day, J. F., Thornburg, R. W., Thorpe,

S. R., Baynes, J. W. 1980. *J. Biol. Chem.* 255:2360–65
7. Morell, A. G., Irvine, R. A., Sternlieb, I., Scheinberg, I. H., Ashwell, G. 1968. *J. Biol. Chem.* 243:155–59
8. Hickman, J., Ashwell, G., Morell, A. G., van den Hamer, C. J. A., Scheinberg, I. H. 1970. *J. Biol. Chem.* 245:759–66
9. Ashwell, G., Morell, A. G. 1974. *Adv. Enzymol.* 41:99–128
10. Pricer, W. E. Jr., Ashwell, G. 1971. *J. Biol. Chem.* 246:4825–33
11. Morell, A. G., Scheinberg, I. H. 1972.

Biochem. Biophys. Res. Commun.
48:808–15
12. Hudgin, R. L., Pricer, W. E., Ashwell, G., Stockert, R. J., Morell, A. G. 1974. *J. Biol. Chem.* 249:5536–43
13. Morell, A. G., Gregoriadis, G., Scheinberg, I. H., Hickman, J., Ashwell, G. 1971. *J. Biol. Chem.* 246:1461–67
14. Kornfeld, R., Kornfeld, S. 1976. *Ann. Rev. Biochem.* 45:217–37
15. Kawasaki, T., Ashwell, G. 1976. *J. Biol. Chem.* 251:5292–99
16. Stockert, R. J., Morell, A. G., Scheinberg, I. H. 1977. *Science* 197:667–68
17. Paulson, J. C., Hill, R. L., Tanabe, T., Ashwell, G. 1977. *J. Biol. Chem.* 252:8624–28
18. Kawasaki, T., Ashwell, G. 1976. *J. Biol. Chem.* 251:1296–1302
19. Strickland, D. K., Andersen, T. T., Hill, R. L., Castellino, F. J. 1981. *Biochemistry* 20:5294–97
20. Andersen, T. T., Freytag, J. W., Hill, R. L. 1981. *J. Biol. Chem.* In press
21. Tanabe, T., Pricer, W. E. Jr., Ashwell, G. 1979. *J. Biol. Chem.* 254:1038–43
22. Schwartz, A. L., Marshak-Rothstein, A., Rup, D., Lodish, H. F. 1981. *Proc. Natl. Acad. Sci. USA* 78:3348–52
23. Warren, R., Doyle, D. 1981. *J. Biol. Chem.* 256:1346–55
24. Sawamura, T., Nakada, H., Fujii-Kuriyama, Y., Tashiro, Y. 1980. *Cell Struct. Funct.* 5:133–46
25. Steer, C. J., Kempner, E. S., Ashwell, G. 1981. *J. Biol. Chem.* 256:5851–56
26. Baenziger, J. U., Maynard, Y. 1980. *J. Biol. Chem.* 255:4607–13
27. Sarkar, M., Liao, J., Kabat, E. A., Tanabe, T., Ashwell, G. 1979. *J. Biol. Chem.* 254:3170–74
28. Stockert, R. J., Morell, A. G., Scheinberg, I. H. 1974. *Science* 186:365–66
29. Steer, C. J., Ashwell, G. 1980. *J. Biol. Chem.* 255:3008–13
30. Stowell, C. P., Lee, Y. C. 1980. *Adv. Carbohydr. Chem. Biochem.* 37:225–81
31. Krantz, M. J., Holtzman, N. A., Stowell, C. P., Lee, Y. C. 1976. *Biochemistry* 15:3963–68
32. Stowell, C. P., Lee, Y. C. 1978. *J. Biol. Chem.* 253:6107–10
33. Kawaguchi, K., Kuhlenschmidt, M., Roseman, S., Lee, Y. C. 1981. *J. Biol. Chem.* 256:2230–34
34. Stowell, C. P., Lee, R. T., Lee, Y. C. 1980. *Biochemistry* 19:4904–8
35. Kawaguchi, K., Kuhlenschmidt, M., Roseman, S., Lee, Y. C. 1980. *Arch. Biochem. Biophys.* 205:388–95
36. Regoeczi, E., Hatton, M. W. C., Wong,

K.-L. 1974. *Can. J. Biochem.* 52:155–61
37. Jeanloz, R. W. 1972. In *Glycoproteins,* ed. A. Gottschalk, pp. 565–611. Amsterdam: Elsevier
38. Jamieson, G. A., Jett, M., DeBernardo, S. L. 1971. *J. Biol. Chem.* 246:3686–93
39. Weigel, P. H., Schnaar, R. L., Kuhlenschmidt, M. S., Schmell, E., Lee, R. T., Lee, Y. C., Roseman, S. 1979. *J. Biol. Chem.* 254:10830–38
40. Tolleshaug, H., Berg, T., Nilsson, M., Norum, K. R. 1977. *Biochim. Biophys. Acta* 499:73–84
41. Tolleshaug, H., Ose, T., Berg, T., Wandel, M., Norum, K. R. 1977. In *Kupffer Cells and Other Liver Cells,* ed. E. Wisse, D. L. Knook, pp. 333–41. Amsterdam: Elsevier/North-Holland
42. Tolleshaug, H., Berg, T., Frölich, L., Norum, K. R. 1979. *Biochim. Biophys. Acta* 585:71–84
43. Weigel, P. H. 1980. *J. Biol. Chem.* 255:6111–20
44. Baenziger, J. U., Fiete, D. 1980. *Cell* 22:611–20
45. Tolleshaug, H. 1981. *Int. J. Biochem.* 13:45–51
46. Tolleshaug, H., Berg, T. 1980. *Hoppe-Seylers Z. Physiol. Chem.* 361:1155–64
47. Schwartz, A. L., Rup, D., Lodish, H. F. 1980. *J. Biol. Chem.* 255:9033–36
48. Wall, D. A., Wilson, G., Hubbard, A. L. 1980. *Cell* 21:79–93
49. Stockert, R. J., Haimes, H. B., Morell, A. G., Novikoff, P. M., Novikoff, A. B., Quintana, N., Sternlieb, I. 1980. *Lab. Invest.* 43:556–63
50. Kolb-Bachofen, V. 1981. *Biochim. Biophys. Acta* 645:293–99
51. Goldstein, J. L., Anderson, R. G. W., Brown, M. S. 1979. *Nature* 279:679–85
52. Weigel, P. H. 1980. *J. Cell. Biol.* 87:855–61
53. Regoeczi, E., Taylor, P., Hatton, M. W. C., Wong, K.-L., Koj, A. 1978. *Biochem. J.* 174:171–78
54. Regoeczi, E., Taylor, P., Debanne, M. T., Marz, L., Hatton, M. W. C. 1979. *Biochem. J.* 184:399–407
55. Harford, J., Ashwell, G. 1981. *Proc. Natl. Acad. Sci. USA* 78:1557–61
56. Nakada, H., Sawamura, T., Tashiro, Y. 1981. *J. Biochem.* 89:135–41
57. Pearse, B. M. F. 1976. *Proc. Natl. Acad. Sci. USA* 73:1255–59
58. Anderson, R. G. W., Brown, M. S., Goldstein, J. L. 1977. *Cell* 10:351–64
59. Bridges, K., Harford, J., Klausner, R., Ashwell, G. 1982. *Proc. Natl. Acad. Sci. USA* 79:350–54

60. Willingham, M. C., Pastan, I. 1980. *Cell* 21:67–77
61. Dunn, W. A., Hubbard, A. L., Aronson, N. N. Jr. 1980. *J. Biol. Chem.* 255:5971–78
62. Weigel, P., Oka, J. A. 1981. *J. Biol. Chem.* 256:2615–17
63. Kolset, S. O., Tolleshaug, H., Berg, T. 1979. *Exp. Cell Res.* 122:159–67
64. Gregoriadis, G., Morell, A. G., Sternlieb, I., Scheinberg, I. H. 1970. *J. Biol. Chem.* 245:5833–37
65. LaBadie, J. H., Chapman, K. P., Aronson, N. N. Jr. 1975. *Biochem. J.* 152:271–79
66. Dunn, W. A., LaBadie, J. H., Aronson, N. N. Jr. 1979. *J. Biol. Chem.* 254:4191–96
67. Berg, T., Tolleshaug, H. 1980. *Biochem. Pharmac.* 29:917–25
68. Hubbard, A. L., Stukenbrok, H. 1979. *J. Cell. Biol.* 83:65–81
69. Haimes, H., Stockert, R. J., Morell, A., Novikoff, A. B. 1981. *Proc. Natl. Acad. Sci. USA* 78:6936–39
70. Burger, R. L., Schneider, R. J., Mehlman, C. S., Allen, R. H. 1975. *J. Biol. Chem.* 250:7707–13
71. Thomas, P., Summers, J. W. 1978. *Biochem. Biophys. Res. Commun.* 80:335–39
72. Tolleshaug, H., Chindemi, P. A., Regoeczi, E. 1981. *J. Biol. Chem.* 256:6526–28
73. Pricer, W. E. Jr., Ashwell, G. 1976. *J. Biol. Chem.* 251:7539–44
74. Stockert, R. J., Howard, D. J., Morell, A. G., Scheinberg, I. H. 1980. *J. Biol. Chem.* 255:9028–29
75. Stockert, R. J., Gärtner, U., Morell, A. G., Wolkoff, A. W. 1980. *J. Biol. Chem.* 255:3830–31
76. Tolleshaug, H., Berg, T. 1979. *Biochem. Pharmacol.* 28:2919–22
77. Oka, J., Weigel, P. 1980. *Fed. Proc.* 39:2616
78. Doyle, D., Hou, E., Warren, R. 1979. *J. Biol. Chem.* 254:6853–56
79. Baumann, H., Hou, E., Doyle, D. 1980. *J. Biol. Chem.* 255:10001–12
80. Doyle, D., Baumann, H., Hou, E., Warren, R. 1981. In *International Cell Biology.* New York: Rockefeller Univ. Press. In press
81. Regoeczi, E., Debanne, M. T., Hatton, M. W. C., Koj, A. 1978. *Biochim. Biophys. Acta* 54:372–84
82. Klausner, R. D., Bridges, K., Tsunoo, H., Blumenthal, R., Weinstein, J. N., Ashwell, G. 1980. *Proc. Natl. Acad. Sci. USA* 77:5087–91

83. Blumenthal, R., Klausner, R. D., Weinstein, J. N. 1980. *Nature* 288:333–38
84. Somlyo, A. P., Somlyo, A. V., Friedmann, N. 1971. *Ann. NY Acad. Sci.* 185:108–14
85. Goldman, R., Rottenberg, H. 1973. *FEBS Lett.* 33:233–38
86. Henning, R. 1975. *Biochim. Biophys. Acta* 401:307–16
87. Weigel, P. 1981. *Biochem. Biophys. Res. Commun.* 101:1419–25
88. Regoeczi, E., Hatton, M. W. C., Charlwood, P. A. 1975. *Nature* 254:699–701
89. Lunney, J. K. 1976. *Studies on the regulation of glycoprotein homeostasis.* PhD thesis. Johns Hopkins Univ., Baltimore, Md.
90. Regoeczi, E., Hatton, M. W. C. 1976. *Can. J. Physiol. Pharmacol.* 54:27–34
91. Lunney, J. K., Ashwell, G. 1976. *Proc. Natl. Acad. Sci. USA* 73:341–43
92. Kawasaki, T., Ashwell, G. 1977. *J. Biol. Chem.* 252:6536–43
93. Kuhlenschmidt, T. B., Lee, Y. C. 1980. *Fed. Proc.* 39:1968
94. Drickamer, K. 1981. *J. Biol. Chem.* 256:5827–39
95. Brown, J. L., Roberts, W. K. 1976. *J. Biol. Chem.* 251:1009–14
96. Teichberg, V. I., Silman, I., Beitsch, D. D., Resheff, G. 1975. *Proc. Natl. Acad. Sci. USA* 72:1383–87
97. Barondes, S. H. 1981. *Ann. Rev. Biochem.* 50:207–31
98. Winkelhake, J. L., Nicolson, G. L. 1976. *J. Biol. Chem.* 251:1074–80
99. Stockert, R. J., Morell, A. G., Scheinberg, I. H. 1976. *Biochem. Biophys. Res. Commun.* 68:988–93
100. Hickman, J., Shapiro, L. J., Neufeld, E. F. 1974. *Biochem. Biophys. Res. Commun.* 57:55–61
101. Stahl, P., Schlesinger, P., Rodman, J. S., Doebber, T. 1976. *Nature* 264:86–88
102. Achord, D. T., Brot, F. E., Sly, W. S. 1977. *Biochem. Biophys. Res. Commun.* 77:409–15
103. Brown, T. L., Henderson, L. A., Thorpe, S. R., Baynes, J. W. 1978. *Arch. Biochem. Biophys.* 188:418–28
104. Wilson, G., Eidelberg, M., Michalak, V. 1979. *J. Gen. Physiol.* 74:495–509
105. Schlesinger, P. H., Doebber, T. W., Mandell, B. F., White, R., deSchryver, C., Rodman, J. S., Miller, M. J., Stahl, P. 1978. *Biochem. J.* 176:103–9
106. Steer, C. J., Clarenburg, R. 1979. *J. Biol. Chem.* 254:4457–61
107. Hubbard, A. L., Wilson, G., Ashwell, G., Stukenbrok, H. 1979. *J. Cell. Biol.* 83:47–64

108. Kolb, H., Kolb-Bachofen, V., Schlepper-Schäfer, J. 1979. *Biol. Cell.* 36:301–8

109. Kolb, H., Vogt, D., Herbertz, L., Corfield, A., Schauer, R., Schlepper-Schäfer, J. 1980. *Hoppe-Seylers Z. Physiol. Chem.* 361:1747–50

110. Stahl, P. D., Rodman, J. S., Miller, M. J., Schlesinger, P. H. 1978. *Proc. Natl. Acad. Sci. USA* 75:1399–1403

111. Kawasaki, T., Etoh, R., Yamashina, I. 1978. *Biochem. Biophys. Res. Commun.* 81:1018–24

112. Mizuno, Y., Kozutsumi, Y., Kawasaki, T., Yamashina, I. 1981. *J. Biol. Chem.* 256:4247–52

113. Townsend, R., Stahl, P. 1981. *Biochem. J.* 194:209–14

114. Maynard, Y., Baenziger, J. U. 1981. *Fed. Proc.* 40:1010

115. Prieels, J.-P., Pizzo, S. V., Glasgow, L. R., Paulson, J. C., Hill, R. L. 1978. *Proc. Natl. Acad. Sci. USA* 75:2215–19

116. Hill, R. L., Pizzo, S. V., Imber, M., Lehrman, M., Prieels, J.-P., Glasgow, L. R., Guthrow, C. E., Paulson, J. C. 1980. In *Birth Defects: Orig. Artic. Ser.* 16:85–91

117. Furbish, F. S., Krett, N. L., Barranger, J. A., Brady, R. O. 1980. *Biochem. Biophys. Res. Commun.* 95:1768–74

118. Lehrman, M. A., Hill, R. L. 1981. *Proc. Katzir-Katchalsky Conf., 9th, Israel.* Weizmann, Inst. Sci.

119. Neufeld, E. F., Lim, T. W., Shapiro, L. J. 1975. *Ann. Rev. Biochem.* 44:357–76

120. Neufeld, E. F., Sando, G. N., Garvin, A. G., Rome, L. H. 1977. *J. Supramol. Struct.* 6:95–101

121. Kaplan, A., Achord, D. T., Sly, W. S. 1977. *Proc. Natl. Acad. Sci. USA* 74:2026–30

122. Neufeld, E. F. 1981. In *Lysosomes and Lysosomal Storage Diseases.* ed. J. W. Callahan, J. A. Lowden, pp. 115–29. New York: Raven

123. Sly, W. S., Natowicz, M., Gonzalez-Noriega, A., Grubb, J. H., Fischer, H. D. 1981. See Ref. 122, pp. 131–46

124. Ullrich, K., Mersmann, G., Fleischer, M., von Figura, K. 1978. *Hoppe-Seylers Z. Physiol. Chem.* 359:1591–98

125. Fischer, H. D., Gonzalez-Noriega, A., Sly, W. S., Morré, D. J. 1980. *J. Biol. Chem.* 255:9608–15

126. Sahagian, G. G., Distler, J., Jourdian, G. W. 1981. *Proc. Natl. Acad. Sci. USA* 78:4289–93

127. Willingham, M. C., Pastan, I. H., Sahagian, G. G., Jourdian, G. W., Neufeld, E. F. 1981. *Proc. Natl. Acad. Sci. USA* 78:6967–71

128. Goldwasser, E., Kong, C. K. H., Eliason, J. 1974. *J. Biol. Chem.* 249:4202–6

129. Bocci, V., Pacini, A., Pessina, G. P., Bargigli, V., Russi, M. 1977. *Experimentia* 33:164–66

130. Bose, S., Hickman, J. 1977. *J. Biol. Chem.* 252:8336–37

131. Conway, T. P., Morgan, W. T., Liem, H. H., Müller-Eberhard, U. 1975. *J. Biol. Chem.* 250:3067–73

132. Tatumi, K., Suzuki, Y., Sinohara, H. 1979. *Biochim. Biophys. Acta* 583:504–11

133. Ikekubo, K., Pervos, R., Schneider, A. B. 1980. *Metabolism* 29:673–81

134. Marshall, J. S., Green, A. M., Pensky, J., Williams, S., Zinn, A., Carlson, D. M. 1974. *J. Clin. Invest.* 54:555–62

135. Van Baelen, H., Mannaerts, G. 1974. *Arch. Biochem. Biophys.* 163:53–56

136. Hossner, K. L., Billiar, R. B. 1979. *Biochim. Biophys. Acta* 585:543–53

137. Hatton, M. W. C., Regoeczi, E., Wong, K.-L. 1974. *Can. J. Biochem.* 52:845–53

138. Miller, R. R., Kuhlenschmidt, M. S., Coffee, C. J., Kuo, I., Glew, R. H. 1976. *J. Biol. Chem.* 251:4751–57

139. Gan, J. C. 1979. *Arch. Biochem. Biophys.* 194:149–56

140. Melchers, F. 1973. In *Membrane-Mediated Information,* ed. P. W. Kent, 2:39–56. New York: Elsevier

141. Finbloom, D. S., Magilavy, D. B., Harford, J. B., Rifai, A., Plotz, P. H. 1981. *J. Clin. Invest.* 68:214–24

142. Thomas, P., Hems, D. A. 1975. *Biochem. Biophys. Res. Commun.* 67:1205–9

143. Hildenbrandt, G. R., Aronson, N. N. Jr. 1979. *Biochim. Biophys. Acta* 587:373–80

144. Nelsestuen, G. L., Suttie, J. W. 1971. *Biochem. Biophys. Res. Commun.* 45:198–203

Ann. Rev. Biochem. 1982. 51:555–85
Copyright © 1982 by Annual Reviews Inc. All rights reserved

STEROL BIOSYNTHESIS[1]

George J. Schroepfer, Jr.

Departments of Biochemistry and Chemistry, Rice University, Houston,
Texas 77001

The following abbreviations are used: HMG, 3-hydroxy-3-methylglutaric acid; TLC, thin layer chromatography; GLC, gas-liquid chromatography; HPLC, high pressure liquid chromatography; AY-9944, trans-1,4-bis(2-chlorobenzylaminomethyl)-cyclohexane dihydrochloride.

CONTENTS

INTRODUCTION

In Volume 50 of this series, I reviewed the enzymatic reactions leading to squalene and other isoprenoids in animal cells, and discussed some aspects of the regulation of these biosynthetic reactions (1). Undaunted by the effort of preparing and repeatedly revising that survey, I attempt here to cover the remaining reactions in the biosynthesis of cholesterol. Limited by time, space, and my own capacity for masochism, I restrict this overview mainly to reactions occurring in vertebrates (which in reality largely means reactions studied in rat liver or preparations derived therefrom). I do not therefore cover a number of interesting areas, including sterol biosynthesis

[1]The configuration of the hydrogen at carbon atom 5 in the various sterols mentioned in this paper is α. The designation of the configuration as 5α is omitted throughout the paper to conserve space.

0066-4154/82/0701-0555$02.00

in plants, yeast, and fungi, and other lower forms such as marine organisms. Expansion of knowledge in this last area has been particularly rapid in the last few years, and a complex array of sterols with unusual features (both in the side chain and in the nucleus) has been identified and studied. Djerassi (2) and Goad (3) have presented stimulating reviews on aspects of this subject.

The metabolism of cholesterol and the function(s) of sterols are also excluded from this survey, and the main emphasis is placed on the nature of the intermediates in the biosynthesis of cholesterol in vertebrates. In a sense, therefore, this review represents an update of one written a decade ago (4). During that period, our understanding of the intermediates involved in the overall conversion of lanosterol to cholesterol has advanced significantly, but despite the picture presented in most textbooks and reviews, it is still rudimentary, and the mechanisms of the reactions involved and their regulation are even more poorly understood.

A number of factors contribute to the slow progress in this field. Methods for the separation of the potential intermediates are frequently laborious and often less than completely satisfactory. Significant advances in techniques available for the isolation and characterization of sterols have been made over the past decade, but these have not been universally applied and are frequently used uncritically. The low steady-state concentrations of the potential sterol intermediates in most animal cells compound the difficulties, especially since these compounds frequently must be resolved from the relatively enormous quantities of cholesterol present in most animal cells (and crude enzyme preparations) prior to detailed analysis. The very low solubility of almost all of the concerned compounds in water, and the probable existence of multiple pathways of metabolism, offer additional problems in metabolic studies in this area. Moreover, very few potential intermediates are available commercially. Moderate talent in organic synthesis and a considerable capacity for hard work are therefore required to generate authentic standards for the characterization of potential intermediates and for the preparation of isotopically labeled substrates. Unfortunately, many of the sterols exhibit low chemical stability, which adds the additional burden of repeated chemical synthesis.

A further and major impediment to progress in this field is a consequence of the cellular localization of the enzyme systems involved. All of the enzymes are embedded in membranes and therefore are intractable to many standard procedures for protein purification. With only a few exceptions, truly commendable attempts at purification of key enzymes have brought meager rewards, and knowledge of the individual enzymes in this chain of reactions is extremely rudimentary. For the most part, washed preparations of microsomes represent the most highly purified preparation for most of the individual enzymes (assuming a given reaction is catalyzed by only a single enzyme).

The need for a rigorous definition of the metabolic relationships of sterol intermediates in the biosynthesis of cholesterol has been given additional impetus recently by the suggestion that some oxygenated sterol precursors of cholesterol and oxygenated metabolites of cholesterol itself may serve as important regulators of the synthesis of cholesterol and other essential isoprenoid derivatives and, as a consequence, of cellular replication (1, 5–8). The careful analysis of the enzymatic reactions involved in the transformations of the various sterol intermediates is a clear prerequisite for understanding cellular regulatory processes involving these compounds.

ENZYMATIC CONVERSION OF SQUALENE TO LANOSTEROL

The formation of lanosterol from squalene proceeds via an oxygen-dependent epoxidation to give 2,3-epoxysqualene, which then undergoes enzymatic cyclization to yield lanosterol. Early aspects of the chemistry and biochemistry of this process have been reviewed (9). Bloch and his associates have made substantial progress in the enzymology and mechanism of squalene epoxidase of rat liver (10–13). The reaction requires microsomes, supernatant fraction, a reduced pyridine nucleotide and molecular oxygen (10), and is not inhibited significantly by carbon monoxide or by azide, cyanide, hydroxylamine, EDTA, or o-phenanthroline at a concentration of 1 mM. The supernatant fraction can be replaced by FAD, a phospholipid, and a partially purified protein from the soluble fraction (11). The phospholipid and the protein from the soluble fraction can be replaced by the detergent, Triton X-100 (12). The same detergent also solubilized the microsomal epoxidase activity. Upon chromatography on DEAE-cellulose columns, the solubilized extract gave two components that were inactive by themselves, but when combined gave epoxidase activity (12). In the reconstituted system, FAD, Triton X-100, molecular oxygen, and NADPH were also required. One of the solubilized protein species was believed to be NADPH-cytochrome c reductase, and the other a flavoprotein with an easily dissociable prosthetic group. More recently, Ono et al (14) reported partial purification of squalene epoxidase from rat liver microsomes. Demonstration of significant epoxidase activity required the presence of the partially purified epoxidase, NADPH-cytochrome P-450 reductase, NADPH, and FAD. The requirement for NADPH-cytochrome P-450 reductase, coupled with the commonly (but not invariably) observed requirement for a cytochrome P-450 system in enzymatic epoxidation of olefins, raises the possibility that the epoxidase is a cytochrome P-450 system.

However, the failure of carbon monoxide to inhibit squalene epoxidase activity in microsomal preparations (10), the inability to demonstrate significant amounts of cytochrome P-450 in partially purified preparations of the epoxidase (12, 14), and the apparent lack of impairment of epoxidase activity in yeast mutants deficient in heme biosynthesis (15, 16) do not support such a contention. The precise nature of the epoxidase and its mode of catalysis will no doubt be clarified shortly. The addition of squalene in phosphatidyl serine or phosphatidyl choline liposomes to microsomal preparations has been advocated for studies of the enzymatic conversion of squalene to lanosterol, and of the role of soluble proteins in regulation of the concerned reactions (17). Ferguson & Bloch (13) reported an extensive purification (to ~95% purity) and characterization of a soluble protein from rat liver which increases the activity of microsomal epoxidase; it was designated as supernatant protein factor (SPF). SPF activates not only squalene epoxidase, but also 2,3-oxidosqualene cyclase. However, significant binding of squalene or squalene 2,3-epoxide to the purified SPF was not observed (18, 19). A partially purified preparation of SPF has also been obtained from hog liver (20). The precise role of SPF and its possible identity with another soluble protein, SCP_1, reported (21, 22) to activate the conversion of squalene to lanosterol, are uncertain (13, 19, 23). Possible roles for SPF in stimulating the epoxidase and the cyclase have been discussed (18–20, 24, 25); the idea that SPF acts as a specific carrier for the polyisoprenoid substrates appears uncertain (19). SPF facilitates the transfer of squalene from one microsomal population to another (i.e. from trypsin-digested microsomes to normal microsomes, and the reverse process) (24, 25).

The enzymatic formation of both 2,3-epoxysqualene and 2,3–22,23-diepoxysqualene has been reported in liver (26), yeast (27), and CHO cells in culture (28). Squalene epoxidase activity, but not cyclase activity, was low in mammalian tissues with low capacity for sterol synthesis (rat brain, lung, kidney, muscle, and human placenta) (29). van Tamelen and Heys (30) have studied the substrate specificity of epoxidation in a 9,000 X g supernatant fraction of rat liver, using a number of analogues of squalene and related compounds. The initial reports by van Tamelen and Clayton and by Corey and Bloch and their associates of the enzymatic cyclization of 2,3-epoxysqualene to give lanosterol have been reviewed previously (9). Subsequent studies of the enzymology, mechanism, and specificity of this reaction have been reported (31–43). The enzymatic cyclization of 2,3–22,23-diepoxysqualene in yeast reportedly gives 24,25-epoxylanosterol (27). Incubation of 2,3(S)-22(S),23-diepoxysqualene with a 10,000 X g supernatant fraction of a rat liver homogenate preparation gives efficient formation of 24S,25-epoxycholesterol (44). Under anaerobic conditions, the 24,25-

epoxylanosterol was formed in high yield from the diepoxysqualene substrate. The enzymatic cyclization of 2,3-epoxysqualene is inhibited by 3β-(β-dimethylaminoethoxy)-androst-5-en-17-one (27, 45, 46), by tris(2-diethylaminoethyl)-phosphate hydrochloride (47), and by 4,4,10β-trimethyl-trans-decal-3β-ol (28). A heme-deficient yeast mutant has been isolated that is also deficient in 2,3-epoxysqualene cyclase (16). The enzymatic cyclization of 2,3-epoxysqualene gives, in the appropriate organisms (or preparations derived therefrom), cycloartenol (48, 49), β-amyrin (50, 51), and fusidic acid (52). Rohmer et al (53) found that a cell-free preparation from *Acetobacter pasteurianum* catalyzed the cyclization of (RS)-2,3-epoxysqualene to give 3α-hydroxyhop-22(29)-ene, 3β-hydroxyhop-22(29)-ene, hopane-3α,22-diol, and hopane-3β-22-diol. These findings and others indicated nonstereospecificity with respect to the epoxide function, and that both enantiomers of the squalene epoxide served as substrates for the cyclase in this organism. These findings are in contrast to those reported for the epoxy-squalene cyclase of eukaryotes in which only the (3S)-enantiomer of the squalene epoxide serves as a substrate. This same laboratory (54) also reported that a cell-free system from *Methylococcus capsulatus* catalyzed the conversion of (RS)-2,3-epoxysqualene to lanosterol and its 3α-hydroxy epimer, 3-epilanosterol. Horan et al (55) reported the formation of β-amyrin from a bicyclic derivative of 2,3-epoxysqualene. A large number of studies on the biosynthesis of tetrahymanol, in which cyclization of squalene proceeds directly, rather than via its 2,3-epoxy derivative, have been reviewed (56).

In a notable application of ^{13}C-NMR spectroscopy to the biosynthesis of cholesterol (57) the use of mevalonic acid samples labeled specifically with ^{13}C provided independent evidence for a 1,2-shift of the methyl group to carbon atom 13 of lanosterol in the cyclization of 2,3-epoxysqualene. The chirality of the methyl group of C-6 of mevalonate is retained in the methyl group at C-13 of cholesterol (58), which indicates that the intramolecular migration of this methyl group in the cyclization of 2,3-epoxysqualene to lanosterol occurs without configurational change of the methyl group. A clever application of ^{1}H-NMR provided important evidence that the terminal methyl groups of squalene are derived from C-2 of mevalonic acid, that the enzymatic conversion of squalene to 2,3-epoxysqualene and lanosterol occurs without interchange of the isopropylidene methyls, and that the 4α-methyl group of lanosterol is derived from C-2 of mevalonic acid (59).

The possibility has been considered (60) that the individual microsomal enzymes responsible for the overall conversion of farnesyl pyrophosphate to squalene-2,3-epoxide (and thence to lanosterol) may be so closely associated (spatially or functionally) that asymmetric handling of the squalene may allow its epoxidation at one specific end. This matter has been extensively explored (61) and no evidence was found to support the existence of such a process.

While lanosterol appears to be the primary, if not exclusive, product of the enzymatic cyclization of 2,3-epoxysqualene in liver, its Δ^7-isomer also appears to be a primary product of the cyclization process in skin. Gaylor (62) reported substantial formation of the Δ^7 isomer of lanosterol from labeled squalene in slices of rat skin. While Δ^7 lanosterol was not detected in normal rat skin, its presence in substantial amounts in skins of triparanol-treated rats has been reported (63). The report (64) that microsomes of rat liver failed to catalyze the isomerization of the Δ^8-nuclear double bond of lanosterol to the Δ^7 position suggested that the formation of Δ^7 lanosterol in skin may represent an alternative mode of cyclization of 2,3-epoxysqualene. The observed retention of the 9α hydrogen upon formation of Δ^7 lanosterol from [2-^{14}C,4R-^{3}H] mevalonate is compatible with this possibility (65).

STEROL INTERMEDIATES IN THE BIOSYNTHESIS OF CHOLESTEROL

The enzymatic conversion of lanosterol to cholesterol (Figure 1) requires three general processes: reduction of the Δ^{24}-double bond, removal of the three "extra" methyl groups, and "shift" of the nuclear double bond from the Δ^8 position to the Δ^5 position. We have reviewed some aspects of this matter previously (4, 9), and the reader is referred to these reviews and to those of Bloch (66) and Clayton (67), and to a book by Nes & McKean (68).

The enzymes responsible for the overall conversion of lanosterol to cholesterol (and, in fact, from farnesyl pyrophosphate to cholesterol) are generally considered to be membrane associated and localized in the microsomal fraction of cell-free homogenates. A number of these transformations, for example, the conversion of squalene to cholesterol (69) and 7-dehydrocholesterol to cholesterol (70) are enhanced by the addition of the 100,000 X g supernatant fraction of cell-free homogenates, and considerable effort has been expended on the purification and characterization of these stimulatory factors (11, 13, 21, 71–80). Certain aspects of this research were reviewed in 1974 in this series (81). Although a "carrier role", analogous to

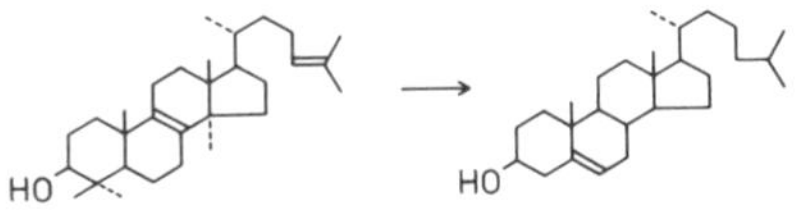

Figure 1 Conversion of lanosterol to cholesterol.

the role of acyl carrier protein in fatty acid synthesis, has been suggested for these stimulatory factors, their physiological significance and mode of action in sterol biogenesis remain unclear. Many recent articles (19, 23, 24, 79, 80, 82, 83) discuss these matters and compare the various stimulatory proteins with recently purified proteins believed to be important in the binding of fatty acids (84, 85) or in the transfer of cholesterol and other lipids from microsomes to mitochondria (82, 86). These matters are not yet clearly defined and need further investigation. Considerable confusion as to the identity (or nonidentity) of the individual protein species isolated in various laboratories exists; their direct comparison (by standard approaches and by modern immunological techniques) should be encouraged to permit further exploration of their physiological significance and mechanisms of action.

In 1972, we presented a list of potential intermediates in the formation of cholesterol from lanosterol that had been detected in animal tissues, or whose enzymatic formation had been reported, and/or whose convertibility to cholesterol had been reported (4). Table 1 is an updated version of this listing. The references listed in 1972 have not been repeated, but are contained in (4). Evidence for the assignment of structure to the isolated sterols was not, in all cases, rigorous. Also listed in the table are a number of sterols that are believed to be derived from agnosterol (4,4,14α-trimethyl-cholesta-7,9(11),24-trien-3β-ol), a sterol long known to occur in sheep wool grease (87) along with lanosterol, and their corresponding 24,25-dihydro derivatives. These and a number of other sterols are listed for completeness; although they are enzymatically convertible to cholesterol, it is unlikely that they play an intermediate role in cholesterol biosynthesis.

Upon inspection of this listing, one cannot but be impressed by the complexity of this area of metabolism. Moreover, it is clear that analysis of sterol mixtures from animal tissues represents a less than simple matter. What follows is a brief consideration of the various chromatographic procedures and spectral methods available for isolation and characterization of sterols.

Chromatography of Sterols

Sterols differing only by the presence or absence of the Δ^{24}-double bond in the side chain can frequently be resolved on silicic acid columns (63, 145), by GLC (146), or, more easily and completely, by TLC (147–149) or column chromatography (127, 147, 150) on silver nitrate impregnated supports. Separation into classes of C_{30}, C_{29}, C_{28}, and C_{27} sterols can be partially achieved on silicic acid columns of high resolution (63). Unfortunately, these columns require long development times. To avoid this problem, a number of investigators have used simple TLC systems or short silicic

Table 1 Potential sterol intermediates in the biosynthesis of cholesterol whose presence in animal tissues has been detected or whose enzymatic formation has been reported and/or whose convertibility to cholesterol has been reported

Compound	Formation or detection in tissues (references)	Convertibility to cholesterol (references)
4,4,14α-trimethyl-cholesta-8,24-dien-3β-ol	4, 88–99	4, 100
4,4,14α-trimethyl-cholesta-7,24-dien-3β-ol	4, 65	4
4,4,14α-trimethyl-cholesta-8,24-dien-3-one		4
4,4,14α-trimethyl-cholest-8-en-3β-ol	4, 88, 89, 91–95, 97–99	4, 100–105
4,4,14α-trimethyl-cholest-8-ene-3β,15β-diol[a]		106
4,4,14α-trimethyl-cholest-8-ene-3β,15α-diol[a]		106
4,4,14α-trimethyl-cholesta-7,9(11)-dien-3β-ol[b]		107, 108
4,4-dimethyl-14α-hydroxymethyl-cholest-8-en-3β-ol		8, 105, 106
4,4-dimethyl-14α-hydroxymethyl-cholest-7-en-3β-ol		4
4,4-dimethyl-14α-formyl-cholest-8-en-3β-ol		8, 105, 106
4,4-dimethyl-14α-formyl-cholest-7-en-3β-ol	104, 109	4
4,4-dimethyl-cholesta-8,24-dien-3β-ol	4, 91	4
4,4-dimethyl-cholesta-7,24-dien-3β-ol	4	
4,4-dimethyl-cholesta-8,14,24-trien-3β-ol	110, 111	
4,4-dimethyl-cholesta-8,14-dien-3β-ol	4, 89, 100–104	4, 100, 101, 105
4,4-dimethyl-cholesta-7,14-dien-3β-ol	103, 104, 112, 113	100
4,4-dimethyl-cholesta-7,9(11), 14-trien-3β-ol[b]		114
4,4-dimethyl-cholesta-7,9(11)-dien-3β-ol[b]	108, 114	108, 114
4,4-dimethyl-cholest-8-en-3β-ol	4, 89, 91, 115	4, 115
4,4-dimethyl-cholest-7-en-3β-ol	4	
4,4-dimethyl-cholest-8(14)-en-3β-ol	4, 109	4, 101, 102, 105, 109, 112
4,4-dimethyl-cholest-8-en-3-one	115	
3β-hydroxy-4β-methyl-cholest-7-en-4α-carboxylic acid	4	
3β-hydroxy-4β-methyl-cholest-7(or 8)-en-4α-carboxylic acid	4	
4α,14α-dimethyl-cholest-7-en-3β-ol	89, 116	
4α,14α-dimethyl-cholest-8-en-3β-ol	89	
4α-methyl-cholesta-8,24-dien-3β-ol	4, 110, 111	
4α-methyl-cholesta-7,24-dien-3β-ol	4	
4α-methyl-cholest-8-en-3β-ol	4, 89	4
4α-methyl-cholest-7-en-3β-ol	4, 89	4

Table 1 *(Continued)*

Compound	Formation or detection in tissues (references)	Convertibility to cholesterol (references)
4β-methyl-cholesta-8,24-dien-3β-ol[c, d]	4	
4β-methyl-cholest-8-en-3β-ol[c, d]	115	115
4α-methyl-cholest-7-en-3-one	4	
4-hydroxymethylene-cholest-7-en-3-one		4
3β-hydroxy-cholest-7-en-4α-carboxylic acid	4	
14α-methyl-cholest-7-en-3β-ol		4, 117, 118
14α-methyl-cholest-8-en-3β-ol	119	
14α-hydroxymethyl-cholest-8-en-3β-ol	119	
14α-hydroxymethyl-cholest-7-en-3β-ol		117, 118, 120
14α-methyl-cholest-7-ene-3β,15β-diol[a]		4, 121, 122
cholesta-8,24-dien-3β-ol	4, 94, 99, 110, 123	4
cholesta-7,24-dien-3β-ol	4, 90, 111, 124–126	4, 127
cholesta-8,14,24-trien-3β-ol	111	
cholesta-8,14-dien-3β-ol	4, 102, 108, 117, 118, 128–131	4
cholesta-7,9(11)-dien-3β-ol[b]	108	
cholesta-7,14-dien-3β-ol	117, 118	4, 132
cholesta-5,7,24-trien-3β-ol	4, 111, 127	4
cholesta-5,24-dien-3β-ol	4, 90, 110, 123, 124, 127	4
cholesta-5,8-dien-3β-ol	133	
cholest-8-en-3β-ol	4, 98, 99, 117, 118, 130, 131	4
cholest-8-ene-3β,6α-diol[a]		4
cholest-8(14)-en-3β-ol	4, 101, 102, 117, 118, 120, 130, 134, 135	4
cholest-8(14)-en-3β-ol-15-one[a]		132
cholest-8(14)-ene-3β,15β-diol		4, 130, 136
cholest-8(14)-ene-3β,15α-diol		4, 130, 136
cholest-8(14)-ene-3β,7α-diol		137
cholest-7-en-3β-ol	4, 90, 98, 101, 102, 117, 118, 120, 124, 127, 130–132, 134, 135, 138, 139	4, 140

Table 1 *(Continued)*

Compound	Formation or detection in tissues (references)	Convertibility to cholesterol (references)
cholest-7-en-3-one	4	
cholest-7-ene-3β,6β-diol[a]		4
cholest-7-ene-3β,6α-diol[a]		4
cholest-7-ene-3β,5α-diol[d]		4, 141
14β-cholest-7-ene-3β,15α-diol[a]		131
14β-cholest-7-en-3β-ol[a]		134
cholestan-3β,8α-diol-7-one[a]		137
cholest-8-ene-3β,7ξ-diols[a]		137
cholestane-3β,7β,8α-triol[a]		137
7α,8α-epoxy-cholestan-3β-ol[a]		137
cholesta-5,7-dien-3β-ol	4, 110, 111, 117, 118, 127, 131, 133, 142–144	4

[a] Very low probability of involvement as an intermediate in the biosynthesis of cholesterol.
[b] Proposed as an intermediate in the enzymatic conversion of agnosterol to cholesterol.
[c] The configuration of the methyl group at C-4 in these sterols appears to be the opposite of that stated (see text).
[d] Low probability of involvement as an intermediate in the biosynthesis of cholesterol.

acid columns, approaches that this reviewer regards as inadequate. The separation of sterols differing only in the number and location of nuclear double bonds is also difficult. While a number of these isomers can be separated by GLC (146), others have essentially the same chromatographic behavior (4, 147). TLC on supports impregnated with silver nitrate is very useful for the separation of some isomers, but others have very similar chromatographic behavior (Δ^0 and $\Delta^{8(14)}$, Δ^8 and Δ^7, $\Delta^{8,14}$ and $\Delta^{7,14}$). More useful separations can be achieved on carefully prepared columns of adsorbents impregnated with silver nitrate (4, 127, 151–153). Two separate column systems have been utilized, one to separate monounsaturated from diunsaturated sterols (and the various diunsaturated sterols from each other), and a second to separate the various monounsaturated sterols from each other (4, 118, 130, 152). Unfortunately, the development times required for these columns are quite long. Recently, an HPLC method which permits the relatively rapid separation of various C_{27} sterol precursors of cholesterol that differ only in the number and location of nuclear double bonds (154, 155) and other HPLC systems to separate various sterols (156–160) have been presented. Unfortunately, the capacity of HPLC systems is low, an undesirable feature for applications to sterol mixtures derived from animal tissues, where cholesterol predominates. A relatively rapid chromatographic procedure for the separation of the various C_{27} sterols, with higher capacity and with comparable resolution to that of the HPLC method, utilizes columns of alumina impregnated with silver nitrate (150, 161) at medium pressure (60–100 psi). Applications of this method have been presented (95, 96, 99, 131, 162). HPLC has also been used to separate various fatty acid esters of cholesterol (163). The chromatographic properties and mass spectral fragmentation of a large number of dioxygenated sterols (C_{27}, C_{28}, and C_{29}) have recently been reported (164).

Infrared Spectroscopy

Infrared spectroscopy continues to play an important role in the routine characterization of synthetic and naturally occurring sterols. Apart from the detection of easily recognizable functional groups, infrared spectroscopy provides exceptionally fine detail, which allows comparison with spectra of sterols of known structure. With instruments of high resolution, most sterols are easily distinguishable from each other in the fingerprint region of the spectra. However, only extremely slight differences are observed in spectra of sterols differing only in the presence or absence of a Δ^{24}-double bond in the side chain (127).

Ultraviolet Spectroscopy

Ultraviolet spectroscopy is of particular value in the study of a,β-unsaturated ketones, and can be used to differentiate between heteroannular and homoannular dienes. Empirical rules have been presented for the calculation of expected values of absorption maxima in these compounds (165 and references therein). A sophisticated treatment of a,β-unsaturated ketones has recently been published (166).

Optical Rotation, Optical Rotatory Dispersion, and Circular Dichroism

The optical rotation can, in favorable cases, provide important structural information for a given sterol. Optical rotatory dispersion provides extremely valuable information with respect to the location and stereochemical orientation of functional groups in steroidal molecules (167, 168 and references therein). Refined treatments of the circular dichroism spectra of steroidal dienes and a,β-unsaturated ketones have recently been presented (169–171).

Mass Spectrometry

Mass spectrometry presents a powerful tool for providing, in many cases, not only the molecular weight of a pure compound, but also important structural information. Due caution should be applied to assignment of structure based upon complex fragmentations of these molecules observed upon electron impact. A number of investigators have utilized such an approach rather uncritically, using only low resolution mass spectrometry. The establishment of the origin of a given ion is not a simple matter, and in most cases requires determination of the exact mass by high resolution studies, metastable ion analyses, and extensive studies of analogous compounds and of isotopically labeled compounds. The reader is referred to examples of such determinations in the work of Djerassi and co-workers (172–177) and others (178, 179). A computer-assisted approach for the interpretation of the mass spectra of sterols has recently been presented and evaluated for a number of marine sterols (179). A particularly novel application of mass spectrometry to sterol metabolism is found in a recent publication by Bjorkhem & Lewenhaupt (180), who were able to demonstrate, after inhalation of $^{18}O_2$ by rats with biliary fistula, its substantial incorporation into cholic acid and chenodeoxycholic acid, but little into biliary cholesterol. Analysis of appropriate fragment ions in the mass spectra of the bile acids permitted estimation of the ^{18}O content at C-3 and C-7 of chenodeoxycholic acid. The combined results indicated preferential use of newly synthesized cholesterol for the synthesis of these bile acids.

Nuclear Magnetic Resonance Spectroscopy

Conventional ^1H-NMR spectroscopy provides a powerful tool in structural elucidation of sterols (181). A great deal of information has been gathered on the effects of various substituents on the chemical shifts of the C-18 and C-19 protons (181–183), which has been particularly useful in structural analyses. Although a considerable amount of valuable structural information can be obtained using 60–100 MHz ^1H-NMR spectrometers, the analyses of spectra recorded with 220 MHz (or higher) spectrometers are considerably simplified (122). However, in the absence of the availability of such instrumentation, extremely valuable information can be obtained at 60–100 MHz through the use of lanthanide shift reagents. Several recent applications of this latter approach in the sterol field (122, 130, 184, 185) include two studies (122, 130) in which the results obtained were unequivocally confirmed by X-ray crystallographic analyses (121, 122, 136). A notable application of ^1H-NMR spectroscopy to the biosynthesis of cholesterol is found in the demonstration of the origin of the 4α-methyl group of lanosterol from C-2 of mevalonic acid (59). Very recently, highly impressive communications have appeared that describe the use of proton two-dimensional J spectroscopy (186, 187) and the combination of nuclear Overhauser difference and spin decoupling-difference nuclear magnetic resonance techniques (187, 188), which allowed most and all, respectively, of the proton resonances of 1-dehydrotestosterone and 11β-hydroxyprogesterone to be resolved and assigned.

^{13}C-NMR also provides a powerful tool for use in structural elucidation of complex molecules and in biosynthetic studies. Key papers deal with assignments of the ^{13}C resonances to individual carbon atoms in cholesterol (57, 189–193), and a review on the ^{13}C-NMR spectra of steroids has appeared (193). The natural abundance ^{13}C-NMR spectra of a number of C_{27} sterol precursors of cholesterol, differing only in the number and location of nuclear double bonds have been studied, and peak assignments for the individual carbon atoms of cholest-8-en-3β-ol, cholest-8(14)-en-3β-ol, cholest-7-en-3β-ol, cholesta-8,14-dien-3β-ol, and cholesta-7,14-dien-3β-ol were made (194). These data have proven useful in the elucidation of the structure of a number of synthetic sterols (195–198) and in the assignment of the ^{13}C-NMR resonance in the spectra of zymosterol (162, 199), ergosta-8,24(28)-dien-3β-ol, and ergosta-8,24(28)-diene-3β,6α-diol (200), of a sterol (isolated from tissues of pregnant rats treated with AY-9944, and their newborn progeny) to which the structure cholesta-5,8-dien-3β-ol was assigned (133), a series of 24-ethyl-Δ^7-sterols from Spanish olive oil (201), and two new $\Delta^{8(14)}$ sterols (4-methylene-24(R)-ethyl-cholest-8(14)-en-3β-ol and 4-methylene-24(S)-ethyl-cholest-8(14)-en-3β-ol) from Red Sea sponges (202). Detailed analyses of the ^{13}C olefinic carbon shieldings of a large number of sterols and related cyclic compounds were made in an attempt to derive some empirical shift rules for these compounds (203). Included were monounsaturated sterols and sterols containing homoannular and heteroannular conjugated diene systems. Also considered were allylic and homoallylic shielding effects on quarternary, methine, methylene, and methyl carbon atom in these systems. Some aspects of this matter have been explored further by Eggert & Djerassi (204) who studied additional monounsaturated androstane and cholestane compounds. Several exceptions to apparent general findings (202) were noted and caution must therefore be used in applying empirical considerations to an unknown unsaturated sterol without supplemental analyses. Batta et al (205) have recently shown that the 25R and 25S diastereoisomers of 5β-cholestane-3α,7α,26-triol can be differentiated by ^{13}C-NMR.

Enzymatic Conversion of Lanosterol to Cholesterol

We have previously suggested criteria for the establishment of a potential intermediary role of a given sterol in the enzymatic conversion of lanosterol to cholesterol (4, 9). These criteria can be briefly summarized as follows: *1.* isolation from tissues in pure form and unequivocal establishment of structure, *2.* demonstration of its enzymatic formation from a known precursor (acetate, mevalonate, squalene, etc), *3.* demonstration of its convertibility to cholesterol, *4.* demonstration of its enzymatic formation from its postulated immediate precursor, and *5.* demonstration of its enzymatic conversion to the next postulated intermediate. The quantitative importance of a given intermediate is much more difficult to establish and is complicated by the existence of multiple pathways and other factors, as discussed in more depth previously (4).

Results obtained over the past two decades indicate that the reduction of the Δ^{24}-double bond of lanosterol can occur at any stage in the overall conversion to cholesterol (4, 9). However, the concentration and/or activity of the "Δ^{24} reductase" appear to vary in different animal tissues and at different stages of development.

Theoretically, nonspecificity with respect to the order of removal of the three extra methyl groups of lanosterol also is possible. However, in contrast to an earlier report (206), subsequent studies (207–211) have indicated that the equatorially oriented 4α-methyl group is specifically removed prior to removal of the 4β-methyl group. It has been commonly assumed that the 14α-methyl group of lanosterol is removed prior to those at carbon atom 4. This belief was based upon the fact that none of the sterols isolated from animal tissues (4, 9) were assigned structures corresponding to $4,14\alpha$-dimethyl sterols or 14α-methyl sterols. However, the reported isolations of $4\alpha,14\alpha$-dimethyl sterols and 14α-methyl sterols from a variety of plant sources and yeasts (157, 212–218), of $4\alpha,14\alpha$-dimethyl-cholest-7-en-3β-ol and $4\alpha,14\alpha$-dimethyl-cholest-8-en-3β-ol from meconium of newborn infants (89), and of $4\alpha,14\alpha$-dimethyl-cholest-7-en-3β-ol from cerebral white matter and optic nerve of rabbit (116) indicate that, at least in some organisms or tissues, the removal of the three extra methyl groups of lanosterol can be initiated by removal of the 4α-methyl group rather than the 14α-methyl group. To my knowledge, a systematic search for the presence of $4\alpha,14\alpha$-dimethyl sterols or 14α-methyl sterols using modern methodology has not been conducted. In a preliminary study, we have been unable to detect the presence of 14α-methyl sterols in rat skin (R. A. Pascal, Jr. and G. J. Schroepfer, Jr., unpublished), an organ in which steady-state concentrations of sterol precursors of cholesterol are unusually high (4).

The removal of the two methyl groups at carbon atom 4 appears to occur as outlined in Figure 2 (207–209, 211, 219–225). The possibility that 4β-methyl -3β- hydroxysterols are discrete intermediates was suggested by the reported isolation of 4β-methyl-cholest-8,24-dien-3β-ol (226, 227) and 4β-methyl-cholest-8-en-3β-ol (115) from animal tissues, and the conversion of the latter sterol to cholesterol by a liver homogenate preparation (115). The chemical synthesis of 4β-methyl-cholest-8-en-3β-ol and a number of related 4α-methyl and 4β-methyl sterols has been completed (228, 229). Comparisons of the reported physical and spectral properties of these synthetic sterols with those of the corresponding isolated sterols, to which the 4β-methyl assignment had been made, indicate that the latter sterols correspond more closely to synthetic 4β-methyl sterols than to 4β-methyl sterols

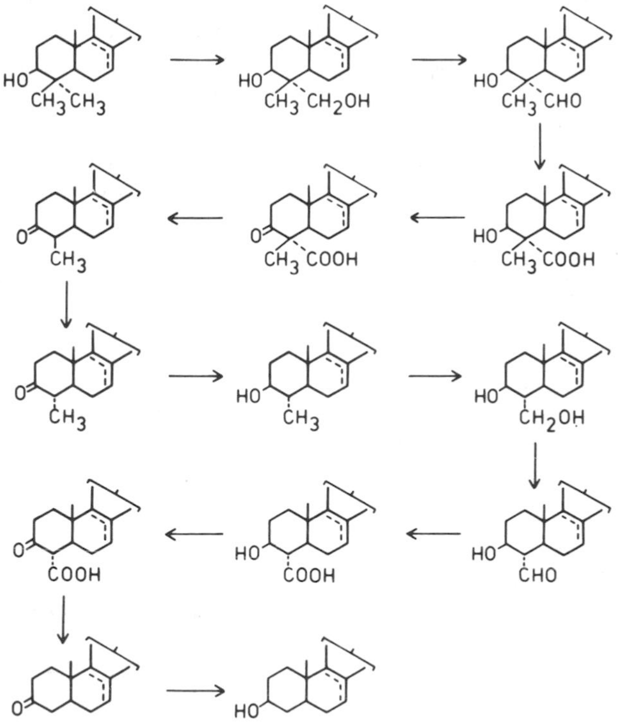

Figure 2 Scheme for the enzymatic removal of the methyl groups at carbon atom 4 of lanosterol and related sterol precursors of cholesterol

(228). The reported isolation of 4β-methyl-stigmasta-7,24(28)-dien-3β-ol (230) has similarly been revised (231) to indicate a 4α orientation of the methyl group at C-4.

Existing evidence does not support a significant role for 4β-methyl-3β-hydroxysterols in the biosynthesis of cholesterol. However, the report by Sharpless et al (208) of the enzymatic formation of both 4α-methyl-cholestan-3β-ol and 4β-methyl-cholestan-3β-ol from 4β-methyl-cholestan-3-one in rat liver homogenate preparations is noteworthy as is the reported (208) more efficient conversion of 4 α-methyl-cholestan-3β-ol to cholestan-3β-ol relative to the 4β-methyl isomer in the same liver preparations. However, the same workers (207) reported that 4β-methyl-4α-hydroxymethyl-cholestan-3β-ol, but not 4α-methyl-4β-hydroxymethyl-cholestan-3β-ol, was convertible to cholestan-3β-ol. During enzymatic conversion of 4,4-dimethyl-[3β-^{18}OH]-cholest-7-en-3β-ol to 4α-methyl-cholest-7-en-3β-ol in rat liver microsomes most of the labeled oxygen is retained (232). Demethylation of a 4α-methyl sterol in an alga occurs with inversion of the 4β

hydrogen to the equatorial 4α position (233, 234). The detailed mechanisms of the enzymatic removal of each of the methyl groups at C-4 are not completely understood. It is commonly assumed that the initial reaction is an oxygen-dependent hydroxylation that yields the corresponding hydroxymethyl derivative (4, 66, 67, 207, 235 and references therein). However, neither the isolation nor the enzymatic formation of a 4α-hydroxymethyl sterol from animal tissues, have been demonstrated. This situation may derive from the fact that the overall enzymatic removal of the C-4 substituent from 4α-hydroxymethyl sterols appears to proceed at a much faster rate than that of the corresponding 4α-methyl sterol (207, 236). Oxygen and a reduced pyridine nucleotide have been reported to be required for the catalysis, by rat liver microsomes, of the overall conversion of 4α-methyl-cholest-7-en-3β-ol to 3β-hydroxy-cholest-7-ene-4α-carboxylic acid (219). Decarboxylation of the latter product requires its conversion to the corresponding 3-keto derivative by an NAD-dependent microsomal enzyme, which has been partially purified (18-fold) from rat liver microsomes (237). Gaylor et al (223) have reported that, under defined conditions, for each equivalent of 4α-hydroxymethyl-cholest-7-en-3β-ol converted to the corresponding 4α-carboxylic acid by rat liver microsomes, two equivalents each of reduced pyridine nucleotide and of oxygen were consumed (223). Cytochrome P-450 appears not to be involved in the reactions concerned in the demethylation at C-4 (91, 238, 239), but a cyanide-sensitive process has been reported to be involved in this overall process in both liver and yeast systems (224, 236, 238–240). NaCN (5 mM) inhibits to a similar extent not only the overall enzymatic (Triton-treated microsomes) removal of the methyl group at C-4 of 4α-methyl-cholest-7-en-3β-ol, but also the enzymatic removal of the hydroxymethyl group of of 4 α-hydroxymethyl-cholest-7-en-3β-ol (236). While cyanide (0.83 mM) reportedly had no effect on the metabolism of lanosterol to zymosterol in a cell-free yeast preparation (241), Aoyama et al (240) recently reported a 50% inhibition of the conversion, by yeast microsomes, of (1, 7, 15, 22, 26, 30-^{14}C) lanosterol to labeled CO_2 by 0.08 mM cyanide. $^{14}CO_2$ formation from the labeled lanosterol in a yeast microsomal system also was inhibited by antibodies to yeast cytochrome b_5 (240). A similar finding with liver microsomes has also been reported (242). Gaylor and co-workers (243) believe the overall enzymatic conversion of a 4α-methyl sterol to the corresponding 4α-carboxylic acid is catalyzed by a single oxidase, 4-methyl sterol oxidase, but in view of the state of the enzymology in this area, this appears premature. Moreover, the question of whether or not two functional path-ways, oxygen-dependent and oxygen-independent, exist for the conversion of the 4α-hydroxymethyl and 4α-formyl intermediates to the corresponding 4α-carboxylic acid should be reconsidered. The enzymatic removal of the hydroxymethyl group at C-4 by Triton-treated microsomes clearly proceeds very much faster in the presence of oxygen than in its absence under the conditions studied (236). These results are important, but extrapolation to intact liver appears premature at best. This situation is analogous to the metabolism of ethanol in liver by two separate systems, one involving an oxidase and the other a dehydrogenase [see (244) for a critical review of a proposed oxygen-dependent, ethanol-metabolizing system in liver]. In the case of the 4α-hydroxymethyl sterols, assessment of the quantitative contributions of the two types of metabolism appears nontrivial at this time.

The evolution of our knowledge concerning possible mechanisms involved in the enzymatic removal of the methyl group at C-14 of lanosterol has been especially interesting. Over 20 years ago, Olson et al (245) reported that 3 mol of labeled carbon dioxide were formed per mole of cholesterol formed from "methyl labeled" lanosterol in a $700 \times g$ supernatant fraction of rat liver homogenate. This finding implied that carbon atom 32 of lanosterol ultimately formed carbon dioxide. The presence of the neighboring Δ^8 double bond of lanosterol was assumed to facilitate removal of C-32 (66). The importance of the nuclear double bond in the demethylation at C-14 was indicated by the subsequent observation that 4,4,14α-trimethyl-cholestan-3β-ol was not converted to cholestan-3β-ol in rat liver preparations, whereas 4,4-dimethyl-cholestan-3β-ol, 4α-methyl-cholestan-3β-ol, 4β-methyl-4α-hydroxymethyl-cholestan-3β-ol, and 4α-hydroxymethyl-cholestan-3β-ol were (207). It has generally been assumed that the initial reaction in the removal of carbon 32 of 14α-methyl sterol precursors of cholesterol involved an oxygen dependent hydroxylation to yield the corresponding hydroxymethyl derivative. Recent studies in liver and yeast systems have indicated an involvement of a cytochrome P-450 system (88, 91, 105, 246, 247). While the enzymatic removal of carbon atom 32 and the enzymatic conversion of a number of Δ^8 and Δ^7 hydroxymethyl sterols to cholesterol have been demonstrated in liver and yeast systems (8, 103–105, 109–113, 117, 120, 135, 248), the isolation and/or clear demonstration of their enzymatic formation from the corresponding 14α-methyl sterols have not been reported to date.

In the middle sixties, it was commonly assumed that the enzymatic removal of each of the three "extra" methyl groups, including the 14α-methyl group, of lanosterol proceeded via hydroxmethyl, aldehyde, and carboxylic acid intermediates. In the case of a Δ^8 (or Δ^7) 14α-methyl sterol, such a sequence would yield a β,γ-unsaturated acid. After reviewing available information regarding nonenzymatic decarboxylations of β,γ-unsaturated acids and stimulated by discussions of this matter by others (249, 250), we (251) and Fried et al (248) noted that an analogous decarboxylation of a Δ^8 or Δ^7 steroidal acid would be expected to yield a $\Delta^{8(14)}$

sterol. The isolation of cholest-8(14)-en-3β-ol from rat skin (153, 252), the demonstration of the occurrence of a number of $\Delta^{8(14)}$ sterols in nature (202, 253–259), and the demonstration of the enzymatic convertibility of a number of $\Delta^{8(14)}$ sterols to cholesterol (101, 153, 248, 251, 260) are compatible with this concept. A modification of this idea, at least as far as the involvement of a C-32 carboxylic acid, was clearly indicated by a report of the inability to detect the formation of the putative carboxylic acid intermediate (109) and, more importantly, by the demonstration (113) that, upon incubation of lanost-7-ene-3β,32-diol with rat liver microsomes, carbon atom 32 was recovered as formic acid and not as carbon dioxide. This finding has been confirmed and extended with the same substrate (103, 104) and with [32-³H]-lanost-8-ene-3β,32-diol (104) in liver systems, with [32-¹⁴C]-24,25-dihydrolanosterol in yeast and liver systems (105, 261), and with [32-³H]-14 α-hydroxymethyl-5α-cholest-7-en-3β-ol in liver systems (117). Incubation of the latter compound with washed microsomes plus NADPH yielded HTO and tritium-labeled formate in approximately equal amounts, a finding in agreement with those of Alexander et al (113) and of Akhtar et al (103, 104) on the corresponding 4,4-dimethyl substituted sterol. [³H]-Formate was metabolized to HTO in the presence of the 10,000 X g supernatant fraction of a rat liver homogenate, but was recovered unchanged upon incubation with washed liver microsomes and NADPH (117). The latter finding accounts for the results observed in a previous study, which indicated the formation of labeled CO_2 from C_{32} derived from "methyl-labeled" lanosterol upon incubation with a 700 X g supernatant fraction of a rat liver homogenate (245). Attempts to demonstrate either the formation of formaldehyde upon enzymatic removal of C-32 from cholesterol precursors or the inhibition of this process by formaldehyde have been unsuccessful (104). While the removal of C-32 of a Δ^8- or Δ^7-32-formyl substituted sterol could provide a ready explanation for the natural occurrence of $\Delta^{8(14)}$ sterols and their possible role in the biosynthesis of cholesterol, this general mode of formation would not, by itself, account for the results of studies (employing mevalonic acid labeled stereospecifically at C-2) that indicated a stereospecific loss of the 15α hydrogen of lanosterol upon enzymatic formation of cholest-7-en-3β-ol, 7-dehydrocholesterol, and cholesterol (262, 263). This finding prompted investigations of the enzymatic conversion of $\Delta^{8,14}$ and $\Delta^{7,14}$ sterols to cholesterol, reactions that have been clearly demonstrated [see Table 1 of this review and Table 2 of (4)]. Although the conversion of cholesta-8,14-dien-3β-ol to cholesta-7,14-dien-3β-ol by washed rat liver microsomes could not be detected (4, 153), the reverse reaction occurred readily under the same conditions (129). These studies were facilitated by the development of a chromatographic method (4, 153) for separation of these isomers. Efficient conversions of the $\Delta^{8,14}$ and $\Delta^{7,14}$

sterols to cholest-7-en-3β-ol and cholesterol in the 10,000 X g supernatant fraction of rat liver homogenate preparations under aerobic conditions were observed (4, 264, 265). Under anaerobic conditions, cholesta-8,14-dien-3β-ol gave cholest-8-en-3β-ol, cholest-7-en-3β-ol, and a very small amount of cholest-8(14)-en-3β-ol (4, 265); cholesta-7,14-dien-3β-ol gave cholest-7-en-3β-ol and a small amount of cholest-8(14)-en-3β-ol (4, 265). The enzymatic reduction of the Δ^{14} double bond of $\Delta^{7,14}$ and $\Delta^{8,14}$ sterols by rat liver microsomes is dependent on the presence of NADPH (266) and is inhibited by AY-9944 (4, 102, 138).

Concern over the appropriateness of studies of the metabolism of 14α-methyl-Δ^7 sterols and 14α-hydroxymethyl-Δ^7-sterols for investigation of the mechanisms involved in the removal of C-32 of cholesterol precursors appears unwarranted in view of the demonstration that the Δ^7 isomer of lanosterol occurs in tissues [Table 1 of this review and Table 2 of (4)]. Furthermore, the isomerization of 14α-methyl-cholest-7-en-3β-ol and 14α-hydroxymethyl-cholest-7-en-3β-ol to give the corresponding Δ^8-sterols has been demonstrated in rat liver microsomes (119). The latter findings are in contrast to those of Gaylor et al (64) who reported that lanost-8-en-3β-ol, lanosta-8,24-dien-3β-ol, lanost-7-en-3β-ol, 14α-methyl-cholest-8-en-3β-ol, and 14α-methyl-cholest-7-en-3β-ol were unchanged upon incubation with rat liver microsomes under anaerobic conditions.

The possibility that 15α-hydroxylation of the 14α-methyl-sterol precursors of cholesterol accounts for the stereospecific loss of the 15α-hydrogen during conversion of lanosterol to cholesterol and provides the source of oxygen incorporated into formic acid during the enzymatic removal of C-32 of 14α-methyl sterols appears unlikely in view of the inability to demonstrate conversion of 14α-methyl-cholest-7-ene-3β,15α-diol to cholesterol in rat liver preparations (121, 122, 267) and the report that the enzymatic conversion, in rat liver homogenate preparations, of the 15α-hydroxy derivative of 24,25-dihydrolanosterol was considerably less efficient than that of 24,25-dihydrolanosterol itself (106). Moreover, 14α-hydroxymethyl-cholest-7-ene-3β,15α-diol is not convertible to cholesterol in rat liver preparations (98).

The enzymatic formation, by rat liver microsomes, of material with the properties of 4,4-dimethyl-14α-formyl-cholest-7-en-3β-ol from the corresponding 14α-hydroxymethyl substrate has been reported (104, 109). Dudowitz and Fried (109) reported the process to occur under anaerobic conditions in the presence of NAD. On the other hand, Akhtar et al (104) reported little or no conversion under these conditions (or with NADP or no added cofactors present), but did report substantial formation of the aldehyde under aerobic conditions in the presence of added NADPH. They also reported little or no formation of formic acid upon incubation of lanost-7-ene-3β,32-diol with liver microsomes under aerobic conditions in the absence of added cofactors, or if NAD was added. In the presence of NADPH, substantial formation of labeled formate was reported. An attempt to isolate the 14α-formyl derivative after incubation of labeled lanost-8-en-3β-ol was unsuccessful (105). Dudowitz & Fried (109) reported that oxygen was required for the enzymatic removal of C-32 of 4,4-dimethyl-14 α-formyl-cholest-7-en-3β-ol in rat liver microsomes. Akhtar et al (104) reported little metabolism of the same compound under aerobic conditions in the presence of added NAD or NADP, or in the absence of added cofactors. In the presence of added NADPH, they reported substantial metabolism of the 14α-aldehyde.

Current evidence indicates that a cytochrome P-450 system is involved, in both yeast and rat liver, in removal of C-32 of 14α-methyl sterols. A reconstituted system consisting of yeast microsomal cytochrome P-450 ($\sim$ 70% pure), cytochrome P-450 reductase (highly purified), NADPH, and oxygen catalyzed the conversion of lanosterol to a product that was assigned the structure 4,4-dimethyl-cholesta-8,14,24-trien-3β-ol on the basis of analysis of its mass spectrum (247). Carbon monoxide inhibited formation of this product about 80%. A single species of cytochrome P-450 appeared to catalyze removal of the 14α-methyl group of lanosterol in yeast by three successive oxygenation steps: formation of a 14α-hydroxymethyl derivative, formation of a 14α-formyl derivative, and finally, oxidative removal of C-32 to give formic acid (247). Subsequently, Gibbons et al (105) reported that oxidative removal of C-32 of lanost-8-en-3β-ol by rat liver microsomes was inhibited by carbon monoxide, but that this inhibitor had no effect on the removal of C-32 from lanost-8-ene-3β,32-diol or lanost-8-en-3β-ol-32-al. These and previous findings strongly suggest involvement of a cytochrome P-450 system in the initial hydroxylation of the 14α-methyl group of lanosterol and related precursors of cholesterol, but that such a system is not involved in the further metabolism of the 14α-hydroxymethyl or the 14α-formyl sterol. Further studies are needed to resolve these important matters. Akhtar et al (103, 104) reported the exclusive formation of 4,4-dimethyl-cholesta-8,14-dien-3β-ol   and   4,4-dimethyl-cholesta-7,14-dien-3β-ol as the initial C-29 products of the enzymatic (liver microsomes) removal of C-32 of lanost-8-ene-3β,32-diol and of lanost-7-ene-3β,32-diol, respectively. These findings are unexpected in view of the demonstration that liver microsomes catalyze the isomerization of the Δ^7 double bond of 14α-hydroxymethyl (and 14α-methyl)sterols to the Δ^8 position (119) and the isomerization of the $\Delta^{7,14}$-diene system of 5α-cholesta-7,14-dien-3β-ol to give the corresponding $\Delta^{8,14}$-diene (129). Moreover, incubation of lanost-7-ene-3β,32-diol with rat liver microsomes, oxygen, and NAD (with or without NADPH) gave a different product, 4,4-dimethyl-cholest-8(14)-en-3

β-ol (109, 248). The latter compound was characterized by GLC and TLC on silver nitrate impregnated plates. Unfortunately, the full details of this work were never presented. More recently, we have observed efficient conversion of 14α-methyl-cholest-7-en-3β-ol to cholesterol by the 10,000 $\times$ g supernatant fraction of rat liver homogenate preparations (117, 118). Other labeled compounds isolated and characterized were cholesta-8,14-dien-3β-ol, cholesta-7,14-dien-3β-ol, cholest-8(14)-en-3β-ol, cholest-8-en-3β-ol, cholest-7-en-3β-ol, and cholesta-5,7-dien-3β-ol. Small amounts of polar sterols also were formed, as well as fatty acid esters of a number of sterols, including the incubated substrate, cholesta-8,14-dien-3β-ol, cholesta-7,14-dien-3β-ol, cholest-8-en-3β-ol, cholest-7-en-3β-ol, and cholesterol. These results illustrate the value of the use of 14α-methyl sterols, rather than 4,4,14α-trimethyl sterols, as substrates, and the application of chromatographic methods, which permit resolution of the various steroidal products. The formation of the $\Delta^{8(14)}$, $\Delta^{8,14}$, and $\Delta^{7,14}$ sterols (along with the more well-established Δ^8, Δ^7, and $\Delta^{5,7}$ sterols) is worthy of note.

The metabolism of 14α-hydroxymethyl-cholest-7-en-3β-ol has also been studied. This sterol was efficiently converted to cholesterol in the 10,000 $\times$ g supernatant fraction of rat liver homogenate preparations (117). Cholest-8(14)-en-3β-ol was virtually the sole product isolated after incubation of this 14α-hydroxymethyl sterol with washed rat liver microsomes supplemented with NAD (135). Incubations with washed microsomes, NAD, and an NADPH-generating system under aerobic conditions gave cholesterol and cholest-7-en-3β-ol as the major C_{27} monohydroxysterol products (120). More importantly, under both aerobic and anaerobic conditions, incubations of the 14α-hydroxymethyl sterol in the absence of added cofactors yielded total C_{27} monohydroxy sterol products in amounts comparable to those observed in the cofactor supplemented, aerobic incubations. These findings are in conflict with those of Akhtar et al (104) and Dudowitz & Fried (109), who found that molecular oxygen was required for the conversion of 14α-hydroxymethyl sterols to the 32-norsteroidal products. Moreover, cholest-8(14)-en-3β-ol was essentially the only C_{27} monohydroxysterol formed under these conditions (120). The variable requirement for pyridine nucleotide cofactors in washed microsome systems might arise from the presence of tightly bound pyridine nucleotide in the microsomes. Accordingly, a series of aerobic incubations of 14α-hydroxymethyl-cholest-7-en-3β-ol were carried out using rat liver microsomes that had been treated with Triton WR-1339. This treatment has previously been reported to provide microsomal preparations that show an absolute requirement for added NAD in the oxidative demethylation of 4,4-dimethyl-cholest-7-en-3 β-ol (268). With such microsomes, essentially no C_{27} monohydroxysterols were formed from the 14α-hydroxymethyl sterol; addition of NAD, or

NAD plus an NADPH generating system, restored the ability of the Triton-treated microsomes to catalyze the removal of C-32. With NAD alone, cholest-8(14)-en-3β-ol was essentially the only C_{27} monohydroxysterol product; with NAD and an NADPH generating system, cholesterol, cholest-7-en-3β-ol, and an unidentified diene sterol (distinct from cholesta-8,14-dien-3β-ol, cholesta-7,14-dien-3β-ol, or cholesta-5,7-dien-3β-ol) were formed in addition to the $\Delta^{8(14)}$ sterol. These and previous (117) findings clearly indicate that enzymes exist in rat liver microsomes capable of removing the C-32 of 14α-hydroxymethyl-cholest-7-en-3β-ol to yield cholest-8(14)-en-3β-ol and formic acid, and that removal of C-32 from 14α-hydroxymethyl-cholest-7-en-3β-ol to yield cholest-8(14)-en-3β-ol can proceed under anaerobic conditions. The findings provide a ready explanation for the existence of $\Delta^{8(14)}$ sterols in nature (153, 202, 252–259), an occurrence that is not accommodated by other proposals for the enzymatic removal of C-32 of 14α-methyl sterols (103–105, 245, 269).

At this point, it is appropriate to consider the metabolism of cholest-8(14)-en-3β-ol and other $\Delta^{8(14)}$ sterols. The enzymatic conversion of cholest-8(14)-en-3β-ol (153, 251) and other $\Delta^{8(14)}$ sterols (101, 104, 109, 112, 248, 260) to cholesterol in rat liver preaparations is well established. The metabolism of these sterols is dependent on the presence of molecular oxygen (4, 109, 153, 248, 251), and a requirement for NADPH has been suggested (109). Moreover, an inhibition of the metabolism of cholest-8(14)-en-3β-ol by carbon monoxide has been indicated (102). This sterol is converted to the corresponding $\Delta^{8,14}$ sterol in rat liver homogenate preparations (128), but neither it nor other $\Delta^{8(14)}$ sterols are metabolized under anaerobic conditions. These results suggest that molecular oxygen, NADPH (102, 109), and possibly a cytochrome P-450 system are involved in the conversion of a $\Delta^{8(14)}$-sterol to the corresponding $\Delta^{8,14}$-sterol. The possibility that a $\Delta^{8(14)}$-15-hydroxysterol is an intermediate in the latter conversion has been considered; although formation of $\Delta^{8(14)}$-15-hydroxysterols has not been reported, 5α-cholest-8(14)-ene-3β,15α-diol and 5α-cholest-8(14)-ene-3β,15β-diol are efficiently converted to cholesterol in liver preparations (4, 130, 136, 270, 271). Moreover, washed rat liver microsomes catalyze the conversion of each of the $\Delta^{8(14)}$-3β,15-dihydroxysterols to cholesta-8,14-dien-3β-ol (4, 130). The overall conversion of each of the two $\Delta^{8(14)}$-3β,15-dihydroxysterols to cholesterol proceeds, successively, via cholesta-8,14-dien-3β-ol, cholest-8-en-3β-ol, cholest-7-en-3β-ol, and cholesta-5,7-dien-3β-ol (4, 130, 270). Both cholest-8(14)-ene-3β, 15α-diol and its 15β-hydroxy epimer serve as substrates for this process. Interconversion of the two epimeric diols could occur either via the $\Delta^{8(14)}$-15-ketosterol (which serves as a substrate for cholesterol formation in this system (132)), or by direct epimerization of the 15-hydroxyl of a $\Delta^{8(14)}$-3

β,15-dihydroxysterol (4, 130). However because of the location of the label (at C-15) in the two $\Delta^{8(14)}$-3β,15-dihydroxysterols, the former possibility has been considered unlikely (130, 132). An enzyme system in rat liver microsomes that catalyzes the direct epimerization of an allylic hydroxyl function of a steroid has been reported (272).

Several attmepts in which a "trap" has been utilized have failed to demonstrate formation and accumulation of a $\Delta^{8(14)}$ sterol from various sterol precursors of cholesterol. Such results have been interpreted as indicating that $\Delta^{8(14)}$ sterols are not intermediates in the biosynthesis of cholesterol (101, 104, 112, 260). For reasons presented previously (4), such negative results are difficult to interpret. Gibbons (101) reports that 4,4-dimethyl-cholesta-8,14-dien-3β-ol is a better substrate for the formation of C_{27} sterol in liver microsomal preparations than the corresponding 4,4-dimethyl-$\Delta^{8(14)}$ sterol, but this does not exclude a possible intermediary role of the $\Delta^{8,14}$ sterol in the conversion of the $\Delta^{8(14)}$-sterol to cholesterol. Reports that lanost-8-en-3β-ol is converted more efficiently than 4,4-dimethyl-$\Delta^{8(14)}$-sterols to C_{27} sterols (101, 104, 112) should be noted in evaluating the quantitative importance of $\Delta^{8(14)}$-sterols in the biosynthesis of cholesterol. However, such comparisons should be viewed with caution since the binding affinities of the sterols to different proposed carrier proteins and their accessibility to the concerned enzymes, which may exist as multienzyme complexes, are unknown.

Aerobic metabolism of [7-^{3}H]14α-formyl-cholest-7-en-3β-ol in washed rat liver microsomes in the presence of an NADPH-generating system, reportedly yields substantial formation of labeled cholest-8(14)-ene-3β,7α-diol, and an intermediate role for this compound in the conversion to cholesta-7,14-dien-3β-ol and cholesterol was proposed (269). Independent proposals for oxygen-dependent removal of C-32 of 14α-hydroxymethyl and 14α-formyl substituted sterol precursors of cholesterol have been advanced (103–105, 247).

The combined results presented above indicate considerable confusion, not only with respect to the mechanisms involved in the enzymatic removal of C-32 of 14α-methyl and 14α-hydroxymethyl sterol precursors of cholesterol, but also with regard to the identity and metabolism of the C-32 norsteroidal products. Further studies of these matters are clearly indicated. However, the existence of multiple pathways for these processes appears highly probable. For example, metabolism of the 14α-hydroxymethyl sterols by two separate types of processes, i.e. oxidase-type and dehydrogenase-type, is certainly possible.

Specific inhibitors of reactions involved in the removal of C-32 of 14α-methyl sterol precursors of cholesterol may prove valuable in extending our knowledge of these enzymatic reactions. 14α-Ethyl-cholest-7-ene-

$3\beta,15\alpha$-diol is a potent inhibitor of the metabolism of lanosterol and 24,25-dihydrolanosterol in CHO-K1 cells (94, 99) and in cell-free homogenates of rat liver (95). Similarly, 14α-hydroxymethyl-cholest-7-ene-$3\beta,15\alpha$-diol is a potent inhibitor of the metabolism of lanosterol and 24,25-dihydrolanosterol in cell-free preparations of rat liver (98). While the mechanism of action of these inhibitors, which also suppress the level of HMG-CoA reductase activity (7, 99, 273, 274), has not been determined, both are structural analogues of lanosterol and may affect the proposed cytochrome P-450-dependent hydroxylation of the 14α-methyl group of lanosterol. Triparanol, 1-[p-(β-diethylaminoethoxy)-phenyl]-1-(p-tolyl)-2-(p-chloro-phenyl)ethanol; a fungicide, S-n-butyl-S-p-tert-butylbenzyl-N-3-pyridyldithiocarbonimidate; and triaminol, α-(2,4-dichlorophenyl)-α-phenyl-5-pyrimidine methanol, inhibit 14α-demethylation in lower forms (214, 216, 275). Triaminol also inhibits 14α-demethylation in rat liver enzyme preparations (92). The modes of action of these compounds have not been established.

A possible relationship exists between the metabolism of lanosterol and HMG-CoA reductase activity (276–278). 14α-Ethyl-cholest-7-ene-$3\beta,15\alpha$-diol not only inhibits lanosterol metabolism but is also potent in lowering the levels of HMG-CoA reductase activity in cells (99, 273, 274). However, a mutant of the CHO-K1 cell, in which HMG-CoA reductase activity is resistant to this inhibitor, is not resistant to its effects on the metabolism of lanosterol and 24,25-dihydrolanosterol (99), which suggests that the inhibition of lanosterol metabolism by the 15-oxygenated sterol does not cause the suppression of HMG-CoA reductase activity in these cells. Moreover, high concentrations of lanosterol and 24,25-dihydrolanosterol have little effect on sterol synthesis from labeled acetate in primary cultures of fetal mouse liver cells (279). Metabolites of 14α-methyl sterol precursors of cholesterol (i.e. 14α-hydroxymethyl and 14α-formyl derivatives) are, however, potent inhibitors of sterol synthesis and lower the levels of HMG-CoA reductase in cultured mammalian cells (1, 6–8).

The overall conversion of cholesta-8,24-dien-3β-ol (and of its 24,25-dihydroanalogue) to cholesterol involves an isomerization that yields the corresponding Δ^7 sterol, which, in an oxygen-dependent reaction, is converted to the corresponding $\Delta^{5,7}$ sterol. Reduction of the Δ^7 bond of the latter sterol yields the Δ^5 sterol (9). In most tissues, enzymatic reduction of the Δ^{24} double bond can apparently occur at any stage in the overall conversion of a Δ^8 sterol to the Δ^5 sterol (9, 127). The $\Delta^8 \rightarrow \Delta^7$ isomerization in liver involves a loss of the 7β-hydrogen and the introduction of solvent hydrogen at C-9 (151 and references therein). In the case of sterols lacking substitution at C-14, the reverse reaction has been detected, but only with great difficulty (140, 280, 281). However, the enzymatic isomerization of the Δ^7 bond of Δ^7-14α-hydroxymethyl and $\Delta^{7,14}$ sterols to the corresponding Δ^8 and $\Delta^{8,14}$ sterols can be readily demonstrated (119, 129). The activity of the $\Delta^8 \rightarrow \Delta^7$ isomerase is higher in microsomes prepared from cholestyramine-fed rats than in microsomes from control rats (125). The isolation of yeast mutants deficient in the $\Delta^8 \rightarrow \Delta^7$ isomerase is also noteworthy (199, 282, 283). 14α-Hydroxymethyl-cholest-6-ene-3β,15α-diol causes an accumulation of labeled cholest-8-en-3β-ol, derived from [2-^{14}C]mevalonic acid, when incubated with rat liver homogenate preparations (98). This finding suggests that the Δ^6-14α-hydroxymethyl-3β,15α-diol causes a specific inhibition of the $\Delta^8 \rightarrow \Delta^7$ isomerase, a microsomal enzyme activity for which specific inhibitors had not been described previously. Cytochrome b_5 is apparently involved in the enzymatic conversion of cholest-7-en-3β-ol to cholesta-5,7-dien-3β-ol (143). Cholesta-7,24-dien-3β-ol and cholesta-5,24-dien-3β-ol (and its sulfate ester) are major sterols of epididymal tissue and epididymal spermatozoa, respectively (126, 284). The high steady-state concentrations of these sterols in these specialized cells is noteworthy. Ikekawa et al (285) recently reported the isolation and characterization of a new C_{26} sterol (22-trans-27-norcholesta-5,22-dien-3β-ol) from the urine of a female child with congenital adrenal hyperplasia. Prior to this report, the reported occurrence of sterols with shortened side chains was limited to marine organisms.

CONCLUDING REMARKS

I have attempted to review selected areas critical to the understanding of the enzymatic reactions involved in the overall conversion of squalene to cholesterol. A few topics have been covered in depth because of their obvious importance or in an attempt to define or clarify conflicting observations or interpretations of experimental findings. For reasons noted earlier, research in this area of biochemistry is not simple and requires both a recognition of the inherent complexities and a great deal of hard work. Definition of the individual reactions involved in the biosynthesis of cholesterol from squalene is not simply a matter of crossing t's and dotting i's. It is only with such knowledge that significant advances can be expected in understanding regulation of the biosynthesis of cholesterol, its precursors, and their metabolites and the functions of these compounds in normal and diseased states.

ACKNOWLEDGMENTS

The support of the National Institutes of Health (HL-15376 and HL-22532) and the Robert A. Welch Foundation is gratefully acknowledged. Thanks are also due to Professor Carl Djerassi for preprints of several manuscripts.

Literature Cited

1. Schroepfer, G. J. Jr. 1981. *Ann. Rev. Biochem.* 50:585–621
2. Djerassi, C. 1981. *Pure Appl. Chem.* 53:873–90
3. Goad, L. J. 1981. *Pure Appl. Chem.* 51:837–52
4. Schroepfer, G. J. Jr., Lutsky, B. N., Martin, J. A., Huntoon, S., Fourcans, B., Lee, W.-H., Vermilion, J. 1972. *Proc. R. Soc. London Ser. B* 180:125–46
5. Kandutsch, A. A., Chen, H. W., Heiniger, H.-J. 1978. *Science* 201:498–501
6. Schroepfer, G. J. Jr., Pascal, R. A. Jr., Shaw, R., Kandutsch, A. A. 1978. *Biochem. Biophys. Res. Commun.* 83:1024–31
7. Schroepfer, G. J. Jr., Parish, E. J., Pascal, R. A. Jr., Kandutsch, A. A. 1980. *J. Lipid Res.* 21:571–84
8. Gibbons, G. F., Pullinger, C. R., Chen, H. W., Cavenee, W. K., Kandutsch, A. A. 1980. *J. Biol. Chem.* 255:395–400
9. Frantz, I. D. Jr., Schroepfer, G. J. Jr. 1967. *Ann. Rev. Biochem.* 36:691–726
10. Yamamoto, S., Bloch, K. 1970. *J. Biol. Chem.* 245:1670–74
11. Tai, H.-H., Bloch, K. 1972. *J. Biol. Chem.* 247:3767–73
12. Ono, T., Bloch, K. 1975. *J. Biol. Chem.* 250:1571–79
13. Ferguson, J. B., Bloch, K. 1977. *J. Biol. Chem.* 252:5381–85
14. Ono, T., Takahashi, K., Odani, S., Konono, H., Imai, Y. 1980. *Biochem. Biophys. Res. Commun.* 96:522–28
15. Bard, M., Woods, R. A., Haslam, J. M. 1974. *Biochem. Biophys. Res. Commun.* 56:324–30
16. Gollub, E. G., Liu, K.-P., Dayan, J., Adlersberg, M., Sprinson, D. B. 1977. *J. Biol. Chem.* 252:2846–54
17. Morin, R. J., Srikantaiah, M. V. 1980. *J. Lipid Res.* 21:1143–47
18. Nakamura, M., Sato, R. 1979. *Biochem. Biophys. Res. Commun.* 89:900–6
19. Caras, I. W., Friedlander, E. J., Bloch, K. 1980. *J. Biol. Chem.* 255:3575–80
20. Lin, L.-F. H. 1980. *Biochemistry* 19:5135–40
21. Srikantaiah, M. V., Hansbury, E., Loughran, E. D., Scallen, T. J. 1976. *J. Biol. Chem.* 251:5496–505
22. Gavey, K. L., Scallen, T. J. 1978. *J. Biol. Chem.* 253:5476–83
23. Noland, B. J., Arebalo, R. E., Hansbury, E., Scallen, T. J. 1980. *J. Biol. Chem.* 255:4282–89
24. Kojima, Y., Friedlander, E. J., Bloch, K. 1981. *J. Biol. Chem.* 256:7235–39
25. Friedlander, E. J., Caras, I. W., Lin, L. F., Bloch, K. 1980. *J. Biol. Chem.* 255:8042–45
26. Corey, E. J., Ortiz de Montellano, P. R., Lin, K., Dean, P. D. G. 1967. *J. Am. Chem. Soc.* 89:2797–98
27. Field, R. B., Holmlund, C. E. 1977. *Arch. Biochem. Biophys.* 180:465–71
28. Chang, T.-Y., Schiavoni, E. S. Jr., McCrae, K. R., Nelson, J. A., Spencer, T. A. 1979. *J. Biol. Chem.* 254:11258–63
29. Astruc, M., Tabacik, C., Descomps, B., Crastes de Paulet, A. 1977. *Biochim. Biophys. Acta* 487:204–11
30. van Tamelen, E. E., Heys, J. R. 1975. *J. Am. Chem. Soc.* 97:1252–53
31. Dean, P. D. G., Ortiz de Montellano, P. R., Bloch, K., Corey, E. J. 1967. *J. Biol. Chem.* 242:3014–19
32. van Tamelen, E. E., Sharpless, K. B., Willett, J. D., Clayton, R. B., Burlingame, A. L. 1967. *J. Am. Chem. Soc.* 89:3920–92
33. Corey, E. J., Ortiz de Montellano, P. R., Yamamoto, H. 1968. *J. Am. Chem. Soc.* 90:6254–55
34. van Tamelen, E. E., Sharpless, K. B., Hanzlik, R., Clayton, R. B., Burlingame, A. L., Wszolek, P. C. 1967. *J. Am. Chem. Soc.* 89:7150–51
35. Barton, D. H. R., Gosden, A. F., Mellows, G., Widdowson, D. A. 1968. *Chem. Commun.* pp. 1067–68
36. Corey, E. J., Lin, K., Yamamoto, H. 1969. *J. Am. Chem. Soc.* 91:2132–34
37. Crosby, L. O., van Tamelen, E. E., Clayton, R. B. 1969. *Chem. Commun.* pp. 532–33
38. Anderson, R. J., Hanzlik, R. P., Sharpless, K. B., van Tamelen, E. E., Clayton, R. B. 1969. *Chem. Commun.* pp. 53–54
39. Yamamoto, S., Lin, K., Bloch, K. 1969. *Proc. Natl. Acad. Sci. USA* 63:110–17
40. Mercer, E. I., Johnson, M. W. 1969. *Phytochemistry* 8:2329–31
41. Tabacik, C., Astruc, M., Descomps, B., Crastes de Paulet, A. 1975. *Biochim. Biophys. Acta* 398:490–95
42. Herin, M., Sandra, P., Krief, A. 1979. *Tetrahedron Lett.* pp. 3103–6
43. Caras, I. W., Bloch, K. 1979. *J. Biol. Chem.* 254:11816–21
44. Nelson, J. A., Steckbeck, S. R., Spencer, T. A. 1981. *J. Biol. Chem.* 256:1067–68
45. Fung, B., Holmlund, C. E. 1976. *Biochem. Pharmacol.* 25:1249–54
46. Sipe, J. D., Holmlund, C. E. 1972. *Biochim. Biophys. Acta* 280:145–60
47. Reid, W. W. 1968. *Phytochemistry* 7:451–52

48. Rees, H. H., Goad, L. J., Goodwin, T. W. 1968. *Tetrahedron Lett.* pp. 723–25
49. Rees, H. H., Goad, L. J., Goodwin, T. W. 1969. *Biochim. Biophys. Acta* 176: 892–94
50. Corey, E. J., Ortiz de Montellano, P. R. 1967. *J. Am. Chem. Soc.* 89:3362–63
51. Barton, D. H. R., Mellows, G., Widdowson, D. A., Wright, J. J. 1971. *J. Chem. Soc. Chem. Commun.* pp. 1142–48
52. Godtfredsen, W. O., Lorek, H., van Tamelen, E. E., Willett, J. D., Clayton, R. B. 1968. *J. Am. Chem. Soc.* 90:208–9
53. Rohmer, M., Anding, C., Ourisson, G. 1980. *Eur. J. Biochem.* 112:541–47
54. Rohmer, M., Bouvier, P., Ourisson, G. 1980. *Eur. J. Biochem.* 112:557–60
55. Horan, H., McCormick, J. P., Arigoni, D. 1973. *J. Chem. Soc. Chem. Commun.* pp. 73–74
56. Caspi, E. 1980. *Acc. Chem. Res.* 13:97–104
57. Popjak, G., Edmond, J., Anet, F. A. L., Easton, N. R. Jr. 1977. *J. Am. Chem. Soc.* 99:931–35
58. Clifford, K. H., Phillips, G. T. 1975. *J. Chem. Soc. Chem. Commun.* pp. 419–20
59. Stone, K. J., Roeske, W. R., Clayton, R. B., van Tamelen, E. E. 1969. *Chem. Commun.* pp. 530–32
60. Etemadi, A. H., Popjak, G., Cornforth, J. W. 1969. *Biochem. J.* 111:445–51
61. Cornforth, J. W., Ross, F. P. 1977. *Proc. R. Soc. London Ser. B* 199:213–30
62. Gaylor, J. L. 1963. *J. Biol. Chem.* 238:1649–55
63. Clayton, R. B., Nelson, A. N., Frantz, I. D. Jr. 1963. *J. Lipid Res.* 4:166–78
64. Gaylor, J. L., Delwiche, C. V., Swindell, A. C. 1966. *Steroids* 8:353–66
65. Hornby, G. M., Boyd, G. S. 1971. *Biochem. J.* 124:831–32
66. Bloch, K. 1965. *Science* 150:19–28
67. Clayton, R. B. 1965. *Q. Rev.* 19:168–200
68. Nes, W. R., McKean, M. L. 1977. *Biochemistry of Steroids and Other Isoprenoids.* Baltimore: Univ. Park Press. 690 pp.
69. Tchen, T. T., Bloch, K. 1957. *J. Biol. Chem.* 226:921–30
70. Kandutsch, A. A. 1962. *J. Biol. Chem.* 237:358–62
71. Ritter, M. C., Dempsey, M. E. 1970. *Biochem. Biophys. Res. Commun.* 38: 921–29
72. Ritter, M. C., Dempsey, M. E. 1971. *J. Biol. Chem.* 246:1536–47
73. Ritter, M. C., Dempsey, M. E. 1973. *Proc. Natl. Acad. Sci. USA* 70:265–69
74. Scallen, T. J., Schuster, M. W., Dhar, A. K. 1971. *J. Biol. Chem.* 246:224–30
75. Scallen, T. J., Srikantaiah, M. V., Seetharam, B., Hansbury, E., Gavey, K. L. 1974. *Fed. Proc.* 33:1733–46
76. Scallen, T. J., Seetharam, B., Srikantaiah, M. V., Hansbury, E., Lewis, M. K. 1975. *Life Sci.* 16:853–74
77. Gaylor, J. L., Delwiche, C. V. 1976. *J. Biol. Chem.* 251:6638–45
78. Gaylor, J. L., Billheimer, J. T., Longino, M. A., Trzaskos, J. M. 1979. *J. Lipid Res.* 20:1045
79. Billheimer, J. T., Gaylor, J. L. 1980. *J. Biol. Chem.* 255:8128–35
80. Dempsey, M. E., McCoy, K. E., Baker, H. N., Dimitriadou-Valiadou, A., Lorsbach, T., Howard, J. B. 1981. *J. Biol. Chem.* 256:1867–73
81. Dempsey, M. E. 1974. *Ann. Rev. Biochem.* 43:967–90
82. Bloj, B., Zilversmit, D. B. 1977. *J. Biol. Chem.* 252:1613–19
83. Gavey, K. L., Noland, B. J., Scallen, T. J. 1981. *J. Biol. Chem.* 256:2993–99
84. Hackney, J. F., Beale, D. 1976. *Biochem. J.* 155:511–21
85. Ockner, R. K., Maning, J. A. 1974. *J. Clin. Invest.* 54:326–38
86. Bloj, B., Hughes, M. E., Wilson, D. B., Zilversmit, D. B. 1978. *FEBS Lett.* 96:87–89
87. Windaus, A., Tschesche, R. 1930. *Hoppe-Seylers Z. Physiol. Chem.* 190: 51–61
88. Gibbons, G. F., Mitropoulos, K. A. 1972. *Biochem. J.* 127:315–17
89. Gustafsson, J.-A., Eneroth, P. 1972. *Proc. R. Soc. London Ser. B* 180:179–86
90. Bojesen, E., Bojesen, I., Capito, K. 1973. *Biochim. Biophys. Acta* 306: 237–48
91. Gibbons, G. F., Mitropoulos, K. A. 1973. *Biochem. J.* 132:439–48
92. Mitropoulos, K. A., Gibbons, G. F., Connell, C. M., Woods, R. A. 1976. *Biochem. Biophys. Res. Commun.* 71:892–900
93. Tint, G. S., Salen, G. 1977. *Metabolism* 26:721–29
94. Miller, L. R., Izumi, A., Pinkerton, F. D. 1980. *Fed Proc.* 39:1907
95. Raulston, D. L., Pajewski, T. N., Miller, L. R., Phillip, B. W., Shapiro, D. J., Schroepfer, G. J. Jr. 1980. *Biochem. Int.* 1:113–19
96. Shecter, I., Fogelman, A. M., Popjak, G. 1980. *J. Lipid Res.* 21:277–83
97. Ciavatti, M., Michel, G., Renaud, S. 1980. *Biochim. Biophys. Acta* 620:297–307

98. Miller, L. R., Pascal, R. A. Jr., Schroepfer, G. J. Jr. 1981. *J. Biol. Chem.* 256:8085–91
99. Pinkerton, F. D., Izumi, A., Andersen, C. M., Miller, L. R., Kisic, A., Schroepfer, G. J. Jr. 1982. *J. Biol. Chem.* 257:1929–36
100. Watkinson, I. A., Wilton, D. C., Munday, K. A., Akhtar, M. 1971. *Biochem. J.* 121:131–37
101. Gibbons, G. F. 1974. *Biochem. J.* 144: 59–68
102. Gibbons, G. F., Mitropoulos, K. A. 1975. *Biochim. Biophys. Acta* 380: 270–81
103. Akhtar, M., Freeman, C. W., Wilton, D. C., Boar, R. B., Copsey, D. B. 1977. *Bioorg. Chem.* 6:473–81
104. Akhtar, M., Alexander, K., Boar, R. B., McGhie, J. F., Barton, D. H. R. 1978. *Biochem. J.* 169:449–63
105. Gibbons, G. F., Pullinger, C. R., Mitropoulos, K. A. 1979. *Biochem. J.* 183:309–15
106. Gibbons, G. F., Mitropoulos, K. A., Pullinger, C. R. 1976. *Biochem. Biophys. Res. Commun.* 69:781–89
107. Rahimtula, A. D., Wilton, D. C., Akhtar, M. 1969. *Biochem. J.* 112:545–46
108. Tavares, I. A. F., Munday, K. A., Wilton, D. C. 1977. *Biochem. J.* 166:11–15
109. Dudowitz, A. E., Fried, J. 1969. *Fed. Proc.* 28:665
110. Ramsey, R. B., Fredericks, M. 1977. *Biochem. Pharmacol.* 26:1161–67
111. Ramsey, R. B., Fredericks, M. 1977. *Biochem. Pharmacol.* 26:1169–73
112. Alexander, K. T. W., Akhtar, M., Boar, R. B., McGhie, J. F., Barton, D. H. R. 1971. *Chem. Commun.* pp. 1479–81
113. Alexander, K., Akhtar, M., Boar, R. B., McGhie, J. F., Barton, D. H. R. 1972. *J. Chem. Soc. Chem. Commun.* pp. 383–85
114. Tavares, I. A. F., Munday, K. A., Wilton, D. C. 1977. *Biochem. J.* 166:17–20
115. Scallen, T. J., Dhar, A. K., Loughran, E. D. 1971. *J. Biol. Chem.* 246:3168–74
116. Adamczewska-Goncerzewicz, Z., Trzebny, W. 1981. *J. Neurochem.* 36: 1378–82
117. Trowbridge, S., Lu, Y. C., Shaw, R., Chan, J., Spike, T. 1975. *Fed. Proc.* 34:560
118. Chan, J. T., Spike, T. E., Trowbridge, S. T., Schroepfer, G. J. Jr. 1979. *J. Lipid Res.* 20:1007–19
119. Pascal, R. A. Jr., Schroepfer, G. J. Jr. 1980. *Biochem. Biophys. Res. Commun.* 94:932–39
120. Pascal, R. A. Jr., Chang, P., Schroepfer, G. J. Jr. 1980. *J. Am. Chem. Soc.* 102:6599–601
121. Spike, T. E., Wang, H.-J., Paul, I. C., Schroepfer, G. J. Jr. 1974. *J. Chem. Soc. Chem. Commun.* pp. 477–78
122. Spike, T. E., Martin, J. A., Huntoon, S., Wang, A. H.-J., Knapp, F. F. Jr., Schroepfer, G. J. Jr. 1978. *Chem. Phys. Lipids* 21:31–58
123. Emmons, G. T., Rosenblum, E. R., Malloy, J. M., McManus, I. R., Campbell, I. M. 1980. *Biochem. Biophys. Res. Commun.* 96:34–38
124. Bojesen, I., Roepstorff, P. 1973. *Biochim. Biophys. Acta* 316:83–90
125. Yamaga, N., Gaylor, J. L. 1978. *J. Lipid Res.* 19:375–82
126. Legault, Y., VandenHeuvel, W. J., Arison, B. H., Bleau, G., Chapdelaine, A., Roberts, K. D. 1978. *Steroids* 32: 649–58
127. Ener, M. A., Frantz, I. D. Jr. 1973. *J. Biol. Chem.* 248:6697–6700
128. Lutsky, B. N., Schroepfer, G. J. Jr. 1971. *Lipids* 6:957–59
129. Hsiung, H. M., Spike, T. E., Schroepfer, G. J. Jr. 1975. *Lipids* 10:623–26
130. Huntoon, S., Fourcans, B., Lutsky, B. N., Parish, E. J., Emery, H., Knapp, F. F. Jr., Schroepfer, G. J. Jr. 1978. *J. Biol. Chem.* 253:775–82
131. Pascal, R. A. Jr., Schroepfer, G. J. Jr. 1980. *J. Biol. Chem.* 255:3565–70
132. Monger, D. J., Parish, E. J., Schroepfer, G. J. Jr. 1980. *J. Biol. Chem.* 255: 11122–29
133. Fumagalli, R., Bernini, F., Galli, G., Anastasia, M., Fiecchi, A. 1980. *Steroids* 35:665–72
134. Galli-Kienle, M., Anastasia, M., Cighetti, G., Manzocchi, A., Galli, G. 1977. *Eur. J. Biochem.* 73:1–6
135. Chang, P., Schroepfer, G. J. Jr. 1977. *Fed. Proc.* 36:816
136. Phillips, G. N., Quiocho, F. A., Sass, R. L., Werness, P., Emery, H., Knapp, F. F. Jr., Schroepfer, G. J. Jr. 1976. *Bioorg. Chem.* 5:1–10
137. Fiecchi, A., Galli-Kienle, M., Scala, A., Galli, G., Paoletti, R., Grossi-Paoletti, E. 1972. *J. Biol. Chem.* 247:5898–904
138. Lutsky, B. N., Hsiung, H. M., Schroepfer, G. J. Jr. 1975. *Lipids* 10:9–11
139. Gibbons, G. F. 1977. *Biochem. Biophys. Res. Commun.* 75:995–1003
140. Scala, A., Galli-Kienle, M., Anastasia, M., Galli, G. 1974. *Eur. J. Biochem.* 48:263–69
141. Alexander, K., Akhtar, M. 1975. *Biochem. J.* 145:345–52
142. Aberhart, D. J., Caspi, E. 1971. *J. Biol. Chem.* 246:1387–92

143. Reddy, V. V. R., Kupfer, D., Caspi, E. 1977. *J. Biol. Chem.* 252:2797–801
144. Esvelt, R. P., DeLuca, H. F., Wichmann, J. K., Yoshizawa, S., Zurcher, J., Sar, M., Stumpf, W. E. 1980. *Biochemistry* 19:6158–61
145. Frantz, I. D. Jr., Mobberley, M. L., Schroepfer, G. J. Jr. 1960. *Progr. Cardiovasc. Diseases* 2:511–18
146. Clayton, R. B. 1962. *Biochemistry* 1: 357–66
147. Galli, G., Grossi-Paoletti, E. 1967. *Lipids* 1:72–75
148. Vroman, H. E., Cohen, C. F. 1967. *J. Lipid Res.* 8:150–52
149. Kammereck, R., Lee, W.-H., Paliokas, A., Schroepfer, G. J. Jr. 1967. *J. Lipid Res.* 8:282–84
150. Pascal, R. A. Jr., Farris, C. L., Schroepfer, G. J. Jr. 1980. *Anal. Biochem.* 101:15–22
151. Lee, W.-H., Kammereck, R., Lutsky, B. N., McCloskey, J. A., Schroepfer, G. J. Jr. 1969. *J. Biol. Chem.* 244:2033–40
152. Lutsky, B. N., Martin, J. A., Schroepfer, G. J. Jr. 1971. *J. Biol. Chem.* 246:6737–44
153. Lee, W.-H., Lutsky, B. N., Schroepfer, G. J. Jr. 1969. *J. Biol. Chem.* 244: 5440–48
154. Thowsen, J. R. 1977. *Fed. Proc.* 36:780
155. Thowsen, J. R., Schroepfer, G. J. Jr. 1979. *J. Lipid Res.* 20:681–85
156. Rees, H. R., Donnahey, L., Goodwin, T. W. 1976. *J. Chromatogr.* 116:281–91
157. Trocha, P. J., Jasne, S. J., Sprinson, D. B. 1977. *Biochemistry* 16:4721–26
158. Hunter, I. R., Walden, M. K., Heftmann, E. 1978. *J. Chromatogr.* 153: 57–61
159. Hansbury, E., Scallen, T. J. 1978. *J. Lipid Res.* 19:742–46
160. Hansbury, E., Scallen, T. J. 1980. *J. Lipid Res.* 21:921–29
161. Pascal, R. A. Jr., Farris, C., Schroepfer, G. J. Jr. 1979. *Fed Proc.* 38:633
162. Taylor, U. F., Kisic, A., Pascal, R. A. Jr., Izumi, A., Tsuda, M., Schroepfer, G. J. Jr. 1981. *J. Lipid Res.* 22:171–77
163. Carroll, R. M., Rudel, L. L. 1981. *J. Lipid Res.* 22:359–63
164. Aringer, L., Nordstrom, L. 1981. *Biomed. Mass Spectrom.* 8:183–203
165. Fieser, L. F., Fieser, M. 1959. *Steroids,* pp. 15–24. New York: Reinhold, 945 pp.
166. Liljefors, T., Allinger, N. L. 1978. *J. Am. Chem. Soc.* 100:1068–73
167. Eliel, E. L. 1962. *Stereochemistry of Carbon Compounds,* pp. 398–433. New York: McGraw-Hill. 486 pp.
168. Djerassi, C. 1960. *Optical Rotatory Dispersion,* New York: McGraw-Hill. 293 pp.
169. Gawronski, J., Liljefors, T., Norden, B. 1979. *J. Am. Chem. Soc.* 101:5515–22
170. Kuball, H.-G., Acimis, M., Altschul, J. 1979. *J. Am. Chem. Soc.* 101:20–27
171. Burnett, R. D., Kirk, D. N. 1981. *J. Chem. Soc. Perkin Trans. I,* pp. 1460–68
172. Karliner, J., Budzikiewicz, H., Djerassi, C. 1966. *J. Org. Chem.* 31:710–13
173. Muccino, R. R., Djerassi, C. 1973. *J. Am. Chem. Soc.* 95:8726–33
174. Massey, I. J., Djerassi, C. 1979. *J. Org. Chem.* 44:2448–56
175. Brown, F. J., Djerassi, C. 1980. *J. Am. Chem. Soc.* 102:807–16
176. Brown, F. J., Djerassi, C. 1981. *J. Org. Chem.* 46:954–63
177. Knapp, F. F. Jr., Wilson, M. S., Schroepfer, G. J. Jr. 1976. *Chem. Phys. Lipids* 16:31–59
178. Knapp, F. F. Jr., Schroepfer, G. J. Jr. 1976. *Chem. Phys. Lipids* 17:466–500
179. Lavanchy, A., Varkony, T., Smith, D. H., Grey, N. A. B., White, W. C., Carhart, R. E., Buchanan, B. G., Djerassi, C. 1980. *Org. Mass Spectrom.* 15: 355–66
180. Bjorkhem, I., Lewenhaupt, A. 1979. *J. Biol. Chem.* 254:5252–56
181. Bhacca, N. S., Williams, D. H. 1964. *Applications of NMR Spectroscopy in Organic Chemistry, Illustrations from the Steroid Field,* San Francisco: Holden-Day. 198 pp.
182. Zurcher, R. F. 1963. *Helv. Chim. Acta* 46:2054–88
183. Cohen, A. I., Rock, S. Jr. 1964. *Steroids* 3:243–57
184. Iida, T., Kikuchi, M., Tamura, T., Matsumoto, T. 1979. *J. Lipid Res.* 20: 279–84
185. Iida, T., Tamura, T., Matsumoto, T. 1980. *J. Lipid Res.* 21:326–38
186. Hall, L. D., Sanders, J. K. M., Sukumar, S. 1980. *J. Chem. Soc. Chem. Commun.* pp. 366–68
187. Hall, L. D., Sanders, J. K. M. 1981. *J. Org. Chem.* 46:1132–38
188. Hall, L. D., Sanders, J. K. M. 1980. *J. Chem. Soc. Chem. Commun.* pp. 368–70
189. Reich, H. J., Jautelat, M., Messe, M. T., Weigert, F. J., Roberts, J. D. 1969. *J. Am. Chem. Soc.* 91:7445–54
190. Eggert, H., Djerassi, C. 1973. *J. Org. Chem.* 38:3788–92
191. Bhacca, N. S., Djerassi, C. 1976. *J. Org. Chem.* 41:71–78

192. Joseph-Nathan, P., Mejira, G., Abramo-Bruno, D. 1979. *J. Am. Chem. Soc.* 101:1289–91
193. Blunt, J. W., Stothers, J. B. 1977. *Org. Magn. Reson.* 9:439–63
194. Tsuda, M., Schroepfer, G. J. Jr. 1979. *J. Org. Chem.* 44:1290–93
195. Tsuda, M., Parish, E. J., Schroepfer, G. J. Jr. 1979. *J. Org. Chem.* 44:1282–89
196. Parish, E. J., Tsuda, M., Schroepfer, G. J. Jr. 1979. *Chem. Phys. Lipids* 25:111–24
197. Parish, E. J., Tsuda, M., Schroepfer, G. J. Jr. 1979. *Chem. Phys. Lipids.* 24:167–82
198. Anastasia, M., Fiecchi, A., Gariboldi, P., Galli, G. 1980. *J. Org. Chem.* 45:2528–31
199. Taylor, U. F., Pascal, R. A. Jr., Tsuda, M., Kisic, A. 1980. *Fed. Proc.* 39:1904
200. Pierce, A. M., Pierce, H. D., Unrau, A. M., Oehlschlager, A. C., Woods, R. A. 1979. *Lipids* 14:376–79
201. Itoh, T., Kikuchi, Y., Tamura, T., Matsumoto, T. 1981. *Phytochem.* 20:761–64
202. Kho, E., Imagawa, D. K., Rohmer, M., Kashman, Y., Djerassi, C. 1981. *J. Org. Chem.* 46:1836–39
203. Tsuda, M., Schroepfer, G. J. Jr. 1979. *Chem. Phys. Lipids* 25:49–68
204. Eggert, H., Djerassi, C. 1981. *J. Org. Chem.* 46:5399–5401
205. Batta, A. K., Williams, T. H., Salen, G., Greeley, D. N., Shefer, S. 1980. *Lipid Res.* 21:130–35
206. Gaylor, J. L., Delwiche, C. V. 1963. *Steroids* 4:207–12
207. Sharpless, K. B., Snyder, T. E., Spencer, T. A., Maheshwari, K. K., Guhn, G., Clayton, R. B. 1969. *J. Am. Chem. Soc.* 90:6874–75
208. Sharpless, K. B., Snyder, T. E., Spencer, T. A., Maheshwari, K. K., Nelson, J. A., Clayton, R. B. 1969. *J. Am. Chem. Soc.* 91:3394–96
209. Rahman, R., Sharpless, K. B., Spencer, T. A., Clayton, R. B. 1970. *J. Biol. Chem.* 245:2667–71
210. Nelson, J. A., Kahn, S., Spencer, T. A., Sharpless, K. B., Clayton, R. B. 1975. *Bioorg. Chem.* 4:363–76
211. Miller, W. L., Gaylor, J. L. 1970. *J. Biol. Chem.* 245:5375–81
212. Djerassi, C., Knight, J. C., Wilkerson, D. I. 1963. *J. Am. Chem. Soc.* 85:835
213. Goad, L. J., Williams, B. L., Goodwin, T. W. 1967. *Eur. J. Biochem.* 3:232–36
214. Doyle, P. J., Patterson, G. W., Dutky, S. R., Cohen, C. F. 1971. *Phytochem.* 10:2093–98

215. Atallah, A. M., Nicholas, H. J. 1971. *Steroids* 17:611–18
216. Ragsdale, N. N. 1975. *Biochim. Biophys. Acta* 380:81–96
217. Pierce, A. M., Mueller, R. B., Unrau, A. M., Oehlschlager, A. C. 1978. *Can. J. Biochem.* 56:794–800
218. Itoh, T., Yoshida, K., Yatsu, T., Tamura, T., Matsumoto, T. 1981. *J. Am. Oil Chem. Soc.* 58:545–50
219. Miller, W. L., Gaylor, J. L. 1970. *J. Biol. Chem.* 245:5369–74
220. Swindell, A. C., Gaylor, J. L. 1968. *J. Biol. Chem.* 243:5546–55
221. Hornby, G. M., Boyd, G. S. 1970. *Biochem. Biophys. Res. Commun.* 40:1452–54
222. Bechtold, M. M., Delwiche, C. V., Comai, K., Gaylor, J. L. 1972. *J. Biol. Chem.* 247:7640–56
223. Gaylor, J. L., Miyake, Y., Yamano, T. 1975. *J. Biol. Chem.* 250:7159–67
224. Brady, D. R., Crowder, R. D. 1978. *J. Biol. Chem.* 253:3101–5
225. Crowder, R. D., Brady, D. R. 1979. *J. Biol. Chem.* 254:408–13
226. Sanghvi, A., Balasubramanian, D., Moscowitz, A. 1967. *Biochemistry* 6:869–72
227. Sanghvi, A. 1970. *J. Lipid Res.* 11:124–30
228. Knapp, F. F. Jr., Trowbridge, S. T., Schroepfer, G. J. Jr. 1975. *J. Am. Chem. Soc.* 97:3522–24
229. Knapp, F. F. Jr., Schroepfer, G. J. Jr. 1975. *Steroids* 26:339–57
230. Pyrek, J. St. 1969. *Chem. Commun.* pp. 107–8
231. Pyrek, J. St., Schmidt-Szalowska, A. 1977. *Roekniki Chemii Ann. Soc. Chim. Pobrorum.* 51:951–58
232. Wilton, D. C., Akhtar, M. 1975. *Biochem. J.* 149:233–35
233. Knapp, F. F., Goad, L. J., Goodwin, T. W. 1973. *J. Chem. Soc. Chem. Commun.* pp. 149–50
234. Knapp, F. F., Goad, L. J., Goodwin, T. W. 1977. *Phytochem.* 16:1677–81
235. Gaylor, J. L., Delwiche, C. V. 1973. *Ann. NY Acad. Sci.* 212:122–38
236. Miller, W. L., Brady, D. R., Gaylor, J. L. 1971. *J. Biol. Chem.* 246:5147–53
237. Rahimtula, A. D., Gaylor, J. L. 1972. *J. Biol. Chem.* 247:9–15
238. Gaylor, J. L., Mason, H. S. 1968. *J. Biol. Chem.* 243:4966–72
239. Ohba, M., Sato, R., Yoshida, Y., Nishino, T., Katsuki, H. 1978. *Biochem. Biophys. Res. Commun.* 85:21–27
240. Aoyama, Y., Yoshida, Y., Sato, Y., Susani, M., Ruis, H. 1981. *Biochim. Biophys. Acta* 663:194–202

241. Alexander, K. T. W., Mitropoulos, K. A., Gibbons, G. F. 1974. *Biochem. Biophys. Res. Commun.* 60:460–67
242. Fukushima, H., Grinstead, G. F., Gaylor, J. L. 1981. *J. Biol. Chem.* 256:4822–26
243. Brady, D. R., Mattingly, T. W., Gaylor, J. L. 1976. *Anal. Biochem.* 70:413–23
244. Ingelman-Sundberg, M., Johansson, I. 1981. *J. Biol. Chem.* 256:6321–26
245. Olson, J. A., Lindberg, M., Bloch, K. 1957. *J. Biol. Chem.* 226:941–56
246. Aoyama, Y., Yoshida, Y. 1978. *Biochem. Biophys. Res. Commun.* 82:33–38
247. Aoyama, Y., Yoshida, Y. 1978. *Biochem. Biophys. Res. Commun.* 85:28–34
248. Fried, J., Dudowitz, A., Brown, J. W. 1968. *Biochem. Biophys. Res. Commun.* 32:568–74
249. Richards, J. H., Hendrickson, J. B. 1964. *The Biosynthesis of Steroids, Terpenes, and Acetogenins,* p. 313. New York: Benjamin. 416 pp.
250. Knight, J. C., Klein, P. D., Szczepanik, P. A. 1966. *J. Biol. Chem.* 241:1502–8
251. Lee, W.-H., Schroepfer, G. J. Jr. 1968. *Biochem. Biophys. Res. Commun.* 32:635–38
252. Lutsky, B. N., Schroepfer, G. J. Jr. 1969. *Biochem. Biophys. Res. Commun.* 35:288–93
253. Zalkow, L. H., Cabat, G. A., Chetty, G. L., Ghosal, M., Keen, G. 1968. *Tetrahedron Lett.* pp. 5727–29
254. Barton, D. H. R., Harrison, D. M., Moss, G. P., Widdowson, D. A. 1970. *J. Chem. Soc.* pp. 775–85
255. Bouvier, P., Rohmer, M., Benveniste, P., Ourisson, G. 1976. *Biochem. J.* 159:267–71
256. Withers, N. W., Goad, L. J., Goodwin, T. W. 1979. *Phytochem.* 18:899–901
257. Kokke, W. C. M. C., Bohlin, L., Fenical, W., Djerassi, C. 1981. *Phytochem.* 20:127–34
258. Bohlin, L., Kokke, W. C. M. C., Fenical, W., Djerassi, C. 1981. *Phytochemistry* 20:2397–2401
259. Teshima, S.-I., Patterson, G. W. 1981. *Comp. Biochem. Physiol.* 69:175–81
260. Fiecchi, A., Scala, A., Cattabeni, F., Grossi-Paoletti, E. 1970. *Life Sci.* 9:1201–5
261. Mitropoulos, K. A., Gibbons, G. F., Reeves, B. E. A. 1976. *Steroids* 27:821–29
262. Canonica, L., Fiecchi, A., Galli-Kienle, M., Scala, A., Galli, G., Grossi Paoletti, E., Paoletti, R. 1968. *J. Am. Chem. Soc.* 90:3597–98
263. Gibbons, G. F., Goad, L. J., Goodwin, T. W. 1968. *Chem. Commun.* pp. 1458–60
264. Lutsky, B. N., Schroepfer, G. J. Jr. 1968. *Biochem. Biophys. Res. Commun.* 33:492–96
265. Lutsky, B. N., Schroepfer, G. J. Jr. 1970. *J. Biol. Chem.* 245:6449–55
266. Akhtar, M., Rahimtula, A. D., Watkinson, I. A., Wilton, D. C., Munday, K. A. 1969. *Chem. Commun.* pp. 149–50
267. Martin, J. A., Huntoon, S., Schroepfer, G. J. Jr. 1970. *Biochem. Biophys. Res. Commun.* 39:1170–74
268. Miller, W. L., Kalafer, M. E., Gaylor, J. L., Delwiche, C. V. 1967. *Biochemistry* 6:2673–78
269. Galli-Kienle, M., Anastasia, M., Cighetti, G., Galli, G., Fiecchi, A. 1980. *Eur. J. Biochem.* 110:93–105
270. Schroepfer, G. J. Jr., Fourcans, B., Huntoon, S., Lutsky, B. N., Vermilion, J. 1971. *Fed. Proc.* 30:1105
271. Huntoon, S., Schroepfer, G. J. Jr. 1970. *Biochem. Biophys. Res. Commun.* 40:476–80
272. Hampl, R., Starka, L. 1969. *J. Steroid Biochem.* 1:47–56
273. Schroepfer, G. J. Jr., Parish, E. J., Kandutsch, A. A. 1977. *J. Am. Chem. Soc.* 99:5494–96
274. Schroepfer, G. J. Jr., Parish, E. J., Tsuda, M., Raulston, D. L., Kandutsch, A. A. 1979. *J. Lipid Res.* 20:994–98
275. Kato, T., Kawase, Y. 1976. *Agric. Biol. Chem.* 40:2379–88
276. Spence, J. T., Gaylor, J. L. 1977. *J. Biol. Chem.* 252:5852–58
277. Havel, C., Hansbury, E., Scallen, T. J., Watson, J. A. 1979. *J. Biol. Chem.* 254:9573–82
278. Ortiz de Montellano, P. R., Beck, J. P., Ourisson, G. 1979. *Biochem. Biophys. Res. Commun.* 90:897–903
279. Kandutsch, A. A., Chen, H. W. 1973. *J. Biol. Chem.* 248:8408–17
280. Wilton, D. C., Rahimtula, A. D., Akhtar, M. 1969. *Biochem. J.* 114:71–73
281. Yabusaki, Y., Nishino, T., Ariga, N., Katsuki, H. 1979. *J. Biochem.* 85:1531–37
282. Barton, D. H. R., Corrie, J. E. T. Widdowson, D. A., Bard, M., Woods, R. A. 1974. *J. Chem. Soc. Perkin Trans. I* pp. 1326–33
283. Barton, D. H. R., Gunatilaka, A. A. L., Jarman, T. R., Widdowson, D. A., Bard, M., Woods, R. A. 1975. *J. Chem. Soc. Perkin Trans. I* pp. 88–92
284. Bleau, G., VandenHeuvel, W. J. A. 1974. *Steroids* 24:549–56
285. Ikekawa, N., Fujimoto, Y., Ishiguro, M., Suwa, S., Hirayama, Y., Mizunuma, H. 1979. *Science* 204:1223–24

Ann. Rev. Biochem. 1982. 52:587–616
Copyright © 1982 by Annual Reviews Inc. All rights reserved

BACTERIORHODOPSIN AND RELATED PIGMENTS OF HALOBACTERIA

Walther Stoeckenius and Roberto A. Bogomolni

Department of Biochemistry and Biophysics, and Cardiovascular Research
Institute, University of California, San Francisco, California 94143

CONTENTS

Perspectives and Summary

Bacteriorhodopsin (bR) is a light energy–transducing pigment so far found only in halobacteria, prokaryotes that occur in natural brines where the NaCl concentration is at or near saturation. Its chromophore, retinal, is bound via a protonated Schiff base to the ϵ-amino group of a lysine residue in the 26,000-mol wt protein, which generates a broad absorption band in the green region of the spectrum. Bacteriorhodopsin is a transmembrane protein that aggregates to form crystalline patches in the membrane. The patches preserve their structure when isolated and are known as the purple membrane (pm). When bR absorbs light it ejects protons from the cell thus generating an electrochemical gradient across the cell membrane that may reach 280–300 mV. This proton gradient directly drives some metabolic processes, e.g. ATP synthesis, and probably also locomotion. Most of the absorbed energy, however, is apparently first converted from a proton into

587

0066-4154/82/0701-0587$02.00

Na^+ and K^+ electrochemical gradients, which due to the high salt concentrations present have a high energy storage capacity, and can regenerate the protonmotive force (PMF) or directly drive other energy-requiring processes, e.g. amino acid uptake.

The crystalline arrangement of bR in pm has allowed determination of its tertiary structure at ~7 Å resolution. The polypeptide chain crosses the membrane seven times, forming a tight crescent-shaped cluster of α-helical segments oriented with their long axis roughly normal to the plane of the membrane. The last 3 to 6 amino acids of the N terminus are exposed on the exterior surface of the membrane, while the last 17–24 amino acids of the C terminus are accessible from the aqueous phase on the cytoplasmic side.

In the dark, pm consists of 13-*cis* and all-*trans* retinal-contining bR conformers in equal amounts. Both components undergo cyclic photoreactions, but a branching pathway leads from the 13-*cis* into the all-*trans* cycle, and the back reaction is very slow, so that in the light the 13-*cis* component rapidly disappears. Only the all-*trans* photoreaction cycle translocates protons across the membrane. At least four intermediates have been identified in the cycle; their rise and decay constants range from 10^{11} to 10^2 sec^{-1}. A transient deprotonation of the Schiff base and all-*trans* to 13-*cis* isomerization of the retinal occur during the cycle as well as protein conformational changes.

Bacteriorhodopsin obviously is the simplest and best understood of all known active transport systems and offers the best opportunity to explore the relevant molecular mechanisms in detail. It is, therefore, not surprising that work on bR has been expanding rapidly in the past few years and has become one of the most active fields in life science research (1). Several recent reviews, some partially redundant, cover the subject comprehensively (2–7). However, new results have not only expanded our knowledge but also seriously challenged some of our earlier conclusions. The complete amino acid sequence of bR is now known, and independent attempts by two groups to fit it to a structural model have yielded essentially the same arrangements. The location of retinal in the structure has been determined by neutron diffraction, but its assumed binding site at Lys 41 had to be revised. If a single binding site can be assigned, all evidence now points to Lys 216.

It has become ever more obvious that the originally proposed unidirectional photoreaction cycle with four to five intermediates in addition to bR, while correct in its main features, is insufficient to explain a wide variety of observations. Many modifications have been proposed, but none of them fits all data satisfactorily. Early experiments on isolated pm in low salt to determine the number of protons translocated in one cycle yielded values of $S_{H^+} \simeq 1.0$. More recent values from intact cells, cell envelopes, and lipid vesicles are closer to $S_{H^+} = 2.0$ and even slightly higher. If a fixed stoichiometry is assumed $S_{H^+} = 2.0$ is the best available estimate, but a sliding stoichiometry dependent on environmental conditions is a strong possibility.

Several laboratories have begun to explore the genetics of halobacteria with modern techniques. An interesting interaction between plasmid and chromosome has emerged and a very active insertion mechanism has been shown to cause the unusually high frequency of 10^{-2} to 10^{-4} with which spontaneous mutations of gas vacuole formation and carotenoid and bacterio-opsin synthesis occur. A base sequence has been obtained for the bacterio-opsin gene.

Bacteriorhodopsin also appears to function as signal transducer for phototactic responses, and together with a second blue-absorbing retinal pigment it allows the bacteria to accumulate in light with a strong green component. Recent work on this response shows that action spectra do not necessarily correspond to the absorption spectra of the pigments. More important, a third retinal pigment, P_{588} or halorhodopsin, has been discovered in the same cells, which strongly resembles bR but apparently mediates a light-driven Na^+ ejection from the cells. Its role in the energy metabolism of the cells is still controversial, but it apparently functions as a receptor pigment for phototaxis in the long wavelength region of the visible spectrum.

We concentrate here on these newer developments and cover the literature published since our 1979 review (as necessary) but not comprehensively.

The Structure of Bacteriorhodopsin

The amino acid sequence analysis of bR has been completed independently by two research groups (8, 9). Bacteriorhodopsin contains no histidine and consists of 247 or 248 amino acids; of these 70% are hydrophobic and distributed in clusters. Differences between the two published sequences are minor; however, Khorana et al find after Trp 137 a second Trp, which is missing in Ovchinnikov's sequence. (We use Khorana's residue numbers throughout this review.) Using partially purified mRNA as a template a partial cDNA transcript corresponding to the N-terminal region of bR has been synthesized. It shows the expected codons for amino acids 1 to 8 and for an additional 13 amino acids preceding the N-terminus (10), which fit the partial sequence of a presumed precursor of bR determined by Dellweg & Sumper (11). Recently, the complete base sequence for the bacteriorhodopsin gene has been published (11a). It does not contain intervening sequences and the standard genetic code is used. One additional codon (for aspartate) is present at the COOH-terminus and no promoter could be identified in the 360 nucleotides preceding the initiation codon AUG.

Ovchinnikov et al have rearranged their earlier suggested chain folding, so that Lys 41, which according to Bridgen & Walker's results (12) forms the Schiff base with retinal, is now close to the cytoplasmic rather than the external side of the membrane (8). This brings their proposed chain folding into agreement with the topography of proteolytic cleavage sites observed by Gerber et al (13). Subsequently, several other groups have published similar suggestions (7, 14–18). Only one recent proposal is significantly different, placing Lys 41 in a separate short helical segment oriented parallel to the membrane surface (19). The arrangement we prefer is shown in Figure 1. The sequences linking the α-helical segments on the membrane surface are generally assumed to be short. The sequence connecting helices B and C on the external surface, which is 10–14 amino acid residues long, is apparently the longest. While proteolytic cleavage sites (8, 13) and chemical labels (20–23) lend some experimental support, these arrangements are only tentative. Their main purpose should be to suggest new experiments, which would permit rejection or refinement.

Michel et al (24) have recrystallized pm into a two-dimensional ortho-rhombic lattice that gives diffraction data with an in-plane resolution of ~ 6.5 Å and, because it is centrosymmetric, facilitates the assignment of phases. Most importantly, the in-plane projection of the structure has experimentally confirmed the assumption that the high density area containing the cluster of seven rods in the Unwin—Henderson map corresponds to one bR molecule. A much higher resolution in-plane projection was obtained by low-dose, low-temperature electron microscopy and the combination of image data from different crystal domains (25). The individual areas were not simply aligned translationally, but were also separately corrected for differences in focus and specimen motion, and weighted so that poorer data contributed less to the final image. The resulting map has a resolution of 3.7 Å (Figure 2). It confirms the general features of the low resolution map, and in the lipid regions shows peaks, with the expected 5.6-Å spacing of lipid chains and apparent aromatic amino acid side chains protruding from the α-helices. However, the height of the latter requires that they be attributed to stacking of several side chains from adjacent helices, which is consistent with data from UV spectroscopy (26). A large negative peak between helices 5 and 6 next to the largest putative side chain peak, indicates a roughly cylindrical 45-Å long space nearly devoid of matter.

To resolve the structure in the third dimension the specimen must be tilted. The images become more difficult to analyze as the tilt angle in-

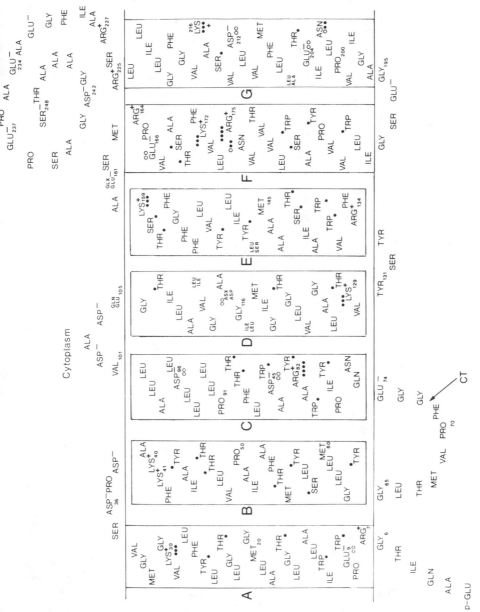

Figure 1 The amino acid sequence of bR according to Khorana et al (9) with differences in the sequence of Ovchinnikov et al (8) indicated below the Khorana sequence. Trp 138 is missing from Ovchinnikov's sequence. Potential proton donors (·) and acceptors (o) and charged groups (−, +) are marked. The arrangement in α-helical segments and linking sequences is taken from Agard & Stroud (17). The arrow marks the chymotrypsin cleavage site.

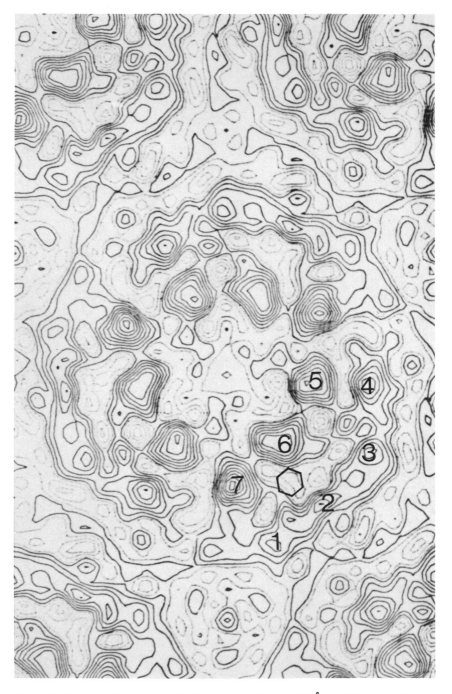

Figure 2 In-plane electron scattering density map of pm at 3.7 Å resolution according to Hayward & Stroud (25) viewed from the cytoplasmic side. The retinal position of King et al (33) is marked by a hexagon. Helix positions are numbered as referred to in the text.

creases, and the resolution obtained in the Henderson—Unwin structure across the membrane is only ~14 Å. The limited degree of tilt possible also results in a distortion of the reconstructed image. Agard & Stroud (17) have developed a method to correct for this distortion and applied it to the data collected by Henderson & Unwin. The reconstructed image as expected shows an increase in the length of the α-helix rods from 35Å to 45 Å, and the taper at their ends has been removed. More important, four regions of density connecting the rods on one surface of the membrane and one such connection on the opposite surface emerge. These apparently represent the portions of the polypeptide chain connecting the transmembrane helices and/or the nonhelical C- and N-terminal segments. Since the sidedness of the model is now known (27, 28), Agard & Stroud could identify the surface with four connecting regions as the extracellular side, and because only three links between helices should exist there, one presumably represents the N-terminal segment. The dense region on the cytoplasmic surface could either be a linking segment or the C-terminal tail. If this interpretation is accepted only 5 likely arrangements for the polypeptide chain in the map remain out of the possible 10080, when no restrictions are applied.

A further restriction for alignment of the amino acid sequence with the density map arises from the retinal-binding site. When Bridgen & Walker's (12) retinyl peptide was matched with the sequence it apparently identified Lys 41 as the binding site. However, a recent reinvestigation in several laboratories showed that the apparent linkage site varies with reduction conditions. Rapid reduction in the light results in linkage to Lys 216 only, and some investigators accept this as the correct site (18, 29, 30, 30a). However, slower reductions in the dark, after proteolysis by chymotrypsin, yield linkages to both Lys 41 and Lys 216 (18). Ovchinnikov et al (31, 31a) found, in the dark, binding to only Lys 41, and in the light, binding also to another lysine residue in the large chymotryptic fragment (amino acids 72–248). which has recently been identified as Lys 216 (Lys 215 in their sequence) (31a). They interpret their observations as evidence for a transient migration of the Schiff base during the photoreaction cycle from Lys 41 to a Lys residue at position 129 or beyond in the sequence. While this is a possible interpretation of the data, the evidence is by no means conclusive. Reduction in all cases required unphysiological conditions, takes at least many minutes, and could be due to an intermediate in a minor, functionally unimportant side reaction. Even the correct identification of Lys 41 in the retinyl peptide has been questioned (29). Most of the evidence favors Lys 216 as the native binding site. If it can be shown, however, that the retinal, when linked to Lys 41 after reduction, is still in its original noncovalent binding site (18), it then follows that Lys 41, located near the

N terminus, and Lys 216, near the C terminus, must lie close together in the tertiary structure. This would eliminate three of the five folding schemes allowed by the observed linking sequences of Agard & Stroud (17) and leave only the two shown in Figure 3.

The position of retinal has also been determined by neutron diffraction. At the resolution obtained from profile diffraction, the β-ionone ring can be located close to the center of the membrane (32). The polyene chain, which forms an angle of $\sim24°$ with the membrane plane, is not resolved, but probably extends from the ring position toward the cytoplasmic surface, because both Lys 41 and Lys 216 in all proposed models are closer to it. The in-plane difference map for pm substituted with deuterated retinal has also been calculated using the phases and intensity distributions from electron microscopy. It shows three larger peaks on a rather noisy background (33). Their position is consistent with the two most likely models selected by Katre et al (18) but at the present resolution cannot be used to distinguish between them.

Engelman et al (15) have proposed a folding of the polypeptide chain based on quite different considerations. After selecting seven segments of the sequence as the most probable transmembrane α helices, they calculated their scattering densities and sought the best match with the densities in the map, simultaneously trying to minimize the length of connecting links and the number of non-neutralized charged groups in the membrane interior. Their two most preferred arrangements are identical to the two models selected by Katre et al (18), on the basis of chromophore location, from the five arrangements that are possible, based on the linking sequences observed (17).

There is further experimental evidence that favors these two models. Engelman & Zaccai (34) have obtained neutron diffraction data from purple

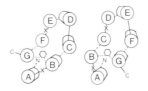

Figure 3 The two most likely arrangements of the bR polypeptide chain superimposed on a simplified presentation of the scattering density map given in Figure 2, using the criteria of Katre et al (18) to eliminate three of the five possible arrangements of Agard & Stroud (17). Letters in the circles correspond to the α-helices labeling in Figure 1. The linking sequences on the external side (- - - - -) and cytoplasmic side (———) are taken from (17); linking sequences missing in their structure are marked X. The positioning of the carboxyl- (C) and amino-terminals (N) is arbitrary except that they are linked to the correct α helix. The retinal position (hexagon) is linked by dotted lines to the helices bearing Lys 41 and Lys 216.

membrane substituted with deuterated valine that show that the protein folded in this manner has α helices with a hydrophobic outer surface in contact with the lipid and the hydrophilic groups pointing inward. An aqueous channel or large pocket inside the protein can be excluded on the basis of data from neutron diffraction in ^2H_2O (35); however, the density of exchangeable hydrogens is higher in the center of the protein, which is consistent with the assumption that the hydrophilic groups are concentrated there (36).

Thus, while attempts at independent structure refinement from several laboratories are in good agreement, some discrepancies must also be noted. Sigrist & Zahler (37) report that phenylisothiocyanate at neutral pH exclusively and nearly stoichiometrically reacts with Lys 216 with only minor effects on spectroscopic properties and function of bR. This observation is difficult to reconcile with all recent studies of the retinal-binding site. Renthal (23) has dansylated Lys 40 (or 41) and from fluorescent energy transfer measurements concludes that the minimal distance to the chromophore is 29 Å; this is clearly inconsistent with binding of retinal at Lys 41. Dansylation of Asp 102 or 104 yields a very short distance to the chromophore, which is also inconsistent with both most favored models discussed above. Kouyama et al (38) have used energy transfer in partially reduced pm to localize the chromophore in the membrane plane. Their most favored position is ~12 Å closer to the in-plane center of the molecule than the position obtained by King et al. Glaeser et al (39) contend that the density distribution over the area attributed to helices 5 to 7 in the high resolution map (Figure 2) is not compatible with a predominantly α helical conformation. They attribute it to mostly β sheet, and adduce additional evidence from CD and IR spectra. While the argument based on CD data is not convincing, the IR spectra show some unusual features that need explanation. Rothschild & Clark (40) already noted an ~10 cm^{-1} blue shift of the amide I band to ~1665 cm^{-1} and ascribed it to a slight distortion of the α helices. The strong shoulder on the low frequency side of the 1665 cm^{-1} peak emphasized in Glaeser et al's argument disappears in Rothschild & Clark's spectra when they subtract the 1645 cm^{-1} water peak. A much weaker shoulder, however, is visible in their spectra of membrane dried from D_2O, and may indicate the presence of a small amount of β structure. The IR work clearly needs to be extended. IR polarization data are consistent with the average α-helix orientation as deduced from X-ray data and electron microscopy (41).

Schreckenbach et al (42, 43) have probed the retinal-binding site stereochemistry by observing the spectral effects of chromophore reconstitution with retinal and retinal analogues. They deduced that three or more groups with pKs of 3.8, 4.5, and >10.5 interact with retinal in the formation of the chromophore, and that energy transfer occurs from tryptophan. There are more extensive data on energy transfer from aromatic amino acids (44, 45). Kalisky et al find only one unquenched tryptophan and conclude that this is consistent with the chain folding in Engelman's most preferred model (15) and King et al's chromophore position (33). Nakanishi's group has reconstituted the chromophore with retinal analogues containing shortened chains of conjugated double bonds (46). The observed shifts of the visible absorption maximum fit a chromophore model with one negatively charged group 3.0 Å distant from the Schiff base and another 3.5 Å from the C5 atom of the β-ionone ring. The latter could be provided by Asp 212 or, in Engelman's most likely model, by Asp 296.

Further information on the structure can obviously be expected from several different approaches. Chemical modification of amino acid side chains and their effects on the structure and function have been extensively used. Most of these efforts suffer from the undetermined location of the modified amino acids in the sequence and/or the three-dimensional structure. A three-dimensional high resolution structure would, of course, solve the problem. It will, however, be extremely difficult to obtain by electron microscopy; and three-dimensional bR crystals recently obtained (47, 48) unfortunately have failed, so far, to give high resolution X-ray diffraction patterns.

The Photoreaction Cycle

The available spectroscopic data are insufficient to determine unequivocally the number of photoreaction cycle intermediates, their kinetic constants, absorption spectra, and connectivity. Models must, therefore, be assumed, and criteria must be found to test them. The model proposed originally by Lozier et al (49) used four observable ground state intermediates, K_{590}, L_{550}, M_{410}, and O_{640} arranged in alphabetical order in a unidirectional, unbranched cycle, with the possibility that an additional poorly resolved intermediate, N_{530}, also existed. However, it was recognized at the time that this simple model did not fit all experimental data satisfactorily (49). While most investigators still recognize the four main ground state intermediates K, L, M, and O, additional ones located in the main path have been described by some, e.g. J_{625} preceding K (50), and X between L and M (51, 52). Numerous branches have been introduced into the cycle that bypass one or more of the main intermediates, e.g. M and O (53, 54), and/or introduce new ones, e.g. P between O and bR (55). Rapidly established equilibria between intermediates within the main sequence or in a side branch have also been suggested. A frequently used feature are two spectroscopically similar M intermediates, as already postulated by Slifkin & Caplan (56); they may occur in series or in parallel (see 57–59).

A detailed discussion of all these modifications would far exceed the space alloted for this review, and the more substantial contributions have been covered recently by Ottolenghi (6). Most of the investigators consider only one or two steps in the photoreaction cycle, and their interpretations have more or less serious shortcomings. For instance, it is often assumed without further justification that absorbance changes at a given wavelength versus time after excitation measure concentration changes of only one intermediate. Some investigators have concluded that O cannot arise from M because its rise time is faster than the M decay, whereas this is entirely possible if the O decay is faster than the M decay. Moreover, the ratio of these two time constants may change from >1 to <1 with a change in temperature or other environmental conditions. Furthermore photoselection effects may be mistaken for concentration changes if the rotational diffusion time of membrane fragments is of the same order of magnitude as the observed photocycle reaction. Fully light-adapted preparations are usually considered to comprise a homogenous population of bR molecules containing all-*trans* retinal; however, aside from the possibility that spectroscopically indistinguishable conformational substates exist that yield nonexponential kinetics at low temperature, recent extraction experiments have detected some 13-*cis* retinal in fully light-adapted bR, (60–63), so that intermediates of the 13-*cis* retinal complex may contribute to the observed absorbance changes. Furthermore, at very high light intensities, 13-*cis* retinal-containing bR is generated ("dark adaptation by light") (64), and this may affect the results obtained in the kinetic resonance Raman experiments (65). Finally, Keszthelyi and his collaborators have recently discovered changes in light scattering intensity during the photoreaction cycle that are large enough to affect the absorbance measurements (L. Keszthelyi, personal communication).

The most comprehensive study of the photoreaction cycle published so far (66) takes most but not all of these complications into account. It uses a comprehensive new set of data, establishes a series of criteria and a systematic procedure to develop and test cycle models, and employs a new nonlinear least squares analysis. This approach again yields roughly the same K, L, M, and O intermediates recognized earlier, but shows that neither a unidirectional cycle, nor simple branching from any intermediate back to bR satisfactorily fits the data. Neither does the assumption of two M intermediates with different kinetics, unless the spectra are strongly temperature dependent.

The transient isomerization from all-*trans* to 13-*cis* retinal during the bR_{570} photoreaction cycle, first postulated from chromophore extraction data of the M intermediate (66a), has been confirmed (66b) and extended to the L intermediate with improved extraction techniques (66c). More importantly, earlier conflicting interpretations of the Raman spectra, were settled in favor of a 13-*cis* configuration for M, using a 15 deutero-retinal chromophore (66d). The most recent time-resolved UV and resonance Raman spectra indicate that isomerization to the 13-*cis* configuration is already complete within the life time of K (66e–66g), as had been postulated earlier from theoretical considerations (reviewed in 12).

The functional significance of this apparent complexity of the photoreaction cycle is difficult to understand and one wonders whether it could be an artifact due to experimental condition. At least in one case it has been conclusively shown that nonphysiological conditions highly modify the cycle. A branching reaction from L back to bR is insignificant at room temperature, but at very low temperature becomes dominant and inhibits formation of M (53, 54). Moreover, the generally assumed uniformity and stability of pm preparations, according to our experience and anecdotal reports from other laboratories, cannot be taken for granted. Conflicting observations, for instance the opposite pH dependence of the above-mentioned branching reaction at L reported by Kalisky et al (54) and Iwasa et al (53) could probably be avoided with better defined pm preparations. A beginning has been made in this direction (67). No comprehensive study of the photoreaction cycle under physiological conditions, e.g. in intact cells, has been reported so far. It is, of course, technically difficult, and the effects of electrochemical gradients on the cycle kinetics [(68–70); Z. Dancshazy, S. L. Helgerson, and W. Stoeckenius, in preparation] must be taken into account. However, such a study should yield physiologically more significant data, and may result in a simpler cycle model.

Nevertheless, recent work has confirmed and extended some important but tentative earlier results. Not only K and M but all of the main intermediates show photoreactions back to bR (57, 71, 72); the 13-*cis* conformation of the M chromophore has been confirmed by resonance Raman spectroscopy (73), and new evidence that the 13-*cis* configuration is already present in K (74) and L (62) has been presented.

Recent observations on transient absorbance and fluorescence changes in the near UV are consistent with the earlier interpretation (75) that they indicate deprotonation of one tyrosine residue during the L → M transition (45, 76–78), and low-temperature data indicate that tyrosine deprotonation is a prerequisite for M formation (54). Interestingly, a kinetic component of the absorbance change at 297 nm does not correspond to any observed in the visible range (77), which demonstrates that kinetic UV spectroscopy can reveal aspects of the photoreaction cycle that are not seen, or seen only with difficulty, in the visible range. A similar kinetic component is, however, observed with resonance Raman spectroscopy and has been attributed to an intermediate X between L and M (51, 52). Further interesting results may also be expected from kinetic IR spectroscopy, which is beginning to be used (W. Mantele, F. Siebert and W. Kreutz, personal communication).

Purple membrane in aqueous suspension orients in an electric field (79–82). This effect has been used to measure simultaneously the kinetics of charge displacements and absorbance changes during the photocycle (83). The light-induced photocurrents consist of a fast, unresolved negative component (which would correspond to movement of a positive charge toward the cytoplasmic side of the membrane) and a series of "positive" electric transients in the microsecond and millisecond range, which closely follow the kinetics of photocycle absorbance changes from L to bR. Similar electric transients had already been observed by Skulachev's group (84, 84a) and others (85–89) in planar film systems. The time resolved charge displacements have been attributed to proton movements within the protein (83–84a). As one would therefore expect, illumination of M leads to a reversal of the charge displacement observed in the L → M transition (90, 91). If this interpretation is correct it has some interesting consequences for the possible models of proton transport. While some of the proton movements through long permanent proton wires (92–94) would likely be synchronized with some of the absorbance changes (93), we would rather propose that short segments of wires (or other translocation paths) are established and broken as the molecule proceeds through the reaction cycle (78). However, if two protons are translocated in one cycle (see following section) they would have to move simultaneously through the same or through two identical pathways. This would argue against the widely held belief that the Schiff base forms part of the translocation path.

The Stoichiometry of Proton Pumping

We have recently discussed in more detail available data on the number of protons translocated in one photoreaction cycle (95). Early results from flash spectroscopy with pm suspended at slightly alkaline pH and low ionic strength (49) indicated that less than one proton transiently appears in the solution for every bR molecule cycling. Its release is somewhat slower than the rise of M_{412}, and its uptake parallels the reformation of bR. Corrected for the time overlap of uptake and release the stoichiometry (S_{H+}) becomes 1.0. However, subsequent experiments in several laboratories showed that the magnitude and even the sign of the pH changes observed may depend strongly on reaction conditions, and that the stoichiometry could be significantly higher.

Convincing evidence for a higher stoichiometry was obtained independently by three groups working with intact cells (96–98). In cell suspensions at neutral or slightly acidic pH, close to two protons are ejected from the cell in every photocycle ($S_{H^+} \simeq 2.0$). The value of $S_{H^+} = 1.0$ reported earlier for comparable conditions (99) is due to the high quantum efficiency assumed. The results agree with the later work if the now generally accepted quantum efficiency of 0.25 to 0.30 is used. Renard & Delmelle, however, found that S_{H^+} declines at alkaline pH, reaching 1.0 between pH 7.5 and 8.0 (98), while the quantum efficiency for cycling remains unchanged (100).

Envelope vesicles also yield S_{H^+} values near 2.0 (96) as do bR-containing egg lecithin vesicles; this yield is independent of ionic strength between 10 and 100 mM KCl (97). Our own data [(101); W. Stockenius and R. Bogomolni unpublished results] with asolectin vesicles give lower values of $S_{H^+} \simeq 1.25$. However, with this system it is difficult to obtain reliable, quantitative data and the values could easily be in error. A much better model system should be the planar films of large area that incorporate pm fragments in known concentration and orientation into a lipid bilayer (85). Unfortunately, high leakage currents so far severely limit its usefulness, and the reported S_{H^+} value of ~2 at pH 5.8 is due to an arithmetical error; the actual value is 0.2 (J. I. Korenbrot, personal communication).

In pm suspensions, pumped protons are not readily distinguishable from protons that may be released or taken up on the same side of the membrane due to light-induced conformational changes of the protein (Bohr protons). However, S_{H^+} has been measured over a much wider range of conditions in pm suspensions than in other systems. When the H^+/M or S_{H^+} values found differ, Bohr protons are usually assumed to account for the discrepancy. If, as some of the experiments with closed systems indicate, Bohr protons do not contribute significantly to the pH changes observed, a variable stoichiometry of the proton pump must be considered. Alternatively, a variable overlap in time of the release and uptake of protons from the membrane could account for observed proton concentration changes smaller than those seen in closed systems.

Ort & Parson (102), in flash experiments, observed that S_{H^+} increased from close to 1.0 to nearly 2.0 between 10 mM and 200 mM NaCl or KCl, and then leveled off. Changes in pH between 6.0 and 8.75 (at 200 mM KCl) had no effect. A decreased S_{H^+} value at high light intensity was attributed to photoreversal of an early intermediate. The stoichiometry and ionic strength effect have recently been confirmed (97).

Most of the other available data result from photosteady state ΔpH measurements made in the laboratories of S. R. Caplan and B. Hess. From Caplan's measurements at 1 M KCl and low light intensity H^+/M values as high as 20 (103) and 500 (104) can be calculated. However, the authors express their data as H^+/bR, which yields values near 2.0, and assume that photocycling of only a few bR molecules causes a cascade of conformational changes throughout the membrane, which expose more and more acid groups in all bR molecules. The work has been reviewed recently (5). This explanation implies that flash experiments are not directly comparable to photosteady-state experiments, a view also held by Hess et al (105, 106). However, their highest H^+/M values observed at low light intensity and high ionic strength do not exceed 3 under essentially the same conditions, where Caplan's group reports values of 20 or more. Our own photosteady-state experiments, at least for the more limited range of conditions so far explored, fail to confirm the observations on which Caplan and Hess base their interpretations (95). We do not need to invoke Bohr protons, and the pH changes observed are compatible with the stoichiometry of transported protons observed in closed systems.

Do the three factors that appear to affect the H^+/M or S_{H^+} values in pm suspensions; ionic strength, light intensitiy, and pH actually affect the stoichiometry of the pump? If we assume a maximal S_{H^+} of 2, as indicated by the experiments with closed systems, the lower values obtained at higher light intensity could be explained by light-induced unproductive back reactions of intermediates, which are known to occur (see section on the photoreaction cycle). Light intensity dependence of H^+/M measurements may seem to preclude this explanation, because M_{412} presumably does not absorb the actinic light used. It is, however, known that M_{412} can back-react to short circuit the transport, and illumination of other intermediates or light absorption by neighboring molecules may affect this reaction. Alternatively, photoproducts of intermediates may contribute to the absorbance at the wavelength where M_{412} is measured, so that the M concentration is overestimated.

It has been shown by Kuschmitz & Hess (106) that in first approximation the ionic strength effect fits the Gouy-Chapman theory if a surface charge density of 0.005–0.0025 $e^-/\text{Å}^2$ and light-activated dissociable surface groups are assumed. A difference in the pk of proton-binding groups of pm suspensions in the light and in the dark supports this assumption (105). Tsuji & Neumann (107, 108) have shown that application of a high voltage pulse to pm suspensions resulting in transmembrane potentials in the physiological range generates transient protonation changes that are comparable in extent and kinetics to those seen during the photoreaction cycle, but are not accompanied by corresponding absorbances changes of the chromophore. However, if Bohr protons are excluded, a stoichiometry of 3.0 must be assumed for Hess's results. Another possible explanation that does not require changes in pump stoichiometry is based on a theory developed by Kell (109). Based on electrode theory he proposed that near a charged surface a potential well may exist that would prevent protons from escaping into the aqueous phase, but allows a rapid diffusion in the plane of the surface. This might allow protons released on the external surface to rebind without appearing in the medium.

Most interesting is the pH effect. Not only do the photosteady state H^+/M values depend strongly on pH, they can even become negative, i.e. light-dependent accumulation of M causes alkalization of the medium. The transition from positive to negative values typically occurs between pH 5 and 6 at room temperature or below, but at elevated temperature it shifts to higher pH values (110). The most extensive pH range has been tested by Takeuchi et al (111) who showed maximum proton binding near pH 4.0 and maximum release near pH 9.0, with a sharp decrease at lower and higher pH values. However, significant changes in the M concentration occur below pH 5.0 and above pH 8.0, and Takeuchi et al do not calculate the H^+/M ratios. In our opinion the changes reported are entirely compatible with the assumption that the kinetics of proton release and uptake are strongly dependent on pH, and Bohr protons need not be invoked. However, as the experiments with whole cells indicate, a variable (e. g. pH-dependent) stoichiometry may have to be taken into account. A number of investigators have used chemically or physically modified membrane for similar experiments, e.g. acetylated membrane or high guanidinium-HCl concentrations (111), partial delipidation and lipid substitution (112), or addition of valinomycin and beauvericin (113). So far these results have contributed little in resolving the stoichiometry problem. Recently Ebrey's group has reported that enzymatic removal of part or all of the 17 C-terminal amino acids reduces the stoichiometry with little or no effect on the photocycle kinetics. They also suggest that this may occur spontaneously on prolonged storage of pm and may account for some of the inconsistencies in the results (67).

The stoichiometry can also be determined by measuring charge movements upon flash illumination of oriented pm. Suspensions oriented in an electric field yielded a net movement of one elementary charge over 100 Å for each molecule cycling (83). The authors originally concluded that two protons were moving across the 50-Å thick membrane. More recent work, however, suggests that this may not be the correct interpretation (L. Keszthelyi, personal communication).

In summary, at present most experiments are compatible with a maximal stoichiometry of 2.0. Experimental results concerning the postulated participation of Bohr protons are ambiguous. The stoichiometry may be lower at higher pH values, higher light intensity, and lower ionic strength.

Very closely related to the stoichiometry problem are the conflicting interpretations given for the biphasic rise and decay of light-induced pH changes in suspensions of cell envelopes or lipid vesicles containing bR. The even more complex kinetics seen in cell suspensions and some envelope preparations may be due in part to the presence of another light-driven ion pump, halorhodopsin, and gated ion channels (see below). In cell envelope and vesicle preparations, Caplan's group has attributed the fast phase of the pH rise to Bohr protons and only the slow phase to H^+ translocation. We attribute both processes to proton transport and assume that the observed components are due to the rate-limiting movement of other, charge-compensating ions across the membrane. No significant new data from either group have been published since Eisenbach & Caplan's (5) and our (4) reviews. However, a theoretical analysis, tested on bR-containing lipid vesicles under a variety of conditions, satisfactorily accounts for the fast phase as due to pumped protons and exclude a contribution from Bohr protons (114–116).

Monomeric Bacteriorhodopsin

Preparations of bR monomers have proved valuable in studies of protein-protein and lipid-protein interactions in pm. However, the important question whether or not the aggregation of bR into the crystalline lattice of pm patches is of functional significance remains unanswered. In addition to Triton X-100 solubilization (117) several new techniques have been used to prepare apparently functional monomers of bR. The commonly used test for the monomeric state is absence of the exciton band couple near 570 nm in CD spectra. It has the advantage of being simple and non-destructive and should theoretically be able to detect the presence of specific aggregates as small as dimers or trimers. Unfortunately, a rigorous evaluation of its sensitivity faces difficulties, and a quantitative method has not been developed. The earlier rotational diffusion time measurements or freeze-fracture electron micrographs cannot prove or disprove the presence of such small aggregates (61, 118). However, with their most recent rotational diffusion measurements, Heyn et al distinguish between monomers and small aggregates (119, 120) and Cherry & Godfrey (121) using the same technique estimate that not more than 5% of the protein can be aggregated. The monomeric state in the work discussed below is deduced mainly from the CD spectra, which taken together with the recent rotational diffusion measurements constitute good evidence that most, if not all, of the bR is indeed present in the monomeric state (121).

Dencher & Heyn (123) showed that bR can be solubilized with β-octylglucoside. However, at least with some commercially available preparations, solubilization is less complete, and the solubilized protein is less stable in β-octylglucoside than in Triton X-100 (see also 24). Khorana's group has developed a technique that allows rapid exchange of Triton X-100 against other detergents, and apparently preserves the monomeric state and completely removes the endogenous lipids (124, 125). Lipid vesicles containing bR oriented "inside out" are obtained after mixing of the delipidated, solubilized bR with detergent-lipid micelles and dialysis. Even completely denatured bacterioopsin or its two chymotryptic fragments can be recombined and reconstituted with lipid and retinal to yield vesicles capable of light-driven proton translocation (125, 126). Alternatively, Triton X-100- or β-octylglucoside-solubilized pm can be dialyzed with or without added lipid to form vesicles. In lipid vesicles with a high lipid-to-protein ratio, bR exists in the monomeric state above the gel-to-liquid phase transition temperature, T_c, and aggregates into a lattice below it. The lipids used are mainly dimyristoyl-and dipalmitoyl-phosphatidylcholine (DMPC and DPPC) (61, 118, 119). The minimal lipid-to-protein ratio at which complete disaggregation occurs probably depends on the lipid species and is close to 80:1 (mole lipid/mole protein) in DMPC or DPPC. However, these vesicles usually contain undetermined amounts of residual native lipid, and the degree of bR orientation is not known. Khorana's technique presumably yields nearly perfectly oriented lipid vesicles from completely delipidated bR (124). Unfortunately, no data on the aggregation state of bR in these vesicles are available.

Monomeric bR, whether in detergent micelles or lipid bilayers, typically shows a small blue shift of 5–10 nm, a decrease in absorbance, and a reduction in the extent of light adaptation (60, 61, 118, 123). The decreased red shift of λ_{max} upon illumination correlates well with the increased amount of residual 13-*cis* retinal extractable from fully light-adapted monomers. The extent of light adaptation was found to depend on the wavelength of actinic light. This is consistent with the assumption that light adaptation is caused by the decay of an intermediate in the 13-*cis* bR photocycle to an all-*trans* bR cycle intermediate, and that in monomeric bR (but not in intact pm) the back reaction occurs at a significant rate. Experiments with lipid vesicles above and below T_c indicated that this effect is due to protein-protein interaction rather than to a change in the physical state of the lipid environment (61, 120).

No careful study of the bR photoreaction cycle in the monomeric state has been published so far. Only a few observations, mainly concerning the rise and decay of the M intermediate, can be found in the literature (120, 123). It is not clear to what extent the monomeric state as opposed to the lipid or detergent environment is responsible for the relatively small differences found. However, the M decay in DMPC or DPPC vesicles is about six times slower than in pm, independent of the aggregation state of bR [(120); W. Stoeckenius and S. B. Hwang, unpublished observations]. In the case of lipid vesicles the problem is further complicated by the observation that membrane potential and proton gradients affect the photocycle kinetics [(69, 127); Dancshazy, Helgerson & W. Stoeckenius, in preparation). Nevertheless, it has been shown as conclusively as one can hope for, that monomeric bR is capable of proton translocation (128, 128a). Whether or not its efficiency is comparable to that of the aggregated form as claimed remains to be seen. One would expect a lower efficiency on the basis of the reduced extent of light adaptation, because only the *trans* retinal-containing bR translocates protons (58). Even if the relative efficiency in model systems above and below T_c is the same, this is of questionable significance unless it can also be shown that the absolute efficiency is comparable to that of pm. For most lipid vesicle systems it seems to be much lower, which may, of course, be due to misorientation of bR in these vesicles. Only Govindjee et al (97) report a quantum efficiency for H^+ translocation and reaction cycle kinetics essentially the same as found in pm, but the state of bR in their lipid vesicles is not known. Nevertheless, while the evidence for equal efficiency in proton translocation by monomers in their natural environment cannot be inferred from the lipid vesicle results available, there is no strong evidence against it.

A conclusive answer to this question may be required to understand the physiological role of lattice structure in pm. When retinal synthesis is partially blocked in growing cells by nicotine, monomeric bR and bacterioopsin are present in a brown membrane fraction, which is the precursor of pm (129). Addition of retinal leads to bR lattice formation. In contrast to an earlier conclusion (130) energy is apparently not necessary for this process (26, 131, 132), but part of the bR remains in nonaggregated form (129, 133). We are not aware of any study to determine the amount of monomeric bR that may be present in the cell membrane of *H. halobium* under typical laboratory growth conditions. In cells from natural brines we have found the total amount of bR to be very small and apparently present mainly, if not exclusively, in the form of monomers [(134); W. Stoeckenius and R. Bogomolni, unpublished observations]. Since the optimal amount of bR for a cell under a given set of environmental conditions is not known, a physiological role of monomeric bR cannot be excluded. At our present level of knowledge one could even argue that aggregation of bR into pm is largely a laboratory artifact or represents a storage form of bR. However, recent observations on a halobacteria bloom in the Dead Sea suggest that pm may also occur in natural environments (134a).

Phototaxis

The changes in swimming direction of bacteria induced by chemical or light stimuli are commonly called chemotaxis and phototaxis. We follow this custom, rather than use the more elaborate terminology developed by sensory physiologists. While *H. halobium* shows chemotactic responses (135,

136) they have not yet been investigated very thoroughly and are beyond the scope of this review. They have, however, already proved to be very useful in the selection of high motility strains and mutants with impaired light-sensing.

The original description of phototaxis in *Halobacterium halobium* i.e. that increases in blue and decreases in green light intensity cause reversal of the swimming direction [for recent reviews see (137, 138)], identifies only part of the response. Actively growing cells show a spontaneous reversal frequency, which is modulated by changes in light intensity. Increases in the intensity of blue light increase, and decreases in this intensity decrease, the reversal frequency. The inverse is true for green light. The effects of simultaneous changes of blue and green light intensity add or subtract depending on their direction. After a step change in light intensity the cells adapt and gradually return to the original reversal frequency (135). The observation that both photosystems PS_{370} and PS_{565}, named for their apparent action spectra maxima, respond to both increases and decreases in light intensity has interesting consequences. Because the cells integrate the signals from the repellent system PS_{370} and the attractant system PS_{565}, the observed action spectra do not correspond to the absorption spectra of the photosensory pigments but to their difference spectrum, weighted by a still unknown factor. If the absorption spectra overlap substantially, as seems likely, the receptor pigment absorption spectra cannot be deduced from the difference spectrum (135).

The problem of identifying PS_{370} is further complicated by the observation that the cell membrane carotenoids can function as accessory pigments for PS_{370} (139). Identification of the receptor molecules will probably require mutants that have lost either one or the other of the sensory pigments. If light barriers are used for selection, the newly recognized sensitivity of both systems to decreases and increases in light intensity must be taken into account. Both systems require retinal (135, 139, 140). That retinal actually forms the chromophore of the receptors has been shown by substituting 3,4 dehydroretinal (retinal$_2$). This shifts the action spectra maxima 15–20 nm toward longer wavelength as expected from the red-shift of retinal$_2$ compared to retinal (141).

It has generally been assumed that the photoreceptor in PS_{565} is bR. Two recent observations show this to be an oversimplification at best. The *H. halobium* mutants, ET-15 and ET-15S, lack bR (142), but still have the full complement of retinal-dependent light responses, with an action spectrum peak near 590 nm, which closely fits the action spectrum for Na^+ pump activity. The bR-containing parent strain ET-1001 shows a broader action spectrum maximum extending from 560 to 590 nm (143). The maximum near 590 nm points to P_{588}, which is present in all strains as the attractant photoreceptor. Given the problems still associated with the apparent action spectra, no definitive conclusion seems possible at present. However, bR is clearly not required for either the photoattractant or photorepellent response. When present, it usually exists in much higher concentration than P_{588} and could function either as an antenna pigment for P_{588} or be an independent, much less efficient photoreceptor. This observation also casts some doubt on the proposed simple hypotheses that a change in membrane potential triggers both the attractant and repellent response (144) or that proton motive force governs the attractant response (145). If that were the case, bR should be much more effective as a photoreceptor than P_{588}, since it is present in much higher concentration, or one would have to postulate an unreasonably high stoichiometry for the sodium pump.

Light-dependent protein methylation changes very similar to the methylation changes that govern chemosensory adaptation have also been observed in *H. halobium* (146, 147). It is not known whether or not the also observed light- and retinal-dependent protein phosphorylation changes (148) are involved in the sensory transduction mechanism. They appear to be regulated by $\Delta \mu_{H^+}$ (148a).

Genetics

A beginning has been made in the study of *Halobacterium* genetics and the results have turned out to be of general interest. Contrary to earlier conclusions, the large satellite bands observed in density gradients of whole cell DNA have been shown to consist mainly of multiple copies of large plasmids (149, 150). In *H. halobium* 15% of the total DNA with an A + T content of 42%, vs 32% for the main band, consists mainly of identical circular plasmid DNA molecules (pHHI) of 100×10^6 mol wt. A similar plasmid with extensive homologies is present in *H. cutirubrum*. The apparently related species *H. salinarium, H. capanicum,* and *H. tunisiensis,* all heavily pigmented rods that also produce purple membrane, do not contain plasmids. However, they carry sequences homologous to pHHI sequences in their chromosome (151).

It has long been known that certain easily observable characteristics in rod-shaped extreme halophiles, such as pigment and gas vacuole formation, are lost spontaneously with frequencies as high as 10^{-2} to 10^{-4}. Reversions also occur, but often with somewhat lower frequencies. These mutations are easily detected by inspection of colonies on agar plates. The loss of pm production is not easily detected. It can, however, be observed with reasonable sensitivity when a bacterioruberin-negative mutant is used and the change in colony color is not masked by the main carotenoid pigment. Loss of pm occurs spontaneously with a frequency of $\sim 10^{-4}$. The mutants can be separated on retinal-containing growth medium, where retinal-negative, opsin-positive cells will form bacteriorhodopsin (151, 152).

To analyze the mechanism causing this high variability, gas vacuole formation has mostly been used but the high variability of other traits are, presumably, caused by the same mechanism. Contrary to expectations, the high frequency loss of gas vacuoles is not due to loss of the plasmid. Instead, multiple insertions into plasmid DNA have been found in these mutants and excisions in their revertants. Moreover, the first insertion appears to facilitate further insertions, so that hypervariable strains are generated. The loss of opsin synthesis correlates with insertions on several locations in the plasmid, but retinal synthesis does not seem to be affected by changes in the plasmid (152). It seems that in halobacteria, insertion mechanisms into plasmid and possibly also into chromosome DNA are used to generate an unusually high genetic variability. How it benefits these organisms in their unique natural environment remains to be explored.

P_{588} or Halorhodopsin

A new retinal-containing pigment resembling bR, which appears to mediate electrogenic, light-driven sodium extrusion, has recently been identified. It appears to be present in all strains of H. halobium so far investigated and has been referred to in the literature as NaP (153), P_{588} (154), or halo-rhodopsin (155). Two brief reviews have recently appeared (156, 156a).

Illumination of cells rich in bR causes one or more transient alkaliniza-tions of the medium, finally followed by a sustained acidification. The transient proton inflows have been attributed to gated coupling of the primary proton electrochemical gradient to energy-requiring cellular pro-cesses, ATP synthesis (157) and/or Na^+ ejection (158–161). More recently, a new, triphenyltin-sensitive proton channel has been described (162) that accounts for part of the initial proton inflow. However, not all of the proton inflow could be explained by these mechanisms. In envelope vesicle prepara-tions, which normally show only light-induced acidification, Kanner & Racker (163) observed that low concentrations of uncouplers cause a sus-tained alkalinization. A red mutant strain of R_1, R_1 mr, containing reduced amounts of bR showed only light-induced alkalinization even after treat-ment with hydroxylamine, which destroys bR (164). In envelope vesicles of this strain the H^+ influx was increased by uncouplers but abolished by valinomycin in the presence of K^+ (165). This suggested that light gener-ated a membrane potential by translocating an ion other than H^+, which then drove the uptake of protons. A number of additional observations further confirmed this explanation and identified Na^+ as the most likely candidate for the primary species transported. Lindley & MacDonald also demonstrated an increased Na^+ efflux under illumination and concluded that a light-driven Na^+ pump exists in these cells (153, 165).

Lanyi and his collaborators confirmed and further extended these findings and showed, in agreement with the model proposed, that in the presence of a proton ionophore the pH gradient balances the inside negative electric potential. Action spectra indicated that in R_1mR, bR significantly contributed to generation of the electric membrane potential but that the uncoupler-enhanced contribution was mediated by a pigment absorbing maximally around 585 nm (166); the same is true for a slowly sedimenting fraction of R_1 envelope vesicles (167).

In nicotine-inhibited R_1mR cells, the light responses were reestablished by addition of all-*trans* retinal and the action spectra showed maxima near 590 nm. Matsuno-Yagi & Mukohata, therefore, concluded that the pigment that mediates the light-induced alkalinization is a retinyl complex with an absorption maximum between 580 and 600 nm. R_1mR also shows an increase in ATP levels under illumination (168). However, the interpretation that the new pigment mediates ATP synthesis is inconclusive, because the strain still contains significant amounts of bR, as shown by Greene & Lanyi (166).

Independently, Weber and collaborators isolated bR-deficient mutant strains from *H. halobium* NRL that showed light-induced alkalinization enhanced by the addition of uncouplers. A white mutant strain, ET-15, which contains less than three molecules of bR per cell (142), shows an action spectrum for the uncoupler-enhanced light-induced alkalinization with a maximum near 585 nm. This strongly suggests that the mutant contains a pigment similar or identical to the light-driven Na^+ pump that causes the light-dependent uptake of protons in R_1mR and R_1.

Direct evidence for the existence of a pigment associated with light-driven passive proton uptake was obtained by using difference spectroscopy after addition of retinal to hydroxylamine-bleached preparation of ET-15 envelope vesicles (169). The regenerated pigment absorbs maximally at 588 nm, does not exhibit the light adaptation typical of bR, and its spectrum shifts ~ 40 nm toward the blue around pH 9.6 in the dark. The pigment-mediated ion translocations are abolished by heating to 75°C for five minutes and, under identical conditions, the pigment is much less sensitive to hydroxylamine attack than bR (156, 168, 169). Work with mutants has provided evidence that bR and P_{588} are probably coded by different genes, since they can be separated by independent mutational events (142, 170).

The content of P_{588} in ET-15 is approximately 10^4 molecules/cell. Flash spectroscopy of ET-15 envelope vesicles reveals transient absorbance changes different from those of bR. Moreover, in distinction to bR, P_{588} shows a reversible 23 nm blue shift when exposed to low ionic strength. The stable forms at high (P_{588}^s) and low (P_{565}^w) sodium chloride concentration undergo different photoreaction cycles (154, 171) (shown in Figure 4). The effect of salt on the spectrum of P_{588} and its photochemical activity at low ionic strength have been confirmed recently, and it was also shown that the retinal isomer extracted from unirradiated halorhodopsin is predominantly all-*trans,* while the 13-*cis* content increases on formation of a photointermediate at ~ 400 nm (176). R. A. Bogomolni and H. J. Weber (unpublished observations) have identified a new intermediate state, P_{680}^s, in the photocycle of P_{588}^s. This form, which seems analogous to the O intermediate of the bR_{570} photocycle, is tentatively positioned between P_{380}^s and P_{588}^s in the scheme. When the pigment is labeled with radioactive retinal and the Schiff base is reduced, SDS-polyacrylamide gel electrophoresis shows the radioactivity associated with a protein band which runs slightly faster than bR (J. K. Lanyi and D. Oesterhelt, personal communication).

The biological significance of this pigment is not well understood. It has been shown that P_{588} may function as a phototactic receptor (143) in motile strains derived from ET-15 cells (see section on phototaxis,) but this is not necessarily its only function. A significant role in maintaining cell's Na^+

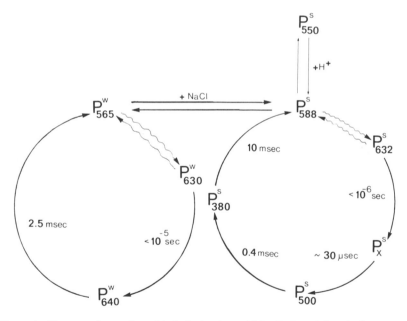

Figure 4 Photoreaction cycles of halorhodopsin at high (P_{588}) and low ionic strength (P_{565}^w) are shown according to (142). Wavy arrows indicate photoreactions, solid arrows thermal reactions. Thermal intermediates are labeled with subscripts indicating maximum wavelengths in the difference spectrum ($P_t - P_{588}$) and are arranged in tentative unbranched sequences. Transition half-times shown are at $\sim20°$C and pH ~8. Intermediate P_x^s has been proposed to account for a gap in the flash kinetic data between the unresolved depletion signal at wavelengths longer than 580 nm and the 30 μsec absorbance increase at 500 nm.

and K^+ gradients is unlikely. Halobacteria have developed an efficient Na^+ ejection mechanism via an electrogenic Na^+/H^+ antiporter (172). This mechanism accounts for most of the light-driven sodium extrusion in cell envelope vesicles that contain both P_{588} and bR. The same Na^+/H^+ antiporter appears to mediate efficient Na^+ ejection in ET-15 vesicles if a proton motive force (PMF) is generated by respiration (173). The electric potential generated by P_{588} should increase PMF and generate ATP. This mechanism, however, cannot be sustained by the cell without detrimental acidification of the cytoplasm, unless a mechanism for proton extrusion exists (142). Under strictly anaerobic conditions, the light-induced increase in ATP level is transient. However, if the electroneutral OH^-/Cl^- antiporter triphenyltin is added, ATP synthesis is sustained (174). [It is not clear why in an earlier publication the same group found that ATP synthesis under similar conditions is sustained longer without additions and that triphenyltin inhibits membrane potential generation (168, 175).]

However, a role for P_{588} in the bioenergetics of the cell cannot be excluded. It has an optimum for generating a membrane potential at pH 8.0 (166). At this pH bR becomes very inefficient in active H^+ extrusion, both through slowdown of its photoreaction cycle and decrease in the stoichiometry (see section on the stoichiometry of proton pumping), and PMF is provided entirely by the electric potential. Further investigations on the functioning of the different transport systems under these conditions should be of interest. However, regardless of its physiological function, P_{588} will probably play a fundamental role in our understanding of active transport. It is very similar to bR and certainly also has the significant advantage that its activity is controlled by light and can be followed by kinetic spectroscopy; yet it does not share bR's one potential disadvantage, that active proton transport may prove to be a special case and the principles involved may not be applicable to other ion transport mechanisms.

ACKNOWLEDGMENTS

Work from the authors' laboratory reported in this review was supported by NIH grants GM 27057 (R. A. Bogomolni and W. Stoeckenius) and GM 28767 (R. A. Bogomolni), and by NASA grant NSG 7151. We are grateful to numerous colleagues who sent us manuscripts prior to publication and/ or read this review in early versions and gave us their comments. Special thanks go to Dr. R. H. Lozier, who also gave generously of his time in compiling the computerized reference list and helping to edit the manuscript.

Literature Cited

1. Garfield, E. 1980. *Curr. Contents* 23(40):5–12
2. Lanyi, J. K. 1978. *Microbiol. Rev.* 4:682–706
3. Bayley, S. T., Morton, R. A. 1978. *CRC Crit. Rev. Microbiol.* 6:151–205
4. Stoeckenius, W., Lozier, R., Bogomolni, R. 1979. *Biochim. Biophys. Acta* 505:215–78
5. Eisenbach, M., Caplan, S. R. 1979. *Curr. Top. Membr. Transp.* 12:165–250
6. Ottolenghi, M. 1980. *Adv. Photochem.* 12:97–200
7. Stoeckenius, W. 1980. *Acc. Chem. Res.* 13:337–44
8. Ovchinnikov, Y. A., Abdulaev, N. G., Feigina, M. Y., Kiselev, A. V., Lobanov, N. A. 1979. *FEBS Lett.* 100: 219–24
9. Khorana, H. G., Gerber, G. E., Herlihy, W. C., Gray, C. P., Anderegg, R. J., Nihei, K., Biemann, K. 1979. *Proc. Natl. Acad. Sci. USA* 76:5046–50
10. Chang, S. H., Majumdar, A., Dunn, R., Makabe, O., RajBhandary, U. L., Khorana, H. G., Ohtsuka, I., Tanaka, T., Taniyama, Y. O., Ikehara, M. 1981. *Proc. Natl. Acad. Sci. USA* 78:3398–402
11. Dellweg, H. G., Sumper, M. 1980. *FEBS Lett.* 116:303–6
11a. Dunn, R., McCoy, J., Simsek, M., Majumdar, A., Chang, S. H., RajBhandary, U. L., Khorana, H. G. 1981. *Proc. Natl. Acad. Sci. USA* 78:6744–48
12. Bridgen, J., Walker, I. D. 1976. *Biochemistry* 15:792–98
13. Gerber, G. E., Gray, C. P., Wildenauer, D., Khorana, H. G. 1977. *Proc. Natl. Acad. Sci. USA* 74:5426–30
14. Packer, L., Tristram, S., Herz, J. M., Russell, C., Borders, C. L. 1979. *FEBS Lett.* 108:243–48
15. Engelman, D. M., Henderson, R., McLachlen, A. D., Wallace, B. A. 1980. *Proc. Natl. Acad. Sci. USA* 77:2023–27
16. Wallace, B. A., Henderson, R. 1980. In *Electron Microscopy at Molecular Dimensions* ed. W. Baumeister, W. Vogell, pp. 57–60. Berlin: Springer
17. Agard, D., Stroud, R. M. 1982. *Biophys. J.* In press
18. Katre, N., Wolber, P., Stroud, R. M., Stoeckenius, W. 1981. *Proc. Natl. Acad. Sci. USA* 78:4068–72
19. Walker, J. E., Carne, A. F., Schmitt, H. W. 1979. *Nature* 278:653–54
20. Dumont, M. E., Wiggins, J. W., Hayward, S. B. 1981. *Proc. Natl. Acad. Sci. USA* 78:2947–51
21. Lemke, H. D., Oesterhelt, D. 1981. *Eur. J. Biochem.* 115:595–604

22. Harris, G., Renthal, R., Tuley, J., Robinson, N. 1979. *Biochim. Biophys. Res. Commun.* 91:926–31
23. Renthal, R. 1981. *Biophys. J.* 33:173a
24. Michel, H., Oesterhelt, D., Henderson, R. 1980. *Proc. Natl. Acad. Sci. USA* 77:338–42
25. Hayward, S., Stroud, R. 1981. *J. Mol. Biol.* 151:491–517
26. Papadopoulos, G. K., Cassim, J. Y. 1981. *Photochem. Photobiol.* 33:455–66
27. Hayward, S. B., Grano, D. A., Glaeser, R. M., Fisher, K. A. 1978. *Proc. Natl. Acad. Sci. USA* 75:4320–24
28. Henderson, R., Jubb, J. S., Whytock, S. 1978. *J. Mol. Biol.* 123:259–74
29. Lemke, H. D., Oesterhelt, D. 1981. *FEBS Lett.* 28:255–60
30. Bayley, H., Huang, K. S., Radhakrishnan, R., Ross, A. H., Takagaki, Y., Khorana, H. G. 1981. *Proc. Natl. Acad. Sci. USA* 78:2225–29
30a. Mullen, E., Johnson, A. H., Akhatar, M. 1981. *FEBS Lett.* 130:187–93
31. Ovchinnikov, Yu. A., Abdulaev, N. G., Tsetlin, V. I., Zakis, V. I. 1980. *Bioorg. Chem. USSR* 6:1427–29
31a. Ovchinikov, Yu. A., 1981. In *Chemiosmotic Proton Circuits in Biological Membranes*, ed. V. P. Skulachev, P. C. Hinkle, pp. 311–20. Reading, Mass: Addison-Wesley
32. King, G. I., Stoeckenius, W., Crespi, H., Schoenborn, B. 1979. *J. Mol. Biol.* 130:395–404
33. King, G. I., Mowery, P. C., Stoeckenius, W., Crespi, H. L., Schoenborn, B. P. 1980. *Proc. Natl. Acad. Sci. USA* 77:4762–30
34. Engelman, D. M., Zaccai, G. 1980. *Proc. Natl. Acad. Sci. USA* 77:5894–98
35. Zaccai, G., Gilmore, D. J. 1979. *J. Mol. Biol.* 132:181–91
36. Rogan, P. K., Zaccai, G. 1981. *J. Mol. Biol.* 145:281–84
37. Sigrist, H., Zahler, P. 1980. *FEBS Lett.* 113:307–11
38. Kouyama, T., Kimura, Y., Kinosita, K., Ikegami, A. 1982. *J. Mol. Biol.* 153:337–59
39. Glaeser, R. M., Jap, B. K., Maestre, M. F., Hayward, S. B. 1981. *Biophys. J.* 33:218a
40. Rothschild, K. J., Clark, N. A. 1979. *Science* 204:311–12
41. Rothschild, K. J., Clark, N. A. 1979. *Biophys. J.* 25:473–88
42. Schreckenbach, T., Walckhoff, B., Oesterhelt, D. 1978. *Biochemistry* 17: 5353–59

43. Schreckenbach, T., Walckhoff, B., Oesterhelt, D. 1978. *Photochem. Photobiol.* 28:205–11
44. Kalisky, O., Feitelson, J., Ottolenghi, M. 1981. *Biochemistry* 20:205–9
45. Fukomoto, J. M., Hopewell, W. D., Karvaly, B., El-Sayed, M. A. 1981. *Proc. Natl. Acad. Sci. USA* 78:252–55
46. Nakanishi, K., Balogh-Nair, V., Arnoldi, M., Tsujimoto, K., Honig, B. 1980. *J. Am. Chem. Soc.* 102:7945–47
47. Michel, H., Oesterhelt, D. 1980. *Proc. Natl. Acad. Sci. USA* 77:1283–85
48. Henderson, R. 1980. *J. Mol. Biol.* 139:99–109
49. Lozier, R. H., Bogomolni, R. A., Stoeckenius, W. 1975. *Biophys. J.* 15:955–62
50. Dinur, U., Honig, B., Ottolenghi, M. 1981. *Photochem. Photobiol.* 33:523–27
51. Marcus, M. A., Lewis, A. 1978. *Biochemistry* 17:4722–35
52. Lewis, A. 1979. *Phil. Trans. R. Soc. London Ser. A* 292:315–27
53. Iwasa, T., Tokunaga, F., Yoshizawa, T. 1980. *Biophys. Struct. Mech.* 6:253–70
54. Kalisky, O., Ottolenghi, M., Honig, B., Korenstein, R. 1981. *Biochemistry* 20:649–55
55. Gillbro, T. 1978. *Biochim. Biophys. Acta* 504:175–86
56. Slifkin, M. A., Caplan, S. R. 1975. *Nature* 253:56–58
57. Hess, B., Kuschmitz, D. 1977. *FEBS Lett.* 74:20–24
58. Lozier, R. H., Niederberger, W., Ottolenghi, M., Sivorinovsky, G., Stoeckenius, W. 1978. In *Energetics and Structure of Halophilic Microorganisms* ed. S. R. Caplan, M. Ginzburg, pp. 123–41. New York: Elsevier/North-Holland Biomed. Press
59. Ohno, K., Takeuchi, Y., Yoshida, M. 1981. *Photochem. Photobiol.* 33:573–78
60. Casadio, R., Gutowitz, H., Mowery, P., Taylor, M., Stoeckenius, W. 1979. *Biochim. Biophys. Acta* 590:13–23
61. Casadio, R., Stoeckenius, W. 1980. *Biochemistry* 19:3374–81
62. Tsuda, M., Glaccum, M., Nelson, B., Ebrey, T. G. 1980. *Nature* 287:351–53
63. Groenendijk, F. W. T., DeGrip, W. J., Daemen, F. J. M. 1980. *Biochim. Biophys. Acta* 617:430–38
64. Sperling, W., Carl, P., Rafferty, C. N., Kohl, K. D., Dencher, N. A. 1979. *FEBS Lett.* 97:129–32
65. Stockburger, M., Klusmann, W., Gatterman, H., Massig, G., Peters, R. 1979. *Biochemistry* 18:4886–900
66. Nagle, J. F., Parodi, L. A., Lozier, R. H. 1982. *Biophys. J.* In press

66a. Pettei, M. J., Yudd, A. P., Nakanishi, K., Henselman, R., Stoeckenius, W. 1977. *Biochem.* 16:1955–59
66b. Tsuda, M., Glaccum, M., Nelson, B., Ebrey, T. G. 1980. *Nature* 287:351–53
66c. Mowery, P. C., Stoeckenius, W. 1981. *Biochemistry* 20:2302–6
66d. Braiman, M., Mathies, R. 1980. *Biochemistry* 19:6421–28
66e. Kuschmitz, D., Hess, B. 1982. *FEBS Lett.* In press
66f. Braiman, M., Mathies, R. 1982. *Proc. Natl. Acad. Sci. USA* 79:403–7
66g. Hsieh, C. L., Nagumo, M., Nicol, M., El Sayed, M. A. 1981. *J. Phys. Chem.* 85:2714–17
67. Govindjee, R., Ohno, K., Ebrey, T. G. 1981. *Am. Soc. Photobiol., 9th Ann. Meet. Abstr.* MPM-Cl, p. 83
68. Hellingwerf, K. J., Schuurmans, J. J., Westerhoff, H. V. 1978. *FEBS Lett.* 92:181–86
69. Quintanilha, A. T. 1980. *FEBS Lett* 117:8–12
70. Westerhoff, H. V., Hellingwerf, K. J., Arents, J. C., Scholte, B. J., van Dam, K. 1981. *Proc. Natl. Acad. Sci. USA* 78:3554–58
71. Hurley, J. B., Becher, B., Ebrey, T. G. 1978. *Nature* 272:87–88
72. Kalisky, O., Lachish, U., Ottolenghi, M. 1978. *Photochem. Photobiol.* 28:261–63
73. Braiman, M., Mathies, R. 1980. *Biochemistry* 19:5421–28
74. Braiman, M., Mathies, R. 1982. *Proc. Natl. Acad. Sci. USA* 79:403–7
75. Bogomolni, R. A., Stubbs, L., Lanyi, J. K. 1978. *Biochemistry* 17:1037–41
76. Bogomolni, R. A., Renthal, R., Lanyi, J. K. 1978. *Biophys. J.* 21:183a
77. Hess, B., Kuschmitz, D. 1979. *FEBS Lett.* 100:334–40
78. Bogomolni, R. A. 1980. In *Bioelectrochemistry*, ed. H. Keyzer, F. Gutmann, pp. 83–95. New York: Plenum
79. Shinar, R., Druckmann, S., Ottolenghi, M., Korenstein, R. 1977. *Biophys. J.* 19:1–5
80. Druckmann, S., Ottolenghi, M. 1981. *Biophys. J.* 33:263–68
81. Keszthelyi, L. 1980. *Biochim. Biophys. Acta* 598:429–36
82. Kimura, Y., Ikegami, A., Ohno, K., Saigo, S., Takeuchi, Y. 1981. *Photochem. Photobiol.* 33:435–39
83. Keszthelyi, L., Ormos, P. 1980. *FEBS Lett.* 109:189–93
84. Drachev, L. A., Kaulen, A. D., Skulachev, V. P. 1978. *FEBS Lett.* 87:161–67
84a. Skulachev, V. P. 1978. In *Membrane*

Proteins ed. P. Nicholls et al, pp. 49–59. Oxford: Pergamon
85. Korenbrot, J. I., Hwang, S. B. 1980. *J. Gen. Physiol.* 76:649–82
86. Fahr, A., Lauger, P., Bamberg, E. 1981. *J. Membr. Biol.* 60:51–62
87. Bamberg, E., Fahr, A. 1980. *Ann. NY Acad. Sci.* 358:324–27
88. Trissl, H. W., Montal, M. 1977. *Nature* 266:655–57
89. Hong, F. T., Montal, M. 1979. *Biophys. J.* 25:465–72
90. Dancshazy, Z., Drachev, L. A., Ormos, P., Nagy, K., Skulachev, V. P. 1978. *FEBS Lett.* 96:59–63
91. Ormos, P., Dancshazy, Z., Keszthelyi, L. 1980. *Biophys. J.* 31:207–14
92. Nagle, J. F., Morowitz, H. J. 1978. *Proc. Natl. Acad. Sci. USA* 75:298–302
93. Nagle, J. F., Mille, M. 1981. *J. Chem. Phys.* 74:1367–72
94. Merz, H., Zundel, G. 1981. *Biochem. Biophys. Res. Commun.* 101:540–46
95. Stoeckenius, W., Lozier, R., Bogomolni, R. 1981. See Ref. 31a, pp. 283–309
96. Bogomolni, R. A., Baker, R. A., Lozier, R. H., Stoeckenius, W. 1980. *Biochemistry* 19:2152–59
97. Govindjee, R., Ebrey, T. G., Crofts, A. R. 1980. *Biophys. J.* 30:231–42
98. Renard, M., Delmelle, M. 1980. *Biophys. J.* 32:993–1006
99. Hartmann, R., Sickinger, H. D., Oesterhelt, D. 1977. *FEBS Lett* 82:1–6
100. Renard, M., Delmelle, M. 1981. *FEBS Lett.* 128:245–48
101. Lozier, R. H., Niederberger, W., Bogomolni, R. A., Hwang, S. B., Stoeckenius, W. 1976. *Biochim. Biophys. Acta* 440:545–56
102. Ort, D. R., Parson, W. W. 1979. *Biophys. J.* 25:341–53
103. Klemperer, G., Eisenbach, M., Garty, H., Caplan, S. R. 1978. See Ref. 58, pp. 291–96
104. Caplan, S. R., Eisenbach, M., Garty, H. 1978. See Ref. 58, pp. 49–66
105. Hess, B., Korenstein, R., Kuschmitz, D. 1978. In *Transport by Proteins,* ed. G. Blauer, H. Sund, pp. 187–98, Berlin de Gruyter, New York:
106. Kuschmitz, D., Hess, B. 1981. *Biochemistry* 20:5950–65
107. Tsuji, K., Neumann, E. 1981. *Int. J. Biol. Macromol.* 3:231–42
108. Tsuji, K., Neumann, E. 1981. *FEBS Lett.* 128:265–68
109. Kell, D. B. 1978. *Biochim. Biophys. Acta* 549:55–99
110. Garty, H., Klemperer, G., Eisenbach,

M., Caplan, S. R. 1977. *FEBS Lett.* 81:238–42
111. Takeuchi, Y., Ohno, K., Yoshida, M., Nagano, K. 1981. *Photochem. Photobiol.* 33:587–92
112. Bakker, E. P., Caplan, S. R. 1978. *Biochim. Biophys. Acta* 503:362–79
113. Avi-Dor, Y., Rott, R., Schnaiderman, R. 1979. *Biochim. Biophys. Acta* 545:15–23
114. Arents, J. C., van Dekken, H., Hellingwerf, K. J., Westerhoff, H. V. 1981. *Biochemistry* 20:5114–23
115. Westerhoff, H. V., Arents, J. C., Hellingwerf, K. J. 1981. *Biochim. Biophys. Acta* 637:69–79
116. Arents, J. C., Hellingwerf, K. J., van Dam, K., Westerhoff, H. V. 1981. *J. Membr. Biol.* 60:95–104
117. Reynolds, J. A., Stoeckenius, W. 1977. *Proc. Natl. Acad. Sci. USA* 74:2803–4
118. Cherry, R. J., Muller, U., Henderson, R., Heyn, M. P. 1978. *J. Mol. Biol.* 121:283–98
119. Heyn, M. P., Cherry, R. J., Dencher, N. A. 1981. *Biochemistry* 20:840–48
120. Heyn, M. P., Kohl, K.-D., Dencher, N. A. 1981. In *Proc. Int. Symp. Vectorial Reactions in Electron and Ion Transport in Mitochondria and Bacteria,* ed. F. Palmieri, et al, pp 237–44, Amsterdam: Elsevier-North Holland Biomed Press
121. Cherry, R. J., Godfrey, R. E. 1981. *Biophys. J.* 36:257–76
122. Deleted in proof
123. Dencher, N. A., Heyn, M. P. 1978. *FEBS Lett.* 96:322–26
124. Huang, K. S., Bayley, H., Khorana, H. G. 1980. *Proc. Natl. Acad. Sci. USA* 77:323–27
125. Huang, K. S., Bayley, H., Liao, M. J., London, E., Khorana, H. G. 1981. *J. Biol. Chem.* 256:3802–9
126. Lind, C., Hojeberg, B., Khorana, H. G. 1981. *J. Biol. Chem.* 256:8298–305
127. Hellingwerf, K. J., Arents, J. C., Scholte, B. J., Westerhoff, H. J. 1979. *Biochim. Biophys. Acta* 547:561–82
128. Dencher, N. A., Heyn, M. P. 1979. *FEBS Lett.* 108:307–10
128a. Bamberg, E., Dencher, N. A., Fahr, A., Heyn, M. P. 1981. *Proc. Natl. Acad. Sci. USA* 78:7502–6
129. Hwang, S. B., Tseng, Y. W., Stoeckenius, W. 1981. *Photochem. Photobiol.* 33:419–27
130. Sumper, M., Reitmeier, H., Oesterhelt, D. 1976. *Angew. Chemie Int. Ed. Engl.* 15:187–94
131. Hiraki, K., Hamanaka, T., Mitsui, T., Kito, Y. 1978. *Biochim. Biophys. Acta* 536:318–22

132. Hiraki, K., Hamanaka, T., Mitsui, T., Kito, Y. 1981. *Photochem. Photobiol.* 33:429–33
133. Peters, R., Peters, J. 1978. See Ref. 103, pp. 315–321
134. Stoeckenius, W. 1981. *J. Bacteriol.* 148:352–60
134a. Oren, A., Shilo, M. 1981. *Arch. Microbiol.* 130:185–7
135. Spudich, J. L., Stoeckenius, W. 1979. *Photochem. Photobiophys.* 1:43–53
136. Schimz, A., Hildebrand, E. 1979. *J. Bacteriol.* 140:749–53
137. Hildebrand, E. 1978. In *Taxis and Behaviour* ed. G. Hazelbauer, pp. 37–73. London: Chapman & Hall
138. Lenci, F., Colombetti, F. 1978. *Ann. Rev. Biophys. Bioeng.* 7:341–61
139. Dencher, N. A., Hildebrand, E. 1979. *Z. Naturforsch. Teil C* 34:841–47
140. Dencher, N. A. 1978. See Ref. 58, pp. 67–88
141. Sperling, W., Schimz, A. 1980. *Biophys. Struct. Mech.* 6:165–69
142. Weber, H. J., Bogomolni, R. A. 1981. *Photochem. Photobiol.* 33:601–8
143. Narurkar, V., Spudich, J. L. 1981. *Biophys. J.* 33:218a
144. Wagner, G., Geissler, G., Linhardt, R., Mollwo, A., Vonhof, A. 1980. In *Plant Membrane Transport: Current Conceptual Issues*, pp. 641–44. Amsterdam: Elsevier/North Holland Biomed. Press
145. Baryshev, V. A., Glagolov, A. N., Skulachev, V. P. 1981. *Nature* 292:338–40
146. Spudich, J. L., Stoeckenius, W. 1980. *Fed. Proc.* 39:1972
147. Schimz, A. 1981. *FEBS Lett.* 125:205–7
148. Spudich, J. L., Stoeckenius, W. 1980. *J. Biol. Chem.* 255:5501–3
148a. Spudich, E. N., Spudich, J. L. 1981. *J. Cell. Biol.* 91:895–900
149. Weidinger, G., Klotz, G., Goebel, W. 1979. *Plasmid* 2:377–86
150. Simon, R. D. 1978. *Nature* 273:314–17
151. Pfeifer, F., Weidinger, G., Goebel, W. 1981. *J. Bacteriol.* 145:369–74
152. Pfeifer, F., Weidinger, F., Goebel, W. 1981. *J. Bacteriol.* 145:375–81
153. MacDonald, R. E., Greene, R. V., Clark, R. D., Lindley, E. V. 1979. *J. Biol. Chem.* 254:11831–38
154. Bogomolni, R. A., Weber, H. J. 1980. *Fed. Proc.* 39:1846
155. Mukohata, Y., Matsuno-Yagi, A., Kaji, Y. 1980. In *Saline Environment* ed. H. Morishita, M. Masui, pp. 31–37. Osaka, Japan: Organ. Comm. Jpn. Conf. Halophilic Microorg.
156. Lanyi, J. K. 1980. *J. Supramol. Struct.* 13:83–92
156a. MacDonald, R. E. 1981. See Ref. 31a, pp. 321–35
157. Bogomolni, R. A., Baker, R. A., Lozier, R. H., Stoeckenius, W. 1976. *Biochim. Biophys. Acta* 440:68–88
158. Lanyi, J. K., MacDonald, R. E. 1977. *Fed. Proc.* 36:1824–27
159. Eisenbach, M., Cooper, S., Garty, H., Johnstone, R., Rottenberg, H., Caplan, S. R. 1977. *Biochim. Biophys. Acta* 465:599–613
160. Bogomolni, R. A. 1977. *Fed. Proc.* 36:1833–39
161. Wagner, G., Hartmann, R., Oesterhelt, D. 1978. *Eur. J. Biochem.* 89:169–79
162. Helgerson, S. L., Stoeckenius, W. 1981. *Fed. Proc.* 39:1849
163. Kanner, B. I., Racker, E. 1975. *Biochem. Biophys. Res. Commun.* 64:1054–61
164. Matsuno-Yagi, A., Mukohata, Y. 1977. *Biochem. Biophys. Res. Commun.* 78:237–43
165. Lindley, E. V., MacDonald, R. E. 1979. *Biochem. Biophys. Res. Commun.* 88:491–99
166. Greene, R. V., Lanyi, J. K. 1979. *J. Biol. Chem.* 254:10986–94
167. Greene, R. V., MacDonald, R. E., Perrault, G. J. 1980. *J. Biol. Chem.* 255:3245–47
168. Matsuno-Yagi, A., Mukohata, Y. 1980. *Arch. Biochem. Biophys.* 199:297–303
169. Lanyi, J. K., Weber, H. J. 1980. *J. Biol. Chem.* 255:243–50
170. Weber, H. J., Bogomolni, R. A. 1982. *Methods Enzymol.* In press
171. Bogomolni, R. A., Belliveau, J. W., Weber, H. J. 1981. *Biophys. J.* 33:217a
172. Lanyi, J. K., MacDonald, R. E. 1976. *Biochemistry* 15:4608–14
173. Luisi, B. F., Lanyi, J. K., Weber, H. J. 1981. *FEBS Lett.* 117:354–58
174. Mukohata, Y., Kaji, Y. 1981. *Arch. Biochem. Biophys.* 206:72–76
175. Mukohata, Y., Kaji, Y. 1981. *Arch. Biochem. Biophys.* 208:615–17
176. Ogurusu, T., Maeda, A., Sasaki, N., Yoshizawa, T. 1981. *J. Biochem.* 90:1267–1273
177. Rothschild, K. J., Marrero, H. 1982. *Proc. Natl. Acad. Sci. USA.* In press
178. Rothschild, K. J., Zagarski, M., Canton, B. 1981. *Biochem Biophys. Res. Commun.* 103:483–89
179. Bagley, K., Dollinger, G., Eisenstein, L., Singh, A. K., Zimanyi, L. 1982. *Proc. Natl. Acad. Sci. USA.* In press
180. Sapienza, C., Doolittle, W. F., 1982. *Nature* 295:385
181. Spudich, E. N., Spudich, J. L. 1982. *Proc. Natl. Acad. Sci. USA.* In press

ADDED IN PROOF   Since this paper was submitted some interesting new results have come to our attention.

Fourier transform infrared difference spectra of bR and photocycle intermediates K and M show mostly differences in chromophore vibrations and are in excellent agreement with the corresponding resonance Raman spectra (177–179). They confirm that the Schiff base is protonated in bR and K is unprotonated in M. They are further consistent with models for the proton translocation mechanism, which assume that in the bR to K transition the Schiff base proton moves away from a counterion. They may also indicate protonation of a charge-perturbed carboxyl group in the M state.

C. Sapienza and W. F. Doolittle (180) have recently demonstrated that *H. halobium* contains many families of highly mobile, repeated base sequences, found clustered and dispersed throughout the genome. This complements the results on the unusually high genetic variability in *H. halobium* and related halobacterium species apparently caused by insertions (149–152) and again emphasizes the differences between halobacteria and the eubacteria. Hybridization with *H. Volcanii* DNA shows that at least some of these repeated sequences are more strongly conserved than unique sequence DNA.

The spectral transistion $P^w_{565} \rightarrow P^s_{588}$ is anion specific (H. J. Weber, M. E. Taylor, and R. A. Bogomolni, unpublished observation). Chloride is more effective than bromide and neither NO_3^-, ClO_4^-, acetate, nor $PO_4 H_2^-$ are active.

B. Schobert and J. K. Lanyi reported at the 82nd Annual Meeting of the American Society of Microbiology, Atlanta 1982, that halorhodopsin is an inwardly directed chloride ion pump rather than an outwardly directed sodium ion pump. The previously observed sodium ion transport was apparently caused by the presence of bacteriorhodopsin and the Na^+/H^+ antiporter in the vesicle preparations used.

E. N. and J. L. Spudich (181) have developed a procedure for isolating light energy conversion-deficient mutants. One of these apparently lacks both bR and P_{588}. Its phototaxis, however, is intact, and the presence of a photoactive pigment in its membrane can be demonstrated which cycles much more slowly than bR or P_{588} (R. A. Bogomolni, E. N. Spudich, and J. L. Spudich, unpublished observations).

Ann. Rev. Biochem. 1982.51:617–54

SnRNAs, SnRNPs, AND RNA PROCESSING

Harris Busch, Ramachandra Reddy, Lawrence Rothblum, and Yong C. Choi[1]

Department of Pharmacology, Baylor College of Medicine, Houston, Texas 77030

CONTENTS

[1]We are grateful to all those who sent us preprints of their work, and to our colleagues in this and other laboratories for suggestions and comments on the manuscript. Work in our laboratory has been supported by grants from Cancer Center Program Grant 10893, awarded by DHEW; the Bristol-Myers Fund; the Michael E. DeBakey Medical Foundation; the Pauline Sterne Wolff Memorial Foundation; the Taub Foundation; and The William S. Farish Fund.

0066-4154/82/0701-0617$02.00

PERSPECTIVES AND SUMMARY

On rare occasions, convergences occur in science that lead to a clarification of several seemingly unrelated lines of investigation. Such a series of events has linked the small nuclear RNAs (snRNAs) and small nuclear ribonucleoproteins (snRNPs) (reviewed in 1–10) to splicing of premessenger RNA (pre-mRNA) transcripts (11–18) and to autoimmune diseases (19–23). From these different avenues of research, the hypothesis emerged that one function of the U-snRNP class of snRNP particles is to guide the elimination of intervening sequences or introns (IVS) of pre-mRNA by binding to splice sites of the IVS in heterogeneous nuclear RNP (hnRNP) particles. The mechanism of excision of the IVS, and ligation of the useful portions of the mRNA are not yet known, nor is it clear if the snRNPs are involved in transport of mRNA to the cytoplasm and incorporation into polysomes (12) as their primary function.

Several groups (11–13) working on viral and eukaryotic mRNA have suggested such a role in splicing for snRNP particles (Figure 1). Lerner et al (12) evaluated the hypothesis based upon calculations of numbers of hydrogen bonds between U1 RNA and IVS cleavage sites, common features of the splice junctions, and details of interactions of small and large nuclear particles. The rigorous testing of the hypothesis should provide a satisfactory evaluation of the concept and the biochemical mechanisms involved in splicing. There are serious technical problems in developing assay systems for bona fide processing of hnRNA. Now that sequencing and characterization of the U-snRNAs are essentially complete, the U-snRNPs are being considered as possible macromolecular aggregates with specific cleavage and ligase functions. When these particles are available in native states, it will be essential to devise processing reactions utilizing pre-mRNA to test their potential functions.

SMALL RNAs

Historical Aspects

In early studies (24) on the nucleolar U-snRNA species an RNA fraction was found in the 4-8S region of a sucrose gradient that had different base composition from that of the rRNA or tRNA. Due to its high content of

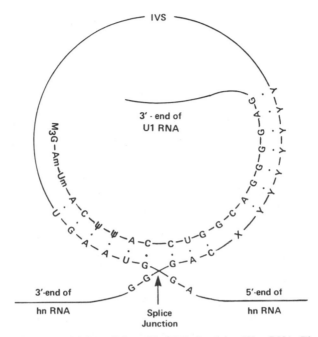

Figure 1 A proposed model for splicing of hnRNA, involving U1 snRNA. The model is derived from others (12, 13). A similar model involving viral precursor RNA and viral snRNA has been proposed (11). The consensus sequences for the splice junctions are from (12, 15) and the sequence of U1 RNA (8, 63, 118) were reported earlier.

uridylic acid it was designated U-RNA (24–29). There was initial concern that such small RNA species could be degradation products of higher-molecular-weight RNA as had been the case for "chromosomal RNA" and other small RNA products (30, 31). To control nuclear RNases is difficult partly because a number of them have specific processing functions; they differ from cytoplasmic contaminants in that they are released during isolation of nuclear subfractions. Some of these problems have been overcome using improved methods (32, 33) and RNase inhibitors.

By 1966, snRNA had been observed (34–36), but specific RNA species had not been characterized although Reich et al (37) had identified a small viral RNA. Subsequently, Pene et al (38) isolated 5.8S RNA, and other snRNAs were identified (39–51).

After studies were made on RNase contents of nuclear preparations, we chose the Novikoff hepatoma as an optimal tissue for study (32). When "RNase free" citric acid nuclei were employed (33), single species of highly purified U1 and U2 snRNAs could be isolated (25). The purification, terminal nucleoside analysis, high uridylic acid content and evidence that U-snRNA's are specifically localized were reported (25–29, 49). U3 RNA was found only in the nucleolus and not in the nucleoplasm (25–29). These results were consistent with those using gel electrophoresis (49, 51). In the absence of chemical evidence, it was not clear whether the gel bands were the same products we had isolated or were degradation products of other RNA species. It was also not apparent how many were isomeric structures like U1A and U1B RNA (4,8).

In other studies, the U3 RNA was purified by a number of techniques. Nine lines of evidence proved that the RNAs were not breakdown products of other RNA species. One line of evidence (49, 24) was that the turnover rates were very low by comparison to other species, particularly rRNA.

Despite the reproducibility of the isolation procedures, the only satisfactory way to show that U-snRNAs are unique species, and not specific cleavage products or artifacts of degradation has been to determine their sequences.

Conservation Through Evolution

Small RNAs are present in viruses and in prokaryotic and eukaryotic cells, but their gel electrophoretic patterns are different. After snRNAs were found in human cells (40, 41, 43–45, 49–54); rat tissues (24–29, 54, 55); Chinese hamster (50); and mouse (41, 42, 56, 57), systematic studies concluded that snRNAs are present in all vertebrates (49–59). snRNAs are independent of cell type and malignancy (49, 61). U1, U2, and U3 RNAs appear to be highly conserved (6, 41, 42, 62, 63). Sequence analysis of U1 RNA structures of chicken, rat, and human tissues show it to be over 95% conserved. There are only two nucleotide differences between rat and human U1 RNAs (63). Notable similarities have been found for the sequence of U1 RNA of *Drosophila* (63a) with other U1 RNA sequences.

Detailed studies (39, 41, 42) on nonvertebrates, e.g. cockroach, meal worm, blowfly, sea urchins, *Tetrahymena,* and *Chironymus tentans* revealed that they all contain small RNAs, but that their mobility on gels is different from vertebrate snRNAs. U1, U2, U3 RNAs of Dictyostelium contain a cap structure similar to that found in rat U-snRNAs (64, 64a).

The U3 RNA sequence of rat (65) is homologous to that of U3 RNA of *Dictyostelium* (66), which indicates the similarity in structure of U-snRNAs of vertebrates and invertebrates. The characteristics of small RNAs are shown in Table 1. Table 2 lists the major small RNAs, their 5'-termini, chain lengths, and their alternate nomenclature.

Most recently, we have found that dinoflagellates, which are considered to be eukaryotes that evolved 3 billion years ago, also contain U1-U6 snRNAs or very similar capped molecular species (R. Reddy, D. Spector, D. Henning, and H. Busch, unpublished results). Diener (66a) has suggested that base-pairing of U1 RNA to intron "ancestors" may be part of a mechanism for viroid formation.

Table 1   General characteristics of U SnRNAs

Size range 90–400 nucleotides

Copies per cell 1×10^6 molecules for the most abundant RNAs (e.g. U1 RNA); 1×10^4 to 5×10^5 for other RNAs

Half-lives U snRNAs are stable; half-lives of up to one cell cycle

Specific localization localized to specific subcellular compartments (e.g. U3 RNA in nucleolus; U1, U2, U4, U5, U6, and La 4.5 RNAs in nucleoplasm; and Y RNAs in cytoplasm)

Polymerases capped snRNAs U1 to U6 synthesized by pol II, and other small RNAs synthesized by pol III

Cap structure U1 to U6 snRNAs are capped, U1 to U5 RNAs contain a trimethyl-guanosine cap, and U6 RNA has a different cap

RNP particles all U snRNAs exist in RNP particles

Associated with precursor RNAs The U snRNAs are associated with precursor hnRNAs and pre-rRNA

Synthesis of Small RNAs

Transfer RNA (tRNA) and 5S RNA are synthesized by RNA polymerase III (67–69); 5.8S RNA is synthesized as part of 45S ribosomal RNA precursor by RNA polymerase I (70–72). RNA polymerase II, which synthesizes hnRNA (73), also synthesizes U-snRNA (74–78). The sensitivity of U1 and U2 RNA biosynthesis to α-amanitin in whole cel systems (73a–76) and in cell-free systems (78), and to 5-6-dichloro-1-β-D-ribofuranosyl benzimidazole (77) also indicates that polymerase II is responsible for U1 and U2 RNA biosynthesis. snRNAs may be synthesized by RNA polymerase I (52), but there is no confirmation of this. 7S RNA, 7-3 RNA (52, 79), La 4.5 RNA (80) and Y RNAs (81) have been reported to be synthesized by polymerase III. Thus it appears that capped small U1–U6 RNAs are synthesized by polymerase II, and other small RNAs are synthesized by polymerase III.

Precursors of snRNAs

There are precursors of U1 and U2 RNA that are slightly larger than the mature U-snRNA's. They are detected in the cytoplasmic fraction within 10 min of [³H]uridine labeling (51, 82, 82a). Precursors of U1 and U2 RNAs have been analyzed by fingerprinting (83). Since these RNAs are capped, the larger size of the precursors is presumed to reflect elongation at 3' ends. Salditt-Georgieff et al (84) found $m_3^{2,2,7}G$ in the cap I structures isolated from the <750-base long nuclear RNA fraction of Chinese hamster ovary cells labeled briefly with [methyl-³H]methionine. These data suggest that longer precursor molecules for snRNAs exist than were previously reported. Tamm et al (77) also found evidence for long precursors for

Table 2 Termini and chain lengths of snRNAs

RNA	5′ terminus	Chain length	Alternative nomenclature
U snRNA			
U1	m$_3$GpppAmUmA	165	D
U2	m$_3$GpppAmUmC	188–189	C
U3	m$_3$GpppAmAG	210–214	A, D2
U4	m$_3$GpppAmGmC	142–146	F
U5	m$_3$GpppAmUmA	116–118	G′, 5S III
U6	XpppGUG	107–108	H1, 4.5 III
Other snRNA			
tRNA	pN	~80	
5S	pppN	121	G
5.8S	pN	158	E
La 4.5	pppG	90–94	
La 4.5 I	pppG	98–99	H2
7S	pppG	294–295	L
VAI, II	pppG	157–160	
7–1	pppN	~260	M
7–2	pppN	~290	M
7–3	pppN	~300	K
8S	pppN	~400	
Y1–Y3	pppN	~100	

snRNAs but did not find trimethylguanosine in hnRNAs longer than 1000 nucleotides. U1 and U2 RNAs have been reported to be derived from transcription units that may be as long as 5 kb (85). Precursors to U4, U5, or U6 RNAs have not been reported. As yet none of the precursors has been sequenced.

Genes Coding for SnRNA

DNA clones prepared for U1 RNA include chicken U1 RNA (78), human U1 RNA (86–89), and rat U1 RNA (90). The clones are prepared in the usual cloning vectors and analyzed for sequence homology to the reported U1 RNA. No IVS were found in any of the U1, U2, or U3 genes studied; in addition, the genes are not clustered as reported earlier, but rather are dispersed throughout the genome (66, 78, 86–90). Genes for the U1 RNA and tRNA are found in proximity in the same Eco R1 fragment; the gene sequence is identical to the structure of human U1 RNA (89). For other "pseudogenes," the sequences do not match precisely with those of the U1 RNA (86–88). These pseudogenes are dispersed throughout the genome in an approximate ratio of 10 : 1 compared to the true genes (86); some of them are flanked by 16–19-nucleotide long direct repeats (87), which are: AGAAACAGGCTTTTGC for U1 pseudogene, TAAATAATCAG-GATGGAA for U2 pseudogene, and TAAAATGCTAATTATCCAA for U3 pseudogene. Genes for U2 RNA have been isolated from human (86) and mouse (R. C. Huang, personal communication). The mouse U2 RNA gene sequence is identical to the mouse U2 RNA sequence (8, 19) and does not contain any IVS.

The first snRNA gene to be isolated and sequenced was the U3 RNA gene from *Dictyostelium* (66). The gene coding for *Dictyostelium* U3 RNA is identical in sequence to *Dictyostelium* RNA. Recently, genes for U6 RNA have been cloned and sequenced along with their flanking regions (89a, 89b). One had a TATAAT 31 nucleotides upstream of the coding region. The flanking regions had heterogeneity immediately outside the genes. There is uncertainty about the presence of a Hogness or TATAAA box in small RNA genes. If there is, it is further upstream from the cap site than the 25 bp of most mRNA genes.

Other features of sequenced eukaryotic genes, such as the putative adenylation signal (AATAAA, 25 bp prior to the 3' end of the RNA) and the termination signal (TTT, at the 3' end of the RNA) are not found for the U1 gene (78). As expected, the adenylation signal is missing, since U1 RNA is not polyadenylated. It is interesting that the termination signal is missing, since the presence of this signal seems to be a general feature of genes transcribed by RNA polymerases II and III (78). The lack of a Hogness box and of the putative termination signal (78) is consistent with the possibility that U1 RNA and other snRNAs are transcribed as part of a larger primary transcript.

Are the genes for the isolated and sequenced snRNAs the same as those transcribed in vivo? This is, at present, difficult to answer for any individual member of a multigene family.

Early studies estimated the number of genes coding for U1, U2, and U3 RNAs by DNA-RNA hybridization (91, 92). In baby hamster cells, Engberg et al (91) have suggested that 2000 genes code for each snRNA. Marzluff et al (92) have analyzed mouse genome sequences for snRNAs and have suggested that there are 100–2000 copies per genome. However, these values are much higher than the approximately 10 determined by cloning studies, and probably include many pseudogenes. The genes for U-snRNAs are unusual in not having a defined TATAAA box, IVS, and the usual termination signals.

Recently, Nojima and Kornberg (unpublished) isolated genes and pseudogenes for mouse U1 and U2 snRNAs. In *Xenopus* oocytes, the U2 RNA gene was transcribed with apparent fidelity but the pseudogenes were inactive.

Subcellular Localization

Several studies have been made (24–29, 49, 51, 76, 82, 93–96) to find the subcellular localization of different small RNAs (Figure 2). The methods used to fractionate the nuclei from cytoplasm include aqueous (21–25, 49, 51) and nonaqueous methods (76, 93, 96), and the citric acid method (95). The results obtained by different investigators for the location of small RNAs vary slightly; however, the data point to the localization of small RNAs as shown in Figure 2. The U- RNPs (U1, U2, U4, U5, and U6) are nucleoplasmic, as shown by indirect immunofluorescence using specific antibodies directed against these RNPs (20). Similar studies show La 4.5 and 4.5I RNPs to be located in the nucleus and Y RNPs to be located in the cytoplasm (81).

Association of snRNAs with Precursor RNAs

Studies made to further locate the snRNAs show that all capped snRNAs are associated with precursor RNAs. U3 RNA is hydrogen bonded to nucleolar 28S and 32S RNAs, and this bonding is stable to treatment with phenol and SDS at 25° (97). U3 was the first snRNA to be shown associated

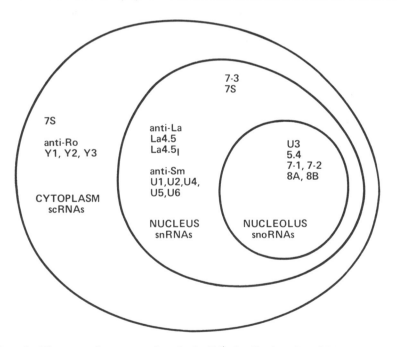

Figure 2 Diagrammatic representation of subcellular localization of small RNAs of rat. The three categories of small RNAs, based on subcellular localization, are small cytoplasmic RNAs (scRNAs), small nuclear RNAs (snRNAs), and small nucleolar RNAs (snoRNAs).

with precursor RNAs. The nucleoplasmic U1 and U2 RNAs are found in hnRNP particles (98) as are other snRNAs (99–112a). U4, U5, and U6 RNAs are also bound to hnRNP particles (111). Other studies show U6 RNA in perichromatin granules (113, 114), 7S RNA bound to mRNA in polysomes (115), 7S and 7-3 RNA in polysomes (116), La 4.5 RNA hydrogen bonded to hnRNA (117), 7-3 RNA in the nuclear matrix (117a), and 7-1 and 8S RNA hydrogen bonded to nucleolar 28S RNA (95).

SEQUENCES OF UsnRNAs

U1 RNA

The initial sequence data of U1 RNA (see Table 1 for alternative nomenclature), which is in the highest concentration in the nucleus (Figure 3) revealed that its 5' cap contained the trimethyl G residue (118, 119). This was the first report of a cap structure on any RNA species, and was followed by studies on cap of mRNA and viral RNA (120, 121).

The U1 RNA sequence contains a large number of uridine residues many of which are clustered (118). Gel analysis of 4-8S RNA separated U1 RNA into two species (122), and the faster moving U1A RNA was sequenced (118). Using rapid sequencing methods (63, 78, 123a) a number of modifications were made of the original sequence of U1 RNA (Figure 3). The U1A RNA from the chicken, rat, and HeLa cells is highly conserved (63). Of the species examined chicken (63), rat, (8, 63) human, and insects (12) contain only one species of U1 RNA. Only mouse contains two U1 RNAs, which differ in the center of the molecule (19). The two forms of U1 RNA found (122) in Novikoff hepatoma appear to be isomers. In addition to establishing its unique structure, U1 RNA sequence analysis led to a concept of hydrogen bonding to specific pre-mRNA consensus IVS sequences and to analysis

$pm_3^{2,2,7}G$

p	10	20	30	40	50

pAUACΨΨACC UGGCAGGGGA GAUACCAUGA UCACGAAGGU GGUUUUCCCA
　　　　　　　　　　　　　　　　　C　　　　　　G C

```
              60          70          80          90         100
GGGCGAGGCU  UAUCCAUUGC  ACUCCGGAUG  UGCUGACCCC  UGCGAUUUCC
              C    CC      m            G
```

```
             110         120         130         140         150
CCAAAUGCGG  GAAACUCGAC  UGCAUAAUUU  GUGGUAGUGG  GGGACUGCGU
       U
```

```
             160    165
UCGCGCUCUC  CCCUG_OH
       U
```

U1 RNA

Figure 3 Nucleotide sequence of U1 RNA of Novikoff hepatoma (118, 123A). or rat brain (63). Human (63), and Chicken (63) nucleotide variants are shown below the main sequence.

of the secondary structures of U1 RNA, and later to analysis of homologies with other U-snRNAs. Multiple secondary structures are possible with similar stability numbers (123). With the techniques developed for specific immunoprecipitation of the U1 snRNP particles, it is possible to conduct direct studies on the conformational state of the U1 RNA in snRNP particles (124). When T1 RNase digestion is followed by immunoprecipitation of the U1 snRNP particles, only one major site specific cleavage is observed, at position 107. The additional fragments produced by RNase A cleavage permit a more detailed analysis of the RNA structure within the particles. The structure derived is in good agreement with one hypothetical structure determined by analysis of stability numbers (123). For these analyses, the computer program of Korn et al (125) was invaluable.

U2 RNA

The most remarkable feature of the U2 snRNA (7, 122, 126; Figure 4) is the large number of modifications in its 5' end (126). Detailed analysis of the 5' cap (127) produced both enzymatic and chemical data that unequivocally established the presence of the 5' trimethyl G and pyrophosphate linkage to the remainder of the molecule. The 5' structure of the U2 RNA (127) was the first for which detailed chemical evidence was provided. Among the unanswered questions raised in this study was whether the turnover rate of trimethyl G is the same or different from the remainder of the molecule and whether, as suggested by Fredrickson et al (74, 76), this snRNA molecule is involved in synthesis of hnRNA.

The presence of 2'-O-methylated nucleotides and the remarkable number of ψ residues on the 5' end of U2 RNA clearly indicates its chemical distinction from U1 RNA, and further that there are an extraordinary number of post-transcriptional modifications. No special functions of U2 RNA are known. Like U1 RNA, U2 RNA is in an Sm protein containing snRNP, which is in the extranucleolar chromatin.

$$pm_3^{2,2,7}G$$

p	10	20	30	40	50

p AUCGC$\psi\psi$CU   CGGCCψUUUG   GCUAAGAUCA $^{6m}_{m}$GUGψAGψAψ CψG$\psi\psi$CU$_{m}$AU

| 60 | 70 | 80 | 90 | 100 |

CAGUψUAAψA   UCUGAUACGU   CCUCUAUCCG   AGGACAAUAU   AψUAAAUGGA

| 110 | 120 | 130 | 140 | 150 |

UUUUUGGAAC   UAGGAGUUGG   AAUAGGAGCU   UGCUCCGUCC   ACUCCACGCA
                        A

| 160 | 170 | 180 | 189 |

UCGACCUGGU   AUUGCAGUAC   CUCCAGGAAC   GGUGCACC(A)$_{OH}$

U2 RNA

Figure 4   Nucleotide sequence of U2 RNA of Novikoff hepatoma (8, 122, 126, 126a).

U3 RNA

U3 RNA (1,128) is of interest because of its specific localization in the nucleolus and its hydrogen bonding to nucleolar 35S and 28S RNA (97). This RNA was first purified by molecular sieve chromatography (26). U3 RNA is associated with protein, and is uniquely present in nucleoli (29). However, no significant concentration of U1 and U2 RNA is present in nucleoli. In addition, there is heterogeneity of the U3 RNA species (29).

The hydrogen bonding of U3 RNA to the 28S and 35S RNA is not stoichiometric and accordingly one suggestion is that U3 RNA is involved in the processing of 28S RNA (97).

Two of the three U3 RNA species found in Novikoff hepatoma cells have been sequenced (65, 129), and the minor differences in their sequences are shown in Figure 5. The third species of U3 RNA has not been completely sequenced, but it appears to be a minor variant of U3B RNA. The U3 RNA species are capped with trimethylguanosine (119, 130). Despite the two insertion/deletions in each RNA, U3A and U3B RNA have identical lengths of 214 nucleotides (129). When these two RNAs are aligned for maximum homology there are 17 base substitutions, including purine → purine, pyrimidine → pyrimidine, and purine → pyrimidine substitutions. The U3 RNA of HeLa cells consists mainly of one species that is similar but not identical to U3 RNA of Novikoff hepatoma cells (61).

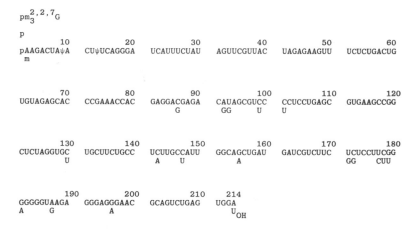

Figure 5 Nucleotide sequence of U3 RNA of Novikoff hepatoma (65, 129). For Dictyostelium U3 RNA, see (66).

$$pm_3^{2,2,7}G$$

p	10	20	30	40	50

p$_{mm}^{A}$GC$^\Psi$UUG$_{m}^{C}$G CAGUGGCAGU AUCGUAGCCA AUGAGGUU$_{A}^{U}$A UCCGAGGCGC

	60	70	80	90	100

GAUUAUUGCU AAUUG$_{m}^{A}$AAAC UU$^\Psi$UCCCAA$^\Psi$ ACCCCGCCAU GACGACUUGA
                                              G   C

$_{m}^{6}$ 110   120   130   140   145

AAUAUAGUCG GCAUUGGCAA UUUUUGACAG UCUCUACGGA GACUG(G)$_{OH}$

Sequence of U4 RNA

Figure 6 Nucleotide sequence of U4 RNA of Novikoff hepatoma (131) or rat (132). Mouse (133), human (132), and chicken (132) nucleotide variants are shown below the main sequence.

U4 and U5 RNA

The U4 RNA is another nucleoplasmic RNA that has important homologies to U1 RNA (131, 132). Its sequence (131–133) contains the same $m_3^{2,2,7}$ G cap of other U snRNAs and it contains a ψ near the 5' cap (Figure 6). In addition, position 63 is 2'-O-methylated and position 99 is m^6A. At present it appears that U4 RNA and possibly U5 RNA (Figure 7) are closely related to U1 RNA (132).

When U5 RNA was first purified, it was referred to as 5S RNA III (47). It has at least two (8) or more subspecies. Of the U1 to U6 RNAs, U5 RNA is most enriched in uridine (35%); it is capped with trimethylguanosine. Its complete sequence has been defined (7, 8, 134, 135; Figure 7).

U6 RNA

U6 RNA is associated with purified perichromatin granules (PCG) (113, 114). These dense granules are specifically juxtaposed to chromatin (114, 136). They appear in electron micrographs as dense cores surrounded by a white halo. Because of their association with newly synthesized mRNA

$$pm_3^{2,2,7}G$$

p	10	20	30	40	50

pAUACUCUGG UUUCUCUUCA GAUCGUAUAA AUCUUUCGCC UUU$^\Psi$AC$^\Psi$AAA
mm   m   m   m

	60	70	80	90	100

GAU$^\Psi$UCCGUG GAGAGGAACA ACUCUGAGUC UUAAACCAAU UUUUUGAGGC
                            U   U

	110	118

CUUGUCUUGG CAAGGCU(A)$_{OH}$
UC CUCCAA

Figure 7 Nucleotide sequence of U5 RNA of Novikoff hepatoma (8) or rat (134). Mouse (135), human (134), and chicken (134) nucleotide variants are shown below the main sequence.

```
                                                         6
         10            20            30            40      m    50
XpppGUGCUCGCU  UCGGCAGCAC  AUAUACUAAA  AΨUGGAACGA  ΨACAGAGAAG
                                                              m

         60            70       2              90           100
                                m
AUUAGCAUGG  CCCCUGCGCA  AGGAUGACAC  GCAAAUΨCGU  GAAGCGUUCC
    mm          m  mm       m        m

        108
AUAUUUU(U)OH
```

Figure 8 Nucleotide sequence of U6 RNA of Novikoff hepatoma (137), and mouse (138).

(114, 136), they could be "carriers" or processing elements in or near the functional chromatin. Accordingly, U6 RNA and U6 RNP may be the first U-snRNPs to combine with newly synthesized hnRNP.

The structure of U6 RNA (137, 138; Figure 8) differs from that of the other U snRNAs in two major respects. First, the cap does not contain trimethyl G. It is not known what structure is linked to the nucleotide chain, but it is not a normal nucleotide.

Second, U6 RNA has several clusters of modified nucleotides in the center of the molecule (Figure 8); in other U snRNAs most modifications are in the 5' third of the structure. The U6 RNA contains several 2'-O-methylated nucleotides in addition to m^6A and m^2G. This molecule appears to be highly hydrogen bonded (137). Interestingly, the U1 to U5 RNAs have Am as their first nucleotide in the RNA sequence but U6 RNA has an unmodified Gp. When nucleotide sequences of the U snRNAs are analyzed, significant homologies are found (8, 138a), the most striking of which are between U1 and U4 RNA (Figure 9). Since U1, U2, and U4 to U6 RNPs contain one or more common proteins, these homologous regions may serve as binding sites for these protein(s). Similar homologies between U snRNAs

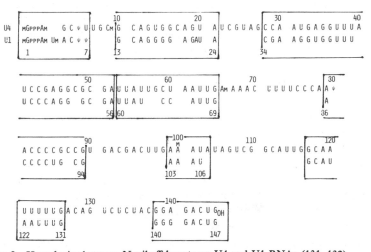

Figure 9 Homologies between Novikoff hepatoma U4 and U1 RNAs (131, 132).

have also been reported by Krol et al (132). Although earlier studies (6, 58) concluded that U snRNAs of different tissues are similar, their sequences show that there may be minor differences in snRNAs of different tissues (134). U5 RNA of chicken brain is one nucleotide shorter on the 3' end compared to chicken liver U5 RNA. Whether this difference is due to differential processing or due to expression of different genes is not known.

snRNA IN RIBONUCLEOPROTEINS

Isolation of snRNPs

Recognizing that snRNAs are probably functional as RNPs, efforts have been made to isolate U snRNPs (28, 40, 59); Raj et al (139) first isolated U1 and U2 RNPs and showed that they contain approximately 10 proteins. There are reported chromatographic methods for isolation of U1 and U2 RNPs (140). Partial purifications of small RNPs have also been reported (111, 113, 139–141).

The first attempts to isolate U-snRNPs made use of Sepharose gel filtration and sucrose density gradient centrifugation, using nuclear extracts (139). Other methods include molecular sieving, ammonium sulfate fractionation, DEAE-cellulose and phosphocellulose chromatography, CsCl density centrifugation, and affinity chromatography (142) with anti-snRNP antibodies.

Since some patients with autoimmune diseases produce antibodies directed against classes of small RNPs (17), several investigators have used antibody columns to isolate small RNPs (142, 143). Although the antibodies are highly specific, the problem appears to be recovery of the RNP particles in a functional form from the antibody columns. The immunological methods offer one approach to purification of the U snRNP particles, but it is difficult to recover undegraded particles.

Proteins Associated with snRNPs

In initial studies on the snRNP proteins, 11 proteins were identified on 2-D gels (139). These have molecular weights of 50,000–70,000 in the fractions that contained the U1 and U2 snRNP particles. However, it is not certain that all of these proteins are associated with the particles. Lerner et al (17) found 7 proteins in the immunoprecipitated particles, but these had low molecular weights of 10,000–30,000. These and larger proteins were found in snRNPs (142). This discrepancy is now being investigated to see whether the small proteins represent subunits or degradation products of the larger proteins or the larger proteins represent aggregates of more native species. Of the proteins associated with capped snRNPs, a 70,000-mol wt protein consisting of several 13,000-mol wt subunits, has been suggested to be the

Sm antigen (144, 145). The 8,000–14,000-mol wt proteins are tightly bound to the U1 RNA, since they do not dissociate in 0.5% sarkosyl (111). The antigenic protein associated with La-RNPs is a 68,000-mol wt protein (146).

Functions of U-snRNPs

The U snRNPs containing U1, U2, U4, U5, and U6 RNA are located in the nucleoplasm. Their function could be in processes relating to "splicing" of hnRNA in hnRNPs (11–13). The first clue to the role of the snRNPs came from reports that U1, U2, and other U snRNPs cosedimented with hnRNP particles (98). The second came from studies showing that some snRNAs are found hydrogen bonded to hnRNAs (99–110). Lerner et al (12) analyzed all the available IVS sequence near the splice consensus; sequences near the splice junctions have been derived by others (reviewed in 15–18).

Mechanisms of splicing must account for Breathnach and Chambon's "rule" that splice points have common sequences (GU . . . AG) (15). This "consensus structure" has been a crucial idea in the development of hypotheses relating to cleavage of hnRNA. Two questions emerged from studies with hnRNA: How does the cell make sense of the final product and why are mistakes relatively uncommon. When the sequences of known small RNAs are analyzed, the 5' end of U1 RNA has good complementarity to consensus sequences (Figure 1) (12).

The evidence for the involvement of U1 RNP in splicing of hnRNA is as follows: (a) the complementarity between conserved consensus sequences and U1 RNA; (b) U1 RNPs lacking the 5' terminus sequence are not associated with hnRNPs (12); (c) splicing of adenovirus hnRNA in HeLa cell nuclei is inhibited when they are preincubated with anti-Sm or anti-RNP antibodies (147). When the nuclei are incubated with anti-Ro or anti-La antibodies, splicing is not inhibited. Since other cellular functions, e.g. polyadenylation, are not inhibited, and since anti-RNA or anti-Sm antibodies react with and presumably inactivate the U snRNPs, it is likely that U snRNPs are required for splicing.

To determine whether IVS are removed in one or multiple steps and, if the latter, whether there is polarity, Avvedimento et al (148) analyzed the processing of collagen hnRNA. Several splicing steps are required for the removal of one IVS, and the polarity is 3' → 5'. This multistep removal of the IVS has been shown to be correlated to complementarity between U1 RNA and sites within the IVS.

In addition to these lines of evidence the data, e.g. the abundance of U1 RNP, which correlates positively with the prevalence of splicing, the presence of U1 RNP in hnRNP where splicing takes place, and the higher amounts of U1 RNP in rapidly dividing cells, are consistent with a role for

U1 RNP in splicing hnRNA. However, yeast and mitochondria, where splicing of hnRNA takes place, do not appear to have snRNAs (6). If so, U1 RNP–mediated splicing may not be the only mechanism for splicing of hnRNAs. Splicing of tRNA and rRNA precursors does not appear to involve a small RNA (15–19).

It seems likely that other snRNAs (U2, U4, U5, and U6) are also involved in processing of hnRNA, since they are associated with hnRNPs. Ohshima et al suggested that U2 RNA ensures specificity of splicing (148b). Daskal et al (113–144) have shown that of the snRNAs, only U6 RNA is associated with perichromatin granules. The U6 RNA may be involved in a nuclear function different from that of U2, U4, or U5 RNAs, possibly in formation of a precursor particle in the chromatin. Pederson & Bhorjee (149) have reported that snRNAs are covalently linked to DNA but this finding has not been confirmed.

Now that the sequences and secondary structures of the U- snRNA species have been defined, the primary tasks are the analyses of the U snRNP structures and their interactions with hnRNP structures. The molecular weight of the proteins of the snRNP particles vary depending upon the method of isolation of (19, 139). With one exception (139), there are no 2-D gel studies on the proteins of the snRNP particles. There are probably up to 20 as yet unidentified proteins associated with the U-snRNPs.

The major problem in the analysis of U snRNP proteins has been the development of suitable purification methods for the particles. When the particles are isolated on a sucrose density gradient, there are many contaminants. When chromatographic or immunochemical methods are used (140–142), the "core particles" have lost enzymes and structural elements. Thus, the problem of "nativeness" is a crucial one both from the point of view of retention of the active elements of the particles and prevention of contamination by nonparticulate structures of the nucleus.

It is important to obtain highly purified U-snRNPs that have specific nuclease functions, and possess ligase activity. This search could be implemented if (a) a proper assay system were available and (b) if the particles could be isolated in a native, uncontaminated form. Almost any defined spliceable hnRNA such as globin hnRNA (hnRNP) or ovomucoid hnRNA (hnRNP) would be satisfactory. Several studies are now in progress with cell free systems (158–161).

In addition to fidelity, there is a problem of rate of reaction. Electron microscopic studies (N. Domae, W. Schreier, R. Reddy, and H. Busch, unpublished) suggest that there is a dynamic equilibrium between the free, small particles in the nucleus and the snRNP-hnRNP complexed particles (143). In cells producing large amounts of hnRNP, such as growing and dividing cells, there are millions of snRNPs interacting with the hnRNP particles, many of which have multiple cleavage sites. The lack of tight binding of the U snRNP-hnRNP complexes has been noted by several groups (98–109). Many of the snRNPs are unattached, as are many hnRNP particles. At 37°, the hnRNP-snRNP complexes are unstable, which suggests that relatively weak binding forces exist between these two (102). When snRNP-hnRNP complexes are examined by electron microscopy, usually only one snRNP particle is found per hnRNP particle (143). From studies with ovomucoid RNA (162), the snRNP particles rapidly associate and dissociate from hnRNP particles randomly cleaving one IVS at a time.

Enzymatic Mechanisms of Precursor RNA Processing

Abelson (16) has pointed out that a variety of RNases are candidates for enzymes that specifically cleave hnRNA during its processing. Thus far only RNase III, P, Q (an exonuclease), T_2, and a few other enzymes exhibit specificity for tRNA precursor cleavage and for processing viral pre-mRNA.

The cleavage sites recognized by the RNases are dependent on RNA conformation (*a*) inherent in the nucleotide sequences (16, 227) and (*b*) resulting from bound protein in RNP. The RNA-directed signal is recognized by RNase III for 16S rRNA and 23S rRNA of the 30S pre-rRNA, and by RNase P and RNase "E" which recognize tRNA and 5S rRNA, respectively (16, 227). The RNP directed signal is recognized by RNase M5 for the maturation of p5S, and by RNase M16 for the maturation of p16S rRNA in *E. coli* (227). RNase III–like enzymes are present in eukaryotes (16, 280), but their functions are not known. The enzyme from chick embryos is associated with RNA, which is essential for 45S pre-rRNA processing (281).

In mitochondrial systems (278–290; see review 291) some special reactions have been found. Manipulation by mutation has been described for pre-mRNA of apocytochrome *b* from yeast mitochondria (287–290). Although mitochondrial snRNA has not been found, mitochondria carry out processing similar to the nuclear pre-mRNA. The processing pathways for pre-mRNA of apocytochrome *b* involve 34S pre-mRNA cleavage into 18S mRNA by 1. five splicing steps of which two independent steps are required to remove one of the four IVS, and 2. a preferred pathway composed of two sequences of events without an obligatory order (287–290). COB-mutants have two different processing pathways: 1. the preferred pathway in which a mutational block of splicing is defective but there is a bypass to subsequent steps, and 2. the obligatory pathway in multiple blocks, which can occur with polar effects in which the upstream mutations block not only one splicing site but also downstream splicing sites. There is also evidence for IVS coding for proteins, which may be involved in splicing (287–290).

As yet, no specific RNA cleavage enzyme or ligase has been found associated with U snRNP particles. The only small RNP shown to be involved in processing of precursor RNAs is RNase P (150–157). This endoribonuclease generates the 5' termini of mature tRNA molecules. The endonuclease consists of a 20,000-mol wt basic protein and a 300-nucleotide long RNA; neither is active alone, but they can be reconstituted to yield an active enzyme (155). The enzyme recognizes the structural conformation of its substrates, tRNA precursor molecules, rather than specific nucleotide sequences. The *E. coli* enzyme is best characterized, including fingerprinting of its RNA component (156). Similar activities have been reported in yeast, chick embryos (153), and human KB cells (157).

These RNases vary in their specificity for conformational states, specific attack sites, and interrelationships with specific proteins, but none has base sequence specificity as a requirement for cleavage. From the consensus sequences involved in the splicing of pre-mRNA, specific base sequences should be recognized by the splicing reactions of the snRNPs and their associated enzymes. Until "native" highly purified U snRNPs are available, it will not be possible to establish whether enzymes similar to RNase P are tightly bound to these particles or whether they are soluble factors in the nucleoplasm.

OTHER SMALL RNAs

La and Y RNPs

Anti-La antibodies precipitate distinct sets of snRNPs that are different from U snRNPs (81, 163). The mouse or rat RNAs are described below.

La 4.5 RNA was first shown (117) to be a group of RNAs 90–100 nucleotides long, hydrogen bonded to poly A–containing nuclear or cytoplasmic RNA from cultured Chinese hamster ovary cells. This RNA binds to nuclear precursor RNAs, presumably to their Alu family sequences, and these hybrids can be reconstituted (117). The RNA is also found in some RNA viruses (80) and is heterogenous at the 3' end in having varying numbers (80) of uridine residues. The sequence of La 4.5 RNA in mouse and hamster has been determined (164) as in Figure 10. A portion of this RNA, 14 nucleotides long, is present in the 300-nucleotide long highly repetitive Alu family DNA, and also in hnRNA (165). Many viruses have sequences common to this RNA (165).

Another RNP precipitated by anti-La antibodies contains 4.5S RNA I (81), the first snRNA to be sequenced (166; Figure 11). Rat 4.5I RNA has

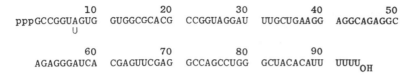

```
        10           20           30           40           50
pppGCCGGUAGUG  GUGGCGCACG  CCGGUAGGAU  UUGCUGAAGG  AGGCAGAGGC
           U

        60           70           80           90
 AGAGGGAUCA  CGAGUUCGAG  GCCAGCCUGG  GCUACACAUU  UUUU_OH
```

Figure 10 Nucleotide sequence of La 4.5 RNA of mouse and hamster (164). Microheterogeneity was found at nucleotide 7.

a 50% sequence homology to La 4.5 RNA (R. Reddy, D. Henning, and H. Busch, unpublished results). La protein (antigen), which is bound to La 4.5 and 4.5I, is present in VA-RNP (163) and EBV-RNP (167). A different set of La RNAs is precipitated from human cells (81). The structure of human La RNAs is not known. Fingerprint analysis indicates La RNAs are less conserved than U snRNAs (81).

Y RNPs, first described to be antigenically nonoverlapping (163) are a subclass of La RNPs (81). Like La RNAs, Y RNAs are not conserved between mouse and human cell lines. Mouse cells contain two Y RNAs (Y1 and Y2) and human cells contain four to five Y RNAs (Y1 to Y5). All these Y RNAs are distinct RNAs with few structural similarities (81). The Novikoff hepatoma contains three Y RNAs (Y1–Y3) and Y2 and Y3 RNA have similar but not identical fingerprints (H. Busch, R. Reddy, D. Henning, and Y. C. Choi, unpublished results). Complementarity between one loop of 4.5S RNA$_I$ and the TATAT box was noted (167a), but its relevance to function is unknown.

7S RNA and Other Small RNAs

7S RNA, first found in RNA viruses (168) is a host coded, conserved RNA species (169). The sequence of 7S RNA from Novikoff hepatoma has been determined (170; Figure 12). 7S RNA is the only cellular small RNA that hybridizes to Alu family DNA sequences (171). 7S RNA is 85% homologous to the La 4.5 RNA and long stretches of Alu family DNA sequences. The homologous sequences between Alu family DNA sequences, mouse B1 sequences, 7S RNA, and La 4.5 RNA are shown in Figure. 13. The significance of these homologies is not clear; however, it is of interest that the DNA sequences at the origins of replication correspond to portions of the

```
        10           20           30           40        U 50
pppGGCUGGAGAG  AUGGC(UC)AGC  CGUUAAAGGC  UAGGCUCACA  ACCAAAAAUA

        60           70           80           90        98
 UAAGAGUUCG  GUUCCCAGCA  CCCACGGCUG  UCUCUCCAGC  CACCUUUUU(U)_OH
```

Figure 11 Nucleotide sequence of La 4.5 I RNA of Novikoff hepatoma (166) modified. Microheterogeneity was found at nucleotide 47.

homologous sequences [(165); see also the review by W. R. Jelinek and C. W. Schmid in this volume]. These may be transposon related.

In addition to these small RNAs several other small RNAs include: RNA 7–1, 7–2 (95), 7–3 (51), and 8S (97), and RNase P RNA (156). There appear to be no additional abundant RNAs in higher eukaryotes other than the RNAs described in the size range of 90–400 nucleotides. However, RNAs with less than 10,000 copies per cell are difficult to detect. RNAs of the size 8–18S have been found and partially characterized (172, 173). Other studies have described RNA species specific to certain stages of tissue development (174, 175). Cloned Alu DNA fragments are transcribed in vitro by polymerase III to yield specific small RNAs. There is evidence that these RNAs exist in vivo (176–177a). 8S RNA contains 5.8S rRNA.

VIRAL RNAs

In 1966, Reich et al (37) compared the chemical and metabolic characteristics of a novel small RNA of adenovirus 2–infected KB cells with those of 5S RNA. They concluded that the former was synthesized during viral infection. Subsequently, other species of low-molecular-weight RNAs were identified in cells infected with SV-40 (178, 179), Epstein Barr (EB) virus (167), vesicular stomatitis virus (VSV) (180, 181), or VSV-defective interfering particles (182, 183). Virions of the retroviruses accumulate discrete classes of host small RNA (168, 169, 184); one class, tRNA, serves as a primer for viral genome replication (185). Small RNA molecules, both viroid and satellite, function as the causative agents of many plant diseases (186, 187), although the mechanisms of infection and propogation are not completely understood.

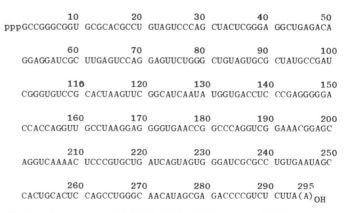

```
           10          20          30          40          50
     pppGCCGGGCGGU GCGCACGCCU GUAGUCCCAG CUACUCGGGA GGCUGAGACA

           60          70          80          90         100
     GGAGGAUCGC UUGAGUCCAG GAGUUCUGGG CUGUAGUGCG CUAUGCCGAU

          110         120         130         140         150
     CGGGUGUCCG CACUAAGUUC GGCAUCAAUA UGGUGACCUC CCGAGGGGGA

          160         170         180         190         200
     CCACCAGGUU GCCUAAGGAG GGGUGAACCG GCCCAGGUCG GAAACGGAGC

          210         220         230         240         250
     AGGUCAAAAC UCCCGUGCUG AUCAGUAGUG GGAUCGCGCC UGUGAAUAGC

          260         270         280         290         295
     CACUGCACUC CAGCCUGGGC AACAUAGCGA GACCCCGUCU CUUA(A)OH
```

Figure 12   Nucleotide sequence of 7S RNA of Novikoff hepatoma (170).

```
7S RNA (rat)    pppGCCGGGC.. .GGUG.CGCA CGCCUGUAGU CCCAGCUACU CGGGAGGCUG AGACAGGCAG AUCGCUUGAG UCCAGGAGUU
                           10         20         30         40         50         60         70         80

Blur 8          AGGCUGGG.AG UGGUGGCUCA CGCCUGUAAU CCCAGAAUUU UGGGAGGCCA AGGCAGGCAG AUACCUGAG  GUCAAGAGUU
(Human Alu)     GGCA.       UGAUGGCAAG UGCCUGUAAU CCCAGCUACU UGGGAGGCUG ACGAAGGAGA AUUGCUUAAA CCUGGA
                           150        160        170        180        190        200        210

B1 (mouse)      CCGGGCA.    UGGUGGUGCA UGCCUUCAAU CCCAGC.ACU CGGGAGGCUG AGGCAGGCGG AUUUC.UGAG UUCGAGCCCA
                           10         20         30         40         50         60         70

La 4.5S (mouse) pppGCCGG.UAG UGGUGGCCCA CGCCGGUAGG AUUUGCUCA. .AGGAGGCAG AGGCAGAGGG AUCACACGAG UUCGAGGCCA
                           10         20         30         40         50         60         70
```

Figure 13 Homologies between human Alu DNA, mouse B1 DNA, rat 7S RNA, and mouse La 4.5 RNA. The sequences in rat 7S RNA, mouse B1 DNA, and mouse La 4.5S RNA are underlined where they are identical to the human Alu sequence.

Adenovirus Associated Small RNAs: VA-RNA

In addition to a 9S mRNA, the adenovirus 2 Ad-2 genome codes for two species of small RNA (188–190). The major viral-associated RNAs, VA RNA$_I$ and VA RNA$_{II}$, are primary transcripts of the viral genome (188–194). They are not products of the processing of larger precursors or degradation products (188, 195, 196). Their genes map close to position 30 on the r-strand of the adenovirus 2 genome, and are transcribed by RNA polymerase III (192, 193, 197). These genes have been transcribed accurately in vitro (191, 195).

Initial questions as to the number of species (189, 190) have been resolved. Mathews & Petterson (188) have investigated the small RNAs of the nuclei and cytoplasm of Ad 2–infected HeLa cells. By fingerprint analysis there are only two major species of small RNAs unique to those cells. Both of these hybridize to adenovirus 2 DNA. Analysis of the structures of the small RNA synthesized in vitro by nuclei of Ad 2–infected KB cells has led to the identical conclusion (197). The 156-nucleotide RNA synthesized in vitro is identical to VA RNA$_I$ sequenced by Ohe & Weissman (196), and the 140-nucleotide virus-coded RNA described by Mathews (198) and synthesized in vitro. Two other species of RNAs 140 nucleotides long also exist and one is probably a degradation product of VA RNA$_I$. The 200-nucleotide long VA RNA (V200) previously identified (193) is a run-through product of the VA RNA$_I$ gene toward a second termination site approximately 40 nucleotides downstream. Most of the fragments of the fingerprint of the V200 RNA can be predicted by the sequence of the VA$_I$ RNA gene (199, 200).

Heterogeneity of VA RNA$_I$ has also been reported by Vennstrom et al (201) who identified a 5' terminus of (pp)pApGpCp, the major 5' terminus of (pp)pGGGGC, and the 3' terminal heterogeneity of CUCCUU(U) in approximately equal molar amounts. The heterogeneities of the two major VA RNAs are better understood when correlated with the nucleotide sequence of the genes (202; Figure 14). The heterogeneity of the 5' sequence can be explained on the basis that RNA polymerase III can initiate at either of two positions in VA RNA$_I$ (Figure 14). The 3' heterogeneity results from termination at any of several contiguous thymidine residues at the 3' ends of both transcripts. From analysis of the sequences of the genes, it appears that the two genes have arisen from the duplication of an ancestral gene, with subsequent divergences that resulted in two RNAs with different secondary structure, and possibly more than one function (202). Recent studies on the variants of the VA RNAs appear to confirm this hypothesis (202a).

VA RNA$_I$ is found in preparations of hnRNA of Ad 2–infected cells (188), which is consistent with formation of an hnRNA-VA-RNA complex. In 1979, Murray & Holliday (11) proposed that VA RNA$_I$ could function as a splicer RNA. They predicted that 19 bases of VA RNA$_I$ could hybridize across a single splice point of the hexon pre-mRNA. In 1980, Mathews (203) reported that the VA RNAs could hybridize in vitro to unfractionated late virus mRNA and to cDNA of this mRNA, but not to cloned portions of the natural gene. The hybrids were constructed between filter bound DNA and RNA in solution. Thus the failure to observe hybrids to portions of the cloned natural genes may have been due to artificial constraints. These data were consistent with a role for VA RNA in splicing, or with the proposal that VA RNA could bind to spliced RNA and play a role in a subsequent metabolic step in mRNA metabolism. It was subsequently determined that VA RNA could be precipitated from extracts of Ad 2– infected HeLa cells by anti-La antibody, which suggested that VA-RNA is in an snRNP (12).

Two recent findings suggest that VA RNA is not involved in splicing viral hnRNA. As noted earlier, Yang et al (147) found that incubation of Ad 2–infected HeLa cell nuclei with anti-La antibody, does not prevent the processing of adenovirus mRNA. They found that anti-Sm and anti-RNP antibodies, which did not interact with the VA-RNA snRNP, inhibited the processing of VA-mRNA. If either small VA RNA is involved in splicing, it would have to associate with the hnRNA soon after polyadenylation. However, VA RNA is not found associated with the nascent adenovirus hnRNA (204, 205). The data argue against a role for Ad 2 VA RNA in processing adenovirus hnRNA, but do not preclude a role for these RNAs

VA RNA I

```
ACCGUGCAAA  AGGAGAGCCU  GUAAGCGGGC  ACUCUUCCGU  GGUCUGGUGG  AUAAAUUCGC  AAGGGUAUCA

UGGCGGACGA  CCGGGGUUCG  AACCCCGGAU  CCGGCCGUCC  GCCGUGAUCC  AUGCGGUUAC  CGCCCGCGUG

UCGAACCCAG  GUGUGCGACG  UCAGACAACG  GGGGAGCGCU  CCUUUUUGGCU  UCCUUCCAGG  CGCGGCGGCU

GCUGCGCUAG  CUUUUUUGGC  CACUGGCCGC  GCGCGGCGUA  AGCGGUUAGG  CUGGAAAGCG  AAAGCAUUAA
```
VA RNA II
```
GUGGCUCGCU  CCCUGUAGCC  GGAGGGUUAU  UUUCCAAGGG  UUGAGUCGCA  GGACCCCGG   UUCGAGUCUC

GGGCCGGCCG  GACUGCGGCG  AACGGGGGUU  UGCCUCCCCG  UCAUGCAAGA  CCCCGCUUGC  AAAUUCCUCC

GGAAACAGGG  ACGAGCCCCU  UUUUUGCUUU  UCCCAGAUGC  AUCCGGUGCU  GCGGCAGACG  CG
```

Figure 14 Nucleotide sequence of the VA RNA genes of Adenovirus-2 (202). The sequence is of the anticoding strand and is given as the RNA. The genes for the VA RNAs are underlined. VA$_{200}$ is underlined with a broken line extending to VA RNA$_1$. Alternative initiation and termination sites are shown with arrowheads.

in either transport of processed Ad 2 mRNA molecules from the nucleus or disruption of the metabolism of host cell hnRNA (206). Experiments with mutants of Ad 2, which have VA RNA genes mutated in vitro, suggest that these genes are essential for viability (206a). Recently, T. Shenk, (personal communication) has obtained evidence for a role of VA-RNA in translation as an inhibitor of host mRNA.

Other Viral RNAs

SIMIAN VIRUS-40 (SV-40) ASSOCIATED SMALL RNA   A small RNA is synthesized late in the simian virus lytic infection (178, 207). Although it is synthesized from the SV-40 genome, it is not required for viral viability. This RNA is complementary to the 91 codon alternative reading frame of the 3' terminal portion of the early genome (179, 208). A third splice occurs in SV40 early mRNA late in the lytic cycle (209) coincident with the synthesis of the SV-40 associated small RNA, and suggests that the SV-40 snRNA plays an indirect role in this splicing event. An snRNA synthesized in SV-40 infected cells is capable of affecting transcription, but it has not been isolated (210).

EB VIRUS ASSOCIATED RNA   Recently, Lerner et al (167) reported on two small RNAs encoded by the virus genome that are synthesized in large amounts in infected cells. They are precipitated with anti-La antibody, which indicates that they are in nuclear RNPs. These two RNAs, which have recently been sequenced (210a), are approximately 180 nucleotides long; they have a 5' pppA and lack poly(A).

RNAs ASSOCIATED WITH STANDARD AND DEFECTIVE VSV PARTI-
CLES   When cells are infected with standard and defective interfering (DI) VSV, or when the particles are incubated in vitro, a 47-nucleotide long RNA is synthesized (183, 211, 212). This RNA has been sequenced (212, 213); pppAp is the 5' end and it lacks poly(A) on the 3' end. This RNA could play a part in regulating the balance between RNA transcription and replication (214, 215).

Standard VSV synthesizes a plus-strand leader RNA that differs from the RNA synthesized by DI particles (180, 216). This RNA is synthesized when the virus associated, RNA-dependent RNA polymerase initiates transcription on the viral genome. The RNA is complementary to the 3' end of the genomic RNA and is apparently cleaved from the growing RNA chain before transcription of the viral mRNAs. Three other small RNAs, in addition to the 47-nucleotide leader sequence, are synthesized during in vitro transcription of VSV virions (181). These RNAs have chain lengths of 28, 42, and 70 nucleotides, contain (p) ppAA at their 5' terminal, and are not polyadenylated. The 42- and 28-nucleotide long RNAs contain 5' terminal sequences identical to those of the N and NS mRNAs (181). The authors hypothesize that these leader sequences are synthesized concurrently by multiple initiations of transcription, and that they are then elongated sequentially, which results in the synthesis of the mRNAs coding for the viral proteins N, NS, M, b, and L.

RETROVIRUS ASSOCIATED SMALL RNAs   Retrovirus virions contain several low-molecular-weight RNAs as well as the viral genomic RNA (217–219), none of which are viral encoded. However, they represent a discrete subclass of the complement of host small RNA (220). The association of tRNA with the viral genome and its function as the primer for initiation of the viral RNA-dependent DNA polymerase is well documented (220–224). The functions of the other viral associated low-molecular-weight RNAs are unknown. The 4.5S RNA species associated with Moloney leukemia virus and spleen focus–forming virus (80) differ from the host cell species in that they contain as many as thirty extra uridylate residues at their 3' termini. The 4.5S RNA associated with the genome of spleen focus forming virus is composed of more than thirty components (80) that vary in lengths of the poly(U). The viral genomic RNA is associated with the larger molecules (about 5S), and the cytoplasmic RNAs of the T3-K-1 cells are associated with the smaller (close to 4.5S) RNAs. When Moloney-murine leukemia virions, or spleen focus forming virions are isolated from vertebrate cells, other than cells derived from rodents, 4.5S RNA species are not found in the viral genomes (225). These data substantiate the finding that the 4.5S RNA molecules associated with the viral genome are provided by the host. Harada et al (225) have compared the 4.5S RNA species in the virions with those species that have been previously characterized. They concluded that the 4.5S RNA associated with the viral genome is not $4.5S_I$ RNA. The virus associated RNA is likely to be the same RNA as that found in snRNP recognized by anti-La antibody.

PRE-rRNA PROCESSING

Progress has been made in the study of pre-rRNA processing (15, 16, 226–230). Complete or nearly complete nucleotide sequences have been determined for many different rDNAs including those of human and mouse mitochondria, *Zea mays* chloroplast, *E. coli,* yeast, *Xenopus,* rat, mouse, and human. Some of these contain IVS that are cleaved out of the final rRNA products.

Table 3 shows examples of the evolutionary diversity of rDNA, which is characterized by various sizes, types, and arrangements and by the presence of spacers such as IVS, insertions, or gaps (230–247). Despite these wide evolutionary diversities there are common features:

(a) Conservation of gene polarity: 5'-16-18S RNA → 23-28S rRNA-3'.
(b) Prokaryotic genes contain 5S rDNA at the 3' end, interspersed spacer tDNAs, and smaller size high-molecular-weight rRNA.
(c) Eukaryotic genes have separated 5S rDNA and tDNA from rDNAs. They contain 5.8S rDNA and higher-molecular-weight rDNAs. The loss of 5S rRNA colinearity occurs in yeast (248) and *Dictyostelium* (249), which contain 5S rDNA in strands opposite to the coding rDNA strand.
(d) Chloroplast genes in higher plants contain 4.5S rDNA.
(e) Insect genes (Dipteran) have disjoined 5.8S rDNA (5.8 rRNA$_\alpha$ + 5.8S rRNA$_\beta$).

Initial Transcripts and Processing Intermediates

Pre-rRNA is generally large, and in some instances there are several precursors that are processed differently. Extensive studies have been made on 45S

Table 3 Diversity of rRNA genes (rDNA)

Species	Compartment	Order of gene	Pre-rRNA	Reference
Rat, mouse, man	Nucleolar	18S–5.8S–28S	45S	231
Xenopus	Nucleolar	18S–5.8S–28S	40S	230
Drosophila	Chromosomal	18S–5.8S–2S–28S$_\alpha$– Gap (and Insert)–28S$_\beta$	38S	230
Yeast	Chromosomal	17S–5.8S–26S	37S	231
Physarum	Extrachromosomal	19S–5.8S–28S$_a$–IVS$_a$– 28S$_b$–IVS$_b$–28S$_c$	40S	232
Tetrahymena	Extrachromosomal	17S–5.8S–26S$_a$–IVS$_a$– 28S$_b$	>35S	233
Leishmania	Extrachromosomal	16S–5.8S–26S$_\alpha$–Gap– 26S$_\beta$		234
Mouse, man	Mitochondrial	tRNA–12S–tRNA– 16S–tRNA		235, 236
Yeast	Mitochondrial	15S–tRNA–21S$_a$– IVS–21S$_b$		237
Neurospora	Mitochondrial	17S–tRNAs–25S$_a$– IVS–25S$_b$–tRNAs		238, 239
Zea mays, spinach	Chloroplastic	16S–tRNA–23S– 4.5S–5S		240–242
Euglena	Chloroplastic	16S–tRNAs–23S–5S		243, 244
Chlamydomonas	Chloroplastic	16S–tRNA–7S–3S– 23S$_a$–IVS–23S$_b$–5S		245, 246
Prokaryotes	Chromosomal	16S–tRNA–23S–5S– tRNA	30S	247

pre-rRNA of mammals, 40S of *Xenopus,* and 38S of *Drosophila.* The 35S pre-rRNA from *Tetrahymena* contains one IVS, and 40S pre-rRNA from *Physarum* contains two IVS (See Table 3). Mechanisms that process the pre-rRNA include specific cleavages (226), 5'-end trimming as found in 32S pre-rRNA of yeast (228) and 41S pre-rRNA from mammals (226), and splicing, as observed in Physarum (232).

LARGE CLEAVAGE INTERMEDIATES   At the initial cleavage stage, a set of specific sites at spacer junctions are cleaved. Initial cleavages of 30S pre-rRNA of *E. coli* produce tRNAs, p5S rRNA, p23S rRNA, and p16S rRNA (250). 32S pre-rRNA from yeast produces p17S rRNA (18S) and p26 rRNA (29S) (231, 248). In some organisms, the spacers between 18S rRNA and 28S rRNA have two or more defined cleavage sites that are not cleaved in a defined order. The alternate sites result in alternate pathways (226, 251); for example, in mammals 41S → 18S + 36S, and 41S → 20S + 32S (226).

Table 3 shows there are many variants of pre-28S (23S) rRNA among organisms, some of which undergo splicing reactions (Table 3).

Eukaryotic pre-rRNA ("32S") contains 5.8S rRNA at the 5' end. Similarly, pre-26S rRNA ("29S") of yeast has 5.8S rRNA at its 5' ends. The 29S pre-rRNA, is the precursor of 5.8S rRNA and 26S rRNA (252). Chloroplast pre-23S rRNA (1.65×10^6 daltons) from spinach is a precursor of both 4.5S rRNA and 28S rRNA (23S) (242).

Splicing occurs in mitochondrial and cholorplast pre-28S rRNA, both of which contain an IVS. Pre-21S rRNA ($5.1–5.4 \times 10^3$ bases) from yeast mitochondria also contains an IVS (1.2×10^3 bases) as does pre-25S rRNA (253) from *Neurospora* mitochondria (2.3×10^3 bases) (254). Prokaryotic pre-23S rRNA (p23S) from *E. coli* contains short spacers, 114 bases on the 5' end, and 71 bases on the 3' end (255).

PRE-18S (16S) rRNA   The pre-16S rRNA from *E. coli* (p16S) contains two flanking sequences, 146 bases on the 5' end and 43 bases on the 3' end (256). *E. coli* RNase III⁻ and P⁻ mutants produce a 20S pre-16S rRNA that contains ppp-leader sequences-16S-tRNA-tRNA (257). The precursor from yeast is 18S or p17S rRNA (252). In mammalian cells, pre-18S rRNA is 20S which has an identical 5'-end to 18S rRNA (226, 258).

MATURE rRNAs AND THEIR FORMATION (18S OR 16S rRNA)   Complete 18S or 16S rRNA sequences have been determined for human mitochondria (954 bases) (235), mouse mitochondria (236), *Zea mays* chloroplast (240), *E. coli* (247), yeast (259), and *Xenopus* (260). There is a 75% homology of mitochondrial and chloroplast 12–16S rRNA to *E. coli,* a substantial homology of yeast to *E. coli,* and greater than 70% homology of *Xenopus* 18S rRNA to yeast 17S rRNA. A universal homology has been observed in the 3' end portion containing $-m_2^6Am_2^6A-$. Eukaryotic 18S rRNA contains sequences with base modifications, including m^6A, m^7G and AIB $-m'\psi$ (259–261).

In addition to the primary structures, the secondary structures have been defined for prokaryotic 16S rRNA (262, 263). For more than 100 species, there are 37 helices, variable and conserved sequences, and a universal sequence (GCCGUAAACGAUG) located in positions 821–879.

28S (23S) rRNA   One form of 28S rRNA is a single continuous strand and the other is two discontinuous strands designated as 28S $rRNA_\alpha$ (the 5'-end portion) and 28S $rRNA_\beta$ (the 3' end portion).

The continuous 28S rRNA is derived from precursors with or without splicing. Splicing at one level of pre-rRNA has been demonstrated in *Physarum* (232) and *Tetrahymena* (233) and at another in pre-28S of yeast mitochondria (253). The 21S rRNA of yeast mitochondria is derived from pre-21S rRNA ($5.1–5.4 \times 10^3$ bases) by two alternative pathways: (*a*) 3'-end processing (800 bases) followed by splicing (1.2×10^3 bases) and (*b*) splicing followed by the 3'-end processing (253).

The discontinuous 28S $rRNA_\alpha$ and 28S $rRNA_\beta$ are derived from the nearly "mature" 28S rRNA (28S $rRNA_\alpha$-gap-28S $rRNA_\beta$) or the "newly synthesized 28S rRNA" by cleavages involved in the elimination of the "gap" (approximately 500 bases) (251, 264). This type of rRNA is observed in insects and thought to result from hidden breakage of newly synthesized 28S rRNA (264). Some prokaryotic ribosomal subunits contain a 14S rRNA and a 16S rRNA instead of the usual 23S rRNA (265).

Complete sequences of 28S or 23S rRNA are known for human mitochondria (1159 bases) (235), mouse mitochondria (1589 bases) (236), *Zea mays* chloroplast (241), and *E. coli* (2904 bases) (247). Partial sequences have been obtained for yeast (266) and *Xenopus* (267). There appears to be greater than 70% homology of mitochondria and chloroplast RNA to *E. coli* RNA.

Secondary structures have been constructed for 23S rRNAs from prokaryotes, chloroplast 23 rRNA from *Z. mays,* and mitochondrial 16S rRNA from human and mouse (268). 7 domains (I–VII) are observed; four are conserved (II, III, V, VI). It has been proposed that the domain V or VI is the site of the IVS in other organisms.

5.8S rRNA   One group theorizes that 5.8S rRNA resulted from the mutations of 5S rRNA from *E. coli* (269) while others propose that it evolved from the 5' end of 23S rRNA (270, 271). In yeast, 5.8S rRNA is cleaved from 7S RNA produced by 3' processing (252).

There are two forms of 5.8S rRNAs, the normal 5.8S rRNA (160 bases) and a cleaved form consisting of 5.8S rRNA$_\alpha$ (130 bases) and 5.8S rRNA$_\beta$ (30 bases), sometimes referred to as "normal" 5.8S rRNA (α) and 2S rRNA (β). The cleaved form has been observed in *Drosophila* (272) and *Sciara* (273), where pre-5.8S rRNA contains a spacer of 28 bases; the 28 base spacer is cleaved out during maturation. It has been suggested that the cleaved 5.8S rRNA is hydrogen bonded to a cleaved form of 28S rRNA$_{\alpha,\beta}$ (274). Recently, Erdmann compiled the 5.8S rRNA structures from 12 different species (275).

4.5S rRNA   4.5S rRNA (80–100 bases) is a small rRNA found in chloroplasts of higher plants. Its origin has been suggested to be by 3' fragmentation of progenitor to 23S rRNA from *E. coli* (271, 253). The primary sequences and structural homologies have been described for a number of higher plants (271, 276–279). In tobacco (279) and *Zea mays* (241) the linkage is 23S rRNA - spacer - 4.5S (103 bases)-spacer (256 bases)-5S rRNA (121 bases).

Role of snRNAs in rRNA Processing

U3 RNA and 8S RNA are two of the snRNAs located in the nucleolus. U3 and 8S RNAs are hydrogen bonded to precursor ribosomal RNAs. (51, 97) It has been suggested that U3 RNA or 8S RNAs have complementarity to regions where specific cleavage of 45S and other ribosomal RNA precursors are cleaved to yield mature 5.8, 18, and 28S RNAs.

Some evidence supports a role for nucleolus specific snRNAs in processing pre-rRNA (51, 97). Two types of small RNAs are associated with nucleolar 28S RNA. One, 5.8S rRNA, is associated with 28S RNA in an equimolar ratio, and the other snRNAs, U3 RNA and 8S snRNA, are in a molar ratio of 0.3 (97). An additional 7–1 snRNA is associated with nucleolar 28S RNA (95). Two species of nucleolar 28S RNAs may exist, 28S:5.8S± snRNAs (U3, 7S-1, and 8S). In addition, 32S pre-28S RNA is associated with U3 snRNA (97). Accordingly, U3 snRNA may exist as a stable complex with both nucleolar 28S RNA and 32S RNA, and may be an important component in processing.

Pre-rRNA Splicing

Direct evidence for splicing mechanisms has been obtained from in vitro systems using *Tetrahymena* nuclei or nucleoli. The initial ribosomal RNA transcript is cleaved at two positions 407 nucleotides apart and spliced to form a 35S pre-rRNA (282, 283). The excised linear IVS (407 bases) forms a circular product (282, 283). The double splice system of Physarum has neither a simultaneous attack nor a preferential site in splice order (232). To define the junction point, rRNA/IVS/rRNA, a number of structural determinations have been made. The systems studies include Tetrahymena (IVS = 407 bases, splice junctions TCTCT/AATTG——AATCG/TAAG) (284), Chlamydomonas = 870 bases, splice junctions CGT/AAA—— AGG/CGT (246), and yeast mitochondria (IVS = 1143 bases, splice junction GGATA/ATT——TGA/ACAGG) (285, 286). The splicing mechanism for the RNA polymerase I system differs from those of the RNA polymerase II and III systems (15). The functions of excised IVS are not known. The yeast mitochondrial IVS may be an mRNA because its structure contains the initiator codon AUG, codons for 235 amino acids, and the terminator codon UAA (285).

CONCLUSION AND PROSPECTS

Small RNAs discovered about 15 years ago are metabolically stable, conserved through evolution, and localized to specific subcellular compartments. The sequencing of small RNAs has resulted in the discovery of cap structures, and the nucleotide sequences of all the six capped snRNAs are defined, in some cases for several species. The discovery that patients with autoimmune diseases produce antibodies directed against small RNPs, and the hypothesis that U1 RNA may be involved in splicing hnRNAs, has

Table 4 Suggested functions of small RNAs

Function	References
Splicing of hnRNA	11–13
Processing tRNA precursors	150–157
Processing of ribosomal precursors	97, 281
Intranuclear or nucleocytoplasmic transport	6, 12
DNA replication	165, 295
Modulators of transcription	6, 97
Stimulator of transcription	210, 296
Genetic reprogramming	294
Chromosome organization	294
Part of nuclear or chromatin structure	51, 149
Structural role in hnRNP particle	102
Induces embryonic heart differentiation	174
Control of translation	297, 298
Acts as incompatability factor	300
Control of cell division	299
Involved in crossing-over during meosis	301

brought small RNAs to the attention of many investigators. All the capped snRNAs appear to be synthesized by polymerase II, and other small RNAs by polymerase III. Genes having identical sequences to U1, U2, and U3 RNA have been isolated and, interestingly, the human genome appears to contain more pseudogenes for small RNAs than real genes.

Table 4 lists the suggested functions for small RNAs. Apparently, the different small RNAs have different functions that are varied and diverse. There is conclusive evidence for the involvement of an RNA component as part of RNase P in processing of tRNA precursors.

We have to repeat that the role of snRNAs in the processing of hnRNA is only a hypothesis, with which, however, some experimental evidence accumulated to date is consistent. There are reports that suggest that the model is incomplete, and alternative splicing mechanisms may exist that do not utilize snRNA. The results of Spritz et al (292) indicate that regions other than consensus regions of introns can affect processing. The possibility exists that the processing site and its interaction with snRNA may be affected by the secondary structure(s) of the RNA(s) involved. Accordingly, all processing sites may not be identical. Naora (293) reported that a better match between U1 RNA and the splice junction of insulin pre-mRNA could be made with our reported model for the secondary structure of U1 RNA (124), sequences of the exon, and the intron portions of insulin pre-mRNA. In addition, mitochondria have not been found to contain small RNA molecules, yet mitochondrial messengers are spliced. Clearly, as in the biochemists epitaph "more work needs to be done," particularly to evaluate a transport function.

Literature Cited

1. Busch, H., Ro-Choi, T. S., Prestayko, A. W., Shibata, H., Crooke, S. T., El-Khatib, S. M., Choi, Y. C. Mauritzen, C. M. 1971. *Perspect. Biol. Med.* 15: 117–39
2. Busch, H., Choi, Y. C., Daskal, I., Inagaki, A., Olson, M. O. J., Reddy, R., Ro-Choi, T. S., Shibata, H., Yeoman, L. C. 1972. *Gene Transcription in Reproductive Tissue. Karolinska Symposia on Research Methods in Reproductive Endocrinology,* ed. E. Diczfalusy, Stockholm: Karolinska Inst. pp. 36–66
3. Ro-Choi, T. S., Busch, H. 1974. *Cell Nucleus* 3:151–208
4. Busch, H. 1976. *Perspect. Biol. Med.* 19:549–67
5. Naora, H. 1977. In *The Ribonucleic Acids,* ed. P. R. Stewart, D. S. Lethan, pp. 61–71. New York/Heidelberg/Berlin: Springer

6. Hellung-Larsen, P. 1977. *Low Molecular Weight RNA Components in Eukaryotic Cells.* Copenhagen: FADL's Forlag
7. Choi, Y. C., Ro-Choi, T. S. 1980. *Cell Biol.* 3:609–67
8. Reddy, R., Busch, H. 1981. *Cell Nucleus* 8:261–306
9. Weinberg, R. A. 1973. *Ann. Rev. Biochem.* 42:329–54
10. Zieve, G. 1981. *Cell* 25:296–97
11. Murray, V., Holliday, R. 1979. *FEBS Lett.* 106:6–7
12. Lerner, M. R., Boyle, J. A., Mount, S. M., Wolin, S. L., Steitz, J. A. 1980. *Nature* 283:220–24
13. Rogers, J., Wall, R. 1980. *Proc. Natl. Acad. Sci. USA* 77:1877–79
14. Roberts, R. 1980. *Nature* 283:132–33

648 BUSCH ET AL

15. Breathnach, R., Chambon, P. 1981. *Ann. Rev. Biochem.* 50:349–83
16. Abelson, J. 1979. *Ann. Rev. Biochem.* 48:1035–69
17. Crick, F. 1979. *Science* 204:264–71
18. Sharp, P. A. 1981. *Cell* 23:643–46
19. Lerner, M. R., Steitz, J. A. 1979. *Proc. Natl. Acad. Sci. USA* 76:5495–99
20. Lerner, E. A., Lerner, M. R., Janeway, C. A. Jr., Steitz, J. A. 1981. *Proc. Natl. Acad. Sci. USA* 78:2737–41
21. Lerner, M. R., Steitz, J. A. 1981. *Cell* 25:298–300
22. Douvas, A., Tan, E. M. 1981. *Cell Nucleus* 8:369–87
23. Tan, E. M. 1979. *Cell Nucleus* 7:457–77
24. Muramatsu, M., Hodnett, J. L., Busch, H. 1966. *J. Biol. Chem.* 241:1544–47
25. Hodnett, J. L., Busch, H. 1968. *J. Biol. Chem.* 243:6334–42
26. Nakamura, T., Prestayko, A. W., Busch, H. 1968. *J. Biol. Chem.* 243: 1368–75
27. Moriyama, Y., Hodnett, J. L., Prestayko, A. W., Busch, H. 1969. *J. Mol. Biol.* 39:335–49
28. Prestayko, A. W., Busch, H. 1968. *Biochim. Biophys. Acta* 169:332–37
29. Prestayko, A. W., Tonato, M., Lewis, C., Busch, H. 1971. *J. Biol. Chem.* 246:182–87
30. Artman, M., Roth, J. S. 1971. *J. Mol. Biol.* 60:291–301
31. Heyden, H. W., Zachau, H. G. 1971. *Biochim. Biophys. Acta* 232:651–60
32. Chakravorthy, A., Busch, H. 1967. *Cancer Res.* 27:789–92
33. Higashi, K., Adams, H. R., Busch, H. 1966. *Cancer Res.* 2196–2201
34. Sporn, M. B., Dingman, C. W. 1963. *Biochim. Biophys. Acta* 68:389–400
35. Rosset, R., Monier, R. 1963. *Biochim. Biophys. Acta* 68:653–56
36. Hadjiolov, A. A., Venkov, P. V., Tsanev, R. 1966. *Anal. Biochem.* 17:263–67
37. Reich, P. R., Forget, B. G., Weissman, S. M., Rose, J. A. 1966. *J. Mol. Biol.* 17:428–39
38. Pene, J. J., Knight, E., Darnell, J. E. 1968. *J. Mol. Biol.* 33:609–23
39. Egyhazi, E., Daneholt, B., Edstrom, J. E., Labert, B., Ringborg, J. 1969. *J. Mol. Biol.* 44:517–32
40. Enger, M. D., Walters, R. A. 1970. *Biochemistry* 9:3551–62
41. Hellung-Larsen, P., Frederiksen, S. 1972. *Biochim. Biophys. Acta* 262:290–307
42. Hellung-Larsen, P., Frederiksen, S. 1977. *Comp. Biochem. Physiol. B* 58: 273–81

43. Larsen, C. J., Galibert, F., Lelong, J. C., Boiron, M. 1967. *CR Acad. Sci. D* 264:1523–26
44. Larsen, C. J., Galibert, F., Hampe, A., Boiron, M. M. 1968. *CR Acad. Sci.* 267:110–13
45. Larsen, C. J., Galibert, F., Hampe, A., Boiron, M. 1969. *Bull. Soc. Chim. Biol.* 51:649–68
46. Loening, U. E. 1967. *Biochem. J.* 102:251–57
47. Ro-Choi, T. S., Reddy, R., Henning, D., Busch, H. 1971. *Biochem. Biophys. Res. Commun.* 44:963–72
48. Ro-Choi, T. S., Moriyama, Y., Choi, Y. C., Busch, H. 1970. *J. Biol. Chem.* 245:1970–77
49. Weinberg, R. A., Penman, S. 1968. *J. Mol. Biol.* 38:289–304
50. Zapisek, W. F., Saponara, A. G., Enger, M. D. 1969. *Biochemistry* 8:1170–81
51. Zieve, G., Penman, S. 1976. *Cell* 8:19–31
52. Zieve, G., Benecke, B. J., Penman, S. 1977. *Biochemistry* 16:4520–25
53. Galibert, F., Lelong, J. C., Larsen, C. J., Boiron, M. 1967. *Biochim. Biophys. Acta* 142:89–98
54. Dingman, C. W., Peacock, A. 1968. *Biochemistry* 7:659–67
55. Moriyama, Y., Ip, P., Busch, H. 1970. *Biochim. Biophys. Acta* 209:161–70
56. Fredriksen, S., Tonnesen, T., Hellung-Larsen, P. 1971. *Arch. Biochem. Biophys.* 142:238–46
57. Fredriksen, S., Hellung-Larsen, P. 1972. *Exp. Cell Res.* 71:289–92
58. Rein, A., Penman, S. 1969. *Biochim. Biophys. Acta* 190:1–9
59. Rein, A. 1971. *Biochim. Biophys. Acta* 232:306–13
60. Yazdi, E., Gyorkey, F. 1971. *J. Natl. Cancer Inst.* 47:765–70
61. Nohga, K., Reddy, R., Busch, H. 1981. *Cancer Res.* 41:2215–20
62. Lerner, M. R., Boyle, J., Mount, S., Weliky, J., Wolin, S., Steitz, J. A. 1982. In *RNA Polymerase, tRNA and Ribosomes: Their Genetics and Evolution,* ed. S. Osawa, H. Ozeki, H. Uchida, T. Yura. Japan. In press
63. Branlant, C., Krol, A., Ebel, J. P., Lazar, E., Gallinaro, H., Jacob, M., Sri-Widada, J., Jeanteur, P. 1980. *Nucleic Acids Res.* 8:4143–54
63a. Mounts, S., Steitz, J. A. 1982. *Nucleic Acids Res.* 9:6351–68
64. Wise, J. A., Weiner, A. M. 1981. *J. Biol. Chem.* 256:956–63
64a. Takeishi, K., Kaneda, S. 1981. *J. Biochem.* 90:299–308

65. Reddy, R., Henning, D., Busch, H. 1979. *J. Biol. Chem.* 254:11097–11105
66. Wise, J. A., Weiner, A. M. 1980. Cell 22:109–18
66a. Diener, T. O. 1981. *Proc. Natl. Acad. Sci. USA* 78:5014–15
67. Marzluff, W. F., Murphy, E. C. Jr., Huang, R. C. C. 1974. *Biochemistry* 13:3689–96
68. McReynolds, L., Penman, S. 1974. *Cell* 1:139–45
69. Weinmann, R., Roeder, R. G. 1974. *Proc. Natl. Acad. Sci. USA* 71:1790–94
70. Roeder, R. G., Rutter, W. J. 1970. *Proc. Natl. Acad. Sci. USA* 65:675–82
71. Chesterton, C. J., Butterworth, P. H. W. 1971 *FEBS Lett.* 12:301–8
72. Reeder, R. H., Roeder, R. G. 1972. *J. Mol. Biol.* 70:433–41
73. Zybler, E. A., Penman, S. 1971. *Proc. Natl. Acad. Sci. USA* 68:2861–65
73a. Ro-Choi, T. S., Raj, N. B. K., Pike, L. M., Busch, H. 1976. *Biochemistry* 15:3823–28
74. Frederiksen, S., Hellung-Larsen, P., Gram-Jensen, E. 1978. *FEBS Lett.* 87:227–31
75. Gram-Jensen, E., Hellung-Larsen, P., Frederiksen, S. 1979. *Nucleic Acids Res.* 6:321–30
76. Eliceiri, G. L. 1980. *J. Cell Physiol.* 102:199–207
77. Tamm, I., Kikuchi, T., Darnell, J. E., Salditt-Georgieff, M. 1980. *Biochemistry* 19:2743–48
78. Roop, D. R., Kristo, P., Stumph, W. E., Tsai, M. J., O'Malley, B. W. 1981. *Cell* 23:671–80
79. Reichel, R., Benecke, B. 1980. *Nucleic Acids Res.* 8:225–33
80. Harada, F., Kato, N., Hoshino, H. O. 1979. *Nucleic Acids Res.* 7:909–18
81. Hendrick, J. P., Wolin, S. L., Rinke, J., Lerner, M. R., Steitz, J. A. 1981. *Mol. Cell. Biol.* 1:1138–49
82. Eliceiri, G. L. 1974. *Cell* 3:11–14
82a. Frederiksen, S., Hellung-Larsen, P. 1975. *FEBS Lett.* 58:374–78
83. Eliceiri, G. L., Sayavedra, M. S. 1976. *Biochem. Biophys. Res. Commun.* 72:507–12
84. Salditt-Georgieff, M., Harpold, M., Chen-Kiang, S., Darnell, J. E. Jr. 1980. *Cell* 19:69–78
85. Eliceiri, G. L. 1979. *Nature* 279:80–81
86. Denison, R., Van Arsdell, S., Bernstein, L., Weiner, A. 1981. *Proc. Natl. Acad. Sci. USA* 78:810–14
87. Van Arsdell, S. W., Denison, R. A., Bernstein, L. B., Weiner, A., Manser, T., Gesteland, R. F. 1981. *Cell* 26:11–17

88. Manser, T., Gesteland, R. F. 1981. *J. Mol. Appl. Genet.* 1:117–25
89. Lund, E., Dahlberg, J. E., Buckland, R., Cooke, H. 1982. *Proc. 10th ICN-UCLA Symp. Mol. Cell Biol.* 109:
89a. Hayashi, K. 1981. *Nucleic Acids Res.* 9:3379–88
89b. Ohshima, Y., Okada, N., Tani, T., Itoh, Y., Itoh, M. 1981. *Nucleic Acids Res.* 9:5145–58
90. Alonso, A., Krieg, L., Winter, H., Sekeris, C. E. 1980. *Biochem. Biophys. Res. Commun.* 95:148–55
91. Engberg, J., Hellung-Larsen, P., Frederiksen, S. 1974. *Eur. J. Biochem.* 41:321–28
92. Marzluff, W. F., White, E. L., Benjamin, R., Huang, R. C. C. 1975. *Biochemistry* 14:3715–24
93. Frederiksen, S., Flodgard, H., Hellung-Larsen, P. 1981. *Biochem. J.* 193:743–48
94. Gunning, P. W., Austin, L., Jeffrey, P. L. 1979. *J. Neurochem.* 32:1725–36
95. Reddy, R., Li, W.-Y., Henning, D., Choi, Y. C., Nohga, K., Busch, H. 1981. *J. Biol. Chem.* 256:8452–59
96. Gurney, T., Eliceiri, G. 1980. *J. Cell Biol.* 87:398–403
97. Prestayko, A. W., Tonato, M., Busch, H. 1971. *J. Mol. Biol.* 47:505–15
98. Sekeris, C. E., Niessing, J. 1975. *Biochem. Biophys. Res. Commun.* 62:642–50
99. Sekeris, C. E., Guialis, A. 1981. *Cell Nucleus* 8:247–59
100. Northemann, W., Klump, H., Heinrich, P. C. 1979. *Eur. J. Biochem.* 99:447–56
101. Northemann, W., Scheurlen, M., Gross, V., Heinrich, P. C. 1977. *Biochem. Biophys. Res. Commun.* 76:1130–37
102. Zieve, G., Penman, S. 1981. *J. Mol. Biol.* 145:501–23
103. Jacob, M., Devilliers, G., Fuchs, J-P., Gallinaro, H., Gattoni, R., Judes, C., Stevenin, J. 1981. *Cell Nucleus* 8:193–246
104. Deimel, B., Louis, C., Sekeris, C. E. 1977. *FEBS Lett.* 73:80–84
105. Gallinaro, H., Jacob, M. 1981. *Biochim. Biophys. Acta* 652:109–20
106. Gallinaro, H., Jacob, M. 1979. *FEBS Lett.* 104:176–82
107. Guimont-Ducamp, C., Sri-Widada, J., Jeanteur, P. 1977. *Biochimie* 59:755–58
108. Howard, E. F. 1978. *Biochemistry* 17:3228–36
109. Seifert, H., Scheurlen, M., Northemann, W., Heinrich, P. C. 1979. *Biochim. Biophys. Acta* 564:55–66

110. Flytzanis, C., Alonso, A., Louis, C., Krieg, L., Sekeris, C. E. 1978. *FEBS Lett.* 96:201–6
111. Brunel, C., Sri-Widada, J., Lelay, M.-N., Jeanteur, P., Liautard, J.-P. 1981. *Nucleic Acids Res.* 9:815–30
112. Maxwell, E. S., Maundrell, K., Scherrer, K. 1980. *Biochem. Biophys. Res. Commun.* 97:875–82
112a. Calvet, J. P., Pederson, T. 1981. *Cell* 26:363–70
113. Daskal, Y., Komaromy, L., Busch, H. 1980. *Exp. Cell Res.* 126:39–46
114. Daskal, Y. 1981. *Cell Nucleus* 8:117–37
115. Walker, T. A., Pace, N. R., Erikson, R. L., Erikson, E., Behr, F. 1974. *Proc. Natl. Acad. Sci. USA* 71:3390–94
116. Gunning, P. W., Beguin, P., Shooter, E. M., Austin, L., Jeffrey, P. L. 1981. *J. Biol. Chem.* 256:6670–75
117. Jelinek, W., Leinwand, L. 1978. *Cell* 15:205–14
117a. Miller, T. E., Huang, C. Y., Pogo, A. O. 1978. *J. Cell Biol.* 76:692–704
118. Reddy, R., Ro-Choi, T. S., Henning, D., Busch, H. 1974. *J. Biol. Chem.* 249:6486–94
119. Ro-Choi, T. S., Reddy, R., Choi, Y. C., Raj, N. B., Henning, D. 1974. *Fed. Proc.* 33:1548
120. Rottman, F., Shatkin, A., Perry, R. 1974. *Cell* 3:1977–79
121. Shatkin, A. J. 1976. *Cell* 9:645
122. Shibata, H., Reddy, R., Henning, D., Ro-Choi, T. S., Busch, H. 1974. *Mol. Cell. Biochem.* 4:3–19
123. Krol, A., Branlant, C., Lazar, E., Gallinaro, H., Jacob, M. 1981. *Nucleic Acids Res.* 9:841–58
123a. Reddy, R., Henning, D., Busch, H. 1981. *Biochim. Biophys. Res. Commun.* 98:1076–83
124. Epstein, P., Reddy, R., Busch, H. 1981. *Proc. Natl. Acad. USA* 78:1562–66
125. Korn, L. J., Queen, C. L., Wegman, N. M. 1977. *Proc. Natl. Acad. Sci. USA* 74:4401–5
126. Shibata, H., Ro-Choi, T. S., Reddy, R., Choi, Y. C., Henning, D., Busch, H. 1975. *J. Biol. Chem.* 250:3909–20
126a. Reddy, R., Henning, D., Epstein, P., Busch, H. 1981. *Nucleic Acids Res.* 9:5645–59
127. Ro-Choi, T. S., Choi, Y. C., Henning, D., McCloskey, J., Busch, H. 1975. *J. Biol. Chem.* 250:3921–28
128. Busch, H., Smetana, K. 1970. In *The Nucleolus,* pp. 285–314. New York: Academic
129. Reddy, R., Henning, D., Busch, H. 1980. *J. Biol. Chem.* 255:7029–33
130. Reddy, R., Ro-Choi, T. S., Henning, D., Shibata, H., Choi, Y. C., Busch, H. 1972. *J. Biol. Chem.* 247:7245–50
131. Reddy, R., Henning, D., Busch, H. 1981. *J. Biol. Chem.* 256:3532–38
132. Krol, A., Branlant, C., Lazar, E., Gallinaro, H., Jacob, M. 1981. *Nucleic Acids Res.* 9:2699–2716
133. Kato, N., Harada, F. 1981. *Biochem. Biophys. Res. Commun.* 99:1477–85
134. Krol, A., Gallinaro, H., Lazar, E., Jacob, M., Branlant, C., 1981. *Nucleic Acids. Res.* 9:769–88
135. Kato, N., Harada, F. 1981. *Biochem. Biophys. Res. Commun.* 99:1468–76
136. Puvion, E., Moyne, G. 1981. *Cell Nucleus* 8:59–115
137. Epstein, P., Reddy, R., Henning, D., Busch, H. 1980. *J. Biol. Chem.* 255:8901–6
138. Harada, F., Kato, N., Nishimura, S. 1980. *Biochem. Biophys. Res. Commun.* 95:1332–40
138a. Busch, H., Reddy, R., Henning, D., Epstein, P. 1980. In *International Cell Biology,* ed. H. Schweiger, pp. 47–52. Berlin: Springer
139. Raj, N., Ro-Choi, T. S., Busch, H. 1975. *Biochemistry* 14:80–87
140. Fuchs, J. P., Jacob, M. 1979. *Biochemistry* 18:4202–8
141. Kinlaw, C. S., Berget, S. M. 1981. *Fed. Proc.* 40:1765
142. White, P. J., Gardner, W. D., Hoch, S. O. 1981. *Proc. Natl. Acad. Sci. USA* 78:626–30
143. Liew, T. S., Tan, E. M. 1979. *Arth. Rheum.* 22:635
144. Takano, M., Golden, S. S., Sharp, G. C., Agris, P. 1981. *Biochemistry* 20:5929–36
145. Takano, M., Agris, P. F., Sharp, G. C. 1981. *Clin. Res.* 28:559A
146. Teppo, A. M., Gripenberg, M., Kurk, P., Baklein, K., Helve, T., Wegelius, O. 1981. *Biochim. Biophys. Acta.* In press
147. Yang, V. W., Lerner, M., Steitz, J. A., Flint, S. J. 1981. *Proc. Natl. Acad. Sci. USA* 78:1371–75
148. Avvedimento, V. E., Vogeli, G., Yamada, Y., Maizel, J. E., Pastan, I., Crombrugghe, B. 1980. *Cell* 21:689–96
148a. Dickson, A., Ninomiya, Y., Bernard, P., Pesciotta, M., Parsons, J., Green, G., Eikenberry, E. F., de Crombrugghe, B., Vogeli, G., Pastan, I., Fietzek, P., Olsen, B. R. 1981. *J. Biol. Chem.* 16:8407–15
148b. Ohshima, Y., Itoh, M., Okada, N., Miyata, T. 1981. *Proc. Natl. Acad. Sci. USA* 78:4471–74

149. Pederson, T., Bhorjee, J. S. 1979. *J. Mol. Biol.* 128:451–80
150. Altman, S. 1978. *Int. Rev. Biochem.* 17:19–44
151. Altman, S., Bowman, E., Garber, R., Kole, R., Koski, R., Stark, B. C. 1980. *tRNA: Biological Aspects,* pp. 71–82. Cold Spring Harbor, NY: Spring Harbor Lab.
152. Bothwell, A. L. M., Garber, R. L., Altman, S. 1976. *J. Biol. Chem.* 251:7709–16
153. Bowman, E. J., Altman, S. 1980. *Biochim. Biophys. Acta* 613:439–47
154. Garber, R. L., Siddiqui, M. A. Q., Altman, S. 1978. *Proc. Natl. Acad. Sci. USA* 75:635–39
155. Kole, R., Altman, S. 1979. *Proc. Natl. Acad. Sci. USA* 76:3795–99
156. Kole, R., Baer, M. F., Stark, B. C., Altman, S. 1980. *Cell* 19:881–87
157. Koski, R., Bothwell, A., Altman, S. 1976. *Cell* 9:101–16
158. Blanchard, J. M., Weber, J., Jelinek, W., Darnell, J. E. Jr. 1978. *Proc. Natl. Acad. Sci. USA* 75:5344–48
159. Harada, H., Igarashi, T., Muramatsu, M. 1980. *Nucleic Acids Res.* 8:587–99
160. Goldenberg, C. J., Raskas, H. J. 1980. *Biochemistry* 19:2719–23
161. Manley, J-L., Sharp, P., Gefter, M. 1979. *Proc. Natl. Acad. USA* 76:160–64
162. Tsai, M. J., Ting, A. C., Nordstrom, J. L., Zimmer, W., O'Malley, B. W. 1980. *Cell* 22:219–30
163. Lerner, M., Boyle, J., Hardin, J. A., Steitz, J. A. 1981. *Science* 211:400–2
164. Harada, F., Kato, N. 1980. *Nucleic Acids Res.* 1273–85
164a. Haynes, S. R., Toomey, T. P., Leinwand, L., Jelinek, W. R. 1981. *Mol. Cell. Biol.* 1:573–83
165. Jelinek, W. R., Toomey, T. P., Leinwand, L., Duncan, C. H., Biro, P. A., Choudary, P. V., Weissman, S. M., Rubin, C. M., Houck, C. M., Deininger, P. L., Schmid, C. W. 1980. *Proc. Natl. Acad. Sci. USA* 77:1398–402
166. Ro-Choi, T. S., Reddy, R., Henning, D., Takano, T., Taylor, C., Busch, H. 1972. *J. Biol. Chem.* 247:3205–22
167. Lerner, M. R., Andrews, N. C., Miller, G., Steitz, J. A. 1981. *Proc. Natl. Acad. Sci. USA* 78:805–9
167a. Gojobori, T., Nei, M. 1981. *J. Mol. Evol.* 17:245–50
168. Bishop, J. M., Levinson, W., Sullivan, D., Farrshier, L., Quintrell, N., Jackson, J. 1970. *Virology* 42:927–37
169. Erikson, E., Erikson, R. L., Henry, B., Pace, N. R. 1973. *Virology* 53:40–46
170. Li, W-Y., Reddy, R., Henning, D., Epstein, P., Busch, H. 1982. *J. Biol. Chem.* In press
171. Weiner, A. M. 1980. *Cell* 22:209–18
172. Savage, H., Grinchishin, V., Fang, W., Busch, H. 1974. *Physiol. Chem. Phys.* 6:113–26
173. Benecke, B., Penman, S. 1977. *Cell:* 12:939–46
174. Deshpande, A. K., Jakowlew, S. B., Arnold, H., Crawford, P. A., Siddiqui, M. A. Q. 1977. *J. Biol. Chem.* 252:6521–27
175. Gunning, P. W., Shooter, E. M., Austin, L., Jeffrey, P. L. 1981. *J. Biol. Chem.* 256:6663–69
176. Pan, J., Elder, J. T., Duncan, C. H., Weissman, S. M. 1981. *Nucleic Acids Res.* 9:1161–70
177. Duncan, C. H., Jagadeeswaran, P., Wang, R., Weissman, S. M. 1981. *Gene.* In press
177a. Haynes, S. R., Jelinek, W. R. 1981. *Proc. Natl. Acad. Sci. USA* 78:6130–34
178. Hutchinson, M. A., Hunter, T., Eckhart, W. 1979. *Cell* 15:65–77
179. Alwine, J. C., Khoury, G. 1980. *J. Virol.* 36:701–8
180. Colonno, R. J., Banerjee, A. K. 1976. *Cell* 8:197–204
181. Testa, D., Chands, P. K., Banerjee, A. K. 1980. *Cell* 21:267–75
182. Emerson, S. V., Dierks, P. M., Parson, J. T. 1977. *J. Virol.* 23:708–16
183. Rao, D. D., Huang, A. S. 1978. *Proc. Natl. Acad. Sci. USA* 76:3742–45
184. Peters, G. G., Harada, F., Dahlberg, J. E., Panet, A., Haseltine, W. A., Baltimore, D. 1977. *J. Virol.* 21:1031–42
185. Harada, F., Sawyer, R. C., Dahlberg, J. E. 1975. *J. Biol. Chem.* 250:3487–97
186. Diener, T. O. 1981. *Sci. Am.* 244:66–73
187. Koper, J. M., Tousignant, M. E., Lot, H. 1976. *Biochem. Biophys. Res. Commun.* 72:1237–43
188. Mathews, M. B., Pettersson, U. 1978. *J. Mol. Biol.* 119:293–328
189. Soderlund, H., Pettersson, U., Vennstrom, B., Philipson, L., Mathews, M. B. 1976. *Cell* 7:585–93
190. Raska, K., Schuster, L. M., Varrichio, F. 1976. *Biochem. Biophys. Res. Commun.* 69:79–84
191. Weinmann, R., Grendler, T. G., Raskas, H., Roeder, R. G. 1976. *Cell* 7:557–66
192. Weill, P. A., Segall, J., Harris, B., Ng, S. Y., Roeder, R. G. 1979. *J. Biol. Chem.* 254:6163–73
193. Weinmann, R., Raskas, H. J., Roeder, R. G. 1974. *Proc. Natl. Acad. Sci. USA* 71:3426–30

194. Price, R., Penman, S. 1972. *J. Virol.* 9:621–26
195. Ohe, K., Weissman, S. M., Cooke, N. 1969. *J. Biol. Chem.* 244:5320–32
196. Ohe, K., Weissman, S. M. 1979. *J. Biol. Chem.* 245:6991–7009
197. Harris, B., Roeder, R. G. 1978. *J. Biol. Chem.* 253:4120–27
198. Mathews, M. B. 1975. *Cell* 6:223–29
199. Celma, M. L., Pan, J., Weissman, S. M. 1977. *J. Biol. Chem.* 252:9032–42
200. Pan, J., Celma, M. L., Weissman, S. M. 1977. *J. Biol. Chem.* 252:9047–54
201. Vennstrom, B., Pettersson, U., Philipson, L. 1978. *Nucleic Acids Res.* 5:195–204
202. Akusjarvi, G., Mathews, M. B., Andersson, P., Vennstrom, B., Petterson, U. 1980. *Proc. Natl. Acad. Sci. USA* 77:2424–28
202a. Mathews, M. B., Grodzicker, T. 1981. *J. Virol.* 38:849–62
203. Mathews, M. B. 1980. *Nature* 285:575–77
204. Blanchard, J. 1980. *Biochem. Biophys. Res. Commun.* 97:524–29
205. Gallinaro, H., Gattoni, R., Stevenin, J., Jacob, M. 1980. *Biochem. Biophys. Res. Comm.* 95:20–26
206. Beltz, G., Flint, S. J. 1979. *J. Mol. Biol.* 131:353–73
206a. Shenk, T., Fowlkes, D. M., Thimmappaya, B., Weinberger, C. 1981. *Fed. Proc.* 40:1755
207. Alwine, J. C., Dhar, R., Khoury, G. 1980. *Proc. Natl. Acad. Sci. USA* 77:1379–83
208. Weissman, S. M. 1980. *Mol. Cell. Biochem.* 35:29–38
209. Mark, D. F., Berg, P. 1980. *Cold Spring Harbor Symp. Quant. Biol.* 44:55–62
210. Krause, M. O., Ringuette, M. J. 1977. *Biochem. Biophys, Res. Commun.* 76:796–803
210a. Rosa, M. D., Gottlieb, E., Lerner, R. M. and Steitz, J. A. 1981. *Mol. Cell Biol.* 1:785–96
211. Emerson, S. V., Dierks, P. M., Parsons, J. T. 1977. *J. Virol.* 23:708–16
212. Schubert, M., Keene, J. D., Lazzarini, R. A., Emerson, S. V. 1978. *Cell* 15:103–12
213. Semler, B. L., Perrault, J., Abelson, J., Holland, J. J. 1978. *Proc. Natl. Acad. Sci. USA* 75:4704–8
214. Rao, D. D., Huang, A. S. 1980. *J. Virol.* 36:756–65
215. Leppert, M., Rittenhouse, L., Perrault, J., Summers, D. F., Kolakofsky, D. 1979. *Cell* 18:735–48
216. Collonno, R. J., Banerjee, A. K. 1978. *Cell* 15:93–101

217. Bonar, R. A., Sverak, L., Bolognesi, D. P., Langlors, A. J., Beard, D., Beard, J. W. 1967. *Cancer Res.* 27:1138–57
218. Duesberg, P. H. 1968. *Proc. Natl. Acad. Sci. USA* 60:1511–18
219. Bishop, M. M., Levinson, W. E., Quintrell, N., Sullivan, D., Fanshier, L., Jackson, J. 1970. *Virology* 42:182–95
220. Sawyer, R. C., Harada, F., Dahlberg, J. E. 1974. *J. Virol.* 13:1302–11
221. Dahlberg, J. E., Sawyer, R. C., Taylor, J. M., Faras, A. J., Levinson, W. E., Goodman, H. M., Bishop, J. 1974. *J. Virol.* 13:1126–33
222. Harada, F., Taylor, J. M., Levinson, W. E., Bishop, J. M., Goodman, H. M. 1974. *J. Virol.* 13:1134–42
223. Harada, F., Peters, G. G., Dahlberg, J. E. 1979. *J. Biol. Chem.* 254:10979–85
224. Peters, G. G., Glover, C. 1980. *J. Virol.* 35:31–40
225. Harada, F., Kato, N., Hoshino, H. 1979. *Nucleic Acids Res.* 7:909–17
226. Perry, R. P. 1976. *Ann. Rev. Biochem.* 45:605–29
227. Schlessinger, D. 1980. *Ribosomes, Structure, Function and Genetics,* ed. G. Chambliss et al. pp. 767–80. Baltimore: Univ. Park Press
228. Planta, R. J., Meyerink, J. H. 1980. See Ref. 227, pp. 872–87
229. Boyington, J. E., Gillham, N. W., Lambowitz, A. M. 1980. See Ref. 227, pp. 903–49
230. Long, E. O., Dawid, I. B. 1980. *Ann. Rev. Biochem.* 49:727–64
231. Busch, H., Rothblum, L., eds. 1982. *The Cell Nucleus* Vols. 9, 10. In press
232. Gubler, U., Wyler, T., Seebeck, T., Braun, R. 1980. *Nucleic Acids Res.* 8:2647–64
233. Engberg, J., Din, N., Eckert, W. A., Kaffenberger, W., Pearlman, R. E. 1980. *J. Mol. Biol.* 142:289–313
234. Leon, W., Fouts, D. L., Manning, J. 1978. *Nucleic Acids Res.* 5:491–504
235. Anderson, S., Bankier, A. T., Barrell, B. G., de Bruijn, M. H. L., Coulson, A. R., Drouin, J., Eperon, I. C., Nierlich, D. P., Roe, B. A., Sanger, F., Schreier, P. H., Smith, A. J. H., Staden, R., Young, I. G. 1981. *Nature* 290:457–65
236. Van Etten, R. A., Walberg, M. W., Clayton, D. A. 1980. *Cell* 22:157–70
237. Locker, J., Morimoto, R., Synenki, R. M., Rabinowitz, M. 1980. *Curr. Genet.* 1:163–72
238. Hahn, U., Lazarus, C. M., Lunsdorf, H., Kuntzel, H. 1979. *Cell* 17:191–200
239. Heckman, J. E., Yin, S., Alzner-DeWeerd, B., RajBhandary, U. L. 1979. *J. Biol. Chem.* 254:12694–700

240. Schwarz, Z., Kossel, H. 1980. *Nature* 283:739–42
241. Edwards, K., Kossel, H. 1981. *Nucleic Acids. Res.* 9:2853–69
242. Hartley, M. R. 1979. *Eur. J. Biochem.* 96:311–20
243. Gray, P. W., Hallick, R. B. 1978. *Biochemistry* 17:284–89
244. Orozco, E. M. Jr., Rushlow, K. E., Dodd, J. R., Hallick, R. B. 1980. *J. Biol. Chem.* 255:10397–11003
245. Malnoe, P., Rochaix, J.-D. 1978. *Mol. Gen. Genet.* 166:269–75
246. Allet, B., Rochaix, J.-D. 1979. *Cell* 18:55–60
247. Brosius, J., Dull, T. J., Sleeter, D. D., Noller, H. F. 1981. *J. Mol. Biol.* 148:107–27
248. Planta, R. J., Retel, J., Klootwijk, J., Meyerink, J. H., DeJonge, P., Van Keulen, H., Brand, R. C. 1977. *Biochem. Soc. Trans.* 5:462–66
249. Maizels, N. 1976. *Cell* 9:431–38
250. Lund, E., Dahlberg, J. E. 1977. *Cell* 11:247–62
251. Long, E. O., Dawid, I. B. 1980. *J. Mol. Biol.* 138:873–78
252. Veldman, G. M., Brand, R. C., Klootwijk, J., Planta, R. J. 1980. *Nucleic Acids Res.* 8:2907–20
253. Merten, S., Synenki, R. M., Locker, J., Christianson, T., Rabinowitz, M. 1980. *Proc. Natl. Acad. Sci. USA* 77:1417–21
254. Mannella, C. A., Collins, R. A., Green, M. R., Lambowitz, A. M. 1979. *Proc. Natl. Acad. Sci. USA* 76:2635–39
255. Bram, R. J., Young, A. A., Steitz, J. A. 1980. *Cell* 19:393–402
256. Dahlberg, A. E., Dahlberg, J. E., Lund, E., Tokimatsu, H., Rabson, A. B., Calvert, P. C., Reynold, F., Zahalak, M. 1978. *Proc. Natl. Acad. Sci. USA* 75:3598–602
257. Gegenheimer, P., Apirion, D. 1980. *J. Mol. Biol.* 143:227–57
258. Egawa, K., Choi, Y. C., Busch, H. 1971. *J. Mol. Biol.* 56:565–77
259. Rubtsov, P. M., Musakhanov, M. M., Zakharyev, V. M., Krayev, A. S., Skryabin, K. G., Bayev, A. A. 1980. *Nucleic Acids Res.* 8:5779–94
260. Salim, M., Maden, B. E. H. 1981. *Nature* 291:205–8
261. Choi, Y. C., Busch, H. 1978. *Biochemistry* 17:2551–60
262. Woese, C. R., Magrum, L. J., Gupta, R., Siegel, R. B., Stahl, D. A., Kop, J., Crawford, N., Brosius, J., Gutell, R., Hogan, J. J. 1980. *Nucleic Acids. Res.* 8:2275–94
263. Goltz, C., Brimacombe, R. 1980. *Nucleic Acids Res.* 8:2377–96
264. Jordan, B. R. 1975. *J. Mol. Biol.* 98:277–80
265. MacKay, R. M., Zablen, L. B., Woese, C. R., Doolittle, W. F. 1979. *Arch. Microbiol.* 123:165–72
266. Bayev, A. A., Georgiev, O. J., Hadjiolov, A. A., Nikolaev, N., Skryabin, K. G., Zakharyev, V. M. 1981. *Nucleic Acids Res.* 9:789–99
267. Moss, T., Boseley, P. G., Birnstiel, M. L. 1980. *Nucleic Acids Res.* 8:467–85
268. Branlant, C., Krol, A., Machatt, M. A., Pouyet, J., Ebel, J.-P., Edwards, K., Kossel, H. 1981. *Nucleic Acids Res.* In press
269. Erdmann, V. A. 1976. *Prog. Nucleic Acid Res. Mol. Biol.* 18:45–90
270. Nazar, R. N. 1980. *FEBS Lett.* 119:212–14
271. Gerbi, S. A., Gourse, R. L., Clark, C. G. 1982. *Cell Nucleus* In press
272. Pavlakis, G. N., Jordan, B. R., Wurst, R. M., Vournakis, J. N. 1979. *Nucleic Acids Res.* 7:2213–38
273. Jordan, B. R., Latil-Dammotte, M., Jourdan, R. 1980. *Nucleic Acids Res.* 8:3565–773
274. Ishikawa, H. 1979. *Biochem. Biophys. Res. Commun.* 90:417–24
275. Erdmann, V. A. 1981. *Nucleic Acids Res.* 9:r25–r42
276. Bowman, C. M., Dyer, T. A. 1979. *Biochem. J.* 183:605–13
277. MacKay, R. M. 1981. *FEBS Lett.* 123:17–18
278. Wildeman, A. G., Nazar, R. N. 1980. *J. Biol. Chem.* 255:11896–11900
279. Takaiwa, F., Sugiura, M. 1980. *Mol. Gen. Genet.* 180:1–4
280. Ferrai, S., Yehle, C. O., Robertson, H. D. 1980. *Proc. Natl. Acad. Sci. USA* 77:2395–99
281. Denoya, C., Costa Giomi, P. C., Scodeller, E. A., Vasquez, C., La Torre, J. L. 1981. *Eur. J. Biochem.* 115:375–83
282. Carin, M., Jensen, B. F., Jentsch, K. D., Leer, J. C., Nielsen, O. F., Westergaard, O. 1980. *Nucleic Acids Res.* 8:5551–66
283. Brabowski, P. J., Zaug, A. J., Cech, T. R. 1981. *Cell* 23:467–76
284. Wild, M. A., Sommer, R. 1980. *Nature* 283:693–94
285. Dujon, B. 1980. *Cell* 20:185–97
286. Bos, J. L., Osinga, K. A., Vander Horst, G., Hecht, N. B., Tabak, H. F., Van Ommen, G. J. B., Borst, P. 1980. *Cell* 20:207–14
287. Haid, A., Grosch, G., Schmelzer, C., Schweyen, R. J., Kaudewitz, F. 1980. *Curr. Genet.* 1:155–61
288. Halbreich, A., Pajot, P., Foucher, M.,

Grandchamp, C., Slonimski, P. 1980. *Cell* 19:321–29
289. Van Ommen, G. J., Boer, P. H., Groot, G. S. P., DeHaan, M., Roosendaal, E., Grivell, L. A., Haid, A., Schweyen, R. J. 1980. *Cell* 20:173–83
290. Schmelzer, C., Haid, A., Grosch, G., Schweyen, R. J., Kaudewitz, F. 1981. *J. Biol. Chem.* 256:7610–19
291. Lewin, B. 1980. *Cell* 22:645–46
292. Spritz, R. A., Jagadeeswaran, P., Chowdary, P. V., Biro, P. A., Elder, J. T., DeRiel, J. K., Manley, J. L., Gefter, M. L., Forget, B. G., Weissman, S. M. 1981. *Proc. Natl. Acad. Sci. USA* 78:2455–59
293. Naora, H., Deacon, N. J. 1981. *Differentiation* 18:125–32

294. Goldstein, L., Wise, G. E., Ko, C. 1977. *J. Cell Biol.* 73:322–31
295. Pogo, A. O. 1981. *Cell Nucleus,* 8:337–67
296. Kanehisa, T., Oki, Y., Ikuta, K. 1974. *Biochim. Biophys. Acta* 277:584–89
297. Bester, A. J., Kennedy, D. S., Heywood, S. M. 1975. *Proc. Natl. Acad. Sci. USA* 72:1523–27
298. Rao, M. S., Blackstone, M., Busch, H. 1977. *Biochemistry* 16:2756–62
299. Howard, E., Stubblefield, E. 1971. *Exp. Cell Res.* 70:460–62
300. Tomizawa, J., Itoh, T., Selzer, G., Som, T. 1981. *Proc. Natl. Acad. Sci. USA* 78:1421–25
301. Hotta, Y., Stern, H. 1981. *Cell.* 27:309–19

Ann. Rev. Biochem. 1982. 52:655–93

CHEMICAL MUTAGENESIS

B. Singer[1] and J. T. Kuśmierek[2]

Department of Molecular Biology and Virus Laboratory, University of California, Berkeley, California 94720

CONTENTS

Perspectives and Summary

Mutagenesis is an inescapable biological event and not necessarily deleterious. Certainly in the context of Darwinian evolution, survival of the fittest implies that mutation can confer a genetic advantage, as well as disadvantage. However, the mutations we observe are usually detected as errors. For example, the frequency of human genetic disorders is extraordinarily high (1). The commonly held hypothesis that this is due to environmental chemicals and radiation is only partially true, since the intrinsic stability of nucleic acids themselves can be responsible for a high level of mutation (see section on mutations resulting from thermodynamic properties.)

[1]This work was supported by Grant CA 12316 from the National Cancer Institute and Grant PCM-7921649 from the National Science Foundation.

[2]Permanent address: Institute of Biochemistry and Biophysics, Polish Academy of Sciences, Warsaw 02532.

The classic definition of mutation is a heritable change in the genetic material. In order to be expressed, replication must occur, although mutation can also be lethal, and in fact, mutagenic efficiency is the ratio between mutational and lethal events. This ratio varies enormously in different systems. When transforming DNA is treated with a direct acting mutagen in vitro and then tested in vivo, cell toxicity and repair are not major considerations. However in bacterial or mammalian cells, mutagenesis involves metabolic activation (or deactivation), various repair systems, the state of the cell, and the effect of the mutagen on the physiology of the cell. For this reason it is useful to critically analyze mutagenesis data studied in different systems over many years, and not just that obtained from recently developed short-term tests.

In the last few years, many new chemical modifications of nucleic acids[3] have been elucidated and hypotheses have been advanced regarding their possible involvement in mutagenesis. Unfortunately, for many known mutagens, only a portion of the likely products has been isolated. This means that even when specific mispairing mechanisms are possible or have been shown, e.g. in transcription, direct mutagenesis by such chemicals may be the result of as yet uncharacterized products. Model experiments with several mutagens have revealed the important role of tautomerism and rotation about the glycosyl bond in both chemical and spontaneous mutagenesis. For other derivatives, the observed mutagenesis appears to be the result of the generation of apurinic or apyrimidinic sites.

Any covalently bound substituent or adduct may be seen as a point mutation or as frameshift, deletion, or insertion mutations. We focus on these mutagenic events, and do not consider intercalating agents, which are generally frameshift mutagens, UV and other radiation damage, or the effects of base analogues.

What is a Mutagen?

Although mutation has a single definition, measurement is difficult and largely qualitative. The researcher is limited in the type of mutational events that may be observed by phenotypic changes, quite apart from mutations that either do not result in amino acid replacements or result in a replacement that does not affect the function of a protein.

Nevertheless rough quantitative comparisons of mutagenic efficiency can be made within a given system, but in no way are such data absolute, because a "good" mutagen in one system e.g. hydroxylamine (HA) (2–6),

---

[3]Single letter abbreviations for nucleic acid bases are used to denote the base moiety in a nucleoside, nucleotide, or polynucleotide. Chemical reactions of free bases differ from those of nucleosides and are not included, since they do not generally represent reactions taking place in nucleic acids.

may be a "nonmutagen" in another (7, 8), even though metabolic activation is not required. A good mutagen, e.g. N-methyl-N'-nitro-N-nitrosoguanidine (MNNG) (7, 9–11), may also be extremely toxic to a particular cell so that, for equal survival, the test is performed at 10^{-5} the concentration of a related chemical with low cytotoxicity, e. g. ethylmethanesulfonate (EtMS) (11). Without correction for the amounts used, it would appear that EtMS is a potent mutagen. Some comparisons of mutagenicity in various systems, normalized for the amount of mutagen, are listed in Table 1. This table can not and does not relate mutagenicity to the absolute amount of modification. Methylating agents are about 20 times as reactive as ethylating agents (15), but to make the problem of extent of reaction more difficult, the various alkylating agents in Table 1 have varying chemical stabilities. The nitrosoureas decompose at pH 7, 37°, with a $t_{\frac{1}{2}} \simeq 20$ min (16), while the nitrosoguanidines are quite stable, with a $t_{\frac{1}{2}}$ varying from 1–20 h depending on the buffers used (17, 18), except in the presence of thiols, which can decrease the $t_{\frac{1}{2}}$ to $\simeq$ 1–2 min (17). Since thiols are present in cells it is likely that one of the decomposition products of MNNG is actually the reactive molecule after the initial alkylation. MNNG, in contrast to N-methyl-N-nitrosourea (MeNU) or methylmethanesulfonate (MeMS), reacts with the amino groups of proteins to form guanidino derivatives (19–21), which may represent 100-fold the DNA alkylation (22). MNNG has an extraordinarily high mutagenicity, not only in the systems in Table 1, but also in intracellular phage λ (22). In contrast, when λ was treated with MNNG extracellularly, the mutation frequency was low. These data led to the conclusion (22) that mutagenicity is not determined solely by the nature of the base alteration, but is highly dependent on the simultaneous chemical modification of some other cellular factors. This, and a dependence on conformation (23), may explain why MNNG is a poor mutagen in T4 and S13 phages, *Haemophilus influenzae* DNA, and TMV RNA (23–25).

As can also be seen in Table 1, the nitrosoureas are much less mutagenic than the nitrosoguanidines, yet the distribution of alkyl derivatives in nucleic acids is the same for both types of mutagens, (see section on chemical reactions of simple alkylating agents). This again points to the difficulty in assessing both mutation and the chemical event(s) that are responsible.

Before the advent of sophisticated mutagen testing involving revertants at specific loci, testing was primarily done by measuring forward mutations of in vitro mutagenized phage, transforming DNA, seeds, TMV RNA, or *Drosophila*. Except in the *Drosophila,* no metabolic or enzymatic activation of chemicals occurred, and there was no toxicity to consider other than degradation of the nucleic acid. With these techniques, mutagenesis could theoretically be related to specific chemical modification of a nucleic acid,

Table 1 Mutagenic efficiency of alkylating agents[a]

	Test system													
Reagent[b]	TA 100 (7)[c]		TA 100 (10)[d]		HGPRT⁻TG^R V79 (11)[e]		TG^R V79 (12)[f]		8 aza G V79 (13)[g]		8 aza G V79 (14)[g]		HGPRT CHO (14a, 14b)[h]	
MeMS	4	*0.63*	32	*350*	9	*1.25*	1.1	*15*	10	*5.7*	4.8	*46*	3.8	*140*
MeNU	28	*4.4*	43	*470*	210	*38*	40	*400*					7.5	*243*
MNNG	8,600	*1,375*	8,700	*≈10^5*	3.8 × 10^6	*30*	1,560	*70*	330	*79*	1,160	*125*	1,480	*200*
EtMS	1	*0.16*	1	*≈11*	1	*33*	1	*140*	1	*17.5*	1	*107*	1	*1,550*
EtNU	7	*1.1*	8	*≈85*	165	*44*	11	*300*					1.25	*550*
ENNG	2,200	*350*			1.4 × 10^6	*23*			720	*19.8*			216	*522*

[a] The first number in each column represents ratios of mutants obtained compared to EtMS. Data are calculated on the basis that for each series the concentration of each alkylating agent is identical. In many instances these figures differ markedly from those, shown in italics, reported by the authors who measured mutation at equal toxicity or survival, which in some cases requires the use of very low reagent concentration. One additional factor, not included in the calculations, is the difference in reactivity between methylating and ethylating agents, which gives the impression that the ethylating agents are less mutagenic than the corresponding methylating agents. If the extent of alkylation is taken into consideration, ethylating agents are almost always more mutagenic than the analogous methylating agent.

[b] For definitions of these abbreviations see pages 657 and 669.

[c] In italics are revertants per nanomole.

[d] In italics are revertants per micromole.

[e] In italics are revertants/10^6 survivors.

[f] In italics are mutants/10^5 cells at D$_{37}$ dose.

[g] In italics are mutants/10^5 survivors.

[h] In italics are mutants/10^6 survivors.

even though the expression of mutation involved replication in vivo. Nevertheless, our view of what constitutes a mutagen was greatly influenced by these early studies. For example, deamination by nitrous acid and a change in tautomerism by hydroxylamine can only be significantly produced under in vitro conditions where low pH or high salt are permissible.

When nucleic acids were treated in vitro, the observed mutagenicity and resultant amino acid exchanges (26) appeared to correlate with either a base change (i. e. by deamination) or a tautomeric shift (i. e. by hydroxylamine or methoxyamine). This simple explanation of mutation as a function of a single base change can not be extended to in vivo systems, where the reagent must penetrate the cell wall and may need to be activated, and the nucleic acid may be repaired by several pathways that themselves may be mutagenic (i. e. induced error-prone repair). Under these conditions it is not surprising that the measurement of mutation is greatly affected by the test system.

For example, 2-acetylaminofluorene (AAF) is not considered mutagenic in *Drosophila* (27), but in silkworm oocytes it is (28). The difference can not be explained on the basis of differing metabolic activation but may be determined by the genetic makeup. Mutagenesis by aflatoxin B_1 (AFB_1) and benzo(a)pyrene [B(a)P] is dependent on the cell type used for activation. Mutation in V79 cells is positive for B(a)P and negative for AFB_1, when rat fibroblasts furnish the activating enzyme, and the opposite is true when rat liver cells are used (29). MeMS, in the Ames assay, is the only alkylating agent tested that was positive in Salmonella TA 100 but negative in Salmonella TA 1535 (30). MNNG was mutagenic in both TA 100 and TA 1535 (30).

When reagents that are reactive only on single-stranded nucleic acids are used to mutate double-stranded nucleic acids in vivo, the results are equivocal. Such mutagens include formaldehyde, chloroacetaldehyde, bisulfite, hydroxylamine, and methoxyamine. Most, if not all, mutagens are much more reactive in single-stranded nucleic acids, so that these regions are probably preferentially modified in replicating DNA.

It is difficult to elucidate the mechanism by which a specific chemical induces mutants only by counting revertants. Multiple chemical events are occurring; only one locus is being selected. As a minimum, mechanistic studies require mapping and possibly sequencing of each revertant locus to exclude suppression.

Bearing in mind all of these caveats, the following sections focus on positive findings. Chemicals that have been found mutagenic in any test are considered if there are significant data on their chemical reactions. For many simple mutagens, data are presented on the possible consequences of modification on transcription.

Mutations Resulting from Thermodynamic Properties

Hydrolytic reactions, which occur in nucleic acids under physiological conditions, cause deamination (31, 32), depurination (33, 34), and depyrimidination (35). The rate of deamination of C → U is considerably greater than A → HX. Depurination of G or A is much greater than depyrimidination of C or T. As shown in Table 2, all the reaction rates are lower in double-stranded DNA than in single-stranded. In particular, deamination is increased more than two orders of magnitude in single-stranded DNA (34).

Deamination, if unrepaired, is a direct mutagenic event (43, 44), and stable base pair can be formed. The loss of a base can lead to strand scission (45). If this does not occur, the decreased information in the template is likely to affect fidelity in replication (46). This may occasionally be a genetically fixed event (47).

The innate physicochemical properties of nucleosides include tautomerism, ionization, and *syn-anti* rotation of bases around the glycosyl bond. All of these are reversible events, but at any given moment their frequency is calculated to be on the order of 10^{-1} to 10^{-5}, as shown in Table 2. It should be emphasized that these data are derived from model experiments and not

Table 2 Thermodynamic characteristics of nucleosides leading to mutation

Thermodynamic characteristics	Events/day/10^{10} bases (rat liver cell)	
	Double stranded[a]	Single stranded[a]
Depurination	$\cong 10^4$	$\cong 4 \times 10^4$
Depyrimidination	$\cong 5 \times 10^2$	$\cong 2 \times 10^3$
Deamination	$\cong 1.7 \times 10^2$	$\cong 4.3 \times 10^4$
	Equilibrium constant (mole^{-1} sec^{-1})	
	Double stranded	Single stranded[b] (calculated from monomer)
All tautomeric shifts[c]	f	$\cong 10^{-4}$
Total ionization[d]	f	$\cong 10^{-2}$–10^{-5}
Base rotation[e]	f	$\cong 10^{-1}$–10^{-2}

[a] From Lindahl (36) and Hartman (37). Singlestrand represents about 1–3% of doublestrand.
[b] Data extrapolated from physical constants for bases, nucleosides, and nucleotides, including model compounds.
[c] Topal & Fresco (38, 39) and references therein.
[d] From published pK values (40, 41).
[e] Calculated by Haschemeyer & Rich (42).
[f] Not determined.

from polynucleotides. It is likely that *syn-anti* rotation is greatly inhibited in a polymer, but the values of the tautomerism and ionization constants may be similar in polymers to those obtained for model compounds. In transcription, the incoming NTP base or exocyclic group may rotate (48, 49), and in addition, the base could be in the rare tautomeric form or could be ionized. Thus, the substrate must be considered, as well as the template, in producing mutation.

The possible changes in base pairing that result from dynamic properties are speculative. Most of our information is on tautomerism and comes from theoretical calculations, but there also exist recent experimental data suggesting that the rare tautomers are involved in mispairs (50). The occasional occurrence of a purine-pyrimidine mispair with one base in a rare tautomeric state was originally postulated by Watson & Crick (51) to explain spontaneous transitions. Later, Topal & Fresco (38) extended this hypothesis to purine transversions in which the template base is in the rare tautomer state and the incoming purine nucleoside triphosphate is the *syn* rotamer. Pyrimidine-pyrimidine oppositions can not be explained by direct hydrogen bonds, but appear to occur as a result of stacking forces.

Up to this point we have discussed nonenzymatic recognition in the formation of a base pair with the proper geometry to fit a double helix. Obviously, no base pair would be formed with an apurinic or apyrimidinic site. The fact that transcription occurs in depurinated nucleic acids must be a function of base selection by the enzyme (46). In the absence of enzyme discrimination, we would expect that improper base incorporation would occur with great frequency, particularly considering the sum of both hydrolytic and physicochemical events in a DNA (Table 2).

The present data on error rate in synthesis in vitro using synthetic deoxypolynucleotides and *Escherichia coli* DNA polymerase I is $\simeq 10^{-4}$–10^{-6} (52). Using a natural DNA, ϕX174, and DNA polymerases, fidelity is higher, and the error rate is on the order of 10^{-6}–10^{-7} (53, 54). Drake (55) has summarized the rate of forward mutation in various genomes to be 2×10^{-8} (DNA phages) to 7×10^{-11} (*Drosophila*) per base pair replicated. The difference between in vitro and in vivo fidelity must be ascribed to the many correcting steps in replication. The subject of fidelity is dealt with in detail in the review by Loeb & Kunkel in this volume.

The use of the word spontaneous to describe mutations of nucleic acids is unnecessarily vague. All the mechanisms by which chemicals or radiation are presumed to cause mutation occur at a lesser level in an untreated genome as the result of the inherent nature of the bond involved. Thus, spontaneous mutation might more properly be termed background mutation.

Chemical Reactions

SIMPLE NONALKYLATING AGENTS   This class of reagents includes formaldehyde (HCHO), hydroxylamine (H_2N-OH, HA;) methoxyamine ($H_2N-O-CH_3$, MHA), nitrous acid (HNO_2), bisulfite (HSO_3^-), and hydrazine (H_2N-NH_2, HZ). The reagents are shown in Figure 1 and the variety of products in Figures 2–4.

Formaldehyde was first studied in the 1940s and found to be mutagenic in a variety of systems, but interest lagged when the chemistry was difficult to elucidate and fashions in mutation research changed (56). It is now clear that formaldehyde reacts reversibly with amino groups of nucleic acids (57, 58), as has long been known for proteins. With nucleic acids the reaction is strongly dependent on conformation, so that formaldehyde binding has been used as a probe of nucleic acid double-strandedness (59–62). In single-stranded nucleic acids, the hydroxymethyl derivatives of A, G, and C have been isolated (63) (Figure 2). A second and slower reaction results in the formation of $-CH_2-$ cross-links, which were first shown by Feldman in formaldehyde treated RNA (58). The unequivocal demonstration of cross-linking in DNA was provided by Shapiro and co-workers (64). Until their isolation of five methylene cross-links (G-G, A-A, A-G, A-C, G-C) it was not clear that cytidine could also react in that manner (Figure 4a–e). Although cross-links have been considered to have genetic effects, Auerbach et al (56) are dubious that the mutagenicity of formaldehyde is directly due to these reactions, rather than resulting from secondary effects on cellular components.

Hydroxylamine [a presumed intermediate in nitrate reduction in vivo (65)], and methoxyamine, are among the few mutagens that can directly cause a change in base pairing by favoring a tautomeric shift (66). Although hydroxylamine can react with A, C, and U, but not T, the reaction with U, which degrades the nucleoside to ribosylurea, is very slow, and significant only at pH 10 (67). The reaction of cytidine or adenosine at pH 5–6

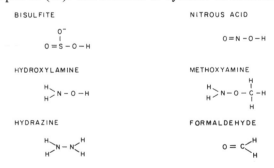

Figure 1   Structural formulas of simple nonalkylating mutagens.

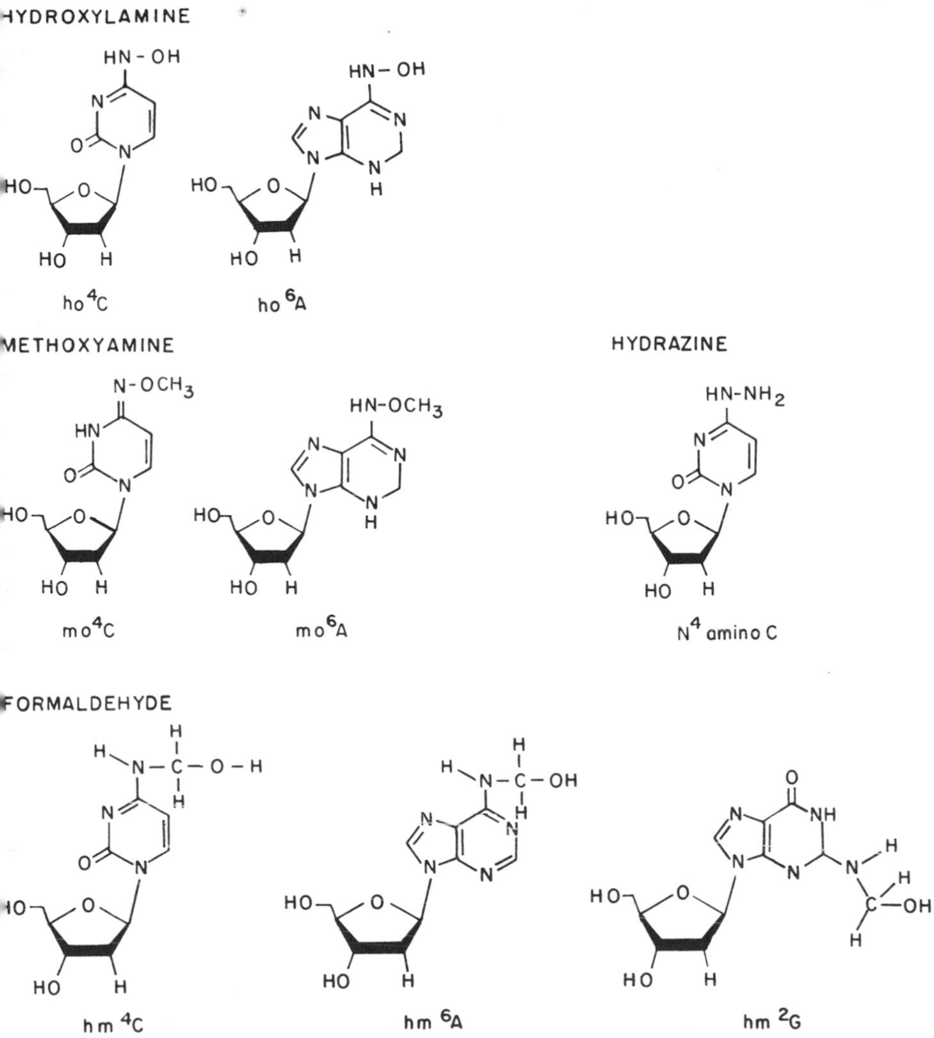

Figure 2 Reaction products of nonalkylating agents with nucleic acids and nucleotides.

with hydroxylamine or methoxyamine leads to replacement of the amino group by a hydroxyamino (–NHOH) or methoxyamino group (–NH-O-CH$_3$) (Figure 2). Saturation of the 5,6 double bond of C also occurs, but the major reaction product is the amino substituted derivative (68). In the case of 5-hydroxymethyl-C, only the N^4 substitution occurs, at a rate about 1/10 that of the reaction with cytidine (69). Reactivity is greatly decreased in double-stranded nucleic acids and it is assumed that the mutagenic reaction in DNA is with single-stranded segments (15). Reaction with adenosine is extremely slow, but in vivo N^6-hydroxy A or N^6-methoxy A,

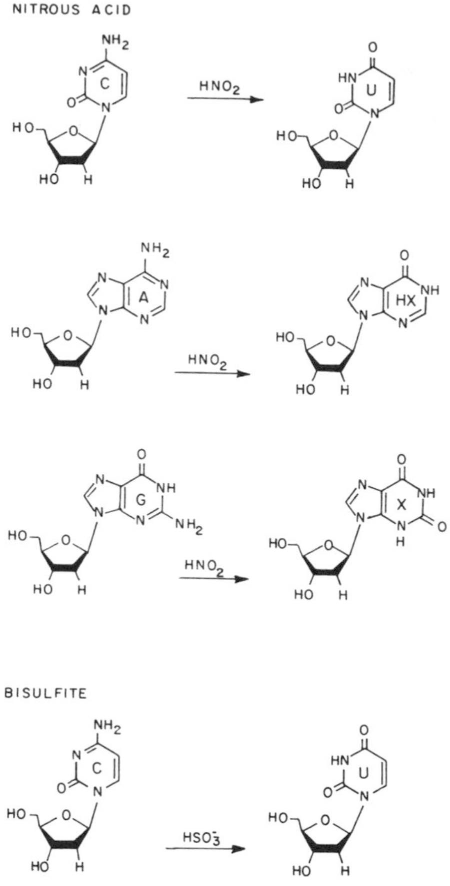

Figure 3   Deamination reactions of nucleic acids.

(as well as the cytidine derivatives) may contribute to mutation (70, 71). Transforming DNA (3, 5, 72), phages T4 (2) and S_D (73), TMV RNA (23), herpes simplex virus DNA (6), and foot and mouth disease virus (74) have all been highly mutated in vitro, but mutation is difficult to demonstrate when hydroxylamine and methoxyamine are directly tested under physiological conditions in bacteria or cells (7, 8).

Nitrous acid, the classical oxidative deamination reagent ($-NH_2 \rightarrow -OH$), was once considered as the "perfect" mutagen. Deamination of of C $\rightarrow$ U, A $\rightarrow$ HX, or G $\rightarrow$ X changed base pairing in a predictable manner (75) (Figure 3). However, it was recognized very early that cross-links were formed in HNO_2-treated *Bacillus subtilis* transforming DNA (76), and cross-linked diguanyl and guanyladenine were later isolated (77) (Figure 4

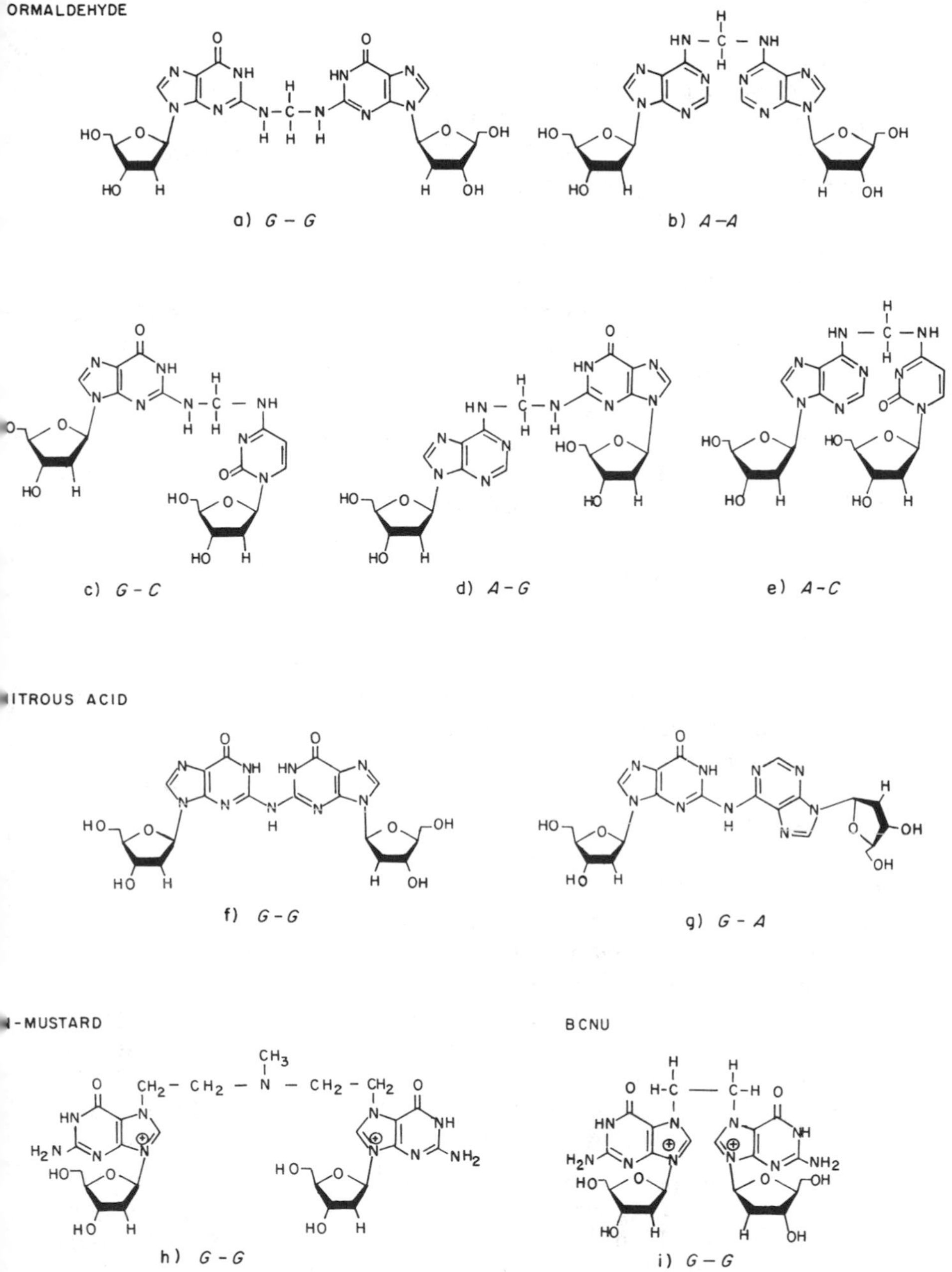

Figure 4 Cross-links identified in polynucleotides and nucleic acids. (*a–e*) Chaw et al (64); (*f,g*) Shapiro et al (77); (*h*) Chun et al (142); (*i*) Tong & Ludlum (181) and Tong et al (175). For simplification in drawing, some of the sugar residues are not stereochemically correct, but are shown as the authors drew them. All structures contain D-deoxyribose.

f,g). Kinetic data have recently led Frankel et al (78) to surmise that still another chemical reaction must occur to account for their mutagenesis data. Again, as in the case of HA and MHA, the conditions required for generating HNO_2 from a precursor such as sodium nitrate, or nitrite are inappropriate for mutation studies with cells. However, the optimal pH for $NO_2^- \rightarrow HNO_2$ is known to exist in the human stomach.

Bisulfite (HSO_3^-) is produced endogenously in mammals as an intermediate in the catabolism of sulfur-containing amino acids (79). Bisulfite undergoes a number of reactions with nucleic acids (79), but the observed mutations (4) are likely to be the result of the deamination of C $\rightarrow$ U (80, 81) (Figure 3). Both Shapiro (80) and Hayatsu (81) studied the reactions of bisulfite with nucleosides or DNA and agree that only C is modified, but there are multiple products as a consequence of the mechanism by which C is deaminated (79). These include, as intermediates, the addition product of bisulfite to the 5,6 double bond, which leads by deamination to the analogous dihydrouracil derivative, and, following the slow removal of bisulfite, to the final product, uracil. Protein-nucleic acid cross-links can be formed by a transamination reaction involving the ε-amino group of lysine (or a terminal –NH₂) and the 4–NH₂ of cytidine. Such cross-links have been reported to occur in bisulfite treated S_D phage (82, 83) and turnip yellow mosaic virus (84). The reaction with cytidine in situ can stop at the intermediate product, 5,6-dihydro-6-sulfopyrimidine, presumably as the result of shielding of the amino group of cytidine by protein (82).

5-Methylcytosine (m^5C) found in DNA, is also deaminated by bisulfite, but at a rate two orders of magnitude slower than C (85). In contrast to deamination of C $\rightarrow$ U, deamination of m^5C in DNA can lead to T, which, not being repaired (as is U by a uracil glycosylase), consequently would cause GC $\rightarrow$ AT transition. m^5C deamination could be the cause of some "hotspots" for spontaneous transition mutations (86), although it is doubtful if the formation of T is the only mutational event. In T-even bacteriophages, where 5-hydroxymethylcytidine (hm^5C) and its glucosylated derivative replace cytidine, reaction with bisulfite does not result in deamination, but rather gives 5-hydroxymethylenesulfonate as a product (87, 88).

Except for this reaction with hm^5C, bisulfite reaction with C has been detected only in single-stranded nucleic acids (89, 90). This general specificity for a nonhydrogen bonded base suggests that the C–6 of the 5,6 double bond, to which bisulfite must add as the first step in deamination, may be sterically unavailable to the bulky SO_3^- ion. Because of its single-strand specificity, bisulfite has been used successfully to determine secondary structure (91–94) and as a site-specific mutagen (95). Cytidine was deaminated in a selected single-stranded section of SV40 DNA, and the gap was filled using DNA polymerase I. A GC $\rightarrow$ AT transition was detected following in vivo replication (95).

Hydrazine is a widely used industrial chemical that has been termed a weak mutagen in vivo (96, 97), but is undoubtedly mutagenic when used to modify transforming DNA in vitro (2, 3). The major reaction at pH 6 is with cytosine, and results in a displacement at N^4 to generate N^4-amino C (98, 99) (Figure 2). Data obtained with monomeric compounds indicate that N^4-amino C would seldom lead to changed base pairing since it has a $K_t \simeq 30$ in favor of the amino form (like C), as contrasted with N^4-hydroxy C, which is predominately in the imino form (like U) with $K_t \simeq 10$ (66). Studies of hydrazine reactions with other pyrimidine bases indicate that the pyrimidine ring of T and U is rapidly degraded (96, 100), so that in DNA, apyrimidinic sites might well be generated.

Additional complications are known to affect the evaluation of the mutagenicity of formaldehyde, hydroxylamine, and hydrazine: all three chemicals are highly reactive with molecular oxygen, forming peroxy radicals, which by additional reactions form peroxides. It should also not be overlooked that the six mutagens discussed in this section also react more or less readily with proteins and cause nucleic acid-protein cross-links. Thus, although each of these mutagens can form nucleic acid derivatives capable of changed base pairing, such events certainly are not necessarily responsible for the observed mutations.

SIMPLE ALKYLATING AGENTS   A large group of mutagens are simple alkylating agents (Figure 5), although they differ greatly in their

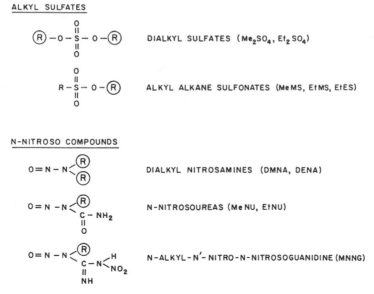

Figure 5 Structural formulas of simple alkylating agents. Only the dialkylnitrosamines require metabolic activation.

mutagenicity (for examples see Table 1). The apparent greater mutagenic efficiency of methylating agents as compared to the analogous ethylating agent is not true if the extent of alkylation is considered (8, 101), since ethylating agents, on a molar basis, are about 1/20 as reactive with nucleic acids as are methylating agents. For these reasons we have been working for a number of years to elucidate the chemical reactions of nucleic acids with good mutagens (e.g. N-nitroso compounds) as compared to poor mutagens (e.g. alkyl sulfates) and of methylating agents as compared to ethylating agents. Some of the studies relevant to this review have been published in a series of papers and reviews (15, 102–119), and are therefore only briefly summarized here.

All oxygens and nitrogens in polynucleotides (except the nitrogen attached to the sugar) can be alkylated in aqueous solution at neutral pH. The structures of the derivatives that have been characterized are shown in Figure 6. However, the exocyclic amino groups and N-1 of G (Figure 6, last two columns) have been found to react with ethylating agents only in in

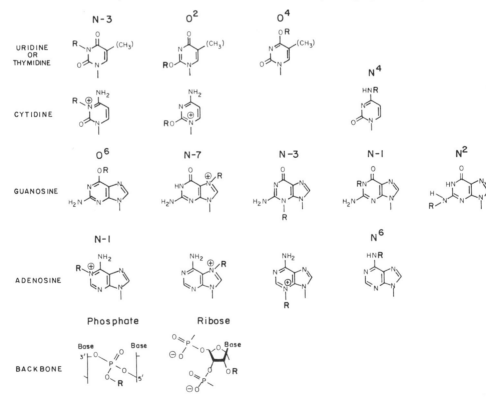

Figure 6   Sites of reaction of simple alkylating agents with nucleic acids or polynucleotides in neutral aqueous solution.

vitro model experiments (102–104, 114, 118, 120). The metabolically activated aromatic amines and polycyclic hydrocarbons differ markedly in this respect, as is discussed in the section on their chemistry.

The order of reactivities of alkylating agents with nucleic acids, both in vitro and in vivo is methyl > ethyl > higher homologues. Alkyl sulfates react almost entirely with nitrogen, while N-nitroso compounds (including metabolically activated nitrosamines) react primarily with oxygen. Alkylalkane sulfonates react in a manner intermediate between these two groups. Ethylating agents in general are relatively more reactive toward oxygen than analogous methylating agents. Approximately 80% of the reaction of N-ethyl-N-nitrosourea (EtNU) with DNA in vivo (119) or in mammalian cell culture (116, 121) is on oxygen, with the ethyl phosphotriester representing about 50–60% of all ethylation products, O^6-EtG and O^2-EtT 7–9% each, and O^4-EtT and O^2-EtC 2–4% each. In RNA, the 2' oxygen of ribose is also alkylated (112). Although the mutagenicity differs (Table 1), the sites and proportion of alkylation of nucleic acids in vitro is the same for reaction with nitrosoureas and nitrosoguanidines (113).

The differences in products formed between single-stranded and double-stranded nucleic acids are small, except for substitutions at the N-1 of A and N-3 of C, which are much less reactive in double-stranded nucleic acids (114, 122, 123). This is to be expected, inasmuch as both of these positions are involved in hydrogen bonding. However, there is enough thermal denaturation at 37° (60, 62, 124) to allow significant alkylation of these positions even in a double-stranded polynucleotide (123, 125, 126). In contrast to the base-paired nitrogens, the extent of reaction of oxygens is not a function of strandedness, since the O^6 of G, and O^4 of T, and the O^2 of C all possess a reactive unbonded electron pair, while the O^2 of T is not hydrogen bonded.

The N-alkyl-N-nitrosamines are mutagens (127) that require metabolic activation before they can act as alkylating agents (128, 129). Both the rate of activation and the corresponding mutagenicity are increased as the pH decreases. The active alkylating species is believed to be the same as for directly acting agents, that is, an alkyl carbonium cation. The chemical reactions of dimethylnitrosamine and diethylnitrosamine with DNA in vivo are, within experimental variation, identical to the corresponding nitrosoguanidine and nitrosourea reactions (B. Singer and D. Grunberger, in preparation). Nitrosamines are not only metabolized in vivo but are also biosynthesized in mammals from nitrite and secondary or tertiary amines (130–132). Biosynthesis and/or metabolism are enhanced by lowering the pH (133–134); ascorbate appears to inhibit biosynthesis (135).

The cyclic alkylating agents, which comprise a number of unrelated mutagens (Figure 7), owe their alkylating activity to a reactive unstable ring structure. Of these agents, the earliest to be studied were the N- and S-mustards, the former of which (HN-2) is also used therapeutically. Brookes & Lawley, in a pioneering series of studies, found that the mustards react with N-7 of G, and both a monoadduct (Figure 8*a,b*) and a cross-linked adduct were formed in vitro and in vivo (136, 137). Later, derivatives resulting from reaction of the N-1 of A and N-3 of C were identified (138), so that it now appears that S-mustard resembles alkyl sulfates in its reactions except that, being bifunctional, inter- and intrastrand cross-linking also occur (139–141). N-mustard also cross-links through the N-7 of G (142) (Figure 4*h*). In mammalian cells, HN-2 appears to form DNA-protein cross-links as well (143). Phosphoramide mustard, which is a mutagen formed by metabolism of the chemotherapeutic agent, cyclophosphoramide, reacts with guanosine to form a highly unstable N-7 derivative (144).

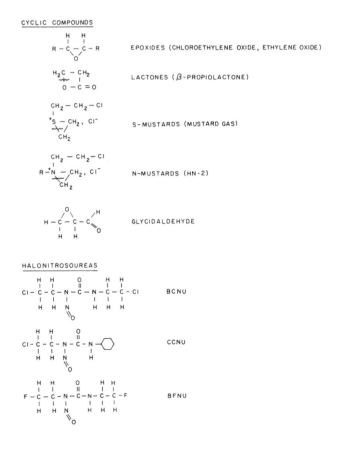

Figure 7   Structural formulas of cyclic alkylating agents and halonitrosoureas.

β-Propiolactone (βPL) (Figure 7) also reacts with the N-7 of G and N-1 of A (145–147), and may form intramolecular cross-links (148). There is less information on other reactions, although it might be predicted that the N-3 of A could also be modified. In contrast to N- and S-mustard, which are inefficient mutagens (12, 149), βPL is mutagenic in many test systems (10, 150).

Aliphatic epoxides such as ethylene oxide and propylene oxide (Figure 7), widely used as sterilants and in industry, are weakly mutagenic (8, 151, 152). The epoxides resemble typical alkylating agents and react with DNA and RNA at the N-7 of G and the N-1 and N-3 of A, to form hydroxyethyl or hydroxypropyl derivatives (153). No O^6–alkyl derivative of G was detected using propylene oxide.

Chloroethylene oxide (Figure 7) and its open-ring derivative, chloroacetaldehyde, are mutagenic metabolites of vinyl chloride (154–157), and their reactions with nucleosides and nucleic acids have been studied intensively. They form the fluorescent cyclic etheno derivatives, $1,N^6$-etheno A (ϵA), $3,N^4$-etheno C (ϵC) (158–163), and $1,N^2$-etheno G (ϵG) (164) (Figure 8j,k,l), although the latter has only been prepared from guanosine. It is likely that the reaction is initially on the ring nitrogen (163, 165), followed by a rapid formation of a cyclic hydrate (166–168). However, the dehydration to the etheno compound is a relatively slow step, particularly for the C derivative (168), and in nucleic acids both the hydrated and the dehydrated cyclic derivatives may exist (J. T. Kusmierek and B. Singer, in preparation). It has been suggested on the basis of in vivo experiments that the N-7 of G also reacts (169). The rate of chloroacetaldehyde reaction is greatly decreased in double-stranded polynucleotides (170). Glycidaldehyde (Figure 7) is a related mutagen forming a Δ-imidazoline ring derivative of guanosine (171, 172) (Figure 8m). Similar reactions with other bases may yet be found.

Another group of biologically important alkylating agents, the halonitrosoureas (Figure 7), includes agents with both antineoplastic and mutagenic activity (173). There are two interesting facets concerning their neutral aqueous reaction with nucleic acids: 1,3-bis-[2-chloroethyl]-1-nitrosourea (BCNU) appears to react with DNA in three separate ways to form haloethyl, hydroxyethyl, and aminoethyl derivatives, all at the N-7 of G (174) (Figure 8c,d,e). Cyclic ethano derivatives of A (Figure 8i) and C (Figure 8h), derived from haloethyl reaction at the N-1 of A and N-3 of C, are also formed (175, 176). In addition, O^6-hydroxyethyl G (175, 176) (Figure 8f) and 3-hydroxyethyl C (174) (Figure 8g) have been isolated from DNA. Although not all these derivatives have been characterized from bis-[2-fluoroethyl]-nitrosourea (BFNU) or 1-[2-chloroethyl]-3-cyclohexyl-1-nitrosourea (CCNU) treated DNA, model experiments indicate that the same positions are reactive (174, 177–179). The bifunctionality of haloni-

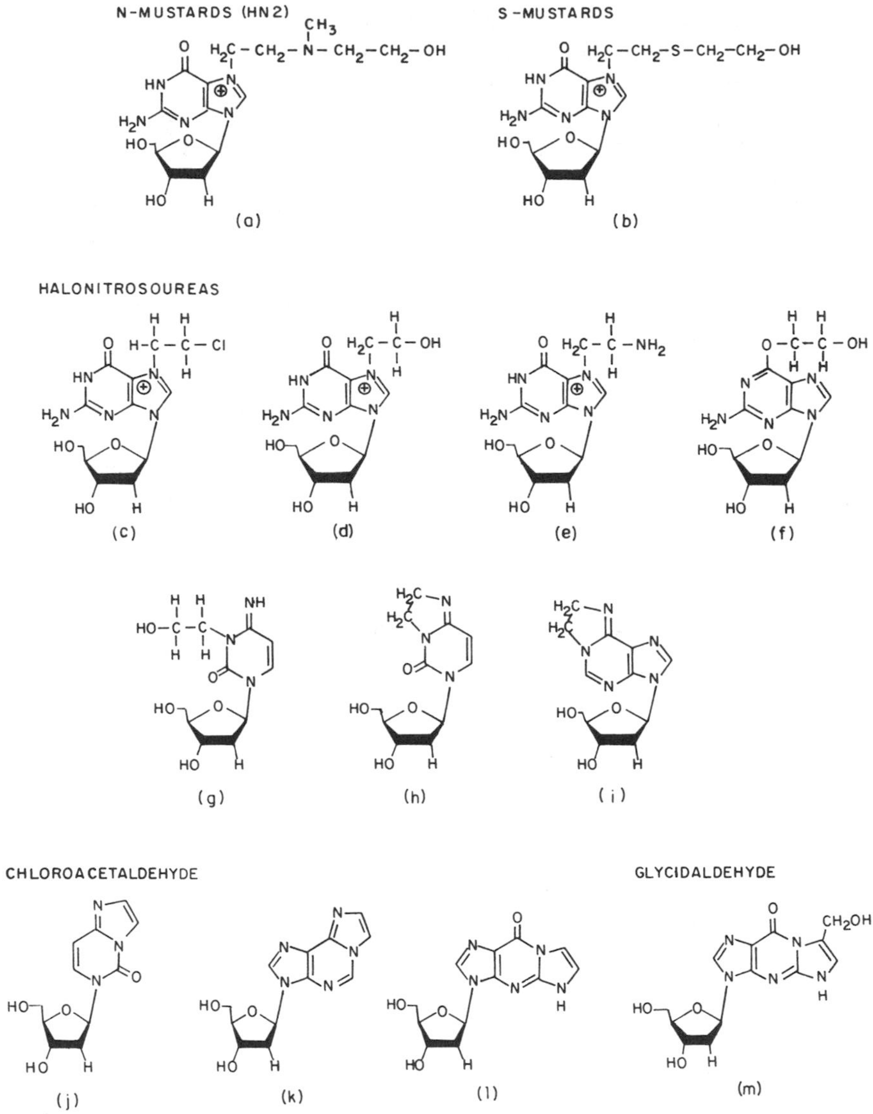

Figure 8 Products identified from reaction of bifunctional alkylating agents with nucleic acids. (*a*) Brookes & Lawley (136); (*b*) Brookes & Lawley (137); (*c*) Tong et al (175); (*d,e*) Gombar et al (174); (*f*) Tong et al (175); (*g,h*) Gombar et al (174); (*i*) Tong & Ludlum (176); (*j*) Barrio et al (158), Wang et al (162), Green & Hathway (160), and Laib & Bolt (161); (*k*) Barrio et al (158), Wang et al (163), Green & Hathway (160), and Laib & Bolt (161); (*l*) Sattsangi et al (164); (*m*) Goldschmidt et al (171) and Van Duuren & Loewengart (172).

trosoureas is similar to that of the N- and S-mustards in forming cross-links (148, 180–182) (Figure 4*i*). Only the BCNU diguanyl cross-link through the N-7 of G has been isolated (181), but both in vitro and in vivo, additional cross-linking appears to occur. One strong possibility involves initial reaction with O^6 of G (175, 183).

In the previous section on nonalkylating mutagens, it was shown that most of the reactions changed base-pairing, which would, if expressed, be mutagenic. In contrast, alkylating agents, while reacting at many sites, do not appear to modify the amino groups in nucleic acids. Most of the reactions of mutagenic alkylating agents would be presumed to block base-pairing. All four O-alkyl derivatives (O^6-alkyl G, O^4-alkyl T, O^2-alkyl T, and O^2-alkyl C) might specifically change base-pairing, since they can form a wobble pair if the alkyl group is *anti* to the Watson-Crick side (183a). However, data on transcription of polynucleotides containing a variety of modified nucleosides indicate that any alkyl substitution on the Watson-Crick side leads to nonspecific misincorporation (see section on biochemical effects of modified nucleosides).

CHEMICAL REACTIONS OF METABOLICALLY ACTIVATED ARO-MATIC MUTAGENS   Metabolic activation (184–190) is a requisite for the mutagenicity (10, 184, 186, 191–195) of these compounds. The identification of biologically active metabolites is too large a subject for this review, which is restricted to clearly identified products of reaction of such metabolites with nucleic acid components. In almost all cases, the adducts have been found in vivo. Although the terminology uses the name of the proximate mutagen, it should be understood that this actually refers to a metabolite.

The aromatic amines (Figure 9) include 2-acetylaminofluorene (AAF), N-hydroxy-1-naphthylamine (1-NA), N-hydroxy-2-naphthylamine (2-NA), N,N-dimethyl-4-aminoazobenzene (DAB; butter yellow), 4-nitroquinoline-1-oxide (4-NQO), and the pyrolysis products from proteins, exemplified by 2-amino-6-methyl-dipyrido-[1,2-a : 3',2'-d]imidazole (Glu-P-1). A new type of mutagenic compound, isolated from broiled fish or beef (IQ and Me-IQ) is a substituted quinoline (196). While the structures of the mutagens have been rigorously determined, the site(s) of nucleic acid reaction are not yet known. It can be seen from Figure 10 that the sites of reaction, as far as they have been identified, are not the same for each amine, nor are they reminiscent of the common alkyl derivatives. Except for 1-NA (197, 198) and 4-NQO (199), all react at the N^2 or the C-8 of G (189, 200–206). Two adducts of 1-NA with O^6 of G are found (198), while 2-NA and 4-NQO react at the N^6 of A (199, 202). There is indirect evidence that

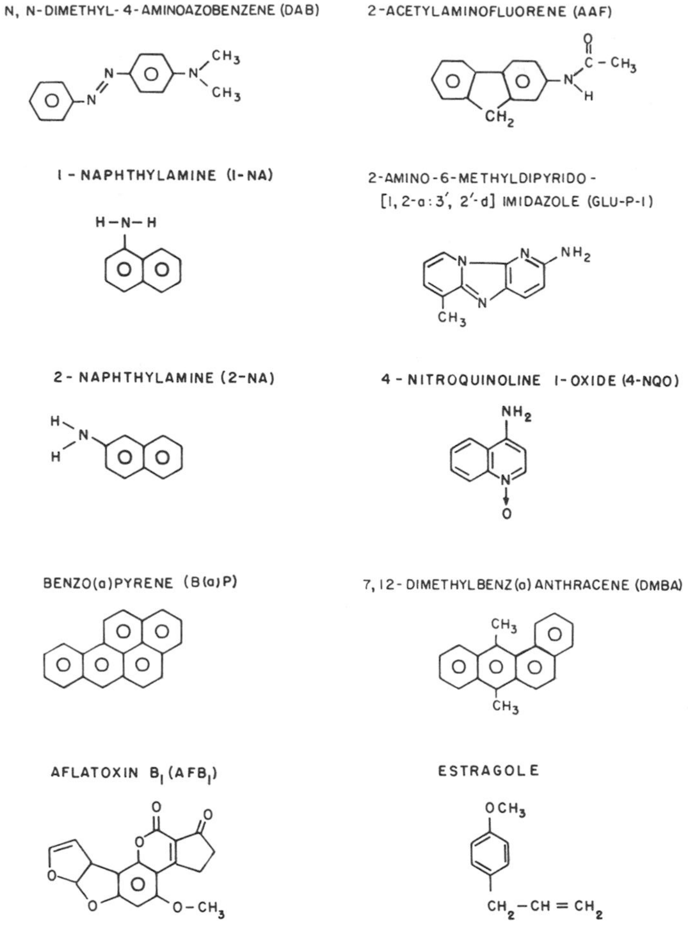

Figure 9 Structural formulas, as proximate mutagens, of aromatic amines, polyaromatic hydrocarbons, and natural product mutagens. All require metabolic activation.

AAF may react at the N-7 of G (207). The reaction of AAF with the C-8 of G destabilizes the normal *anti* conformation of the nucleoside in favor of the *syn* conformation; this is termed "base displacement" (208, 209). However, in vivo, an unacetylated derivative is found that is in the normal *anti* conformation (210).

Another group of metabolically activated aromatic mutagens is represented by the polyaromatic hydrocarbons (PAH) (Figure 9), which include B(a)P, 7,12-dimethylbenz(a)anthracene (DMBA), and 7-bromomethyl-benz(a)anthracene (7-BMBA) (formula not shown). The most studied of this group, B(a)P, reacts with the N^2 of G (211–215) (Figure 11*a*), N^6 of

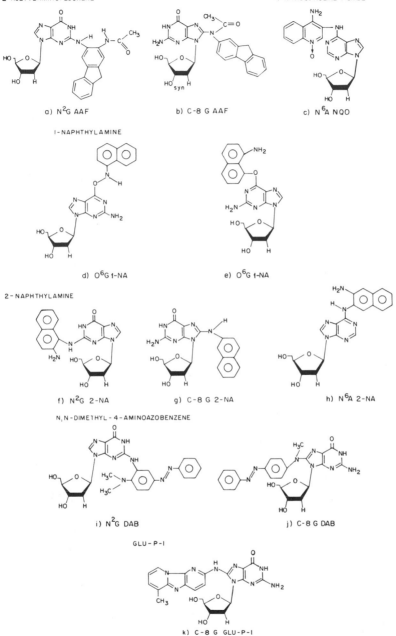

2-ACETYLAMINOFLUORENE

a) N²G AAF

b) C-8 G AAF

4-NITROQUINOLINE-1-OXIDE

c) N⁶A NQO

1-NAPHTHYLAMINE

d) O⁶G 1-NA

e) O⁶G 1-NA

2-NAPHTHYLAMINE

f) N²G 2-NA

g) C-8 G 2-NA

h) N⁶A 2-NA

N,N-DIMETHYL-4-AMINOAZOBENZENE

i) N²G DAB

j) C-8 G DAB

GLU-P-I

k) C-8 G GLU-P-I

Figure 10 Identified products of reaction of metabolites of aromatic amines with nucleic acids. (*a*) Westra et al (200); (*b*) Kriek et al (201); (*c*) Kawazoe et al (199); (*d,e*) Kadlubar et al (198); (*f,g,h*) Kadlubar et al (202); (*i,j*) Beland et al (203) and Tarpley et al (204); (*k*) Hashimoto et al (189, 205) and Imamura et al (206).

BENZO (a) PYRENE

a) N²G B(a)P   b) N-7 G B(a)P   c) N⁶A B(a)P *trans*

DIMETHYLBENZ (a) ANTHRACENE

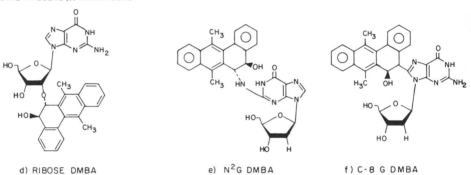

d) RIBOSE DMBA   e) N²G DMBA   f) C-8 G DMBA

7- BROMOMETHYBENZ (a) ANTHRACENE

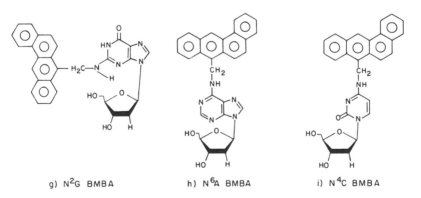

g) N²G BMBA   h) N⁶A BMBA   i) N⁴C BMBA

Figure 11 Identified reaction products of metabolites of polyaromatic hydrocarbons with nucleic acids. (*a*) Jeffrey et al (211, 212), Koreeda et al (213), Weinstein et al (214), and Nakanashi et al (215); (*b*) King et al (217); (*c*) Jeffrey et al (216); (*d*) Kasai et al (223); (*e*) Jeffrey et al (221); (*f*) Nakanishi et al (222); (*g,h,i*) Dipple et al (218), and Rayman & Dipple (219).

A (216) (Figure 11c), and N-7 of G (217) (Figure 11b). The N-7 G adduct has been difficult to characterize, since the glycosyl bond of the adduct in DNA is extraordinarily labile [$t_{1/2} \simeq 3$ h (217) compared to $\simeq 155$ h for 7-MeG (117)]. 7-BMBA reacts with all the exocyclic amino groups (218, 219) (Figures 11g,h,i). The crystal structure of the 7-BMBA N^6-A adduct shows that the nucleoside is *syn* and the hydrocarbon is perpendicular to the base (220). DMBA appears to be quite different in its reactions. In DNA, an N^2-G adduct is found (221) (Figure 11e), but in RNA both the C-8 of G (222) (Figure 11f) and 2'-O of ribose (223) (Figure 11d) react.

Aflatoxin (Figure 9), a fungal product, forms a labile N-7 G derivative (224–227) (Figure 12) with a $t_{1/2} \simeq 8$ h at pH 7.3 (228). At least ten minor compounds, now under study, are also formed (229, 330). Another natural product mutagen, estragole (Figure 9), reacts at the N^2 of G to form three isomers, as well as at the N^6 of A (231) (Figure 12). The same reactions have been reported for safrole, a closely related mutagen (232) (formulas not shown).

The above summary does not indicate whether the identified derivative is a major product or whether a product was looked for and not found, with one exception: B(a)P does not react with phosphodiesters (233). In many cases, too little is known of in vivo reactions to judge whether an identified product represents the major site of reaction. It does appear that, in contrast to all other mutagens, the N^2 and C-8 of G are modified. There are a sufficient number of examples to predict that N-7 derivatives of G can be formed, but the lability of two such derivatives so far characterized indicates that the major consequence of such adduct formation is an apurinic site. The reported chain breakage of AAF-DNA is suggested to be the result of a highly labilized modified guanine (186).

Metabolites of aromatic mutagens are alkylating agents, which makes it likely that all oxygens and nitrogens will react to some degree (see previous section). This premise is supported by the fact that the reported sites of reaction include the N^2, O^6, and N-7 of G, the N^4 of C, the N^6 of A, and the 2'-O of ribose. On the basis of the reported reactivities of the N-7 of G, it is expected that the N-3 of A also reacts. However, the chemical lability of the glycosyl bond of such derivatives is so high that they would be difficult to detect except as apurinic sites (234, 235).

Parameters of Base Pairing

Base mispairing is the formation of planar, hydrogen bonded pairs of bases other than G-C and A-T(U). Base mispairs are thermodynamically allowed, at least in the moment of formation. This direct mispairing is the simplest kind of mutational mechanism. Such pairs can be considered mispairs only in a biological sense, in that they do not yield a faithful copy of the template.

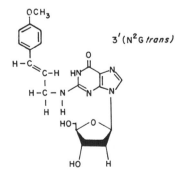

Figure 12 Identified products of reaction with nucleic acids of metabolites of aflatoxin B_1 [Martin & Garner (224), Essigmann et al (225), Lin et al (226), and Autrup et al (227)]; estragole [Phillips et al (231)]; and safrole [Phillips et al (232)].

However, physicochemically the interaction may be correct. Even in the absence of chemical modification, base mispairing can occur as the result of innate properties of common nucleosides, such as tautomerism, ionization, and *syn-anti* rotation around the glycosyl bond.

Various interactions, termed wobble (236), have been demonstrated by UV, IR, NMR, and Raman spectra, and by hybridization techniques, etc., in tailor-made oligonucleotides or polynucleotides containing mismatched bases (237–242). The unequivocal existence of stable G-U wobble pairs in nucleic acids has been shown only in tRNAs (243–245). Wobble mispairs postulated in deoxypolynucleotides on the basis of experimental data, include G-T (246), T-T (247, 248), and C-T (249). A-A and C-C pairs that are not observed require additional changes (*anti → syn;* ionization). Additional wobble pairs without tautomerism are described in codon-anticodon interactions (250) and in RNA hairpins (251–255). These pairs have a variable stability, but usually permit maximization of the frequency of the other common base pairs. In naturally occurring DNA and RNA wobble pairs have a very low probability. This reflects both the low stability of the pair and the discrimination of the transcribing enzyme(s). Poly[d(A-T)] has often been used in fidelity studies of DNA synthesis. The highest level of fidelity, as assessed by GMP misincorporation, is on the order of $1/10^6$ bases synthesized (52). This would not provide evidence for a G-T wobble occurring during DNA synthesis at a greater frequency than expected on the basis of mispairing by the rare tautomeric form.

This discussion of mispairing at the monomer and polymer level may not be completely applicable to the situation in a replication complex, where geometrical and other restrictions may not permit some of the mispairs. Another unknown factor is the level of discrimination of enzymes involved in the replication process.

TAUTOMERISM   A classic mispair is the one where one base is in the rare tautomeric form (51). Spectral methods indicate that common tautomers are predominant both in solution (256–260) and in the vapor phase (261). Evidence for the rare tautomer of common bases is not available, because spectroscopic methods can only detect a few percent of this tautomer under ideal conditions. Thus K_t values for monomers must be derived from the pK values of fixed tautomers. These are on the order of 10^{-4} to 10^{-5} (257, 262).

In contrast, at the polymer level the rare tautomers of common bases in mispairs can be detected using UV spectroscopy (50). Hydrogen bonding schemes have been proposed for A-C and I-U base mispairs in helices dominated by A-U or I-C pairs. Each mispair contains one of its bases in the rare tautomeric form. Which of the bases in a mispair is in the unfavored form depends upon the stacking interactions of these bases with their nearest neighbor Watson-Crick pairs, according to Fresco et al (50). In this case the enhancement of energetically unfavorable tautomers is at the expense of the stacking interactions. Similar effects of the neighbor bases on mispairing are shown for 2-aminopurine (263) which, in transcription, acts more like G when in a C polymer than in an A polymer (264). Even more distant neighbors may also exert an effect (265).

The substitution of the 4-amino group in cytidine by hydroxylamine or methoxyamine is an example of a reaction causing a tautomeric shift: the imino tautomer is predominant in water solution (Figure 13) and is the only form observed in nonpolar solutions (66). Model experiments on possible base pairing have also been reported. 1-methyl-N^4-methoxycytosine forms 1:1 complexes with a 9-substituted adenine, thus behaving like U. No interaction between 1-substituted N^4-methoxycytosine and 9-ethylguanine was observed in nonpolar solution, which is in line with the observed tautomerism favoring the imino form (266). The association constants of 1-methyl-N^4-methoxycytosine and its 5-methyl derivative with adenosine in chloroform solutions are identical (267). This is contrary to expectation, since the 5-methyl substituent should, based on analogous experiments

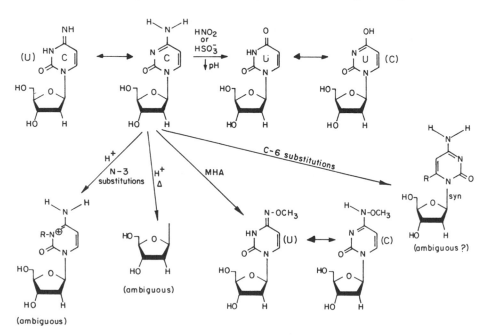

Figure 13   Various reactions of cytidine leading to changes in base pairing.

(268), sterically hinder rotation of the exocyclic N^4-methoxylamino group, so that it is constrained to the *syn* form with respect to the ring N(3). This would prevent Watson-Crick base pairing (although monomer interactions are not restricted to Watson-Crick pairing). Nevertheless, experimentally, Watson-Crick pairing occurs in poly(U,N^4-methoxy C) complexed with poly(A) (269).

An example of a "frozen" tautomer of a purine base is O^6-alkyl guanine. In the enol form, this modified base can form a wobble pair with U, which would be a mutagenic event. However, there is no evidence for any complex formation between O^6-methyl G and U, C, or A in chloroform at the monomer level (267).

Base tautomerism, which does occur, is difficult to measure directly and unequivocally, but this property of N^4-hydroxycytidine has been used in elegant experiments on site directed mutagenesis (270–273). Polymerases can utilize N^4-hydroxy CTP in the absence of CTP (271, 274). Once incorporated, this base can mispair when it is in the imino form. Thus C → T transitions have been recovered in vivo (275).

IONIZATION   Ionization is a function of pH, in contrast to tautomerism, which is an innate property of neutral molecules. Under physiological conditions the degree of ionization, that is the ratio of charged to uncharged species, ranges from 10^{-2} to 10^{-5}, depending on the pK of the nucleoside (40, 41). The resultant protonation or deprotonation can cause mispairing in the same way as tautomerism (Figure 13).

SYN-ANTI ROTATION   Rotation of a base around the glycosyl bond increases the ways in which changed base-base interaction can occur. The purine-purine base pair in which one purine is in the rare tautomeric form and the other is in the *syn* conformation has been postulated by Topal & Fresco (38) to explain some spontaneous transversions. Purine nucleosides have long been considered capable of existing in the *syn* conformation (42, 276). Recently, certain pyrimidine nucleosides [m^6dU (277) and pseudouridine (278)] have also been shown to exist as the *syn* rotamer. Although in crystals the nucleosides and nucleotides are usually found in either *anti* or *syn* conformations, there is one report that crystals can contain two molecules in the asymmetric cell unit, one *syn,* the other *anti* (279). A new finding of the significance of rotamers involves alternating d(G,C), which forms a double-stranded left-handed helix in which G is *syn* (280–285), in striking contrast to the more common right-handed DNA helix with bases in the *anti* orientation.

In solution, where there is freedom of rotation, the evaluation of the population of *syn* and *anti* rotamers of nucleosides and nucleotides varies from more than 50% of *syn* (286, 287) to 70–100% of *anti* (288). This wide variation in rotamers can be attributed to at least two factors. There are differences in the experimental techniques and also in the theoretical models used for evaluation of experimental data. In addition, the energy differences between the two rotamers may be very low, so that minor experimental conditions would have a large effect on the forms observed. Even in the case of the sterically constrained 8-bromoadenosine, this energy difference is on the order of only 1 Kcal/mol (289). This means that in enzymatic reactions the proper conformation can be easily adopted and maintained, particularly in a helix.

In has been suggested that since a modified nucleotide such as m^5C promotes the formation of left-handed helix (290), a possible consequence of chemical reactions could be rotation of the base around the glycosyl bond (291–294). A further consequence would be an increased availability for reaction of normally unavailable or buried sites on nucleotides. Although these proposals are based on the behavior of synthetic oligonucleotides, such DNA structures have been postulated to play a role in mutagenesis.

Biochemical Effects of Modified Nucleosides

The precise mechanism by which a mutagen exerts its biological effect has not been defined in any system, inasmuch as there can be both direct and indirect effects at every stage in replication. However, assuming that a modified nucleoside plays some role in biological expression, several types of experiments, in which one component is modified, are useful as models of point mutation. These include translation of messengers, measurement of codon-anticodon interaction, hybridization of polynucleotides, and synthesis of polynucleotides using natural and synthetic templates. When the modified nucleoside can form at least two hydrogen bonds, templates and messengers are active. In hybridization studies, secondary structure can prevent this bonding even though it is theoretically possible. A recent review by Singer & Kröger (295) covers these interactions.

Relatively few nucleic acid derivatives have been studied in terms of their ability to direct transcription. Mutagenic modification of a polynucleotide leads to at least two products and usually more (see previous section on chemical reactions). Particularly in the case of the large aromatic metabolites, not all products have been identified. It does, however, appear that transcription is inhibited by the adducts formed by metabolites of AAF and B(a)P. In these instances, transcription is terminated at, or near, the modified base in ϕX174 DNA, using DNA polymerase I (296–300).

The effect of mutagens that add small aliphatic groups to nucleosides can be assessed by copolymerizing a single modified derivative with an unmodified carrier (the major component), which then serves as a template for transcription. With few exceptions (264, 301, 302) (Table 3), these experiments have utilized polyribonucleotides and DNA-dependent RNA polymerase in the presence of Mn^{2+}. Although it would be preferable for model transcription studies to use polydeoxynucleotides or site-specific insertion of the modified derivative in natural DNAs, the problems and labor to synthesize such models is great compared to the relative ease of preparing defined ribopolynucleotides. There is also an advantage in using ribopolynu-

Table 3 Summary of effect of nucleoside modification on transcription of ribopolynucleotide templates[a]

Basepairing unchanged	Ambiguous behavior[b]	Changed pairing only
N^4-acetyl C	N^4-hydroxy C	N^4-methoxy C
N^6-methyl A	N^6-methoxy A	Iso-C (2-amino-
N^6-hydroxyethyl A	N^4-methyl C	2 deoxy U)
N^6-isopentenyl A	N^2-methyl G	Xanthosine
5-halo, methyl, hydroxyl U or C	O^2-alkyl U	
2'-O-methyl A or C or U	O^4-alkyl U	
	O^6-alkyl G[c]	
7-methyl G	1-methyl A	
2-thio U or C	3-methyl C[c]	
4-thio U	3-methyl U	
Iso-A (3-ribosyla- denine)	1,N^6-etheno A	
	3,N^4-etheno C[c, d]	
	2-aminopurine[e]	

[a] Most of these data were published in a different form by Singer & Kröger. Additional data are in references in the text.

[b] The term ambiguous behavior is used for two classes of modified bases. The first class can form hydrogen bonds with two different base residues (N^4-hydroxy C, N^6-methoxy A, 2-aminopurine). All the remaining modifications lead to misincorporation of more than two bases, usually without stable hydrogen bonds. This latter type of ambiguity is not completely random and is consistent for each modification. Experiments in which the effect of the presence of Mn^{2+} versus Mg^{2+} were studied, indicate that this divalent ion does not change the pattern of misincorporation.

[c] Misincorporation of ribonucletides also from deoxypolynucleotide templates containing a single modification.

[d] Reaction of chloroacetaldehyde with poly(rC) or poly(dC) leads to two products (168). Transcription of these modified polymers without prior dehydration to a single product (ϵC) does not represent the specific misincorporations directed by ϵC. Unpublished data (J. T. Kuśmierek and B. Singer) indicate that misincorporations resulting from the presence of the hydrated intermediate differ markedly from those of ϵC.

[e] Misincorporation of deoxynucleotides from deoxypolynucleotide template.

cleotides and an RNA polymerase: the lack of proofreading and repair maximizes any errors so that polymers with a very small proportion of the modified nucleoside can be used. This minimizes the effect of changed secondary structure on transcription. Although for the purpose of relating in vitro transcription data to biological systems, the presence of a very small amount of modified base is preferred, in some cases homopolymers capable of forming hydrogen bonds were successfully transcribed. This requires that the stacking interactions resemble those of the unmodified polynucleotide.

The data in Table 3 are arranged in three categories: modifications having no discernable effect on base pairing, modifications causing ambiguous incorporation of more than one nucleoside, and modifications completely changing base pairing through a change in hydrogen bonding.

Modifications studied that are on the non-Watson-Crick side or on the sugar have no effect on base pairing (e.g. C-5 of pyrimidines, 2'-O, N-7 of G, and iso A) (303–308). All three pyrimidine oxygens can be changed to sulfur without significant distortion of hydrogen bonding (308, 309). Although modification of exocyclic amino groups can lead to any of the three categories in Table 3, if the substituent in the polymer is *anti* to the Watson-Crick side, normal pairing occurs. In transcription, this is the case for N^4-acetyl-C and the three N^6-modified A derivatives (307, 308).

Derivatives show ambiguous behavior when an essential Watson-Crick site is blocked (N-1 of purines or N-3 of pyrimidines), and no more than a single hydrogen bond can be formed (301, 304, 307). Transcription however is not terminated at these sites. This is also true of the etheno derivatives, ϵC and ϵA (Figure 8), in which a bulky, planar substituent blocks two Watson-Crick sites. The inability to form a specific base pair then leads to misincorporation (310).

Another reason for ambiguity is rotation of a substituent on an exocyclic amino group into the Watson-Crick side, which then sterically interferes with base pairing and eliminates a possible hydrogen bond. We infer from transcription data that N^4-hydroxy C not only acts like U due to the favored tautomer and *anti* conformation, but the hydroxyamino group can rotate *syn,* which results in some ambiguity (304, 308). N^6-Methoxy A can, as a consequence of tautomerism, act like G as well as like A, but not like U or C (310a). The experimental results indicate that there is a shift in the tautomeric equilibrium, but the methoxyamino substituent lies *anti* in transcription.

The methyl group in both N^4-methyl C and N^2-methyl G (308) prefers to lie *anti,* but also rotates. The O-alkyl derivatives of (O^6-alkyl G, O^2-alkyl T, O^4-alkyl T) appear to have the substituent *syn,* on the basis of transcription data [(302, 311, 312); B. Singer, unpublished], but the changed electron distribution can also be responsible. The crystal structure of O^6-methyl G also indicates that the methyl group is *syn* (313).

The extent of transcription of these derivatives in the system used in this laboratory is approximately equal to the amount of modified nucleoside in the polymer. Thus the misincorporations observed are not rare events. The high level of misincorporation that occurs in the RNA polymerase system is not likely to be of the same magnitude with DNA polymerases that have editing functions and different discrimination. Nevertheless, if a misincorporation is possible, it may occur in any system.

Three derivatives show changed base pairing that can be explained on a chemical basis. N^4-Methoxy C, which is formed by reaction with methoxyamine, is in the imino form (66) and thus substitutes for U (308). This complete tautomeric shift is unusual and is reflected not only in transcription data (308), but is also supported by physical studies of the double-stranded product in which N^4-methoxy C is basepaired with A (269). The behavior of the two other derivatives with changed base pairing is also predictable, since iso C, a model compound, and xanthosine (deaminated G) have the hydrogen bonding equivalence of U (308) (Table 3).

Based on polynucleotide studies on the possible mutational effects of modified nucleosides, some general conclusions can be drawn:

1. The bulky aromatic substituents, regardless of their position on a nucleoside, will interfere with replication simply because of size. However, the rapid depurination of some of these products may exert a mutagenic effect through error-prone repair of the apurinic site. In addition, bypass of the polymerase could lead to frameshifts. This is the observed type of mutation by these chemicals in the Ames test.

2. Smaller alkyl derivatives, with substituents affecting hydrogen-bonding, if unrepaired, are likely to cause misincorporation and give random point mutations. All of the O-alkyl derivatives formed by the highly mutagenic N-nitroso compounds would be predicted to be mutagenic on this basis.

3. Mutation by most nonalkylating agents can be ascribed to a change in a base by deamination (HNO_2, HSO_3^-), or a change in the preferred tautomer in the modified derivative (HA, MHA, HZ). The mechanism of formaldehyde mutation is still unclear.

How It Works For One Nucleoside

In this last section we choose cytidine as a model for a variety of direct mutagenic reactions, particularly those occurring under physiological conditions. Throughout this review we emphasize that both complicated organic chemistry and physical chemistry can produce the same mutagenic end point. Figure 13 illustrates a series of known reactions, some of which have been studied as to their effect in replication, and all of which have been studied in model systems. It is apparent that tautomerism is always of importance, whether due to modification or simply as a normal property. The reactions causing the greatest ambiguity are those where Watson-Crick hydrogen-bonding sites are blocked, or unavailable due to ionization, glycosyl bond cleavage, or base rotation.

The reactions of cytidine illustrated in Figure 13 represent the maximum amount of information we possess for any nucleoside. It is unfortunate that such simple mechanistic pathways can not be described for most mutagens. It is likely that mutagens producing large derivatives cause mutations through indirect mechanisms, such as error-prone repair pathways. Other indirect mechanisms may exist for even simple reagents, as a result of their interaction with a variety of cellular components, and not only DNA.

Literature Cited

1. Neel, J. V. 1979. In *Banbury Report,* ed. V. K. McElheny, S., Abrahamson, 1:7–26. Cold Spring Harbor, NY: Cold Spring Harbor Lab.
2. Freese, E., Freese, E. B. 1966. *Radiat. Res. Suppl.* 6:97–140
3. Bresler, S. E., Kalinin, V. L., Perumov, D. A. 1968. *Mutat. Res.* 5:1–14
4. Mukai, F., Hawryluk, I., Shapiro, R. 1970. *Biochem. Biophys. Res. Commun.* 39:983–88
5. Kimball, R. F., Hirsch, B. F. 1975. *Mutat. Res.* 30:9–20
6. Chu, C.-T., Parris, D. S., Dixon, R. A. F., Farber, F. E., Schaffer, P. A. 1979. *Virology* 98:168–81
7. McCann, J., Choi, E., Yamasaki, E., Ames, B. N. 1975. *Proc. Natl. Acad. Sci. USA* 72:5135–81
8. Hemminki, K., Falck, K., Vainio, H. 1980. *Arch. Toxicol.* 46:277–85
9. Bresler, S. E., Kalinin, V. L., Sukhodolova, A. T. 1972. *Mutat. Res.* 15:101–12
10. Bartsch, H., Malaveille, C., Camus, A. M., Martel-Planche, G., Brun, G., Hautefeuille, A., Sabadie, N., Barbin, A., Kuroki, T., Drevon, C., Piccoli, C., Montesano, R. 1980. In *Molecular and Cellular Aspects of Carcinogen Screening Tests,* ed. R. Montesano, H. Bartsch, L. Tomatis, pp. 179–241. Lyons, France: IARC Sci. Publi. No. 27
11. Hodgkiss, R. J., Brennard, J., Fox, M. 1980. *Carcinogenesis* 1:175–87
12. Suter, W., Brennand, J., McMillan, S., Fox, M. 1980. *Mutat. Res.* 73:171–81

13. Peterson, A. R., Peterson, H., Heidelberger, C. 1979. *Cancer Res.* 39:131–38
14. Chu, E. H. Y., Malling, H. V. 1968. *Proc. Natl. Acad. Sci. USA* 61:1306–12
14a. Couch, D. B., Hsie, A. W. 1978. *Mutat. Res.* 57:209–16
14b. Couch, D. B., Forbes, N. L., Hsie, A. W. 1978. *Mutat. Res.* 57:217–24
15. Singer, B. 1975. *Prog. Nucleic Acid Res. Mol. Biol.* 15:219–84, 310–12
16. Druckrey, H., Preussmann, R., Ivankovic, S., Schmahl, D. 1967. *Z. Krebsforsch.* 69:103–201
17. Lawley, P. D., Thatcher, C. J. 1970. *Biochem. J.* 116:693–707
18. LaPolla, J. P., Harris, C. M., Vary, J. C. 1972. *Biochem. Biophys. Res. Commun.* 49:133–38
19. Nagao, M., Yokoshima, T., Hosoi, H., Sugimura, T. 1969. *Biochim. Biophys. Acta* 192:191–99
20. Nagao, M., Hosoi, H., Sugimura, T. 1971. *Biochim. Biophys. Acta* 237:369–77
21. Yoda, K., Sakiyama, S., Fujimura, S. 1978. *Biochim. Biophys. Acta* 521:677–88
22. Yamamoto, K., Kondo, S., Sugimura, T. 1978. *J. Mol. Biol.* 118:413–30
23. Singer, B., Fraenkel-Conrat, H. 1969. *Prog. Nucleic Acid Res. Mol. Biol.* 9:1–30
24. Singer, B., Fraenkel-Conrat, H. 1969. *Virology* 39:395–99
25. Herriott, R. M. 1971. In *Chemical Mutagens,* ed. A. Hollaender, 1:175–217. New York: Plenum

26. Singer, B., Fraenkel-Conrat, H. 1974. *Virology* 60:485–90
27. Fahmy, O. G., Fahmy, M. J. 1972. *Cancer Res.* 32:550–57
28. Tazima, T. 1980. In *Chemical Mutagens* ed. F. J. deSerres, A. Hollaender, 6:203–38. New York: Plenum
29. Langenbach, R., Freed, H. J., Raveh, D., Huberman, E. 1978. *Nature* 276: 277–80
30. Ames, B. N., McCann, J., Yamasaki, E. 1975. *Mutat. Res.* 31:347–64
31. Lindahl, T., Nyberg, B. 1974. *Biochemistry* 13:3405–10
32. Karran, P., Lindahl, T. 1980. *Biochemistry* 19:6005–11
33. Shapiro, R., Kang, S. 1969. *Biochemistry* 8:1806–10
34. Lindahl, T., Nyberg, B. 1972. *Biochemistry* 11:3610–18
35. Lindahl, T., Karlstrom, O. 1973. *Biochemistry* 12:5151–54
36. Lindahl, T. 1979. *Prog. Nucleic Acids Mol. Biol.* 22:135–92
37. Hartman, P. E. 1980. *Environ. Mutagenesis* 2:3–16
38. Topal, M. D., Fresco, J. R. 1976. *Nature* 263:285–89
39. Topal, M. D., Fresco, J. R. 1976. *Nature* 263:289–93
40. Dunn, D. B., Hall, R. H. 1975. In *Handbook of Biochemistry and Molecular Biology Nucleic Acids*, ed. G. D. Fasman, 1:65–215. Cleveland, Ohio: CRC. 3rd ed.
41. Singer, B. 1975. See Ref. 40, pp. 409–47
42. Haschemeyer, A. E. V., Rich, A. 1967. *J. Mol. Biol.* 27:369–84
43. Duncan, B. K., Miller, J. H. 1980. *Nature* 287:560–61
44. Hayakawa, H., Sekiguchi, M. 1978. *Biochem. Biophys. Res. Commun.* 83: 1312–18
45. Lindahl, T., Andersson, A. 1972. *Biochemistry* 11:3618–22
46. Shearman, C. W., Loeb, L. A. 1979. *J. Mol. Biol.* 128:197–218
47. Schaaper, R. M., Shearman, C. W., Loeb, L. A. 1981. *Proc. Am. Assoc. Cancer Res.* 22:92
48. Engel, J. D., von Hippel, P. H. 1978. *J. Biol. Chem.* 253:927–34
49. Ward, D. C., Cerami, A., Reich, E., Acs, G., Altwerger, L. 1969. *J. Biol. Chem.* 244:3243–50
50. Fresco, J. R., Broitman, S., Lane, A.-E. 1980. In *Mechanistic Studies of DNA Replication and Genetic Recombination*, ed. B. Alberts, C. F. Fox, pp. 753–68. New York: Academic
51. Watson, J. D., Crick, F. H. C. 1953. *Nature* 171:964–67
52. Agarwal, S. S., Dube, D. K., Loeb, L. A. 1979. *J. Biol. Chem.* 254:101–6
53. Fersht, A. R. 1979. *Proc. Natl. Acad. Sci. USA* 76:4946–50
54. Kunkel, T. A., Loeb, L. A. 1980. *J. Biol. Chem.* 255:9961–66
55. Drake, J. W. 1969. *Nature* 221:1132
56. Auerbach, C., Moutschen-Dahmez, M., Moutschen, J. 1977. *Mutat. Res.* 39:317–62
57. Fraenkel-Conrat, H. 1954. *Biochim. Biophys. Acta* 15:308–9
58. Feldman, M. Ya. 1973. *Prog. Nucleic Acid Res. Mol. Biol.* 13:1–49
59. von Hippel, P. H., Wong, K. Y. 1971. *J. Mol. Biol.* 61:587–613
60. Lukashin, A. V., Vologodskii, A. V., Frank-Kamenetskii, M. D., Lyubchenko, Y. L. 1976. *J. Mol. Biol.* 108: 665–82
61. McGhee, J. D., von Hippel, P. H. 1977. *Biochemistry* 16:3267–76
62. McGhee, J. D., von Hippel, P. H. 1977. *Biochemistry* 16:3276–93
63. McGhee, J. D., von Hippel, P. H. 1975. *Biochemistry* 14:1281–96
64. Chaw, Y. F. M., Crane, L. E., Lange, P., Shapiro, R. 1980. *Biochemistry* 19: 5525–31
65. Brock, T. D. 1979. *Biology of Microorganisms*, p. 428. New Jersey: Prentice-Hall. 802 pp. 3rd ed.
66. Brown, D. M., Hewlins, M. J. E., Schell, P. 1968. *J. Chem. Soc. C*, pp. 1925–29
67. Kochetkov, N. K., Budowsky, E. I. 1969. *Prog. Nucleic Acid Res. Mol. Biol.* 9:403–38
68. Fraenkel-Conrat, H., Singer, B. 1972. *Biochim. Biophys. Acta* 262:264–68
69. Brown, D. M., Hewlins, M. J. E. 1968. *J. Chem. Soc. C*, pp. 1922–24
70. Freese, E. B. 1966. *Mutat. Res.* 5:299–301
71. Budowsky, E. I., Sverdlov, E. D., Spasokukotskaya, T. N., Koudelka, J. A. 1975. *Biochim. Biophys. Acta* 300: 1–13
72. Freese, E. B., Freese, E. 1964. *Proc. Natl. Acad. Sci. USA* 52:1289–97
73. Budowsky, E. I., Krivisky, A. S., Preobrazhenskaya, E. S. 1979. *Mutat. Res.* 59:285–89
74. Maes, R., Mesquita, J. 1970. *Arch. ges. Virusforsch.* 29:77–82
75. Schuster, H., Schramm, G. 1958. *Z. Naturforsch. Teil B* 136:698–704
76. Litman, R. M. 1961. *J. Chim. Phys.* 58:997–1007
77. Shapiro, R., Dubelman, S., Feinberg, A., Crain, P., McCloskey, J. 1977. *J. Am. Chem. Soc.* 99:302–3

78. Frankel, A. D., Duncan, B. K., Hartman, P. E. 1980. *J. Bacteriol.* 142:335–38
79. Shapiro, R. 1982. In *Induced Mutagenesis: Molecular Mechanisms and their Implication for Environmental Problems*, ed. C. W. Lawrence, L. Prakash, F. Sherman. New York: Plenum. In press
80. Shapiro, R., Servis, R. E., Welcher, M. 1979. *J. Am. Chem. Soc.* 92:422–24
81. Hayatsu, H., Wataya, Y., Kai, K., Iida, S. 1970. *Biochemistry* 9:2858–65
82. Sklyadneva, V. B., Chekanovskaya, L. A., Nikolaeva, I. A., Tikchonenko, T. I. 1979. *Biochim. Biophys. Acta* 565:51–66
83. Sklyadneva, V. B., Shie, M., Tikchonenko, T. I. 1979. *FEBS Lett.* 107:129–33
84. Ehresmann, B., Briand, J.-P., Reinbolt, J., Witz, J. 1980. *Eur. J. Biochem.* 108:123–29
85. Wang, R. Y. H., Gehrke, C. W., Ehrlich, M. 1980. *Nucleic Acids Res.* 8:4777–90
86. Coulondre, C., Miller, J. H., Farabaugh, P. J., Gilbert, W. 1978. *Nature* 274:775–80
87. Shiragami, M., Hayatsu, H. 1976. *Nucleic Acids Res.* 2:s31–32
88. Hayatsu, H., Shiragami, M. 1979. *Biochemistry* 18:632–37
89. Shapiro, R., Law, D. C. F., Weisgras, J. M. 1972. *Biochem. Biophys. Res. Commun.* 49:358–63.
90. Shapiro, R., Braverman, B., Louis, J. B., Servis, R. E. 1973. *J. Biol. Chem.* 248:4060–64
91. Chambers, R. W., Aoyagi, S., Furukawa, Y., Zawadzka, H., Bhanot, O. S. 1973. *J. Biol. Chem.* 248:5549–51
92. Piper, P. W., Clark, B. F. C. 1974. *Nucleic Acids Res.* 1:45–51
93. Iserentant, D., Fiers, W. 1979. *Eur. J. Biochem.* 102:595–604
94. Mills, D. R., Kramer, F. R., Dobkin, C., Nishihara, T., Cole, P. E. 1980. *Biochemistry* 19:228–36
95. Shortle, D., Nathans, D. 1978. *Proc. Natl. Acad. Sci. USA* 75:2170–74
96. Kimball, R. F. 1977. *Mutat. Res.* 39:111–26
97. Parodi, S., DeFlora, S., Cavanna, M., Pino, A., Robbiano, L., Bennicelli, C., Brambilla, G. 1981. *Cancer Res.* 41:1469–82
98. Brown, D. M., McNaught, A. D., Schell, P. 1966. *Biochem. Biophys. Res. Commun.* 24:967–71
99. Brown, D. M., Osborne, M. R. 1971. *Biochim. Biophys. Acta* 247:514–18

100. Brown, D. M., 1974. In *Basic Principles in Nucleic Acid Chemistry*, ed. P. O. P. Ts'o, 2:1–90. New York: Academic
101. Thielmann, H. W., Schroeder, C. H., O'Niell, J. P., Brimer, K., Hsi, A. W. 1979. *Chem.-Biol. Interactions* 26: 233–43
102. Singer, B. 1972. *Biochemistry* 11: 3939–47
103. Sun, L., Singer, B. 1974. *Biochemistry* 13:1905–13
104. Singer, B., Sun, L., Fraenkel-Conrat, H. 1974. *Biochemistry* 13:1913–20
105. Singer, B., Fraenkel-Conrat, H. 1975. *Biochemistry* 14:772–82
106. Sun, L., Singer, B. 1975. *Biochemistry* 14:1795–1802
107. Singer, B., Sun, L., Fraenkel-Conrat, H. 1975. *Proc. Natl. Acad. Sci. USA* 72:2232–36
108. Singer, B. 1975. *Biochemistry* 14: 4353–57
109. Singer, B. 1976. *FEBS Lett.* 63:85–88
110. Kusmierek, J. T., Singer, B. 1976. *Biochim. Biophys. Acta* 422:420–31
111. Kusmierek, J. T., Singer, B. 1976. *Nucleic Acids Res.* 4:989–1000
112. Singer, B., Kusmierek, J. T. 1976. *Biochemistry* 15:5052–57
113. Singer, B. 1976. *Nature* 264:333–39
114. Singer, B. 1977. *J. Toxicol. Environ. Health* 2:1279–95
115. Singer, B., Kröger, M., Carrano, M. 1978. *Biochemistry* 17:1246–50
116. Singer, B., Bodell, W. J., Cleaver, J. E., Thomas, G. H., Rajewsky, M. F., Thon, W. 1978. *Nature* 276:85–88
117. Singer, B. 1979. *J. Natl. Cancer Inst.* 61:1329–39
118. Singer, B., Bodell, W. J. 1979. *Biochemistry* 18:2860–63
119. Singer, B., Spengler, S., Bodell, W. 1981. *Carcinogenesis.* 10:1069–73
120. Farmer, P. B., Foster, A. B., Jarman, M., Tisdale, M. J. 1973. *Biochem. J.* 135:203–13
121. Swenson, D. H., Harbach, P. R., Trzos, R. J. 1980. *Carcinogenesis* 1:931–36
122. Singer, B. 1982. In *Molecular and Cellular Mechanisms of Mutagenesis*, ed. J. F. Lemontt, W. M. Generoso. New York: Plenum. In press
123. Jensen, D. E., Reed, D. J. 1978. *Biochemistry* 17:5098–5107
124. Mandal, C., Kallenbach, N. R., Englander, S. W. 1979. *J. Mol. Biol.* 135:391–411
125. Bodell, W. J., Singer, B. 1979. *Biochemistry* 18:2860–63
126. Beranek, D. T., Weis, C. C., Swenson, D. H. 1980. *Carcinogenesis* 1:595–606
127. Neale, S. 1976. *Mutat. Res.* 32:229–66

128. Montesano, R., Bartsch, H. 1976. *Mutat. Res.* 32:179–228
129. Pegg, A. E. 1977. *Adv. Cancer Res.* 25:195–269
130. Lijinsky, W., Greenblatt, M. 1972. *Nature New Biol.* 237:177–78
131. Taylor, H. W., Lijinsky, W. 1975. *Cancer Res.* 35:812–15
132. Mirvish, S. M. 1975. *Toxic. Appl. Pharmacol.* 31:325–51
133. Guttenplan, J. B. 1980. *Carcinogenesis* 1:439–44
134. Hsia, C. C., Sun, T. T., Wang, Y. Y., Anderson, L. M., Armstrong, D., Good, R. A. 1981. *Proc. Natl. Acad. Sci. USA* 78:1878–81
135. Guttenplan, J. B. 1978. *Cancer Res.* 38:2018–22
136. Brookes, P., Lawley, P. D. 1960. *Biochem. J.* 77:478–84
137. Brookes, P., Lawley, P. D. 1961. *Biochem. J.* 80:496–503
138. Brookes, P., Lawley, P. D. 1963. *Biochem. J.* 89:138–44
139. Flamm, W. G., Bernheim, N. J., Fishbein, L. 1970. *Biochim. Biophys. Acta* 224:657–59
140. Kircher, M., Fleer, R., Ruhland, A., Brendel, M. 1979. *Mutat. Res.* 63:273–89
141. Ross, W. E., Ewig, R. A. G., Kohn, K. W. 1978. *Cancer Res.* 38:1502–6
142. Chun, E. H. L., Gonzales, L., Lewis, F. S., Jones, J., Rutman, R. J. 1969. *Cancer Res.* 29:1184–94
143. Ewig, R. A. G., Kohn, K. W. 1977. *Cancer Res.* 37:2114–22
144. Mehta, J. R., Przybylski, M., Ludlum, D. P. 1980. *Cancer Res.* 40:4183–86
145. Roberts, J. J., Warwick, G. P. 1963. *Biochem. Pharmacol.* 12:1441–42
146. Maté, U., Solomon, J. J., Segal, A. 1977. *Chem.-Biol. Interactions* 18:327–36
147. Chen, R., Mieyal, J. J., Goldthwait, D. A. 1981. *Carcinogenesis* 2:73–80
148. Kubinski, H., Szybalski, E. H. 1975. *Chem.-Biol. Interactions* 10:41–55
149. Corbett, T. H., Heidelberger, C., Dove, W. F. 1970. *Molec. Pharmacol.* 6:667–79
150. Brusick, D. J. 1977. *Mutat. Res.* 39:241–56
151. Fishbein, L. 1969. *Ann. NY Acad. Sci.* 163:869–94
152. Ehrenberg, L. 1979. In *Banbury Report,* ed. V. K. McElheny, S. Abrahamson, 1:157–90. Cold Spring Harbor, NY: Cold Spring Harbor Lab.
153. Lawley, P. D., Jarman, M. 1972. *Biochem. J.* 126:893–900
154. Barbin, A., Bresil, H., Croisy, A., Jacquignon, P., Malaveille, C., Montesano, R., Bartsch, H. 1975. *Biochem. Biophys. Res. Commun.* 67:596–603
155. Huberman, E., Bartsch, H., Sachs, L. 1975. *Int. J. Cancer* 16:639–44
156. McCann, J., Simmon, V., Streitwieser, D., Ames, B. N. 1975. *Proc. Natl. Acad. Sci. USA* 72:3190–93
157. Bartsch, H., Malaveille, C., Barbin, A., Planche, G. 1979. *Arch. Toxicol.* 41:249–77
158. Barrio, J. R., Secrist, J. A. III, Leonard, N.J. 1972. *Biochem. Biophys. Res. Commun.* 46:597–604
159. Secrist, J. A. III, Barrio, J. R., Leonard, N. J., Weber, G. 1972. *Biochemistry* 11:3499–3506
160. Green, T., Hathaway, D. E. 1978. *Chem.-Biol. Interactions* 22:211–24
161. Laib, R. J., Bolt, H. M. 1978. *Arch. Toxicol.* 39:235–40
162. Wang, A. H.-J., Barrio, J. R., Paul, I. C. 1976. *J. Am. Chem. Soc.* 98:7401–8
163. Wang, A.H.-J., Dammann, L. G., Barrio, J. R., Paul, I. C. 1974. *J. Am. Chem. Soc.* 96:1205–13
164. Stattsangi, P. D., Leonard, N. J., Frihart, C. R. 1977. *J. Org. Chem.* 42:3292–96
165. Biernat, J., Ciesiolka, J., Górnicki, P., Adamiak, R. W., Krzyzosiak, W. J., Wiewiórowski, M. 1978. *Nucleic Acids Res.* 5:789–804
166. Barrio, J. R., Sattsangi, P. D., Gruber, B. A., Dammann, L. G., Leonard, N. J. 1976. *J. Am. Chem. Soc.* 98:7408–14
167. Sattsangi, P. D., Barrio, J. R., Leonard, N. J. 1980. *J. Am. Chem. Soc.* 102:770–74
168. Krzyzosiak, W. J., Biernat, J., Ciesiolka, J., Gulewicz, K., Wiewiórowski, M. 1981. *Nucleic Acids Res.* 9:2841–51
169. Osterman-Golkar, S., Hultmark, D., Segerback, D., Calleman, C. J., Gothe, R., Ehrenberg, L., Wachmeister, C. A. 1977. *Biochem. Biophys. Res. Commun.* 76:259–66
170. Kimura, K., Nakanishi, M., Yamamoto, T., Tsuboi, M. 1977. *J. Biochem.* 81:1699–1703
171. Goldschmidt, B. M., Blazej, T. P., Van Duuren, B. L. 1968. *Tetrahedron Lett.* 13:1583–86
172. Van Duuren, B. L., Loewengart, G. 1977. *J. Biol. Chem.* 252:5370–71
173. Bradley, M. O., Sharkey, N. A., Kohn, K. W., Layard, M. W. 1980. *Cancer Res.* 40:2719–25
174. Gombar, C. T., Tong, W. P., Ludlum, D. B. 1980. *Biochem. Pharmacol.* 29:2639–43
175. Tong, W. P., Kirk, M. C., Ludlum, D.

B. 1981. *Biochem. Biophys. Res. Commun.* 100:351–57
176. Tong, W. P., Ludlum, D. B. 1981. *Proc. Am. Assoc. Cancer Res.* 22:29
177. Tong, W. P., Ludlum, D. B. 1978. *Biochem. Pharmacol.* 27:77–81
178. Tong, W. P., Ludlum, D. B. 1979. *Biochem. Pharmacol.* 28:1175–79
179. Ludlum, D. B., Tong, W. P. 1978. *Biochem. Pharmacol.* 27:2391–94
180. Kohn, K. W. 1977. *Cancer Res.* 37: 1450–54
181. Tong, W. P., Ludlum, D. B. 1981. *Cancer Res.* 41:380–82
182. Lown, J. W., McLaughlin, L. W., Chang, Y.-M. 1978. *Bioorganic Chem.* 7:97–110
183. Erickson, L. C., Laurent, G., Sharkey, N. A., Kohn, K. W. 1980. *Nature* 288:727–29
183a. Singer, B. 1980. In *Carcinogenesis: Fundamental Mechanisms and Environmental Effects,* ed. B. Pullman, P.O.P. Ts'o, H. Gelboin, pp. 92–102. Dorndrecht, Holland: Reidel
184. Nagao, M., Sugimura, T. 1976. *Adv. Cancer Res.* 23:131–69
185. Moschel, R. C., Baird, W. M., Dipple, A. 1977. *Biochem. Biophys. Res. Commun.* 76:1092–98
186. Brookes, P. 1977. *Mutat. Res.* 39: 257–84
187. Miller, E. C. 1978. *Cancer Res.* 38:1479–96
188. Kriek, E., Westra, J. G. 1979. In *Chemical Carcinogens and DNA,* ed. P. L. Grover, 2:1–28. Boca Raton, Fla: CRC
189. Hashimoto, Y., Shudo, K., Okamoto, T. 1980. *Biochem. Biophys. Res. Commun.* 92:971–76
190. Selkirk, J. K. 1980. In *Carcinogenesis: Modifiers of Chemical Carcinogenesis,* ed. T. J. Slaga, 5:1–31 New York: Raven
191. Huberman, E., Aspiras, L., Heidelberger, C., Grover, P. L., Sims, P. 1971 *Proc. Natl. Acad. Sci. USA* 68:3195–99
192. Maher, V. M., McCormick, J. J. 1978. In *Polycyclic Hydrocarbons and Cancer,* ed. H. V. Gelboin, 2:137–59 New York: Academic
193. Wood, A. W., Chang, R. L., Huang, M.-T., Levin, W., Lehr, R. E., Kumar, S., Thakker, D. R., Yagi, H., Conney, A. H. 1980. *Cancer Res.* 40:1985–89
194. Newbold, R. F., Brookes, P., Harvey, R. G. 1979. *Int. J. Cancer* 24:203–9
195. Stark, A. A. 1980. *Ann. Rev. Microbiol.* 34:235–62
196. Kasai, H., Yamaizumi, Z., Wakabayashi, K., Nagao, M., Sugimura, T., Yokoyama, S., Miyazawa, T., Ni-

shimura, S. 1980. *Chem. Lett.* pp. 1391–94
197. Kadlubar, F. F., Miller, J. A., Miller, E. C. 1978. *Cancer Res.* 38:3628–38
198. Kadlubar, F. F., Melchior, W. B., Flammang, T. J., Gagliano, A. G., Yoshida, H., Geacintov, N. E. 1981. *Cancer Res.* 41:2168–74
199. Kawazoe, Y., Araki, M., Huang, G. F., Okamoto, T., Tada, M., Tada, M. 1975. *Chem. Pharm. Bull. (Tokyo)* 23: 3041–43
200. Westra, J. G., Kriek, E., Hittenhausen, H. 1976. *Chem.-Biol. Interactions* 15: 149–64
201. Kriek, E., Miller, J. A., Juhl, U., Miller, E. C. 1967. *Biochemistry* 6:177–82
202. Kadlubar, F. F., Unruh, L. E., Beland, F. A., Straub, K. M., Evans, F. E. 1980. *Carcinogenesis* 1:139–50
203. Beland, F. A., Tullis, D. L., Kadlubar, F. F., Straub, K. M., Evans, F. E. 1980. *Chem.-Biol. Interactions* 31:1–17
204. Tarpley, W. G., Miller, J. A., Miller, E. C. 1980. *Cancer Res.* 40:2493–99
205. Hashimoto, Y., Shudo, K., Imamura, M., Okamoto, T. 1980. *Nucleic Acids Res. Symp. Ser.* 8:s109–12
206. Imamura, M., Takeda, K., Shudo, K., Okamoto, T., Nagata, C., Kodama, M. 1980. *Biochem. Biophys. Res. Commun.* 96:611–17
207. Drinkwater, N. R., Miller, E. C., Miller, J. A. 1980. *Biochemistry* 19: 5087–92
208. Grunberger, D., Weinstein, I. B. 1979. *Prog. Nucleic Acid Res. Mol. Biol.* 23:106–49
209. Grunberger, D., Weinstein, I. B. 1979. In *Chemical Carcinogenesis and DNA,* ed. P. L. Grover, 1:60–93. Boca Raton Fla: CRC
210. Evans, F. E., Miller, D. W., Beland, F. A. 1980. *Carcinogenesis* 1:955–59
211. Jeffrey, A. M., Jennette, K. W., Blobstein, S. H., Weinstein, I. B., Beland, F. A., Harvey, R. G., Kasai, H., Miura, I., Nakanishi, K. 1976. *J. Am. Chem. Soc.* 98:5714–15
212. Jeffrey, A. M., Weinstein, I. B., Jennette, K. W., Grzeskowiak, K., Nakanishi, K., Harvey, R. G., Autrup, H., Harris, C. 1977. *Nature* 269:348–50
213. Koreeda, M., Moore, P. D., Yagi, H., Yeh, H. J. C., Jerina, D. M. 1976. *J. Am. Chem. Soc.* 98:6720–22
214. Weinstein, I. B., Jeffrey, A. M., Jennette, K. W., Blobstein, S. H., Harvey, R. G., Harris, C., Autrup, H., Kasai, H., Nakanishi, K. 1976. *Science* 197:592–95

215. Nakanishi, K., Kasai, H., Cho, H., Harvey, R. G., Jeffrey, A. M., Jennette, K. W., Weinstein, I. B. 1977. *J. Am. Chem. Soc.* 99:258–60
216. Jeffrey, A. M., Grzeskowiak, K., Weinstein, I. B., Nakanishi, K., Roller, P., Harvey, R. G. 1979. *Science* 206:1309–11
217. King, H. S. W., Osborne, M. R., Brookes, P. 1979. *Chem.-Biol. Interactions* 24:345–53
218. Dipple, A., Brookes, P., Mackintosch, D. S., Rayman, M. P. 1971. *Biochemistry* 10:4323–30
219. Rayman, M. P., Dipple, A. 1973. *Biochemistry* 12:1538–42
220. Carrell, H. L., Glusker, J. P., Moschel, R. C., Hudgins, W. R., Dipple, A. 1981. *Cancer Res.* 41:2230–34
221. Jeffrey, A. M., Blobstein, S. H., Weinstein, I. B., Beland, F. A., Harvey, R. G., Kasai, H., Nakanishi, K. 1976. *Proc. Natl. Acad. Sci. USA* 73:2311–15
222. Nakanishi, K., Komura, H., Miura, I., Kasai, H., Frenkel, K., Grunberger, D. 1980. *JCS Chem. Commun.* pp. 82–83
223. Kasai, H., Nakanishi, K., Frenkel, K., Grunberger, D. 1977. *J. Am. Chem. Soc.* 99:8500–2
224. Martin, C. N., Garner, R. C. 1977. *Nature* 267:863–65
225. Essigmann, J. M., Croy, R. G., Nadzan, A. M., Busby, W. F. Jr., Reinhold, V. N., Buchi, G., Wogan, G. N. 1977. *Proc. Natl. Acad. Sci. USA* 74:1870–74
226. Lin, J-K., Miller, J. A., Miller, E. C. 1977. *Cancer Res.* 37:4430–38
227. Autrup, H., Essigmann, J. M., Croy, R. G., Trump, B. F., Wogan, G. N., Harris, C. C. 1979. *Cancer Res.* 39:694–98
228. Wang, T. V., Cerutti, P. 1980. *Biochemistry* 19:1692–98
229. Wogan, G. N., Croy, R. G., Essigmann, J. M., Bennett, R. A. 1980. In *Carcinogenesis: Fundamental Mechanisms and Environmental Effects*, eds. B. Pullman, P. O. P. Ts'o, H. Gelboin, Dordrecht, Holland: D. Riedel Publ. Co. 13:179–91
230. Cerutti, P. A., Wang, V. T., Amstad, P. 1980. See Ref. 229, pp. 465–77
231. Phillips, D. H., Miller, J. A., Miller, E. C., Adams, B. 1981. *Cancer Res.* 41:176–86
232. Phillips, D. H., Miller, J. A., Miller, E. C., Adams, B. 1981. *Cancer Res.* 41:2664–71
233. Gamper, H. B., Bartholomew, J. C., Calvin, M. 1980. *Biochemistry* 19:3948–56
234. Fujii, T., Saito, T., Nakasaka, T. 1980.

J. Chem. Soc. Chem. Commun, pp. 758–59
235. Saito, T., Fujii, T. 1979. *J. Chem. Soc. Chem. Commun.,* p. 135
236. Crick, F. H. C. 1966. *J. Mol. Biol.* 19:548–55
237. Lomant, A. J., Fresco, J. R. 1975. *Prog. Nucl. Acids. Res. Mol. Biol.* 15:185–218
238. Gillam, S., Waterman, K., Smith, M. 1975. *Nucleic Acids Res.* 2:625–34
239. Dodgson, J. B., Wells, R. D. 1977. *Biochemistry* 16:2367–74
240. Ackermann, T., Gramlich, V., Klump, H., Knable, T., Schmid, E. D., Seliger, H., Stulz, J. 1979. *Biophysical Chem.* 10:231–38
241. Romaniuk, P. J., Hughes, D. W., Gregoire, R. J., Bell, R. A., Neilson, T. 1979. *Biochemistry* 18:5109–16
242. Wallace, R. B., Shaffer, J., Murphy, R. F., Bonner, J., Hirose, T., Itakura, K. 1979. *Nucleic Acids Res.* 6:3543–57
243. Quigley, G. J., Seeman, N. C., Wang, A. H. J., Suddath, F. L., Rich, A. 1975. *Nucleic Acids Res.* 2:2329–41
244. Sussman, J. L., Kim, S. H. 1976. *Biochem. Biophys. Res. Commun.* 68:89–96
245. Ladner, J. E., Jack, A., Robertus, J. D., Brown, R. S., Rhodes, D., Clark, B. F. C., Klug, A. 1975. *Nucleic Acids Res.* 2:1629–37
246. Early, T. A., Olmsted, J., Kearns, D. R., Lezius, A. G. 1978. *Nucleic Acids Res.* 5:1955–70
247. Cornelis, A. G., Haasnoot, J. H. J., den Hartog, J. F., deRooij, M., van Boom, J. H., Cornelis, A. 1979. *Nature* 281:235–36
248. Haasnoot, C. A. G., den Hartog, J. H. J., de Rooij, J. F. M., van Boom, J. H., Altona, C. 1980. *Nucleic Acids Res.* 8:169–81
249. Hall, Z. W., Lehman, I. R. 1968. *J. Mol. Biol.* 36:321–33
250. Grosjean, H. J., de Henau, S., Crothers, D. M. 1978. *Proc. Natl. Acad. Sci. USA* 75:610–14
251. Fiers, W., Contreras, R., Duerinck, F., Haegeman, G., Iserentant, D., Merregaert, J., MinJou, W., Molemans, F., Raeymaekers, A., van den Berghe, A., Volckaert, G., Ysebaert, M. 1976. *Nature* 260:500–7
252. Branlant, C., Krol, A., Machatt, M. A., Ebel, J. P. 1979. *FEBS Lett.* 107:177–81
253. Carbon, P., Ebel, J. P., Ehresmann, C. 1981. *Nucleic Acids Res.* 9:2325–33
254. Delihas, N., Anderson, J., Sprouse, H. M., Dudock, B. 1981. *Nucleic Acids Res.* 9:2801–5

255. Stiegler, P., Carbon, P., Zucker, M., Ebel, J. P., Ehresmann, C. 1981. *Nucleic Acids Res.* 9:2153–72
256. Miles, H. T. 1981. *Proc. Natl. Acad. Sci. USA* 47:791–802
257. Katritzky, A. R., Waring, A. J. 1962. *J. Chem. Soc.,* pp. 1540–44
258. Psoda, A., Shugar, D. 1971. *Biochim. Biophys. Acta* 247:507–13
259. Evans, F. E., Sarma, R. H. 1974. *J. Mol. Biol.* 89:249–53
260. Stolarski, R., Remin, M., Shugar, D. 1977. *Z. Naturforsch. Teil C* 32:894–900
261. Nowak, M. J., Szczepaniak, K., Barski, A., Shugar, D. 1978. *Z. Naturforsch. Teil C* 33:878–83
262. Kulikowski, T., Shugar, D. 1978. *Nucleic Acids Res. Spec. Publ.* 4:7–10
263. Janion, C. 1980. In *DNA-Recombination Interactions and Repair,* ed. S. Zadrazil, J. Sponar, Oxford: Pergamon
264. Watanabe, S. M., Goodman, M. F. 1981. *Proc. Natl. Acad. Sci. USA* 78:2864–68
265. Conkling, M. A., Koch, R. E., Drake, J. W. 1980. *J. Mol. Biol.* 143:303–15
266. Brown, D. M., Hewlins, M. J. E. 1969. *Nature* 221:656–57
267. Psoda, A., Kierdaszuk, B., Pohorille, A., Geller, M., Kusmierek, J. T., Shugar, D. 1981. *Int. J. Quantum Chem.* 20:543–54
268. Shoup, R. R., Miles, H. T., Becker, E. D. 1972. *J. Phys. Chem.* 76:64–70
269. Spengler, S., Singer, B. 1981. *Biochemistry* 20:7290–94
270. Flavell, R. A., Sabo, D. L., Bandle, E. F., Weissmann, C. 1974. *J. Mol. Biol.* 89:255–72
271. Sabo, D. L., Domingo, E., Bandle, E. F., Flavell, R. A. Weissmann, C. 1977. *J. Mol. Biol.* 112:235–52
272. Taniguchi, T., Weissmann, C. 1978. *J. Mol. Biol.* 118:533–65
273. Müller, W., Weber, H., Meyer, F., Weissmann, C. 1978. *J. Mol. Biol.* 124:343–58
274. Budowsky, E. I., Sverdlov, E. D., Spasokukotskaya, T. N. 1972. *Biochim. Biophys. Acta* 287:195–210
275. Janion, C., Glickman, B. W. 1980. *Mutat. Res.* 72:43–47
276. Berthod, H., Pullman, B. 1971. *Biochim. Biophys. Acta* 232:525–606
277. Birnbaum, G. I., Hruska, F. E., Niemczura, W. P. 1980. *J. Am. Chem. Soc.* 102:5586–90
278. Neumann, J. M., Bernassau, J. M., Gueron, M., Tran-Dinh, S. 1980. *Eur. J. Biochem.* 108:457–63
279. Watenpaugh, K., Dow, J., Jensen, L. H., Furberg, S. 1968. *Science* 159:206–7
280. Pohl, F. M., Jovin, T. M. 1972. *J. Mol. Biol.* 67:375–96
281. Wang, A. H. J., Quigley, G. J., Kolpak, F. J., Crawford, J. L., van Boom, J. H., van der Marel, G., Rich, A. 1979. *Nature* 282:680–86
282. Wang, A. H. J., Quigley, G. J., Kolpak, F. J., van der Marel, G., van Boom, J. H., Rich, A. 1981. *Science* 211:171–76
283. Arnott, S., Chandrasekaran, R., Birdsall, D. L., Leslie, A. G. W., Ratliff, R. L. 1980. *Nature* 283:743–45
284. Crawford, J. L., Kolpak, F. J., Wang, A. H. J., Quigley, G. J., van Boom, J. H., van der Marel, G., Rich, A. 1980. *Proc. Natl. Acad. Sci. USA* 77:4016–20
285. Klysik, J., Stirdivant, S. M., Larson, J. E., Hart, P. A., Wells, R. D. 1981. *Nature* 290:672–77
286. Tran-Dinh, S., Guschlbauer, W. 1975. *Nucleic Acids Res.* 2:873–86
287. Chachaty, C., Langlet, G. 1976. *FEBS Lett.* 68:181–86
288. Stolarski, R., Dudycz, L., Shugar, D. 1980. *Eur. J. Biochem.* 108:111–21
289. Stolarski, R., Pohorille, A., Dudycz, L., Shugar, D. 1980. *Biochim. Biophys. Acta* 610:1–19
290. Behe, M., Felsenfeld, G. 1981. *Proc. Natl. Acad. Sci. USA* 78:1619–23
291. Grunberger, D., Nelson, J. H., Cantor, C. R., Weinstein, I. B. 1970. *Proc. Natl. Acad. Sci. USA* 66:488–94
292. Sage, E., Leng, M. 1980. *Proc. Natl. Acad. Sci. USA* 77:4597–4601
293. Sage, E., Leng, M. 1981. *Nucleic Acids Res.* 9:1241–50
294. Santella, R. M., Grunberger, D., Weinstein, I. B., Rich, A. 1981. *Proc. Natl. Acad. Sci. USA* 78:1451–55
295. Singer, B., Kröger, M. 1979. *Prog. Nucleic Acid Res. Mol. Biol.* 23:151–94
296. Braverman, B., Shapiro, R, Szer, W. 1975. *Nucleic Acids Res.* 2:501–7
297. Grunberger, D., Weinstein, I. B. 1971. *J. Biol. Chem.* 246:1123–28
298. Moore, P., Strauss, B. S. 1979. *Nature* 278:664–66
299. Moore, P. D., Rabkin, S. D., Strauss, B. S. 1980. *Nucleic Acids Res.* 8:4473–84
300. Yamaura, I., Rosenberg, B. H., Cavalieri, L. F. 1981. *Chem.-Biol. Interactions* 37:171–80
301. Ludlum, D. B. 1971. *Biochim. Biophys. Acta* 247:412–18
302. Mehta, J. R., Ludlum, D. B. 1978. *Biochim. Biophys. Acta* 521:770–78

303. Ludlum, D. B. 1970. *J. Biol. Chem.* 245:477–82
304. Fraenkel-Conrat, H., Singer, B. 1971. In *Biological Effects of Polynucleotides*, ed. R. F. Beers Jr., W. Braun, pp. 13–19. New York: Springer
305. Means, G. E., Fraenkel-Conrat, H. 1971. *Biochim. Biophys. Acta* 247:441–48
306. Gerard, G. F., Rottman, F., Boezi, J. A. 1972. *Biochem. Biophys. Res. Commun.* 46:1095–1101
307. Kröger, M., Singer, B. 1979. *Biochemistry* 18:3492–3500
308. Singer, B., Spengler, S. 1981. *Biochem-istry* 20:1127–32
309. Kröger, M., Singer, B. 1979. *Biochemistry* 18:91–95
310. Spengler, S., Singer, B. 1981. *Nucleic Acids Res.* 9:365–73
310a. Singer, B., Spengler, S. 1982. *FEBS Lett.* In press
311. Gerchman, L. L., Ludlum, D. B. 1973. *Biochim. Biophys. Acta* 308:310–16
312. Singer, B., Fraenkel-Conrat, H., Kusmierek, J. T. 1978. *Proc. Nat. Acad. Sci. USA* 75:1722–26
313. Cook, W. J., Gartland, G. L., Bugg, C. E. 1980. *Acta Crystallogr.* B36:2467–70

Ann. Rev. Biochem. 1982. 51:695–726

THE MOLECULAR BIOLOGY
OF TRYPANOSOMES

Paul T. Englund, Stephen L. Hajduk, and Joan C. Marini

Department of Physiological Chemistry, Johns Hopkins School of Medicine, Baltimore, Maryland 21205

CONTENTS

PERSPECTIVES AND SUMMARY

Trypanosomes and their close relatives are parasitic protozoa that cause some of mankind's major diseases. These diseases, which include African trypanosomiasis (sleeping sickness), South American trypanosomiasis (Chagas' disease), and leishmaniasis (kala azar) debilitate millions of people, primarily in developing countries (1, 2). In addition, bovine trypanosomiasis makes a major part of Africa unsuitable for cattle grazing and consequently contributes to human malnutrition. Since few satisfactory treatments are available for any of these diseases, there is a desparate need for study of these parasites. Fortunately, many of these species are extremely interesting in their own right, and in addition some can be easily

695

0066-4154/82/0701-0695$02.00

and safely manipulated in the laboratory. For these reasons, the trypanosomes make attractive subjects for research by biochemists and molecular biologists as well as by medical scientists. One purpose of this review is to encourage more investigation of trypanosomes.

We discuss two subjects concerning the molecular biology of trypanosomes. These subjects are unrelated to each other but each represents a unique property of these parasites. First is the mitochondrial DNA, known as kinetoplast DNA, which is found in all species of these parasites. This DNA consists of thousands of DNA circles within the cell's single mitochondrion. Remarkably, these circles are not free but are interlocked together to form a single massive network. Two kinds of circles are present. There are 25–50 large circles, known as maxicircles, which like other mitochondrial DNAs contain genes essential for mitochondrial biogenesis. In addition there are five to ten thousand small circles known as minicircles. The minicircles within a network are usually heterogeneous in sequence and are usually not transcribed. Their function is not yet known. In this review we emphasize the structure and genetic function of kinetoplast DNA and also its novel mode of replication.

The second topic of this review is antigenic variation in African trypanosomes. African trypanosomes are bloodstream parasites, and antigenic variation is the process by which they evade the immune system of their animal host. Each parasite is completely covered with a dense surface coat composed of a unique protein known as the variant surface glycoprotein. Recognition of this surface coat by the host immune system results in killing of most of the infecting parasites. However, a few trypanosomes escape by undergoing antigenic variation. In this process they synthesize a different variant surface glycoprotein and therefore replace their old coat with a new one. Although cells with this new coat also soon fall victim to the immune system, some survive by undergoing antigenic variation again. This process continues almost indefinitely because each parasite has the potential to synthesize more than 100 different variant surface glycoproteins. We review current knowledge of antigenic variation, including recent discoveries of its genetic basis, and the molecular properties of the variant surface glycoprotein.

AN OVERVIEW OF TRYPANOSOME BIOLOGY

Trypanosomes are flagellated protozoa of the order kinetoplastidae and the family trypanosomatidae (3, 4). The trypanosomatids of major importance to human health include the African trypanosomes, of which *Trypanosoma brucei* is the prototype, the South American trypanosomes, of which *Trypanosoma cruzi* is of central importance, and several species of *Leishmania*. The family is much broader than these medically important species.

Other trypanosomatids infect many other species of animals, some even infect plants, and some, such as those in the genus *Crithidia,* are parasites only of insects. Some relatives of the family trypanosomatidae (the bodonidae) are free living and therefore are not parasites at all. [See (4–6) for a definitive description of the biology of trypanosomes.]

The Life Cycle of Trypanosoma brucei

The trypanosomatids have evolved a remarkable variety of life cycles, although in most cases at least one stage of the life cycle occurs in an insect vector. In the animal host some species dwell in the bloodstream and others intracellularly. As an example, the following is a brief description of the life cycle of *Trypanosoma brucei,* a parasite of the mammalian bloodstream and lymphatic system. Its life cycle is neither typical nor unique, but it serves as background for our subsequent discussion.

There are three subspecies of *Trypanosoma brucei* that are virtually identical biochemically and morphologically but differ in host range and virulence (4). All three are found only in equatorial Africa. *Trypanosoma brucei rhodesiense* and *Trypanosoma brucei gambiense* cause the acute and chronic forms of African sleeping sickness in humans. *Trypanosoma brucei brucei* is noninfectious to humans, at least in part because it is lysed by high density lipoprotein in human serum (7), but it is a cause of trypanosomiasis in cattle and other mammals. It is the latter subspecies that has been most intensively studied.

T. brucei is transmitted by the tsetse fly *Glossina,* within which it undergoes important developmental changes (8, 9; Figure 1). The tsetse ingests parasites when taking a blood meal from an infected animal. In the insect midgut the trypanosomes then differentiate into procyclic forms. The procyclics have lost the variant surface glycoprotein that formed the surface coat in the animal bloodstream forms (10), they are noninfectious to the animal host, and they undergo drastic changes in metabolism (described briefly below). After about three weeks, the parasites migrate to the insect salivary gland where they develop ultimately into metacyclic forms. The metacyclic trypanosomes morphologically resemble the animal bloodstream form. They are infectious to animals and they express the variant surface glycoprotein (11, 12). When the fly bites another animal, the parasites present in the saliva are injected into the blood where they quickly develop into long slender bloodstream forms. These cells divide rapidly by binary fission, alter their metabolic pathways, and continue to express variant surface glycoproteins. They begin to undergo antigenic variation. As the infection proceeds, the long slender *T. brucei,* provided they are "pleomorphic," differentiate further into short stumpy bloodstream forms (8, 9, 13). These forms are nondividing (8), and it is generally believed that they are preadapted to life in the insect. After the tsetse takes its blood meal

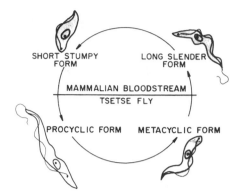

Figure 1 Abbreviated version of the *T. brucei* life cycle. It does not show the epimastigote form that is a precursor of the metacyclic form in the fly salivary gland. See (115) for a complete version. The shaded trypanosomes contain the surface coat, which is exchanged during antigenic variation.

from the infected animal, the short stumpy trypanosomes complete the life cycle by transforming into procyclics in the insect midgut (14).

Whereas pleomorphic strains can exist as either long slender forms or short stumpy forms, some laboratory strains of *T. brucei* are monomorphic (4, 15). These strains, consisting exclusively of long slender trypanosomes, grow to very high parasitemias and usually kill the host within a few days. They do not generate significant numbers of short stumpy forms, and they infect the tsetse with difficulty if at all (15). They survive in the laboratory only by syringe passage. Pleomorphic strains become monomorphic after extensive syringe passage (4, 16).

The different developmental stages of *T. brucei* differ dramatically in their metabolism (17, 18). We mention here only their mitochondrial metabolism as this is controlled in part by genes in the kinetoplast DNA. In the procyclic forms, present in the fly midgut, the mitochondria are fully active and have abundant cristae. These cells actively respire using a cyanide-sensitive electron transport system and have Krebs cycle enzymes (19). In contrast, the long slender bloodstream forms have completely suppressed many mitochondrial functions. Their mitochondrial volume is reduced and they have very few cristae. They are completely deficient in cytochromes and Krebs cycle enzymes, and their ATP-generating mechanism depends exclusively on glycolysis (20, 21). They do retain some mitochondrial functions, however. For example, they regenerate NAD^+ from NADH in a reaction involving a mitochondrial glycerol-3-phosphate oxidase (22, 23).

Laboratory Manipulations of Trypanosomes

T. brucei is easily grown in the laboratory, and cells in two of its developmental stages are available in large quantities. The blood stream forms can be isolated from the blood of infected rodents. Parasitemias exceeding 10^9 cells/ml of blood are readily obtainable, and the trypanosomes can be quantitatively separated from blood cells by chromatography on DEAE cellulose (24). After isolation the parasites can be kept in liquid medium for several hours for the purpose of labeling with biosynthetic precursors (25–27). However, they will not undergo significant cell division in this medium. The bloodstream forms will multiply in culture indefinitely, provided bovine fibroblasts are added as feeder cells, but these cultures will reach densities of only about 5×10^6 cells/ml (28, 29). These culture forms retain the ability to undergo antigenic variation (30). Trypanosomes that are similar if not identical to insect procyclic forms can also be grown in culture at 27°C to densities of about 5×10^7 cells/ml (31). The culture medium can be innoculated with bloodstream forms of pleomorphic strains, and transformation to procyclics will occur under appropriate conditions. Transformation in the reverse direction, from procyclic to bloodstream forms, is not straightforward; it usually requires incubation of the cultured procyclics with explants of tsetse salivary glands (31). *T. brucei* can be easily cloned. A single parasite, either a bloodstream form or a metacyclic, will successfully infect a laboratory rodent. Procyclic trypanosomes will develop colonies on agarose plates (32).

A major hindrance to research on trypanosomatids, including *T. brucei,* has been the absence of genetic techniques. Until very recently there was not even a suggestion of a sexual cycle in these parasites, but now there are two indications of the presence of genetic exchange. First, an analysis of enzyme polymorphisms in 17 strains of *T. brucei,* collected in Uganda during a two-year period, strongly suggests that the parasites are diploid and that they undergo mating and recombination (33). Second, genetic recombination has been detected between drug resistant strains of *Crithidia fasciculata* (J. Glassberg, L. Miyazaki, and M. Rifkin, personal communication). Further study of these processes, and the availability of mutant strains, could make an enormous contribution to research on the subjects described in this review.

KINETOPLAST DNA

The first indication of the existence of kinetoplast DNA came about 75 years ago with the discovery of a structure near the base of the trypanosome's flagellum that stained brightly with basic dyes. This structure, at one time presumed to be involved in cell motility, was named the kinetoplast. More recently, electron microscopy of thin sections show that the kinetoplast is a disk-like structure that resides within the matrix of the cell's single mitochondrion. Isolation of this structure in the early 1970s revealed that it is a massive network consisting of thousands of interlocked DNA circles

(35–38) (Figure 2). Each cell contains only a single network that, depending on the species, comprises 10–25% of the total cellular DNA. As stated above, the networks consist of two types of DNA circles. The minicircles comprise about 95% of the mass of a network and the maxicircles make up the remainder. Maxicircles carry genes similar to those on mitochondrial DNA in other eukaryotic cells (see below). However, the function of minicircles and the reason for the network structure are still not known. What follows is the current view on the structure of kinetoplast DNA, its possible functions, and its novel mode of replication. Other reviews cover in more detail the early work on kinetoplast DNA (39–44).

Network Structure

Electron micrographs of isolated networks show massive sheets of DNA, but, because of the dense crowding, it is difficult to draw many conclusions about structural details (Figure 2). However, each loop in the network must be part of an individual DNA circle and the circles must be interlocked, as the entire structure can be decatenated, using type II topoisomerases, to form individual minicircles and maxicircles (45).

The maxicircles are threaded through the network presumably by multiple catenations. Occasionally they can be seen as long loops at the network periphery (46). It is clear that maxicircles do not form an essential skeletal framework on which minicircles are threaded (47), because maxicircles can be removed from networks by cleavage with an appropriate restriction enzyme without affecting network structure. Furthermore, some mutant trypanosomes (see below), are completely deficient in maxicircles but have otherwise normal minicircle networks (48, 49).

Isolated networks, except for those undergoing replication, are remarkably uniform in size and shape. Despite their enormous size they sediment as homogeneous species, with a s_{app} of up to 5000S (51, 52). Their molecular weights are $\sim 10^{10}$ and networks from most species contain 5000–10000 minicircles and 25–50 maxicircles (53). Isolated networks range in diameter from about $5 \mu m$ for African trypanosomes to about 15 μm for *Crithidia*. Techniques for isolation of networks usually take advantage of their high sedimentation coefficient. Networks can be easily purified from nuclear DNA by differential centrifugation of deproteinized cell lysates (54). From some species it is easy to obtain a milligram of pure intact networks.

Within the mitochondrial matrix the huge network is condensed in diameter over tenfold and appears in the form of a disk (39). Nothing is known about the forces that constrain the network into this highly organized structure. In all trypanosomatids the disk is situated near and perpendicular to the basal body of the flagellum, but the significance of this proximity is also not yet known.

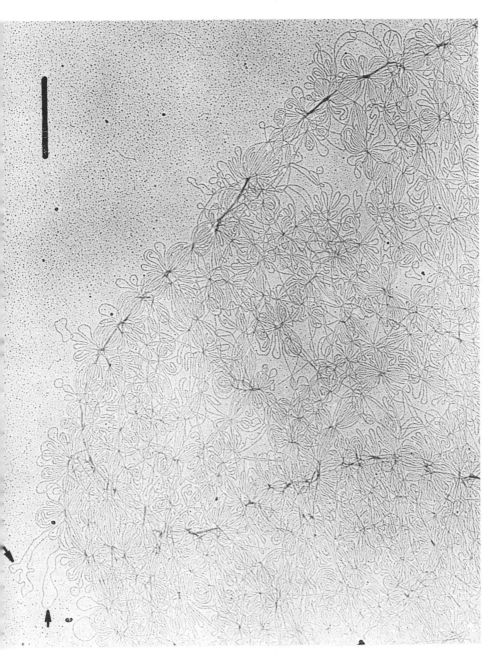

Figure 2 Electron micrograph of the edge of a kinetoplast DNA network from *Crithidia fasciculata*. All of the small loops are minicircles and the large loops (arrows) are maxicircles. The bar is 1 μm.

Networks similar to that shown in Figure 2 are present in all species of trypanosomatid that contain a functioning mitochondrion. This fact implies that networks were formed early in the evolution of these protozoa and that they play an important role in the life of these cells (55). No other cells, outside the order kinetoplastidae, are known to contain DNA networks.

Minicircles

Depending on the species, minicircles range in size from about 700 to 2500 base pairs. Although the thousands of minicircles within a network are all the same or very nearly the same size, they are usually heterogeneous in sequence (56, 57). The degree of minicircle sequence heterogeneity varies in different species. Both the heterogeneity of minicircles within a network and the virtual absence of sequence homology between minicircles of different species imply that minicircle sequences evolve very rapidly (58).

SEQUENCE ORGANIZATION   Among the best characterized minicircle sequences are those from the African trypanosomes *T. brucei* and *T. equiperdum*. Although these species are very closely related, their minicircle sequence heterogeneity differs dramatically. *T. brucei* minicircle heterogeneity is more extreme than that found in any other trypanosomatid (59–61b); in contrast, *T. equiperdum* minicircles are homogeneous in sequence (62). Minicircles in both species are about 1 kb. Reassociation analysis of *T. brucei* minicircles shows that there are 250–300 different minicircle sequence classes in a network (59–61). This value indicates that there are on the average roughly 20 copies of each type of minicircle within a network, although hybridization studies indicate that some species are more abundant than others (61). Analysis of two *T. brucei* minicircles that have been cloned and sequenced (63) reveals a constant and a variable region within these molecules. The constant region has 122 base pairs in which all except 8 are identical in the two minicircles. Comparison of the two variable regions shows that they do not have any extensive homology. Studies of other *T. brucei* minicircles by reassociation kinetics and by electron microscope heteroduplex analysis raises the possibility that the common sequence may be found in many and possibly all minicircles in the network (59–61b). The homogeneous minicircle found in *T. equiperdum* networks has also been sequenced (64). Remarkably, it contains a sequence that is virtually identical to the 122-base pair constant sequence in the two *T. brucei* minicircles.

Minicircles in other trypanosomatids appear to be intermediate in heterogeneity between *T. brucei* and *T. equiperdum*. Sequence analysis of several cloned minicircles from *Leishmania tarentolae* networks shows that these molecules also have constant sequences and variable sequences, although the former differs almost completely from those found in *T. brucei* and *T. equiperdum* [(32, 65) D. E. Arnot, personal communication]. The constant sequence found in minicircles may represent a replication origin; however, the replication origins in mitochondrial DNAs from different animal species are the least-conserved part of those molecules.

TRANSCRIPTION   In an effort to understand the function of minicircles, there has been a search for their transcripts. Several laboratories were unable to detect minicircle transcripts in *T. brucei* (either bloodstream or procyclic forms), *Crithidia luciliae,* or *Leishmania tarentolae* (32, 50, 66, 67), which strongly implies that minicircles are not transcribed. However, in another trypanosomatid, *C. acanthocephali,* a minicircle transcript has been reported (68). This transcript is complementary to only one of the minicircle strands, and, based on electron microscopy of DNA-RNA hybrids, is about 240 nucleotides long. Otherwise, this transcript has not been characterized in detail. It seems likely that transcription of minicircles is not a regular occurrence in trypanosomatids, although, because of the contrasting findings in *C. acanthocephali,* the subject deserves more investigation. There remains a possibility that transcription of minicircles may occur only during certain stages of the life cycle.

OTHER PROPERTIES OF MINICIRCLES   The apparent absence of minicircle transcripts in most trypanosomatids together with the absence of long open reading frames in their sequences (63–65) make it likely that minicircles serve a noncoding function. Further support for this comes from the fact that minicircle sequences, in most species, evolve exceedingly rapidly. If the minicircle sequences do have a noncoding function, the nature of that function and the selective forces that act on the sequences are completely unknown.

If sequences within minicircles are not conserved, there may be structural features within them that are conserved. An unusual structural feature has been found in fragments of *L. tarentolae* minicircles, and preliminary evidence suggests that these properties are also found in minicircles from other species. These fragments migrate anomalously during electrophoresis on polyacrylamide gels (69). For example, one *L. tarentolae* fragment, which is 494 base pairs long, migrates on a 12% polyacrylamide gel as if it were 1380 base pairs. This anomaly is not due to modification of the DNA, as it is also observed on a cloned fragment, and therefore it must be the nucleotide sequence itself that dictates an unusual conformation in the molecule (J. C. Marini, unpublished observations). This conformation must hinder passage of the molecule through the gel.

Maxicircles

Maxicircles have a structure and coding function that appear to be not unlike those of mitochondrial DNA in other eukaryotic cells. Their unique feature is that they are not free in the mitochondrial matrix, but instead are threaded through a network of thousands of minicircles (46, 70). The network contains roughly 25–50 maxicircles, all of which appear to have the same sequence. They do not have sequences in common with minicircles. Depending on the species, maxicircles range in size from 20–38 kb, and about 80% of their base pairs are AT. Maxicircles can be removed from networks by a single cleavage with an appropriate restriction enzyme. The linearized maxicircles can be easily purified (71) and their fragments can be cloned (61a, 61b, 71a) (K. Stuart, personal communication). [See (32, 42, 61) for a more detailed discussion of maxicircles.]

TRANSCRIPTION   Preliminary studies suggest that maxicircle transcripts in several trypanosomatids resemble those from other eukaryotic mitochondrial DNAs. The major transcripts are ribosomal RNAs (50, 66, 67). There also is a family of polyadenylated RNAs that presumably are messengers for a small number of mitochondrial proteins (32, 50) (K. Stuart, personal communication). No maxicircle-coded tRNAs have been discovered. The ribosomal RNAs are 9 and 12S (67). They are present in roughly equimolar amounts, and their sizes, in *T. brucei,* are only about 500 and 1000 nucleotides. They are the smallest ribosomal RNAs known, and their small size may be a factor in the apparent instability of the mitochondrial ribosomes. The size and sequence of the 9 and 12S ribosomal RNAs are conserved in *T. brucei, Leishmania, Phytomonas,* and *Crithidia* (32, 49, 72). From direct sequence analysis of the *T. brucei* maxicircle rRNA genes, these molecules also have been found to share some homology with human mitochondrial and *E. coli* rRNA (58). So far no maxicircle translation products have been detected. However, a cloned *T. brucei* maxicircle fragment directs the synthesis of two polypeptides in *E. coli* minicells, and these may be related to authentic maxicircle gene products (61b).

MUTANTS   A powerful approach to the study of maxicircle function that has not been fully exploited, depends on the fact that *T. brucei,* when growing in the animal bloodstream, suppresses the biogenesis of its mitochondrion (9). Therefore the bloodstream forms apparently do not need many of the maxicircle gene products. This discovery has led to the isolation and characterization of a number of conditionally lethal maxicircle mutants (41, 49, 61, 62, 73). Since cells with defective maxicircle genes will grow in the animal bloodstream, but not in the insect vector, they are designated

I⁻ (meaning insect⁻) (74). I⁻ strains can be passaged in the laboratory by syringe. However, some I⁻ strains, such as *T. equiperdum* or *T. evansi*, exist in nature. These close relatives of *T. brucei* are not carried by the tsetse but instead are transmitted either venereally or in the saliva of vampire bats. They do not undergo a cycle of differentiation (4). Some strains of *T. equiperdum* and *T. evansi* are completely deficient in maxicircles but otherwise have normal minicircle networks (49), whereas another *T. equiperdum* strain has a maxicircle with a large deletion (75). Some *T. brucei* I⁻ strains have maxicircles that appear to be indistinguishable from normal maxicircles (49, 76). These maxicircles may have point mutations that inactivate essential genes. Some I⁻ strains are not only deficient in maxicircles, but are also completely deficient in minicircles (61, 62, 75, 77). These dyskinetoplastic cells sometimes arise spontaneously, but they can also be induced by agents such as acridine or ethidium bromide (41, 78). No mutant cells, however, have been found that have maxicircles but not a minicircle network.

MAXICIRCLE FUNCTION IN *T. BRUCEI* DEVELOPMENT   An exciting prospect for future investigation concerns the genetic regulation of mitochondrial biogenesis in *T. brucei*. As described earlier, *T. brucei* bloodstream forms, but not the procyclics, are completely deficient in cytochrome oxidase, cytochrome *c*, and other essential components of electron transport and oxidative ATP synthesis. Some of these components are probably products of nuclear genes and some are probably coded by maxicircles. It will be of great interest to study the shift in gene expression as the trypanosomes transform from bloodstream forms to procyclics. Preliminary experiments have revealed that maxicircle transcripts are reduced only five to ten fold in the bloodstream forms when compared to the levels found in the procyclics (50). This reduction does not seem adequate to ensure a complete suppression of mitochondrial function, and therefore other regulatory mechanisms may be involved.

Kinetoplast DNA Replication

The replication of the kinetoplast raises problems not found with any other DNA. In fact, it is difficult to imagine how a network containing thousands of minicircles and a handful of maxicircles, all of which are interlocked, could replicate at all. Yet the network replicates quickly and with precision. During the mitochondrial S phase, which in some species is only one hour, the cell's single network replicates to form two (79–81). These daughter networks, each identical to the parent, are subsequently passed on to the two progeny cells during cell division. In this section we describe the current view of kinetoplast DNA replication. We emphasize studies on *C.*

fasciculata but the available evidence suggests that replication may proceed by a similar mechanism in all trypanosomatids. [See (44, 82, 83) for a more detailed discussion.

Most studies of kinetoplast DNA replication have focused on the minicircles, the major components of the network. Each minicircle in a network replicates once per generation by a semi-conservative mechanism (84, 85). However, they do not replicate while attached to the network. Instead, they are individually released from it so that they can replicate as free minicircles (86). The existence of free minicircles implies that there must be specific enzymes that release these molecules from the network for the purpose of replication and reattach the progeny. Although the "releasing enzyme" and the "reattachment enzyme" have not yet been purified it is likely that they are type II topoisomerases. Enzymes of this kind can efficiently catenate and decatenate duplex DNA circles (87), and in fact the HeLa and T4 phage Type II topoisomerases will completely decatenate kinetoplast DNA (45, 88). All details of the model for kinetoplast DNA replication (see following paragraph and Figure 3) are not rigorously proven, but they are consistent with virtually all of the available data (52, 85, 86, 89, 90).

THE REPLICATION SCHEME   At the beginning of the replication process, in the G1 phase of the cell cycle, the network contains thousands of minicircles that are all covalently closed. This kind of network is designated Form I. When the S phase begins, individual minicircles are released from the network by the hypothetical releasing enzyme for the purpose of replication. Each released minicircle, which probably is covalently closed, then replicates by a Cairns mechanism to form two progeny, which are probably nicked. The hypothetical reattachment enzyme then links these progeny back onto the network. Because reattachment occurs at the network periphery, the replicating network contains a central zone of unreplicated covalently closed minicircles and a peripheral zone of nicked minicircles that have already been replicated and reattached. As replication proceeds, the central zone shrinks, the peripheral zone enlarges, and, because two minicircles are reattached for each one removed, the whole network grows in size. At any time during the S phase there are probably several hundred free minicircles that have been released from the network for the purpose of replication. Finally, when the S phase ends, all minicircles in the network have undergone replication and all are nicked. The network, designated Form II, is double size and often appears to be preparing to divide in two. It resembles either an elongated ellipse or a two-lobed structure when viewed in the electron microscope. The final step in the process, which occurs sometime during the G2 phase, includes both the completion of division of the double size network and the covalent closure of all the

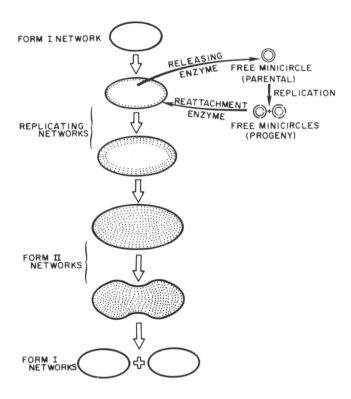

Figure 3 Model for replication of a kinetoplast DNA network. The dotted region of the networks contains nicked minicircles. See text for discussion. [Reprinted from (82) with permission.]

minicircles. The products are two Form I networks, each identical to the parent, which segregate into the progeny cells at cell division.

Nicking of minicircles probably plays a central role in regulating the replication of networks. There must be some mechanism that ensures that each minicircle replicates once and only once per generation, and it may be the nicks that label the molecule as having already completed replication. It could be the releasing and the reattachment enzymes themselves that distinguish the nicked and covalently closed minicircles. The releasing enzyme must act only on those minicircles that have not yet undergone replication and it may recognize them by the fact that they are covalently closed. In contrast, the reattachment enzyme must act on those free minicircles that have completed replication, and it may recognize them because they are nicked. It then reattaches these molecules to the network periphery.

The release and reattachment of minicircles may be completely random processes. Therefore, the two progeny of a minicircle may, with equal probability, end up in separate networks at cell division, or in the same one (69). In the latter case, one of the daughter networks and all of its descendants will no longer contain progeny of the minicircle being considered. The other daughter network will contain two copies. This random distribution of minicircle progeny could contribute to the rapid evolution of minicircle sequences. It could also account for the fact that minicircles of identical sequence are not clustered within a network but instead are widely if not randomly distributed throughout the structure (91).

MAXICIRCLE REPLICATION   Virtually nothing is known about maxicircle replication, including whether it takes place free of the network. If it does, *random* reattachment of maxicircle progeny could present a serious problem. Since there is only one mitochondrion per cell, containing a network with only 25–50 maxicircles, the random reattachment of maxicircle progeny would often result in unequal distribution of these progeny to the daughter network. After only a few generations some cells could completely lose maxicircles, and with them some genes essential to mitochondrial biogenesis. Therefore, if maxicircles are released from the network for replication, it is likely that they are reattached in a way that ensures their equal distribution to progeny cells at cell division.

In *T. brucei* networks, maxicircles are especially prominent as long loops extending from the network periphery. They are usually distributed uniformly around the network periphery, but in double size Form II networks, they are clustered near the middle of the network near the cleavage point (89, 92, 93). Furthermore, some unit size networks look as if they had just been formed by cleavage of one of these double structures, as they have all of their maxicircles clustered on one side (89). One interpretation of this is that the maxicircle progeny are lined up on the Form II network in a way that ensures their equal segregation to the daughter networks. One important function of the minicircle network could be that of a structural framework to facilitate maxicircle segregation, but there is no supporting evidence for this speculation.

Prospectives for Research in Kinetoplast DNA

Future studies on kinetoplast DNA will focus on the characterization of maxicircle genes, changes in maxicircle transcription during the parasite life cycle, and replication. Some clues to the function of minicircles and networks may also be obtained from study of cells with mutationally defective networks or from cells with unconventional networks. An example of the latter is in *Herpetomonas ingenoplastis* (94) (an insect parasite). Its networks probably contain maxicircles, as they have 36-kb circles that cross-hybridize with *T. brucei* maxicircles. However, they do not contain minicircles. Instead, they have thousands of larger circles that are heterogeneous in size (16–23 kb). The nonparasitic bodonids may also have unconventional networks. Although their kinetoplast DNA has not yet been isolated, its appearance in thin sections resembles that in *H. ingenoplastis* (5).

There are two medically important reasons for studying kinetoplast DNA. First, an increased understanding of its biochemistry may lead to new concepts of chemotherapy against these parasites. None of the host organisms of these parasites has DNA that resembles kinetoplast DNA, and therefore it may be possible to specifically inhibit its replication or expression. This inhibition should result in the selective killing of pathogens such as *Leishmania donovani* or *Trypanosoma cruzi,* which appear to require kinetoplast gene products in all stages of their life cycle. In contrast, the animal bloodstream forms of *T. brucei* do not have active mitochondria and therefore might not be killed by inhibitors of kinetoplast function.

Second, characterization of kinetoplast DNA sequences appears to be useful in identification of strains of these parasites for diagnostic or epidemiological purposes. Many of the *Trypanosoma* or *Leishmania* strains are morphologically and biochemically very similar, and more tools are needed to distinguish strains that may differ in virulence. Progress has already been made in distinguishing strains by kinetoplast DNA hybridization or by comparing restriction digests of minicircles (55, 65, 95–95b). Maxicircles or nuclear DNA may also have sequences useful for strain characterization (96).

The major concern about kinetoplast DNA is its overall function. Whereas maxicircles appear to be typical mitochondrial DNAs, nothing is known about the function of minicircles or the reason for the network structure. This frustrating problem of function does not apply to antigenic variation which we discuss in the remainder of this review. Although there are many mysteries about the mechanism of antigenic variation, its function is very clear.

ANTIGENIC VARIATION IN AFRICAN TRYPANOSOMES

Background

The optimum effect of a parasite on its host, from the parasite's point of view, is not to kill the host quickly but instead to cause a chronic infection. In a chronic infection the parasite will have won itself more time in which to be passed on to other hosts, and therefore it will have increased its own likelihood of survival. A chronic infection is possible only if the parasite can resist the host's defense mechanisms. The African trypanosomes resist, very efficiently, through antigenic variation.

Over 70 years ago it was found that African sleeping sickness is characterized by relapsing parasitemia (97). Bloodstream trypanosomes grow to high levels and then disappear. In about 10 days they return and grow again to high levels. Then they disappear again. This cycle of recrudescense and remission seems to continue indefinitely until the patient finally dies. Similar cycles of infection are found in cattle suffering from trypanosomiasis.

It soon became apparent that the parasites in each relapse were antigenically distinct from those that had come earlier and those that followed (98). Antiserum specific for one form would agglutinate homologous but not heterologous populations of trypanosomes. These parasites have evolved a system for altering their antigens, which presumably are localized on their surfaces, for the purpose of evading the immune system of the host. The devastating impact of this evasion was noted at the beginning of this chapter. It has frustrated all attempts to develop effective vaccines against these parasites.

Antigenic variation is a property of the African trypanosomes, which cause human and bovine trypanosomiasis. The most important of these species are *T. brucei, T. vivax,* and *T. congolense.* Other trypanosomatids, such as *T. cruzi* and *Leishmania,* apparently do not undergo antigenic variation and have developed other defenses against the host immune system.

It is only in recent years that we have begun to develop an understanding of antigenic variation at the molecular level. In 1969 electron microscope studies revealed that all bloodstream forms of *T. brucei* are coated with a unform 12–15-nm electron dense surface layer just outside the plasma membrane (11). This coat, which seems to cover the entire cell including the flagellum, is also found on the infective metacyclic forms from the tsetse salivary gland, but is absent on the procyclic forms present in the fly midgut. Its importance to antigenic variation is revealed by its uniform and specific staining with ferritin-labeled, variant specific antibodies (99). Biochemical studies, reported in 1975, showed that the coat is composed of $\sim 10^7$ molecules of a unique glycoprotein with a mol wt of about 65,000 (100). These molecules probably form a monolayer just outside the plasma membrane. This one glycoprotein, which may be the only protein exposed on the cell surface, carries all of the antigenic determinants of the living trypanosome. Although it is highly immunogenic itself, its function is probably to shield all subsurface molecules from attack by host antibodies. Presumably antigenic variation involves shedding of the coat and replacing it with one composed of a new and different glycoprotein. Remarkably, the proteins isolated from different variants differ dramatically in amino acid sequence (101, 102).

A distinctive nomenclature has been developed to describe the phenomenon of antigenic variation (103). Variable antigen type (VAT) refers to a trypanosome expressing a particular antigen. The antigen is usually called the variant surface glycoprotein (VSG). Many different VATs, all derived from a trypanosome clone, form a serodeme. It is not known how large the VAT repertoire in a serodeme is, although, as will be shown later, the number is probably finite. In one courageous study of a *T. equiperdum* serodeme, 101 distinctive VATs were detected, and the repertoire could easily run to several hundred or more (104). Not only do we not know the number of VATs per serodeme, but we also do not know either the number or the geographical distribution of different serodemes that exist in nature. Different serodemes usually have different repertoires of VATs, although serologically related VATs are occasionally found in different serodemes (105). These VATs are known as iso-VATs or isotypes. Isotypes have VSGs with similar but probably not identical amino acid sequences (106, 107).

VATs are often given a name such as AnTat 1.30, MITat 1.2, or ILTat 1.4. AnTat means Antwerp Trypanozoon antigen type (103). The number 1.30 refers to serodeme 1 (from Antwerp) and VAT 30. Similarly, MITat and ILTat refer to serodemes developed at the Molteno Institute in Cambridge, England, and at the International Laboratory for Research on Animal Diseases (ILRAD) in Nairobi.

In the following section we review in more detail the phenomenon of antigenic variation, the structure of the antigen, and finally the recent experiments aimed at elucidating the genetic basis of antigenic variation. Unless otherwise indicated, all experiments were done on *T. brucei*. However, there are also important papers describing antigenic variation in *T. congolense* (108–114). Several other reviews of antigenic variation have been published recently (115–121).

The Phenomenon of Antigenic Variation

EARLY VIEWS   Early studies of antigenic variation led to the widespread belief that animals bitten by an infected tsetse would receive trypanosomes carrying a unique antigen type (122–125). Multiplication of these trypanosomes, expressing the "basic antigen," would result in the first parasitemia. Subsequent variants were thought to appear in a similar order in any infection. Furthermore, if any variant recycled through the tsetse the whole sequence would start over again. The basic antigen always came first. This simple scheme raised expectations that a vaccine against trypanosomiasis could be developed easily. It was known that vaccination with a unique purified antigen would protect against homologous trypanosomes (100, 126, 127). Therefore, immunization with the basic antigen, and perhaps with several of the antigens that appear early in the infection, should provide protection against the disease. Unfortunately, the situation turned out to be much more complex.

MORE RECENT VIEWS   A more precise description of a trypanosome infection was made possible by the techniques of immunofluorescense and complement-dependent immune lysis (128). These techniques, and the availability of monospecific antisera, have made possible the identification of the VATs in any population. Application of these techniques to metacyclic trypanosomes obtained from tsetse salivary glands led to three important and unexpected discoveries. First, the metacyclics are not of a single antigen type but instead are heterogenous (12, 128a). It is not yet known how many different metacyclic VATs exist in the saliva of a tsetse, but three different VATs make up only 40% of the AnTat metacyclics (129, 130). Second, a comparison of metacyclics from different serodemes revealed that each has a distinct set of metacyclic VATs (124, 130). Finally, the relative proportions of the three AnTat metacyclic VATs is always the same no matter what VAT has been used to infect the tsetse (130). In this sense the cycle does seem to start over in the fly, as had been originally proposed. However, there is not a basic antigen type. Instead, there is a characteristic repertoire of metacyclic VATs. Although the relative proportion of different metacyclic VATs does seem to remain constant throughout the duration of infection of a tsetse, there remains a possibility that all the different metacyclic VATs arose from a common precursor by antigenic variation within the fly's salivary gland.

Immunofluorescence studies have also shown that VAT heterogeneity is found in the first parasitemia that develop during the 10 days after a tsetse bite (124, 125, 131, 132). Studies of the AnTat trypanosomes revealed that the first parasites detectable in the bloodstream are antigenically the same as the infecting metacyclics (129). These VATs begin to disappear about 5 days after infection. Antigenic variants derived from these early VATs then multiply to form the first major parasitemia occurring about 10 days after infection. About 20 different VATs are present in this population. The VAT used to infect the tsetse always seems to be one of the major VATs (> 10% of the total) present in the first parasitemia (133).

The realization of the complexity of a tsetse-induced infection has diminished expectations for a vaccine against trypanosomiasis. Not only is there extensive VAT heterogeneity both in the metacyclic trypanosomes and in the early bloodstream forms, but also the tsetse-transmitted trypanosomes are especially prone to rapid antigenic variation (12, 124). However, it may be possible to identify all of the metacyclic VATs in a serodeme. Then a vaccine consisting of the appropriate VSGs might provide protection against trypanosomes belonging to that particular serodeme (134).

INFECTIONS WITH A TRYPANOSOME CLONE   Because of the complexity of tsetse-induced infections, there have been many studies of infections induced by a single trypanosome (104, 128, 135, 136). Provided that the infecting parasite is from a population not recently derived from the fly, the infection is characterized by a fairly homogeneous first parasitemia. This parasitemia generally consists of a single major VAT (the homotype, which may constitute more than 99% of the population) and some minor VATS (the heterotypes, which are sometimes present at a frequency of less than 1/10,000). Host antibody will eliminate the homotype, and one or more of the heterotypes will multiply to form the next parasitemia (128). In the second and subsequent parasitemias, the population of trypanosomes may include two or more major VATs.

The availability of homogeneous populations of VATs from infections such as these has made possible the production of monospecific antisera and the isolation of pure VSGs. In addition, studies of these infections have shown clearly that the sequence of expression of different VSGs is not rigorously conserved in different infections, although there is a strong tendency for some to appear early and for others to appear late (104, 128, 136). It is obviously important to the parasite that some VSGs are kept in reserve for later stages of the infection. If VSG expression was completely random, then all different VATs might appear at a low level at the beginning of the infection and the host might be able to eliminate them all.

THE ROLE OF ANTIBODY   Nothing is known about the factors that trigger antigenic variation. Until recently it was suspected that host antibody might induce this process, although it is now clear from several experiments that this view is incorrect. Antigenic variation occurs even during the early stages of infection before an immune response has developed (128), it occurs in immunosuppressed animals (133), and it occurs in vitro in the absence of antibody (30). Therefore, it is likely that antibody plays a selective rather than an inductive role in antigenic variation. It eliminates homotypes and therefore allows one of the heterotypes to emerge.

The Variant Surface Glycoprotein

Antigenic variation is expressed through the VSG. Early studies revealed that VSGs isolated from different VATs differ dramatically in isoelectric point, carbohydrate content, amino acid composition, and amino acid sequence of the N-terminal region (100–102, 137). These differences clearly explain the different serological specificities of these molecules. However, because all of the different VSGs seem capable of forming morphologically similar surface coats and since they presumably bind to the membrane through similar linkages, it was expected that these molecules would share important common structural features. These common features might be concentrated near the C terminus of the VSG, because it is this part of the molecule that was thought to be buried within the surface coat near the plasma membrane (118).

RECENT STRUCTURAL STUDIES   Some common structural features have in fact been revealed by recent amino acid sequencing studies and by carbohydrate analyses. The former have been obtained directly or have been deduced from sequences of cloned cDNAs prepared from VSG mRNAs (107, 138–141). Two VSGs have been sequenced in part by both methods (139–141). From these recent structural studies it is possible to draw three important conclusions.

First, carbohydrate analysis on six different VSGs revealed oligosaccharides attached directly to the C-terminal residues (141). In some VSGs this residue is serine whereas in others it is either aspartic acid or asparagine. Surprisingly, in the latter cases cDNA sequence analysis indicated that the residue is aspartic acid rather than asparagine (107, 138, 139). Glycosylation of an aspartic acid is unprecedented, although the possibility remains that the aspartic acid was converted to asparagine in a post-translational modification. This novel glycosyl group is discussed in the following section.

Second, the initial translation product has sequences that are not present in the mature VSG (139, 140). Comparison of protein and cDNA sequences of VSGs 117 and 221 of the MITat serodeme revealed a sequence of predominantly hydrophobic amino acids that is removed from the C terminus of the VSG either during a processing step or possibly during isolation of the glycoprotein. This sequence is 23 residues in the case of VSG 117 and 17 residues in the case of VSG 221. Removal of these sequences exposes as the new C terminus the glycosylated aspartic acid (or possibly asparagine) in VSG 117 and the glycosylated serine in VSG 221 (139, 140). Inspection of cDNA sequences for other VSGs suggests that very similar hydrophobic C-terminal tails may be present on the initial translation product of those VSGs as well (107, 138). The function of these hydrophobic tails is not yet known. The initial translation products of VSGs 117 and 221 also contain N-terminal sequences, not present on the mature VSG, that resemble the signal sequence found on most secretory proteins (139, 140).

Third, important sequence homologies exist in the C-terminal region of the mature VSGs (107, 138, 139, 141). In several different VSGs from three different serodemes, there is a striking conservation of several hydrophobic and hydrophilic regions near the C terminus. Furthermore, the position of several half-cystine residues is conserved within the C-terminal region, and in some of the VSGs there is conservation of a substantial fraction of the amino acids in the hundred residues nearest the C terminus (138). These results suggest a common three-dimensional structure in the C-terminal region of the VSGs that may be stabilized by disulfide bridges. This conserved conformation may correspond to the C-terminal domain that was deduced earlier from tryptic fragmentation of several VSGs (102). Tryptic digestion of VSGs results in an N-terminal fragement with a mol wt of 40,000–50,000 and a C-terminal fragment(s) usually in the range of 10,-000–17,000 mol wt. These fragments are assumed to derive from two independent structural domains within the VSG.

CARBOHYDRATE ANALYSIS   Carbohydrate content varies from 7 to 17% in different VSGs (137). The sugars present in purified preparations of these proteins are mannose, galactose, and glucosamine. Galactosamine, mannosamine, and sialic acid are absent in the *T. brucei* VSGs studied so far. Recent structural and biosynthetic studies indicate that there are two kinds of carbohydrate chains on VSGs. Some are conventional asparagine-linked oligosaccharides and the others (see previous section), are attached by an unknown linkage to the C-terminal serine, aspartic acid (or asparagine) (141, 142). The oligosaccharides linked to internal asparagines are incorporated at the time of protein synthesis in a reaction inhibited by tunicamycin (26, 27). The C-terminal oligosaccharides are synthesized later in a reaction that seems resistant to tunicamycin (26). VSG, which is synthesized in the presence of tunicamycin and therefore lacks oligosaccharide linked to internal asparagines, appears to be transported normally to the cell surface (26, 142a). However, it may be more sensitive than normal VSG to proteolytic degradation (26). The C-terminal oligosaccharide confers immunological cross-reactivity to the VSG (see the following section).

IMMUNOLOGICAL CROSS-REACTIVITY   It has been known for many years that antibodies specific for the surface coat of living trypanosomes are variant-specific and do not cross-react with heterologous trypanosomes. However, in 1978 it was reported that antibodies against a *purified* VSG will react with almost any heterologous VSG tested (143–145). Cross-reactivity was detected even between VSGs isolated from *T. brucei* and *T. congolense* (143), and between different VSGs from a strain of *T. equiperdum* (145a). Because of the absence of cross-reactivity on living trypanosomes it was concluded that the cross-reacting determinants of the VSG are not accessible on the surface of the trypanosome, but instead must be buried within the interior of the coat. Recent experiments indicate that the cross-reacting determinants consist primarily, if not exclusively, of carbohydrate (141).

Cross-reactivity of a VSG is eliminated by periodate oxidation (146) and is not found on nonglycosylated VSGs synthesized in a cell-free ribosomal system (147, 161). A study of tryptic or pronase peptides from several VSGs indicate that it is the C-terminal glycopeptides that are immunologically cross-reactive (141). These glycopeptides contain mannose, glucosamine, and galactose, but their structure has not yet been determined.

ORGANIZATION OF VSG IN THE SURFACE COAT   It is not yet known how the VSGs are organized on the cell surface. One can assume that they form a monolayer. Since the coat is 12–15 nm thick and the VSGs are at most only 65,000 mol wt, these proteins must be highly asymmetric molecules. The N-terminal region, which is highly variable in amino acid sequence and probably contains the antigenic determinants detectable on the living trypanosome, presumably resides on the exterior surface (118). The more conserved C-terminal region, together with the cross-reacting oligosaccharide, resides in the interior of the coat near the plasma membrane. The purified VSG exists as a dimer in solution (147a, 147b), but it is not known if VSG dimers are also the fundamental structural unit of the surface coat. Treatment of living trypanosomes with a cross-linking reagent resulted in the formation of VSG oligomers, but did not result in joining of VSG to any other protein (147b). It is not yet known how VSGs are attached to the trypanosome's plasma membrane. These molecules bind tightly to the living cell, yet are released quickly when the cell is disrupted by mechanical or osmotic treatment (100). Detergent is not required for the release or solubilization of the VSG, although a small fraction of the VSG usually resists solubilization unless detergent is added (26,100). Several interesting suggestions have been made to explain the linkage of VSG to plasma membrane, although so far none has been tested (26, 109, 118, 138, 139). Eventually it will be necessary to explain how the coat can be removed when the trypanosome enters the tsetse fly and how it can be efficiently exchanged during antigenic variation.

EASE OF VSG PURIFICATION   Structural studies on VSGs are greatly facilitated by the ready availability of these proteins. Several milligrams of VSG can be obtained from trypanosomes in the blood of a single rat, and the purification scheme is simple and reliable (100, 147c). The availability of abundant amounts of VSG, from different VATs, should soon lead to more extensive knowledge of their physicochemical properties and a crystallographic determination of their three-dimensional structure.

The Genetic Basis of Antigenic Variation

During the last two or three years several exciting discoveries have provided important clues as to how antigenic variation may work.

POSSIBLE GENETIC MECHANISMS Three distinct mechanisms have been considered. First, each trypanosome could have a single VSG gene that continually changes by mutation and therefore gives rise to continually changing gene products. This model was discarded for many reasons, some of which will be clear from the discussion that follows. Second, antigenic variation could involve the rearrangement of chromosomal sequences to create new VSG genes. This mechanism could be related to that which accounts for the coding of immunoglobulins. Although an enormous number of genes could be created by genomic sequence fusions, it is unlikely that the trypanosome needs such a versatile system. Genes for only several hundred different VSGs should be sufficient to ensure the parasite's survival. According to the third mechanism, the genes for all of the VSGs in the repertoire are present intact within the trypanosome's chromosome. Only one of these genes is expressed at a time, and antigenic variation simply involves the shutting off of one gene and the turning on of another. The presence of all of the VSG genes intact on the chromosome should not create too great a load for the trypanosome to carry. Even a thousand VSG genes could involve less than 10% of the total genome. The third mechanism has turned out to be more or less correct, although as we shall see antigenic variation does involve genomic rearrangements. We describe this mechanism in more detail in the following sections.

HYBRIDIZATION EXPERIMENTS The study of the genetic basis of antigenic variation has depended on the availability of cloned cDNAs prepared from VSG mRNAs. The use of these cDNAs as probes in hybridization experiments has led to three important discoveries. First, in Southern hybridizations every probe tested hybridized with a set of genomic fragments from every VAT tested that belonged to the same serodeme (148–151). These and other experiments have proven that the gene for a VSG is present in a VAT whether or not that gene is expressed. Second, in Northern hybridizations, a cDNA probe hybridized with a unique RNA only from VATs expressing the VSG that corresponded to the probe (149, 151, 151a). This RNA was absent in other VATs. Therefore, although many VSG genes are present in the trypanosome's chromosome, only the one that codes for the cell's VSG, is transcribed. Third, in Southern hybridizations, a cDNA probe sometimes hybridized to different sized genomic restriction fragments in different VATs (148–151). This result indicates that a particular VSG gene may be located at different chromosomal sites in different VATs, and it implies that antigenic variation involves genomic rearrangements.

Two distinct patterns of genomic rearrangement have been detected. In the simpler case, one copy of the VSG gene being probed is found on the same restriction fragment(s) in every VAT tested. This "basic copy" is thought to be at the same location on the chromosome whether or not the trypanosome expresses the gene. However, in cells expressing the gene there is an extra copy found on another restriction fragment(s). This extra copy is called the "expression-linked copy," and it is thought that this copy of the gene is transcribed (149). Expression-linked copies have been reported for some VSG genes in MITat, AnTat, IsTat, LiTat, and ILTat trypanosomes (149–151, 151b), (J. Young, P. Majiwa, and R. Williams, personal communication). In a more complex case, some copies of the VSG gene being probed are found on different sized restriction fragments from a number of different VATs. Copies are found on different sized fragments even when comparing VATs that do not express the gene, and there is no obvious correlation of sequence rearrangement with gene expression. These freewheeling genomic rearrangements were originally reported for some VSG genes in ILTat trypanosomes (148), although they probably also occur in some VATs in other serodemes. These two kinds of genomic rearrangement are discussed in more detail in subsequent sections.

THE ORGANIZATION OF VSG GENES   So far a number of VSG genes have been mapped either by probing Southern blots of genomic DNA or by analysis of genomic fragments cloned in λ phage or cosmid vectors (121, 152–158). No introns have been detected in several VSG genes that have been studied (156–158). Little is known about the linkage of VSG genes on the trypanosome's chromosome, although in the case of the ILTat trypanosomes it seems that sequentially expressed VSG genes are not closely linked (121). Based on Southern hybridization studies, some of the VSG genes appear to be present in two or more copies, and if hybridization is conducted at low stringency even more copies are detected [(152, 154, 157); J. Young, J. Donelson, R. Williams, personal communication]. This fact indicates that there are families of VSG genes that have similar but not identical sequences. Further insight into the relationship of these genes has come from the use of cDNA probes that correspond to either the 3' or the 5' half of the mRNA sequence (154). Probes specific for the 3' sequences generally detect multiple bands on Southern blots of genomic DNA (150, 154, 157). This result is not unexpected, as this part of the probe carries coding information for the C-terminal region, which is known to be conserved in different VSGs. The probe specific for the 5' sequences, which code for the highly variable N-terminal region of the VSG, usually hybridizes with much fewer bands [(150, 154, 157); J. Young, J. Donelson, R. Williams, personal communication]. However, a 3' probe does not detect all of the VSG genes, and 3' probes derived from different cDNAs appear to hybridize with different sets of fragments. This result implies that among all of the VSG genes in the trypanosome's chromosome, there are a number of different families. Members of a family are related by the similarity of their 3'-

terminal sequences. Some of the genomic sequences recognized by the probe may not be functional VSG genes but instead may be pseudogenes or evolutionary fossils incapable of coding for a VSG. The overall conclusion drawn from these experiments is that the VSG genes evolved by gene duplication followed by divergence of the many copies. The 5' ends of the genes, corresponding to the N-terminal region of the proteins, were allowed to diverge much more rapidly than the 3' ends (154, 155, 157).

TRANSCRIPTION OF VSG GENES   As mentioned earlier, antigenic variation is under transcriptional control. Only one kind of VSG transcript is found in each cell, and it corresponds to the particular VSG gene being expressed. In VATs that do not have an expression-linked copy it is not known which copy of the VSG gene undergoes transcription. In those that do have the expression-linked copy, it is likely that it is transcribed. Three kinds of evidence support this view. First, there is a correlation between the presence of an expression-linked copy and transcription of the gene. The extra copy of the gene disappears when the cell undergoes antigenic variation (149–151), and it also disappears, together with the surface coat, when the cell transforms into a procyclic form (150). Second, if the cell's chromatin is treated with DNase I, the expression-linked copy is selectively degraded (156). DNase sensitivity is characteristic of transcriptionally active genes (159). Finally, the mRNA sequence seems to match the sequence of the expression-linked copy but it differs slightly from that of the basic copy (158). This interesting observation is discussed in a following section.

The mRNA for a VSG gene constitutes nearly 10% of the cell's total polyadenylated RNA and because of its abundance is easily isolated either by immunochemical fractionation of polyribosomes or by hybridization to a cloned cDNA (147, 160–165). These molecules are usually about 2 kb, which is more than adequate to code for a protein of 500–600 amino acid residues. The sequences of the 3' nontranslated regions of several of these mRNAs have been deduced from cDNA sequences. Although mRNAs transcribed from gene families in other organisms generally have 3' nontranslated sequences that are strongly divergent, these sequences in VSG mRNAs are remarkably conserved (107, 138–140). They differ by single base changes, small insertions, or deletions. Their possible significance is discussed in the following section.

EXPRESSION-LINKED COPIES OF VSG GENES   The appearance of an expression-linked copy of a VSG gene seems to involve duplication of the basic copy of the gene and transposition of the duplicate to a new location in the chromosome where it is transcribed. There are several kinds of evidence to support this model. First, the appearance of the expression-linked copy is not accompanied by loss of any other copies, which implies duplication of the gene (149–151). Second, mapping experiments indicate that the nucleotide sequences flanking the expression-linked copy differ from those flanking the basic copy (154, 156, 158). Third, in VATs 117 and 118 in the MITat serodeme which have been studied in detail, the VSG genes appear to be transposed to the same chromosomal site, as determined by comparative mapping of the sequences flanking their expression-linked copies (158). Finally, as described in the previous section, it is likely that the expression-linked copy of the gene undergoes transcription. All of these results tentatively indicate that the chromosome may contain a VSG gene expression site to which sequences corresponding to the basic copy of certain VSG genes can be transposed. Antigenic variation would involve insertion of a new gene sequence into the expression site and removal of one already there. This mechanism may be related to the "cassette mechanism," which has been proposed to account for mating type switches in yeast (166).

Recent studies have provided possible clues to the mechanism of duplication-transposition. Although mapping and sequencing studies show that the basic copy and the expression-linked copy of a VSG gene are identical in structure over most of their sequences, it is now clear that they have important differences near their 3' ends (158, 167). These non-homologies have been detected by mapping of hybrids of the VSG mRNA and cloned basic copy DNA (154, 158), by comparison of the restriction enzyme cleavage maps of the basic copy and the expression-linked copy (158, 167), and by comparison of nucleotide sequences of the cloned basic copy and cDNA prepared from VSG mRNA (158). (Genomic clones of expression-linked copies are not yet available but they are assumed to have the same sequence as the cDNA.) In the case of VSG gene 117, the sequence of the mRNA differs from that of the basic copy in the 100–150 nucleotides that precede the poly(A) tail (158). This sequence contains coding information for part of the VSG's C-terminal hydrophobic tail as well as the 3' nontranslated sequence of the mRNA. Therefore, duplication-transposition of the VSG basic copy sequence must involve an alteration of its 3' terminus.

Although the 3' end of the mRNA differs from the corresponding region of the basic copy gene, the two sequences do not differ very much (158). The differences could be accounted for by a few base changes and some small insertions and deletions. In fact, the basic copy of the gene for VSG 117 has what appears to be a fully functional 3' terminus. The basic copy gene could code for a VSG with a C-terminal hydrophobic tail very similar to that coded by the mRNA, and it has a termination codon followed by a sequence similar to the 3' nontranslated sequence on the mRNA. Two possible explanations have been suggested for the presence of a proper 3' terminus on the basic copy gene (158). One is that the basic copy is itself transcribed under some circumstances to produce a functional gene product. A second and more interesting possibility is that transposition may be to a site that already contains a sequence that could serve as a functional 3' terminus. Transposition could involve homologous recombination of the 3' end of the basic copy sequence with the similar sequence in the expression site. Therefore, the basic copy's 3' end would be replaced by a new one. This homologous recombination may occur anywhere within the 150 nucleotides at the 3' end of the basic copy gene. This mechanism clearly explains why mRNAs for different VSGs have very similar 3' nontranslated sequences. These conserved sequences would be required for transposition rather than for function of the mRNA.

Nothing is known about the events that occur at the 5' end of the gene during transposition except that a sequence that flanks the basic copy gene on its 5' side seems to be co-transposed along with the gene to its expression site. In the VSG genes studied so far, this sequence is probably $\sim$ 1 kb (154, 156, 167).

OTHER KINDS OF SEQUENCE REARRANGEMENT   Not all VSG genes have expression-linked copies. Some of the ILTat VSG genes, and probably some VSG genes in other trypanosomes, fall into this category. cDNA probes specific for VSG genes expressed in ILTat 1.2, 1.3, and 1.4 trypanosomes characteristically hybridize to two or more copies of the gene in all ILTat VATs tested (121, 148, 152, 153). Some of these copies are found on restriction fragments that vary in size in different VATs, whereas others are found on fragments that do not vary. These apparent rearrangements are not restricted to VATs expressing the gene, as they were found in all VATs tested. There is no correlation of sequence rearrangement with gene expression.

Mapping has revealed that the rearrangements involve large deletions or insertions beyond the 3' terminus of the gene (121, 152, 153). There does not seem to be gene duplication in VATs that express the gene, and there is no evidence for transposition.

Not all ILTat VSG genes behave in this way, as the ILTat 1.1 gene has an expression-linked copy similar to those described in the previous section (J. Young, P. Majiwa, and R. Williams, personal communication). Further study will be needed before the mechanism of activation of the other ILTat VSG genes is understood. Based on our present knowledge, it seems likely that they are regulated by a mechanism completely different from those that have expression-linked copies.

CONCLUSION

The techniques of molecular biology have already revealed much about the genetic processes of trypanosomes. They are contributing to our understanding of unique DNA structures and novel mechanisms of gene regulation. They have not yet had an impact on the diseases caused by these parasites, but it is likely that they will.

ACKNOWLEDGMENTS

We are grateful to many colleagues for sending us unpublished manuscripts, and we thank John Donelson, John Young, George Cross, Jerry Hart, and Ken Marcu for comments on an early draft of this review. Research in the authors' laboratory was supported by NIH Grant GM-27608-13 and WHO Grant 790184. S. L. Hajduk was supported by a fellowship from the Rockefeller Foundation and J. C. Marini by Medical Scientist Training Program Grant GM-7309. We thank Shirley Metzger for preparation of the manuscript and Ann Hajduk for art work.

Literature Cited

1. Walsh, J. A., Warren, K. S. 1979. *N. Engl. J. Med.* 301:967–74
2. Goodman, H. C. 1978. *Ann. Immunol.* 129C:267–74
3. Vickerman, K. 1976. In *Biology of the Kinetoplastida*, ed. W. H. R. Lumsden, D. A. Evans, 1:1–34. London: Academic
4. Hoare, C. A. 1972. *The Trypanosomes of Mammals. A Zoological Monograph.* Oxford: Blackwell
5. Vickerman, K., Preston, T. M. 1976. See Ref. 3, pp. 35–130
6. McGhee, R. B., Cosgrove, W. B. 1980. *Microbiol. Rev.* 44:140–73
7. Rifkin, M. R. 1978. *Proc. Natl. Acad. Sci. USA* 75:3450–54
8. Robertson, M. 1912. *Proc. R. Soc. Ser. B.* 85:527–39
9. Vickerman, K. 1965. *Nature* 208:762–66
10. Barry, J. D., Vickerman, K. 1979. *Exp. Parasitol.* 48:313–24
11. Vickerman, K. 1969. *J. Cell Sci.* 5:163–93
12. LeRay, D., Barry, J. D., Vickerman, K. 1978. *Nature* 273:300–2
13. Wijers, D. J. B. 1957. *Nature* 180:391–92
14. Wijers, D. J. B., Willett, K. C. 1960. *Ann. Trop. Med. Parasitol.* 54:341–50
15. Ashcroft, M. T. 1960. *Ann. Trop. Med. Parasitol.* 54:44–53
16. Fairbain, H., Culwick, A. T. 1947. *Ann. Trop. Med. Parasitol.* 41:26–29
17. Bowman, I. B. R., Flynn, I. W. 1976. See Ref. 3, pp. 435–76
18. Gutteridge, W. E., Coombs, G. H. 1977. *Biochemistry of Parasitic Protozoa.* Baltimore: Univ. Park Press
19. Bienen, E. J., Hammadi, E., Hill, G. C. 1981. *Exp. Parasitol.* 51:408–17
20. Fulton, J. D., Spooner, D. F. 1959. *Exp. Parasitol.* 8:137–62
21. Hill, G. C. 1976. *Biochim. Biophys. Acta* 456:149–93
22. Grant, P. T., Sargent, J. R. 1960. *Biochem. J.* 76:229–37
23. Opperdoes, F. R., Borst, P., Spits, H. 1977. *Eur. J. Biochem.* 76:21–28
24. Lanham, S. M., Godfrey, D. G. 1970. *Exp. Parasitol.* 28:521–34
25. Taylor, D. W., Cross, G. A. M. 1977. *Parasitology* 74:47–60
26. Rovis, L., Dube, D. K. 1981. *Mol. Biochem. Parasitol.* 4:77–93
27. Strickler, J. E., Patton, C. L. 1980. *Proc. Natl. Acad. Sci. USA* 77:1529–33
28. Hirumi, H., Doyle, J. J., Hirumi, K. 1977. *Science* 196:992–94
29. Hill, G. C., Shimer, S. P., Caughey, B., Sauer, L. S. 1978. *Science* 202:763–65
30. Doyle, J. J., Hirumi, H., Hirumi, K., Lupton, E. N., Cross, G. A. M. 1980. *Parasitology* 80:359–69

31. Cunningham, I. 1977. *J. Protozool.* 24:325–28
32. Simpson, L., Simpson, A. M., Kidane, G., Livingston, L., Spithill, T. W. 1980. *Am. J. Trop. Med. Hyg.* 29:1053–63
33. Tait, A. 1980. *Nature* 287:536–38
34. Deleted in proof
35. Riou, G., Paoletti, C. 1967. *J. Mol. Biol.* 28:377–82
36. Laurent, M., Steinert, M. 1970. *Proc. Natl. Acad. Sci. USA* 66:419–24
37. Renger, H. C., Wolstenholme, D. R. 1970. *J. Cell Biol.* 47:689–702
38. Simpson, L., da Silva, A. 1971. *J. Mol. Biol.* 56:443–73
39. Simpson, L. 1972. *Int. Rev. Cytol.* 32:139–207
40. Borst, P., Fairlamb, A. H. 1976. In *Biochemistry of Parasites and Host-Parasite Relationships,* ed. H. Van den Bossche, pp. 169–91. Amsterdam: North-Holland
41. Hajduk, S. L. 1978. *Prog. Mol. Subcell. Biol.* 6:158–200
42. Borst, P., Hoeijmakers, J. H. J. 1979. *Plasmid* 2:20–40
43. Barker, D. C. 1980. *Micron* 11:21–62
44. Englund, P. T. 1980. In *Biochemistry and Physiology of Protozoa,* ed. M. Levandowsky, S. H. Hutner, 4:333–83. New York: Academic. 2nd ed.
45. Marini, J. C., Miller, K. G., Englund, P. T. 1980. *J. Biol. Chem.* 255:4976–79
46. Steinert, M., Van Assel, S. 1975. *Exp. Cell Res.* 96:406–9
47. Weislogel, P. O., Hoeijmakers, J. H. J., Fairlamb, A. H., Kleisen, C. M., Borst, P. 1977. *Biochim. Biophys. Acta* 478:167–79
48. Fairlamb, A. H., Weislogel, P. O., Hoeijmakers, J. H. J., Borst, P. 1978. *J. Cell Biol.* 76:293–309
49. Borst, P., Hoeijmakers, J. H. J. 1979. *Extrachromosomal DNA, ICN-UCLA Symp. Mol. Cell Biol* 15:515–31
50. Hoeijmakers, J. H. J., Snijders, A., Janssen, J. W. G., Borst, P. 1981. *Plasmid* 5:329–50
51. Riou, G. F., Gutteridge, W. E. 1978. *Biochimie* 60:365–79
52. Englund, P. T. 1978. *Cell* 14:157–68
53. Borst, P. 1976. In *Handbook of Biochemistry and Molecular Biology,* ed. G. D. Fasman, 2:375–78. Cleveland, Ohio:CRC. 3rd ed.
54. Laurent, M., Van Assel, S., Steinert, M. 1971. *Biochem. Biophys. Res. Commun.* 43:278–84
55. Steinert, M., Van Assel, S., Borst, P., Newton, B. A. 1976. In *The Genetic Function of Mitochondrial DNA,* ed. C.

Saccone, A. M. Kroon, pp. 71–81. Amsterdam: North-Holland
56. Riou, G., Yot, P. 1975. *C. R. Acad. Sci. Ser. D.* 280:2701–4
57. Kleisen, C. M., Borst, P. 1975. *Biochim. Biophys. Acta* 407:473–78
58. Borst, P., Hoeijmakers, J. H. J., Hajduk, S. L. 1981. *Parasitology* 82:81–93
59. Steinert, M., Van Assel, S. 1980. *Plasmid* 3:7–17
60. Donelson, J. E., Majiwa, P. A. O., Williams, R. O. 1979. *Plasmid* 2:572–88
61. Stuart, K., Gelvin, S. R. 1980. *Am. J. Trop. Med. Hyg.* 29:1075–81
61a. Brunel, F., Davison, J., Merchez, M., Borst, P., Weijers, P. J. 1979. In *DNA-Recombination Interactions and Repair,* ed. S. Zandrazil, J. Sponar, pp. 45–54. Oxford: Pergammon
61b. Brunel, F., Davison, J., Thi, V. H., Merchez, M. 1980. *Gene* 12:223–34
62. Riou, G. F., Saucier, J.-M. 1979. *J. Cell Biol.* 82:248–63
63. Chen, K. K., Donelson, J. E. 1980. *Proc. Natl. Acad. Sci. USA* 77:2445–49
64. Barrois, M., Riou, G., Galibert, F. 1981. *Proc. Natl. Acad. Sci. USA* 78:3323–27
65. Barker, D. C., Arnot, D. E., Butcher, J. 1982. *UNDP/World Bank/WHO Workshop on Biochemical Characterization of Leishmania.* Washington DC: In press
66. Hoeijmakers, J. H. J., Borst, P. 1978. *Biochim. Biophys. Acta* 521:407–11
67. Simpson, L., Simpson, A. M. 1978. *Cell* 14:169–78
68. Fouts, D. L., Wolstenholme, D. R. 1979. *Nucleic Acids Res.* 6:3785–804
69. Challberg, S. S., Englund, P. T. 1980. *J. Mol. Biol.* 138:447–72
70. Kleisen, C. M., Weislogel, P. O., Fonck, K., Borst, P. 1976. *Eur. J. Biochem.* 64:153–60
71. Simpson, L. 1979. *Proc. Natl. Acad. Sci. USA* 76:1585–88
71a. Masuda, H., Simpson, L., Rosenblatt, H., Simpson, A. M. 1979. *Gene* 6:51–73
72. Cheng, D., Simpson, L. 1978. *Plasmid* 1:297–315
73. Hajduk, S. L., Cosgrove, W. B. 1979. *Biochim. Biophys. Acta* 561:1–9
74. Opperdoes, F. R., Borst, P., de Rijke, D. 1976. *Comp. Biochem. Physiol. B* 55:25–30
75. Frasch, A. C. C., Hajduk, S. L., Hoeijmakers, J. H. J., Borst, P., Brunel, F., Davison, J. 1980. *Biochim. Biophys. Acta* 607:397–410
76. Hajduk, S. L., Vickerman, K. 1981. *Mol. Biochem. Parasitol.* 4:17–28

77. Riou, G., Baltz, T., Gabillot, M., Pautrizel, R. 1980. *Mol. Biochem. Parasitol.* 1:97–105
78. Riou, G. F., Belnat, P., Benard, J. 1980. *J. Biol. Chem.* 255:5141–44
79. Steinert, M., Van Assel, S. 1967. *Arch. Int. Physiol. Biochim.* 75:370–71
80. Cosgrove, W. B., Skeen, M. J. 1970. *J. Protozool.* 17:172–77
81. Simpson, L., Braly, P. 1970. *J. Protozool.* 17:511–17
82. Englund, P. T., Marini, J. C. 1980. *Am. J. Trop. Med. Hyg.* 29:1064–69
83. Englund, P. T., Hajduk, S. L., Marini, J. C., Plunkett, M. L. 1982. In *Mitochondrial Genes,* ed. P. Slonimski, P. Borst, G. Attardi. New York: Cold Spring Harbor Lab. In press.
84. Manning, J. E., Wolstenholme, D. R. 1976. *J. Cell Biol.* 70:406–18
85. Simpson, L., Simpson, A. M., Wesley, R. D. 1974. *Biochim. Biophys. Acta* 349:161–72
86. Englund, P. T. 1979. *J. Biol. Chem.* 254:4895–900
87. Gellert, M. 1981. *Ann. Rev. Biochem.* 50:879–910
88. Miller, K. G., Liu, L. F., Englund, P. T. 1981. *J. Biol. Chem.* 256:9334–39
89. Hoeijmakers, J. H. J., Weijers, P. J. 1980. *Plasmid* 4:97–116
90. Brack, C., Delain, E., Riou, G. 1972. *Proc. Natl. Acad. Sci. USA* 69:1642–46
91. Kleisen, C. M., Borst, P., Weijers, P. J. 1976. *Eur. J. Biochem.* 64:141–51
92. Borst, P., Fairlamb, A. H., Fase-Fowler, F., Hoeijmakers, J. H. J., Weislogel, P. O. 1976. See Ref. 55, pp. 59–69
93. Steinert, M., Van Assel, S., Steinert, G. 1976. See Ref. 40, pp. 193–202
94. Hajduk, S. L., Hoeijmakers, J. H. J., Borst, P., Vickerman, K. 1982. *J. Cell Biol.* In press
95. Morel, C., Chiari, E., Plessman Camargo, E., Mattei, D. M., Romanha, A. J., Simpson, L. 1980. *Proc. Natl. Acad. Sci. USA* 77:6810–14
95a. Arnot, D. E., Barker, D. C. 1981. *Mol. Biochem. Parasitol.* 3:47–56
95b. Borst, P., Fase-Fowler, F., Gibson, W. C. 1981. *Mol. Biochem. Parasitol.* 3:117–31
96. Borst, P., Fase-Fowler, F., Frasch, A. C. C., Hoeijmakers, J. H. J., Weijers, P. J. 1980. *Mol. Biochem. Parasitol.* 1:221–46
97. Massaglia, M. A. 1907. *CR Acad. Sci.* 145:687–89
98. Lourie, E. M., O'Connor, J. R. 1937. *Ann. Trop. Med. Parasitol.* 31:319–40
99. Vickerman, K., Luckins, A. G. 1969. *Nature* 224:1125–26
100. Cross, G. A. M. 1975. *Parasitology* 71:393–417
101. Bridgen, P. J., Cross, G. A. M., Bridgen, J. 1976. *Nature* 263:613–14
102. Johnson, J. G., Cross, G. A. M. 1979. *Biochem. J.* 178:689–97
103. 1978. *Bull. WHO* 56:467–80
104. Capbern, A., Giroud, C., Baltz, T., Mattern, P. 1977. *Exp. Parasitol.* 42:6–13
105. Van Meirvenne, N., Magnus, E., Vervoort, T. 1977. *Ann. Soc. Belge Med. Trop.* 57:409–23
106. Vervoort, T., Barbet, A. F., Musoke, A. J., Magnus, E., Mpimbaza, G., Van Meirvenne, N. 1981. *Immunology* 44:223–32
107. Matthyssens, G., Michiels, F., Hamers, R., Pays, E., Steinert, M. 1981. *Nature* 293:230–33
108. Rovis, L., Barbet, A. F., Williams, R. O. 1978. *Nature* 271:654–56
109. Richards, F. F., Rosen, N. L., Onodera, M., Bogucki, M. S., Neve, R. L., Hotez, P., Armstrong, M. Y. K., Konigsberg, W. H. 1981. *Fed. Proc.* 40:1434–39
110. Rosen, N. L., Onodera, M., Hotez, P. J., Bogucki, M. S., Elce, B., Patton, C., Konigsberg, W. H., Cross, G. A. M., Richards, F. 1981. *Exp. Parasitol.* 52:210–18
111. Onodera, M., Rosen, N. L., Lifter, J., Hotez, P. J., Bogucki, M. S., Davis, G., Patton, C. L., Konigsberg, W. H., Richards, F. F. 1981. *Exp. Parasitol.* 52:427–39
112. Bogucki, M. S., Onodera, M., Rosen, N. L., Lifter, J., Hotez, P. J., Konigsberg, W. H., Richards, F. F. 1982. *Exp. Parasitol.* 53:1–10
113. Nantulya, V. M., Doyle, J. J., Jenni, L. 1980. *Parasitology* 80:123–31
114. Risse, N. J., Reinwald, E., Rautenberg, P. 1980. In *The Host-Invader Interplay,* ed. H. Van den Bossche, pp. 245–48 Amsterdam: Elsevier/North-Holland Biomed. Press
115. Vickerman, K. 1974. *CIBA Found. Symp.* 25:53–80
116. Gray, A. R., Luckins, A. G. 1976. In *Biology of the Kinetoplastida,* ed. W. H. R. Lumsden, D. A. Evans, pp. 493–542. London: Academic
117. Doyle, J. J. 1977. In *Immunity to Blood Parasites in Animals and Man,* ed. L. H. Miller, J. A. Pino, J. J. McKelvey, pp. 31–63. New York: Plenum
118. Cross, G. A. M. 1978. *Proc. R. Soc. London Ser. B* 202:55–72
119. Vickerman, K. 1978. *Nature* 273:613–17

120. Turner, M. J. 1980. In *The Molecular Basis of Microbial Pathogenicity,* ed. H. Smith, J. J. Skekel, M. J. Turner, pp. 138–58. Weinheim: Chemie GmbH

121. Marcu, K. B., Williams, R. O. 1981. In *Genetic Engineering,* ed. J. K. Setlow, A. Hollaender, 3:129–55. New York: Plenum

122. Gray, A. R. 1965. *J. Gen. Micro.* 41:195–214

123. Gray, A. R. 1975. *Trans. R. Soc. Trop. Med. Hyg.* 69:131–38

124. Jenni, L. 1977. *Acta Trop.* 34:35–41

125. Hudson, K. M., Taylor, A. E. R., Elce, B. J. 1980. *Parasitol. Immunol.* 2:58–69

126. Baltz, T., Baltz, D., Pautrizel, R., Richet, C., Lamblin, G., Degand, P. 1977. *FEBS Lett.* 82:93–96

127. Holms, P. H. 1980. In *Vaccines Against Parasites, Symposia of the British Society for Parasitology,* ed. A. E. R. Taylor, R. Muller, 18:75–105. London: Blackwell Sci.

128. Van Meirvenne, N., Janssens, P. G., Magnus, E. 1975. *Ann. Soc. Belge Med. Trop.* 55:1–23

128a. Esser, K. M., Schoenbechler, M. J., Gingrich, J. B., Diggs, C. L. 1981. *Fed. Proc.* 40:1011

129. Barry, J. D., Hajduk, S. L., Vickerman, K., LeRay, D. 1979. *Trans. Roy. Soc. Trop. Med. Hyg.* 73:205–8

130. Hajduk, S. L., Cameron, C., Barry, J. D., Vickerman, K. 1981. *Parasitology* 83:595–607

131. LeRay, D., Barry, J. D., Easton, C., Vickerman, K. 1977. *Ann. Soc. Belge Méd. Trop.* 57:369–81

132. Stanley, H. A., Honigberg, B. M., Cunningham, I. 1979. *Z. Parasitenkd.* 58:141–49

133. Hajduk, S. L., Vickerman, K. 1981. *Parasitology* 83:609–21

134. Jenni, L., Brun, R. 1981. *Trans. R. Soc. Trop. Med. Hyg.* 75:150–1

135. McNeillage, G. J. C., Herbert, W. J., Lumsden, W. H. R. 1969. *Exp. Parasitol.* 25:1–7

136. Miller, E. N., Turner, M. J. 1981. *Parasitology* 82:63–80

137. Johnson, J. G., Cross, G. A. M. 1977. *J. Protozool.* 24:587–91

138. Rice-Ficht, A. C., Chen, K. K., Donelson, J. E. 1981. *Nature* 294:53–57

139. Boothroyd, J. C., Cross, G. A. M., Hoeijmakers, J. H. J., Borst, P. 1980. *Nature* 288:624–26

140. Boothroyd, J. C., Paynter, C. A., Cross, G. A. M., Bernards, A., Borst, P. 1981. *Nucleic Acids Res.* 9:4735–43

141. Holder, A. A., Cross, G. A. M. 1981. *Mol. Biochem. Parasitol.* 2:135–50

142. Cross, G. A. M., Holder, A. A., Allen, G., Boothroyd, J. C. 1980. *Am. J. Trop. Med. Hyg.* 29:1027–32

142a. Strickler, J. E., Patton, C. L. 1982. *Mol. Biochem Parasitol.* In press

143. Barbet, A. F., McGuire, T. C. 1978. *Proc. Natl. Acad. Sci. USA* 75:1989–93

144. Cross, G. A. M. 1979. *Nature* 277:310–12

145. Cross, G. A. M. 1979. *J. Gen. Microbiol.* 133:1–11

145a. Labastie, M. C., Baltz, T., Richet, C., Giroud, C., Duvillier, G., Pautrizel, R., Degand, P. 1981. *Biochem. Biophys. Res. Commun.* 99:729–36

146. Barbet, A. F., McGuire, T. C., Musoke, A. J., Hirumi, H. 1979. In *Pathogenicity of Trypanosomes,* ed. G. Losos, A. Chouinard, pp. 38–43. Ottawa: Int. Dev. Res. Centre

147. Hoeijmakers, J. H. J., Borst, P., Van den Burg, J., Weissmann, C., Cross, G. A. M. 1980. *Gene* 8:391–417

147a. Auffret, C. A., Turner, M. J. 1981. *Biochem. J.* 193:647–50

147b. Strickler, J. E., Patton, C. L. 1982. *Experimental Parasitol.* 53:117–32

147c. Strickler, J. E., Mancini, P. E., Patton, C. L. 1978 *Experimental Parasitol.* 46:262–76

148. Williams, R. O., Young, J. R., Majiwa, P. A. O. 1979. *Nature* 282:847–49

149. Hoeijmakers, J. H. J., Frasch, A. C. C., Bernards, A., Borst, P., Cross, G. A. M. 1980. *Nature* 284:78–80

150. Pays, E., Van Meirvenne, N., Le Ray, D., Steinert, M. 1981. *Proc. Natl. Acad. Sci. USA* 78:2673–77

151. Agabian, N., Thomashow, L., Milhausen, M., Stuart, K. 1980. *Am. J. Trop. Med. Hyg.* 29:1043–49

151a. Pays, E., Debronche, M., Lheureux, M., Vervoort, T., Block, J., Gannon,F., Steinert, M. 1980. *Nucleic Acids Res.* 8:5965–81

151b. Pays, E., Lheureux, M., Vervoort, T., Steinert, M. 1981. *Mol. Biochem. Parasitol.* 4:349–57

152. Williams, R. O., Young, J. R., Majiwa, P. A. O., Doyle, J. J., Shapiro, S. Z. 1980. *Am. J. Trop. Med. Hyg.* 29:1037–42

153. Williams, R. O., Young, J. R., Majiwa, P. A. O., Doyle, J. J., Shapiro, S. Z. 1980. *Cold Spring Harbor Symp. Quant. Biol.* 45:945–49

154. Borst, P., Frasch, A. C. C., Bernards, A., Van der Ploeg, L. H. T., Hoeijmakers, J. H. J., Arnberg, A. C., Cross, G. A. M. 1980. *Cold Spring Harbor Symp. Quant. Biol.* 45:935–43

155. Deleted in proof

156. Pays, E., Lheureux, M., Steinert, M. 1981. *Nature* 292:265–67
157. Davison, J., Brunel, F., Merchez, M., Ha Thi, V. 1982. *Gene.* 17:101–6
158. Bernards, A., Van der Ploeg, L. H. T., Frasch, A. C. C., Borst, P., Boothroyd, J. C., Coleman, S., Cross, G. A. M. 1981. *Cell* 27:497–505
159. Weintraub, H., Groudine, M. 1976. *Science* 193:848–56
160. Williams, R. O., Marcu, K. B., Young, J. R., Rovis, L., Williams, S. 1978. *Nucleic Acids Res.* 5:3171–82
161. Lheureux, M., Vervoort, T., Van Meirvenne, N., Steinert, M. 1979. *Nucleic Acids Res.* 7:595–609

162. Shapiro, S. Z., Young, J. R. 1981. *J. Biol. Chem.* 256:1495–98
163. Eggitt, M. J., Tappenden, L., Brown, K. N. 1977. *Parasitology* 75:133–41
164. Merritt, S. C. 1980. *Mol. Biochem. Parasitol.* 1:151–66
165. Cordingly, J. S., Turner, M. J. 1980. *Mol. Biochem. Parasitol.* 1:129–37
166. Hicks, J. B., Strathern, J. N., Herskowitz, I. 1977. In *DNA Insertion Elements, Plasmids, and Episomes,* ed. A. I. Bukhari, J. A. Shapiro, S. L. Adhya, pp. 457–62. New York: Cold Spring Harbor Lab.
167. Pays, E., Lheureux, M., Steinert, M. 1981. *Nucleic Acids Res.* 9:4225–38

Ann. Rev. Biochem. 1982. 51:727–61

MOLECULAR MECHANISMS IN GENETIC RECOMBINATION

David Dressler and Huntington Potter

Department of Biochemical Sciences, Princeton University, Princeton, New Jersey 08540 and Department of Microbiology, Harvard School of Public Health, Boston, Massachusetts 02115

PROSPECTUS AND SUMMARY

Our picture of the molecular events that occur during genetic recombination is much more complete and satisfying than the description that was available only two or three years ago. Through an interesting fusion of data involving classical genetic experiments in fungi, the analysis of recombination intermediates recovered from intact cells, and the study of enzymes produced by genes known to be involved in recombination in *Escherichia coli,* an increasingly consistent pattern has emerged.

0066-4154/82/0701-0727$02.00

One of the unifying themes of the last two decades has been provided by a recombination intermediate envisioned by Robin Holliday in 1964 (1, 1a). Genetic studies in fungi had set conditions that a recombination model had to satisfy if it were to be applicable to the most highly ordered form of genetic recombination known—the eukaryotic meiosis. Holliday proposed a model that satisfied these conditions and used, as its centerpiece, an intermediate in which the two recombining chromosomes were covalently held together at a region of homology through a crossover connection formed by the reciprocal exchange of two of the four strands in the participating DNA molecules. Although the specific intermediate suggested by Holliday has been retained in much of our present thinking about recombination, the original Holliday model has undergone several modifications as a consequence of both theoretical and experimental advances. The relatively simple ways in which the interactions between DNA strands could be thought about in 1964 (relying entirely on Watson and Crick base-pairing but lacking any specific consideration of the role of enzymes) could not anticipate all of the complexities of the recombination process. And, as genetic studies in fungi became more elaborate and insightful, refinements and extensions of the mechanism for forming the Holliday intermediate were required. But in spite of the fact that the original model has evolved over a period of 15 years with contributions from several sources, the Holliday structure itself has been retained as the preferred theoretical intermediate.

The Holliday intermediate acquired a physical reality in the 1970s when DNA molecules engaged in recombination in *E. coli* were recovered from cells and found to have a crossover connection of the type predicted by Holliday. This, in effect, helped unify the study of recombination. Whereas Holliday had relied upon and sought to explain recombination in eukaryotes, the finding in *E. coli* of the central recombination intermediate of the model meant that perhaps prokaryotes and eukaryotes share a common mechanism for the exchange of genetic information. Thus all of the experience and technical expertise available in prokaryotic systems could be brought to bear on the study of recombination with the expectation that the findings would be of more general applicability.

In the last few years, studies with *E. coli* cell lysates and with the purified products of genes known to be involved in genetic recombination, especially the recA protein, have deepened our understanding of the recombination process to the level of enzymology. In fact, it is now possible to discuss a coherent model for recombination in which each step is specified in terms of a known enzyme and a specific biochemical reaction mechanism. One of the intermediate stages in this biochemical pathway is the formation of a Holliday-type crossover connection between the two recombining DNA molecules.

This review begins with a description of the Holliday model and then traces its evolution from a geneticist's blackboard diagram into a detailed enzymological model understood at the level of specific reaction mechanisms.[1]

A convenient place to begin then is with a discussion of the model first elaborated by Holliday, because, in a prototype way, this model addresses the major questions that must be solved in understanding the recombination process.

THE PROTOTYPE HOLLIDAY MODEL

A prototype model that uses the Holliday recombination intermediate is shown in Figure 1 (1, 8–10). Two homologous double helices are aligned, and in each the positive strands (or, alternatively, the negative strands) are nicked open in a given region. The free ends thus created leave the complementary strands to which they had been hydrogen-bonded and become associated instead with the complementary strands in the homologous double helix (Figure 1 *a–f*). The result of this reciprocal strand exchange is to establish a tentative physical connection between the two DNA molecules that are going to recombine. This linkage can be made stable through a process of DNA repair, which in this case can be as simple as the formation of two phosphodiester bonds by the enzyme DNA ligase (Figure 1*e*).

Holliday envisioned nicks in the recombining chromosomes as promoting the strand exchange reactions. However, if homologous pairing of the two DNA molecules can occur prior to breakage and reunion, then an enzyme with nicking-closing (or topoisomerase) activity could be used to carry out the genome fusion reaction (11–13). In this case, the strand breakage and reunion events would be accomplished in a concerted reaction still leading, however, to the structure shown in Figure 1*e*.

The structure shown in Figure 1*e* is the Holliday recombination intermediate. Sigal & Alberts (14) have shown, by building a space-filling model of DNA, that in this intermediate, steric hindrance is minimal and virtually all of the bases can be paired. Thus, whatever the energy barrier to forming the intermediate, it is not an inherently strained or unstable structure.

---

[1]We concentrate on recent studies of generalized DNA recombination at the level of molecular biology and enzymology. Several other excellent reviews are available that deal with related topics. These include Clark's review of the genes involved in bacterial recombination (2), Hotchkiss's comparative analysis of various types of models for recombination (3), Nash's review of the site-specific recombination system of phage λ (4), Fox's discussion of recombination in prokaryotes (5), Stahl's review of the role of special DNA sites in recombination (6), and Radding's 1978 general review of the field (7). Indeed the second half of the Cold Spring Harbor Symposium in 1978 (Volume 43) was devoted to specific research papers on recombination.

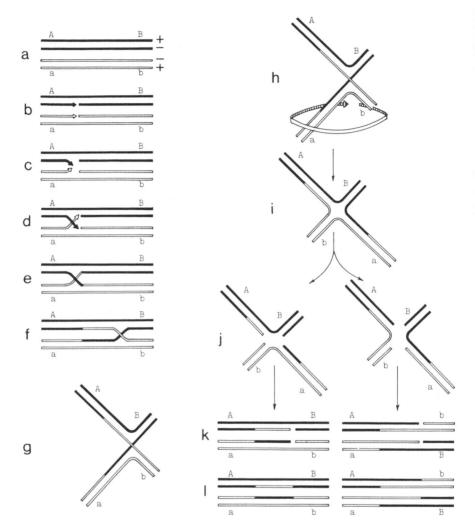

Figure 1   A prototype model for genetic recombination using the Holliday intermediate.

While the Holliday intermediate is expected to be stable, it need not be static. A continuing reciprocal strand exchange by the two polynucleotide chains involved in the crossover can occur (in conjunction with the rotation of the four double-helical arms around their cylindrical axes), allowing the point of linkage between the two DNA molecules to move to the right or to the left (Figure 1*e, f*). This dynamic property of the Holliday structure can lead to the development of regions of heteroduplex DNA during recombination and, when mismatched bases occur, to the formation of areas of double helix that are genetically heterozygous. It was, in fact, the observation of genetic heterozygosity in the recombinant chromosomes of haploid fungal spores by Kitani, Olive, and El-ani (15; see also 16–21) that helped motivate the proposal of this model by Holliday.

The fluidity of the Holliday structure and the rate of cross-bridge migration have been studied theoretically by Meselson (22) and experimentally by Warner and his colleagues (23, 24). They conclude that this rate is high enough to allow the rapid formation of regions of hybrid DNA under physiological conditions.

The remainder of Figure 1 shows the maturation mechanism proposed for the Holliday intermediate. Because of its structural symmetry, it is expected that the intermediate can be processed in either of two related ways, to give rise to two different pairs of recombinant chromosomes (25; see also 9, 14). This symmetry in structure, and the dual maturation potential that results, are most easily appreciated if one rotates the intermediate into another planar configuration (as in Figure 1*i*). Then, cleavage on an east-west axis, or on a north-south axis, leads to the release of unit-size DNA molecules in which potentially heterozygous regions exist, and in which the flanking genes may either remain in their original linkage or, with equal probability, emerge in a reciprocally recombinant linkage (Figure 1 *l*). Genetic distances are measured from the 50% of the cases in which the flanking genes are exchanged. The remaining maturation events are silent from the point of view of traditional recombination, except for the occurrence of heterozygous DNA that remains as a footprint of the former crossover connection.

This model or one related to it, is attractive as a general mechanism for recombination because it can account for the genetic properties of recombinant chromosomes that emerge from the most highly ordered form of gene exchange known, the eukaryotic meiosis. As studied in fungi, where all four of the recombinant chromosomes formed in a single meiosis are preserved together and therefore can be comparatively analyzed, the following general conclusions have emerged:

1. Recombination proceeds with a net conservation of genetic material: for every two chromosomes that enter the recombination process, two emerge (26).
2. Recombinant chromosomes are produced in reciprocal pairs, i.e. not just in the population as a whole, but during individual crossover events (26; see also 27).
3. A region of genetic heterozygosity is frequently found in the immediate area of the chromosome where a recombination event has occurred (15–21). This is detected when a haploid fungal spore divides to give rise

immediately to two different lines of progeny. For example, in a mating of *A arg$^+$ B* with *a arg$^-$ b,* among recombinant (*Ab* or *aB*) spores, several per hundred, although haploid, will contain information for both the *arg$^+$* and *arg$^-$* phenotypes and, upon the first round of semiconservative DNA replication, will give rise to both *arg$^+$* and *arg$^-$* progeny.

4. Lastly, if one selectively analyzes spores that show such genetic heterozygosity, the genes on either side of the heterozygous region are found, with equal probability, either to occur in a recombinant linkage, or to have retained their original linkage [for reviews see Kitani, Olive & El-ani and Hurst, Fogel & Mortimer (15, 16, 20)]. That is, a chromosome that divides to give both *arg$^+$* and *arg$^-$* progeny will show either the parental linkage (*AB* or *ab*) or the recombinant linkage (*Ab* or *aB*) for the flanking genes.

The prototype recombination mechanism shown in Figure 1 accounts for all of these genetic findings. That recombinant chromosomes are produced in reciprocal pairs in a process that does not involve the destruction of any of the participating DNA molecules is evident from the overall drawing. The formation of regions of genetically heterozygous DNA is seen in panels *a–f,* where the crossed strand exchange develops and the crossover bridge then moves laterally along the DNA. Lastly, the finding of an equal likelihood for the parental or recombinant linkage for the flanking genes on either side of a region of genetic heterozygosity is understood in terms of the structural symmetry of the crossover connection and the dual maturation alternatives shown in panels *i–l.* Other models for recombination differ in that they produce only one recombinant at a time [for instance, the model of Boon & Zinder, (28)], or piece together one new recombinant chromosome while discarding the rest of the participating DNA in the form of fragments [for instance, the model of Broker & Lehman, (29)].

The prototype Holliday model was derived entirely from a consideration of genetic data obtained with fungi. Further support for one aspect of the model—the development of regions of heterozygous DNA in recombinant chromosomes—has been obtained in studies with prokaryotic viruses; for example, λ bacteriophages that have undergone recombination (17–19, 21). However, the dispersion of the immediate products of recombination in these and virtually all prokaryotic and eukaryotic systems precludes a genetic analysis that would bear upon the important questions of reciprocity and net DNA conservation. [For the one exception, see (27).] It is nonetheless taken as axiomatic by workers in the field that the genetic data obtained with fungi, and their resulting implications, will be applicable to a wider range of plant and animal systems.

Physical Evidence for the Existence of the Holliday Recombination Intermediate

While the prototype Holliday model was proposed on the basis of genetic studies in eukaryotes, it received its first biochemical support from studies carried out with small DNA molecules in prokaryotic systems. Basically these experiments involved the isolation of recombination intermediates formed under the influence of the recombination system of *E. coli.*

The viruses S13 (30, 31), ϕX 174 (32, 33), and λ (34, 35), and plasmids related to ColE1 (10, 36, 37) have proved to be especially useful in the search for intermediates in the recombination process. These small DNA molecules can be isolated without breakage and then analyzed biochemically; moreover, because, under appropriate conditions, they do not encode their own recombination systems but utilize the recombination apparatus of the host cell, their behavior provides a window onto a larger view.

In considering the structure of recombining DNA in simple organisms, it is necessary to take into account that the chromosomes of prokaryotes, viruses, plasmids, and cellular organelles are generally circular. We have thus far discussed recombination intermediates in terms of linear DNA molecules, such as those found in eukaryotic chromosomes. However, if a recombination intermediate were to be formed between two DNA circles, one would expect the composite molecule to assume the appearance of a figure-8. For instance, Figure 2 diagrams the structure that would result if two DNA circles were to undergo the general type of strand nicking and exchange events proposed by Holliday.

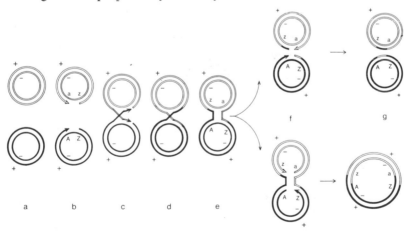

Figure 2 Circular DNA molecules undergoing a reciprocal strand invasion to form a figure eight structure.

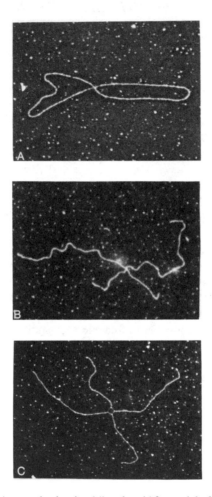

Figure 3 Electron micrographs showing (*A*) a plasmid figure eight form, (*B*) a chi form, and (*C*) a chi form in which the single-stranded connections can be seen in the region of the crossover.

OBSERVATION OF FIGURE-8s Dimeric DNA molecules having the shape of a figure-8 were first found among the viral DNA from S13 and ϕX infected cells by Doniger, Thompson, Warner & Tessman (30, 31) and by Benbow, Zuccarelli & Sinsheimer (32, 33) (for a review, see 10, 37). In subsequent studies, figure-8s were also found in intracellular λ DNA (34), among plasmid DNA isolated from *E. coli* (10), and for the yeast 2 μ circle (38). An electron micrograph of a plasmid figure-8 molecule is shown in Figure 3*A*. Inasmuch as these forms appeared to be composed of two interacting monomer circles, they immediately became candidates for intermediates in genetic recombination.

OBSERVATION OF CHI FORMS   The figure-8, although a highly suggestive structure, is inherently ambiguous. This geometry could represent two genomes covalently held together at a region of DNA homology (as in Figure 2C), but two alternative interpretations are also possible. The figure-8 could result from two monomer circles interlocked like links in a chain, or from a double-length circle that accidently overlaps itself in the middle. To remove this ambiguity, the circular DNA molecules were opened with a restriction enzyme prior to electron microscopy (10, 23). An enzyme was chosen that would cut monomeric DNA rings once, at a unique site, to generate unit-size rods. Upon enzyme digestion both interlocked monomer rings and double-length circles should be cleaved into two separable, unit-size rods. On the other hand, if the figure-8 represents two DNA circles covalently connected at a region of homology, then the enzyme would be expected to leave the fusion point intact and to convert the figure-8 into a bilaterally symmetrical, dimeric structure shaped like the Greek letter chi.

Figure 3B is an electron micrograph of a chi-shaped molecule (and may be compared with Figure 1e). It is representative of more than 1500 chi forms found in a study of plasmid DNA recovered from wild-type E. coli (10). After restriction enzyme cleavage, such dimeric forms were observed at a frequency of about 1%, amidst a simple background of unit-length rods. This illustrates the particular power of the electron microscope: it allowed one to carefully observe thousands of individual molecules and to analyze, with a high level of confidence, fused structures that were occurring at a low, but physiological frequency.

A characteristic feature of the chi forms is that they always display a special symmetry: the point of contact between the two unit-size genomes occurs so as to divide the molecule into a structure with two pairs of equal-length arms (10, 23) (see Figure 3B). Because the fusion point (the potential crossover) is doubly equidistant from a defined base sequence (the restriction enzyme cleavage site), it necessarily occurs at a region of DNA homology. The fusion point is observed with equal probability at numerous and perhaps all locations along the DNA molecule, that is, at varying distances from the restriction enzyme cutting site (10).

The finding of chi forms indicated that one could recover from intact cells genomes stably fused together by an interaction that occurs at a region of DNA homology. In addition to the plasmid chi forms discussed above, such molecules have also been observed among the DNA of phage G4 by Thompson, Camien & Warner (23), phage λ by Valenzuela & Inman (34), the yeast 2 μ circle by Bell & Byers (40), and adenovirus by Wolgemuth & Hsu (41).

THE CROSSOVER POINT   The ability to observe chi forms in the electron microscope encouraged enquiry about the fine structure of the crossover point. If the nature of the DNA strand connections could be determined, it would allow a direct comparison of the experimentally observed chi forms with the diagrammatic structures shown in Figure 1. However, when the Holliday form occurs in the planar configuration shown in Figure 1e, it is expected that the two polynucleotide chains forming the crossover will overlie one another, obscuring the nature of the crossover connection. But if the intermediate were to assume the planar configuration shown in Figure 1i, then the region of the crossover would occur in a potentially more open state, with a more analyzable geometry. In this case one would expect to see four double helical segments emerging from a ring of connecting single-stranded DNA (as in Figure 1i). In fact it has been possible to observe chi forms with just this structure (10, 36).

The electron micrographs in Figures 3C and 4 are examples of chi forms in which the polynucleotide strand connections in the crossover region are visible. The double-helical DNA has become partially unwound in the crossover region during spreading for the electron microscope. The single strands connecting the four arms of the recombining molecules are easily distinguished, and one can conclude that the point of contact is indeed a crossover connection involving DNA strands and is not due for instance, to a synaptic protein holding the two DNA molecules together. The observation of this DNA strand substructure in the crossover region allows these molecules to be correlated exactly with the planar representation of the Holliday intermediate shown in Figure 1i.

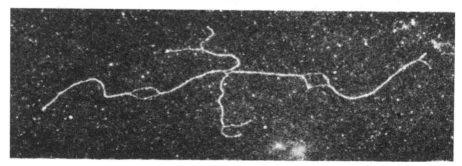

Figure 4   A chi form that has been prepared for electron microscopy in the presence of a high concentration of formamide to cause partial denaturation of the DNA double helix at regions rich in A-T base pairs. The single-stranded connections in the crossover region can be seen. Compare with Figure 1i.

It is predicted that when a Holliday structure has assumed the planar configuration in which the crossover region is open, the equal-length arms will always occur in a *trans* configuration, as in Figure 1*i*. This prediction has been fulfilled in all of the open molecules that have been observed (10) (compare Figure 3*C* and 4 with Figure 1*i*).

EFFECT OF THE RECA LOCUS   In preparations of plasmid DNA recovered from wild-type cells, chi forms were observed with frequencies ranging from 1% to 3%. However, examination of material obtained from recombination-deficient (recA⁻) strains revealed few structures larger than monomers and no chi forms.

To summarize, the DNA molecules recovered from intact cells, shown in Figures 3 and 4, are interpreted as representing the two planar configurations of the Holliday intermediate as diagrammed in Figure 1, *g* and *i*. The fact that such molecules are only recovered from recA⁺ cells serves as the basis for the conclusion that they are involved in physiological recombination (10).

Monomer and Multimer Circles as Products of Genetic Recombination

When viral and plasmid DNA molecules are recovered from wild-type cells they are found to consist of a set of monomer and multimer size DNA rings (40, 41). Not only are chi forms and figure-8s the product of genetic recombination, but so, in fact, are all of these multimeric species. This is known from an experiment in which purified monomer rings are transfected in parallel into recA⁺ and recA⁻ cells. In recA⁻ cells, the monomers can only replicate to form more monomers. But in recA⁺ cells, they are able to regenerate the full spectrum of monomer and multimer size plasmid genomes (36, see also 35, 42). An analogous result holds for the transfection of multimers (dimers, trimers, or tetramers). In recA⁻ cells, the entering plasmids can only replicate to form more molecules of the same size class, but in recA⁺ cells, all size plasmids are produced.

The monomer to multimer interconversion can be understood in terms of the maturation of recombination intermediates formed between two unit-size circles. For instance, as shown in Figure 2 *D–G*, one of the alternative maturation pathways for the Holliday crossover connection in the figure-8 (the east-west cut) regenerates the two parental monomer circles. However, the use of the second maturation pathway (the north-south cut) directly converts the recombination intermediate into a dimer-size circle. In this case, a subsequent *intramolecular* recombination event between two homologous areas within the dimer is required to regenerate monomer plasmids and produce unit-size recombinant DNA molecules [for further discussion see (36)].

These data support the overall reciprocal recombination between DNA circles as a two-stage process leading first, in a bimolecular reaction, to a multimer-size structure, then, second, in an intramolecular recombination event that breaks the multimer apart into two smaller circles in which the original genes have been recombined into a new configuration. (In this process, the dimer DNA circle can be viewed as a long-lived intermediate as compared to the short-lived figure-8.)

The ability to trace the flow of genetic information from the monomer to the multimer state and back again under the guidance of the recA recombination system provides an in vivo assay for recombination that is distinct from the traditional bacterial conjugation assay. In the monomer-multimer interconversion, one is observing recombination between two covalently intact DNA molecules, in contrast to the less demanding strand assimilation reaction that can result in bacterial mating after a segment of single-stranded DNA is transferred from the donor to the recipient. The monomer-multimer interconversion assay therefore allows one to study recombination reactions that should be more closely analogous to those that occur between intact chromosomes. Thus far this assay has been used to confirm the requirement for recA in the overall recombination process and to strongly suggest the need for recF (43) and possibly for topoisomerase I, when the interacting molecules are covalently intact. The monomer-multimer interconversion assay, employing as it does, defined DNA substrates, should also be useful because specific DNA sites which may be promotors (or inhibitors) of recombination, can be placed in the plasmid, and their influence on multimer formation assessed.

Initiation Mechanisms for Forming the Holliday Recombination Intermediate

While figure-8s and chi forms offer physical evidence in support of the existence of the Holliday intermediate, the data do not directly bear on the question of how this structure is formed. No evidence exists to indicate that the precise type of equivalent nicking and strand exchange events shown in Figure 1 *a–d* are correct. Several alternative initiation mechanisms, distinct from the one proposed by Holliday, have therefore been considered, and two representative examples are shown in Figures 5 and 6.

The Proposal of Meselson & Radding

An impetus for considering alternative initiation mechanisms arose from the need to accommodate new genetic data obtained in fungi. Increasingly sophisticated studies in fungi by Hurst, Fogel, Mortimer, Rossignol and

Hastings and their colleagues showed that the regions of heterozygous DNA in a finished pair of recombinant chromosomes are seldom perfectly reciprocal (16, 44, 45; see also 46). In overview, the region of heterozygous DNA in one progeny recombinant chromosome generally encompasses more genetic markers (is longer) than the heterozygous region on the reciprocally recombinant chromosome, which is in contrast to the prediction inherent in the diagram in Figure 1 a–e.

Meselson & Radding (47) therefore proposed a two-step initiation mechanism for the formation of the Holliday structure (Figure 5) (see also 3, 48). They sought to explain the dissimilarity in the heterozygous regions by suggesting that, initially, only one of the two participating double helices is broken open. After nicking, DNA synthesis would occur on this double helix and lead to the displacement of a single-stranded DNA tail. The displaced tail would then invade the second double helix at a region of homology, forming the first heteroduplex region (Figure 5 a–d.) This process would eventually provoke a reciprocal nicking and strand invasion, which would occur at a slightly different place and give rise to a second, shorter heteroduplex region (Figure 5 e–g). After the filling-in of any resultant gaps and the trimming of any remaining tails, this two-stage initiation mechanism would yield a Holliday intermediate with asymmetric heteroduplex regions, at the outset of recombination (Figure 5h).

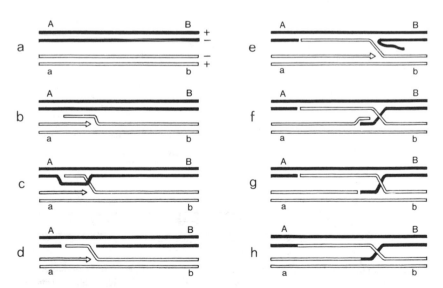

Figure 5 The Meselson-Radding initiation mechanism, which couples DNA synthesis to the formation of the Holliday recombination intermediate.

The Interwrapping Model

A final prototype proposal for genome fusion leading to the formation of a Holliday-type structure is shown in Figure 6. In this case, as distinct from the Meselson-Radding mechanism, DNA pairing precedes strand breakage. Genome fusion occurs when two DNA molecules undergo a localized denaturation at homologous places, and the exposed pairs of positive and negative strands become interwrapped (Figure 6 *a–c* [for discussion see

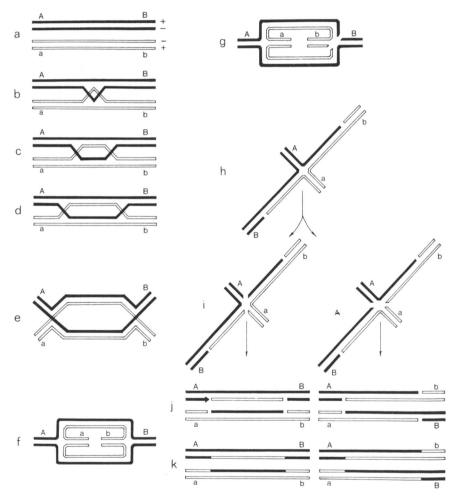

Figure 6   A prototype model for DNA recombination in which the recombination intermediate is formed by the interwrapping of complementary DNA strands after localized denaturation of two double helices at a region of homology. DNA synthesis is not involved in this initiation.

(43); see also (49–51)]. [Conceivably, a four-stranded DNA structure, not initially requiring DNA denaturation, is also possible and could be used to establish homologous pairing (11–13, see also 108).] After the initial contact, strand interwrapping is established and extended for several hundred base pairs, with transient nicks being introduced by a nicking-closing enzyme (51, 52) to allow the rotation necessary for helix formation and further interwrapping. The structure shown in Figure 6d results and is characterized by two crossover connections of the type proposed by Holliday, one at either end of the interwrapped segment (compare Figure 6d with Figure 1e).

Although the interwrapped structure is twofold more complex, it is not fundamentally different from the single crossover structure shown in Figure 1. One imagines that such an intermediate may be matured (Figure 6 g–i) by the same type of strand-nicking events that have been proposed for the Holliday-type recombination intermediate. In this type of model, dissimilarities in the heterozygous regions could be created not at the beginning of recombination but at the maturation stage when the recombining chromosomes are broken apart and there is a potential for nick translation at the borders of the heterozygous regions (Figure 6j arrow).

A convenient way to distinguish the various models is to realize that the two models using free ends of DNA to initiate strand transfer result in a single Holliday crossover junction, while the prototype interwrapping model that employs a topoisomerase initially results in a double Holliday crossover.

The essential point is that several distinct types of initiation mechanism could be used to form a recombination intermediate with the basic crossover connection envisioned by Holliday.

In Vitro Studies of Genetic Recombination

Having some confidence in the structure of the intermediate stages involved in recombination, one is prepared to attempt the transition from the level of molecular biology to the mechanistically more specific level of enzymology. In the end, this will allow recombination to be understood as a series of steps through which known enzymes process two DNA molecules, to achieve an exchange of genetic information. There are two traditional approaches that can provide this type of understanding. The first involves making cell extracts in which it is anticipated that genetic recombination might occur. These extracts can then be fractionated, and the enzymes responsible for each step in the reaction purified and studied. This is the approach that proved successful in elaborating the enzymology of DNA replication (52–55). First, extracts were made that could carry out the entire replication process. The extracts were then fractionated so that individual proteins could be identified and characterized. Although these proteins were able to perform their functions in conjunction with the entire system, they often had no activity by themselves and could not have been studied as isolated proteins.

Cell extracts have, in fact, been shown to be active in carrying out DNA recombination (56–61). For instance, when monomer-sized plasmid DNA molecules are added to cell lysates of recombination-proficient *E. coli,* it is observed that during the incubation, an increasing number of monomer rings are converted to figure-8s and higher multimers (56). A second assay system in *E. coli* involves the observation in cell extracts of the reverse reaction: the dimer → monomer conversion (57).

As was the case for the in vivo figure-8s, it was possible to show that the molecules formed in vitro by the unfractionated cell extract consisted of two genomes held together at a region of DNA homology. That is, the figure-8s could be converted into chi forms by appropriate restriction enzyme cleavage. At least some of the figure-8s were held together by covalent strand connections in the region of the crossover, because the molecules could withstand treatment with alkali. The fact that plasmids were not fused with high efficiency in extracts made from recombination-deficient (recA⁻) cells indicated that the fusion was likely to represent a first step in recombination (56, 57). However, the exact biochemical roles of the enzymes responsible for the recombination could not be deduced so long as they were present with numerous other proteins in the cell extract. For instance, it could not be determined whether the recA protein was playing a direct role in the genome fusion reaction, or whether it functioned indirectly by regulating the synthesis or activity of other cellular proteins. In order to answer such questions, a cell extract has to be fractionated into functionally distinct recombination proteins and this has not yet been achieved.

There is an alternative, though potentially more risky approach to the method of observing recombination in a cell extract and then fractionating the extract that has thus far proved more successful. This alternative is to purify the products of genes known to be involved in or related to recombination from genetic analysis, and study their properties in hopes of finding a partial reaction that an isolated protein can carry out. One then attempts to relate this activity to the overall process of recombination.

Figure 7   A free single strand is assimilated into an intact duplex circle, which leads to the formation of a D-loop. (This figure is explained on page 745.)

Using this approach, four purified proteins have been found to be of particular interest with respect to generalized genetic recombination: the product of the *recA* gene, the product of the *recB* and *recC* genes, single-stranded DNA-binding protein, and, more tentatively, DNA topoisomerase I. Studies of the action of these proteins, either working alone or in conjunction with one or more of the others, have provided support for a coherent enzymological model for recombination. What follows is a description of the analysis of these proteins and an elaboration of the way in which the study of the reactions they are able to catalyze in vitro has led to the evolution of this prototypic enzymatic model.

BIOCHEMICAL ANALYSIS OF THE recA PROTEIN

Mutations in the *recA* gene of *E. coli* have been shown by Clark and his colleagues to reduce genetic recombination more than 1000-fold, as measured in bacterial conjugation and phage transduction studies (2, 62). A similar effect for the *recA* locus is observed when the monomer → dimer interconversion assay is employed (36, 42, 57).

The biochemical identification, and the subsequent purification to homogeneity of the product of the *recA* gene has been instrumental in elucidating its role in the recombination process.

In independent studies, the *recA* gene product was identified in cell extracts of *E. coli* by McEntee (63), by Gudas & Mount (64) by Emmerson & West (65) and by Little & Kleid (66). These researchers demonstrated that mutations in the *recA* gene altered the electrophoretic mobility or the isoelectric point of a particular 40,000-mol wt protein previously known as Protein X. This protein had first been observed some years earlier (67, 68) as a polypeptide whose synthesis was markedly increased by a number of agents that adversely affect DNA metabolism (for example UV light, nalidixic acid, bleomycin, or mitomycin c). The normal level of Protein X (recA protein) is about 2000 molecules per cell (69); the induced level is about 50,000 molecules per cell, which corresponds to about 6% of the total cellular protein (K. McEntee, personal communication). The ability to follow the recA protein physically on gels after column chromatography provided the first effective, albeit difficult, method for purifying the product of the recA gene (71). [The subsequent finding, discussed below, of an ATPase activity associated with the recA protein has greatly aided purification, and now milligram amounts of recA can readily be purified from a few liters of cells (72–77).]

The RecA Protein as a Protease

Astonishingly, the first physiological role discovered for the recA protein was completely unrelated to genetic recombination. Roberts & Roberts knew from the work of Brooks & Clark (78) that the treatment of lysogenic recA$^+$ strains of *E. coli*, with UV light or mitomycin c, would cause λ prophages to enter into lytic growth. They then found that this induction resulted from the inactivation of the λ repressor through its cleavage into two pieces (79). Extracts from recA$^+$ cells contained an activity that was able to accomplish the cleavage of exogenously added λ repressor, provided ATP was added as a cofactor (80). Pursuing this result, Roberts, Roberts & Craig used the λ repressor cleavage reaction to purify the enzyme in the extract that was responsible, and found a continued enrichment for the same 40,000-mol wt protein that had previously been identified as recA (81).

Thus, for some years the only physiological function associated with the recA protein was that of a protease. How could this result be understood? It was interpreted in terms of an earlier suggestion that the product of the *recA* gene might exert its pleiotropic effects by functioning as a regulatory protein (82)—in this case by using its protease activity to activate proenzymes, or inactivate the repressors for various genes including those involved in genetic recombination and DNA repair. Indeed the recA protein does function as a regulatory protease. It cleaves not only the repressors of the phages λ and P22, but also the protein made by the *lex* gene, which is the repressor for the *recA* gene itself and several other genes involved in DNA metabolism (for instance *uvrA, sfiA, himA, umuC*) (83–85; for review see 86, 86a).

Because of its clearly demonstrated activity as a protease, a key concern became whether the recA protein had any direct role in catalyzing the strand rearrangements that must occur during recombination. Kobayashi & Ikeda addressed this question by infecting *E. coli* carrying a temperature-sensitive lesion in the *recA* gene with two types of mutant λ phages (87). In the presence of inhibitors of both RNA synthesis and protein synthesis (to prevent the production of any new recombination enzymes), recombinant viruses were nonetheless produced, as detected by in vitro packaging. However, recombinant phage were only produced at the permissive temperature for the recA protein. Thus, in addition to its regulatory role as a protease, it was clear that the recA protein also participates directly in genetic recombination. It is, however, not understood to this day why a single protein should be able to catalyze reactions as apparently disparate as proteolysis and DNA strand rearrangement, or what physiological or evolutionary significance this implies.

Interaction of the RecA Protein with DNA

The first clue as to the way in which recA participates in recombination came, oddly enough, from the study of the λ repressor cleavage reaction. During this analysis, it was discovered that the more highly purified recA became, the less active it was as a protease. It then transpired that a random DNA or RNA oligonucleotide was essential for the cleavage reaction. In the earlier stages of purification, this essential cofactor had been serendipitously supplied by a small contaminating amount of polynucleotide phosphorylase in the recA preparation, which polymerized a small amount of ADP in the ATP preparation (72). The rationale for the use of an oligonucleotide as an activator for the protease function of recA can be readily understood. Under conditions of abnormal DNA metabolism, the cell accumulates single-stranded regions in its DNA and oligonucleotide fragments (for example through the excision of damaged bases). Phage λ is sensitive to this signal and takes advantage of the host recA protein to inactivate its self-imposed repression system and enter into a cycle of lytic growth. One may even speculate that the λ repressor evolved from a cellular repressor that is subject to recA protease control.

The ability of the recA protein to interact with single-stranded nucleic acid was further explored by the Ogawas and their colleagues. They showed that single-stranded DNA would stimulate recA to display a strong ATPase activity, even in the absence of a protein that was a potential cleavage substrate (71). If recA was able to interact with DNA in a way that led to the hydrolysis of ATP, perhaps this was a clue to its direct role in recombination. Several laboratories therefore embarked on a direct study of the recA protein. A logical starting place was to further explore the interaction of recA with single-stranded DNA.

DNA Reannealing Promoted by the RecA Protein

RecA protein is able to promote the rapid renaturation of complementary single strands from randomly sheared viral DNA, hydrolyzing ATP in the process (73). When recA protein was withheld from the reaction, DNA reannealing was at least 50 times slower. Many recombination models (including those discussed in Figures 1, 5, and 6) include a step in which a single-stranded element is annealed to its complementary strand in a recipient double helix. Therefore this DNA-annealing result might be related to an early step in recombination.

To explore this possibility more systematically Shibata, DasGupta, Cunningham & Radding (74), and McEntee, Weinstock & Lehman (88) took the important step of introducing the use of more defined DNA substrates. They asked whether a free single strand could be induced by the recA protein to anneal to its complementary strand in an intact recipient double helix. Indeed the recA protein was found to catalyze just such a strand uptake reaction, and to form a structure called a D-loop. In the D-loop a triple-stranded area exists where the linear single strand has partially invaded the recipient double helix and hydrogen bonded to its complementary strand, leaving the other, noncomplementary strand, in an unpaired state.

The ability of recA protein to promote the annealing of a single strand to a recipient double helix (the D-loop formation reaction) is shown schematically in Figure 7.

D-loop formation serves well as a model reaction mirroring a potential first step in recombination: the invasion of one double helix by a single strand from a second double helix. In some situations, this may be the essence of recombination, for example the transformation of bacteria by exogenously added DNA (89). Also, in bacterial mating, a single strand is transferred from the male to the female cell and could, in principle, be directly incorporated into the recipient chromosome (90). However, recombination is normally viewed as occurring between two double-stranded DNA molecules; we return to this situation later. For the present, we continue to focus on the model reaction represented by the formation of D-loops, since this allows us to explore the biochemistry of strand assimilation in a simplified system.

Mechanism of Uptake of Single-Stranded DNA into Duplex DNA Mediated by RecA Protein

While it may appear simple, the formation of the D-loop is a complex reaction and raises several questions about the mechanism by which the recA protein works. Among these are the following:

1. How does the recA protein bring together the single-stranded and duplex DNA molecules that are going to undergo genome fusion?
2. How does the recA protein induce the recipient double helix to take up the invading single strand?
3. How does the recA protein guide the search for homology between the two reacting DNA species?
4. How does ATP supply energy for this reaction?

Binding of RecA Protein to DNA

In a reaction in which the recA protein is carrying out genome fusion (for instance, D-loop formation), the amount of recA protein needed is directly proportional only to the amount of single-stranded (not double-stranded) DNA in the system (91–93). In fact, the amount of recA protein required is stoichiometric. About one 40,000-mol wt monomer for every 5 bases of single-stranded DNA is needed to optimize the strand assimilation reaction (91, 92, 94). The problem then is to explain how the recA protein interacts with single-stranded and double-stranded DNA to promote genome fusion.

RecA will bind to both single- and double-stranded DNA, but in different and complex ways. In both cases, the product is a protein-DNA complex that can be assayed by retention on a membrane filter. The recA protein binds most easily to single-stranded DNA (95; see also 75, 91). ATP is not required for this binding reaction, yet when ATP is added, recA rapidly hydrolyzes it while cycling on and off the single-stranded DNA. The complex therefore appears to be dynamic, rather than stable.

In contrast, under physiological conditions, the binding of recA protein to double-stranded DNA requires ATP and is at least 100 times slower than the binding to single-stranded DNA (95). The greater affinity of recA for single-stranded DNA can be understood in physiological terms as reflecting the need to keep the vast excess of intact double-stranded DNA in the cell from titrating recA protein away from the sites where it is to function, i.e. in single-stranded regions involved or in strand assimilation reactions leading to recombination or DNA repair.

Throughout these studies on the interaction of recA and DNA, the non-(or weakly) hydrolyzable analogue of ATP—ATP-γ-S—has been particularly useful in defining the role of the high energy cofactor by trapping the recA-DNA complexes at stable intermediate stages (88, 91, 96–98). Some aspects of the overall interaction of recA and DNA appear to reflect what may be called credit card energetics (99, 100). The binding of the protein to its substrate occurs at no initial cost in terms of energy, but this indebtedness must be redeemed later as the protein hydrolyzes ATP to recover its former conformation (or state of freedom).

While recA protein binds only weakly to double-stranded DNA, this binding is strongly stimulated by the addition of (homologous or non-homologous) single-stranded DNA to the reaction mixture (91). However, competition binding studies indicate that the binding sites on recA for double-stranded and single-stranded DNA are identical or overlapping (95). This would appear to rule out models for genome fusion in which recA protein has two facets, one each for binding single-stranded and double-stranded DNA. A model that would be consistent with these results is that, under physiological conditions, single-stranded DNA is required to activate recA protein, perhaps by serving as a scaffold for the formation of a recA filament. Such a filament could involve a bilayer of recA formed by protein-protein interactions. One side of the bilayer would bind to single-stranded DNA while the other side would be prepared to bind to double-stranded DNA.

In overview, the binding of recA protein to single-stranded DNA would appear to be the first step in initiating a strand uptake reaction.

Unwinding of the Recipient DNA Molecule by RecA Protein

Since the recipient double helix must become partially denatured to accept the invading single strand, an important question is whether the recA protein can carry out this partial denaturation reaction. Cunningham,

Shibata, DasGupta & Radding assayed such unwinding by binding recA protein to nicked duplex rings in the presence of ATP-γ-S and then covalently sealing the DNA-recA protein complexes with ligase (101). If, in binding, the recA protein had caused a localized unwinding of the DNA helix, the ligated product would show a change in linking number (superhelicity), which would be reflected as a change in electrophoretic mobility upon gel analysis. While the recA protein would bind to intact double-stranded DNA, there was no indication that the DNA was unwound. However, when single-stranded DNA (homologous or nonhomologous) was added to this basic reaction mixture, unwinding was observed.

A straightforward interpretation of this and of the previous results of the binding of recA to single-stranded DNA, is that the recA protein has two functions: binding to single-stranded DNA and then partially denaturing the recipient duplex DNA molecule. This interpretation, based on the DNA-binding properties of recA, could be further supported if mutant recA proteins were to be found that were defective in binding to either single-stranded or double-stranded DNA, or in forming protein-protein interactions. Such mutational studies on the *recA* gene (71, 102) and the development of techniques for site-specific mutagenesis (103, 104) should make this possible.

The Search for Homology

If strand assimilation is to occur, the single strand must, somehow, find its complement in the recipient double helix. This search for homology is perhaps the most central problem facing the recA protein, since the hydrogen-bonding surfaces of the recipient double helix are already satisfied and thus unavailable to the searching single strand. One might have thought that the first stable complex that could be formed between recA, single-stranded DNA, and double-stranded DNA would be a finished hydrogen-bonded structure such as a D-loop. However in an experiment in which the strand-assimilation reaction was carried out in the presence of ATP-γ-S rather than ATP, it was found that recA was able to mediate the formation of a filter-bindable protein-DNA complex (91, 92). Because it dissociated in the presence of a protein-denaturing detergent, this complex must lack the region of hydrogen-bonded base pairs characteristic of a D-loop and instead be held together only by the binding of the two DNA species to the recA protein. The essential point is that this finding indicates the existence of a pre-D-loop intermediate. Evidently, the failure to hydrolyze ATP-γ-S prevents the recA protein from functioning beyond the formation of an initial complex. Only if the reaction is carried out in the presence of the hydrolyzable cofactor, ATP, or if single-stranded DNA-binding protein is added (which enhances the efficiency of the reaction with ATP-γ-S) (92), can the initial, nonspecific complex be converted into a (soap-stable) D-loop.

The above experiments suggest a mechanism in which the initial interaction between the searching single strand and the recipient double helix occurs randomly at a region of nonhomology to form a nonspecific triple-stranded complex. The search for homology is then carried out as the recA protein reiteratively forms such complexes and hydrolyzes ATP, or perhaps as the recA protein translates the two DNA molecules with respect to each other. Only when the single- and double-stranded DNA are complexed at homologous places can a stable D-loop be formed.

Enzymological Analysis of the Strand Uptake Reaction

The evolution of our understanding of the recA protein has occurred in several stages. The single-strand assimilation reaction represented by the formation of a D-loop is only the first and simplest type of recombination reaction that can be catalyzed by recA. In the discussion below we turn to the use of increasingly complex DNA substrates to explore more fully the types of genome fusion reactions that this remarkable protein can carry out, which lead ultimately to the formation of a Holliday structure.

One pair of model DNA substrates has proved particularly useful in working out the mechanism of recA action. These are a single-stranded circle and a homologous duplex rod. The single-stranded circle can be derived from a virus such as ϕX or M-13 and the homologous duplex rod can be produced by restriction enzyme cleavage of the double-stranded replicative form of the virus. West, Cassuto & Howard-Flanders, Das Gupta, Shibata, Cunningham & Radding and Cox & Lehman have shown that, upon incubation with these two substrates, the recA protein is able to promote the strand rearrangements leading to the pairing of the single-stranded circle with its complementary strand in the linear duplex rod (105–108) (Figure 8).

This reaction can be understood by analogy with the formation of a D-loop. Recall that in D-loop formation, recA protein can be thought of as binding first to single-stranded DNA, and then partially unwinding the potential target duplex molecule. Following this line of reasoning with the current pair of substrates, the recA protein is viewed as binding to the

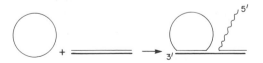

Figure 8   A single-stranded circle and a linear duplex rod pair under the influence of the recA protein; one strand from the duplex rod becomes base-paired with the circle.

single-stranded circle, which thus becomes the donor (or attacking) strand. The resulting recA/single-strand complex then forms a nonspecific triple-stranded structure with the target molecule (the duplex rod), repeatedly unwinds the duplex rod, and initiates the search for homology. If a homologous alignment is created between the single-stranded circle and a region in the interior of the duplex rod, only limited hydrogen bonding can occur (due to topological constraints on strand interwrapping) and the association is unstable. However, when the homologous alignment happens to involve the single-stranded circle and one end of the duplex rod, strand transfer can occur, with the circle becoming intertwined with its complementary strand. If the reaction proceeds to completion, the final products are a nicked duplex ring and a free unit-length single strand.

Using the single-stranded circle and the homologous duplex rod as substrates, three specific biochemical questions can be asked regarding the mechanism by which the recA protein promotes strand uptake.

The Polarity of RecA-Mediated Strand Transfer

A single-stranded circle and a duplex rod can be used to show that the strand transfer catalyzed by recA has a directionality. Since only one strand in the duplex rod is complementary to the single-stranded circle, the target molecule has only one 5' and one 3' end that are potentially capable of hydrogen bonding with the donor circle. If the donor strand (the recA-covered single-stranded circle) enters the recipient duplex with a polarity, tracking in 5' → 3' or 3' → 5' as it becomes associated with its complementary strand in the target duplex, then nonhomologous DNA at one end of the linear duplex or the other should block the reaction. West, Cassuto & Howard-Flanders, Kahn, Cunningham, DasGupta & Radding, and Cox & Lehman have, in fact, shown that the reaction will only occur when a 3' end in the duplex is complementary and thus available to the entering circular single-stranded DNA (109, 110, 110a). Inasmuch as the overall reaction is seen to involve the annealing of the single-stranded circle (which, because it is coated with recA protein, is considered to be the donor molecule) with the 3' end of its complement in the duplex rod (the recipient molecule), it is clear that the assimilating strand tracks in with a 5' to 3' polarity. One can thus picture the strand rearrangements leading to the formation of crossover connections as involving the unidirectional entering of donor DNA strands in a manner that parallels the unidirectional reactions leading to the formation of biological polymers in general.

The Formation of Heterozygous DNA

We have seen that the recA protein can catalyze genome fusion reactions between perfectly complementary DNA molecules. However, in nature, the genetic role of recombination involves the interaction between two slightly different genomes. And, in fact, as discussed at the beginning of this review, a region of genetically heterozygous DNA is a characteristic footprint in a chromosome that has undergone a recombination event (15–21). This is why the formation of heteroduplex DNA has been built into the Holliday model and indeed into most recombination models. In all of these models, the heteroduplex regions play a crucial role in the specific step of forming the crossover connection. Mechanistically they allow strand continuities to be surrendered in a stepwise fashion so that the DNA molecules remain intact throughout the recombination process and no double-stranded breaks need ever be created. Furthermore, even if it were never to lead to a stable crossover connection, the formation of genetically heterozygous DNA in and of itself is important, as each heterozygous region is able, if not repaired, to generate, through subsequent cycles of semiconservative replication, chromosomes with new combinations of genes.

If regions of heteroduplex DNA are to be created during recombination, there must be a mechanism for taking advantage of the complementarity of DNA to form a stable crossed-strand exchange without allowing an occasional mismatched base pair to interfere with the formation of the crossover connection.

DasGupta & Radding (111) have asked whether the recA protein is able to carry out its strand assimilation reactions when this requires the formation of heterozygous DNA. That is, can the strand-transfer reaction proceed through an area of nonhomology to form a joint molecule containing mismatched bases. By using duplex linear and single-stranded circular substrates from closely related but not identical phages, they were able to show that a few single base pair mismatches were not enough to inhibit the formation of heteroduplex DNA by the recA protein. However, three mismatches in a row were sufficient to stop strand transfer (as was gross nonhomology). In general, a high concentration (for instance 10%) of single mismatches in a short distance was able to inhibit the continued reaction. Thus, the recA protein promotes the transfer of a DNA strand unless an area of significant nonhomology is encountered. In such a situation, the fusion points in the resultant joint molecules tend to build up at the border between homologous and nonhomologous DNA, as if they became trapped in the area of nonhomology, or as if the strand transfer process were driven by the recA protein only in one direction, rather than being the result of a random walk.

Extensive Heteroduplex DNA Formation Requires Continued RecA Action

The original blackboard recombination models assumed that once the invading single strand had established an initial region of hydrogen bonding with its recipient target strand, further tracking-in of the donor would occur by the process of branch migration. As originally defined, this is the name given to a sequence of events in which two identical DNA strands, competing for the same complementary partner, move back and forth making and breaking the same number of hydrogen bonds so that there is no net energy change in the system (see Figure 1 *d,e*). As applied to genetic recombination, once the initial strand transfer has been made, continued assimilation (or retreat) of the invading strand would occur as a simple one-dimensional random walk, understandable as passive branch migration. This view must now be modified for reactions in the presence of recombination enzymes. Cox & Lehman have shown that even after the initial strand transfer step, strand assimilation continues to be actively driven by the recA protein and requires the continual hydrolysis of ATP (107). In their experiment the strand assimilation catalyzed by the recA protein is subdivided into two distinct reactions. Using as model DNA substrates a single-stranded circle and a homologous duplex rod, it was possible to demonstrate that the first stage of the reaction results in a short (300–500 base-pair) region of heteroduplex DNA formed between the single-stranded circle and one end of the duplex rod. This initial phase of the reaction can occur rapidly in the presence of either ATP or ATP-γ-S plus single-stranded DNA-binding protein, which indicates that extensive ATP hydrolysis is not required. In the presence of ATP-γ-S, however, further strand assimilation does not occur. In contrast, when ATP is available, the heteroduplex region is slowly extended in a second phase of the reaction so that, eventually, an entire single strand from the duplex rod becomes associated with the single-stranded circle. The addition of ATP-γ-S, or the removal of the ATP enzymatically, immediately abolishes further branch migration.

These results favor a mechanism for active branch migration in which the recA protein binds cooperatively to areas in the crossover region in the recombining molecules to form a type of "protein-DNA filament." More recA protein then adds onto one end of the filament to drive the branch migration.

RecA-Mediated Interaction Between Duplex DNA Molecules

The most recent step in exploring the reactions that are catalyzed by the recA protein has involved the use of double-stranded DNA substrates. This study is, of course, motivated by the fact that recombination generally occurs between two duplex DNA molecules: for instance, the monomer-dimer interconversion for DNA circles in *E. coli* and, in eukaryotes, the recombination that occurs at meiosis.

West and Howard-Flanders and their colleagues (75) and Cunningham and Radding and their colleagues (112) have found that duplex rings containing one or more single-stranded regions or gaps 60–700 nucleotides in length will efficiently interact with supercoiled DNA (or each other), when provided with recA protein and ATP. If there was only one gap per duplex ring, the products generally had the appearance of a figure 8; if there were many gaps, the product consisted of huge complexes containing large numbers of monomers (75). This result suggested that each gap can serve as a focal point for a DNA fusion reaction. In fact, this experiment provides the strongest evidence for the assumption adopted in this review that homologous DNA pairing is initiated when recA protein binds to a single-stranded DNA element.

What precisely happens after the recA protein binds? In the reaction between a gapped duplex ring and a supercoil, it appears that the recA protein promotes the transfer of a free end in the gapped molecule into an intact double helical area in the recipient molecule to form a D-loop type structure. Hydrogen bonds are broken on both the donor and recipient double helix as a single strand is transferred between them.

The above result demonstrates that the recA protein can transfer a single strand to a recipient duplex molecule even when that single strand starts out hydrogen bonded in a double helical state, provided the strand to be transferred is next to a gap.

What then is the role for the gap, for it is necessary—a nicked DNA duplex is not a substrate for strand transfer by the recA protein. Moreover, data from West, Cassuto & Howard-Flanders can be interpreted to show that the single-stranded DNA in the gap must be homologous to the recipient duplex, in order to potentiate the transfer of an adjacent free end (106). One satisfying interpretation is that the search for homology is carried out by the single-stranded DNA in the gap (after the recA protein has bound there). Once the search is completed, a free end adjacent to the gap can be transferred from one double helix to the other.

Two-Strand Reciprocal Exchanges Mediated by RecA Protein

Thus far we have discussed only reactions in which recA promotes the transfer of one DNA strand from a donor molecule to a recipient molecule. However, the formation of a complete crossover connection as visualized in the prototype Holliday model and the variations on this theme require that there be two reciprocal strand invasions (Figures 1, and 5, 6). Recent experiments have begun to explore the capacity of recA to carry out such reciprocal exchanges.

RecA protein is able to reciprocally exchange DNA strands between two double helices. That is, as a strand from a donor double helix invades a recipient double helix, the displaced strand in the recipient molecule can invade the first double helix in much the same way that Holliday and Meselson & Radding envisioned the initial step in recombination. To demonstrate this reaction, West, Cassuto & Howard-Flanders (106) used, as substrates, a single-stranded circle with a small fragment annealed to it and a duplex rod that was homologous to the double-stranded area on the circle but was slightly longer (see Figure 9). When the recA protein and ATP were added to these reactants, the two duplex areas exchanged strands. Apparently, the single-stranded circle first began to interact with the end of the duplex rod. Then, when sufficient transfer had occurred so that the displaced strand of the duplex rod contained unpaired bases that were homologous to the duplex area on the single-stranded circle, a reciprocal strand transfer was initiated involving the end of the annealed fragment. The final result was that the single-stranded circle now had acquired one strand of the input duplex rod while losing its original annealed fragment to the other strand of the duplex rod. A double (reciprocal) strand exchange had taken place (106).

During the period when both strand exchanges were occurring simultaneously, the arrangement of strands at the crossover point acquired the exact conformation of a Holliday structure. Using a longer, linear duplex and a longer duplex region on a single strand of DNA, DasGupta, Wu, Shibata, Kahn, Cunningham & Radding (113) succeeded in trapping the molecules in the process of the double-strand exchange and showed that the strand substructure had exactly the geometry of the chi forms observed in vivo and discussed above. This observation has been confirmed, and an electron micrograph in which the strand substructure in the crossover can be visualized is shown in Figure 10.

An Auxiliary Role for DNA-Binding Protein

Although the recA protein alone can catalyze a variety of DNA strand-exchange reactions, including those that will give rise to a Holliday type crossover connection, the question remains what role other cellular proteins may have in the overall recombination process. Three *E. coli* proteins appear likely to play such a role in conjunction with the recA protein.

The *E. coli* single-stranded DNA-binding protein has an even greater affinity for single-stranded DNA (114) than recA protein. And, in fact, single-stranded DNA-binding protein will effectively compete with recA for this interaction. When DNA-binding protein is added to the various reac-

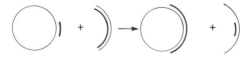

Figure 9 A more complex version of the reaction shown in Figure 8. The recA protein catalyzes a reciprocal strand transfer in which one DNA strand from the duplex rod and one strand from the circle exchange places.

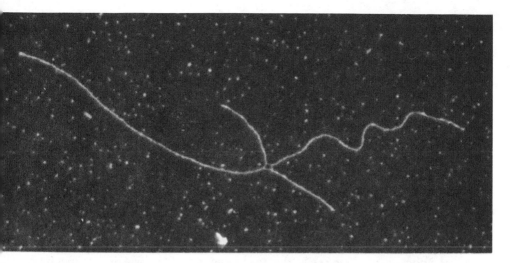

Figure 10 A chi form made *in vitro* by the recA protein acting on two homologous DNA rods with short terminal gaps.

tions with recA, it dramatically improves the efficient use of recA and ATP; less recA protein is required and less ATP is hydrolyzed to generate an equivalent level of strand exchange (92, 115).

A Potential Role for Topoisomerase

For all of the reactions so far discussed—the formation of D-loops, the assimilation of a single-stranded circle into the end of a duplex rod, the fusion of gapped and supercoiled duplex circles, and finally the reciprocal strand exchange leading to a Holliday structure—there is an apparent need for a free end on one of the reacting molecules. This requirement appears to reflect the topological problem of forming a Watson-Crick double helix, rather than an essential feature of the reaction. Once a DNA double helix is unwound by the recA protein, a single strand with a free end can easily assimilate into the duplex, wrapping around its complementary strand to form a stable helix with one turn for every ten base pairs. However, if no free end is available, very few base pairs can be formed before the interacting molecules are topologically constrained from further pairing. Only a limited interaction is possible. The partial interaction of complementary single-stranded circles can be made stable and completed by the action of a DNA relaxation enzyme or topoisomerase, which allows the complementary strands to extend their interaction until a covalently closed double helical DNA ring is generated (51, 52). Models for recombination have been proposed that utilize enzyme-catalyzed interwrapping of DNA strands to fuse duplex DNA molecules into a double Holliday junction (as in Figure 6). Indeed, when Radding and his colleagues add a topoisomerase—the ω protein from *E. coli*—to a basic reaction mixture containing recA, ATP, single-stranded circles, and supercoiled duplex rings, the single-stranded circles become stably interwrapped with (one strand of) the duplex ring at a region of homology (116). The ω protein, by creating a transient break and solving the topological problem of interwrapping, had obviated the need for a permanent free DNA end during the strand transfer reaction.

A Role for the RecBC Enzyme

Above all of the seeming complexity of the reactions catalyzed by the recA protein, one common feature stands out—the requirement that a region of single-stranded DNA exist in one of the participating molecules. During recombination in living cells, this requirement could in principle be met by single-stranded gaps due to DNA repair (117, 118), by single-stranded regions in the replication forks (119, 120), or by DNA synthesis leading to the displacement of a 5'-ended, single-stranded tail as in Figure 5 (3, 47, 48). However, single-stranded DNA might also be provided by the action of a protein known to be involved in recombination from genetic studies, the product of the *E. coli recB* and *recC* genes.

The recBC enzyme is the multisubunit protein product of the *recB* and *recC* genes of *E. coli* (121). Mutations in these genes were found by Emmerson & Howard-Flanders (122) to reduce recombination to about 1% of the wild-type level, a substantial reduction, though not nearly as great as that caused by recA mutations [(which may imply that other proteins can at least partially substitute for recBC (see 2)]. The recBC enzyme was first identified as a potent exo- and endonuclease driven by the hydrolysis of ATP (123–125). Recently recBC has been shown to have another activity on DNA that may be more relevant to its role in recombination. When recBC enzyme is incubated with double-stranded DNA and ATP in the presence of single-stranded DNA binding protein, its nuclease activities are suppressed (125). Under these conditions, the recBC enzyme primarily functions to unwind the duplex molecule to expose single-stranded regions (126, 127). Telander-Muskavitch & Linn (128) observed such single-stranded regions coated with the single-stranded DNA-binding protein either at the end of the duplex, giving the molecule a Y shaped appearance, or in the middle of the duplex, or both. When the denatured regions occur in the middle of the molecule, the duplex regions to either side are apparently held together in one continuous rod with the single-stranded loops taking the appearance of two "rabbit ears" in the electron microscope (128,

129). A study of the kinetics of formation of these molecules by Taylor & Smith has led to the following model for the action of recBC under physiological conditions (129): the "head" of the enzyme binds to one strand at the end of a duplex molecule and begins to unwind the double helix at a rate of about 300 nucleotides/second. At the same time, the "tail" of the enzyme releases its hold on the single strand at a rate of about 200 nucleotides/second. The difference in the two rates—100 nucleotides/second —results in a growing loop of single-stranded DNA ahead of the separated strands as the enzyme travels down the duplex molecule. The single-stranded DNA in the loop formed by the recBC enzyme and in the two released tails are coated with DNA binding protein. Then, if the complementary parts of the two tails reanneal, the rabbit eared structure results.

The exposed single-stranded regions that are a consequence of the movement of the recBC enzyme down the DNA molecule offer themselves as potential donor strands in recombination. Coating of these DNA elements by recA protein could be the first stage in activating the single-stranded DNA to invade a double helix (see also 130). Guided by the recA protein, the coated loops of DNA would search for a region of homology in a recipient duplex. When homologous alignment is achieved, a DNA topoisomerase would be expected to be able to interwrap the DNA strands to form a stable recombination intermediate as in Figure 6.

CONCLUDING REMARKS

Considering all of the above data, it is possible to deduce a coherent model for recombination that attempts to specify each step of the process in biochemical terms. Although the model must be viewed as tentative, it does integrate much of the experimental data gathered over the last few years in studies of *E. coli* proteins known or likely to be involved in recombination. These proteins are the products of recA, recBC, the single-stranded DNA binding protein, and a topoisomerase (nicking-closing enzyme).

The biochemical model begins with the enzyme product of the *E. coli* genes *recB* and *recC*. this multisubunit enzyme would initiate genetic recombination by travelling internally down a DNA double helix and creating a region of local denaturation in which several hundred base pairs of positive and negative strand DNA are held apart so that their hydrogen bonding surfaces are exposed. In this reaction the recBC enzyme works with the single-stranded DNA-binding protein (or possibly recA itself), whose role is to stabilize the separated strands. This is the DNA molecule that will initiate recombination by serving as the donor. One of its exposed single-stranded regions will "attack" a second DNA molecule.

It is at this stage that the recA protein functions. It binds to the exposed single-stranded DNA in the donor molecule, and the resulting recA-DNA complex then begins to explore other, intact double-helical molecules, searching for a region of homology. For this purpose the recA protein is assumed to have two functions: one is to bind to the single-stranded DNA in the donor molecule, while the other is to partially denature the recipient duplex. The localized denaturation of the target molecule occurs repeatedly, accompanied by the hydrolysis of ATP, until a region of homology is found and the single-stranded area of the donor is able to hydrogen bond to its complementary strand in the recipient.

The initial contact between the donor and recipient is likely to involve only a limited amount of hydrogen bonding. Unless there is a free end in either the donor or recipient DNA molecule, the two strands will not be able to wind around each other and form a long stretch of Watson-Crick double helix. This problem, however, can be solved by a nicking-closing enzyme, which recognizes the underwound character of the DNA in the region where strand contact has been made and introduces a transient nick that allows the two strands to fully interwrap. If all four strands remain covalently intact, a second strand interwrapping between the two as yet uninvolved strands in the donor and recipient molecules must occur, again facilitated by a nicking-closing enzyme. In this case, the final structure formed would initially look like the interwrapped recombination intermediate shown in Figure 6d. There are two Holliday-type crossover connections separated by regions of heterozygous DNA. However, if any of the interacting strands should become nicked during the recA pairing reaction (perhaps by the nuclease activities of the recBC enzyme), the strand rotations necessary to form a crossover connection can occur more directly, and a structure with a single Holliday connection results. That is, the events that start as in Figure 6 switch to those of Figures 1 or 5.

The one remaining problem is to break the two recombining DNA molecules apart. No bacterial recombination gene has yet been identified that is specifically associated with the maturation of recombination intermediates. But recent studies show the existence of such trans-nicking functions (as in Figure 1j) in the site-specific recombination systems coded for by some transposons (131, 132) and by phage λ (P. Shu and A. Landy, personal communication). Further, the product of the phage T4 gene 49 is an endonuclease that can recognize a Holliday crossover junction in vitro and perform the trans nicking leading to the maturation of this recombination intermediate [G. Mosig; R. Weisberg, B. Kemper, K. Mizuuchi and J. Hays, personal communications; (133)]. DNA ligase then seals the nicks to complete the maturation process and regenerate intact duplex molecules. Presumably, generalized genetic recombination will utilize a similar enzyme.

Literature Cited

1. Holliday, R. 1964. *Genet. Res.* 5:282–303
1a. Whitehouse, H. L. K. 1963. *Nature* 199:1034–40
2. Clark, A. J. 1973. *Ann. Rev. Genet.* 7:67–85
3. Hotchkiss, R. D. 1974. *Ann. Rev. Microbiol.* 28:445–68
4. Nash, H. A. 1977. *Curr. Topics Microbiol. Immunol.* 78:171
5. Fox, M. S. 1978. *Ann. Rev. Genet.* 12:47–68
6. Stahl, F. W. 1979. *Ann. Rev. Genet.* 13:7–24
7. Radding, C. M. 1978. *Ann. Rev. Biochem.* 47:847–80
8. Holliday, R. 1968. In *Replication and Recombination of Genetic Material,* ed. W. Peacock, R. Brock, p. 157–74, Canberra: Australian Acad. Sci.
9. Holliday, R. 1974. *Genetics* 78:273–85
10. Potter, H., Dressler, D. 1976. *Proc. Natl. Acad. Sci. USA* 73:3000–4
11. McGavin, S. 1971. *J. Mol. Biol.* 55:293–98
12. Wilson, J. H. 1979. *Proc. Natl. Acad. Sci. USA* 76:3671–45
13. Kikuchi, Y., Nash, H. A. 1979. *Proc. Natl. Acad. Sci. USA* 76:3760–64
14. Sigal, N., Alberts, B. 1972. *J. Mol. Biol.* 71:789–93
15. Kitani, Y., Olive, L. S., El-ani, A. S. 1962. *Am. J. Bot.* 49:697–706
16. Hurst, D. D., Fogel, S., Mortimer, R. K. 1972. *Proc. Natl. Acad. Sci. USA* 69:101–5
17. Russo, V. E. A. 1973. *Mol. Gen. Genet.* 122:353
18. White, R. L., Fox, M. S. 1974. *Proc. Natl. Acad. Sci. USA* 71:1544–48
19. Enea, V., Zinder, N. D. 1976. *J. Mol. Biol.* 101:25–38
20. Fogel, S., Hurst, D. D. 1967. *Genetics* 57:455–81
21. Lam, S. T., Stahl, M. M., McMilin, K. D., Stahl, F. W. 1974. *Genetics* 77:425–33
22. Meselson, M. S. 1972. *J. Mol. Biol.* 71:795–98
23. Thompson, B. J., Camien, M. N., Warner, R. C. 1976. *Proc. Natl. Acad. Sci. USA* 73:2299–2303
24. Warner, R. C., Fishel, R. A., Wheeler, F. C. 1979. *Cold Spring Harbor Symp. Quant. Biol.* 43:957–68
25. Emerson, S. 1969. In *Genetic Organization: A Comprehensive Treatise,* ed. E. Caspari, A. Ravin, 1:267. New York: Academic
26. Pascher, A. 1918. *Ber. Dtsch. Bot. Ges.* 36:136–68
27. Sarthy, P. V., Meselson, M. S. 1976. *Proc. Natl. Acad. Sci. USA* 73:4613–17
28. Boon, T., Zinder, N. D. 1971. *J. Mol. Biol.* 58:133–51
29. Broker, T. R., Lehman, I. R. 1971. *J. Mol. Biol.* 60:131–49
30. Doniger, J., Warner, R. C., Tessman, I. 1973. *Nature New Biol.* 242:9–12
31. Thompson, B. J., Escarmis, C., Parker, B., Slater, W. C., Doniger, J., Tessman, I., Warner, R. C. 1975. *J. Mol. Biol.* 91:409–19
32. Benbow, R. M., Zuccarelli, A. J., Shafer, A., Sinsheimer, R. L. 1974. In *Mechanisms in Recombination,* ed. R. Grell, p. 3. New York: Plenum
33. Benbow, R. M. Zuccarelli, A. J., Sinsheimer, R. L. 1975. *Proc. Natl. Acad. Sci. USA* 72:235–39
34. Valenzuela, M. S., Inman, R. B. 1975. *Proc. Natl. Acad. Sci. USA* 72:3024–28
35. Hobom, G., Hogness, D. S. 1974. *J. Mol. Biol.* 88:65–87
36. Potter, H., Dressler, D. 1977. *Proc. Natl. Acad. Sci. USA* 74:4168–72
37. Warner, R. C., Tessman, I. 1978. In *The Single-Stranded DNA Phages,* ed. D. T. Denhardt, D. Dressler, p. 417. Cold Spring Harbor, New York: Cold Spring Harbor Lab.
38. Bell, L., Byers, B. 1979. *Proc. Natl. Acad. Sci. USA* 76:3445–49
39. Wolgemuth, D. J., Hsu, M. T. 1980. *Nature* 287:168–71
40. Rush, M. G., Warner, R. C. 1968. *Cold Spring Harbor Symp. Quant. Biol.* 33:459–66
41. Roth, T. F., Helinski, D. R. 1967. *Proc. Natl. Acad. Sci. USA* 58:650–57
42. Bedbrook, J. R., Ausubel, F. M. 1976. *Cell* 9:707–16
43. Potter, H., Dressler, D. 1979. *Cold Spring Harbor Symp. Quant. Biol.* 43:969–85
44. Fogel, S., Mortimer, R. K., Lusnak, K., Tavares, F. 1979. *Cold Spring Harbor Symp. Quant. Biol.* 43:1325–41
45. Rossignol, J. L., Paquette, N., Nicolas, A. 1979. *Cold Spring Harbor Symp. Quant. Biol.* 43:1343–52
46. Hastings, P. J. 1975. *Ann. Rev. Genet.* 9:129–44
47. Meselson, M. S., Radding, C. M. 1975. *Proc. Natl. Acad. Sci. USA* 72:358–61
48. Catcheside, P. J., Angel, T. 1974. *Aust. J. Biol. Sci.* 27:219–29
49. Cross, R. A., Lieb, M. 1967. *Genetics* 57:549–60
50. Sobell, H. M. 1975. *Proc. Natl. Acad. Sci. USA* 72:279–83

51. Champoux, J. J. 1977. *Proc. Natl. Acad. Sci. USA* 74:5328–32
52. Kirkegaard, K., Wang, J. C. 1978. *Nucleic Acids Res.* 5:3811–20
53. Tomizawa, J. 1978. In *DNA Synthesis: Present and Future*, ed. I. Molineux, M. Kohiyama, p. 797. New York: Plenum
54. Schekman, R., Weiner, A., Kornberg, A. 1974. *Science* 231:170–73
55. Wickner, S., Hurwitz, J. 1974. *Proc. Natl. Acad. Sci. USA* 71:4120–24
56. Potter, H., Dressler, D. 1978. *Proc. Natl. Acad. Sci. USA* 75:3698–3702
57. Kolodner, R. 1980. *Proc. Natl. Acad. Sci. USA* 77:4847–51
58. Roeder, G. S., Sadowski, P. D. 1979. *Cold Spring Harbor Symp. Quant. Biol.* 43:1023–32
59. Ogawa, H., Araki, H., Tsujimoto, Y. 1979. *Cold Spring Harbor Symp. Quant. Biol.* 43:1033–41
60. Attardi, D. G., Mattoccia, E., Tocchini-Valentini, G. P. 1977. *Nature* 270:754–56
61. Benbow, R. M., Krauss, M. R. 1977. *Cell* 12:191–204
62. Clark, A. J., Margulies, A. D. 1965. *Proc. Natl. Acad. Sci. USA* 53:451–59
63. McEntee, K. 1977. *Proc. Natl. Acad. Sci. USA* 74:5275–79
64. Gudas, L. J., Mount, D. W. 1977. *Proc. Natl. Acad. Sci. USA* 74:5280–84
65. Emerson, P. T., West, S. C. 1977. *Mol. Gen. Genet.* 155:77–85
66. Little, J. W., Kleid, D. G. 1977. *J. Biol. Chem.* 25:6251–52
67. Gudas, L. J., Pardee, A. B. 1975. *Proc. Natl. Acad. Sci. USA* 72:2330–34
68. Inouye, M., Pardee, A. B. 1970. *J. Biol. Chem.* 245:5813–19
69. Paoletti, C. as cited in Cox, M. M., Lehman, I. R. 1981. *Proc. Natl. Acad. Sci. USA* 78:3433–37
70. Deleted in proof
71. Ogawa, T., Wabiko, H., Tsurimoto, T., Horii, T., Masukata, H., Ogawa, H. 1978. *Cold Spring Harbor Symp. Quant. Biol.* 43:909–15
72. Roberts, J. W., Roberts, C. W., Craig, N. L., Phizicky, E. M. 1978. *Cold Spring Harbor Symp. Quant. Biol.* 43:917–29
73. Weinstock, G. M., McEntee, K., Lehman, I. R. 1979. *Proc. Natl. Acad. Sci. USA* 76:126–30
74. Shibata, T., DasGupta, C., Cunningham, R. P., Radding, C. M. 1979. *Proc. Natl. Acad. Sci. USA* 76:1638–42
75. West, S. C., Cassuto, E., Mursalim, J., Howard-Flanders, P. 1980. *Proc. Natl. Acad. Sci. USA* 77:2569–73
76. Cox, M. M., McEntee, K., Lehman, I. R. 1981. *J. Biol. Chem.* 256:4676–78
77. Shibata, T., Cunningham, R. P., Radding, C. M. 1981. *J. Biol. Chem.* 256:7557–64
78. Brooks, K., Clark, A. J. 1967. *J. Virol.* 1:283–93
79. Roberts, J. W., Roberts, C. W. 1975. *Proc. Natl. Acad. Sci. USA* 72:147–51
80. Roberts, J. W., Roberts, C. W., Mount, D. W. 1977. *Proc. Natl. Acad. Sci. USA* 74:2283–87
81. Roberts, J. W., Roberts, C. W., Craig, N. L. 1978. *Proc. Natl. Acad. Sci. USA* 75:4714–18
82. Witkin, E. 1976. *Bacteriol. Rev.* 40:869–907
83. Little, J. W., Edmiston, S. H., Pacelli, L. Z., Mount, D. W. 1980. *Proc. Natl. Acad. Sci. USA* 77:3225–29
84. Kenyon, C. J., Walker, G. C. 1980. *Proc. Natl. Acad. Sci. USA* 77:2819–23
85. Fogliano, M., Schendel, P. 1981. *Nature* 289:196–98
86. Gottesman, S. 1981. *Cell* 23:1–2
86a. Howard-Flanders, P. 1981. *Sci. Am.* 245:72–80
87. Kobayashi, I., Ikeda, H. 1978. *Molec. Gen. Genet.* 166:25–29
88. McEntee, K., Weinstock, G. M., Lehman, I. R. 1979. *Proc. Natl. Acad. Sci. USA* 76:2615–19
89. Fox, M. S. 1966. In *Macromolecular Metabolism*, p. 163–96. Boston: Little, Brown
90. Vapnek, D., Rupp, W. D. 1971. *J. Mol. Biol.* 60:413–24
91. Shibata, T., Cunningham, R. P., DasGupta, C., Radding, C. M. 1979. *Proc. Natl. Acad. Sci. USA* 76:5100–4
92. McEntee, K., Weinstock, G. M., Lehman, I. R. 1980. *Proc. Natl. Acad. Sci. USA* 77:857–61
93. Shibata, T., DasGupta, C., Cunningham, R. P., Williams, J. G. K., Osber, L., Radding, C. M. 1981. *J. Biol. Chem.* 256:7565–72
94. Cassuto, E., West, S. C., Mursalim, J., Conlon, S., Howard-Flanders, P. 1980. *Proc. Natl. Acad. Sci. USA* 77:3962–66
95. McEntee, K., Weinstock, G. M., Lehman, I. R. 1981. *J. Biol. Chem.* 256:8835–44
96. Weinstock, G. M., McEntee, K., Lehman, I. R. 1981. *J. Biol. Chem.* 256:8829–34
97. Weinstock, G. M., McEntee, K., Lehman, I. R. 1981. *J. Biol. Chem.* 256:8845–49
98. Weinstock, G. M., McEntee, K., Lehman, I. R. 1981. *J. Biol. Chem.* 256:8850–55

99. Sugino, A., Higgins, N. P., Brown, P. O., Peebles, C. L., Cozzarelli, N. R. 1978. *Proc. Natl. Acad. Sci. USA* 75:4838–42
100. Hill, T. L. 1969. *Proc. Natl. Acad. Sci. USA* 64:267–74
101. Cunningham, R. P., Shibata, T., DasGupta, C., Radding, C. M. 1979. *Nature* 281:191–95
102. McEntee, K., Epstein, W. 1977. *Virology* 77:306–18
103. Shortle, D., Nathans, D. 1978. *Proc. Natl. Acad. Sci. USA* 75:2170–74
104. Weber, H., Taniguchi, T., Müller, W., Meyer, F., Weissmann, C. 1979. *Cold Spring Harbor Symp. Quant. Biol.* 43:669–77
105. DasGupta, C., Shibata, T., Cunningham, R. P., Radding, C. M. 1980. *Cell* 22:437–46
106. West, S. C., Cassuto, E., Howard-Flanders, P. 1981. *Proc. Natl. Acad. Sci. USA* 78:2100–4
107. Cox, M. M., Lehman, I. R. 1981. *Proc. Natl. Acad. Sci. USA* 78:3433–37
108. West, S. C., Cassuto, E., Howard-Flanders, P. 1981. *Nature* 290:29–33
109. West, S. C., Cassuto, E., Howard-Flanders, P. 1981. *Proc. Natl. Acad. Sci. USA* 78:6149–53
110. Kahn, R., Cunningham, R. P., DasGupta, C., Radding, C. M. 1981. *Proc. Natl. Acad. Sci. USA* 78:4786–90
110a. Cox, M. M., Lehman, I. R. 1981. *Proc. Natl. Acad. Sci. USA* 78:6018–22
111. DasGupta, C., Radding, C. M. 1981. *Proc. Natl. Acad. Sci. USA* In press
112. Cunningham, R. P., DasGupta, C., Shibata, T., Radding, C. M. 1980. *Cell* 20:223–35
113. DasGupta, C., Wu, A. M., Kahn, R., Cunningham, R. P., Radding, C. M. 1981. *Cell* 25:507–16
114. Sigal, N., Delius, H., Kornberg, T., Gefter, M. L., Alberts, B. 1972. *Proc. Natl. Acad. Sci. USA* 69:3537–41
115. Shibata, T., DasGupta, C., Cunningham, R. P., Radding, C. M. 1980. *Proc. Natl. Acad. Sci. USA* 77:2606–10
116. Cunningham, R. P., Wu, A., Shibata, T., DasGupta, C., Radding, C. M. 1981. *Cell* 24:213–23
117. West, S. C., Cassuto, E., Howard-Flanders, P. 1981. *Nature* 294:659–62
118. Rupp, W. D., Wilde, C., Reno, D., Howard-Flanders, P. 1971. *J. Mol. Biol.* 61:25–44
119. Inman, R. B., Schnös, M. 1971. *J. Mol. Biol.* 56:319–25
120. Wolfson, J., Dressler, D. 1972. *Proc. Natl. Acad. Sci. USA* 69:2682–86
121. Tomizawa, J. I., Ogawa, H. 1972. *Nature New Biol.* 239:14–16
122. Emmerson, P. T., Howard-Flanders, P. 1967. *J. Bacteriol.* 93:1729–31
123. Barbour, S. D., Clark, A. J. 1970. *Proc. Natl. Acad. Sci. USA* 65:955–61
124. Oishi, M. 1969. *Proc. Natl. Acad. Sci. USA* 64:1292–99
125. Goldmark, P. J., Linn, S. 1970. *Proc. Natl. Acad. Sci. USA* 67:434–41
126. MacKay, V., Linn, S. 1976. *J. Biol. Chem.* 251:3716–19
127. Rosamond, J., Telander, K. M., Linn, S. 1979. *J. Biol. Chem.* 254:8646–52
128. Telander-Muskavitch, K. M., Linn, S. 1980. *J. Supramol. Struct.* 4:375 (Suppl.)
129. Taylor, A., Smith, G. R. 1981. *Cell* 22:447–57
130. Williams, J. G. K., Shibata, T., Radding, C. M. 1981. *J. Biol. Chem.* 256:7573–82
131. Reed, R. R. 1981. *Cell* 25:713–19
132. Reed, R. R., Grindley, N. D. F. 1981. *Cell* 25:721–28
133. Mizuuchi, K., Kemper, B., Hays, J., Weisberg, R. A. 1982. *Cell.* In press

Ann. Rev. Biochem. 1982. 51:763–93

SYNTHESIS OF THE YEAST
CELL WALL AND ITS REGULATION[1]

Enrico Cabib and Rowena Roberts[2]

National Institute of Arthritis, Diabetes, Digestive and Kidney Diseases,
Bethesda, Maryland 20205

Blair Bowers[2]

National Institute of Heart, Blood and Lung, Bethesda, Maryland 20205

CONTENTS

[1]The US Government has the right to retain a nonexclusive royalty-free licence in and to any copyright covering this paper.

[2]We are indebted to Drs. N. Elango, M. P. Fernandez, W. B. Jakoby, and V. Notario for critical reading of the manuscript.

PERSPECTIVES AND SUMMARY

The cells of plants, bacteria, and fungi, in contrast to those of animals, are encapsulated in a sturdy wall. This structure has attracted interest mainly for two reasons. From the scientific standpoint, since the wall determines the shape of the cell, it is a convenient model for the study of morphogenesis. With regard to practical applications, the cell wall often is the main distinguishing feature between the cells of a parasite, such as a bacterium or fungus, and those of its host. Thus, it can be used as a specific target for antibacterial and antifungal agents.

Budding yeasts are well suited for the study of cell wall formation. They are single-cell organisms, as easy to grow and to maintain as bacteria. The mode of growth, however, is quite different from that of bacteria. During the budded stage of the cell cycle, only the bud along with its wall is growing whereas the mother cell is essentially quiescent. This makes it easy to distinguish the new wall from the old one, as well as to correlate the growth of the wall with the position in the cell cycle by monitoring the size of the bud relative to the mother cell. Because the wall encloses the cell, the mechanisms that control cell size must be intimately related with, or identical to those that control cell wall synthesis. Thus, one may learn about the control of cell size by studying cell wall regulation.

Morphologically, the cell wall of budding yeasts presents a fairly uniform structure with the exception of the septal region (the bud scar region after cell separation), where one can distinguish primary and secondary septa. Synthesis of a precursor structure to the primary septum begins at bud emergence, as a ring around the constriction between mother and daughter cell. Later, the septum grows centripetally to close the communication between the two cells. Secondary septa are added subsequently.

Chemically, the composition of yeast cell walls is relatively simple. The wall consists, almost in its entirety, of homopolysaccharides of glucose, mannose, and N-acetylglucosamine, only one of which, the mannose polymer appears to be linked to a relatively small protein moiety. Of these polysaccharides, a β-glucan, containing mainly $1 \rightarrow 3$ and $1 \rightarrow 6$ linkages, is the main structural component. Another polysaccharide, chitin, constitutes the primary septum, thereby providing a specific marker for septum formation.

Since enzymes that catalyze at least partial synthesis of the three main types of cell wall polysaccharides from sugar nucleotides have been found, it is now possible to study the formation of the cell wall and its control at the molecular level. In this way, it was discovered that chitin synthetase, an enzyme attached to the plasma membrane, exists in a latent form that can be transformed into active enzyme by proteolysis. It is assumed that this or a similar reaction is involved in the triggering of primary septum formation. $\beta(1 \rightarrow 3)$Glucan synthetase, the enzyme responsible for the formation of the most important structural polysaccharide of the cell wall, is also associated with the plasma membrane and is strongly activated by certain nucleotides, especially GTP, at micromolar concentrations. Probably, these activators function in the regulation of cell wall growth. Finally, mannan, a protein-linked polysaccharide that presumably functions as a cement in the wall structure, is formed through a complex pathway that includes lipid-linked intermediates and is quite similar to that of mammalian glycoprotein synthesis. The isolation of mutants with defects in mannan structure has aided considerably in the elucidation of the biosynthetic pathway. These studies highlight another important feature of yeast, i.e. its suitability for genetic manipulation, which greatly enhances its usefulness.

Antibiotics have been found that inhibit the synthesis of yeast cell wall polysaccharides. These include polyoxins for chitin, papulacandin and aculeacin for glucan, and tunicamycin for mannan. Although these inhibitors are not suitable for therapeutic applications, their existence clearly supports the rationale that compounds able to interfere with cell wall formation may be useful in the control of infections caused by pathogenic yeasts and fungi.

This chapter is concerned with the cell wall of budding yeasts, with an emphasis on the research carried out in recent years. For reviews of earlier work or of related topics, see (1–6).

YEAST CELL WALL

Chemistry

Yeast cell walls are constructed almost entirely of two classes of polysaccharides: (a) polymers of mannose covalently linked to peptides, or mannoproteins, and (b)polymers of glucose, or glucans. The glucans and mannoproteins occur in roughly equal amounts in the wall. A third sugar polymer of N-acetylglucosamine, chitin, is characteristically a major component of fungal walls, but is present only in minor amounts (about 1%) in the yeasts.

Mannoproteins are easily extracted from intact cells by autoclaving in citrate buffer at neutral pH or by dilute alkali at lower temperatures. The water soluble mannoproteins obtained by such procedures are a heterogeneous group, exhibiting a range of molecular weights, sugar content, and molecular charge. They include at least two functionally distinct types of molecules: exoenzymes, e.g. invertase and acid phosphatase, and structural components of the wall. The chemistry of yeast mannoproteins has been reviewed by Ballou (7). The molecules are complex (Figure 1). In *Saccharomyces* mannan, a core $\beta(1 \rightarrow 6)$ mannose chain of 15–17 units is linked to an asparagine of the peptide through a di-N-acetylchitobiose unit. A much longer outer chain, on the order of 100 to 200 mannosyl residues, has branching side chains, as does the inner core. Phosphodiester bonds link some of the side chains. Shorter mannose chains are attached directly to serine and threonine residues of the peptides (8, 9). The structure shown in Figure 1 is derived from studies on *Saccharomyces cerevisiae*. The mannoproteins of other genera of yeasts have not been defined in comparable detail and appear to be more complex. They differ in the nature of the linkages, the branching of the side chains and in the presence of nonmannose sugars in the side chains (7). The peptides of the mannoproteins have not been studied.

In addition to the linkages mentioned, others may be present in the mannoprotein, because mannan is depolymerized by mild alkaline treat-

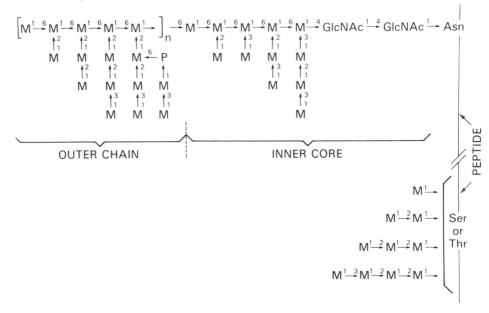

Figure 1 Structure of mannoprotein, according to Ballou (7). All linkages are α, except those between the two N-acetylglucosamine residues and between the mannose and N-acetylglucosamine, which are of the β type. In the outer chain, the lateral branches are representative of those found in the polysaccharide, but their order and frequency may vary. *M* stands for mannose.

ment (10) and the disintegration of the wall of intact cells by lytic enzymes is greatly facilitated by pretreatment with sulfhydryl compounds (11). Perhaps a thioester linkage would accomodate both results, because it would be very sensitive to alkali and may be susceptible to substitution by sulfhydryl groups.

Different classes of yeast glucans can be distinguished by their solubility characteristics. The extractability of glucans varies, however, depending on the use of intact cells or isolated cell walls as starting material. In the former case, treatment with alkali solubilizes very little glucan (12). Subsequent heating with acetic acid extracts a glucan that contains predominantly $\beta(1 \rightarrow 6)$ with some $\beta(1 \rightarrow 3)$ linkages and appears to be highly branched (12, 13). The residue, now soluble in dilute alkali (12, 14), is predominantly $\beta(1 \rightarrow 3)$-linked glucan with a small number of $\beta(1 \rightarrow 6)$ linkages, probably at branch points (14). When cell walls, rather than intact cells, are extracted with alkali at room temperature, up to 20% of the cell wall can be solubilized (15, 16). The liberated material appears to be a glucan of the predominantly $\beta(1 \rightarrow 3)$ type (16). The implications of these results for the cell wall structure are discussed in the next section.

In contrast to the complex glucans and mannans, the third polysaccharide wall component, chitin, is a linear polymer of $\beta(1 \rightarrow 4)$-linked N-acetylglucosamine units of well-characterized structure (17, 18). Chitin is polymorphic and exists in three crystalline forms designated α, β and γ. Chitin in yeast walls appears to be α-chitin (19–21). In α-chitin the chains are antiparallel in alignment and oriented so that hydrogen bonding is maximal (18). As a consequence, chitin is highly resistant to chemical extraction. Chitin-containing structures are normally left as residues after complete removal of the glucans and mannoproteins by extensive chemical extractions (22, 23).

Structure

The microarchitecture of the yeast wall is not well defined, partly because the bulk of the components appear to be structurally amorphous and partly because the polymers seem to interweave to a considerable degree. Thin-section electron microscopy shows differentially staining "layers" in the wall, especially in certain fixatives (24, 25). These differences in affinity for electron stains cannot be reliably related to differences in chemical composition among the layers, but suggest areas enriched with different components. Similar layering is not visible in freeze-fracture replicas of yeast cell walls (26). The rarity of other than cross-fractures is consistent with a highly cohesive and cross-linked structure for the whole depth of the wall. The surface of the cross-fracture is disappointingly featureless. It appears granular, with a greater density of "granules" appearing in the inner half of the wall, which again indicates differences in component disposition in the wall.

The presence of a fine net of microfibrils (7.5–10 nm thick) has been documented in the inner layer of wall fragments by negative staining after digestion of amorphous glucan by highly purified $\beta(1 \to 6)$ and $\beta(1 \to 3)$ glucanases from *Bacillus circulans* (27). This microfibrillar material in normal walls is thought to be a $\beta(1 \to 3)$ linked glucan by analogy with fibrillar nets, of larger dimensions, that are formed by protoplasts allowed to regenerate walls. The regenerated nets were shown by X-ray diffraction and chemical analysis to be crystalline $\beta(1 \to 3)$ glucan (28–30). The fibrillar nets produced by protoplasts are a more highly aggregated, unbranched form of the glucan than is found in the normal wall, and are soluble in dilute alkali. An alkali insoluble residue from protoplast nets resembles microfibrils found in normal walls (29). The microfibrils revealed by treatment of normal walls with *B. circulans* glucanases are removed and the wall structure is totally disintegrated after digestion with an endo $\beta(1 \to 3)$ glucanase isolated from *Schizosaccharomyces versatilis,* which suggests that the fibrillar network is the major structural scaffolding of the wall (27, 31). This notion is supported by the finding that even untreated wall can be completely disrupted by treatment with certain $\beta(1 \to 3)$ glucanases (32, 33).

Since mannan-specific antibodies agglutinate intact cells (7) and concanavalin A binds to the cell surface (34, 35), mannoproteins clearly are exposed at the external surface of the wall. There is no evidence that such glycoproteins form any crystalline or fibrillar structures, although electron micrographs often show a brush-like appearance of the outermost edge of the wall. Attempts have been made to examine the distribution of mannan within the wall by several different techniques of thin section cytochemistry (36–38). False positives can be controlled for, but false negatives may be due to loss of reactive sites during the preparative procedure or to masking by other wall components. Given this caveat, the cytochemistry confirms that mannan is present at the surface and extends some distance into the wall. Mannan-specific staining is also seen next to the plasmalemma at the inner surface of the wall (37, 38). The latter staining may be due to mannan-containing exoenzymes that are presumably concentrated in the periplasmic space (39, 40). Interestingly, no effect on cell wall structure could be detected in three classes of mutants that produce mannoproteins lacking the mannose side chains (41). A fourth class of mutants that retain the core structure of the mannan but lack most of the outer chain do show defects in wall morphology as well as growth abnormalities. The specific biochemical lesion in this group of mutants has not been identified (42).

More than 90% of the chitin in normal yeast wall is found in the region of the bud scars (Figure 2). The chitin is distributed in the form of an annulus with an external thinner rim and a thin central plate (22, 23, 43). The central plate forms the primary septum between the mother cell and the completed bud. The remainder of the chitin is dispersed over the whole cell wall (38, 43). In *S. cerevisiae,* the chitin in the bud scar appears as small lozenge-shaped particles in negatively stained preparations (44). The location of chitin within the intact wall can be inferred from thin-section micrographs, because chitin is resistant to all electron stains and appears electron

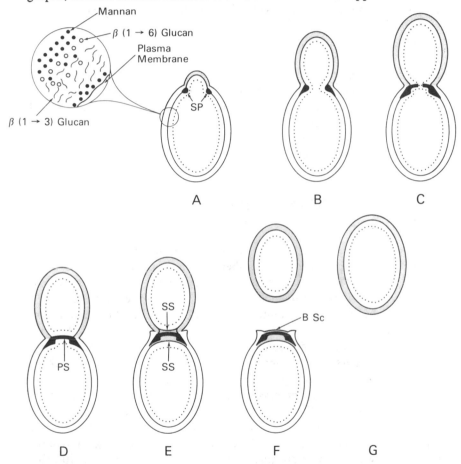

Figure 2 Scheme of the cell cycle of budding yeasts. The insert *at left* depicts a tentative structure of the cell wall, at higher magnification. As explained in the test, the β(1,6)glucan contains some β(1 → 3) linkages and the reciprocal is true for the β(1 → 3)glucan. The possibility of linkages between mannan molecules has also been mentioned. In A to G, the old wall has been represented as empty and the new one is stippled; chitin is filled in black and the plasma membrane is shown as a dotted line. SP, septal primordia; PS, primary septum; SS, secondary septa; BSc, bud scar.

lucent against a variably staining matrix (45). Recently, specific staining of chitin has been achieved in thin sections by using wheat germ agglutinin adsorbed to gold marker particles (38); this lectin specifically binds to N-acetylglucosamine. The specific label confirmed the localization deduced by chemical extraction procedures.

The bud scar region seems to be the only differentiated region of the yeast cell wall. Except for the concentration of chitin in this region, other wall components appear to be chemically identical to those found in other parts of the wall (43, 46).

Figure 2 diagrams the tentative wall structure of *S. cerevisiae*. In this scheme, the predominantly $\beta(1 \rightarrow 6)$glucan has been drawn near the cell surface, where it may serve as a barrier against extraction of the $\beta(1 \rightarrow 3)$glucan. The location is suggested by the need for removal of the $\beta(1 \rightarrow 6)$polysaccharide from intact cells prior to solubilization of $\beta(1 \rightarrow 3)$glucan. As mentioned in the preceding section, extraction of $\beta(1 \rightarrow 3)$glucan is much easier when starting with broken cell walls; presumably material from the barrier-free inner layer can gain access to the exterior through cracks in the wall.

CELL WALL AND SEPTUM FORMATION DURING THE YEAST CELL CYCLE

In the vegetative cycle of budding yeasts, the development of a new cell is signaled by the appearance of a protuberance on the surface of a preexisting cell (Figure 2A). The protuberance, a bud, is about 1 μm wide at the base and its cell wall is apparently continuous with that of the mother cell (47). At the constriction between mother cell and bud, a ring of electron-lucent material develops at the inner surface of the wall (47; Figure 2A). This material has been identified as chitin on the basis of its fluorescence with certain brighteners (48, 49) and the absence of fluorescence when chitin synthesis is inhibited with polyoxin D (49). Chitin rings also have been isolated from cells in the process of producing their first bud (50). Subsequent to its emergence, the bud and its wall grow to a size somewhat smaller than that of the mother cell (see next section), while nuclear division and separation take place (Figure 2B). The plasma membrane invaginates at the junction between the two dividing cells, and concomitantly the chitin ring grows centripetally (Figure 2C) until a complete primary septum is formed (Figure 2D). Secondary septa, which appear to contain mannan and glucan like the remainder of the wall (43, 46), are laid down on both sides of the primary septum (Figure 2E). The secondary septum made by the mother cell is morphologically different and slightly thicker than that made by the daughter cell. Finally, the cells separate in such a fashion that the chitinous primary septum remains with the mother cell; the resulting raised crater-like structure forms the bud scar (2; Figure 2F). The mother cell is now ready to bud again, but the new cell must attain full size before budding (Figure 2G).

CELLULAR ASPECTS OF CELL WALL BIOSYNTHESIS

The biosynthesis of the cell wall is governed by the preexisting cell wall structure and the environmental condition in addition to the genetic information. Interaction among these factors, which results in cell wall synthesis at the appropriate time and cellular location, is then reflected in the morphogenesis of the yeast cell. Thus, observation of the growth process should provide clues to general mechanisms of cell wall biosynthesis and its regulation. Any model for cell wall biosynthesis should be consistent with cellular morphogenesis.

An analysis of the growth of budding yeast cells elicits a number of questions pertaining to the biosynthesis of the cell wall and its regulation. One might ask first how and when the site for bud formation is chosen and what the mechanism is for ensuring that one bud is formed during one division cycle. Second, during bud enlargement, what controls exist to ensure that synthesis occurs in the appropriate location and with integration of the components so as to produce a cell with the proper shape? Third, how is septum formation temporally and spatially coordinated with the biosynthesis of other cell wall components and with the final events in the mitotic cell cycle? Finally, what processes regulate the growth of a virgin (previously unbudded) yeast cell during the period between septum completion and bud initiation? Investigation of in vivo biosynthesis of yeast cell walls by direct observation of the growth process, by the use of inhibitors and mutants, and by examination of the relationship of events in the cell cycle to each other have provided some insight into these questions.

Initiation of Cell Wall Synthesis at Bud Emergence

Probably one of the most crucial points in the yeast cell cycle for the regulation of cell wall synthesis is the initiation of a new bud (Figure 2A). Several lines of evidence suggest that the beginning of the cell division cycle and subsequent bud initiation are integrated with growth and cell volume. Virgin yeast cells of S. cerevisiae normally begin this process only when the cell has reached a size equal to that of the mother cell (51–53). This critical cell size, however, varies according to the growth conditions (54–57). Furthermore, small sized cells produced in poor growth medium increase in size when shifted to a richer medium and only produce a bud when they have attained the larger size (52, 54, 55). Also, the size (volume) of a cell at initiation of budding may be altered by mutation. Such small-cell mutants of *S. cerevisiae* have been isolated; they grow at the same rate as wild type but initiate cell division at a smaller size than the wild-type strain (58). Thus, cell size at the initiation of the cell cycle is controlled both by growth rate and the genetic information.

The time at which bud formation begins, relative to other events in the cell cycle, may shift depending on the growth rate (59). These results indicate that bud formation and initiation of cell wall synthesis need not be correlated with the initiation of DNA synthesis. However, the process of bud initiation does correlate with duplication of the spindle plaque (centriole) and separation of the duplicated plaques. Budded cells have not been observed in which the spindle plaque had not duplicated; a temperature-sensitive mutant in which the spindle plaque is not duplicated at the restrictive temperature does not bud at this temperature (60). Furthermore, when separation of the duplicated spindle plaque is inhibited by a mutation, mutant cells may form more than one bud per cell, which implies that plaque separation exerts a negative control over bud initiation (61).

Controls must exist, not only for the time at which a bud is formed, but also for the site of bud initiation. In general, it appears that bud formation in *S. cerevisiae* occurs in an ordered fashion at the poles of the cell, the area of maximum cell curvature. In haploid strains the first bud initiated by a virgin cell is located at the pole of the birth scar, whereas the first bud formed by a diploid cell is usually located at the pole of the cell distal to the birth scar (62). The mechanism for bud localization can be disrupted. In cells growing at slow growth rates (57, 63) and also in cells of cell division cycle mutants affected in *cdc* 24 (64), budding may occur at abnormal sites, which suggests that the bud site selection may be controlled by a single gene.

For some time it was thought that one of the processes required for bud initiation was the formation of a chitin ring. In cells with a very tiny bud, a ring of chitin, as detected by fluorescent staining, was observed; in some cases, the ring appeared to be formed before a bud was observed (48). However, it now seems that the chitin ring is not obligatory for bud initiation and growth, because cells in which chitin synthesis is inhibited by polyoxin D are capable of bud emergence, although not of septum formation (49). Until recently, the determination of bud formation was also thought to require and to be directed by microtubules. Extranuclear microtubules have been observed radiating into the newly formed bud; they are proximal to vesicles (60), which could be involved in cell wall lysis and synthesis. An obligate role of microtubules in the regulation of budding and cell wall synthesis is now questionable because bud formation occurs in cells in which no microtubules were detected after treatment with the microtubule inhibitor, methyl benzimidazol-2-yl-carbamate (65).

Cell Wall Synthesis during Bud Enlargement

REGULATION OF SYNTHESIS   Autoradiographic analysis of cells in asynchronous populations has demonstrated that, during bud growth, glucan and mannan are probably synthesized continuously (66). Similar observations have also been reported when synthesis was studied biochemically in synchronous cultures (67, 68). The continuous synthesis of mannan and glucan during bud growth could be the result of either the constant activity of stable synthetic enzymes or the continuous formation of the active but labile enzymes. Results of studies of in vivo cell wall formation during inhibition of protein synthesis suggest that the enzymes involved in both mannan and glucan synthesis are quite stable in the cell (69, 70). Additionally, the RNAs that code for the protein portion of mannan seem to decay only slowly (70), which indicates that the protein necessary for mannan may also be readily available in the cell.

Although the enzymes necessary for the construction of the major components of the bud cell wall are available during bud formation, it has become clear that their activity is responsive to changes in the environment. Addition of compounds such as diphenylamine (71) and 2-deoxy-D-glucose (2-DG) (72, 73) to the culture medium may result in inhibition of the synthesis of one cell wall component but may stimulate the synthesis of another. Changes in cell wall composition and structure have been noted under conditions that change the lipid composition of the membrane (74–76). Nutritional conditions appear to affect the ratio of glucan to mannan, the phosphorus content, and the percentage of protein in the cell wall, whereas changes in growth rate effect little variation in cell wall composition with regard to glucan, mannan, or protein (77). Alterations in cell wall composition resulting from different nutritional conditions appear to be reflected in variation in cell shape and cell wall ultrastructure (77). Furthermore, cells are responsive to changes in the concentration of small molecules. Walker & Duffus (78) have shown that magnesium deficiency results in buds of abnormal shape and often accelerates the cells into division. Perhaps the most appropriate mechanism for control of such responses is at the level of effectors for modulation of cell wall synthetases, which is discussed in a later section.

LOCALIZATION OF SYNTHESIS   Localization of cell wall synthesis during the growth of a bud has been demonstrated with fluorescent stains and autoradiography (79–82). In general, the cell wall of the growing bud is not derived from material in the mother cell wall but rather is synthesized de novo (79, 81, 82). Mannan incorporation into the cell wall, as detected by autoradiographic analysis, occurs uniformly over the surface of the very young bud, but becomes localized to the bud apex during subsequent growth of the bud (82). Alkali insoluble cell wall material, probably glucan, is also deposited at the tip of the bud distal to the mother cell (80). In addition, the deposition of acid phosphatase, an extracellular mannoprotein enzyme, is limited to the bud when cells are derepressed for synthesis of the enzyme (83). Therefore, it appears that construction of a bud results from synthesis of new wall material and incorporation of this material mostly in the portion of the bud distal to the mother cell. Nevertheless, the exact means by which cell wall synthesis is localized to generate a bud with an ellipsoidal shape is unclear (1, 4).

INSERTION OF THE NEW CELL WALL   The new wall is incorporated into the preexisting wall architecture, i.e. the mode of synthesis of the individual components must provide integration with the older material. Two mechanisms for cell wall synthesis have been proposed. One states that cell wall polymers, which at the growing points may not be completely solidified by crosslinks, slide apart, providing an area for insertion of new material. The second explanation is that new building blocks are inserted at growth sites in which bonds in recently formed cell wall are cleaved (lysed) (84). Some insight into the insertion mechanism developed during studies of the inhibition of cell wall synthesis with 2-DG. 2-DG inhibits either mannan or glucan synthesis, depending upon the carbon source provided for growth (85). This was also demonstrated with protoplasts in which 2-DG inhibited both fibrillar glucan formation and mannoprotein complex formation (86). When growing cells are exposed to 2-DG or 2-deoxy-2-fluoro-D-glucose, lysis of the cell and dissolution of the walls at the growing points were observed (84, 87, 88). Cell lysis has also been observed when cells are treated with aculeacin A, another inhibitor of cell wall synthesis (89). In contrast to the lysis of cells observed in the presence of inhibitors of cell wall synthesis, exposure of cells to the glucanase inhibitor δ-gluconolactone results in the arrest of yeast cell growth induced by auxin (90). Thus, in the absence of continued cell wall growth but with continuation of cellular growth, lysis of the cell wall occurs, which demonstrates the presence of lytic enzymes capable of cell wall dissolution. In view of these results lysis of wall material accompanied by synthesis and insertion of new material seems the more likely mechanism for growth.

FUNCTION OF cdc GENES   The mechanism by which cell wall growth during the period of bud formation is limited to cell wall synthesis in the bud appears to be related to the function of the genes defined by temperature-sensitive cell division cycle (cdc) mutants (91). Mutants blocked in spindle plaque duplication, DNA synthesis, bud initiation, or nuclear division generally continue cellular enlargement during arrest of the cell cycle

(52). Growth at the restrictive temperature does not appear to be limited to the bud. Instead, both mother cell and bud enlarge, which suggests that the interruption of the cell cycle results in disruption of the mechanism that localizes cell wall synthesis in the bud. This aspect of localization of cell wall synthesis has been more extensively studied with mutants defective in the gene *cdc* 24. Incubation of unbudded cells of the mutants at the restrictive temperature results in the arrest of bud formation and continued growth of unbudded cells, which indicates that cellular enlargement can be separated from bud formation (64, 92). In these mutant cells it appears that cell wall material as well as extracellular mannoproteins, e.g. acid phosphatase, are deposited randomly over the cell surface (64, 83, 92). Thus, in addition to its involvement with the selection of the bud site, the *cdc* 24 gene product appears to function in the localization of cell wall synthesis. Furthermore, because the *cdc* 24 product appears to affect cell shape (64) it may regulate in some way the balance between different synthetic enzymes and/or between synthesis and lysis of cell wall material.

Septum Formation

RELATIONSHIP TO OTHER CELL CYCLE EVENTS   The final event in the yeast cell cycle is the formation of a septum. The dependence of septum formation upon completion of other cell cycle events has largely been elucidated through observations with the *cdc* mutants of *S. cerevisiae*. Mutants arrested in DNA synthesis and nuclear division are unable to construct a septum, which suggests that these events are necessary (91). Such an arrangement, the dependence of septation upon DNA synthesis and nuclear division, ensures that a "cell" is not delimited without a nucleus. Septum formation also apparently requires the construction of a bud. The phenotype of mutant *cdc* 24 strains at the restrictive temperature supports this assertion. Unbudded cells of such mutants, when shifted to high temperature, enlarge and execute both DNA synthesis and nuclear division, but never construct a septum between the two daughter nuclei (91, 93). Thus, bud formation, in addition to DNA synthesis and nuclear division, is necessary for septum formation.

The *cdc* mutants in which incubation at the restrictive temperature allows bud formation but blocks cytokinesis are also unable to construct a normal septum (94). This dependence is logical in view of observations that suggest that cytokinesis and deposition of the primary septum are coincident. Correct execution of the final step in the septation process, deposition of the secondary septum, requires previous construction of the primary septum. When construction is inhibited, material for the secondary septum appears to be synthesized but is laid down abnormally (45), therefore the primary septum probably functions as a template for secondary septum synthesis. The final two processes in the completion of cell division, cytokinesis and septum formation, seem to be set in motion quite early in the cell cycle. The execution point (the time at which the gene product functions) for the *cdc* mutants unable to undergo cytokinesis may occur as early as the first 8% of the cycle (94). Determination of the time at which the synthesis of chitin, the constituent of the primary septum, is initiated suggests that chitin synthesis begins close to the initiation of bud formation, and continues through the cell cycle to septation (49, 95). In summary, the initial events involved in septum formation probably occur concomitant with bud emergence, whereas the subsequent construction of the septum depends on bud formation, the completion of DNA synthesis and nuclear division, and cytokinesis.

CHITIN SYNTHESIS   Chitin synthesis itself can be separated from the construction of the primary septum, i.e. even though the primary septum is not constructed, chitin synthesis may nevertheless occur. Two such situations have been described (64, 92, 96). Arrest of haploid cells at the initiation of the cell cycle by exposure to the mating pheromone results in the deposition of three times more chitin in these cells than in untreated cells (96). Such synthesis, which occurs at the elongating tip of the cell, may be accompanied by increased amounts of chitin synthetase zymogen and active chitin synthetase (96). Additionally, in cell-cycle arrested cells defective in the *cdc* 24 gene, chitin (three times as much as in normal cells) is deposited randomly over the surface of unbudded cells (64, 92). Under these abnormal conditions, chitin synthesized at sites other than the septum may reflect the displacement of the normal guidance system that activates the zymogen form of chitin synthetase and results in the formation of the primary septum.

Cell Wall Synthesis after Cytokinesis

Because division in *S. cerevisiae* is asymmetrical under many growth conditions, cell wall synthesis must continue during the G1 interval of the cell cycle, i.e. the time between completion of mitosis and initiation of DNA synthesis, so that the newly formed daughter attains a size equal to that of the mother cell (Figure 2*F, G*). Under growth conditions in which division is nearly symmetrical (fast growth rates), increase in cell size during G1 probably is only 10% of the total growth. As the rate of growth decreases, however, the size of the daughter cell, which at birth may be only about 50% of the size of the mother cells, increases significantly during the unbudded period (56). The response of the cell to growth conditions is effected in G1. This notion is supported by studies of cells shifted from poor to rich growth medium in which deficiency in size is compensated for during G1 (52, 54, 55). Cell wall synthesis during G1, sometimes described as maturation of the bud, appears to be different from that observed prior to cell division. In fact, the decrease in the rate of synthesis of glucan and mannan coincident with septation and prior to bud initiation may reflect this shift in cell wall growth (97). Newly made cell wall material synthesized by unbudded cells appears to be inserted randomly over the cell surface (80, 82). This delocalization of cell wall deposition may involve both glucan and mannan (80, 82), and possibly chitin (43), synthesis. It is possible that during this time the system that directs localized wall deposition is nonfunctional, which would allow random wall synthesis.

Remarks

The results discussed in this section suggest points at which cell wall growth and septation are controlled and their interrelations with other events in the cell cycle. They do not supply information about the nature of the regulatory mechanisms. In the next section the enzymatic steps in the biosynthesis of cell wall components are considered together with their possible contribution to the understanding of cell wall regulation.

ENZYMATIC SYNTHESIS OF CELL WALL POLYSACCHARIDES

Chitin

PROPERTIES AND REGULATION OF CHITIN SYNTHETASE   Particulate preparations of chitin synthetase were initially obtained from *Neurospora crassa* (98), and later from *Saccharomyces* (95, 99) and many other fungi (100–120). The substrate is UDP-N-acetylglucosamine and the stoichiometry of the reaction, as measured with the yeast enzyme, is:

$$2n \text{ UDP-N-}\alpha\text{-D-GlcNAc} \rightarrow 2n \text{ UDP} + (\text{GlcNAc } \beta(1 \rightarrow 4)\text{GlcNAc})_n.$$

A common property of preparations from many different sources is the stimulation by divalent cations, usually Mg^{2+} and Mn^{2+} (an absolute requirement for the yeast enzyme), and by free N-acetylglucosamine (98, 99, 102, 105, 107–109, 111–113, 116). On the other hand, polyoxins, antibiotics produced by *Streptomyces cacaoi* (121), act as powerful competitive inhibitors of chitin synthetase (99, 103, 106, 122), with K_i values in the micromolar range.

A key feature of chitin synthetase, first detected in the yeast enzyme (95), is its occurrence in a latent state. With adequate protection against proteolytic action, fresh preparations are essentially inactive. Enzymatic activity can be elicited by incubation with any of a number of proteases (95, 123). A similar behavior is displayed by many fungal synthetases (113–118, 124–128), with the possible exception of an enzyme from *Coprinus* (129).

The crypticity of particulate chitin synthetase is not due to enclosure within a vesicle. After solubilization with digitonin, the enzyme continues to require protease treatment for activity (130). It has been postulated, therefore, that most of chitin synthetase, in the native state, exists in a zymogen form (95, 130). This is an empirical definition. It may well be that proteolytic activation occurs by destruction of an inhibitor tightly bound to the enzyme. The point cannot be settled until highly purified preparations of the enzyme are available.

Solubilization of chitin synthetase allowed some further probing into its properties. A modest purification of the solubilized and proteolytically activated enzyme was achieved by chromatography on Sepharose 6B. The enzyme eluted as a large protein, possibly an aggregate, of molecular weight between 5×10^5 and 1×10^6 (130). The activity of purified preparations is affected by lipids with acidic phospholipids acting as stimulators; the most effective is phosphatidylserine. In contrast, free unsaturated fatty acids, but not their saturated counterparts, strongly inhibit the enzyme (130). Sterols, which are major constituents of yeast membranes, may also interact with chitin synthetase. Thus, mutants of *Candida albicans* with a low ergosterol content exhibit increased activity of the synthetase (131), and polyene antibiotics, which bind to sterols, were reported to affect in various ways the activity of the *Mucor* enzyme (132). It is doubtful that these effects are directly related to the normal regulation of the enzyme in vivo. They may only reflect that the synthetase, a membrane protein, normally exists in a lipid environment and is sensitive to its variations.

Although the enzyme activity is affected by lipids, there is no evidence that a lipid intermediate, of the type involved in glycoprotein synthesis (133), participates in the formation of chitin. Addition of dolichylphosphate to the solubilized enzyme has no effect on the reaction (130), and tunicamycin, a potent inhibitor of the transfer of N-acetylglucosamine-1-phosphate from UDP-N-acetylglucosamine to dolichylphosphate, depresses the rate only slightly (130). A competitive inhibitory effect of tunicamycin on the chitin synthetase of *Neurospora* was reported (134), but here too the effect is small, which suggests that it may be due to weak competition with substrate by virtue of the uridine-like moiety of tunicamycin (135).

The solubilized synthetase from yeast is only slightly stimulated by free N-acetylglucosamine. Incubation of the enzyme in the presence of only substrate and Mg^{2+} leads to the formation of chitin, which precipitates form solution (130). There is, therefore, no evidence that the reaction requires a primer. Examination of the reaction product by electron microscopy shows chitin in the form of rod-like particles about 60 nm long and 9 nm wide, very similar in appearance to those seen in isolated septa by negative staining (44, 130).

SUBCELLULAR DISTRIBUTION OF CHITIN SYNTHETASE   Chitin synthetase is a particulate enzyme that can be obtained either by lysis of protoplasts or by mechanical disruption of intact cells, followed by centrifugation. To ascertain the intracellular localization of the synthetase, Duran et al (136), adapting a procedure previously used by Scarborough (137) for *Neurospora,* coated yeast protoplasts with Concanavalin A. The lectin appears to cross-link the plasmalemma and reinforce it, so that osmotic lysis of the protoplasts sets free the membrane in a single piece or in a few large fragments. Subsequent centrifugation of the lysate in a Renografin gradient led to isolation of membranes in a discrete band. Most of the chitin synthetase in the lysate was associated with this band. When purified membranes were treated with trypsin to activate the synthetase, and then incubated with UDP-N-acetylglucosamine, the synthesized chitin remained attached to the membrane, as detected either with a fluorescent brightener or by autoradiography (138). This result suggests that the enzyme catalyzing the formation of the polysaccharide is also located on the plasma membranes. Fractionation of lysates from *Candida albicans* protoplasts led to a distribution of chitin synthetase identical to that found for *Saccharomyces* (118). On the other hand, Schekman & Brawley (96), who omitted the concanavalin A step, reported only 60% of the enzyme in the plasma membrane fraction of *S. cerevisiae,* as identified with protoplast surface markers, and the remainder in a lighter fraction.

In the experiments described above, extracts were prepared by osmotic lysis of yeast protoplasts. Quite different results were reported after fractionation of extracts obtained by disrupting intact cells with glass beads. With the latter procedure, Bartnicki-Garcia, Bracker, and their co-workers were able to isolate a fraction of small (40–70 nm in diameter) granules, rich in chitin synthetase, which they termed chitosomes (125, 139). Although chitosomes were initially obtained from *Mucor rouxii,* they were subsequently detected in extracts from *Saccharomyces* and several other fungi (117). Because of the drastic conditions used in their preparation, it is difficult to decide whether chitosomes exist as such in the cell. If they do exist, they may be carriers of chitin synthetase between the endoplasmic reticulum and the plasma membrane (see 5).

Although some chitin synthetase may be attached to different organelles, the evidence discussed above indicates that the bulk of the enzyme in yeast is associated with the plasma membrane. An important question is whether the enzyme spans the membrane or is specifically located on one of its two sides. Since the substrate of chitin synthetase, UDP-N-acetylglucosamine, is formed inside the cell and the plasma membrane is impermeable to phosphoric esters, one would expect the nucleotide to interact with the enzyme on the cytoplasmic face of the membrane. This sidedness of chitin synthetase was confirmed by the finding that a brief treatment of the external face of protoplasts with glutaraldehyde did not affect the enzyme (136). As soon as the inner face was exposed by lysis, however, glutaraldehyde irreversibly inactivated the synthetase. Since the product of the synthetase reaction, chitin, is found in vivo outside the plasma membrane, the synthesizing system must be able, not only to tranfer N-acetylglucosamine from the sugar nucleotide to a growing chitin chain, but also to extrude the polysaccharide through the membrane. It is not clear whether the synthetase alone can achieve this result or whether other membrane components are required.

Glucan

Low levels of glucose incorporation from UDP-glucose into an insoluble polysaccharide were detected (140) with toluene-permeabilized yeast cells. Soon afterwards, reports appeared on the formation of glucan in cell-free extracts (141–143). In these studies, UDP-glucose was found to give rise to a $\beta(1 \rightarrow 3)$-linked product, but Balint et al (141) claimed that use of GDP-glucose as substrate resulted in a $\beta(1 \rightarrow 6)$-linked glucan. In all these cases the incorporations into product only amounted to a few percent of the added substrate. Much greater activity was reported (144) with a particulate preparation from S. cerevisiae similar to those previously used for chitin synthetase. The reason for the increased incorporation appears to reside in the factors present in the reaction mixture. Thus, the activity is enhanced by bovine serum albumin, by glycerol and, especially, by a nucleoside triphosphate. Under optimal conditions, up to 40% of the substrate can be converted into product in 20 min at 30°C.

As ascertained by chemical and enzymatic methods, the product is a linear $\beta(1 \rightarrow 3)$glucan, with an average chain length of 60–80 glucose units. Therefore it is much smaller than the glucan found in the cell wall and lacks $\beta(1 \rightarrow 6)$ branches. Reduction of the terminal glucose with sodium borohydride followed by acid hydrolysis led to the isolation of sorbitol labeled with ^{14}C as was the UDP-glucose used as substrate (144). It appears, therefore, that the enzyme catalyzes formation of glucan chains de novo, rather than addition of glucose to an endogenous primer.

The stoichiometry of the reaction can be written as:

$$2n \text{ UPD-}\alpha\text{-D-Glc} \rightarrow 2n \text{ UDP} + [\text{Glc } \beta(1 \rightarrow 3)\text{Glc}]_n$$

Papulacandin B and aculeacin A, which inhibit glucan synthesis in vivo (145, 146), depress somewhat synthetase activity, but at much higher concentrations than those effective on intact cells or protoplasts (H. Kawai and E. Cabib, unpublished observations).

The stimulation by nucleotides, which may be involved in the regulation of the synthetase in vivo, was studied in detail (147). The most effective stimulator is GTP, which increases the enzyme activity severalfold, with an $S_{0.5}$ (concentration for half-maximal stimulation) of 0.2 μM. ITP is active at a similar concentration, whereas the $S_{0.5}$ for ATP is about 100-fold higher. Another difference between ATP and GTP is that the effect of the former is inhibited by EDTA, whereas that of GTP is stimulated by the complexing agent, probably due to protection against contaminating phosphatases. Because of the difference in conditions required for optimal stimulation, it was not possible to ascertain whether the effect of the two nucleotides was additive.

Stimulation of glucan synthetase is not restricted to the nucleoside triphosphates. GDP is also active at a somewhat higher concentration, and even inorganic pyrophosphate and higher polyphosphates stimulate, although with a much larger $S_{0.5}$ (147a). On the other hand, esterification of the terminal phosphate of a nucleoside triphosphate with a methyl group or with a nucleoside abolishes the activation. In fact, these esters competitively inhibit the effect of stimulators (147a). These results suggest that the enzyme, or a regulatory subunit thereof, has an activation site that interacts with the pyrophosphate group, and a binding site for the nucleotide moiety of the activator. This concept, together with the structural requirements for an activator, is depicted in Figure 3.

In the course of these studies, it was noted that analogues of ATP and GTP, with imino or methylene groups substituting for the α,β or the β,γ oxygen, or with a sulfur atom attached to the terminal phosphate, were able to stimulate the enzyme with an $S_{0.5}$ usually similar to that of the parent nucleotides. Since the effect of these analogues was, in most cases, unaffected by EDTA, it was possible to use adenosine and guanosine derivatives in the same reaction mixture. In this way, it was shown that the effects of adenosine-β,γ-imino-triphosphate and of guanosine-β,γ-imino-triphosphate are not additive, which suggests that adenosine and guanosine nucleotides interact with the same domain on the enzyme. This notion was confirmed by the finding that the inhibitory nucleotide esters mentioned above compete with the imino analogues of both adenosine and guanosine nucleotides, yielding the same K_i values for both type of compounds.

With nucleotide derivatives containing imino or methylene groups, the phosphate attached to such groups cannot participate in an enzymatic transfer reaction. Since both α,β and β,γ imino and methylene analogues

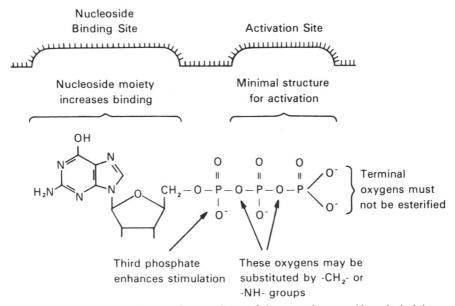

Figure 3 Structural requirements for an activator of glucan synthetase and hypothetical sites of interaction with the enzyme. GTP is represented as a model compound.

are active, stimulation of the enzyme occurs without transfer of either or both of the two terminal phosphates of nucleoside triphosphates, or of their nucleotidyl moiety. Thus, simple binding of the appropriate compound to the enzyme is probably sufficient for stimulation. This does not mean, however, that a chemical reaction with the formation of covalent bonds cannot take place with the parent nucleotides, ATP or GTP. In fact, this possibility was suggested by experiments in which the particulate preparation was incubated at 30°C with ATP, after which the nucleotide was eliminated by dilution and centrifugation. In a subsequent assay the enzyme was found to be active in the absence of nucleotide (147). The effect depended on time and temperature of incubation, which implies the participation of some enzymatic process. The nature of the process is unknown. There is some evidence that the activation can be reversed by incubation in the presence of Mg^{2+} ions (147), but the instability of the synthetase at 30°C interfered with further progress in this line of experimentation.

In contrast with chitin synthetase, glucan synthetase is not activated by treatment with proteases. Incubation with trypsin only leads to loss of activity (144). Conversely, chitin synthetase is not stimulated by the nucleotides active in the glucan system (E. Cabib, unpublished observations).

The subcellular distribution of glucan synthetase was studied by the same methods used for chitin synthetase, with identical results (144). It appears, therefore, that both enzymes are positioned on the inner face of the plasma membrane.

The localization of glucan synthetase and its regulation by guanosine nucleotides and their analogues resemble closely those of the adenylate cyclase system of mammalian cells (148). Further work is required for establishing whether this is mere coincidence or an indication of a common regulatory mechanism.

Mannan

In contrast to chitin and glucan, mannan is formed by a very complex pathway, similar to that found in the synthesis of mammalian glycoproteins. Mannan is the only component of the yeast cell wall known to be of glycoprotein nature, as discussed above. The carbohydrate moiety of other internal and periplasmic glycoproteins from yeast has been found to match the structure of either one portion or the whole of wall mannan (149, 150). In the in vitro studies to be discussed below, it has been tacitly assumed that the biosynthetic pathway is the same in all these cases, since there is presently no way of knowing which specific glycoprotein is being formed with the various enzyme preparations used. It should also be kept in mind that mannan, as already mentioned (Figure 1), contains both short mannosyl chains linked to serine and/or threonine as well as a large polysaccharide consisting of an outer chain and a core region linked to asparagine through a diacetylchitobiosyl bridge. The biosynthesis of the two moieties proceeds by different pathways (5, 151, 152; Figure 4).

It was originally observed that GDP-mannose is the precursor for all the mannosyl groups found in mannan outer chain (153). This notion has been supported and extended to all other mannosyl residues by subsequent studies. Before reaching the polysaccharide, however, many of the mannosyl residues are processed through a lipid intermediate (Figure 4), as first reported by Tanner (154). This finding was confirmed (155) and the lipid was later identified as dolichol (156–158) as in the case of mammalian glycoproteins (133).

Dolichyl-phosphate-mannose is formed from GDP-mannose and dolichyl-phosphate in a reaction inhibited by diumycin (159, 160). The responsible enzyme has been solubilized and found to require a divalent cation (161). In the synthesis of short oligosaccharide chains linked to hydroxyaminoacids, dolichyl-phosphate-mannose is the donor for the first mannosyl group (162–164). The enzyme that catalyzes this reaction has also been solubilized and reported to require Mg^{2+} or Mn^{2+} for activity (161). Subsequent mannosyl groups of the short chains are directly transferred from GDP-mannose (162, 164).

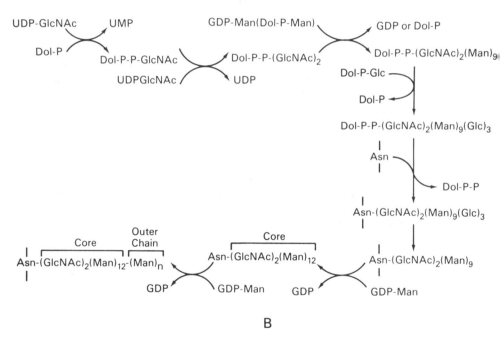

Figure 4 Pathway of mannan biosynthesis. *A,* short oligosaccharide attached to serine or threonine. *B,* large polysaccharide attached to asparagine. Dol, dolichol; P, phosphate. Adapted from (5).

The first step in the formation of the asparagine-linked polysaccharide also begins with dolichyl-phosphate, but proceeds by the transfer of N-acetylglucosamine-1-phosphate from UDP-N-acetylglucosamine to the lipid in the presence of Mg^{2+} or Mn^{2+} (165, 166; Figure 4*B*). This is the specific step blocked by tunicamycin (167), an antibiotic that inhibits synthesis of glycoproteins in yeast (168, 169) as well as in animals (133). A second N-acetylglucosamine is added from UDP-N-acetylglucosamine (165, 166; Figure 4B) in a reaction inhibited by bacitracin (170). The oligosaccharide bound to dolichyl-pyrophosphate grows by subsequent addition of nine mannosyl residues (171–173). Of these, the first four are probably transferred from GDP-mannose and the following five from dolichyl-phosphate-mannose (174). Finally, three glucose residues are added in succession (173, 175) to yield dolichyl-P-P-GlcNAc$_2$ Man$_9$Glc$_3$. The precursor for the glucosyl groups is UDP-glucose (173), but dolichyl-phosphate-glucose is probably an intermediate (176, 177). The addition of glucose to the oligosaccharide-lipid appears to be the signal for transfer of the sugar moiety "en bloc" to acceptor proteins, as judged from pulse-chase experiments (175), an idea that is supported by results in vitro (178, 179). In detergent solubilized preparations, the transfer reaction, which required Mn^{2+}, was about 25-fold faster with glucosylated than with non-glucosylated oligosaccharide-lipid (178). Sharma et al (179) report no transfer at all with dolichyl-pyrophosphate-GlcNAc$_2$-Man$_9$. Surprisingly, the mannosyl-free compound, dolichyl-pyrophosphate-GlcNAc$_2$, readily transfers its chitobiose moiety to protein (166, 180). This compound was used in model reactions with synthetic peptides as acceptors (181). The results confirmed the need for the sequence Asn-X-Thr(Ser) in a sugar acceptor, as previously postulated for animal glycoproteins (182). Such a requirement was not found for transfer of mannose from dolichyl-phosphate-mannose to serine-containing peptides (181).

After transfer of the oligosaccharide to protein, the next step is the hydrolysis of the three glucose residues (175). An enzyme capable of removing the terminal glucose and, more slowly, the penultimate residue, has been partially purified from yeast (183). The core region of the protein-bound oligosaccharide must now be completed, after which the outer chain is added (see Figure 4B). In these latter steps, all the mannosyl residues appear to be donated directly by GDP-mannose (152). The enzymes responsible for the transfer have been solubilized and found to specifically require Mn^{2+} (152, 184), as previously found for the particulate preparation (153). At this stage of the synthesis, different types of linkages, i.e. $\alpha(1 \rightarrow 6)$, $\alpha(1 \rightarrow 2)$, $\alpha(1 \rightarrow 3)$, and 6-phosphorylmannosyl are formed (Figure 1). To expedite study of the synthesis of these linkages, either mannose or small mannose oligosaccharides, sometimes in the reduced form, were used as acceptors (184–187). These studies were facilitated by the isolation of mutants with a defective mannan (for review see 7, 188). Some of these mutants lack all side branches of the outer chains, i.e. they possess only the $\alpha(1 \rightarrow 6)$ backbone; others lack either $\alpha(1 \rightarrow 2)$-linked or $\alpha(1 \rightarrow 3)$-linked mannose units. The *mnn*4 mutant does not contain the terminal mannosylphosphoryl groups. In general, the results in vitro coincide with those obtained in vivo,i.e. mutants in whose mannan a certain linkage is missing also lack the corresponding transferase (184, 187, 189). An exception is mutant *mnn*2, which produces a mannan lacking all side chains; it should be devoid of $\alpha(1 \rightarrow 2)$ mannosyl transferase, but shows normal activity of the enzyme and normal synthesis of outer chain in extracts (152, 189).

Based on the results obtained with *mnn* mutants, the picture emerges of a stepwise construction of the outer chain, beginning with the $\alpha(1 \rightarrow 6)$backbone and continuing with successive addition of lateral branches; different transferases are probably responsible for each type of linkage.

Whereas the structure and biosynthetic pathway of the carbohydrate portion of cell wall mannan are fairly well established, little is known about the protein moiety. Blocking of protein synthesis with cycloheximide inhibits formation of mannan in intact cells (70, 190) or its secretion by protoplasts (191), presumably by preventing formation of acceptor for the carbohydrate moiety.

As for the localization of mannan synthesis, autoradiographic evidence suggests that the polysaccharide is made intracellularly (192). By analogy with mammalian glycoproteins (193) one would expect the transfer of glucosylated oligosaccharide to occur onto a nascent protein chain in the endoplasmic reticulum. Further processing would take place in the Golgi complex or its yeast equivalent, followed by transfer to the wall, perhaps by means of secretory vesicles. Studies of intracellular distribution (194–199) indicate the presence of mannosyl transferases both in internal and plasma membranes, but neither the subcellular organelles nor the transferases were well characterized. Better information is accumulating from studies of temperature-sensitive yeast mutants blocked in the secretion of certain enzymes of mannoprotein nature. At the nonpermissive temperature, several of these mutants accumulate intracellular organelles containing the secretory enzymes (200, 201). Cytological studies of double mutants suggest the pathway: endoplasmic reticulum → Golgi-like vesicles → secretory vesicles → exocytosis through the plasma membrane for the mannoproteins (201).

ENZYMATIC DEGRADATION OF CELL WALL POLYSACCHARIDES

As mentioned above there are indications that growth of the cell wall may require breakage of some bonds while others are formed. Enzymes that degrade $\beta(1 \rightarrow 3)$glucan were found intracellularly (202), in the periplasmic space (203, 204) and in the growth medium (203). According to one report (205), the intracellular enzyme was sequestered in vesicles. Protoplasts were also shown to excrete glucanases into the medium (203). It has been claimed

(205) that the activity of a vesicle-bound exoglucanase fraction increased just before budding. Del Rey et al (206) were unable to confirm these results, and detected instead a doubling of exoglucanase activity after DNA duplication. In the same laboratory, it was found that a mutant lacking exoglucanase grows normally (207). Thus, there is no clear evidence that links the activity of known glucanases to cell wall growth.

Recently, an endochitinase was isolated from *S. cerevisiae* (207a). This enzyme, which is associated with mannan, is located partly within intracellular vesicles and partly in the periplasmic space. The accumulation in vesicles may be a step prior to secretion, since protoplasts secrete part of the enzyme into the medium (207b). Because of its location, it is possible that the role of this enzyme consists in catalyzing a partial hydrolysis of chitin in the primary septum, prior to cell separation.

HOW IS THE CELL WALL PUT TOGETHER?

The ultimate goal of the research reviewed here is to understand, in molecular terms, the mechanisms by which the wall is constructed and its synthesis is regulated. In concluding the chapter, it is worthwhile to assess how far along this road the studies of the preceeding pages have led us.

Of the events associated with cell wall formation, the most precisely localized in time and space is probably the initiation of the primary septum. It is also the simplest from the chemical standpoint since the product is exclusively the homopolysaccharide chitin. The enzyme responsible for its formation, chitin synthetase, appears to be uniformly distributed on the plasma membrane in a latent state. Since the area involved in chitin formation, i.e. an annular zone between mother cell and bud, is a tiny fraction of the total cell surface, it seems reasonable to conclude that only a small amount of chitin synthetase zymogen is activated at each budding event. This would explain why most of the enzyme is found in a cryptic state even in a rapidly growing culture. The enzyme responsible for activation was thought to be the vacuole-bound protease B (95, 123, 208, 209), but this notion had to be abandoned, because mutants totally lacking the protease show no impairment in cell division (210–212). This does not necessarily mean that the basic idea is incorrect. The latent character of chitin synthetase has now been confirmed in a large number of fungi, as discussed above, and in all such cases activity could be elicited by treatment with an appropriate protease. This does not preclude the possibility that activation of chitin synthetase zymogen may take place by a nonproteolytic mechanism (5, 44), but it does make it unlikely. Preliminary results indicate the presence in *Saccharomyces* of a membrane-bound enzymatic activity, probably a protease, that activates chitin synthetase (M. P. Fernandez, J. V. Correa, and E. Cabib, unpublished results).

The unraveling of mechanisms involved in the regulation of chitin synthesis might be greatly facilitated by the availability of conditional mutants, but efforts to isolate such mutants have been unsuccessful (M. L. Slater and E. Cabib, unpublished results).

Whatever the chemical mechanism of zymogen activation may be, it remains to be seen how the localization of the activation is achieved. It seems inescapable that some intracellular vesicle or organelle must participate in this process in order to limit the activated area, but the existence of such a directional apparatus can be defined only by default: the apparatus seems to have broken down in *Saccharomyces* mutant *cdc* 24, and budding is blocked while deposition of chitin occurs over the whole cell surface (92).

The formation of lateral cell walls is not as sharply localized as that of the septum, but remains restricted to the growing bud surface. An increase in the cell surface means extension not only of the cell wall but also of the underlying plasma membrane. This point was forcefully brought out by Field & Schekman (83) who monitored new surface growth by derepressing acid phosphatase, a periplasmic enzyme, and following its localization. They showed, under a variety of conditions, that surface growth and secretion were coupled. Furthermore, temperature-sensitive mutants blocked in secretion are also blocked in surface growth (83). Some of these mutants accumulate a large number of intracellular vesicles, rich in secretory enzymes (201). These results suggest that both plasma membrane and periplasmic enzymes are added by a process of exocytosis (201), which has important consequences for glucan and mannan deposition. As mentioned, glucan synthetase appears to be bound to the plasma membrane. The membranes of mother cell and bud are continuous, yet glucan is laid down only on the bud surface. A possible explanation of this apparent paradox could be that the new membrane, added by exocytosis of vesicles, carries glucan synthetase in an activated form similar to that obtained in vitro by incubation of the enzyme with nucleotides (147). At a later stage, when the bud has attained its maximum size, the enzyme would be inactivated. In view of the effect of Mg^{2+} ions on inactivation of the synthetase in vitro (147), it is attractive to speculate that the increase in Mg^{2+} concentration observed just before cell division (78) might have a role in this process.

As for mannan, it may be assumed that its fate is similar to that of the periplasmic secretory mannoproteins, such as invertase and acid phosphatase. The evidence from secretory mutants that has been reviewed suggests that the synthesis of all these glycoproteins begins in the endoplasmic reticulum and is completed in Golgi-like bodies, followed by transport to the cell surface in secretory vesicles. Regulation of the process is probably intimately related to that of protein and membrane formation, about which little is known. Thus, although many details of the complex chemical and intracellular pathway of mannan synthesis have been elucidated in recent years, we can say practically nothing about its regulation.

Whereas our knowledge of this field is so fragmentary that surprises probably await us in the future, present results suggest that the biosynthesis of each component of the cell wall is regulated by a different mechanism; partial proteolysis for chitin; interaction with nucleotides for glucan; control of glycoprotein biosynthesis for mannan. This situation was certainly unexpected and would not probably have been detected if work had not been carried out at the subcellular and enzymatic level. The localization of biosynthesis also seems to vary according to the cell wall component: whereas $\beta(1 \rightarrow 3)$glucan and chitin are probably manufactured directly at the plasma membrane, mannan appears to be formed intracellularly and later transported to the surface in vesicles. Nothing is known about the final assembly of the cell wall components once they have crossed the plasma membrane. Yet, some new linkages are probably formed at that stage, as noted in the discussion of mannan structure.

Undoubtedly, the studies of the past ten years or so have considerably increased our understanding of the formation of the yeast cell wall and have provided promising models for morphogenesis in general. Among such models of general significance is the notion that enzymes involved in the biosynthesis of structural cell wall components can be subjected to strict posttranslational regulation. Yet, essential portions of the picture are absent. Nowhere is our ignorance more complete than in the area of those directional systems that carry with great precision vesicles and other organelles to selected targets in the cell. This is certainly one of the most important, if difficult, topics for future efforts.

Literature Cited

1. Cabib, E. 1975. *Ann. Rev. Microbiol.* 29:191–214
2. Cabib, E., Ulane, R., Bowers, B. 1974. *Curr. Top. Cell. Regul.* 8:1–32
3. Gooday, G. W. 1977. *J. Gen. Microbiol.* 99:1–11
4. Farkas, V. 1979. *Microbiol. Rev.* 43:117–44
5. Cabib, E., Shematek, E. M. 1981. In *Biology of Carbohydrates,* ed. V. Ginsburg, P. W. Robbins, 1:51–90. New York: Wiley
6. Cabib, E. 1981. In *Encyclopedia of Plant Physiology,* ed. W. Tanner, F. A. Loewus, 13B:395–415. Heidelberg: Springer.
7. Ballou, C. E. 1976. *Adv. Microb. Physiol.* 14:93–158
8. Ballou, C. E., Raschke, W. C. 1974. *Science* 184:127–34
9. Sentandreu, R., Northcote, D. H. 1969. *Carbohydr. Res.* 10:584–85
10. Nakajima, T., Ballou, C. E. 1974. *J. Biol. Chem.* 249:7679–84
11. Duell, E. A., Inoue, S., Utter, M. F. 1964. *J. Bacteriol.* 88:1762–73
12. Bacon, J. S. D., Farmer, V. C., Jones, D., Taylor, I. F. 1969. *Biochem. J.* 114:557–67
13. Manners, D. J., Masson, A. J., Patterson, J. C., Bjorndal, H., Lindberg, B. 1973. *Biochem. J.* 135:31–36

14. Manners, D. J., Masson, A. J., Patterson, J. C. 1973. *Biochem. J.* 135:19–30
15. Eddy, A. A., Woodhead, J.S. 1968. *FEBS Lett.* 1:67–68
16. Fleet, G. H., Manners, D. J. 1976. *J. Gen. Microbiol.* 94:180–92
17. Muzzarelli, R. A. A. 1977. *Chitin,* pp. 1–309. New York: Pergamon
18. Minke, R., Blackwell, J. 1978. *J. Mol. Biol.* 120:167–82
19. Beran, K., Holan, Z., Baldrian, J. 1972. *Folia Microbiol.* 17:322–30
20. Houwink, A. L., Kreger, D. R. 1953. *Antonie van Leeuwenhoek J. Microbiol. Serol.* 19:1–24
21. Ruiz-Herrera, J., Bartnicki-Garcia, S. 1974. *Science* 186:357–59
22. Bacon, J. S. D., Davidson, E. D., Jones, D., Taylor, I. F. 1966. *Biochem. J.* 101: C36–C38
23. Cabib, E., Bowers, B. 1971. *J. Biol. Chem.* 246:152–59
24. Djaczenko, W., Cassone, A. 1971. *J. Cell Biol.* 52:186–90
25. Cassone, A. 1973. *Experientia* 29: 1303–5
26. Moor, H., Muhlethaler, K. 1963. *J. Cell Biol.* 17:609–28
27. Kopecka, M., Phaff, H. J., Fleet, G. H. 1974. *J. Cell Biol.* 62:66–77
28. Svoboda, A., Necas, O. 1970. In *Yeast Protoplasts, Proc. 2nd Int. Symp. Yeast Protoplasts,* ed. O. Necas, A. Svoboda, pp. 211–15. Brno: J. E. Purkyne Univ. Press
29. Kreger, D. R., Kopecka, M. 1975. *J. Gen. Microbiol.* 92:207–20
30. Kreger, D. R., Kopecka, M. 1978. *J. Gen. Microbiol.* 108:269–74
31. Phaff, H. J. 1971. In *The Yeasts,* ed. A. H. Rose, J. S. Harrison, 2:135–210. New York: Academic
32. Bacon, J. S. D., Gordon, A. H., Jones, D., Taylor, I. F., Webley, D. M. 1970. *Biochem. J.* 120:67–78
33. Kitamura, K., Kaneko, T., Yamamoto, Y. 1974. *J. Gen. Appl. Microbiol.* 20: 323–44
34. Horisberger, M., Bauer, H., Bush, D. A. 1971. *FEBS Lett.* 18:311
35. Tkacz, J. S., Cybulska, E. E., Lampen, J. O. 1971. *J. Bacteriol.* 105:1–5
36. Mundkur, B. 1960. *Exp. Cell Res.* 20:28–42
37. Tronchin, G., Poulain, D., Biquet, J. 1979. *Arch. Microbiol.* 123:245–49
38. Horisberger, M., Volanthen, M. 1977. *Arch. Microbiol.* 115:1–7
39. Arnold, W. N. 1973. *Physiol. Chem. Physics* 5:117–23
40. Arnold, W. N., Garrison, R. G. 1981. *Curr. Microbiol.* 5:57–60
41. Ballou, C. E., Kern, K. A., Raschke, W. C. 1973. *J. Biol. Chem.* 248:4667–73
42. Ballou, L., Cohen, R. E., Ballou, C. E. 1980. *J. Biol. Chem.* 255:5986–91
43. Molano, J., Bowers, B., Cabib, E. 1980. *J. Cell Biol.* 85:199–212
44. Cabib, E., Duran, A., Bowers, B. 1979. In *Fungal Walls and Hyphal Growth,* ed. J. H. Burnett, A. P. J. Trinci, pp. 189–201. Cambridge: Cambridge Univ. Press
45. Bowers, B., Levin, G., Cabib, E. 1974. *J. Bacteriol.* 119:564–75
46. Bush, D. A., Horisberger, M. 1973. *J. Biol. Chem.* 248:1318–20
47. Marchant, R., Smith, D. G. 1968. *J. Gen. Microbiol.* 53:163–69
48. Hayashibe, M., Katohda, S. 1973. *J. Gen. Appl. Microbiol.* 19:23–39
49. Cabib, E., Bowers, B. 1975. *J. Bacteriol.* 124:1586–93
50. Vrsanska, M., Kratky, Z., Biely, P., Machala, S. 1979. *Z. Allg. Mikrobiol.* 19:357–62
51. Hartwell, L. H., Unger, M. W. 1977. *J. Cell Biol.* 75:422–35
52. Johnston, G. C., Pringle, J. R., Hartwell, L. H. 1977. *Exp. Cell Res.* 105:79–98
53. Carter, B. L. A., Jagadish, M. N. 1978. *Exp. Cell Res.* 112:15–24
54. Johnston, G. C., Ehrardt, C. W., Lorincz, A., Carter, B. L. A. 1979. *J. Bacteriol.* 137:1–5
55. Lorincz, A., Carter, B. L. A. 1979. *J. Gen. Microbiol.* 113:287–95
56. Tyson, C. B., Lord, P. G., Wheals, A. E. 1979. *J. Bacteriol.* 138:92–98
57. Lord, P. G., Wheals, A. E. 1980. *J. Bacteriol.* 142:808–18
58. Sudbery, P. E., Goodey, A. R., Carter, B. L. A. 1980. *Nature* 288:401–4
59. Rivin, C. J., Fangman, W. L. 1980. *J. Cell Biol.* 85:96–107
60. Byers, B., Goetsch, L. 1975. *J. Bacteriol.* 124:511–23
61. Hereford, L. M., Hartwell, L. H. 1974. *J. Mol. Biol.* 84:445–61
62. Streiblova, E. 1970. *Can. J. Microbiol.* 16:827–31
63. Thompson, P. W., Wheals, A. E. 1981. *J. Gen. Microbiol.* 121:401–9
64. Sloat, B. F., Adams, A., Pringle, J. R. 1981. *J. Cell Biol.* 89:395–405
65. Quinlan, R. A., Pogson, C. I., Gull, K. 1980. *J. Cell Sci.* 46:341–52
66. Biely, P., Kovarik, J., Bauer, S. 1973. *Arch. Mikrobiol.* 94:365–71
67. Sierra, J. M., Sentandreu, R., Villanueva, J. R. 1973. *FEBS Lett.* 34: 285–90

68. Hayashibe, M., Abe, N., Matsui, M. 1977. *Arch. Microbiol.* 114:91–92
69. Elorza, M. V., Sentandreu, R. 1969. *Biochem. Biophys. Res. Commun.* 36: 741–47
70. Elorza, M. V., Lostau, C. M., Villanueva, J. R., Sentandreu, R. 1976. *Biochim, Biophys. Acta* 454:263–72
71. Phai, L. D., Reuter, G. 1976. *Z. Allg. Mikrobiol.* 16:197–99
72. Poole, R. K., Lloyd, D. 1973. *Arch. Mikrobiol.* 88:256–72
73. Biely, P., Kratky, Z., Kovarik, J., Bauer, S. 1971. *J. Bacteriol.* 107:121–29
74. Challinor, S. W., Power, D. M., Tonge, R. J. 1964. *Nature* 203:250–51
75. Power, D. M., Challinor, S. W. 1969. *J. Gen. Microbiol.* 55:169–76
76. Hanson, B. A., Lester, R. L. 1980. *J. Bacteriol.* 142:79–89
77. McMurrough, I., Rose, A. H. 1967. *Biochem. J.* 105:189–203
78. Walker, G. M., Duffus, J. H. 1980. *J. Cell Sci.* 42:329–56
79. Chung, K. L., Hawirko, R. Z., Isaac, P. K. 1965. *Can. J. Microbiol.* 11:953–57
80. Johnson, B. F., Gibson, E. J. 1966. *Exp. Cell Res.* 41:580–91
81. Tkacz, J. S., Lampen, J. O. 1972. *J. Gen. Microbiol.* 72:243–47
82. Farkas, V., Kovarik, J., Kosinova, A., Bauer, S. 1974. *J. Bacteriol.* 117:265–69
83. Field, C., Schekman, R. 1980. *J. Cell Biol.* 86:123–28
84. Johnson, B. F. 1968. *J. Bacteriol.* 95:1169–72
85. Kratky, Z., Biely, P., Bauer, S. 1975. *Eur. J. Biochem.* 54:459–67
86. Farkas, V., Svoboda, A., Bauer, S. 1969. *J. Bacteriol.* 98:744–48
87. Johnson, B. F. 1968. *Exp. Cell Res.* 50: 692–94
88. Biely, P., Kovarik, J., Bauer, S. 1973. *J. Bacteriol.* 115:1108–20
89. Miyata, M., Kitamura, J., Miyata, H. 1980. *Arch. Microbiol.* 127:11–16
90. Shimoda, C., Yanagishima, N. 1971. *Physiol. Plant.* 24:46–50
91. Hartwell, L. H., Culotti, J., Pringle, J. R., Reid, B. 1974. *Science* 183:46–51
92. Sloat, B. F., Pringle, J. R. 1978. *Science* 200:1171–73
93. Hartwell, L. H., Mortimer, R. K., Culotti, J., Culotti, M. 1973. *Genetics* 74:267–86
94. Hartwell, L. H. 1971. *Exp. Cell Res.* 69:265–76
95. Cabib, E., Farkas, V. 1971. *Proc. Natl. Acad. Sci. USA* 68:2052–56
96. Schekman, R., Brawley, V. 1979. *Proc. Natl. Acad. Sci. USA* 76:645–49
97. Biely, P. 1978. *Arch. Microbiol.* 119: 213–14
98. Glaser, L., Brown, D. H. 1957. *J. Biol. Chem.* 228:729–42
99. Keller, F. A., Cabib, E. 1971. *J. Biol. Chem.* 246:160–66
100. Jaworski, E. G., Wang, L. C., Carpenter, W. D. 1965. *Phytopathology* 55:1309–12
101. Porter, C. A., Jaworski, E. G. 1966. *Biochemistry* 5:1149–54
102. Plessmann Camargo, E., Dietrich, C. P., Sonneborn, D., Strominger, J. L. 1967. *J. Biol. Chem.* 242:3121–28
103. Ohta, N., Kakiki, K., Misato, T. 1970. *Agr. Biol. Chem.* 34:1224–34
104. McMurrough, I., Flores-Carreon, A., Bartnicki-Garcia, S. 1971. *J. Biol. Chem.* 246:3999–4007
105. McMurrough, I., Bartnicki-Garcia, S. 1971. *J. Biol. Chem.* 246:4008–16
106. Hori, M., Eguchi, J., Kakiki, K., Misato, T. 1974. *J. Antibiot.* 27:260–66
107. Jan, Y. N. 1974. *J. Biol. Chem.* 249: 1973–79
108. Gooday, G. W., Rousset-Hall, A. 1975. *J. Gen. Microbiol.* 89:137–45
109. Rousset-Hall, A., Gooday, G. W. 1975. *J. Gen. Microbiol.* 89:146–54
110. Moore, P. M., Peberdy, J. F. 1975. *Microbios* 12:29–39
111. Peberdy, J. F., Moore, P. M. 1975. *J. Gen. Microbiol.* 90:228–36
112. Moore, P. M., Peberdy, J. F. 1976. *Can. J. Microbiol.* 22:915–21
113. Lopez-Romero, E., Ruiz-Herrera, J. 1976. *Antonie van Leeuwenhoek J. Microbiol. Serol.* 42:261–76
114. Archer, D. B. 1977. *Biochem. J.* 164: 653–58
115. Ryder, N. S., Peberdy, J. F. 1977. *FEMS Microbiol. Lett.* 2:199–201
116. Ryder, N. S., Peberdy, J. F. 1977. *J. Gen. Microbiol.* 99:69–76
117. Bartnicki-Garcia, S., Bracker, C. E., Reyes, E., Ruiz-Herrera, J. 1978. *Exp. Mycol.* 2:173–92
118. Braun, P. C., Calderone, R. A. 1978. *J. Bacteriol.* 133:1472–77
119. Vermeulen, C. A., Raeven, M. B. J. M., Wessels, J. G. H. 1979. *J. Gen. Microbiol.* 114:87–97
120. Selitrennikoff, C. P. 1979. *Biochim. Biophys. Acta* 571:224–32
121. Isono, K., Asahi, K., Suzuki, S. 1969. *J. Am. Chem. Soc.* 91:7490–505
122. Endo, A., Kakiki, K., Misato, T. 1970. *J. Bacteriol.* 104:189–96
123. Ulane, R. E., Cabib, E. 1976. *J. Biol. Chem.* 251:3367–74
124. Ruiz-Herrera, J., Bartnicki-Garcia, S. 1976. *J. Gen. Microbiol.* 97:241–49

125. Ruiz-Herrera, J., Lopez-Romero, E., Bartnicki-Garcia, S. 1977. *J. Biol. Chem.* 252:3338–43
126. Isaac, S., Ryder, N. S., Peberdy, J. F. 1978. *J. Gen. Microbiol.* 105:45–50
127. Van Laere, A. J., Carlier, A. R. 1978. *Arch. Microbiol.* 116:181–84
128. Arroyo-Begovich, A., Herrera, J. R. 1979. *J. Gen. Microbiol.* 113:339–45
129. Gooday, G. W. 1979. In *Fungal Walls and Hyphal Growth,* ed. J. H. Burnett, A. P. J. Trinci, pp. 203–23. Cambridge: Cambridge Univ. Press
130. Duran, A., Cabib, E. 1978. *J. Biol. Chem.* 253:4419–25
131. Pesti, M., Campbell, J. M., Peberdy, J. F. 1981. *Curr. Microbiol.* 5:187–90
132. Rast, D. M., Bartnicki-Garcia, S. 1981. *Proc. Natl. Acad. Sci. USA* 78:1233–36
133. Parodi, A. J., Leloir, L. F. 1979. *Biochim. Biophys. Acta* 559:1–37
134. Selitrennikoff, C. P. 1979. *Arch. Biochem. Biophys.* 195:243–44
135. Takatsuki, A., Kawamura, K., Okina, M., Kodama, Y., Ito, T., Tamura, G. 1977. *Agr. Biol. Chem.* 41:2307–9
136. Duran, A., Bowers, B., Cabib, E. 1975. *Proc. Natl. Acad. Sci. USA* 72:3952–55
137. Scarborough, G. A. 1975. *J. Biol. Chem.* 250:1106–11
138. Duran, A., Cabib, E., Bowers, B. 1979. *Science* 203:363–65
139. Bracker, C. E., Ruiz-Herrera, J., Bartnicki-Garcia, S. 1976. *Proc. Natl. Acad. Sci. USA* 73:4570–74
140. Sentandreu, R., Elorza, M. V., Villanueva, J. R. 1975. *J. Gen. Microbiol.* 90:13–20
141. Balint, S., Farkas, V., Bauer, S. 1976. *FEBS Lett.* 64:44–47
142. Lopez-Romero, E., Ruiz-Herrera, J. 1977. *Biochim. Biophys. Acta* 500:372–84
143. Lopez-Romero, E., Ruiz-Herrera, J. 1978. *Antonie van Leeuwenhoeck J. Microbiol. Serol.* 44:329–39
144. Shematek, E. M., Braatz, J. A., Cabib, E. 1980. *J. Biol. Chem.* 255:888–94
145. Baguley, B. C., Rommele, G., Gruner, J., Wehrli, W. 1979. *Eur. J. Biochem.* 97:345–51
146. Mizoguchi, J., Saito, T., Mizuno, K., Hayano, K. 1977. *J. Antibiot.* 30:308–13
147. Shematek, E. M., Cabib, E. 1980. *J. Biol. Chem.* 255:895–902
147a. Notario, V., Kawai, H., Cabib, E. 1982. *J. Biol. Chem.* In press
148. Helmreich, E. J. M., Zenner, H. P., Pfeuffer, T., Cori, C. F. 1976. *Curr. Top. Cell. Regul.* 10:41–87
149. Smith, W. L., Ballou, C. E. 1974. *Biochemistry* 13:355–61
150. Hashimoto, C., Cohen, R. E., Zhang, W-J., Ballou, C. E. 1981. *Proc. Natl. Acad. Sci. USA* 78:2244–48
151. Farkas, V., Bauer, S., Vagabov, V. M. 1976. *Biochim. Biophys. Acta* 428:583–90
152. Parodi, A. J. 1979. *J. Biol. Chem.* 254:8343–52
153. Behrens, N. H., Cabib, E. 1968. *J. Biol. Chem.* 243:502–9
154. Tanner, W. 1969. *Biochem. Biophys. Res. Commun.* 35:144–50
155. Sentandreu, R., Lampen, J. O. 1971. *FEBS Lett.* 14:109–13
156. Tanner, W., Jung, P., Behrens, N. H. 1971. *FEBS Lett.* 16:245–48
157. Sentandreu, R., Lampen, J. O. 1972. *FEBS Lett.* 27:331–34
158. Jung, P., Tanner, W. 1973. *Eur. J. Biochem.* 37:1–6
159. Babczinski, P. 1980. *Eur. J. Biochem.* 112:53–58
160. Bretthauer, R. K., Wu, S., Irwin, W. E. 1973. *Biochim. Biophys. Acta* 304:736–47
161. Babczinski, P., Haselbeck, A., Tanner, W. 1980. *Eur. J. Biochem.* 105:509–15
162. Babczinski, P., Tanner, W. 1973. *Biochem. Biophys. Res. Commun.* 54:1119–24
163. Bretthauer, R. K., Wu, S. 1975. *Arch. Biochem. Biophys.* 167:151–60
164. Sharma, C. B., Babczinski, P., Lehle, L., Tanner, W. 1974. *Eur. J. Biochem.* 46:35–41
165. Lehle, L., Tanner, W. 1975. *Biochim. Biophys. Acta* 399:364–74
166. Reuvers, F., Habet-Willems, C., Reinking, A., Boer, P. 1977. *Biochim. Biophys. Acta* 486:541–52
167. Lehle, L., Tanner, W. 1976. *FEBS Lett.* 71:167–70
168. Kuo, S.-C., Lampen, J. O. 1974. *Biochem. Biophys. Res. Commun.* 58:287–95
169. Kuo, S.-C., Lampen, J. O. 1976. *Arch. Biochem. Biophys.* 172:574–81
170. Reuvers, F., Boer, P., Steyn-Parve, E. P. 1978. *Biochem. Biophys. Res. Commun.* 82:800–4
171. Lehle, L., Tanner, W. 1978. *Biochim. Biophys. Acta* 539:218–29
172. Parodi, A. J. 1978. *Eur. J. Biochem.* 83:253–59
173. Trimble, R. B., Maley, F., Tarentino, A. L. 1980. *J. Biol. Chem.* 255:10232–38
174. Lehle, L. 1980. *Eur. J. Biochem.* 109:589–601

175. Parodi, A. J. 1979. *J. Biol. Chem.* 254: 10051–60
176. Parodi, A. J. 1976. *FEBS Lett.* 71: 283–86
177. Parodi, A. J. 1977. *Eur. J. Biochem.* 75: 171–80
178. Trimble, R. B., Byrd, J. C., Maley, F. 1980. *J. Biol. Chem.* 255:11892–95
179. Sharma, C. B., Lehle, L., Tanner, W. 1981. *Eur. J. Biochem.* 116:101–8
180. Lehle, L., Tanner, W. 1978. *Eur. J. Biochem.* 83:563–70
181. Bause, E., Lehle, L. 1979. *Eur. J. Biochem.* 101:531–40
182. Neuberger, A., Marshall, R. D. 1969. In *Carbohydrates and Their Role,* ed. H. W. Schulz, R. F. Cain, R. W. Wrastad, pp. 115–32. Westport, Conn: Avi
183. Kilker, R. D., Saunier, B., Tkacz, J. S., Herscovics, A. 1981. *J. Biol. Chem.* 256:5299–303
184. Nakajima, T., Ballou, C. E. 1975. *Proc. Natl. Acad. Sci. USA* 72:3912–16
185. Schutzbach, J. S., Ankel, H. 1971. *J. Biol. Chem.* 246:2187–94
186. Farkas, V., Vagabov, V. M., Bauer, S. 1976. *Biochim. Biophys. Acta* 428: 573–82
187. Karson, E. M., Ballou, C. E. 1978. *J. Biol. Chem.* 253:6484–92
188. Ballou, C. E. 1974. *Adv. Enzymol.* 40: 239–70
189. Farkas, V., Bauer, S. 1976. *Folia Microbiol.* 21:459–64
190. Sentandreu, R., Lampen, J. O. 1970. *FEBS Lett.* 11:95–99
191. Farkas, V., Svoboda, A., Bauer, S. 1970. *Biochem. J.* 118:755–58
192. Kosinova, A., Farkas, V., Machala, S., Bauer, S. 1974. *Arch. Microbiol.* 99: 255–63
193. Hubbard, S. C., Ivatt, R. J. 1981. *Ann. Rev. Biochem.* 50:555–83
194. Cortat, M., Matile, P., Kopp, F. 1973. *Biochem. Biophys. Res. Commun.* 53: 482–89
195. Ruiz-Herrera, J., Sentandreu, R. 1975. *J. Bacteriol.* 124:127–33
196. Larriba, G., Elorza, M. V., Villanueva, J. R., Sentandreu, R. 1976. *FEBS Lett.* 71:316–20
197. Lehle, L., Bauer, F., Tanner, W. 1977. *Arch. Microbiol.* 114:77–81
198. Marriot, M., Tanner, W. 1979. *J. Bacteriol.* 139:565–72
199. Welten-Verstegen, G. W., Boer, P., Steyn-Parve, E. P. 1980. *J. Bacteriol.* 141:342–49
200. Novick, P., Schekman, R. 1979. *Proc. Natl. Acad. Sci. USA* 76:1858–62
201. Novick, P., Field, C., Schekman, R. 1980. *Cell* 21:205–15
202. Abd-el-al, A. T. H., Phaff, H. J. 1968. *Biochem. J.* 109:347–60
203. Farkas, V., Biely, P., Bauer, S. 1973. *Biochim. Biophys. Acta* 321:246–55
204. Villa, T. G., Notario, V., Villanueva, J. R. 1975. *Arch. Microbiol.* 104:201–6
205. Cortat, M., Matile, P., Wiemken, A. 1972. *Arch. Mikrobiol.* 82:189–205
206. Del Rey, F., Santos, T., Garcia-Acha, I., Nombela, C. 1979. *J. Bacteriol.* 139:924–31
207. Santos, T., Del Rey, F., Conde, J., Villanueva, J. R., Nombela, C. 1979. *J. Bacteriol.* 139:333–38
207a. Correa, J. U., Elango, N., Polacheck, I., Cabib, E. 1982. *J. Biol. Chem.* In press
207b. Elango, N., Correa, J. U., Cabib, E. 1982. *J. Biol. Chem.* In press
208. Cabib, E., Ulane, R. 1973. *Biochem. Biophys. Res. Commun.* 50:186–91
209. Cabib, E., Ulane, R., Bowers, B. 1973. *J. Biol. Chem.* 248:1451–58
210. Wolf, D. H., Ehmann, C. 1978. *FEBS Lett.* 92:121–24
211. Wolf, D. H., Ehmann, C. 1979. *Eur. J. Biochem.* 98:375–84
212. Zubenko, G. S., Mitchell, A. P., Jones, E. W. 1979. *Proc. Natl. Acad. Sci. USA* 76:2395–99

Ann. Rev. Biochem. 1982. 51:795–812

ENZYME THERAPY: PROBLEMS AND SOLUTIONS

John S. Holcenberg[1]

Department of Pharmacology and Toxicology, Medical College of Wisconsin, Milwaukee, Wisconsin 53226

CONTENTS

Perspectives and Summary

Enzymes are unique therapeutic agents. Small amounts of these biological catalysts can rapidly produce large, very specific effects at physiological pH and temperature. Their large molecular size limits their distribution in the body after parenteral administration, but allows for local therapeutic effects.

These properties have led to numerous applications, first as crude topical and oral preparations with proteolytic and hydrolytic activity, and later as highly purified enzymes for the treatment of cancer, clotting disorders, genetic defects, inflammation, digestive problems, drug toxicities, and kidney failure.

Native enzymes are often unsuitable for these therapeutic uses because of their antigenicity, rapid inactivation, and limited distribution in the body. It is rarely possible to obtain sufficient quantities of the desired enzyme from the same species. Foreign enzyme proteins can elicit an immune response that alters the clearance rate or enzyme activity, or causes severe allergic reactions in the host. After intravenous injection of an enzyme, its activity

[1]This work is supported by USPHS Grant CA 20061, the Burroughs Wellcome Fund, and the Midwest Athletes Against Childhood Cancer.

0066-4154/82/0701-0795$02.00

in plasma decreases exponentially with time due to distribution to other fluids and tissues and to proteolysis and excretion. Distribution appears to be related to molecular size, charge, and glycosylation. The rate of degradation of some enzymes appears to be related to surface change or accessibility of free amino groups.

In the past decade, many techniques have been developed to improve the therapeutic properties of enzymes. These include 1. soluble chemical modifications, 2. insoluble chemical modifications or binding to surfaces, and 3. encapsulation in biodegradable or inert materials. Since cofactors and several enzymes can be coupled or encapsulated together, these techniques may expand therapeutic uses of enzymes to more complex reactions and reaction sequences.

Enzyme therapy has been extensively reviewed in several recent books and chapters (1–4). This chapter concentrates on the special problems and solutions in the use of enzyme therapy in treatment of cancer, genetic disorders, clotting problems, and toxic reactions. Each of these areas presents unique problems for the development of therapeutic enzymes.

Aminohydrolases have been used in cancer chemotherapy to deplete specific amino acids required for tumor cell growth. Preparations with asparaginase and glutaminase activity are used in the treatment of human acute lymphocytic leukemia. Such therapy requires prolonged depletion of the target amino acid in the fluids bathing the tumor cells. Since the amino acid concentration in these fluids equilibrates with blood, an extracorporeal enzyme reactor could deplete the amino acid. However, currently available reactors are not efficient enough to achieve the degree of depletion produced by the injection of soluble enzymes. Injected enzymes must have prolonged activity in circulation and elicit little immune response. Soluble adducts with polyethylene glycol, dextran, amino acids, and succinic anhydride have been shown to improve these properties with several antitumor enzymes.

The treatment of genetic diseases with enzymes that replace the defective or absent ones presents several special problems. In many cases, the enzyme activity is absent in the lysosomes of a particular organ. Consequently, the replacement enzymes must be glycoproteins containing residues that interact with specific receptors on the diseased cells, be incorporated into the lysosomes, and retain activity. Furthermore, these enzymes need to be nonantigenic, since they must be administered chronically. Recent work has found suitable animal models for these genetic diseases, has identified the cellular uptake mechanisms of these enzymes, and has investigated encapsulation and other means to target the enzymes to specific tissues.

Both the absence of clotting factors and excessive clotting can be treated with enzymes. Most clotting factors are highly specific proteolytic enzymes.

Absence of an individual factor can be treated by infusion of crude or purified preparations. Chemically modified preparations are needed to decrease antigenicity and prolong the activity of some clotting factors. Excessive clot formation in vital organs can be treated by activation or addition of fibrinolytic enzymes. The major problems of fibrinolytic therapy are excess depletion of fibrinogen, inability to direct the enzymes to the clot, and lack of large amounts of the human enzymes. Recent advances are the commercial production of human urokinase from tissue culture cells and the discovery of a more specific human tissue plasminogen activator.

Removal of circulating levels of endogenous and exogenous toxic materials may be greatly facilitated by specific enzymes. For this use, the enzymes must be incorporated into a biocompatible device that can be rapidly attached to the circulation or implanted in an appropriate site.

Cancer Chemotherapy

USES AND PROBLEMS   The major impetus for enzyme therapy of cancer came from the observation by Broome that the antilymphoma factor (4a) in guinea pig serum was asparaginase (5). Subsequently, bacterial sources were found for this enzyme and clinical trials showed antileukemic activity (6). Asparaginase was a particularly exciting drug because its cytotoxicity was based on a biochemical difference between sensitive and resistant cells. Sensitive tumor cells had insufficient asparagine synthetase activity to support growth in media lacking this amino acid. When the circulating levels of asparagine were depleted by asparaginase treatment, the sensitive cells were unable to grow or survive. Most normal cells have sufficient constitutive or inducible levels of asparagine synthetase to survive depletion of asparagine in the surrounding media or circulation (6a).

An important concept from this work with asparaginase was that tumor cells could have a masked requirement for a nutrient. Under normal conditions, these tumor cells do not need to synthesize asparagine, since it is readily available in the diet. Sensitive cells are killed only when the circulating level of asparagine is depleted by the enzyme.

Tumor cells may have other masked deficiencies in enzymes that catalyze the synthesis of usually nonessential nutrients. Treatment with enzymes that deplete the nutrient would unmask the requirement and selectively kill the deficient cells. This therapeutic strategy depends on the discovery of the hidden requirement and the development of suitable enzymes that can degrade the nutrient. These enzymes should have the following characteristics: 1. low substrate K_m, 2. optimal activity at physiological pH, 3. stability in blood and other body fluids, 4. slow clearance from circulation, 5. no inhibition by high concentrations of its products, 6. no cofactor require-

ments or prosthetic groups that can easily dissociate from the enzyme, and
7. low endotoxin contamination.

All nonessential amino acids are candidates for this approach. A review
of the biosynthetic pathways and requirements for amino acids in animals
and tissue culture shows that certain tumor cells require glutamine, arginine, cysteine, or citrulline for growth and survival (7). These amino acids
have single biosynthetic pathways that are altered in some tumor cells.
Glutaminase-asparaginase enzymes have antitumor activity in animals and
man (8, 9). The major therapeutic problems of the current glutaminase-
asparaginase enzymes are their antigenicity and short half-life, and their
toxicity, which consists of negative nitrogen balance and neurotoxicity.
Chemical modification of *Acinetobacter* glutaminase-asparaginase with succinic anhydride lowers the isoelectric point from 8.2 to about 5 and prolongs
the half-life in man from less than 1 hour to greater than 1 day (8, 10, 11).
Preliminary studies indicate that amino acid supplements can reverse the
negative nitrogen balance in man and animals (12). The mechanism of
central nervous system toxicity is unknown. It is not due to glutamine
depletion in the cerebrospinal fluid, since the levels of this amino acid
actually rise in the fluid. Other possible causes of this toxicity are altered
levels of glutamine metabolites, glutamate, or ammonia in the brain or
cerebrospinal-fluid (8). Techniques to alter the antigenicity are outlined at
the end of this section.

Effective arginine-degrading enzymes are currently being studied. The
most promising enzymes are arginine deiminase from *Streptococcus faecalis,*
mammalian arginase coupled to polyethylene glycol, and arginine decarboxylase (7). At present, cysteine- or citrulline-degrading enzymes are not
available with the properties needed for effective enzyme therapy.

The other nonessential amino acids do not appear to be good candidates
for this therapeutic approach mainly because they are synthesized by multiple pathways. Serine may be an exception, since one of its two biosynthetic
routes, the glycine-serine interconversion, can be inhibited by dihydrofolate
reductase inhibitors.

Essential amino acids are required by both host and tumor. Therefore,
enzymatic or dietary depletion of these amino acids for prolonged periods
would be highly toxic to the host. Deprivation of selected essential amino
acids for 2–3 weeks is relatively well tolerated and may have pronounced
antitumor effects. For example, treatment of mice with indolyl-3-alkane-α
-hydroxylase from a *Pseudomonas* soil-isolate lowered tryptophan concentrations in plasma, brain, and lungs and inhibited the growth of a variety
of tumors (13). The biological activity of this heme-containing enzyme
indicates that certain oxidases can function as drugs in vivo. In contrast,
a tyrosine phenol-lyase from *Erwinia herbicola* is not a very effective antitu-

mor agent largely because it requires pyridoxal phosphate, which can be stripped from the enzymes in vivo (14). If this enzyme could be chemically modified to stabilize the binding of pyridoxal phosphate, it might be an effective therapeutic agent.

Phenylalanine ammonia lyase has produced variable antitumor effects in animals. Methionine is not completely essential, since some tissues can form the amino acid from homocysteine. Some tumor cells lack this pathway. A methionine-degrading enzyme from *Clostridium sporogenes* has been isolated, but it has a relatively high substrate K_m and requires pyridoxal phosphate (15). Other methionase enzymes with better properties are currently being studied (16).

Vitamins, sugars, lipids, nucleic acids, and peptides are other circulating compounds that may be required in higher concentrations by tumor cells than normal cells and could be targets of enzyme depletion. To date, enzymes have been utilized to deplete only folates and ribonucleic acids.

Folates can be degraded by oxidative deamination, hydroxylation, and hydrolysis of the peptide bond linking glutamate to the pteroic acid derivatives. The carboxypeptidases that catalyze this glutamate hydrolysis have been studied in great detail (17). They depleted folate in the media and inhibited the growth of a number of tumor cell lines in tissue culture. The inhibition was reversed by the addition of the end products of folate coenzyme–requiring reactions: thymidine, hypoxanthine, serine, and glycine. In vivo studies in animals and man have shown only partial depletion of folate levels and only modest antitumor effect. These disappointing results may be due to rapid clearance of the enzyme, which could be improved by chemical modification.

These enzymes may have a more important use in the rescue of patients from the toxicity of the antifol, methotrexate. Methotrexate inhibits dihydrofolate reductase, the enzyme that catalyzes the reduction of folate to the tetrahydrofolate level. Administration of high-dose methotrexate, followed in 12–36 hr by treatment with the reduced folate, ^{5}N-formyltetrahydrofolate (leucovorin), has increased the therapeutic index of methotrexate for the treatment of several human tumors. Carboxypeptidase G hydrolyzes methotrexate to 4-amino-^{10}N-methylpteroate, which is a poor inhibitor of dihydrofolate reductase. Carboxypeptidase G rescue has several advantages over leucovorin. The enzyme effect is irreversible, while leucovorin competes with methotrexate for cellular uptake. Repeated doses of leucovorin may increase the reduced folate levels and lessen the effectiveness of methotrexate, while the enzyme accentuates the methotrexate effect by depleting folate stores. The limited distribution of the enzyme may also be an advantage. For example, Abelson and co-workers (18) showed in a patient with a brain tumor that carboxypeptidase G was excluded from the cerebrospinal fluid by the blood brain barrier. Systemic administration of the enzyme hydrolyzed methotrexate in the blood and not in the cerebrospinal fluid, which allowed for prolonged effect of the antifol in the brain.

Many different ribonuclease (RNase) enzymes are present in animal and human plasma, urine, and tissues (19). Most of these enzymes are small enough to be eliminated by the kidneys probably through glomerular filtration, tubular reabsorption, and destruction. The clearance of endogenous RNase is about 4% of the clearance of creatinine in man. Exogenously administered RNase has been shown to have variable antitumor and antiviral effects depending on the source and molecular weight of the enzyme. The native bull semen and dimerized pancreatic RNases (molecular weights 27,000–29,000) have more activity toward double-stranded RNA, are taken up by cells at a faster rate, and have much more antitumor activity than the monomeric pancreatic enzyme. Recently, bovine pancreatic RNase was tested in the USSR for the treatment of tick-borne encephalitis. Patients who receive the enzyme every 4 hr over 5–6 days in doses sufficient to elevate the plasma enzyme activity by 2- to 3-fold recovered faster than those receiving gamma globulin. The enzyme has now been introduced in that country as a new antiviral drug for this disease (19). It is not known whether the dimeric enzyme would be more active.

Enzymes have also been used to alter the cell surface of tumor cells (20). In vitro treatment with *Vibrio cholerae* neuraminidase can modify the physical, biological, and immunological properties of neoplastic cells without altering their viability. Injection of these treated cells can induce a specific immune response and improve the survival of animals inoculated with untreated tumor cells. Clinically, neuraminidase-treated cells have been used to immunize patients to their tumors, and neuraminidase has been injected directly into superficial tumors. The results to date are difficult to interpret and reproduce.

Enzymes do not have to be specific for the surface components on tumor cells if the cells are treated in vitro and then injected. In vivo treatment requires some method to target the enzyme to the tumor cell or requires the selection of an enzyme that recognizes a unique structure on tumor cells. We need a much greater understanding of the tumor cell surface and the action of enzymes that alter this surface before this therapy can be optimally used.

SOLUTIONS   One solution to the problems with a particular enzyme is to find another native enzyme that is easier to purify, has less endotoxin, is less immunogenic, and has better kinetic and biological properties. For example, *E. coli* asparaginase elicits antibodies that often prevent its prolonged use in man. *Erwinia* asparaginase has similar properties but does not cross-react immunologically with the *E. coli* enzyme. It can be substituted when this problem occurs. Both *E. coli* and *Erwinia* asparaginase can hydrolyze glutamine to a limited extent. Since the glutamine concentration in plasma is ten times the asparagine concentration, glutamine can compete for the active site and lessen the enzymatic activity toward asparagine. In addition, the depletion of glutamine may account for some of the toxicity of these enzymes (6, 7). Both of these problems can be overcome by an asparaginase enzyme without glutaminase activity. Such an enzyme from *V. succinogenes* has been shown to have antitumor activity and be less immunosuppressive than *E. coli* asparaginase (21).

Soluble chemical modifications, encapsulation, and binding to surfaces have been used to improve the biological properties and reduce the antigenicity of antitumor enzymes.

The plasma half-life of *E. coli* and *Erwinia* asparaginase preparations have been increased by deamination, acylation, and carbodiimide reactions with free amino groups (22–24). In most cases, the plasma half-life increased as the isoelectric point of the protein decreased. We showed that succinylation and glycosylation with glycopeptides of *Acinetobacter* glutaminase-asparaginase greatly increase the half-life in mice, rats, rabbits, and man (8, 11). The antigenicity of this enzyme was not altered by these modifications.

Work with synthetic peptides, both free and coupled to proteins, has shown that certain polyamino acids can decrease the immune response to these foreign polymers (25). Uren & Ragin (26) applied this technique to *E. coli* and *Erwinia* asparaginases by reaction of these enzymes with DL-alanine-N-carboxyanhydride. The polyalanated preparations had prolonged half-lives, were less immunogenic in mice, and cross-reacted with antibodies developed to the native enzymes 300- and 500-fold less.

Other conjugates can also decrease antigenicity. Poly(N-vinylpyrrolidone) conjugated to β-D-N-acetyl hexosaminidase A also increased the half-life and decreased the reactivity to antibodies developed to the native enzyme (27). In addition, the soluble adduct between serum albumin and Hog uricase failed to elicit antibodies or react with antibodies to the native enzyme.

Coupling dextrans of various molecular weight to proteins can also increase their half-life in circulation. The effect of this reaction on immunogenicity has not been studied in detail (27, 28).

The attachment of polyethylene glycol to proteins often produces a prolongation of plasma half-life and a diminution or elimination of immunogenicity (27). Usually polyethylene glycol with a mol wt of 5000 is activated to 2-O-methoxypolyethylene glycol-4,6-dichloro-S-triazine and reacted with the enzyme at controlled temperature and pH (29). Polyethylene glycol conjugates of beef liver arginase and glutaminase-asparaginase from an *Achromobacter sp.* had longer circulating half-lives, no immunogenicity, and increased anitumor effect than the native enzymes. Unfortunately, the activity of some antitumor enzymes is rapidly lost by the coupling reaction: these include phenylalanine ammonia-lyase, *E. coli* and *V. succinogenes* asparaginase, and *Acinetobacter* glutaminase-asparaginase.

These soluble modifications also can increase the stability of the enzymes to heat and protease inactivation. In several cases, the substrate K_m has been increased and the pH optimum shifted. For example, Schmer & Roberts (30) showed that the tryptophan-degrading enzyme, indolyl-3-alkane-α-hydroxylase has a pH optimum of 3.5, with low activity at neutral pH. Coupling with either polyacrylic or polymaleic acids increased the pH optimum and increased the activity at pH 7 threefold.

Enzymes have also been entrapped in microspheres of synthetic or natural materials (31). Chang and co-workers (32) studied encapsulation in various biostable, synthetic materials with techniques that allowed a wide range of pore size and cell diameter. The major advantages of these cells are their ultrathin membranes and high ratio of surface area to volume. The usual pore size permits exchange of permeable small molecules but not of the enzyme or antibodies. These microspheres have several disadvantages. 1. The kinetic properties of the enzyme are usually decreased by inactivation or by slow diffusion of substrates and products across the membrane. 2. The cells can invoke an inflammatory response if injected subcutaneously or intraperitoneally. This reaction leads to a coating of the spheres with fibrin or connective tissue, which further decreases the rate of diffusion of substrate and products. 3. Injection of the spheres intravenously can plug capillary beds (32).

Biodegradable microspheres have been made of lipids (liposomes) and albumin. These cells are usually entrapped and degraded by the reticuloendothelial system. Nevertheless, they potentially can be targeted to other cells by surface attachment of specific antibodies, F(ab)$_2$ fragments, lectins, glycoproteins, lipids, or ligands. The composition of lipids in liposomes has been varied to alter their charge, permeability, rate of degradation, and uptake by certain tissues. A recent exciting approach is the use of lipids with phase transitions above physiological temperature. This composition allows release of the contents of the liposomes at sites by local heating (33). Biodegradable microspheres have two major disadvantages: 1. they are unstable in plasma and tissues and 2. the membrane lipids may actually increase the immune response to a foreign protein (34).

Red blood cell ghosts have several advantages as carriers of enzymes. When prepared from animals of the same species and blood type, they are immunologically compatible. They have prolonged bioaction, substrate selectivity for membrane transport, and the potential for tissue targeting.

Proteins can be entrapped in red blood cells by hypotonic hemolysis and resealing in isotonic salt solutions, glycol- or glycerol-induced osmotic lysis, and electric hemolysis in isotonic solutions. The last method appears to have the greatest potential, since the resealed ghosts are most stable in vivo after this procedure (31).

The immobilization of therapeutic enzymes has been extensively studied (2, 35). Coupling methods include absorption, ionic binding, cross-linking, entrapment, and covalent attachment. Table 1 summarizes the advantages and disadvantages of each method. For prolonged antitumor activity, the immobilized enzyme must be accessible to substrates and products, not leach from the surface, and not interact with antibodies. The absence of leaching should greatly decrease or eliminate antibody formation and prevent immediate hypersensitivity types of allergic reactions. These requirements are met by the cross-linking, entrapment, and covalent attachment techniques.

The only example of cross-linking of antitumor enzymes is in pancreatic ribonuclease, which has improved antitumor activity as the dimer (19).

Table 1 Comparison of immobilizing methods[a]

Method	Advantages	Disadvantage
Absorption	Simple, inexpensive. Very nonspecific binding. Easy to regenerate, Any substrate size.	Binding affected by pH, ionic strength, and temperature. Continuous leakage of enzyme. Low capacity.
Ionic binding	Gentle, reversible. Very nonspecific binding. Inexpensive.	Binding affected by pH, ionic strength and temperature. Limited applicability. Too nonspecific. Low capacity.
Cross-linking	Many reagents available. No enzyme leaching. High capacity.	Diffusion of substrate becomes limiting. Rubbery consistency.
Entrapment	Large quantity/volume. Can vary size thickness and hardness. Biocompatible and/or biodegradable.	Diffusion limited. Only low-molecular-weight substrates. Loss of activity from solvents.
Covalent attachment	Most stable. Many carriers and coupling techniques. Stable; little leakage of enzyme.	Costly. Least gentle. Many coupling reagents are toxic.

[a] Adapted from (35).

Entrapment in microcapsules is discussed above. Enzymes also can be entrapped within the interstitial spaces of cross-linked water insoluble polymers like polyacrylamide, polyvinyl alcohol, polyacrylics, and polystyrene. For example, Horvath and co-workers (36) covalently bound asparaginase to a polycarboxylic gel layer attached to the inner wall of small-bore nylon tubes. The tubes rapidly depleted asparaginase in a liver perfusion system.

The kinetics of asparagine hydrolysis by this system were a complex mixture of diffusional effects, partitioning, and Michaelis-Menten that varied with the age of the tubes. Theoretical approaches to the kinetics of solid-supported enzymes have been developed (37–39). In general, at lower substrate concentrations and flow rates, the reactions are largely controlled by the rate of diffusion. At high substrate concentrations and flow rates, the diffusion layer is negligible and the apparent K_m approaches the true K_m for the immobilized enzyme. This true K_m value will equal that for the native enzyme in solution if the cross-linking has not altered the characteristics of the enzyme.

The most commonly used method of immobilization is by covalent attachment to a support that is insoluble, slightly soluble, or readily soluble. Chemical methods for coupling include diazonium reactions with tyrosine hydroxyl groups, isocyanate or isothiocyanate reactions with amino groups, acid azide reactions with a variety of groups, and cyanogen bromide, maleic anhydride, Woodward's Reagent K, carbodiimides, alkylation, cyanuric chloride, and halogen derivative reactions (35).

Binding to the surface of a polymer allows for greater exposure of the enzyme to its substrate than entrapment. However, it also permits hydrolysis of the covalent linkage, which results in leaching of the enzyme. The cleaved enzyme can elicit an immune response that can inactivate both the free and bound enzyme (40). Furthermore, hydrolysis of the covalent linkage makes it impossible to know whether a therapeutic effect is due to the bound or free enzyme (41). An additional problem is the deposition of fibrin and plasma proteins on the surface, which can impair the diffusion of substrate and product to the enzyme.

These problems with covalently attached enzymes can be overcome by developing more stable chemical bonds, coating the enzyme with materials like polyethylene glycol to decrease immunogenicity, using more biocompatible surfaces, and impregnating the surface with heparin or fibrinolytic enzymes.

Enzyme Therapy in Genetic Diseases

USES AND PROBLEMS   Enzyme replacement is the logical approach for the treatment of genetic diseases that are caused by defined enzyme defects. This is especially true for lysosomal storage diseases where the pathology is due to a lack of a degradative enzyme in this subcellular organelle (42). Tissue culture studies with fibroblasts from patients with various lysosomal storage diseases have shown that exogenous enzymes can enter the cell, gain access to the accumulated substrates, retain intracellular activity for many days, and normalize substrate turnover (43). These in vitro corrections have been accomplished with fibroblasts from individuals deficient in various glycosphingolipidases, mucopolysaccharidases, glycogenases, glycoproteinases, glucoasminidase, and glucuronidase.

Early clinical trials with crude enzymes were not successful because of the following problems: 1. short half-life of enzyme activity in circulation and in the cell, 2. inability to target the enzymes to the specific tissue or subcellular site, 3. difficulty in monitoring the clinical results of the therapy, and 4. immunological complications (43). Subsequently, human trials have used highly purified, well-characterized enzymes, usually isolated from human placenta, spleen, and urine. These enzymes include β-hexosaminidase A, α-galactosidase, β-glucosidase, α-glucosidase, and glucocerebrosidase (43, 44). The plasma half-lives were usually less than 20 min, but enzyme activity was demonstrated in liver biopsies, and accumulated substrate was catabolized.

When β-hexosaminidase was administered intravenously to a patient with a G_{M2} gangliosidosis characterized by accumulation in the central nervous system, the enzyme appeared in liver, but not in cerebrospinal fluid or biopsied brain tissue. Administration of enzyme into the cerebrospinal fluid also did not alter the accumulation of ganglioside in the brain. These results show that the blood brain barrier prevents enzyme treatment of disorders with severe central nervous system involvement.

In two lysosomal storage diseases, Fabry and Gaucher's type I, the accumulated material may equilibrate sufficiently with plasma so that circulating replacement enzyme can gradually deplete the tissue stores. In other cases, the enzyme must gain access to the substrate within the lysosomes of particular tissues. Animal studies have shown rapid uptake of these enzymes by liver but not by other tissues. Since these diseases are characterized by accumulation in neurons, endothelium, fibroblasts, reticuloendothelial system, muscle, or cartilage, methods are needed to target the enzymes away from the liver and to these sites.

SOLUTIONS   In the last decade, considerable information has developed on specific glycoprotein uptake systems of various mammalian cells and tissues. These recognition markers include β-galactose for hepatocytes, kidney, spleen, thymus, heart, and lung; mannose-6-phosphate for fibroblasts; low density lipoproteins for fibroblasts, vascular endothelium, and smooth muscle; α-mannose and N-acetylglucosamine for reticuloendothelial cells; fucose for hepatocytes; and an as yet uncharacterized group on β-hexosaminidase A for brain synaptic plasma membranes (43, 44). The following strategies are being studied to use these systems to deliver enzyme to cells other than the liver: 1. selective removal of carbohydrates on the enzyme to expose the desired recognition marker; 2. coupling of the desired recognition marker to the enzyme; 3. selection of a specific isozyme with high uptake for the desired tissue; and 4. use of nontoxic carbohydrates, oligosaccharides, or glycopeptides to selectively block enzyme uptake by the liver.

The problem of entry of enzymes into the brain may also be solvable. Recent studies show that the blood brain barrier can be transiently opened by carotid artery injection of hypertonic solutions or hyperbaric oxygen (44). During this period, enzyme injected intraarterially may enter the central nervous system.

Another approach is the entrapment of the enzyme in liposomes or red blood cells and then the targetting of them to the pathologic organs. Desnick and co-workers (45) showed that negatively and positively charged liposomes containing β-glucuronidase were rapidly cleared by the liver, with a retention of activity in the liver for up to 11 days. Some activity was detected in spleen and kidney, but no activity was found in bone marrow or brain. A recent report shows the enzymes may be incorporated into the brain by use of liposomes containing sulfatide, phosphatidylcholine, and cholesterol (46). Preliminary studies have shown that liposomes can be targeted to specific cells by surface incorporation of antitarget-site antibodies (47). Moreover the distribution of liposomes potentially can be altered by their physical size, lamellar structure, and lipid composition and by the surface incorporation of drugs, ligands, and lectins.

The liposomes have two major disadvantages for enzyme therapy. 1. They appear to enhance immune response to the entrapped proteins and 2. positively charged liposomes disrupt intracellular organelles (43). An immune response will occur if the exogenous enzyme is from another species or if the host makes no enzyme protein rather than an inactive enzyme. The soluble chemical modifications discussed above may overcome this immunogenicity. Careful selection of lipids should prevent the tissue damage.

Entrapment in red blood cells not only can target enzymes to the liver and spleen, but also can decrease any immune response. Potentially, materials can be coupled to the red cell membrane to target these natural carriers to other tissues. For example, Beutler et al (48) has used Rh antibody–coated human erythrocytes containing β-glucosidase to increase the uptake of these cells by the reticuloendothelial system in patients with Gaucher's disease.

A new solution is suggested by the report of direct transfer of a lysosomal enzyme from lymphoid cells to deficient fibroblasts in tissue culture (49).

Such a transfer might be produced in vivo from histocompatible normal lymphoid cells. Permanent correction of the enzyme defect is possibly by transplanting HLA-identical bone marrow stem cells, skin, liver, or kidneys. As techniques develop for genetic engineering, the normal gene can perhaps be introduced into the patient's own cells prior to transplantation.

Another major advance has been the immobilization of all four enzymes of the urea cycle and inorganic pyrophosphatase in a fibrin fiber formed from fibrinogen by the action of thrombin and transglutaminase (50). This multienzyme system was able to carry out the urea cycle with much greater efficiency than the soluble enzymes. It could be used for the treatment of any of the genetic defects in the urea cycle. Moreover, it illustrates a technology that can be applied to the enzyme therapy of many other complex metabolic pathways.

Enzyme Therapy in Clotting Disorders

USES AND PROBLEMS   Inherited or acquired defects in the glycoproteins required for clot formation can lead to severe hemorrhaging. Normal clotting, initiated after a vascular injury, includes the deposition of platelets, activation of Factor XII, and release of a tissue factor. These events lead to a cascade of activation of proteolytic factors that eventually cleave fibrinogen to fibrin and cross-link the fibrin into an insoluble matrix. Finally, activation of the fibrinolytic system causes the dissolution of the hemostatic plug.

The most common congenital defects are hemophillia A (lack of Factor VIII coagulant), hemophilia B (lack of Factor IX), and Von Willebrand's factor (lack of part of Factor VIII complex). Acquired defects occur in fibrinogen and Factor V due to liver disfunction, and in Factors II, VII, IX, and X due to vitamin K deficiency, drugs, or liver dysfunction.

Historically, fresh-frozen plasma or freeze-dried Cohn Fraction I concentrates were used for rapid replacement of these factors. Recently, cryoprecipitates and other partially purified fractions of blood have become commercially available. These preparations have greatly improved the chances of survival and the quality of life of patients with hemophilia. Nevertheless, the following problems limit their effectiveness: 1. The crude preparations contain many unneeded, antigenic materials, and may contain hepatitis virus; 2. Some preparations have activated factors that can cause excessive clotting: 3. Certain factors (particularly VII and VIII) have short half-lives, and 4. Antibodies may develop after prolonged use (51).

Enzymes and enzyme activators also have been used clinically to accelerate the dissolution of clots in pulmonary emboli, myocardial infarction, cerebral infarction, and occlusion of lower limb vessels (52, 52a). Streptokinase and urokinase activate plasminogen to plasmin, a serine protease that hydrolyzes a wide variety of proteins including fibrin at lysyl and arginyl residues. Streptokinase must combine with plasminogen to become an activator. Urokinase is a serine protease that directly cleaves plasminogen. Plasminogen gains specificity by being absorbed to the clot and being activated at the surface or within the matrix of the clot. Plasma contains various protease inactivators that prevent excessive plasmin proteolysis.

A problem with streptokinase is its lack of specificity toward plasminogen within clots. Consequently, plasma plasminogen is reduced to near zero levels at therapeutic concentrations of streptokinase. Since continued clot lysis requires incorporation of plasminogen in new clots, this depletion of plasminogen may prevent sustained thrombolytic action. In addition, excessive free plasmin activity may hydrolyze fibrinogen to degradation products that interfere with clot formation and lead to hemorrhage.

Urokinase has more specificity for plasminogen and fibrin. Thus, clot lysis can be produced with minor decrease in plasma plasminogen concentration. An activator has been isolated from tissue extracts that has even higher affinity for fibrin. In animal studies of experimental pulmonary emboli, this protein had a much more selective effect on clot lysis than the other activators (53).

The major problems with urokinase are its rapid clearance ($t_{1/2} = 9-16$ min) and limited availability. It is a human enzyme and therefore not antigenic. Streptokinase is readily available but is antigenic to man. Due to prior infections with microorganisms containing streptokinase, many people have antibodies to this protein. The antibody inactivates streptokinase and decreases its plasma half-life from 80 to 18 min. For treatment, sufficient streptokinase must be infused to bind the antibody before a therapeutic effect is seen. Fortunately, acute, life-threatening allergic reactions are rare (52).

Two enzymes isolated from snake venoms, Ancrod and Batroxobin, catalyze the cleavage of fibrinopeptide A and the formation of a weak, easily lysed clot. Injection of either enzyme results in a fall in plasma fibrinogen, an increase in fibrin degradation products, a decrease in plasminogen, and no change in fibrinolytic activity. Interest in these defibrinating enzymes arose from reports that subjects bitten by the Malayan pit viper had very low fibrinogen levels but little toxicity. In clinical trials, Ancrod was as effective as heparin for prevention of deep vein clots (52). A disadvantage is the antigenicity of these enzymes. Neutralizing antibodies develop after 4–6 weeks of treatment.

SOLUTION   Most clotting factors have been extensively purified and characterized. This knowledge should be applied to develop large scale purification methods for Factors IX and VIII coagulant and Von Willebrand's Factor. Introduction of the appropriate genes in bacteria may also be useful in large scale production of the proteins, although coupling of the correct polysaccharides may be needed for full activity. Soluble modifications of Factor VIII may extend its biological half-life and decrease its antigenicity.

Soluble chemical modifications of plasminogen activators should be able to prolong their half-life in circulation and decrease any interaction with antibodies. Unfortunately, efforts to date have led to loss of activity (52). Characterization of the specific binding sites for tissue plasminogen activator (53) on fibrin will, we hope, lead to the targeting of other activators to fibrin and not fibrinogen.

Urokinase has been made more available by the development of tissue culture methods for its large scale production from embryonic kidney cells (54). This technique may be the way to produce other human enzymes to replace deficiencies of clotting factors and for other genetic disorders.

Enzyme Therapy in Toxic Reactions

USES AND PROBLEMS   The usual treatment of poisoning consists of prevention of further absorption, achieved by emesis, gastric lavage and oral charcoal; administration of an antidote; enhancement of excretion; and supportive care. Occasionally, those measures are supplemented by exchange of peritoneal fluid or by hemodialysis. Dialyzable poisons, drugs, and endogenous toxins include barbiturates, salicylates, carbon tetrachloride, lithium, amphetamines, ethylene glycol, most antibiotics, methanol, bromides, ammonia, uric acid, bilirubin, and urea (55). The efficiency of dialysis depends on the concentration gradient between blood and the dialysis fluid. Thus, toxins that are bound to tissues or plasma proteins will be poorly dialyzed. An alternative has been perfusion of blood through resins or charcoal to absorb certain toxins.

Enzymes should be ideal agents for removal of toxins since they usually have high affinity to compete with plasma proteins and tissue sites for the toxin. This approach depends on the discovery of an appropriate enzyme, the elimination of any toxic products of the enzymatic reaction, and the avoidance of allergic reactions to the enzymes.

There are few example of enzyme therapy in toxic reactions. Carboxypeptidase G has been used to reverse methotrexate effects (17). Uricase is currently being used in Europe to treat high levels of uric acid in patients with leukemia. Many toxins are candidates for enzyme degradation. These include poorly dialyzed drugs like digitalis glycosides, benzodiazepams, phenothiazines, glutethimide, methaqualone, and tricyclic antidepressants.

Specific enzymes may be able to degrade some of the unusual peptide and protein toxins from mushrooms, insects, and snakes.

SOLUTIONS Soluble enzymes may require frequent administration because of rapid clearance and may produce allergic reactions. Chemical modifications can improve these properties. For example, uricase from two sources was nonimmunogenic when coupled to polyethylene glycol and was superior to native enzyme in lowering blood uric acid levels in animals (27).

Enzymes have been encapsulated or coupled to surfaces to provide stable reactors for the removal of toxins. Chang (56) pioneered this approach by the use of microencapsulted catalase to treat H_2O_2-induced lesions in acatalasemic mice and microencapsulated urease to treat high urea levels in renal failure. In the case of urease, one product, ammonia, had to be removed because of its potential toxicity. A compact artifical kidney was devised that consisted of microcapsules containing urease, an ammonia absorbant, and activated charcoal, which removed other uremic metabolites. This system was effective because of the ultrathin membrane (400 Å) and large surface area. For example, 30 ml of 100-micron diameter microcapsules had a larger total surface area and could exchange small molecules at least 100 times faster than a conventional artificial kidney. This system requires regeneration or replacement of the absorbers and resins. Chang's studies showed that the microspheres were biocompatible and the reactor did not destroy blood cells.

Schmer entrapped urease in red blood cell ghosts bound to a sephadex matrix with Concanavalin A and glutaraldehyde (31). This system rapidly hydrolyzed urea but had poor flow characteristics in vivo.

A simpler approach to the enzymatic treatment of uremia has been proposed and tested by Setala (57). He utilized intact, nonpathogenic bacteria that can metabolize most of the toxic materials in patients with uremia. The lyophilized bacteria are encapsulated in 25-micron microspheres, packed into enteric capsules, and administered orally. The microspheres released in the small intestine degrade the toxins within the gut lumen. Since the intestinal tract is in equilibrium with the blood, the toxins are gradually cleared from the body.

Uricase has been encapsulated in red blood cells or bound to the membranes of a dialysis unit (35, 58). Both systems rapidly removed uric acid. Further development of such reactors has been hampered by their cost and the increasing efficiency of standard dialysis techniques in removing small, water-soluble toxins. Future efforts should be applied to the use of enzyme reactors to degrade materials that are not ordinarily dialyzable.

Literature Cited

1. Holcenberg, J. S., Roberts, J. eds. 1981. *Enzymes as Drugs.* New York: Wiley-Intersci. 455 pp.
2. Chibata, I., 1978. *Immobilized Enzymes.* New York: Wiley-Intersci. 284 pp.
3. Holcenberg, J. S., Roberts, J., 1977. *Ann. Rev. Pharmacol. Toxicol.* 17:97–116
4. Cooney, D. A., Rosenbluth, R. J., 1974. *Adv. Pharmacol. Chemother.* 12:185–289
4a. Kidd, J. G. 1953. *J. Exp. Med.* 98:565–81
5. Broome, J. D., 1963. *J. Exp. Med.* 118:99–120
6. Capizzi, R. L., Cheng, Y.-C. 1981. See Ref. 1, pp. 1–44
6a. Horowitz, B., Madras, B., Meister, A., Old, L., Boyse, E., Stockert, E. 1968. *Science* 160:533–35
7. Holcenberg, J. S. 1981. See Ref. 1, pp. 22–57
8. Holcenberg, J. S., Borella, L. D., Camitta, B. M., Ring, B. J., 1979. *Cancer Res.* 39:3145–51
9. Mitta, S., Chou, F. C., Roberts, J., Steinherz, P., Miller, D., Tan, C., 1980. *Proc. Am. Assoc. Cancer Res.* 21:143
10. Warrell, R. P., Chou, T.-C., Gordon, C., Tan, C., Roberts, J., Sternberg, S. S., Philips, F. S., Young, C. W., 1980. *Cancer Res.* 40:4546–51
11. Holcenberg, J. S., Scher, G., Teller, D. C., Roberts, J., 1975. *J. Biol. Chem.* 4165–70
12. Kien, C. L., Holcenberg, J. S. 1981. *Cancer Res.* 41:2056–62
13. Roberts, J. 1981. See Ref. 1, pp. 63–76
14. Meadows, G. G., Digiovanni, J., Minor, L., Elmer, G. W., 1976. *Cancer Res.* 36:167–71
15. Kreis, W., 1979. *Cancer Treat. Rep.* 63:1069–72
16. Carr, K. A., Arduino, M. J., 1981. *Bios* 52:13–22
17. Kalghatgi, K. K., Bertino, J. R. 1981. See Ref. 1, pp. 77–102
18. Abelson, H. T., Ensminger, W., Rosowski, A., Uren, J., 1978. *Cancer Treat. Rep.* 62:629–33
19. Levy, C. C., Karpetsky, T. P. 1981. See Ref. 1, pp. 103–66
20. Jelsema, C. L., Killion, J. J., Winkelhake, J. L. 1981. See Ref. 1, pp. 259–312
21. Durden, D. L., Distasio, J. A. 1981. *Int. J. Cancer* 27:59–65
22. Wagner, O., Irions, E., Arens, A., Bauer, K., 1969. *Biochem. Biophys. Res. Commun.* 37:383–92
23. Rutter, D. A., Wade, H. E., 1971. *Br. J. Exp. Pathol.* 52:610–14
24. Hare, L. E., Handschumacher, R. E., 1973. *Mol. Pharmacol.* 9:534–41
25. Sela, M., 1966. *Adv. Immunol.* 5:29–129
26. Uren, J. R., Ragin, R. C., 1979. *Cancer Res.* 39:1927–33
27. Abuchowski, A., Davis, F. F. 1981. See Ref. 1, pp. 367–83
28. Foster, R. L., Wileman, T., 1979. *J. Pharm. Pharmacol.* 31 (Suppl.): 37P
29. Abuchowski, A., van Es, T., Palcuk, N. C., Davis, F. F., 1977. *J. Biol. Chem.* 252:3578–81
30. Schmer, G., Roberts, J., 1979. *Cancer Treat. Rep.* 63:1123–26
31. Schmer, G., Holcenberg, J. S. 1981. See Ref. 1, pp. 385–94
32. SiuChong, E. D., Chang, T. M. S., 1974. *Enzyme* 18:218–39
33. Weinstein, J. N., Magin, R. L., Yatvin, M. B., Zaharko, D. S., 1979. *Science* 204:188–91
34. Hudson, L. D. S., Fiddler, M. B., Desnick, R. J., 1979. *J. Pharmacol. Exp. Ther.* 208:507–14
35. Weetall, H. H., Cooney, D. A. 1981. See Ref. 1, pp. 395–443
36. Horvath, C., Sardi, A., Woods, J. A., 1973. *J. Appl. Physiol.* 34:181–87
37. Kobayashi, T., Laidler, K. J., 1973. *Biochim. Biophys. Acta* 302:1–12
38. Bunting, I. P. S., Laidler, K. J., 1974. *Biotechnol. Bioengin.* 16:119–34
39. Kuchel, P. W., Roberts, D. V., 1974. *Biochim. Biophys. Acta* 364:181–92
40. Cooney, D. A., Weetall, H. H., Long, E., 1975. *Biochem. Pharmacol.* 24:503–15
41. Sampson, D., Han, T., Hersh, L. J., Murphy, G. P., 1974. *J. Surg. Oncol.* 6:39–48
42. Desnick, R. J., Bernlohr, R. W., Krivit, W., eds. 1973. *Enzyme Therapy and Genetic Diseases, Birth Defects* Orig. Article Ser. 9(2). Baltimore: Williams & Wilkins. 233 pp.
43. Grabowski, G. A., Desnick, R. J. 1981. See Ref. 1, pp. 167–208
44. Gregoriadis, G., Dean, M. F., 1981. *Nature* 278:603–4
45. Desnick, R. J., Fiddler, M. B., Thorpe, S. R., Steger, L. D. 1977. *Biomedical Applications in Immobilized Enzymes and Proteins,* ed. T. M. S. Chang, pp. 227–44. New York: Academic
46. Naoi, M., Yagi, K., 1980. *Biochem. Int.* 1:591–6
47. Gregoriadis, G., 1974. *Enzyme Replacement Therapy of Lysosomal Storage Diseases,* ed. J. M. Tager, J. M. Hooghwinkel, W. T. Daoms, pp. 131–48. Amsterdam: North-Holland

48. Beutler, E. L., Dale, G. L., Kuhl, W. 1980. See Ref. 46, pp. 369–81
49. Olsen, I., Dean, M. F., Harris, G., Muir, H. 1981. *Nature* 291:244–7
50. Inada, Y., Tazawa, Y., Attygalle, A., Saito, Y. 1980. *Biochem. Biophys. Res. Commun.* 96:1586–91
51. Lazerson, J. 1981. See Ref. 1, pp. 241–58
52. Fletcher, A. P., Alkjaersig, N. K. 1981. See Ref. 1, pp. 209–40
52a. Verstraete, M. 1978. *Semin. Hematol.* 15:35–54
53. Matsuo, O., Rijken, D. C., Collen, D. 1981. *Nature* 291:589–90
54. Barlow, G. H. 1976. *Methods Enzymol.* 45:239–44
55. Rumack, B. H., Peterson, R. G. 1980. Cassarett and Doull's Toxicology: *The Basic Science of Poisons,* ed. J. Doull, C. D. Klaassen, M. O. Amdur, pp. 677–98. New York: MacMillan, 2nd ed.
56. Chang, T. M. S. 1973. *Birth Defects: Orig. Artic. Ser.* 9:66–76
57. Setala, K. 1979. *Clin. Nephrol.* 11:156–66
58. Ihler, G., Lantzy, A., Purpura, J., Glew, R. H. 1975. *J. Clin. Invest.* 56:595–602

Ann. Rev. Biochem. 1982. 51:813–44

REPETITIVE SEQUENCES IN EUKARYOTIC DNA AND THEIR EXPRESSION

Warren R. Jelinek

Department of Biochemistry, New York University Medical Center, New York, 10016

Carl. W. Schmid

Department of Chemistry, University of California at Davis, California 95616

CONTENTS

0066-4154/82/0701-0813$02.00

PERSPECTIVES AND SUMMARY

It has been more than ten years since it was realized that eukaryotic DNAs contain different sequence classes whose constituent members can be broadly classified as (*a*) unique (approximately one copy per haploid genome), (*b*) moderately repetitive ($\sim 10^3$–10^5 copies per haploid genome), or (*c*) highly repetitive ($\sim 10^6$ copies per haploid genome) (81). Since then the structure and function of the DNAs in each of these classes has been extensively investigated. The unique DNA contains, but is not necessarily composed exclusively of, protein encoding sequences, although some protein coding genes are also represented more than once per haploid genome. The highly repetitive DNA consists of clustered repetitions of relatively short sequence units that are thought not to be interspersed with other sequence types. They are generally thought to be structural components of chromosomes, residing mainly at centromeric and telemeric positions, but their functions are unknown. The moderatively repetitive sequences can be divided into two categories, long and short. They are interspersed with unique sequences. Some long dispersed repeats have structures similar to proretroviruses, and it is possible that retroviruses evolved from them. They are mobile, in some instances causing recognizable mutations at their sites of insertion into chromosomal DNA. They are transcribed into discretely sized RNA molecules that have modified 3' and 5' ends, characteristic of cellular and viral mRNAs. The short dispersed repeats can be categorized into different sequence families with different numbers of individual family members. Some families contain as many as 5×10^5 members, while others contain relatively few. Although structural analyses of cloned short repeats suggest they are mobile DNA elements, further studies are required to demonstrate their mobility. They are transcribed as part of mRNA precursor molecules, and some are also transcribed as discretely sized RNAs by RNA polymerase III. These repeats have alternately been suggested as sequence recognition elements involved in intricate pathways of gene expression or as parasitic DNAs that confer no phenotypic advantage on the cells in which they reside. Substantial information has been gathered concerning their structures and their associations with protein encoding sequences both in DNA and in RNA transcripts, but their functions have yet to be determined.

INTRODUCTION

Dispersed Repeats

Although dispersed repetitive sequences are characteristic of most eukaryotic DNAs (1–6), no single unifying description of their arrangement can be applied to all. However, the DNAs of such diverse species as sea urchins,

Xenopus, and humans exhibit a pattern of dispersed repeats, known as the "short period" interspersion pattern, in which 100–300-base pair (bp) long repeated sequences are interspersed with longer, single copy sequences of approximately 1000–2000 bp (2, 7, 8). Approximately 50–80% of the DNA from such organisms exhibit the short-period interspersion pattern (3–7, 9–15). Its existence in a portion of an organism's DNA does not preclude the presence of other types of sequence arrangements within the same DNA. For example, in addition to the short-period interspersed repeats, eukaryotic DNAs also contain interspersed repeated sequences that are significantly longer than 100–300 bp (16), long tandem repeats that are not interspersed with single copy sequences (reviewed in 17, 18), and long single copy sequences that are not interspersed with short repeated sequences (19). Long repeats in sea urchin DNA may contain short repetitive elements (20); possibly in some instances the latter are derived exclusively from the former. In the DNAs of some organisms a significant proportion of the short-period interspersed repeats occur as inverted repeated pairs (8, 21–24) with an arrangement and spacing similar to that of the total repeats within the same DNA, and are thus an organizational subset of the total short-period interspersed repeats.

Not all eukaryotic DNAs exhibit the short-period interspersion pattern. As the first documented exception, *Drosophila* DNA was found to have a long-period interspersion pattern with approximately 5000-bp long repeated sequences dispersed among single copy sequences that might be as long as 35,000 bp (25–27). Recent evidence has demonstrated that these long repeats, even though they consist of different sequence families, share a common structure in which the approximately 5000-bp repeats are flanked on either side by shorter direct repeats of approximately 255–400 bp (27–35). This arrangement of shorter direct repeats flanking long dispersed repeated sequences is also found in yeast DNA (36–39) and is characteristic of the structure of integrated retrovirus DNAs both in birds (40–46) and mammals (47–54), and resembles the structures of known bacterial transposons (55, 56). Chicken DNA has an intermediate interspersion pattern consisting of approximately 2000 nucleotide repeats spaced by approximately 4500 nucleotides of single copy sequences (57). In some instances dispersed repeats within chicken DNA have a clustered organization in which members of different repeated sequence families are arranged in arrays as long as 2×10^4 bp (58) that may be extensively methylated (59). These repeats differ from the short-period interspersed repeat sequences in that, except for their ends, the sequences neighboring any one repetitive family member are members of other repeated sequence families rather than single copy sequences. This type of clustering of different repeat sequences, as well as scrambling of different repeats within different clusters has also been observed in *Drosophila* (60) and sea urchin DNAs (61–63), and may occur to a greater or lesser degree in all DNAs that contain repetitive sequences. Plant DNAs also have a diversity of sequence arrangements [briefly reviewed in (64)]. As one especially noteworthy example, pea DNA has a general organization in which approximately 300-bp long single copy sequences are interspersed with approximately 4000-bp long repeated sequences (65).

The DNA sequence organization of the aquatic fungus *Achlya* illustrates the opposite extreme of sequence arrangement, with interspersed repetitive sequences so sparse that it is impossible to measure their lengths directly (66, 67). As limiting values, the interspersed repeats in *Achlya* are longer than 27,000 bp and are spaced by single copy sequences that are longer than 135,000 bp. Most of these repeated sequences can be ascribed to known multigene families such as ribosomal genes (66). Thus, various patterns of dispersed repeats can be distinguished in the DNAs of a variety of different eukaryotic organisms, and although the short-period interspersion pattern of dispersed repeats exists in the DNAs of the majority of organisms examined, it appears not to be an obligatory attribute of eukaryotic DNAs. There is no reason to believe that the different arrangements of dispersed repeats perform common functions in DNAs of the various organisms in which they are present. It is, however, plausible to imagine that the interspersed repeats in organisms with similar DNA sequence organizations may serve common functions and/or reflect a common dispersal mechanism. Because short-period interspersed repeats are present in most eukaryotic DNAs examined, and among single copy sequences that contain protein coding sequences (68), they possibly function in the control of gene expression (2, 69, 70) and therefore have been extensively investigated [for references through 1979 see (1)]. As discussed below, they are active in transcription, being represented in HnRNA and mRNA, and at least one family of dispersed repeats in mammals has been shown to direct the transcription of discretely sized low-molecular-weight RNA molecules by RNA polymerase III (see below). Despite considerable effort the biological function(s), if any, of these dispersed repeats remain obscure.

Clustered Repeats

In addition to dispersed repeats, eukaryotic DNAs contain clustered repeats, some of which are composed of simple sequences with repeat units no more than six bp long repeated as many as 10^6–10^7 times (71), while others are hundreds of bp long and are repeated millions of times per haploid genome (reviewed in 17). Because of their structures, many of these repeated sequence clusters have buoyant densities in cesium chloride different from the majority of an organism's DNA and can be observed as satellites to the main band DNA. In general, these simple, tandemly repeated sequence arrays are present in centromeric and telemeric hetero-chromatin and are normally transcriptionally quiescent. Recently, however, satellite DNAs in the newts *Triturus crestatus carnifex* (72, 73) and *Notopthalmus viridescens* (74, 75) have been shown to be transcribed during the lampbrush chromosome stage of oogenesis. This transcription may result from the failure of normal transcription termination from upstream histone genes, among which this satellite DNA sequence is interspersed (74, 75). It is unknown whether such RNA transcripts are produced merely by "accident" due to failure of transcription termination or whether their production serves some biological function during the developmental stage at which they are synthesized. Both ribosomal (reviewed in 18) and 5s RNA genes (76, 77), and histone genes in some (reviewed in 78) but not all organisms (79) are also examples of long tandem arrays of complex repeated sequences, some portions of which are transcriptionally active and represent multigene families in which the copy numbers per haploid genome vary between a few hundred and many thousands. In some eukaryotes, copies of both ribosomal and histone genes are not associated with their major gene clusters (80) and are possibly dispersed to new chromosomal locations by a mechanism that could disperse essentially any (transcribed) DNA sequence to different chromosomal positions (see below). Such a mechanism might eventually disperse enough copies of a multigene family so that the original gene cluster could be deleted without any drastic phenotypic effects on an organism.

SHORT DISPERSED REPEATS

Number of Different Repeat Families

The number of different families of short interspersed repeats is an important factor in considering their biological functions. Prior to the introduction of newer techniques, the renaturation rate kinetics for the formation of duplex DNA provided the only estimates of this number (81). There are, however, a number of complications associated with the interpretation of renaturation kinetics of whole genomic DNA. First, repeated sequences are best described as families of related but individually distinct members that share sufficient homology to cross-hybridize (81). Mutational divergence within a family of related sequences reduces the cross-renaturation rate of members of such a family as compared to the renaturation rate of perfectly complementary sequences (82). Second, renaturation rate studies of genomic DNA do not distinguish between repeated components that are clustered and those that are dispersed. Moreover, repeated sequences usually exhibit a distribution of repetition frequencies so that it is often impossible to identify discrete components according to their renaturation rates when total genomic DNA is used. Third, the renaturation rates of DNA depend on the length of complementary sequences, which, in the case of dispersed repeats, also depends on the unknown details of their sequence organization (83). Within these limitations, renaturation rate studies of genomic DNA have been interpreted to mean that different repetitive sequence families containing between 10 and 10,000 members are present in a variety of eukaryotic DNAs (reviewed in 84). The availability of cloned DNA fragments makes it possible to test this interpretation with greater precision than before.

Kline et al (85) examined 18 different cloned *Strongylocentrotus purpuratus* repetitive sequences by hybridization with total *S. purpuratus* DNA. The copy numbers of these cloned repeats vary between 3 and 12,500 per haploid genome. The average divergence of the cloned repeats closely matched that of the total short repetitive sequence fraction. For 13 of the 18 cloned repeats there were no detectable highly divergent sequence relatives, while 5 of the 18 did show divergence within their sequence families. The apparent copy number of one cloned repeat depended strongly on the stringency of the hybridization conditions. Under stringent conditions, there were apparently 20 copies per haploid genome, whereas under nonstringent conditions the apparent copy number exceeded 400 (61). Similar analyses with different cloned members of the same repeat family suggest the presence of at least 10–20 distinct subfamily sequences in the *S. purpuratus* genome (61). In marked contrast to this result, hybridization studies of the major dispersed repeat sequence family in human DNA, the *Alu* family of dispersed repeats, which has over 300,000 members, does not reveal subfamilies of sequences (86).

The biological significance of the conservation of some families and the divergence of others is not understood, but could be explained if the members of some families have been dispersed to new chromosomal locations from a single progenitor sequence more recently, if some sequences are under stronger selective pressures than others to remain undiverged, or if the members of some sequence families are periodically replaced with another member of the same family (87, 88). To extend the issue of sequence diversity within a single family of repeat sequences to its limit, the question can be raised whether a single nucleotide difference between two otherwise identical members of a repeat family would imply functional subfamilies. Alternatively, extensive sequence divergence among different members of a repeat family might be tolerated without disrupting the biological function of individual family members if some crucial (possibly short) region of sequence were kept relatively constant by selection. In this case individual family members might perform identical functions, but standard hybridization analyses would divide them into meaningless subfamilies with respect to their biological roles.

The Human Alu Family of Dispersed Repeats

Like sea urchin DNA, mammalian DNAs also contain short-period dispersed repeats. However, the short-period repeats in both primates and rodents contain a single prominent family of sequences that accounts for approximately one of every two to three of the dispersed repeats that contribute to the short-period interspersion pattern. Because of its ubiquity, this family is currently the most thoroughly described dispersed repeat sequence in the mammalian genome (see below). Studies of the renaturation rate kinetics of denatured human DNA, while not conclusive, initially suggested that the human interspersed repetitious sequences might consist mainly of one or only a few prominent sequence families (89). This possibility was tested further by allowing denatured human DNA to reanneal such that only repetitive sequences would reform duplexes. This DNA was then freed of unreacted single-stranded DNA by treatment with S1 nuclease and separated by agarose gel electrophoresis. A distinct band of double-stranded DNA approximately 300-bp long was visible in the gel superimposed on a background of essentially random lengths of duplex DNA. 60% of the mass of this crude preparation of 300-bp repetitive DNA could be cleaved at a common site by the restriction endonuclease *Alu* I, which suggests that it might be composed predominantly of a single "*Alu* family" of sequences (90). Similar results were obtained when 300-bp inverted repeated DNA sequences were isolated and used as the substrate for *Alu* I restriction endonuclease digestions (90).

The *Alu* sequence family accounts for a minimum of $\sim$ 3–6% of the human genome, a value independently determined by two different methods. The mass yield of this sequence family following S1 nuclease digestion of renatured DNA was 3%, which indicates approximately 300,000 copies per haploid human DNA complement (90). Renaturation rate studies indicated approximately 500,000 or more *Alu* family members per haploid human genome (86, 89). 500,000 copies of this sequence would comprise $\sim$ 6% of the mass of the human genome. The abundance of this family is sufficient to account for a large proportion of all the interspersed 300-nucleotide repeats in human DNA. A random distribution of at least 300,-000 copies of this sequence throughout the entire genome would give an average spacing of 8,000 nucleotides (2.5×10^9 bp per haploid human genome/3×10^5 *Alu* copies) between *Alu* family members [see (91) for human genome size]. If, instead of being distributed over the entire genome, *Alu* family members were distributed over only 60% of the human genome, as has been described for the distribution of most of the short-period interspersed repeats in a variety of eukaryotic DNAs (3–15), then the average spacing between *Alu* family members would be approximately 5,000 bp. The average distance between all short-period interspersed repeats in the human genome is approximately 2,200 bp (7, 8). Thus, it follows that one of approximately every two and a half interspersed repeats could be an *Alu* family member.

As an example of the interspersion frequency of *Alu* members, the 56,000 bp of human DNA containing the epsilon, A-gamma, G-gamma, delta, and beta globin genes also contain seven *Alu* family members (92, 93), with an average spacing of approximately 8000 bp. However, the spacing is not constant, but varies between the different *Alu* family members. Recently, the structure of a human *onc* gene (c-sis) has been determined. This 12 kb gene contains three *Alu* family members located in two intervening sequences (94). To confirm the high frequency and wide distribution of *Alu* family members in the human genome, 100 randomly selected clones bearing human DNA fragments approximately 15,000–20,000-bp long were screened by hybridization with purified, radiolabeled *Alu* DNA sequences. 94 of the clones hybridized with the radiolabeled *Alu* sequence probe (W. R. Jelinek, unpublished). Similar results were reported by Tashima et al (95). In an analogous experiment, 75% of unselected clones bearing African Green Monkey DNA fragments were found to hybridize with a purified human *Alu* sequence (G. Grimaldi, T. McCutchen, and M. Singer, unpublished). Thus, analyses of both randomly selected, cloned DNA samples and defined cloned DNA fragments containing known genes confirm that *Alu* family members are both highly repeated and widely dispersed throughout human and monkey DNAs, and comprise the most abundant family of dispersed repeats.

Non-Alu Family Dispersed Repeats

Although *Alu* family members account for a major fraction of the interspersed repeats in human, and, as described below, in other mammalian DNAs as well, they do not account for all of these repeats. Shen & Maniatis (96) reported the presence of five distinct families of interspersed repeated sequences in the rabbit beta-like globin gene cluster, and Haigwood et al (97) demonstrated the presence of at least three different dispersed repeat sequence families that differ from the *Alu* sequence (M. H. Edgell, personal communication) in the mouse beta and beta-like gene cluster, although an *Alu*-type sequence is also present downstream from the beta major gene (98). In addition to seven *Alu* family members, the human beta-like globin gene cluster also contains a 6400-bp repeated sequence located downstream from the beta gene, not obviously related to the *Alu* sequence, that is also interspersed throughout human DNA approximately 3000–4800 times (16). This component comprises ∼ 1% of the mass of the human genome. A similar, but not identical sequence is located upstream from the G-gamma gene and has structures similar to the long terminal repeats of proretroviral DNA and mobile DNA elements of both *Drosophila* and yeast (100).

The DNAs from a number of different rodents contain a repeat originally described as a 1.3-kb *Eco* RI fragment (101, 102), but more recently shown to be at least 3 kb long (88), that is dispersed, and present about 20,000 times per haploid genome. This sequence corresponds to at least 2–3% of the mass of the rodent genome. A repeat sequence originally identified in a fragment of mouse ribosomal gene, nontranscribed spacer DNA is also scattered throughout the mouse genome (103, 104). The mammalian genome contains between 100 and 2000 loci complementary to the low-molecular-weight RNAs U1, U2, and U3 (105, 106). Recently, Van Arsdell et al (107) identified human genomic clones containing pseudogenes for these RNA species and presented evidence consistent with the notion that they are also dispersed in the human genome. Similar conclusions concern sequences homologous to RNA U6 (108). Of 15 cloned DNA fragments prepared from 300-bp S1 nuclease–released human repeats, two have been identified as non-*Alu* family members that also may be dispersed through-out human DNA. Preliminary results suggest that one of these non-*Alu* families corresponds to as much as $\sim$ 3% of the mass of the human genome (C. W. Schmid and colleagues, unpublished). Little else is known about these non-*Alu* interspersed repeats, but their presence indicates that mammalian DNAs contain multiple families of dispersed repeats in addition to the ubiquitous *Alu* family.

A significant mass fraction of the dispersed repeats in the mammalian genome have now been identified in a small number of different families of sequences. With regard to human DNA, the *Alu* family, and the 6400-bp family and the one non-*Alu* family sequence described above together comprise perhaps as much as 9% of the genomic mass. The mass fraction of the genome attributable solely to dispersed repeats is difficult to measure precisely. For human DNA this fraction falls in the broad range of 15–30% of the genome (7, 89). Accordingly, three families of sequences account for one third or more of the mass of dispersed repeats in human DNA. That similar considerations apply to the rodent genome are strongly reinforced by the reported presence of the 3-kb sequence family, which, as described above, comprises an additional 2–3% of rodent DNA.

Sequence Conservation of the Alu Family

Because the individual members of a family of repeated DNA sequences are similar but not identical to one another, it is necessary to represent the overall base sequence of such a family as the consensus or most frequent sequence of its members. The consensus sequence of the human *Alu* family has been estimated by two methods. *Alu* family members are so abundant and so highly conserved in sequence that it was possible to determine a partial consensus sequence of the entire, crude 300-bp DNA fraction liberated by S1 nuclease from human DNA that was denatured and subsequently renatured, such that only repetitious sequences would reform duplexes (109). As a second approach, this 300-bp DNA fraction was cloned, and the sequences of ten individual clones were determined (109, 111). The consensus sequence derived from these ten clones is shown in Figure 1. The nucleotide sequence of individual cloned *Alu* family members differ by an average of only 10% from this consensus sequence (110). Most of these sequence variations are individual point mutations that appear to be randomly distributed throughout the 300-nucleotide sequence.

The sequences of four other human *Alu* family members have been determined from genomic clones containing an insulin gene (112), the epsilon globin gene (113), the G-gamma globin gene (114), and the delta globin gene (114). In all four of these clones, the *Alu* sequence was also approximately 300-bp long and differed by only approximately 10% from the consensus sequence given in Figure 1. These sequences allowed one end of the *Alu* repeat to be precisely defined (the left end as indicated in Figure 1). The other end of the *Alu* repeat is less precisely defined. It contains an A-rich sequence that is not conserved among different *Alu* family members, but maintains the general structure $[N(A)_n]_m$, where N represents any nucleotide (or in some instances more than one nucleotide), n is usually less than 20, and m is usually less than 10. In some instances this pattern of internal repeats in the A-rich sequence is interrupted by other nucleotides (see below and Figure 3).

Like human DNA, rodent DNA has a short-period interspersion pattern dominated by a major family of interspersed repetitious sequences (24, 115, 116). However, unlike the human *Alu* sequence, which is approximately 300-bp long, the rodent equivalent sequence is only $\sim$ 130-bp long (115, 116). The human *Alu* sequence is an imperfect dimer formed of two directly repeated, approximately 130-nucleotide monomer units with a 31-bp insertion in the second monomer that is missing from the first monomer unit. The plus (+) overlining in Figure 1 indicates a dAMP-rich sequence that defines the end of the first monomer unit. Krayev et al (115) have determined the nucleotide sequences of three dispersed repeats from the mouse genome, the so-called B1 repeat sequence. In Figure 1, the consensus sequence derived by these investigators is compared to that of human *Alu* and that derived by Haynes et al (116) for one type of *Alu*-equivalent sequence, the type 1 sequence, in Chinese hamster DNA. The nucleotide sequence of an African Green Monkey DNA fragment found as an insert in an SV40 viral DNA has also been determined (117) and also is compared with the human *Alu* sequence. The mouse sequence is 129-bp long and the CHO sequence is 134-bp long. They both display homology with one monomer unit of the dimeric human *Alu* sequence. The total length of the second human monomer agrees more closely than the first with the length of the rodent sequence, which contains a region (indicated by the tilde overlining

```
                                 + + + + + + + + + + + + + + + + + +
HUMAN    GGTGAAACCCCGTCTCTACTAAAAATACAAAAATTAGCCGGGCGT  GGTGCGCGCGCCTGTAATCCCAGCTACTCGGGAGGCTGAGGCAGGAGAATCGC
MONKEY   A.........................G..........  T..T.  ....TAT.A...T.G...TT.G..............................T.
MOUSE    ......................A........  ...AT....T.G.......A...........A........C.G..TT.
CHO      ..A...A.T.....CA.A..T. G.........T. G.............A....G.A........C.G..TT.
                                                  T


HUMAN    TTGAACCCAGGAGGTGGAGGTTGCAGTGAGCCGAGATCGCGCCACTGCACTCCAGCCTGG                                    GCA
MONKEY   .......AC.........T...T.....A..A....AA...T..T.............                                        ..G

MOUSE    ..GTT.GA. .          .......TCTTTC AGAGT    GAGTTCCAGGACACCAGGCTA
CHO      ....GTT.AA...        .......TCTACCAGAGTTCCTGAGTT CAAGACA   GGCTA .
             C                                     T
             T                                     A


HUMAN    ACAGAGCCGAGACTCCATCTC A-Rich Sequence
MONKEY   ......A
MOUSE    ......  .A..C.TG...  A-Rich Sequence
CHO      ......  .A..C.TG...  A-Rich Sequence
```

Figure 1 Comparison of the human *Alu* consensus sequence, an African Green Monkey *Alu*-equivalent sequence, the mouse B1 consensus sequence, and the Chinese hamster *Alu*-equivalent sequence.

The top line is the consensus sequence of the human *Alu* repeat extending from the first nucleotide of the repeat to the first nucleotide before the 3' A-rich region as determined from sequences in (108, 110, 111, 114). All other sequences are compared to the human sequence. The second line represents the sequence of a fragment of African Green Monkey DNA found as an insert in an SV40 viral genome (117). The third line represents the mouse B1 consensus sequences determined by Krayev et al (115). The fourth line represents the Chinese hamster type 1 *Alu*-equivalent sequence (116). A dot at any position indicates the same nucleotide as in the human *Alu* consensus sequence. A letter indicates the nucleotide present at each position that differs from that in the human *Alu* consensus sequence. Occasionally a blank space has been inserted to facilitate the alignment of the sequences. The plus (+) overlining indicates the end of the first monomer unit of the human *Alu* consensus sequence. The asterisk (*) overlining indicates a nine-base sequence that is perfectly conserved in the three consensus sequences and in the single African Green Monkey sequence. The second monomer of the human *Alu* sequence has a 31-bp insert not present in the first monomer unit. This insert is immediately to the left of the 9-nucleotide conserved sequence and is indicated by dashed overlining (–). The mouse and Chinese hamster sequences each have a 32-bp insert not represented in the human or monkey sequences. This 32-bp sequence is located immediately to the right of the 9-nucleotide conserved sequence and is indicated by tilde overlining (~).

of the mouse sequences in Figure 1) corresponding approximately in length but not in sequence to the insert in the second monomer of the primate repeat (indicated by the dash overlining in Figure 1). The second monomer of the human *Alu* sequence, the monkey sequence, the mouse B1 consensus sequence, and the Chinese hamster *Alu*-equivalent consensus sequences all agree well up to the position of the insert in the human sequence. The human and monkey sequences beyond the insert agree with the rodent sequences for nine bases (indicated by the asterisks over the human sequence), which form a perfect inverted repeat with a single C residue at the position of symmetry. The homology between the rodent and human sequences again breaks down at the position indicated by the tilde overlining of the mouse sequence, but resumes beyond this region and extends to the ends of the primate and rodent repeats.

Thus, by nucleotide sequence comparisons it is easy to recognize that the primate and rodent repeats belong to the same sequence family. By the criteria of molecular hybridization, however, they should be considered different families, since they cross-hybridize rather poorly. This poor cross-hybridization has been used to advantage in the detection of small amounts of human DNA in the presence of excess rodent DNA following transformation of rodent cells with human DNA sequences (118, 119). The important issue is whether these sequences effect a common function in the different species in which they reside. If so, then considerable sequence divergence might be tolerated within a repeat sequence family without disrupting function, and the inability of these dispersed repeats to cross-hybridize may be an irrelevant criterion in the evaluation of their biological significance. The 157-bp monkey DNA sequence shown in Figure 1 probably does not represent the entire monkey *Alu* equivalent family sequence, since S1-nuclease digestion of renatured repetitive monkey or galago DNA releases 300-bp long duplex DNA fragments analogous in size to the human *Alu* sequence (120). The predominant length of *Alu* family members in primates is therefore approximately 300 bp, while the rodent equivalent is approximately 130-bp long and extensively homologous in sequence to one of the monomers of the primate dimer. Presumably these rodent and primate repeats are descendents of a common ancestral sequence that has been well-preserved during recent evolution. 300-bp dimeric *Alu* family members may be present in rodent DNA or monomeric *Alu* family members in primate DNAs, but this is not yet known.

These observations raise a number of questions concerning the evolutionary origins of the primate dimer. Did a single dimer arise once, followed by its dispersion to approximately 300,000 different locations in the primate genome, or did dispersion of monomers occur first, followed by their conversion to dimers (or first dimers followed by conversion to monomers)?

This would have had to occur many times to convert all monomers to dimers (or dimers to monomers), possibly by replacement (87, 88). Are the monomers in rodents and the dimers in primates continually being dispersed to new chromosomal locations? Does the same mechanism that disperses monomers also disperse dimers and other non *Alu* repeats? An answer to this last question is suggested below, but the first two remain unresolved.

A Second Alu-Type Dispersed Repeat in Rodent DNAs

A second type of dispersed repeat family, termed the rodent type 2 *Alu* equivalent family, has been detected in Chinese hamster and rat DNAs (116, 121, 122). The nucleotide sequence of four members of this family are given in Figure 2. The first 61 residues correspond to residues 47 through 107 of the type 1 rodent *Alu* equivalent family sequence. The next 96 residues are not obviously *Alu* type 1 rodent equivalent sequences nor primate *Alu* sequences, although they do have some short regions of sequence in common with them. The two members from CHO DNA represented in Figure 2 were found in two randomly selected genomic clones, while the two rat type 2 *Alu* equivalent family members were found juxtaposed to one another to form a dimeric structure in the second intron of the growth hormone gene (122). Like type 1 *Alu* equivalent family members, type 2 family members also have one end precisely defined in all examples examined and the other end imprecisely defined by an A-rich sequence that exhibits extensive sequence homology among different members of the repeat sequence family, but is not identical in sequence in different family members. Thus it appears that subsequences of one type of dispersed repeat can become associated with other sequences to form "composite" repeat sequences that are also dispersed throughout chromosomal DNA.

Dispersal of Short Repeats

The ubiquity of the short-period interspersed repeats in eukaryotic DNAs, and in particular of *Alu* family members in mammalian DNAs, raises concerns about the mechanisms of dispersal of these repeats throughout eukaryotic genomes. Figure 3 gives the sequences of the A-rich region (roman type) as well as the structures (boldface type) surrounding eleven *Alu* or *Alu*-equivalent family members from four different mammalian species. Each is flanked on either side by direct repeats ranging in size from 8–20 bp. The sequences of these flanking repeats are not conserved among different *Alu* family members, but are unique to each *Alu* sequence. By analogy with the direct repeats flanking known bacterial transposons or insertion sequences and mobile DNA elements in eukaryotic DNAs, the

```
HAMSTER    GGCTGGAGAGATGGCTC GA GGTTAA GAGCA  CCAACTGCTGTCTTCCAGAGGTCCTGAGTTCAATTCCCAGCAACCACATGGTGGCTCACAACAATCTATAATGAGATCT
HAMSTER    .........A...   ....TG....A.............................T...............C...CG.T.....C..
RAT        ........AGT.. ....C..G........A...............................C...G..A......C
RAT        ........AGC.. ....GC..G................................................C...G..A........
4.5sI   ppp ........AGCC....A.. ..TAGGCTCACAACCAAAAATATAA  ......GG.........C....GGCTGCTCTCCAGCCACCTTTTT-OH
RNA

HAMSTER    GGTGCCCTCTTCTGGTGTGCAGATATATGGAAG CAGAA  TGTTG TATACATAATAATAATAAAATCTT  AAAAAA
HAMSTER    .........C....... ......  .......TTT.........
RAT        A.........TCT  ...A...CTACAG...ACT...  ....C......  ...T......AAACAAAAACGG
RAT        A.........ATCT  ...A...CTACAG...ACT...T  .......  ...T......ACAAAACAAAAACAAAAACAAAA
```

Figure 2 Comparison of four type 2 *Alu*-equivalent sequences from Chinese hamster and rat DNA with the rat 4.5sI RNA sequence. The top line is the sequence of one Chinese hamster type 2 *Alu*-equivalent sequence to which the three other DNA sequences and the sequences of the 4.5sI RNA are compared. A dot at any position indicates the same nucleotide as in the Chinese hamster type 2 *Alu*-equivalent sequence shown in the first line, and a letter at any position indicates a nucleotide different from that on the first line. The continuous overlining indicates a sequence present in the type 2 *Alu*-equivalent sequence that is also present in the rat 4.5sI RNA sequence, but "scrambled" in position with respect to one another. The sequences were determined as follows: Chinese hamster DNA sequence (116, 121), rat DNA sequence (122), and rat 4.5sI RNA sequence (R. Reddy, P. C. Hennig, and H. Busch, unpublished).

2.	AAGATTCACTTGTTTAG	. . human alu . . .	A_{12}GAGAGATTGATTGA$_2$**AAGATTCACTTGTTTAG**
3.	AAATGGATGGAGAC	. . . . human alu .	A_{14}GA$_3$GA$_3$GA$_4$GA$_5$GA$_6$GA$_3$**AAATGGATGTAGAAC**
4.	GTTTAGATAAG	. . . human alu .	A_{25}**GTTTAGATAAA**
5.	AAAAGAAACTTGGAAAGAG	. . type 2 cho alu . .	A_2TA$_3$TA$_3$TA$_4$TCTTA$_7$**AAAAGGAAACTTGGAAAGGA**
6.	AAAGATGCCCCGCTACAG	. . type 2 cho alu .	A_2TA$_3$TA$_3$TA$_4$TCTTTTTA$_4$**AAAAGATGCTCAG**
7.	AACATACTAATTTTG	. . type 1 cho alu .	A_4CA$_2$**AACTATAATTTTG**
8.	GTCAGCC	. . type 1 cho alu . .	TGA$_5$CCA$_5$GA$_7$GA$_5$GA$_5$GA$_3$GTTCCAGGCCAGTCAG
9.	AGCTCATGAATGAAG	. . type 1 cho alu .	CCA$_5$CA$_3$TCA$_4$CCAGACAGGCACAGCCCC**AGCCCAT**
10.	GAGACAACAAATCAGAG	. . type 1 mouse alu .	A_7CCA$_3$CCA$_3$CCA$_3$CCA$_6$CC**GAGACAACAAATCAAAT**
11.	GAGTAATGACAGAGAG	. . type 2 rat alu .	A_4CA$_4$CGG
	. . . .		ACA$_4$CA$_5$CA$_5$CA$_4$**CAGTAATGACAGAGAG**
12.	AGAAACAGGCTTTTCGC .	. human U1.101 DNA .	A_{14}GCA$_2$GA$_3$**AGAAACAGGCTTTT-GC**
13.	TAAATAATCAGGATGGA	. . human U2.13 DNA .	**TAAATAATCAGGATGGAA**
14.	TAAAATGCTAATTATCCAA	. human U3.5 DNA . .	**TAAAATGCTAATTATCCA**
15.	TAGAGTGC	. human U6 DNA . . .	A_{18}**TTGAGGC**

DIRECT REPEAT • alu • A-RICH SEQUENCE • DIRECT REPEAT

Figure 3 Comparison of the sequences flanking *Alu* family members in three different mammalian species and snRNA pseudogenes in humans.

The sequences of 11 *Alu* family members and 4 U-series snRNA pseudogenes are compared. The dotted line represents the *Alu* sequence or the U-series snRNA homologous sequence. The sequences were all determined from cloned genomic DNA fragments as follows: *1.* an *Alu* family member located downstream from a human insulin gene (112); *2–4.* human *Alu* family members located in the beta and beta-like gene cluster (113, 114); *5 and 6.* Chinese hamster type 2 (116, 121); *7–9.* Chinese hamster type 1 (116) *Alu*-equivalent family members located in random genomic clones; *10.* a mouse type 1 *Alu*-equivalent sequence from a random genomic clone (115); *11.* two rat *Alu*-equivalent sequences juxtaposed to one another in an intron of the rat growth hormone gene (122); *12–14.* RNAs U1, U2, and U3 pseudogenes from human DNA (107); and *15.* a human U6 (pseudo)gene (108).

direct repeats flanking *Alu* family members may have resulted from the duplication of a unique DNA sequence at the target site of *Alu* insertion into chromosomal DNA (55, 56, 123). If so, then the *Alu* sequence has probably dispersed throughout the mammalian genome by a mechanism related in some aspects to that used by known mobile DNA elements.

Alu and *Alu*-like sequences are not the only short dispersed repeats bordered on one side by an A-rich region and flanked by direct repeats. Van Arsdell et al (107) demonstrated that a pseudogene for U1 RNA in the human genome also has these structures (Example 12 in Figure 3). Likewise, pseudogenes for U2 and U3 RNAs (107) are flanked by short direct repeats (Examples 13 and 14 in Figure 3), and Hayashi (108) determined the sequence around one example of a human U6 gene that also has short, directly repeated flanking sequences (Example 15 in Figure 3). These observations suggest that the mechanism responsible for the dispersion of *Alu* family members throughout mammalian DNAs may have dispersed other (perhaps all) interspersed repeat sequences as well. In this context, E. M. Stephenson and J. G. Gall (unpublished) observed a histone H4 pseudogene in *Notopthalamus viridescens* DNA flanked on either side by short direct repeats of 8 bp with the 5' flanking direct repeat immediately abutting the first nucleotide of the mRNA encoded sequence. To investigate this possibility further, it will be of interest to examine the flanking sequences of a number of different short-period interspersed repeats in the DNA of a variety of different species.

LONG DISPERSED REPEATS

Structural Resemblance to Proretroviral DNAs

The long-period dispersed repeats in *Drosophila* DNA and the Ty1 sequences in yeast DNA are mobile DNA elements (see below) analogous in structure to proretroviruses (see Table 1). They resemble integrated retroviral DNA in that each is approximately 5-kbp–7-kbp long and is flanked on either side by approximately 300–600-bp long direct repeats, which in yeast are known as the delta sequences (27–39) and are presumed to correspond to the long terminal repeats (LTR) at the ends of integrated retrovirus DNAs (40–45). The entire structure is flanked by short direct repeats of approximately five nucleotides (29–32, 35, 37–39, 40–45, 47, 48, 50–54, 124), one copy of which is thought to be generated from the other during the insertion event at the target site of insertion into chromosomal DNA (55, 56, 123). The members of a dispersed repeat family in mouse DNA known as the VL30 gene (125–127) are also flanked by direct repeats of approximately 400 bp (126). Although no sequence homology has been found between these dispersed repeats and known retroviruses, the repeats do produce 30s RNAs that are efficiently packaged into virions by pseudo-type infection (125, 127). Like the long-period repeats in *Drosophila* and yeast, the VL30 genes may be mobile elements in mouse DNA. Young (128, 129) has estimated that ~ 16–17% of *Drosophila* DNA is composed of long dispersed repeats that can be grouped into ~ 70 different sequence families whose members range in copy number between ~ 3 and 100 per haploid genome. The Ty1 sequence is present ~ 35 times in yeast DNA, while the delta sequences are present about 100 times and occur at some chromosomal positions unassociated with the remainder of Ty1 DNA (36).

In both *Drosophila* (32–35, 128–132) and yeast (36–39, 133–137) these dispersed repeats are thought to be mobile, because they are not always present at the same chromosomal locations. Different species of *Drosophila* have different copy numbers of the different repeated families, and different strains of the same species have members of the same repeat family in different chromosomal positions (32–35, 128–133). Likewise, the Ty1 sequence in yeast is present at different chromosomal locations in different strains (36–39, 133–137). Sequence alterations associated with one Ty1 element have been observed during the propagation of a single yeast clone under laboratory conditions during the course of a month (36), which suggests that these repeats may disrupt chromosomal DNA at a relatively rapid rate. In addition, a region of yeast DNA enriched in delta sequences shows a high frequency of sequence rearrangements between different yeast strains (36), presumably due to relatively frequent homologous recombination between various of the delta sequences (134). Both Ty1 (133–137) and the *Drosophila* repeat family known as copia (132) have been implicated in the generation of mutations at their sites of insertion into chromosomal DNA, Ty1 is associated with the loss of activity of adjacent genes (38, 133, 134) and with an increase in the expression of adjacent genes (135–137). Likewise, the insertion of retrovirus DNAs into chromosomal DNA is mutagenic (138) and is thought in some instances to initiate or enhance transcription from adjacent genes (139) that may be involved in tumorogenesis.

Structural Differences Between Long and Short Dispersed Repeats and their Flanking Sequences

Both the long-period dispersed repeats, as exemplified by the copia sequence and related *Drosophila* long repeats, the yeast Ty1 sequence and proretrovirus sequences, and the short-period dispersed repeats as exemplified by human *Alu,* the rodent type 1 and type 2 *Alu*-equivalent sequences, U1, U2, and U3 pseudogenes, and a U6 gene appear to have been (and probably continue to be) dispersed by transposition. However, structural differences between these sequences suggest that the details of their dispersal mechanisms must differ. The long terminal repeats of proretroviruses, copia, and similar *Drosophila* long-period repeats and the yeast delta sequences are all demarcated by short inverted repeats that range in size from 2 to 17 bp that are bordered by the dinucleotides TG . . . CA in all examples examined (reviewed in 123, 140; Table 1). Bacterial insertion sequences and transposable elements have similar inverted repeats that are required for transposition (55, 56), and by analogy these structures are assumed to be necessary for the mobility of the eukaryotic long-period dispersed repeats. Neither human *Alu* family members, rodent type 1, type 2 *Alu*-equivalent family members, U1, U2, U3, nor the U6 pseudogenes have these inverted repeat structures at their ends. The short direct repeats flanking the long dispersed repeats in *Drosophila* and yeast and proretroviral DNAs in avian and mammalian cells are only 4–6 bp long (29–32, 35, 37–39, 40–45, 47, 48, 50–54, 124; Table 1) while those that flank *Alu* and *Alu*-equivalent sequences and the dispersed U-RNA pseudogenes are ~ 8–20 bp long (Figure 3). Thus, the long- and short-period interspersed repeats differ by a number of properties that imply differences in the mechanisms by which they disperse throughout chromosomal DNAs. The short-period dispersed repeats are themselves approximately equal in length to the LTRs that flank the long dispersed repeats. Are they the LTRs for as yet unidentified transposable sequences in mammalian DNA?

TRANSCRIPTION OF DISPERSED REPEATS

Alu Family Members Within and Near RNA Polymerase II Transcription Units

It has been recognized for many years that hnRNA molecules like their DNA templates contain interspersed repetitive sequences [(141–145) and references cited below]. In some molecules these repeats occur more than

Table 1 Comparison of long-period dispersed repeats and proretrovirus DNAs

DNA element	Element size (bp)	LTR size (bp)	IR at LTR ends?	Sequence of LTR ends	TATA Box and AATAAA in LTR	Number of bp duplicated at target	Refs
Dm mdg 1	7200	422	yes	TGT...ACA	yes	4	31
Dm mdg 3	5600	268	yes	TGT...CAG	yes	5	30
Dm copia	5600	276	yes	TGT...ACA	yes	5	28, 29
Yeast Ty1	5600	334	yes	TGT...TCA	yes	5	37, 124
Molony murine sarcoma provirus	5900	588	yes	TGT...ACA	yes	4	48, 49
Molony murine leukemia provirus	8800	515	yes	TGA...TCA	yes	4	47, 50, 51
Avian spleen necrosis provirus	8300	569	yes	TGT...ACA	yes	5	40

once in inverted orientation and can be isolated as "fold back" or inverted repeated double-stranded hnRNA (dshnRNA) segments (146–151). Comparisons of the sequences of major RNAse T1 oligonucleotides from the dshnRNA of cultured human cells (151) with the human *Alu* family DNA sequence confirmed that *Alu* family members are heavily represented in hnRNA molecules (109, 152), comprising as much as 18%–25% of the mass of HeLa Cell hnRNA (150, 153). Likewise, HeLa cell cytoplasmic, polyribosomal-associated, poly(A)-terminated RNA molecules also contain these sequences, but at a considerably lower frequency per molecule than in hnRNA (153, 154). Similar observations have also been made for Chinese hamster [(24); W. R. Jelinek, unpublished] and mouse (147, 148) nuclear and cytoplasmic RNA molecules.

Two genes that have been examined in detail contain *Alu* family members in intervening sequences and presumably so do their primary transcription products. The rat growth hormone gene contains two type 2 rodent *Alu* equivalent family members in one intervening sequence (122), and a human *onc* gene (c-sis) contains three *Alu* family members located in two intervening sequences (94). Some *Alu* family members are therefore located within RNA polymerase II transcription units. Apparently they are more frequently represented in regions of hnRNA transcripts that are removed during processing or turnover than in the regions that are conserved and transported to the cytoplasm (153, 154). On the other hand, *Alu* family members do not appear to be located exclusively within RNA transcription units. The seven *Alu* family members in the human beta and beta-like globin gene cluster are located between pairs of genes that are coordinately active in transcription during human development (92, 155), and possibly define the endpoints of some functional or evolutionary DNA units. Whether transcription extends into any of these *Alu* sequences awaits further mapping data to define the transcription units for the beta and beta-like globin genes, particularly in view of the observation that transcription of the mouse beta major globin gene extends well beyond the poly(A) addition site (156). From data currently available there appears to be no easily distinguishable relationship between the positions of *Alu* family members and RNA polymerase II transcription units in genomic DNA.

Developmental and Tissue-Specific Transcription of Short Dispersed Repeats

Like the *Alu* family sequence, short-period interspersed repeats in sea urchins are also transcribed as part of long RNA molecules. Costantini et al (157) demonstrated that approximately 80% of the short repeats in the sea urchin genome had copies present in oocyte RNA, as did approximately 35% of the long repeat sequences. Using nine different cloned repeat DNA sequences these authors also demonstrated that both strands of each repeat were present approximately equally in the oocyte RNA; there are similar findings for intestine and gastrula nuclear RNAs (158). No simple relationship could be found between the amount of the different repeats present in the RNA and their copy frequencies in genomic DNA. Some repeats present in relatively low copy numbers in DNA were represented more abundantly in RNA than other repeats that are present in higher copy numbers in the DNA. Further analyses indicated that most of the single copy egg poly(A)$^+$ RNAs were covalently linked to transcripts of a subset of the total short repetitive sequences present in genomic DNA (159). Thus, transcript prevalence of these repeats appears to be independent of genomic reiteration frequency. According to these observations egg poly(A)$^+$ RNAs could be grouped into several hundred sets, with each set containing single copy sequences covalently linked to members of the same repeat sequence family (159). Furthermore, the abundance of transcripts from some of the repeat sequence families in intestine nuclear RNA differed from that in gastrula stage nuclear RNA, which suggests there is selected accumulation of the different interspersed repeats in different tissues. The mechanism by which these different abundances of different repeat transcripts are achieved is unknown, and could either be due to differential transcription and/or to differential stability of the different larger RNAs of which they are a part. Polyribosomal RNA could not be demonstrated to contain transcripts of these repeat sequence families (158), an observation consistent with the finding that in mammalian cells the abundance of *Alu* family sequences is considerably reduced in polyribosomes, compared to nuclear RNAs (153, 154).

The observations that short period interspersed repeats are close to protein encoding sequences in DNA and that their transcripts are found in nuclear RNA molecules have been used to argue that they function either in DNA to regulate transcription or in RNA to regulate the processing of primary transcription products to mRNA molecules (2, 69, 70, 148, 149, 151–153). However, repetitive sequences are widespread in DNA, and in some instances are located within RNA transcription units, and in addition different cells express different specific classes of genes. It is, therefore, just as easy to argue that tissue-specific patterns of transcription of different families of repetitious sequences are the consequence, not the cause of differential gene transcription. In this regard it is notable that two related vitellogenin genes of *Xenopus laevis,* whose expression are coordinately controlled by estrogen, have different patterns of repetitive sequences within their intervening sequences (160). One interpretation of this is that the presence of the dispersed repeated sequences is not required for the coordinated expression of these two genes. Current data are insufficient to decide whether the short-period dispersed repeats have an active role in gene expression.

Discrete Cellular Low-Molecular-Weight RNAs Homologous to Alu Family Sequences

Recent evidence suggests that *Alu* family members are transcribed as discrete short RNA molecules, presumably by RNA polymerase III. Jelinek & Leinwand (161) initially identified a discrete low-molecular-weight RNA, known as 4.5s RNA, that could be purified from Chinese hamster ovary cells because it formed base-paired duplexes with poly(A)-terminated hnRNA and mRNA molecules. The nucleotide sequence of the 4.5s RNA isolated from mice, Syrian hamsters, and Chinese hamsters has been determined (116, 162). The molecule is 96 residues long, terminated by pppGp at its 5' end and by a short oligo(U) sequence of variable length at its 3' end, both of which are characteristic of RNA polymerase III transcription products. It contains a 26-residue long region at its 5' end that has 65% homology with the rodent *Alu* consensus sequence, followed by a 10-residue long sequence that has no homology with the *Alu* sequence, followed in turn by a 50-residue long sequence that contains the 3' end of the RNA and has 88% homology with the rodent *Alu* consensus sequence (Figure 4). The gene(s) for the 4.5s RNA can thus be considered as a member of the rodent *Alu* family of repetitive sequences that has diverged from the consensus or most frequent rodent *Alu* sequence. Presumably the sequence homology between the 4.5s RNA and one strand of *Alu* family members allows it to base pair with hnRNA molecules, since they contain transcripts of both strands of the *Alu* family sequence in high abundance (24, 115, 147–151, 153). Whether this association between the 4.5s RNA and the poly(A)-terminated hnRNA and cytoplasmic RNA molecules occurs within living cells is unknown. The 4.5s RNA has not been detected in human cells.

However, another discrete low-molecular-weight RNA present in HeLa cells and in those from a variety of different eukaryotic species, the 7s RNA, could base pair with human *Alu* family members. There is a recent determination of the nucleotide sequence of the rat 7s RNA (W. Y. Li, R. Reddy, T. Henning, P. Epstein, and H. Busch, unpublished). The molecule is 295 residues long and shows extensive sequence homology with the human *Alu* family sequence and with the type 1 rodent *Alu* equivalent sequence as demonstrated in Figure 4. Like the 4.5s RNA, the 7s RNA can be considered an *Alu* family member with a sequence somewhat divergent from the consensus or most frequent *Alu* sequence. Ribonuclease T1 oligonucleotide fingerprint analyses indicate that murine and avian 7s RNAs contain many similar oligonucleotides, and thus the 7s RNA has been conserved during recent evolution (164). If the sequence homology between *Alu* family members and the 7s RNA is biologically significant then avian species are expected to have *Alu* related sequences in their DNAs. In this respect it is germane to note that Czernilofsky et al (165) have identified *Alu*-like sequences in avian sarcoma virus RNA and Breathnach & Chambon (166) refer to the presence of an *Alu* related sequences in intron C of the chicken conalbumin gene. *Alu* sequences are thus likely to be present in the DNAs of a variety of nonmammalian vertebrates as well as in mammalian DNAs.

The nucleotide sequence of a second example of another low-molecular-weight RNA, known as $4.5s_I$ RNA, originally identified in rat cells by RoChoi et al (167) has recently been determined (R. Reddy, P. C. Hennig, and H. Busch, unpublished). The molecule is 99 residues long and shows considerable homology with the type 2 rodent *Alu* equivalent sequence, although its sequence is "scrambled" with respect to the type 2 repeat, as indicated by the continuous overlining in Figure 2. The biological functions of the $4.5s$, the $4.5s_I$, and the $7s$ RNAs are not currently known. Since each appears to be a single molecular sequence rather than a group of closely related sequences as are *Alu* family members, each must be encoded by a single gene or group of invariant genes all of which are related in sequence to one another and to *Alu* family members.

The $4.5s$ RNA contains two short regions of sequence that align with sequences found within other low molecular weight RNAs known to be transcribed by RNA polymerase III (168, 169). This group of RNAs includes the Adenovirus type 2 VA-1 and VA-2 RNAs (170, 171), the Epstein-Bar Virus EBER-1 and EBER-2 RNAs (172), the various tRNAs (168, 169), and as described below, RNAs transcribed in vitro from human *Alu* family members. Lerner et al (173, 174) demonstrated that ribonucleoprotein particles containing the adenovirus VA RNAs, the Epstein-Bar Virus EBER RNAs, and the cellular $4.5s$ RNA are all recognized by the La-specific group of autoantibodies produced by certain patients with systemic lupus erythematosis. The determinant of the antigen/antibody reaction is thought to be the protein with which these RNAs interact, since the purified RNAs themselves are not antigenically reactive. Thus, on the basis of protein interactions and nucleotide sequence comparisons, the $4.5s$ RNA can be classified as a member of a group of RNAs consisting at least in part of the low-molecular-weight RNAs transcribed by RNA polymerase III from two viral DNAs. Presumably these low-molecular-weight RNAs perform some required but as yet unidentified function during the viral life cycle. Perhaps the $4.5s$ RNA performs an analogous cellular function.

Alu Family Members are RNA Polymerase III Transcription Units

Alu family members are themselves transcribed in vitro by RNA polymerase III. Originally, Duncan et al (175), and subsequently Fritsch et al (155) demonstrated that two DNA sequences in the human beta and beta-like

```
                                            T
                                            C
CHO       CCAGGCCATTGGTGGCACACACCTTTAGTCCCAGC ACTCAGGAGGCAGAGGCAGGAGGATCACTTGAGTTCAAGAGCCAG
MOUSE       .G.... ......TG..TG.........G................C.....TT...........G.. ......
C 4.5s    pppG..G.TTG....CG...G..GG.... GATTTG CTGA..............AG...........AC......G.... .....
M 4.5s    pppG..G.TAG....CG...G..GG.... GATTTG CTGA..............AG...........AC......G.... .....
R 7s      pppG..G... C....CG...G..G...........T...G.......T...A...A......CG......C.AG......
                                                                        ___
                                                                        186 NT

                    A
                    T
CHO       CCTGGTCTACCAGAGTTCCTGAGTT CAAGCCA   GGCTATACAGAGAAACCCTGTCT  A-Rich Sequence
MOUSE       .....T..G.........C..G.A..CCAG.......  ...............  A-Rich Sequence
C 4.5s      .....G.....C.TTTT-OH
M 4.5s      .....G.....C.TTTTT-OH
R 7s        .....GC.A..T...                      .....C....  .....TTA(A)-OH
```

Figure 4 Comparison of the Chinese hamster and mouse type 1 *Alu*-equivalent sequence with the mouse and Chinese hamster 4.5s RNA and the rat 7s RNA sequences.

The top line is the consensus sequence for the Chinese hamster type 1 *Alu*-equivalent sequence (116). The mouse type 1 *Alu*-equivalent sequence [also known as the B1 sequence, (115)] is compared to it, as are the mouse (162) and Chinese hamster (116) 4.5s RNA sequences and the rat 7s RNA sequence (W. Y. Li, R. Reddy, T. Hennig, P. Epstein and H. Busch, unpublished). A dot at any position indicates the same nucleotide as in the Chinese hamster consensus sequence. A blank space indicates a deleted nucleotide and a letter indicates a nucleotide different from the consensus nucleotide. In the RNA sequences the letter *T* has been used in place of a *U* for easier comparison with the DNA sequences.

globin gene region can be transcribed in vitro by RNA polymerase III to yield two discrete low-molecular-weight RNA molecules, one of 515 residues and the other of 575 residues. The transcription templates for these RNAs are *Alu* family members (114, 176). Likewise, in a cloned DNA fragment from an unknown region of the human genome, an *Alu* family member directed the transcription of a discrete low-molecular-weight RNA by RNA polymerase III in the cell free transcription system (154). Detailed mapping revealed that the start site for transcription was close to or coincident with the first nucleotide of the *Alu* sequence (114). Transcription proceeded through the *Alu* sequence and terminated beyond the A-rich region at the 3' side of the *Alu* sequence somewhere in a unique DNA sequence having two regions of four adenylic acid residues, which resulted in an oligo-uridylate sequence at the 3' end of the RNA. Although human *Alu* family members are composed of two monomer units similar to one another in nucleotide sequence, transcription in the cell-free system has only been observed to begin within the first monomer. Elder et al (154) identified the sequence GAGTTCPuAGACC [also noted by (164, 165)] in the first monomer unit as the most likely internal "control" region for RNA polymerase III transcription of human *Alu* family members. A modified second copy of this sequence is present in the second monomer unit of the human *Alu* sequence, but is interrupted by the 31-base pair insert that is absent from the first monomer unit (see above and Figure 1). By analogy to the internal control region of *Xenopus* 5s DNA (177), this region in the *Alu* sequence lies downstream from the transcription initiation site. Each *Alu* family member produces one or only a few transcription products of defined length. However, the RNA transcripts from different *Alu* family members differ in length from one another, presumably because transcription termination occurs in different unique sequences beyond different *Alu* family members (155, 175). Assuming this is also true in vivo, the entire family of approximately 300,000 *Alu* members would produce a heterogeneous set of RNAs having the shared *Alu* sequence at their 5' ends, but unique sequences at their 3' ends.

Like human *Alu* family members, members of the rodent type 2, but not the type 1 Chinese hamster *Alu*-equivalent sequence family, serve as templates for the transcription of discrete low-molecular-weight RNAs by RNA polymerase III in vitro, and analogous RNAs have been isolated from live cells (121). Transcription begins at the first nucleotide of the type 2 sequence, proceeds through the repeat, and terminates within an oligo T sequence encoded in the A-rich regions of the repeat (121). In some type 2 sequences this transcription termination sequence is absent, and transcription continues beyond the A-rich region into a unique sequence beyond the repeat. In this respect type 2 *Alu*-equivalent family members act like the first half of the human dimeric *Alu* sequence, while type 1 *Alu*-equivalent family members act like the second half, neither of which appear to have an initiation site for RNA polymerase III (121, 154). The function of these RNA transcripts remains undetermined, but they perhaps play a role in the dispersal of *Alu* sequences throughout mammalian DNA, possibly through a reverse transcript intermediate (107, 178) that subsequently becomes inserted into chromosomal DNA by a mechanism related to that used by known transposable elements and insertion sequences (55, 56), but this is only speculation. If it is proven, this mechanism could explain: why some oncogenes lack intervening sequences that are present in cellular genes from which they are thought to be derived (179–181), why an alpha globin pseudogene lacks both intervening sequences characteristic of most globin genes (182, 183), and why a histone H4 pseudogene in *Notopthalamus viridescens* has flanking direct repeats of nine base pairs (E. Stephenson and J. Gall, unpublished).

Transcription of Long Dispersed Repeats

Long-period dispersed repeats are also transcriptionally active. Copia and similar DNA elements in *Drosophila* were first noticed because they produced discretely sized RNA molecules that account for substantial proportions of the poly(A)-terminated RNA fraction (27, 184, 185). These RNAs contain a 5' "cap" structure (32), and are presumed to be synthesized by RNA polymerase II. Transcription begins within the 5'-long terminal repeat and extends into the 3'-long terminal repeat (H. E. Schwartz, T. J. Lockett and M. W. Young, unpublished); the same is true for the delta sequences bordering the yeast Ty1 element (186). Like the LTRs of proretroviruses, these repeats also contain the canonical TATA box (28–31, 37, 124), presumed to signal transcription initiation 25–30 bp downstream from it (187), and an AATAAA sequence that is thought to play a role in poly(A) addition approximately 20 residues downstream from it. Ty1 sequences in yeast are also heavily represented in poly(A)-terminated RNA (186), and presumably their transcription products have the same structure as copia RNAs, since their DNAs are homologous in structure to copia DNA.

Because of the structural resemblance between copia and Ty1 DNA and their RNA transcripts, and integrated retroviral DNA and retroviral RNA (Table 1), it seems likely that these long-period interspersed repeats were dispersed throughout the genomes of *Drosophila* and yeast, respectively, by a mechanism similar to retroviral insertion into chromosomal DNA that possibly included reverse transcription of RNA followed by integration of the DNA copies into cellular DNA. In this respect it is germane to note that extrachromosomal circular copies of copia DNA have been isolated from cultured *Drosophila* cells (188). However, their origin remains unclear. They could be self-replicating plasmids, reverse transcription products of copia RNA, or excision products from the genome produced by homologous recombination between the two long terminal repeats. Whatever their origins, they do bear a striking structural resemblance to the unintegrated circular DNAs of avian and mammalian retroviruses (41–43). Thus, both copia and other long-period dispersed repeats of *Drosophila* and the Ty1 sequences in yeast could be considered proretroviruses, possibly of cellular origin, that are produced without any extracellular virions. Conversely, retroviruses may have evolved from mobile genetic elements (40–54, 140, 189).

CONCLUDING REMARKS

Whether dispersed repeats can perform functions useful to an organism at their sites of insertion remains unknown. The constancy of structure of the primate beta-related gene cluster (190), which includes a number of *Alu* family members (92, 93, 155), suggests they might have a cellular function and have therefore been preserved at this chromosomal location during recent primate evolution. However, it could be counter-argued that once a dispersed repeat sequence has inserted itself at a new chromosomal location it is difficult to remove it and so it remains in position even though it performs neither useful nor deleterious functions there. Without any selective pressure such an element would be free to accumulate mutations and thus diverge from its original sequence. Insertion at some chromosomal loci could profoundly disrupt normal physiological functions, resulting in drastic changes in phenotype (38, 132–134, 135, 137, 138, 191). In some instances this could lead to cell death if an important function were disrupted, while in other instances it could confer selective advantage, as in the case of retrovirus insertion followed by increased expression of adjacent genes. If such increased expression promotes cell growth, the cells that grow fast would have an advantage because they would produce more of themselves than other cells. Ultimately, however, they might kill the organism and, thus, themselves as well. In still other instances insertion might be advantagous to an organism if the inserted sequences performed a useful cellular function at the inserted site, and selection pressure might then act to preserve such an inserted element. The frequency of insertion at new chromosomal loci has different implications depending on whether such insertions occur during the lifetime of an individual organism or on an evolutionary time scale. Except in yeast (36), there is currently no estimate of this frequency, but it could happen frequently in the somatic cells of metazoans and remain undetected unless drastic phenotypic changes such as uncontrolled growth (i.e. cancer) resulted.

A number of possible functions have been suggested for *Alu* family members (152), but no conclusive evidence has yet been presented to confirm these suggestions. It is particularly noteworthy that *Alu* family members contain a G-rich sequence extensively homologous to a sequence at or near the origin of papovavirus DNA replication (152). Accordingly, *Alu* family members have been suggested as origins of DNA replication, and there is some corroborating evidence for this (192). However, considerably more experimentation is required to confirm this suggestion. Like the functions of other dispersed repeats, that of the *Alu* family remains in question. It has even been suggested that dispersed repeats serve no phenotypically useful functions for the organisms in which they reside, but that they are parasitic and serve only to perpetuate themselves (193, 194). If this is true, they may still have profound effects on cellular physiology because of mutations caused by their movement to new chromosomal locations.

The long-period and short-period dispersed repeats that have been studied are either transcription units for discretely sized RNAs themselves, and/or they are extensively homologous to discretely sized RNAs that are transcribed elsewhere. For the proretroviruses, the function of the RNAs are known. They are the viral genome and mRNAs. For copia and similar long-period dispersed repeats in *Drosophila* and the Ty1 sequence yeast, the function of the RNAs is less clear. Copia is translated in vitro into a number of proteins of 18,000–51,000 daltons (195), and thus copia RNAs may be mRNAs. The functions of the RNAs transcribed from and homologous to human *Alu* family members and the rodent *Alu*-equivalent sequences (i.e. 4.5s, $4.5s_I$, and 7s) as well as the U2, U3, and U6 RNAs, which are also homologous to dispersed repeats, are unknown. The U1 RNA has been implicated in RNA:RNA splicing (196, 197) and thus may serve a cellular function unrelated to its postulated role in the dispersal of the U1 pseudogenes (107). Do other families of dispersed repeats also have discrete RNAs homologous to them or transcribed from them? If so, might they be instrumental in the dispersion of the repeats throughout an organism's genome? We currently do not know the answers to these questions, but at least we do know enough to raise them.

Literature Cited

1. Lewin, B. 1980. *Gene Expression,* p. 529. New York: Wiley. 1160 pp. 2nd ed.
2. Davidson, E. H., Britten, R. J. 1973. *Q. Rev. Biol.* 48:565–613
3. Bonner, J., Garrard, W. T., Gottesfeld, J., Holmes, D. S., Sevall, J. S., Wilkes, M. 1973. *Cold Spring Harbor Symp. Quant. Biol.* 38:303–10
4. Chamberlain, M. E., Britten, R. J., Davidson, E. H. 1975. *J. Mol. Biol.* 96:317–33
5. Goldberg, R. B., Crain, W. R., Ruderman, J. V., More, G. P., Higgins, T. R.,
   Gelfand, R. C., Galau, G. A., Britten, R. J., Davidson, E. H. 1975. *Chromosoma* 51:225–51
6. Davidson, E. H., Gallau, G. A., Angerer, R. C., Britten, R. J. 1975. *Chromosoma* 51:253–59
7. Schmid, C. W., Deininger, P. L. 1975. *Cell* 6:345–58
8. Deininger, P. L., Schmid, C. W. 1976. *J. Mol. Biol.* 106:773–90
9. Firtel, R. A., Kindel, K. 1975. *Cell* 5:401–11

10. Graham, D. E., Neufeld, B. R., Davidson, E. H., Britten, R. J. 1974. *Cell* 1:127–37
11. Efstratiadis, A., Crain, E. T., Britten, R. J., Davidson, E. H., Kafatos, F. C. 1976. *Proc. Natl. Acad. Sci. USA* 73:2289–93
12. Angerer, R. C., Davidson, E. H., Britten, R. J. 1975. *Cell* 6:29–39
13. Pearson, W. R., Wu, J-R., Bonner, J. 1978. *Biochemistry* 17:51–59
14. Davidson, E. H., Hough, B. R., Amenson, C. S., Britten, R. J. 1973. *J. Mol. Biol.* 77:1–23
15. Zimmerman, J. L., Goldberg, R. B. 1977. *Chromosoma* 59:227–52
16. Adams, J. W., Kaufman, R. E., Kretschner, P. J., Harrison, M., Nienhuis, A. W. 1980. *Nucleic Acids Res.* 8:6113–28
17. Brutlag, D. 1980. *Ann. Rev. Genet.* 14:121–44
18. Long, E. D., Dawid, I. B. 1980. *Ann. Rev. Biochem.* 49:727–66
19. Wyman, A. R., White, R. 1980. *Proc. Natl. Acad. Sci. USA* 77:6754–58
20. Chaudhari, N., Craig, S. P. 1979. *Proc. Natl. Acad. Sci. USA* 76:6101–5
21. Wilson, D. A., Thomas, C. A. 1974. *J. Mol. Biol.* 84:115–38
22. Dott, P. J., Chuang, C. R., Saunders, G. F. 1976. *Biochemistry* 15:4120–25
23. Jelinek, W. R. 1977. *J. Mol. Biol.* 115:591–601
24. Jelinek, W. R. 1978. *Proc. Natl. Acad. Sci. USA* 75:2679–83
25. Manning, J. E., Schmid, C. W., Davidson, N. 1975. *Cell* 4:141–55
26. Crain, W. R., Eden, F. C., Pearson, W. R., Davidson, E. H., Britten, R. J. 1976. *Chromosoma* 56:309–26
27. Finnegan, D. J., Rubin, G. M., Young, M. W., Hogness, D. S. 1978. *Cold Spring Harbor Symp. Quant. Biol.* 42:1053–63
28. Levis, R., Dunsmuir, P., Rubin, G. 1980. *Cell* 21:581–88
29. Dunsmuir, P., Brorein, W. J., Simon, M. A., Rubin, G. M. 1980. *Cell* 21:575–80
30. Bayev, A. A., Krayev, A. S., Lyubomirskaya, N. V., Ilyin, Y. V., Skryabin, K. G., Georgiev, G. P. 1980. *Nucleic Acids. Res.* 8:3263–73
31. Kulguskin, V. V., Ilyin, Y. V., Georgiev, G. P. 1981. *Nucleic Acids. Res.* 9:3451–64
32. Rubin, G. M., Brorein, W. J., Dunsmuir, P., Flavell, A. J., Levis, R., Strobel, E., Toole, J. J., Young, E. 1981. *Cold Spring Harbor Symp. Quant. Biol.* 45:619–28

33. Young, M. W., Schwartz, H. E. 1981. *Cold Spring Harbor Symp. Quant. Biol.* 45:629–40
34. Georgiev, G. P., Ilyin, Y. V., Chmeliauskaite, V. G., Ryskov, A. P., Kramerov, D. A., Skryabin, K. G., Krayev, A. S., Lukaniidin, E. M., Grigoryan, M. S. 1981. *Cold Spring Harbor Symp. Quant. Biol.* 45:641–54
35. Tchurikov, N. A., Ilyin, Y. V., Skryabin, K. G., Ananiev, E. V., Baayev, A. A., Krayev, A. S., Zelentsova, E. S., Kulguskin, V. V., Lyubomirskaya, N. V., Georgiev, G. P. 1981. *Cold Spring Harbor Symp. Quant. Biol.* 45:655–65
36. Cameron, J. R., Loh, E. Y., Davis, R. W. 1979. *Cell* 16:739–51
37. Farnbauch, P. J., Fink, G. R. 1980. *Nature* 286:352–56
38. Fink, G., Farabaugh, P., Roeder, G., Chaleff, D. 1981. *Cold Spring Harbor Symp. Quant. Biol.* 45:575–80
39. Eibel, H., Gafner, J., Stotz, A., Philippsen, P. 1981. *Cold Spring Harbor Symp. Quant. Biol.* 45:609–17
40. Shimotohno, K., Mizutani, S., Temin, H. M. 1980. *Nature* 285:550–54
41. Guntaka, R. V., Richards, O. C., Shank, P. R., Kung, H-J., Davidson, N., Fritsch, E., Bishop, J. M., Varmus, H. E. 1976. *J. Mol. Biol.* 106:337–57
42. Shank, P. R., Hughes, S. H., Kung, H-J., Majors, J. E., Quintrell, N., Guntaka, R. V., Bishop, J. M., Varmus, H. E. 1978. *Cell* 15:1383–95
43. Highfield, P. E., Rafield, L. F., Gilmer, T. M., Parsons, J. T. 1980. *J. Virol.* 36:271–9
44. Majors, J. E., Swanstrom, R., DeLorbe, W. J., Payne, G. S., Hughes, S. H., Ortiz, S., Quintrell, N., Bishop, J. M., Varmus, H. E. 1981. *Cold Spring Harbor Symp. Quant. Biol.* 45:731–8
45. Shimotohno, K., Temin, H. 1981. *Cold Spring Harbor Symp. Quant. Biol.* 45:719–30
46. Skalka, A., Ju, G., Hishinuma, F., DeBona, P. J., Astrin, S. 1981. *Cold Spring Harbor Symp. Quant. Biol.* 45:739–46
47. Shoemaker, C., Goff, S., Gilboa, E., Paskind, M., Mitra, S., Baltimore, D. 1980. *Proc. Natl. Acad. Sci. USA* 77:3932–36
48. Dhar, R., McClements, W. L., Enquist, L. W., Vande Woude, G. F. 1980. *Proc. Natl. Acad. Sci. USA* 77:3937–41
49. Benz, E. W., Wydo, R. M., Nadal-Ginard, B., Dina, D. 1980. *Natuure* 288:665–9

50. Sutcliffe, J. G., Sninnick, T. M., Verma, I. M., Lerner, R. A. 1980. *Proc. Natl. Acad. Sci. USA* 77:3302–6
51. VanBereren, C., Goddard, J. G., Berns, A., Verma, I. M. 1980. *Proc. Natl. Acad. Sci. USA* 77:3307–11
52. McClements, W. L., Dhar, R., Blair, D. G., Enquist, L., Oskarsson, M., Vande Woude, G. F. 1981. *Cold Spring Harbor Symp. Quant. Biol.* 45:699–705
53. Sutcliffe, J. G., Shinnick, T. M., Lerner, R. A. 1981. *Cold Spring Harbor Symp. Quant. Biol.* 45:707–10
54. Shoemaker, C., Goff, S., Gilboa, E., Paskind, M., Mitra, S., Baltimore, D. 1981. *Cold Spring Harbor Symp. Quant. Biol.* 45:711–17
55. Shapiro, J. A. 1979. *Proc. Natl. Acad. Sci. USA* 76:1933–37
56. Calos, M. P., Miller, J. H. 1980. *Cell* 20:579–95
57. Eden, F. C., Hendrick, J. P., Gottlieb, S. S. 1978. *Biochemistry* 17:5113–21
58. Eden, F. C., Burns, A. T. H., Goldberger, R. F. 1980. *J. Biol. Chem.* 225:4843–53
59. Eden, F. C., Musti, A. M., Sobieski, D. A. 1981. *J. Mol. Biol.* 148:129–51
60. Wensink, P. C., Tabata, S., Pachl, C. 1979. *Cell* 18:1231–46
61. Scheller, R. H., Anderson, D. M., Posakony, J. W., McAllister, L. B., Britten, R. J., Davidson, E. H. 1981. *J. Mol. Biol.* 149:15–39
62. Posakony, J. W., Scheller, R. H., Anderson, D. M., Britten, R. J., Davidson, W. H. 1981. *J. Mol. Biol.* 149:41–67
63. Anderson, D. M., Scheller, R. H., Posakony, J. W., Trabert, S., McAllister, L. B., Beall, C., Britten, R. J., Davidson, E. H. 1981. *J. Mol. Biol.* 145:5–28
64. Murray, M. G., Palmer, J. D., Cuellar, R. E., Thompson, W. F. 1979. *Biochemistry* 18:5259–66
65. Murray, M. G., Cuellar, R. E., Thompson, W. F. 1978. *Biochemistry* 17:5781–90
66. Hudspeth, M. E. S., Goldberg, R. B., Timberlake, W. E. 1977. *Proc. Natl. Acad. Sci. USA* 74:4332–36
67. Pellegrini, M., Timberlake, W. E., Goldberg, R. B. 1981. *Mol. Cell. Biol.* 1:136–43
68. Davidson, E. H., Hough, B. R., Klein, W. H., Britten, R. J. 1975. *Cell* 4:217–38
69. Britten, R. J., Davidson, E. H. 1969. *Science* 165:349–57
70. Davidson, E. H., Britten, R. J. 1979. *Science* 204:1052–59
71. Gall, J. G., Cohen, E. H., Atherton, D. D. 1973. *Cold Spring Harbor Symp. Quant. Biol.* 38:417–21
72. Varley, J. M., Mcgregor, H. C., Nardi, I., Andrews, C., Erba, H. P. 1980. *Chromosoma* 80:289–307
73. Varley, J. M., Mcgregor, H. C., Erba, H. P. 1980. *Nature* 283:686–88
74. Diaz, M. O., Barsacchi-Pilone, G., Mahon, K. H., Gall, J. G. 1981. *Cell* 24:649–59
75. Stephenson, E. C., Erba, H. P., Gall, J. G. 1981. *Cell* 24:639–47
76. Fedoroff, N. V., Brown, D. D. 1978. *Cell* 13:701–16
77. Miller, J. R., Cartwright, E. M., Brownlee, G. G., Federoff, N. V., Brown, D. D. 1978. *Cell* 13:717–25
78. Kedes, L. H. 1979. *Ann. Rev. Biochem.* 48:837–70
79. Heintz, N., Zernik, M., Roeder, R. G. 1981. *Cell* 24:661–68
80. Childs, G., Maxson, R., Cohn, R. H., Kedes, L. 1981. *Cell* 23:651–63
81. Britten, R. J., Kohn, D. E. 1968. *Science* 161:529–40
82. Bonner, T. I., Brenner, D. J., Neufeld, D. R., Britten, R. J. 1973. *J. Mol. Biol.* 81:123–35
83. Wetmer, J. G., Davidson, N. 1968. *J. Mol. Biol.* 31:349–70
84. Davidson, E. H. 1976. *Gene Activity in Early Development.* New York: Academic. 452 pp. 2nd ed.
85. Kline, W. H., Thomas, T. L., Lai, C., Scheller, R. H., Britten, R. J., Davidson, E. H. 1978. *Cell* 14:889–900
86. Rinehart, F. P., Ritch, T. G., Deininger, P. L., Schmid, C. W. 1981. *Biochemistry* 20:3003–10
87. Dover, G., Coen, E. 1981. *Nature* 290:731–32
88. Brown, S. D. M., Dover, G. 1981. *J. Mol. Biol.* 150:441–66
89. Houck, C. M., Rinehart, F. P., Schmid, C. W. 1978. *Biochem. Biophys. Acta* 518:37–52
90. Houck, C. M., Rinehart, F. P., Schmid, C. W. 1979. *J. Mol. Biol.* 132:289–306
91. Lewin, B. See Ref. 1, p. 959
92. Fritsch, E. F., Lawn, R. M., Maniatis, T. 1980. *Cell* 19:959–72
93. Coggins, L. W., Grindlay, G. J., Vass, J. K., Slater, A. A., Montague, P., Stinson, M. A., Paul, J. 1980. *Nucleic Acids Res.* 8:3319–33
94. Della Fovira, R., Gelman, E. P., Gallo, R. C., Wong-Stall, F. 1981. *Nature* 292:31–35
95. Tashima, M., Calabretta, B., Torelli, G., Scofield, M., Maizel, A., Saunders,

G. 1981. *Proc. Natl. Acad. Sci. USA* 78:1508–12

96. Shen, C. K. J., Maniatis, T. 1980. *Cell* 19:379–91

97. Haigwood, N. L., Jahn, C. L., Hutchinson, C. A. III, Edgell, M. H. 1981. *Nucleic Acids Res.* 9:1133–50

98. Toomey, T. P. 1981. *A small nuclear RNA molecule homologous to a repetitive DNA sequence in mammalian cells.* PhD thesis. Rockefeller Univ., New York. 80 pp.

99. Deleted in proof

100. Jagadeeswaran, P., Pan, J., Spritz, R. A., Duncan, C. H., Biro, P. A., Tuan, D., Forget, B. G., Weissman, S. M. 1982. *Nucleic Acids Res.* In press

101. Horz, W., Hess, I., Zachau, H. G. 1974. *Eur. J. Biochem.* 45:501–10

102. Sheau-Mei, C., Schildkraut, C. C. 1980. *Nucleic Acids Res.* 8:4075–90

103. Arnheim, N. P., Seperack, P., Banerji, J., Lang, R. B., Miesfeld, R., Marcu, K. B. 1980. *Cell* 22:179–85

104. Marcu, K. B., Arnheim, N., Banerji, J., Penncavage, N. A., Seperack, P., Lang, R., Miesfeld, R., Harris, L., Greenberg, R. 1981. *Cold Spring Harbor Symp. Quant. Biol.* 45:899–911

105. Marzluff, W. F., White, E. L., Benjamin, R., Huang, R. C. C. 1975. *Biochemistry* 14:3715–24

106. Engberg, J., Hellung-Larsen, P., Frederiksen, S. 1974. *Eur. J. Biochem.* 41:321–28

107. Van Arsdell, S. W., Denison, R. A., Bernstein, L. B., Weiner, A. M., Manser, T., Gesteland, R. F. 1981. *Cell* 26:11–17

108. Hayashi, K. 1981. *Nucleic Acids Res.* 9:3379–88

109. Rubin, C. M., Houck, C. M., Deininger, P. L., Friedmann, T., Schmid, C. W. 1980. *Nature* 284:372–74

110. Deininger, P. L., Jolly, D. J., Rubin, C. M., Friedmann, T., Schmid, C. W. 1981. *J .Mol. Biol.* 151:17–33

111. Deininger, P. L., Jolly, D. J., Rubin, C. M., Houck, C. M., Friedmann, T., Schmid, C. W. 1981. In *Mechanistic Studies of DNA Replication,* ed. B. Alberts, C. F. Fox. New York: Academic

112. Bell, G. I., Pictet, R., Rutter, W. J. 1980. *Nucleic Acids Res.* 8:4091–4109

113. Baralle, F. E., Shoulders, C. C., Goodbourn, S., Jeffreys, A., Proudfoot, N. J. 1980. *Nucleic Acids Res.* 8:4393–404

114. Duncan, C. H., Jagadeeswaran, P., Wang, R. R. C., Weissman, S. M. 1981. *Gene* 13:185–96

115. Krayev, A. S., Kramerov, D. A., Skryabin, K. G., Ryskov, A. P., Bayev,

A. A., Georgiev, G. P. 1980. *Nucleic Acids Res.* 8:1201–5

116. Haynes, S. R., Toomey, T. P., Leinwand, L., Jelinek, W. R. 1981. *Mol. Cell Biol.* 1:573–83

117. Dhruva, B. R., Shenk, T., Subramanian, K. N. 1980. *Proc. Natl. Acad. Sci. USA* 77:4514–18

118. Gusella, J. F., Keys, C., Varsanyi-Breiner, A., Kao, F-T., Jones, C., Puck, T. T., Housman, D. 1980. *Proc. Natl. Acad. Sci. USA* 77:2829–33

119. Murray, M. J., Shilo, B-Z., Shih, C., Cowing, D., Hsu, H. W., Weinberg, R. A. 1981. *Cell* 25:355–61

120. Houck, C. M., Schmid, C. W. 1981. *J. Mol. Evol.* 17:148–51

121. Haynes, S. R., Jelinek, W. R. 1981. *Proc. Natl. Acad. Sci. USA* 78:6130–34

122. Page, G. S., Smith, S., Goodman, H. M. 1981. *Nucleic Acids. Res.* 9:2087–104

123. Flavell, A. 1981. *Nature* 289:10–11

124. Gafner, J., Philippsen, P. 1980. *Nature* 286:414–18

125. Howk, R. S., Troxler, D. H., Lowy, D., Duesberg, P. H., Scolnick, E. M. 1978. *J. Virol.* 25:115–23

126. Keshet, E., Shaul, Y. 1981. *Nature* 289:83–85

127. Scolnick, E. M., Vaas, W. C., Howk, R. S., Duesberg, P. H. 1979. *J. Virol.* 29:964–72

128. Young, M. W. 1979. *Proc. Natl. Acad. Sci. USA* 76:6274–78

129. Young, M. W. 1981. In *Genetic Engineering: Principles and Methods,* ed. J. K. Stelow, A. Hollander, 3:109–28. New York: Plenum

130. Strobel, E. P., Dunsmuir, P., Rubin, G. M. 1979. *Cell* 17:429–39

131. Potter, S. S., Brorein, W. J. Jr., Dunsmuir, P., Rubin, G. M. 1979. *Cell* 17:415–27

132. Ghering, W. J., Paro, R. 1980. *Cell* 19:897–904

133. Chaleff, D. T., Fink, G. R. 1980. *Cell* 21:227–37

134. Roeder, G. S., Fink, G. R. 1980. *Cell* 21:239–49

135. Williamson, V. M., Young, E. T., Ciriacy, M. 1981. *Cell* 23:605–14

136. Errede, B., Cardillo, T. S., Sherman, F., Dubois, E., Deschamps, J., Wiame, J. M. 1980. *Cell* 22:427–36

137. Errede, B., Cardillo, T. S., Weaver, G., Sherman, F. 1981. *Cold Spring Harbor Symp. Quant. Biol.* 45:593–602

138. Varmus, H. E., Quintrell, W., Ortiz, S. 1981. *Cell* 25:23–36

139. Neel, B. G., Hayward, N. S., Robinson, H. L., Fang, J., Astrin, S. M. 1981. *Cell* 23:323–34

140. Temin, H. M. 1980. *Cell* 21:599–600
141. Darnell, J. E., Balint, R. 1970. *J. Cell Physiol.* 76:349–61
142. Hough, B. R., Davidson, E. H. 1972. *J. Mol. Biol.* 70:491–509
143. Holmes, D. S., Bonner, J. 1974. *Proc. Natl. Acad. Sci. USA* 71:1108–12
144. Molloy, G. R., Jelinek, W. R., Salditt, M., Darnell, J. E. 1974. *Cell* 1:43–53
145. Smith, M. J., Hough, B. R., Chamberlain, M. E., Davidson, E. H. 1974. *J. Mol. Biol.* 85:103–25
146. Jelinek, W. R., Darnell, J. E. 1972. *Proc. Natl. Acad. Sci. USA* 69:2537–41
147. Ryskov, A. P., Farashyan, V. R. Georgiev, G. P. 1972. *Biochem. Biophys. Acta* 262:568–72
148. Ryskov, A. P., Saunders, G. F., Farashyan, V. R., Georgiev, G. P. 1973. *Biochem. Biophys. Acta* 312:152–64
149. Jelinek, W. R., Molloy, G. R., Fernandez-Munoz, R., Salditt-Georgieff, M., Darnell, J. E. 1974. *J. Mol. Biol.* 82:361–70
150. Fedoroff, N. V., Wellauer, P. K., Wall, R. 1977. *Cell* 10:597–610
151. Robertson, H. D., Dickson, E., Jelinek, W. R. 1977. *J. Mol. Biol.* 115:571–89
152. Jelinek, W. R., Toomey, T. P., Leinwand, L., Duncan, C. H., Biro, P. A., Choudary, P. V., Weissman, S. M., Rubin, C. M., Houck, C. M., Deininger, P. L., Schmid, C. W. 1980. *Proc. Natl. Acad. Sci. USA* 77:1398–402
153. Jelinek, W. R., Evans, R., Wilson, M., Salditt-Georgieff, M., Darnell, J. E. 1978. *Biochemistry* 17:2776–83
154. Elder, J. T., Pan, J., Duncan, C. H., Weissman, S. M. 1981. *Nucleic Acids Res.* 9:1171–89
155. Fritsch, E. F., Shen, C. K. J., Lawn, R. M., Maniatis, T. 1981. *Cold Spring Harbor Symp. Quant. Biol.* 45:761–75
156. Hofer, E., Darnell, J. E. 1981. *Cell* 23:585–93
157. Costantini, F. D., Scheller, R. H., Britten, R. J., Davidson, E. H. 1978. *Cell* 15:173–87
158. Scheller, R. H., Costantini, F. D., Kozlowski, M. R., Britten, R. J., Davidson, E. H. 1978. *Cell* 15:189–203
159. Costantini, F. D., Britten, R. J., Davidson, E. H. 1981. *Nature* 287:111–7
160. Ryffel, G. U., Muellener, D. B., Wyler, T., Wahli, W., Weber, R. 1981. *Nature* 291:429–31
161. Jelinek, W. R., Leinwand, L. 1978. *Cell* 15:205–14
162. Harada, F., Kato, N. 1981. *Nucleic Acids Res.* 8:1273–85
163. Weiner, A. M. 1980. *Cell* 22:209–18
164. Erikson, E., Erikson, R. L., Henry, B., Pace, N. R. 1973. *Virology* 53:40–46
165. Czernilofsky, A. P., Levinson, A. D., Varmus, H. E., Bishop, J. M., Tischer, E., Goodman, H. M. 1980. *Nature* 287:198–203
166. Breathnach, R., Chambon, P. 1981. *Ann. Rev. Biochem.* 50:349–83
167. RoChoi, T. S., Reddy, R., Henning, D., Takano, T., Taylor, C. W., Busch, H. 1972. *J. Biol. Chem.* 247:3205–22
168. Fowlkes, D. M., Shenk, T. 1980. *Cell* 22:405–13
169. Shenk, T. 1981. *Curr. Top. Microbiol. Immunol.* In press
170. Reich, P. R., Forget, B. G., Weissman, S. M., Rose, J. A. 1966. *J. Mol. Biol.* 17:428–39
171. Soderlund, H., Pettersson, U., Vennstrom, B., Philipson, L., Mathews, M. B. 1976. *Cell* 7:585–93
172. Rosa, M. D., Gottlieb, E., Lerner, M. R., Steitz, J. A. 1981. *Mol. Cell. Biol.* 1:785–96
173. Lerner, M. R., Boyle, J. A., Hardin, J. A., Steitz, J. A. 1981. *Science* 211:400–2
174. Lerner, M. R., Andrews, N. C., Miller, G., Steitz, J. A. 1981. *Proc. Natl. Acad. Sci. USA* 78:805–9
175. Duncan, C., Biro, P. A., Choudary, P. V., Elder, J. T., Wang, R. R. C., Forget, B. G., DeRiel, J. K., Weissman, S. M. 1979. *Proc. Natl. Acad. Sci. USA* 76:5095–99
176. Pan, J., Elder, J. T., Duncan, C. H., Weissman, S. M. 1981. *Nucleic Acids Res.* 9:1151–70
177. Sakonju, S., Bogenhagen, D. F., Brown, D. D. 1980. *Cell* 19:13–25
178. Jagadeeswaran, P., Forget, B. G., Weissman, S. M. 1981. *Cell* 26:141–42
179. Goff, S. P., Gilboa, E., Witte, O. N., Baltimore, D. 1980. *Cell* 22:777–85
180. Franchini, G., Even, J., Sherr, C. J., Wong-Stall, F. 1981. *Nature* 290:154–57
181. Bishop, J. M. 1981. *Cell* 23:5–6
182. Rishioka, Y., Leder, A., Leder, P. 1980. *Proc. Natl. Acad. Sci. USA* 77:2806–9
183. Vanin, E. F., Goldberg, G. I., Tucker, P. W., Smithies, O. 1980. *Nature* 286:222–26
184. Rubin, G. M., Finnegan, D. J., Hogness, D. S. 1976. *Prog. Nucleic Acid Res. Mol. Biol.* 19:221–26
185. Young, M. W., Hogness, D. S. 1977. In *Eukaryotic Genetics Systems, ICN-UCLA Symp. Mol. Cell. Biol.*, ed. G. Wilcox, 8:315–21. New York: Academic

186. Elder, T. R., John, T. P., Stinchcomb, D. T., Davis, R. W. 1981. *Cold Spring Harbor Symp. Quant. Biol.* 45:581–84
187. Goldberg, M. 1978. PhD thesis. Stanford Univ.
188. Flavell, A. J., Ish-Horowicz, D. 1981. *Nature* 292:591–95
189. Temin, H. M. 1971. *J. Natl. Cancer Inst.* 46:3–7
190. Barrie, P. A., Jeffreys, A. J., Scott, A. F. 1981. *J. Mol. Biol.* 149:319–36
191. McClintock, B. 1967. *Carnegie Inst. Wash. Yearbook* 65:568–78
192. Taylor, J. H., Watanabe, S. 1981. In *Structure and DNA-Protein Interactions of Replication Origins,* ICN-UCLA

Symp. Mol. Cell. Biol., ed. D. A. Ray, C. F. Fox, 22:597–606. New York: Academic. 623 pp.
193. Doolittle, W. F., Sapienza, C. 1980. *Nature* 284:601–3
194. Orgel, L. E., Crick, F. H. C. 1980. *Nature* 284:604–7
195. Flavell, A. J., Ruby, S. W., Toole, J. J., Roberts, B. E., Rubin, G. M. 1980. *Proc. Natl. Acad. Sci. USA* 77:7101–11
196. Lerner, M. R., Boyel, J. A., Mount, S. M., Wolin, S. L., Steitz, J. A. 1980. *Nature* 283:220–24
197. Rogers, J., Wall, R. 1980. *Proc. Natl. Acad. Sci. USA* 77:1877–79

Ann. Rev. Biochem. 1982. 51:845–68
copyright © 1982 Annual Reviews Inc. All rights reserved

THE BIOLOGY AND MECHANISM OF ACTION OF NERVE GROWTH FACTOR

Bruce A. Yankner and Eric M. Shooter

Department of Neurobiology, Stanford University School of Medicine, Stanford, California 94305

CONTENTS

PERSPECTIVES AND SUMMARY

In order to understand the role of a nerve growth factor, we must first define what is meant by nerve growth. The nervous system grows not only by cellular hyperplasia and hypertrophy but also by the ramification of processes with the formation of specific synaptic connections. This latter feature sets apart the development of the nervous system in its requirement for precisely oriented growth. Such a constraint has been built into the mechanism of action of nerve growth factor (NGF) and distinguishes it from other peptide growth hormones. Regulation of neuronal development can be considered in terms of the control of cell death, differentiation, and orienta-

845

tion. These biological functions of NGF and their mechanisms of action are the subject of this review. The chemistry, biosynthesis, and measurement of NGF have been described elsewhere (1–5).

Considerable regression of neurons by cell death during development has been well established (6). One proposal is that cell death may function as a sorting out process in the establishment of specific synaptic connections (7). Another suggestion is that the redundancy of neurons in the embryo may confer a selective or developmental advantage (8). Neuronal cell death may be modulated by the availability of diffusible factors. Several cell types of neural crest origin require NGF for survival both in vivo and in vitro during development. The exposure of developing nerve terminals to NGF appears to be important for neuronal cell viability. The initial source of NGF in vivo may be the innervated target organ, which could supply the neuronal cell body with NGF by retrograde axonal transport.

The outgrowth of neuritic processes must be regulated in its orientation in order to form the synaptic connections appropriate to the functioning nervous system. The chemotactic effect of NGF is manifested by the rapid reorientation of the neuritic growth cone to a gradient of NGF. One possible mechanism for this transient response may involve rapid changes in intracellular cyclic nucleotide levels and the efflux of intracellular calcium.

The differentiation of neurons requires a considerable synthetic effort in order to construct and maintain an extensive neurite arborization. The synthesis of RNA and protein is regulated by NGF. Although continuous protein synthesis appears to be required for neurite extension, the necessity for RNA synthesis is dependent on the substratum to which the cell adheres and the degree of prior exposure to NGF. The translocation of the NGF-receptor complex to intracellular structures such as the nuclear membrane may be important in the mechanism of action.

THE ROLE OF NGF IN THE DEVELOPMENT OF THE NERVOUS SYSTEM

NGF and Neuronal Survival

Perhaps the most dramatic example of the biological importance of NGF is the neuronal cell death observed following the administration of NGF antibody to the immature animal. Several days of injection of antibody to NGF results in the irreversible degeneration of the paravertebral and prevertebral sympathetic ganglia (9). The first ultrastructural changes observed are the disorganization of the nucleolus and the disruption of the nuclear membrane followed by a longer-term general cellular disintegration (10, 11). These changes are almost completely reversible on addition of a sufficient dose of NGF within 48 hr after antibody administration. The same effects of antibody to NGF were observed in complement-deficient mice (12), which suggests that these degenerative changes are due to NGF deprivation rather than complement-mediated cytotoxicity.

Susceptibility to immunosympathectomy gradually decreases with increasing age (13, 14). In the adult rat, injection of NGF antibody or autoimmunization against NGF results in decreased neuronal size and protein content as well as a disproportionate drop in the activities of tyrosine hydroxylase and dopamine β hydroxylase (15). The latter may reflect the degeneration of adrenergic nerve terminals (16). These effects were in large part, but not completely reversed as antibody titers dropped. This is consistent with the finding (17) that even in the adult animal, immunization against NGF results in a 30–40% cell loss in the superior cervical ganglion (SCG).

Injection of antibody to NGF during the postnatal period does not affect the sensory neurons of the dorsal root ganglion (DRG), although these cells require NGF for survival in culture (9, 18). Although it is possible that the injected antibody was somehow physically excluded from sensory neurons, it seems more likely that the requirement for NGF in culture may be substituted *in vivo* by other factors or by innervation. As demonstrated by Varon and his colleagues, nonneuronal cells can substitute for NGF in maintaining the survival of sensory neurons in culture (19, 20). Although sensory neurons are unaffected by NFG antibody postnatally, prenatal dependence on NGF in vivo has shown by the degeneration of these cells after exposure to maternal NFG antibody in utero (21). In a similar fashion, exposure of 17-day-old rat fetuses to NGF antibody results in degeneration of chromaffin cell precursors in the adrenal medulla (22). These cells are unaffected postnatally. In addition to demonstrating the physiological relevance of NGF, the NGF antibody is useful as a probe of the stage of ontogenesis in which a neuron is dependent on the hormone.

During normal development, there is a considerable decrease in the number of sympathetic neurons in the SCG. Administration of NGF to 6-day-old rats prevented this decrease and actually brought about an increase in the number of neurons in the ganglion (23). Using tritiated thymidine autoradiography, it was demonstrated in the rat that neuronal cell division is complete before birth (24). The increased number of neurons observed after administration of NGF appears to reflect a more rapid rate of neuronal maturation and the prevention of normal cell death (24, 25). An increased number of mitoses has been observed in rat sympathetic and chick sensory ganglia treated with NGF, which appears to be due to enhanced proliferation of nonneuronal cells (24). Since isolated nonneuronal cells do not respond to NGF (26), it is possible that their increased mitoses in the ganglion is secondary to NGF-stimulated neurite outgrowth.

The explanted SCG requires NGF for survival in culture during the postnatal period with the exception of a restricted period from day 14–18 in the mouse (27). Sympathethic neurons separated from nonneuronal cells also depend on NGF for survival in culture (28–30). Withdrawal of NGF after short-term culture results in the rapid degeneration of virtually all cells (30). In contrast, withdrawal of NGF after 7 weeks of culture followed by 10 days in the absence of NGF results in the loss of only about 50% of the neurons. This gradual reduction in the dependence on NGF for survival with age in culture is consistent with the decreased degenerative response to NGF antibody observed in the adult animal.

Communication between the periphery and the neuron appears to be necessary for its survival during early development. Axotomy of the SCG by either surgical interruption or chemical destruction with 6-hydroxy-dopamine results in the degeneration of the corresponding cell bodies during the postnatal period (31, 32). This damage could be prevented by administration of NGF (33, 34). Axotomy was most effective during the early postnatal period. Its effects on neuronal degeneration gradually decreased with increasing age until they were no longer apparent by 3 weeks of age in the rat (32, 33). The local control of neuronal survival and neurite extension was investigated in vitro by an ingenious multichamber culture system in which the cell bodies of sympathetic neurons were separated from the periphery by a fluid-impermeable barrier (35). Neuronal survival was maintained by NGF in the central chamber containing the cell bodies, but neurites did not penetrate the barrier unless NGF was also present in the peripheral chamber. Once neurites had extended into the peripheral chamber, NGF could be withdrawn from the cell bodies in the central chamber, and as long as NGF was still in contact with neurites in the peripheral chamber neuronal survival was maintained.

A mechanism for delivery of NGF from the periphery to the neuronal cell body became apparent with the discovery of retrograde axonal transport (36). Injection of ^{125}I-NGF into the anterior eye chamber of the rat resulted in the subsequent appearance of label in the neuronal cell bodies of the SCG a considerable distance away. The label was mostly intact βNGF as determined by SDS-polyacrylamide gel electrophoresis and recognition by NGF antibody (36, 37). The SCG on the side of the injected eye showed intense labeling of a small number of neurons (presumably those innervating the iris) whereas the SCG from the noninjected side showed diffuse labeling throughout the ganglion (36). The difference between the two sides was maximal 16 hr after injection and could be completely abolished by prior axotomy on the injected side. Intraocular injection of colchicine or vinblastine partially blocked retrograde transport, which suggests a role for microtubules in this process. Uptake of NGF by nerve terminals was found to be hormone-specific and receptor-mediated and was not quantitatively affected by simultaneous uptake and transport of a variety of lectins and toxins (37). Moreover, the uptake of NGF was not stimulated by electrical activity as is the nonspecific uptake of other macromolecules (38). The dose response relationship for retrograde transport of NGF suggests two uptake systems; one with high affinity and low capacity and another with intermediate affinity and intermediate capacity (37). This is consistent with the finding of two affinities for receptors on sensory neurons, sympathetic neurons, and PC12 cells. The relationship of these two receptor types will be discussed in the section on the cell surface receptor.

The subcellular localization of retrogradely transported NGF was determined by electron microscopic autoradiography of ^{125}I-NGF and cytochemical analysis of horseradish peroxidase-conjugated NGF (39, 40). In axons, labeled NGF was found to be distributed in small vesicles and larger structures confined by smooth membranes. In the neuronal cell body, NGF was primarily found in multivesicular and dense bodies (presumed to be lysosomes), vesicles, and smooth endoplasmic reticulum, 14 hr after injection. About 3% of the labeled NGF was observed in the nucleus at this time (39). A similar subcellular distribution was determined for retrogradely transported tetanus toxoid (41). The final destination of internalized NGF was not conclusively established in these studies because they were not carried out over a time course sufficient for the loss of labeled NGF from the cell body. Therefore, subsequent redistribution of NGF to other sites in the cell (possibly by vesicular fusion) could not be ruled out. This is especially relevant to the translocation of NGF to the nucleus, which has been demonstrated in the PC12 cell line (42). It was found that nuclear translocation is a very slow process and accounts for only a small fraction of the total NGF associated with the cell; it may therefore be difficult to distinguish it from background using morphological methods (5).

The physiological significance of retrograde axonal transport has been surmised from experiments in which the intraocular injection of ^{125}I-NGF was followed by an increase in ganglionic tyrosine hydroxylase (TH) activity, to a greater extent on the injected than on the noninjected side (43). Axotomy of the SCG on the injected side eliminated the difference in TH induction and prevented the selective accumulation of ^{125}I-NGF. Hendry (44) found that identified neurons innervating the iris were selectively enlarged following intraocular injection of NGF and increased their ganglionic TH activity.

The sensory neurons of the chick dorsal root ganglion have also been demonstrated to retrogradely transport NGF from the periphery (45–47). Although the response of these cells to NGF is known to be restricted to early development, retrograde transport was also observed in the adult (46).

Retrograde axonal transport of NGF has been observed in cholinergic neurons of the chick and rat ciliary ganglion (47) and in selective cells of the rat brain (48), even though these cells have no known response to NGF. Therefore, retrograde transport per se does not seem to be adequate to elicit a biological response to NGF. Its presence in these cells may, however, be a remnant of an earlier period of NGF-responsiveness during development.

Several mechanistic roles have been suggested for the action of retrogradely transported NGF. The effects of NGF are perhaps mediated by second messengers, which can be released by the NGF-receptor complex after its transfer from the nerve terminal to the cell body (49). Alternatively the direct transfer of the hormone-receptor complex to intracellular membranes may be a key step in the mechanism of action (5). Evidence in support of this model is discussed below in the section on internalization.

In summary, the importance of NGF in maintaining neuronal viability during ontogeny has been well established for several cell types of neural crest origin. Considering the overabundance of cells in the early nervous system, it may be more meaningful to think of NGF as regulating the degree of cell death, since a limited quantity in the periphery will determine the ultimate number of neurons in the ganglion.

NGF and Neuronal Differentiation

The first evidence for an effect of NGF on neuronal differentiation was provided by experiments in which sarcoma tissue cocultured with sympathetic or sensory ganglia brought about the pronounced outgrowth of nerve fibers (50, 51). Neurite outgrowth from the chick dorsal root ganglion has become the most commonly used bioassay for NGF (18). The ganglion is comprised of two neuronal populations, the mediodorsal (MD) and ventrolateral (VL) neurons. These are distinguishable during the embryonic period by their morphological features and rates of development (52). The effects of NGF appear to be limited primarily to the MD neurons (18). NGF stimulates the development of the neurotransmitter substance P in sensory neurons (53–55). Neurite outgrowth in response to NGF is accompanied by pronounced changes at the ultrastructural level including an increase in the size of the nucleolus, rough endoplasmic reticulum, and golgi, an increased quantity of microfilaments and microtubules, and the organization of the latter into neuritic processes (56–58).

Several tumor-derived continuous cell lines respond to NGF. Human and murine neuroblastoma cell lines bind NGF and respond by neurite extension (59–63). Human melanoma cell lines have receptors for NGF and appear to respond by increased cell survival in culture (64). Recently, a continuous cell line (PC12) has been cloned from an NGF-responsive rat pheochromocytoma (65, 66). The PC12 cell line has become one of the most commonly used systems in studies of the mechanism of action of NGF. PC12 cells proliferate in serum-containing culture in the absence of NGF, and resemble chromaffin cells in their capacity to synthesize, store, and release catecholamines (67). In response to NGF, PC12 cells eventually stop dividing and extend neuritic processes (66). This is accompanied by the acquisition of several characteristics of sympathetic neurons including electrical excitability, responsiveness to acetylcholine, and the capacity to form synaptic connections (58, 67, 68). In addition, the specific induction of choline acetyltransferase and acetylcholinesterase has been demonstrated (68–72). Tyrosine hydroxylase is not induced even at high concentrations of NGF (66, 70). Addition of low concentrations of dexamethasone to NGF, however, results in the induction of tyrosine hydroxylase in PC12 (73). This may be analogous to the induction of tyrosine hydroxylase in the superior cervical ganglion observed after addition of NGF in culture, since the neurons of the ganglion may have been previously exposed to glucocorticoids in vivo. Withdrawal of NGF from PC12 results in the reversal of its differentiated changes, with disintegration of neuritic processes and recommencement of cell proliferation (66).

The PC12 cell line permits study of the initial response to NGF and as such is analogous to the embryonic neuroblast in vivo. This system provides several experimental advantages over primary neurons, such as the availability of a large and homogeneous population of cells for biochemical studies. In addition, PC12 cells respond to NGF without requiring it for survival or maintenance, thus permitting well-controlled experiments. A caveat, however, in the use of a tumor-derived cell line as a model system is that it may not always precisely mimic its physiological counterpart.

The effects of NGF on neurotransmitter synthesis have received considerable attention. Changes have been demonstrated in the enzyme activities of tyrososine hydroxylase (74), dopamine β-hydroxylase (74), choline acetyltransferase (68, 69), and phenylethanolamine-N-methyl transferase (75). The most thoroughly studied enzyme has been tyrosine hydroxylase (TH). Evidence for the induction of TH in response to NGF comes from (1) local administration of NGF to the SCG in vivo (2), systemic administration of NGF, and (3) direct addition of NGF to ganglia or neurons in culture.

Intraocular injection of NGF was found to result in a preferential increase in both total TH activity and average neuronal cell diameter in the SCG on the injected side (43, 44). In the latter study, a slow-release cellulose linkage of NGF was used in order to minimize leakage into the systemic circulation (44). It was not ascertained in these studies whether the increase in TH activity was specific or secondary to the general growth effect of NGF manifested by the increase in neuronal size. Increased specific activity of TH has been demonstrated, however, in the SCG following systemic administration of NGF (74). In contrast to the characteristic effects of NGF on neuronal growth and survival, the induction of TH is independent of the age of the animal (4). The pituitary-adrenocortical axis is stimulated by NGF, as manifested by the enhanced production of ACTH and glucocorticoids (76). In the cultured SCG, NGF and dexamethasone each bring about an increase in the specific activity of TH and are synergistic together (77). The effects of the two agents can be distinguished by their sensitivities to inhibitors of RNA synthesis (78). Induction of TH by dexamethasone is blocked by α-amanitin and actinomycin D, whereas the response to NGF is not. Enzyme induction by both NGF and dexamethasone is blocked by cycloheximide. If the induction of TH in the SCG following systemic administration of NGF were a composite of a direct effect of NGF and an indirect effect due to glucocorticoids, then treatment of the animal with an inhibitor of RNA synthesis should block the glucocorticoid effect and result in only partial inhibition of TH induction. However, treatment of the animal with actinomycin D completely blocks the induction of TH (79, 80). The induction of TH by NGF in vivo may therefore be exclusively a result of the enhanced glucocorticoid production following stimulation of the pituitary-adrenocortical axis. Nevertheless, a direct effect of NGF on TH induction in the SCG has been observed in culture (77–81). Interestingly, this effect could only be demonstrated at supraphysiological concentrations of NGF (29, 76, 77, 82). Enzyme induction is not significant in the concentration range of NGF receptor occupancy and is maximal at concentrations 10- 30-fold higher than that required to saturate the low affinity NGF receptor (see section on the cell surface receptor). In contrast, stimulation by NGF of neuronal survival growth and neurite extension is observed in the concentration range of receptor occupancy (1, 29, 83). Although the NGF dose-response for TH induction was shifted to greater sensitivity by the addition of glucocorticoids, it remained in a supraphysiological concentration range (81). In a cloned pheochromocytoma cell line (PCG-2), it was also found that very high concentrations of NGF are required to induce TH (84). In contrast, much lower, physiological levels of epidermal growth factor (EGF) are sufficient to induce TH in this cell line. The effect of NGF on TH induction in culture may therefore reflect cross-reactivity of NGF at high concentrations with the receptor for EGF or another peptide hormone.

It has been shown that the adrenergic or cholinergic character of a sympathetic neuron can be determined by factors in the periphery (85). NGF, however, was not found to influence the choice of neurotransmitter, even though it maintained neuronal survival and stimulated total neurotransmitter synthesis (86). An interesting counterpoint to this permissive function of NGF is its effect on the developing adrenal medulla. Prenatal injection of NGF was shown to produce massive transformation of chromaffin cell precursors to sympathetic neurons (22). The treatment was ineffective after the second postnatal week. Despite this aberrant effect of exogenous NGF, prenatal injection of antibody to NGF resulted in widespread degeneration of chromaffin cell precursors. These cells therefore depend on NGF during development, but the amount or duration of exposure to the hormone is critical in determining their differentiated fate. This may also reflect a balance between NGF and other hormones, such as glucocorticoids (87). Injection of NGF into the prenatal animal has also been shown to increase the number of noradrenergic neurons in the gut and in ectopic locations (88). It is apparent that NGF can influence the differentiated fate of cells of neural crest origin in some but not all instances.

Neurite Orientation

The first evidence that NGF could orient as well as stimulate neurite outgrowth came from the sarcoma transplantation experiments of Levi-Montalcini & Hamburger (50, 51). Sarcoma tissue grafted onto the chorioallantoic membrane of a chick embryo was observed to "attract" sympathetic nerve fibers into the lumen of the veins of the host. In vitro, sympathetic ganglia were cocultured with explants of tissues that are normally sparsely or densely innervated in vivo. When both types of tissue were present, the ganglia were observed to extend neurites more rapidly and profusely toward the tissue that is more densely innervated in vivo (89–91). A dramatic demonstration in vivo of the chemotactic activity of NGF was provided by the results of NGF injection into the brain of the neonatal rat (92). Following repeated injection for a 7–10 day period, nerve fibers originating from sympathetic ganglia were observed to enter the spinal cord via the ventral and dorsal spinal roots and ascend the dorsal columns to the site of injection in the brainstem. Discontinuation of NGF administration resulted in the gradual disappearance of the ectopic nerve fibers. The growth of nerve fibers in the direction of the highest concentration of NGF has also been demonstrated in culture (93–96). When exposed to a local concentration gradient of NGF, growth cones of chick sensory neurites were observed to rapidly turn and grow toward the NGF source (95, 96). The possible mechanism of this chemotactic response is discussed in the section on cyclic AMP and phosphorylation.

Another level of control of neurite orientation has been found in neurite bundling or fasciculation (97). Chick sensory neurons exposed to a steep but transient NGF gradient exhibited considerable assymmetry in the subsequent outgrowth of neurites from ganglia or cell aggregates, but not from single cells. The effect was partially inhibited by antibody against the neural cell adhesion molecule (CAM), which also decreased the average diameter

of neurite fascicles. It was suggested that the initial directional signal provided by an NGF gradient can be amplified and preserved by interactions
among neurites.

THE MECHANISM OF ACTION

Cyclic AMP (cAMP) and Phosphorylation

Cyclic AMP is one of the most ubiquitous and pleiotypic regulatory molecules in nature. Its function as an intracellular "second messenger" has been
demonstrated for several hormones. Its place in the mechanism of action
of NGF, however, has been a source of considerable controversy. Initial
studies on chick sensory neurons could not demonstrate changes in intracellular cAMP levels after the addition of NGF (98). Although some sprouting
of neurites from ganglia was observed following the addition of Bt_2 cAMP,
this was felt to be morphologically different from the NGF response and
could be mimicked by the butyrate moiety alone. Evidence supporting a role
for cAMP in the response to NGF was provided by studies on PC12 cells
(99, 100). It was observed that PC12 cells plated in the presence of NGF
showed a rapid increase in cell-cell and cell-substratum adhesion. This was
accompanied by a rapid rise in the intracellular concentration of cAMP,
which was maximal by 7 min and back down to baseline by 22 min.
Moreover, the addition of cAMP or its analogues increased cell adhesion
and neurite extension. Similar effects were observed for agents that raised
intracellular cAMP levels, such as theophylline and cholera toxin (100).
Membrane changes correlating with these effects were suggested by the
increased agglutinability of PC12 cells by the lectins phytohemagglutinin
and ricin after exposure to NGF or cAMP. An effect was also found on the
rate of calcium efflux from the cell (100), which has been disputed (101).
Increased cAMP levels in response to NGF have been reported for the
cultured SCG and DRG (102–104). Some reports contradict this effect of
NGF on cAMP levels (105–107).

Regulation of cell function by cAMP is in many cases dependent on the
stimulation of cAMP-dependent protein kinase. It has been found that
NGF and cAMP each stimulate the phosphorylation of nuclear proteins in
sympathetic ganglia, one of which may be histone H1 (108, 109). A detailed
analysis of the proteins phosphorylated in response to NGF has been performed in PC12 cells (110). NGF, cholera toxin, and cAMP were shown
to stimulate phosphorylation of the same set of proteins including tyrosine
hydroxylase, ribosomal protein S6, histones H1 and H3, and the nonhistone
chromosomal HMG 17 protein. The degree of phosphorylation of histone
H2A is reduced. EGF and insulin stimulate the phosphorylation of different

subgroups of these proteins, although it is known from other systems that these phosphorylations are cAMP-independent (111, 112). Moreover, it was determined that combinations of saturating levels of EGF, insulin, or NGF were additive in phosphorylation of the ribosomal protein S6, which suggests different mechanisms for the stimulation due to each hormone (110). Combinations of cholera toxin or cAMP with NGF were not additive, however, but actually less than stimulation with NGF alone. It was concluded from these results that cAMP and NGF stimulate phosphorylation by the same mechanism. The same pattern of NGF-mediated phosphorylation was preserved even after 3 days of incubation with NGF. This is puzzling if it were to be explained solely by a cAMP stimulation, since it was found that cAMP levels are only transiently elevated in response to NGF, returning to baseline after 22 min (99). In addition to the proteins described above, it has been found that NGF can also stimulate the phosphorylation of microtubule-associated proteins (G. E. Landreth, C. Richter-Landsberg, and E. M. Shooter, unpublished). It is interesting to note that plasma membrane proteins are not phosphorylated after the addition of NGF to either the intact cell or isolated membranes (W. H. Wilson and E. M. Shooter, unpublished).

Is NGF-induced neurite extension-mediated by cAMP? Recent comparison of the morphological responses of PC12 cells to NGF and cAMP revealed fundamental differences (113). Both agents promote neurite extension, but the combination of saturating levels of each has an additive or greater effect. Moreover, it was found that neurite extension due to cAMP is unaffected by inhibitors of RNA synthesis, unlike the response to NGF alone. One effect of NGF on PC12 cells is to impart the capacity for rapid regeneration of a disrupted neurite. This capacity is dependent on the time of exposure to NGF (114). cAMP alone could not impart this regenerative capacity nor did cAMP accelerate its acquisition when combined with NGF (115). The responses of PC12 cells to NGF and cAMP were also examined by time lapse cinematography (115). Addition of NGF was followed by rapid membrane protrusive activity, but neurite extension did not appear for many hours. Addition of cAMP, on the other hand, was followed within 10 min by the sprouting of neurites. These neurites elongated for the next 4–6 hr, but not subsequently, in contrast to NGF-induced neurites, which continued to elongate for days after initiation. Furthermore, removal of cAMP resulted in the rapid retraction of neurites (half-life of 30 min) in contrast to the very stable NGF-induced neurites (half-life of about 24 hr). Taken together, these data suggest that NGF does not induce neurite extension solely through a cAMP-dependent mechanism. Whereas NGF mediates the formation of a stable neurite, cAMP induces a rapidly reversible structure.

Which aspects of NGF action may be mediated through cAMP? As discussed under neurite orientation, dorsal root ganglion neurites rapidly reorient their direction of growth in response to a gradient of NGF (95, 96). A similar chemotactic response is observed with cAMP, cGMP, and phosphodiesterase inhibitors (96). Although calcium by itself is ineffective, when combined with a calcium ionophore it is chemotactic. Dantrolene, which inhibits the release of intracellular calcium from skeletal muscle (116), blocks the chemotactic responses to both NGF and cAMP (96). Transient elevation of cAMP in the neurite, which leads to the extrusion of intracellular calcium, may reorient the neurite in a rapid and reversible fashion. This may account for the rapid response of the growth cone to local concentration gradients of NGF in contrast to the more long-term establishment of a stable neurite.

Active Transport and Metabolism

One of the earliest effects of NGF on its target cells is the enhanced uptake of metabolites including nucleotides, amino acids, and sugars (117, 118). Rapid changes in the Na content of sympathetic and sensory ganglionic cells have also been observed (119, 120). In the absence of NGF, intracellular levels of Na rise, but upon addition of NGF they are rapidly depleted. This was found to be due to increased extrusion of intracellular Na rather than to changes in membrane permeability. Extrusion of intracellular Na was almost immediate upon presentation of NGF, whereas a delay of several minutes preceded the uptake of metabolites such as deoxyglucose. Active transport systems for many metabolites are Na^+-dependent. On this basis, Skaper & Varon (120) have proposed that NGF may regulate the uptake of metabolites by altering the Na concentration gradient across the membrane.

The uptake of metabolites may be important in the well-documented role of NGF in maintaining the viability of its target cells. It has been shown that NGF can substitute for serum in maintaining the survival of PC12 cells in culture (121). Dissociation of this effect from neurite outgrowth was demonstrated with an inhibitor of RNA synthesis, which blocks neurite outgrowth without affecting cell survival. NGF-stimulated uptake of α-aminoisobutyric acid has been demonstrated in PC12 and is also independent of RNA synthesis (122). The degree of uptake correlates with the occupancy by NGF of its cell surface receptors (D. Kedes and E. M. Shooter, unpublished). Additional evidence for the dissociation of effects on neurite outgrowth and cell survival is evident from the action of insulin on PC12 cells. Insulin, like NGF, can substitute for serum in maintaining the survival of PC12 cells in culture, but unlike NGF does not promote neurite outgrowth (B. A. Yankner and E. M. Shooter, unpublished). Insulin also stimulates the uptake of α-aminoisobutyric acid in PC12 cells in serum-free culture (D. Kedes and E. M. Shooter, unpublished).

The stimulation of RNA metabolism by NGF is suggested by the increased nucleolar size (56), increased RNA polymerase activity (123), and elevated cellular RNA content (124) following administration of NGF. Early studies on the effects of NGF on RNA and protein synthesis in ganglionic cells were hampered by the requirement of these cells for NGF in order to survive in culture. Therefore, enhancement of a process by NGF could not be distinguished from maintenance in comparison to a degenerating control. Putative increased labeling of the RNA of chick sensory neurons after the addition of NGF was found to actually reflect decreased labeling by the degenerating controls (117, 125–127). Sympathetic ganglia cultured in the presence of NGF and high concentrations of actinomycin D showed only minimal perturbation of neurite outgrowth (126). The interpretation that neurite outgrowth in response to NGF did not require RNA synthesis turned out to be a simplification that was clarified by studies on the PC12 cell line. Neurite outgrowth from PC12 cells upon initial exposure to NGF is blocked by inhibitors of RNA synthesis (128). However, when PC12 cells are pretreated with NGF for 1 to 2 weeks and then have their neurites mechanically removed, these neurites regenerate within 24 hr in the presence of NGF unaffected by inhibitors of RNA synthesis. The interpretation offered by Burstein & Greene (128) was that initiation of neurite outgrowth requires an RNA synthesis–dependent buildup of necessary precursors, whereas the regeneration of neurites can utilize the preformed pool without the necessity for ongoing RNA synthesis. The apparent transcription-independent neurite outgrowth from ganglionic cells treated with NGF in culture may then be explained by the prior exposure of these cells to NGF in vivo. Several responses to NGF that are more rapid than neurite outgrowth require ongoing transcription. These include the induction of acetylcholinesterase in PC12 cells (71, 72) and ornithine decarboxylase in the SCG (129), embryonic DRG (130), and PC12 cells (107, 131). Both of these effects can be dissociated from the neurite outgrowth response (72, 131).

In contrast to the transient requirement for transcription, ongoing translation appears to be necessary for the neurite outgrowth response to NGF. Cycloheximide has been shown to block NGF-mediated neurite outgrowth in chick sensory ganglia without loss of cell viability (126, 127, 132, 133). NGF has been reported to increase the incorporation of labeled amino acids into tyrosine hydroxylase (80, 134). Selective induction of enzymes such as choline acetyltransferase, acetylcholinesterase, and ornithine decarboxylase has not been rigorously demonstrated to result from the selective modulation of protein synthesis. Using two-dimensional gel electrophoresis, the patterns of protein synthesis before and after the addition of NGF to PC12 cells were determined (135). Several quantitative changes were noted, although the synthesis of new proteins following addition of NGF could not be detected. Another study using a similar protocol, but with computerized gel scanning, reported significant quantitative changes in the synthesis of as much as 30% of the proteins analyzed (136). Enhanced incorporation of labeled sugars into two glycoproteins has also been observed in PC12 cells in response to NGF (135).

Insight into the functional significance of the transcription requirement for NGF-mediated neurite outgrowth was provided by Mizel & Bamburg (133). Chick sensory and sympathetic ganglia cultured on untreated tissue culture plates require both RNA and protein synthesis for NGF-mediated neurite outgrowth. When ganglia are cultured in plasma clots or on tissue culture plates coated with collagen or polylysine, RNA synthesis is no longer completely required for NGF-mediated neurite outgrowth, although the response is still dependent on protein synthesis. These results suggest that part of the transcription requirement may be for the production of substratum-associated proteins. In support of this idea, it was found that ganglia cultured on plates previously exposed to other ganglia will extend neurites in the presence of actinomycin D (133). Recently, it was found that PC12 cells plated on basement membrane produced by corneal epithelial cells will extend neurites transiently in the absence of NGF (D. Gospodarowicz, personal communication). The active components of the basement membrane involved in this response appear to be proteoglycans (L. Reichart, personal communication).

The Cell Surface Receptor

Rapid effects following the addition of NGF, such as alterations in cell surface and neurite morphology (137, 138) and stimulated uptake of metabolites (see previous section) are almost certainly direct consequences of the binding of NGF to its cell surface receptors. Receptors for NGF have been demonstrated in sympathetic ganglia (139, 140), dorsal root sensory ganglia (83, 140, 141), brain (142, 143), PC12 pheochromocytoma (42, 144–146), and cultured melanoma cells (64, 147). Loss of responsiveness to NGF in embryonic chick sensory ganglia during development was shown to be paralleled by a loss of cell surface receptors for NGF (148). Binding has been assayed with ^{125}I-labeled NGF by rapid contrifugation (141), filtration (139), and polyethylene glycol precipitation (149). A serological cytotoxicity assay has also been described (143) that detects the high affinity NGF receptor (described below). The biologically active β subunit of the 7S NGF complex has been used in binding studies (5). It was found that the 7S NGF complex, maintained intact by an excess of the α and λ subunits, does not bind to receptors on dorsal root ganglia (150). In addition, when the complex is chemically cross-linked it is biologically inactive (151). It has been reported, however, that 7S NGF is almost as potent as the β subunit in stimulating responses in cultured cells (29, 152). The complex must therefore dissociate under these conditions. A possible mechanism could be the depletion of zinc from the 7S complex, which would considerably reduce its stability (153).

The NGF receptor from adult rabbit superior cervical ganglia has been solubilized with detergents and partially isolated by chromatography on Sepharose and sucrose density gradient centrifugation (154, 155). The protein was found to be highly assymmetric, with a molecular weight of 135,000 and a sedimentation coefficient of $4.3s$. A larger form of the receptor was also observed that was considered to represent an aggregate of the $4.3s$ moiety. Recently it was found that certain lectins inhibit the binding of ^{125}I-NGF to the solubilized receptor primarily by a reduction in receptor affinity, which suggests that the receptor is a glycoprotein (156).

The binding characteristics of NGF receptors on sensory and sympathetic neurons have been extensively studied (83, 139–141). Binding of NGF to the intact cell was shown to be specific and saturable (141). In contrast, binding to disrupted membranes is nonsaturable and displays multiple equilibrium dissociation constants (140, 157). This heterogeneity of receptor affinity was attributed to negative cooperativity, a phenomenon in which increasing ligand concentration results in a compensatory decrease in receptor affinity (158). The development of a method for the preparation of ^{125}I-labeled NGF that retained full biological activity and displayed low nonspecific binding permitted the identification of a class of NGF receptors with high affinity. Using this technique, Sutter and his colleagues (83) demonstrated two types of receptors on intact sensory neurons; a high affinity ($K_d = 2 \times 10^{-11}$ M) and a lower affinity ($K_d = 1.4 \times 10^{-9}$ M) receptor. When binding of ^{125}I-NGF was determined on membrane preparations of dorsal root ganglia, two binding sites were also found with steady state and kinetic properties similar to those on intact cells (157). In addition, a lower affinity binding site ($K_d \cong 10^{-6}$ M) was revealed that was not detected on intact cells. Disruption of the cell appears to expose low affinity intracellular binding sites, which accounts for the apparent nonsaturability of NGF binding to membranes derived from sympathetic and sensory ganglia (140, 157). The difference in the affinities of the two receptor sites on the cell surface is reflected in their rates of dissociation (83). Addition of an excess of unlabeled NGF to intact sensory neurons after the binding of ^{125}I-NGF has reached equilibrium, results in a rapid dissociation ($t_{1/2}$ of seconds) followed by a slower dissociation ($t_{1/2} \cong 30$ min). The quantity of label dissociating in each of these two components is dependent on the loading dose of ^{125}I-NGF and consistent with the expected occupancy of two independent receptors based on their affinities. If the receptors were interconvertible by negative cooperativity, then their saturation with unlabeled NGF should convert and maintain all the receptors in the lower affinity form and give a monophasic dissociation of ^{125}I-NGF regardless of the loading dose. This was not observed (83). An alternative explanation of the data is suggested by binding studies on PC12 cells (145). Because of the differential dissociation of NGF from high and low affinity receptors, a brief wash of the cells with unlabeled NGF selectively removes ^{125}I-NGF bound to low affinity receptors and thus allows an assessment of both types of binding. Using this technique, it was demonstrated in PC12 cells that ^{125}I-NGF binds initially to low affinity receptors and that high affinity receptors appear only after a short time lag (145). The continuous generation of high from low affinity receptors is observed after ^{125}I-NGF is removed from the medium after it has initiated binding. It is unlikely that this could be because of ^{125}I-NGF dissociating-off low affinity and rebinding to high affinity receptors because of the dilution of dissociated ^{125}I-NGF. These results are consistent with competition assays that suggest that only the low affinity receptor is present on PC12 cells not previously exposed to NGF (159). The nature of the receptor conversion is not known, but it may reflect the clustering of receptors described for other hormones (160, 161) and recently for NGF (162).

The data of Landreth & Shooter (145) were recently repeated with a different interpretation (163). These authors suggest that the two binding sites on PC 12 cells represent two independent classes of receptors with differing rates of association and dissociation but with the same affinity ($K_d = 2 \times 10^{-10}$ M). Although a time lag was not detected in the binding of NGF to the high affinity receptor, their assay was performed at a 30-fold higher NGF concentration than that used by Landreth & Shooter (145). At this concentration of NGF, receptor conversion may be too rapid to resolve (G. E. Landreth and E. M. Shooter, unpublished). In addition, it was shown that tryptic digestion of NGF receptors prior to the addition of ^{125}I-NGF could be adjusted to selectively destroy low affinity receptors, leaving high affinity receptors intact (163). This was presented as evidence against receptor conversion and in support of preexistence of the high affinity receptor. With the correct titration of tryptic digestion, however, the number of low affinity receptors can be reduced to a low enough level so that the subsequent binding of NGF might result in the complete conversion of low to high affinity receptors as observed.

It has been suggested that high affinity receptors for NGF are associated with the cytoskeleton (163). This was based on the resistance of the receptors to extraction with nonionic detergents. Direct interaction of the receptor with the cytoskeleton, however, has not been demonstrated. The cell surface localization of these binding sites for NGF was inferred from the observation that treatment with azide and deoxyglucose or incubation at low temperature only partially inhibited binding. Although these treatments have been found to inhibit hormone internalization in some systems (164), the unknown degree of inhibition in PC12 makes these results difficult to interpret. Studies with labeled antibody to NGF suggest that high affinity receptors are localized to the cell surface but that a fraction of NGF binding measured as high affinity actually represents internalized NGF (B. A. Yankner and E. M. Shooter, unpublished). The temperature dependence of the high affinity receptor (42) accounts for one report of NGF binding to PC12 cells in which only low affinity receptors ($K_d = 2.9 \times 10^{-9}$ M) were detected at low temperature (144).

The long-term regulation of high and low affinity receptors has recently been studied in PC12 cells (5). After ^{125}I-NGF was added to PC12 cells and receptor binding had reached an apparent steady state, the occupied low affinity receptors were selectively destroyed by tryptic digestion leaving the occupied high affinity receptors intact. Subsequently, NGF-bound high affinity receptors were depleted from the cell surface by internalization to a much greater extent than controls in which low affinity receptors were also present (B. A. Yankner and E. M. Shooter, unpublished). A possible explanation of these results is that the conversion of low to high affinity receptors is continuous and functions to replenish high affinity receptors depleted from the cell surface by internalization, as illustrated in Figure 1.

The Nuclear Receptor

Detergent extraction of the cell does not solubilize all of the NGF binding sites, which led to the discovery of intracellular NGF receptors in the nucleus (165). Receptors for NGF have been demonstrated in nuclei isolated from chick dorsal root ganglia (165) and PC12 pheochromocytoma cells (42). Nuclei were isolated from PC12 cells by a rigorous method involving mechanical shearing and buoyant density centrifugation (42). These purified nuclei were free of plasma membrane and lysosomal enzyme markers, and morphological examination by electron microscopy showed that the preparation was free of contamination by membranes, cytoplasmic organelles, and cytoskeletal filaments. Two bindings sites for NGF were detected on isolated nuclei with different affinities ($K_d = 8 \times 10^{-11}$ M and 9×10^{-9} M). The two receptor types were quantitatively recovered in isolated nuclear membrane but could not be detected on chromatin. The nuclear receptors exhibit specificity for NGF and are enriched in NGF target tissues.

In contrast to receptors in isolated plasma membranes, the receptors in the nucleus are resistant to extraction with nonionic detergents (42, 165). In the intact PC12 cell, over 90% of the specifically bound ^{125}I-NGF, which is detergent-resistant, could be recovered in the purified nucleus (5). The nuclear-bound label represents intact β NGF as determined by SDS-polyacrylamide gel electrophoresis. It has recently been reported that nonnuclear detergent-resistant binding in many fold higher levels than the nuclear component can be detected in PC12 (163). This component, which is lost in the detergent treatment for extraction of the nuclear component (42), is maintained by gentle extraction in a calcium-free buffer system and was hypothesized to be cytoskeletal in origin (163). The nuclear component of cell-bound NGF does not itself appear to be associated with the cytoskeleton, as judged by its resistance to mechanical shearing and cytoskeletal disrupting drugs (5).

In order to determine the relationship of the cell surface and nuclear receptors for NGF, NGF-nonresponsive mutants of the PC12 cell line were isolated (B. A. Yankner and E. M. Shooter, unpublished). Colonies of PC12 cells, which continued to divide despite long-term exposure to NGF, were selected and cloned. One spontaneously occurring mutant, NR2, is deficient in both high and low affinity cell surface receptors for NGF. The higher affinity nuclear receptor is also lacking, despite an unaltered number of low affinity nuclear receptors. Thus it is possible that the high affinity nuclear

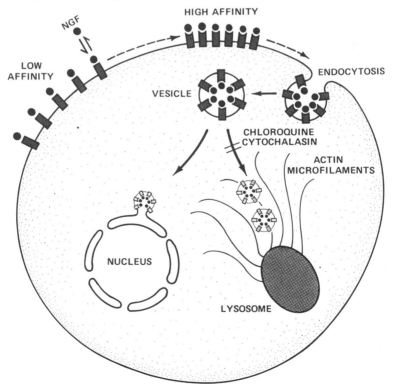

Figure 1   Steps in the binding, internalization, and intracellular transport of NGF.

receptor is translocated to the nucleus from the cell surface. This is consistent with the increased number of nuclear receptors observed after the exposure of PC12 cells to NGF (42).

The possible significance of intracellular receptors for peptide hormones has only recently been considered because of the long-standing belief that peptide hormones act solely through second messengers generated at the cell surface. Nuclear receptors have now been demonstrated for insulin (166, 167), EGF (168, 169), and gonadotropins (170). The nuclear receptors for NGF, insulin and EGF share several properties including localization to the nuclear membrane, resistance to detergent extraction, and binding characterized by a biphasic Scatchard curve (42, 167, 168).

Internalization

The internalization of peptide hormones by receptor-mediated endocytosis has been well established (164). Internalization of NGF in vivo by retrograde axonal transport has already been discussed in the section on NGF and neuronal survival. It was recently demonstrated by fluorescence intensification microscopy that NGF receptors on PC12 cells and sensory neurons cluster and are internalized in vesicles in a fashion similar to that observed for EGF and insulin (160, 162). The downregulation of cell surface receptors for NGF and their degradation in lysosomes has been demonstrated in PC12 cells (5). Degradation of NGF can be inhibited by disruption of the actin cytoskeleton with cytochalasins B and D (5). Cytochalasin does not, however, inhibit the binding of NGF to low and high affinity surface receptors or prevent internalization. Thus it appears that the interaction with microfilaments is in the interior of the cell and may function to guide endocytic vesicles to lysosomes, as illustrated in Figure 1.

Inhibition of lysosomal degradation does not appear to interfere with the initial neurite outgrowth response to NGF in PC12 cells, which suggests that the degradation products of NGF and its receptors may not be a part of the mechanism of action (5). Nevertheless, the lysosomal pathway may play an important role in regulating the rapidity and degree of the hormonal response. This is apparent in the time course of nuclear translocation. The initial appearance of NGF in the nucleus is preceded by a time lag of several hours during which the lysosomal degradation of NGF is pronounced. Chloroquine interferes with the lysosomal pathway (171) and eliminates the time lag in nuclear translocation (5). The commitment of the cell to neurite outgrowth in response to NGF also exhibits a time lag similar to that observed for nuclear translocation. This requirement for continuous incubation with hormone is apparently modulated by the degree of lysosomal processing and may regulate the stimulus for differentiation in vivo by rendering transient pulses of NGF ineffective. Translocation of NGF to the lysosomes and the nucleus appear to represent parallel, competitive pathways. Additional pathways to other intracellular organelles may also exist. The importance of NGF in the nucleus for the biological response is evident from studies of neurite regeneration in PC12 cells. Following exposure to NGF, PC12 cells acquire the capacity to rapidly regenerate a mechanically disrupted neurite (114). In a subclone of the PC12 cell line, it was observed that neurite regeneration could occur even after the removal of NGF and its plasma membrane receptors by exhaustive tryptic digestion (5). The capacity for neurite regeneration decayed, however, with the dissociation of NGF from the nucleus.

The biochemical determination of ^{125}I-NGF in the nucleus is very sensitive, but complicated by the condition of the cells in culture and the generation of intracellular nonspecific binding during cellular subfractionation. Initial reports of 30–60% of total NGF binding in the nucleus in sympathetic neurons and PC12 cells were probably overestimated (42, 172). Carefully controlled studies on monolayer cultures of PC12 cells have revealed that about 5% of the specifically bound ^{125}I-NGF is in the nucleus after 24 hr, and 15% after 6 days of incubation (5). The subtlety of this effect may account for the difficulty of its detection using morphological methods (40). One report of morphological confirmation of NGF in the nucleus using light microscopic autoradiography is questionable because of the low resolution of the technique (173). The slow accumulation of NGF in the nucleus contrasts with its rapid association with lysosomes (5). As described above, however, the association of NGF and/or its receptor with the nucleus closely parallels the effects of the hormone on neurite outgrowth.

NGF has been introduced directly into the cytoplasm of PC12 cells by fusion with NGF-loaded erythrocyte ghosts without effect (173a). In addition, NGF antibodies introduced into the cytoplasm did not prevent the neurite outgrowth response to NGF. These results suggest that NGF, in its intact form, does not have a cytoplasmic locus of action. However, it is well known that internalized peptide hormones are contained within membrane-bounded compartments and are not accessible to the cytoplasm.

An important question relates to the mechanism by which regulatory functions of a peptide hormone could be mediated by its endocytosis. Binding to the plasma membrane receptor can activate an enzymatic activity such as a protein kinase or adenyl cyclase. Although there is no evidence that NGF or other peptide hormones are released into the cytoplasm, the redistribution of the hormone-receptor complex to intracellular membranes could be mediated by the fusion of endocytic vesicles. The effector activity might then be stimulated by the hormone-receptor complex in a new milieu. It has been demonstrated that only one molecule of ricin or diphtheria toxin needs to be internalized and transferred to the rough endoplasmic reticulum in order to profoundly alter protein synthesis (174, 175). In a similar fashion, a very small proportion of the hormone and its plasma membrane receptors may need to be redistributed to intracellular membranes in order to modulate specific regulatory processes.

CONCLUDING REMARKS

The targets of NGF action in the peripheral nervous system have been well characterized. The sites of synthesis of the protein, however, are largely unknown. The identification of small quantities of NGF may be facilitated in the future by the use of recombinant DNA technology in order to probe for the NGF mRNA. This will permit the definitive test of the theory that innervated end-organs supply the ganglionic neurons with NGF.

Understanding the mechanism of action of NGF, or any peptide hormone, depends on understanding its receptor. The purification of the receptor and its reconstitution into membranes will be necessary in order to elucidate the molecular signals generated by hormone binding. An interesting question is whether similar signals, such as protein phosphorylation, are utilized by different hormones and if so, what sort of fine-tuning determines the specificity of the response. As pointed out at the beginning of this article, there are several levels of complexity in the action of NGF, including its capacity to orient as well as stimulate neuronal cell growth and differentiation. We have discussed the mechanism of action in terms of multiple biochemical pathways. How such diversity of function is incorporated in a single hormone and its receptor is a biological puzzle just beginning to be unraveled.

Literature Cited

1. Mobley, W. C., Server, A. C., Ishii, D. N., Riopelle, R. J., Shooter, E. M. 1977. *N. Engl. J. Med.* 297:1096–1104, 1149–58, 1211–18
2. Bradshaw, R. A. 1978. *Ann. Rev. Biochem.* 47:191–216
3. Greene, L. A., Shooter, E. M. 1980. *Ann. Rev. Neurosci.* 3:353–402
4. Thoenen, H., Barde, Y.-A. 1980. *Physiol. Rev.* 60:1284–335
5. Shooter, E. M., Yankner, B. A., Landreth, G. E., Sutter, A. 1981. *Rec. Progr. Horm. Res.* 37:417–46
6. Oppenheim, R. W. 1981. In *Studies in Developmental Neurobiology: Essays in Honor of Viktor Hamburger,* ed. W. M. Cowan, pp. 74–133. Oxford Univ. Press
7. Hughes, A. 1965. *J. Embryol. Exp. Morphol.* 13:9–34
8. Hamburger, V. 1975. *J. Comp. Neurol.* 160:535–46
9. Levi-Montalcini, R., Booker, B. 1960. *Proc. Natl. Acad. Sci. USA* 46:384–91
10. Sabatini, M. T., Pellegrino de Iraldi, A., DeRobertis, E. 1965. *Exp. Neurol.* 12:370–83
11. Levi-Montalcini, R., Caramia, F., Angeletti, P. U. 1969. *Brain Res.* 12:54–73
12. Goedert, M., Otten, U., Schäfer, T. H., Schwab, M., Thoenen, H. 1980. *Brain Res.* 20:399–409
13. Levi-Montalcini, R., Angeletti, P. U. 1966. *Pharmacol. Rev.* 18:619–28
14. Goedert, M., Otten, U., Thoenen, H. 1978. *Brain Res.* 148:264–68
15. Otten, U., Goedert, M., Schwab, M., Thibault, J. 1979. *Brain Res.* 176:79–90
16. Bjere, B., Wiklund, L., Edwards, D. C. 1975. *Brain Res.* 92:257–58
17. Gorin, P. D., Johnson, E. M. Jr. 1980. *Brain Res.* 198:27–42
18. Levi-Montalcini, R., Angeletti, P. U. 1968. *Physiol. Rev.* 48:534–69
19. Burnham, P. A., Raiborn, C., Varon, S. 1972. *Proc. Natl. Acad. Sci. USA* 69:3556–60

20. Varon, S., Raiborn, C., Burnham, P. A. 1974. *J. Neurobiol.* 5:355–71
21. Johnson, E. M., Gorin, P. D., Brandeis, L. D., Pearson, J. 1980. *Science* 210: 916–18
22. Aloe, L., Levi-Montalcini, R. 1979. *Proc. Natl. Acad. Sci. USA* 76:1246–50
23. Hendry, I. A., Campbell, J. 1976. *J. Neurocytol.* 5:351–60
24. Hendry, I. A. 1967. *Rev. Neurosci.* 2: 149–94
25. Zaimis, E. 1971. *J. Physiol.* 216:65–66
26. McCarthy, K. D., Partlow, L. W. 1976. *Brain Res.* 114:415–26
27. Coughlin, M. D., Bayer, D. M., Black, I. B. 1977. *Proc. Natl. Acad. Sci. USA* 74:3438–42
28. Greene, L. A. 1977. *Dev. Biol.* 53:96–105
29. Chun, L. Y., Patterson, P. H. 1977. *J. Cell Biol.* 75:694–704
30. Chun, L. Y., Patterson, P. H. 1977. *J. Cell Biol.* 75:705–11
31. Angeletti, P. U., Levi-Montalcini, R. 1970. *Proc. Natl. Acad. Sci. USA* 65: 114–21
32. Hendry, I. A. 1975. *Brain Res.* 90: 235–44
33. Hendry, I. A. 1975. *Brain Res.* 94: 87–97
34. Levi-Montalcini, R., Aloe, L., Mugnaini, E., Oesch, F., Thoenen, H. 1975. *Proc. Natl. Acad. Sci. USA* 72:595–99
35. Campenot, R. B. 1977. *Proc. Natl. Acad. Sci. USA* 74:4516–19
36. Hendry, I. A., Stoeckel, K., Thoenen, H., Iversen, L. L. 1974. *Brain Res.* 68: 103–21
37. Dumas, M., Schwab, M. E., Thoenen, H. 1979. *J. Neurobiol.* 10:179–97
38. Stoeckel, K., Dumas, M., Thoenen, H. 1978. *Neurosci. Lett.* 10:61–64
39. Iversen, L. L., Stoeckel, K., Thoenen, H. 1975. *Brain Res.* 88:37–43
40. Schwab, M. E. 1977. *Brain Res.* 130: 190–96
41. Schwab, M. E., Thoenen, H. 1977. *Brain Res.* 122:459–74
42. Yankner, B. A., Shooter, E. M. 1979. *Proc. Natl. Acad. Sci. USA* 76:1269–73
43. Paravicini, U., Stoeckel, K., Thoenen, H. 1975. *Brain Res.* 84:279–91
44. Hendry, I. A. 1977. *Brain Res.* 134: 213–23
45. Stoeckel, K., Schwab, M., Thoenen, H. 1975. *Brain Res.* 89:1–14
46. Stoeckel, K., Schwab, M., Thoenen, H. 1975. *Brain Res.* 99:1–16
47. Max, S. R., Schwab, M., Dumas, M., Thoenen, H. 1978. *Brain Res.* 159: 411–15
48. Ebbott, S., Hendry, I. 1978. *Brain Res.* 139:160–63
49. Thoenen, H., Schafer, T., Heumann, R., Schwab, M. 1981. *INSERM Eur. Symp., Horm. Cell Regul.* 5:15–34
50. Levi-Montalcini, R., Hamburger, V. 1951. *J. Exp. Zool.* 116:321–62
51. Levi-Montalcini, R., Hamburger, V. 1953. *J. Exp. Zool.* 123:233–78
52. Carr, U. M., Simpson, S. B. 1978. *J. Comp. Neurol.* 182:727–40
53. Otten, U., Goedert, M., Mayer, N., Lembeck, F. 1980. *Nature* 287:158–59
54. Kessler, J. A., Black, I. B. 1980. *Proc. Natl. Acad. Sci. USA* 77:649–52
55. Kessler, J. A., Black, I. B. 1981. *Brain Res.* 208:135–45
56. Angeletti, P. U., Levi-Montalcini, R., Caramia, F. 1971. *J. Ultrastruct. Res.* 36:24–36
57. Schwab, M., Thoenen, H. 1975. *Cell Tissue Res.* 158:543–53
58. Tischler, A. S., Greene, L. A. 1978. *Lab. Invest.* 39:77–89
59. Waris, T., Rechardt, L., Waris, P. 1973. *Experientia* 29:1128–29
60. Kolber, A. R., Goldstein, M. N., Moore, B. W. 1974. *Proc. Natl. Acad. Sci. USA* 71:4203–7
61. Hermetet, J. C., Ciesielsk-Treska, J., Mandel, P. 1972. *CR Soc. Biol.* 166: 1120–33
62. Revoltella, R., Bertolini, L., Pediconi, M., Vigneti, E. 1974. *J. Exp. Med.* 140:437–51
63. Revoltella, R. P., Butler, R. H. 1980. *J. Cell Physiol.* 104:27–33
64. Fabricant, R. N., DeLarco, J. E., Todaro, G. J. 1977. *Proc. Natl. Acad. Sci. USA* 75:565–69
65. Tischler, A. S., Greene, L. A. 1975. *Nature* 258:341–42
66. Greene, L. A., Tischler, A. S. 1976. *Proc. Natl. Acad. Sci. USA* 73:2424–28
67. Dichter, M. A., Tischler, A. S., Greene, L. A. 1977. *Nature* 268:501–4
68. Schubert, D., Hienemann, S., Kidakoro, Y. 1977. *Proc. Natl. Acad. Sci. USA* 74:2579–83
69. Greene, L. A., Rein, G. 1977. *Nature* 268:349–51
70. Edgar, D. H., Thoenen, H. 1978. *Brain Res.* 154:186–90
71. Rieger, F., Schelanski, M. L., Greene, L. A. 1980. *Dev. Biol.* 76:238–43
72. Greene, L. A., Rukenste, A. 1980. *J. Biol. Chem.* 256:6363–67
73. Otten, U., Towbin, M. 1980. *Brain Res.* 193:304–8
74. Thoenen, H., Angeletti, P. U., Levi-Montalcini, R., Kettler, R. 1971. *Proc. Natl. Acad. Sci. USA* 68:1598–1602

75. Luizzi, A., Fappen, F. H., Kopin, I. J. 1977. *Brain Res.* 138:309–15
76. Otten, U., Baumann, J. B., Girard, J. 1979. *Nature* 282:413–14
77. Otten, U., Thoenen, H. 1976. *Brain Res.* 111:438–41
78. Rohrer, H., Otten, U., Thoenen, H. 1978. *Brain Res.* 159:436–39
79. Stoeckel, K., Solomon, F., Paravicini, U., Thoenen, H. 1974. *Nature* 250: 150–51
80. MacDonnell, P. C., Tolson, N., Guroff, G. 1977. *J. Biol. Chem.* 252:5859–63
81. Otten, U., Thoenen, H. 1977. *J. Neurochem.* 29:69–75
82. Nagaiah, K., MacDonnell, P., Guroff, G. 1977. *Biochem. Biophys. Res. Commun.* 75:832–37
83. Sutter, A., Riopelle, R. J., Harris-Warrick, R. M., Shooter, E. M. 1979. *J. Biol. Chem.* 254:5972–82
84. Goodman, R., Herschman, H. R. 1978. *Proc. Natl. Acad. Sci. USA* 75:4587–90
85. Patterson, P. H. 1978. *Ann. Rev. Neurosci.* 1:1–17
86. Chun, L. Y., Patterson, P. H. 1977. *J. Cell Biol.* 75:712–18
87. Unsicker, K., Krisch, B., Otten, U., Thoenen, H. 1978. *Proc. Natl. Acad. Sci. USA* 75:3498–502
88. Kessler, J. A., Cochard, P., Black, I. B. 1979. *Nature* 280:141–42
89. Chamley, J. H., Goller, I., Burnstock, G. 1973. *Dev. Biol.* 31:362–79
90. Chamley, J. H., Dowel, J. J. 1975. *Exp. Cell Res.* 90:1–7
91. Ebendal, T., Jacobson, C. O. 1977. *Exp. Cell Res.* 105:379–87
92. Levi-Montalcini, R. 1976. *Prog. Brain Res.* 45:235–56
93. Charlwood, K. A., Lamont, D. M., Banks, B. E. C. 1972. In *Nerve Growth Factor and Its Antiserum,* ed. E. Zaimis, J. Knight, pp. 102–7 London: Athlone
94. Letorneau, P. C. 1978. *Dev. Biol.* 66: 183–96
95. Gundersen, R. W., Barrett, J. N. 1979. *Science* 206:1079–80
96. Gundersen, R. W., Barrett, J. N. 1980. *J. Cell Biol.* 87:546–54
97. Rutishauser, U., Edelman, G. M. 1980. *J. Cell Biol.* 87:370–78
98. Frazier, W. A., Ohlendorf, L. E., Boyd, L. F., Aloe, L., Johnson, E. M., Ferrendelli, J. A., Bradshaw, R. A. 1973. *Proc. Natl. Acad. Sci. USA* 70:2448–52
99. Schubert, D., Whitlock, C. 1977. *Proc. Natl. Acad. Sci. USA* 74:4055–58
100. Schubert, D., LaCorbiere, M., Whitlock, C., Stallcup, W. 1978. *Nature* 273: 718–23
101. Landreth, G., Cohen, P. 1980. *Nature* 283:202–4
102. Nikodijevic, B., Nikodijevic, O., Yu, M. W., Pollard, H., Guroff, G. 1975. *Proc. Natl. Acad. Sci. USA* 72:4769–71
103. Narumi, S., Fiyita, T. 1978. *Neuropharmacol.* 17:73–76
104. Skaper, S. D., Bottenstein, J. E., Varon, S. 1979. *J. Neurochem.* 43:1845–51
105. Otten, U., Hatanaka, H., Thoenen, H. 1978. *Brain Res.* 140:385–89
106. Lakshmanan, J. 1978. *Brain Res.* 157: 173–77
107. Hatanaka, H., Otten, U., Thoenen, H. 1978. *FEBS Lett.* 92:313–16
108. Yu, M. W., Hori, S., Tolson, N., Huff, K., Guroff, G. 1978. *Biochem. Biophys. Res. Commun.* 81:941–45
109. Yu, M. W., Tolson, N. W., Guroff, G. 1980. *J. Biol. Chem.* 255:10481–92
110. Halegoua, S., Patrick, J. 1980. *Cell* 22: 571–81
111. Carpenter, G., King, L., Cohen, S. 1979. *J. Biol. Chem.* 254:4884–91
112. Nimmo, H. G., Cohen, P. 1977. *Adv. Cyclic Nucleotide Res.* 8:145–266
113. Gunning, P. W., Landreth, G. E., Bothwell, M. A., Shooter, E. M. 1981. *J. Cell Biol.* 89:240–45
114. Greene, L. A. 1977. *Brain Res.* 133: 350–53
115. Gunning, P. W., Landreth, G. E., Letorneau, P. C., Shooter, E. M. 1981. *J. Neurosci.* 1:1085–90
116. Van Winkle, W. D. 1976. *Science* 191: 1130–31
117. Horin, Z.-I., Varon, S. 1977. *Brain Res.* 124:121–23
118. Skaper, S. D., Varon, S. 1979. *Brain Res.* 163:89–100
119. Skaper, S. D., Varon, S. 1979. *Biochem. Biophys. Res. Commun.* 88:563–68
120. Skaper, S. D., Varon, S. 1980. *Brain Res.* 197:379–89
121. Greene, L. A. 1978. *J. Cell Biol.* 78: 747–55
122. McGuire, J. C., Greene, L. A. 1979. *J. Biol. Chem.* 254:3363–67
123. Huff, K., Lakshmanan, J., Guroff, G. 1978. *J. Neurochem.* 31:599–606
124. Gunning, P. W., Landreth, G. E., Layer, P., Ignatius, M., Shooter, E. M. 1981. *J. Neurosci.* 1:368–79
125. Angeletti, P. U., Gandini-Attardi, D., Toschi, G., Salvi, M. I., Levi-Montalcini, R. 1965. *Biochim. Biophys. Acta* 95:111–20
126. Partlow, L. M., Larrabee, M. G. 1971. *J. Neurochem.* 18:2101–18
127. Burnham, P. A., Varon, S. 1974. *Neurobiol.* 4:53–70

128. Burstein, D. E., Greene, L. A. 1978. *Proc. Natl. Acad. Sci. USA* 75:6059–63
129. MacDonnell, P. C., Nagaiah, K., Lakshmanan, J., Guroff, G. 1977. *Proc. Natl. Acad. Sci. USA* 74:4681–84
130. Lakshmanan, J. 1979. *Biochem. J.* 178:248–48
131. Greene, L. A., McGuire, J. C. 1978. *Nature* 276:191–94
132. Ludueno, M. A. 1973. *Dev. Biol.* 33:268–84
133. Mizel, S. B., Bamburg, J. R. 1976. *Dev. Biol.* 49:20–28
134. Max, S. R., Rohrer, H., Otten, U., Thoenen, H. 1978. *J. Biol. Chem.* 253:8013–15
135. McGuire, J. C., Greene, L. A., Furano, A. V. 1978. *Cell* 15:357–65
136. Garrels, J. I., Schubert, D. 1979. *J. Biol. Chem.* 254:7978–85
137. Connolly, J. L., Greene, L. A., Viscarello, R., Riley, W. D. 1979. *J. Cell Biol.* 82:820–27
138. Griffin, C. G., Letorneau, P. C. 1980. *J. Cell Biol.* 86:156–61
139. Banerjee, S. P., Snyder, S. H., Cuatrecasas, P., Greene, L. A. 1973. *Proc. Natl. Acad. Sci. USA* 70:2519–23
140. Frazier, W. A., Boyd, L. F., Bradshaw, R. A. 1974. *J. Biol. Chem.* 249:5513–19
141. Herrup, K., Shooter, E. M. 1973. *Proc. Natl. Acad. Sci. USA* 70:3384–88
142. Frazier, W. A., Boyd, L. F., Pulliam, M. W., Szutowicz, A., Bradshaw, R. A. 1974. *J. Biol. Chem.* 249:5918–23
143. Zimmerman, A., Sutter, A., Samuelson, J., Shooter, E. M. 1978. *J. Supramol. Struct.* 9:351–61
144. Herrup, K., Thoenen, H. 1979. *Exp. Cell Res.* 121:71–78
145. Landreth, G. E., Shooter, E. M. 1980. *Proc. Natl. Acad. Sci. USA* 77:4751–55
146. Calissano, P., Shelanski, M. L. 1980. *Neuroscience* 5:1033–39
147. Sherwin, S. A., Sliski, A. H., Todaro, G. J. 1979. *Proc. Natl. Acad. Sci. USA* 76:1288–92
148. Herrup, K., Shooter, E. M. 1975. *J. Cell Biol.* 67:118–25
149. Banerjee, S. P., Cuatrecasas, P., Snyder, S. H. 1976. *J. Biol. Chem.* 251:5680–85
150. Harris-Warrick, R. M., Bothwell, M. A., Shooter, E. M. 1980. *J. Biol. Chem.* 255:11284–89
151. Stach, R. W., Shooter, E. M. 1980. *J. Neurochem.* 34:1499–1505
152. Varon, S., Nomura, J., Shooter, E. M. 1968. *Biochemistry* 7:1296–1303
153. Pattison, M. E., Dunn, M. F. 1975. *Biochemistry* 14:2733–39
154. Costrini, N. V., Bradshaw, R. A. 1979. *Proc. Natl. Acad. Sci. USA* 76:3242–45
155. Costrini, N. V., Kogan, M., Kukreja, K., Bradshaw, R. A. 1979. *J. Biol. Chem.* 254:11242–46
156. Costrini, N. V., Kogan, M. 1981. *J. Neurochem.* 36:1175–80
157. Riopelle, R. J., Klearman, M., Sutter, A. 1980. *Brain Res.* 199:63–77
158. De Meyts, P., Roth, J., Neville, D. M. Jr., Gavin, J. R. III, Lesnick, M. A. 1973. *Biochem. Biophys. Res. Commun.* 55:154–61
159. Cohen, P., Sutter, A., Landreth, G., Zimmerman, A., Shooter, E. M. 1980. *J. Biol. Chem.* 255:2949–54
160. Schlessinger, J., Shechter, Y., Willingham, M. C., Pastan, I. 1978. *Proc. Natl. Acad. Sci. USA* 75:2659–63
161. McKanna, J. A., Haigler, H. T., Cohen, S. 1979. *Proc. Natl. Acad. Sci. USA* 76:5689–93
162. Levi, A., Shechter, Y., Neufeld, E. J., Schlessinger, J. 1980. *Proc. Natl. Acad. Sci. USA* 77:3469–73
163. Schechter, A. L., Bothwell, M. A. 1981. *Cell* 24:807–74
164. Goldstein, J. L., Anderson, R. G. W., Brown, M. S. 1979. *Nature* 279:679–85
165. Andres, R. Y., Jeng, I., Bradshaw, R. A. 1977. *Proc. Natl. Acad. Sci. USA.* 74:2785–89
166. Goldfine, I. D., Smith, G. J., Wong, K. Y., Jones, A. L. 1977. *Proc. Natl. Acad. Sci. USA* 74:1368–72
167. Vigneri, R., Goldfine, I. D., Wong, K. Y., Smith, G. J., Pezzino, V. 1978. *J. Biol. Chem.* 253:2098–2103
168. Johnson, L. K., Vlodavsky, I., Baxter, J. D., Gospodarowicz, D. 1980. *Nature* 287:340–43
169. Savion, N., Vlodavsky, I., Gospodarowicz, D. 1981. *J. Biol. Chem.* 256:1149–54
170. Rao, C. V., Mitra, S. 1979. *Biochim. Biophys. Acta* 584:454–66
171. Libby, P., Bursztajn, S., Goldberg, A. C. 1980. *Cell* 19:481–91
172. Johnson, E. M., Andres, R. Y., Bradshaw, R. A. 1978. *Brain Res.* 150:319–31
173. Marchisio, P. C., Naldtini, L., Calissano, P. 1980. *Proc. Natl. Acad. Sci. USA* 77:1656–60
174. Heumann, R., Schwab, M., Thoenen, H. 1981. *Nature* 292:838–40
175. Yamaizumi, M., Mekada, E., Uchida, T., Okada, Y. 1978. *Cell* 15:245–50
176. Eiklid, K., Olsnes, S., Pihl, A. 1980. *Exp. Cell Res.* 126:321–26

Ann. Rev. Biochem. 1982. 51:869–900
Copyright © 1982 by Annual Reviews Inc. All rights reserved

INITIATION FACTORS
IN PROTEIN BIOSYNTHESIS[1]

Umadas Maitra, Evan A. Stringer,[2] and Asok Chaudhuri[3]

Department of Developmental Biology and Cancer, Albert Einstein College of
Medicine, Bronx, New York 10461

CONTENTS

[1]*Abbreviations* The following abbreviations are used: IF-1, IF-2, and IF-3: prokaryotic
initiation factors 1, 2, and 3, respectively; eIF-1, eIF-2, eIF-3, eIF-4A, eIF-4B, eIF-4C, eIF-5,
eIF-6: eukaryotic initiation factors 1, 2, 3, 4A, 4B, 4C, 5, and 6 respectively; $m^7GpppXm$,
7-methyl guanosine-(5') triphosphate (5')2-0 methylated nucleoside; m^7GDP, 7-methyl guanosine 5'-diphosphate; CBP, Cap binding protein.

[2]Present Address: Department of Experimental Cell Biology, Mount Sinai School of Medicine, New York, NY 10029

[3]Present address: Rockefeller University, New York, NY 10021.

0066-4154/82/0701-0869$02.00

PERSPECTIVES AND SUMMARY

The ribosome-mediated translation of mRNA into protein from aminoacyl-tRNA substrates can be divided into three phases: chain initiation, elongation, and termination. Accumulated evidence shows that these phases are physiologically discrete, with specific proteins required for each step. Since chain initiation is the subject of this review, we confine our discussions to this process. For excellent current reviews of the mechanism of the other two steps, elongation and termination, see (1–4).

Initiation of translation of mRNA into protein consists of a series of discrete reactions during which the ribosomal subunits interact with mRNA and with methionyl initiator tRNA in such a way that the ribosomes bind at a specific site of an mRNA that contains the initiation codon AUG corresponding to the beginning of a cistron. During this process, methionyl initiator tRNA, which provides the N-terminal amino acid residue of all nascent polypeptide chains, is also positioned on the ribosome by base-pairing between the anticodon of the initiator tRNA with the initiation codon of mRNA. The ribosomal chain initiation complex thus formed is ready to form a peptide bond with an incoming aminoacyl-tRNA during the elongation phase of protein synthesis. A set of several specific protein molecules, called initiation factors, and GTP direct the various steps leading to the formation of a functional ribosomal chain initiation complex.

Because the prokaryotic system is better defined and continues to provide a useful model system for further work in eukaryotic initiation, we first briefly summarize the saliant features of our current understanding of the mechanism of polypeptide chain initiation in prokaryotes and later discuss in greater detail the eukaryotic systems.

Since the publication of reviews on protein synthesis (5), (6), several excellent review articles dealing specifically with initiation of protein synthesis have been published (7–9). The present article provides a comprehensive review of the role of initiation factors in protein biosynthesis by giving both a retrospective view of the field and an appraisal of the current developments.

A. INITIATION OF PROTEIN SYNTHESIS IN PROKARYOTES

I. Initiation Reaction

Most of our knowledge in this field stems from in vitro work with *Escherichia coli.* In *E. coli,* the initiation of protein synthesis proceeds with the formation of a primary 30S initiation complex consisting of 30S ribosomal subunits, mRNA, GTP, and formyl methionyl-tRNA (fMet-tRNA$_f$). mRNA and fMet-tRNA$_f$ are bound first to the 30S ribosomal subunit, with subsequent addition of the 50S subunit to form the 70S initiation complex. Evidence for this mechanism has been reviewed in a number of recent articles (7–9). Formation of the initiation complex requires GTP, Mg^{2+}, NH$_4^+$, and three protein factors, designated as IF-1, IF-2, and IF-3. These protein factors are loosely associated with 30S ribosomes and are readily dissociated by treatment with 1 M NH$_4$Cl. Isolation of factors in electrophoretically homogeneous form in several laboratories during the last decade (10–24) has permitted their detailed characterization and facilitated the understanding of the way they act at the individual steps in 70S initiation complex formation. The overall initiation reaction can be written as follows:

$$\text{fMet-tRNA}_f + \text{mRNA} + 30S + 50S \xrightarrow[\text{GTP}]{\text{IF1, IF2, IF3}} [\text{fMet-tRNA}_f\text{-70S-mRNA}] + \text{GDP} + \text{P}_i$$

All three factors act catalytically in the overall process of initiation. Concomitant with the formation of the 70S initiation complex, GTP is hydrolyzed to GDP + P$_i$ (25–29).

II. Physicochemical Properties of E. Coli Initiation Factors

Extensive physicochemical characterization of *E. coli* initiation factors have been carried out. IF-1 and IF-3 are relatively heat-stable basic protein molecules, each consisting of a single polypeptide chain (15–24). The reported M_r values of the two proteins are 8,900–9,400 for IF-1 (16, 17, 20, 22, 24) and 21,000–23,500 for IF-3 (16–18, 20, 23, 24, 30). IF-2, on the other hand, is a relatively heat-labile acidic protein (14, 17, 25, 31). Two forms of IF-2, designated IF-2a (M_r = 90,000–118,000) and IF-2b (M_r = 82,000–90,000) have been described (24, 32–35). Both forms are nearly equally active in all biochemical properties involving IF-2. Later studies have convincingly demonstrated that IF-2b is an artifact of the isolation procedure; its origin is due to limited proteolysis of IF-2a (24, 32). The amino acid composition of all three initiation factors have been determined (22, 24) and IF-1 has been crystallized (22).

III. Studies of the Individual Partial Reactions Leading to 70S Initiation Complex Formation

The detailed individual partial reactions leading to the 70S initiation complex formation via the intermediate formation of a 30S initiation complex are summarized below. The exact sequences of some of the steps are still being debated.

1. DISSOCIATION OF 70S RIBOSOMES   It is now generally accepted that following termination of mRNA translation, the 70S ribosomes are released from the polysomal complex and are then in equilibrium with their subunits. At physiological Mg^{2+} concentration, the rate of dissociation is slow, and nearly all ribosomes exist as tight 70S-couples (36, 37). Since initiation of protein synthesis starts on the 30S-particle, which subsequently joins the 50S-subunit, dissociation of 70S-ribosomes is necessary to ensure a pool of free 30S- and 50S-subunits.

It has been clearly demonstrated (18, 30, 31, 38–45) that this dissociation process is mediated by the combined action of IF-1 and IF-3 as follows:

$$70S \; \frac{K_1}{\overline{\overline{K_2}}} \; 50S + 30S \; \frac{+ \; IF\text{-}3}{(IF\text{-}1)} \; 50S + 30S \cdot IF\text{-}3$$

In this reaction, IF-3 does not significantly change the rate of dissociation of 70S-ribosomes but rather acts primarily as an antiassociation factor (31, 42, 46–51). It is the initiation factor, IF-1, that directly aids in the dissociation process by increasing the rate of dissociation into free 50S and 30S (31,42,46–50). IF-3, on the other hand, binds to free 30S subunits, which originate by dissociation of 70S ribosome. IF-1 may also aid in the process (45). The 30S·IF-3 complex is unable to reassociate with 50S subunits and thus, in effect, displaces the equilibrium toward free subunits (46–48).

2. FORMATION OF A 30S·IF-1·IF-2·IF-3 COMPLEX   The first step in the formation of the 30S initiation complex involves cooperative binding of three initiation factors to a 30S subunit to form a 30S·IF-1·IF-2·IF-3 particle to which mRNA and fMet-$tRNA_f$ subsequently bind to form the 30S initiation complex. IF-2 specifically binds to 30S subunits in the absence of all other components of initiation. This association is stabilized by IF-1 and IF-3 (52–55). Under these conditions, one molecule of IF-2 is bound per active 30S subunit (53). IF-3 also binds to 30S independent of other components of initiation. However, this binding is also stabilized by IF-1 and IF-2 (56, 57). Stable binding of IF-1 to 30S is also obtained in the presence of IF-2 and IF-3 (58, 59). All three factors bind contiguously near the 3' end of 16S RNA on the 30S subunit at adjacent or overlapping sites that are located at the interface in the 70S ribosome. The protein neighbor-

hood of this 30S region have been identified (43, 60–66).

3. FORMATION OF 30S INITIATION COMPLEXES   Following binding of the three initiation factors to 30S ribosomes, the initiator fMet-tRNA$_f$ and mRNA specifically bind to the 30S·IF-1·IF-2·IF-3 particle. The actual order of addition of fMet-tRNA$_f$ and mRNA to such a 30S complex is, however, uncertain. Evidence suggests that in the in vitro system fMet-tRNA$_f$ and mRNA may bind independently of one another to the 30S complex (44, 56, 57, 67–72). However, in vitro experimental data (56, 57) on the formation of a complete 30S initiation complex are compatible with the binding of mRNA preceding that of fMet-tRNA$_f$.

The 30S·IF-3·IF-2·IF-1 complex binds to mRNA at the site that includes the initiation codon, AUG or GUG (73, 74). The mechanism of recognition of the initiation site of a cistron in a natural mRNA by 30S ribosome is an important event in initiation of protein synthesis and is dealt with later in this section (see ii).

Formation of the 30S initiation complex can be depicted by the following reaction:

$$30S·IF-1·IF-2·IF-3 + mRNA + GTP + fMet-tRNA_f$$
$$\rightarrow [30S·IF-1·IF-2·mRNA·GTP·fMet-tRNA_f] + IF-3$$

The 30S initiation complex can be isolated as a stable intermediate by Sephadex G-100 gel filtration or by sucrose gradient centrifugation.

GTP acts as a steric effector in this reaction, permitting stable association of IF-2 with ribosomes at a relatively low concentration of the factor (53, 54, 75–77). In the presence of excess IF-2, however, 30S initiation complexes can be formed even in the absence of GTP (76, 77). The nucleotide is not hydrolyzed at this stage and remains bound to the 30S complex (75, 76). One mole each of GTP and fMet-tRNA$_f$ is bound per mole of IF-2 bound to the 30S initiation complex. Thus in formation of the latter IF-2 acts stoichiometrically. IF-1 is also bound to the complex (21), while IF-3 is released from the 30S ribosomes during binding of fMet-tRNA$_f$ (56, 57).

Although all three initiation factors are required for maximal binding of fMet-tRNA$_f$ to ribosomes, IF-2 plays the central role in the selection of fMet-tRNA$_f$ for binding to 30S ribosomes. Its requirement is absolute. While the function of IF-3 is to aid 30S ribosomes in initiation site selection in natural mRNA, the major role of IF-1 at the 30S level is stabilization of binding of IF-2 to 30S initiation complex (52, 53, 78), which is necessary because 70S-dependent GTPase activity of IF-2 provides a means by which GTP can be removed during the junction of the 30S complex with the 50S subunits to form a 70S complex (78). It is well documented that the removal of GTP is necessary for formation of an active 70S initiation complex (see section A III 4 for a detailed discussion). IF-1 has also been implicated in the stabilization and increased affinity of mRNA binding to ribosomes (79, 80).

i. Role of IF-2 in specific recognition of fMet-tRNA$_f$ IF-2 is known to be the factor responsible for binding of initiator tRNA to ribosomes, since this factor alone is capable of binding fMet-tRNA$_f$ to isolated 30S subunits when AUG is used as messenger. The exact steps involved in this process remain unclear. Highly purified IF-2 is unable to form stable complexes with either GTP or fMet-tRNA$_f$ under physiological conditions (34), although weak interactions of IF-2 with GTP occur in the absence of initiator tRNA and of IF-2 with fMet-tRNA$_f$ in the absence of GTP (26, 81–84). Highly purified IF-2 forms a GTP-independent binary complex with fMet-tRNA$_f$ in the absence of free Mg^{2+} ions (83, 85–87). Although the role of this complex in the initiation mechanism is not yet understood, measurement of binary complex allowed Sundari et al (86) to determine which structural features of the initiator tRNA are recognized by IF-2. These workers have found that IF-2 will not bind to free fMet or to a short fMet oligonucleotide, but will bind strongly to any tRNA structure covalently attached to an N-blocked methionine group. The *E. coli* initiator and noninitiator methionine tRNAs (which have many differences in primary structure), fMet-tRNA$_f^{Met}$ molecules containing numerous structural modifications (86), and N-blocked eukaryotic initiator tRNAs all bound strongly to the factor. The work of Clark & Marcker (88) showed the importance of tRNA structure in ribosome binding; at higher Mg^{2+} concentration, there is no requirement of initiation factors and formylation for binding of Met-tRNA$_f^{Met}$ to ribosomal P-site. Under these conditions, Met-tRNA$_f^{Met}$ binds with a much higher affinity than other aminoacyl-tRNAs, including Met-tRNA$_m^{Met}$, which shows that the ribosome itself has an inherent specificity for the tRNA portion of the structure. At lower Mg^{2+} concentrations, the factor-independent interaction is weakened, and formylation of initiator tRNA further decreases its affinity for the ribosomes (86, 89). Thus, under physiological conditions, little binding of fMet-tRNA$_f^{Met}$ occurs in the absence of IF-2. These results suggest that the role of IF-2 in the formation of ribosomal initiation complexes is to stabilize the interaction of the formylated initiator tRNA with the ribosome.

In certain bacterial strains, e.g. *Streptococcus faecalis* (90–92) and an *E. coli* mutant (93) that can grow in a defined medium in the complete absence of folate, protein synthesis is initiated in the absence of formylation of

Met-tRNA$_f^{Met}$. It was shown that the ribothymidine (rT) present in the sequence GTΨC in loop IV of tRNA$_f^{Met}$ was replaced by uridine (due to the absence of methylation) (90–93). Presumably this single alteration of the structure of tRNA$_f^{Met}$ allows unformylated Met-tRNA$_f^{Met}$ to bind more efficiently in response to initiation codons than the corresponding unformylated Met-tRNA$_f^{Met}$ formed in folate-plus cells.

ii. The Recognition of Initiation Sites in Natural mRNA by Ribosomes, and the Role of Initiation Factors in the Process The 30S ribosomal subunit possesses the properties necessary for specifically recognizing and selecting initiation sites in a natural mRNA and IF-3 aids the 30S ribosomes in this selection process (for review see 73, 74). The central question is what are the unique structural features of the region of mRNA surrounding the initiation AUG codons that distinguish them from noninitiating internal AUG codons?

Every mRNA molecule or cistron has one "ribosome-binding site" for each of its independently synthesized polypeptide products. Shine & Dalgarno (94) and Steitz and her collaborators (74, 95) have shown that virtually all ribosome-binding sites have an AGGAGGU(3') sequence (or a close derivative) at a relatively similar position on the 5' side of the initiation codon of a cistron. This sequence has been shown to base-pair to a pyrimidine-rich region at the 3' end of 16S rRNA chains (—-GAUCACCUCCUUA$_{OH}$3'). This pairing positions the initiating AUG codon so that it can bind to the anticodon of an initiator tRNA.

Several features of the initiation site sequences are particularly noteworthy. First, comparison of the ribosome-binding sites of a wide variety of mRNAs with the presumed complementary region at the 3' end of 16S RNA indicates heterogeneity of the initiation signals (73, 74). The number of nucleotides capable of forming true Watson-Crick base pairs with the 3'-OH terminal oligonucleotide end of 16S RNA varies from three to nine. It has been postulated that the greater the number of nucleotides in the ribosome-binding site of mRNA involved in base-pairing with the 3' end of 16S RNA, the higher is the efficiency with which ribosomes bind to initiator sites (73, 74) and the less the dependency on IF-3 for initiation complex formation (96, 97). Another distinguishing feature of different initiation sites is the variability of the distance covered by the initiation triplet and polypurine stretch involved in base-pairing to 16S RNA. This distance is also likely to influence the formation and stability of the mRNA-ribosome complex (73, 74).

Shine & Dalgarno's suggestion of a purine-rich ribosome binding site has been validated by a body of experimental evidence, the most important of which is the direct demonstration (95) of the formation of a RNA·RNA hybrid between the polypurine stretch of an RNA (R17 A-cistron RNA fragment) and the pyrimidine rich 3' end of 16S RNA during initiation complex formation. [For a detailed discussion see (73 and 74)]. More recently, Jay et al (98) have reported that a chemically synthesized icosadeoxyribonucleotide containing the several features found in prokaryotic mRNAs ribosome binding sites can direct the formation of a functional 70S initiation complex in a reaction dependent on both IF-2 and IF-3.

It should be emphasized, however, that since in some cases the length of the mRNA region complementary to 16S RNA 3' end can be as short as 3 base pairs, the possibility exists that in a natural mRNA, there may exist several such noninitiating regions that are characterized by having an AUG triplet preceded by an appropriate polypurine stretch 3–4 base pairs long. These AUG triplets may not be true initiation codons corresponding to the beginning of a cistron. Several such "false" potential initiation sites are, in fact, found in MS2 RNA (99). These regions are usually "buried" in RNA secondary structures and are thus unavailable for base-pairing with 16S RNA or with initiator tRNA. It is likely that mRNA uses secondary and tertiary structures to prevent ribosome recognition of noninitiator AUG triplets present in a potential ribosome binding site.

What function do IF-3 and other initiation factors and ribosomal proteins involved in initiation play in this recognition process?

Data based on cross-linking experiments have shown that IF-3 and also IF-2 and IF-1 bind near the 3' end of 16S RNA during 30S initiation complex formation (43, 61, 66, 100). The protein neighborhood organized around this region of 16S RNA consists of S7, S1, S11, S12, S18, and S21 (59, 64, 100, 101). Since this region of 16S RNA is involved in the recognition of the mRNA initiation site it is likely that these ribosomal proteins together with IF-3 function in the formation of a 30S initiation complex with mRNA.

Dahlberg & Dahlberg (100) have proposed that in the free 30S subunit, the 16S rRNA may assume a conformation in which a major portion of the pyrimidine-rich sequence is in a base-paired hairpin loop structure (102), thereby making it inaccessible for base-pairing with the "ribosome binding site" of an mRNA. It has been suggested that S1 and IF-3, by binding to this region, may function to hold the 3' end of 16S RNA in an unpaired conformation thereby making this region accessible for hydrogen bonding with the ribosome binding sites in mRNA. In agreement with this, it has been observed that S1 has the property of unwinding a variety of stacked or helical single-stranded polynucleotides, with polypyrimidines appearing to be the preferred substrates over polypurines (74, 103).

It has also been proposed (65, 104) that in 70S couples, the 3' ends of both 16S and 23S RNA are paired into each other. This gives a molecular basis for association of 50S and 30S subunits in 70S couples. By binding to the 3' terminus of 16S RNA, IF-3 could destabilize base-pairing between these two RNAs and thus promote formation of free 50S and 30S·IF-3 particles. In addition, the binding of IF-3 may result in modification of overall conformation of the 30S subunit suitable for base-pairing with mRNA.

4. FORMATION OF 70S INITIATION COMPLEXES AND RECYCLING OF IF-2   As mentioned above, the complete 30S initiation complex contains bound fMet-tRNA$_f$, intact GTP, mRNA, IF-2, and IF-1. Addition of 50S subunits to the 30S initiation complex results in the joining of 50S subunits to the 30S initiation complex to form the 70S initiation complex that contains bound fMet-tRNA$_f$ and mRNA in a 1:1 molar ratio (5–9). During the 50S subunit joining reaction, GTP originally present in the 30S initiation complex is hydrolyzed by ribosome-bound IF-2 to GDP and P_i. IF-2 (and IF-1), GDP, and P_i are released from the ribosomal complex. Subsequent to this release, fMet-tRNA$_f$ present in 70S ribosomal complex becomes active in peptide bond formation (5–9). This reaction is conveniently measured by the ability of 70S-bound fMet-tRNA$_f$ to react with puromycin. The subunits joining reaction can, therefore, be written as follows:

$$[30S·mRNA·IF-2·IF-1·fMet-tRNA_f·GTP] + 50S \rightarrow$$
$$[mRNA·70S·fMet-tRNA_f] + IF-2 + GDP + P_i + IF-1$$

i. Function of GTP hydrolysis   The function of GTP hydrolysis in the formation of 70S initiation complexes has been investigated in detail. As stated in section A III 3, IF-2 acts stoichiometrically during 30S initiation complex formation and remains bound to the complex. During GTP-dependent formation of a 70S initiation complex, IF-2 is released from the ribosome and thus acts catalytically in initiation reactions. In contrast, when GTP is replaced by nonhydrolyzable analogue, GDPCP, IF-2 remains bound to the 70S ribosomal complex (20, 26, 52–54, 75, 76).

To examine whether the free energy of GTP hydrolysis is required for a functional 70S initiation complex formation, Dubnoff et al. (76) made use of the observation that GTP bound in the 30S initiation complexes is unstable and is easily removed by incubation of the 30S initiation complex for 15 min at 37° followed by Sephadex G-100 gel filtration. This procedure yields 30S complexes depleted of GTP but containing fMet-tRNA$_f$. Such GTP-deficient 30S complexes can still accept 50S subunits to form 70S complexes active in peptidyl transfer. Since the GTP removed from the complex remained intact, it was clearly demonstrated that the energy of hydrolysis was not involved in forming puromycin reactive 70S complexes.

Furthermore, GDPCP can replace GTP in the 30S initiation complexes, but addition of 50S subunits results in a 70S complex that is inactive in subsequent peptide bond formation (20, 52–54, 75, 105). However, if GDPCP is removed from the 30S complex before the addition of 50S subunits, the ability of the subsequent 70S complex to participate in peptide bond formation is restored (76). Thus, the behavior of 30S complexes formed either with GTP or GDPCP is identical. In either case, once the 30S complex forms, the continued presence of the nucleotide prevents fMet-tRNA from participating in subsequent peptide bond formation. Removal of the nucleotide alters the configuration of the complex to place fMet-tRNA in a reactive state.

These results show that the energy of hydrolysis of GTP is not required for correct positioning of fMet-tRNA$_f$ in the donor site of the 70S initiation complex. Rather, GTP hydrolysis is used to release IF-2 and bound GTP from the 70S complex, thereby activating or "unblocking" fMet-tRNA$_f$ so that peptide bonds can be formed. It has been postulated (105) that the hydrolysis of GTP and/or the release of IF-2 bring about a conformational change of tRNA$^{Met}_f$ so that the fMet moiety and the 3' end of tRNA$_f$ is able to interact with peptidyl transferase. This step was termed "accommodation".

ii. Function of IF-1   In addition to requirements for hydrolysis of GTP, the other component of the initiation system that is involved in catalytic recycling of IF-2 is IF-1 (27, 54, 76, 78). In the absence of IF-1, 70S initiation complexes formed in a GTP-dependent reaction contain bound IF-2 even after GTP hydrolysis has occurred. Addition of IF-1 to such an isolated 70S complex causes quantitative release of IF-2 (54, 78, 106). There is evidence to rule out the presence of a stable IF-2·GDP complex on the surface of the 70S ribosome following GTP hydrolysis and the need for any exchange reactions between IF-1 and guanine nucleotides in effecting the release of IF-2 (78). IF-2 remains on the 70S initiation complexes after the release of guanine nucleotides and can be liberated solely by addition of IF-1 (78).

In summary, it appears that GTP acts as a steric effector in initiation permitting association of IF-2 to 30S ribosomes. The hydrolysis of GTP is necessary to convert one steric effector GTP, to another, GDP, which has a low affinity for the ribosomal complex. GDP readily dissociates from the ribosomes which allows IF-1 to effect the release of IF-2 in a simple interaction with IF-2-bound 70S ribosomal complex.

iii. The Requirement for Initiation Factors in the Joining of 30S Initiation Complexes to 50S Subunits In eukaryotic initiation reactions, the joining of the 60S subunit to the 40S initiation complex requires the participation of a specific protein factor (eIF-5) (see section on eukaryotic initiation reactions). In order to see if the subunit joining reactions in bacterial polypeptide chain initiation require the presence of initiation factors, Stringer et al (78) prepared 30S initiation complexes with poly(U,G) as mRNA in the absence of both IF-3 and IF-1. (IF-1 is not absolutely required for binding of fMet-tRNA$_f$ to 30S subunits, while with synthetic mRNA, IF-3 is also not required.) IF-2 and GTP present in the 30S complexes were dissociated by incubation and removed by sucrose gradient centrifugation to give 30S complexes still containing fMet-tRNA$_f$ but lacking initiation factors and GTP. Such complexes were able to form 70S initiation complexes that were active in peptidyl transfer. Hence once fMet-tRNA$_f$ is bound to the 30S subunit neither initiation factors nor GTP need be present when junction with 50S subunits occurs to yield an active 70S complex.

These results also suggest that an intimate relationship exists between IF-2 and GTP and that continued presence of IF-2 in 30S initiation complexes is only required for active 70S complex formation when GTP is present. In vivo, the only means available for removal of GTP (and hence the subsequent reactivity of fMet-tRNA$_f$ in the peptidyl transfer reaction), is its hydrolysis by the 70S-dependent IF-2-GTPase. Therefore, the presence of IF-2 on the 30S complex at the time of junction with 50S subunit is essential. Hence, an important role of IF-1 at the level of 30S complex formation is also established; it ensures that IF-2 remains bound to the 30S complex throughout the initiation reaction so that it can hydrolyze GTP and facilitate the formation of an active 70S complex.

iv. Ribosomal Sites Involved in GTP Hydroloysis
It has been demonstrated that 50S ribosomal proteins L7/L12 and protein L11 are involved in IF-2-dependent GTP hydrolysis (29, 107–109). IF-2 presumably binds (in 70S complex) at or near L7/L12 site, since the 70S ribosome-bound IF-2 can be cross-linked to L7/L12 (110). The same proteins have been shown to be involved in EF-G-and EF-Tu-catalyzed hydrolysis of GTP during elongation reactions and in RF-1- and RF-2-catalyzed termination reactions (1–4). However, proteins L7/L12 may not be an intrinsic part of the active site of GTPase. Binding of IF-2 or of elongation factors, EF-G and EF-Tu to the ribosomal site containing proteins L7/L12 may cause a change in the configuration of a separate site (the GTPase site) of 50S subunit in such a way that the site becomes active in GTP hydrolysis.

IV. CONCLUDING REMARKS ON PROKARYOTIC INITIATION

The evidence presented above justifies writing the following sequence of reactions leading to the formation of an active 70S·initiation complex:

$$70S \xrightarrow{+ \text{ IF-3, IF-1}} 50S + [30S \cdot IF-3 \cdot IF-1]$$

$$[30S \cdot IF-3 \cdot IF-1] + IF-2 \rightarrow [30S \cdot IF-2 \cdot IF-1 \cdot IF-3] \xrightarrow{+ \text{ mRNA}}$$

$$[30S \cdot mRNA \cdot IF-2 \cdot IF-1 \cdot IF-3]$$

$$[30S \cdot mRNA \cdot IF-2 \cdot IF-1 \cdot IF-3] + \text{fMet-tRNA}_f \text{ GTP} \rightarrow$$

$$[30S \cdot mRNA \cdot IF-2 \cdot IF-1 \cdot \text{fMet-tRNA}_f \cdot GTP] + IF-3$$

30S·Initiation Complex

$$[30S \cdot mRNA \cdot IF-2 \cdot IF-1 \cdot \text{fMet-tRNA}_f \cdot GTP] + 50S \rightarrow$$

$$[mRNA \cdot 70S \cdot \text{fMet-tRNA}_f] + IF-2 + GDP + P_i + IF-1$$

Active 70S Initiation Complex

B. EUKARYOTIC INITIATION REACTIONS

I. Components Involved in Initiation Complex Formation

Initiation of protein synthesis in both eukaryotes and prokaryotes follows the same basic mechanism. The initiator tRNA is bound to the small ribosomal subunit in a GTP-dependent reaction to form a ribosomal subunit complex containing mRNA, GTP, and initiator tRNA. This complex then joins to the larger ribosomal subunit to form an initiation complex capable of translating mRNA to form new polypeptide chains. The main characteristics of the eukaryotic as contrasted to the prokaryotic system are:

1. The ribosomes are larger;
2. More protein initiation factors are involved;
3. The methionyl initiator tRNA is not formylated;
4. There is a 7-methyl guanosine (5') triphosphate (5') 2-O methylated nucleoside residue ($m^7G(5')pppNp^m$) known as "cap" structure present in most eukaryotic mRNAs.
5. The mRNAs are monocistronic in nature.

The increase in order and complexity of both the eukaryotic ribosomes and initiation factors most likely reflects the need for translational control resulting from the stability of eukaryotic mRNAs.

1. INITIATION FACTORS   The rabbit reticulocyte system has been most widely used in the study of eukaryotic peptide chain initiation and, to date, at least eight initiation factors have been isolated (111–121). Each of the factors, eIF-1, eIF-2, eIF-3, eIF-4A, eIF-4B, eIF-4C, eIF-4D, and eIF-5 (122) was isolated from the crude mixture of ribosomal salt-wash proteins on the basis of their ability to stimulate the translation of natural mRNA (globin mRNA) (118–121). A highly purified and fractionated system was used, with completely defined aminoacyl-tRNAs and the two elongation factors EF-1 and EF-2, as well at ATP and GTP. Subsequently, the ability of each of the isolated initiation factors to promote the sequential binding of radioactive Met-tRNA$_f$ and mRNA to the small ribosomal subunit (40S), and the joining of the larger ribosomal subunit (60S) to form a functional 80S initiation complex was studied.

There is still disagreement on the number and functions of the initiation factors needed for initiation complex formation. Several (111–121) agree that there are at least eight factors involved, and they have purified all initiation factors except eIF-4B to apparent homogeneity. The data clearly indicate that, when natural mRNA is used, only eIF-2, eIF-3, and eIF-5 are absolutely essential for the formation of functional 80S initiation complexes (118–121). eIF-1, eIF-4A, eIF-4B, eIF-4C, and eIF-4D are not essential but merely give a two to three fold stimulation of complex formation. When AUG is used in place of natural mRNA as template, only eIF-2 and eIF-5 are essential for 80S initiation complex formation, while eIF-3 gives only twofold stimulation of initiation complex formation (121, 123). These results show that the binding and recognition of natural mRNA requires eIF-3 and the hydrolysis of ATP and is stimulated by the factors eIF-4A,-4B, and-4C (119). This is analogous to the situation in prokaryotes where the binding of natural mRNA has an absolute requirement for IF-3, while binding of synthetic mRNA or the triplet AUG does not. The eukaryotic initiation reaction in vitro can be considered to proceed in at least four discrete steps. These are:

1. Formation of an eIF-2·GTP·Met-tRNA$_f$ ternary complex;
2. Transfer of ternary complex to the 40S ribosomal subunit;
3. Binding of mRNA to the 40S complex;
4. Joining of the 40S initiation complex to the 60S subunit.

These steps are schematically represented in Figure 1.

The above scheme, adapted from that of Trachsel et al (119) shows only the pathway by which the initiation complex is assembled and at which steps the initiation factors participate. The most serious criticism of the results obtained from in vitro assays is that apart from eIF-5, no evidence

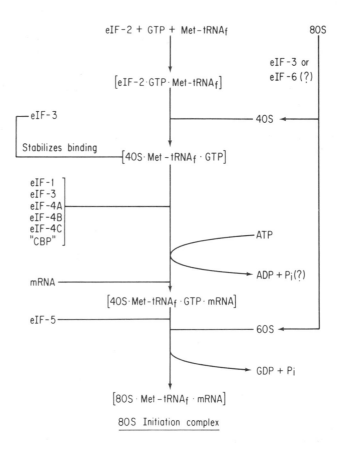

Figure 1   Schematic representation for the assembly of 80S initiation complex formation.

has been presented to demonstrate that these factors are catalytically reutilized in the initiation reaction. Recycling of initiation factors is an essential feature of such a system, and analysis of the results to date suggests that a component (or components) is limiting or lacking.

(2) ANCILLARY FACTORS   In addition to the factors discussed above, there are reports of other protein factors involved in the initiation reactions:

i. Co-eIF-2A, Co eIF-2B, and Co-eIF-2C   These protein factors have been reported to control the formation of ternary complexes (124). Co-eIF-2A, which has a molecular mass of 20,000 daltons, stabilizes ternary complex formation and increases the rate of Met-tRNA$_f$ binding to 40S ribosomal subunits (125–128). A physiological role for Co-eIF-2A was demonstrated when antisera to homogeneous preparations of the factor strongly inhibited protein synthesis in rabbit reticulocyte lysates (129).

Co-eIF-2B (ternary complex dissociation factor, TDF) is a large complex of 12–15 polypeptides with a molecular weight of 450,000 (130, 131). As well as catalyzing the dissociation of the ternary complex it is involved in the AUG-directed binding of Met-tRNA$_f$ to 40S ribosomal subunits, and it appears to be similar to eIF-3.

Co-eIF-2C, a protein fraction purified from high-salt wash of reticulocyte ribosomes (261), promotes ternary complex formation by eIF-2 in the presence of Mg^{2+} and also stimulates eIF-2-dependent binding of Met-tRNA$_f$ to 40S ribosomes in the presence of AUG.

ii. eIF-2-Stimulatory Protein   Salt-washed proteins from the ribosomes of *Artemia salina* and rabbit reticulocytes contain a factor called eIF-2 stimulatory protein (ESP) that enhances the ability of eIF-2 to form ternary and 40S initiation complexes. The stimulation is particularly marked at low concentration of eIF-2 (133, 134). It has been proposed that ESP stabilizes a binary complex of eIF-2 and GTP and is released when Met-tRNA$_f$ joins to form a ternary complex (133):

eIF-2 + GTP $\rightleftharpoons$ eIF-2·GTP

eIF-2·GTP + ESP $\rightleftharpoons$ eIF-2·GTP·ESP

eIF-2·GTP·ESP + Met-tRNA$_f$ $\rightleftharpoons$ eIF-2·GTP·Met-tRNA$_f$ + ESP

Complexes between eIF-2 and ESP have not been detected either in the presence or absence of GTP, which suggests their interaction must be very weak (134). However, it has recently been demonstrated that highly purified eIF-2 does not bind GTP and reports of eIF-2·GTP complexes probably result from contamination of GDP in commercial GTP preparations (123). Until ESP can be shown to stabilize a binary complex between eIF-2 and GTP labeled in the γ phosphate of the nucleotide, its role remains unclear.

iii. mRNA-Inhibition Counteracting Factor   mRNA strongly inhibits ternary complex formation by binding to eIF-2 and dissociating its subunits (123, 135, 136). Protein fractions capable of preventing this inhibition have been isolated from *A. salina* (137) and rabbit reticulocytes (125). There is evidence that the factor isolated from rabbit reticulocytes may be identical with Co-eIF-2A (125).

iv. eIF-6   A factor with a mol wt of 23,000–25,500 isolated from wheat germ (138–140) and from calf liver extracts (141), appears to be responsible for maintaining a pool of 40S and 60S ribosomal subunits. Studies with ^{14}C-labeled eIF-6 indicate that the protein acts primarily as an antiassociation factor by binding to 60S ribosomal subunits to prevent them from interacting with 40S ribosomal subunits (140, 141). eIF-6 has very little 80S ribosomal dissociation activity (139, 141).

II. Physical Properties of Initiation Factors

With the exception of eIF-2 and eIF-3, all the initiation factors consist of single polypeptide chains ranging in molecular weight from 15,000 (eIF-1) to about 160,000 (eIF-5) (111–121). Initially, analysis of purified eIF-2 from rabbit reticulocytes by polyacrylamide gel electrophoresis in the presence of sodium dodecyl sulfate showed that eIF-2 consisted of three nonidentical polypeptide chains with molecular weights in the range of 55,000–52,000, 50,000–48,000, and 38,000–35,000 (116, 118, 142, 143). However, the literature contains conflicting reports on the numbers and stoichiometry of the different polypeptide chains (136, 144–147). Stringer et al demonstrated that eIF-2 purified from either calf liver or rabbit reticulocyte lysates is composed of one 48,000- and one 38,000-dalton polypeptide corresponding to a M_r of 86,000 for the native factor (148,149). Using a different experimental approach, Mitsue et al (150) confirmed that the active factor consists of two subunits.

eIF-3 has been shown to be a large complex protein consisting of numerous polypeptide subunits with a total molecular mass greater than 500,000 daltons and a sedimentation value of between 15S and 18S (116, 119, 143, 151, 152). The number and stoichiometry of the polypeptides comprising eIF-3 have still not been conclusively determined although none of them seem to correspond to any of the ribosomal proteins (143). Meyer et al (153) have observed that initiation factors eIF-2, eIF-3, and eIF-5 may be cleaved into different active forms by protease acting either in the intact cell or during the factor isolation and purification. Addition of the protease inhibitor, phenylmethanesulfonyl fluoride, prevents some but not all of the proteolysis.

III. Partial Reactions Involved in the Assembly of 80S Initiation Complex Formation

1. GENERATION OF EUKARYOTIC RIBOSOMAL SUBUNIT POOL   Since initiation of protein synthesis occurs on separated subunits, a mechanism must exist for maintaining a pool of 40S and 60S subunits or for generating them from 80S ribosomes. It has been reported that initiation factor, eIF-3 ($M_r > 500,000$) isolated from rabbit reticulocyte lysates and ascites cells binds to 40S ribosomal subunits and prevents them from reassociating with 60S subunits (antiassociation activity) (255). Ribosome dissociation factor

activity has also been reported in extracts of native ribosomal subunits of rat liver (171). The rat liver ribosome dissociation activity was later resolved into two molecular species: eIF-3_H (M_r = 500,000–700,000) and eIF-3_L (M_r = 51,000) by glycerol density gradient centrifugation (256). These two protein fractions have not been purified further, and their relationship to known initiation factors remain unclear. In contrast to these results from the mammalian systems, recent work from the wheat germ system shows that a small protein, designated eIF-6 (M_r = 23,000) maintains a pool of ribosomal subunits by binding to 60S subunits and prevents them from reassociating with 40S ribosomal subunits (139, 140). The wheat germ eIF-3, a large protein complex of 11 polypeptides ranging in molecular weights from 25,000 to 120,000, fails to prevent subunits reassociation or to dissociate 80S ribosomes (257). In agreement with these later reports, it has recently been demonstrated that ribosomal subunit antiassociation activity in calf liver extracts is contained in a single protein of molecular weight 25,500 (141). There is no evidence for the presence of any other protein factor that contained ribosomal subunit antiassociation or 80S ribosome dissociating activity in calf liver extracts. The properties of the calf liver 25,500-dalton antiassociation factor is similar to wheat germ eIF-6 (141). Therefore, it is tempting to speculate that mammalian ribosome antiassociation activity is contained in the eIF-6 protein. The presence of ribosome antiassociation activity in high-molecular-weight eIF-3 may be due to association of an eIF-6-like protein with eIF-3, which the other investigators have failed to resolve from eIF-3.

2. TERNARY COMPLEX FORMATION eIF-2 specifically binds Met-tRNA$_f$ in the presence of GTP to form a ternary complex of eIF-2:-GTP·Met-tRNA$_f$ in which the molar ratio of GTP to Met-tRNA$_f$ is 1:1 (148, 149, 154–156). Nonhydrolyzable analogues of GTP can substitute for GTP in ternary complex formation, which shows that GTP plays an allosteric role in the complex formation and that GTP hydrolysis is not involved (155). The formation of a stable binary complex of eIF-2· Met-tRNA$_f$ has been reported (114). Formation of a transient eIF-2·GTP complex has been claimed to be an intermediate step in the formation of the ternary complex (114, 134), but recent studies of nucleotide binding to eIF-2 (123) have shown that eIF-2 does not form a stable complex with [γ-^{32}P]GTP. The factor, however, has a potent GDP binding activity (157, 158). Using isolated subunits of eIF-2 molecules, it was demonstrated (145) that the 48,000-dalton subunits possess Met-tRNA$_f$ and mRNA binding activities, while the smaller subunit possesses GDP binding activity. Cross-linking data also suggest that Met-tRNA$_f$ binds to the larger subunit (159).

3. TRANSFER OF THE TERNARY COMPLEX TO THE 40S RIBOSOMAL
SUBUNIT   The next step in the initiation reaction is the transfer of
eIF-2·GTP·Met-tRNA$_f$ ternary complex to 40S subunit. It has been demonstrated that all three components of the ternary complex remain bound
to subunits in 1:1:1 ratio (160, 161). However, the specific requirements
for transfer of ternary complex to 40S ribosomal subunits are not completely resolved. Several groups have observed that Met-tRNA$_f$, in the
presence of eIF-2 and GTP, binds to 40S subunits in the absence of initiation codon AUG or mRNA. This binding is stabilized by eIF-3 and eIF-1
(119, 121). Both eIF-3 and the ternary complex (eIF-2·GTP·Met-tRNA$_f$)
can bind independently to the 40S subunit, and each of the components
enhances the binding of the other. All of the polypeptides of eIF-2 and eIF-3
participate in this binding (161). Several other groups have also reported
that the transfer reaction requires an additional initiation factor fraction for
maximum binding of Met-tRNA$_f$ to 40S subunits (131, 156, 162). Majum-
der et al have, however, reported that the binding of Met-tRNA$_f$ to 40S
subunits is dependent on the presence of AUG codon (124, 130), and these
observations have been confirmed (149, 163).

4. BINDING OF mRNA   The observation that Met-tRNA$_f$ (together with
eIF-2 and GTP) binds to 40S subunits in the absence of mRNA led to the
suggestion that Met-tRNA$_f$ binding to the 40S ribosomal subunit is a stringent prerequisite for subsequent binding of mRNA. As a consequence of
this proposed mechanism, the binding of mRNA to the ribosome in the
absence of Met-tRNA$_f$ would be unlikely even in the presence of all the
other components of initiation (119, 121, 156, 164).

eIF-2 is able to bind a wide variety of RNA species, including mRNA,
rRNAs, and tRNAs, but it shows the highest affinity for eukaryotic
mRNAs (123, 165–168). Numerous reports claim eIF-2 plays a role in the
recognition and binding of mRNA (165–168). However, the exact mechanism by which eIF-2 aids in the recognition process is unclear. eIF-3, like
its prokaryotic counterpart IF-3, is essential for the translation of natural
mRNA (119, 121, 169). The factor has been shown to bind to the 40S
subunit independent of other components of initiation, including ATP and
GTP (161, 170, 171). The mechanism by which eIF-3 facilitates the translation of mRNA is not clear, but it has been suggested that when it binds to
the 40S subunit, it forms a mRNA recognition and binding site (169).

Natural mRNA contains extensive secondary structures and it has been
estimated that in the case of globin mRNA 70% of the bases are arranged
in base pairs (172). Because the secondary structure of mRNA is a conformation with a high energy barrier, the structure has to be unwound or
relaxed in order to expose the codons for translation. There is evidence that
eIF-3 may be involved in this unwinding process (169).

In addition to eIF-2 and eIF-3, three other initiation factors have been shown to have a stimulatory role in the binding of mRNA. Trachsel et al observed that Messenger binding was promoted by the factors eIF-4A, eIF-4B, and ATP (119). The omission of either of these factors resulted in a significant reduction of mRNA binding without any reduction in the amount of Met-tRNA$_f$ bound. eIF-4C has a less pronounced effect. In contrast, Benne et al have reported that omission of eIF-4A, eIF-4B, and eIF-4C caused only a marginal reduction in the binding of mRNA (121). These factors had little effect on the synthesis of methionyl-puromycin when either the triplet AUG or globin mRNA was used as messenger. A significant reduction in globin synthesis was however, observed by both groups when eIF-4A, eIF-4B, and eIF-4C were omitted from the assays. The precise mechanism of interaction and role of these factors are not clear at present.

5. THE ROLE OF THE 5' CAP IN INITIATION OF PROTEIN SYNTHESIS

The majority of eukaryotic mRNAs contain a cap structure of the general form $m^7GpppX^m_p$ at their 5' ends. There is abundant evidence that the cap moiety enhances the efficiency of initiation. The enzymatic addition of the cap structure to the 5' end of mRNA isolated from a system in which the capping process had been inhibited, increased the affinity of the mRNA for the ribosomal subunits (173). Chemical analogues of the cap structure, such as m^7GMP, m^7GDP and m^7GpppN^m (where N is any nucleotide) inhibited both the ribosome binding of capped mRNAs and their translation in cell-free extracts (174–183). This suggests that there was either a specific site on the ribosome or a factor that recognized capped mRNA and bound it to the ribosome.

The 5' cap structure may play an important role in stabilizing the binding of mRNA to 40S ribosomal subunits. As discussed previously (see section A III 3 ii) there is strong evidence that a region at the 3' end of 16S ribosomal RNA base-pairs with the polypurine stretch located on the 5' side of the AUG initiation codon of an mRNA (94, 95). Analysis of mRNA and rRNA from eukaryotes has not revealed any comparable complementary sequences. Recently Nakashima et al demonstrated the interaction between sequences at the 5' end of capped mRNA and the 3' end of 18S RNA in both 40S and 80S initiation complexes by the use of photochemical cross-linking (184). It is not clear if the cross-linking is restricted to the 3' end of the rRNA or if it occurs at many sites throughout the molecule, but it is clear that the interaction is not the same as the "Shine & Dalgarno" base-paired structure. It has been postulated that the evolution of the cap structures in higher cells may facilitate ribosome binding of mRNA and replace the mRNA:rRNA interaction that occurs in prokaryotes.

6. CAP-BINDING PROTEIN   Since the cap structure is a common feature in eukaryotic mRNA and its role in enhancing the efficiency of translation of mRNA has been established, attention was focused on the initiation factors to see if one of them could specifically recognize capped mRNA.

Initial reports that eIF-4B was the factor that specifically recognized capped mRNA (180) were discounted when Sonenberg & Shatkin by a cross-linking technique identified the protein that recognizes the cap structure of mRNA (168). This protein was purified from rabbit reticulocytes by affinity chromatography on a m^7GDP-Sepharose column and was shown to have a mol wt of 24,000 (185, 186). The purified protein exhibited functional activity by stimulating translation of capped mRNAs in extracts of HeLa cells. The translation of noncapped mRNA was unaffected (185, 186). In agreement with this finding, it was observed that monoclonal antibodies prepared against cap binding protein inhibited translation of capped mRNAs, while translation of uncapped mRNAs was unaffected (187). An attractive model for the role of cap-binding protein suggests that it first binds to the 5' end of the mRNA to form a protein·mRNA complex which, by virtue of the affinity of cap binding protein for eIF-3, then binds to 40S ribosomal subunits containing eIF-3. After initiation complex formation, the cap-binding protein is released for recycling, and in its interaction with the 5' end of the mRNA it is replaced by eIF-3 (188).

7. RECOGNITION OF THE INITIATOR CODON AUG IN EUKARYOTIC mRNAs BY 40S RIBOSOMAL SUBUNITS   Besides having a cap structure at their 5' end, eukaryotic mRNAs are generally monocistronic. The importance of the cap in ensuring that translation begins at the 5' end of the message was illustrated by Rosenberg & Paterson (189, 190). They showed that, following the addition of the cap structure to the 5' end of the polycistronic bacteriophage or bacterial mRNAs, only the 5'-proximal cistrons of the polycistronic mRNAs were translated in an in vitro eukaryotic protein synthesizing system. That naturally occurring uncapped mRNA can effectively compete with capped mRNA suggests that, even though the m^7G terminus has a major role in promoting initiation of most messages, an additional mechanism must operate in the recognition of mRNA.

Kozak & Shatkin (191) and Kozak (192) have postulated a "scanning model" for the recognition of the initiation codons in mRNAs by the 40S preinitiation complexes. The model postulates that a 40S ribosomal subunit binds at the 5' terminus of a message regardless of the sequence and subsequently advances along the mRNA until it encounters the first AUG triplet, whereupon it halts and the 60S subunit joins it to form the 80S initiation complex. Recognition of the AUG codon as a "stop signal" would presumably involve base-pairing with the anticodon of Met-tRNA$_f$, which is bound to the 40S complex prior to message attachment. The model also postulates that although the presence of a 5' cap structure enhances the efficiency of 40S binding, it seems to be the 5' end of an RNA molecule as such, and not the m⁷G moiety that comprises the primary recognition signal.

Several features characteristic of eukaryotic initiation are compatible with the scanning model. First, it has been observed that circularization of an mRNA chain by phage T_4 RNA ligase abolishes the ability of RNA molecules to form initiation complexes with eukaryotic ribosomes (193). In contrast, *E. coli* ribosomes were able to form initiation complexes with both linear and circularized mRNA molecules (193). These results suggest that the presence of a free, exposed 5' terminus on a RNA molecule is a stringent requirement for recognition by eukaryotic ribosomes. The critical requirement for a nearby 5' terminus probably accounts for the inability of eukaryotic ribosomes to initiate at internal sites in polycistronic transcripts (194). Second, except for only a few messages (195, 196) it has been observed that translation in an eukaryotic system always begins at the AUG codon closest to the 5' end of the mRNA, irrespective of the surrounding nucleotide sequences. There is also variability in the location of the first AUG codon within the 5'-proximal regions of different mRNAs (195). In this connection, the elegant genetic analysis of yeast iso-1-cytochrome *c* mutants by Sherman et al (197) are particularly noteworthy. In their analysis of second-site revertants obtained after mutational inactivation of the normal AUG initiator codon, these investigators observed that generation of a new AUG codon anywhere within a region spanning 37 nucleotides and presumably at any site preceeding and following the site of the normal initiation codon restored translation of iso-1-cytochrome *c* in vivo. Finally, there is experimental evidence that 40S ribosomes migrate from the 5' end of an mRNA to the first AUG codon, where initiation of translation begins (191, 198). In the presence of any number of initiation inhibitors (edeine, fluoride, etc), which inhibit joining of 60S subunits with the 40S initiation complexes, a large number of 40S ribosomes was found to accumulate in the 5'-flanking region preceeding initiator AUG codon of an mRNA. These results have been explained on the basis of movement of 40S ribosomes from the 5' end of the mRNA toward the initiator codon. Presumably, the presence of initiation inhibitors facilitates detection of the normal process (191, 198).

It should, however, be emphasized that there are a number of eukaryotic messages in which translation does not begin at the first AUG codon [see (260) for a more detailed discussion]. Therefore, it is apparent that the mechanism by which ribosomes select the initiating AUG codon in an mRNA is governed by other factors not considered in the model postulated by Kozak.

8. THE ROLE OF ATP IN THE BINDING OF mRNA   The requirement for ATP in eukaryotic initiation was first shown in the wheat germ system (199, 200). Later, it was shown that omission of ATP, or its substitution with the nonhydrolyzable analogue, 5'-adenylyl methylene diphosphonate, causes a strong reduction in the binding of mRNA to 40S ribosomal subunits (119). This suggests that the hydrolysis of ATP may be an important step in the binding of mRNA but the details are not clear.

It was discussed in section B III 4 that natural mRNA contains extensive secondary structures that must be relaxed in order to allow translation to occur (172). This unwinding process is involved in overcoming an energy barrier, and ATP may be required to facilitate the process. eIF-3, which has been implicated in the unwinding process may interact with ATP, but there is no evidence to support this. A possible requirement for ATP arises from the "scanning" model outlined above. The migration of the 40S ribosomal preinitiation complex from the 5' end of the mRNA to the distal AUG initiation codon may require energy, and this could be provided by the hydrolysis of ATP.

9. THE FORMATION OF 80S INITIATION COMPLEXES   The junction of the 40S initiation complex with the 60S ribosomal subunit results in the formation of the 80S initiation complex. Unlike the prokaryotic system, this step requires the participation of a specific joining factor called eIF-5 (112, 118, 123, 201, 202). This factor has been purified from rabbit reticulocyte lysates and has been shown to be a single polypeptide chain ranging in size from 125,000 to 160,000 daltons. eIF-5 has no effect on the formation of 40S initiation complexes, but its requirement is absolute in the formation of 80S complexes. The joining reaction catalyzed by eIF-5 requires the hydrolysis of GTP. Nonhydrolyzable analogues of GTP, such as 5'-guanyl-yl methylene diphosphonate or the corresponding imido analogues are inhibitory [(119), A. Chaudhuri and U. Maitra, unpublished results.] During the joining reaction, eIF-5 mediates the release of initiation factors eIF-2 and eIF-3 from the 40S initiation complex, provided GTP is present in a hydrolyzable form (160, 202). In the absence of 60S ribosomal subunits, eIF-5 caused a reduction in the amount of Met tRNA$_f$ bound to the 40S initiation complexes. It was proposed that, when eIF-2 and eIF-3 are released from the 40S initiation complex, a metastable complex containing only Met-tRNA$_f$ results, which, in the presence of 60S ribosomal subunits, is active in the formation of 80S complexes. Once eIF-2 and eIF-3 were released from the 40S initiation complex, no further requirement for eIF-5 could be demonstrated either in the junction with the 60S ribosomal subunit or in the formation of puromycin-reactive 80S initiation complexes (202). Chaudhuri et al, using pure 40S ribosomal initiation complexes isolated by

sucrose density centrifugation, have also recently shown that 60S subunits are not required in the reaction causing the destabilization of the 40S ribosomal initiation complexes (A. Chaudhuri and U. Maitra, unpublished results). The 40S initiation complexes made in the absence of GTP were destabilized by eIF-5. This casts doubt on the claim that GTP hydrolysis is necessary for the destabilization reaction. When GTP was present, only the hydrolyzable form allowed the reaction to proceed (A. Chaudhuri and U. Maitra, unpublished results).

Several investigators have demonstrated that eIF-5 was able to hydrolyze the GTP in the 40S initiation complexes without the 60S ribosomal subunit being present (202, 204). Whether the eIF-2 in the 40S initiation complex plays a cooperative role with eIF-5 in the hydrolysis of GTP is not known.

Two additional initiation factors, eIF-4C (17,000 daltons) and eIF-4D (15,000 daltons), have been reported to take part in the joining reaction. It has been suggested that eIF-4C is partly responsible for the formation of Met-tRNA$_f$·80S complexes in the absence of mRNA binding. It is thought that eIF-4C stabilizes Met-tRNA$_F$·40S complexes, possibly by changing the conformation of the 40S subunits to allow them to interact with the 60S ribosomal subunits (119, 121, 205). eIF-4D has no effect on the binding of Met-tRNA$_f$ or mRNA to 40S or 80S ribosomes. The main role of eIF-4D seems to be in ensuring that Met-tRNA$_f$ is bound in the correct position on 80S initiation complexes, as it gives a three-fold stimulation of methionyl puromycin synthesis (121). It is possible that eIF-4C, eIF-4D, and eIF-5 may interact and enhance each other's activities. eIF-4C and eIF-4D may enhance the purported catalytic activity of eIF-5 (119) and all three factors may be necessary for maximum puromycin reactivity.

IV. The Role of Initiation Factors in Regulation of Translation

1) PROTEIN KINASES SPECIFIC FOR eIF-2   In rabbit reticulocyte lysates, protein synthesis is dependent on the presence of heme, the prosthetic group of hemoglobin. The rate of protein synthesis suddenly drops to a low level and almost ceases after a few minutes of incubation in the absence of heme. This cessation of synthesis was shown to be due to the activation of an inhibitor from an inactive precursor or proinhibitor. The active form of the inhibitor, designated heme-controlled inhibitor (HCI) or heme-controlled repressor (HCR) has been purified and found to possess a cAMP-independent protein kinase activity that ultilizes ATP to phosphorylate the 38,000-mol wt (α) subunit of eIF-2 (206–210). A detailed account of this and other translational inhibitors have been reviewed (206, 207).

Treatment of hemin-containing rabbit reticulocyte lysates with oxidized glutathione (211–213) or with double-stranded RNA (206, 213, 214) also causes inhibition of translation, due to activation of a translational inhibitor (dsI). Addition of double-stranded RNA to interferon-treated cells also activates the dsI (215–219). Purified dsI also possesses a cAMP-independent protein kinase activity that phosphorylates the α subunit of eIF-2 (206, 213, 214). Both HRI and dsI seem to phosphorylate the same site (220, 221).

2) THE MECHANISM OF INHIBITION OF eIF-2 ACTIVITY BY PHOSPHO-RYLATION   After considerable work, the details of how phosphorylation of eIF-2 causes inhibition of translation are still not clear. Phosphorylation of eIF-2 with HCI and ATP does not prevent it from forming ternary complexes or 40S and 80S initiation complexes (160, 203, 222). In addition, release of eIF-2 following 80S initiation complexes also seem to be unaffected by phosphorylation of eIF-2. A protein factor recently identified in reticulocyte lysates, termed stabilizing protein factor (SF), stabilizes ternary complex formation by eIF-2 in the presence of Mg^{2+} (223). SF is necessary for HCI and ATP to inhibit eIF-2 activity (223). The results of Gupta and his co-workers are similar in that phosphorylated eIF-2 is unable to interact with either co-eIF-2C, which stimulates ternary complex formation in the presence of 1 mm Mg^{2+}, or with co-eIF-2B, which dissociates ternary complexes in the presence of 5 mm Mg^{2+} (224, 225). deHaro and Ochoa have observed that their eIF-2 stimulatory protein (ESP) stimulates ternary complex formation by unmodified eIF-2 and not by phosphorylated eIF-2 (133, 226). Recently Ohtsuki et al have shown that, when both eIF-2 and a 65,000-dalton ribosomal protein are phosphorylated by interferon-induced kinase (dsI), phosphorylated eIF-2 is no longer able to make ternary complexes or transfer Met-tRNA to 40S initiation complexes when the phosphorylated 65,000-dalton protein is also present (227).

The reports that inhibition due to HCI could be overcome by the addition of excess eIF-2 (228–230) have been invalidated by the observations that a protein factor, isolated from both supernatant and ribosomes of rabbit reticulocytes, could overcome the inhibition of translation in heme-deficient lysates. eIF-2, free of this factor, had no effect (231–236). The mode of action of rescue factor (RF) is not clear. RF, purified from supernatant, does not dephosphorylate phosphorylated eIF-2, nor does it prevent phosphorylation of eIF-2 by HRI in a fractionated in vitro system (234). A phosphatase activity that dephosphorylates phosphorylated eIF-2 has recently been purified from reticulocyte lysates (237).

It is clear from these results that phosphorylation of eIF-2 has little effect on the action of the purified factor in ternary complex formation or in 40S initiation complex formation in vitro. The inhibitory effects appear to result from the inability of the phosphorylated factor to interact with additional protein factors that enhance or modulate, but do not seem to be essential for the functioning of eIF-2. Whether these additional factors are true initiation factors, cofactors, or ribosomal proteins is still not known.

Recent reports (238) suggest that alterations of the redox state of eIF-2 should also be considered in the regulation of eIF-2 activity. It has been proposed that oxidation of the thiol groups of eIF-2 may cause conformational changes in the protein molecule that inhibit recycling of phosphorylated eIF-2 in the initiation reaction thus causing shut-off of protein synthesis.

3. INITIATION FACTORS INVOLVED IN DISCRIMINATION AND SELECTION OF mRNAs Differential rates of initiation of translation of mRNAs may play an important role in regulation of protein synthesis in animal cells. Recent studies (259) on translation of reovirus mRNAs in SC-1-infected cells clearly show that the major reovirus mRNAs are poor initiators of translation relative to host mRNAs and that this is probably due to low affinity of reoviral mRNAs for an initiation component. This indicates that competition among mRNAs for binding to initiation component(s) may be an important step in regulating the rate at which they are translated. Several reports stemming from the original observations of Wigle & Smith (239) show that eIF-4A and eIF-4B may be involved in the preferential translation of one species of mRNA over another (240–243). Different mRNAs may have different affinities for these factors or the ribosomal binding sites that affect the rate at which they are translated (244, 245). In cell-free translating systems, optimal translation of a particular species of mRNA in a mixture of mRNA molecules can be obtained by increasing the concentration of eIF-4A (117, 242, 246) or eIF-4B (240, 243, 247). eIF-4B preparations frequently contain cap-binding protein as a contaminant, thus discriminating activity may be mediated through the cap-binding protein and not through eIF-4B.

Picornavirus infection inhibits the translation of host cell mRNA at the level of initiation without causing degradation or modification of mRNA. Extracts from poliovirus-infected HeLa cells do not translate capped vesicular stomatitis virus (VSV) mRNA, while uncapped poliovirus mRNA is translated with a high efficiency (246, 248). Initially it was found that preparations of eIF-4B could restore the translation of VSV mRNA (249, 250), but the "restoring activity" has subsequently been purified from the salt-wash proteins of rabbit reticulocytes and shown to be associated with a protein complex that included cap binding protein and several high-molecular-weight polypeptides (251, 258). Thus it appears that polio virus may inhibit cellular protein synthesis by inactivation of the protein involved in capped mRNA recognition. The mode of inactivation is not known.

Heywood et al (252) have separated eIF-3 from embryonic chick muscle into "core" and "discriminatory" polypeptides. The core eIF-3 is capable of translating myosin and rabbit globin mRNA with equal efficiency, but addition of the discriminatory polypeptides results in preferential translation of myosin mRNA (253, 254). Modulation of the activity of these discriminatory protein could provide a means by which the translation of specific mRNA molecules or classes of mRNAs could be altered during differentiation.

V. CONCLUSION

It is clear from the above discussion that although there are basic similarities between the process of initiation in prokaryotes and eukaryotes, the details of the steps involved are much more complex in eukaryotes.

In the prokaryotic system, a series of discrete reactions can be defined in which mRNA, ribosome, initiator tRNA, and protein factors interact to give rise to a functional 70S-polypeptide chain initiation complex that can engage in peptide bond formation. This is in contrast to the situation in eukaryotes where, even though the overall pathway for the formation of functional 80S-initiation complexes is known, the details of the mechanisms of action of the initiation factors are far from clear. Some basic facts still have to be established in the eukaryotic system. It is still not clear how many initiation factors are necessary for formation of functional 80S-initiation complexes and whether the factors are reused in a catalytic fashion. Recycling of initiation factors is an essential feature of such a system and present evidence suggests that a component (or components) is limiting or lacking in the in vitro system.

ATP is not required at any step during initiation complex formation in the prokaryotic system, but in eukaryotes, formation of 40S-initiation complexes with natural mRNA requires the hydrolysis of ATP. The function of ATP hydrolysis is not known, but it is presumably involved in the recognition of mRNAs by the 40S-ribosomes. It is also not known how eukaryotic ribosomes select the correct initiation site on a mRNA molecule with such precision that translation can start at the AUG codon corresponding to the beginning of a cistron.

It is clear that the mode of action of various regulatory elements that control eukaryotic initiation cannot be elucidated until the details of the molecular mechanisms of the initiation factors are understood.

ACKNOWLEDGMENT

The authors experimental work described in this review was supported by National Institutes of Health Grant GM-15399. The authors wish to thank Mr. Pradip Raychaudhuri for considerable help during the preparation of the manuscript.

Literature Cited

1. Miller, D. M., Weissbach, H. 1977. In *Molecular Mechanisms of Protein Biosynthesis,* ed. H. Weissbach, S. Pestka, pp. 323–73. New York: Academic
2. Brot, N. 1977. See Ref. 1, pp. 375–411
3. Harris, R. J., Pestka, S. 1977. See Ref. 1, pp. 413–42
4. Caskey, C. T. 1977. See Ref. 1, pp. 443–65
5. Haselkorn, R., Rothman-Denes, L. B. 1973. *Ann. Rev. Biochem.* 42:397–438
6. Lengyel, P. 1974. In *Ribosome,* ed. M. Nomura, A. Tissières, P. Lengyel, pp. 13–52. Cold Spring Harbor, New York: Cold Spring Harbor Lab.
7. Ochoa, S., Mazumder, R. 1974. *The Enzymes* 10:1–51
8. Revel, M. 1977. See Ref. 1, pp. 245–321
9. Grunberg-Manago, M., Gros, F. 1977. *Progr. Nucl. Acid. Res. Mol. Biol.* 20: 209–84
10. Iwasaki, K., Sabol, S., Wabha, A. J., Ochoa, S. 1968. *Arch. Biochem. Biophys.* 125:542–47
11. Revel, M., Herzberg, M., Becarevic, A., Gros, F. 1968. *J. Mol. Biol.* 33:231–49
12. Maitra, U., Dubnoff, J. S. 1968. *Fed. Proc.* 27:398
13. Chae, Y. B., Mazumder, R., Ochoa, S. 1969. *Proc. Natl. Acad. Sci. USA* 62: 1181–88
14. Herzberg, M., Lelong, J. C., Revel, M. 1969. *J. Mol. Biol.* 44:297–308
15. Hershey, J. W. B., Remold-O&Donnell, E., Kolakofsky, D., Dewey, K. F., Thach, R. E. 1971. *Methods Enzymol.* 20:235–47
16. Dubnoff, J. S., Maitra, U. 1971. *Methods Enzymol.* 20:248–61
17. Dubnoff, J. S., Lockwood, A. H., Maitra, U. 1972. *Arch. Biochem. Biophys.* 149:528–40
18. Sabol, S., Sillero, M. A. G., Iwasaki, K., Ochoa, S. 1970. *Nature* 228:1269–73
19. Revel, M., Aviv, H., Groner, Y., Pollack, Y. 1970. *FEBS Lett.* 9:213–17
20. Wabha, A. J., Miller, M. J. 1974. *Methods Enzymol.* 30:3–18
21. Hershey, J. W. B., Dewey, K. F., Thach, R. E. 1969. *Nature* 222:944–47
22. Lee-Huang, S., Sillero, M. A. G., Ochoa, S. 1971. *Eur. J. Biochem.* 18: 536–43
23. Dondon, J., Godefroy-Colburn, T., Graffe, M., Grunberg-Manago, M. 1974. *FEBS Lett.* 45:82–87
24. Hershey, J. W. B., Yanov, J., Johnston, K., Fakunding, J. L. 1977. *Arch. Biochem. Biophys.* 182:626–38
25. Kolakofsky, D., Dewey, K. F., Hershey, J. W. B., Thach, R. E. 1968. *Proc. Natl. Acad. Sci. USA* 61:1066–70
26. Lelong, J. C., Grunberg-Manago, M., Dondon, J., Gros, D., Gros, F. 1970. *Nature* 226:505–10
27. Chae, Y. B., Mazumder, R., Ochoa, S. 1969. *Proc. Natl. Acad. Sci. USA* 63: 828–33
28. Dubnoff, J. S., Maitra, U. 1972. *J. Biol. Chem.* 247:2876–83
29. Lockwood, A. H., Maitra, U., Brot, N., Weissbach, H. 1974. *J. Biol. Chem.* 249: 1213–18
30. Dubnoff, J. S., Maitra, U. 1971. *Proc. Natl. Acad. Sci. USA* 68:318–23
31. Naaktgeboren, N., Roobol, K., Voorma, H. O. 1977. *Eur. J. Biochem.* 72:49–56
32. Kolakofsky, D., Dewey, K., Thach, R. E. 1969. *Nature* 223:694–97
33. Fakunding, J. L., Traugh, J. A., Traut, R. R., Hershey, J. W. B. 1972. *J. Biol. Chem.* 247:6365–67
34. Miller, M. J., Wabha, A. J. 1973. *J. Biol. Chem.* 248:1084–90
35. Krauss, S. W., Leder, P. 1975. *J. Biol. Chem.* 250:3752–58
36. Davis, B. D. 1971. *Nature* 231:153–57
37. Noll, M., Hapke, B., Schreier, M. H., Noll, H. 1973. *J. Mol. Biol.* 75:281–94
38. Subramanian, A. R., Davis, B. D., Beller, R. J. 1969. *Cold Spring Harbor Symp. Quant. Biol.* 34:223–30
39. Subramanian, A. R., Davis, B. D. 1970. *Nature* 228:1273–75
40. Albrecht, J., Stap, F., Voorma, H. O., Van Knippenberg, P. H., Bosch, L. 1970. *FEBS Lett.* 6:297–301
41. Miall, S. H., Tamaoki, T. 1972. *Biochemistry* 11:4826–30

42. Godefroy-Colburn, T., Wolfe, A. D., Dondon, J., Grunberg-Manago, M., Dessen, P., Pantaloni, D. 1975. *J. Mol. Biol.* 94:461–78

43. Baan, R. A., Duijfjes, J. J., van Leerdam, E., Van Knippenberg, P. H., Bosch, L. 1976. *Proc. Natl. Acad. Sci. USA* 73:702–6

44. Noll, M., Noll, H. 1972. *Nature New Biol.* 238:225–28

45. Dottavio-Martin, D., Suttle, D. P., Ravel, J. M. 1979. *FEBS Lett.* 97: 105–10

46. Sabol, S., Ochoa, S. 1971. *Nature New Biol.* 234:233–36

47. Pon, C. L., Friedman, S. M., Gualerzi, C. 1972. *Mol. Gen. Genet.* 116:192–98

48. Thibault, J., Chestier, A., Vidal, D., Gros, F. 1972. *Biochimie* 54:829–35

49. Sabol, S., Meier, D., Ochoa, S. 1973. *Eur. J. Biochem.* 33:332–40

50. Gottlieb, M., Davis, B. D., Thompson, R. C. 1975. *Proc. Natl. Acad. Sci. USA* 72:4238–42

51. Chaires, J. B., Pande, C., Wishnia, A. 1981. *J. Biol. Chem.* 256:6600–7

52. Lockwood, A. H., Sarkar, P., Maitra, U. 1972. *Proc. Natl. Acad. Sci. USA* 69: 3602–5

53. Fakunding, J. L., Hershey, J. W. B. 1973. *J. Biol. Chem.* 248:4206–12

54. Benne, R., Naaktgeboren, N., Gubbens, J., Voorma, H. O. 1973. *Eur. J. Biochem.* 32:372–80

55. Chu, J., Mazumder, R. 1974. *FEBS Lett.* 40:335–38

56. Vermeer, C., Boon, J., Talens, A., Bosch, L. 1973. *Eur. J. Biochem.* 40: 283–93

57. Vermeer, C., deKievit, R. J., van Alphen, W. J., Bosch, L. 1973. *FEBS Lett.* 31:273–76

58. Grunberg-Manago, M., Godefroy-Colburn, T., Wolfe, A. D., Dessen, P., Pantaloni, D., Springer, M., Graffe, M., Dondon, J., Kay, A. 1973. In *Regulation of Transcription and Translation in Eukaryotes,* ed. E. Bautz, pp. 213–49. Berlin/New York: Springer

59. Benne, R., Ebes, F., Voorma, H. O. 1973. *Eur. J. Biochem.* 38:265–73

60. Langberg, S., Kahan, L., Traut, R. R., Hershey, J. W. B. 1977. *J. Mol. Biol.* 117:307–19

61. Bollen, A., Heimark, R. L., Cozzone, A., Traut, R. R., Hershey, J. W. B. 1975. *J. Biol. Chem.* 250:4310–14

62. Hawley, D. A., Slobin, L. I., Wahba, A. J. 1974. *Biochem. Biophys. Res. Commun.* 61:544–50

63. Van Duin, J., Kurland, C. G., Dondon,

J., Grunberg-Manago, M. 1975. *FEBS Lett.* 59:287–90

64. Heimark, R. L., Kahan, L., Johnston, K., Hershey, J. W. B., Traut, R. R. 1976. *J. Mol. Biol.* 105:219–30

65. Van Duin, J., Kurland, C. G., Dondon, J., Grunberg-Manago, M., Branlant, C., Ebel, J. P. 1976. *FEBS Lett.* 62:111–14

66. Czernilofsky, A. P., Kurland, C. G., Stoffler, G. 1975. *FEBS Lett.* 58:281–84

67. Greenshpan, H., Revel, M. 1969. *Nature* 224:331–35

68. Van der Hofstad, G. A. J. M., Foekens, J. A., Van den Elsen, P. J., Voorma, H. O. 1976. *Eur. J. Biochem.* 66:181–92

69. Leder, P., Nau, M. 1967. *Proc. Natl. Acad. Sci. USA* 58:774–81

70. Noll, M., Noll, H. 1974. *J. Mol. Biol.* 90:237–51

71. Jay, G., Kaempfer, R. 1974. *Prod. Natl. Acad. Sci. USA* 71:3199–3203

72. Jay, G., Kaempfer, R. 1975. *J. Biol. Chem.* 250:5742–48

73. Steitz, J. A. 1978. In *Biological Regulation and Control,* ed. R. Goldberger, 1:349–99. New York/London: Plenum

74. Steitz, J. A., Sprague, K. U., Steege, D. A., Yuan, R. C., Laughrea, M., Moore, P. B., Wabha, A. J. 1977. In *Nucleic Acid-Protein Recognition,* ed. H. J. Vogel, pp. 491–508. New York: Academic

75. Thach, S., Thach, R. E. 1971. *Nature New Biol.* 229:219–21

76. Dubnoff, J. S., Lockwood, A. H., Maitra, U. 1972. *J. Biol. Chem.* 247: 2884–94

77. Mazumder, R. 1972. *Proc. Natl. Acad. Sci. USA* 69:2770–73

78. Stringer, E. A., Sarkar, P., Maitra, U. 1977. *J. Biol. Chem.* 252:1739–44

79. Mazumder, R. 1971. *FEBS Lett.* 18: 64–66

80. Groner, Y., Revel, M. 1971. *Eur. J. Biochem.* 22:144–52

81. Mazumder, R., Chae, Y. B., Ochoa, S. 1969. *Proc. Natl. Acad. Sci. USA* 63:98–103

82. Rudland, P. S., Whybrow, W. A., Clark, B. F. C. 1971. *Nature New Biol.* 231:76–78

83. Groner, Y., Revel, M. 1973. *J. Mol. Biol.* 74:407–10

84. Petersen, H. U., Kruse, T. A., Worm-Leonhard, H., Siboska, G. E., Clark, B. F. C., Boutorin, A., Remy, P., Ebel, J. P., Dondon, J., Grunberg-Manago, M. 1981. *FEBS Lett.* 128:161–65

85. Majumdar, A., Bose, K. K., Gupta, N. K., Wabha, A. J. 1976. *J. Biol. Chem.* 251:137–40

86. Sundari, R. M., Stringer, E. A., Schul-

man, L. H., Maitra, U. 1976. *J. Biol. Chem.* 251:3338–45

87. Van der Hofstad, G. A. J. M., Foekens, J. A., Bosch, L., Voorma, H. O. 1977. *Eur. J. Biochem.* 77:69–75

88. Clark, B. F. C., Marcker, K. A. 1966. *J. Mol. Biol.* 17:394–406

89. Leder, P., Bursztyn, H. 1966. *Cold Spring Harb. Symp. Quant. Biol.* 31:297–301

90. Delk, A. S., Rabinowitz, J. C. 1974. *Nature* 252:106–9

91. Samuel, C. E., Rabinowitz, J. C. 1974. *J. Biol. Chem.* 249:1198–1206

92. Delk, A. S., Rabinowitz, J. C. 1975. *Proc. Natl. Acad. Sci. USA* 72:528–30

93. Baumstark, B. R., Spremulli, L. L., RajBhandary, U. L., Brown, G. M. 1977. *J. Bact.* 129:457–71

94. Shine, J., Dalgarno, L. 1974. *Proc. Natl. Acad. Sci. USA* 71:1342–46

95. Steitz, J. A., Jakes, K. 1975. *Proc. Natl. Acad. Sci. USA* 72:4734–38

96. Berrisi, H., Groner, Y., Revel, M. 1971. *Nature New Biol.* 234:44–47

97. Vermeer, C., van Alphen, W., van Knippenberg, P. H., Bosch, L. 1973. *Eur. J. Biochem.* 40:295–308

98. Jay, E., Seth, A. K., Jay, G. 1980. *J. Biol. Chem.* 255:3809–12

99. Fiers, W., Contreras, R., Duerinck, F., Haegeman, G., Iserentant, D., Merregaert, J., Min Jou, W., Molemans, F., Raeymaekers, A., Van den Berghe, A., Volckaert, G., Ysebaert, M. 1976. *Nature* 260:500–7

100. Dahlberg, A. E., Dahlberg, J. E. 1975. *Proc. Natl. Acad. Sci. USA* 72:2940–44

101. Mackeen, L. A., Kahan, L., Wabha, A., Schwartz, I. 1980. *J. Biol. Chem.* 255:10526–31

102. Ehresmann, C., Stiegler, P., Mackie, G. A., Zimmermann, R. A., Ebel, J. P., Fellner, P. 1975. *Nucleic Acids Res.* 2:265–78

103. Szer, W., Hermoso, J. M., Boublik, M. 1976. *Biochem. Biophys. Res. Commun.* 70:957–64

104. Branlant, C., Sri Widada, J., Krol, A., Ebel, J. P. 1976. *Nucleic Acids Res.* 3:1671–87

105. Thach, S. S., Thach, R. E. 1971. *Proc. Natl. Acad. Sci. USA* 68:1791–95

106. Sarkar, P., Stringer, E. A., Maitra, U. 1974. *Proc. Natl. Acad. Sci. USA* 71:4986–90

107. Fakunding, J. L., Traut, R. R., Hershey, J. W. B. 1973. *J. Biol. Chem.* 248:8555–59

108. Kay, A., Sander, G., Grunberg-Manago, M. 1973. *Biochem. Biophys. Res. Commun.* 51:979–86

109. Mazumder, R. 1973. *Proc. Natl. Acad. Sci. USA* 70:1939–42

110. Heimark, R. L., Hershey, J. W. B., Traut, R. R. 1976. *J. Biol. Chem.* 251:7779–84

111. Merrick, W. C., Anderson, W. F. 1975. *J. Biol. Chem.* 250:1197–1206

112. Merrick, W. C., Kemper, W. M., Anderson, W. F. 1975. *J. Biol. Chem.* 250:5556–62

113. Kemper, W. M., Berry, K. W., Merrick, W. C. 1976. *J. Biol. Chem.* 251:5551–57

114. Safer, B., Adams, S. L., Anderson, W. F., Merrick, W. C. 1975. *J. Biol. Chem.* 250:9076–82

115. Safer, B., Anderson, W. F., Merrick, W. C. 1975. *J. Biol. Chem.* 250:9067–75

116. Safer, B., Adams, S. L., Kemper, W. M., Berry, K. W., Lloyd, M., Merrick, W. C. 1976. *Proc. Natl. Acad. Sci. USA* 73:2584–86

117. Staehelin, T., Trachsel, H., Erni, B., Boschetti, A., Schreier, M. H. 1975. *FEBS Proc. Meet.* 10:309–23

118. Schreier, M. H., Erni, B., Staehelin, T. 1977. *J. Mol. Biol.* 116:727–53

119. Trachsel, H., Erni, B., Schreier, M. H., Staehelin, T. 1977. *J. Mol. Biol.* 116:755–67

120. Safer, B., Anderson, W. F. 1978. *Crit. Rev. Biochem.* 5:261–90

121. Benne, R., Hershey, J. W. B. 1978. *J. Biol. Chem.* 253:3078–87

122. Anderson, W. F., Bosch, L., Cohn, W. E., Lodish, H., Merrick, W. C., Weissbach, H., Wittmann, H. G., Wool, I. G. 1977. *FEBS Lett.* 76:1–10

123. Chaudhuri, A., Stringer, E. A., Valenzuela, D., Maitra, U. 1981. *J. Biol. Chem.* 256:3988–94

124. Majumdar, A., Dasgupta, A., Chatterjee, B., Das, H. K., Gupta, N. K. 1979. *Methods Enzymol.* 60:35–52

125. Roy, R., Ghosh-Dastidar, P., Das, A., Yaghmai, B., Gupta, N. K. 1981. *J. Biol. Chem.* 256:4719–22

126. Dasgupta, A., Das, A., Roy, R., Ralston, R., Majumdar, A., Gupta, N. K. 1978. *J. Biol. Chem.* 253:6054–59

127. Dasgupta, A., Majumdar, A., George, A. D., Gupta, N. K. 1976. *Biochem. Biophys. Res. Commun.* 71:1234–41

128. Ghosh-Dastidar, P., Giblin, D., Yaghmai, B., Das, A., Das, H. K., Parkhurst, L., Gupta, N. K. 1980. *J. Biol. Chem.* 255:3826–29

129. Ghosh-Dastidar, P., Yaghmai, B., Das, A., Das, H. K., Gupta, N. K. 1980. *J. Biol. Chem.* 255:365–68

130. Gupta, N. K., Chatterjee, B., Chen, Y. C., Majumdar, A. 1975. *J. Biol. Chem.* 250:853–62

131. Majumdar, A., Roy, R., Das, A., Dasgupta, A., Gupta, N. K. 1977. *Biochem. Biophys. Res. Commun.* 78:161–69
132. Das, A., Gupta, N. K. 1977. *Biochem. Biophys. Res. Commun.* 78:1433–41
133. de Haro, C., Ochoa, S. 1979. *Proc. Natl. Acad. Sci. USA* 75:2713–16
134. de Haro, C., Ochoa, S. 1979. *Proc. Natl. Acad. Sci. USA* 76:2163–64
135. Kaempfer, R., Rosen, H., Israeli, R. 1978. *Proc. Natl. Acad. Sci. USA* 75:640–54
136. Barrieux, A., Rosenfeld, M. 1978. *J. Biol. Chem.* 253:6311–14
137. Malathi, V. G., Mazumder, R. 1979. *Biochem. Biophys. Res. Commun.* 89:585–90
138. Russell, D. W., Spremulli, L. L. 1978. *J. Biol. Chem.* 253:6647–49
139. Russell, D. W., Spremulli, L. L. 1979. *J. Biol. Chem.* 254:8796–800
140. Russell, D. W., Spremulli, L. L. 1980. *Arch. Biochem. Biophys.* 201:518–26
141. Valenzuela, D., Chaudhuri, A., Maitra, U. 1981. *Fed. Proc.* 40:1749
142. Tahara, S., Traugh, J., Safer, B., Merrick, W. C. 1976. *Fed. Proc.* 35:1515
143. Benne, R., Wong, C., Leudi, M., Hershey, J. W. B. 1976. *J. Biol. Chem.* 251:7675–81
144. Spremulli, L. L., Walthall, B. J., Lax, S. R., Ravel, J. M. 1977. *Arch. Biochem. Biophys.* 178:565–75
145. Barrieux, A., Rosenfeld, M. 1977. *J. Biol. Chem.* 252:3843–47
146. Lloyd, M., Osborne, J. C., Merrick, W. C. 1977. *Fed. Proc.* 36:869
147. Baan, R. A., Keller, P. B., Dahlberg, A. E. 1981. *J. Biol. Chem.* 256:1063–66
148. Stringer, E. A., Chaudhuri, A., Maitra, U. 1979. *J. Biol. Chem.* 254:6845–48
149. Stringer, E. A., Chaudhuri, A., Valenzuela, D., Maitra, U. 1980. *Proc. Natl. Acad. Sci. USA* 77:3356–59
150. Mitsui, K., Datta, A., Ochoa, S. 1981. *Proc. Natl. Acad. Sci. USA* 78:4128–32
151. Emanulov, I., Sabatini, D., Lake, J., Freienstein, C. 1978. *Proc. Natl. Acad. Sci. USA* 75:1389–93
152. Spremulli, L. L., Walthall, B. J., Lax, S. R., Ravel, J. M. 1979. *J. Biol. Chem.* 254:143–48
153. Meyer, L., Brown-Luedi, M., Corbett, S., Tolan, D. R., Hershey, J. W. B. 1981. *J. Biol. Chem.* 256:351–56
154. Ranu, R. S., Wool, I. G. 1976. *J. Biol. Chem.* 251:1926–35
155. Dettman, G. L., Stanley, W. M. 1972. *Biochim. Biophys. Acta* 287:124–33
156. Levin, D. H., Kyner, D., Acs, G. 1973. *Proc. Natl. Acad. Sci. USA* 70:41–45
157. Stringer, E. A., Chaudhuri, A., Maitra, U. 1977. *Biochem. Biophys. Res. Commun.* 76:586–92
158. Walton, G. M., Gill, G. N. 1976. *Biochem. Biophys. Acta* 418:195–203
159. Nygard, O., Westermann, P., Hultin, T. 1980. *FEBS Lett.* 113(1):125–28
160. Trachsel, H., Staehelin, T. 1978. *Proc. Natl. Acad. Sci. USA* 75:204–8
161. Peterson, D. T., Merrick, W. C., Safer, B. 1979. *J. Biol. Chem.* 254:2509–16
162. Pinphanichakarn, P., Kramer, G., Hardesty, B. 1976. *Biochem. Biophys. Res. Commun.* 73:625–31
163. Smith, K. E., Richards, A. C., Arnstein, H. R. V. 1976. *Eur. J. Biochem.* 62:243–55
164. Darnbrough, C., Legon, S., Hunt, T., Jackson, R. 1973. *J. Mol. Biol.* 76:379–403
165. Kaempfer, R., Hollender, R., Abrams, W., Israeli, R. 1978. *Proc. Natl. Acad. Sci. USA* 75:209–13
166. Di Segni, G. D., Rosen, H., Kaempfer, R. 1979. *Biochemistry* 18:2847–54
167. Hellerman, J. G., Shafritz, D. A. 1975. *Proc. Natl. Acad. Sci. USA* 72:1021–25
168. Sonenberg, N., Shatkin, A. J. 1978. *J. Biol. Chem.* 253:6630–32
169. Ilan, J., Ilan, J. 1976. *J. Biol. Chem.* 251:5718–25
170. Benne, R., Hershey, J. W. B. 1976. *Proc. Natl. Acad. Sci.* 73:3005–9
171. Thompson, H. A., Sadnik, I., Scheinbuks, J., Maldave, K. 1977. *Biochemistry* 16:2221–30
172. Holder, J. W., Lingrel, J. B. 1975. *Biochemistry* 14:4209–15
173. Muthukrishnan, S., Moss, B., Cooper, J. A., Maxwell, E. S. 1978. *J. Biol. Chem.* 253:1710–15
174. Filipowicz, W., Furuichi, Y., Sierra, J. M., Muthukrishnan, S., Shatkin, A. J., Ochoa, S. 1976. *Proc. Natl. Acad. Sci. USA* 73:1559–63
175. Asselbergs, F. A. M., Peters, W., Van Venrooij, W. J., Bloemendal, H. 1978. *Eur. J. Biochem.* 88:483–88
176. Canaani, D., Revel, M., Groner, Y. 1976. *FEBS Lett.* 64:326–31
177. Groner, Y., Grasfeld, H., Littaur, U. Z. 1976. *Eur. J. Biochem.* 71:281–93
178. Hickey, E. D., Weber, L. A., Baglioni, C. 1976. *Proc. Natl. Acad. Sci. USA* 73:19–23
179. Roman, R. J. D., Booker, S. N., Marcus, A. 1976. *Nature* 260:359–60
180. Shafritz, D. A., Weinstein, J. A., Safer, B., Merrick, W. C., Weber, L. A., Hickey, E. D., Baglioni, C. 1976. *Nature* 261:291–94
181. Suzuki, H. 1977. *FEBS Lett.* 79:11–14

182. Weber, L. A., Feman, E. R., Hickey, E. D., Williams, M. C., Baglioni, C. 1976. *J. Biol. Chem.* 251:5657–62
183. Willems, M., Wieringa, B. E., Mulder, J., Geert, A. B., Gruber, M. 1979. *Eur. J. Biochem.* 93:469–79
184. Nakashima, K., Darzynklewicz, E., Shatkin, A. J. 1980. *Nature* 286:226–30
185. Sonenberg, N., Rupprecht, K. M., Hecht, S. M., Shatkin, A. J. 1979. *Proc. Natl. Acad. Sci. USA* 76:4345–49
186. Sonenberg, N., Trachsel, H., Hecht, S. M., Shatkin, A. J. 1980. *Nature* 285:331–33
187. Sonenberg, N., Skup, D., Trachsel, H., Millward, S. 1981. *J. Biol. Chem.* 256:4138–41
188. Sonenberg, N., Morgan, M. A., Testa, D., Colonno, R. J., Shatkin, A. J. 1979. *Nucleic Acid Res.* 7:15–29
189. Paterson, B. M., Rosenberg, M. 1979. *Nature* 279:692–96
190. Rosenberg, M., Paterson, B. M. 1979. *Nature* 279:696–701
191. Kozak, M., Shatkin, A. J. 1978. *J. Biol. Chem.* 253:6568–77
192. Kozak, M. 1978. *Cell* 15:1109–23
193. Kozak, M. 1979. *Nature* 280:82–85
194. Smith, A. E., Kamen, R., Mangel, W. F., Shure, H., Wheeler, T. 1976. *Cell* 9:481–87
195. Kozak, M. 1980. *Cell* 22:7–8
196. Hagenbüchle, O., Tosi, M., Schibler, U., Bovey, R., Wellauer, P. K., Young, Y. A. 1981. *Nature* 289:643–46
197. Sherman, F., Stewart, J. W., Schweingruber, A. M. 1980. *Cell* 20:215–22
198. Kozak, M. 1980. *Cell* 22:459–67
199. Marcus, A. 1970. *J. Biol. Chem.* 245:955–61
200. Marcus, A. 1970. *J. Biol. Chem.* 245:962–66
201. Benne, R., Brown-Luedi, M. L., Hershey, J. W. B. 1978. *J. Biol. Chem.* 253:3070–77
202. Peterson, D. T., Safer, B., Merrick, W. C. 1979. *J. Biol. Chem.* 254:7730–35
203. Deleted in proof
204. Odom, O. W., Kramer, G., Henderson, A. B., Pinphanichakarn, P., Hardesty, B. 1978. *J. Biol. Chem.* 253:1807–16
205. Thomas, A., Goumans, H., Voorma, H. O., Benne, R. 1980. *Eur. J. Biochem.* 107:39–45
206. Farrell, P. J., Balkow, K., Hunt, T., Jackson, R. J., Trachsel, H. 1977. *Cell* 11:187–200
207. Ochoa, S., de Haro, C. 1979. *Ann. Rev. Biochem.* 48:549–80
208. Levin, D., Ranu, R. S., Ernst, V., London, I. M. 1976. *Proc. Natl. Acad. Sci. USA* 73:3112–16
209. Kramer, G., Cimadevilla, J. M., Hardesty, B. 1976. *Proc. Natl. Acad. Sci. USA* 73:3078–82
210. Gross, M., Mendelewski, J. 1977. *Biochem. Biophys. Res. Commun.* 74:559–69
211. Kosower, N. S., Vanderhoff, G. A., Benerefe, B., Hunt, T., Kosower, E. M. 1971. *Biochem. Biophys. Res. Commun.* 45:816–21
212. Clemens, M. J., Safer, B., Merrick, W. C., Anderson, W. F., London, I. M. 1975. *Proc. Natl. Acad. Sci. USA* 72:1286–90
213. Levin, D. H., Ernest, V., London, I. M. 1977. *Fed. Proc.* 36:868
214. Levin, D. H., London, I. M. 1978. *Proc. Natl. Acad. Sci. USA* 75:1121–25
215. Lebleu, B., Sen, G. C., Shaila, S., Cabrer, B., Lengyel, P. 1976. *Proc. Natl. Acad. Sci. USA* 73:3107–11
216. Zilberstein, A., Federman, P., Shulman, L., Revel, M. 1976. *FEBS Lett.* 68:119–24
217. Roberts, W. K., Hovanessian, A., Brown, R. E., Clemens, M. J., Kerr, I. M. 1976. *Nature* 264:477–80
218. Cooper, J. A., Farrell, P. J. 1977. *Biochem. Biophys. Res. Commun.* 77:124–31
219. Lewis, J. A., Falcoff, E., Falcoff, R. 1978. *Eur. J. Biochem.* 86:497–509
220. Ernst, V., Levin, D. H., Leroux, A., London, I. M. 1980. *Proc. Natl. Acad. Sci. USA* 77:1286–90
221. Gross, M., Rynning, J., Knish, W. M. 1981. *J. Biol. Chem.* 256:589–92
222. Benne, R., Salimans, M., Goumans, H., Amesz, H., Voorma, H. O. 1980. *Eur. J. Biochem.* 104:501–9
223. Ranu, R. S., London, I. M. 1979. *Proc. Natl. Acad. Sci. USA* 76:4128–32
224. Ranu, R. S., London, I. M., Das, A., Dasgupta, A., Majumdar, A., Ralston, R., Roy, R., Gupta, N. K. 1978. *Proc. Natl. Acad. Sci. USA* 75:745–49
225. Das, A., Ralston, R. O., Grace, M., Roy, R., Ghosh-Dastidar, P., Das, H. K., Yaghmai, B., Palmieri, S., Gupta, N. K. 1979. *Proc. Natl. Acad. Sci. USA* 76:5076–79
226. de Haro, C., Datta, A., Ochoa, S. 1978. *Proc. Natl. Acad. Sci. USA* 75:243–47
227. Ohtsuki, K., Nakamura, M., Koika, T., Ishida, N., Baron, S. 1980. *Nature* 287:65–67
228. Clemens, M. J., Henshaw, E. C., Rahmimoff, H., London, I. M. 1974. *Proc. Natl. Acad. Sci. USA* 71:2946–50
229. Kaempfer, R. 1974. *Biochem. Biophys. Res. Commun.* 61:591–97

230. Ranu, R. S., Levin, D. H., Delaunay, J., Ernest, V., London, I. M. 1976. *Proc. Natl. Acad. Sci. USA* 73:2720–24
231. Gross, M. 1975. *Biochem. Biophys. Res. Commun.* 67:1507–15
232. Gross, M. 1976. *Biochim. Biophys. Acta* 447:445–49
233. Ralston, R., Das, A., Dasgupta, A., Roy, R., Palmieri, S., Gupta, N. K. 1978. *Proc. Natl. Acad. Sci. USA* 75:4858–62
234. Ralston, R., Das, A., Grace, M., Das, H. K., Gupta, N. K. 1979. *Proc. Natl. Acad. Sci. USA* 76:5490–94
235. Amesz, H., Goumans, H., Haubrich-Morree, T., Voorma, H., Benne, R. 1979. *Eur. J. Biochem.* 98:513–20
236. Siekierka, J., Mitsui, K., Ochoa, S. 1981. *Proc. Natl. Acad. Sci. USA* 78: 220–23
237. Crouch, D., Safer, B. 1980. *J. Biol. Chem.* 255:7918–24
238. Jagus, R., Safer, B. 1981. *J. Biol. Chem.* 256:1324–29
239. Wigle, D. T., Smith, A. E. 1973. *Nature New Biol.* 242:136–40
240. Nudel, U., Lebleu, B., Revel, M. 1973. *Proc. Natl. Acad. Sci. USA* 70:2139–44
241. Kabat, D., Chappell, M. R. 1977. *J. Biol. Chem.* 252:2684–90
242. Blair, G. E., Dahl, H. H. M., Truelsen, E., Lelong, J. C. 1977. *Nature* 265: 651–53
243. Golini, F., Thach, S. S., Birge, C. H., Safer, B., Merrick, W. C., Thach, R. E. 1976. *Proc. Natl. Acad. Sci. USA* 73:3040–44
244. Lodish, H. F. 1974. *Nature* 251:385–88
245. Lodish, H. F. 1976. *Ann. Rev. Biochem.* 45:39–72
246. Ehrenfeld, E., Lund, H. 1977. *Virology* 80:297–308
247. Lawrence, C., Thach, R. E. 1974. *J. Virol.* 14:598–610
248. Doyle, S., Holland, J. 1972. *J. Virol.* 9:22–28
249. Helentjaris, T., Ehrenfeld, E. 1978. *J. Virol.* 26:510–21
250. Kaufmann, Y., Goldstein, E., Penman, S. 1976. *Proc. Natl. Acad. Sci. USA* 73:1834–38
251. Trachsel, H., Sonenberg, N., Shatkin, A. J., Rose, J. K., Lelong, K., Bergman, J. E., Gordon, J., Baltimore, D. 1980. *Proc. Natl. Acad. Sci. USA* 77:770–74
252. Heywood, S. M., Kennedy, D. S., Bester, A. J. 1974. *Proc. Natl. Acad. Sci. USA* 71:2428–31
253. Gette, W. R., Heywood, S. M. 1979. *J. Biol. Chem.* 254:9879–85
254. Heywood, S. M., Kennedy, D. S. 1979. *Arch. Biochem. Biophys.* 192:270–81
255. Trachsel, H., Staehelin, T. 1979. *Biochim Biophys. Acta* 565:305–14
256. Jones, R. L., Sadnik, I., Thompson, H. A., Moldave, K. 1980. *Arch. Biochem Biophys.* 199:277–85
257. Checkley, J. W., Cooley, L., Ravel, J. M. 1981. *J. Biol. Chem.* 256:1582–86
258. Tahara, S. M., Morgan, M. A., Shatkin, A. J. 1981. *J. Biol. Chem.* 256:7691–94
259. Walden, E. W., Godefroy-Colburn, T., Thach, R. E. 1981. *J. Biol. Chem.*
260. Kozak, M. 1981. *Current Topics in Microbiol. and Immunol.* ed. A. J. Shatkin 93:81–123
261. Das, A., Bagchi, M., Roy, R., Ghosh-Dastidar, P., Gupta, N. K. 1982. *Biochem. Biophys. Res. Commun.* 104: 89–98

Ann. Rev. Biochem. 1982. 51:901–34

EUKARYOTIC DNA REPLICATION: VIRAL AND PLASMID MODEL SYSTEMS

Mark D. Challberg and Thomas J. Kelly

Department of Molecular Biology and Genetics, Johns Hopkins University
School of Medicine, Baltimore, Maryland 21205

CONTENTS

PERSPECTIVES AND SUMMARY

The complex events that occur during the replication of eukaryotic chromosomes are not yet understood. By analogy with recent experience in prokaryotic systems, it is expected that much of the basic information about chromosomal replication will be obtained by studying the replication of the

0066-4154/82/0701-0901$02.00

genomes of viruses and other extrachromosomal elements. The same analogy suggests that rapid progress in this area will require the detailed analysis of in vitro systems that faithfully carry out the replication of such genomes. There are a number of potentially useful viral and plasmid model systems for studying eukaryotic DNA replication. Each has certain unique virtues for illuminating particular aspects of the replication process. This review focuses on those systems that have so far proven most amenable to experimental attack.

Among the eukaryotic viruses with linear genomes the two best characterized groups are the parvoviruses and the adenoviruses. The parvoviruses have some of the properties that have made the small coliphages such attractive models for studying prokaryotic DNA replication. Their genomes are single-stranded and have a limited genetic capacity. It seems likely that the parvoviruses rely almost completely upon the cellular DNA replication machinery for their multiplication. The genomes of the adenoviruses are double-stranded and considerably more complex than those of the parvoviruses. At least two of the proteins required for adenovirus DNA replication are encoded by the viral genome, but others, including DNA polymerase(s), are contributed by the host cell. The linearity of the parvovirus and adenovirus genomes imposes the need for special replicative mechanisms to carry out the synthesis of the sequences at the 5' ends of progeny DNA molecules. This is a consequence of the facts that the known DNA polymerases require preformed primers to initiate synthesis, and that they catalyze chain elongation exclusively in the 5' to 3' direction. The parvoviruses utilize a self-priming mechanism that operates at self-complementary terminal nucleotide sequences. The adenoviruses appear to employ a novel mechanism in which a protein serves to prime DNA synthesis at the 5' termini. One or both of these mechanisms may be relevant to the replication of the ends of eukaryotic chromosomes. Parvovirus and adenovirus DNA replication involve only a continuous mode of DNA synthesis such as occurs at the leading strand of the chromosomal replication fork. All daughter strands grow from their 5' toward their 3' termini, and thus, can be elongated without the need for the synthesis and subsequent joining of Okazaki fragments. These considerations suggest that the enzymology of parvovirus and adenovirus DNA replication may be significantly simpler than that of chromosomal replication. This apparent simplicity makes these two viral systems an attractive place to begin the detailed biochemical dissection of eukaryotic DNA replication. In the case of adenovirus there exists a soluble in vitro system that accurately carries out the initiation and elongation of viral DNA chains.

The circular genomes of the papovaviruses and the yeast plasmids represent more complete (albeit more complex) model systems for studying eukaryotic DNA replication. Both are organized by cellular histones into a periodic array of nucleosomes similar in structure to those of cellular chromatin. Thus, it should be possible to utilize these systems to analyze the role of histones in the replication process and to study the mechanisms by which newly synthesized DNA is assembled into chromatin. The initiation of papovavirus DNA replication may be closely analogous to the initiation of chromosomal replicons. Replication begins within a unique origin on the viral chromosome and proceeds bidirectionally. The initiation reaction requires a highly specific interaction between the origin sequence and a virus-coded initiation protein, the T antigen. The nature of this interaction is currently being probed by both genetic and physical approaches. The analysis of initiation of yeast plasmid replication is not yet so far advanced. However, segments of yeast DNA that confer the ability to replicate autonomously in the yeast cells have been identified and sequenced. Both papovaviruses and yeast plasmids probably rely exclusively on host proteins for all of the steps in DNA replication subsequent to initiation. Moreover, the available data suggest that the events that occur during fork movement are indistinguishable from those that occur during cellular DNA replication. Both continuous (on the leading strand) and discontinuous (on the lagging strand) DNA synthesis take place at each fork. The latter involves a complex series of events that begins with the synthesis of RNA primers and ends with the covalent joining of Okazaki fragments to growing nascent DNA strands. As in the case of the parvoviruses and adenoviruses, the study of the enzymology of initiation and chain elongation is in its infancy. However, an in vitro system from yeast that replicates exogenously added plasmid DNA, has recently been developed. The existence of this system together with the availability of yeast mutants with defects in DNA replication, should allow a combined genetic-biochemical approach to the isolation and characterization of yeast replication proteins.

PARVOVIRUS DNA REPLICATION

As the name implies the parvoviruses are the smallest and least complex of the DNA animal viruses. Their genomes are nonpermuted, linear, single-stranded DNA molecules containing 4500–5500 nucleotides. The parvoviruses that infect vertebrates have been classified into two major groups: the nondefective or autonomous parvoviruses (e.g. H-1, minute virus of mice, or Kilham rat virus) and the defective parvoviruses (adenoassociated viruses). Multiplication of members of the latter group requires coinfection with a helper virus such as adenovirus or herpes virus. A number of excellent reviews of the biology of the parvoviruses have been published (1–6).

Autonomous parvoviruses

GENOME STRUCTURE   Although the genomes of the autonomous parvoviruses are largely single-stranded, both the 3' and 5' termini are folded
into duplex structures (7–12). The precise structure of the 5' terminus is not
yet clear because only limited nucleotide sequence information is available.
However, on the basis of nuclease protection experiments, the size of the
5'-terminal duplex region has been estimated at 70–130 base pairs (9, 13).
The nucleotide sequences of the 3' termini of several rodent parvoviruses
(MVM, KRV, H-1, and H-3) have recently been determined (12). The
sequences show a high degree of self complementarity within the first 115
or 116 nucleotides. In each case, maximum base pairing (102 of 115 bases
paired) is achieved by a Y-shaped hairpin structure. Analysis of the cleavage
products produced by digestion of parvovirus DNA with single-strand-
specific endonucleases strongly suggests that this structure or one very
similar to it actually exists in solution. The nucleotide sequences of the
3'-terminal hairpins of the rodent parvoviruses are highly conserved, differing in only two or three nucleotides (12). The 3'-terminal sequences of the
human (LuIII) and bovine parvoviruses differ considerably from those of
the rodent viruses, but both can be folded into a similar Y-shaped hairpin
structure (12), as can the termini of the defective parvovirus genomes (see
below). Thus, while the significance of this novel topological feature of the
termini is not yet clear, the fact that it is maintained in spite of great
differences in the absolute nucleotide sequence of the termini suggests that
it has an important function. The possible role of the 3'- and 5'-terminal
hairpin structures in parvovirus DNA replication is discussed below.

DNA REPLICATION *IN VIVO*   Because of the small size of the autonomous parvovirus genome, it is likely that most of the functions required for
viral DNA replication are provided by the host cell. A large number of
studies have demonstrated that these viruses multiply best in actively dividing cells (14–23). When synchronized or partially synchronized cell populations are employed as hosts, the shortest latent periods are obtained when
the infection is initiated in early S phase (16, 17, 20, 21, 23). Under these
conditions the onset of viral DNA replication occurs in late S phase (20,
22, 23) and is apparently dependent upon proteins synthesized at that time
(22). These observations have been interpreted as indicating that the early
steps of parvovirus DNA replication require one or more cellular DNA
replication proteins that are transiently expressed during S phase.
    The replication of autonomous parvovirus DNA takes place in the nu-
cleus of the infected cell. As in the case of bacteriophage with single-
stranded genomes, the first step in DNA replication is the conversion of the

infecting parental single strands to duplex replicative form (RF) molecules. The resulting parental RFs then undergo replication to produce progeny RFs, which accumulate in the nucleus and serve as templates for the synthesis of new viral genomes. The basic structural features of RFs have been deduced from analysis of labeled intracellular DNA isolated either from cells infected with ^{32}P-labeled virions (24, 25) or cells exposed to a short pulse of radioactive precursor (25–31). In either case the major duplex form detected is approximately unit length and is referred to as the monomer RF. Analysis of the products of restriction endonuclease digestion has provided evidence that the termini of monomer RFs have two possible structures: a structure in which the terminal sequences are folded back to form a hairpin and a structure in which the terminal sequences are in an extended duplex configuration (25; see Figure 1). In the majority of monomer RFs with terminal hairpins, the two complementary strands are covalently linked, as indicated by their ability to "snap back" after denaturation (25–31). A number of investigators have observed dimeric and higher oligomeric RF forms in parvovirus-infected cells (25–28, 30, 32). The dimer RFs consist of linear structures in which two monomeric units are joined in a "tail to tail" configuration (i.e. with alternating viral and complementary strands.) In a significant fraction of cases the DNA strands of the individual monomeric units that make up a dimer RF are linked by covalent bonds. The structure of higher oligomeric RF molecules has not been extensively studied; however, the majority of these forms yield unit length single strands upon denaturation, which indicates that they contain single-stranded nicks spaced approximately one genome apart (25). In addition to the fully duplex RF molecules, parvovirus-infected cells contain molecules that are partially duplex and partially single stranded (26, 29, 33, 34). These species are preferentially labeled during short pulses and presumably represent the metabolically active replicative intermediates (RIs). The results of pulse-chase experiments are consistent with the hypothesis that RIs are precursors to both RF molecules and progeny single strands (29). The detailed structure of RIs has not yet been determined. Recently, Revie et al (35) obtained evidence that a protein of molecular weight 60,000–70,000 is covalently linked to the 5' ends of each strand in the duplex intracellar forms of parvovirus H-1 DNA. The role of this protein, if any, in viral DNA replication is not known.

The origins of DNA replication appear to reside at the ends of the parvovirus genome. Naturally occurring mutants with extensive internal deletions have been detected in virus populations propagated at high multiplicities of infection (36). The smallest of these retain only about 250 nucleotides from each end of the genome, yet appear to replicate efficiently in the presence of coinfecting wild-type virus. This suggests that the termini con-

I. SS → RF

II. RF → SS

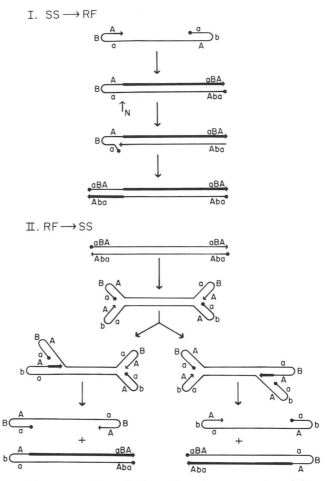

Figure 1. Model for AAV DNA replication; schematic representation of the two basic replicative processes. *I* Conversion of single strands to duplex RF molecules. DNA synthesis is initiated by self-priming at the 3' terminus and proceeds to the end of the template, thereby producing a monomer RF whose complementary strands are covalently linked via a terminal hairpin. A nick (*N*) is introduced at a site opposite the original 3' end creating a new 3'-OH primer terminus. DNA synthesis initiated at this site generates a monomer RF with extended termini. *II,* Production of single strands by displacement synthesis. A monomer RF undergoes hairpin rearrangement to produce a rabbit eared structure. DNA synthesis is initiated at either 3' terminus by a self-priming mechanism. Synthesis of a new daughter strand is accompanied by displacement of one of the parental strands. The products of displacement synthesis are a unit length single strand and a monomer RF with a hairpin terminus. The former can either be converted to an RF by the pathway shown in *I* or encapsidated in a progeny virion. The latter can be processed further by pathway *I* to yield a monomer RF with extended termini.

Complementary nucleotide sequences are represented by upper and lower case letters. Arrowheads denote 3' termini and dots denote 5' termini. [Adapted from (49, 59, 61)].

tain all of the critical *cis* functions required for DNA replication. The presence of terminal hairpin structures in both mature viral DNA and intracellular RF DNA suggests that initiation of DNA synthesis may involve a self-priming mechanism. This possibility is discussed more fully below.

Analysis of the size distribution of nascent parvovirus DNA labeled during pulses as short as 20 sec has revealed no evidence for the 4–5S segments (Okazaki fragments) observed in papovavirus or chromosomal DNA replication (37). Thus, it seems most likely that the synthesis of both strands of parvovirus DNA proceeds from one end of the genome to the other by a completely continuous mechanism. It also seems likely that chain elongation is carried out by one or more cellular DNA polymerases, perhaps in concert with other cellular replication proteins. The levels of DNA polymerase α activity in infected cells is closely correlated with the rate of viral DNA synthesis (38). Moreover, viral DNA synthesis in nuclei or nuclear lysates derived from parvovirus-infected cells is sensitive to aphidicolin, an inhibitor of DNA polymerase α, and to anti-DNA polymerase α antibody (39, 40). The recent finding that viral DNA synthesis in isolated nuclei is inhibited by 2'–3'-dideoxythymidine triphosphate has led to the suggestion that DNA polymerase γ may also play a role in parvovirus DNA replication (40).

The question of whether parvovirus DNA replication requires the participation of any virus-coded proteins is not yet settled, largely due to insufficient genetic analysis of these viruses. Studies of temperature-sensitive mutants and naturally occurring deletion mutants indicate a possible requirement for a viral protein(s) in RF and progency strand replication (41, 42); however, further work is required to confirm this.

As alluded to briefly above, parvovirus DNA replication in vitro has been observed in isolated nuclei and nuclear lysates from virus-infected cells (39–40). These systems continue DNA synthesis on replication intermediates initiated in vivo, but do not initiate new rounds of DNA synthesis. Recently, it has been reported that a soluble extract from uninfected human cells can carry out the conversion of single-stranded H-1 DNA to the double-stranded monomer RF form (43). Each of these systems may prove useful in elucidating some of the biochemical mechanisms of chain elongation.

Defective Parvoviruses (AAV)

GENOME STRUCTURE   In contrast to the autonomous parvoviruses, whose virions normally contain a preponderance of DNA strands with a single polarity, the adenovirus-associated viruses (AAV) encapsidate DNA strands of plus and minus polarity at roughly equal frequencies (44). The purified viral DNA molecules form single-stranded circles when incubated under annealing conditions, which indicates the presence of an inverted terminal repetition (45). Detailed nucleotide sequence analysis of the termini (46) has demonstrated that the inverted terminal repetition is 145 nucleotidess in length. The first 125 nucleotides display extensive self-complementarity and can be folded into a Y-shaped hairpin structure in which 107 of 125 nucleotides are base paired. The topology of this hairpin structure is similar to that present at the 3' termini of rodent autonomous parvoviruses, although the terminal nucleotide sequences of the two groups of viruses are quite different. Interestingly, the 125 nucleotides that comprise the putative hairpin structure can exist in either of two sequence orientations at each end of the genome. The two possible sequence orientations, termed flip and flop, are inverted complements of each other. As discussed below these two sequence orientations are thought to be generated during the replication of the AAV genome.

REPLICATION *IN VIVO*   Like the autonomous parvoviruses, AAV replicates in the nucleus of the infected cell. The early steps in infection (attachment, penetration, and uncoating) can proceed in the absence of helper virus, but viral DNA replication and RNA synthesis require the presence of coinfecting adenovirus (the natural helper) or herpes virus (47). The human AAV, the best characterized group, will replicate in cultured cells derived from a variety of animal species as long as the helper virus can also replicate in the same cells (48).

The intracellular forms of AAV DNA have been partially characterized, and the available data suggest that their structures are quite similar to those of the autonomous parvoviruses. Double-stranded monomer and dimer RF molecules have been identified by sedimentation analysis of intracellular DNA (49–51). A substantial fraction of the duplex molecules are capable of spontaneous renaturation following alkaline denaturation and neutralization, which indicates that they contain covalently linked complementary strands. Restriction enzyme analysis shows that the termini of RF molecules can exist as either a hairpin structure in which the two complementary strands are linked or as an extended structure in which the two strands are unlinked (50, 51).

The origin and direction of DNA chain elongation during AAV DNA replication has been deduced from a study of the distribution of incorporated radioactivity in newly synthesized strands (52). Each of the two complementary strands of monomer RF isolated after a short pulse of ^{3}H thymidine shows a unimodal distribution of label, with the highest specific activity near the 3' end and the lowest specific activity near the 5' end. This result strongly suggests that the origin of replication of each strand is at or near its 5' terminus and that the overall direction of chain growth is 5' to 3'. As in the case of the autonomous parvoviruses, sedimentation studies of nascent AAV strands have not revealed evidence for Okazaki fragments. Thus, the synthesis of AAV DNA probably occurs by a completely continuous mechanism.

Little is known about the nature of the proteins involved in AAV DNA replication. In principal such proteins could be coded by the AAV genome, the genome of the helper virus, or by the genome of the host cell. Although it seems likely that cellular replication proteins will prove to play a major role in the replication process none has been implicated to date. Suggestive evidence for a role for AAV capsid protein in the production of progeny genomes has been obtained in studies with L-canavanine, an arginine analogue that inhibits the accumulation of VP3, the major AAV capsid protein (53). In the presence of the inhibitor progeny AAV single strands fail to accumulate, but RF synthesis apparently proceeds normally. This is consistent with the hypothesis that accumulation of progeny genomes is coupled to the synthesis of virus capsids; however, further direct verification of this is needed. A number of attempts have been made to define the adenovirus gene function(s) required for the helper activity. Two lines of evidence indicate that expression of one or more early adenovirus genes is sufficient to allow replication of AAV. First, certain adenovirus mutants that do not express late genes have been shown to be competent helpers (54–57). Second, AAV multiplication has been demonstrated in cells microinjected with early adenovirus mRNA (58). Recently, in an extension of the microinjection experiments, W. Richardson and H. Westphal (personal communication) have shown that mRNA encoded by adenovirus early region E4 is sufficient to provide helper activity, but that the regulation of E4 expression is complex, requiring the prior expression of early regions E1a and E2. This is consistent with previous findings that adenovirus mutants with lesions in E1a and E2 display reduced helper activity (53, 56, 57). The precise nature of the E4 gene product responsible for helper activity and its mechanism of action are not yet known.

Models for parvovirus DNA replication

Although information concerning the structure and kinetic interrelationships of parvovirus replication intermediates is rather limited, the available data are consistent with the following general scheme (49, 59) (illustrated in Figure 1 for the case of AAV). The infecting parental single-stranded genome is first converted to a double-stranded RF molecule (Figure 1 (I,

SS→RF). DNA synthesis is then initiated at a terminus of the RF and a new daughter strand is synthesized in the 5' to 3' direction. During this process one strand of the RF serves as template and the other is displaced (Figure 1 *II.*, RF→SS). The displaced single strand has one of two possible fates, perhaps depending on availability of capsid proteins. It may be converted to an RF molecule by a mechanism analogous to that which operates on the original parental genome. Alternatively, it may be encapsidated to form the genome of a progeny virion.

As discussed briefly above, the initiation of parvovirus DNA replication appears to involve a self-priming mechanism, at least in part. The evidence for this is most compelling in the case of the SS→RF reaction. Model studies in vitro have demonstrated that the 3' terminus of the viral genome is capable of serving as primer for complementary strand synthesis by several different DNA polymerases (7). Moreover, the observation that the two strands of monomer RF molecules isolated from infected cells are often covalently joined via a terminal hairpin strongly suggests that a similar process occurs in vivo (25–31, 49). Initiation of displacement synthesis (RF →SS) may also involve self-priming (26, 49, 60). This would require rearrangement of the duplex DNA at the end of an RF molecule to form a "rabbit-eared" structure as shown in Figure 1. Since such a rearrangement would not be expected to occur spontaneously, it has been suggested that it is mediated by specific protein-DNA interactions (60). This mechanism, although plausible, does not yet have any direct experimental support.

Assuming that initiation of parvovirus DNA replication involves extension from a 3' hairpin terminus, it follows that there must exist a mechanism for resolving the hairpin and completing the synthesis of the terminus. Most of the proposed mechanisms for accomplishing this task are based upon a model originally proposed by Cavalier-Smith to explain the replication of the termini of eukaryotic chromosomes (61). In this model (see Figure 1) a newly synthesized duplex RF molecule containing a terminal hairpin is specifically nicked at a point opposite the original 3' end of the hairpin structure, creating a new 3'OH primer terminus. DNA synthesis is then initiated at that point and proceeds 5' to 3', generating an RF molecule with an extended terminus. A characteristic feature of this (and related) models is the generation of a new terminal nucleotide sequence that is the inverted complement of the original terminal sequence. Thus, the finding that the ends of AAV DNA molecules exist in two possible sequence orientations constitutes strong circumstantial evidence in favor of the scheme depicted in Figure 1. The presence of dimeric and oligomeric RF molecules in parvovirus infected cells can be accounted for by simply postulating that the resolution of terminal hairpin structures by the Cavalier-Smith mechanism is not strictly coupled to DNA synthesis. The longer molecular forms would be generated when DNA synthesis proceeds through a terminal hairpin prior to its resolution (49, 59, 60). Such structures could be converted back to monomer RFs by operation of the site-specific nicking/DNA synthesis activities at internal sites (49).

The basic features of the replication of the defective and the autonomous parvovirus genomes are likely to be quite similar, although there clearly must be differences in detail. In the case of the defective parvovirus, AAV, it seems probable that initiation of displacement synthesis occurs with roughly the same frequency at the two ends of RF molecules, since the nucleotide sequences in the immediate neighborhood of the putative initiation sites are indistinguishable. This would lead to the production of equal numbers of progeny genomes with plus and minus polarity, as is actually observed. In addition the continued operation of the Cavalier-Smith mechanism on progeny RF molecules would be expected to randomize the distribution of terminal sequence orientations such that the orientation at one end of the viral genome would be uncorrelated with that at the other end. This has also been verified experimentally (62). In the case of the autonomous parvoviruses, the two ends of the viral genome are not related in nucleotide sequence. Thus, the rates of initiation of displacement synthesis could be quite different at the two ends of an RF molecule. This may explain the observation that the progeny genomes found in autonomous parvovirus virions are predominantly of one polarity. The autonomous pavoviruses also differ from AAV in that only one sequence orientation is present at the 3' ends of progeny genomes (12) instead of the two predicted by the Cavalier-Smith model (the status of the 5' termini is not clear, since the nucleotide sequence data is not complete). There are a number of explanations for this finding, all of which require some modification of the model depicted in Figure 1. For example, it is possible that replicative intermediates with one of the two possible 3'-terminal sequence orientations are replicated preferentially or that progeny single strands with the appropriate 3' sequence orientation are encapsidated preferentially. Finally, it remains possible that the autonomous parvoviruses utilize an initiation mechanism other than self-priming. Further work is required to determine which, if any, of these possibilities is correct.

The model shown in Figure 1 may have more general application. There is good reason to believe that eukaryotic chromosomes contain single linear DNA molecules (63). Moreover, Forte & Fangman (64) have recently obtained evidence that the ends of yeast chromosomes are closed by hairpin loops. The similarity of this structure to the intracellular forms of parvovirus DNA raises the interesting possibility that these viruses have borrowed a cellular mechanism that normally operates to replicate chromosomal termini.

ADENOVIRUS DNA REPLICATION

In size and genetic complexity the adenoviruses occupy an intermediate position among the animal viruses. Their genomes are nonpermuted linear, double-stranded DNA molecules containing 35,000–45,000 base pairs. During the past several years the adenoviruses (especially the human sero-types) have served as extremely valuable models for probing the basic mechanisms of gene expression and DNA replication in eukaryotic cells (65–68).

Genome Structure

Adenovirus genomes, like those of the parvoviruses, have specialized termi-nal structures that appear to play a central role in DNA replication. The early observation that adenovirus single strands are capable of forming unit length circles when incubated under annealing conditions indicated the presence of an inverted terminal repetition (69, 70). Recently, the terminal nucleotide sequences of several adenoviruses have been determined (71–75). From analysis of these sequences the following general points can be made: (*a*) The inverted terminal repetitions extend to the very ends of the viral genome. (*b*) In every case the nucleotide sequences at the two ends of the genome are precisely the same. (*c*) The sizes of the inverted terminal repetitions are different in different adenoviruses, ranging from 103 base pairs (Ad2 and Ad5) to 162 base pairs (Ad12). (*d*) The sequences of the inverted terminal repetitions of different adenoviruses show a high degree of homology especially in the first 50 base pairs. The most striking region of homology is a sequence of 14 base pairs (nucleotides 9–22) that is con-served in all adenoviruses examined to date including representatives of three human subgroups and the simian adenovirus SA7 (74).

A second novel structural feature of the termini of adenovirus DNA is the presence of a covalently bound terminal protein (76–83). The terminal protein is linked to the 5' terminus of each adenovirus DNA strand, and, in the case of adenovirus type 5, has a molecular weight of 55,000 (82, 83). Recently, the protein-DNA linkage has been shown to consist of a phos-phodiester bond between the β-OH of a serine residue in the terminal protein and the 5'-OH of the terminal deoxycytidine residue in the DNA (84). Interestingly, removal of the terminal protein greatly reduces the specific infectivity of isolated adenovirus DNA (79). The basis for this effect is not understood, but one reasonable possibility, suggested by recent in vitro studies, is that the terminal protein on the parental DNA strands may facilitate initiation of viral DNA replication (see below).

In infected cells the adenovirus terminal protein is synthesized in the form of a larger (M_r=80,000) precursor protein that is subsequently cleaved to produce the mature 55-kd species (85–87). Analysis by hybrid-selected translation techniques indicates that the 80-kd precursor is encoded by the viral genome (86). The 5' termini of nascent adenovirus DNA strands synthesized in vitro or in vivo are attached exclusively to the precursor form of the terminal protein (85, 87). The processing event that results in conversion to the 55-kd form appears to occur as a late step in virion assembly (86, 87).

Replication In Vivo

The basic structural features of adenovirus replication intermediates have been defined by electron microscopy (88–90) and other physical methods (91–98). Understanding of the structure of the intermediates has led in turn to a reasonably clear picture of the pathways involved in adenovirus DNA replication. These are summarized in the model shown in Figure 2 (90). Replication can be initiated at either end of the double-stranded parental genome. Following each initiation event, a daughter strand is synthesized in the 5' to 3' direction with concomitant displacement of the parental strand of the same polarity, thereby producing a branched (or type I) replication intermediate. The frequency of initiations at the two ends of the viral genome is approximately the same, and multiple initiation events on the same replicating molecule are common. As indicated in Figure 2, there are two possible pathways for the synthesis of the viral DNA strand complementary to the displaced parental strand. In the major pathway, displace-

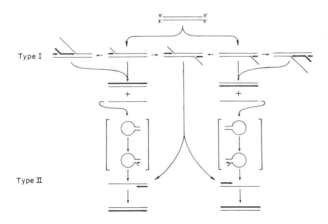

Figure 2.   Model for adenovirus DNA replication. See text for detailed discussion. [Adapted from (90)].

ment synthesis proceeds to completion, resulting in the formation of a daughter duplex and the release of a free single strand. The latter then serves as a template for DNA synthesis initiated at or near its 3' end. This results in the formation of a linear (or type II) replication intermediate. The minor pathway is a consequence of the fact that displacement synthesis can be initiated at either end of the genome. Thus, there is a low, but finite probability that before completion of a round of displacement synthesis initiated at one end of the genome, a second initiation will occur at the opposite end. In such doubly initiated molecules, both parental strands serve as templates for daughter strand synthesis. Two type II replication intermediates are produced when the oppositely moving displacement forks meet.

All initiation events that occur during adenovirus DNA replication take place at the 3' terminus of the template DNA strand (90, 91, 93, 97). Since the nucleotide sequences at the 3' termini of the l and r strands of adenovirus DNA are identical it seems probable that all initiation events proceed by the same molecular mechanism. (If, for example, the single strands released by displacement synthesis were to cyclize by hybridization of their self-complementary terminal sequences as suggested in Figure 2 (94) the ends of all intracellular forms of adenovirus DNA, both duplex and single-stranded, would have an identical structure.) The nature of the initiation mechanism has been the subject of much recent work and is discussed in a later section of this review.

It is apparent from the model described above that there is no requirement, at least in principle, for a discontinuous mode of DNA synthesis during any phase of adenovirus DNA replication. All nascent daughter strands grow from their 5' termini toward their 3' termini, and thus, can be elongated without the need for the synthesis and subsequent joining of Okazaki fragments. There is, in fact, considerable experimental support for the view that adenovirus DNA replication proceeds by a completely continuous mechanism. Electron microscopic studies of viral replication intermediates have shown that internal initiations on displaced single strands seldom, if ever, occur (90, 99). Moreover, experiments in which the free 3'-OH groups of replication intermediates were labeled using terminal transferase demonstrated that each nascent strand contains only one such group and that it is located at the growing end of the strand (90). Finally, analysis of the size distribution of nascent strands pulse-labeled in vitro or in vivo has generally failed to reveal evidence for Okazaki fragments (100–108).

Replication Proteins

Two virus-coded proteins have been directly implicated in adenovirus DNA replication. The first is the adenovirus terminal protein described above, which has been shown to be required for adenovirus DNA replication in vitro, and probably plays an important role in the initiation reaction (see below). The second virus-coded protein known to be involved in viral DNA replication is a 72-kd DNA binding protein (DBP) (109, 110). Like the terminal protein, the adenovirus DBP is the product of an early adenovirus gene (111, 112). In fact, recent evidence indicates that both proteins are expressed from the same transcriptional unit (86). The DBP binds cooperatively and tightly to single-stranded DNA from a variety of sources (109, 113–116). At saturation approximately one molecule of DBP is bound for every seven nucleotides (114). Hydrodynamic measurements and electron microscopic observations indicate that the protein holds the DNA chains in an extended configuration (114). Thus, the DNA binding properties of adenovirus DBP are similar in some respects to those of prokaryotic single-stranded DBPs, such as bacteriophage T4 gene 32 protein. However, attempts to demonstrate that the adenovirus DBP lowers the melting temperature of DNA have so far been unsuccessful (114, 115), and the effects of the protein on the kinetics of renaturation have not been determined. Interestingly, careful filter-binding studies have demonstrated that the adenovirus DBP also binds to a limited extent to double-stranded DNA (113, 115). The binding occurs exclusively at the ends of linear duplex molecules, as indicated by restriction endonuclease cleavage analysis of DBP-DNA complexes.

Electron microscopic studies have demonstrated that the single-stranded regions of type I and type II adenovirus replication intermediates are completely coated by DBP (99). Direct evidence that the protein is involved in viral DNA replication has been obtained by analysis of the properties of the adenovirus mutant, H5ts125, which produces a thermolabile DBP in infected cells (117). At the restrictive temperature no detectable viral DNA replication occurs in H5ts125-infected cells (118, 119). Moreover, when H5ts125-infected cultures are maintained at the permissive temperature until after the onset of viral DNA replication and then shifted to the nonpermissive temperature, the rate of viral DNA synthesis declines to very low levels within one hour (119). The precise role of the DBP in viral DNA replication is not yet clear. Recent in vitro studies suggest that the DBP is required for chain elongation (120, 121), but an additional role in the initiation process, suggested by previous in vivo studies (117), has not been excluded.

It is generally believed that cellular proteins play major roles in adenovirus DNA replication. It seems certain, for example, that one or more cellular polymerases is required for viral DNA chain elongation, since no new DNA polymerase activity has been detected in adenovirus-infected cells (122, 123). The rate of adenovirus DNA replication in isolated nuclei or subnuclear replication complexes is reduced by inhibitors of DNA polymerase α (aphidicolin) and DNA polymerase γ (dideoxythymidine triphosphate), which suggests that both enzymes are involved in viral DNA replication (124, 126). However, this conclusion is tentative, since it rests upon unproven assumptions concerning the specificity of the inhibitors in virus-infected cells. The recent development of in vitro replication systems should allow rapid progress in defining both viral and cellular replication proteins.

A number of studies have addressed the question of whether adenovirus DNA in infected cells is complexed with cellular histones. Analysis of the fate of ^{32}P-labeled adenovirus has shown that after uncoating takes place, at least a fraction of the parental viral DNA molecules become organized into nucleoprotein structures that have some of the properties of cellular chromatin (127, 128). However, it is not clear whether these molecules participate in transcription and/or replication. In fact, two lines of evidence strongly suggest that the bulk of the metabolically active intracellular viral DNA is not packaged into nucleosomes. First, direct electron microscopic visualization of intracellular adenovirus DNA molecules (including those involved in replication and transcription) reveals that the duplex viral DNA has a relatively smooth contour, in contrast to the cellular chromatin where the typical nucleosomal pattern is evident (99, 129). In addition, direct analysis by SDS polyacrylamide gel electrophoresis indicates that histones are not a prominent constituent of purified replication complexes (99). These studies leave open the question of whether duplex viral DNA is complexed with nonhistone cellular proteins or with viral proteins.

Replication In Vitro

The first in vitro studies of adenovirus DNA replication were performed with isolated nuclei or subnuclear replication complexes (102, 107, 130–132). The replication observed in these systems represents a continuation of synthesis on replicative intermediates formed in vivo. Nascent strands can be completed in vitro, but de novo initiation of new DNA chains apparently does not occur. More recently, a soluble enzyme system capable of replicating exogenously added adenovirus DNA has been described (133, 134). This system carries out both the initiation and elongation of viral DNA chains. The in vitro replication reaction requires nuclear extract from adenovirus-infected cells, adenovirus DNA-55 kd protein complex as template, the four deoxynucleoside triphosphates, Mg^{2+}, and ATP. Replication does not require ribonucleoside triphosphates other than ATP. Serveral lines of evidence indicate that the DNA synthesis observed in this in vitro system closely resembles adenovirus DNA replication in vivo (85, 133–137). First, daughter strand synthesis is initiated at either terminus and proceeds exclusively in the 5' to 3' direction. Second, replication takes place by a displacement mechanism on type I replicative intermediates identical in structure to those observed in vivo. Third, replication is dependent upon the presence of factors unique to virus-infected cells, including the adenovirus 72-kd DNA binding protein. Fourth, the rate of DNA chain elongation in vitro (about 1200 nucleotides per minute) is comparable to the in vivo rate. Finally, the DNA synthesized in vitro consists of long adenovirus DNA strands that are hydrogen bonded, but not covalently linked, to the input DNA template. A significant fraction of the newly synthesized DNA strands approximate full length.

Efficient DNA replication in vitro requires the presence of the terminal protein on the input DNA template (133, 134). If the terminal protein is removed by proteolysis or treatment with alkali, only a very limited repair-like reaction is observed. The function of the terminal protein on parental strands is not yet clear, although it has been suggested that it may serve to facilitate initiation of DNA replication, perhaps via protein-protein interactions with factors present in the infected-cell extracts (133, 134). The 5' termini of nascent DNA strands synthesized in vitro are also linked to terminal protein. These terminal protein molecules may play a direct role in initiation of replication, perhaps during the priming reaction (82, 85).

Initiation of Replication

In considering the mechanisms that might be responsible for priming the synthesis of adenovirus DNA strands at the molecular termini, it is evident that the linearity of the viral genome imposes certain constraints. In particular, classical RNA priming schemes are not directly applicable, because, after removal of the putative primer, the original problem of generating the DNA sequence at the extreme 5' terminus is still present. A number of models for initiation of adenovirus DNA replication that involve self-priming are possible in principle, but none has significant experimental support. All such models predict that newly synthesized DNA is covalently linked to the parental DNA template at least transiently; this has never been observed during adenovirus DNA replication (100–107). In addition detailed sequence analysis of the terminal regions of the adenovirus genome has not revealed the sequence arrangments predicted by hairpin priming at terminal palindromes (71–75). Thus, recent speculations concerning the mechanism of initiation have centered on the novel idea that adenovirus DNA replication may be primed by protein (82). In one version of this idea (85) the primary initiation event is the formation of an ester linkage between the α-phosphoryl group of dCTP and the β-OH of a serine residue in the 80-kd terminal protein precursor. It is envisioned that this event takes place as part of a concerted reaction in which a free 80kd terminal protein associates with the terminus of the parental genome. Such an association might be mediated by interactions with specific terminal nucleotide sequences and/or interactions with the terminal protein attached to the parental DNA. (The latter possibility is suggested by the observation that deproteinized DNA is not an effective template for adenovirus DNA replication in vitro.) As a result of these interactions the 3'-OH of the dCMP residue, which becomes covalently linked to the 80kd protein, is positioned in such a way that it can serve as a primer for subsequent chain elongation.

Experimental support for this model has recently been obtained by analysis of viral DNA replication in the in vitro system that is dependent upon exogenous template. It has been demonstrated that the 5' ends of nascent adenovirus DNA strands synthesized in this system are linked to the 80-kd terminal protein by a phosphodiester bond between the β-OH of a serine residue in the protein and the 5'-OH of the terminal cytidine residue in the DNA (85). Moreover, it has been possible to study the protein-nucleotide joining reaction in the absence of DNA chain elongation by incubation of infected-cell extracts with [α-^{32}P] dCTP as the only deoxynucleoside triphosphate (138, 139, 284). Under these conditions a covalent complex between the 80-kd protein and dCMP can be detected. Formation of the complex requires the presence of template (adenovirus DNA-55-kd terminal protein complex) and ATP. The reaction is specific for dCTP. Most, if not all, of the 80-kd protein-dCMP complex that is formed is bound by noncovalent bonds to the input DNA template. The results of pulse-chase experiments suggest that 80-kd protein-dCMP complexes can be incorporated into newly synthesized adenovirus DNA upon addition of the other three triphosphates. These results are consistent with the basic elements of the protein priming model. It should now be possible to test other predictions of the model by further analysis of the protein-nucleotide joining reaction, including its reconstitution from purified components.

PAPOVAVIRUS DNA REPLICATION

The genomes of the papovaviruses, simian virus 40 (SV40) and polyoma virus, represent simple models for a single mammalian cell replicon. The genomes are circular, duplex DNA molecules of about 5.5 kb. Both within virions and within the nucleus of infected cells, the viral DNA is condensed by cellular histones into a "minichromosome" (140–142). The complete nucleotide sequences of polyoma and SV40 are known, (143–145) and the structure and genetic organization of these viruses has been studied extensively (reviewed in 146–148).

Initiation

Papovavirus DNA replication begins at a unique site and proceeds bidirectionally (149–151). Initiation requires an interaction between a specific nucleotide sequence and the product of the viral A gene (T antigen). In spite of the fact that the initiation of papovavirus replication in vitro has never been conclusively demonstrated, considerable insight into the process has been gained by genetic analysis of the viral replication functions and by studies with purified T antigen. The exact nucleotide with which replication begins is not known. The site of initiation has, however, been located on the viral genome by electron microscopic analysis of replication intermediates (149, 150), and by analysis of the distribution of radioactivity in pulse-labeled viral DNA (151). In addition, limits on the boundaries of the origin have been deduced from studies of viable deletion mutants and evolutionary variants (152–158). The most striking structural feature in the origin region is a 27-bp perfect palindrome, flanked on one side by 17 A/T base pairs (159–161). It is likely that this structure is important in initiation, since it has been shown that both deletions and base substitution mutations within (or within a few base pairs of) the 27-bp palindrome cause cis-acting defects in viral DNA replication (162–168).

The initiation of papovavirus DNA replication also requires a viral gene product. Temperature-sensitive mutants that carry a lesion in the coding region for the viral T antigen (tsA mutants) fail to initiate DNA replication at the nonpermissive temperature (169, 170). The SV40 T antigen is an 82-kd phosphoprotein located primarily in the nucleus of cells infected by SV40 (171–175). T antigen, and the structurally related D2 hybrid protein (produced by the adenovirus SV40 hybrid, AD2$^+$D2), bind specifically to SV40 DNA or SV40 chromatin at or near the origin of replication (176–184). Tjian has shown that the D2 protein binds sequentially to three tandemly arranged binding sites in the origin region (181). At low concentrations, the D2 protein binds to a site (site I) located between the initiation codon for early proteins and the 27-bp palindrome. At higher protein concentrations it binds to an adjacent site (site II) centered on the palindrome, and at highest concentrations, the D2 protein binds to a third site (site III) located adjacant to site II toward the coding region for late proteins. Methylation-protection experiments have shown that the D2 protein makes symmetrical (or nearly symmetrical) contacts in the major groove at each binding site (181). The pentanucleotide sequence GCCTC is present twice in site I and four times in site II; it is probable that this sequence represents at least part of the recognition sequence for T antigen.

It seems likely that the interaction between T antigen and binding site II is necessary for the initiation of viral DNA replication. As mentioned, deletions and base substitutions within the palindrome cause cis-acting defects in DNA replication. These defects have been correlated with the failure of T antigen to bind to the mutant DNA (168, 185). Moreover, conditionally lethal mutations within binding site II can be suppressed by a second mutation within the coding region for T antigen (186). It is of interest to note that several independent second-site supressor mutants have missense changes within the same short region of the T antigen coding sequence. Each of these T antigens supresses a variety of different origin mutations; thus, mutations within this region of T antigen may relax the stringency of the interaction between T antigen and DNA (R. Margolskee and D. Nathans, personal communication).

The possible role of the binding of T antigen to site I in DNA replication is much less clear. Mutants in which part or all of site I is deleted are viable (165, 187); thus binding to site I is not essential for initiation. On the other hand, it has been reported that viral DNA lacking most of site I replicates somewhat less efficiently than wild-type DNA (168). This effect may be an indirect consequence of cooperative interactions between T antigen molecules, i.e. binding of T antigen to site I may stimulate binding at site II (168, 188).

The events that serve to establish bidirectional replication following the binding of T antigen to the origin are completely unknown. One approach to this problem has been to survey purified T antigen for various enzymatic activities. T antigen copurifies with an ATPase activity (189–191) that is not dependent on the presence of DNA, although it is stimulated two to ten fold by poly(dT) (189, 191). It has been suggested that T antigen may use the free energy derived from the hydrolysis of ATP to promote local unwinding of the DNA helix at the origin as a prelude to the initiation of DNA synthesis. While this remains a reasonable suggestion, no T antigen-promoted DNA unwinding has yet been demonstrated. The association of T antigen with a protein kinase activity has also been reported (190). It is not yet clear whether this activity is intrinsic to the T antigen protein or is due to a contaminant.

Although the papovaviruses represent useful models for studying the mechanisms by which DNA replication can start at internal sites, there may be limitations to the analogy between viral and cellular initiation. First, replication of viral DNA differs fundamentally from that of cellular DNA in that multiple rounds of replication occur during a single cell cycle. Second, while the initiation of chromosomal replication is clearly regulated both spatially and temporally, there is as yet no direct evidence that it occurs at specific nucleotide sequences. Indeed, Harland & Laskey have recently reported that circular fragments of SV40 or polyoma DNA that lack the viral origin undergo one round of replication after injection into *Xenopus laevis* eggs (192, 193). Based on these results, it was suggested that the viral origin sequence and the T antigen protein do not have cellular analogues; rather, they constitute a device for circumventing the cellular mechanism that limits replication to one round per cell cycle. Harland & Laskey point out that the *Xenopus* egg may represent a special case, since it replicates its DNA much faster than adult cells. On the other hand these results dramatize the need for further characterization of the sites where chromosomal replication begins.

Chain Elongation

Several subcellular systems have been described that are capable of limited papovavirus DNA synthesis in vitro. These include unfractionated cell lysates (194), isolated nuclei (195), and purified complexes of replicating viral chromatin and associated replication proteins (196–198). These systems are capable, to varying degrees, of chain elongation on viral genomes that have initiated replication in vivo. Analysis of these systems, combined with studies of viral replication in vivo, have yielded a fairly detailed description of the events that occur during the process of elongating viral DNA. To date, the in vitro systems have been relatively refractory to attempts at fractionation and reconstitution (196); thus, very little is known about the enzymology of the replication process. A brief description of the biochemistry of the events at the replication fork is given below. An extensive review on this subject has recently been published (199).

Both strands of the viral genome are synthesized at each growing fork. Since all DNA polymerases synthesize DNA exclusively in the 5' to 3' direction, one strand (the lagging strand) must grow in a direction opposite that of the movement of the fork. It is generally accepted that this is accomplished by repeated initiations of short nascent strands (Okazaki fragments) on the lagging side of the fork (200). Nascent viral DNA in vivo and in vitro contains a significant fraction of short chains, ranging in size from about 40 nucleotides to 300 nucleotides, with an average length of about 135 nucleotides (201–204). Consistent with their postulated role as intermediates in DNA synthesis, these short chains are incorporated into longer DNA chains shortly after their synthesis.

It is generally believed that Okazaki fragments are initiated by the synthesis of a short oligoribonucleotide primer. In the case of papovavirus DNA replication, this idea is supported by the finding that a significant fraction of the Okazaki fragments synthesized in vitro contain an oligoribonucleotide at their 5' end (205–207). The primer oligonucleotides have a 5'-ribopurine triphosphate and are of a unique size (about 10 nucleotides), but do not have a unique sequence (206–208). All four ribo- and deoxyribonucleotides are present at the RNA-DNA junction in approximately equal amounts (209, 210). Recent evidence suggests that there are preferred sites for the initiation of Okazaki fragment synthesis on SV40 DNA (211), but the significance of this observation is not yet clear. Reichard & Eliasson (212) have shown that, like the *E. coli* primase (DNA G protein) (213), the enzyme system responsible for synthesizing the primer RNA is not absolutely specific for ribonucleotides. By controlling the relative concentrations of ribo- and deoxyribonucleoside triphosphates in vitro, it is possible to obtain primer oligonucleotides containing internal deoxynucleotides. This observation, and the fact that the synthesis of primers in vitro is not inhibited by α-ananitin (212), suggest that none of the three known cellular RNA polymerases is responsible for primer synthesis.

Before an Okazaki fragment can be incorporated into a long DNA chain, the RNA primer must be removed and the newly created gap must be filled in by a DNA polymerase. Little is known about the enzymes that carry out these steps. It is clear, however, that at least one of the factors involved is only loosely associated with replicating molecules and normally is found in the cytoplasmic fraction following disruption of infected cells (214, 215). Thus the replication of viral DNA either in washed nuclei or in isolated chromatin complexes is much less efficient than in whole cell lysates, and yields viral DNA containing unjoined Okazaki fragments (204). These Okazaki fragments do not contain RNA primers, so it is likely that the defect does not involve primer excision (204). The defect(s) can be completely corrected by the addition of a cytoplasmic extract from uninfected host cells (214, 215), or by an extract of *Drosophila* embryos (216). These observations form the basis of a complementation assay for the missing factors. Although purification attempts are now only at a preliminary stage, recent results suggest that the factors present in *Drosophila* embryos can be separated by gel filtration into two fractions, neither of which is DNA polymerase α (216) (see below).

A good deal of attention has been given to the question of whether Okazaki fragments are intermediates in the synthesis of both viral DNA strands or of only the lagging strand. There is now agreement that short nascent fragments are derived predominantly from the lagging side of the fork (217–220). Thus, it appears that the replication of papovavirus DNA is semidiscontinuous, i.e. the leading strand grows by an essentially continuous process and the lagging strand is synthesized discontinuously. There is evidence that RNA primer synthesis may occur on the leading side of the fork at a low frequency (220, 221), but it is doubtful that this is an obligatory feature of leading strand synthesis.

There is little direct evidence as to which, if any, of the known cellular DNA polymerases carries out the synthesis of DNA chains during papovavirus replication. Several indirect lines of evidence suggest that DNA polymerase α is the predominant replicative enzyme. Both DNA polymerase α and γ are associated with purified viral replication complexes; polymerase β is not found in detectable amounts (222–224). Aphidicolin, which inhibits purified polymerase α but not polymerase γ (225), is a potent inhibitor of SV40 replication in vivo and in vitro (226). Conversely, 2' 3'-dideoxythymidine triphosphate, which specifically inhibits purified polymerase γ, has little or no effect on SV40 replication in vitro (222, 226–228). Finally, the replication activity of isolated replication complexes that have been treated with N-ethylmaleimide can be restored by the addition of purified polymerase α (226). These experiments provide strong circumstantial support for an important role for polymerase α in viral replication; however, since the effect of inhibitors on purified DNA polymerases can be profoundly influenced by the assay system (229), a definitive answer will probably await the purification of the replicative polymerase on the basis of a functional assay.

Assembly of Viral Chromatin

It is clear that when the two parental strands of the viral genome are unwound at the replication fork, there must be some alteration in nucleosome structure at that point. It is also evident that newly synthesized DNA must be rapidly organized by histones into nucleosomes since (a) the histone to DNA ratio of replicating chromatin is not detectably different from that of mature viral chromatin (141) and (b) electron microscopic studies have failed to detect regions in replicating chromatin that are deficient in nucleosomes (230, 231). The fate of the histones on parental DNA has been examined by analyzing viral chromatin replicated in the presence of cyclohexamide, which effectively blocks histone synthesis (232). Nucleosome monomers were isolated from purified viral chromatin after micrococcal nuclease digestion. Hybridization analysis revealed that the nascent DNA in these nucleosomes was derived predominately from the leading side of the replication fork. This observation is consistent with a conservative model for histone segregation: at the replication fork, parental histones associate with the daughter duplex derived from the leading side of the fork, and new histones associate with the daughter duplex on the lagging side of the fork.

Although, as mentioned, the assembly of nascent DNA into nucleosomes appears to be rapid, Okazaki fragments have a structure that more closely resembles naked DNA than chromatin (233). It thus seems likely that the daughter duplex on the lagging side of the fork is usually assembled into chromatin after the Okazaki fragments are ligated to the 5' ends of long nascent DNA chains. It should be noted in this context that because the length of Okazaki fragments is approximately the same as the length of DNA contained in a nucleosome, several investigators have suggested that chromatin structure might dictate one or more of the details of Okazaki fragment metabolism (232–236). This remains an attractive but untested idea.

Termination of Replication and Segregation of Daughter Molecules

The segregation of the two circular sibling DNA molecules at the termination of papovavirus replication represents a good model for the merger of two replicons in a long eukaryotic chromosome. In the case of SV40, the rate of movement of each fork in a replicating molecule is approximately equal, and replication terminates when the two forks meet at a point on the genome diametrically opposite the origin (149, 151). Termination does not require a specific nucleotide sequence (237–239). The molecular events that result in two separated daughter molecules are not known, although it is clear that a transient discontinuity must be introduced into at least one of the parental strands. Unit length circular molecules containing a covalently continuous parental strand and a daughter strand with a short gap of about 50 nucleotides in the termination region are present at a low level (1–2% of viral molecules) in SV40-infected cells (240–242). It has also been reported recently that catenated circular dimers are present in relatively high amounts (10% of viral DNA molecules) (243). Pulse-chase experiments suggest that both of these structures might represent intermediates in the formation of covalently closed cirular monomer length genomes (240, 243, 244). Sundin & Varshavsky have proposed a simple model consistent with these results (243). In this model, replication at each of the two forks proceeds until the forks meet, at which point the two parental strands are no longer held together by hydrogen bonds. The resulting structure is a catenated circular dimer with a small gap in each of the nascent strands. Replication is completed by gap filling and ligation to form two covalently closed circles. The catenated circles can be resolved into monomer circles by one of the known mammalian topoisomerases, which are capable of introducing transient single-strand (topoisomerase I) or double-strand (topoisomerase II) breaks into DNA molecules (reviewed in 245). Late replication intermediates (about 90% replicated) have been reported to accumulate during SV40 DNA replication (211, 246). The relationship of these intermediates to the termination process is not yet clear.

YEAST PLASMID DNA REPLICATION

Yeast is rapidly emerging as one of the potentially most useful systems for studying eukaryotic DNA replication. There are several reasons for this. First, the genetics of DNA replication in yeast is well advanced. Over 32 complementation groups of temperature-sensitive mutants with defects in the control of the cell division cycle have been defined [cdc mutants; (247–249)]; several of the cdc mutations have been shown to affect DNA replication. Second, two classes of extrachromosomal DNA have been identified that appear to represent useful models for studying yeast chromosomal replication: the 2 μm plasmid, which occurs naturally in most yeast strains (250, 251), and chimeric plasmids (produced by in vitro recombination), which contain segments of yeast chromosomal DNA that confer the ability to replicate autonomously in yeast cells (252–256). Each of these plasmids can be isolated in large quantities in intact form. Finally, and perhaps most importantly, in vitro systems that replicate exogenously added plasmid DNA have recently been developed (257–259).

Autonomously Replicating Sequences

Auxotrophic mutants of yeast can be complemented by transformation with chimeric plasmids containing the corresponding wild-type yeast gene (260). The frequency of transformation is normally low, but is increased by about 1000-fold if the plasmid is capable of autonomous replication (252, 254, 255). Using high frequency transformation as an assay, segments of DNA that allow the autonomous replication of any attached DNA have been isolated from the chromosomal DNA both of yeast and a number of other eukaryotes (252–256, 261). It seems likely that at least some of these DNA segments contain origins of chromosomal DNA replication. There has been no direct demonstration as yet, however, that replication initiates within an autonomously replicating sequence either during chromosome replication or during the replication of plasmid DNA in vivo.

The frequency of autonomously replicating sequences in the yeast genome averages about 1 per 30 kb (256). Cloned segments of yeast DNA that contain an autonomously replicating sequence can be divided into two classes: segments that contain only single-copy sequences and segments that contain repetitive sequences in addition (256). One family of the repetitive class has been analyzed in some detail (262). Several independent clones containing *Sal* I fragments of yeast chromosomal DNA were isolated on the basis of homology to a prototype autonomously replicating sequence. All of the plasmids isolated in this way were able to replicate autonomously. Since the frequency of occurrence of autonomously replicating sequences in randomly cloned *Sal* I fragments is one in four, it seems likely that in this family the ability to replicate autonomously is tightly associated with the repetitive sequence. It is possible that the repetitive autonomously replicating sequences represent origins that are coordinately controlled, although a great deal more work is needed to verify this hypothesis.

Tschumper & Carbon (263) have determined the complete nucleotide sequence of two independent autonomously replicating sequences. *ars*-2 has been localized to a 100-bp segment of DNA near the *Arg*-4 gene. The *ars*-2 sequence does not have significant homology with any known animal, viral or prokaryotic replication origin. Its only obvious structural feature is a sequence of 10 A/T base pairs with a dyad axis of symmetry. *ars*-1 has been located on a 1.5-kb segment of unique sequence near the *Trp*-1 gene. The 1.5-kb segment contains three closely spaced 8–10-bp sequences in common with the *ars*-2 sequence; however, these common sequences are located several hundred base pairs away from sequences known to be required for the function of *ars*-1. The significance of this homology is thus not clear. Again, a great deal more data will have to be collected in order to determine what, if any, functional relationships exist between the various autonomously replicating sequences.

One additional property of these sequences is of note. Plasmids carrying autonomously replicating sequences are unstable when the cells are grown under nonselective conditions. It has been reported, however, that hybrid plasmids containing both an autonomously replicating sequence and sequences from the centromere of a yeast chromosome are stable mitotically and segregate 2+:2– during meiosis (262, 264, 265). Such plasmids thus represent ideal model systems for studying the mechanisms of control of segregation of yeast chromosomes.

The 2-μm Plasmid

The yeast 2-μm plasmid is a circular DNA molecule of about 6 kb present at a level of about 50–100 copies per cell in most yeast strains (250, 251, 266). The function of this plasmid in yeast is not known, but its structure is well defined (267–271). The plasmid consists of 2 unique segments of DNA (2.7 and 2.3 kb) separated by two copies of a 599-bp sequence that are perfect inverted repeats of each other. Reciprocal intramolecular recombination between the two inverted repeat sequences produces two intracellular forms of the plasmid that differ in the orientation of the unique segments relative to each other. The complete nucleotide sequence of the plasmid has recently been published (272).

The genetic organization of the 2 μm plasmid is not yet well understood. The nucleotide sequence contains three open reading frames with the potential for coding for 48-kd, 37-kd, and 33-kd polypeptides (272). Poly-A containing RNA transcripts spanning the three open reading frames have been identified in yeast strains that harbor the plasmid (273). The corresponding proteins have not yet been identified. An analysis of hybrid plasmids produced by in vitro recombination has shed some light on the possible functions encoded by the plasmid genome (274). A plasmid with an insertion into the small unique segment interconverts between the two possible forms only when transfected into a yeast strain carrying an endogenous wild-type plasmid. It thus seems likely that the plasmid encodes a diffusible product that facilitates intramolecular recombination within the inverted repeat segments. Similarly, another hybrid plasmid missing one inverted repeat and most of the unique regions is maintained at a lower copy number in strains not carrying an endogenous plasmid. On the basis of this finding, it has been suggested that the plasmid produces a trans-acting product that increases the efficiency with which it is replicated; of course, other interpretations are possible.

The intracellular location of the plasmid is not clear. The fact that it is transcribed into RNA (270, 273, 275), that it is organized by cellular histones into a chromatin-like structure (276, 277), and that its replication is under the control of nuclear genes (278), all point to the idea that the plasmid must reside within the nucleus for at least part of the cell cycle. However, nuclei isolated by conventional methods contain less than 5–10% of the total cellular plasmid (266). It has recently been reported that in growing cells, but not stationary-phase cells, the plasmid is associated with a rapidly sedimenting subnuclear structure termed the folded chromosome (279). Thus, it seems likely that the plasmid maintains some association with the nucleus, but the nature of this association is not known.

The replication of 2μm plasmid DNA in vivo has not been studied extensively, but a few important facts are known. In synchronized cells, plasmid replication occurs exclusively during early S phase (280). Replication does not occur at the restrictive temperature in cells bearing the cdc 28, cdc 4, and cdc 7 mutations (278), all of which prevent the passage of cells from G1 phase to S phase (248, 249). Thus, it seems certain that one or more nuclear genes expressed only during S phase are required for plasmid replication. Since only a small segment of the plasmid is absolutely required for autonomous replication (see below) it is likely that plasmid replication is carried out largely, if not exclusively, by nuclear replication proteins. Density-shift experiments have suggested that each of the nearly 100 copies of the plasmid per cell replicates once during each cell division cycle (280). Thus, the replication of the plasmid must be responsive to the cellular control mechanism that limits chromosomal DNA replication to one round per cell division.

The plasmid DNA appears to replicate via "Cairns" type intermediates, which are observable in the electron microscope (281). It seems likely that replication begins at a unique origin. Using high frequency transformation of the leu-2 gene as an assay, Broach & Hicks (274) have identified a cis-acting region of the plasmid genome that is essential for autonomous replication. The region is located near the junction of one of the inverted repeat sequences and the large unique region. The sequence of the 599-bp inverted repeat near the junction contains an extensive sequence of near perfect dyad symmetry, although the significance of this structure in DNA replication is not clear. The precise location of the origin of plasmid replication is not yet known, but preliminary electron microscopic studies and analysis of plasmid replication in vitro are consistent with the idea that the origin is within the region required for autonomous replication [C. Newlon, personal communication, cited in (274)].

Plasmid DNA Replication In Vitro

Recently, in vitro systems capable of replicating exogenously added plasmid DNA have been described (257, 258). These systems hold a great deal of promise for the eventual biochemical analysis of the proteins involved in yeast DNA replication. By the limited tests available, the replication that occurs in these in vitro systems appears to be similar to that which occurs in vivo. Replication initiates in vitro at one or more unique origins (see below) on the plasmid DNA and proceeds bidirectionally with formation of Cairns-type replication intermediates. In extracts prepared from the temperature-sensitive mutants cdc 28, cdc 4, cdc 7, and cdc 8, plasmid DNA replication is diminished when the cells are grown at the nonpermissive temperature (257, 258); moreover, replication in extracts prepared from the mutant cdc 8 grown at the permissive temperature is thermolabile relative to that observed with wild-type extracts (258). Thus it appears that plasmid replication in vitro depends on at least some of the gene products known to be required for plasmid replication in vivo, and in vitro complementation assays should be a viable approach to the purification of replication proteins.

A number of hybrid plasmids have been tested for their ability to serve as templates for DNA replication in vitro (258). All plasmids containing the putative in vivo origin of 2μm circle replication were replicated in vitro with approximately the same efficiency as authentic 2μm DNA. Plasmids lacking the 2μm origin were replicated less efficiently. Curiously, although the E. coli vector plasmid pBR322 was essentially inactive as a template for replication in vitro, the plasmid pMB9 was replicated with an efficiency of about 30% relative to the 2-μm plasmid. Moreover, an electron microscopic study of intermediates formed during the replication of a hybrid plasmid containing both 2-μm DNA and pMB9 DNA indicated the following: approximately 60% of initiation events occurred at or near the putative 2-μm origin; 25% occurred at a specific site within the pMB9 sequences, and the remainder were distributed randomly around the plasmid. Since pMB9 DNA will not replicate autonomously in yeast cells, it appears that the requirement for a specific origin sequence is less stringent in vitro than in vivo.

Very little is currently known about yeast replication proteins. Two DNA polymerases (DNA polymerase I and DNA polymerase II) have been identified (282). Both purified enzymes are sensitive to the drug aphidicolin (283). A yeast mutant expressing a DNA polymerase I that is resistant to aphidicolin has been isolated (283). Plasmid replication both in vivo and in extracts prepared from this mutant is relatively resistant to aphidicolin. Thus it seems very likely that the yeast DNA polymerase I is the predominant replicative polymerase.

CONCLUSIONS

Many of the general features of the replication of animal virus genomes and yeast plasmids are now understood. In addition, the number of potentially useful model systems may well be extended by the use of recombinant DNA techniques to isolate nonviral, autonomously replicating elements in higher eukaryotes. The exploitation of these systems should now make it possible to move from largely descriptive studies to more detailed analyses of the enzymology of eukaryotic DNA replication.

Literature Cited

1. Ward, D., Tattersall, P., eds. 1978. *Replication of Mammalian Parvoviruses.* Cold Spring Harbor, NY: Cold Spring Harbor Lab. 547 pp.
2. Berns, K. I., Hauswirth, W. W. 1982. In *Organization and Replication of Viral DNA,* ed. A. S. Kaplan. Cleveland: CRC
3. Siegl, G. 1976. *Virol. Monogr.* 15:1–30
4. Rose, J. A. 1974. *Comprehensive Virol.* 3:1–61
5. Berns, K. I. 1974. *Curr. Top. Microbiol. Immunol.* 65:1–30
6. Toolan, H. W. 1972. *Prog. Exp. Tumor Res.* 16:410–30
7. Bourguignon, G. J., Tattersall, P. J., Ward, D. C. 1976. *J. Virol.* 20:290–306
8. Salzman, L. A. 1977. *Virology* 72:12–16
9. Chow, M. B., Ward, D. C. 1978. See Ref. 1, pp. 205-17
10. Lavelle, G., Mitra, S. 1978. See Ref. 1, pp. 219-29
11. Salzman, L. A., Fabisch, P. 1979. *J. Virol.* 30:946–950
12. Astell, C. R., Smith, M., Chow, M. B., Ward, D. C. 1979. *Cell* 17:691–703
13. Mitra, S. 1980. *Ann. Rev. Genet.* 14:347–97
14. Tennant, R. W., Layman, K. R., Hand, R. E. 1969, *J. Virol.* 4:872–78
15. Tennant, R. W., Hand, R. E. Jr. 1970. *Virology* 42:1054–63
16. Hampton, E. G. 1970. *Can. J. Microbiol.* 16:266–68
17. Tattersall, P. 1972. *J. Virol.* 4:586–90
18. Siegl, G., Hallauer, C., Novak, A. 1972. *Arch. Gesamte. Virusforsch.* 36:351–62
19. Salzman, L. A., White, W. L., McKerlie, M. L. 1972. *J. Virol.* 10:573–77
20. Siegl, G., Gautschi, M. 1973. *Arch. Gesamte. Virusforsch.* 40:105–18
21. Rhode, S. L. III. 1973. *J. Virol.* 6:856–61
22. Rhode, S. L. III. 1974. *J. Virol.* 14:791–801
23. Parris, D. S., Bates, R. C. 1976. *Virology* 73:72–78
24. Salzman, L. A., White, W. 1973. *J. Virol.* 11:299–305
25. Ward, D. C., Dadachanji, D. 1978. See Ref. 1, pp. 297-314
26. Tattersall, P., Crawford, L. V., Shatkin, A. J. 1973. *J. Virol.* 12:1446–56
27. Gunther, M., May, P. 1976. *J. Virol.* 20:86–95
28. Rhode, S. L. III. 1977. *J. Virol.* 21:694–712
29. Li, A. T., Lavelle, G. C., Tennant, R. W. 1978. See Ref. 1, pp. 341–54
30. Hayward, G. S., Bujard, H., Gunther, M. 1978. See Ref. 1, pp. 327–40
31. Siegl, G., Gautschi, M. 1978. See Ref. 1, pp. 315–25

32. Mayor, H. D., Jordan, L. E. 1976. *J. Gen. Virol.* 30:337–42
33. Lavelle, G., Li, A. T. 1977. *Virology* 76:464–67
34. Siegl, G., Gautschi, M. 1976. *J. Virol.* 17:841–53
35. Revie, D., Tseng, B. Y., Grafstrom, R. H., Goulian, M. 1979. *Proc. Natl. Acad. Sci. USA* 76:5539–43
36. Faust, E. A., Ward, D. C. 1979. *J. Virol.* 32:276–92
37. Tseng, B. Y., Grafstrom, R. H., Revie, D., Oertel, W., Goulian, M. 1979. *Cold Spring Harbor Symp. Quant. Biol.* 43:263–70
38. Bates, R. C., Kuchenbuch, C. P., Patton, J. T., Stout, E. R. 1978. See Ref. 1, pp. 367-82
39. Pritchard, C., Stout, E. R., Bates, R. C. 1981. *J. Virol.* 37:352–62
40. Kolleck, R., Tseng, B. Y., Goulian, M. 1982. In press
41. Rhode, S. L. III. 1976. *J. Virol.* 17:659–67
42. Rhode, S. L. III. 1978. See Ref. 1, pp. 279-96
43. Kolleck, R., Goulian, M. 1981. *Proc. Natl. Acad. Sci. USA* 78:6206–10
44. Rose, J. A., Berns, K. I., Hoggan, M. D., Koczot, F. J. 1969. *Proc. Natl. Acad. Sci. USA* 64:863–69
45. Koczot, F. J., Carter, B. J., Garon, C. F., Rose, J. A. 1973. *Proc. Natl. Acad. Sci. USA* 74:560–63
46. Lusby, E., Fife, K. H., Berns, K. I. 1980. *J. Virol.* 34:402–9
47. Rose, J. A., Koczot, F. J. 1972. *J. Virol.* 10:1–18
48. Casto, B. C., Atchison, R. W., Hammon, W. McD. 1967. *Virology* 32:52–59
49. Straus, S. E., Sebring, E. D., Rose, J. A. 1976. *Proc. Natl. Acad. Sci. USA* 73:742–46
50. Straus, S. E., Sebring, E. D., Rose, J. A. 1978. See Ref. 1, pp. 243-55
51. Hauswirth, W. W., Berns, K. I. 1979. *Virology* 93:57–68
52. Hauswirth, W. W., Berns, K. I. 1977. *Virology* 78:488–99
53. Myers, M. W., Carter, B. J. 1981. *J. Biol. Chem.* 256:567–70
54. Handa, H., Shiroki, K., Shimojo, H. 1975. *J. Gen. Virol.* 29:239–42
55. Straus, S. E., Ginsberg, H. S., Rose, J. A. 1976. *J. Virol.* 17: 140–48
56. Ostrove, J. M., Berns, K. I. 1980. *Virology* 104:502–5
57. Myers, M. W., Laughlin, C. A., Jay, F. T., Carter, B. J. 1980. *J. Virol.* 35:65–75
58. Richardson, W. D., Carter, B. J., Westphal, H. 1980. *Proc. Natl. Acad. Sci. USA* 77:931–35
59. Berns, K. I., Hauswirth, W. W. 1978. See Ref. 1, pp. 13–32
60. Tattersall, P., Ward, D. C. 1976. *Nature* 263:106–9
61. Cavalier-Smith, T. 1974. *Nature* 250: 467–70
62. Lusby, E., Bohenzky, R., Berns, K. I. 1981. *J. Virol.* 37:1083–86
63. Kavenoff, R., Zimm, B. H. 1973. *Chromosoma* 41:1–27
64. Forte, M. A., Fangman, W. L. 1979. *Chromosoma* 72:131–50
65. Flint, S. J. 1980. In *DNA Tumor Viruses*, ed. J. Tooze, pp. 383-442. Cold Spring Harbor, NY: Cold Spring Harbor Lab. 958 pp.
66. Flint, S. J., Broker, T. R. 1980. See Ref. 65, pp. 443-546
67. Ziff, E. B. 1980. *Nature* 287:491–96
68. Kelly, T. J. Jr. 1981. See Ref. 2. In press
69. Wolfson, J., Dresser, D. 1972. *Proc. Natl. Acad. Sci. USA* 69:3054–57
70. Garon, C. F., Berry, K. W., Rose, J. A. 1972. *Proc. Natl. Acad. Sci. USA* 69:2391–95
71. Steenbergh, P. H., Maat, J., van Ormondt, H., Sussenbach, J. S. 1977. *Nucleic Acids Res.* 4:4371–89
72. Arrand, J. R., Roberts, R. J. 1979. *J. Mol. Biol.* 128:577–94
73. Dijkema, R., Dekker, B. M. M. 1979. *Gene* 7:15
74. Tolun, A., Alestrom, P., Pettersson, U. 1979. *Cell* 17:705–13
75. Shinagawa, M., Padmanabhan, R. 1980. *Proc. Natl. Acad. Sci. USA* 77: 3831–35
76. Robinson, A. J., Younghusband, H. B., Bellett, A. J. D. 1973. *Virology* 56:54–69
77. Robinson, A. J., Bellett, A. J. D. 1974. *Cold Spring Harbor Symp. Quant. Biol.* 39:523–31
78. Brown, D. T., Westphal, M., Burlingham, B. T., Winterhoff, U., Doerfler, W. 1975. *J. Virol.* 16:366–87
79. Sharp, P. A., Moore, C., Haverty, J. L. 1976. *Virology* 75:442–56
80. Keegstra, C. S., van Wielink, P. S., Sussenbach, J. S. 1977. *Virology* 76:444–47
81. Padmanabhan, R., Padmanabhan, R. V. 1977. *Biochem. Biophys. Res. Commun.* 80:955–64
82. Rekosh, D. M. K., Russell, W. C., Bellett, A. J. D., Robinson, A. J. 1977. *Cell* 11:283–95
83. Carusi, E. A. 1977. *Virology* 76:380–94
84. Desiderio, S. V., Kelly, T. J. 1981. *J. Mol. Biol.* 145:319–37
85. Challberg, M. D., Desiderio, S. V., Kelly, T. J. Jr. 1980. *Proc. Natl. Acad. Sci. USA* 77:5105–9

86. Stillman, B. W., Lewis, J. B., Chow, L. T., Mathews, M. B., Smart, J. E. 1981. *Cell* 23:497–508
87. Challberg, M. D., Kelly, T. J. Jr. 1981. *J. Virol.* 38:272–77
88. Sussenbach, J. S., van der Vliet, P. C., Ellens, D. J., Jansz, H. S. 1972. *Nature New Biol.* 239:47–49
89. Ellens, D. J., Sussenbach, J. S., Jansz, H. S. 1974. *Virology* 61:427–42
90. Lechner, R. L., Kelly, T. J. Jr. 1977. *Cell* 12:1007–20
91. Weingartner, B., Winnacker, E. L., Tolun, A., Pettersson, U. 1976. *Cell* 9:259–68
92. Sussenbach, J. S., Tolun, A., Pettersson, U. 1976. *J. Virol.* 20:532–34
93. Sussenbach, J. S., Kuijk, M. G. 1978. *Virology* 84:509–17
94. Daniell, E. 1976. *J. Virol.* 19:685–95
95. Lavelle, G., Patch, C., Khoury, G., Rose, J. 1975. *J. Virol.* 16:775–82
96. Flint, S. J., Berget, S. M., Sharp, P. A. 1976. *Cell* 9:559–71
97. Horwitz, M. S. 1976. *J. Virol.* 18:307–15
98. Tolun, A., Pettersson, U. 1975. *J. Virol.* 16:759–66
99. Kedinger, C., Brison, O., Perrin, F., Wilhelm, J. 1978. *J. Virol.* 26:364–71
100. van der Eb, A. J. 1973. *Virology* 51:11–23
101. Horwitz, M. S. 1971. *Virology* 8:675–83
102. Brison, O., Kedinger, C., Wilhelm, J. 1977. *J. Virol.* 24:423–43
103. Yamashita, T., Arens, M., Green, M. 1977. *J. Biol. Chem.* 252:7940–46
104. Bellett, A. J. D., Younghusband, H. B. 1972. *J. Mol. Biol.* 72:691–709
105. Pearson, G. D. 1975. *J. Virol.* 16:17–26
106. Vlak, J. M., Rozijn, T. H., Sussenbach, J. S. 1975. *Virology* 63:168–75
107. Winnacker, E. L. 1975. *J. Virol.* 15:744–58
108. Ariga, H., Shimojo, H. 1979. *Biochem. Biophys. Res. Commun.* 87:588–97
109. van der Vliet, P. C., Levine, A. J. 1973. *Nature New Biol.* 246:170–74
110. Levine, A. J., van der Vliet, P. C., Rosenwirth, B., Rabek, J., Frenkel, G., Ensinger, M. 1975. *Cold Spring Harbor Symp. Quant. Biol.* 39:559–66
111. van der Vliet, P. C., Levine, A. J., Ensinger, M., Ginsberg, H. S. 1975. *J. Virol.* 15:348–54
112. Lewis, J. B., Atkins, J. F., Baum, P. R., Solem, R., Gesteland, R. F., Anderson, C. W. 1976. *Cell* 7:141–51
113. Schechter, N. M., Davies, W., Anderson, C. W. 1980. *Biochemistry* 19:2802–10
114. van der Vliet, P. C., Keegstra, W., Jansz, H. S. 1978. *Eur. J. Biochem.* 86:389–98
115. Fowlkes, D. M., Lord, S. T., Linne, T., Pettersson, U., Philipson, L. 1979. *J. Mol. Biol.* 132:163–80
116. Nass, K., Frenkel, G. D. 1980. *J. Virol.* 35:314–19
117. van der Vliet, P. C., Sussenbach, J. S. 1975. *Virology* 67:415–26
118. Ensinger, M. J., Ginsberg, H. S. 1972. *J. Virol.* 10:328–39
119. van der Vliet, P. C., Sussenbach, J. S. 1975. *Virology,* 67:415–26
120. van der Vliet, P. C., Landberg, J., Jansz, H. S. 1977. *Virology* 80:98–110
121. Horwitz, M. S. 1978. *Proc. Natl. Acad. Sci. USA* 75:4291–95
122. de Jong, A., van der Vliet, P. C., Jansz, H. S. 1977. *Biochem. Biophys. Acta* 476:156–65
123. Bolden, A., Aucker, J., Weissbach, A. 1975. *J. Virol.* 16:1584–92
124. Abboud, M. M., Horwitz, M. S. 1979. *Nucleic Acids Res.* 6:1025–39
125. van der Vliet, P. C., Kwant, M. M. 1978. *Nature* 276:532–34
126. Longiaru, M. L., Ieda, J. E., Jarkovsky, Z., Horwitz, S. B., Horwitz, M. S. 1979. *Nucleic Acids Res.* 6:3369–86
127. Tate, V. E., Philipson, L. 1979. *Nucleic Acids Res.* 6:2769–85
128. Sergeant, A., Tigges, M. A., Raskas, H. J. 1979. *J. Virol.* 29:888–98
129. Beyer, A. L., Bouton, A. H., Hodge, L. D., Miller, O. L. 1981. *J. Mol. Biol.* 147:269–98
130. Sussenbach, J. S., van der Vliet, P. C. 1972. *FEBS Lett.* 21:7–15
131. Kaplan, L. M., Kleinman, R. E., Horwitz, M. S. 1977. *Proc. Natl. Acad. Sci. USA* 74:4425–29
132. Yamashita, T., Arens, M., Green, M. 1975. *J. Biol. Chem.* 250:3273–79
133. Challberg, M. D., Kelly, T. J. Jr. 1979. *Proc. Natl. Acad. Sci. USA* 76:655–59
134. Challberg, M. D., Kelly, T. J. Jr. 1979. *J. Mol. Biol.* 135:999–1012
135. Stillman, B. W. 1981. *J. Virol.* 37:139–47
136. Kaplan, L. M., Hiroyoshi, A., Hurwitz, J., Horwitz, M. S. 1979. *Proc. Natl. Acad. Sci. USA* 76:5534–39
137. Reiter, T., Futterer, J., Weingartner, B., Winnacker, E. 1980. *J. Virol.* 35:662–71
138. Lichy, J. H., Horwitz, M. S., Hurwitz, J. 1981. *Proc. Natl. Acad. Sci. USA* 78:2678–82
139. Challberg, M. D., Ostrove, J. M., Kelly, T. J. Jr. 1982. *J. Virol.* 41:265–70
140. Griffith, J. D. 1975. *Science* 187:1202–3

141. Cremisi, C., Pignatti, P. F., Croissant, O., Yaniv, M. 1976. *J. Virol.* 17:204–11
142. Shelton, E. R., Wassarman, P. M., DePamphilis, M. L. 1978. *J. Mol. Biol.* 125:491–514
143. Reddy, V. B., Thimmappaya, B., Dhar, R., Subramanian, K. N., Zain, B. S., Pan, J., Ghosh, P. K., Celma, M. L., Weissman, S. M. 1978. *Science* 200:494–502
144. Fiers, W. R., Contreras, G., Haegeman, G., Rogiers, R., Van de Voorde, A., van Heuverswyn, H. J., Volckaert, G., Ysebaert, M. 1978. *Nature* 273:113–20
145. Soeda, E., Arrand, J. R., Smolar, N., Walsh, J. E., Griffin, B. E. 1980. *Nature* 283:445–53
146. Kelly, T. J. Jr., Nathans, D. 1977. *Adv. Virus Res.* 21:85–126
147. Fareed, G. C., Davoli, D. 1977. *Ann. Rev. Biochem.* 46:471–522
148. Tooze, J. ed. 1980. See Ref. 65. 958 pp.
149. Fareed, G. C., Garon, C. F., Salzman, L. A. 1972. *J. Virol.* 10:484–91
150. Crawford, L. V., Robbins, A. K., Nicklin, P. M. 1974. *J. Gen. Virol.* 25:133–42
151. Danna, K. J., Nathans, D. 1972. *Proc. Natl. Acad. Sci. USA* 69:3097–3100
152. Lai, C. -J., Nathans, D. 1974. *J. Mol. Biol.* 89:179–93
153. Shenk, T. 1977. *J. Mol. Biol.* 113:503–15
154. Gutai, M. W., Nathans, D. 1978. *J. Mol. Biol.* 126:259–74
155. Subramanian, K. N., Shenk, T. 1978. *Nucleic Acids Res.* 5:3635–42
156. Magnusson, G., Berg, P. 1979. *J. Virol.* 32:523–29
157. Wells, R. D., Hutchinson, M. A., Eckhart, W. 1979. *J. Virol.* 32:517–22
158. Bendig, M. M., Folk, W. R. 1979. *J. Virol.* 32:530–35
159. Soeda, E., Arrand, J. R., Smolar, N., Griffin, B. E. 1979. *Cell* 17:357–70
160. Yang, R. C., Wu, R. 1979. *Science* 206:456–62
161. Seif, I., Khoury, G., Dhar, R. 1979. *Cell* 18:963–77
162. Shortle, D., Nathans, D. 1978. *Proc. Natl. Acad. Sci. USA* 75:2170–74
163. Shortle, D., Nathans, D. 1979. *Cold Spring Harbor Symp. Quant. Biol.* 43:663–68
164. Shortle, D., Nathans, D. 1979. *J. Mol. Biol.* 131:801–17
165. DiMaio, D., Nathans, D. 1980. *J. Mol. Biol.* 140:129–42
166. Shenk, T. 1978. *Cell* 13:791–98
167. Gluzman, Y., Frisque, R. J., Sambrook, J. 1979. *Cold Spring Harbor Symp. Quant. Biol.* 44:293–300
168. Myers, R. M., Tjian, R. 1980. *Proc. Natl. Acad. Sci. USA* 77:6491–95
169. Tegtmeyer, P. 1972. *J. Virol.* 10:591–98
170. Chou, J. Y., Martin, R. G. 1975. *J. Virol.* 15:145–56
171. Tegtmeyer, P. 1975. *Cold Spring Harbor Symp. Quant. Biol.* 39:9–17
172. Schaffhausen, B. S., Silver, J. E., Benjamin, T. L. 1978. *Proc. Natl. Acad. Sci. USA* 45:79–84
173. Tegtmeyer, P., Rundell, K., Collins, J. K. 1977. *J. Virol.* 21:647–57
174. Pope, J. H., Rowe, W. P. 1964. *J. Exp. Med.* 120:121–30
175. Rapp, F., Kitahara, T., Butel, J. S., Melnick, J. L. 1964. *Proc. Natl. Acad. Sci. USA* 52:1138–42
176. Hassell, J. A., Lukanidin, E., Fey, G., Sambrook, J. 1978. *J. Mol. Biol.* 120:209–47
177. Reed, S. I., Ferguson, J., Davis, R. W., Stark, G. R. 1975. *Proc. Natl. Acad. Sci. USA* 72:1605–9
178. Jessel, D., Hudson, J., Landau, T., Tenen, D., Livingston, D. M. 1975. *Proc. Natl. Acad. Sci. USA* 72:1960–64
179. Jessel, D., Landau, T., Hudson, J., Lalor, T., Tenen, D., Livingston, D. M. 1976. *Cell* 8:535–45
180. Tjian, R. 1978. *Cell* 13:165–79
181. Tjian, R. 1978. *Cold Spring Harbor Symp. Quant. Biol.* 43:655–62
182. Shalloway, D., Kleinberger, T., Livingston, D. M. 1980. *Cell* 20:411–22
183. Persico-DiLauro, M., Martin, R. G., Livingston, D. M. 1977. *J. Virol.* 24:451–60
184. Reiser, J., Renart, J., Crawford, L. V., Stark, G. R. 1980. *J. Virol.* 33:78–87
185. McKay, R. D. G., DiMaio, D. 1981. *Nature* 289:810–13
186. Shortle, D. R., Margolskee, R. F., Nathans, D. 1979. *Proc. Natl. Acad. Sci. USA* 76:6128–31
187. DiMaio, D., Nathans, D. 1982. *J. Mol. Biol.* In press
188. Myers, R. M., Rio, D. C., Roberts, A. K., Tjian, R. 1981. *Cell* 25:373–84
189. Tjian, R., Robbins, A., Clark, R. 1979. *Cold Spring Harbor Symp. Quant. Biol.* 44:103–11
190. Griffin, J. D., Spangler, G., Livingston, D. M. 1979. *Cold Spring Harbor Symp. Quant. Biol.* 44:113–22
191. Giacherio, D., Hager, L. P. 1979. *J. Biol. Chem.* 254:8113–16
192. Harland, R. M., Laskey, R. A. 1980. *Cell* 21:761–71
193. Laskey, R. A., Harland, R. M. 1981. *Cell* 24:283–84
194. DePamphilis, M. L., Beard, P., Berg, P. 1974. *J. Biol. Chem.* 250:4340–47

195. Winnacker, E.-L., Magnusson, G., Reichard, P. 1972. *J. Mol. Biol.* 72:523–37
196. Su, R. T., DePamphilis, M. L. 1978. *J. Virol.* 28:53–65
197. Edenberg, H. J., Waqar, M. A., Huberman, J. 1976. *Proc. Natl. Acad. Sci. USA* 73:4392–96
198. Gourlie, B. B., Krauss, M. R., Buckler-White, A. J., Benbow, R. M., Pigiet, V. 1981. *J. Virol.* 38:805–14
199. DePamphilis, M. L., Wassarman, P. M. 1980. *Ann. Rev. Biochem.* 49:627–66
200. Kornberg, A. 1980. *DNA Replication,* Ch. 11. San Francisco: Freeman
201. Fareed, G. C., Salzman, N. P. 1972. *Nature New Biol.* 238:274–79
202. Magnusson, G., Pigiet, V., Winnacker, E.-L., Abrams, R., Reichard, P. 1973. *Proc. Natl. Acad. Sci. USA* 70:412–15
203. Francke, B., Hunter, T. 1974. *J. Mol. Biol.* 83:99–121
204. Anderson, S., DePamphilis, M. L. 1979. *J. Biol. Chem.* 254:11495–11504
205. Hunter, T., Francke, B. 1974. *J. Mol. Biol.* 83:123–30
206. Kaufmann, G., Anderson, S., DePamphilis, M. L. 1977. *J. Mol. Biol.* 116:549–67
207. Eliasson, R., Reichard, P. 1978. *J. Biol. Chem.* 253:7469–75
208. Reichard, P., Eliasson, R., Soderman, G. 1974. *Proc. Natl. Acad. Sci. USA* 71:4901–5
209. Pigiet, V., Eliasson, R., Reichard, P. 1974. *J. Mol. Biol.* 84:197–216
210. Anderson, S., Kaufmann, G., DePamphilis, M. L. 1977. *Biochemistry* 16:4990–98
211. Tapper, D. P., DePamphilis, M. L. 1980. *Cell* 22:97–108
212. Reichard, P., Eliasson, R. 1978. *Cold Spring Harbor Symp. Quant. Biol.* 43:271–77
213. Rowen, L., Kornberg, A. 1978. *J. Biol. Chem.* 253:770–74
214. Otto, B., Reichard, P. 1975. *J. Virol.* 15:259–67
215. DePamphilis, M. L., Berg, P. 1975. *J. Biol. Chem.* 250:4348–54
216. Mastrome, G., Eliasson, R., Reichard, P. 1981. *J. Mol. Biol.* 153:627–44
217. Hunter, T., Francke, B., Bacheler, L. 1977. *Cell* 12:1021–28
218. Perlman, D., Huberman, J. A. 1977. *Cell* 12:1029–43
219. Kaufmann, G., Bar-Shavit, R., DePamphilis, M. L. 1978. *Nucleic Acids Res.* 5:2535–45
220. Narkhammar-Meuth, M., Kowalski, J., Denhardt, D. T. 1981. *J. Virol.* 39:21–30
221. Narkhammar-Meuth, M., Eliasson, R., Magnusson, G. 1981. *J. Virol.* 39:11–20
222. Edenberg, H. J., Anderson, S., DePamphilis, M. L. 1978. *J. Biol. Chem.* 253:3278–80
223. Otto, B., Fanning, E. 1978. *Nucleic Acids Res.* 5:1715–28
224. Tsubota, Y., Waqar, M. A., Burke, J. F., Milavetz, B. I., Evans, M. J., Kowalski, D., Huberman, J. A. 1979. *Cold Spring Harbor Symp. Quant. Biol.* 43:693–704
225. Ikegami, S., Taguchi, T., Okashi, M. 1978. *Nature* 275:458–59
226. Krokan, H., Schaffer, P., DePamphilis, M. L. 1979. *Biochemistry* 18:4431–43
227. Waqar, M. A., Evans, M. J., Huberman, J. A. 1978. *Nucleic Acids Res.* 5:1933–46
228. Van der Vliet, P. C., Kwant, M. M. 1978. *Nature* 276:532–34
229. Ikeda, J. E., Longiaru, M., Horwitz, M. S., Hurwitz, J. 1980. *Proc. Natl. Acad. Sci. USA* 77:5827–31
230. Seidman, M. M., Garon, C. F., Salzman, N. P. 1978. *Nucleic Acids Res.* 5:2877–93
231. Cremisi, C., Chestier, A., Yaniv, M. 1977. *Cold Spring Harbor Symp. Quant. Biol.* 42:409–16
232. Seidman, M. M., Levine, A. J., Weintraub, H. 1979. *Cell* 18:439–50
233. Herman, T. M., DePamphilis, M. L., Wassarman, P. M. 1979. *Biochemistry* 18:4563–71
234. Herman, T. M., DePamphilis, M. L., Wassarman, P. M. 1981. *Biochemistry* 20:621–30
235. Hewish, D. R. 1976. *Nucleic Acids Res.* 3:69–78
236. Rosenberg, B. H. 1976. *Biochem. Biophys. Res. Commun.* 72:1384–91
237. Lai, C.-J., Nathans, D. 1975. *J. Mol. Biol.* 97:113–18
238. Griffin, B. E., Fried, M. 1975. *Nature* 256:175–79
239. Brockman, W. W., Gutai, M. W., Nathans, D. 1975. *Virology* 66:36–52
240. Fareed, G. C., McKerlie, M. L., Salzman, N. P. 1973. *J. Mol. Biol.* 74:95–111
241. Laipis, P. J., Sen, A., Levine, A. J., Mulder, C. 1975. *Virology* 68:115–23
242. Chen, M. C. Y., Birkenmeier, E., Salzman, N. P. 1976. *J. Virol.* 17:614–21
243. Sundin, O., Varshavsky, A. 1980. *Cell* 21:103–14
244. Jaenisch, R., Levine, A. J. 1973. *J. Mol. Biol.* 73:199–212
245. Gellert, M. 1981. *Ann. Rev. Biochem.* 50:879–910

246. Seidman, M., Salzman, N. P. 1979. *J. Virol.* 30:600–9
247. Hartwell, L. H., Culotti, J., Pringle, J. R., Reid, B. J. 1974. *Science* 183:46–51
248. Hereford, L. M., Hartwell, L. H. 1974. *J. Mol. Biol.* 84:445–61
249. Hartwell, L. H. 1976. *J. Mol. Biol.* 104:803–17
250. Sinclair, J. H., Stevens, B. J., Sanghavi, P., Rabinowitz, M., 1967. *Science* 156:1234–37
251. Livingston, D. M. 1977. *Genetics* 86:73–84
252. Beggs, J. D. 1978. *Nature* 275:104–8
253. Stinchcomb, D. T., Struhl, K., Davis, R. W. 1979. *Nature* 282:39–43
254. Struhl, K., Stinchcomb, D. T. Scherer, S. S., Davis, R. W. 1979. *Proc. Natl. Acad. Sci. USA* 76:1035–39
255. Hsiao, C.-L., Carbon, J. 1979. *Proc. Natl. Acad. Sci. USA* 76:3829–33
256. Chan, C. S., Tye, B.-K. 1980. *Proc. Natl. Acad. Sci. USA* 77:6329–33
257. Jazwinski, S. M., Edelman, G. M. 1979. *Proc. Natl. Acad. Sci. USA* 76:1223–27
258. Kojo, H., Greenberg, B. D., Sugino, A. 1982. *Proc. Natl. Acad. Sci. USA* 78:7261–65
259. Scott, J. F. 1980. *ICN-UCLA Symp. Mol. Cell. Biol.* 19:379–88
260. Hinnen, A., Hicks, J. B., Fink, G. R. 1978. *Proc. Natl. Acad. Sci. USA* 75:1929–33
261. Stinchcomb, D. T., Thomas, M., Kelly, J., Selker, E., Davis, R. W. 1980. *Proc. Natl. Acad. Sci. USA* 77:4559–63
262. Chan, C. S., Maine, G., Tye, B.-K. 1982. *ICN-UCLA Symp. Mol. Cell. Biol.* 21: In press
263. Tschumper, G., Carbon, J. 1981. *ICN-UCLA Symp. Mol. Cell. Biol.* 22:489–500
264. Clarke, L., Carbon, J. 1980. *Nature.* 287:504–9
265. Hsiao, C.-L., Carbon, J. 1981. *Proc. Natl. Acad. Sci. USA* 78:3760–64

266. Clark-Walker, G. D., Miklos, G. L. G. 1974. *Eur. J. Biochem.* 41:359–65
267. Hollenberg, C. P., Degelmann, A., Kustermann-Kuhn, B., Royer, H. D. 1976. *Proc. Natl. Acad. Sci. USA* 73:2072–76
268. Guerineau, M., Grandchamp, C., Slonimski, P. P. 1976. *Proc. Natl. Acad. Sci. USA* 73:3030–34
269. Beggs, J. D., Guerineau, M., Atkins, J. F. 1976. *Molec. Gen. Genet.* 148:287–94
270. Gubbins, E. J., Newlan, C. S., Kann, M. D., Donelson, J. E. 1977. *Gene* 1:185–207
271. Cameron, J. R., Philippsen, P., Davis, R. W. 1977. *Nucleic Acids Res.* 4:1429–48
272. Hartley, J. L., Donelson, J. E. 1980. *Nature* 286:860–64
273. Broach, J. R., Atkins, J. F., McGill, C., Chow, L. 1979. *Cell* 16:827–39
274. Broach, J. R., Hicks, J. B. 1980. *Cell* 21:501–8
275. Guerineau, M. 1977. *FEBS Lett.* 80:426–28
276. Livingston, D. M., Hahne, S. 1979. *Proc. Natl. Acad. Sci. USA* 76:3727–31
277. Nelson, R. G., Fangman, W. L. 1979. *Proc. Natl. Acad. Sci. USA* 76:6515–19
278. Livingston, D. M., Kupfer, M. M. 1977. *J. Mol. Biol.* 116:249–60
279. Taketo, M., Jazwinski, S. M., Edelman, G. M. 1980. *Proc. Natl. Acad. Sci. USA* 77:3144–48
280. Zakian, V. A., Brewer, B. J., Fangman, W. L. 1979. *Cell* 17:923–34
281. Petes, T. D., Williamson, D. H. 1975. *Cell* 4:249–253
282. Chang, L. M. S. 1977. *J. Biol. Chem.* 252:1873–80
283. Sugino, A., Kojo, H., Greenberg, B. D., Brown, P. O., Kim, K. C. 1981. *ICN-UCLA Symp. Mol. Cell. Biol.* 22:529–53
284. Pincus, S., Robertson, W., Rekosh, D. 1981. *Nucleic Acids Res.* 9:4919–37

Ann. Rev. Biochem. 1982. 51:935–71

SUBUNIT COOPERATION AND ENZYMATIC CATALYSIS[1]

Charles Y. Huang, Sue Goo Rhee, and P. Boon Chock

Laboratory of Biochemistry, National Heart, Lung, and Blood Institute, National Institutes of Health, Bethesda, Maryland 20205

CONTENTS

Perspectives and Summary

Subunit structure is a common feature of macromolecular organization. It is now well recognized that most intracellular enzymes are composed of subunits. During the past two decades, our knowledge of the relationship between subunit interaction and enzymatic activity has improved tremendously, but many challenging problems still confront the protein chemist and the enzymologist. The core of current research efforts may be viewed as the quest for the functional significance of quaternary structure. Why do enzymes contain subunits? What are the inherent advantages of polymeric enzymes? How do the polypeptide chains communicate? In what way does subunit interaction affect enzymatic catalysis? The questions that can be raised are almost endless, and the complexities involved in them defy simple answers. However, these questions can be addressed in terms of existing concepts and recent developments.

One general benefit of subunit aggregation is that it permits the protein to maintain a more favorable surface-to-volume ratio, to shield certain hydrophobic regions, and to achieve structural symmetry (1, 2). Furthermore, the self-recognition of subunits in the association process may serve as a safeguard against errors of genetic translation. These possibilities may explain why some oligomeric enzymes apparently lack regulatory properties and display normal kinetic patterns.

The most notable advantage of polymeric enzymes is the regulation of catalytic activity through the cooperative effects of subunits. The prevailing models of Monod et al (2) and Koshland et al (3) have been successfully applied, e.g. by Schachman's and Koshland's groups, to the analysis of cooperative systems. Several modified versions have appeared, including one that incorporates negative cooperativity into the Monod-type model (4). Although from a kinetic standpoint a multisubunit enzyme is not required for the cooperative type of kinetics, since it can be generated by a monomeric enzyme that exists in several conformations at steady state or utilizes alternative pathways in a multisubstrate reaction, the fact that regulatory enzymes, almost without exception, are polymeric structures, argues strongly for the role of subunit interaction in cooperativity. The potential for achieving sophisticated control undoubtedly is greater in a polymeric, rather than a monomeric, enzyme. However, most of the present models are based on ligand-promoted conformational equilibria. While they adequately describe the change of enzymatic activity in response to metabolite concentration, they often fail to link intersubunit cooperation to the catalytic mechanism per se. An exception is the work of Berhhard and co-workers on sturgeon muscle glyceraldelhyde-3-phosphate dehydrogenase. They were able to show a direct catalytic role for the concerted transitions between alternative enzyme subunit structures.

Another advantage of subunit complexation is the combination of chemically or functionally different proteins into a single complex such that the overall catalytic or regulatory efficiency is enhanced or the substrate specificity is altered or broadened. Examples for these cases are found in tryptophan synthase (5), aspartate transcarbamoylase (6), lactose synthetase (7), and anthranilate synthetase (8). The activation of a number of enzymes by the calcium-binding protein calmodulin is another prominent example of how protein-protein interaction can be instrumental in the modulation of a wide range of cellular processes (9, 10).

A more specific role for subunit cooperation has been proposed by Lazdunski and by Boyer. The "flip-flop" model of Lazdunski (11, 12) was originally developed to explain the Michaelis-Menten type of kinetics displayed by enzymes exhibiting *strong* negative cooperativity[2] or half-of-the-sites reactivity. The "alternate-site" or "binding-change" mechanism advanced by Boyer (13) is a more general one that can account for various cases of negative cooperativity. With these models, ligand-induced subunit cooperation becomes an integral part of the catalytic process. The binding of substrate to one subunit is coupled to the acceleration of a catalytic step or to the facilitation of release of a product. An "exchange" of ligands is thus made possible through site-site interactions. The importance of this type of mechanism is that it may operate in other systems in which a tightly bound effector must be rapidly displaced in order to accomplish effective metabolic control with respect to time.

In recent years, new theories and diagnostic procedures have been formulated, and more experimental approaches, notably the hybridization technique, isotopic exchange, NMR, use of structural analogues such as alternative substrate and competitive inhibitor, and transient kinetic methods, have been applied to the study of structural and kinetic aspects of polymeric enzymes. As a consequence, the correlation between subunit interaction and enzymatic catalysis, at least in some instances, is now more firmly established.

Introduction

The correlation between subunit-subunit interactions and catalysis has been demonstrated for various enzymes over the past years. In most cases, however, the interactions involved are the ligand-promoted interactions among identical or similar polypeptide chains. The effects of such interactions on catalysis are often classified into two types: K_m (or K_d) effect and V (or k_{cat}) effect. In reality, these two effects cannot be cleanly separated. First, the Michaelis constant usually contains k_{cat} as one of its components. Second k_{cat} generally reflects the rate constant of the rate-limiting step, which may be the release of a tightly bound product rather than the conversion of the central complex; that is, the k_{cat} effect may be principally a product K_d effect. Third, the chemical equilibrium of an enzyme-catalyzed reaction as defined by the Haldane relationship clearly indicates that a change in K_m often must be accompanied by a change in V. If the binding of ligand to one subunit affects the K_d of another subunit, it may also affect its k_{cat}. Many cooperative phenomena are discussed in the literature in terms of K_d effect only, because binding rather than kinetic data are used.

[2]With "strong" negative cooperativity, Michaelian kinetics or normal binding isotherm will be observed. Deviations from normal behavior are seen only at extremely high ligand concentration.

When kinetic data are used, it is assumed that the enzyme-ligand equilibria are rapidly attained and the k_{cat} remains invariant. These assumptions certainly would not hold in a steady-state situation.

For some time the main feature of cooperative effects arising from subunit interaction has been considered to be the control of enzymatic activity as a function of substrate or effector concentration. Thus, the significance of positive cooperativity is thought to be manifested in the sigmoidal velocity versus substrate concentration curve, which shows how enzymatic activity can be dramatically changed by a relatively small fluctuation in ligand level; and the significance of negative cooperativity is how the enzyme can be made insensitive to a large fluctuation of a ligand. Recently, more attention has been given to the role of ligand-promoted subunit interaction on the catalytic mechanism. For example, both the flip-flop mechanism of Lazdunski and the alternative-site mechanism of Boyer stress how the binding of a ligand to a second subunit can be linked to the acceleration of a catalytic step on the first subunit. In this chapter, we concentrate on the developments connecting subunit interaction to enzymatic catalysis and the tools that have been employed to examine such relationships. Many related topics have been expertly reviewed; for instance, quaternary structure (14, 15); cooperative effects (16, 17); negative cooperativity and half-of-the-sites reactivity (11, 12, 18, 19); multienzyme complexes (5); and protein-protein interaction and enzymatic activity with emphasis on the self-associating enzymes (20). Our review is illustrative rather than comprehensive; we chose examples that focus on particular problems or democratic the use of a particular experimental approach.

Current Models

This section discusses current models describing the relationship between subunit interactions and enzymatic catalysis and defines some terminologies.

POSITIVE AND NEGATIVE COOPERATIVITY   Positive cooperativity is a phenomenon identifiable by a sigmoidal v vs S plot or a concave upward double reciprocal $(1/v$ vs $1/S)$ plot. Originally the term positive cooperativity meant that the binding of a ligand to one site promotes the subsequent binding of ligand to the other sites. In contrast, negative cooperativity (or anticooperativity) describes the phenomenon where the binding of one ligand impairs the subsequent binding of other ligands. The double reciprocal plot is concave downward. In a Hill plot, log $(v/V -v)$ vs log S, a Hill coefficient (slope of such a plot at the inflection point) greater or smaller than one is often taken as an indication of positive or negative cooperativity. The model of Monod, Wyman & Changuex (2) assumes that an enzyme having identical and noninteracting subunits arranged in a symmetrical manner can exist in two conformational states, T and R, which have different ligand binding affinities. A key feature of the Monod model is the conservation of symmetry, i.e. the T and R transition is a concerted one such that all the subunits in a given state are equivalent. The cooperative phenomenon, however, arises from the preexisting $T \rightleftharpoons R$ equilibrium and is an apparent cooperativity. The two-state Monod model is a limiting case that can be extended to accommodate more complicated situations. For instance, there is no reason why it cannot be extended to three or more configurational states. As noted by Frieden the dissociation-association of monomer-polymer is analogous to the Monod model except that the two conformational forms in equilibrium have a different number of subunits (21). Application of the model to kinetic, rather than binding, studies requires that the enzyme-ligand and the $T \rightleftharpoons R$ transition (or monomer $\rightleftharpoons$ oligomer reaction) equilibrate rapidly relative to the assay time. If the transition is slow so that the ligand-promoted transition does not occur to an appreciable extent during the course of assay for enzymatic activity, the system will behave like a mixture of two enzymes (or like a single enzyme if it initially exists predominantly in one state). A mixture of two enzymes with different kinetic characteristics acting on the same substrate may give rise to a concave downward double reciprocal plot resembling negative cooperativity. The original Monod model based on rapid equilibria, however, cannot generate negative cooperativity (22). A more recent proposal by Viratelle and Seydoux (4) called pseudoconservative transition is a modified Monod model that permits the two-state model to accommodate both positive and negative cooperativities. In this model, at least one of the states contains two classes of binding sites (or subunits) with distinct affinities for the same ligand (pairwise asymmetry). In other words, for a polymeric enzyme with n subunits, the $T_n \rightleftharpoons R_n$ transition is modified to a $T_n \rightleftharpoons R_{n/2}R'_{n/2}$ transition. The negative cooperativity, again, is an apparent one arising from the fact that the R form of the enzyme is an equimolar "mixture" of two different enzyme species.

The model of Koshland et al (3) is based on the induced-fit theory. Mathematically, this model is similar to those developed by Adair (23) and Pauling (24). In the simplest case, the subunits are initially identical and the binding of ligand to one subunit affects the subsequent binding of ligand to the remaining subunits. The model is therefore often referred to as the sequential model as opposed to the concerted model of Monod et al. It can be readily adapted to the case where the subunits are initially nonidentical (preexisting asymmetry). Since the binding of a ligand to one subunit can either improve or impair the binding affinities for the remaining subunits, positive, negative, and "mixed" cooperativities may result.

Apparent cooperative effects can occur for other reasons. The models described above are applicable to treatment of kinetic data only if rapid equilibrations of enzyme-ligand and subunit conformational states are assumed. Also, these models rarely consider multisubstrate reactions in which the mechanism of substrate addition becomes a factor. For example, rapid-equilibrium treatment of ordered binding of substrates A and B will result in an equation with only A and AB terms (the B-containing terms are missing), which may not correctly describe the mechanism. In a steady-state situation, multisubstrate reactions can give rise to power terms higher than the number of enzyme subunits. In addition, a monomeric enzyme may exhibit cooperative-type kinetics because of alternative binding or reaction pathways (25, 26). A number of enzymes have been shown to display positive cooperativity only in the presence of an allosteric (noncompetitive) inhibitor. With the Monod model, such a phenomenon has been interpreted as the enzyme existing predominantly in the R state and the presence of the inhibitor regenerating the cooperative effect by shifting the conformational equilibrium toward the T state. However, if the enzyme-substrate-inhibitor complex is active, the presence of inhibitor will create an alternative reaction pathway in a steady-state situation, thereby giving rise to apparent cooperativity. As has been mentioned in the perspectives and summary section, enzyme isomerization can likewise lead to apparent cooperative effects. Apparent negative cooperativity can also arise from preexisting asymmetry of enzyme subunit structure, substrate activation, and the presence of more than one enzyme form (enzymes acting on the same substrate or a mixture of native and partially denatured enzyme etc).

Differentiation of models of cooperative interaction based on the shape of the saturation curve is virtually impossible. In a few special cases, however, such a differentiation is possible. With the Monod model, the allosteric transition constant, L, in the presence of a ligand becomes $L[(1+ S/K_T)/(1+ S/K_R)]^n$, where K_R and K_T are the ligand dissociation constants for the R and T states, respectively. Thus, with the concerted mechanism, the enzyme can be totally converted to the R form even though not all the binding sites are occupied. The total conversion of aspartate transcarbamoylase to the R state by the binding of 3–4 mol of the bisubstrate analogue N-(phosphonacetyl)-L-aspartate to the six available sites is one such example (27). Under favorable conditions, distinction between the simple Monod model and the Koshland model or between true negative cooperativity and preexisting asymmetry also is possible provided a noncooperative, competitive ligand is available (28).

THE FLIP-FLOP AND ALTERNATE-SITE MODELS   The models for cooperative effects presented above emphasize the regulatory aspects of polymeric enzymes such that subunit cooperation becomes synonymous with allosteric control. Attempts to tie subunit interaction to a catalytic mechanism have been made by Lazdunski and Boyer. The flip-flop model was first advanced by Lazdunski (11, 29) to account for the Michaelian kinetics observed with many multisubunit enzymes, especially those displaying extreme negative cooperativity. The oligomeric enzyme is considered functionally to be a polydimer. The binding of substrate to the first subunit causes a strong anticooperativity such that the affinity of the second subunit for substrate is greatly decreased. A chemical event on the first subunit then revives the substrate-binding capacity for the second subunit, and the binding energy is utilized to facilitate the catalysis on the first subunit. The two subunits thus complete the flip-flop cycle by serving alternately as the catalytic site. The interdependence of the two subunits for each other's catalytic activity also serves to justify the need for subunit structure. The basic mechanism of Boyer's alternate-site model (13) is similar to the flip-flop model. A binding or a chemical event that occurs on one subunit can facilitate the product release on another subunit. However, variations of Boyer's model permit the simultaneous binding of substrate to both subunits and are capable of explaining the phenomena of general negative cooperativity (30). These mechanisms have been examined by alternative substrate (31), quenched-flow (32), and isotopic exchange techniques (33) and are discussed in greater detail in later sections.

Subunit Cooperation in Catalysis

ALKALINE PHOSPHATASE   Strong negative cooperativity has been observed for both inorganic phosphate (P_i)-binding and catalysis of (See Footnote 2) alkaline phosphatase isolated from either *Escherichia coli* (12, 34–36) or mammalian tissues (36, 37). The molecular characteristics of these enzymes are very similar. However, the *E. coli* enzyme has been more extensively investigated because of its availability in large quantity. *E. coli* alkaline phosphatase is a dimeric enzyme consisting of two identical subunits (38). Symmetry in subunit arrangement is indicated by X-ray crystallography (39), and by the fact that only a single ^{113}Cd NMR resonance was observed for the $E \cdot Cd_2$ complex (40). Alkaline phosphatase is a Zn^{2+} metalloenzyme (41), but Zn^{2+} can be substituted by a variety of metal ions such as Co^{2+}, Cd^{2+}, Mn^{2+}, Cu^{2+}, or Ni^{2+} (42–45). Currently, it is well established that at alkaline pH the enzyme can bind four equivalents of Zn^{2+} and two equivalents of Mg^{2+}, while only two Zn^{2+} per dimer are required for the enzymatic activity (46, 47). The catalysis involves the formation of a covalent intermediate E-P, the phosphoryl enzyme (48, 49). Most P_i-binding data show that one mole of phosphate is bound to each mole of dimer, with a dissociation constant of $\sim 10^{-6}$ M. However, at high phosphate concentration, the binding of a second equivalent of P_i could be detected (12). Studies of E-P formation from enzyme and $^{32}P_i$ also showed that the Zn^{2+} enzyme at low pH or the Cd^{2+} enzyme at pH 6.5 forms only one mole of E-P per mole of dimer (50–53). The stoichiometry is consistent with the results of most rapid kinetic studies at low pH, which show a burst of one equivalent of product formed per dimer during the presteady-state phosphorylation of the enzyme (53–59). At alkaline pH, rapid kinetic data also show a burst amplitude equivalent to the formation of one mole of product per dimer during the presteady-state phase (32), but a ^{31}P NMR study shows that the detectable enzyme-bound phosphate is the noncovalent complex, $E \cdot P$ (36). ^{31}P NMR technique was also used (60) to demonstrate that the stoichiometry of phosphate binding is dependent on the number of divalent metal ions bound to the enzyme. For example, when two metal ions, either Zn^{2+} or Cd^{2+}, bound to the enzyme, only 1 mol of phosphate was found noncovalently bound to the Zn^{2+} enzyme and 1 mol of covalently bound phosphate was formed with the Cd^{2+} enzyme at pH 8.0. In contrast, enzyme containing four Zn^{2+} or four Cd^{2+} plus two Mg^{2+} ions is capable of binding 2 mol of phosphate to form $E \cdot P_{i_2}$ or $E-P_2$ respectively. This was further confirmed by the observation that addition of two Cd^{2+} ions to the apodiphosphoryl enzyme, prepared by treatment of the four Cd^{2+} enzyme with a cheleting agent, results in dephosphorylation of one of the active sites. This provides a reasonable explanation for the variation in phosphate binding stoichiometry reported for this enzyme (61). To investigate the mechanism of the metal ion-dependent "half-site" binding activity, Otvos & Armitage (40) used ^{113}Cd NMR to show that when the first two Cd^{2+} were added to the apodimer, a single ^{113}Cd resonance is obtained, and phosphorylation of the Cd^{2+} enzyme is accompanied by the migration of one Cd^{2+} from the nonphosphorylated subunit to the phosphorylated subunit, thus leaving half of the subunits devoid of metal ion and hence incapable of binding phosphate. This substrate-induced cofactor migration is in accord with the observed negative cooperativity of binding reported for this enzyme. In the case of $Cd_4 Mg_2 E-P_2$, the ^{113}Cd chemical shift shows that the diphosphoryl enzyme is asymmetrical (36).

In order to resolve the apparent paradox between the anticooperative binding and the observed Michaelian kinetics, Lazdunski and his co-workers (11, 29) proposed a flip-flop mechanism in which a catalytic role is conferred to the idle subunit of this dimer in the overall reaction. Specifically, binding of the first substrate molecule causes a conformational change such that the affinity of the other subunit for substrate is greatly reduced. Phosphorylation of the enzyme then leads to the regaining of substrate-binding capacity for the other subunit, and the binding energy is utilized to facilitate the dephosphorylation process. The two subunits on the dimeric enzyme thus complete the flip-flop cycle by serving alternately as the catalytic site. With the flip-flop model one would expect a normal Michaelis-Menten kinetic response, because the second power terms in the initial rate equation for the flip-flop model can be cancelled to yield a Michaelis-Menten type of rate expression. Because in the presence of an alternative substrate an additional reaction pathway is generated in the flip-flop model such that noncancelable second power terms are present, alternative substrates have been employed to examine the mechanism. The flip-flop model is compared with the simple model proposed for this enzyme (35, 51, 53) in which the alternative substrate simply behaves like a competitive inhibitor. Results of such experiments using CMP as alternative substrate and 6-bromo-2-hydroxy-3-naphthoyl-O-anisidine phosphate (NASBIP) as substrate yielded a linear competitive inhibition pattern (31). When the alternative substrate is present at constant ratios to the substrate, linear intersecting double reciprocal plots are predicted (62) for the flip-flop model. Instead, parallel plots consistent with the simple model were obtained for AMP as substrate and CMP as alternative substrate, or 4-methylumbelliferyl phosphate (4-MUP) as substrate and AMP as alternative substrate (31). In addition, nonlinear, noncompetitive inhibition by the product P_i is predicted for the flip-flop model, but linear competitive inhibition was observed (31). Furthermore, k_{cat} determined at pH 8.0 and 25° in 0.1 M Tris·HCl is 27 sec^{-1}, which agrees well with the "off-rate" of P_i, $\sim$ 25 sec^{-1}, determined by Hull et al (61) using the NMR technique under identical conditions. If the k_{cat} and the off-rate of phosphate are of the same magnitude, then the substrate-facilitated pathway is unnecessary, unless binding of substrate will facilitate the conversion of the covalently bound phosphate to the noncovalently bound E·P_i complex. This possibility has been investigated by studying the dephosphorylation of the phosphoryl enzyme prepared at pH 5.7 in the presence or absence of various substrates and P_i using the pH jump technique (32). The results showed that the dephosphorylation step is not rate limiting in the catalytic cycle at pH 8.3, and that the presence of substrates or inhibitor has no effect on this rate. The rate constant determined for the dephosphorylation step agrees well with that reported by Aldridge et al (63). Lazdunski et al (29) had reported that a competitive inhibitor, p-chloroanilidophosphonate, can increase the dephosphorylation rate. This discrepancy can be accounted for by the fact that the later measurement was performed at pH 5.0 where the released phosphate can recombine with the enzyme to form a stable phosphorylated intermediate. Thus the observed enhancement in the dephosphorylation rate by the competitive inhibitor is likely derived from the binding competition between the inhibitor and phosphate. Kelly et al (64) have shown that an analogue of inorganic pyrophosphate, methylene diphosphonate, or imidodiphosphate at low concentration stimulates but at high concentration inhibits the rate of hydrolysis of p-nitrophenyl phosphate catalyzed by either the $E.\ coli$ or bovine intestine enzyme. The observed stimulation at low analogue concentration appears to support the flip-flop model if the analogue binds to the second site. However, a flip-flop model predicts that such activation would exhibit nonlinear double reciprocal plots ($1/v$ vs $1/S$). Since the plots for this activation are linear and the effect of the analogue is simply the improvement of substrate affinity (64), it suggests that the analogue binds to an allosteric site other than the substrate binding site. All the evidence presented so far, together with the lack of substrate effect on either the rate of dephosphorylation or the rate of phosphate dissociation, indicates that the flip-flop mechanism is not valid for alkaline phosphatase.

Since additional binding of Mg^{2+} to the Zn^{2+} enzyme results in an increase in the affinity of P_i at the second site (60), high levels of Mg^{2+} were used to explore its catalytic role. It was found that at a saturating Mg^{2+} level, with p-nitrophenyl phosphate as substrate, the K_m for the second site was reduced by more than 50-fold, but that Mg^{2+} exerts essentially no effect on the low K_m site (65). As a consequence, the difference between the two K_m values was suppressed from 270 to 5 μM by Mg^{2+} ions. In other words, Mg^{2+} can practically desensitize the strong negative cooperativity of $E.\ coli$ alkaline phosphatase. These data indicate that strong negative cooperativity, at least in the case of alkaline phosphatase, can provide a regulatory role. It provides a means for a rapid surge or decrease in enzymatic activity through changes in the concentration of an effector such as Mg^{2+} ions.

SUCCINYL COENZYME A SYNTHETASE   Succinyl-CoA synthetase catalyzes the reversible formation of succinyl-CoA from succinate and CoA with the concomitant cleavage of ATP to ADP and P_i (66, 67). The overall reaction is believed to proceed via three partial reactions involving a phosphorylated enzyme intermediate (68) and a tightly bound succinyl phosphate (69).

$$E + ATP \rightarrow E\text{-}P + ADP$$
$$E\text{-}P + \text{succinate} \rightarrow E\cdot\text{succinyl phosphate}$$
$$E\cdot\text{succinyl phosphate} + CoA \rightarrow E + \text{succinyl-CoA} + P_i \qquad 1.$$

Succinyl-CoA synthetase from $E.\ coli$ is a tetrameric enzyme with an $\alpha_2\beta_2$ structure (mol wt, 140,000; mol wt of α subunit, 30,000; mol wt of β subunit, 39,000) and contains two active sites, each of which is arranged at the interface of α and β subunits (67, 70, 71). The α subunit binds ATP and contains the active-site histidine residue that is phosphorylated as a catalytic intermediate (68, 72), while the β subunit contains sites for attachment of the substrates succinate and CoA (70, 71). Half-of-the sites reactivity was observed in this phosphohistidine intermediate formation: only one of the two identical α subunits could be phosphorylated at any one time (73, 74). The phosphorylation at one subunit induces a conformational change in the other subunit such that its susceptibility to attack by a variety of proteases decreases dramatically (74).

In this connection, Boyer and co-workers applied a new approach for the detection of substrate modulation of catalytic steps to this enzyme to assess whether *alternating site cooperativity* between two catalytic sites occurs (75). This method involved measurement of the effect of variation in substrate concentration on the extent of oxygen exchange between [^{18}O] P_i and succinate per molecule of ATP cleaved during steady-state succinyl-CoA synthesis. With the *E. coli* enzyme, which has an $\alpha_2\beta_2$ structure, a pronounced increase in oxygen exchange per ATP cleaved was observed, while variation in the CoA concentration did not have any effect. However, with the pig heart enzyme, which has an $\alpha\beta$ structure and possesses only one catalytic site, no modulation of oxygen exchange by ATP concentration was observed. These results led them to suggest that attachment of ATP (binding or phosphorylation of histidine residue) to one catalytic subunit of the *E. coli* enzyme promotes catalytic events at the other site. Their suggestion was further substantiated by the ^{31}P-NMR spectrum obtained by Vogel & Bridger (76). The addition of succinate alone to the monophosphorylated enzyme caused neither the decrease of the phosphohistidyl resonance, nor the formation of a new peak due to succinyl phosphate. Further addition of ATP to this already phosphorylated enzyme was necessary before a succinyl-phosphate resonance appeared. If a nonhydrolyzable analogue of ATP, AMPPCP, was used, however, no change in the phosphohistidyl resonance was produced. Therefore, it was concluded that catalysis of succinyl phosphate production does not occur at one active site unless phosphorylation by ATP occurs in the other half of the molecule. Utilizing hybrid enzymes ($\alpha'\alpha\beta_2$), Bridger and co-workers also showed that phosphorylation at one site also promotes fast release of succinyl-CoA from another site (67) and that the half-of-the-sites phosphorylation is not a consequence of permanent asymmetry, but a consequence of true *alternating sites cooperativity* (77).

Similar half-of-the-sites reactivity has been observed with malate thiokinase isolated from *Pseudomonas* MA (78, 79). Some structural similarity is also evident. Malate thiokinase, which catalyzes the reversible formation of malyl-CoA or succinyl-CoA from ATP, CoA, malate, or succinate, is an octameric enzyme with an $\alpha_4\beta_4$ structure (mol wt of α subunit = 34,000; mol wt of β subunit = 42,000).

MITOCHONDRIAL ATPase   Mitochondrial ATPase catalyzes the terminal step in the synthesis of ATP and P_i during oxidative phosphorylation.

In addition it is involved in virtually every ATP-dependent partial reaction of oxidative phosphorylation (80). This enzyme complex is composed of five different kinds of subunit, and exists in the form of $\alpha_n\beta_n\gamma_m\delta_m\epsilon_m$ where n and m are equal to or larger than 2 and 1, respectively (80). Currently, it is believed that the value of n is 3 and the catalytic site is located on the β subunit (33, 80–82). The enzyme complex is known to bind at least 5 mol of AMPPNP per mole of protein and three of these nucleotides bind to the "tight site" on the α subunits (83). The catalytic role of the tight nucleotide site is not known, even though it has been shown that binding of one AMPPNP or ADP to one of the tight sites results in a substantial inhibition of the ATPase activity (84, 85). Based on the observed slow off-rate of ADP or 1,N⁶-ethanoadenosine diphosphate from the tight nucleotide sites on the α subunits of chloroplast ATPase, Cantley & Hammes (86) ruled out the possibility that these tight nucleotide sites are catalytic sites for ATPase activity. Using the acid based–induced synthesis of ATP (87) and rapid mixing and quenching techniques (88), Boyer et al (13) demonstrated that the tightly bound ATP detectable in isolated submitochondrial particles is not an intermediate in the net synthesis of ATP. This evidence argues against a direct involvement of the tightly bound nucleotide in the synthesis and hydrolysis of ATP catalyzed by the β subunits. Nevertheless, these tightly bound nucleotides are involved in some way in the function of the enzyme. In addition, restoration of the enzyme activity has been shown to require the recombination of α, β, and γ subunits in the case of the *E. coli* enzyme (89).

The following observations suggest cooperative interaction between the catalytic subunits:

1. Nonlinear double reciprocal plots of MgATP concentration versus the rate of ATP hydrolysis were observed with the rat liver mitochondrial ATPase (90). These data indicate the existence of two active sites exhibiting negative cooperativity. Apparent negative cooperativity was also observed for inner membrane vesicle of rat liver mitochondria when the ATPase activity was monitored as a function of MgATP concentration in Tris·Cl buffer (91). However, this deviation from the normal Michaelian kinetics can be overcome by using Tris·bicarbonate buffer. The V in the latter is twofold higher than that obtained in the Tris·Cl system, and the K_m is about the same as the low K_m value determined for the Tris·Cl buffer (91). Thus it appears that the bicarbonate anion is capable of desensitizing the negative cooperativity of this ATPase. Similar results were also reported for the purified enzyme (91). On the other hand, steady-state kinetics of ATP hydrolysis catalyzed by either membrane-bound or soluble mitochondrial ATPase from bovine heart followed a normal Michaelian kinetic pattern when the reaction was studied in the presence of 0.1 M Cl⁻ anion (92).

2. Modification of a single tyrosine residue on only one of the β subunits with 4-chloro-7-nitrobenzofurazan leads to complete inactivation of both ATPase and oxidative phosphorylation activity of the bovine heart mitochondrial enzyme (81). Similarly covalent modification of a single tyrosyl residue on the chloroplast ATPase prevents ADP and AMPPNP binding at the catalytic site and abolishes the ATPase activity (86). Based on this observation and the assumption that two β subunits are involved in catalysis, Cantley & Hammes (86, 93) have proposed a half-of-the-sites reactivity for the chloroplast ATPase. In addition, modification of the arginine residues also results in inactivation of the beef heart enzyme (94). The incorporation of the first two molecules of arginine reagent, phenylglyoxal, led to 50% inhibition, and the inactivation of the remainder of activity required the incorporation of up to 16 mol of phenylglyoxal per mole of enzyme. Since two molecules of phenylglyoxal is known to condense with one arginine (95), the incorporation of the first two phenylglyoxal corresponds to the modification of a single ariginyl residue. The observed biphasic inactivation by phenylglyoxal is consistent with either subunit interaction or preexisting assymmetry.

In addition to the evidence described above, which suggests the possibility of subunit interaction being involved in mitochondrial and chloroplast ATPase catalysis, Boyer and co-workers (33) reported that: (a) when soluble mitochondrial ATPase hydrolyzes ATP, the extent of water oxygen incorporation into each P_i formed increases markedly as ATP concentration is lowered (13, 30, 96). This suggests that enzyme-bound ATP reversibly forms enzyme-bound ADP and P_i and the bound ADP and P_i cannot be released until ATP binds to a low affinity site. A similar observation was also reported for submitochondrial particles (97). (b) During ATP hydrolysis catalyzed by submitochondrial particles, removal of ADP by pyruvate kinase inhibits $P_i \rightleftharpoons$ ATP and ATP $\rightleftharpoons H_2O$ exchanges (98). This suggests that ADP binding is required for ATP release. (c) During net oxidative phosphorylation, removal of medium ATP inhibits the medium $P_i \rightleftharpoons H_2O$ exchange even though intermediate ATP $\rightleftharpoons H_2O$ exchange continues (99), as if ATP is required for P_i release. In addition, a decrease of either ADP or P_i concentration during synthesis of ATP by submitochondrial particles greatly increases the extent of water-oxygen incorporation into each ATP formed (99), as if binding of P_i and ADP can facilitate the release of ATP (33) (d) Rapid kinetic study of acid base–induced synthesis of ATP catalyzed by chloroplast thylakoid membrane showed that an ATP tightly bound to the isolated membranes is a transient intermediate in the catalytic sequence for ATP synthesis (100). Based on these observations, Boyer (33, 98, 99) proposed an energy-linked binding-change mechanism, sometimes referred to as the binding-change or alternating site model, to explain the experimental data. In this mechanism, during net oxidative phosphorylation, ATP is formed at one site but is transitorily tightly bound and not released until ADP and P_i bind at a second site. The binding in turn converts the tight ATP site to a "loose" ATP site and simultaneously the loose ADP and P_i sites are converted to tight sites. Under conditions of net ATP hydrolysis, the tightly bound ATP at one site is hydrolyzed to form tightly bound ADP and P_i, and binding of the second ATP to the loose site results in the conversion of the tight ADP and P_i sites to loose sites and loose ATP sites to tight sites. In essence, this proposed mechanism is similar to the flip flop model of Lazdunski except that in Boyer's model the second substrate molecule can bind to low affinity site while in Lazdunski's case, the low affinity site is converted to a high affinity site prior to substrate binding. Therefore, Lazdunski's model would predict Michaelian kinetics while Boyer's model could explain the nonlinear double reciprocal plot of MgATP concentration versus the velocity of ATP hydrolysis reported by Ebel & Lardy (90). In both mechanisms, product(s) release is rate limiting in the catalytic cycle, and binding of the second substrate molecule facilitates the rate of products release. Using the fact that purified ATPase is known to bind one mole of phosphate very tightly per mole of enzyme, and the off-rate of the bound phosphate is slow $[t_{\frac{1}{2}} \simeq 2$ min (101, 102)], Hutton & Boyer demonstrated that this slow phosphate off-rate is greatly accelerated by the addition of ATP, and to a lesser extent ADP (96). This observation is in harmony with the binding-change mechanism proposed for ATPase. A similar mechanism has also been proposed by Adolfken & Moudrianakis (103).

Using an ATP analog, 2',3'-O-(2,4,6-trinitrophenyl)ATP (TNP-ATP) Grubmeyer & Penefsky (104) showed that beef heart mitochondrial ATPase contains two high affinity sites for TNP-ATP and the two sites are equally effective in catalyzing the hydrolysis of TNP-ATP to TNP-ADP and P_i. The K_m for TNP-ATP hydrolysis is about 1000 times lower than that for ATP, and the V is about 600 times slower. Results of the inhibition studies indicate that TNP-ATP binds to the catalytic site of the enzyme. Binding studies show that the difference in binding affinity for the first as compared to the second TNP-ATP bound is about two to three orders of magnitude. The values of K_d were determined in the absence of Mg^{2+}, and these values (the first one is too tight to be measured) are very much lower than the single K_m value determined in the presence of Mg^{2+}. In addition, during prolonged incubation of the enzyme with a high concentration of TNP-ATP in Mg^{2+} buffer, the enzyme appeared to pick up a third TNP-ATP. The rate of TNP-ATP hydrolysis was found to be very slow when the ratio of enzyme to substrate was raised such that only site 1 was occupied by the substrate (105). However, the hydrolysis of site 1–bound TNP-[γ-^{32}P]ATP was accelerated by 15- to 20-fold when sufficient TNP-

ATP, GTP, ATP, or ITP was added to the reaction mixture to fill site 2. This observation, together with the fact that the two sites on the enzyme catalyze the hydrolysis of TNP-ATP at the same rate (104) and the likelihood that the two catalytic sites are located on two identical subunits (106), is consistent with catalytic site cooperativity in the mechanism of action of mitochondrial ATPase. The acceleration of the rate of hydrolysis was also observed with ATP as substrate, but the enhancement factor is significantly lower than that with TNP-ATP as substrate (105). This is in agreement with the fact that the enzyme binds TNP-ATP much tighter than ATP, and the addition of excess nucleotide can significantly replace the bound [γ-^{32}P]ATP. Based on the fact that much of the bound substrate, either TNP-ATP or ATP, remained unhydrolyzed, Grubmeyer & Penefsky (105) proposed that product release is not rate limiting and the binding of ATP to the second catalytic site facilitates the hydrolysis of ATP at site 1. This explanation is inconsistent with the observation that the extent of water oxygen incorporation into each P_i formed increases markedly with the lowering of ATP concentration (30, 96, 97). However, all the data can be accommodated by a composite mechanism (Scheme 1).

In this modified binding-change scheme, rapid equilibrium in Step *3* now greatly favors the E' $_{<ATP}$ form. In the absence of high ATP concentration, the hydrolysis proceeds via steps *1 → 2 → 3 → 4* where Step *4* is very slow. When ATP concentration is sufficiently high to saturate the second site, ATP hydrolysis proceeds via the substrate-facilitated pathways, namely, *3 → 5 → 8 → 9* and *6 → 7 → 8 → 9*. In this scheme, the equilibrium that is expected to favor the formation of ADP and P_i over ATP is reversed through enzyme-induced stabilization of E'$_{<ATP}$ over E"$<$ADP,P_i, and the formation of the E'$_{<ATP}^{\cdot ATP}$ species at high ATP concen-

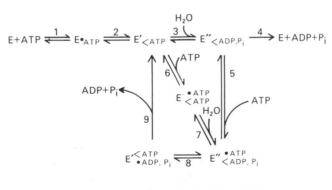

< TIGHTLY BOUND LIGAND
• LOOSELY BOUND LIGAND

Scheme 1

tration would explain the observed negative cooperative kinetic pattern (90) and the binding stoichiometry reported by Grubmeyer & Penefsky (104). In essence, there exists a significant amount of evidence in support of the alternating-site mechanism in the case of the mitochondrial ATPase system. However, to clearly demonstrate the catalytic mechanism of this enzyme system, more work is needed to elucidate the role of the tight nucleotide site and of the other subunits involved.

The alternative-site mechanism has also been proposed for the other two better studied membrane ATPases, namely, (Na^+, K^+)-ATPase (107–111) and the Ca^{2+}-ATPase (112–114) of sarcoplasmic reticulum. However, current data on the former show that only a single nucleotide binding site exists per functional unit of the enzyme, and the observed high and low affinity ATP sites are derived from the interconversion between two identical enzyme forms (11, 115, 116). In the case of Ca^{2+}-ATPase, evidence shows that the low affinity ATP site appears to be an allosteric site on the monomer (117–119). However, further work is needed to rule out the possible role of subunit aggregates in the catalytic process.

AMINOACYL tRNA SYNTHETASE   Many aminoacyl tRNA synthetases exhibit negatively cooperative binding and half-of-the-sites reactivity (120, 121). For convenience, we discuss them as a group; however, it is not our intent to suggest that all the synthetases function in a negatively cooperative manner. These enzymes catalyze the binding of amino acids to the 3' termini of transfer RNA. The reaction involves first the activation of an amino acid to form an aminoacyl adenylate, followed by the transfer of the amino acid to a specific tRNA (121–125) as shown in Equation 2

$$E + AA + ATP \underset{PP_i}{\overset{}{\rightleftarrows}} E \cdot AA\text{–}AMP \xrightarrow{\text{tRNA}} AA\text{–}tRNA + E + AMP \qquad 2.$$

where AA-AMP denotes aminoacyl adenylate and AA an amino acid. This mechanism is supported by the fact that (a) in the absence of tRNA, the enzyme catalyzes the formation of aminoacyl adenylate with a rate equal to 20 times or more the rate of aminoacylation reaction (123, 126, 127). (b) The rate for the transfer of amino acid from aminoacyl adenylate to the tRNA, e.g. Ile from Ile-AMP to $tRNA^{Ile}$, is found to be the same as the k_{cat} for the aminoacylation of the tRNA under similar conditions. In addition, when isoleucyl-tRNA synthetase was incubated with $[^{14}C]$-isoleucine, tRNA^{Ile}, and [γ^{32}P]ATP, a "burst" of pyrophosphate formation was detected, but no burst was observed for the formation of Ile-tRNA^{Ile} (128).

Among the synthetases, one of the clear examples for subunit interaction is the tyrosyl-tRNA synthetase from *Bacillus stearothermophilus*. This enzyme is composed of two identical subunits of molecular weight 44,000 (129). The X-ray data, with a resolution of 2.7 Å, show that the two subunits are symmetrically arranged in the dimer (130). The catalytic site on each subunit is situated 18 Å from the twofold axis, such that any small ligand bound at the first site would have minimal direct effect on the second site. However, substrate binding studies show that the dimeric enzyme binds only one L-tyrosine (131). In the presence of L-Tyr, ATP, and pyrophosphatase (to remove pyrophosphate such that the reaction would favor the formation of E·AA-AMP), it was found, either by the gel filtration method (131) or by monitoring the amplitude of the reaction burst (132), that only one tyrosyl adenylate was formed per dimer. However, once the tyrosyl adenylate was formed, and prior to its release, the enzyme was capable of binding a second ATP and Tyr. This is evident by the fact that Tyr enhances the rate of hydrolysis of E^{Tyr}·Tyr-AMP complex and that ATP exerts a strong synergistic effect on the K_d of Tyr and the enhancement factor. Based on the observation that with saturating concentrations of Tyr and ATP, ATP hydrolyzed twice as fast as E·Tyr-AMP, Fersht (120) suggested that E·(Tyr-AMP)$_2$ was formed. This interpretation is complicated by the fact that ATP hydrolysis was detected in the absence of Tyr, thus the threefold enhancement by Tyr could be due to a separate reaction. Nevertheless, all these data suggest that binding of the second Tyr and ATP facilitates the release or the breakdown of tyrosyl adenylate. The net result of this effect on the overall reaction (Equation 2) is minimal, since the rate of aminoacylation is very much faster than the rate of E·Tyr-AMP hydrolysis (120, 121, 133). Based on the finding that binding of the second Tyr and ATP induces the E·Tyr-AMP complex to a new conformational state, Fersht et al (134) suggested that this additional substrate binding would lead to stabilization of the high energy intermediate, Tyr-AMP.

Binding of tRNA^{Tyr} to tyrosyl-tRNA synthetase also exhibits negative cooperativity. In the presence of Mg^{2+}, two equivalents of tRNA^{Tyr} can bind to each dimer with a binding affinity ten times higher for the first site compared with that for the second site (131). However, data obtained from equilibrium gel filtration in the absence of Mg^{2+} show that the enzyme only binds one equivalent of tRNA^{Tyr} (135). Similar binding patterns were also obtained with *E. coli* tyrosyl-tRNA synthetase, a dimeric enzyme. Both *E. coli* and *B. stearothermophilus* enzymes exhibit biphasic initial-rate kinetics as a function of Tyr concentration (135, 136). The kinetics of tRNA^{Tyr}-synthetase interaction has also been reported (137, 138). The results show that tRNA binds to the synthetase by a two-step mechanism; namely, a nearly diffusion controlled initial binding step followed by a conformational transition. In fact the two-step tRNA binding mechanism has also been observed for other synthetases such as the yeast Ser-tRNA synthetase (137–139) and Phe-tRNA synthetase (140) systems. This mechanism has been proposed as a means for the enzyme to discriminate noncognate tRNAs. The initial binding step has a broad specificity, while the conformational change step would discriminate the noncognate tRNA. This idea is supported by the fact (140) that only one relaxation was detected when yeast Phe-tRNA synthetase was reacted with *E. coli* tRNA^{Tyr}, whereas a two-step reaction was observed with yeast tRNA^{Phe}.

Other synthetases that have been reported to exhibit negative cooperativity in either substrate binding or catalysis include the monomeric enzymes such as valyl-tRNA synthetase from *B. stearothermophilus* (120, 123, 141) and isoleucyl-tRNA synthetase from *E. coli* (128, 142); the dimeric enzymes (α_2) such as seryl-tRNA synthetase from yeast (143) and methionyl-tRNA synthetase from *E. coli* (144, 145); and the tetrameric enzymes ($\alpha_2\beta_2$), such as phenylalanyl-tRNA synthetase from *E. coli* (146, 147) and yeast (148, 149). Monomeric isoleucyl enzyme is capable of binding two equivalents of valine, ATP (128), and tRNA^{Val} (150). The steady-state data obtained for isoleucyl tRNA synthetase seem to support a mechanism in which the binding of a second isoleucine occurs between the isoleucyl adenylate formation and the aminocylation of tRNA (142, 151). Fersht (120) reported that another monomeric enzyme, valyl-tRNA synthetase from *B. stearothermophilus,* can also bind 2 mol each of valine and ATP per molecule. However, studies on the yeast enzyme (152) showed that the monomeric enzyme possesses only one catalytic site and binds one valine. This enzyme contains two ATP sites; one binds the AMP moiety of ATP and the other is involved in the interaction with the 3'-terminal adenosine of tRNA.

In investigations of the rate-limiting step for the aminoacylation reactions, the rate constants used for comparison in many cases were determined from partial reactions in the absence of a certain substrate(s). Binding of tRNA causes a decrease in the rate of adenylate formation such that it contributes significantly to the rate-limiting step for the Tyr enzyme (128, 153), which suggests that the rate-limiting step is either the aminoacylation process or Tyr-AMP formation (Equation 2). This is in contrast to the belief that the dissociation rate of aminoacyl-tRNA from the enzyme is the rate-determining step (154–156). Similarly, it was found (157) that the dissociation rate of Val-tRNA and Ile-tRNA from Val-specific and Ile-specific enzymes, respectively, is not the rate-determining step for the aminoacylation. In addition, the rate-limiting step for the *E. coli* phenylalanenyl enzyme is a conformational change of an enzyme-product complex subsequent to tRNA aminoacylation (133). Although it has been suggested that binding of amino acid can facilitate the dissociation rate of E·AA-tRNA complex in the partial reaction studied (154, 157), the significance of this effect in the overall reaction and the possible involvement of a flip-flop type of mechanism remain to be demonstrated.

GLUTAMINE SYNTHETASE   Glutamine synthetase (GS) catalyzes the biosynthesis of glutamine from glutamate in the presence of divalent cations (158, 159). There is strong evidence that this biosynthetic reaction proceeds sequentially: an enzyme-bound intermediate, γ-glutamyl phosphate, forms from ATP and glutamate, followed by phosphate displacement by ammonia to give P_i and glutamine (159).

E. coli glutamine synthetase (mol wt 600,000) consists of 12 identical subunits arranged in the form of two stacked eclipsed hexagons. Each subunit can exist in an adenylylated form ($GS_{\overline{12}}$) in which AMP is covalently attached to a specific tyrosine residue or in an unadenylylated form ($GS_{\overline{0}}$) (158). The latter is active in glutamine synthesis only when Mg^{2+} is present, while catalysis of glutamine synthesis by $GS_{\overline{12}}$ requires the presence of Mn^{2+}.

Heterologous subunit interactions between adenylylated and unadenylylated subunits in hybrid enzyme forms have been demonstrated (160). When the K_m of glutamate for the biosynthetic reaction catalyzed by adenylylated subunit alone was measured in the presence of Mn^{2+}, significantly different K_m values were obtained for the glutamine synthetase with different states of adenylylation.

Recently, evidence for possible homologous subunit interaction was obtained when the extent of oxygen exchange between [^{18}O]glutamate and phosphate per molecule of glutamine formed was evaluated (at various NH_4^+ concentrations) (161). This allows calculation of the minimum number of reaction reversals in which bound glutamine is converted to bound glutamate prior to release of glutamine. At high NH_4^+ no detectable reversals occurred. However, at low NH_4^+, over 15 reversals of bound glutamine formation occurred. In contrast to the effect seen with NH_4^+, ATP concentration did not modulate the oxygen exchange.

These results possibly indicate that NH_4^+ binding to the catalytic site of one subunit promotes catalytic events at another site (161). However, as Bild & Boyer indicate, their results do not eliminate other explanations.

Another indication that NH_4^+ may play an important role in catalytic cooperativity was that double-reciprocal plots of initial velocity data with varied NH_4^+ concentration are not linear at subsaturating glutamate concentration, which indicates a negatively cooperative interaction between

NH_4^+ and enzyme (162). Several lines of evidence for homologous subunit interaction come from the interaction of the enzyme with methionine-sulfoximine, a glutamate analogue. Glutamine synthetases from mammalian and other sources are inhibited by L-methionine-S-sulfoximine, one of the four possible stereoisomers of methionine-sulfoximine (159, 163, 164). The inhibition is initially competitive with glutamate, followed by the phosphorylation of methionine-sulfoximine by ATP, which leads to the irreversible inhibition due to the formation of tightly bound products ADP and methionine-sulfoximine phosphate. The noncovalent binding of methionine-sulfoximine and ADP is so tight that they remain bound except under extreme conditions like acid treatment below pH 4 or boiling (165). The formation of methionine-sulfoximine phosphate and ADP is comparable to the formation of γ-glutamyl phosphate and ADP during the catalytic cycle. In addition, L-methionine-R-sulfoximine, which does not inactivate glutamine synthetase irreversibly in the presence of ATP, was found to inhibit reversibly (166).

An extreme case of negative cooperativity was observed with rat liver glutamine synthetase, which has eight subunits. Complete irreversible inhibition of this enzyme was attained by the binding of only about 4 mol of methionine-sulfoximine phosphate per octamer (167). However, a possible flip-flop mechanism with half-of-the-sites reactivity was eliminated, based on kinetic results obtained with the pig brain enzyme (168). Detailed kinetic studies on the irreversible inhibition of *E. coli* glutamine synthetase in the presence of excess ATP and L-methionine-S-sulfoximine revealed that the apparent first order rate constant of irreversible inhibition decreases progressively from the expected first order rate; thus an inactivated subunit possibly retards the reactivity of its neighboring subunit (166, 169). Fully inactivated ($>$ 95%) glutamine synthetase contains 9–10 mol of methionine-sulfoximine phosphate and ADP per dodecamer. This full inactivation caused by substoichiometric binding of methionine-sulfoximine phosphate has been substantiated (164, 166).

The binding of both S- and R-isomers of L-methionine-sulfoximine, separately or together, to glutamine synthetase have been monitored by various optical techniques (166, 169–172). The UV spectral change produced by the S-isomer binding to Mn·unadenylylated GS was linearly related to the saturation function of this isomer (172). The binding isotherm (166, 169, 171) obtained from fluorescence titration yielded negatively cooperative binding for the S-isomer binding to $MgGS_{\overline{0}}$, as indicated by its Hill coefficient, $n_H = 0.65$, for the S-isomer binding to $MnGS_{\overline{12}}$ ($n_H = 0.7$), and for the R-isomer binding to $MnGS_{\overline{12}}$ ($n_H = 0.87$); but it yielded positively cooperative binding for the R-isomer binding to $MgGS_{\overline{0}}$ ($n_H = 1.3$). Similar results were also obtained from spectrophotometric (170, 172) and calorimetric (173) titration data. Further indication of strong subunit interaction comes from the studies on the glutamine synthetase inactivated irreversibly but partially by the S-isomer and ATP. The partially inactivated enzyme showed increased K_m for NH_4^+ but decreased K_m for ATP in the biosynthetic reaction (F. C. Wedler, personal communication). Furthermore, when the partiallyinactivated enzyme was dissociated by treating with EDTA and 5,5'-dithio-bis-(2-nitrobenzoic acid), (DTNB), oligomers with even numbers of subunits (4, 6, 8, 10, and 12) were formed (165). These results indicate that the inactive subunits in a dodecamer restrict the conformational flexibility of adjacent active subunits by preventing partial unfolding and by blocking the reaction of certain –SH groups with DTNB thereby preventing disruption of the bonding contacts between subunits at those locations.

GLYCERALDEHYDE-3-PHOSPHATE DEHYDROGENASE   Glyceraldehyde-3-phosphate dehydrogenase (GPDH) catalyzes reversibly the oxidative phosphorylation of D-glyceraldehyde-3-phosphate to 1,3-diphosphoglycerate according to the two following consecutive reaction schemes:

$$RCHO + NAD^+ + E\text{-}SH \rightleftharpoons E\text{-}SCOR + NADH + H^+$$

D-glyceralde-   enzyme   acyl enzyme   3a.
hyde-3-phosphate

$$E\text{-}SCOR + HPO_4^{2-} \xrightleftharpoons{NAD^+} E\text{-}SH + RCO\text{-}OPO_3^{2-}$$

                           3b.

1,3-diphosphoglycerate

GPDH's isolated from a variety of sources (yeast, rabbit muscle, lobster muscle, sturgeon muscle, and Bacillus) are tetrameric (mol wt 145,000), are made up of chemically identical monomers containing one uniquely reactive SH group per subunit, show a long, virtually identical sequence around the active site cysteine residue, and show strong sequence homologies in general (174).

Coenzyme binding   In the case of the yeast enzyme, it was shown that the binding of NAD^+ occurs in a positively cooperative fashion under certain conditions, e.g. at pH 8.5 and 40°C (175). Using a combination of equilibrium and rapid kinetic techniques, Kirshner and co-workers (175, 176) have shown that the binding of NAD^+ to the yeast enzyme can be described by the concerted model of monod et al (2, 18). According to this proposal, a relatively slow T to R conversion occurs on addition of NAD^+ to the apoenzyme, which exists in about 98% T form. The T form is enzymatically inactive and has a lower affinity for NAD^+ (k_d 2 X 10^{-3} M compared with 10^{-4} M). Later studies (177) showed that the binding of NADH is hyperbolic in contrast to that of NAD^+. Thus, NADH binds equally well to both R and T forms. The rationalization for these observations is that NADH, which does not possess the quaternary pyridinium ring, cannot stabilize the R form. Consequently, the enzyme is active in the reverse reaction only in the presence of NAD^+, which converts the enzyme to the R form.

Equilibrium studies of NAD^+ binding to the yeast enzyme at pH 8.5 (178, 179) showed positively cooperative behavior. But the overall nature of the curves suggested a complex binding isotherm that manifests positive cooperativity at low degrees of saturation, but does not reach the 100% saturation expected of positive cooperativity. Koshland and his co-workers explained this complex binding isotherm by postulating a ligand-induced sequential model in which subunit interactions lead to a mixture of positive and negative cooperativity. Alternatively, the same data have been interpreted (4) as arising from a pseudoconservative transition in which one tetrameric conformation has functional pairwise asymmetry.

Although a possibility that the mixed cooperativity might be due to the presence of multispecies or partially denatured enzyme (affinity heterogeneity) has been carefully examined by previous workers (179), when the yeast enzyme was rapidly purified by affinity chromatography (180), a binding isotherm was observed that shows simple positive cooperativity and saturation at 4 NAD^+ molecules bound per tetramer. Recently, Niekamp et al (181) reported that the data obtained by various physical techniques for NAD^+ binding to the yeast GPDH can be equally well fitted on the basis of either a sequential model or a concerted model.

In contrast to the yeast enzyme, GPDH's isolated from muscles of rabbit (182–184), lobster (185), and sturgeon (186) and from *B. stearothermophilus* (187) bind NAD^+ in a negatively cooperative manner. Negatively cooperative binding was first demonstrated (183, 185) by showing that NAD^+ binds to the tetrameric rabbit muscle enzyme with decreasing affinity (see 1 and 2 in Table 1). Pointing out that this progressive decrease cannot be described by the statistical relationship predicted from the concerted model, they proposed a sequential model in which the binding of one NAD^+ molecule to one promoter affects the conformation of a second promoter. The sequential nature of the ligand-induced changes was further demonstrated by the observation that the binding of the first coenzyme molecule to the tetramer induces the largest conformational changes, as indicated by the sulfhydryl group reactivity (183, 188), optical properties (184), and thermodynamic parameters (189), and subsequent binding steps contribute progressively less to the total change.

Table 1 Dissociation constants for NAD^+ binding to rabbit muscle glyceraldehyde-3-phosphate dehydrogenase

	K_1	K_2	K_3	K_4	Experimental conditions
1. Conway & Koshland (183)[a]	$<10^{-11}$	$<10^{-9}$	3×10^{-7}	2.6×10^{-5}	3°C; pH = 8.5; equilibrium dialysis
2. DeVijlder & Slater (182)	$<5 \times 10^{-8}$	$<5 \times 10^{-8}$	4×10^{-6}	3.5×10^{-5}	20°C; pH = 8.2; ultracentrifugation
3. Price & Radda (191)	—	—	1.7×10^{-6}	3.4×10^{-5}	25°C; pH = 8.2; fluorescence
4. Bell & Dalziel (192)	1×10^{-8}	9×10^{-8}	4.1×10^{-6}	3.6×10^{-5}	25°C; pH = 7.6; $K_1 K_2$ fluorescene; $K_3 K_4$ from UV
5. Scheek & Slater (193)	— 2.8×10^{-8}	— 2.8×10^{-8}	9×10^{-7} 2.7×10^{-7}	9×10^{-7} 2.7×10^{-7}	25°C; gel filteration 15°C; fluorescence
6. Henis & Levitzki (198)	1.1×10^{-7}	3.5×10^{-7}	5.9×10^{-7}	5.4×10^{-6}	25°C; pH = 7.5; equilibrium dialysis in the presence of ADP ribose

[a] Numbers in parentheses refer to the literature cited section.

On the other hand, strictly equal contributions of each bound coenzyme molecule (NAD^+ or NADH) to the total signal changes have been observed by various optical techniques for GPDH's from rabbit muscle (190–194) and sturgeon muscle (186, 195), even when binding heterogeneity among sites was observed. Due to both the difficulties in evaluating extremely tight binding constants and the dissociable nature of GPDH tetramer, it has also been controversial whether the binding affinities of NAD^+ decrease by an order of 10^6 in successive steps as originally reported by Conway & Koshland (see column 1 in Table 1). When GPDH was purified rapidly (193) from rabbit muscle, the negative interaction calculated from the NAD^+ binding was much smaller than those reported previously (182, 183, 192), and only two different classes of binding constants were obtained (5 in Table 1).

The binding of NAD^+ to sturgeon muscle GPDH also occurs at four sites per tetramer, and the affinity decreases as the extent of bound NAD^+ increases. This heterogeneity in NAD^+ binding affinity could be adequately described by assuming two independent pairs of binding sites (186, 195). Similarly, the binding of NADH seemed to involve two distinct classes of binding sites (195). Based on the binding studies in conjunction with the half-of-the-sites reactivity observed with the sulfhydryl group, Seydoux et al proposed that the negative cooperativity is due to two independent pairs of binding sites preexistent in the tetramer. In any event, as has been previously pointed out, it is virtually impossible to deduce the molecular mechanism of ligand binding solely from the shape of the binding isotherm

(28). Recent reinvestigation of NAD^+ binding to rabbit muscle GPDH (197, 198) indicates that the progressive decrease of binding affinity observed as the extent of bound NAD^+ increase is not as dramatic as reported earlier (Table 1, 1 and 2 vs 6) and that the binding isotherm of ϵ-NAD, an analogue of NAD, to the enzyme can be suitably fitted by both the preexistent asymmetry model and a model involving ligand-induced conformational changes.

X-ray crystallographic data also do not distinguish between the preexisting asymmetry and the sequential model; while the electron density map of lobster muscle holoenzyme shows asymmetric assembly of four subunits (199), the *B. stearothermophilus* holoenzyme, whose structure and properties are very similar to those of the muscle enzyme, is found to possess a precise 222 symmetry (200).

In an effort to develop a new method for delineating the mechanism of the negative cooperativity of coenzyme binding, Henis & Levitzki (197, 198) studied the binding of NAD and ϵ-NAD to rabbit muscle GPDH in the absence and presence of other noncooperative, competitive ligands such as acetyl pyridine-adeninedinucleotide, adenosine diphosphoribose (ADP-ribose), ATP, ADP, and AMP. The preexistent asymmetry model and the ligand-induced sequential model predict very different results (28). The former model, which does not allow for intersubunit interactions, predicts that the presence of a noncooperatively competing ligand will shift the binding curve of the cooperative ligands (NAD or ϵ-NAD) to higher concentrations, without affecting the cooperativity of its binding. On the other hand, in the sequential model, the binding of a noncooperative ligand will alter the cooperativity in the binding of the primary ligand.

The Hill coefficient (n_H) at 50% saturation by ϵ-NAD, a low affinity substrate for GPDH, was evaluated as a measure of cooperativity. The strong negative cooperativity exhibited by ϵ-NAD in the absence of a competing ligand ($n_H = 0.64$) was found to be abolished in the presence of acetyl pyridine adenine dinucleotide ($n_H = 1.00$) and strongly weakened by ATP ($n_H = 0.89$), ADP ($n_H = 0.93$), and AMP ($n_H = 0.88$), but was not affected by the addition of ADP-ribose ($n_H = 0.65$). The Hill coefficients for NAD^+ binding in the presence of noncooperatively competing ligands were also variable; $n_H = 0.58$ with ADP-ribose and $n_H = 0.88$ with ATP. (In the absence of a competing ligand, n_H could not be evaluated because the first two NAD^+ molecules bind too tightly.) Therefore, Henis & Levitzki concluded that the negative cooperativity in NAD^+ and ϵ-NAD^+ binding involves ligand-induced conformational changes (197, 198). In addition, the dissociation constants of all four NAD^+ molecules were determined in the presence of ADP-ribose, which reduced the binding affinity of ϵ-NAD but did not change the cooperativity of ϵ NAD binding. Subsequently, assuming that ADP-ribose also does not influence the cooperativity of NAD binding, four intrinsic dissociation constants for NAD were evaluated (Table 1, 6). Furthermore, Henis & Levitzki (197, 198) proposed that the results from this competition study corroborate their previous hypothesis (184, 201) that the structure of the pyridine moiety of the coenzyme analogues plays a role in orienting the adenine moiety at the adenine subsite, thereby affecting the cooperativity in the binding of the coenzyme analogue, which is mediated through the adenine subsites.

GPDH is a dissociable oligomeric enzyme (202). It has also been reported that the dissociated species are catalytically active (198, 203) and that they bind the coenzyme more tightly than the nondissociated tetramer (203). In this connection, it was suggested that the negative cooperativity is due to the rapid equilibrium between monomer-dimer-tetramer (203). However, this rapid equilibrium cannot account for all heterogeneous binding affinities observed, because under the experimental conditions used in most of the binding studies, GPDH is in its tetrameric form (198, 202). Recently results (204) indicate that the catalytically active species in pig muscle GPDH is a tetramer over the enzyme concentration range of 10^{-8}–10^{-4} M.

Half-of-the-Sites reactivity and coenzyme binding to acyl-enzyme intermediate Another allosteric property, exhibited by GPDH from different sources, is half-of-the-sites reactivity with respect to alkylation or acylation by certain sulfhydryl reagents, including the substrate or substrate analogues. GPDH contains four especially reactive sulfhydryl groups per tetrameric molecule, which react with substrates to form a thiol ester acyl-enzyme intermediate in the catalytic cycle. When a chromophoric pseudosubstrate, β-(2-furyl)acryloyl phosphate (FAP), was incubated with muscle holo-GPDH, only two equivalents of the acyl group (FA group) could be introduced per tetramer, unless forcing conditions were used to achieve a higher stoichiometry (186, 205). Thus, Bernhard and his co-workers assumed a "dimer of dimers" structure for the tetrameric molecule. The rabbit muscle apo-GPDH also shows half-of-the-sites reactivity toward alkylating reagents such as fluorodinitrobenzene and p-fluoro-m,m-dinitrophenyl sulfone (206, 207). However, not all sulfhydryl reagents cause the half-of-the-sites reactivity, and the yeast GPDH differs from the muscle enzymes in reaction with alkylating and acylating reagents. For example, the yeast enzyme shows half-of-the-sites reactivity toward iodoacetic acid and iodoacetamide (208), whereas these reagents show strict all-of-the-sites reactivity (single exponential time curve for four-SH groups) in muscle enzymes (207, 209).

The character of a "dimer of dimer" is also evident in the thermal denaturation of the oligomeric enzyme. Both in plant species and in muscle enzymes, the catalytic activity drops rapidly to 50%, and subsequently declines at a much slower rate for the remaining activity (210, 211).

As in the case of coenzyme binding, a controversy has been going on for many years as to whether this half-of-the-sites reactivity is a consequence of preexistent asymmetry, ligand-induced subunit interaction, or concerted transition (18, 174, 196). MacQuarrie & Bernhard (209, 212) proposed a procedure for distinguishing induced from preexistent asymmetry in quaternary structure of the rabbit muscle GDPH as depicted in Equation 4:

$$FAP \,+\, E(SH)_4 \,\rightarrow\, FA_2E(SH)_2 \xrightarrow{RX} FA_2E(SR)_2 \xrightarrow{P_i} (HS)_2E(SR)_2$$

$$\xrightarrow{FAP} (FA)(SH)E(SR)_2 \qquad\qquad 4.$$

First, the di(2-furylacryloyl)enzyme, $FA_2E(SH)_2$, was prepared from the holoenzyme utilizing its half-of-the-sites reactivity with FAP. Subsequently, the two remaining sites were alkylated with iodoacetate (RX) utilizing its all-of-the-sites reactivity. The acyl groups were then removed from this diacyl-dialkyl enzyme by phosphorylysis. When the resultant dialkyl enzyme was tested for reaction stoichiometry with FAP, only one could be reintroduced. This result cannot be explained in terms of an induced model, and indeed, can only be explained by a preexistent asymmetry model if there is a subunit rearrangement (209).

Unlike the rabbit muscle enzyme, when similar studies were carried out with the yeast apoGPDH (208), the results supported the induced model. In this experiment, the apoenzyme was allowed to react with 2 mol of fluorodinitrobenzene (FDNB) to produce the dimodified enzyme, $(DNP)_2E(SH)_2$ (DNP denotes dinitrophenyl), which lacks catalytic activity and has greatly decreased –SH reactivity. The remaining sites were then modified with iodoacetamide by forcing conditions, so that all four sites are alkylated. The DNP groups were then removed by thiolysis to regenerate a disubstituted enzyme, $(HS)_2E(SR)_2$, which was unreactive toward further sulfhydryl reagent and inactive in enzyme assay, as predicted by the induced model. According to the preexistent asymmetry model, the removal of the DNP groups from the reactive and active subunits should have regenerated enzymatic activity and rapid–SH reactivity.

A half-of-the-sites effect is also observed in yeast GPDH for the substrate 1,3-diphosphoglycerate. Detailed studies showed (213) that acylation of the first two sites with the substrate decreases the rate of acylation of the third site, and that excess substrate accelerates the hydrolysis of the 3-phosphoglycerolyl enzyme both in the presence and absence of NAD. This acceleration by excess acyl phosphate was not seen with acetyl phosphate or FAP, even though a half-of-the-sites reactivity was observed with both. This observation provides support for the concept that the yeast GPDH operates through a type of flip-flop mechanism (11, 12) for acyl phosphate hydrolysis.

In the case of lobster and sturgeon muscle GPDH, all four sites can be acylated, apparently with equal reactivity, by 1,3-diphosphoglycerate or glyceraldehyde-3-phosphate in the presence of NAD^+ (186, 214, 215), whereas the reaction of pseudosubstrate FAP leads to half-of-the-sites reactivity. However, under the steady-state process, the properties of the four acyl groups fall into two distinct classes. From the transient kinetics of the reductive dephosphorylation of 1,3-diphosphoglycerate (reverse direction of Equation 3), Kellershohn & Seydoux (215) demonstrated explicitly that a half-acylated enzyme species predominates during the catalytic process of sturgeon GPDH. When the reductive dephosphorylation was initiated by adding NADH and 1,3-diphosphoglycerate to the unacylated, coenzyme-free, apoenzyme, a well-defined lag phase was observed. This lag phase is independent of the enzyme and NADH concentration but shows a hyperbolic decrease as the concentration of 1,3-diphosphoglycerate increases. When the reaction was initiated with the preacylated enzyme, this lag time became shorter. The lag time disappeared completely when half of the four available sites of the tetrameric enzyme had been acylated. Further preacylation of the enzyme induced a new transient phase with a pseudoburst (negative lag time). This "negative lag" phase was due to a temporary hyperactivation of the enzymic reaction, which disappeared after several turnovers, but not after a single turnover. Since the oxidized coenzyme NAD^+, which is a product of the overall reaction, activates the acylation reaction by 10^2 to $\sim 10^4$-fold (215, 216), the positive lag phase disappeared when the two tight-binding sites were saturated by NAD^+. Upon saturation of the two remaining loose sites, the negative lag phase appeared. The final steady-state of the reaction, however, was not dependent on the initial degree of preacylation or NAD^+ binding. These observations led Kellershohn & Seydoux to conclude that GPDH contains two distinct pairs of acylation sites. The preacylation of the first pair of sites did allow immediate attainment of the steady-state rate. Further preacylation of the second pair of sites led to a hyperactivation phase, which was replaced by the normal steady state after a few turnovers. Likewise, in the phosphorolysis (Equation 3b) of the tetra-3-phosphoglyceroyl·enzyme, two of the four acyl groups were very rapidly phosphorylized, whereas the other two acyl groups were phosphorylized at the slower steady-state velocity (211). It was also pointed out that the nonequivalence among sites does not provide an increase of the catalytic efficiency of tetrameric GPDH. In fact, the fully acylated enzyme appeared to be about twice as active as the diacylated enzyme, which is clearly not in agreement with the hypothesis (217) that oligomeric enzymes gain catalytic efficiency through a type of flip-flop mechanism.

Under physiological conditions, muscle GPDH exists principally as the acyl enzyme (211), because the rate limiting step in the oxidative-phosphorylation is the phosphorolysis of the acyl enzyme (214, 218, 219) and ATP stabilizes this conformational species (211). Therefore, it is important to consider the effect of coenzyme binding not only on the unsubstituted enzyme, but also on the stoichiometrically significant enzyme-substrate covalent intermediates.

In order to obviate measurements of the effect of coenzyme on the chemical bonding properties of the acyl-enzyme linkage, Bernhard and his coworkers (34, 40, 48–50) took advantage of the chromophoric properties of the $FA_2E(SH)_2$. This acyl-enzyme, like the physiologically important one, undergoes phosphorolysis, arsenolysis, and reduction by NADH. All catalytic reactions involving acyl enzyme exhibit an absolute requirement for bound NAD^+ (205, 216, 220). NAD^+ is a required cocatalyst even for the reduction of the acyl enzyme by NADH. Binding of NAD^+ at the two nonacylated sites of the FA_2-GPDH is required for catalytic reduction with NADH at the acyl sites (221). This is a direct demonstration of a functional role for subunit interaction. The thiol ester bond of the acyl enzyme is activated by NAD^+ and NADH for the deacylation reaction (phosphorolysis or arsenolysis) and reduction reaction, respectively (205, 220, 222). In FA_2-GPDH, this activation is accompanied by large acyl-spectral shifts: a "red shift" with NAD^+ and a "blue shift" with NADH. Only the electronically perturbed acyl group participates in the chemical reaction involving acyl groups, and the deacylation reactions, and the spectral shifts (both red and blue) show biphasic kinetics. For the deacylation reaction, the relative amplitude of the fast and slow phases depend on NAD^+ concentration, but are independent of the nature and concentration of the acyl acceptor. The amplitude of the fast phase of NAD^+-induced spectral change in apoenzyme is equal to that of the fast phase in phosphorolysis (or arsenolysis) at low NAD^+ concentration. This biphasicity strongly suggests that in the apoacyl enzyme there are already two different conformational states: one leading to an instantaneous interaction and spectral perturbation by NAD^+, and another that binds NAD^+ without spectral perturbation, but isomerizes slowly to the spectrally perturbed and catalytically active acylenzyme conformation. Based upon these observations, Malhotra & Bernhard (211, 222) proposed a model involving concerted transition between two conformations; (in this model FA_2GPDH was considered as a dimer of dimer and each circle or square represents asymmetric monoacylated dimers (see Scheme 2).

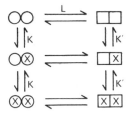

Scheme 2 A model for glyceraldehyde-3-phosphate dehydrogenase. $\bigcirc\bigcirc$, FA_2E; $\square\square$, FA_2E'; X, NAD^+ or NADH; L, allosteric constant; and K and K', binding constants of coenzyme.

In this model, NAD^+ binds preferentially to FA_2E' over FA_2E (K = 7 mM^{-1}, K' = 33 mM^{-1}). The NAD^+ binding to FA_2E' causes a rapid spectral perturbation, and the resulting (NAD·FA) (FA)E' or (NAD·FA) (NAD ·FA)E' undergoes rapid deacylation in the presence of phosphate or arsenate. The slow spectral perturbation and slow deacylation reaction are due to the required transition from FA_2E to FA_2E' before NAD^+ binding. The concerted model predicts a positively cooperative binding isotherm for NAD^+, but no obvious cooperativity was observed for the NAD^+ binding to acyl enzyme (222). This was attributed to the near unity Hill coefficient (n_H = 1.11), which could be evaluated from data obtained by transient kinetics (222).

Unlike NAD^+, NADH preferentially binds to FA_2E over FA_2E', and the former species gives rise to the blue-shifted acyl spectrum upon binding of NADH. Consequently, it is expected that the kinetic pattern of spectral shifts obtained by adding NAD^+ and NADH be complementary: the amplitude of the fast phase in one is equal to that of the slow phase in the other. Bernhard and co-workers conclude that the concerted transitions between alternative enzyme structures are involved directly in the catalysis and are the rate-limiting processes for both forward and reverse reactions.

ASPARTATE TRANSCARBAMOYLASE Aspartate transcarbamoylase (ATCase) from *E. coli* is an enzyme that can be adequately described by the model of Monod et al. The unique structure of this enzyme permits the full utilization of the hydridization technique to study its subunit organization, bonding strength, intersubunit communication, and concerted transition.

ATCase exhibits positive cooperativity with respect to substrate concentration and is subject to feedback inhibition by CTP (223). Cooperative binding is also observed with the substrate analogue succinate in the presence of carbamoyl phosphate, and the feedback inhibition can be reversed by ATP. The enzyme contains two catalytic trimers (C subunits) that are cross-linked noncovalently by three regulatory dimers (R subunits) (6). Each R subunit binds a zinc ion, and contains a binding site for the allosteric inhibitor CTP or the activator ATP. X-ray crystallographic studies reveal no preexisting asymmetry in the arrangement of the subunits (224, 225).

The C and R subunits can be separated (226). The isolated C subunit is fully active in catalyzing the formation of carbamoyl aspartate from carbamoyl phosphate and aspartate but displays no sigmoidal dependence on aspartate concentration. This by itself is an indication of the crucial role played by the R subunits in the mediation of the allosteric transitions of the intact enzyme.

Evidence for a concerted transition in ATCase comes mainly from two types of experiments. The first type involves the demonstration of the conversion of the enzyme to a particular conformational state at less than saturation levels of active site ligands. As has been mentioned previously, the concerted transition mechanism predicts that, in the presence of a ligand that binds preferentially to one of the conformational states, total conversion of the enzyme to that state may result prior to full occupancy of the ligand binding sites. Strictly speaking, however, total conversion to a given final conformational state merely shows that there is a preexisting equilibrium between several enzyme forms; it does not necessarily indicate a concerted transition where all the functionally identical polypeptide chains assume the same configuration. Thus, the second type of experiment involving the reconstruction of hybrid enzymes from native and modified C and R subunits is needed to complete the proof. With ATCase, it was observed that the binding of active-site ligands led to an increase in the effective hydrodynamic volume of the molecule (a maximal decrease of 3.5% in the sedimentation coefficient) and a sixfold enhancement of the chemical reactivity of 24 sulfhydryl groups on the R subunits (227). These changes reach 50% completion when only 15% of the sites is filled by succinate. The effect of the bisubstrate analogue, N-(phosphoacetyl)-L-aspartate (PALA) is more dramatic. Binding of PALA to ATCase is very tight (K_d on the order of 10^{-8} M) such that stoichiometric binding occurs under usual experimental conditions (228). In the presence of about four equivalents of PALA, maximal changes in sedimentation coefficient and thiol group reactivity are attained (229, 230). At a PALA-to-enzyme ratio of 3:1, biphasic first-order plots were observed for the reaction with p-mercuribenzoate (PMB), which can be resolved into two components corresponding to the reaction rates of the fully liganded (R state) and unliganded (T state) enzymes. The inhibitor CTP and the activator ATP were shown to stabilize the T and R conformations, respectively. In addition, UV spectral changes of the protein caused by PALA binding was completed only when all six equivalents of PALA were bound. This suggests that the gross conformational changes represented by a decrease in the sedimentation coefficient and an increase in sulfhydryl reactivity can be distinguished from local changes reflected by the spectral shift. When the Zn^{2+} ion on the R subunits are replaced by Ni^{2+}, ATCase retains the allosteric properties of the Zn^{2+} enzyme and exhibits absorption bands at 360 and 440 nm due to charge transfer transitions (231). Binding of PALA to this enzyme species gives rise to two different bands at 390 and 465 nm. Again, complete spectral changes are observed when about four equivalents of PALA are bound, thereby demonstrating that the conformational changes due to PALA binding to the C subunits are transmitted to the R subunits to effect a concerted transition.

Further support for concerted transition in ATCase is provided by studies on hybrids constructed from native R subunits and mixtures of active and pyridoxylated, inactive C chains (232). Two hybrids, one containing two active sites in one C subunit and none in the other, and one containing one active site each in both C subunits, exhibit identical kinetic behavior in terms of V and sigmoidal dependence on aspartate concentration. The ligand-promoted changes in sedimentation coefficients and reactivity toward PMB resemble those of the native enzyme. That active sites located on the same or different C chains behave in the same manner indicates the absence of discrete "cooperative units" within the enzyme molecule. Consequently, the allosteric transition must be fully concerted. Schachman and co-workers (233) were able to calculate an internally consistent set of parameters for ATCase based on the two-state model of Monod et al. From kinetic and physical measurements, they estimated the allosteric transition constant L ($= T/R$) to be 250 in the absence of any ligand. Binding of aspartate, succinate, and PALA to the T state is at least 20 times weaker than for the R state. The preferential binding of CTP to the T state shifts the equilibrium to 1250.

There are observations that contradict the concerted model. The binding of CTP to the R subunits has been reported to be negatively cooperative (234–236). Apparent half-of-the-sites saturation has also been observed for the binding of carbamoyl phosphate to the enzyme in the absence of succinate (237) and the binding of succinate in the absence of carbamoyl phosphate (238). However, technical problems such as the preparation of the enzyme, the growth medium of the E. coli cells (238), the decomposition of carbamoyl phosphate, and the poor affinity of succinate may account for some of the variations in binding data (233).

The hybridization technique has been used to study the subunit arrangement, the propagation of conformational changes from one polypeptide change to another, and the bonding strength between different chains. Rapid mixing of native C and R subunits with succinylated R subunits (R_s) resulted in four hybrid forms ($R_sR_sR_s$, R_sR_sR, R_sRR, RRR) that differed in electrophoretic mobility (239). These results substantiated earlier evidence that the six R chains are arranged as three dimeric units. The presence of succinylated R subunits also reduced the cooperative effect of aspartate binding. As mentioned previously, the communication between R and C chains is also evident in experiments performed with enzymes with Ni^{2+}-substituted R subunits (231).

Nitration of tyrosyl residue on the C subunit permits the construction of hybrid ATCase with a sensitive spectral probe (240, 241). The hybrid containing native R subunits and active nitrated C chains exhibits the characteristic regulatory properties of the native enzyme. Addition of CTP shifts the enzyme toward the T state, while ATP causes a shift toward the R conformation as judged by the change in the sedimentation coefficient and the altered kinetic pattern (241). The R and T states are found to be reflected in the decrease or increase of absorbance at 430 nm. Subsequently, the nitrated C chain was made inactive by pyridoxylation. Unlike the nitrated C chain, which displays a decrease at 430 nm on binding succinate in the presence of carbamoyl phosphate, the pyridoxylated, nitrated C chain shows no spectral perturbation by these ligands. With the hybrid containing one native and one inactive nitrated C chain, binding of succinate to the native chain causes an increase in the absorbance of the nitrotyrosyl groups of the unliganded C chain at 430 nm (242). This increase in absorbance as a function of succinate concentration coincides with the decrease in sedimentation coefficient and the increase in sulfhydryl-group reactivity of the hybrid. The results demonstrate clearly that the effects of ligand binding is communicated to the unliganded subunit, and the resultant gross conformational change is accompanied by an alteration in the tertiary structure of the unliganded C chains.

The effect of ligand binding on the strength of the intersubunit bonds in ATCase has been studied by using a less stable enzyme-like complex that lacks one R subunit (243, 244). This species disproportionates in solutions of low ionic strength to form native enzyme and free C subunits. The slow step in the disproportionation process is the rupture of two bonds between the C and R chains, which can be measured by an electrophoretic technique (243). Upon binding of the bisubstrate analogue PALA, the rate of disproportionation is enhanced 300-fold. This rate increase indicates that the ligand-promoted transition from the constrained to the relaxed conformations involves a weakening of the intersubunit bonds corresponding to about 1.7 kcal/mole.

Concluding remarks

The enzyme systems reviewed in the foregoing sections provide examples for various mechanisms involving subunit interaction and catalysis and for the diverse approaches employed to study these interactions and mechanisms. The flip-flop model was originally developed on the basis of strong negative cooperativity exhibited by alkaline phosphatase. It is now believed that this enzyme does not operate by such a mechanism. In the case of alkaline phosphatase, therefore, the significance of strong negative cooperativity is regulatory rather than catalytic. At high levels of Mg^{2+}, the "idled" subunit on alkaline phosphatase is revitalized, presumably to meet metabolic needs. Negative cooperativity is also evident in the aminoacyl tRNA synthetase, glutamine synthetase, and GPDH systems. Direct involvement of subunit interaction in catalysis is well documented in studies with succinyl CoA synthetase and mitochondrial ATPase. Thus, in these cases, the significance of negative cooperativity is a catalytic one, being related to the alternate-site mechanism. Sturgeon muscle GPDH and *E. coli* aspartate transcarbamoylase, on the other hand, are convincing examples for the concerted transition models. The propagation of ligand-promoted conformational change from one polypeptide chain to another is elegantly illustrated by the work of Schachman and co-workers on ATCase. In the case of sturgeon muscle GPDH, the concerted transition is shown to be the rate-limiting step in catalysis. In short, evidence accumulated in the past two decades tells us that subunit interactions are vital to enzyme regulation and catalysis and that the mechanisms involved are sufficiently different that they need not conform to a certain general model.

Literature Cited

1. *Subunit Structure of Proteins,* 1964. *Brookhaven Symp. in Biol.* No. 17, pp. 174–83. Wash. DC: US Dept Commerce
2. Monod, J., Wyman, J., Changeux, J.-P. 1965. *J. Mol. Biol.* 12:88–118
3. Koshland, D. E. Jr., Némethy, G., Filmer, D. 1966. *Biochemistry* 5:365–85
4. Viratelle, O. M., Seydoux, F. J. 1975. *J. Mol. Biol.* 92:193–205
5. Reed, L. J., Cox, D. J. 1970. *Enzymes* 1:213–40
6. Schachman, H. K. 1972. In *Protein-Protein Interactions,* ed. R. Jaenicke E. Helmreich, pp. 17-54. Berlin: Springer
7. Brodbeck, V., Denton, W. L., Tanahashi, N., Ebner, K. E. 1967. *J. Biol. Chem.* 242:1391–97
8. Ito, J., Yanofsky, C. 1969. *J. Bacteriol.* 97:734–42
9. Wang, J. H., Waisman, D. M. 1979. *Curr. Top. Cell. Regul.* 15:47–107
10. Klee, C. B., Crouch, T. H., Richman, P. B. 1980. *Ann. Rev. Biochem.* 49:489–515
11. Lazdunski, M. 1972. *Curr. Top. Cell. Regul.* 6:267–310
12. Lazdunski, M. 1974. *Prog. Bioorg. Chem.* 3:81–140
13. Boyer, P. D., Gresser, M., Vinkler, C., Hackney, D., and Choate, G. 1977. In *Structure and Function of Energy-Transducing Membranes,* ed. K. van Dam and B. F. van Gelder, pp. 261-74. Amsterdam: Elsevier
14. Reithel, F. J. 1963. *Adv. Protein Chem.* 18:124–226
15. Klotz, I. M., Langerman, N. R., Darnell, D. W. 1970. *Ann. Rev. Biochem.* 39:25–62
16. Koshland, D. E. 1970. *Enzymes* 1: 341–96
17. Wyman, J. 1972. *Curr. Top. Cell. Regul.* 6:209–26
18. Levitzki, A., Koshland, D. E. Jr. 1976. *Curr. Top. Cell. Regul.* 10:1–40
19. Levitzki, A. 1978. *Quantitative Aspects of Allostery in Molecular Biology, Biochemistry and Biophysics,* Vol. 28. Heidelberg: Springer
20. Frieden, C. 1971. *Ann. Rev. Biochem.* 40:653–96
21. Frieden, C. 1967. *J. Biol. Chem.* 242: 4045–52
22. Dalziel, K., Engel, P. C. 1968. *FEBS Lett.* 1:349–52
23. Adair, G. S. 1925. *J. Biol. Chem.* 63: 529–45
24. Pauling, L. 1935. *Proc. Natl. Acad. Sci. USA* 21:186–91
25. Ferdinand, W. 1966. *Biochem. J.* 98: 278–83
26. Sweeney, J. R., Fisher, J. B. 1968. *Biochemistry* 7:561–65
27. Howlett, G. J., Blackburn, M. N., Compton, J. G., Schachman, H. K. 1977. *Biochemistry* 16:5091–99
28. Henis, Y. I., Levitzki, A. 1979. *Eur. J. Biochem.* 102:449–65
29. Lazdunski, M., Petitclerc, C., Chappelet, D., Lazdunski, C. 1971. *Eur. J. Biochem.* 20:124–39
30. Choate, G., Hutton, R. L., Boyer, P. D. 1979. *J. Biol. Chem.* 254:286–90

31. Bale, J. R., Chock, P. B., Huang, C. Y. 1980. *J. Biol. Chem.* 255:8424–30
32. Bale, J. R., Huang, C. Y., and Chock, P. B. 1980. *J. Biol. Chem.* 255:8431–36
33. Boyer, P. D. 1979. In *Membrane Bioenergetics,* ed. C. P. Lee, G. Schatz, L. Ernster, pp. 461-79. Reading, Penn: Addison-Wesley
34. Simpson, R. T., Vallee, B. L. 1970. *Biochemistry* 9:954–58
35. Reid, T. W., Wilson, I. B. 1971. *Enzyme* 4:373–415
36. Coleman, J. E., Chlebowski, J. F. 1979. *Adv. Inorg. Biochem.* 1:1–66
37. Fernley, F. N. 1971. *Enzyme* 4:417–47
38. Rothman, F., Byrne, R. 1963. *J. Mol. Biol.* 6:330–40
39. Hanson, A. W., Applebury, M. L., Coleman, J. E., Wyckoff, H. W. 1970. *J. Biol. Chem.* 245:4975–76
40. Otvos, J. D., Armitage, I. M. 1980. *Biochemistry* 19:4031–43
41. Plocke, D. J., Levinthal, C., Vallee, B. L. 1962. *Biochemistry* 1:373–78
42. Lazdunski, C., Petitclerc, C., Lazdunski, M. 1969. *Eur. J. Biochem.* 8:510–17
43. Chappelet, D., Lazdunski, C., Petitclerc, C., Lazdunski, M. 1970. *Biochem. Biophys. Res. Commun.* 40:91–96
44. Vallee, B. L., Ulmer, D. D. 1972. *Ann. Rev. Biochem.* 41:91–128
45. Lazdunski, C., Chappelet, D., Petitclerc, C., Leterrier, F., Douzou, P., Lazdunski, M. 1970. *Eur. J. Biochhem.* 17:239–45
46. Anderson, R. A., Bosron, W. F., Kennedy, F. S., Vallee, B. L. 1975. *Proc. Natl. Acad. Sci. USA* 72:2989–93
47. Bosron, W. F., Anderson, R. A., Falk, M. C., Kennedy, F. S., Vallee, B. L. 1977. *Biochemistry* 16:610–14
48. Engstrom, L., Agren, G. 1958. *Acta Chem. Scand.* 12:357
49. Barrett, H., Butler, R., Wilson, I. B. 1969. *Biochemistry* 8:1042–47
50. Applebury, M. L., Johnson, B. P., Coleman, J. E. 1970. *J. Biol. Chem.* 245:4968–76
51. Schwartz, J. H. 1963. *Proc. Natl. Acad. Sci. USA* 49:871–78
52. Lazdunski, C., Petitclerc, C., Chappelet, D., Lazdunski, M. 1969. *Biochem. Biophys. Res. Commun.* 37:744–49
53. Chlebowski, J. F., Coleman, J. E. 1976. *J. Biol. Chem.* 251:1207–16
54. Ko, S. H. D., Kézdy, F. J. 1967. *J. Am. Chem. Soc.* 89:7139–40
55. Trentham, D. R., Gutfreund, H. 1968. *Biochem. J.* 106:455–60
56. Fernley, H. N., Walder, P. G. 1969. *Biochem. J.* 111:187–94
57. Halford, S. E., Lennette, D. A., Schlesinger, M. J. 1972. *J. Biol. Chem.* 247:2095–2101
58. Chlebowski, J. F., Coleman, J. E. 1974. *J. Biol. Chem.* 249:7192–202
59. Halford, S. E., Schlesinger, M. J. 1974. *Biochem. J.* 141:845–52
60. Otvos, J. D., Armitage, I. M., Chlebowski, J. F., Coleman, J. E. 1979. *J. Biol. Chem.* 254:4707–13
61. Hull, W. E., Halford, S. E., Gutfreund, H., Sykes, B. D. 1976. *Biochemistry* 15:1547–61
62. Huang, C. Y. 1977. *Arch. Biochem. Biophys.* 184:488–96
63. Aldridge, W. N., Barman, T. E., Gutfreund, H. 1964. *Biochem. J.* 92:236–56
64. Kelly, S. J., Sperow, J. W., Butler, L. G. 1974. *Biochemistry* 13:3503–5
65. Chock, P. B., Bale, J. R., Huang, C. Y. 1980. *Fed. Proc.* 39:1822
66. Bridger, W. A. 1974. *Enzymes* 10:581–606
67. Bridger, W. A. 1981. *Can. J. Biochem.* 59:1–8
68. Bridger, W. A., Millen, W. A., Boyer, P. D. 1968. *Biochemistry* 7:3608–16
69. Nishimura, J. S., Meister, A. 1965. *Biochemistry* 4:1457–62
70. Pearson, P. H., Bridger, W. A. 1975. *J. Biol. Chem.* 250:8524–29
71. Collier, G. E., Nishimura, J. S. 1978. *J. Biol. Chem.* 253:4938–43
72. Bridger, W. A. 1971. *Biochem. Biophys. Res. Commun.* 42:948–54
73. Ramaley, R. F., Bridger, W. A., Moyer, R. W., Boyer, P. D. 1967. *J. Biol. Chem.* 242:4287–98
74. Moffet, F. J., Wang, T., Bridger, W. A. 1972. *J. Biol. Chem.* 247:8139–44
75. Bild, G. S., Janson, C. A., Boyer, P. D. 1980. *J. Biol. Chem.* 255:8109–15
76. Vogel, H. J., Bridger, W. A. 1982. *J. Biol. Chem.* In press
77. Wolodko, W. T., O'Connor, M. D., Bridger, W. A. 1981. *Proc. Natl. Acad. Sci. USA* 78:2140–44
78. Elwell, M., Hersh, L. B. 1979. *J. Biol. Chem.* 254:2434–38
79. Hersh, L. B., Peet, M. 1981. *J. Biol. Chem.* 256:1732–37
80. Penefsky, H. S. 1979. *Adv. Enzymol.* 49:223–80
81. Ferguson, S. J., Lloyd, W. T., Radda, G. K. 1975. *Eur. J. Biochem.* 54:127–33
82. Wagenvoord, R. J., van der Kraan, I., Kemp, A. 1977. *Biochim. Biophys. Acta* 460:17–24
83. Garrett, N. E., Penefsky, H. S. 1975. *J. Biol. Chem.* 250:6640–47

84. Garrett, N. E., Penefsky, H. S. 1975. *J. Supramol. Struct.* 3:469–78
85. Schuster, S. M., Ebel, R. E., Lardy, H. A. 1975. *J. Biol. Chem.* 250: 7848–53
86. Cantley, L. C. Jr., Hammes, G. G. 1975. *Biochemistry* 14:2968–75
87. Jagendorf, A. T., Uribe, E. 1966. *Proc. Natl. Acad. Sci. USA* 55:170–77
88. Yamamoto, T., Tonomura, Y. 1975. *J. Biochem. (Tokyo)* 77:137–46
89. Futai, M. 1977. *Biochem. Biophys. Res. Commun.* 79:1231–37
90. Ebel, R. E., Lardy, H. A. 1975. *J. Biol. Chem.* 250:191–96
91. Pedersen, P. L. 1976. *J. Biol. Chem.* 251:934–40
92. Hammes, G. G., Hilborn, D. A. 1971. *Biochim. Biophys. Acta* 233:580–90
93. Cantley, L. C. Jr. Hammes, G. G. 1975. *Biochemistry* 14:2976–81
94. Marcus, F., Schuster, S. M., Lardy, H. A. 1976. *J. Biol. Chem.* 251: 1775–80
95. Takahashi, K. 1968. *J. Biol. Chem.* 243:6171–79
96. Hutton, R. L., Boyer, P. D. 1979. *J. Biol. Chem.* 254:9990–93
97. Russo, J. A., Lamos, C. M., Mitchell, R. A. 1978. *Biochemistry* 17:473–80
98. Kayalar, C., Rosing, J., Boyer, P. D. 1977. *J. Biol. Chem.* 252:2486–91
99. Hackney, D. D., Boyer, P. D. 1978. *J. Biol. Chem.* 253:3164–70
100. Smith, D. J., Boyer, P. D. 1976. *Proc. Natl. Acad. Sci. USA* 73:4314–18
101. Penefsky, H. S. 1977. *J. Biol. Chem.* 252:2891–99
102. Kashara, M., Penefsky, H. S. 1978. *J. Biol. Chem.* 253:4180–87
103. Adolfsen, R., Moudrianakis, E. N. 1976. *Arch. Biochem. Biophys.* 172: 425–33
104. Grubmeyer, C., Penefsky, H. S. 1981. *J. Biol. Chem.* 256:3718–27
105. Grubmeyer, C., Penefsky, H. S. 1981. *J. Biol. Chem.* 256:3728–34
106. Senior, A. E., Fayle, D. R. H., Downie, J. A., Gibson, F., Cox, G. B. 1979. *Biochem. J.* 180:111–18
107. Stein, W. D., Lieb, W. R., Karlish, S. J. D., Eilam, Y. 1973. *Proc. Natl. Acad. Sci. USA* 70:275–78
108. Robinson, J. D. 1974. *Biochim. Biophys. Acta* 341:232–47
109. Froehlich, J. P., Albers, R. W., Koval, G. J., Goebel, R., Berman, M. 1976. *J. Biol. Chem.* 251:2186–88
110. Cantley, L. C., Gelles, J., Josephson, L. 1978. *Biochemistry* 17:418–27
111. Cantley, L. C. 1981. *Curr. Top. Bioenerg.* 11:201–37
112. Froehlich, J. P., Taylor, E. W. 1975. *J. Biol. Chem.* 250:2013–21
113. Eckert, K., Grosse, R., Levitsky, D. O., Kurzmin, A. V., Smirnow, V. N., Repke, K. R. H. 1977. *Acta Biol. Med. Germ.* 36:K1–K10
114. Verjovski-Almeida, S., Inesi, G. 1979. *J. Biol. Chem.* 254:18–21
115. Smith, R. L., Zinn, K., Cantley, L. C. 1980. *J. Biol. Chem.* 255:9852–59
116. Moczydlowski, E. G., Fortes, P. A. G. 1981. *J. Biol. Chem.* 256:2346–66
117. Dupont, Y. 1977. *Eur. J. Biochem.* 72: 185–90
118. Taylor, J. S., Hattam, D. 1979. *J. Biol. Chem.* 254:4402–7
119. Ariki, M., Boyer, P. D. 1980. *Biochemistry* 19:2001–4
120. Fersht, A. R. 1975. *Biochemistry* 14: 5–14
121. Holler, E. 1978. *Angew Chem. Int. Ed. Engl.* 17:648–56
122. Schimmel, P. R., Söll, D. 1979. *Ann. Rev. Biochem.* 48:601–48
123. Mehler, A. H., Chakraburtty. 1971. *Adv. Enzymol.* 35:443–501
124. Berg, P. 1956. *J. Biol. Chem.* 222: 1025–34
125. Hoagland, M. B. 1955. *Biochim. Biophys. Acta* 16:288–89
126. Baldwin, A. N., Berg, P. 1966. *J. Biol. Chem.* 241:839–45
127. Bartmann, P., Hanke, T., Holler, E. 1975. *Biochemistry* 14:4777–86
128. Fersht, A. R., Kaethner, M. M. 1976. *Biochemistry* 15: 818–23
129. Koch, G. L. E. 1974. *Biochemistry* 13: 2307–12
130. Irwin, M. J., Nyborg, J., Reid, B. R., Blow, D. M. 1976. *J. Mol. Biol.* 105: 577–86
131. Bosshard, H. R., Koch, G. L. E., Hartley, B. S. 1975. *Eur. J. Biochem.* 53: 493–98
132. Fersht, A. R., Ashford, J. S., Bruton, C. J., Jakes, R., Koch, G. L. E., Hartley, B. S. 1975. *Biochemistry* 14:1–4
133. Holler, E. 1976. *J. Biol. Chem.* 251:7717–19
134. Fersht, A. R., Mulvey, R. S., Koch, G. L. E. 1975. *Biochemistry* 14:13–18
135. Jakes, R., Fersht, A. R. 1975. *Biochemistry* 14:3344–50
136. Krajewska-Grynkiewicz, K., Buonocore, V., Schlesinger, S. 1973. *Biochim. Biophys. Acta* 312: 518–27
137. Pingoud, A., Boehme, D., Riesner, D., Kownatzki, R., Maass, G. 1975. *Eur. J. Biochem.* 56:617–22
138. Riesner, D., Pingoud, A., Boehme, D., Peters, F., Maass, G. 1976. *Eur. J. Biochem.* 68:71–80
139. Rigler, R., Pachmann, U., Hirsch, R.,

Zachau, H. G. 1976. *Eur. J. Biochem.* 65:307–15

140. Krauss, G., Riesner, D., Maass, G. 1976. *Eur. J. Biochem.* 68:81–93
141. Koch, G. L. E., Boulanger, Y., Hartley, B. S. 1974. *Nature* 249:316–20
142. Moe, J. G., Piszkiewicz, D. 1979. *Biochemistry* 18:2804–10
143. Krauss, G., Pingoud, A., Boehme, D., Riesner, D., Peters, F., Maass, G. 1975. *Eur. J. Biochem.* 55:517–29
144. Blanquet, S., Iwatsubo, M., Waller, J.-P. 1973. *Eur. J. Biochem.* 36:213–26
145. Blanquet, S., Dessen, P., Iwatsubo, M. 1976. *J. Mol. Biol.* 103:765–84
146. Kosakowski, H. M., Böck, A. 1971. *Eur. J. Biochem.* 24:190–200
147. Fayat, G., Blanquet, S., Dessen, P., Batelier, G.,Waller, J.-P. 1974. *Biochimie* 56:35–41
148. Berther, J.-M., Mayer, P., Dutler, H. 1974. *Eur. J. Biochem.* 47:151–63
149. Fasiolo, F., Remy, P., Pouget, J., Ebel, J.-P. 1974. *Eur. J. Biochem.* 50:227–36
150. von der Haar, F., Cramer, F. 1978. *Biochemistry* 17:3139–45
151. Moe, J. G., Piszkiewicz, D. 1979. *Biochemistry* 18:2810–4
152. Kern, D., Giegé, R., Ebel, J.-P. 1981. *Biochemistry* 20:5156–69
152. Fersht, A. R., Jakes, R. 1975. *Biochemistry* 14:3350–56
154. Yarus, M., Berg, P. 1969. *J. Mol. Biol.* 42:171–89
155. Helene, C., Brun, F., Yaniv, M. 1971. *J. Mol. Biol.* 58:349–65
156. Eldred, E. W., Schimmel, P. R. 1972. *Biochemistry* 11:17–23
157. Lövgren, T. N. E., Pastuszyn, A., Loftfield, R. B. 1976. *Biochemistry* 15:2533–40
158. Stadtman, E. R., Ginsburg, A. 1974. *Enzymes* 10:755–807
159. Meister, A. 1974. *Enzymes* 10:699–754
160. Denton, M. D., Ginsburg, A. 1970. *Biochemistry* 9:617–32
161. Bild, G. S., Boyer, P. D. 1980. *Biochemistry* 19:5774–79
162. Meek, T. D., Villafranca, J. J. 1980. *Biochemistry* 19:5513–19
163. Manning, J. M., Moore, S., Rowe, W. B., Meister, A. 1969. *Biochemistry* 8:2681–85
164. Weisbrod, R. E., Meister, A. 1973. *J. Biol. Chem.* 248:3997–4002
165. Maurizi, M. R., Ginsburg, A. 1982. *J. Biol. Chem.* In press
166. Rhee, S. G., Chock, P. B., Wedler, F. C., Sugiyama, Y. 1981. *J. Biol. Chem.* 256:644–48
167. Tate, S. S., Lew, F. Y., Meister, A. 1972. *J. Biol. Chem.* 247:5312–21

168. Jaenicke, L., Jésior, J. C. 1978. *FEBS Lett.* 90:115–18
169. Rhee, S. G., Chock, P. B., Wedler, F. C., Sugiyama, Y. 1979. *Abstr. Int. Congress Biochem., 11th,* Toronto, Canada
170. Shrake, A., Whitley, E. J. Jr., Ginsburg, A. 1980. *J. Biol. Chem.* 255: 581–89
171. Rhee, S. G., Ubom, G. A., Hunt, J. B., Chock, P. B. 1981. *J. Biol. Chem.* 256: 6010–16
172. Shrake, A., Ginsburg, A., Wedler, F. C., Sugiyama, Y. 1982. *J. Biol. Chem.* In press
173. Gorman, E. G., Ginsburg, A. 1982. *J. Biol. Chem.* In press
174. Harris, J. I., Waters, M. 1976. *Enzymes* 8:1–50
175. Kirschner, K., Eigen, M., Bittman, R., Voight, B. 1966. *Proc. Natl. Acad. Sci. USA* 56:1661–67
176. Kirschner, K., Gallego, E., Schuster, I., Goodall, D. 1971. *J. Mol. Biol.* 58: 29–50
177. Ellenrieder, G. von, Kirschner, K., Schuster, I. 1972. *Eur. J. Biochem.* 26: 220–36
178. Cook, R. A., Koshland, D. E. Jr. 1970. *Biochemistry* 9:3337–42
179. Mockrin, S. C., Byers, L., Koshland, D. E., Jr. 1975. *Biochemistry* 14:5428–37
180. Gennis, L. S. 1976. *Proc. Natl. Acad. Sci. USA* 73:3928–32
181. Niekamp, C. W., Sturtevant, J. M., Velick, S. F. 1977. *Biochemistry* 16: 436–44
182. DeVijlder, J. J. M., Slater, E. C. 1968. *Biochim. Biophys. Acta* 167:23–34
183. Conway, A., Koshland, D. E. Jr. 1968. *Biochemistry* 7:4011–23
184. Schlessinger, J., Levitzki, A. 1974. *J. Mol. Biol.* 82:547–61
185. DeVijlder, J. J. M., Boers, W., Slater, E. C. 1969. *Biochim. Biophys. Acta* 191: 214–20
186. Seydoux, F., Bernhard, S., Pfenninger, O., Payne, M., Malhotra, O. P. 1973. *Biochemistry* 12:4290–300
187. Allen, G. A., Harris, J. I. 1975. *Biochem. J.* 151:747–49
188. Teipel, J., Koshland, D. E., Jr. 1970. *Biochim. Biophys. Acta* 198:183–91
189. Velick, S. F., Baggott, J. P., Sturtevant, J. M. 1971. *Biochemistry* 10: 779–86
190. Boers, W., Oosthüizen, C., Slater, E. C. 1971. *Biochim. Biophys. Acta* 250:35–46
191. Price, N. C., Radda, G. K. 1971. *Biochim. Biophys. Acta* 235:27–31
192. Bell, J. E., Dalziel, K. 1975. *Biochim. Biophys. Acta* 391:249–58
193. Scheek, R. M., Slater, E. C. 1978. *Biochim. Biophys. Acta* 526:13–24

194. Scheek, R. M., Berden, J. A., Hoogh-iemstra, R., Slater, E. C. 1979. *Biochim. Biophys. Acta* 569:124–34
195. Kekemen, N., Kellershohn, N., Seydoux, F. 1975. *Eur. J. Biochem.* 57:69–78
196. Herzfeld, J., Ichiye, T., Jung, D. 1981. *Biochemistry* 20:4936–41
197. Henis, Y. I., Levitzki, A. 1980. *Proc. Natl. Acad. Sci. USA* 75:5055–59
198. Henis, Y. I., Levitzki, A. 1980. *Eur. J. Biochem.* 112:59–73
199. Moras, D., Olsen, K. W, Sabesan, M. N., Buehner, M., Ford, G. C., Rossmann, M. G. 1975. *J. Biol. Chem.* 250:9137–62
200. Biesecker, G., Harris, J. I., Thierry, J. C., Walker, J. E., Wonacott, A. J. 1977. *Nature* 266:328–33
201. Henis, Y. I., Levitzki, A., Gafni, A. 1979. *Eur. J. Biochem.* 97:519–28
202. Hoagland, V. D., Teller, D. C. 1969. *Biochemistry* 8:594–602
203. Keleti, T., Batke, J., Ovádi, J., Janesik, V., Bartha, F. 1977. *Adv. Enzyme Regul.* 15:233–65
204. Vas, M., Lakatos, S., Hajdu, J., Friedrich, P. 1981. *Biochimie* 63:89–96
205. Malhotra, O. P., Bernhard, S. A. 1968. *J. Biol. Chem.* 243:1243–52
206. Levitzki, A. 1973. *Biochem. Biophys. Res. Commun.* 54:889–93
207. Levitzki, A. 1974. *J. Mol. Biol.* 90:451–58
208. Stallcup, W. B., Koshland, D. E. Jr. 1973. *J. Mol. Biol.* 80:41–62
209. MacQuarrie, R. A., Bernhard, S. A. 1971. *J. Mol. Biol.* 55:181–92
210. Malhota, O. P., Srinivasan, K., Srivastava, D. K. 1978. *Biochim. Biophys. Acta* 526:1–12
211. Malhotra, O. P., Bernhard, S. A., Seydoux, F. 1981. *Biochimie* 63:131–41
212. Bernhard, S. A., MacQuarrie, R. A. 1973. *J. Mol. Biol.* 74:74–78
213. Stallcup, W. B., Koshland, D. E., Jr. 1973. *J. Mol. Biol.* 80:77–91
214. Trentham, D. R. 1971. *Biochem. J.* 122:71–77
215. Kellershohn, N., Seydoux, F. J. 1979. *Biochemistry* 18:2465–70
216. Trentham, D. R. 1971. *Biochem. J.* 122:59–69
217. Hill, T. L., Levitzki, A. 1980. *Proc. Natl. Acad. Sci. USA* 77:5741–45
218. Furfine, C. S., Velick, S. F. 1965. *J. Biol. Chem.* 240:844–55
219. Bloch, W., MacQuarrie, R. A., Bernhard, S. A. 1971. *J. Biol. Chem.* 246:780–90
220. Malhotra, O. P., Bernhard, S. A. 1973. *Proc. Natl. Acad. Sci. USA* 70:2077–81
221. Schwendimann, B., Ingbar, D., Bernhard, S. A. 1976. *J. Mol. Biol.* 108:123–38
222. Malhotra, O. P., Bernhard, S. A. 1981. *Biochemistry* 20:5529–38
223. Gerhart, J. C., Pardee, A. B. 1962. *J. Biol. Chem.* 237:891–96
224. Nakae, T., Nikaido, M. 1971. *J. Biol. Chem.* 246:4386–96
225. Wiley, D. C., Evans, D. R., Warren, S. G., McMurray, C. H., Edwards, B. F. P., Franks, W. A., Lipscomb, W. N. 1971. *Cold Spring Harbor Symp. Quant. Biol.* 36:285–90
226. Gerhart, J. C., Schachman, H. K. 1965. *Biochemistry* 4:1054–62
227. Gerhart, J. C., Schachman, H. K. 1968. *Biochemistry* 7:538–52
228. Jacobson, G. R., Stark, G. R. 1973. *J. Biol. Chem.* 248:8003–14
229. Howlett, G. J., Schachman, H. K. 1977. *Biochemistry* 16:5077–83
230. Blackburn, M. N., Schachman, H. K. 1977. *Biochemistry* 16:5084–91
231. Johnson, R. S., Schachman, H. K. 1980. *Proc. Natl. Acad. Sci. USA* 77:1995–99
232. Gibbons, I., Ritchey, J. M., Schachman, H. K. 1976. *Biochemistry* 15:1324–30
233. Howlett, G. J., Blackburn, M. N., Compton, J. G., Schachman, H. K. 1977. *Biochemistry* 16:5091–99
234. Conway, A., Koshland, D. E., Jr. 1968. *Biochemistry* 7:4011–23
235. Winlund, C. C., Chamberlin, M. J. 1970. *Biochem. Biophys. Res. Commun.* 60:43–49
236. Klotz, I. M., Hunston, D. L. 1977. *Proc. Natl. Acad. Sci. USA* 74:4959–63
237. Rosenbusch, J. P., Griffin, J. H. 1973. *J. Biol. Chem.* 248:5063–66
238. Suter, P., Rosenbusch, J. P. 1976. *J. Biol. Chem.* 251:5986–91
239. Nagel, G. M., Schachman, H. K. 1975. *Biochemistry* 14:3195–3203
240. Kirschner, M. W., Schachman, H. K. 1973. *Biochemistry* 12:2987–97
241. Hensley, P., Schachman, H. K. 1979. *Proc. Natl. Acad. Sci. USA* 76:3732–36
242. Yang, Y. R., Schachman, H. K. 1980. *Proc. Natl. Acad. Sci. USA* 77:5187–91
243. Subramani, S., Bothwell, M., Gibbons, I., Yang, Y. R., Schachman, H. K. 1977. *Proc. Natl. Acad. Sci USA* 74:3777–81
244. Subramani, S., Schachman, H. K. 1980. *J. Biol. Chem.* 255:8136–43

AUTHOR INDEX

(Names appearing in capital letters indicate authors of chapters in this volume.)

SUBJECT INDEX

A

Acetylaminofluorene
  mutagenicity of, 659
Acetylcholine
  AcChR and, 518
Acetylcholinesterase
  selective induction of, 857
Acetylglucosaminides
  asialoglycoprotein binding
    and, 536
Achyla DNA
  sequence arrangements in,
    816
ACTH
  melanocyte-stimulating
    hormone and, 149
  nerve growth factor and, 852
Actin
  discovery of, 14
Actinomycin D
  tyrosine hydroxylase
    induction and, 852
Active transport
  nerve growth factor and,
    856–58
  see also Bacteriorhodopsin
Aculeacin
  yeast cell wall polysaccharide
    synthesis inhibition and,
    765
Aculeacin A
  glucan synthesis in vivo and,
    781
Acute lymphocytic leukemia
  enzyme therapy and, 796
Adenine
  deaminated
    removal from DNA, 64
Adenosine triphosphate
  see ATP
Adenosylmethionine
  DNA alkylation and, 62, 67
Adenosylmethionine synthetase
  paramagnetic metal ion
    coupling in
  EPR spectroscopy and,
    390
Adenovirus 2 snRNA 638-40
  nucleotide sequence of, 639
Adenovirus DNA replication,
    912–18
  in vitro, 916–17
  in vivo, 913–14
  initiation of, 917–18
  virus-coded proteins and,
    914–16
Adenoviruses
  genome structure of, 912–13

Adrenocorticotropic hormones,
    148–50
Affinity chromatography
  AcChR purification and,
    492–93
  asialoglycoprotein binding
    activity and, 533–34
Aflatoxin
  mutagenicity of, 677
Aflatoxin B1
  mutagenicity of, 659
  reaction products of, 678
African green monkey DNA
  nucleotide sequence of, 822
Aging
  DNA synthesis fidelity and,
    449–50
  mutagenesis and, 429
Aliphatic oxides
  mutagenicity of, 671
Alkaline phosphatase
  catalysis of, 941–44
  NMR spectroscopy and,
    369–74
Alkalophilic bacteria
  proton-motive force and,
    203–4
Alkyl guanidines
  AcChR and, 516–17
Alkyl sulfates
  mutagenicity of, 669
Alkylating agents
  mutagenic efficiency of, 658
  mutagenicity of, 451, 657,
    667–73
  reaction products of, 671
  reaction sites of, 668
  S-methylcysteine and, 80
  structural formulas of, 667,
    670
Alternaria alternata toxins
  properties of, 319–20
Alternaria phytotoxins
  properties of, 326–27
Alternaria spp
  phytotoxin production and,
    311
Alternaric acid
  phytotoxicity of, 326–27
Alternariol methylether
  phytotoxicity of, 326–27
Alternariolide, 319
Amanitin, 310
  tyrosine hydroxylase
    induction and, 852
Amantidine
  AcChR and, 515
Amines
  aromatic
    mutagenicity of, 673–77

protein breakdown and, 344
Amino acid sequences
  bacteriorhodopsin, 591
  keratin polypeptides and, 225
  of *E. coli* ribosomal proteins,
    156–60
Amino acids
  cancer chemotherapy and,
    796, 798–99
  protein breakdown and, 337
Aminoacyl tRNA synthetase
  catalysis and, 950–53
Aminohydrolases
  cancer chemotherapy and,
    796
Aminoisobutyric acid
  insulin-stimulated uptake of,
    856–57
  nerve growth
    factor-stimulated uptake
    of, 856
Aminopurine
  DNA fidelity and
    mutagenesis and, 451
  mutagenicity of, 439
Aminouracil
  DNA repair enzyme
    inhibition and, 65
Ammonia
  protein breakdown and, 344
Ammonia toxicity
  enzyme therapy and, 809
Amphetamine toxicity
  enzyme therapy and, 809
Amyrin
  enzymatic cyclization of
    epoxysqualene and, 559
Anaerobiosis
  reticulocytic protein
    breakdown and, 346
Ancrod
  clot dissolution and, 808
Anesthetics
  AcChR and, 513–17
Angiotensin-converting enzyme
  assay of, 294–95
  inhibitors of, 301–4
  localization of, 295–96
  properties of, 297–301
  purification of, 297
Angiotensinases
  angiotensin conversion and,
    284
Anthranilate synthetase
  catalytic activity regulation
    and, 936
Antibiotic toxicity
  enzyme therapy and, 809
Antibiotics

CUMULATIVE INDEXES

Contributing Authors, Volumes 47–51

CHAPTER TITLES, VOLUMES 47–51

1054 CHAPTER TITLES